38. $\displaystyle\int u \sin u \, du = \sin u - u \cos u + C$

39. $\displaystyle\int u \cos u \, du = \cos u + u \sin u + C$

40. $\displaystyle\int u^n \sin u \, du = -u^n \cos u + n \int u^{n-1} \cos u \, du$

41. $\displaystyle\int u^n \cos u \, du = u^n \sin u - n \int u^{n-1} \sin u \, du$

FORMS INVOLVING $\sqrt{u^2 \pm a^2}$

42. $\displaystyle\int \sqrt{u^2 \pm a^2}\, du = \frac{u}{2}\sqrt{u^2 \pm a^2} \pm \frac{a^2}{2} \ln\left|u + \sqrt{u^2 \pm a^2}\right| + C$

43. $\displaystyle\int \frac{du}{\sqrt{u^2 \pm a^2}} = \ln\left|u + \sqrt{u^2 \pm a^2}\right| + C$

FORMS INVOLVING $\sqrt{a^2 - u^2}$

44. $\displaystyle\int \sqrt{a^2 - u^2}\, du = \frac{u}{2}\sqrt{a^2 - u^2} + \frac{a^2}{2} \sin^{-1}\frac{u}{a} + C$

45. $\displaystyle\int \frac{\sqrt{a^2 - u^2}}{u}\, du = \sqrt{a^2 - u^2} - a \ln\left|\frac{a + \sqrt{a^2 - u^2}}{u}\right| + C$

EXPONENTIAL AND LOGARITHMIC FORMS

46. $\displaystyle\int ue^u \, du = (u - 1)e^u + C$

47. $\displaystyle\int u^n e^u \, du = u^n e^u - n \int u^{n-1} e^u \, du$

48. $\displaystyle\int u^n \ln u \, du = \frac{u^{n+1}}{n+1} \ln u - \frac{u^{n+1}}{(n+1)^2} + C$

49. $\displaystyle\int e^{au} \sin bu \, du = \frac{e^{au}}{a^2 + b^2}(a \sin bu - b \cos bu) + C$

50. $\displaystyle\int e^{au} \cos bu \, du = \frac{e^{au}}{a^2 + b^2}(a \cos bu + b \sin bu) + C$

INVERSE TRIGONOMETRIC FORMS

51. $\displaystyle\int \sin^{-1} u \, du = u \sin^{-1} u + \sqrt{1 - u^2} + C$

52. $\displaystyle\int \tan^{-1} u \, du = u \tan^{-1} u - \frac{1}{2} \ln(1 + u^2) + C$

53. $\displaystyle\int \sec^{-1} u \, du = u \sec^{-1} u - \ln\left|u + \sqrt{u^2 - 1}\right| + C$

54. $\displaystyle\int u \sin^{-1} u \, du = \frac{1}{4}(2u^2 - 1) \sin^{-1} u + \frac{u}{4}\sqrt{1 - u^2} + C$

55. $\displaystyle\int u \tan^{-1} u \, du = \frac{1}{2}(u^2 + 1) \tan^{-1} u - \frac{u}{2} + C$

56. $\displaystyle\int u \sec^{-1} u \, du = \frac{u^2}{2} \sec^{-1} u - \frac{1}{2}\sqrt{u^2 - 1} + C$

57. $\displaystyle\int u^n \sin^{-1} u \, du = \frac{u^{n+1}}{n+1} \sin^{-1} u - \frac{1}{n+1} \int \frac{u^{n+1}}{\sqrt{1 - u^2}} \, du \quad \text{if } n \neq -1$

58. $\displaystyle\int u^n \tan^{-1} u \, du = \frac{u^{n+1}}{n+1} \tan^{-1} u - \frac{1}{n+1} \int \frac{u^{n+1}}{1 + u^2} \, du \quad \text{if } n \neq -1$

59. $\displaystyle\int u^n \sec^{-1} u \, du = \frac{u^{n+1}}{n+1} \sec^{-1} u - \frac{1}{n+1} \int \frac{u^{n+1}}{\sqrt{u^2 - 1}} \, du \quad \text{if } n \neq -1$

OTHER USEFUL FORMULAS

60. $\displaystyle\int_0^{\infty} u^n e^{-u} \, du = n! \quad \text{if } n \geq 0$

61. $\displaystyle\int_0^{\infty} e^{-au^2} \, du = \frac{1}{2}\sqrt{\frac{\pi}{a}} \quad \text{if } a > 0$

62. $$\int_0^{\pi/2} \sin^n u \, du = \int_0^{\pi/2} \cos^n \, du = \begin{cases} \dfrac{1 \cdot 3 \cdot 5 \cdots (n-1)}{2 \cdot 4 \cdot 6 \cdots n} \cdot \dfrac{\pi}{2} & \text{if } n \text{ is an even integer and } n \geqq 2 \\[2ex] \dfrac{2 \cdot 4 \cdot 6 \cdots (n-1)}{3 \cdot 5 \cdot 7 \cdots n} & \text{if } n \text{ is an odd integer and } n \geqq 3 \end{cases}$$

Differential Equations & Linear Algebra

Differential Equations & Linear Algebra

Second Edition

Jerry Farlow
University of Maine, Orono, Maine

James E. Hall
Westminster College, New Wilmington, Pennsylvania

Jean Marie McDill
California Polytechnic State University, San Luis Obispo, California

Beverly H. West
Cornell University, Ithaca, New York

PEARSON
Prentice Hall

Harlow, England • London • New York • Boston • San Francisco • Toronto • Sydney • Singapore • Hong Kong
Tokyo • Seoul • Taipei • New Delhi • Cape Town • Madrid • Mexico City • Amsterdam • Munich • Paris • Milan

Library of Congress Cataloging-in-Publication Data

Differential equations and linear algebra / J.S. Farlow . . . [et al.]. — 2nd ed.
p. cm.
ISBN 0-13-186061-5
1. Differential equations. 2. Algebras, Linear. I. Farlow, Jerry.
QA371.D4436 2007
515′.35—dc22

2006034955

Vice President and Editorial Director, ECS: *Marcia J. Horton*
Senior Editor: *Holly Stark*
Editorial Assistant: *Jennifer Lonschein*
Executive Managing Editor: *Vince O'Brien*
Managing Editor: *David A. George*
Production Editor: *Winifred Sanchez*
Director of Creative Services: *Paul Belfanti*
Creative Director: *Juan Lopez*
Art Director: *Jonathan Boylan*
Cover and Interior Designer: *511 Design*
Art Editor: *Thomas Benfatti*
Manufacturing Manager: *Alexis Heydt-Long*
Manufacturing Buyer: *Lisa McDowell*
Senior Marketing Manager: *Tim Galligan*

About the Cover: HSB Turning Torso in Malmö, Sweden. Designed by photographer and artist Santiago Calatrava, the building officially opened on August 27, 2005. The tower reaches a height of 623 feet (190 meters) with 54 stories. Upon completion, it was the tallest building in Scandinavia.

Pearson Prentice Hall
Pearson Education, Inc.
Upper Saddle River, NJ 07458

Printed in the United States of America

10 9 8 7 6 5

ISBN: 0-13-186061-5

Pearson Education Ltd., *London*
Pearson Education Australia Pty. Ltd., *Sydney*
Pearson Education Singapore, Pte. Ltd.
Pearson Education North Asia Ltd., *Hong Kong*
Pearson Education Canada, Inc., *Toronto*
Pearson Educación de Mexico, S.A. de C.V.
Pearson Education—Japan, *Tokyo*
Pearson Education Malaysia, Pte. Ltd.
Pearson Education, Inc., *Upper Saddle River, New Jersey*

Contents

Preface to the Second Edition

This text is a response to departments of mathematics (many at engineering colleges) that have asked for a combined course in differential equations and linear algebra. It differs from other combined texts in its effort to stress the modern qualitative approach to differential equations, and to merge the disciplines more effectively.

The advantage of combining the topics of linear algebra and differential equations is that the linear algebra provides the underlying mathematical structure, and differential equations supply examples of function spaces in a natural fashion. In a typical linear algebra course, students ask frequently why vectors spaces other than R^n are of interest. In a combined course based on this text, the two topics are interwoven so that solution spaces of homogenous linear systems and solution spaces of homogeneous linear differential equations appear together quite naturally.

Differential Equations

In recent years, the emphasis in differential equations has moved away from the study of closed-form transient solutions to the qualitative analysis of steady-state solutions. Concepts such as equilibrium points and stability have become the focus of attention, diminishing concentration on formulas.

In the past, students of differential equations were generally left with the impression that all differential equations could be "solved," and if given enough time and effort, closed-form expressions involving polynomials, exponentials, trigonometric functions, and so on, could always be found. For students to be left with this impression is a mathematical felony in that even simple-looking equations such as

$$dy/dt = y^2 - t \quad \text{and} \quad dy/dt = e^{ty^2}$$

do not have closed-form solutions. But these equations *do* have *solutions*, which we can see graphically in Figures 1 and 2.

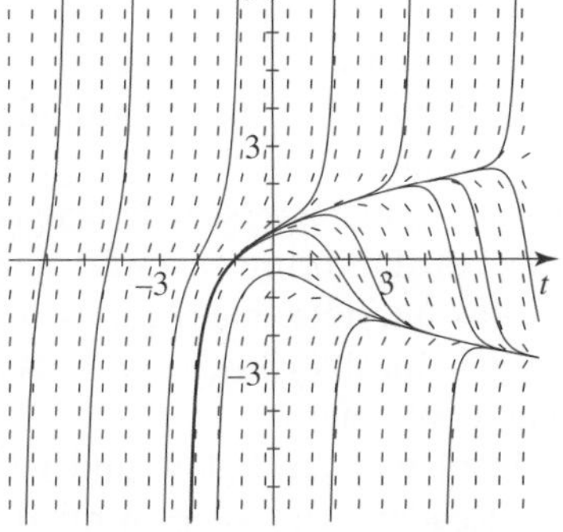

FIGURE 1 $dy/dt = y^2 - t$.

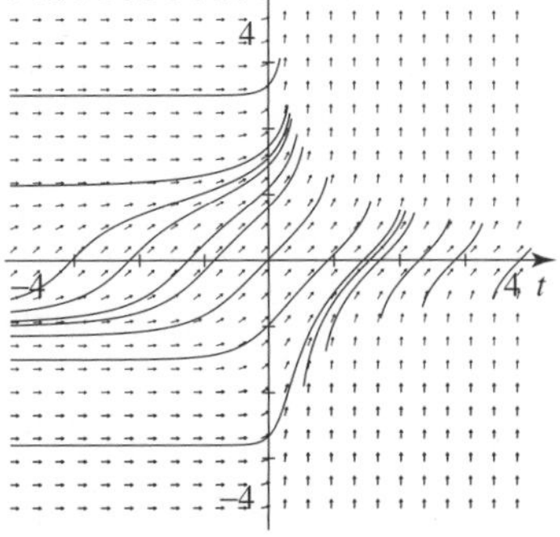

FIGURE 2 $dy/dt = e^{ty^2}$.

In the traditional differential equations course, students spent much of their time grinding out solutions to equations, without real understanding for the solutions or the subject. Nowadays, with computers and software packages readily available for finding numerical solutions, plotting vector and directional fields, and carrying out physical simulations, the student can study differential equations on a more sophisticated level than previous students, and ask questions not contemplated by students (or teachers) in the past. Key information is transmitted instantly by visual presentations, especially when students can watch solutions evolve. We use graphics heavily in the text and in the problem sets.

We are *not* discarding the quantitative analysis of differential equations, but rather increasing the qualitative aspects and emphasizing the links.

Linear Algebra

The visual approach is especially important in making the connections with linear algebra. Although differential equations have long been treated as one of the best applications of linear algebra, in traditional treatments students tended to miss key links. It's a delight to hear those who have taken those old courses gasp with sudden insight when they *see* the role of eigenvectors and eigenvalues in phase portraits.

Throughout the text we stress as a main theme from linear algebra that the general solution of a linear system is the solution to the associated homogeneous equation plus any particular solution.

Technology

Before we discuss further details of the text, we must sound an alert to the technological support provided, in order that the reader not be short-changed.

First is the unusual but very effective resource *Interactive Differential Equations* (IDE) that we helped to pioneer. The authors, two of whom were on the development team for IDE, are concerned that IDE might be overlooked by the student and the instructor, although we have taken great pains to incorporate this software in our text presentation. IDE is easily accessed on the internet at

`www.aw-bc.com/ide`

Interactive Differential Equations (IDE) This original collection of very useful interactive graphics tools was created by Hubert Hohn, at the Massachusetts College of Art, with a mathematical author team of John Cantwell, Jean Marie McDill, Steven Strogatz, and Beverly West, to assist students in really understanding crucial concepts. You will see that a picture (especially a dynamic one that you control) is indeed worth a thousand words. This option greatly enhances the learning of differential equations; we give pointers throughout the text, to give students immediate visual access to concepts.

The 97 "tools" of IDE bring examples and ideas to life. Each has an easy and intuitive point/click interface that requires *no* learning curve. Students have found this software very helpful; instructors often use IDE for short demonstrations that immediately get a point across.

Additional detail is given below under Curriculum Suggestions, and in the section "To the Reader." But keep in mind that IDE is designed as a valuable aid to understanding, and is *not* intended to replace an "open-ended graphic DE solver."

Second, students *must* be able to make their own pictures, with their own equations, to answer their own questions.

We do not want to add computing to the learning load of the students—we would far rather they devote their energy to the mathematics. All that is needed is an ability to draw direction fields and solutions for differential equations, an occasional algebraic curve, and simple spreadsheet capability. A graphing calculator is sufficient for most problems.

A complete computer algebra system (CAS) such as *Derive, Maple, Mathematica*, or *MATLAB* is more than adequate, but not at all necessary. We have found, however, that a dedicated "graphic ODE solver" is the handiest for differential equations, and have provided a good one on the text website:

`www.prenhall.com/farlow`

ODE Software (Dfield and Pplane) John Polking of Rice University has created our "open-ended graphic DE solver," originally as a specialized front-end for MATLAB, but now has a stand-alone Java option on our website. *Dfield* and *Pplane* provide an easy-to-use option for students and avoid the necessity of familiarity or access to a larger computer algebra system (CAS).

Finally, many instructors have expressed interest in projects designed for the more powerful CAS options; we will make a collection available on the text website cited above.

CAS Computer Projects Professor Don Hartig at California Polytechnic State University has designed and written a set of Computer Projects utilizing Maple®, for Chapters 1–9, which will be added to the text website, to provide a guide to instructors and students who might want to use a computer algebra system (CAS). These can be adapted to another CAS, and other projects may be added.

Differences from Traditional Texts

Although we have more pages explicitly devoted to differential equations than to linear algebra, we have tried to provide all the basics of both that either course syllabus would normally require. But merging two subjects into one (while at the same time enhancing the usual quantitative techniques with qualitative analysis of differential equations) requires streamlining and simplification. The result should serve students well in subsequent courses and applications.

Some Techniques De-Emphasized Many of the specialized techniques used to solve small classes of differential equations are no longer included within the confines of the text, but have been retired to the problem set. The same is true for some of the specialized techniques of linear algebra.

Dynamical Systems Philosophy We focus on the long-term behavior of a system as well as its transient behavior. Direction fields, phase plane analysis, and trajectories of solutions, equilibria, and stability are discussed whenever appropriate.

Exploration Problems for nontraditional topics such as bifurcation and chaos often involve guided or open-ended exploration, rather than application of a formula to arrive at a specific numerical answer. Although this approach is not traditional, it reflects the nature of how mathematics advances. This experimental stage is the world toward which students are headed; it is essential that they learn how to do it, especially how to organize and communicate about the results.

Problem Sets Each problem set involves most or all of the following:

- traditional problems for hand calculation (and understanding of techniques)
- additional traditional techniques
- graphical exercises (drawing, matching) to gain understanding of different representations
- real world applications
- some open-ended questions or exploration
- suggested journal entries (writing exercises)

Writing in Mathematics In recent years, the "Writing Across the Curriculum" crusade has spread across American colleges and universities, with the idea of learning through writing. We include "Suggested Journal Entries" at the end of each problem set, asking the student to write something about the section. The topics suggested should be considered simply as possible ideas; students may come up with something different on their own that is more relevant to their own thinking and evolving comprehension. Another way to ask students to keep a scholarly journal is to allow five minutes at the end of class for each student to write and outline what he or she does or does not understand. The goal is simply to encourage writing about mathematics; the degree to which it raises student understanding and performance can be amazing! Further background is provided in the section "To the Reader."

Historical Perspective We have tried to give the reader an appreciation of the richness and history of differential equations and linear algebra through footnotes and the use of "Historical Notes," which are included throughout the book. They can also be used by the instructor to foster discussions on the history of mathematics.

Applications We include traditional applications of differential equations: mechanical vibrations, electrical circuits, biological problems, biological chaos, heat flow problems, compartmental problems, and many more.

Many sections have applications at the end, where an instructor can choose to spend extra time. Furthermore, many problems introduce new applications and ideas not normally found in a beginning differential equations text: reversible systems, adjoint systems, Hamiltonians, and so on, for the more curious reader.

The final two chapters introduce related subjects that suggest ideal follow-up courses.

- Discrete Dynamical Systems: Iterative or difference equations (both linear and nonlinear) have important similarities and differences from differential equations. The ideas are simple, but the results can be surprisingly complicated. Subsections are devoted to the discrete logistic equation and its path to chaos.
- Control Theory: Although one of the most important applications of differential equations is control theory, few books on differential equations spend any time on the subject. This short chapter introduces a few of the important ideas, including feedback control and the Pontryagin maximum principle.

Obviously you will not have time to look at every application—being four authors with different interests and specialties, we do not expect that. But we would suggest that you choose to spend some time on what is closest to *your* heart, and in addition become aware of the wealth of other possibilities.

Changes in the Second Edition

Our goal has been to make the connection between the topics of differential equations and linear algebra even more obvious to the student. For this reason we have emphasized solution spaces (of homogeneous linear systems in Chapter 3, and homogenous linear differential equations in Chapters 4 and 6), and stressed the use of the Superposition Principle and the Nonhomogeneous Principle in these chapters.

Many of the changes and corrections have been in response to many helpful comments and suggestions by professors who have used the first edition in their courses. We have greatly benefited from their experience and insight.

The following major changes and additions are part of the new edition. We have also trimmed terminology, reorganized, and created a less-cramped layout, to clarify what is important.

Chapter 1: First-Order Differential Equations The introductory Section 1.1 has been rewritten and now includes material on direct and inverse variation. Section 1.2 has been expanded with many more problems specific to the qualitative aspects of solution graphs. Runge-Kutta methods for numerical approximations are part of Section 1.4.

Chapter 2: Linearity and Nonlinearity We have tried to make a clearer presentation of the basic structure of linearity, and have added a few new applications.

Chapter 3: Linear Algebra This chapter contains several examples of homogenous DE solution spaces in the section on vector spaces. The applications of the Superposition Principle and Nonhomogeneous Principle have been emphasized throughout. The issue of linear independence has been amplified.

Chapter 4: Higher-Order Linear Differential Equations This chapter emphasizes solution spaces and linear independence of solutions. We have added a new section on variation of parameters.

Chapter 5: Linear Transformations The generalized eigenvector section has been expanded.

Chapter 6: Linear Systems of Differential Equations This chapter has been expanded to include more information on the sketching of phase portraits, particularly for systems with nonreal eigenvalues. Two new sections have been added, on the matrix exponential and on solutions to nonhomogeneous systems of DEs, including the undetermined coefficients and variation of parameters methods (found in Sections 8.1 and 8.2 in the first edition). New examples of applications to circuits, coupled oscillators, and systems of tanks have been added.

Chapter 7: Nonlinear Systems of Differential Equations This is the big payoff that shows the power of all that comes before. This chapter has been extended to include as Section 7.5 the chaos material on forced nonlinear systems (Section 8.5 in the first edition).

Chapter 8: Laplace Transforms This chapter has been reorganized so that it only contains material on Laplace Transforms. The material has been expanded to four sections and a new section on Systems of Laplace Transforms has been added.

Chapter 9: Discrete Dynamical Systems This chapter has been updated to reflect new research by Samer Habre of the Lebanese American University and one of the authors.

Chapter 10: Control Theory and the Appendices These have been slightly revised to be consistent with the improved organization of this edition.

Problems Some 500 problems have been added to the approximately 2,000 problems of the first edition. A problem number in color signifies that there is a brief answer at the back of the book.

Curriculum Suggestions

Scientific and technological programs at many colleges and universities are increasingly squeezed by new courses required by the rapid advances in their fields. Consequently the amount of time for mathematical support courses is being pinched. Students are expected to acquire more skills in fewer courses. One result of this pinch at a number of institutions has been the appearance of an integrated course in linear algebra and differential equations, with a target audience of students having had a smattering of matrix algebra in precalculus or discrete mathematics and a minimal introduction to differential equations in calculus. It is for these students that this material is designed. The presentation is informal. Theory is presented visually and intuitively. Problem-solving skills are emphasized. Contemporary computer support is integrated into the text and problems.

The text is intended as a sophomore–junior level course in differential equations and linear algebra, for students majoring in science, engineering, and mathematics who have taken a one-year course in calculus. For a one-semester course, most instructors would probably rather concentrate on Chapters 1 through 8, skipping sections such as those on chaos (Sections 7.4–7.5) or Laplace transforms (Sections 8.3 and 8.4) as appropriate for their students. Two semesters would allow for more complete coverage of all the chapters.

The table below suggests a possible selection of materials for a one-semester course.

Basic Text Organization

	Core	Optional	Enrichment
Chapter 1: First-Order Differential Equations	1.1–1.5		
Chapter 2: Linearity and Nonlinearity	2.1–2.5	2.6	
Chapter 3: Linear Algebra	3.1–3.6		
Chapter 4: Second-Order Linear Differential Equations	4.1–4.5	4.6–4.7	
Chapter 5: Linear Transformations	5.1–5.3	5.4	
Chapter 6: Linear Systems Differential Equations	6.1–6.4	6.5–6.7	
Chapter 7: Nonlinear Systems Differential Equations	7.1–7.2	7.3	7.4–7.5
Chapter 8: Laplace Transforms	8.1–8.4	8.5	
Chapter 9: Discrete Dynamical Systems			9.1–9.3
Chapter 10: Control Theory			10.1–10.3

The basic text can be adapted to many different circumstances.

- Students with strong DE backgrounds (e.g., from reformed calculus courses) can use Chapter 1 as an overview and omit or review Sections 2.2–2.5.
- Students with strong matrix or linear algebra backgrounds (e.g., from courses in finite or discrete mathematics) will be able to omit or simply review Sections 3.1–3.4.
- Courses emphasizing linear algebra should add Appendix LT to the core, and could include Chapter 9.
- Courses emphasizing differential equations with Laplace transforms would include Chapter 8 and should consider Chapter 10.
- Courses emphasizing differential equations with dynamical systems instead of Laplace transforms would cover Chapter 9.
- A pure differential equations course could omit Sections 3.3–3.6, and all of Chapter 5 except Section 5.3, replacing them with additional sections from Chapters 7 and 8, plus Chapter 10.

Notes on the Uses of IDE

Instructors might find useful the following notes from co-author J. M. McDill on her use of the *Interactive Differential Equations* (IDE) software during a differential equations and linear algebra course at California State Polytechnic University at San Luis Obispo. IDE is available at this website:

`www.prenhall.com/ide`

Our course is over one ten-week quarter that meets four days per week. It is very crowded with material. I use a variety of IDE interactive illustrations; however, I use four or five basic labs to help me make it through the material in time.

- On the day following the introduction of the general method for solutions of linear homogeneous equations with constant coefficients, I show the substitution for an unforced mass–spring system and assign Lab 9: Linear Oscillators: Free Response. In this lab the students can vary each of the parameters, investigate the cases for the damped and undamped mass–spring systems, and obtain concrete visual examples of the undamped, underdamped, critically damped, and overdamped cases.
- The day after I introduce the method of undetermined coefficients for nonhomogeneous linear differential equations with constant coefficients, I show the substitution of parameters for a forced mass–spring system and assign Lab 12: Forced Vibrations: An Introduction. In this lab the students can vary parameters and observe that the motion of the mass is the sum of the transient and steady-state solutions by means of the software, which links graphs to a model of the mass–spring system. (I always found the presentation of the graphical material at the blackboard to be time-consuming.) Also, the undamped forced mass–spring system can be investigated by the students. They can assign parameters to produce pure resonance and beats. Even if formal lab time is not available to the students, IDE can be used as a lecture demonstration and/or the appropriate labs can be assigned for homework.
- As a further illustration of the usefulness of the analytical techniques for solving linear differential equations with constant coefficients, I emphasize

the analogy between series circuits and mass–spring systems. I assign Lab 13: Electrical Circuits as homework to illustrate this analogous behavior.

- By the time we approach the materials in Chapter 6, I am feeling somewhat rushed. I cover Sections 6.1–6.3. Then I assign Lab 16: Linear Classification as homework (usually over a weekend) before I lecture on Section 6.4: Stability and Linear Classification. I find that students who have covered the material visually as guided by the lab do not find the lecture so mind-boggling.
- I usually also sneak in Lab 18: Romeo and Juliet, as I introduce the phase plane. I find that it exerts a certain fascination for the students and provides an unforgettable if unconventional connection between the phase plane trajectories and the solution graphs.

About the Cover

The cover of the text incorporates a photo of the "Turning Torso" apartment tower, designed by the famous Spanish architect, Santiago Calatrava, which is a new landmark in southern Sweden. For details, see

`www.turningtorso.com`

`www.bizzbook.com/map/turningtorsocalatrava.html`

This tower is a unique example of the interaction of linear algebra and differential equations in the original structural design. For instance, the twisting of the tower can be modeled by linear transformations, and the stability under wind gusts can be modeled by differential equations and chaos theory.

For some fun, consider the following. Calatrava's turning tower challenged the skills of extreme skydiver Felix Baumgartner, who adjusted to wind conditions and other variables to make a legendary skydive from a helicopter to the top of the tower, then from the top of the tower to the ground. A mathematical model of his jump would require several differential equations, with differing initial conditions. Baumgartner's physical skills and mental skills were clearly up to the requirements (as well as the photographer's skills). See

`www.spoettel.com`

or search for Felix Baumgartner at

`www.youtube.com`

Acknowledgments

We first and foremost gratefully acknowledge the many ideas we have learned from John Hubbard, of Cornell University. He opened for us the new world of possibilities for studying differential equations and dynamical systems with technology.

Later, Hubert Hohn of the Massachusetts College of Art designed simple, elegantly constructed software, *Interactive Differential Equations*, to illustrate the concepts of differential equations. It has been pure pleasure to see how much insight and delight this has brought to ourselves, our students, and our colleagues.

There are many colleagues and users of the first edition who have provided invaluable help in the revision for this new edition. Professor Samer Habre of the Lebanese American University, a reviewer of the first edition, provided many suggestions and insightful accuracy checking of this edition. Jack Girolo, Don Hartig, Jim Delany, Bill Hesselgrave, and Al Jimenez of Cal Poly, San Luis Obispo were generous with their insights and suggestions.

We are indebted as well to the many other colleagues and previous authors whom we have quoted or referenced in the footnotes. Bob Borrelli and Courtney Coleman of Harvey Mudd College, John Cantwell of St. Louis University, Bjorn Felsager of Haslev Gymnasium in Denmark, and Steve Strogatz of Cornell University deserve special mention.

The following reviewers offered excellent advice, suggestions, and ideas as they read early drafts of the manuscript or taught from the text:

Kathy Eppler, Shawna Haider, *Salt Lake Community College*

Garret Etgen, *University of Houston*

Vincent Fatica, *Syracuse University*

Kenneth Goodearl, Martin Scharlemann, Jeffrey Stopple, *University of California at Santa Barbara*

Joel Lehmann, Patrick Sullivan, *Valparaiso University*

Michael Li, *Mississippi State University*

Paul A. Milewski, *University of Wisconsin at Madison*

Linda Patton, Jordi Pirig Suari, *California State Polytechnic University at San Luis Obispo*

Steve Rovnyak, *Louisiana State Technical University*

Donald Solomon, *University of Wisconsin at Milwaukee*

Marty Sternstein, *Ithaca College*

In addition, special appreciation for class testing with continued and detailed critiques goes to reviewers Chris Goodrich, Creighton Preparatory School and Mark Parker, Carroll College. Each reviewer wrote remarkably detailed and enthusiastic reviews, with many concrete suggestions for improvement or expansion. We were not able to incorporate all the suggested new material for this edition, but hope to include more in future editions.

We also thank our many students and colleagues who have tested our ideas, taken time to write to us, and helped to refine many a presentation or problem.

There are over 500 new problems in this edition. The solutions manual is a joint effort of the authors and two major contributors, Bill Hesselgrave and Mike Robertson, Cal Poly, San Luis Obispo. The timeliness and accuracy of the solutions manual is due in great part to these two contributors. Kelly Barber has contributed many months of precision typing to this effort.

Others to whom we offer grateful appreciation are Jim West, whose fantastic dinners have sustained us all during weeks of concentrated working sessions; John Armstrong, our internet expert; Katrina Thomas, who handily produced the artwork we could not and created the authors' website (see the link at the text website); Leah King for her sharp eye at proofreading solutions; and finally, as highly valued support troops, Diane Bonaccorsi, George Miller, Beth Paris, Gillie Waddington, Connie O'Brien, Keith Henderson, Mary Toscano, and Nancy Toscano.

We would like to express our appreciation to the editorial and production staff at Prentice Hall. First we thank George Lobell, who set up the author team, envisioned this project and encouraged us to write a second edition, and also

Sally Yagan, who continued it. And now we especially thank Holly Stark (Senior Editor, Editorial), Vince O'Brien (Executive Managing Editor, Production), David A. George (Managing Editor, Production), and Winifred Sanchez (Project Manager), who saw this edition to completion. We also thank the many other members of the production staff at Prentice Hall who helped in this project: production editor Scott Disanno, art editor Tom Benfatti, and art director Jon Boylan made special contributions. Throughout the production process, the anonymous copyeditors, typists, and compositors did an especially careful and efficient job of producing, reviewing, and responding to our concerns—we are most grateful.

Finally, we acknowledge that we can never adequately express our appreciation to Michelle Klinger for her painstaking manuscript preparation, careful execution of hundreds of graphs and figures, attention to detail, and gentle reminders that this brilliant example or that lengthy discussion were just not clear enough for students. Mikki's care, mathematical knowledge and talent, and incredible typing and graphics skill have been such an integral part of this endeavor.

Until she had to leave us, Mikki's devotion to our project exceeded any reasonable expectation and overcame astounding obstacles; she found (and corrected) more errors than any of us. Furthermore she independently created enormous clarifications, improved organization, and contributes several original new figures for this second edition. We have been fortunate indeed, for without Mikki, we could never have finished the first edition nor embarked on the second.

Whatever errors remain are of course the responsibility of the authors. We would appreciate having these brought to our attention so as to be corrected as soon as possible. We also appreciate any comments and suggestions from students and instructors.

Jerry Farlow farlow@math.umaine.edu
James E. Hall jehall@westminster.edu
Jean Marie McDill jmcdill@calpoly.edu
Beverly H. West bhw2@cornell.edu

To the Reader

And here let me insert a parenthesis to insist on the importance of written exercises. Compositions in writing are perhaps not given sufficient prominence in certain examinations. In the École Polytechnique, I am told that insistence on such compositions would close the door to very good pupils who know their subject and understand it very well, and yet are incapable of applying it in the smallest degree. I said just above that the word understand has several meanings. Such pupils only understand in the first sense of the word, and we have just seen that this is not sufficient to make either an engineer or a geometrician. Well, since we have to make a choice, I prefer to choose those who understand thoroughly.

—Henri Poincaré

Keeping a Scholarly Journal

One cannot help but be impressed with the large number of important English naturalists who lived during the nineteenth century. In addition to Darwin there were Wallace, Eddington, Thompson, Haldane, Fisher, Jevons, Fechner, Galton, and others. One characteristic that permeated the work of these men was the attention to scientific detail. And an important part of that attention to scientific detail was the keeping of detailed journals: Every observation and impression was recorded. These journals provided not only a place for storing data, but also a means for organizing thoughts, exploring relationships, and formulating and testing ideas. *In fact, they learned through writing.*

Journal keeping has declined in popularity in the twentieth century, but in the last few years there has been a renaissance in the "*learning through writing*" movement. People are beginning to realize that writing is an important *learning tool*, not just a means of communication.

This course can provide you, the student, with an opportunity to enhance the study of the material by keeping a journal. Entries can be made either daily or in conjunction with each section of the text. It is useful to date entries and provide them with short descriptive titles. While there are no general rules telling you what to include or how to write it, your instructor may provide you with some specific guidelines. Most problem sets in this text will suggest at least one possible topic. General suggestions for entries include describing a section's most important idea, reflecting on how the current section relates to earlier material, or formulating questions about concepts not yet clearly understood.

The style of such journal entries is strictly "free form," but you will benefit most by writing a modest amount regularly rather than procrastinating and then trying to "catch up." Even if your instructor doesn't require writing regular journal entries, we urge you to try it as a way of enriching your study of differential equations and linear algebra. Develop the habit of *rereading one or two old entries* each time you write a new one. See how your grasp of the material grows.

Using Technology

Technology is what gives the *power* of the modern approach to differential equations. It does not have to be fancy power, but you're losing a lot if you try to do without. Make it a habit to take advantage of the essential tools provided on the text website,

`www.prenhall.com/farlow`

Interactive Differential Equations (IDE)

So much of learning today is visual. The software package IDE is immediately accessible. The point and click interface allows you to see the equations, their graphical solutions and the physical processes that they model linked together in an interactive illustration. You can change parameters and see the results immediately. The use of the mind-eye link will greatly enhance your learning.

As an articulate programmer, Richard L. Willmore, once said

> *The human mind-eye system, especially that part which responds to things in motion, is by far the highest performance supercomputer that most of us can reasonably expect to use.*

Marginal notes at various points of the text show where one of the 97 tools in IDE is useful for grasping important concepts. At any of these points you can use the website to access immediately any particular tool; then you will be able to explore, without any detailed instruction, the concept in question. There is no syntax or typing—just point and click on a graph or move sliders to choose values. Graphs are linked to moving models and color-coded, so much information is communicated automatically, in less than a minute. You can replay anything as often as you like, with the same or different values of the parameters.

You will also find on the IDE site a laboratory workbook with 31 guided explorations, each designed for one (or sometimes two) 50-minute class periods. These IDE labs are conceptual rather than computational, and worksheets are provided to guide students' experimentation and reflection.

The IDE site gives the text of the laboratory workbook as PDF files; student worksheets can simply be printed out. Students write their answers directly on these worksheets. Sample solutions to the lab questions are given in the instructor's version of the manual (available as hard copy from the publisher with course adoptions), in order to speed grading. When an IDE lab is appropriate to a text section, we list it at the end of the exercise set, just before the suggested journal problem(s).

ODE Software (Dfield and Pplane)

In addition to the text and IDE, you will want access to a graphical solver with which you can enter any differential equation, then see the direction field and the solutions that follow it.

John Polking of Rice University provides our dedicated ODE solver:

- *Dfield*, for a single first-order differential equation, $y' = f(t, y)$;
- *Pplane* for a system of two first-order DEs, or a second-order DE.

These Java applets will enable you to make pictures easily for the qualitative analysis of differential equations. That is the key to really understanding the subject. When you start a problem with a picture, you are probably on the fastest track.

In Summary

Keep in mind that learning these subjects requires your *active* involvement:

- When we ask questions, try to *answer* them;
- *Try* IDE when we suggest it;
- Make *pictures* for differential equations and try to *understand* them;
- Keep a *journal* and work on communication as a major skill.

Prologue

In calculus you studied the derivative, a tool to measure the rate of change of a quantity. When you found that $d(Ce^{kt})/dt = kCe^{kt}$, you had learned that $y(t) = Ce^{kt}$ is a solution of the differential equation

$$y' = ky.$$

Every student of calculus works with differential equations, often without using the term explicitly.

In algebra you solved a system of linear equations like

$$\begin{aligned} 3x - 2y + z &= 7, \\ x + y - z &= -2, \\ x - y + 2z &= 6 \end{aligned}$$

and found that $x = 1$, $y = -1$, $z = 2$ provides the unique answer. Though you may not have known it, you were working with linear algebra ideas, for the coefficients in these equations form an array **A** called a matrix,

$$\mathbf{A} = \begin{bmatrix} 3 & -2 & 1 \\ 1 & 1 & -1 \\ 1 & -1 & 2 \end{bmatrix},$$

and the right-hand sides of the equations and the solutions,

$$\begin{bmatrix} 7 \\ -2 \\ 6 \end{bmatrix} \quad \text{and} \quad \begin{bmatrix} 1 \\ -1 \\ 2 \end{bmatrix},$$

are examples of arrays called vectors. Every student of algebra works with matrices and vectors, objects of linear algebra, even if the term is not used explicitly.

Meteorologist Edward Lorenz developed a simplified weather model in 1973 in an attempt to improve forecasting. Using variables x, y, and z to represent position and temperatures, he used the set of differential equations

$$\begin{aligned} x' &= -\sigma x + \sigma y, \\ y' &= Rx - y - xz, \\ z' &= -bz + xy \end{aligned}$$

to model the jet stream. This system can be written in matrix-vector form as

$$\mathbf{u}' = \mathbf{A}\mathbf{u} + \mathbf{f}(\mathbf{u}),$$

where

$$\mathbf{u} = \begin{bmatrix} x \\ y \\ z \end{bmatrix}, \quad \mathbf{A} = \begin{bmatrix} -\sigma & \sigma & 0 \\ R & -1 & 0 \\ 0 & 0 & -b \end{bmatrix}, \quad \mathbf{f} = \begin{bmatrix} 0 \\ -xz \\ xy \end{bmatrix}.$$

This marriage of differential equations and linear algebra provides a powerful language for studying the population of competing species, the behavior of electrical power networks, the equilibria of chemical, biological and economic systems, and many other complex phenomena of current interest.

Some of these systems exhibit dramatic and catastrophic behavior. Buildings and bridges may collapse as a result of earth tremors. Small local changes in weather may have unexpectedly widespread effects. Overloads in one part of an electric power grid can escalate into a regional blackout. Artificial introduction of a predator into an ecosystem can lead to extinction of another species. Industrial pollutants may wipe out a valuable fishery.

Whether modeling the submicroscopic world of elementary physical particles, the motions among the planets or man-made satellites, or the interaction of medications in the bloodstream, systems of algebraic and differential equations enable us to describe the state of a system and the ways in which it changes. In this text we will develop basic concepts of both linear algebra and differential equations and learn how they support and depend upon one another. Together these partners bring mathematics to bear on the world of practical problems.

Get ready for an exciting adventure!

Differential Equations & Linear Algebra

CHAPTER

1 First-Order Differential Equations

Differential equations—the major interface of mathematics with the real world—are the main tool with which scientists make mathematical models of real systems. That is, differential equations have a central role in connecting the power of mathematics with the description of real phenomena.[1]

—John Hubbard, Cornell University

1.1 Dynamical Systems: Modeling

SYNOPSIS: Models are significant investigative tools in the scientific method. These models may be physical or mathematical, continuous or discrete, stochastic or deterministic, scalar or vector. The original population growth model of Thomas Malthus provides a concrete example.

Models

Many centuries ago, people believed that solar eclipses were caused by a dragon trying to swallow the sun and that earthquakes were caused by disgruntled gods in the underworld. These theories were mental constructs that helped them explain the world around them. Human history is a long record of such interesting visions of reality, illustrating an age-old belief that complex phenomena can be understood by comparison with simpler systems based on limited versions of the real thing. Mythical Chinese dragons and Mayan gods of the underworld were religious models. On the other hand, the ancient Greeks, after observing the heavens for many centuries, constructed a model of the physical universe consisting of a large hollow sphere with the earth at its center and the heavenly bodies moving along various paths on its surface. While this model exhibited reasonably good agreement with the data available at the time, it has since been replaced by a more accurate and comprehensive description of the physical universe. Thus we see that different models can serve to answer different questions in different times and circumstances.

[1] This short description, given in testimony before a congressional committee re scientific research funding, was confirmed by the chemist and physicist also making presentations. The event palpably raised the status of mathematics with the politicians.

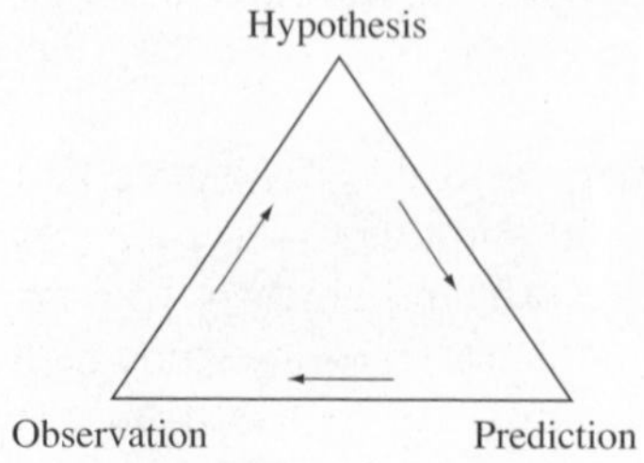

FIGURE 1.1.1 The scientific method: If observation does not verify your prediction, you revise your hypothesis.

Models are the way we understand the world around us. To explain turbulence around an airplane wing, an engineer will talk about the Reynolds number, a physicist about nonlinear resonance, and a mathematician about the stretching and folding of smooth manifolds. Models are a hallmark of the scientific method, that philosophical process by which scientific knowledge is extended and refined. The triangular scheme in Fig. 1.1.1 illustrates the three pillars of the method and the cyclical nature of its evolution. From careful observations, hypotheses are formulated that define the models. Predictions on the basis of the models are then subjected to verification by a new round of observations. If the agreement with observations is imperfect, old hypotheses are modified or new ones formulated, and the process continues. The history of science is a record of continual refinement in the understanding of our world.

A model is not intended to be the "real thing" but represents selected features or aspects of the real thing. Structural engineers build models to simulate enough important properties of bridges, dams, and buildings to predict their performance when stressed by flood or earthquake. Robotics engineers model key actions of joints and muscles.

Types of Models

Continuous
(differential equations)
vs.
Discrete
(iterative equations)

Although modern physics suggests that all systems are ultimately discrete and changes occur in jumps at distinct points in time, the scale of such changes renders them more understandable when treated as changing continuously with time. Thus we treat changing temperature, electric current, and the flow of a fluid as continuous phenomena, not as systems of distinct electrons or molecules. Even a population of fruit flies is treated by the biologist as changing continuously. Most commonly these **continuous-time** systems are modeled by **differential equations**, a major focus of Chapters 1, 2, 4, 6, 7, 8, and 10 of this book.

When the time intervals are longer, changes may be treated as happening in separate jumps: daily, weekly, or yearly variations in economic variables such as stock prices or tax revenues, or ecological variables such as annual populations of animals or plants. For these **discrete-time** or **sampled-data** systems a useful mathematical model is the **iterative equation**, the subject of Chapter 9.

Scalar
(single dependent variable)
vs.
Vector
(multiple dependent variables)

We use **scalar** models when a system is described by a single measurement, such as temperature or pressure, and **vector** models for systems with several varying components, such as latitude and longitude to specify a geographical position. The study of vector or multidimensional systems is facilitated by matrices and other tools of **linear algebra**, the second subject of this book, presented in Chapters 3 and 5.

Dynamical Systems

In this book, we study mathematical models applied to **dynamical systems**: systems that change over time. We will model events such as earthquakes not with underworld gods but with mathematical variables and relationships among them (principally algebraic equations, differential equations, and iterative equations). Our goal may be to understand more completely the system as it is now (How *do* electrons orbit the nucleus of the atom? as Erwin Schrödinger asked) or it may be to predict future states (When *will* California experience "the big one"? Imagine the savings in lives if seismology could have predicted the 2004 tsunami disaster in Southeast Asia.).

The phenomena we study—physical, biological, or social—are found to be in different configurations or **states**, characterized by a set of measurements or observations, and these change or evolve with the passage of time.

Interactive Differential Equations (IDE)

IDE (on the software accompanying this text) contains many models and linked graphs.[2] Graphs and diagrams update as sliders change values; a mouse click activates a blank graph.

The first task is to decide which variables will be used to model the system. Frequently, the refinement process stimulated by the scientific method leads to the addition of variables that were thought (mistakenly) to be insignificant at an earlier stage.

EXAMPLE 1 Cooling Coffee The cup of coffee on your kitchen table is a simple physical system. To understand completely the coffee's interactions with air, cup, or later with your digestive, circulatory, and nervous systems might involve physicists, chemists, biologists, neurologists, or even sociologists and philosophers. But a limited model of some utility can be based on one measurement, the temperature. Using only the temperature, Newton's Law of Cooling (Sec. 2.3) is an example of a differential equation that incorporates the temperature of the surroundings to accurately describe how the coffee temperature changes. ■

Newton's Law of Cooling

The model shows actual temperature data from a cooling cup of coffee. Move sliders for three parameters to fit the model to the data.

The simple temperature model for a cup of coffee becomes serious when applied to calculate how quickly food in a freezer will thaw when the power goes off or how much insulation is needed for architectural spaces in different climates.

Differential Equations and Models

Differential equations relate rates of change to other variables. In many cases the independent variable is time. Systems that change over time are called **dynamical systems**.

Differential Equation

A **differential equation** (DE) is an equation that contains *derivatives* of one or more dependent variables with respect to one or more independent variables.

- An **ordinary differential equation** (ODE) contains only *ordinary* derivatives.
- A **partial differential equation** (PDE) contains *partial* derivatives.

The **order** of a differential equation refers to the highest-order derivative that appears in the equation.

EXAMPLE 2 Classifying Differential Equations

(a) $\dfrac{dy}{dt} = f(t, y)$ is a first-order ODE with independent variable t and dependent variable y.

(b) $\dfrac{d^2y}{dt^2} = f(t, y, y')$ is a second-order ODE with independent variable t and dependent variable y.

[2] Hubert Hohn at the Massachusetts College of Art has a particular talent for using interactive graphics to explain mathematics. Hohn provided the graphics design, programming, and much pedagogical insight for the software package *Interactive Differential Equations* (IDE), developed together with four mathematicians. We will often refer you to IDE and the accompanying exploratory laboratory sessions. The icon in margins and problem sets serves as an alert. IDE can be found at www.aw-bc.com/ide.

(c) $2\frac{d^2y}{dt^2} + y\frac{dy}{dt} + ty^2 = 0$ is also a second-order ODE with independent variable t and dependent variable y.

(d) $\frac{d^5y}{dt^5} - \frac{dy}{dt} = 4yt$ is a fifth-order ODE with independent variable t and dependent variable y.

(e) $\frac{\partial^2 y}{\partial x^2} + \frac{\partial^2 z}{\partial t^2} = xyz$ is a second-order PDE with independent variables x and t and dependent variables y and z. ■

The reason DEs are so important is that often we don't know exactly how variables are related, but we *can* write equations relating their derivatives (rates of change). See the examples that follow.

Constructing Simple First-Order Models

In many instances, we find that the collected data infer that the rate of change of a quantity with respect to time will increase or decrease in a proportionate fashion with some function of the quantity itself and/or other variables. To write a descriptive differential equation, we must include a constant of proportionality k.

EXAMPLE 3 **Constants of Proportionality** Let y be an unknown differentiable function of time. We can express each of the following statements as an equation, using k as a constant of proportionality.

(a) The rate of change of y is *proportional to* y:

$$\frac{dy}{dt} = ky.$$

(b) The rate of change of y is *proportional to* the product of y^2 and t:

$$\frac{dy}{dt} = ky^2t.$$

(c) The rate of change of y is *inversely proportional to* y:

$$\frac{dy}{dt} = \frac{k}{y}.$$

(d) The rate of change of y is *directly proportional to* y^2 and *inversely proportional to* $\sqrt{t}$:

$$\frac{dy}{dt} = k\frac{y^2}{\sqrt{t}}.$$

■

Proportionality:

As in Example 3(d), *proportional to* is sometimes written as *directly proportional to* if the statement indicates both direct and inverse proportionality.

Some Standard First-Order Differential Equation Models

We use first-order differential equations to model many common situations. You are no doubt familiar with at least some of these; they will be discussed in detail in this chapter and the next. The model in Example 4 was first proposed by Malthus in 1798 and is discussed in detail in the next subsection.

EXAMPLE 4 **Exponential Growth** The population P is growing at a rate proportional to the population at any time t:

$$\frac{dP}{dt} = kP, \quad k > 0.$$

■

EXAMPLE 5 **Exponential Decay** Let A be the amount of radioactive material in a sample at any time t. The amount A is decreasing at a rate proportional to the amount at any time t:

$$\frac{dA}{dt} = kP, \quad k < 0.$$

■

Newton's Law of Cooling; Logistic Growth

Use IDE to explore colorful interactive examples of any of these models. As you change the sliders, graphs and diagrams update automatically; a click of the mouse will activate a blank graph.

EXAMPLE 6 **Newton's Law of Cooling or Heating** The rate of change of temperature T of an object is proportional to the difference between the temperature M of the surroundings and the temperature of the object:

$$\frac{dT}{dt} = k(M - T), \quad k > 0.$$

■

EXAMPLE 7 **Logistic Growth** The rate at which a disease is spread (i.e., the rate of increase of the number N of people infected) in a fixed population L is proportional to the product of the number of people infected and the number of people not yet infected:

$$\frac{dN}{dt} = kN(L - N), \quad k > 0.$$

■

EXAMPLE 8 **Voltage Across an Inductor** The voltage drop V is proportional to the rate of change of current I in the inductor:

$$V = L\frac{dI}{dt}.$$

(The proportionality constant in this instance is written as L (instead of k) and called the **inductance**.)

■

The Malthus Model for Population Growth

In 1798 an English clergyman named Thomas Malthus published a paper called "An Essay on the Principles of Population (Growth) as It Affects the Future Improvement of Society."[3] In the paper, Malthus argued that the world's population was growing geometrically ($a, ar, ar^2, \ldots$) but that the world food supply increased only arithmetically ($a, a + d, a + 2d, \ldots$). Because geometric growth outstrips arithmetic growth, he concluded that if there were no reductions due to war and disease, the result would be mass starvation for humanity. Malthus constructed the first mathematical model for population growth; with it he aroused a storm of controversy by aggravating a host of current class, social, and religious issues.

Growth and Decay

Try different values of the rate constant and initial populations. Compare the results.

---

[3] Malthus's paper is included in J. R. Newman, *The World of Mathematics*, vol. 2 (NY: Simon & Schuster, 1956), 1192–1199.

Malthus assumed that the rate of increase of the world's population at any time was proportional to its size at that time (the more people, the more births). If we assume that the world's population $y(t)$ is a continuous function of time t, Malthus's growth principle can be stated by the differential equation

$$\frac{dy}{dt} = ky, \tag{1}$$

where the positive number k is called the **growth** or **rate** constant. (The rate constant is numerically close to the annual percentage growth rate—see Problem 8(a). We will study the rate constant further in Sec. 2.3.) Malthus took $k = 0.03$, corresponding roughly to a 3% annual increase in world population. Malthus's mathematical problem was to determine a function $y(t)$ that satisfied (1) and had the correct value in 1798, represented by the time $t = 0$. Estimating the world population at 0.9 billion, Malthus needed to solve the initial-value problem

Verification:
If $y(t) = 0.9e^{0.03t}$, then

$$\begin{aligned} y'(t) &= 0.027e^{0.03t} \\ &= 0.03\left(0.9e^{0.03t}\right) \\ &= 0.03y(t). \end{aligned}$$

At $t = 0$,

$$y(0) = 0.9e^0 = 0.9.$$

$$\frac{dy}{dt} = 0.03y, \quad y(0) = 0.9, \tag{2}$$

where y is in units of billions of persons.

We will define differential equation and initial-value problem in more detail in Sec. 1.2, but you can verify at once that the exponential function

$$y(t) = 0.9e^{0.03t} \tag{3}$$

satisfies both conditions in (2). This is the Malthusian population prediction.[4] In Fig. 1.1.2 we see how this exponentially growing function will ultimately outstrip any linear food supply curve, however steep.

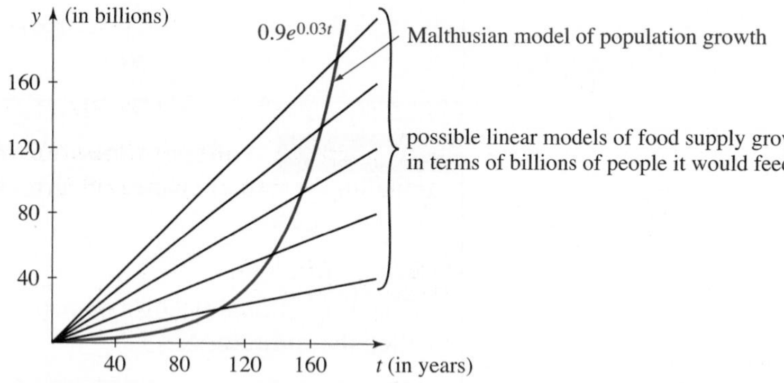

FIGURE 1.1.2 Malthusian population and food growth curves.

How well did the Malthus model work? The data in Table 1.1.1 allow us to compare his predictions with what has actually happened. (The table takes $t = 0$ at 1800 rather than 1798.)

We can see from the data that the Malthus model overestimates the population from the very first decade. After 200 years the prediction is 60 times too high. We must conclude that the Malthusian model oversimplified the real-world system. See Problems 8–10 for some considerations of why. Later in the book we will study some more realistic population models.

---

[4]The original Malthus model is used today only for very simple situations; many better methods are available to forecast populations of everything from whales in the north Pacific to the number of persons in the world with AIDS. We will return to population models in Problems 8–12 and in Secs. 2.5 and 2.6.

Table 1.1.1 Comparison of Malthus model $y(t) = 0.9e^{0.03t}$ and actual world population (in billions).

Year	t	Malthus	Actual	Year	t	Malthus	Actual
1800	0	0.90	0.9	1910	110	24.42	1.8
1810	10	1.21	0.9	1920	120	32.98	1.9
1820	20	1.64	1.0	1930	130	44.52	2.1
1830	30	2.21	1.0	1940	140	60.10	2.3
1840	40	2.99	1.1	1950	150	81.13	2.7
1850	50	4.03	1.2	1960	160	109.53	3.0
1860	60	5.45	1.3	1970	170	147.87	3.5
1870	70	7.35	1.4	1980	180	199.62	4.2
1880	80	9.93	1.5	1990	190	269.49	5.1
1890	90	13.40	1.6	2000	200	363.81	6.0
1900	100	18.09	1.7				

Source: *The Universal Almanac*, 2000.

While the rest of this text is primarily devoted to how to *solve* various equations, you will also gradually build up experience and modeling skills to create appropriate equations for real-world situations.

Higher-Order Differential Equation Models

In later chapters, we will explore higher-order models in two forms: DEs of higher order (Chapter 4) and systems of first-order DEs (Chapter 6 and 7).

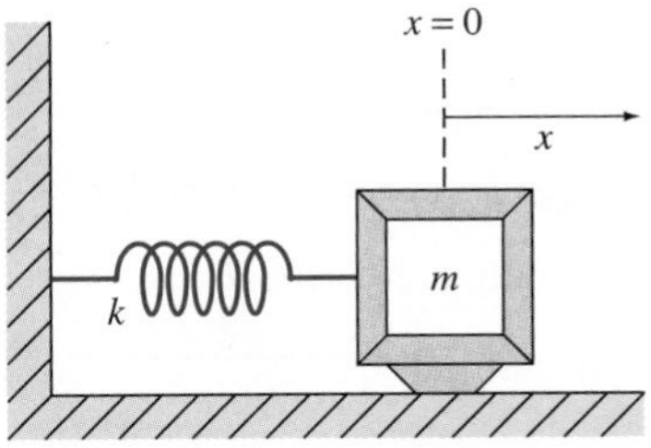

FIGURE 1.1.3 Mass–spring system.

EXAMPLE 9 Hooke's Law The restoring force on a spring, as shown in Fig. 1.1.3, is proportional to the displacement x but opposite in direction:

$$F_{\text{res}} = -kx, \quad k > 0.$$

If friction is negligible, we can assume Newton's First Law of Motion and write

$$m\frac{d^2x}{dt} = -kx. \tag{4}$$

■

EXAMPLE 10 Hooke's Law as a System If we substitute $dx/dt = y$ into the second-order equation (4), we can convert it to an equivalent *system* of first-order equations:

$$\frac{dx}{dt} = y,$$
$$\frac{dy}{dt} = -\frac{k}{m}x.$$

■

Models Using Computer Software

Although the advent of easy-to-use computer software has allowed both the investigation and illustration of more complicated models, many systems were investigated before these tools were available, such as the following.

- During World War I, Italian mathematician Vito Volterra used predator–prey models (multivariable systems of differential equations, for two or more populations) to predict fish populations in the Adriatic Sea. (See Sec. 2.6.)
- The motion of the pendulum and the oscillations of mass–spring system have been studied extensively using analytic methods, some of which will be explored in Chapters 4, 6, and 7.

The following examples are computer models from IDE, the software that accompanies this text.

The Glider

Choose an initial velocity v and angle θ by a click on the graph and watch the path of the glider. Then change the drag coefficient D and repeat to see what changes.

EXAMPLE 11 **Glider** If you have ever played with a balsa-wood glider, you know it flies in a wavy path if you throw it gently and does loop-the-loops if you throw it hard.[5] This quirky behavior is modeled by differential equations from physics and aeronautical dynamics, as illustrated in Fig. 1.1.4.

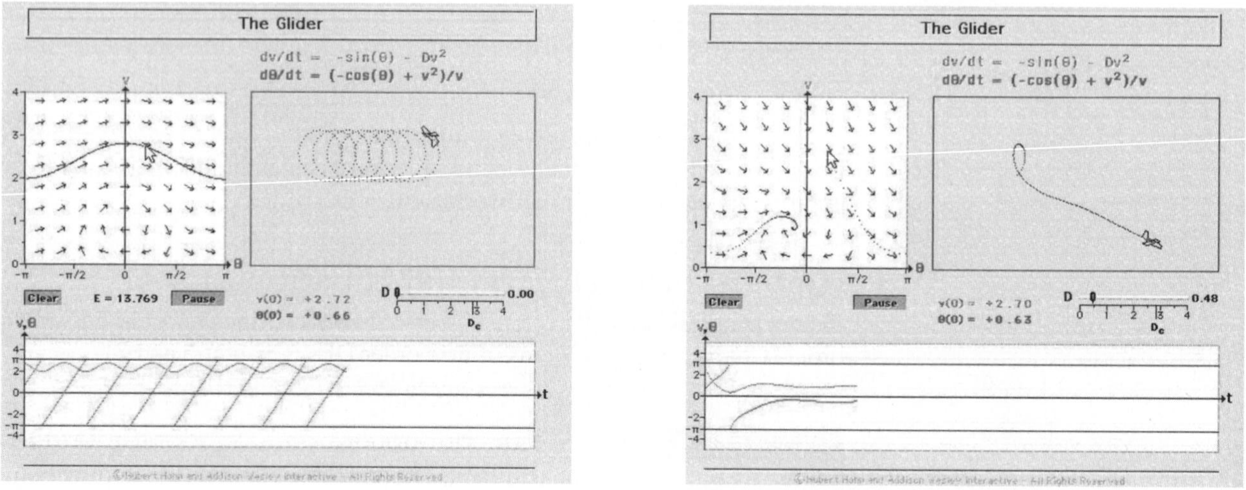

FIGURE 1.1.4 Comparing two paths, with the same initial velocity v and angle θ, but different drag coefficients D.

■

Chemical Oscillator

See the chemical solution change color when you click to choose initial quantities $x(0)$ and $y(0)$ for the two substances being combined.

EXAMPLE 12 **Chemical Oscillator** A revolutionary discovery of the twentieth century was that certain chemical substances (e.g., chlorine dioxide and iodine-malonic acid) can be combined to produce solutions with concentrations that not only change colors over time, but that oscillate between two colors over extended time periods.[6] A system of two differential equations provides a model, illustrated in Fig. 1.1.5. Models for other chemical systems appear in the problems for Section 7.5.

---

[5] Steven Strogatz, Cornell University Professor of Theoretical and Applied Mechanics, suggested the model of the toy glider and provided the analysis given in IDE Lab 19.

[6] Lengyel, Rabai, and Epstein published this result (*Journal of the American Chemical Society*, **112** (1990), 9104). The appropriate differential equations model is discussed in IDE Lab 25 and illustrated by IDE's Chemical Oscillator Tool.

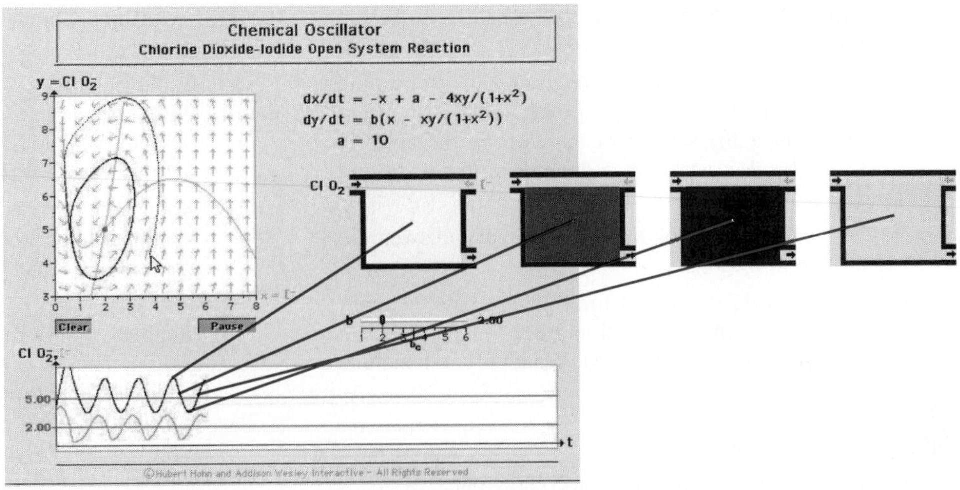

FIGURE 1.1.5 When chlorine dioxide and iodine-malonic acid are combined, the color of the chemical solution actually oscillates over time.

The Glider and Chemical Oscillator examples give a glimpse of what lies ahead as you learn from this text. For now, do not worry about where the differential equations come from, as those answers will evolve as you progress through the text; just observe that the effects of the equations can be graphed and that parameters can be easily changed to observe how the graphs respond. This visual evidence is an enormous aid to understanding a model and using it to predict consequences. Both examples are included in the interactive graphics software package *Interactive Differential Equations* accompanying this text, at www.aw-bc/ide. Try them out to get more of a feel for the action.

Summary

Mathematical modeling allows a scientific approach to solving real-life problems. The Malthus population model gives an example of creating a model, as well as the need to revise after observation fails to match predictions sufficiently.

1.1 Problems

(Problems with blue numbers have brief answers at the back of the book.)

Constants of Proportionality *Write first-order differential equations that express the situations in Problems 1–5.*

1. Let A be the amount of alcohol in someone's bloodstream at time t. The absorption of alcohol by the body is proportional to the amount remaining in the bloodstream at time t.

2. Let A be the amount of radioactive carbon in a fossilized bone. The decay of radioactive carbon in the bone is proportional to the amount present at time t.

3. The increase in the number of people who have heard a rumor in a town of population 20,000 is proportional to the product of the number P who have heard the rumor and the number who have not heard the rumor at time t.

4. A business student invests his money in a risky stock. He finds that his rate of return (dA/dt) is directly proportional to the amount A he has at time t and is inversely proportional to the square root of time t.

5. A student in an engineering course finds the rate of increase of his grade point average G is directly proportional to the number N of study hours/week and inversely proportional to the amount A of time spent on online games.

6. **A Walking Model** Construct a mathematical model to estimate how long it will take you to walk to the store for groceries. Suppose you have measured the distance as 1 mile and you estimate your walking speed as 3 miles per hour. If it takes you 20 minutes to get to the store, what do you conclude about your model?

7. A Falling Model Suppose you drop a ball from the top of a building that is 100 feet tall.

(a) Construct a mathematical model to estimate how long it takes the ball to reach the ground. HINT: An object falling near the surface of the earth in the absence of air friction accelerates downward at the rate of $g = 32.2$ ft/sec^2.

(b) Use calculus to solve the model and answer the question.

(c) After solving the model for your estimate, suppose you actually drop the ball and discover it takes 2.6 seconds to reach the ground. What do you conclude from this result?

8. The Malthus Rate Constant k

(a) If t is time measured in years, show that $k = 0.03$ results in an annual increase (or annual percentage growth rate) of approximately 3% per year, using equation (3).

(b) Plot the world population data in Table 1.1.1 on the same graph as the Malthus model.

(c) How might the Malthus exponential population model be modified to better fit the actual population data in Table 1.1.1? Make an argument for why the exponential function is not unreasonable for these data and identify what must change to fit the given data. This is a qualitative question rather than a quantitative one, so you need not find exact numbers.

9. Population Update World population, now in excess of 6 billion persons, is augmented by 3 persons per second. (In each 2-second period, 9 babies are born, but only 3 persons die.) The annual growth rate was estimated in the 1990s at 1.7%.

(a) Does this agree with the world population predictions by the United Nations in 2004? See Table 1.1.2, and set $t = 0$ in the year 2000. Explain fully.

(b) Calculate the appropriate annual growth rate for each 10-year period given in Table 1.1.2.

Table 1.1.2

Year	2004 Prediction
2000	6,056,000,000
2010	6,843,000,000
2020	7,578,000,000
2030	8,199,000,000

*Source: www.un.org/popin/

10. The Malthus Model Reread the description of the Malthus population model and then discuss the following questions.

(a) With respect to time span, what assumptions did Malthus make in his model of population versus food supply? What conclusions?

(b) Argue whether it is reasonable to extend indefinitely Malthus's exponential population model, as adjusted in Problems 8 and 9. Where or when might it break down?

(c) Malthus's linear food supply models are not elaborated in the text discussion. Argue whether you think any single linear formula could be expected to extend indefinitely. What innovations might cause any food supply graph to change direction or shape as time goes on? What limitations do you see? HINT: Use the following IDE tool.

Logistic Growth
IDE Lab 3 (parts 1, 2, 4) shows successive refinements of the Malthus population model, first to include logistic growth (as in Verhulst's model) and then to add harvesting.

(d) Might Malthus's population model be used to estimate the future numbers of other species of plants and animals?

11. Discrete-Time Malthus Suppose y_n represents the population of the world the nth year after 1800. That is, y_0 is the population in 1800, y_1 is the population in 1801, and so on. From the definition of the derivative as the limit of a difference quotient, we can write

$$\frac{dy}{dt} \approx \frac{y(t+1) - y(t)}{1} = y(t+1) - y(t),$$

so Malthus's differential equation model $dy/dt = 0.03y$ can be *approximated* by $y(t+1) - y(t) = 0.03y$, that is, $y(t+1) = y(t) + 0.03y(t)$. Thus writing y_n for $y(n)$, we obtain[7]

$$y_{n+1} = 1.03y_n. \tag{5}$$

(a) Use this discrete model (5) to estimate the population in the years 1801, 1802, 1803, ..., 1810.

(b) Estimate the world's population in the year 1900 using this discrete model. You might do this with a spreadsheet, using equation (5), or you might develop an algebraic formula by observing a pattern.

(c) Comment on the difference in results comparing this discrete process with Malthus's continuous model in Table 1.1.1. Explain.

12. Verhulst Model In 1840, Belgian demographer Pierre Verhulst modified the Malthus model by proposing that

[7]Equation (5) is called a *recursion* formula or *iterative* formula. Such discrete dynamical systems will be studied in some detail in Chapter 9.

the rate of growth dy/dt of the world's population $y(t)$ should be proportional to $ky - cy^2$; thus

$$\frac{dy}{dt} = ky - cy^2,$$

where both k and c are positive constants. Why do you think such a model may be feasible? What possible interpretations can you give for the term $-cy^2$? HINT: Factor the right-hand side of the differential equation. Such refinements of population growth models will be further explored in Sec. 2.5.

13. **Suggested Journal Entry** Please refer to the compelling argument (in the section "To the Reader") on the benefits of keeping a journal and considering what you have learned from each section. A good way to begin is to expand on one of the models from the text, examples, or problems (with or without an equation), or to do the same for some model you have encountered in another field. Explain your variable(s) and sketch some likely graphs for each with respect to time. Explain in words these graphs and their relationships. Consider and state strengths, weaknesses, and limitations of your rough model.

1.2 Solutions and Direction Fields: Qualitative Analysis

SYNOPSIS: What exactly is a differential equation? What is a solution? How do we approach finding a solution? Since explicit solutions of differential equations and initial-value problems are often unobtainable, we explore methods of finding properties of solutions from the differential equation itself; the principal tool is the geometry of the direction field. We call this approach qualitative analysis.

What Is a Differential Equation?

A **differential equation** (or DE) is an equation containing derivatives; the **order** of the equation refers to the highest-order derivative that occurs. In this chapter we focus on first-order equations that can be written as

$$\frac{dy}{dt} = f(t, y) \quad \text{or} \quad y' = f(t, y). \tag{1}$$

The independent variable t suggests the time scale in dynamical models. The dependent variable y stands for an unknown function constituting a **solution** of the differential equation.

A solution is a function that must satisfy the DE for all values of t. For each DE we are asking:

- Is there any function that satisfies equation (1)?
- Is there more than one function that satisfies equation (1)?

What Is a Solution?

Analytic Definition of Solution

Analytically, $y(t)$ is a **solution** of differential equation (1) if substituting $y(t)$ for y reduces the equation to an identity

$$y'(t) \equiv f(t, y(t))$$

on an appropriate domain for t (such as $(0, \infty)$, $[0, 1)$, or all real numbers).

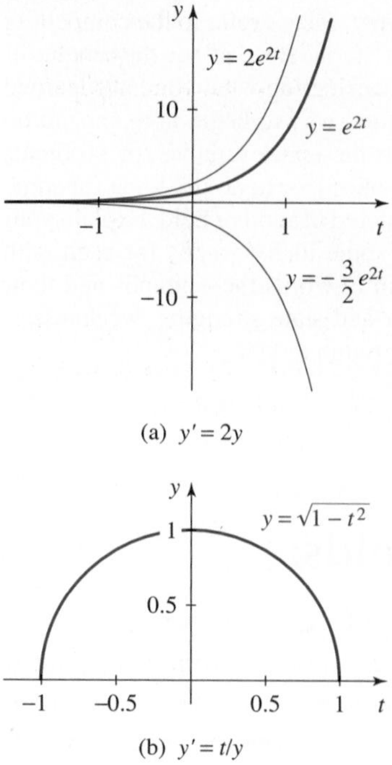

FIGURE 1.2.1 Solution curves for the DEs of Example 1.

EXAMPLE 1 **Solutions of DEs** Verify for each DE that the given $y(t)$ is a solution. (These solutions are graphed in Fig. 1.2.1; you are asked to find other possible solutions in Problems 9 and 10 at the end of the section.)

(a) $y' = 2y$, $y = e^{2t}$. Substituting e^{2t} for y in the DE gives

$$y'(t) = \frac{d}{dt}e^{2t} = 2e^{2t} = 2(e^{2t}) = 2y(t) = f(t, y(t)),$$

true for all t. Thus y is a solution on $(-\infty, \infty)$. Substitution will also show that $2e^{2t}$ and $-3e^{2t}/2$ are solutions. Can you find other solutions for this DE?

(b) $y' = -t/y$, $y = \sqrt{1 - t^2}$. Substitution yields the identity

$$y'(t) = \frac{d}{dt}\left(\sqrt{1 - t^2}\right) = \frac{1}{2}\left(1 - t^2\right)^{-\frac{1}{2}}(-2t) = \frac{-t}{\sqrt{1 - t^2}} = -\frac{t}{y},$$

an identity valid on the interval $(-1, 1)$. Can you find other solutions? ■

Most differential equations have an infinite number of solutions. The simplest differential equation is the one studied in calculus,

$$\frac{dy}{dt} = f(t), \tag{2}$$

where the right-hand side depends only on the independent variable t. The solution is found by "integrating" each side, obtaining

$$y = \int f(t)\,dt + c, \tag{3}$$

where c is an arbitrary constant and $\int f(t)\,dt$ denotes any *antiderivative* of f (that is, any function $F(t)$ such that $F'(t) = f(t)$). The result (3) means that we actually have not one function but many, one for each choice of c. We say that (3) is a **family** of solutions with parameter[1] c.

In general, all solutions of a first-order DE form a *family* of solutions expressed with a single parameter c. Such a family is called the **general solution**. A member of the family that results from a specific value of c is called a **particular solution**.

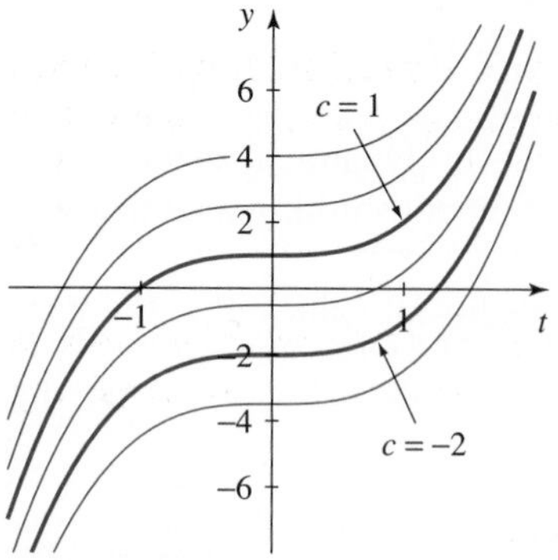

FIGURE 1.2.2 Several solutions $y = t^3 + c$ of $y' = 3t^2$, the DE of Example 2. Two particular solutions are highlighted.

EXAMPLE 2 **Family of Solutions** The *general* solution of $y' = 3t^2$ is

$$y = t^3 + c,$$

where c may be any real value. The *particular* solutions $y = t^3 + 1$ and $y = t^3 - 2$ result from setting $c = 1$ and -2, respectively. Figure 1.2.2 shows the graphs of several members of the general solution of $y' = 3t^2$. ■

Initial-Value Problems

In many models, we will be looking for a solution to the (more general) differential equation that has a specified y-value y_0 at a given time t_0. We call such a specified point an **initial value** by considering that the solution "starts off" from (t_0, y_0).

[1]Parameters represent quantities in mathematics that share properties with variables (the things that change) and constants (the things that do not), and might be called "variable constants." Their nature depends on the context. The "c" in $y = t^3 + c$ is a parameter, whereas y and t are the variables. Another descriptive phrase for parameter that is close to the Greek root of the word is "auxiliary variable."

Initial-Value Problem (IVP)

The combination of a first-order differential equation and an **initial condition**

$$\frac{dy}{dt} = f(t, y), \quad y(t_0) = y_0$$

is called an **initial-value problem**. Its solution will pass through (t_0, y_0).

While a DE generally has a *family* of solutions, an IVP usually has only one. The appropriate value of c for this *particular* solution is found by substituting the coordinates of the specified initial condition into the general solution.

EXAMPLE 3 **Adding an Initial Condition** The function $y(t) = t^3 + 1$ is a solution of the IVP

$$y' = 3t^2, \quad y(0) = 1.$$

Differentiating $y(t)$ confirms that $y'(t) = 3t^2$, and $y(0) = 0^3 + 1 = 1$.

In Fig. 1.2.2 the graph of $y(t) = t^3 + 1$ passes through the point (0, 1); it is the only curve of the family that does. ■

EXAMPLE 4 **Another IVP** Recall the Malthusian population problem

$$\frac{dy}{dt} = 0.03y, \quad y(0) = 0.9,$$

from Section 1.1 (equation (2)). We can see that

$$y(t) = ce^{0.03t} \tag{4}$$

is a one-parameter family of solutions since, for any c,

$$y'(t) = 0.03ce^{0.03t} = 0.03y(t).$$

Substituting the initial condition $y(0) = 0.9$ into the general solution (4) gives $ce^{0.03 \cdot 0} = 0.9$, which implies that $c = 0.9$; the particular solution of the IVP is

$$y(t) = 0.9e^{0.03t}.$$

■

The most common understanding of the phrase *solving a differential equation* is obtaining an *explicit* formula for $y(t)$. However, *solving* may also mean obtaining

- an *implicit* equation relating y and t,
- a *power series representation* for $y(t)$, or
- an appropriate *numerical approximation* to $y(t)$.

More informally, solving may refer to studying a *geometrical representation*, the main emphasis of this section. It is useful first, though, to see how explicit formula solutions work.

Qualitative Analysis

Historically, the main approach to the study of differential equations was the **quantitative** one: the attempt to obtain explicit formulas or power series representations for the solution functions. This notion dominated the thinking of the seventeenth and eighteenth centuries, and a great deal of such work was carried out by Isaac Newton, Gottfried Leibniz, Leonhard Euler, and Joseph Lagrange. Unfortunately, it is relatively rare to be able to find explicit formulas for solutions, not because of lack of skill on the part of researchers but because the family of **elementary functions** (those obtained from algebraic, exponential, logarithmic, and trigonometric functions) is simply too limited to express solutions of most equations we want to solve.[2] Finding power series representations has limitations as well. Even when these classical solutions can be obtained, they often fail to provide much insight into the process being modeled.

Around 1880, the French mathematician Henri Poincaré, working on problems in celestial mechanics, started thinking about investigating the behavior of solutions in a new way.[3] His approach, now called the **qualitative** theory of differential equations, sets aside the search for analytic solutions and studies *properties* of solutions directly from the differential equation itself. In this way one can often demonstrate the existence of constant solutions or of periodic solutions, as well as learn about long-term behavior—limiting values, rates of growth and decay, and chaotic oscillations. While explicit solutions are not ignored when they can be found, the insights obtained by qualitative methods may actually be more valuable.

Direction Fields

The direction field is the most basic and useful tool of qualitative analysis for first-order differential equations. It gives a picture of the family of solutions as a whole, *without ever trying to solve the differential equation explicitly*. We redefine solution as follows:

Slope Fields; Solutions

For different DEs, explore the slopes at different points; click to "stick" a slope line segment. Make your own direction field. Then see how carefully solutions follow the direction field.

Graphical Definition of Solution

A **solution** to a first-order differential equation is a function whose *slope* at each point is specified by the derivative.

The first-order DE $y' = f(t, y)$ provides a formula for the slope y' of the solution at any point (t, y) in the ty-plane. To see what such solution curves look like, we calculate a large number of slopes; a convenient scheme is to use points spaced at regular intervals in the t- and y-directions. Through each of these points we draw a short segment having the slope calculated for that point. A solution curve through such a point will be tangent to the segment there. The collection of segments of proper slope is called a **direction field** (or a **slope field**) for the differential equation.

EXAMPLE 5 **An Informal Example** The differential equation $y' = t - y$ gives a recipe for calculating a slope y' at any point in the ty-plane by substituting its coordinates into the right-hand side of the DE. At the point $A = (2, 1)$, for example, $y' = 2 - 1 = 1$; at point $B = (0.5, 2)$, $y' = 0.5 - 2 = -1.5$.

---

[2]For example, solutions of the relatively simple *pendulum equation* $\theta'' + \frac{g}{L}\sin\theta = 0$ must be expressed in terms of a sophisticated class of functions known as *elliptic integrals*.

[3]J. Henri Poincaré (1854–1912) did fundamental work in celestial mechanics, differential equations, and topology. It has been said that he was the last of the *universal mathematicians*, someone who made major contributions in every area of mathematics. With more and more research being carried out today, it has become impossible for any one person, however brilliant, to be at the forefront of all areas of mathematical research.

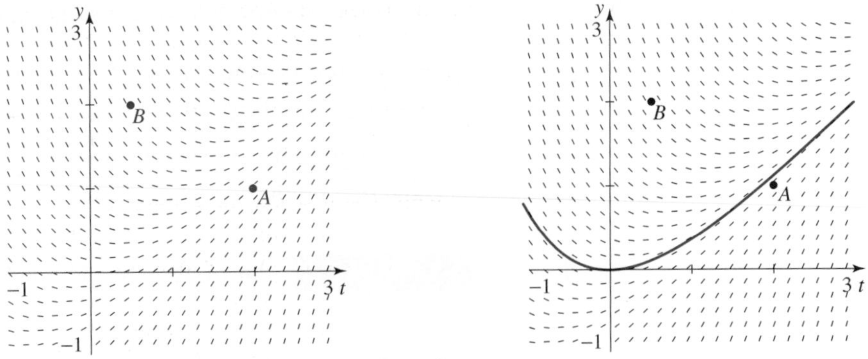

FIGURE 1.2.3 Direction field and solution curve for $y' = t - y$.

Figure 1.2.3 shows a direction field for $y' = t - y$ and one of its solution curves (which includes neither A nor B, but passes between them). From the limited window of Fig. 1.2.3, we might draw several tentative conclusions:

- Solutions seem to be defined for all t (no vertical asymptotes).
- Solutions appear to tend toward the solution we found through (0, 0), but also toward a line with slope 1 as $t \to \infty$.
- Solution curves seem to have an "exponential shape" with one possible exception. (See Problem 11.)
- Solutions appear to tend to ∞ as $t \to \infty$.
- Some solutions (those through points above the line with slope 1 and y-intercept -1) appear to tend to ∞ as $t \to -\infty$ as well.

■

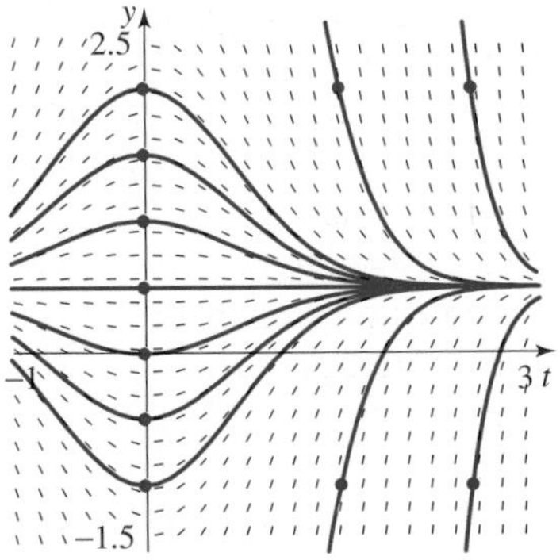

(a) $y' = -2ty + t$

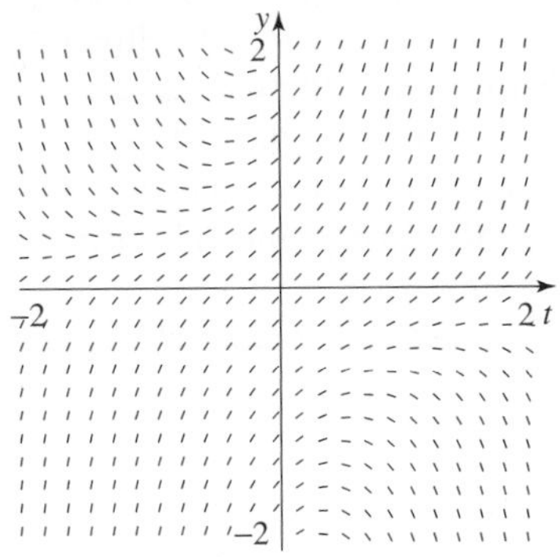

(b) $y' = 2ty + 1$

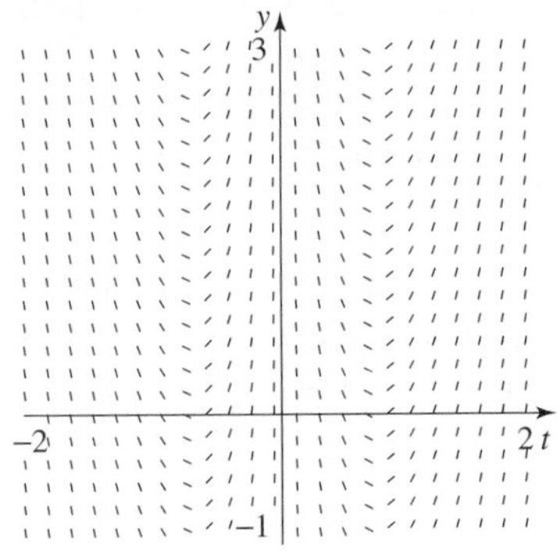

(c) $y' = -\frac{2}{t} + 4t$

FIGURE 1.2.4 Some typical direction fields. Sketch some sample solutions in (b) and (c) to follow the slope marks.

Many computer programs will draw slope fields and solutions for any DE you can enter.[4] Here we want to just focus on concepts, so you probably do not need to call on technology just yet.

Just looking at the direction field immediately gives a feeling for how the solution curves will flow, and can set you on a fast track to understanding the solutions to a given DE. Figure 1.2.4 illustrates several direction fields. We provide a sample set of solutions in (a), for a representative set of initial conditions (dots). Sketch some solutions in (b) and (c), carefully following the direction field. Notice where they converge or diverge, whether and where they approach infinity, and whether they are periodic.

A useful fact to remember from calculus is that if $y' = f(t, y)$, then, by the chain rule,

$$y'' = \frac{d}{dt} f(t, y) = \frac{\partial f}{\partial t}\frac{dt}{dt} + \frac{\partial f}{\partial y}\frac{dy}{dt} = \frac{\partial f}{\partial t} + y'\frac{\partial f}{\partial y},$$

so you can pursue questions of **concavity** of solutions without even solving the DE! (See Problems 15–17.)

[4]On our web site, www.prenhall.com/farlow, we provide an ODE Solver consisting of DFIELD and PPLANE by John Polking of Rice University. (See "To the Reader.")

Recall from calculus that, at any point on a curve,

- $y'' > 0$ means concave up;
- $y'' < 0$ means concave down;
- $y'' = 0$ gives no information about concavity. This result may occur at an inflection point.

EXAMPLE 6 **Concavity** Consider again the equation of Example 5,

$$y' = t - y,$$

for which

$$y'' = 1 - y' = 1 - t + y.$$

We see that

$$\begin{aligned} y'' &= 0 \text{ when } y = t - 1; \\ y'' &> 0 \text{ when } y > t - 1; \\ y'' &< 0 \text{ when } y < t - 1. \end{aligned}$$

Thus solutions are concave up above the line $y = t - 1$, and solutions are concave down below $y = t - 1$. Check these observations against the direction field and solutions drawn in Fig. 1.2.3. By substitution in the DE we can see that in this particular example the line $y = t - 1$ is a particular solution and observe that every other solution lies entirely above or below that line. Our calculations confirm that these solutions have *no* inflection points, because the only points where $y'' = 0$ are not on any other solutions. ■

Equilibria

Equilibrium

For a differential equation, a solution that does not change over time is called an **equilibrium solution**.

For a first-order DE $y' = f(t, y)$, an equilibrium solution is always a horizontal line $y(t) \equiv C$, which can be obtained by setting $y' = 0$.

The symbol $\equiv$ is read as "identically equal" or "always equal."

An equilibrium solution $y(t) \equiv C$ has zero slope. Thus, an easy way to find an equilibrium solution is to set $y'(t) = 0$; any solutions to that algebraic equation that do not depend on t are equilibria.

EXAMPLE 7 **Finding Equilibrium** For the DE

$$y' = -2ty + t,$$

you can *see* (Fig. 1.2.4(a)) the constant solution

$$y(t) \equiv \frac{1}{2}$$

To check this out, we solve $y' = 0$, which confirms that $y(t) \equiv 1/2$ is a solution, and in fact the *only* constant solution for this DE. ■

An important question to ask about any equilibrium solution is: Is it stable?

Stability

For a differential equation $y' = f(t, y)$, an equilibrium solution $y(t) \equiv C$ is called

- **stable** if solutions near it tend toward it as $t \to \infty$;
- **unstable** if solutions near it tend away from it as $t \to \infty$.

Slope Fields

Learn to spot stable and unstable equilibria for a variety of DEs.

Sometimes an equilibrium solution is **semistable**, which means stable on one side and unstable on the other.

EXAMPLE 8 **Observation** The direction field for $y' = y^2 - 4$ drawn in Fig. 1.2.5(a) can help us to discover some general properties of solutions, shown in Fig. 1.2.5(b).

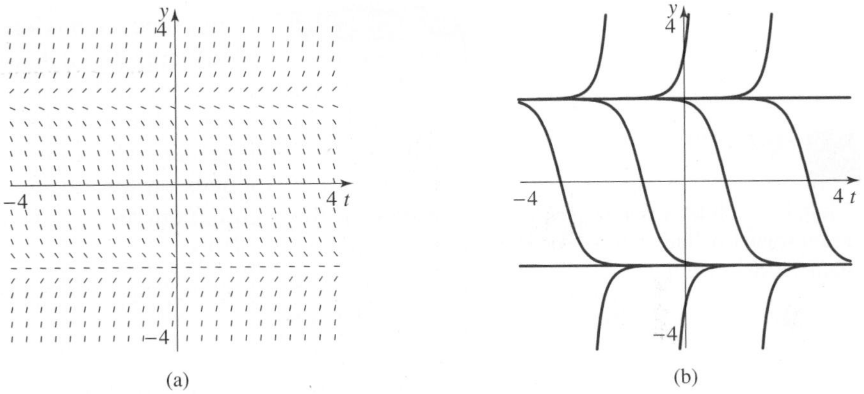

FIGURE 1.2.5 Direction field and some solutions for $y' = y^2 - 4$.

- There are two constant (or **equilibrium**) solutions, $y = -2$ and $y = 2$.
- Solution $y = -2$ is **stable**: solutions near it tend toward it as $t \to \infty$; they "funnel" together.
- Solution $y = 2$ is **unstable**: solutions near it tend away from it, either toward $y = -2$ or toward $+\infty$; these solutions spray apart, the opposite of a funnel.
- The concavity of solutions changes at $y = -2$, $y = 0$, and $y = 2$.
- As $t \to -\infty$, stability of the equilibrium solutions is "reversed."

Can you notice any other useful qualitative properties from the direction field? ■

Isoclines

So far we have plotted the direction field elements at points on a grid. An alternative scheme, useful for plotting fields by hand, is the **method of isoclines**.

Isocline

An **isocline** of a differential equation $y' = f(t, y)$ is a curve in the ty-plane along which the slope is constant. In other words, it is the set of all points (t, y) where the slope has the value c, and is therefore the graph of $f(t, y) = c$.

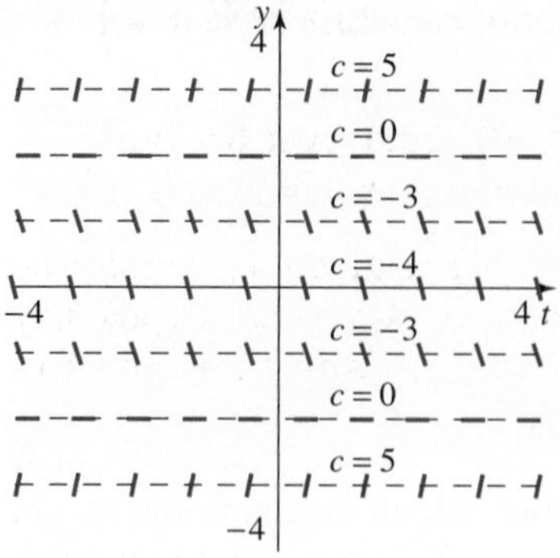

FIGURE 1.2.6 Some isoclines, $y' = c$, and some solutions for $y' = y^2 - 4$ (Example 9).

Once an isocline is determined, all the line elements for points along the curve have the same slope c, making it easy to plot many of them quickly.

EXAMPLE 9 **Simplest Isoclines** For the differential equation $y' = y^2 - 4$ in Example 8, the isoclines have the form $y^2 - 4 = c$ or $y = \pm\sqrt{c+4}$, and are horizontal lines. Even though the isoclines aren't shown in Fig. 1.2.5, the pattern of equal slopes along horizontal lines is easy to see; we have added the isoclines explicitly in Fig. 1.2.6. ■

Isoclines are a handy *guide* to the slopes of solutions for making a quick sketch when you don't have graph paper or computer at hand. Isoclines seldom coincide with a solution, so you don't want to confuse them. Draw them faintly, dashed or in a different color from solutions. In some DEs, isoclines provide a more direct route to analysis of solutions. The following is a favorite example.

Isoclines

Using only a slider and a mouse, sketch some isoclines and *see* how the solutions cross them.

EXAMPLE 10 **Long-Term Behaviors** The direction field for $y' = y^2 - t$ is plotted in Fig. 1.2.7(a) using the method of isoclines. The isoclines are parabolas with equations $c = y^2 - t$, or $t = y^2 - c$, shown for $c = -3, -2, \ldots, 3$.

Try sketching some solutions in the direction field of Fig. 1.2.7(a). Compare with the computer-generated solutions shown in Fig. 1.2.7(b). Try sketching the parabolic isoclines on top of Fig. 1.2.7(b) to see how they line up with points on the solution curves that have the same slopes. (Notice the shorter horizontal axis in (b).)

What observations can you make about these solutions (which actually can never be expressed in "closed form," i.e., as formulas[5])? Your list may include such characteristics as the following:

- There are no constant solutions.
- In forward time (as t gets larger), many solutions seem to "funnel" together and approach $-\infty$ close to the parabola $y = -\sqrt{t}$, while others fly off to $+\infty$ (and may even have vertical asymptotes).

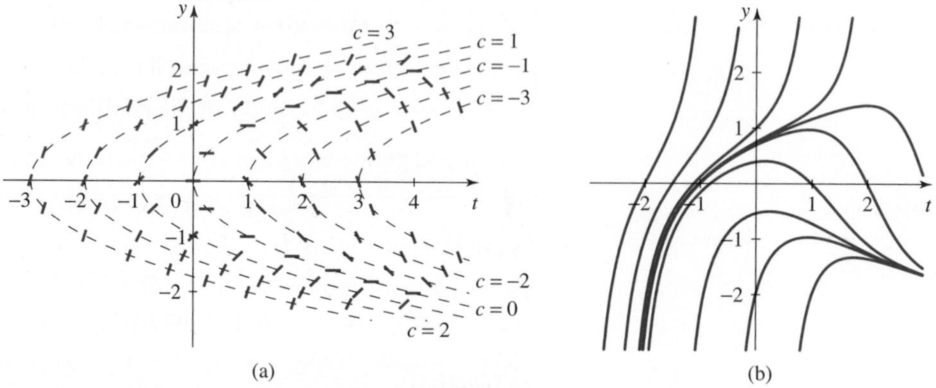

FIGURE 1.2.7 Direction field drawn from isoclines and some solutions for $y' = y^2 - t$.

[5]The proof of this fact, involving differentiable algebra, is far beyond the scope of this course or, in fact, most undergraduate mathematics courses.

- There is a splitting of these two behaviors in forward time close to the parabola $y = +\sqrt{t}$.
- All solutions seem to go to $-\infty$ in backward time, and these backward solutions could also have vertical asymptotes.

Each of the preceding assertions about $y' = y^2 - t$ turns out to be true and can be proven rigorously with qualitative arguments that yield surprisingly quantitative information.[6] Furthermore, this is a case in which isoclines give the clearest information from the direction field. ■

No visual aid is more useful than the direction field for gaining an overview of a first-order differential equation. Although sometimes tedious to plot by hand, direction fields can be produced efficiently by graphing calculator or computer. However, in small-scale calculator or computer pictures, you cannot distinguish small differences in slopes—that is when the ability to add equilibria and isoclines makes your pictures far more valuable.

Not all questions can be answered definitively in all cases from just a direction field. Furthermore, a direction field can mislead if it misses key points. Nevertheless, such a picture can indeed answer many questions and provide guidance on those that remain. We list some questions that a well-drawn direction field may help to answer.

Direction Field Checklist for $y' = f(t, y)$

1. Is the field *well defined* at all points of the *ty*-plane?
2. Does there appear to be a *unique* solution curve passing through each point of the plane?
3. Are there *equilibrium* (constant) solutions? Are such solutions *stable* (nearby solutions are attracted), *unstable* (nearby solutions are repelled), or neither?
4. What is the *concavity* of solutions?
5. Do any solutions appear to "blow up"? That is, do you suspect any *vertical asymptotes*?
6. What is the pattern of the *isoclines*? Do they help visualize behavior?
7. Are there any *periodic* solutions?
8. What is the *long-term behavior* of solutions as $t \to \infty$? as $t \to -\infty$?
9. Does the field have any *symmetries*? What do they tell you about solutions?
10. Can you give any useful *overall description* of the field, with words alone, to someone who has not seen it?

1. *Well defined* means that $f(t, y)$ exists as a single-valued function at the point in question.

2. *Unique* means only one; detailed discussion will be given in Sec. 1.5.

3–9. To explore and clarify these items, see Problems 12–58.

10. Imagine composing a purely verbal e-mail or holding a telephone conversation.

You might consider each of the items in the checklist in relation to Examples 9 and 11, where we answered some but not all of these questions. Try your hand at the others.

[6] A complete reference for these methods and proofs is J. H. Hubbard and B. H. West, *Differential Equations: A Dynamical Systems Approach, Part 1* (TAM 5, NY: Springer-Verlag, 1989), Chapter 1; a subset appears in J. H. Hubbard, "What It Means to Understand a Differential Equation," *College Mathematics Journal* **25**, no. 5 (1994), 380–384.

Summary

Qualitative analysis, based on direction fields and the information they yield about the behavior of solutions to differential equations, provides a significant overview to understanding DEs, especially for the majority of such equations that do not submit to formula solutions. Slopes, equilibria and isoclines are key concepts that we will use throughout the text.

1.2 Problems

Verification *For each differential equation, verify by differentiation and substitution that the given function is a solution.*

1. $y' = y^2 + 4 \quad (|t| < \pi/4); \qquad y = 2\tan 2t$
2. $y' = \dfrac{1}{t}y + t \quad (t > 0); \qquad y = 3t + t^2$
3. $y' = \dfrac{2y}{t} + t \quad (t > 0); \qquad y = t^2 \ln t$
4. $y' - 4ty = 1; \qquad y = \displaystyle\int_0^t e^{-2(s^2-t^2)}\,ds$

IVPs *In Problems 5 and 6 verify that the given function satisfies both the differential equation and the initial condition.*

5. $y' + 3y = e^{-t}, \quad y(0) = -1/2; \quad y = e^{-t}/2 - e^{-3t}$
6. $y' = 2y + 1 - 2t^2, \quad y(0) = 2; \quad y = t + t^2 + 2e^{2t}$

Applying Initial Conditions

7. Verify that $y = ce^{t^2}$ is a solution, for any real c, of $y' = 2ty$. Determine c so that $y(0) = 2$.
8. Verify that the function $y = e^t \cos t + ce^t$ is a solution, for every real c, of the DE $y' - y = -e^t \sin t$. Determine c so that $y(0) = -1$.

Using the Direction Field *For the DEs in Problems 9–11, use the corresponding direction fields to draw some solutions. Try to give the general solutions as formulas, then substitute your guesses into the DE to see if you got them right. Where you did not, explain why.*

9. $y' = 2y$

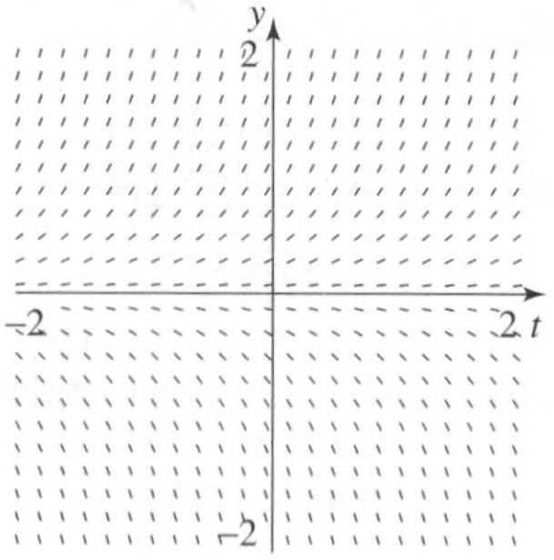

10. $y' = -t/y$

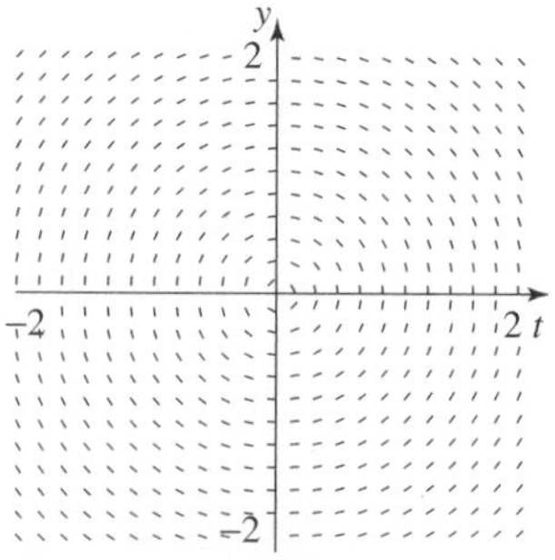

11. $y' = t - y$

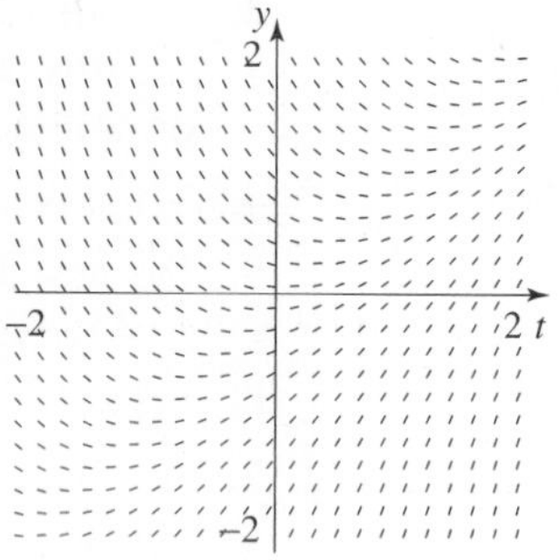

12. **Linear Solution** From the direction field for $y' = t - y$ in Fig. 1.2.3, can you determine the equation of the straight-line solution of the DE? Verify your conjecture.

Stability *For Problems 13–15, sketch the direction fields, then identify the constant or equilibrium solutions and give their stability.*

13. $y' = 1 - y$
14. $y' = y(y + 1)$
15. $y' = t^2(1 - y^2)$

Match Game *Consider the DEs of Problems 16–21.*

(a) *Match each DE with its corresponding direction field in Fig. 1.2.8.*

(b) *State your reasons. What characteristic(s) caused you to make each match?*

16. $y' = 1$ **17.** $y' = y$ **18.** $y' = y/t$

19. $y' = t^2$ **20.** $y' = t^2 + y^2$ **21.** $y' = 1/t$

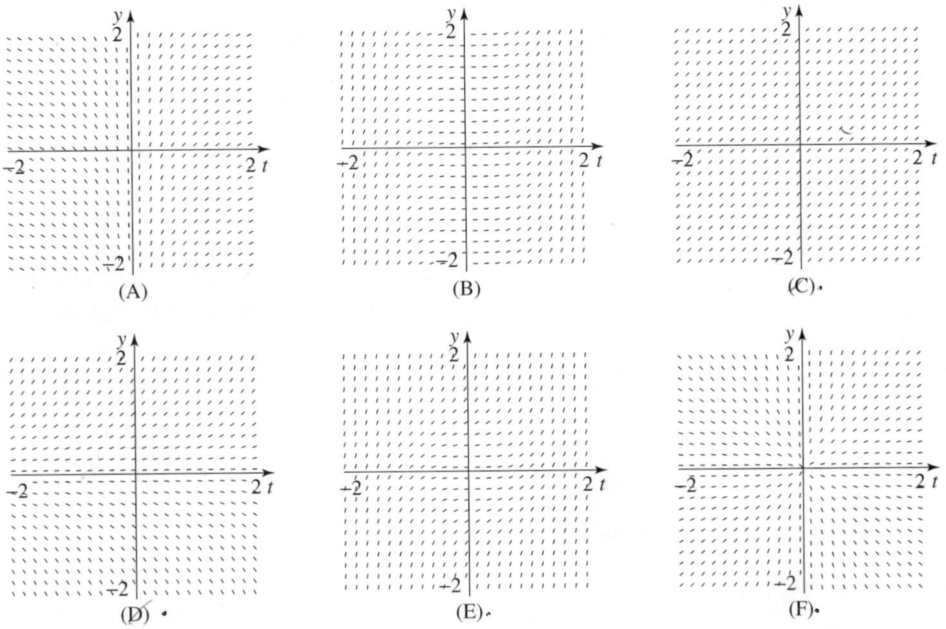

FIGURE 1.2.8 Direction fields that match the equations of Problems 16–21.

Concavity *For Problems 22–24, given with direction fields, calculate y″, determine concavity of solutions and find inflection points. Then sketch some solutions and shade the regions with solutions concave down.*

HINT: *See Figs. 1.2.5, 1.2.7.*

22. $y' = y^2 - 4$

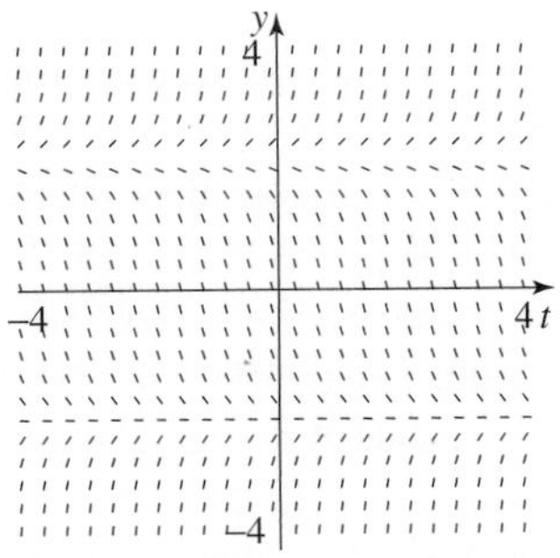

23. $y' = y + t^2$

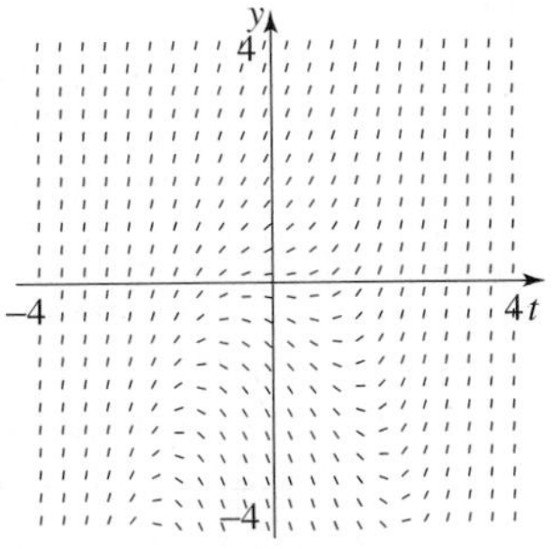

24. $y' = y^2 - t$

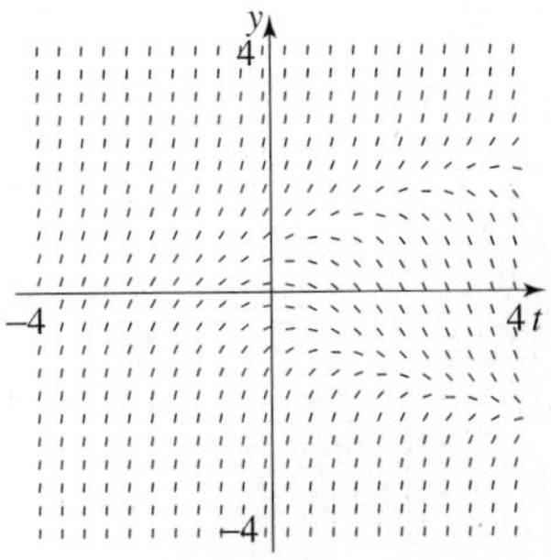

Asymptotes *For each of the DEs in Problems 25–30, tell whether and where you expect vertical asymptotes or oblique asymptotes. Use the direction field to support your case, and state any other arguments. (You are* not *expected to solve the DEs, but you should sketch direction fields for Problems 25–27 sufficiently to answer the asymptote question. For Problems 28–30 direction fields are given.*

25. $y' = y^2$ **26.** $y' = \frac{1}{ty}$ **27.** $y' = t^2$

28. $y' = 2t + y$

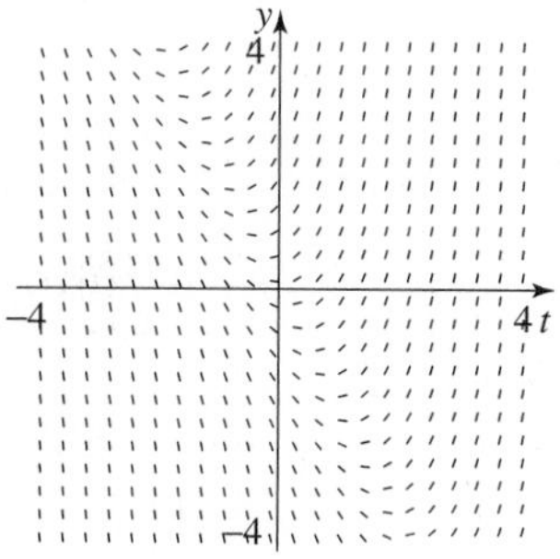

29. $y' = -2ty + t$

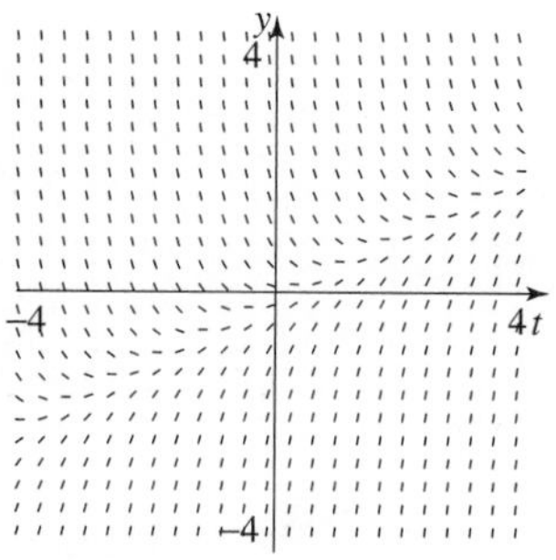

30. $y' = \frac{ty}{t^2 - 1}$

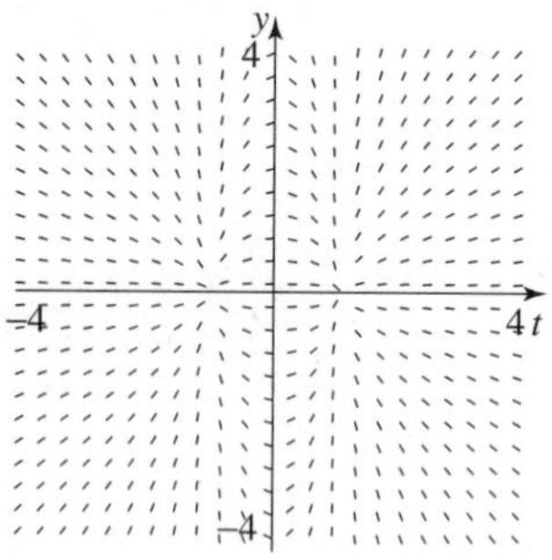

Isoclines *For each of the following DEs in Problems 31–39, lightly sketch by hand several isoclines (it's often helpful to start with $c = 0$, $c = \pm 1$, $c = \pm 2$) and from these sketch a direction field with sample solutions drawn darker.*

Isoclines
Try some examples to get started.

31. $y' = t$ **32.** $y' = -y$

33. $y' = y^2$ **34.** $y' = -ty$

35. $y' = 2t - y$ **36.** $y' = y^2 - t$

37. $y' = \cos y$ **38.** $y' = \sin t$

39. $y' = \cos(y - t)$

Periodicity *Some first-order DEs have periodic solutions or solutions that tend toward an oscillation. For the DEs in Problems 40–46, use isoclines to sketch a direction field and describe whatever sort of periodicity you anticipate in solutions.*

40. $y' = \cos 10t$ **41.** $y' = 2 - \sin t$

42. $y' = -\cos y$ **43.** $y' = \cos 10t + 0.2$

44. $y' = \cos(y - t)$ **45.** $y' = y(\cos t - y)$

46. $y' = \sin 2t + \cos t$

Symmetry *For $y' = f(t, y)$ in Problems 47–52, do the following:*

(a) *Sketch the direction fields and identify visual symmetries.*

(b) *Conjecture how these graphical symmetries relate to algebraic symmetries in $f(t, y)$.*

47. $y' = y^2$ **48.** $y' = t^2$

49. $y' = -t$ **50.** $y' = -y$

51. $y' = \frac{1}{(t+1)^2}$ **52.** $y' = \frac{y^2}{t}$

53. Second-Order Equation Consider the second-order linear differential equation $y'' - y' - 2y = 0$.

(a) Verify that $y = e^{2t}$ is a solution; then check that $y = e^{-t}$ is a solution as well.

(b) Verify that $y = Ae^{2t}$ and $y = e^{2t} + e^{-t}$ are both solutions, where A is any real constant.

(c) Verify that for any constants A and B, a solution is $y = Ae^{2t} + Be^{-t}$.

(d) Determine values for A and B so that the solution of part (c) satisfies both $y(0) = 2$ and $y'(0) = -5$.

Long-Term Behavior *Answer the following questions for the differential equations of Problems 54–59, using the direction fields given as a guide.*

(a) *Are there any constant solutions? If so, what are they?*

(b) *Are there points at which the DE is not defined? How do solutions behave near these points?*

(c) *Are there any straight line solutions? If so, what are they?*

(d) *What can be said about the concavity of solutions?*

(e) *What is the long-term forward behavior of solutions as* $t \to \infty$?

(f) *Where do solutions come from? Look at the long-term backward behavior of solutions as* $t \to -\infty$.

(g) *Do you see asymptotes or periodic solutions?*

54. $y' = t + y$

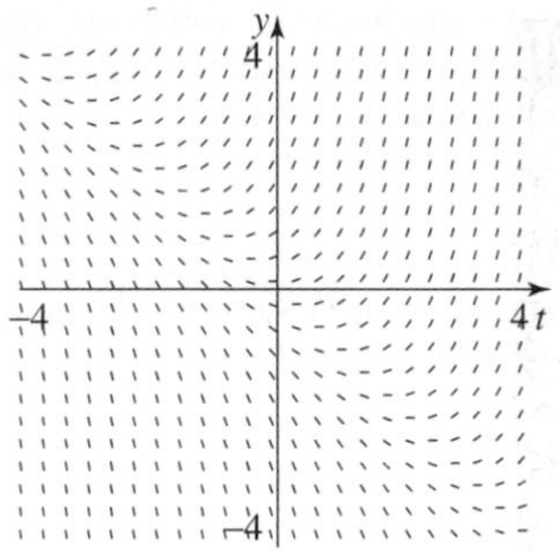

55. $y' = \dfrac{y - t}{y + t}$

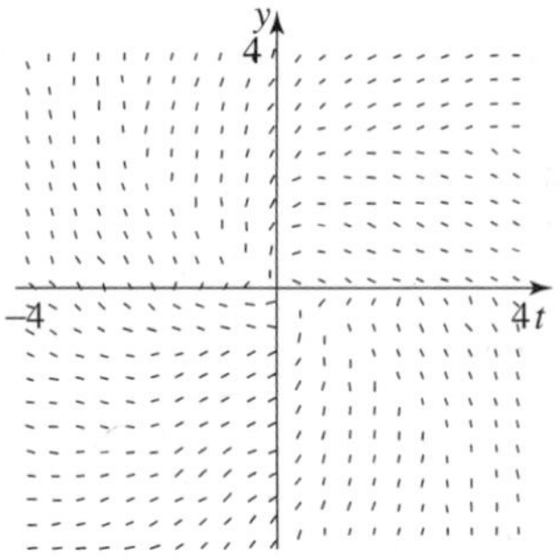

56. $y' = \dfrac{1}{ty}$

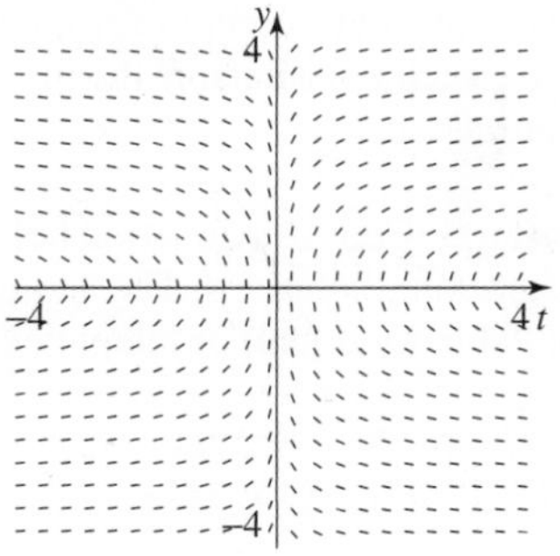

57. $y' = \dfrac{1}{t - y}$

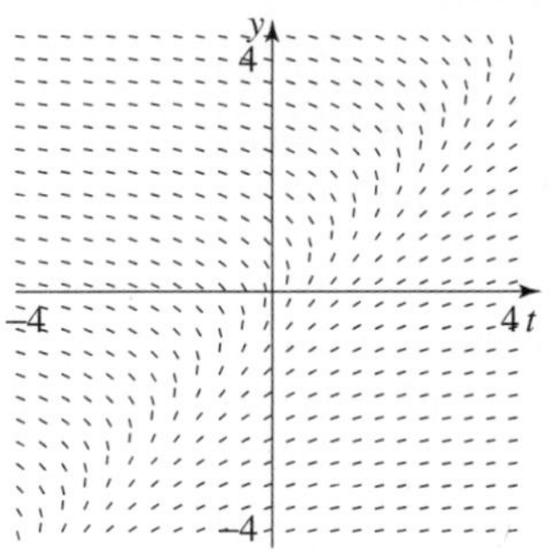

58. $y' = \dfrac{1}{t^2 - y}$

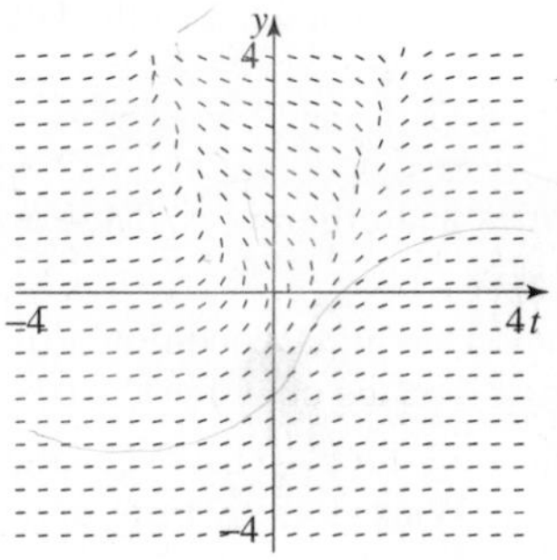

59. $y' = \dfrac{y^2}{t} - 1$

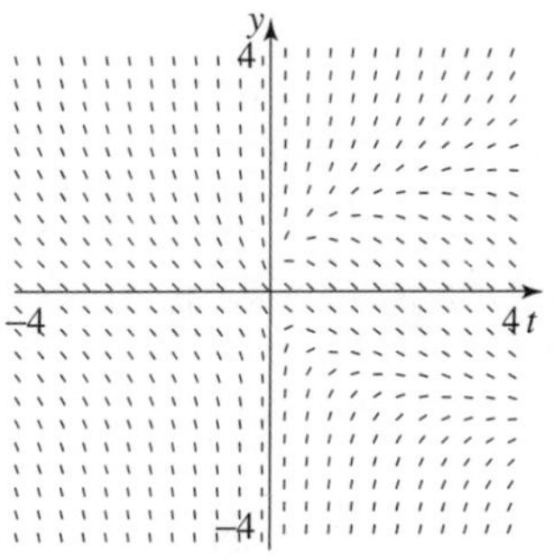

60. Logistic Population Model The simplest case of the logistic model is represented by the DE

$$\frac{dy}{dt} = ky(1 - y),$$

where $k > 0$ is the growth rate constant. Draw a direction field for this equation when $k = 1$. Find the constant solutions. Explain why this model represents limited growth. What happens in the long run (that is, as $t \to \infty$)?

Logistic Growth
Watch what happens in the long run.

61. Autonomy

> **Autonomous Equations**
> When a first-order DE has the form $y' = f(y)$, so the right-hand side doesn't depend on t, the equation is called **autonomous** (which *means* independent of time).

The logistic equation $y' = ky(1 - y)$ in Problem 60 and the equation $y' = y^2 - 4$ in Example 9 are examples of autonomous equations.

(a) List those DEs in Problems 9–17 and 31–39 that are autonomous.

(b) What is the distinguishing property of isoclines for autonomous equations?

62. Comparison Explore the direction fields of the DEs

$$y' = y^2, \quad y' = (y+1)^2, \quad \text{and} \quad y' = y^2 + 1.$$

Describe their similarities and differences. Then answer the following questions:

(a) Suppose each equation has initial condition $y(0) = 1$. Is one solution larger than the other for $t > 0$?

(b) You can verify that $y = 1/(1 - t)$ satisfies the IVP $y' = y^2$, $y(0) = 1$. What does this say about the solution of $y' = y^2 + 1$, $y(0) = 1$?

Coloring Basins

> **Basin of Attraction**
> Suppose $y \equiv c$ is an equilibrium or constant solution of the first-order DE $y' = f(y)$. Its **basin of attraction** is the set of initial conditions (t, y_0) for which solutions tend to c as $t \to \infty$.

An example of shading a basin is shown in Fig. 1.2.9.

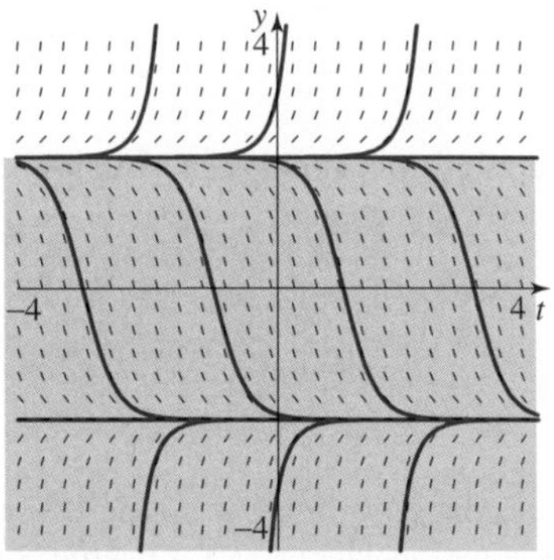

FIGURE 1.2.9 For $y' = y^2 - 4$, the shaded area is the basin of attraction for the stable equilibrium solution $y = -2$.

Determine the basin of attraction for each constant solution of the autonomous equations in Problems 63–66. That is, sketch a direction field and highlight the equilibrium solutions. For each, color the portion of the plane from which solutions are attracted to that equilibrium.

63. $y' = y(1 - y)$

64. $y' = y^2 - 4$

65. $y' = y(y - 1)(y - 2)$

66. $y' = (1 - y)^2$

Computer or Calculator *For the DEs in Problems 67–74, use appropriate software to draw direction fields. Then discuss what you can deduce about their solutions by sketching some representative solutions following the direction field. Include such features as constant and periodic solutions, special solutions to which other solutions tend, and regions in which solutions fly apart from each other.*

> **Slope Fields**
> Direction fields are at your command for some examples quite like these.

67. $y' = y/2$

68. $y' = 2y + t$

69. $y' = ty$

70. $y' = y^2 + t$

71. $y' = \cos 2t$

72. $y' = \sin(ty)$

73. $y' = -\sin y$

74. $y' = 2y + t$

75. Suggested Journal Entry I Discuss the kinds of information about solutions of a DE that may be discovered from studying its direction field. What alternative information is available if you also have explicit formulas for the solutions? If you have such a formula, is the direction field still useful? Explain.

76. Suggested Journal Entry II Choose a direction field that you have made and sketch some solutions. Then compose a verbal description, suitable for e-mail or telephone.

1.3 Separation of Variables: Quantitative Analysis

SYNOPSIS: We study a special class of first-order equations for which $y' = f(t)g(y)$; such equations can be solved analytically by separation of variables using elementary integration. We explore some of the limitations of this quantitative method.

Qualitative-First Rule

A good practice when studying a differential equation is to investigate its qualitative properties as much as possible before trying to get an analytical or numerical solution. The direction field usually gives insight into the behavior of solutions, especially the constant solutions (corresponding to **equilibrium** states) and solutions near them. Many physical systems modeled by differential equations "spend most of their time at or near equilibrium states."

Separable DEs

A **separable** differential equation is one that can be written $y' = f(t)g(y)$.
Constant solutions $y \equiv c$ can be found by solving $g(y) = 0$.

Solving by Separation of Variables: Informal Example

Working informally with differentials, we can quickly solve the differential equation $y' = 3t^2(1 + y)$. We find first that $y(t) \equiv -1$ is a constant (equilibrium) solution. We also see that it is an unstable equilibrium solution, because nearby solutions move away from it. To find the nonconstant solutions, we write

$$\frac{dy}{dt} = 3t^2(1 + y).$$

Divide both sides of the equation by $1+y$ (assuming for the moment that $y \neq -1$), and multiply by dt to get

$$\frac{dy}{1 + y} = 3t^2 dt.$$

Integrating each side,[1] we have

$$\int \frac{dy}{1 + y} = \int 3t^2\, dt,$$

$$\ln |1 + y| = t^3 + c.$$

Exponentiating, we have a family of solutions

$$|1 + y| = e^c e^{t^3},$$

or

$$y = -1 \pm e^c e^{t^3}$$
$$= -1 + ke^{t^3}, \quad k \neq 0.$$

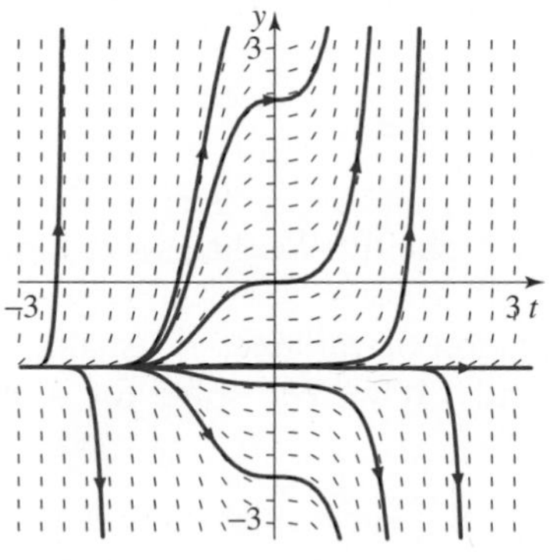

FIGURE 1.3.1 Direction field and solutions to $y' = 3t^2(1 + y)$.

We added the arbitrary constant c in the integration step. The exponential $\pm e^c$ can never be zero, but can be entirely replaced in the solution by a simpler arbitrary constant $k \neq 0$. In this example, since we showed that $y \equiv -1$ is also a solution, we can allow k to be *any* real number. (See Fig. 1.3.1.)

[1] If you are uncomfortable with any of these steps, you can verify the solution by differentiation and substitution as in Sec. 1.2. We will give a formal justification later using the chain rule.

Quantitative Methods

The sad state of affairs in differential equations is that most differential equations do not have nice, explicit formulas for their solutions. The solution of differential equations involves evaluation of indefinite integrals, and every student of calculus knows this isn't automatic. Just remember examples like $\int e^{x^2}\,dx$ and $\int \sin x^2\,dx$.

Slope Fields

Find the separable equations and take a look at their solutions.

There is, however, one important class of first-order differential equations that can be reduced to the evaluation of integrals: those that are *separable*.[2] But the necessary integrations may not be possible. In addition, the result may be an implicit relation between y and t instead of a formula for y in terms of t. Even so, this is an important type of equation, and the **separation of variables** method we illustrated is very useful.[3] And it all works because of the *chain rule!*

Separation of Variables: Why Does It Work?

Suppose $G(y)$ is an antiderivative of $1/g(y)$ and $F(t)$ is an antiderivative of $f(t)$. Suppose also that $y = y(t)$ is a function defined **implicitly** by

$$G(y) = F(t) + c$$

on some t-interval for some real constant c. This means that for appropriate t-values,

$$G(y(t)) = F(t) + c. \tag{1}$$

Then, if y is a differentiable function, we can differentiate equation (1) with respect to t by the chain rule:

Chain Rule:

$$\frac{dG}{dt} = \frac{dG}{dy}\frac{dy}{dt}$$

$$G'(y(t))y'(t) = F'(t).$$

But G and F are antiderivatives of $1/g$ and f, so this can be written as

$$\frac{y'(t)}{g(y(t))} = f(t).$$

Multiplied out, this says that

$$y'(t) = f(t)g(y(t)),$$

which just means that $y(t)$ is a solution of the differential equation $y' = f(t)g(y)$. This explains why our earlier example worked. We wrote $y' = f(t)g(y)$ as $dy/g(y) = f(t)\,dt$, calculated

$$\int \frac{dy}{g(y)} = \int f(t)\,dt,$$

and the result was equation (1), defining solutions implicitly. We then solved for y in terms of t.

So the result of applying the method of separation of variables is usually a one-parameter family of solutions defined implicitly. If we *can* solve for y in terms of t, we usually do.

---

[2]Of course, equations are also separable if they have the form $y' = f(t)/h(y)$, which can be written $y' = f(t)[1/h(y)]$.

[3]As one might suspect, the method goes back almost to the beginning of calculus. Gottfried Leibniz (1646–1716) used it implicitly when he solved the *inverse problem of tangents* in 1691. It was Johann Bernoulli, however, who formulated it explicitly in 1694 and named it in a letter to Leibniz.

Separation of Variables Method for $y' = f(t)g(y)$

Step 1. Set $g(y) = 0$ and solve for equilibrium solutions, if any.

Step 2. Now assume that $g(y) \neq 0$. Rewrite the equation in separated or differential form:

$$\frac{dy}{g(y)} = f(t)\,dt.$$

Step 3. Integrate each side (if integrable):

$$\int \frac{dy}{g(y)} = \int f(t)\,dt + c$$

(obtaining an implicit one-parameter family of solutions).

Step 4. If possible, solve for y in terms of t, getting the explicit solution $y = y(t)$.

Step 5. If you have an IVP, use the initial condition to evaluate c.

EXAMPLE 1 **Separability** Differential equations (a), (b), and (c) are separable. Equation (d) is *not* separable: it resists all efforts to segregate its variables. (What goes wrong when you *try* to separate part (d)?)

(a) $\dfrac{dy}{dt} = -\dfrac{t}{y} \quad \Rightarrow \quad y\,dy = -t\,dt$

(b) $\dfrac{dy}{dt} = t^2 y \quad \Rightarrow \quad \dfrac{1}{y}\,dy = t^2 dt$

(c) $\dfrac{dy}{dt} = y + 1 \quad \Rightarrow \quad \dfrac{1}{y+1}\,dy = dt$

(d) $\dfrac{dy}{dt} = t + y$ (not separable) ■

EXAMPLE 2 **Implicit Solutions** The separable differential equation

$$y' = \frac{t^2}{1 - y^2}$$

is defined for all $y \neq \pm 1$. (What effect does this have on solutions?)

There are no equilibrium solutions because $g(y) = 1/(1 - y^2) = 0$ has no solutions.

Rewrite the equation as

$$(1 - y^2)\,dy = t^2 dt$$

and integrate to obtain

$$y - \frac{y^3}{3} = \frac{t^3}{3} + c.$$

Multiplying through by 3 and letting $3c = k$, we obtain

$$-t^3 + 3y - y^3 = k$$

for the family of implicitly defined solutions.

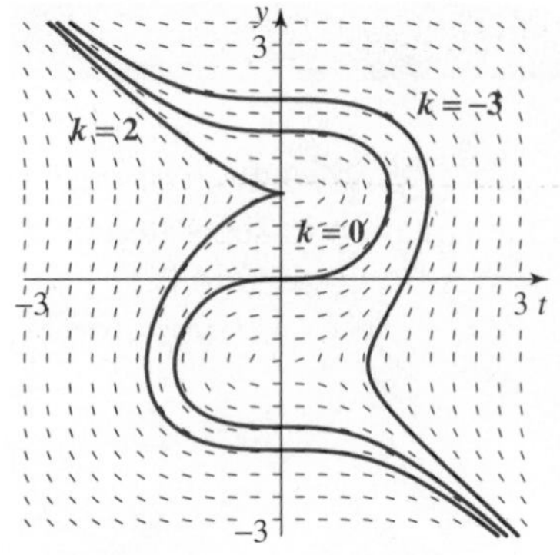

FIGURE 1.3.2 Three curves formed by solutions for $y' = t^2/(1-y^2)$ (Example 2).

The direction field for the equation and solution curves for $k = -3$, $k = 0$, and $k = 2$ are shown in Fig. 1.3.2. Each solution curve is not a single function $y(t)$ but a piecewise combination of several functions. A particular solution to an IVP for this DE would only be *one* of these functions.

Explicit solutions of these cubics for y in terms of t are not feasible, so we can be grateful for the computer drawing. (Section 1.4 will explain how the computer does it.) ■

EXAMPLE 3 Separable IVP Solve the initial-value problem

$$\frac{dy}{dt} = \frac{3t^2+1}{1+2y}, \qquad y(0) = 1. \tag{2}$$

The DE is separable, but there are no equilibrium solutions. Points where $y = -1/2$ must be excluded.

The separation leads to

$$(1+2y)\,dy = (3t^2+1)\,dt.$$

The family of solutions is defined implicitly by

$$y + y^2 = t^3 + t + c.$$

If the curve is to satisfy the initial condition, we must have $y = 1$ when $t = 0$; that is, $1 + 1^2 = 0^3 + 0 + c$, so $c = 2$. The resulting equation,

$$y + y^2 = t^3 + t + 2, \tag{3}$$

can be solved for y by the quadratic formula, giving the *two* solutions

$$y = \frac{1}{2}\left[-1 \pm \sqrt{4t^3 + 4t + 9}\right].$$

Figure 1.3.3 shows them as two branches of the curve given by (3). The (colored) branch with the positive sign contains (0, 1), so that is the solution to the IVP (2). The point where $y = -1/2$ separates the two implicitly defined functions. (Its t-coordinate is approximately -1.06.) At this point the tangent line is vertical and the slope undefined. ■

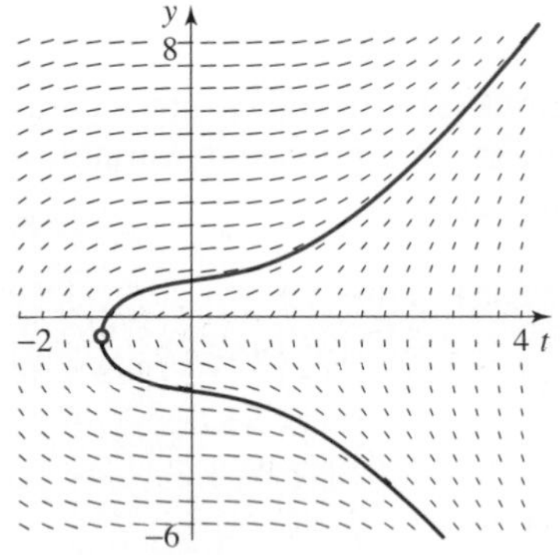

FIGURE 1.3.3 Solution curves for $y' = (3t^2+1)/(1+2y)$, $y(0) = 1$ (Example 3).

EXAMPLE 4 Moving in Circles Solve the IVP

$$y' = -\frac{t}{y}, \qquad y(0) = 1.$$

This separable equation has no equilibrium solutions. The direction field strongly suggests circular solution curves.

Separating variables confirms this. We write

$$y\,dy = -t\,dt.$$

Integrating gives the implicit equation

$$\frac{y^2}{2} = -\frac{t^2}{2} + c$$

for the solution curves.

To have $y = 1$ when $t = 0$ requires that $c = 1/2$, and the corresponding solution curve is the unit circle $t^2 + y^2 = 1$. There are two explicit solutions $y = \pm\sqrt{1-t^2}$; the one with the positive square root contains the initial point (0, 1). Thus the solution to the IVP is a *semi*circle, as shown in Fig. 1.3.4. ■

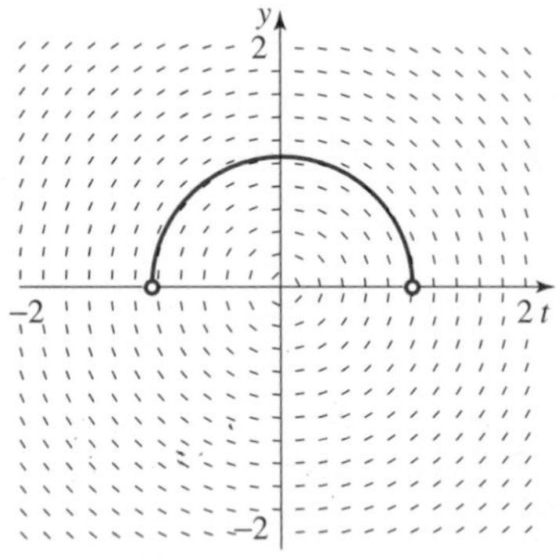

FIGURE 1.3.4 Direction field for $y' = -t/y$, with solution to the IVP of Example 4.

EXAMPLE 5 **The Bell Shape** The initial-value problem

$$y' = -2ty, \qquad y(0) = 1 \tag{4}$$

can be solved by separation of variables to give the general solution of the DE as $y = ke^{-t^2}$. From the initial condition $y(0) = 1$, we determine that $k = 1$, so the IVP solution is simply $y = e^{-t^2}$. This solution is graphed in Fig. 1.3.5. Students who have had statistics will recognize this curve as a multiple of the **normal distribution curve**.

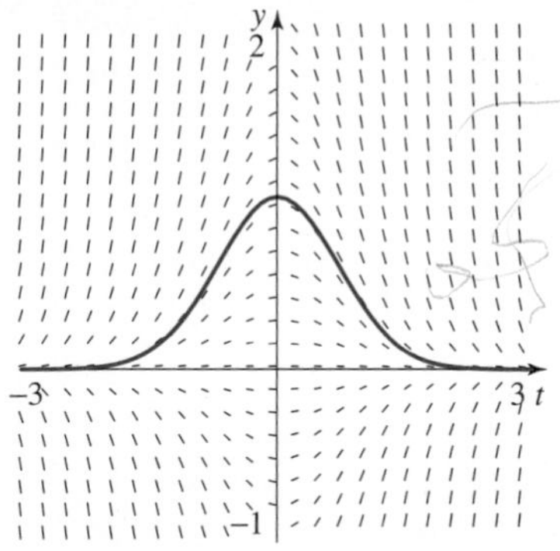

FIGURE 1.3.5 Direction field for $y' = -2ty$ and solution curve $y = e^{-t^2}$ of Example 5.

Summary

We now know one quantitative method, separation of variables, although it applies to only a small minority of first-order equations. While it provides us with some explicit examples for comparison and study, this method is not always satisfactory. Even for a separable equation, we may not be able to evaluate the integrals or to solve for the dependent variable. Limitations like these on quantitative methods cause us to devote substantial effort to the alternatives: qualitative and numerical methods.

1.3 Problems

Separable or Not *Determine whether each equation in Problems 1–10 is separable or not. Write separable ones in separable form. Determine constant solutions, if any.*

1. $y' = 1 + y$
2. $y' = y - y^3$
3. $y' = \sin(t + y)$
4. $y' = \ln(ty)$
5. $y' = e^{t+y}$
6. $y' = \dfrac{y+1}{ty} + y$
7. $y' = \dfrac{e^{t+y}}{y+1}$
8. $y' = t\ln(y^{2t}) + t^2$
9. $y' = \dfrac{y}{t} + \dfrac{t}{y}$
10. $ty' = 1 + y^2$

Solving by Separation *Use separation of variables to obtain solutions to the DEs and IVPs in Problems 11–20. Solve for y when possible.*

11. $y' = \dfrac{t^2}{y}$
12. $ty' = \sqrt{1 - y^2}$
13. $y' = \dfrac{t^2 + 7}{y^4 - 4y^3}$
14. $ty' = 4y$
15. $y' = y\cos t$
16. $4t\,dy = (y^2 + ty^2)dt, \quad y(1) = 1$
17. $y' = \dfrac{1 - 2t}{y}, \quad y(1) = -2$
18. $y' = y^2 - 4, \quad y(0) = 0$
19. $y' = \dfrac{2t}{1 + 2y}, \quad y(2) = 0$
20. $y' = -\dfrac{1 + y^2}{1 + t^2}, \quad y(0) = -1$

Integration by Parts *Problems 21–24 involve a variety of integration techniques. Recall that the formula for* ***integration by parts*** *is $\int u\,dv = uv - \int v\,du$, where the variables u and dv must be assigned carefully. Determine the solutions to the following DEs.*

21. $y' = (\cos^2 y)\ln t$
22. $y' = (t^2 - 5)\cos 2t$
23. $y' = t^2 e^{y+2t}$
24. $y' = tye^{-t}$

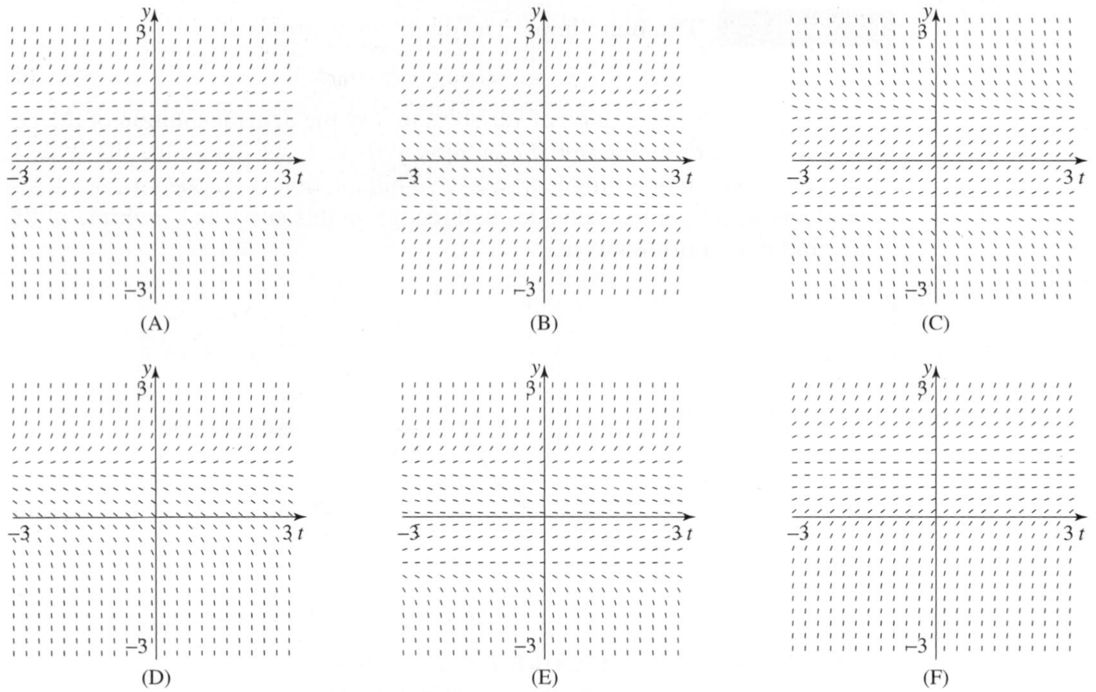

FIGURE 1.3.6 Direction fields that match the equations of Problems 25–30.

Equilibria and Direction Fields *Match each autonomous DE in Problems 25–30 with a direction field in Fig. 1.3.6.* HINT: *First determine the equilibrium solutions and then determine the signs of slopes for appropriate values of y.*

25. $y' = 1 - y^2$ **26.** $y' = y^2 - 1$

27. $y' = y(y-1)(y+1)$ **28.** $y' = (y-1)^2$

29. $y' = (y+1)(y-1)^2$ **30.** $y' = (y^2+1)(y-1)$

Finding the Nonequilibrium Solutions *Solve the DEs in Problems 31–34. Most of these solutions will require the use of the* ***method of partial fractions****. (See Appendix PF for a review of this material.) Write down the equilibrium solutions as well as the nonequilibrium solutions for each DE.* NOTE: *For most of these equations, direction fields are drawn with the previous group of problems.*

31. $y' = 1 - y^2$ **32.** $y' = 2y - y^2$

33. $y' = y(y-1)(y+1)$ **34.** $y' = (y-1)^2$

Help from Technology *For each DE in Problems 35–40, solve analytically to obtain solution curves through the points* $(1, 1)$ *and* $(-1, -1)$*. Then, using an appropriate software package, draw the direction field and superimpose your solution curves onto it.*

35. $\dfrac{dy}{dt} = y$ **36.** $\dfrac{dy}{dt} = \cos t$

37. $\dfrac{dy}{dt} = \dfrac{t}{y^2\sqrt{1+t^2}}$ **38.** $\dfrac{dy}{dt} = y\cos t$

39. $\dfrac{dy}{dt} = \dfrac{2t(y+1)}{y}$ **40.** $\dfrac{dy}{dt} = \sin(ty)$

Making Equations Separable *Many differential equations that are not separable can be made separable by making a proper substitution. One example is the class of first-order equations with right-hand sides that are functions of the combination y/t (or t/y). Given such a DE*

$$\frac{dy}{dt} = f\left(\frac{y}{t}\right),$$

called ***Euler-homogeneous***,[4] *let $v = y/t$. By the product rule, we deduce from $y = vt$ that*

$$\frac{dy}{dt} = v + t\frac{dv}{dt},$$

so the equation becomes

$$v + t\frac{dv}{dt} = f(v),$$

which separates into

$$\frac{dt}{t} = \frac{dv}{f(v) - v}.$$

[4]This use of the term Euler-homogeneous is distinct from the term linear-homogeneous, which will be introduced in Sec. 2.1 and used throughout the text.

Apply this method to solve the Euler-homogeneous DEs and IVPs in Problems 41–44. Plot sample solutions on a direction field and discuss.

41. $\dfrac{dy}{dt} = \dfrac{y+t}{t}$ **42.** $\dfrac{dy}{dt} = \dfrac{y^2+t^2}{yt}, \quad y(1) = -2$

43. $\dfrac{dy}{dt} = \dfrac{2y^4+t^4}{ty^3}$ **44.** $\dfrac{dy}{dt} = \dfrac{y^2+ty+t^2}{t^2}$

Another Conversion to Separable Equations *Given the differential equation $y' = f(at + by + c)$, it can be shown that the substitution $u = at + by + c$, where a, b, and c are constants, will transform the differential equation into the separable equation $u' = a + bf(u)$. Use this substitution to solve the DEs in Problems 45 and 46.*

45. $y' = (y+t)^2$ **46.** $y' = e^{t+y-1} - 1$

47. Autonomous Equations Recall that if the right-hand side is independent of variable t, so that the DE has the form $y' = f(y)$, the equation is called *autonomous*. (See Problem 61 in Sec. 1.2.) Such equations are always separable.

(a) Identify the autonomous equations in Problems 1–20.

(b) How can you recognize the direction field of an autonomous equation?

Slope Fields
Find the autonomous equations. See how this property shows up in the direction fields and solutions. (For comparison, all 12 IDE pictures are printed out in Lab 2, Sec. 5.)

48. Orthogonal Families

Orthogonal Trajectories

When a one-parameter family of curves satisfies a first-order DE, we can find another such family as solution curves of a related DE with the property that *a curve from one family intersects each curve of the other family* **orthogonally** (that is, at right angles: their respective tangent lines are perpendicular). Each family constitutes the set of **orthogonal trajectories** for the other.

Orthogonal Trajectories
See some intriguing pairs of orthogonal trajectories.

For the following questions we use the customary independent variable x instead of t.

(a) Use implicit differentiation to show that the one-parameter family $f(x, y) = c$ satisfies the differential equation $dy/dx = -f_x/f_y$, where $f_x = \partial f/\partial x$ and $f_y = \partial f/\partial y$.

(b) Explain why the curves satisfying $dy/dx = f_y/f_x$ are the orthogonal trajectories to the family in part (a).

(c) In Example 4, we found that the family $x^2 + y^2 = c^2$ of circles with centers at the origin were the solution curves of the separable DE $dy/dx = -x/y$. Use this and part (b) to show that the family of orthogonal trajectories are the straight lines $y = kx$. (See Fig. 1.3.7. These families represent the electric field and equipotential lines around a point charge at the origin.)

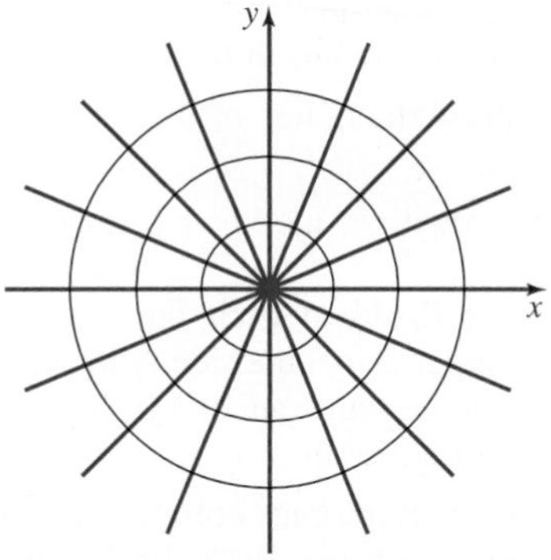

FIGURE 1.3.7 Orthogonal families $y = kx$ and $x^2 + y^2 = c$ for Problem 48(c).

More Orthogonal Trajectories *Use Problem 48 to determine the family of trajectories orthogonal to each given family in Problems 49–51. (See Fig. 1.3.8.)*

49. $y = cx^2$ **50.** $y = c/x^2$ **51.** $xy = c$

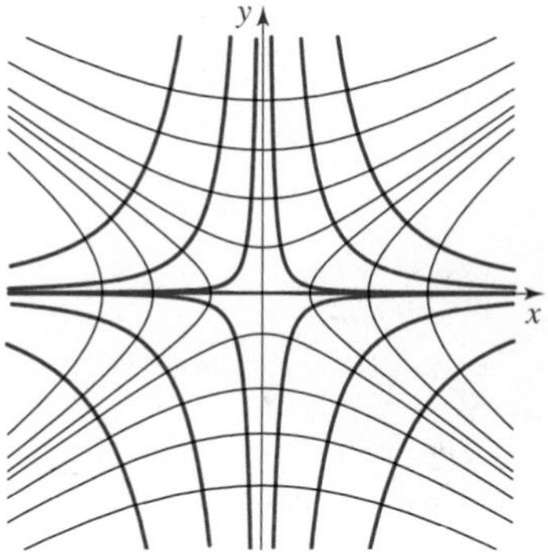

FIGURE 1.3.8 Orthogonal families $y = c/x^2$ (in black) and ? (in color) for Problem 50.

Calculator or Computer *With the help of suitable computer software, for Problems 52–55 graph the families of curves along with their families of orthogonal trajectories.*

52. $y = c$ (horizontal lines)

53. $4x^2 + y^2 = c$ (ellipses)

54. $x^2 = 4cy^3$ (cubics)

55. $x^2 + y^2 = cy$ (coaxial circles)

56. The Sine Function The sine function $\sin x$ has the property that the square of itself plus the square of its derivative is identically equal to one. Find the most general function that has this property.

57. Disappearing Mothball The rate at which the volume of a mothball evaporates from solid to gas is proportional to the surface area of the ball. Suppose a mothball has been observed at one time to have a radius of 0.5 in. and, six months later, a radius of 0.25 in.

(a) Express the radius of the mothball as a function of time.

(b) When will the mothball disappear completely?

58. Four-Bug Problem Four bugs sit at the corners of a square carpet L in. on a side. Simultaneously, each starts walking at the same rate of 1 in/sec toward the bug on its right. See Fig. 1.3.9(a).

(a) Show that the bugs collide at the center of the carpet in exactly L sec. HINT: Each bug always moves in a direction perpendicular to the line of sight of the bug behind it, so the distance between two successive bugs always decreases at 1 in/sec. The bugs always form a square that is shrinking and rotating clockwise.[5]

(b) Using the result from (a), but using *no calculus*, tell how far each bug will travel.

(c) Use differential equations to find the paths of the bugs. Simplify the setup by starting the bugs at the four points $(\pm 1, 0)$ and $(0, \pm 1)$, making $L = \sqrt{2}$. Use Fig. 1.3.9(b) to deduce the relationship $dr \approx -r\,d\theta$, for sufficiently small $d\theta$.

59. Radiant Energy Stefan's Law of Radiation states that the radiation energy of a body is proportional to the fourth power of the absolute temperature T of a body.[6] The rate of change of this energy in a surrounding medium of absolute temperature M is thus

$$\frac{dT}{dt} = k(M^4 - T^4),$$

where $k > 0$ is a constant. Show that the general solution of Stefan's equation is

$$\ln\left|\frac{M+T}{M-T}\right| + 2\tan^{-1}\left(\frac{T}{M}\right) = 4M^3kt + c,$$

where c is an arbitrary constant.

60. Suggested Journal Entry Describe the distinction between quantitative and qualitative analysis. In what ways do they complement one another?

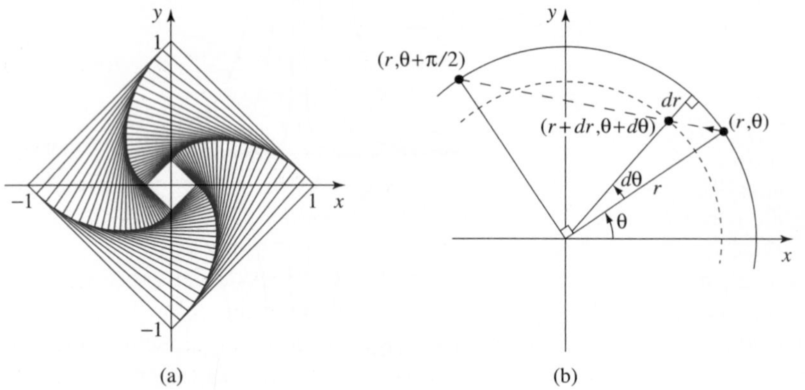

FIGURE 1.3.9 Four-bug problem (Problem 58).

[5] In general, if n bugs are initially at the vertices of a regular n-gon with side L, the time it takes the bugs to collide is $L/[s(1+\cos\alpha)]$, where s is the bugs' speed and α is the interior angle of the polygon. The path of the bug that starts at $(r, \theta) = (1, 0)$ is the logarithmic spiral given in polar coordinates by $r = e^{-(1+\cos\alpha)\theta}$.

[6] Josef Stefan (1835–1893), an Austrian physicist, stated this fact in 1879 as a result of empirical observation of hot bodies over a wide temperature range. Five years later his former student Ludwig Boltzmann (1844–1906) derived the same fact from thermodynamic principles, so it is often called the Stefan–Boltzmann Law.

1.4 Approximation Methods: Numerical Analysis

$y = y + y'\Delta t$
$t = t + \Delta t$

SYNOPSIS: We illustrate the calculation of numerical approximations to the solutions of differential equations, which is especially useful when analytical solutions cannot be found. Euler's method, though crude by modern standards, illustrates the basic ideas behind numerical solutions. A discussion of errors in numerical approximation gives some indication of why caution is needed in its use.

Euler's Method: Informal Example

The direction field for the differential equation $y' = t - y$ is graphed in Fig. 1.4.1. You were asked to draw solution curves on this field in Sec. 1.2, Problem 11. These were curves having *the correct slope* at each point, as given by the direction elements of the field. Let us use this idea to construct an approximate solution to the IVP

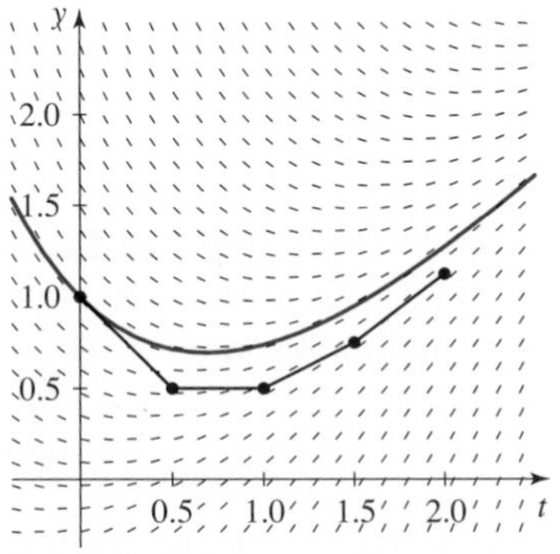

FIGURE 1.4.1 Broken-line approximation and true solution curve of $y' = f(t, y) = t - y$, $y(0) = 1$.

$$y' = f(t, y) = t - y, \quad y(0) = 1.$$

- At the point $(0, 1)$, the field has direction $f(0, 1) = -1$. We follow the line element through $(0, 1)$, down and to the right with slope -1, until we reach $t = 0.5$. The y-value changes by $(-1)(0.5) = -0.5$ (slope times change in t), so we end up at the point $(0.5, 0.5)$.
- Now at this point the slope is $f(0.5, 0.5) = 0.5 - 0.5 = 0$, so we head off to the right on a horizontal segment until we reach $t = 1$, putting us at $(1, 0.5)$. (The y-change was $(0)(0.5) = 0$.)
- Now we calculate a new direction from $f(1, 0.5) = 1 - 0.5 = 0.5$. Moving to the right along a line of slope 0.5 until t equals 1.5 brings us to $(1.5, 0.75)$, since the y-change was $(0.5)(0.5) = 0.25$.
- The slope at $(1.5, 0.75)$ is $f(1.5, 0.75) = 1.5 - 0.75 = 0.75$, so a fourth step lands us on $(2, 1.125)$. (Change in $y = (0.75)(0.5) = 0.375$.)

We are developing a piecewise-linear (or broken-line) approximation to the true solution curve in a step-by-step fashion; the amount t increases each time (0.5) is called the step size. Figure 1.4.1 shows our four steps and the true solution to this IVP superimposed on the direction field.

We have illustrated the use of **Euler's method** for the numerical solution of an IVP.[1] It is the "natural" thing to do. Standing at the initial point, facing in the direction of the line element there, we take a step. Realigning ourselves with the direction field, we take another step. We just "follow our noses" from point to point.

The step-by-step calculation of output or y-values corresponding to a sequence of equally spaced input or t-values, obtaining a sequence of points we can connect with line segments (or curves in some methods), is typical of numerical methods

Slope Fields

You can make such an approximate solution yourself. Click and stick a vector, then start another at its head. Repeat.

[1]Leonhard Euler (1707–1783), a Swiss, was the most prolific mathematician of all time, publishing more than 800 papers, in all branches of mathematics, with careful explanations of his reasoning and lists of false paths he had tried. He remained active throughout his life, despite increasing blindness that began around 1735. By 1771 he had become totally blind, yet still produced half of his work after that point, with help from his sons and colleagues. Euler is responsible for our $f(\,)$ notation for functions, the letter i for imaginary numbers, and the letter e for the base of the natural logarithms. You will hear more about him in Sec. 2.2.

for solving differential equations. After introducing some notation and formulas to streamline the calculations, we will take a look at the errors involved.

Euler's Method: Formal Approach

Consider the initial-value problem

$$y' = f(t, y), \quad y(t_0) = y_0; \tag{1}$$

we will assume for now that the problem has a unique solution $y(t)$ on some interval around t_0.[2] Our goal is to compute approximate values for $y(t_n)$ at the finite set of points $t_1, t_2, \ldots, t_k$, or

$$t_n = t_0 + nh, \quad n = 1, 2, \ldots, k,$$

for a preassigned k. Because

$$t_n - t_{n-1} = (t_0 + nh) - [t_0 + (n-1)h] = h,$$

this common difference h between successive points is called the **step size**. Starting at the initial point (t_0, y_0), where the slope $y' = f(t_0, y_0)$ is given by the line element of the direction field, we move along the tangent line determined by

$$y - y_0 = (t - t_0) f(t_0, y_0) \tag{2}$$

to the approximate solution $y_1(t)$ at t_1. We use the portion of the line between t_0 and t_1 as the first segment of the approximate solution, so equation (2) becomes $y_1 - y_0 = (t_1 - t_0) f(t_0, y_0)$. Because $t_1 - t_0 = h$, we have

$$y_1 = y_0 + hf(t_0, y_0). \tag{3}$$

(Compare this description to the first step in our earlier example.)

Having arrived at (t_1, y_1), we repeat the process by looking along the line element there having slope $f(t_1, y_1)$ (hence this is also called the **tangent-line method**). We move ahead along $y - y_1 = (t - t_1) f(t_1, y_1)$ to point (t_2, y_2), where

$$y_2 = y_1 + hf(t_1, y_1). \tag{4}$$

Continuing this process we obtain the sequence of points (t_n, y_n), where

$$\begin{aligned} y_3 &= y_2 + hf(t_2, y_2), \\ y_4 &= y_3 + hf(t_3, y_3), \\ &\vdots \\ y_k &= y_{k-1} + hf(t_{k-1}, y_{k-1}). \end{aligned}$$

Time Steps

Vary the time steps and compare results.

The resulting piecewise-linear function is the **Euler-approximate** solution to the IVP (1).

---

[2]We will discuss conditions that guarantee such solutions in the next section.

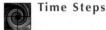

Euler's Method

For the initial-value problem

$$y' = f(t, y), \quad y(t_0) = y_0,$$

use the formulas

$$t_{n+1} = t_n + h, \tag{5}$$

$$y_{n+1} = y_n + hf(t_n, y_n) \tag{6}$$

to compute iteratively the points $(t_1, y_1), (t_2, y_2), \ldots, (t_k, y_k)$, using step size h. The piecewise-linear function connecting these points is the Euler approximation to the solution $y(t)$ of the IVP for $t_0 \le t \le t_k$.

EXAMPLE 1 **Euler at Work** Obtain the Euler-approximate solution of the initial-value problem

$$y' = -2ty + t, \quad y(0) = -1$$

on [0, 0.4] with step size 0.1.

Using the information $f(t, y) = -2ty + t = t(1 - 2y)$, together with

$$t_0 = 0, \quad y_0 = -1, \quad \text{and} \quad h = 0.1,$$

we calculate as follows:

$$t_1 = t_0 + h = 0 + 0.1 = 0.1,$$

$$\begin{aligned} y_1 &= y_0 + hf(t_0, y_0) = y_0 + ht_0(1 - 2y_0) \\ &= -1 + (0.1)(0)[1 - 2(-1)] = -1, \end{aligned}$$

$$t_2 = t_1 + h = 0.1 + 0.1 = 0.2,$$

$$\begin{aligned} y_2 &= y_1 + hf(t_1, y_1) = y_1 + ht_1(1 - 2y_1) \\ &= -1 + (0.1)(0.1)[1 - 2(-1)] = -0.97, \end{aligned}$$

$$t_3 = t_2 + h = 0.2 + 0.1 = 0.3,$$

$$\begin{aligned} y_3 &= y_2 + hf(t_2, y_2) = y_2 + ht_2(1 - 2y_2) \\ &= -0.97 + (0.1)(0.2)[1 - 2(-0.97)] = -0.9112, \end{aligned}$$

$$t_4 = t_3 + h = 0.3 + 0.1 = 0.4,$$

$$\begin{aligned} y_4 &= y_3 + hf(t_3, y_3) = y_3 + ht_3(1 - 2y_3) \\ &= -0.9112 + (0.1)(0.3)[1 - 2(-0.9112)] = -0.826528. \end{aligned}$$

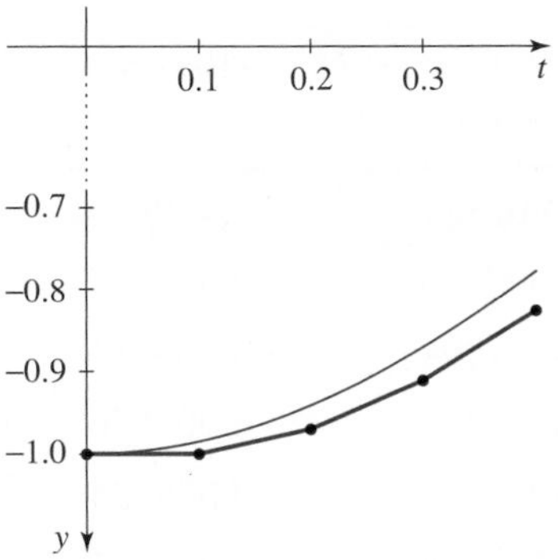

FIGURE 1.4.2 Approximate and actual solutions for $y' = -2ty + t$, $y(0) = -1$, $h = 0.1$.

The exact solution of the IVP is $y(t) = 0.5 - 1.5e^{-t^2}$, which you can confirm by separation of variables or by substitution in the DE. In Figure 1.4.2 and Table 1.4.1 (on the next page) we compare the Euler and true solution values y_n and $y(t_n)$ at the various t steps. Notice that the error grows rapidly.

Spreadsheets:

By the way, it is easy and efficient to carry out the calculations and create displays such as Table 1.4.1 at the same time by using a spreadsheet program such as Excel or Quattro. (See the spreadsheet for Problems 3–10.)

Table 1.4.1 Approximate values y_n for $y' = -2ty+t$, $y(0) = -1$, $h = 0.1$, compared with exact values $y(t_n)$

n	t_n	y_n	$y(t_n)$	Error
0	0.0	−1.000000	−1.000000	0.000000
1	0.1	−1.000000	−0.985075	−0.014925
2	0.2	−0.970000	−0.941184	−0.028815
3	0.3	−0.911200	−0.870897	−0.040303
4	0.4	−0.826528	−0.778216	−0.048312

■

EXAMPLE 2 **That Bell Again** In Example 5 of the previous section we obtained the exact solution $y = e^{-t^2}$ for the initial-value problem $y' = -2ty$, $y(0) = 1$. You can test your understanding of Euler's method by verifying several of the y_n values in Table 1.4.2 for the interval [0, 2] with step size $h = 0.2$. For example,

$$t_6 = t_5 + h = 1.0 + 0.2 = 1.2,$$

$$\begin{aligned} y_6 &= y_5 + hf(t_5, y_5) = y_5 - 2ht_5y_5 \\ &= 0.3993830 - 2(0.2)(1.0)(0.3993830) \\ &= 0.3993830 - 0.1597532 = 0.2396298. \end{aligned}$$

Figure 1.4.3 shows the true solution curve on [0, 2] and the piecewise linear Euler-approximate solution, which is sometimes above and sometimes below the exact solution.

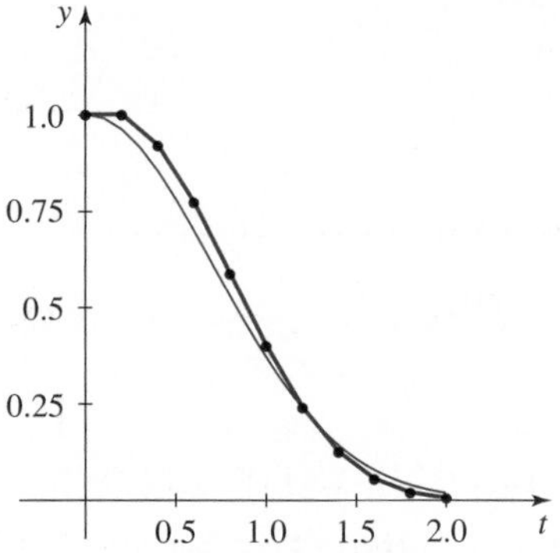

FIGURE 1.4.3 Approximate and actual solutions for $y' = -2ty$, $y(0) = 1$, $h = 0.2$.

Table 1.4.2 Approximate values y_n for $y' = -2ty$, $y(0) = 1$, $h = 0.2$, compared with exact values $y(t_n)$

n	t_n	y_n	$y(t_n)$	Error
0	0.0	1.0000000	1.0000000	0.000000
1	0.2	1.0000000	0.9607894	−0.039211
2	0.4	0.9200000	0.8521437	−0.067856
3	0.6	0.7728000	0.6976763	−0.075124
4	0.8	0.5873280	0.5272924	−0.060036
5	1.0	0.3993830	0.3678794	−0.031504
6	1.2	0.2396298	0.2369277	−0.002702
7	1.4	0.1246075	0.1408584	0.016251
8	1.6	0.0548273	0.0773047	0.022477
9	1.8	0.0197378	0.0391639	0.019426
10	2.0	0.0055265	0.0183156	0.012789

■

Error in Euler's Method

In the two examples, we had analytic solutions for comparison, so the errors in the Euler-approximate values could be obtained easily. What if we have no exact solution—the case of real interest for using numerical methods?

Roundoff error can be monitored by calculating to more decimal places.

There are two kinds of error that arise when Euler's method or any numerical method is used. The first is **roundoff error**, the discrepancy resulting from rounding or chopping numbers at each stage in the computation. In practice, all calculators and computers have limitations in their computational accuracy. Even when calculations are carried out with eight or even sixteen places of precision, roundoff errors can accumulate significantly after many steps. One strategy for monitoring roundoff error is to perform the computations with some given number of places of accuracy, then to repeat the computations using twice as many places. If good agreement results, roundoff error is probably under control.

Discretization Error for Euler's Method: These error formulas just give *bounds* on the discretization errors. Actual errors may be much smaller.

The second kind of error is **discretization error**, the error that results from the *process itself.* In Euler's method, this process is the use of the *linear approximation* of the tangent line instead of the true solution curve in stepping from one value to the next. Using the theory of Taylor series expansions, it can be shown (see Problem 22) that the error in each step is proportional to the square of the step size, h^2. That is,

Local error for a single step

$$|y_i - y(t_i)| \le Ch^2,$$

where the constant C depends on the size of the second derivative of the exact solution. This bound is called the **local discretization error** because it is an estimate for the discrepancy at just one step. After n steps, this error must be multiplied by n. Since n is inversely proportional to h (the more steps, the smaller each one is), the **accumulated** or **global discretization error** is proportional to h.

Global Discretization Error in Euler's Method

If the solution of the IVP $y' = f(t, y)$, $y(t_0) = y_0$ has a continuous second derivative on the interval $[t_0, t_k]$, and y_n is the value of the Euler approximation at t_n, $t_0 < t_1 < \cdots < t_n < \cdots < t_k$, then there exists a constant C such that

Global error for n steps

$$|y_n - y(t_n)| \le Ch, \quad n = 1, 2, \ldots, k, \tag{7}$$

where step size $h = t_n - t_{n-1}$.

Global Error Bound: $\mathcal{O}(h)$ for Euler's Method

We say that the error is of **order one**, corresponding to the first power of h in the estimate (7), and write "$\mathcal{O}(h)$" as a standard abbreviation[3] for Ch.

Comparison of Roundoff and Discretization Errors

We have seen that there is a discretization error e_n resulting from the approximation process, $e_n = y(t_n) - y_n$. But the actual computation gives not y_n but a rounded or chopped version w_n; the roundoff error is $r_n = y_n - w_n$. By the triangle inequality, then, the overall error is really given by

$$\begin{aligned}|y(t_n) - w_n| &= |y(t_n) - y_n + y_n - w_n| \le |y(t_n) - y_n| + |y_n - w_n| \\ &= |e_n| + |r_n|.\end{aligned}$$

[3]The "big-oh" notation $\mathcal{O}(h^p)$ ("order p") is a shorthand to indicate that the long-term behavior of a quantity depending on h, growth or decay, is proportional to the pth power of h.

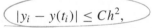

Thus total error is bounded by the sum of the discretization and roundoff errors. It can be shown that roundoff error is inversely proportional to a power of h. These quantities and their sum have graphs as depicted schematically in Fig. 1.4.4.

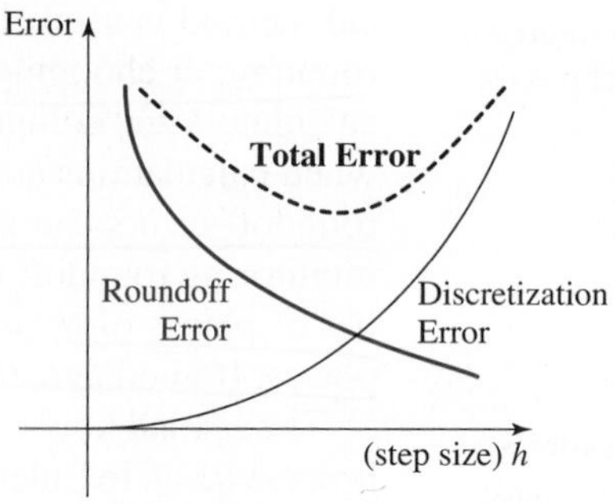

FIGURE 1.4.4 Errors as a function of step size.

Numerical Methods with Step-Size Scaling

Compare errors with step size.

The graph shows that we face a dilemma in choosing a step size: large h means less roundoff error but greater discretization error, while small h reduces discretization error but necessitates more calculations and thus more roundoff error. The graph suggests that there is an optimal choice of h to minimize the total error, but determining this critical step size is more an art than a science. One often resorts to experimentation and experience.

Runge-Kutta Methods

Euler's method advances the approximate solution of a first-order differential equation from point (t_n, y_n) to point (t_{n+1}, y_{n+1}) using only information at the first point. We now introduce two **Runge-Kutta methods**,[4] the second- and fourth-order methods, which look ahead and use the slope field at more than one point.

In the case of the second-order Runge-Kutta method (or midpoint method, sometimes called **midpoint Euler**), we use the slope at (t_n, y_n) to look a *half step* ahead to

$$\left(t_n + \frac{h}{2}, y_n + \frac{h}{2} f(t_n, y_n)\right).$$

We compute the slope at this halfway point and use this "corrected" slope to move from (t_n, y_n) to (t_{n+1}, y_{n+1}), the next point of the approximation.

[4] Carl D. T. Runge (1857–1927) was a German professor of applied mathematics; he devised what is known as the Runge-Kutta method around 1895. Martin W. Kutta (1867–1944) was also a German applied mathematician who made important contributions to the theory of aerodynamics. Proofs of the orders of the Runge-Kutta methods would take more space than we want to spend here; they are found in texts with more detail on numerical approximation. See, for example, J. H. Hubbard and B. H. West, *Differential Equations: A Dynamical Systems Approach, Part 1* (TAM 5, NY: Springer-Verlag, 1989), Sec. 3.3.

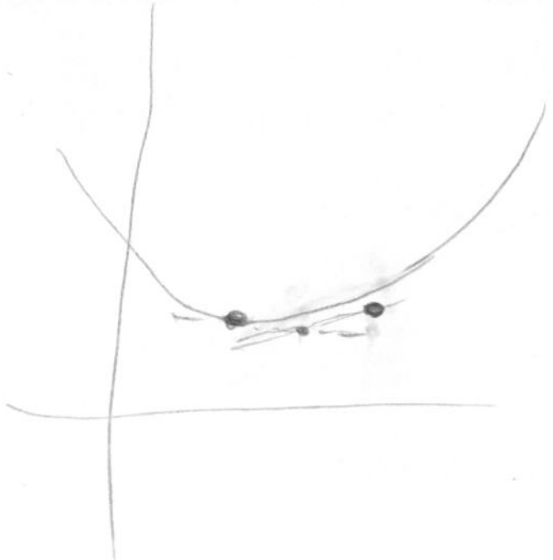

Second-Order Runge-Kutta Method

For the IVP $y' = f(t, y)$, $y(t_0) = y_0$, use the following formulas to compute the points (t_1, y_1), (t_2, y_2), …, of the approximate solution, using step size h:

$$t_{n+1} = t_n + h,$$
$$y_{n+1} = y_n + hk_{n2},$$

where

$$k_{n1} = f(t_n, y_n),$$
$$k_{n2} = f\left(t_n + \frac{h}{2}, y_n + \frac{h}{2}k_{n1}\right).$$

To find k_{n2}, you must first find k_{n1}.

EXAMPLE 3 **Second-Order Runge-Kutta** To obtain the first step of an approximate solution to the IVP

$$y' = t + y, \quad y(0) = 0$$

with step size $h = 0.5$, we calculate as follows:

$$k_{01} = f(t_0, y_0) = t_0 + y_0 = 0 + 0 = 0,$$
$$k_{02} = f\left(t_0 + \frac{h}{2}, y_0 + \frac{h}{2}k_{01}\right) = \left(t_0 + \frac{h}{2}\right) + \left(y_0 + \frac{h}{2}k_{01}\right)$$
$$= (0 + 0.25) + (0 + (0.25)0) = 0.25,$$

$$t_1 = t_0 + h = 0 + 0.5,$$
$$y_1 = y_0 + hk_{02} = 0 + 0.5(0.25) = 0.125.$$

Second-Order Runge-Kutta Approximation:

- Start at $(t_0, y_0) = (0, 0)$.
- Use slope k_{01} for a trial half step and read new slope k_{02} there.
- Return to (t_0, y_0) and use slope k_{02} for a full step, ending at

$$(t_1, y_1) = (0.5, 0.125).$$

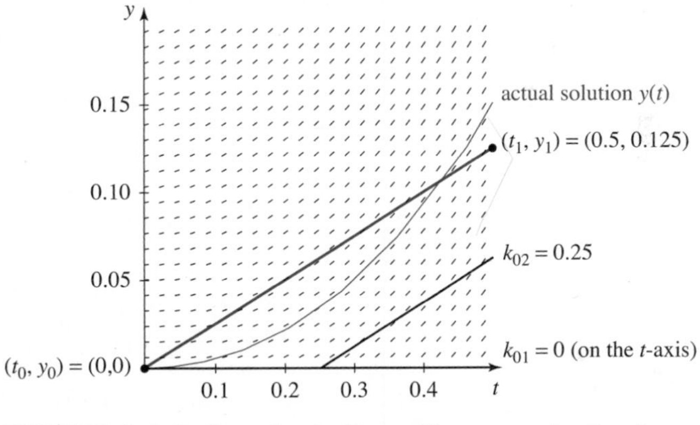

FIGURE 1.4.5 Second-order Runge-Kutta approximation for the first step of Example 3, $y' = t + y$, $y(0) = 0$, $h = 0.5$.

We used a foolishly large step size just to show the process, but even so we see in Fig. 1.4.5 that the approximation is far closer to the solution curve than the zero value that the Euler method would have given for step $h = 0.5$ from $(0, 0)$. ■

The fourth-order Runge-Kutta method adds more refinement to obtain dramatically smaller errors than first- or second-order methods, and is probably the most commonly used method for numerically solving differential equations.[5] Here, one computes the slopes k_i of the direction field at *four* well-chosen points and takes a weighted average of them to obtain a new point.

Fourth-Order Runge-Kutta Method

For the IVP $y' = f(t, y)$, $y(t_0) = y_0$, use the following formulas to compute the points (t_1, y_1), (t_2, y_2), ..., of the approximate solution, using step size h:

$$t_{n+1} = t_n + h,$$
$$y_{n+1} = y_n + \frac{h}{6}(k_{n1} + 2k_{n2} + 2k_{n3} + k_{n4}),$$

where

$$k_{n1} = f(t_n, y_n),$$
$$k_{n2} = f\left(t_n + \frac{h}{2}, y_n + \frac{h}{2}k_{n1}\right),$$
$$k_{n3} = f\left(t_n + \frac{h}{2}, y_n + \frac{h}{2}k_{n2}\right),$$
$$k_{n4} = f(t_n + h, y_n + hk_{n3}).$$

Global Error Bound:

The **order** of a numerical approximation method refers to the bound on its global error:

- $\mathcal{O}(h)$ for Euler's method;
- $\mathcal{O}\left(h^2\right)$ for second-order methods;
- $\mathcal{O}\left(h^4\right)$ for fourth-order methods.

For a fourth-order method, halving the step size reduces the error bound to $1/16$ of its previous value.

Because the averaging of slopes by this Runge-Kutta method is so carefully chosen, the results are amazingly accurate compared with simpler methods. For this reason, many DE solvers use fourth-order Runge-Kutta as the default method. It seems to offer a good balance between discretization and roundoff errors.

EXAMPLE 4 Fourth-Order Runge-Kutta To obtain the first step of an approximate solution to the IVP

$$y' = t + y, \quad y(0) = 0$$

with step size $h = 0.5$, we calculate as follows:

$$k_{01} = f(t_0, y_0) = t_0 + y_0 = 0 + 0 = 0,$$

$$k_{02} = f\left(t_0 + \frac{h}{2}, y_0 + \frac{h}{2}k_{01}\right) = \left(t_0 + \frac{h}{2}\right) + \left(y_0 + \frac{h}{2}k_1\right)$$
$$= (0 + 0.25) + (0 + (0.25)0) = 0.25,$$

$$k_{03} = f\left(t_0 + \frac{h}{2}, y_0 + \frac{h}{2}k_{02}\right) = \left(t_0 + \frac{h}{2}\right) + \left(y_0 + \frac{h}{2}k_2\right)$$
$$= (0 + 0.25) + (0 + (0.25)(0.25)) = 0.31,$$

$$k_{04} = f(t_0 + h, y_0 + hk_{03}) = (t_0 + h) + (y_0 + hk_3)$$
$$= (0 + 0.50) + (0 + 0.15) = 0.655,$$

[5]The fourth-order Runge-Kutta method is what most people think of as *the* Runga-Kutta method, the "workhorse" of numerical methods for solving differential equations. For most equations it is fast and suitably accurate.

$$\text{fourth-order averaged slope } k^* = \frac{0 + 2(0.25) + 2(0.31) + 0.655}{6} = \frac{1.775}{6},$$

$$t_1 = 0 + 0.5,$$

$$y_1 = 0 + 0.5\left(\frac{1.775}{6}\right) \approx 0.15.$$

Fourth-Order Runge-Kutta Approximation:

- Start at $(t_0, y_0) = (0, 0)$.
- Use slope k_{01} for a trial half step and read new slope k_{02} there.
- Return to (t_0, y_0) and use slope k_{02} for *another* trial half-step and read new slope k_{03} there.
- Return to (t_0, y_0) and use slope k_{03} for a trial *full* step and read new slope k_{04} there.
- Calculate the weighted average
$$k^* = \frac{k_{01} + 2k_{02} + 2k_{03} + k_{04}}{6}.$$
- Return to (t_0, y_0) and use the averaged slope k^* for the full step, ending at
$$(t_1, y_1) = (0.5, 0.15).$$

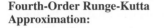

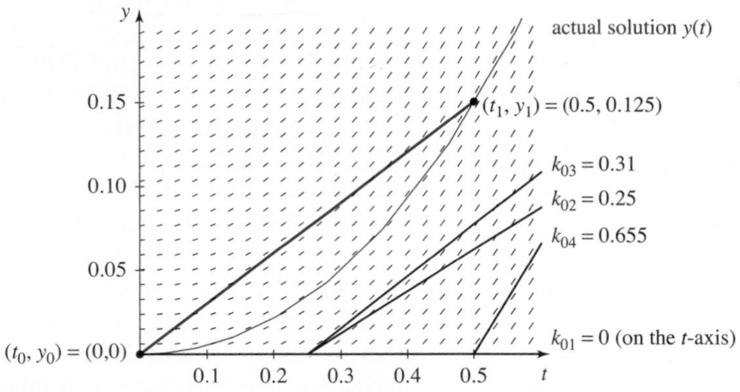

FIGURE 1.4.6 Fourth-order averaging of slopes for a single step of Runge-Kutta approximation for Example 4, $y' = t + y$, $y(0) = 0$, $h = 0.5$.

As Fig. 1.4.6 shows, even after one very large step, the approximation is extremely close to the actual solution curve. Problem 17 gives you a look at the spectacular result of even larger steps. ■

In Table 1.4.3, we tabulate the results of the methods discussed so far for the IVP of Examples 3 and 4. (See Sec. 2.2 for the method of calculation of the exact solution.)

Table 1.4.3 Comparison of numerical approximation for $y' = t + y$, $y(0) = 0$, $h = 0.5$.

Method	One-Step Result	Actual Error	Bound on Global Error
Exact solution	$y(1) = 0.1487\ldots$		
Euler's method	$y_1 = 0$	0.1487...	$h = 0.5$
Second-order Runge-Kutta	$y_1 = 0.125$	0.0237...	$h^2 = 0.25$
Fourth-order Runge-Kutta	$y_1 = 0.15$	0.0013...	$h^4 = 0.0625$

Other Methods

We introduced Euler's method first in this section because it is easy to understand and because its strategy of using the direction field to "snoop ahead" is typical of a number of other methods. In the Runge-Kutta methods, one moves from y_n to y_{n+1} using not just the slope at (t_n, y_n) but also slopes at strategically chosen nearby points, averaged to maximize the accuracy of the approximation.

Another type of averaging strategy looks backward as it works forward. **Multistep methods** base the calculation of y_{n+1} not just on y_n but on two or

Numerical Methods with Step-Size Scaling

Compare five methods on a single initial-value problem.

more previous steps, such as y_{n-1} and y_{n-2}. Such procedures, like the Adams-Bashforth method, require use of a single-step method like Runge-Kutta to get enough points to start.[6] A wide array of methods is available, suited to various particular categories of differential equations. Their implementation and analysis is an important part of the mathematical specialty **numerical analysis**. Formulas get more complicated, but the ideas are simply to achieve better averaging, and computers make that easier.

Variable step size methods, which computers can also efficiently automate, have increased accuracy enormously. A century after Adams and Bashforth, we see methods like Dormand-Prince (the default approximation method in John Polking's ODE solver on the Prentice Hall website).[7] Such variable step size methods adjust the step size at each step. If the possible error gets too large (e.g., when slopes get steep), the calculation will stop and repeat from the last point with a smaller step. The setting for permissible error bound, called **tolerance**, is typically between 10^{-3} and 10^{-12}.

Tolerance is the maximum error the user will accept from a numerical approximation.

There are many other variations on these basic methods, too numerous to describe here, but you will get a glimpse in the exercises. Some are based on the Taylor series expansions from calculus, to go beyond linear approximation. (See Problems 23 and 24.) You will find that Euler and Runge-Kutta methods provide a fine introduction to others. For instance, Richardson's extrapolation is an important way to achieve higher-order accuracy from any given method. (See Problems 25–28.) Computers now enable even further expansion of this extrapolation method, such as the popular Bulirsch-Stoer procedure.[8]

The area of numerical solutions of differential equations is an active area of mathematical research. The efforts to deal with ever more difficult categories of differential equations continue without end. We have given only the flavor of the process that creates the graphic solution pictures.

Summary

While analytical solutions are still useful and appealing and sometimes allow easier determination of the dependence on physical parameters, numerical approximations can serve many of the same purposes for the wide class of models for which analytic solutions are unavailable. Effective numerical procedures are the basis for modern computer graphics packages, which are showing us subtle (and not so subtle) nuances of properties of dynamical systems that were completely unsuspected before the computer age.

---

[6] The **Adams-Bashforth multistep formula** for numerical integration of differential equations is

$$y_{n+1} = y_n + \frac{h}{24}(55y'_n - 59y'_{n-1} + 37y'_{n-2} - 9y'_{n-3}).$$

Its discretization error is proportional to h^5. For details, consult a text on numerical methods. This method was first developed in 1883 by John Couch Adams (1819–1891) and Francis Bashforth (1819–1912) in a book on capillary action. Adams was an English astronomer who mathematically predicted the existence of the planet Neptune in 1846.

[7] The **Dormand-Prince method** simultaneously uses solvers of two different orders (e.g., fourth- and fifth-order Runge-Kutta) at the same set of t_n values (to reduce the number of evaluations of the function). At each step local error is estimated and compared with tolerance, to adjust the step size if necessary. See J. R. Dormand and P. J. Prince, "A Family of Embedded Runge-Kutta Formulae," *Journal of Computational and Applied Mathematics* 6 (1980), 19–26.

[8] The **Bulirsch-Stoer method** is based on stepwise extrapolation, starting with a single step size h and breaking it into smaller step sizes as tolerance demands. Thus, each step consists of many substeps, usually by a modified midpoint method. See Sec. 2.2 of R. Bulirsch and J. Stoer, *Introduction to Numerical Analysis* (NY: Springer-Verlag, 1991).

1.4 Problems

1. **Easy by Calculator** For the IVP $y' = t/y$, $y(0) = 1$,
   (a) Find Euler-approximate solution values at $t = 0.1$, $t = 0.2$, and $t = 0.3$ with $h = 0.1$.
   (b) Repeat (a) with $h = 0.05$.
   (c) Compute an analytic solution $y(t)$, and compare the values of $y(0.2)$ with your results from (a) and (b).

2. **Calculator Again** Consider the IVP $y' = ty$, $y(0) = 1$.
   (a) Use Euler's method to approximate the solution at $t = 1$ with step sizes 1, 1/2, 1/4, and 1/8.
   (b) Solve the problem exactly, and compare the result at $t = 1$ with the approximations calculated in part (a).

Computer Help Advisable *Use of a programmable calculator or computer is advisable for carrying out numerical approximations. A spreadsheet is especially effective in many cases. Euler's method formulas for solving the IVP $y' = t - y$, $y(0) = 1$ on $[0, 2]$ with step size 0.2 are illustrated here.*

	A	B
1	"t_n"	"y_n"
2	0	1
3	= A2 + 0.2	= B2 + 0.2 ∗ (A2 − B2)
4	= A3 + 0.2	= B3 + 0.2 ∗ (A3 − B3)
⋮	⋮	⋮
12	= A11 + 0.2	= B11 + 0.2 ∗ (A11 − B11)

NOTE: *The formulas in line 3 are from equations (5) and (6). Lines 4–12 need not be typed separately; they just repeat the instructions given on line 3, with spreadsheet shortcuts. See Appendix SS for details.*

In Problems 3–10, solve the IVP numerically on the suggested interval, if given, using various step sizes. Compare with values of exact solutions when possible.

3. $y' = 3t^2 - y$, $y(0) = 1$; $[0, 1]$

4. $y' = t^2 + e^{-y}$, $y(0) = 0$; $[0, 2]$

5. $y' = \sqrt{t + y}$, $y(1) = 1$; $[1, 5]$

6. $y' = t^2 - y^2$, $y(0) = 1$; $[0, 5]$

7. $y' = t - y$, $y(0) = 2$

8. $y' = -\dfrac{t}{y}$, $y(0) = 1$

9. $y' = \dfrac{\sin y}{t}$, $y(2) = 1$

10. $y' = -ty$, $y(0) = 1$

11. **Stefan's Law Again** An interesting analysis results from playing with the equation of Stefan's Law (Sec. 1.3, Problem 59). For $dT/dt = k(M^4 - T^4)$, let $k = 0.05$, $M = 3$, $T(0) = 4$.
   (a) Estimate $T(1)$ by Euler's method with step sizes $h = 0.25$, $h = 0.1$.
   (b) Graph a direction field and both multistep approximations from (a). Explain why and how the approximations from (a) take different routes.
   (c) Find an equilibrium solution; relate it to (a) and (b).

12. **Nasty Surprise** Use Euler's method with $h = 0.25$ to approximate the solution of $y' = y^2$, $y(0) = 1$, at $t = 0.25$, $t = 0.50$, $t = 0.75$, and $t = 1$. Verify that the exact solution is $y(t) = 1/(1 - t)$; does this help explain what happened to the Euler approximations?

13. **Approximating e** Obtain an estimate for the value of e by using Euler's method to approximate the solution of the IVP $y' = y$, $y(0) = 1$, at $t = 1$, using smaller and smaller values of h. As h decreases, the approximation for e gets better for a while but will eventually worsen, due to roundoff when h is small enough or calculation coarse enough. If you have a software package that automates Runge-Kutta or other methods, try them and compare the results with Euler.

14. **Double Trouble or Worse** The initial-value problem $y' = y^{1/3}$, $y(0) = 0$, has an infinite number of solutions, two of which are $y(t) = 0$ and $y(t) = (2t/3)^{3/2}$, $t \geq 0$. These solutions are drawn in Fig. 1.4.7; the nonzero solution is tangent to the t-axis at the origin.

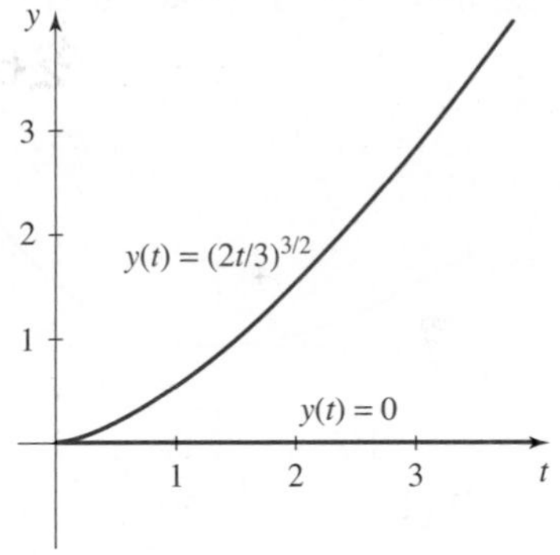

FIGURE 1.4.7 Two solutions of the IVP $y' = y^{1/3}$, $y(0) = 0$, for Problem 14.

   (a) What happens if Euler's method is applied to this problem?
   (b) What happens if the initial condition is changed to $y(0) = 0.01$? If a preprogrammed Euler solution is available, solve on $[0, 6]$ with $h = 0.1$.
   (c) Use the computer to look at the direction field. How does it correlate with your solution in part (b)?

15. **Roundoff Problems** The solution of the IVP $y' = y$, $y(0) = A$, is $y(t) = Ae^t$. If a roundoff error of ε occurs when the value of A is entered, how will this affect the solution at $t = 1$? What about at $t = 10$ and $t = 20$?

16. **Think Before You Compute** We considered the DE $y' = y^2 - 4$ in Example 8 of Sec. 1.2. (The direction field is graphed in Fig. 1.2.5.) What is the result of applying Euler's method (or other methods, if available) to the IVPs

$$y' = y^2 - 4, \; y(0) = 2 \quad \text{and} \quad y' = y^2 - 4, \; y(0) = -2?$$

How accurate are your approximations? What should the solutions be?

Runge-Kutta Method *The fourth-order approximation invented by Runge and Kutta can be surprisingly accurate, even with a ridiculously large step size. To see this, for Problems 17 and 18, use the given step size with the IVP*

$$y' = t + y, \quad y(0) = 0$$

to do the following:

(a) *Compute for a single step the Euler approximation, the second-order Runge-Kutta approximation, and the fourth-order Runge-Kutta approximation.*

(b) *Add the three approximations in part (a) to the graph of the actual solution, as given in Fig. 1.4.8, and describe what you see.*

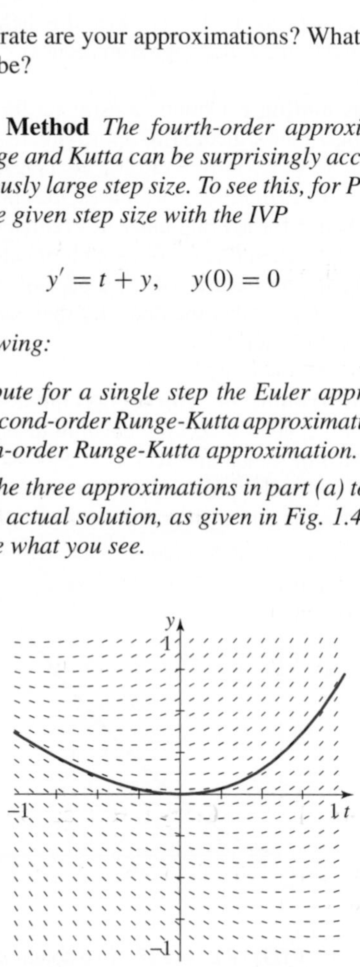

FIGURE 1.4.8 Actual solution of $y' = t + y$, $y(\phi)$, on [−1,1] for Problems 17 and 18, as appropriate.

(c) *Verify that $y(t) = e^t - t - 1$ satisfies the DE, then calculate the numerical values for the actual solution at $y(1)$ or $y(-1)$.*

17. $h = 1.0$

18. $h = -1.0$

Runge-Kutta vs. Euler *Use the Runge-Kutta method to approximate solutions to the following IVPs and compare the results with the results obtained earlier by Euler's method (and with exact values when possible).* You may wish to implement your solutions using a spreadsheet program, *with several more columns for the different slopes k_{ni}.*

19. IVP of Problem 3

20. IVP of Problem 7

21. IVP of Problem 8

22. IVP of Problem 10

23. **Euler's Errors** We will investigate the local discretization error in applying the Euler approximation given by equations (5) and (6).

(a) If $y(t)$ is the exact solution of $y' = f(t, y)$, use the chain rule to calculate $y''(t)$ and explain why it is continuous.

(b) Recall the following from calculus:

Taylor's Theorem
Any continuously infinitely differentiable function can be expanded in polynomial form about a value as follows:

$$f(t) = \sum_{n=0}^{\infty} \frac{f^{(n)}(t_0)}{n!}(t - t_0)^n,$$

where $f^{(n)}(t_0)$ means the nth derivative of $f(t)$ with respect to t, evaluated at t_0.

If the summation is stopped at k instead of ∞, the remainder is

$$R_k = \frac{f^{(k+1)}(t^*)}{(k+1)!}(t - t_0)^{k+1},$$

where t^* is some value of t between t and t_0.

NOTE: Some care must be taken, because the power series expansion of an infinitely differentiable function may converge only on a finite interval around t_0.

Remember that $y(t_{n+1}) = y(t_n + h)$, and deduce that

$$y(t_n + h) = y(t_n) + y'(t_n)h + \frac{1}{2}y''(t_n^*)h^2 \qquad (8)$$

for some t_n^* in the interval (t_n, t_{n+1}).

(c) Subtract equation (6) from equation (8) to conclude that the local discretization error e_{n+1} is given by

$$e_{n+1} = \frac{1}{2}y''(t_n^*)h^2,$$

where we assume that the nth approximation is exact: $y(t_n) = y_n$. Hence, if $|y''(t)| \le M$ on $[t_n, t_{n+1}]$, then $e_{n+1} \le Mh^2/2$.

(d) How small must h be to guarantee that this local discretization error is no greater than some prescribed ε?

24. Three-Term Taylor Series

(a) Replace the so-called two-term Taylor estimate (equation (8) in Problem 23) by the three-term result:

$$y(t_n + h) = y(t_n) + y'(t_n)h + \frac{1}{2}y''(t_n)h^2 + \frac{1}{6}y'''(t_n^*)h^3.$$

Compute the second derivative y'' in terms of f_t and f_y (partial derivatives of f, with respect to t and y, respectively), and deduce the three-term Taylor approximation:

$$y_{n+1} = y_n + hf(t_n, y_n) + \frac{1}{2}h^2[f_t(t_n, y_n) + f_y(t_n, y_n)f(t_n, y_n)].$$

(b) Show that the local discretization error e_{n+1} in this scheme is $\mathcal{O}(h^3)$.

(c) Apply the method to the IVP in Problem 1 and compare the results.

(d) Repeat (c) for Problem 2.

Richardson's Extrapolation *Euler's method gives first-order approximations, but can readily be used to make more accurate higher-order approximations. The basic idea starts with equation (7).*

When Euler's method starts at t_0, the accumulated discretization error in the approximation at $t^* = t_0 + nh$ is bounded by a constant times the step size h, as shown in equation (7). Thus, at t^* the true solution can be written

$$y(t^*) = y_n + Ch + \mathcal{O}(h^2), \tag{9}$$

where y_n is the first-order Euler approximation after n steps. Now repeat the Euler computations with step size $h/2$, so that $2n$ steps are needed to reach t^*. Equation (9) becomes

$$y(t^*) = y_{2n} + Ch/2 + \mathcal{O}(h^2). \tag{10}$$

Subtracting (9) from two times (10) eliminates the Ch term, thus giving a second-*order approximation*

$$y_R(t^*) = 2y_{2n} - y_n. \tag{11}$$

This technique of raising the order of an approximation by using both y_n and the half-step approximation y_{2n} is called ***Richardson's extrapolation****; it can be used with any numerical DE method.*[9]

In Problems 25–28, approximate the solution to the IVP at $t = 0.2$ by Richardson's extrapolation, using Euler's method with $h = 0.1$ and $h/2 = 0.05$. Compare with exact solutions when possible.

25. $y' = y, \quad y(0) = 1$

26. $y' = ty, \quad y(0) = 1$

27. $y' = y^2, \quad y(0) = 1$

28. $y' = \sin ty, \quad y(0) = 1$

29. Integral Equation

(a) Show that the IVP $y' = f(t, y)$, $y(t_0) = y_0$, is equivalent to the **integral equation**

$$y(t) = y_0 + \int_{t_0}^{t} f(s, y(s))\,ds$$

by verifying the following two statements: (i) Every solution $y(t)$ of the IVP satisfies the integral equation; (ii) Any function $y(t)$ satisfying the integral equation satisfies the IVP.

(b) Convert the IVP $y' = f(t)$, $y(0) = y_0$, into an equivalent integral equation as in part (a). Show that calculating the Euler-approximate value of the solution to this IVP at $t = T$ is the same as approximating the right-hand side of the integral equation by a Riemann sum (from calculus) using left endpoints.

(c) Explain why the calculation of part (b) depends on having the right-hand side of the differential equation independent of y.

30. Computer Lab: Other Methods If you have access to software with other methods, choose one or two of Problems 3–10 and make a study (for fixed step size) of different methods. Tell which you think is best and why.

31. Suggested Journal Entry I How do numerical methods complement qualitative and quantitative investigations? Discuss their relative importance in studying differential equations.

32. Suggested Journal Entry II In what ways are the choice of numerical method and choice of step size affected by the particular hardware and software available? Are they also influenced by the particular differential equation studied? Try to give examples to illustrate your discussion.

[9]English scientist and applied mathematician Lewis Fry Richardson (1881–1953) first used this technique in 1927. Its simple approach is both computer efficient and easily extended to provide higher-order approximations. Many variations are popular, particularly for approximating solutions to partial differential equations.

1.5 Picard's Theorem: Theoretical Analysis

SYNOPSIS: We discuss why it is important to consider the problems of existence and uniqueness of solutions of differential equations and initial-value problems. Picard's Theorem gives conditions that guarantee the existence of a unique local solution for an initial-value problem.

Why Study Theory?

When a mathematical model is constructed for a physical system, two reasonable demands are made. First, solutions should exist, if the model is to be useful at all. Second, to work effectively in predicting the *future* behavior of the physical system, the model should produce only one solution for a particular set of *initial conditions*. **Existence** and **uniqueness** theorems help to meet these demands:

- Existence theorems tell us that a model has *at least one* solution;
- Uniqueness theorems tell us that a model has *at most one* solution.

Interestingly enough, such theorems can prove that there is one and only one solution to a problem without actually finding it! This is like proving that there *is* a needle in the haystack even if we cannot actually come up with it!

At first this might not sound worthwhile. Why care if a solution exists if we cannot find it? But knowing in advance that an IVP has a solution and that it is unique tells us that, whether or not we can give an explicit formula, it still makes sense to study properties of the solution or to develop techniques for approximating it.

Issues of existence and uniqueness are not peculiar to differential equations or even to mathematics. Philosophers debate the existence of God while faithful worshippers live their beliefs. Physicists agonize over the existence of subatomic particles while engineers build better mousetraps. The existence of extraterrestrial life (and thus the uniqueness of our own life on earth) intrigues astronomers and astrophysicists while man-made satellites revolutionize communications. Theory and practice operate in distinct ways, yet maintain a useful dialogue.

Existence and uniqueness come up in basic mathematical situations. Sometimes we can demonstrate the existence of a solution by producing it: The number $-3/2$ is a solution of the linear equation $2x + 3 = 0$. (Can you give a convincing argument that this solution is unique?) Things are a little more complicated for the quintic equation $x^5 + x - 1 = 0$, which has only one real root.

EXAMPLE 1 **Arguments for Only One** By applying the Intermediate Value Theorem of calculus to the function

$$f(x) = x^5 + x - 1$$

on the interval [0, 1], we can be sure there is at least one root of $f(x)$ and that it is between 0 and 1: this gives existence. (See Fig. 1.5.1(a).)

Further, since

$$f'(x) = 5x^4 + 1$$

is positive for all x, we know from Rolle's Theorem (also from calculus) that there can't be any other real roots. (See Fig. 1.5.1(b).)

With this information, we can try to approximate the solution numerically without worrying that our effort will be wasted or that we might need to look for more than one answer.

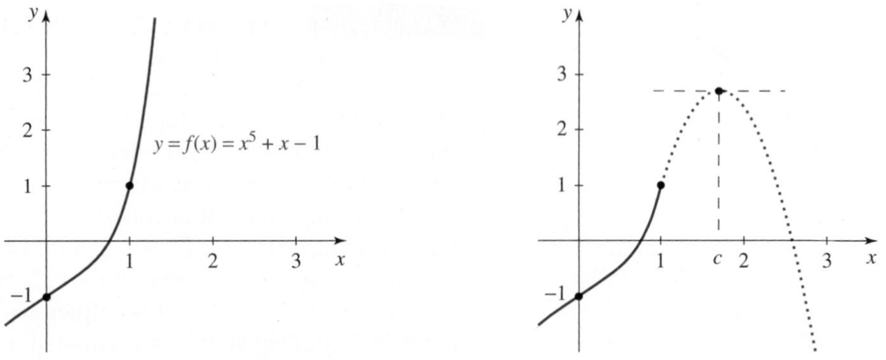

(a) The Intermediate Value Theorem shows that $f(x) = 0$ between 0 and 1.

(b) If f had another zero, f' would vanish between them by Rolle's Theorem.

FIGURE 1.5.1 Existence and uniqueness of solutions to the algebraic equation of Example 1.

What About Differential Equations?

Most meaningful differential equations have solutions. For first-order DEs, the constant of integration usually gives a one-parameter family of them. We will mostly be interested in looking at IVPs. After checking that there are, in fact, solutions to choose from, we will be interested in finding a unique member of the family satisfying the initial condition; that is, the particular solution curve that passes through the initial point. Following the *qualitative-first* rule, it will often be useful to look first at the direction field. Sometimes the information we get this way is pretty obvious. Figure 1.5.2 shows direction fields for

Targets

Look at the direction fields for a variety of DEs. Are these DEs "good," "bad," or "definitionally challenged"?

Fix a target and shoot at it with solutions. Can every target be hit?

$$y' = y \qquad \text{(the } good\text{)}, \tag{1}$$

$$y' = \sqrt{y} \qquad \text{(the } bad\text{)}, \tag{2}$$

$$y' = \sqrt{y} + \sqrt{-y} + \frac{1}{y} \qquad \text{(the } ugly\text{)}. \tag{3}$$

- The field for the first one (Fig. 1.5.2(a)) suggests that there is a unique solution through each point that is defined for all t-values.
- The field in Fig. 1.5.2(b) is good in the upper half-plane. No solutions through (t_0, y_0) exist if $y_0 < 0$, and the situation for initial points on the t-axis needs further investigation.
- Because $y < 0$ makes $\sqrt{y}$ undefined, $y > 0$ is bad for $\sqrt{-y}$, and $y = 0$ is a "no-no" for $1/y$, what is given in equation (3) is really a nondifferential equation, and its direction field is empty.

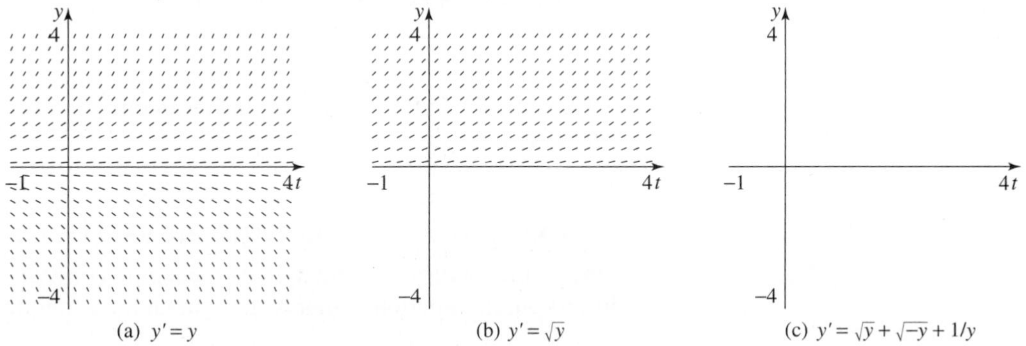

(a) $y' = y$ (b) $y' = \sqrt{y}$ (c) $y' = \sqrt{y} + \sqrt{-y} + 1/y$

FIGURE 1.5.2 Direction fields for good, bad, and ugly differential equations (1), (2), and (3).

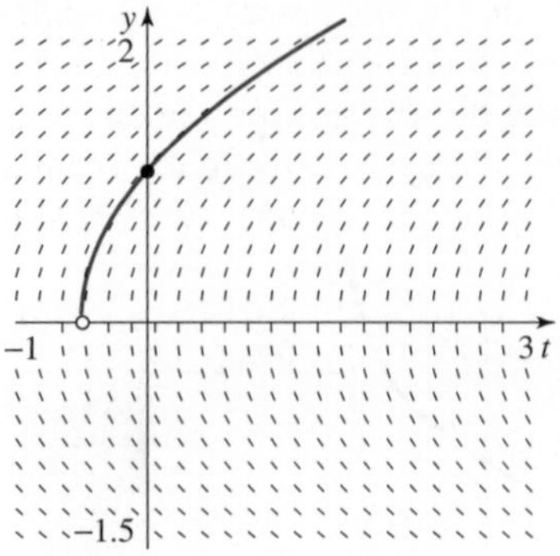

FIGURE 1.5.3 The direction field for $y' = 1/y$ and a solution for $y(0) = 1$ (Example 2).

EXAMPLE 2 **Too Steep** The initial-value problem $y' = 1/y$, $y(0) = 0$, has no solution because the domain of $f(t, y) = 1/y$ does not include the t-axis, where $y = 0$. The direction field in Fig. 1.5.3 suggests that as a solution gets close to the t-axis, its slope "tends to infinity." So we get no solution at all through $y = 0$, but we get a whole family of solutions for $y > 0$, and another whole family of solutions for $y < 0$.

If we change the initial condition and consider the IVP $y' = 1/y$, $y(0) = 1$, the solution $y(t) = \sqrt{2t+1}$ can be obtained by separation of variables; it is defined only on the interval $(-1/2, \infty)$, where $2t + 1 > 0$. Actually the half-parabola $y = \sqrt{2t+1}$ is defined for $t \geq -1/2$, but $y = 0$ at $t = -1/2$, so the differential equation is not satisfied. Despite restricted domains, the picture suggests unique solution curves through points not on the t-axis. ■

Surefire Target

Can some targets be hit by more than one solution?

EXAMPLE 3 **Too Many** From the direction field graphed previously in Fig. 1.5.2(b), we know there are no solutions to

$$y' = \sqrt{y}, \quad y(0) = y_0$$

for y_0 negative, although for y_0 positive the field looks well behaved. Also, the horizontal line elements along the t-axis indicate that the constant function $y = 0$ provides a solution curve through every point $(t_0, 0)$. But it isn't obvious at first whether this is a unique solution, and in fact the direction field alone cannot tell us.

Separating variables gives $y^{-1/2}dy = dt$, so that $2y^{1/2} = t + c$, where the right-hand side can't be negative. Solving for y gives $y = (t + c)^2/4$. Thus the IVP $y' = \sqrt{y}$, $y(0) = 0$, also has the solution

$$y(t) = \begin{cases} 0 & \text{if } t \leq 0, \\ \dfrac{1}{4}t^2 & \text{if } t \geq 0. \end{cases}$$

The solution $y = 0$ is not the only one that passes through the origin! In fact, the situation is even worse. For an arbitrary positive t-value c, the function

$$y(t) = \begin{cases} 0 & \text{if } t < c, \\ \dfrac{1}{4}(t-c)^2 & \text{if } t \geq c \end{cases} \tag{4}$$

is a solution of $y' = \sqrt{y}$, $y(0) = 0$. You are asked to verify this in detail in Problem 21. Figure 1.5.4 shows several such solutions. ■

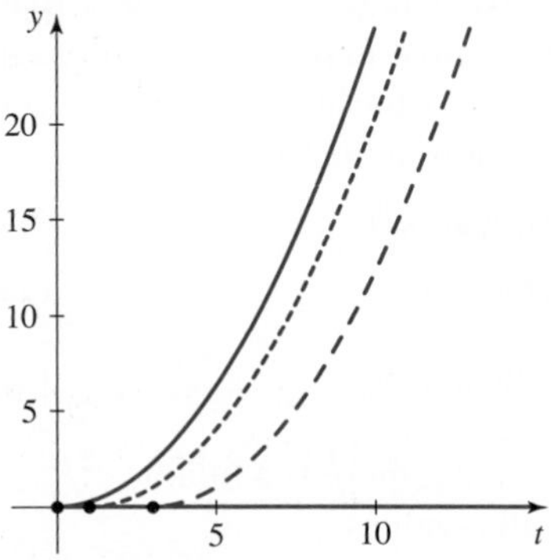

FIGURE 1.5.4 Several solutions of the IVP $y' = \sqrt{y}$, $y(0) = 0$ (Example 3). *One* of these is the constant solution $y(t) = 0$; other solutions that satisfy the DE go along the horizontal axis and then leave along $y = \frac{1}{4}(t - c)^2$ at any $t = c$, shown as a black dot.

Examples 2 and 3 emphasize that a direction field alone does not necessarily alert us to questions of uniqueness.

An Existence and Uniqueness Theorem

Nineteenth-century mathematicians spent much time and effort investigating conditions on the function f that would guarantee a unique solution of the IVP

$$y' = f(t, y), \quad y(t_0) = y_0.$$

One of the most useful results of this kind was obtained by the French mathematician Charles Émile Picard.[1] His theorem requires just that we check the continuity of f and $\partial f/\partial y$, but is only a *local* result: it guarantees a unique solution on *some* t-interval containing t_0—an interval that might turn out to be quite short. What's more, Picard's requirements are what are called *sufficient conditions*: they are more restrictive than absolutely necessary.[2]

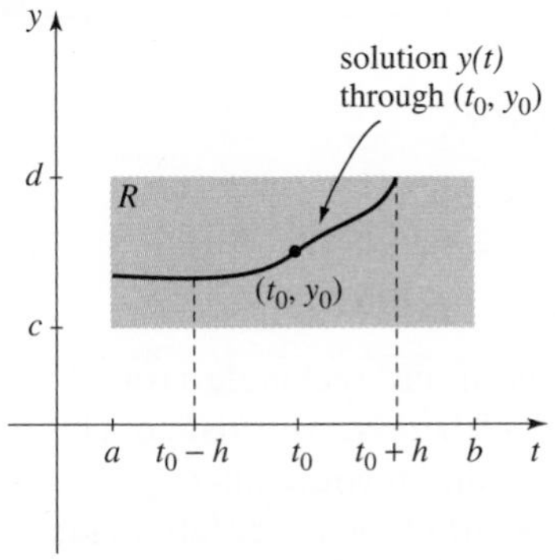

FIGURE 1.5.5 A solution guaranteed by Picard's Theorem.

1. Continuous
   = Solution

2 $f_y(t,y)$ continuous
  unique

> **Picard's Existence and Uniqueness Theorem**
> Suppose that the function $f(t, y)$ is continuous on the region
> $$R = \{(t, y) \mid a < t < b, \ c < y < d\}$$
> and $(t_0, y_0) \in R$. Then there exists a positive number h such that the initial-value problem
> $$y' = f(t, y), \quad y(t_0) = y_0$$
> *has a solution* for t in the interval $(t_0 - h, \ t_0 + h)$.
> If, furthermore, $f_y(t, y)$ is also continuous in R, that solution is *unique*.

For the IVPs of both Examples 2 and 3, there is no open rectangle containing $(0, 0)$ on which f is even defined, let alone continuous.

Figure 1.5.5 suggests how the interval of definition of the solution may be smaller than the interval (a, b) marking the horizontal extent of R. This often happens because the solution curve reaches the top or bottom of R before getting to its left or right end.

> **EXAMPLE 4** **The Obvious** For the simple initial-value problem
> $$y' = 1 + y^2, \quad y(0) = 0,$$
> we have $f(t, y) = 1 + y^2$ and $f_y = 2y$. Both are polynomials and hence are continuous on a rectangle R (any rectangle in fact) containing the initial point $(t_0, y_0) = (0, 0)$. Picard's Theorem guarantees that the initial-value problem has a unique solution on *some* interval (we can't say how large) around $t = 0$.
>
> In this example, it is a simple matter to solve the IVP by separation of variables, which gives $y(t) = \tan t$, and we find that the solution exists on the interval $(-\pi/2, \pi/2)$. The graph is shown in Fig. 1.5.6.
>
> If we asked the broader question, "*What* initial conditions $y(t_0) = y_0$ have a unique solution passing through them?" we would conclude that *all* initial conditions do, because f and f_y are continuous on *any* rectangle R in

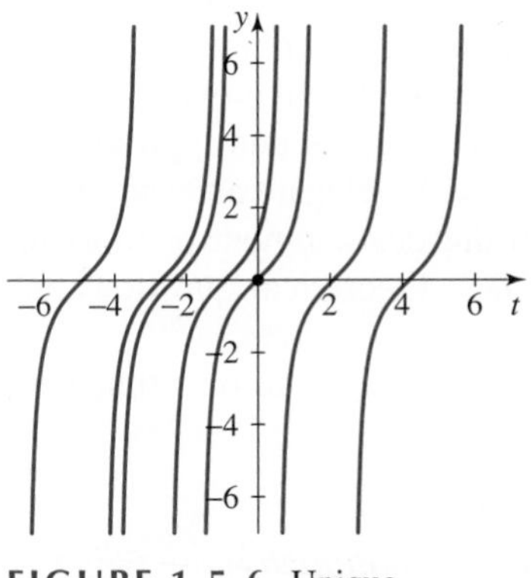

FIGURE 1.5.6 Unique solutions for Example 4.

---

[1] Charles Émile Picard (1856–1941) was one of the greatest mathematicians of the nineteenth century, best known for his contributions in complex variables and differential equations. Before he was 30, Picard was appointed professor at the Sorbonne, an unusual honor. He taught in 1899 at Clark University in Worcester, Massachusetts.

[2] Instead of requiring the continuity of f_y, it is enough to assume that f satisfies a technical requirement (expounded in the mid nineteenth century by Rudolf Otto Sigismund Lipschitz (1832–1903), a German mathematician) that there exists a number K such that $|f(t, y_1) - f(t, y_2)| \leq K|y_1 - y_2|$ for all points (t, y_i) throughout the region of the ty-plane under study. This *Lipschitz condition* includes some functions that do not satisfy the continuity condition. It works for uniqueness because the Lipschitz constant K bounds the rate at which solutions to $y' = f(t, y)$ can pull apart. Further enlightenment on this point is provided in J. H. Hubbard and B. H. West, *Differential Equations: A Dynamical Systems Approach, Part 1* (TAM 5, NY: Springer-Verlag, 1989).

the ty-plane. Again, we can use separation of variables to solve the more general IVP

$$y' = 1 + y^2, \quad y(t_0) = y_0.$$

We obtain the unique solution $y(t_0) = \tan(t + \tan^{-1} y_0 - t_0)$, which exists on an interval around $t = t_0$. ■

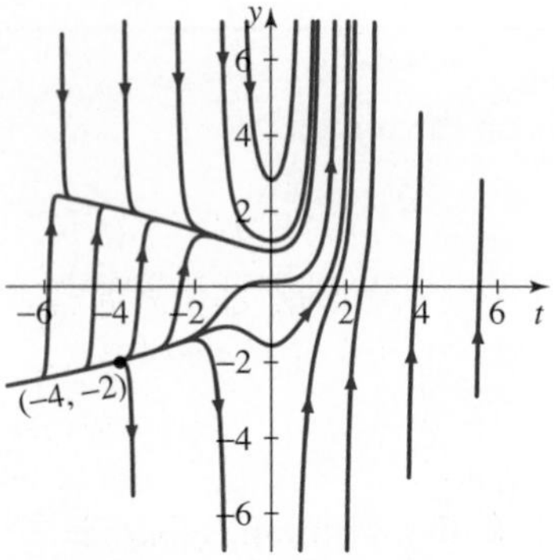

FIGURE 1.5.7 Uniqueness is *not* obvious everywhere for Example 5.

EXAMPLE 5 **The Not So Obvious** For the IVP

$$y' = t^2 + ty^2, \quad y(-4) = -2,$$

we have

$$f = t^2 + ty^2 \quad \text{and} \quad f_y = 2ty.$$

Because both of these functions are continuous in any rectangle around the initial point $(-4, -2)$, the hypothesis of Picard's Theorem is satisfied. Hence we can conclude that the IVP has a unique solution on some interval around $t = -4$. The solutions that seem to be spraying apart in the lower left quadrant of Fig. 1.5.7 are all unique, just very close together. NOTE: You can *see* them as separate instead of merged solutions if you use a computer to zoom in on the area in question. ■

We will not give the proof of Picard's Theorem. One approach is based on converting the IVP to an integral equation (see Problem 29 in Sec. 1.4) and defining a sequence of approximations that tends to the desired solution as a limit.[3] Another approach is to prove that a sequence of Euler approximations for smaller and smaller step sizes converges to the true solution.[4] These proofs offer more than just existence and uniqueness in the abstract. They also give concrete procedures for constructing solutions. (In particular, see Problems 28–38 in this section.)

Most students do not get excited about theoretical topics like existence and uniqueness—they are more interested in how to graph, compute, and approximate. This is too bad, because the theory can be mathematically elegant and satisfying, as well as useful in practice. Theory can help in the development of tools for studying DEs, while new methods may suggest new theoretical questions and even help to resolve them.

Uniqueness

Can you find more than one solution through a given point for $y' = y^{2/3}$ and $y' = y^{4/3}$? What happens near $y = 0$?

Four questions that may be asked about a differential equation often have curiously interrelated answers.

1. Are there any solutions? (*Existence*)
2. How many are there? (*Uniqueness; multiplicity*)
3. What are they like? (*Qualitative theory*)
4. How can we represent them? (*Solution and approximation techniques*)

Each question and its answers can enrich and illuminate the others.

Do not sell theory short! Even though we began this chapter with an emphasis on the third and fourth questions, our discussions really depended on some answers to the first two. We will find that kind of interdependence throughout this book.

---

[3]Such a proof can be found, for example, in M. Braun, *Differential Equations and Their Applications* (NY: Springer-Verlag, 1975).

[4]A proof along these lines can be found in J. H. Hubbard and B. H. West, *Differential Equations: A Dynamical Systems Approach, Part 1* (TAM 5, NY: Springer-Verlag, 1989), 177–178.

Summary

The Picard Theorem gives sufficient conditions for the local existence and uniqueness of solutions of the differential equation $y' = f(t, y)$, and initial-value problems related to it: continuity of f and $\partial f/\partial y$, respectively. The method of proof suggests a procedure for constructing a solution as the limit of a sequence of approximations.

1.5 Problems

Picard's Conditions *For each of Problems 1–8, answer the following questions:*

(a) *Does Picard's Theorem apply to the given IVP? Explain.*

(b) *If your answer to part (a) is yes, is there a largest rectangle for which Picard's conditions hold?*

(c) *If your answer to part (a) is no, are there other initial conditions $y(t_0) = y_0$ for which the answer would be yes? Try describe the set of all such points.*

1. $y' + ty = 1, \quad y(0) = 0$

2. $ty' + y = 2, \quad y(0) = 1$

3. $y' = y^{4/3}, \quad y(0) = 0$

4. $y' = \dfrac{t - y}{t + y}, \quad y(0) = -1$

5. $y' = \dfrac{1}{t^2 + y^2}, \quad y(0) = 0$

6. $y' = \tan y, \quad y(0) = \dfrac{\pi}{2}$

7. $y' = \ln|y - 1|, \quad y(0) = 2$

8. $y' = \dfrac{y}{y - t}, \quad y(1) = 1$

9. Linear Equations The general (and basic) first-order linear differential equation

$$y' + p(t)y = q(t),$$

where p and q are continuous functions on an interval I, will be studied in Sec. 2.2. Show that the Picard conditions apply to any linear DE initial-value problem

$$\begin{aligned} y' + p(t)y &= q(t), \\ y(t_0) &= y_0, \end{aligned}$$

for any t_0 in the interval I.

Eyeballing the Flows *For Problems 10–18, answer the following questions for each of the points A, B, C, and D:*

(a) *Does the differential equation seem to have a unique solution through the point?*

(b) *If yes, on what interval do you think it is defined?*

10.

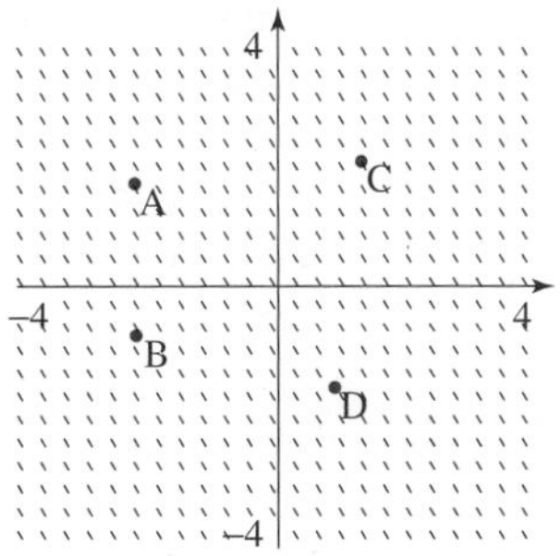

11.

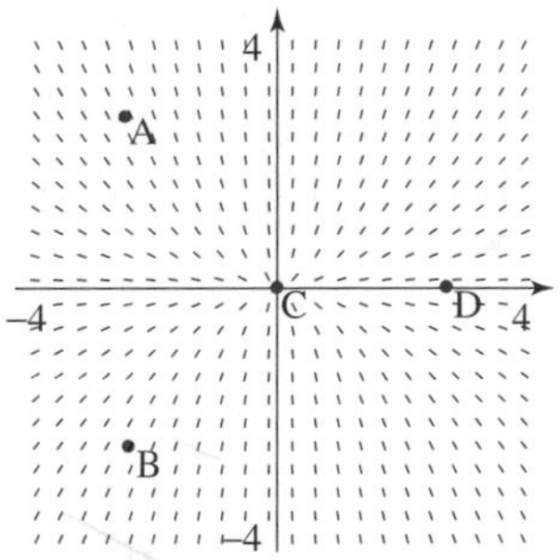

12.

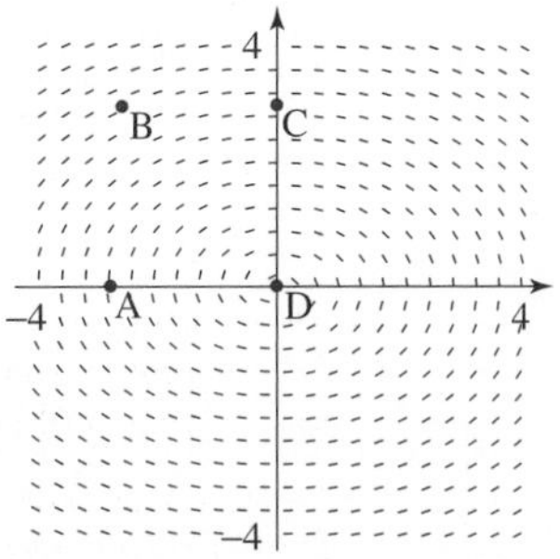

13.

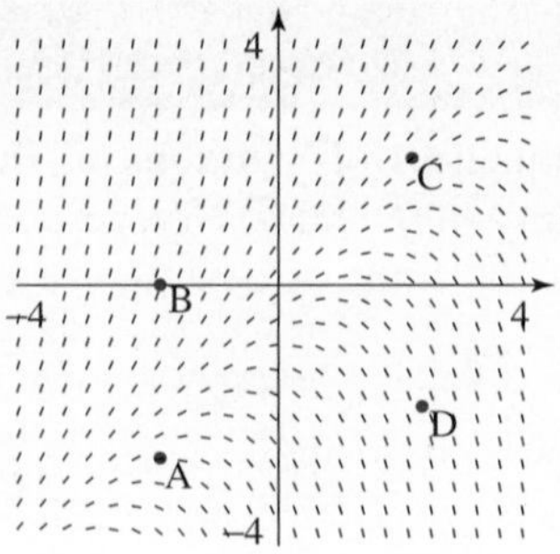

14.

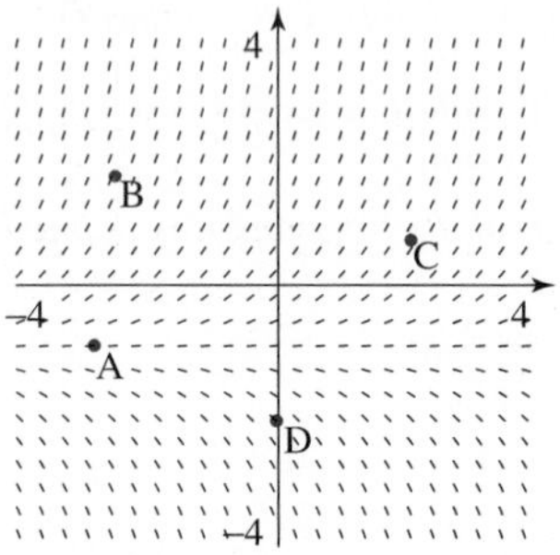

15.

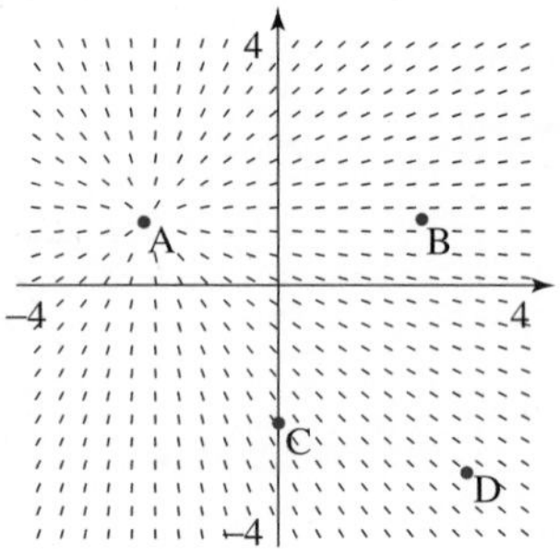

16.

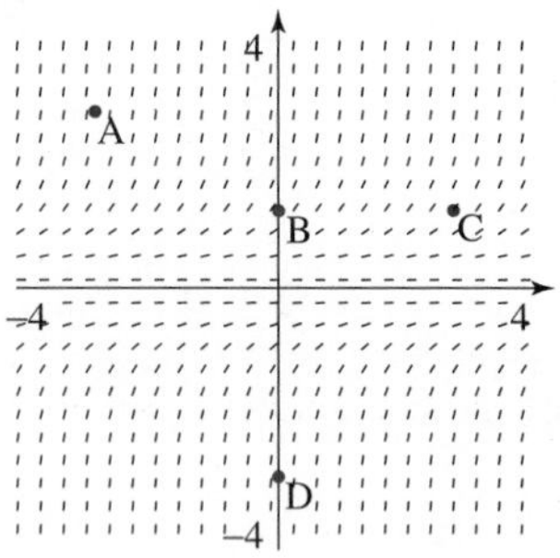

17.

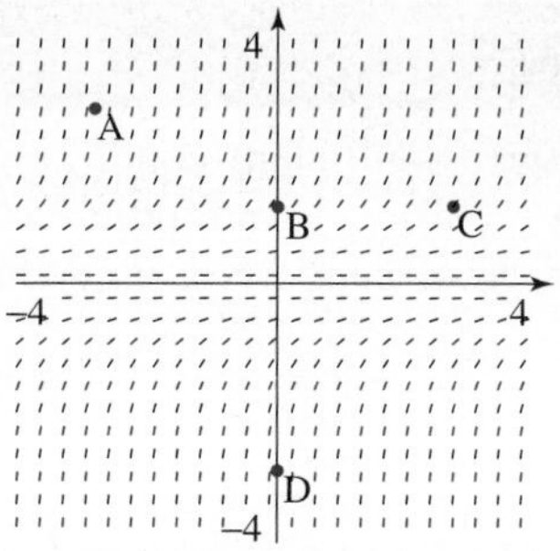

18.

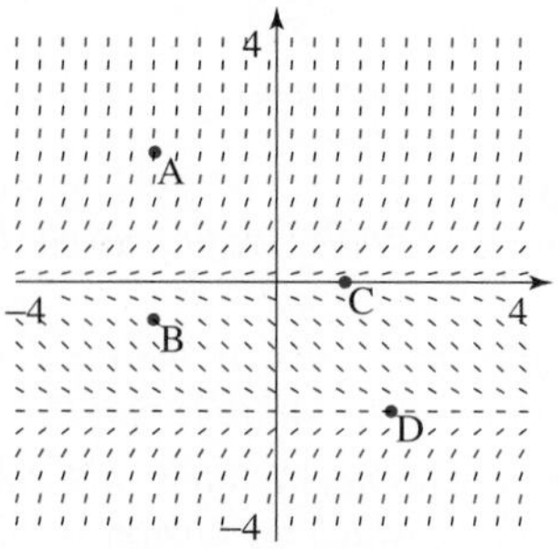

19. Local Conclusions We will investigate the initial-value problem $y' = y^2$, $y(0) = 1$.

(a) Show that Picard's conditions hold. How large a region R can be found?

(b) Draw the direction field and solution.

(c) Solve the IVP using separation of variables. What is the largest t-interval for which this solution is defined? How does it compare with the region R of part (a)?

(d) Generalize the results of parts (a) and (c) to the IVP $y' = y^2$, $y(t_0) = y_0$.

20. Nonuniqueness Show that the IVP $y' = y^{1/3}$, $y(0) = 0$, exhibits nonunique solutions and sketch graphs of several possibilities. What does Picard's Theorem tell you for this problem?

21. More Nonuniqueness For the IVP $y' = \sqrt{y}$, $y(0) = 0$, of Example 3, and for any positive number t_0, show that a solution is given by equation (4). HINT: First show that the DE is satisfied on $(-\infty, t_0)$ and (t_0, ∞). Then verify that the one-sided derivatives agree at $t = t_0$.

22. Seeing versus Believing Look back to Fig. 1.3.1, at the solution in the lower right. Follow it up and backward, toward $y = -1$. Does it actually merge with the solution $y = -1$? Use Picard's Theorem to answer.

23. Converse of Picard's Theorem Fails The conditions of Picard's Theorem may fail at a given point for a differential equation, but the equation may still have a unique solution through the point. (In other words, the *converse* of Picard's Theorem does not hold.)

(a) Show that the uniqueness condition for Picard's Theorem does not apply to the differential equation

$$\frac{dy}{dt} = |y| \tag{5}$$

when $y = 0$.

(b) Find the general solution of the differential equation (5), and verify that the only solution with initial condition $y(0) = 0$ is $y(t) \equiv 0$, thus showing that negation of the hypothesis does *not* insure negation of the conclusion.

24. Hubbard's Leaky Bucket[5] If you are given an empty bucket like the one in Fig. 1.5.8, having a hole in the bottom from which all the water has leaked, can you tell how long ago the bucket was full? Of course not, and the answer can be related to nonunique solutions. The equation describing the height h of the water level comes from Torricelli's Law:

$$\frac{dh}{dt} = -k\sqrt{h},$$

where the positive constant k depends on the size and shape of the bucket and of the hole.

(a) Even before setting a time scale, we can see that Picard's Theorem will not apply. Explain why.

(b) Formulate a general initial condition and solve the DE by separation of variables. Why is it impossible to tell how much time has elapsed? Draw several possible solution curves.

(c) Show that the total emptying time is given by $2\sqrt{h_0}/k$, where h_0 is the initial height of the water in the bucket.

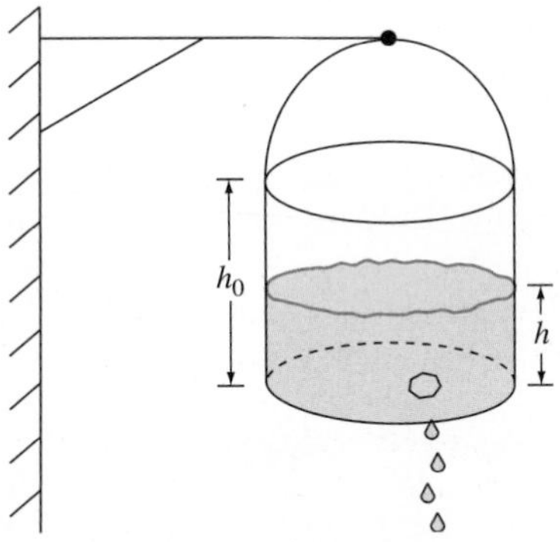

FIGURE 1.5.8 Hubbard's leaky bucket.

25. The Melted Snowball The rate of change dV/dt of the volume V of a melting snowball is proportional to its surface area, so

$$\frac{dV}{dt} = -kV^{2/3}$$

for positive constant of proportionality k.

(a) Explain how the relationship between the surface area and volume of a sphere leads to the power $2/3$.

(b) Suppose that you find a puddle of water that was formerly a snowball; that is, you now have $V = 0$. Can you tell when the snowball melted? Why?

(c) Solve the DE by separation of variables and draw several possible solution curves.

(d) How do your results fit with the Picard Theorem?

26. The Accumulating Raindrop Suppose that a spherical raindrop falling through a moist atmosphere grows at a rate proportional to its surface area.

(a) Explain why

$$\frac{dV}{dt} = kV^{2/3}, \quad k \text{ a positive constant,}$$

models this situation.

(b) Demonstrate nonuniqueness for $dV/dt = kV^{2/3}$, $V(0) = 0$, by constructing several solutions.

27. Different Translations

(a) Show that for arbitrary $a \geq 0$, $y' = y$ has infinitely many solutions that can be written

$$y(t) = e^{(t-a)}.$$

(b) Show that for arbitrary $a \geq 0$, the IVP

$$s' = 2\sqrt{s}, \quad s(0) = 0,$$

has infinitely many solutions

$$s(t) = \begin{cases} 0 & \text{if } t < a, \\ (t-a)^2 & \text{if } t \geq a \end{cases}.$$

(c) Sketch the similar graphs of the two families in (a) and (b). Explain in terms of uniqueness where and why they differ.

[5]Based on an example from J. H. Hubbard and B. H. West, *Differential Equations: A Dynamical Systems Approach, Part 1* (TAM 5, NY: Springer-Verlag, 1989), Sec. 4.2.

Picard Approximations *In Problem 29 of Sec. 1.4 it was shown that the IVP*

$$y' = f(t, y), \quad y(t_0) = y_0,$$

is equivalent to the integral equation

$$y(t) = y_0 + \int_{t_0}^{t} f(s, y(s))\, ds.$$

(Any solution of one is a solution of the other also.) This leads to the following.

> **Picard's Successive Approximations**
>
> *Beginning with a fairly arbitrary first approximation $y_0(t)$ (for example, $y_0(t) \equiv y_0$, the constant function), we define a sequence of approximations $y_1, y_2, \ldots, y_n, \ldots$, using the formula*
>
> $$y_{n+1}(t) = y_0 + \int_{t_0}^{t} f(s, y_n(s))\, ds, \quad n = 0, 1, \ldots.$$

The proof of Picard's Theorem depends on showing both that the sequence $y_0, y_1, y_2, \ldots, y_n, \ldots$, of functions tends to a limit function $y(t)$, and that y is a solution of the integral equation, hence of the initial-value problem.

In Problems 28–31, use the function given as the initial approximation y_0 to the solution of the IVP

$$y' = t - y, \quad y(0) = 1,$$

and generate the first three Picard approximations y_1, y_2, y_3 in each case.

28. $y_0(t) = 1$

29. $y_0(t) = t - 1$

30. $y_0(t) = e^{-t}$

31. $y_0(t) = 1 + t$

32. Computer Lab

(a) Using a computer algebra system to evaluate the necessary integrals, obtain additional Picard approximations for the IVP in Problems 28–31.

(b) Try to guess an analytical solution from these approximations.

Calculator or Computer *Use Picard's Theorem and direction fields to study the following DEs.*

(a) *At what points do the corresponding IVPs fail to have solutions?*

(b) *At what points does uniqueness fail?*

33. $y' = y^{1/4}$

34. $y' = \sin ty$

35. $y' = y^{5/3}$

36. $y' = (ty)^{1/3}$

37. $y' = (y - t)^{1/3}$

38. $y' = 6t^2 - 3y/t$

39. Suggested Journal Entry Discuss the two conditions for existence and uniqueness in Picard's Theorem. Do they seem plausible? Intuitive? The partial derivative f_y measures the change in the slope field in the y-direction, sometimes called the dispersion. Can you relate this to uniqueness?

CHAPTER 2 Linearity and Nonlinearity

If differential equations are the modeling language of the world, then linear algebra must be the grammar rules that bind the language.

—Mark Parker, Carroll College

2.1 Linear Equations: The Nature of Their Solutions

SYNOPSIS: We begin to outline the algebraic structure common to both linear algebraic and linear differential equations. We introduce linear operators and the Superposition Principle of solutions for linear homogeneous equations. We then present the important Nonhomogeneous Principle, which gives the form of the solutions for the general nonhomogeneous linear equation, either algebraic or differential, as the sum of the homogeneous solutions plus any particular solution.

What Is Linear?

First of all, very few physical systems are purely linear. This is unfortunate, since mathematicians have over the years studied linear systems much more extensively than nonlinear systems. But linearity represents an ideal behavior that is seldom attained in the natural world of the scientist or the human-made world of the engineer.

So why study linear systems at all? A standard argument is that they are easy, but you probably will not find that very satisfying. We will emphasize linearity because many nonlinear systems can be approximated by linear systems. Many of the standard models in electrical and mechanical engineering are linear. And many systems behave in a linear manner over a significant portion of their ranges, even though they are nonlinear elsewhere. Most springs behave linearly for small vibrations, for example, and many electrical circuits follow linear patterns for small voltages and currents. And even when systems behave in a nonlinear fashion, they can often be *linearized* (approximated by systems that *are* linear). The occurrences of linearity are widespread and its implications are immense, both in differential equations and in the companion subject of linear algebra, which we will begin to study in detail in Chapter 3.

The concept of *linearity* is central to our investigations. We know that the equation $y = ax + b$ is a linear equation by definition. As our first step we will generalize the properties of this equation to any number of variables.

Linear Algebraic Equation

An equation $F(x_1, x_2, \ldots, x_n) = C$ is **linear** if it is of the form

$$a_1x_1 + a_2x_2 + \cdots + a_nx_n = C, \tag{1}$$

where $a_1, a_2, \ldots, a_n$ and C are constants.

If $C = 0$, the equation is said to be **homogeneous**.

Notice the characteristic look of a linear algebraic equation. Each variable occurs to the first power, and no variables are multiplied.

EXAMPLE 1 **Geometry** The equation $5x + 4y - 3z = 4$ is a linear equation that describes a plane in 3-space. ■

EXAMPLE 2 **Time for Fun?** A student has a monthly entertainment budget of \$50. The expected costs are as follows: \$8 for a movie ticket, \$2 for a video rental, \$7 for a new paperback book, and \$12 for a compact disc. This information is represented by the linear equation

$$8m + 2v + 7b + 12d = 50,$$

where m, v, b, and d represent the number of movie tickets, videos, books, and CDs, respectively, that the student can buy or rent in a month if the entire budget is spent. ■

EXAMPLE 3 **Recognizing Linearity in Algebraic Equations** Examine the following equations to determine which ones are linear equations:

$$4x - 3e^x = 15$$
$$4x - 2y + 3\sqrt{z} = 12$$
$$2x - 3y + 4z + 3 = w$$

It is clear that the term e^x and $\sqrt{z}$ make the first two equations nonlinear. The third equation is linear. ■

We can generalize the concept of a linear equation to a linear differential equation, where each variable is a function of t, and $y, y', y'', \ldots, y^{(n)}$ are the function $y(t)$ and its successive derivatives. The role of the constants is now taken by functions of the independent variable t.

Linear Differential Equation

A differential equation $F(y, y', y'', \ldots, y^{(n)}) = f(t)$ is **linear** if it is of the form

$$a_n(t)\frac{d^ny}{dt^n} + a_{n-1}(t)\frac{d^{n-1}y}{dt^{n-1}} + \cdots + a_1(t)\frac{dy}{dt} + a_0(t)y = f(t), \tag{2}$$

Independent vs. Dependent Variables:

The coefficients $a_n(t)$ need *not* be linear in the independent variable t. The term "linear" for a DE refers only to the *dependent* variable and its derivatives.

where all functions of t are assumed to be defined over some common interval I.

If $f(t) = 0$ over the interval I, the differential equation is said to be **homogeneous**.

In particular, the general first-order linear differential equation can be written

$$y' + p(t)y = f(t);$$

Typically, in **mechanical problems**,

- y is a spatial coordinate;
- t represents time;
- $f(t)$ is a forcing function.

the general second-order linear differential equation is

$$y'' + p(t)y' + q(t)y = f(t).$$

A differential equation that *cannot* be written in the form required by the definition of a linear DE is a *nonlinear* differential equation.

EXAMPLE 4 **Recognizing Linear DEs** The trick for identifying linearity in differential equations is to focus on the *dependent* variable, which is y in these examples. Which of the following DEs are linear?

(a) $y'' + ty' - 3y = 0$ **(b)** $y' + y^2 = 0$ **(c)** $y' + \sin y = 1$
(d) $y' + t^2y = 0$ **(e)** $y' + (\sin t)y = 1$ **(f)** $y'' - 3y' + y = \sin t$

Only two of these equations are not linear, those with y^2 or $\sin y$ terms. All the others are linear in y and its derivatives. ■

Several more differential equations are classified in Table 2.1.1 according to the properties we've just introduced. The concepts of homogeneity and constant coefficients are not applicable to nonlinear equations, and homogeneity requires $f(t) = 0$ *over the entire interval*. For the three DEs that are nonlinear, exactly what makes them so?

Table 2.1.1 Classifying differential equations $F(y, y', y'', \ldots, y^{(n)}) = f(t)$

Differential Equation	Order n	Linear or Nonlinear	Homogeneous or Nonhomogeneous	Coefficients
(a) $y' + ty = 1$	1	*linear*	*nonhomogeneous*	*variable*
(b) $y'' + yy' + y = t$	2	*nonlinear*	—	—
(c) $y'' + ty' + y^2 = 0$	2	*nonlinear*	—	—
(d) $y'' + 3y' + 2y = 0$	2	*linear*	*homogeneous*	*constant*
(e) $y'' + y = \sin y$	2	*nonlinear*	—	—
(f) $y^{(4)} + 3y = \sin t$	4	*linear*	*nonhomogeneous*	*constant*

Nature of Solutions of Linear Equations

Before we actually solve anything, we need to show two basic facts about linear equations, in any setting: **superposition** of solutions to *homogeneous* equations, and a principle for solving *nonhomogeneous* equations.

Let us introduce some simplifying notation, which we will subsequently find very helpful. First we use a single boldface variable to stand for a whole set

Variables:

of variables in a *linear* equation. (This is a *vector* notation to be formalized in Sec. 3.1.)

Algebraic

- For linear algebraic equations of the form (1),

$$\vec{\mathbf{x}} = [x_1, x_2, \ldots, x_n].$$

Differential

- For linear differential equations as in (2),

$$\vec{\mathbf{y}} = [y^{(n)}, y^{(n-1)}, \ldots, y', y].$$

Operators:

We introduce a **linear operator** L as a shorthand to represent an entire operation performed on the variables, as follows:

Algebraic

- For linear algebraic equations of the form (1),

$$L(\vec{\mathbf{x}}) = a_1x_1 + a_2x_2 + \cdots + a_nx_n.$$

Differential

- For linear differential equations as in (2),

$$L(\vec{\mathbf{y}}) = a_n(t)y^{(n)} + a_{n-1}(t)y^{(n-1)} + \cdots + a_1(t)y' + a_0(t)y.$$

EXAMPLE 5 **Operators for Linear DEs** To find the linear operator in a differential equation, isolate all terms in y and its derivatives on the left-hand side. If the left-hand side now has the linear form prescribed by (2), then the linear operator $L(y)$ is exactly that left-hand side of the DE. Thus, in Table 2.1.1 we have the following linear operators for the three linear equations.

(a) $L(\vec{\mathbf{y}}) = y' + ty$ **(d)** $L(\vec{\mathbf{y}}) = y'' + 3y' + 2y$ **(f)** $L(\vec{\mathbf{y}}) = \dfrac{d^4y}{dt^4} + 3y$ ■

Solutions:

Now we can write (1) simply as $L(\vec{\mathbf{x}}) = C$ and (2) as $L(\vec{\mathbf{y}}) = f(t)$.

Algebraic

- A solution of (1) is any $\vec{\mathbf{x}} = [x_1, x_2, \ldots, x_n]$ that satisfies (1).

Differential

- A solution of (2) is any n-times differentiable function y that satisfies (2)—usually just the function y is given, rather than the entire set of its derivatives $\vec{\mathbf{y}} = [y^{(n)}, y^{(n-1)}, \ldots, y', y]$.

Homogeneous Linear Equations:

Useful properties of solutions to *homogeneous* linear equations, both algebraic and differential, are the facts that:

Algebraic *and* **Differential**

- A constant multiple of a solution is also a solution.
- The sum of two solutions is a solution.

We can easily confirm that the operators denoted by L have the following properties, which characterize *linear* operators.

Linear Operator Properties

$$L(k\vec{\mathbf{u}}) = kL(\vec{\mathbf{u}}), \quad k \in \mathbb{R} \tag{3}$$

$$L(\vec{\mathbf{u}} + \vec{\mathbf{w}}) = L(\vec{\mathbf{u}}) + L(\vec{\mathbf{w}}) \tag{4}$$

Differential Operator:

The linear operator used most often is D, where

$$D(y) \equiv \frac{dy}{dt}.$$

Then $y'' + 2y' + 3y$ can be written as

$$L(y) = (D^2 + 2D + 3)y.$$

For the algebraic operator, (3) and (4) can be proved directly by substitution. For the differential operator, (3) and (4) follow from the corresponding properties of the derivative; for example, $(ky)' = ky'$ and $(f + g)' = f' + g'$.

We can state the following, in general:

Superposition Principle for Linear Homogeneous Equations

Let $\vec{\mathbf{u}}_1$ and $\vec{\mathbf{u}}_2$ be any solutions of the *homogeneous linear* equation

$$L(\vec{\mathbf{u}}) = 0.$$

- Their *sum* $\vec{\mathbf{u}} = \vec{\mathbf{u}}_1 + \vec{\mathbf{u}}_2$ is also a solution.
- A *multiple* $\vec{\mathbf{u}} = k\vec{\mathbf{u}}_1$ is a solution for any constant k.

The Superposition Principle follows directly from the properties of a linear operator. We stress the common applicability of this basic principle to both algebraic and linear differential operators, which are illustrated (not proven) in the following example.

EXAMPLE 6 Checking Out Superposition

(a) The points $(1, 3)$ and $2(1, 3) = (2, 6)$ lie on the line

$$y = 3x \quad (\text{or } y - 3x = 0),$$

thus illustrating property (3) with $k = 2$. Property (4) is illustrated by the fact that

$$(1 + 2, 3 + 6) = (3, 9).$$

(b) Likewise the points $(2, 3, 5)$ and $(-4, -5, -9)$ lie in the plane

$$x + y - z = 0,$$

and you can check that the sum $(-2, -2, -4)$ does as well.

(c) Two solutions of

$$y'' - 4y = 0$$

are $y = e^{2t}$ and $y = e^{-2t}$. We can verify that $y = 2e^{2t} + 3e^{-2t}$ is also a solution:

$$\begin{aligned} y' &= 4e^{2t} - 6e^{-2t}, \\ y'' &= 8e^{2t} + 12e^{-2t}, \\ y'' - 4y &= (8e^{2t} + 12e^{-2t}) - 4(2e^{2t} + 3e^{-2t}) = 0. \end{aligned}$$

■

We use the properties of linear operators to obtain another important result.

Nonhomogeneous Principle

Let $\vec{\mathbf{u}}_p$ be any solution (called a particular solution) to a *linear nonhomogeneous* equation

$$L(\vec{\mathbf{u}}) = C \qquad \text{(algebraic)}$$

or

$$L(\vec{\mathbf{u}}) = f(t). \qquad \text{(differential)}$$

Then

$$\vec{\mathbf{u}} = \vec{\mathbf{u}}_h + \vec{\mathbf{u}}_p$$

is also a solution, where $\vec{\mathbf{u}}_h$ is a solution to the *associated homogeneous* equation

$$L(\vec{\mathbf{u}}) = 0.$$

Furthermore, every solution of the nonhomogeneous equation must be of the form $\vec{\mathbf{u}} = \vec{\mathbf{u}}_h + \vec{\mathbf{u}}_p$.

Proof Using property (4) for the linear operator L, we show that the desired sum $\vec{\mathbf{u}} = \vec{\mathbf{u}}_h + \vec{\mathbf{u}}_p$ is indeed a solution:

$$\begin{aligned} L(\vec{\mathbf{u}}) = L(\vec{\mathbf{u}}_h + \vec{\mathbf{u}}_p) &= \underbrace{L(\vec{\mathbf{u}}_h)} + \underbrace{L(\vec{\mathbf{u}}_p)} \\ &= \quad 0 \quad + \; f(t). \end{aligned}$$

Then, to show that every solution has this form, suppose that $\vec{\mathbf{u}}_q$ is also a solution to the linear nonhomogeneous equation and note that $\vec{\mathbf{u}}_q = \vec{\mathbf{u}}_p + (\vec{\mathbf{u}}_q - \vec{\mathbf{u}}_p)$.

We use the properties of a linear operator to show that $\vec{\mathbf{u}}_q - \vec{\mathbf{u}}_p$ is a solution of $L(\vec{\mathbf{u}}) = 0$:

$$\begin{aligned} 0 = L(\vec{\mathbf{u}}_q) - L(\vec{\mathbf{u}}_p) &= L(\vec{\mathbf{u}}_q) + (-1)L(\vec{\mathbf{u}}_p) \\ &= L(\vec{\mathbf{u}}_q) + L(-\vec{\mathbf{u}}_p) \qquad \textit{property (3)} \\ &= L(\vec{\mathbf{u}}_q - \vec{\mathbf{u}}_p). \qquad \textit{property (4)} \end{aligned}$$

Hence, we see that every solution u_q of a nonhomogeneous linear equation is the sum of any single solution u_p plus the solutions of the corresponding homogeneous equation. □

The Superposition and Nonhomogeneous Principles are the fundamental structural theorems for both linear algebra and differential equations, and will be recurrent themes throughout the course. The Nonhomogeneous Principle leads to a three-step solution strategy for linear equations, whether algebraic or differential:

Solving Nonhomogeneous Linear Equations

Step 1. Find all $\vec{\mathbf{u}}_h$ of $L(\vec{\mathbf{u}}) = 0$.

Step 2. Find any $\vec{\mathbf{u}}_p$ of $L(\vec{\mathbf{u}}) = f$.

Step 3. Add them, $\vec{\mathbf{u}} = \vec{\mathbf{u}}_h + \vec{\mathbf{u}}_p$, to get all solutions of $L(\vec{\mathbf{u}}) = f$.

In the following examples, we apply the solution strategy to nonhomogeneous linear differential equations. (See Section 3.2 Example 9 for a nice *algebraic* example.)

EXAMPLE 7 **Nonhomogeneous Linear Differential Equation** Consider the simple DE

$$y' - y = t. \tag{5}$$

Step 1. Solve the associated homogeneous equation $y' - y = 0$, or $y' = y$. A first-order homogeneous linear DE is always separable, so we get

$$y_h = ce^t, \quad \text{for any } c \in \mathbb{R}.$$

Step 2. Confirm, by differentiation and substitution in (5), that

$$y_p = -t - 1$$

is a particular solution. (We will show how to *find* such solutions in Sec. 2.2.)

Step 3. Form the sum, and confirm that $y_h + y_p = ce^t - t - 1$ is a solution.

We can envision combining the y_h and y_p graphs, as shown in Fig. 2.1.1, to obtain the graph of solutions y.

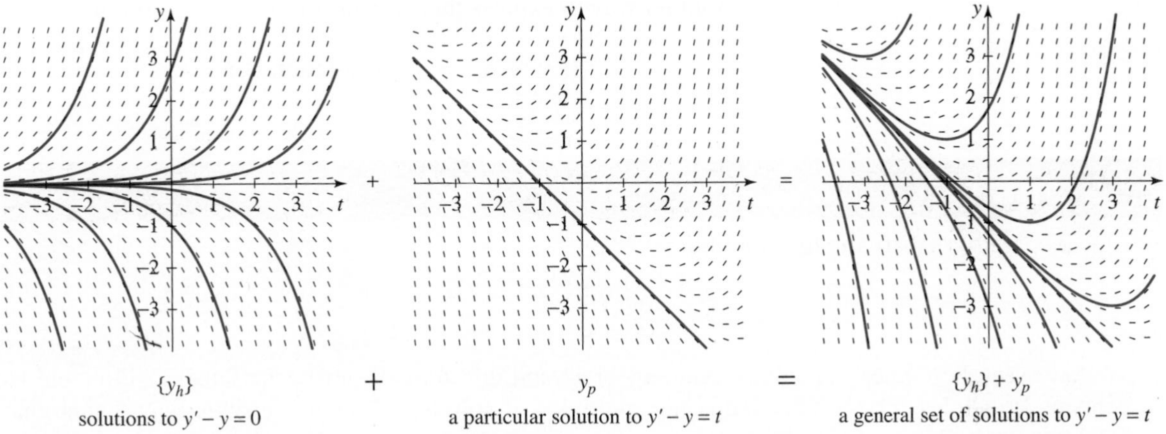

FIGURE 2.1.1 Add to the solutions of the homogeneous equation the particular solution y_p to arrive at the solution set for $y' - y = t$.

Of course, the big question is how do we *find* particular solutions? In the next subsection we will show different methods for obtaining y_p for linear differential equations. But sometimes you can "see" them without heavy machinery (once you know how to look).

EXAMPLE 8 **Example with a Moral** Solve

$$y' + ay = b, \tag{6}$$

where a and b are constants.

Step 1. The corresponding homogeneous equation will soon be (if it is not already) so familiar that you can solve it in your head: $y_h = ce^{-at}$, where c is an arbitrary constant.

Step 2. To find a particular solution y_p, just *look* at the DE and see that the constant function $y_p = b/a$ will work! (You can get the same result with a lot more work by other means.)

Step 3. Thus the general solution to equation (6) is

$$y = \underbrace{ce^{-at}}_{y_h} + \underbrace{\frac{b}{a}}_{y_p}.$$

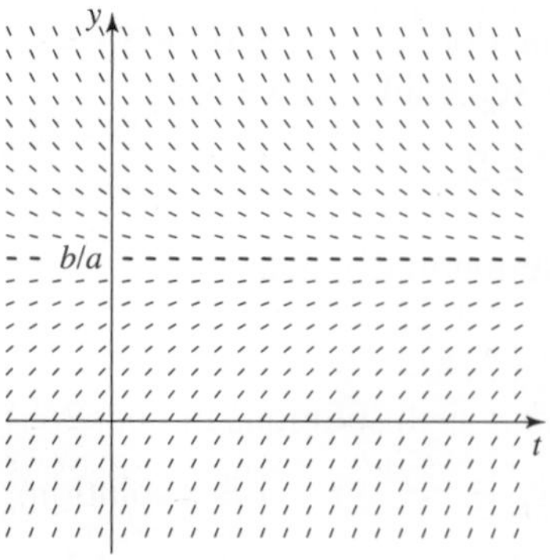

FIGURE 2.1.2 Direction field for $y' + ay = b$.

Another way to approach this problem is to think of the direction field, shown in Fig. 2.1.2. It says all the same things visually—that in the long run solutions tend to b/a, and that at first as t increases, solutions are increasing toward b/a if $y(0) < b/a$, decreasing toward b/a if $y(0) > b/a$. ■

Example 8 shows the wisdom of *thinking* a bit before you calculate. There is no need to go through fancy procedures or formulas, since on the direction field the constant solution is staring you in the face!

If you form the habit of inspecting equations for easy solutions, especially constant solutions, before setting out on a long calculation, you can often obtain the result more quickly and avoid errors. For example, you can find the solution of $y' + 4y = 8$ in your head by thinking "homogeneous solution plus particular solution." Have you got it yet? Here it is: $y = y_h + y_p = ce^{-4t} + 2$.

Another way to express this very useful result is as follows:

$$\text{If } y' + ay = b, \text{ the solution is } y = ce^{-at} + \frac{b}{a}.$$

Summary

Linear operators satisfy the linear properties

$$L(ku) = kL(u),$$
$$L(u + w) = L(u) + L(w).$$

We have seen how linear equations, both algebraic and differential, satisfy the Superposition and Nonhomogeneous Principles. Thus, sums and scalar multiples of solutions to homogeneous linear equations are also solutions, and solutions to nonhomogeneous linear equations have the form

$$u = u_h + u_p.$$

2.1 Problems

Classification *In Problems 1–10, classify the differential equation according to its order, whether it is linear or nonlinear, whether it is homogeneous or nonhomogeneous, and whether its coefficients are constant or variable.*

1. $\dfrac{dy}{dt} + ty^2 = 1$

2. $t\dfrac{dy}{dt} + y = \sin t$

3. $e^t \dfrac{d^2y}{dt^2} + 2\dfrac{dy}{dt} + y = 0$

4. $t\dfrac{d^2y}{dt^2} + \dfrac{dy}{dt} + ty = 1$

5. $\dfrac{d^3y}{dt^3} + \dfrac{d^2y}{dt^2} + \dfrac{dy}{dt} + y = 0$

6. $\dfrac{d^3y}{dt^3} + y = \sin t$

7. $t\dfrac{d^2u}{dt^2} + t\dfrac{du}{dt} + 3u = 1$

8. $\dfrac{d^2w}{dt^2} + w^2\dfrac{dw}{dt} + w = 0$

9. $\dfrac{d^2v}{dt^2} = t^2 v$

10. $\dfrac{d^2y}{dt^2} + y^2 = 0$

11. Linear Operator Notation For the differential equations of Example 4, find the linear operators.

Linear and Nonlinear Operators *An operator is linear if it satisfies the two linear properties (3) and (4); otherwise, it is nonlinear. Which of the following differential operators are linear and which are nonlinear?*

12. $L(y) = y' + 2y$

13. $L(y) = y' + y^2$

14. $L(y) = y' + 2ty$

15. $L(y) = y' - e^t y$

16. $L(y) = y'' + (\sin t)y$

17. $L(y) = y'' + (1 - y^2)y' + y$

Pop Quiz *Solve each of the following equations by inspection in less than 10 seconds. If the equation has an initial condition, find the arbitrary constant in the general solution.*

18. $y' + 2y = 1$

19. $y' + y = 2$

20. $y' - 0.08y = 100$

21. $y' - 3y = 5$

22. $y' + 5y = 1$, $y(1) = 0$

23. $y' + 2y = 4$, $y(0) = 1$

24. Superposition Principle Show that if $y_1(t)$ and $y_2(t)$ are solutions of $y' + p(t)y = 0$, then so are $y_1(t) + y_2(t)$ and $cy_1(t)$ for any constant c.

25. Second-Order Superposition Principle Show that if $y_1(t)$ and $y_2(t)$ are solutions of $y'' + p(t)y' + q(t)y = 0$, then so is $c_1y_1(t) + c_2y_2(t)$ for any constants c_1 and c_2.

Verifying Superposition *For Problems 26–31, verify that the given functions y_1 and y_2 are solutions of the given differential equation, then show that $c_1y_1(t) + c_2y_2(t)$ is also a solution for any real numbers c_1 and c_2.*

26. $y'' - 9y = 0;$ $y_1 = e^{3t};$ $y_2 = e^{-3t}$

27. $y'' + 4y = 0;$ $y_1 = \sin 2t;$ $y_2 = \cos 2t$

28. $2y'' + y' - y = 0;$ $y_1 = e^{t/2};$ $y_2 = e^{-t}$

29. $y'' - 5y' + 6y = 0;$ $y_1 = e^{2t};$ $y_2 = e^{3t}$

30. $y'' - y' - 6y = 0;$ $y_1 = e^{3t};$ $y_2 = e^{-2t}$

31. $y'' - 9y = 0;$ $y_1 = \cosh 3t;$ $y_2 = \sinh 3t$

32. Different Results? Are the solutions y_1 and y_2 given in Problems 26 and 31 really different solutions to the differential equation $y'' - 9y = 0$? Explain.

33. Many From One The function $y(t) = t^2$ is a solution of $y' - \frac{2}{t}y = 0$. Can you find any more solutions? Why do you know these other functions are solutions without substituting them into the equation?

Guessing Solutions *In Problems 34–42, make an educated guess to find a formula for at least one solution y_p of the nonhomogeneous or higher-order equation. After you guess, check your answer by differentiating and substituting in the DE.*

34. $y' + y = e^{-t}$ **35.** $y' + y = e^t$ **36.** $y' - ty = 0$

37. $y' - y = e^t$ **38.** $y'' - a^2y = 0$ **39.** $y' + y/t = t^3$

40. $y'' + y' = 0$ **41.** $y'' + a^2y = 0$ **42.** $y'' - y' = 0$

Nonhomogeneous Principle *For each of the differential equations in Problems 43–46, we give one solution $y(t)$. Verify that $y(t)$ is a solution, then find the rest of the solutions.* HINT: *Use separation of variables to find the homogeneous solutions.*

43. $y' - y = 3e^t,$ $y_p(t) = 3te^t$

44. $y' + 2y = 10\sin t,$ $y_p(t) = 4\sin t - 2\cos t$

45. $y' - \dfrac{2}{t}y = t^2,$ $y_p(t) = t^3$

46. $y' + \dfrac{1}{t+1}y = 2,$ $y_p(t) = \dfrac{t^2 + 2t}{t+1}$

Third-Order Examples *For each of the nonhomogeneous linear DEs in Problems 47 and 48.*

(a) *Verify that the given y_1, y_2, y_3 satisfy the corresponding homogeneous equation.*

(b) *Use the Superposition Principle, with appropriate coefficients, to state the general solution $y_h(t)$ to the corresponding homogeneous equation.*

(c) *Verify that the given $y_p(t)$ is a particular solution to the given nonhomogeneous DE.*

(d) *Use the Nonhomogeneous Principle to write the general solution $y(t)$ to the nonhomogeneous DE.*

(e) *Solve the IVP consisting of the nonhomogeneous DE and the given initial conditions*

47. $y''' - y'' - y' + y = 2t - 1 + 3e^{2t}$

$y_1(t) = e^t,\ y_2(t) = te^t,\ y_3(t) = e^{-t}$

$y_p(t) = 2t + 1 + e^{2t}$

$y(0) = 4,\ y'(0) = 3,\ y''(0) = 4$

48. $y''' + y'' - y' - y = 4\sin t + 3$

$y_1 = e^t,\ y_2 = e^{-t},\ y_3 = te^{-t}$

$y_p = \cos t - \sin t - 3$

$y(0) = 1,\ y'(0) = 2,\ y''(0) = 3$

49. Suggested Journal Entry Write a summary of your concept of the Nonhomogeneous Principle and tell how it applies to the linear DE $2y' + 3y = 5$.

2.2 Solving the First-Order Linear Differential Equation

SYNOPSIS: We solve the general linear differential equation in three steps: find all solutions of the corresponding homogeneous equation; find a particular solution of the nonhomogeneous equation; then add them together.

Euler-Lagrange Two-Stage Method

We saw in Sec. 2.1 that the general solution of the first-order linear equation

$$y' + p(t)y = f(t) \tag{1}$$

has the form $y(t) = y_h + y_p$, where y_h are the solutions of the corresponding homogeneous equation

$$y' + p(t)y = 0 \tag{2}$$

and y_p is any single solution of (1). We solved the corresponding homogeneous equation (2) in Sec. 1.3 using separation of variables, getting a one-parameter family of solutions

$$y_h = ce^{-\int p(t)\,dt}, \tag{3}$$

where c is an arbitrary constant. So now we are halfway finished.

Variation of Parameters:

Change an arbitrary *constant* c to an arbitrary *function* $v(t)$. This general technique will work in other situations and with higher-order equations.

The second step is to find a single solution y_p of the nonhomogeneous equation (1), called the **particular solution,**[1] which we obtain by a clever technique called **variation of parameters**, developed by the French mathematician Joseph Louis Lagrange.[2] Lagrange suspected that the nonhomogeneous solution might be some "modification" of the homogeneous solution because the left-hand side of the DE is the same. So he changed the constant c in the homogeneous solution to a *function* $v(t)$ and tried a solution of the form

$$y_p(t) = v(t)e^{-\int p(t)\,dt}, \tag{4}$$

What is the point of adding homogenous solutions to particular solutions?

calling the unknown function $v(t)$ a *varying parameter*. The idea is to substitute y_p into the nonhomogeneous equation (1) to find out what $v(t)$ must be. So we substitute, differentiate, and simplify:

$$\underbrace{\left[v'e^{-\int p(t)\,dt} + ve^{-\int p(t)\,dt}(-p(t))\right]}_{y_p'} + \underbrace{p(t)v(t)e^{-\int p(t)\,dt}}_{p(t)y_p} = f(t).$$

The happy surprise is that many terms cancel out, and we are left with a result to remember:

$$v'e^{-\int p(t)\,dt} = f(t). \tag{5}$$

Solving (5) for v' gives $v' = f(t)e^{\int p(t)\,dt}$, and this can be integrated to obtain

$$v(t) = \int f(t)e^{\int p(t)\,dt}\,dt.$$

Constants of integration for $\int p(t)dt$ are absorbed by the arbitrary constant in the larger integration, so they can be ignored.

Now, having found $v(t)$, we have determined a particular solution

$$y_p(t) = v(t)e^{-\int p(t)\,dt} = e^{-\int p(t)\,dt}\int f(t)e^{\int p(t)\,dt}\,dt.$$

When we add this particular solution to all solutions y_h of the homogeneous equation, we get the general solution of differential equation (1):

$$y(t) = y_h + y_p = ce^{-\int p(t)\,dt} + e^{-\int p(t)\,dt}\int f(t)e^{\int p(t)\,dt}\,dt. \tag{6}$$

Whew! Equation (6) looks more complicated than it really is because it contains several antiderivatives.

But we seldom use formula (6); we generally solve individual equations just by retracing the steps (3), (4), and (5).

---

[1] The single solution $y_p(t)$ is always called the particular solution although there is nothing *particular* or special about it. It is just any single solution of the nonhomogeneous equation. Even if different people choose different solutions, the difference will be made up by the constants in the solutions.

[2] The two greatest mathematicians of the 18th century, Leonhard Euler (1707–1783) and Joseph Louis Lagrange (1736–1818), were major contributors to the solution of the first-order equation $y' + p(t)y = f(t)$. Euler solved the *homogeneous* equation when $p(t)$ is a constant by introducing the exponential ce^{-pt}, and Lagrange solved the *nonhomogeneous* equation using variation of parameters.

Furthermore, the rather intimidating expression $e^{-\int p(t)\,dt}$ that appears in y_h, y_p, and $v'(t)$ often works out to something much simpler, as in the following example.

EXAMPLE 1 **Variable Coefficient** The IVP

$$y' + \left(\frac{1}{t+1}\right) y = 2, \quad y(0) = 0, \quad t \geq 0 \tag{7}$$

is of the type that occurs in mixing problems in biology and chemistry. (The context will be discussed in Sec. 2.4.)

Step 1. We begin by solving the corresponding homogeneous equation

$$y' + \left(\frac{1}{t+1}\right) y = 0$$

by assuming for a moment that $y \neq 0$, and separating variables, getting the differential form:

$$\frac{dy}{y} = -\frac{dt}{t+1};$$

$$\ln|y| = -\ln(t+1) + c;$$

$$|y| = e^{\ln(t+1)^{-1}+c} = e^c(t+1)^{-1};$$

$$y_h = \frac{\pm e^c}{t+1} = \frac{k}{t+1}.$$

Observe how $e^{-\int p(t)\,dt}$ in Example 1 reduces to $1/(t+1)$. Rather than use the abstract formula, we solved directly for y_h using separation of variables.

NOTE: $e^c > 0$, but you can remove the absolute value sign if you replace $\pm e^c$ by an arbitrary constant k.

Step 2. Using variation of parameters for the particular solution, we try

$$y_p = \frac{v(t)}{t+1},$$

which gives, for the DE (5),

$$\frac{v'(t)}{t+1} = f(t) = 2,$$

or $v'(t) = 2t + 2$. Integrating and choosing the integration constant to be zero, we find $v(t) = t^2 + 2t$. So,

Any solution for $v(t)$ is all we need.

$$y_p = \frac{t^2+2t}{t+1}.$$

Step 3. The general solution to the DE in (7) is given by

$$y(t) = y_h + y_p = \frac{k}{t+1} + \frac{t^2+2t}{t+1}.$$

Step 4. Substituting the initial condition $t = 0$, $y = 0$ into the general solution gives $k = 0$. The solution of the IVP is therefore

$$y(t) = \frac{t^2+2t}{t+1}.$$

For a graphical view, try Problem 21 in this section. ■

These simple steps form the Euler-Lagrange method for solving first-order linear DEs. You will see in Chapters 4 and 6 how this method generalizes to higher-order linear systems.

Should we know the different sorts?

explain?

Euler-Lagrange Method for Solving Linear First-Order DEs

To solve a linear DE

$$\mathbf{y' + p(t)y = f(t),}$$

where p and f are continuous on a domain I, take the following steps.

Step 1. Solve $y' + p(t)y = 0$ (the corresponding homogeneous equation) by separation of variables to obtain the one-parameter family of solutions

$$y_h = ce^{-\int p(t)\,dt},$$

where c is an arbitrary constant.

The DE in Step 2 results from the nonhomogeneous DE by substituting y_h with c replaced by $v(t)$. See equation (5).

Step 2. Solve

$$v'(t)e^{-\int p(t)\,dt} = f(t)$$

for $v(t)$ to obtain a particular solution

$$y_p = v(t)e^{-\int p(t)\,dt}.$$

Step 3. Combine the results of Steps 1 and 2 to form the general solution

$$y(t) = y_h + y_p.$$

Step 4. If you are solving an IVP, only after Step 3 can you substitute the initial condition to find c.

Initial Conditions:

Beware! You *cannot* insert initial conditions for the *nonhomogeneous* problem into solutions of the corresponding *homogeneous* problem.

Another possible approach to solving linear DEs is with an **integrating factor**.

Integrating Factor Method

For first-order (and *only* first-order) linear differential equations

$$y' + p(t)y = f(t), \tag{8}$$

there is a popular alternative way to obtain a particular solution y_p to a nonhomogeneous equation (where $f(t) \neq 0$).[3]

Constant Coefficient: Let us introduce the idea of the integrating factor by first looking at the case where the coefficient $p(t)$ is a constant a, which gives

$$y' + ay = f(t). \tag{9}$$

The idea behind the integrating factor method is the simple observation (also made by Euler) that

$$e^{at}(y' + ay) = \frac{d}{dt}\left(e^{at}y\right), \tag{10}$$

[3]We gave the variation of parameters method first because it generalizes to higher orders and more variables; the integrating factor method does not, but it is often easier to use for first-order equations.

which turns the differential equation (8) into a "calculus problem." To see how this method works, multiply each side of (9) by e^{at}, getting

Key idea:

$$\frac{d}{dt}(e^{at}y) = e^{at}(y' + ay)$$

$$e^{at}(y' + ay) = f(t)e^{at},$$

which, using the fundamental property (10), reduces to

$$\frac{d}{dt}(e^{at}y) = f(t)e^{at}.$$

This equation can now be integrated directly, giving

$$e^{at}y = \int f(t)e^{at}\,dt + c,$$

where c is an arbitrary constant and the integral sign refers to *any* antiderivative of $f(t)e^{at}$. Solving for y gives

$$y(t) = e^{-at}\int f(t)e^{at}\,dt + ce^{-at}. \tag{11}$$

Yes, equation (11) looks suspiciously like a general Euler-Lagrange result (6), but we got there by a different route! You should confirm that this method obtained y_p and y_h in one fell swoop.

Variable Coefficient: We return to the general expression (8) and use an idea motivated by the constant coefficient case:

We seek a function $\mu(t)$, called an integrating factor, that satisfies

$$\mu(t)[y' + p(t)y] = \frac{d}{dt}[\mu(t)y(t)]. \tag{12}$$

We will get a simple formula for $\mu(t)$ if we carry out the differentiation on the right-hand side of (12) and simplify, getting

$$\mu(t)y' + \mu(t)p(t)y = \mu'(t)y + \mu(t)y'.$$

If we now assume that $y(t) \neq 0$, we arrive at

$$\mu'(t) = p(t)\mu(t).$$

But we can find a solution $\mu(t) > 0$ by separating variables,

$$\frac{\mu'(t)}{\mu(t)} = p(t),$$

and integrating,

$$\ln|\mu(t)| = \int p(t)\,dt,$$

so we have an integrating factor

Integrating Factor

$$\mu(t) = e^{\int p(t)\,dt}. \tag{13}$$

NOTE: Since $\int p(t)\,dt$ denotes the *collection* of *all* antiderivatives of $p(t)$, it contains an arbitrary additive constant. Hence, $\mu(t)$ contains an arbitrary *multiplicative* constant. However, since we are interested in finding only one integrating factor, we will pick the multiplicative constant to be one.

Now that we *know* the integrating factor, we simply multiply each side of (8) by the integrating factor (13), giving

$$\mu(t)[y' + p(t)y] = \mu(t)f(t).$$

But, from the property $\mu(t)[y' + p(t)y] = [\mu(t)y]'$, we have

$$[\mu(t)y]' = \mu(t)f(t). \tag{14}$$

We can now integrate (14), getting

$$\mu(t)y(t) = \int \mu(t)f(t)dt + c, \tag{15}$$

and, as long as $\mu(t) \neq 0$, we can solve (15) for $y(t)$ algebraically, getting

$$y(t) = \frac{1}{\mu(t)}\int \mu(t)f(t)dt + \frac{c}{\mu(t)}. \tag{16}$$

We will summarize these results, which give the complete solution (16) to a nonhomogeneous linear DE, of first-order only, without separately finding y_h and y_p.

Integrating Factor Method for First-Order Linear DEs

To solve a linear DE

$$y' + p(t)y = f(t), \tag{17}$$

where p and f are continuous on a domain I:

Step 1. Find the integrating factor $\mu(t) = e^{\int p(t)\,dt}$, where $\int p(t)\,dt$ represents *any* antiderivative of $p(t)$. Normally, pick the arbitrary constant in the antiderivative to be zero. Note that $\mu(t) \neq 0$ for $t \in I$.

Step 2. Multiply each side of the differential equation (17) by the integrating factor to get

$$e^{\int p(t)\,dt}[y' + p(t)y] = f(t)e^{\int p(t)\,dt},$$

which will always reduce to

$$\frac{d}{dt}\left[e^{\int p(t)\,dt}y(t)\right] = f(t)e^{\int p(t)\,dt}.$$

Step 2.5. Check that this is true for your DE and $\mu(t)$ to catch any computational errors.

Step 3. Find the antiderivative of the final equation in Step 2 to get

$$e^{\int p(t)\,dt}y(t) = \int f(t)e^{\int p(t)\,dt}\,dt + c.$$

Step 4. Solve the final equation of Step 3 algebraically for y to get the general solution to equation (17),

$$y = e^{-\int p(t)\,dt}\int f(t)e^{\int p(t)\,dt}\,dt + ce^{-\int p(t)\,dt}, \tag{18}$$

where c is an arbitrary constant, and the integral signs refer to *any* antiderivative of the expressions indicated.

Step 5. If you are solving an IVP, substitute the initial condition in result (18) to evaluate the constant c.

Observe that equation (18) is the same as the solution (6) found by the Euler-Lagrange Method. The integrating factor method simply uses $e^{\int p(t)\,dt}$ more explicitly.

EXAMPLE 2 **Integrating Factor Method** We return to the equation

$$y' - y = t$$

of Sec. 2.1, Example 7, and show how to obtain the solution $y(t)$ by the integrating factor method. We follow the *method* rather than the formulas (after finding μ), which gives a pretty clean calculation.

Step 1. Find the integrating factor:

$$\mu(t) = e^{\int (-1)dt} = e^{-t}.$$

Step 2. Multiply the DE by the integrating factor:

$$e^{-t}(y' - y) = te^{-t},$$

which reduces to

$$\frac{d}{dt}(e^{-t}y) = te^{-t}.$$

Step 2.5. Check that

$$e^{-t}(y' - y) = \frac{d}{dt}(e^{-t}y).$$

Yes, it does, so we proceed.

Step 3. Find the antiderivative:

$$e^{-t}y = \int te^{-t}\,dt = e^{-t}(-t - 1) + c.$$

(Confirm the final step using integration by parts, from calculus.)

Step 4. Solve for y:

$$y(t) = e^t(e^{-t})(-t - 1) + ce^t = -t - 1 + ce^t.$$

■

Transient and Steady-State Solutions

Adding a particular solution of a differential equation to all the homogeneous solutions will not surprise a systems engineer. They think of differential equations in terms of input/output systems. In such a situation, the homogeneous part of the equation represents the "hardwired" portion of the system, while the nonhomogeneous term $f(t)$ represents the input to the system, the part over which the operator has control. For a given input $f(t)$, the solution $y(t)$ is the corresponding output. This process is suggested by the "black box" model in Fig. 2.2.1.

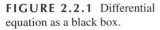

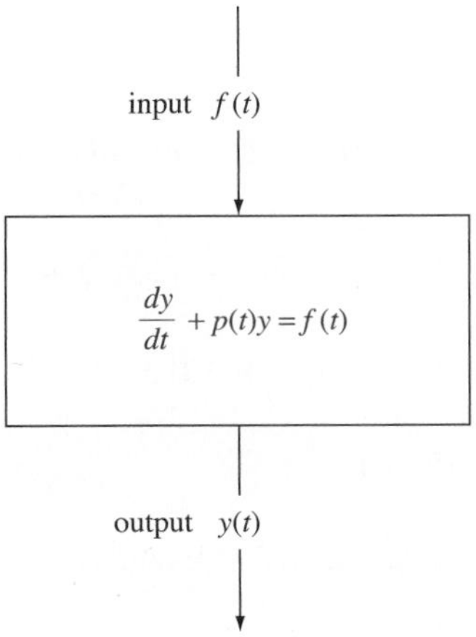

FIGURE 2.2.1 Differential equation as a black box.

Engineers usually seek solutions that do not "blow up" as $t \to \infty$ but rather settle into a constant or periodic solution. This desired long-term behavior is called a **steady-state** solution. On a direction field we see the whole family of solutions attracted to the steady state, while other parts of the solution, called **transients**, die out.

EXAMPLE 3 **Engineering View** The general solution of $y' + y = 2$ is

$$y = y_h + y_p = \underbrace{ce^{-t}}_{\text{transient}} + \underbrace{2}_{\text{steady state}}.$$

Figure 2.2.2 superposes these solutions on the direction field.

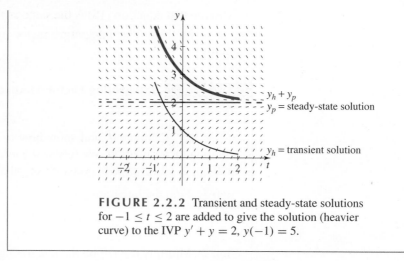

Transient and Steady-State Solutions:

It frequently (but definitely not always) happens that y_h is the transient solution and y_p is the steady-state solution. The key is to check whether $y_h \to 0$ as $t \to \infty$, and whether y_p is constant or periodic.

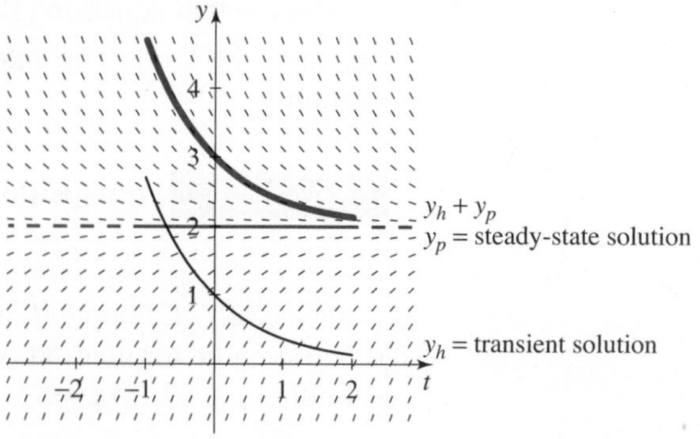

FIGURE 2.2.2 Transient and steady-state solutions for $-1 \le t \le 2$ are added to give the solution (heavier curve) to the IVP $y' + y = 2$, $y(-1) = 5$.

Summary

We have solved the general first-order linear differential equation by two different methods. The Euler-Lagrange method uses two integrations, one for the homogeneous solution and another for the particular solution based on varying a parameter. This separation correlates roughly with the two arbitrarily specified functions in the equation. The two-part structure is similar to that found in solving algebraic equations, and we will find it again when we solve higher-order differential equations. (Variation of parameters will still work for such DEs, but new techniques will be needed for the homogeneous solutions.) A convenient alternative, for linear DEs of first-order only, is the integrating factor method.

2.2 Problems

General Solutions *Find general solutions for the equations given in Problems 1–15.*

1. $\dfrac{dy}{dt} + 2y = 0$

2. $\dfrac{dy}{dt} + 2y = 3e^t$

3. $\dfrac{dy}{dt} - y = 3e^t$

4. $\dfrac{dy}{dt} + y = \sin t$

5. $\dfrac{dy}{dt} + y = \dfrac{1}{1+e^t}$

6. $\dfrac{dy}{dt} + 2ty = t$

7. $\dfrac{dy}{dt} + 3t^2 y = t^2$

8. $\dfrac{dy}{dt} + \dfrac{1}{t}y = \dfrac{1}{t^2}$

9. $t\dfrac{dy}{dt} + y = 2t$

10. $\cos t\dfrac{dy}{dt} + y\sin t = 1$

11. $\dfrac{dy}{dt} - \dfrac{2y}{t} = t^2\cos t$

12. $\dfrac{dy}{dt} + \dfrac{3}{t}y = \dfrac{\sin t}{t^3}$

13. $(1+e^t)\dfrac{dy}{dt} + e^t y = 0$

14. $(t^2+9)\dfrac{dy}{dt} + ty = 0$

15. $\dfrac{dy}{dt} + \dfrac{2t+1}{t}y = 2t$

Initial-Value Problems *For Problems 16–20, find solutions of the given IVPs.*

16. $\dfrac{dy}{dt} - y = 1$, $y(0) = 1$

17. $\dfrac{dy}{dt} + 2ty = t^3$, $y(1) = 1$

18. $\dfrac{dy}{dt} - \dfrac{3}{t}y = t^3$, $y(1) = 4$

19. $\dfrac{dy}{dt} + 2ty = t$, $y(0) = 1$

20. $(1+e^t)\dfrac{dy}{dt} + e^t y = 0$, $y(0) = 1$

21. Synthesizing Facts Reconsider the DE of Example 1,

$$y' + \left(\frac{1}{t+1}\right)y = 2.$$

With a calculator or computer, graph the direction field and some solutions.

(a) Which solution corresponds to the initial value $y(0) = 0$?

(b) Which solution corresponds to the initial value $y(0) = 1$?

(c) Do both behave as you would expect from the algebraic solution $y(t) = (t^2 + 2t + k)/(t + 1)$ given in Example 1? Explain.

(d) Finally, the *entire* line $y = t + 1$ is not the solution to (b), even allowing time to march backward. Where does it stop and why?

Using Integrating Factors *In Problems 22–30, solve each DE by the integrating factor method, Steps 1–4.*

22. $y' + 2y = 0$

23. $y' + 2y = 3e^t$

24. $y' - y = e^{3t}$

25. $y' + y = \sin t$

26. $y' + y = \dfrac{1}{1 + e^t}$

27. $y' + 2ty = t$

28. $y' + 3t^2 y = t^2$

29. $y' + \dfrac{1}{t} y = \dfrac{1}{t^2}$

30. $ty' + y = 2t$

31. Switch for Linearity Solve the nonlinear IVP

$$\frac{dy}{dt} = \frac{1}{t + y}, \quad y(-1) = 0$$

by reinterpreting it with y as the independent variable and t as the dependent variable. HINT: Use the fact that $dy/dt = 1/(dt/dy)$, where $y = y(t)$ and $t = t(y)$ are inverse functions.

32. The Tough Made Easy The differential equation

$$\frac{dy}{dt} = \frac{y^2}{e^y - 2ty}$$

looks impossible to solve analytically, but by treating y as the independent variable and t as the dependent variable, an implicit solution can be found. Carry out this solution. (See the hint for Problem 31.)

33. A Useful Transformation

(a) Use the change of variable $z = \ln y$ to solve the nonlinear DE $dy/dt + ay = by \ln y$, where a and b are constants.

(b) Use the result from (a) to solve $dy/dt + y = y \ln y$.

34. Bernoulli Equation The nonlinear equation

$$\frac{dy}{dt} + p(t)y = q(t)y^\alpha \qquad (19)$$

(where $\alpha \neq 0$, $\alpha \neq 1$) is called a *Bernoulli* equation and can be transformed into a linear equation.[4] It already looks almost linear, except for y^α on the right side.

(a) Divide (19) by y^α and then show that the transformation

$$v = y^{1-\alpha}$$

reduces (19) to a linear equation in v.

(b) Use the transformation in (a) to solve the Bernoulli equation $y' - y = y^3$.

(c) Explain how to solve the given Bernoulli equation when $\alpha = 0$ and when $\alpha = 1$.

Bernoulli Practice *Solve the Bernoulli equations in Problems 35–38. For Problems 39 and 40, use the given initial conditions to solve the given IVP.*

35. $y' + ty = ty^3$

36. $y' - y = e^t y^2$

37. $t^2 y' - 2ty = 3y^4$

38. $(1 - t^2)y' - ty - ty^2 = 0$

39. $y' + \dfrac{y}{t} = \dfrac{1}{ty^2}, \quad y(1) = 2$

40. $3y^2 y' - 2y^3 - t - 1 = 0, \quad y(0) = 2$

41. Ricatti Equation The first-order nonlinear DE

$$y' = p(t) + q(t)y + r(t)y^2$$

is known as the *Ricatti* equation.[5]

(a) Show that if one solution $y_1(t)$ of the Ricatti equation is known, then a more general solution containing an arbitrary constant can be found by substituting $y = y_1(t) + 1/v(t)$ into the DE and requiring $v(t)$ to satisfy $v' = -[q(t) + 2r(t)y_1(t)]\,v - r(t)$, which is a linear equation.

(b) Verify that $y_1(t) = 1$ satisfies the Ricatti equation

$$y' = -1 + 2y - y^2$$

and use this to determine the general solution.

Computer Visuals *For the linear DEs in Problems 42–47:*

(a) *Use an open-ended graphical DE solver to draw a direction field and some solutions.*

(b) *Find the exact solution and relate it to your picture in (a).*

(c) *If there is a steady-state solution, add it to your picture in (a) and highlight it in color. Then identify the transient and steady-state parts of the general solution in (b).*

42. $y' + 2y = t$

43. $y' - y = e^{3t}$

44. $y' + y = \sin t$

45. $y' + y = \sin 2t$

46. $y' + 2ty = 0$

47. $y' + 2ty = 1$

[4] The Bernoullis were a whole family of famous Swiss mathematicians spanning four generations. This equation is named for two brothers, Jakob (1654–1705) and Johann (1667–1748), who worked from 1695–1697 on solving it, in competition with Leibniz.

[5] A specific Ricatti equation was studied by Venetian mathematician Count Jacopo Francesco Ricatti (1676–1754) in 1724 while investigating radii of curves, but Euler in 1760 was the one who discovered that the substitution given in Problem 41(a) gives a *linear* equation in $v(t)$. Nowadays the interest in the Ricatti equation lies in control theory, in which solutions of Ricatti equations provide feedback for control systems.

Computer Numerics *Consider the linear DEs in Problems 48–50, with* $y(0) = 1$.

(a) *Use a computer package with different numerical methods to find* $y(1)$.

(b) *Use the exact solution (as found in the corresponding Problems 42, 43 and 47) to compute* $y(1)$ *and compare with your approximations.*

(c) *Summarize what happens with different methods and different step sizes.*

(d) *Tell which approximations are misleading and why.*

48. $y' + 2y = t$ **49.** $y' - y = e^{3t}$ **50.** $y' + 2ty = 1$

51. Direction Field Detective The direction fields of three DEs are shown in Fig. 2.2.3.

(a) Label the equations linear or nonlinear. Of the two that are linear, which is homogeneous and which is nonhomogeneous?

(b) Explain (algebraically) why the sum of two solutions of a homogeneous linear DE is a solution. Verify that the negative of such a solution is also a solution, so $y = 0$ must be a solution.

(c) Interpret the observations of part (b) for the direction fields in Fig. 2.2.3. Visualize two solutions and then their sum. Does the sum follow the direction field?

52. Recognizing Linear Homogeneous DEs from Direction Fields From your conclusions in Problem 51, decide which of the direction fields in Fig. 2.2.4 represent linear homogeneous equations. Explain your reasons in each case.

53. Suggested Journal Entry Write a summary of your concept of the Euler-Lagrange method for solving linear first-order DEs. Compare Euler-Lagrange with the method of integrating factors.

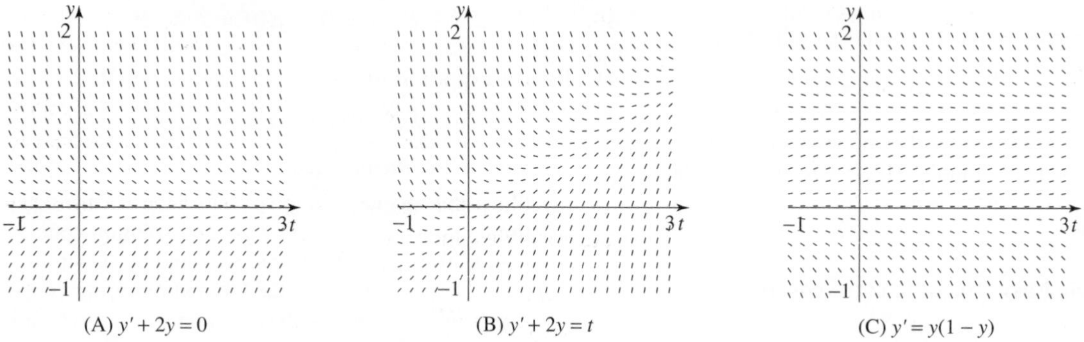

FIGURE 2.2.3 Direction fields for Problem 51.

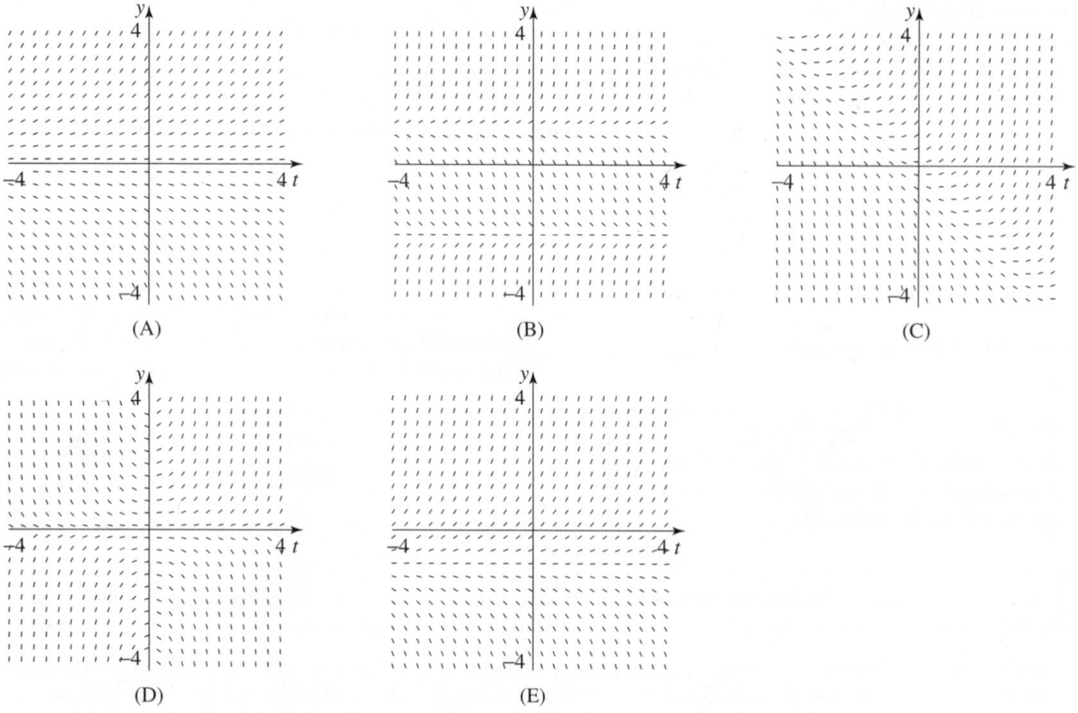

FIGURE 2.2.4 Direction fields for Problem 52.

2.3 Growth and Decay Phenomena

SYNOPSIS: The first-order linear differential equation $y' = ky$, where k is a constant, is an important model for phenomena in science (growth and decay) and finance (interest and annuities).

The Basic Equation and Its Solution

Growth and Decay

Experiment with the parameter k and see how the population responds.

The basic linear differential equation

$$\frac{dy}{dt} = ky \tag{1}$$

wears two hats according to whether the coefficient k is positive or negative. When $k > 0$, k is called the **growth constant** or **rate of growth**, and equation (1) is called the **growth equation**. If, on the other hand, $k < 0$, then we refer to k as the **decay constant** or **rate of decline**, and we call equation (1) the **decay equation**.

We first met this equation in Sec. 1.1: Thomas Malthus used it to estimate world **population growth**. In Sec. 1.2 we verified its solution for $k = 2$, and saw typical solution curves in Fig. 1.2.1(a). In Problem 9 of Sec. 1.2, a part of the direction field was plotted (again for the k-value 2).

If we rewrite (1) in the form $y' - ky = 0$, we can classify it further, using definitions from Sec. 2.1, not only as first-order and linear, but also as homogeneous with constant coefficients.

To solve equation (1) we could invoke the theory of the previous section, but we can instead just separate variables and integrate. We get, for $y \neq 0$,

$$\frac{dy}{y} = k\,dt,$$

$$\ln|y| = kt + C,$$

$$|y| = e^{kt+C} = e^C e^{kt},$$

where constant $e^C > 0$. If we replace this positive constant by an arbitrary constant A, which could be negative, we can replace $|y|$ by y to obtain the general solution of (1) in the form

$$y(t) = Ae^{kt}, \tag{2}$$

for any real constant A. ($A = 0$ yields the constant solution $y = 0$.)

Since $y = A$ when $t = 0$ in equation (2), a solution to the IVP consisting of equation (1) and the initial condition $y(0) = y_0$ is $y(t) = y_0 e^{kt}$.

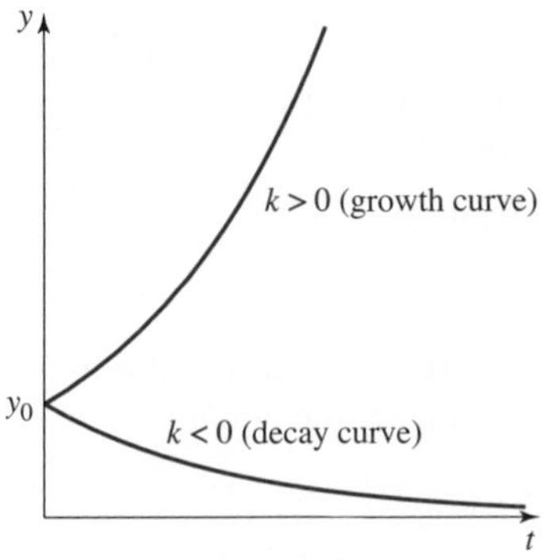

FIGURE 2.3.1 Growth and decay curves $y = y_0 e^{kt}$.

> **Growth and Decay**
> For each k, the solution of the IVP
>
> $$\frac{dy}{dt} = ky, \quad y(0) = y_0 \tag{3}$$
>
> is given by
>
> $$y(t) = y_0 e^{kt}. \tag{4}$$
>
> Solution curves are **growth curves** for $k > 0$ and **decay curves** for $k < 0$. (See Fig. 2.3.1.)

Radioactive Decay

In 1940, a group of boys walking in the woods near the village of Lascaux in France suddenly realized that their dog was missing. They soon found him in a hole too deep for him to climb out. One of the boys was lowered into the hole to rescue the dog and stumbled upon one of the greatest archaeological discoveries of all time. What he discovered was a cave whose walls were covered with drawings of wild horses, cattle, and a fierce-looking beast resembling a modern bull. In addition, the cave contained the charcoal remains of a small fire, and from these remains scientists were able to determine that the cave was occupied 15,000 years ago.

To understand why the remains of the fire are so important to archaeologists, it is important to realize that charcoal is dead wood, and that over time a fundamental change takes place in dead organic matter. All living matter contains a tiny but fixed amount of the radioactive isotope Carbon-14 (or C-14). After death, however, the C-14 decays into other substances at a rate proportional to the amount present. Based on these physical principles, the American chemist Willard Libby (1908–1980) developed the technique of **radiocarbon dating**, for which he was awarded the Nobel prize in chemistry in 1960. The following example illustrates such calculations.

EXAMPLE 1 **Radiocarbon Dating** By chemical analysis it has been determined that the amount of C-14 remaining in samples of the Lascaux charcoal was 15% of the amount such trees would contain when living. The **half-life** of C-14 (the time required for a given amount to decay to 50% of its original value; see Problem 1) is approximately 5600 years. The quantity Q of C-14 in a charcoal sample satisfies the decay equation

$$Q' = kQ. \tag{5}$$

(a) What is the value of the decay constant k?

(b) What is $Q(t)$ at any time t, given the initial amount $Q(0) = Q_0$?

(c) What is the age of the charcoal and, hence, the age of the paintings?

We can answer these questions using the information provided.

(a) The IVP constructed from the DE (5) and initial condition $Q(0) = Q_0$ has the form of (3), so its solution is given by equation (4): $Q(t) = Q_0e^{kt}$ for any time $t \geq 0$. After a half-life of 5600 years, the original amount Q_0 of C-14 will decrease to one-half of Q_0: when $t = 5600$, $Q = Q_0/2$. That is,

$$Q(5600) = Q_0e^{5600k} = \frac{1}{2}Q_0.$$

We solve for k by dividing by Q_0, which gives $e^{5600k} = 1/2$, and taking logs of both sides to obtain $5600k = -\ln 2$. Therefore, we have the decay constant $k = -(\ln 2)/5600 \approx -0.00012378$.

(b) Just substitute the k-value into the general solution:

$$Q(t) = Q_0e^{-t\ln 2/5600} \approx Q_0e^{-0.00012378t}.$$

(c) The final question is to find the t-value that reduces an initial amount Q_0 to 15% of Q_0, so we must solve, for the time t, the equation

$$Q_0e^{-t\ln 2/5600} = 0.15Q_0.$$

Dividing by Q_0 and taking logs we get $-(t\ln 2)/5600 = \ln 0.15$, from which it follows that $t = -(5600\ln 0.15)/\ln 2 \approx 15{,}327$ (years). The decay curve for C-14 is shown in Fig. 2.3.2. ■

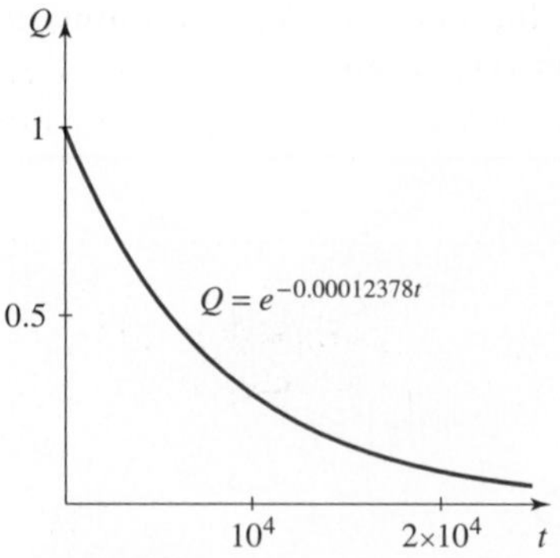

FIGURE 2.3.2 Decay curve for C-14 (half-life of 5600 years).

Compound Interest

The growth equation is a model useful in determining the future value of money. A deposit in a savings account earns interest, which is just a fraction of your deposit added to the total at regular intervals. The fraction is the **interest rate** and is based on a one-year period. Interest of 8% means that 0.08 of the original amount is added after a year, so amount A grows to $A + 0.08A = A(1 + 0.08) = 1.08A$. Table 2.3.1 gives the future value of an initial deposit A_0, year by year, at an interest rate $r = 0.08$, compounded annually.

Table 2.3.1 Annual compound interest

Number of years	Future value of account	Future value of \$100 at 8% annual interest
0	A_0	\$100
1	$A_0(1+r)$	\$100(1.08) = \$108
2	$A_0(1+r)^2$	$\$100(1.08)^2 = \$116.64$
3	$A_0(1+r)^3$	$\$100(1.08)^3 = \$125.97$
⋮	⋮	⋮
N	$A_0(1+r)^N$	$\$100(1.08)^N$

Calculating Future Value:

$A(1) = A_0(1+r)$

$A(2) = A(1)(1+r) = A_0(1+r)^2$

⋮

Now, suppose that the bank pays interest n times per year. The fraction added in each period will be r/n since r is the annual amount, but the process will be carried out nt times over t years. Thus the future value of the account after t years will be

$$A(t) = A_0\left(1 + \frac{r}{n}\right)^{nt}. \tag{6}$$

For example, if the interest rate is $r = 0.06$ and a bank pays interest monthly, the future value of a \$100 deposit after $t = 2$ years is

$$A(2) = \$100\left(1 + \frac{0.06}{12}\right)^{24} = \$100(1.005)^{24} = \$112.72.$$

$y = y_0 e^{kt}$
$\frac{dy}{dt} = ky$

Suppose that the bank makes its payments more and more often: daily, hourly, every minute, every second ... *continuously!* The future value of the account would be determined from the limit

$$A(t) = \lim_{n\to\infty} A_0\left(1 + \frac{r}{n}\right)^{nt} = A_0e^{rt}. \tag{7}$$

We use the calculus fact that $\lim_{n\to\infty}(1 + r/n)^n = e^r$.

We say that the interest computed in this manner is **compounded continuously**, even though no bank could physically make interest payments in a continuous way. The exponential growth rate is a close approximation to compounding daily (that is, for $n = 365$; see Problem 33).

Continuous Compounding of Interest

If an initial amount of A_0 dollars is deposited at an **annual interest rate** of r, **compounded continuously**, the **future value** $A(t)$ of the deposit at time t satisfies the initial-value problem

$$\frac{dA}{dt} = rA, \quad A(0) = A_0 \tag{8}$$

and is therefore given by

$$A(t) = A_0e^{rt}. \tag{9}$$

EXAMPLE 2 **Canarsie Indians** In 1626, so the story goes, the Dutch explorer Peter Minuit paid the Canarsie Indians \$24 for the island of Manhattan, now a borough of New York City. If the Canarsies had had the chance to deposit this money in a savings account that paid at an annual interest rate of 8% compounded continuously, what would have been the value of this account in the year 2000?

To answer this question we calculate that in 2000, $t = 374$. By equation (9), then,

$$A(374) = (\$24)e^{0.08(374)} = \$236{,}756{,}625{,}000{,}000.$$

As an old adage says: Time is money! ■

The Savings Problem

The growth of an account is more interesting when, in addition to the initial deposit and earned interest, **steady contributions** are added as well. This is the situation for systematic savings plans such as retirement accounts and pension funds. If the account earns annual interest at rate r, compounded continuously, and is also continuously increased by deposits that total a dollars per year, the future value of the account satisfies the initial-value problem

$$\frac{dA}{dt} = rA + a, \quad A(0) = A_0. \tag{10}$$

Using methods of the previous section we can solve this nonhomogeneous linear equation to obtain

$$A(t) = A_0 e^{rt} + \frac{a}{r}(e^{rt} - 1). \tag{11}$$

The first term on the right-hand side of equation (11) represents the accumulation due to the initial deposit, while the second term is the amount resulting from the subsequent deposits and the interest that *they* earn. (If $r = 0$, we just get $A(t) = A_0 + ta$ from (10).)

If the quantity a in equations (10) and (11) is negative, this corresponds to systematic withdrawal following the initial deposit. In this case the account value may still grow if withdrawals aren't too large. Larger withdrawals, however, will lead to eventual depletion of the account. (See Problem 31.)

EXAMPLE 3 **Saving Your Cigarette Money** Ravi has just entered college at age 18 and has decided to improve his health and save some money by quitting smoking. He figures he can save \$30 per week in this way. If he deposits this amount in an account paying 10% annual interest compounded continuously, how much will he have in the account when he retires at age 65?

The weekly deposits of \$30 represent a fairly uniform rate of \$1,560 per year (52 weeks at \$30 each). The IVP describing the value of his account is

$$\frac{dA}{dt} = 0.10A + 1560, \quad A(0) = 0,$$

and its solution, by equation (11) or by Euler-Lagrange, is

$$A(t) = 15600(e^{0.1t} - 1).$$

After 47 years the value of his account will be $15600(e^{4.7}-1) = \$1{,}699{,}575.89$. In other words, a regular savings plan of $30 a week for 47 years at this rate yields more than one and a half mllion dollars. ■

Summary

The basic differential equation of growth and decay is one of the most useful models we have, with applications including radioactive decay and the future value of money. As you will see in the problems, other applications pop up everywhere: population studies, blood alcohol levels, light intensity, anesthesiology—and many more. The moral of the section is never to forget your old friend $y' = ky$: *it will serve you well!*

2.3 Problems

1. **Half-Life** The time t_h required for the solution y of the decay problem

$$y' = ky, \quad k < 0, \quad y(0) = y_0$$

to reach one-half of its original value is called the **half-life**.

(a) Find the half-life t_h in terms of the decay rate k.

(b) Show that if the solution has value B at any time t_1, then the solution will have the value $B/2$ at time $t_1 + t_h$.

Growth and decay
Doubling times and half-lives are displayed.

2. **Doubling Time** The time t_d required for the solution y of the growth problem

$$y' = ky, \quad k > 0, \quad y(0) = y_0$$

to reach twice its original value is called the *doubling time*. Find t_d in terms of k.

3. **Interpretation of $1/k$** The reciprocal $|1/k|$ (which has units of time) of the absolute value of the decay constant k in the decay equation $y' = ky$ can be roughly interpreted as the time for y to fall two-thirds of the way from the initial value y_0 to the limiting value 0. Show why this is true and illustrate with a figure. HINT: Evaluate the solution at $t = |1/k|$.

4. **Radioactive Decay** A certain radioactive material is known to decay at a rate proportional to the amount present. Over a 50-year period, an initial amount of 100 grams has decayed to only 75 grams. Find an expression for the amount of material t years after the initial measurement. Calculate the half-life of the material.

5. **Determining Decay from Half-Life** A certain radioactive substance has a half-life of 5 hours. Find the time for a given amount to decay to one-tenth of its original mass.

6. **Thorium-234** Thorium-234 is a radioactive isotope that decays at a rate proportional to the amount present. Suppose that 1 gram of this material is reduced to 0.80 grams in one week.

(a) Find an expression for the amount of Th-234 present at a general time t.

(b) Find the half-life of Th-234.

(c) Find the amount of Th-234 left after 10 weeks.

7. **Dating Sneferu's Tomb** A cypress beam found in the tomb of Sneferu in Egypt contained 55% of the amount of Carbon-14 found in living cypress wood. Estimate the age of the tomb. NOTE: The half-life of C-14 is 5,600 years.

8. **Newspaper Announcement** A 1960 *New York Times* article announced: "Archaeologists Claim Sumerian Civilization Occupied the Tigris Valley 5,000 Years Ago." Assuming the archaeologists used Carbon-14 to date the site, determine the percentage of Carbon-14 found in the relevant samples.

9. **Radium Decay** Radium decays at a rate proportional to the amount present and has a half-life of 1,600 years. What percentage of an original amount will be present after 6,400 years? HINT: This problem is very easy.

10. **General Half-Life Equation** If Q_1 and Q_2 are the amounts of a radioactively decaying substance at time t_1 and t_2, respectively, where $t_1 < t_2$, show that the half-life of the material is given by

$$t_h = \frac{(t_2 - t_1)\ln 2}{\ln(Q_1/Q_2)}.$$

11. **Nuclear Waste** The U.S. government has dumped roughly 100,000 barrels of radioactive waste into the Atlantic and Pacific oceans. The waste is mixed with concrete and encased in steel drums. The drums will eventually rust and seawater will gradually leach the radioactive material from the concrete and diffuse it throughout the ocean. It is assumed that the leached radioactive material would be so diluted that no environmental damage would result.

However, scientists have discovered that one of the pollutants, Americium-241, is sticking to the ocean floor near the drums. Given that Am-241 has a half-life of 258 years, how long will it take for the amount of Am-241 to be reduced to 5% of its initial value?

12. Bombarding Plutonium In 1964, Soviet scientists made a new element with atomic number 104, called simply E104, by bombarding plutonium with neon ions. The half-life of this new element is 0.15 seconds, and it was produced at a rate of 2×10^{-5} micrograms per second. Assuming none was present initially, how much E104 is present after t seconds? HINT: The decay equation must be modified; remember how we obtained equation (10).

13. Blood Alcohol Levels In many states it is illegal to drive with a blood alcohol level greater than 0.10% (one part alcohol per 1,000 parts blood). Suppose someone who was involved in an automobile accident had blood alcohol tested at 0.20% at the time of the accident. Assume that the percentage of alcohol in the blood decreases exponentially at the rate of 10% per hour.

(a) Find the percentage of alcohol in the bloodstream at any value of time t.

(b) How long will it be until this person can legally drive?

14. Exxon Valdez Problem In the tragic 1989 accident of the Exxon ship Valdez that dumped 240,000 barrels of oil into Prince William Sound, the National Safety Board determined that blood tests of Capt. Joseph Hazelwood showed a blood-alcohol content of 0.06%.[1] This testing did not take place until nine hours after the accident. Blood alcohol is eliminated from the system at a rate of about 0.015 percentage points per hour. If the permissible level of alcohol is 0.10%, should the Board determine that the captain could be liable? (We assume that he did not have a drink after the accident.)

15. Sodium Pentathol Elimination Ed is undergoing surgery for an old football injury and must be anesthetized. The anesthesiologist knows Ed will be "under" when the concentration of sodium pentathol in his blood is at least 50 milligrams per kilogram of body weight. Suppose that Ed weighs 100 kg (220 pounds) and that sodium pentathol is eliminated from the bloodstream at a rate proportional to the amount present. If the half-life of the drug is 10 hours, what single dose should be given to keep Ed anesthetized for three hours?

16. Moonlight at High Noon The fact that sunlight is absorbed by water is well known to any diver who has dived to a depth of 100 ft. It is also true that the intensity of light falls exponentially with depth. Suppose that at a depth of 25 ft the water absorbs 15% of the light that strikes the surface. At what depth would the light at noon be as bright as a full moon, which is one three-hundred-thousandth as bright as the noonday sun?

17. Tripling Time The number of bacteria in a colony increases at a rate proportional to the number present. If the number of bacteria doubles in 10 hours, how long will it take for the colony to triple in size?

18. Extrapolating the Past If the number of bacteria in a culture is 5 million at the end of 6 hours and 8 million at the end of 9 hours, how many were present initially?

19. Unrestricted Yeast Growth The number of bacteria in a yeast culture grows at a rate proportional to the number present. If the population of a colony of yeast bacteria doubles in one hour, and if 5 million are present initially, find the number of bacteria in the colony after 4 hours.

20. Unrestricted Bacterial Growth A certain colony of bacteria grows at a rate proportional to the number of bacteria present. Suppose that the number of bacteria doubles every 12 hours. How long will it take this colony to grow to five times its original size?

21. Growth of Tuberculosis Bacteria A strain of tuberculosis bacteria grows at a rate proportional to its size. A researcher has determined that every hour the culture is 1.5 times larger than the hour before, and that initially there were 100 cells present. How many cells are present at any time t?

22. Cat and Mouse Problem On an island that had no cats, the mouse population doubled during the first 10 years, reaching 50,000. At that time the islanders imported several cats, who thereafter ate 6,000 mice every year.

(a) What is the number of mice on the island t years after the arrival of the cats?

(b) How many mice will be on the island 10 years after the arrival of the cats?

(c) Normally, the cats' harvest would not remain constant. What will happen if the cats harvest 10% of the current mouse population each year?

23. Banker's View of *e* A banker once gave the interpretation of the constant e as the value after 10 years of an account earning 10% interest continuously compounded if the initial deposit is one dollar. Explain the merit of this claim.

---

[1]Problem based on an article in the *New York Times* for March 31, 1989. Although the media publicized the captain's blood alcohol level, the actual cause of the accident, in the NTSB's final report, was serious sleep deprivation on the part of the third mate, who was in command at the time but had only slept 6 of the previous 48 hours; W. C. Dement, MD, *The Promise of Sleep* (NY: Delacorte Press, 1999), 52–53.

24. **Rule of 70** In banking circles, the "Rule of 70" states that the time (in years) required for the value of an account to double in value can be approximated by dividing 70 by the annual interest rate (as a percentage, not a decimal). What is the reasoning behind this rule?

25. **Power of Continuous Compounding** In 1820, a William Record of London deposited \$0.50 (or its equivalent in English pounds) for his granddaughter in the Bank of London. Unfortunately, he died before he could tell his granddaughter about the account. One hundred sixty years later, in 1980, the granddaughter's heirs discovered the account. What was the value of the account if the bank paid 6% compounded continuously?

26. **Credit Card Debt** Upon entering college, Meena borrowed the limit of \$5,000 on her credit card to help pay expenses. The credit company charges 19.95% annual interest, compounded continuously. How much will Meena owe when she graduates in four years?

27. **Compound Interest Thwarts Hollywood Stunt** In 1944, a Hollywood publicist decided to dramatize the opening of the movie *Knickerbocker Holiday* by arranging a stunt in which three bottles of whiskey, originally thought to have been given to the Canarsie Indians for the island of Manhattan, were to have been returned to the mayor of New York City plus 8% interest, compounded annually, in bottles. To his horror, just before the gala event, he discovered that the compound interest on the whiskey over a period of 320 years would be more than 100 million bottles. As the agent put it, "The stunt just ain't worth it." Exactly how many bottles of whiskey would need to have been given to New York City's mayor?

28. **It Ain't Like It Used to Be** Sheryl's grandfather told Sheryl that 50 years ago the average cost of a new car was only \$1,000, while today the average cost is \$18,000. What continuous interest rate over the past 50 years would produce this change?

29. **How to Become a Millionaire** Upon graduating from college, Sergei has no money. However, during each year after that, he will deposit $d = \$1{,}000$ into an account that pays interest at a rate of 8% compounded continuously.
    (a) Find the future value $A(t)$ of Sergei's account.
    (b) Find the value for an annual deposit d that would produce a balance of one million dollars when he retires 40 years later.
    (c) If $d = \$2{,}500$, what should be the value of the interest rate r in order for Sergei's balance to be one million dollars after 40 years?

30. **Living Off Your Money** Suppose a rich uncle has left you A_0 dollars, which is invested at rate r compounded continuously. Show that if you make *withdrawals* amounting to d dollars per year (where $d > rA_0$), the time required to deplete the account to zero is

$$\frac{1}{r}\ln\left(\frac{d}{d - rA_0}\right).$$

What happens to the account when the annual withdrawal is not greater than rA_0?

31. **How Sweet It Is** Linda has won the New Jersey megabucks lottery consisting of one million dollars. Suppose that she deposits the money in a savings account that pays an annual rate of 8% compounded continuously. How long will the money last if she makes annual withdrawals of \$100,000?

32. **The Real Value of the Lottery** You must be careful about money. Lottery winners sometimes think they are millionaires when they're not really as rich as they think. Furthermore, there are enormous income taxes to be paid. But in these problems we are just calculating pretax earnings. Suppose that a state lottery's Grand Prize is announced to be one million dollars but that actually the winner is paid \$50,000 each year for the next 20 years. Assuming the state can earn interest on money at 10% over those 20 years, how much is the lottery worth in today's dollars? That is, what does it really cost the state today to set aside funds to cover it? HINT: Denoting time in years by t, solve $A' = 0.10A - 50{,}000$, $A(0) = A_0$; then set $A(20) = 0$ and solve for A_0.

33. **Continuous Compounding** Many banks advertise that they compound interest continuously, meaning that the amount of money $A(t)$ in an account satisfies the DE $A' = rA$, where r is the annual interest rate and t is time in years.
    (a) Show that an interest rate of 8% compounded continuously gives an effective annual interest rate of 8.329%; that is, the yield is the same at 8% compounded continuously as at 8.329% compounded annually.
    (b) Show that an interest rate of r, compounded continuously, gives an effective annual interest rate of $e^r - 1$.
    (c) Compare the effective annual interest rates for 8% compounded continuously and 8% compounded daily.

34. **Good Test Equation for Computer or Calculator** Although the growth equation $y' = ky$ is simple, it is not easy to approximate numerically, particularly over intervals $[0, a)$ for large a. Compare the accuracy of different numerical methods by solving the IVP $y' = y$, $y(0) = 1$, and evaluating the solution at $t = 1$. The exact value of $y(1)$ is e, so all methods can be compared against this value. Try step sizes of 0.1, 0.5, and 1. Although a step size of 1 is enormous, you may be surprised at the accuracy of certain other methods even so—comment on why this might be so.

35. **Your Financial Future** After college you have no money, but you begin to create a retirement account by making

continuous deposits that total $d = \$5{,}000$ per year. Suppose that the account pays interest at an annual rate of 8% compounded continuously. Use a computer or calculator to plot the future value of your account over the next 20 years. Experiment by changing the interest rate and the annual deposits. Which parameter is more important to the future value of your account? To increase your future worth, is it more important to increase your annual deposit or to find an institution that pays a higher rate of interest?

36. Mortgaging a House Kelly and friends buy a house after graduating from college and borrow \$200,000 from the bank to pay for it. Suppose that the bank charges 12% annual interest on the outstanding principle and that Kelly's group plans to make monthly payments of $d = \$2{,}500$ to the bank. Call $A(t)$ the amount of money the group still owes the bank.

(a) What is the initial value problem that describes $A(t)$?

(b) Solve the IVP in part (a).

(c) How long will it take Kelly and friends to pay off the loan?

37. Suggested Journal Entry The future value of a savings account is determined by three things: initial deposit, rate of interest, and length of time on deposit. Discuss the relative importance of these three factors.

2.4 Linear Models: Mixing and Cooling

SYNOPSIS: We use first-order linear differential equations to model mixing and cooling problems and to suggest other applications, such as multiple-compartment models.

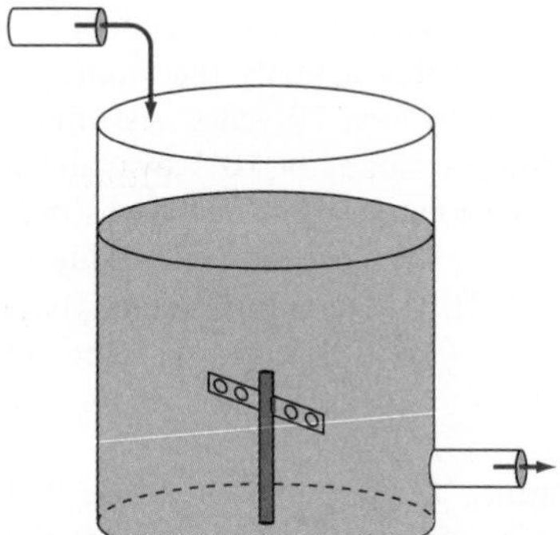

FIGURE 2.4.1 Single-tank configuration, with stirring apparatus.

Mixing Problems

We will start with a simple system consisting of one compartment: some substance flows into a tank, is mixed uniformly with the contents of the tank, and flows out with the mixture. This setting is pictured in Fig. 2.4.1. The goal is to find the amount of the substance in the tank at any time. To do this we let $x(t)$ denote the amount of the substance in the tank at time t. Then x' is the rate of change of this amount; that is, the difference between the rate at which it flows into the tank (RATE IN) and the rate at which it flows out (RATE OUT). This fundamental idea is called the **equation of continuity**.

Mixing Model

If $x(t)$ is the amount of a dissolved substance, then

$$\frac{dx}{dt} = \text{RATE IN} - \text{RATE OUT}, \tag{1}$$

where

$$\begin{aligned} \underset{\uparrow\ [\text{lb/min}]}{\text{RATE IN}} &= \text{(CONCENTRATION IN) (FLOW RATE IN)}, \\ \text{RATE OUT} &= \underset{\uparrow\ [\text{lb/gal}]}{\text{(CONCENTRATION OUT)}}\ \underset{\uparrow\ [\text{gal/min}]}{\text{(FLOW RATE OUT)}}. \end{aligned} \tag{2}$$

The CONCENTRATION variables are just what you think: the *relative* amount, the amount per unit of solution of the mixtures entering and leaving the tank. The FLOW RATE expressions are the rates at which the carrying mixtures enter and leave the tank. Pay careful attention to how the units combine. The following example illustrates these ideas.

EXAMPLE 1 **Brine Mixing** A tank initially contains 50 gal of pure water. A solution containing 2 lb/gal of salt is pumped into the tank at the rate of 3 gal/min. The mixture is stirred constantly and flows out at the same rate of 3 gal/min.

(a) What initial-value problem is satisfied by the amount of salt $x(t)$ in the tank at time t?

(b) What is the actual amount of salt in the tank at time t?

(c) How much salt is in the tank after 20 minutes?

(d) How much salt is in the tank after a long time?

We can answer these questions using the information provided.

(a) To find the IVP we combine the equation of continuity (1) with the rate relationships (2):

$$\begin{aligned}\frac{dx}{dt} &= \text{RATE IN} - \text{RATE OUT}\\ &= (2\text{ lb/gal})(3\text{ gal/min}) - \left(\frac{x}{50}\text{lb/gal}\right)(3\text{ gal/min})\\ &= 6\text{ lb/min} - \frac{3}{50}x\text{ lb/min}.\end{aligned}$$

Since the tank initially contains no salt, $x(0) = 0$. The IVP for question (a) is therefore

$$x' + 0.06x = 6, \quad x(0) = 0. \tag{3}$$

(b) The DE is a linear nonhomogeneous equation of the type studied in Sec. 2.1, Example 8; its solution is

$$x(t) = 100\left(1 - e^{-0.06t}\right). \tag{4}$$

(c) Substitute $t = 20$ into (4) to obtain $x(20) = 100(1 - e^{-1.2}) \approx 69.9$. The amount of salt after 20 minutes is roughly 70 lb.

(d) The amount of salt after a "long time" anticipates that there will be a limiting value. This is given by the "steady-state" term in the solution, the constant value 100. For example, after 2 hours the amount is $x(120) \approx 99.9$ lb.

These results are illustrated by the graph of x in Fig. 2.4.2. ■

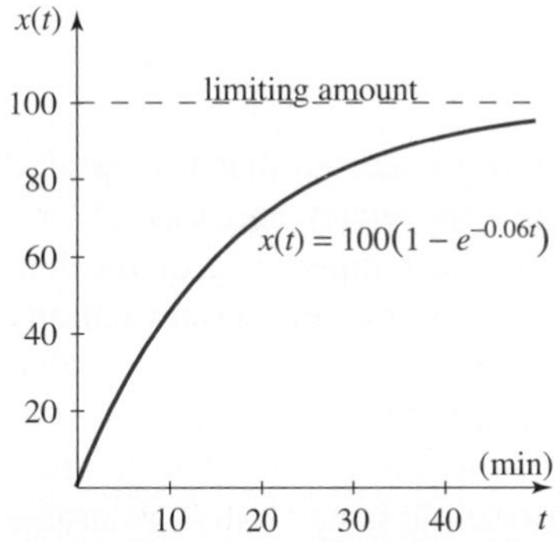

FIGURE 2.4.2 Amount of salt $x(t)$ in Example 1.

EXAMPLE 2 **RATE IN < RATE OUT** Brine containing 1 lb/gal of salt is poured at 1 gal/min into a tank that initially contained 100 gal of fresh water. The stirred mixture is drained off at 2 gal/min.

(a) Until the tank empties, what IVP is satisfied by the amount of salt in it?

(b) What is the formula for this amount of salt?

We tackle these questions as follows.

(a) Call the amount of salt $x(t)$ and write expressions for the flow rates. There is a net outflow of 1 gal/min, so after t minutes the tank contains only $100 - t$ gallons. From equations (2):

$$\begin{aligned}\text{RATE IN} &= (1\text{ lb/gal})(1\text{ gal/min}) = 1\text{ lb/min},\\ \text{RATE OUT} &= \left(\frac{x}{100-t}\text{lb/gal}\right)(2\text{ gal/min})\\ &= \frac{2x}{100-t}\text{ lb/min}.\end{aligned}$$

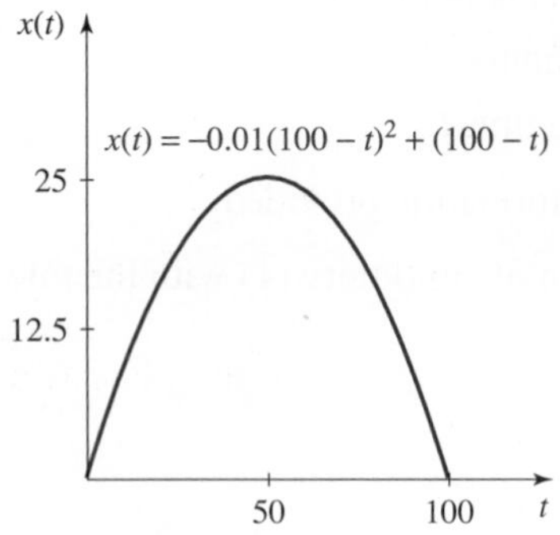

FIGURE 2.4.3 Amount of salt $x(t)$ in Example 2.

Substituting into the continuity equation and using $x(0) = 0$, we obtain

$$\frac{dx}{dt} + \frac{2x}{100 - t} = 1, \quad x(0) = 0 \quad (0 \le t < 100). \tag{5}$$

(b) Finding the salt in the tank as required means solving the IVP (5). The corresponding homogeneous equation has the solution $x_h(t) = c(100 - t)^2$. Using variation of parameters (or the integrating factor method) we find $x_p(t) = 100 - t$. The general solution of the DE is given by

$$x(t) = x_h(t) + x_p(t) = c(100 - t)^2 + (100 - t).$$

Since $x = 0$ when $t = 0$, we find that $c = -0.01$. The amount of salt in the tank after t minutes ($t < 100$) is

$$x(t) = -0.01(100 - t)^2 + (100 - t).$$

This function is graphed in Fig. 2.4.3. ■

More complicated models with more than one compartment are handled in a similar fashion, using an equation of continuity (1) for each compartment. See Problems 10–12, and Sec. 6.2.

Newton's Law of Cooling

Suppose that a steel ball is placed in a pan of boiling water so that it is heated to a temperature of 212°F. The ball is taken from the water and placed in a room where the temperature is a constant 70°F. If the temperature of the ball has dropped to 150°F after 10 minutes, how can we find the temperature at any subsequent time?

If your intuition tells you that the rate of change of the temperature is proportional to the *difference* between the temperature of the ball and the temperature of its surroundings, you are correct.[1] This basic principle is attributed, as are so many other laws, to Sir Isaac Newton (1643–1727).

Newton's Law of Cooling

The rate of change in the temperature T of an object placed in surroundings of uniform temperature M is proportional to the difference between the temperature of the object and the temperature of the surroundings. Mathematically,

$$\frac{dT}{dt} = k(M - T), \tag{6}$$

where $k > 0$ is a constant of proportionality.

Newton's Law of Cooling: Curve Fitting

Vary the parameters k, M, and the initial temperature to fit a solution to real data.

Newton's Law of Cooling says that the temperature of an object will *fall* if its temperature is greater than the surrounding temperature, because $M - T$ is negative and therefore so is T'. On the other hand, if the ambient temperature is greater, $M - T$ is positive and the temperature of the object will *rise*.

[1] We assume surroundings large enough that their constant temperature is not affected by the object studied.

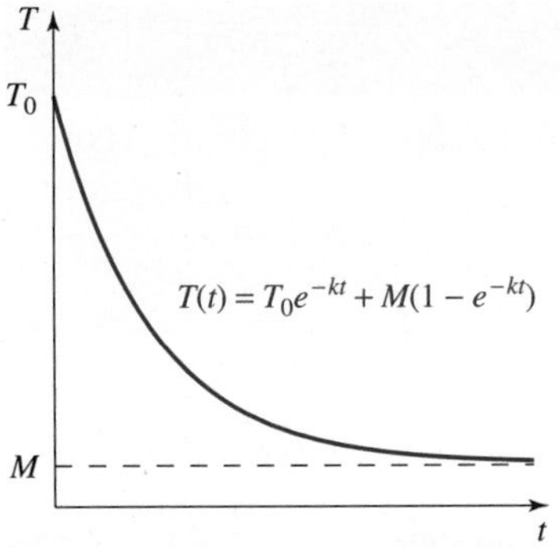

FIGURE 2.4.4 Temperature of object (initial temperature T_0) in room with temperature M.

Consider an object with initial temperature $T(0) = T_0$ placed in surroundings of temperature M. Then $T(t)$ satisfies the IVP

$$\frac{dT}{dt} = k(M - T), \quad T(0) = T_0. \tag{7}$$

This DE can be written as $T' + kT = kM$, so we know from Sec. 2.1 that the solution[2] is

$$T(t) = T_0e^{-kt} + M(1 - e^{-kt}). \tag{8}$$

The temperature $T(t)$ of the object changes exponentially from the initial temperature T_0 to the limiting temperature M. See Fig. 2.4.4.

EXAMPLE 3 **Cool House** At midnight, with the temperature inside your house at 70°F and the outside temperature at 20°F, your furnace breaks down. Two hours later the temperature in your house has fallen to 50°F. We assume that the outside temperature remains at 20°F and consider the following questions:

(a) What IVP is satisfied by the temperature inside the house after midnight?

(b) What formula gives the inside temperature?

(c) At what time will the inside temperature reach 40°F?

Some answers follow.

(a) The IVP requested is provided by Newton's Law of Cooling:

$$T' = k(20 - T), \quad T(0) = 70. \tag{9}$$

(b) To solve the IVP (9), use the general solution (8) to obtain the temperature $T(t) = 70e^{-kt} + 20(1 - e^{-kt})$; therefore,

$$T(t) = 20 + 50e^{-kt}. \tag{10}$$

To find k, we use the temperature given for 2:00 AM. We put $T(2) = 50$ and $t = 2$ into (10) and obtain $50 = 20 + 50e^{-2k}$, from which it follows that $k = -\ln(0.6)/2 \approx 0.255$. Hence our answer to question (b) is

$$T(t) = 20 + 50e^{t\ln(0.6)/2} \approx 20 + 50e^{-0.255t}. \tag{11}$$

The temperature curve is shown in Fig. 2.4.5.

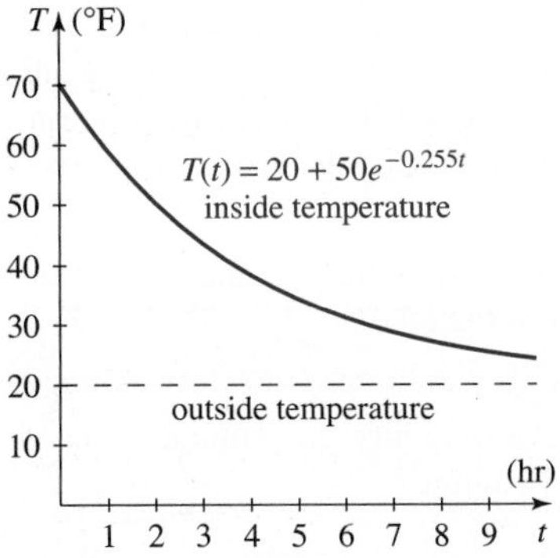

FIGURE 2.4.5 Temperature of the cooling house for Example 3.

(c) Let us reword the final question: For what t-value will $T = 40$? From equation (11), then, we must have $40 = 20 + 50e^{t\ln(0.6)/2}$. Solving, we obtain $t = 2\ln(0.4)/\ln(0.6) \approx 3.592$. Since 3.592 hours is approximately 3 hours and 35 minutes, the temperature of the house will reach 40°F at 3:35 AM. ■

[2] Under the change of variable $y = T - M$, the Newton cooling equation becomes the decay equation; see Problem 14.

Summary

Two simple models for mixing and cooling provide the tools to solve a wide variety of applied problems. They also help build our intuition about differential equations. By rewriting the DE $y' + p(t)y = f(t)$ in the form $y' = f(t) - p(t)y$, we can think of it as representing a mixing problem with inflow $f(t)$ and outflow $p(t)y$. Thus, new problems are illuminated by old ones.

2.4 Problems

1. **Mixing Details** Solve the homogeneous equation
$$\frac{dx}{dt} + \frac{2x}{100 - t} = 0,$$
corresponding to the DE in Example 2, using separation of variables.

2. **English Brine** Initially, 50 lb of salt is dissolved in a tank containing 300 gal of water. A salt solution with 2 lb/gal concentration is poured into the tank at 3 gal/min. The mixture, after stirring, flows from the tank at the same rate the brine is entering the tank.
   (a) Find the amount of salt in the tank as a function of time.
   (b) Determine the concentration of salt in the tank at any time.
   (c) Determine the steady-state amount of salt in the tank.
   (d) Find the steady-state concentration of salt in the tank.
   (e) Use a graphing calculator or computer to sketch the graphs of the future amount of salt in the tank and the concentration of salt in the tank.

3. **Metric Brine** Initially, a 100-liter tank contains a salt solution with concentration 0.5 kg/liter. A fresher solution with concentration 0.1 kg/liter flows into the tank at the rate of 4 liter/min. The contents of the tank are kept well stirred, and the mixture flows out at the same rate it flows in.
   (a) Find the amount of salt in the tank as a function of time.
   (b) Determine the concentration of salt in the tank at any time.
   (c) Determine the steady-state amount of salt in the tank.
   (d) Find the steady-state concentration of salt in the tank.

4. **Salty Goal** At the start, 5 lb of salt are dissolved in 20 gal of water. Salt solution with concentration 2 lb/gal is added at a rate of 3 gal/min, and the well-stirred mixture is drained out at the same rate of flow. How long should this process continue to raise the amount of salt in the tank to 25 lb?

5. **Mysterious Brine** A tank initially contains 200 gallons of fresh water, but then a salt solution of unknown concentration is poured into the tank at 2 gal/min. The well-stirred mixture flows out of the tank at the same rate. After 120 min, the concentration of salt in the tank is 1.4 lb/gal. What is the concentration of the entering brine?

6. **Salty Overflow** A 600-gallon tank is filled with 300 gal of pure water. A spigot is opened and a salt solution containing 1 lb of salt per gallon of solution begins flowing into the tank at a rate of 3 gal/min. Simultaneously, a drain is opened at the bottom of the tank allowing the solution to leave the tank at a rate of 1 gal/min. What will be the salt content in the tank at the precise moment that the volume of solution in the tank reaches the tank's capacity of 600 gal?

7. **Cleaning Up Lake Erie** Lake Erie has a volume of roughly 100 cubic miles, and its equal inflow and outflow rates are 40 cubic miles per year. At year $t = 0$, a certain pollutant has a volume concentration of 0.05%, but after that the concentration of pollutant flowing into the lake drops to 0.01%. Answer the following questions, assuming that the pollutant leaving the lake is well mixed with lake water.
   (a) What is the IVP satisfied by the volume V (in cubic miles) of pollutant in the lake?
   (b) What is the volume V of pollutant in the lake at time t?
   (c) How long will it take to reduce the pollutant concentration to 0.02% in volume?

8. **Correcting a Goof** Into a tank containing 100 gal of fresh water, Wei Chen was to have added 10 lb of salt but accidentally added 20 lb instead. To correct her mistake she started adding fresh water at a rate of 3 gal/min, while drawing off well-mixed solution at the same rate. How long will it take until the tank contains the correct amount of salt?

9. **Changing Midstream** A 1,000-gallon tank contains 200 gal of pure water. (See Fig. 2.4.6.) A brine solution containing 1 lb of salt per gal is flowing into the tank at a rate of 4 gal/sec, and the well-stirred mixture is leaving the tank at the same rate. Let x denote the amount of salt in the tank at time t.
   (a) Set up (but do not solve) the initial-value problem (both DE and initial condition).

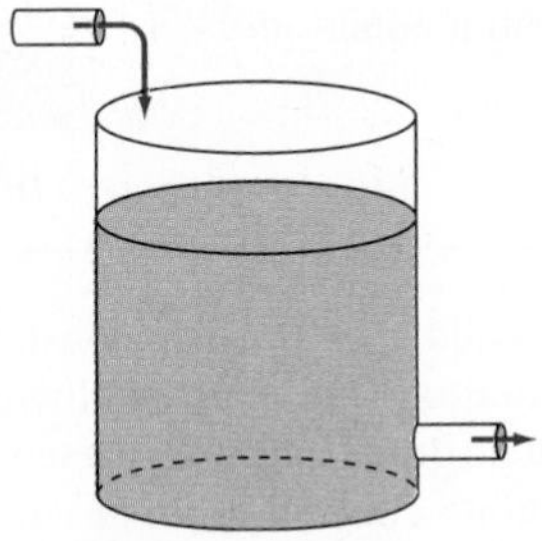

FIGURE 2.4.6 Tank setup for Problem 9.

(b) Suppose that this situation continues for a very long time. What is the equilibrium solution x_{eq}?

(c) After the solution has reached equilibrium, an additional faucet is turned on that supplies brine containing 2 lb/gal at a rate of 2 gal/sec. (See Fig. 2.4.7.) Set up the new initial-value problem, assuming that the clock is restarted when the new faucet is turned on.

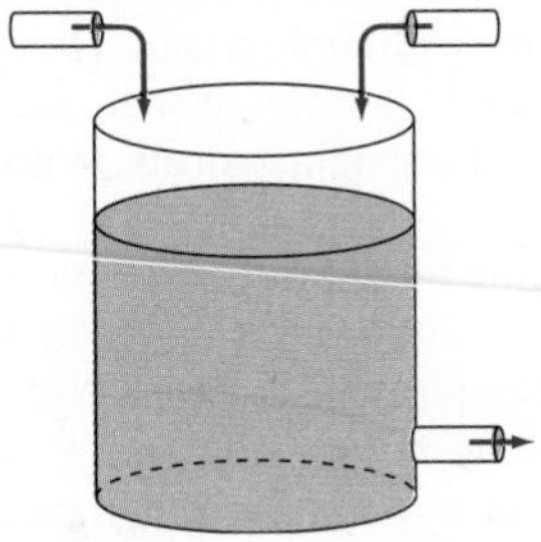

FIGURE 2.4.7 Tank setup for Problem 9 with a second faucet.

(d) How long t_f does it take for the tank to fill completely after the second faucet is turned on?

(e) How much salt is in the tank when $t = t_f$?

(f) Set up the IVP for the amount of salt in the tank for $t > t_f$, assuming that the tank overflows.

10. Cascading Tanks[3] Fresh water is poured at a rate of 2 gal/min into a tank A, which initially contains 100 gal of a salt solution with concentration 0.5 lb/gal. The stirred mixture flows out of tank A at the same rate and into a second tank B that initially contained 100 gal of fresh water. The mixture in tank B is also stirred and flows out at the same rate. (See Fig. 2.4.8.)

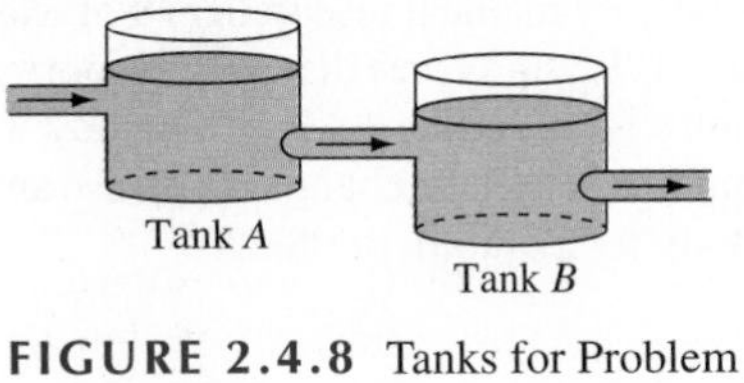

FIGURE 2.4.8 Tanks for Problem 10.

[3]This is an example of a *multiple-compartment* problem.

(a) Determine an IVP satisfied by the amount of salt in tank A.

(b) Find the amount of salt in tank A at any time.

(c) Find the IVP satisfied by the amount of salt in tank B.

(d) Determine the amount of salt in tank B as a function of time.

11. More Cascading Tanks A cascade of several tanks is shown in Fig. 2.4.9. Initially, tank 0 contains 1 gal of alcohol and 1 gal of water, while the other tanks contain 2 gal of pure water. Fresh water is pumped into tank 0 at the rate of 1 gal/min, and the varying mixture in each tank is pumped into the next tank at the same rate. Let $x_n(t)$ denote the amount of alcohol in tank n at time t.

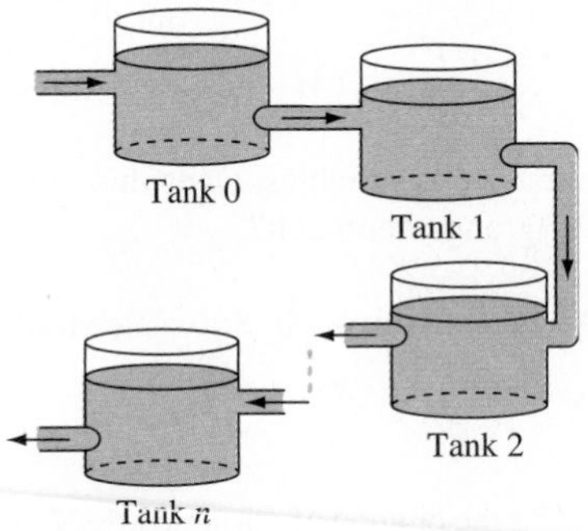

FIGURE 2.4.9 Cascading tanks for Problem 11.

(a) Show that $x_0(t) = e^{-t/2}$.

(b) Show by induction that $x_n(t) = (t^n e^{-t/2})/(n!\,2^n)$, $n = 1, 2, \ldots$.

(c) Show that the maximum value of $x_n(t)$ will be $M_n = n^n e^{-n}/n!$.

(d) Use *Stirling's approximation*, $n! \approx \sqrt{2\pi n}\, n^n e^{-n}$, to show that $M_n \approx (2\pi n)^{-1/2}$.

12. Three-Tank Setup Consider the cascading arrangement of tanks shown in Fig. 2.4.10, with $V_1 = 200$ gal,

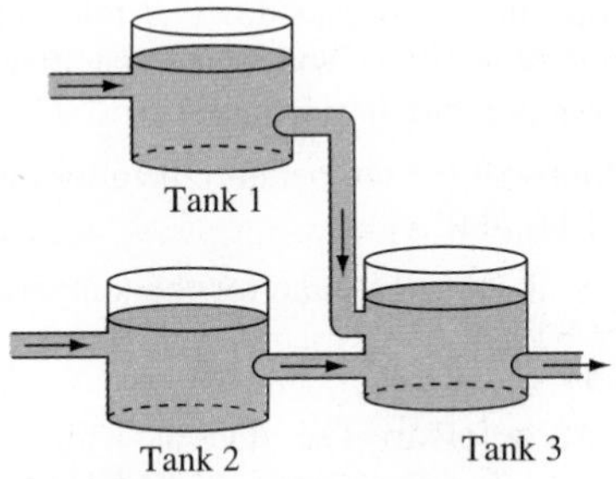

FIGURE 2.4.10 Three-tank setup for Problem 12.

$V_2 = 200$ gal, and $V_3 = 500$ gal as the volumes of brine in the three tanks. Each tank initially contains 20 lb of salt. The inflow rates and outflow rates for tanks 1 and 2 are all 5 gal/sec, but the outflow rate for tank 3 is 10 gal/sec. The flow into tanks 1 and 2 is pure water.

(a) Let $x(t)$, $y(t)$, and $z(t)$ be the amounts of salt in tanks 1, 2, and 3, respectively. Set up the differential equation and the initial conditions that describe how the amounts of salt in tanks 1 and 2 are changing at any time t.

(b) Solve for $x(t)$ and $y(t)$.

(c) Using the results of part (b), solve for $z(t)$.

13. Another Solution Method Instead of using the theory of Sec. 2.1, solve the cooling/heating problem

$$\frac{dT}{dt} = k(M - T), \quad T(0) = T_0,$$

by separation of variables. Does this seem easier than the Euler-Lagrange approach?

14. Still Another Approach Solve Newton's equation

$$\frac{dT}{dt} = k(M - T)$$

by making the change of variable $y = T - M$ to transform it into the decay equation.

15. Using the Time Constant At noon, with the temperature in your house at 75°F and the outside temperature at 95°F, your air conditioner breaks down. Suppose that the time constant $1/k$ for your house is 4 hours.

(a) What will the temperature in your house be at 2:00 PM?

(b) When will the temperature in your house reach 80°F?

NOTE: Engineers often state problems in terms of the *time constant* $1/k$, the reciprocal of the constant of proportionality in Newton's Law of Cooling.

16. Chilling Thought Suppose that it is 70°F in your house when the furnace breaks down at midnight. The outside temperature is 10°F. You notice that after 30 minutes the inside temperature has dropped to 50°F.

(a) What will the temperature be after one hour (that is, at 1:00 AM)?

(b) How long will it take for the temperature to drop to 15°F?

17. Drug Metabolism The rate at which a drug is absorbed into the bloodstream is modeled by the first-order differential equation

$$\frac{dC}{dt} = a - bC(t),$$

where a and b are positive constants and $C(t)$ denotes the concentration of drug in the bloodstream at time t. Assuming that no drug is initially present in the bloodstream, find the limiting concentration of the drug in the bloodstream as $t \to \infty$. How long does it take for the concentration to reach one-half of the limiting value?

18. Warm or Cold Beer? A cold beer with an initial temperature of 35°F warms up to 40°F in 10 minutes while sitting in a room with temperature 70°F. What will the temperature of the beer be after t minutes? After 20 minutes?

19. The Coffee and Cream Problem John and Maria are having dinner, and each orders a cup of coffee. John cools his coffee with some cream. They wait 10 minutes and then Maria cools her coffee with the same amount of cream. The two then begin to drink. Who drinks the hotter coffee?

20. Professor Farlow's Coffee Professor Farlow always has a cup of coffee before his 8:00 AM differential equations class. Suppose the coffee is 200°F when poured from the coffee pot at 7:30 AM, and 15 minutes later it cools to 120°F in a room whose temperature is 70°F. However, Professor Farlow never drinks his coffee until it cools to 90°F. When will the professor be able to drink his coffee?

21. Case of the Cooling Corpse In a murder investigation, a corpse was found by a detective at exactly 8:00 PM. Being alert, he measures the temperature of the body and finds it to be 70°F. Two hours later the detective again measures the temperature of the corpse and finds it to be 60°F. (See Fig. 2.4.11.)

(a) If the room temperature is 50°F, and the detective assumes that the body temperature of the person before death was 98.6°F, at what time did the murder occur?

(b) The commonly assumed 98.6°F was based on an 1861 study in Germany that reported average normal body temperature as 37°C, with only two significant figures. Conversion from Celsius to Fahrenheit gave an unreliable third digit, which has been questioned recently by medical researchers. For example, the result of a 1992 study was that average normal body temperature is more like 98.2°F.[4] By how much does your answer to part (a) change if you use an average normal body temperature of 98.2°F?

---

[4] P. A. Mackowiak, S. S. Wasserman, and M. M. Levine, "A Critical Appraisal of 98.6 Degrees F, the Upper Limit of the Normal Body Temperature, and Other Legacies of Carl Reinhold August Wunderlich," *Journal of the American Medical Association* **268**, **12** (23–30 September 1992), 1578–80.

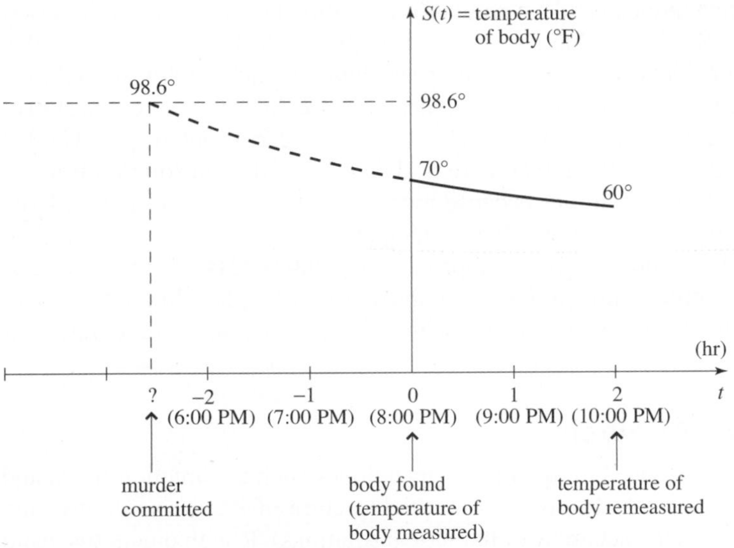

FIGURE 2.4.11 Temperature of the cooling corpse in Problem 20.

22. A Real Mystery At 1:00 PM, Carlos puts into the refrigerator a can of soda, which has been standing out in a room with temperature 70°F. The temperature inside the refrigerator is 40°F. Fifteen minutes later, at 1:15 PM, the temperature of the soda has fallen to 60°F. At some later time Carlos removes the soda from the refrigerator to the room where, at 2:00 PM, the soda temperature is 60°F. At what time did Carlos remove the soda from the refrigerator?

Computer Mixing *Use appropriate computer technology to obtain graphical or numerical solutions for Problems 23 and 24, and describe how the solution behaviors respond to the inflow/outflow comparison. Consider $y(t)$ to be the amount of dissolved substance.*

23. $y' + \dfrac{1}{1-t}y = 2, \quad y(0) = 0.$ (inflow < outflow)

24. $y' + \dfrac{1}{1+t}y = 2, \quad y(0) = 0;$ (inflow > outflow)

25. Suggested Journal Entry Summarize what you have learned thus far about differential equations, and discuss how these tools will be useful in other courses or in your future work.

2.5 Nonlinear Models: Logistic Equation

SYNOPSIS: Nonlinear DEs often fail to have solutions that can be found or expressed as formulas, so we approach them qualitatively. We introduce as significant nonlinear models the logistic equation and the closely related threshold equation, both of which we are also able to solve analytically.

Nonlinear Differential Equations

Consider the following DEs that are not linear:

$$dy/dt = y(1-y), \tag{1}$$

$$dy/dt = \cos(y-t), \tag{2}$$

$$dy/dt = \frac{1}{t^2+y^2}. \tag{3}$$

What options do you have for solving them? This is a good time to review the possible techniques studied so far, both quantitative (analytical or numerical) and qualitative (graphical).

1. Separable
2. Linear

Analytical techniques yield a formula for a solution but cannot do so for *every* first-order DE. So far we have introduced analytical techniques for DEs that are separable or linear. But none of the preceding equations is linear, and only equation (1) is separable. There are more specialized techniques for certain classes of DEs (e.g., see Sec. 1.3, Problems 41–44 and Sec. 2.2, Problems 31–41), but do not be fooled into thinking that there is always an analytical solution formula for every differential equation, because more often than not there is not. Even separable and linear DEs are not always integrable.

A numerical method can give an approximate solution (essentially as close as you like) to any initial-value problem, but that is only a single solution to the DE. What is more, the further you go from the initial condition, the less accurate your numerical solution is likely to be.

Qualitative Analysis

Graphical solutions based on qualitative techniques such as direction fields and isoclines are most likely to give you a quick picture of *all* the solutions (and can also help gauge the accuracy of numerical solutions). Rough qualitative hand sketches can easily be made for equations (1), (2), and (3) from quick calculations of easy isoclines and information on the sign of the slopes, as developed in Sec. 1.2 and shown in Fig. 2.5.1.

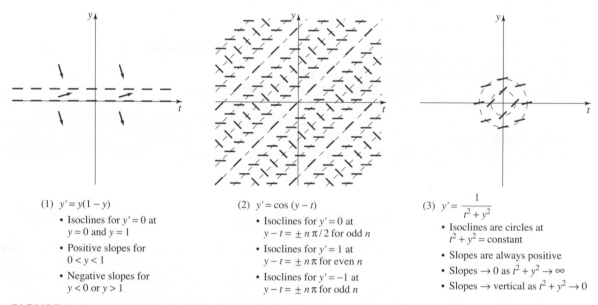

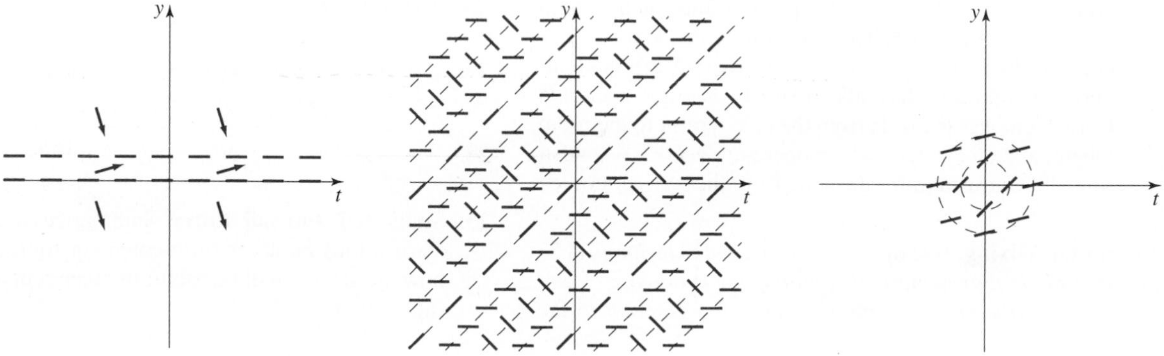

(1) $y' = y(1-y)$
- Isoclines for $y' = 0$ at $y = 0$ and $y = 1$
- Positive slopes for $0 < y < 1$
- Negative slopes for $y < 0$ or $y > 1$

(2) $y' = \cos(y - t)$
- Isoclines for $y' = 0$ at $y - t = \pm n\pi/2$ for odd n
- Isoclines for $y' = 1$ at $y - t = \pm n\pi$ for even n
- Isoclines for $y' = -1$ at $y - t = \pm n\pi$ for odd n

(3) $y' = \dfrac{1}{t^2 + y^2}$
- Isoclines are circles at $t^2 + y^2 = \text{constant}$
- Slopes are always positive
- Slopes $\to 0$ as $t^2 + y^2 \to \infty$
- Slopes $\to$ vertical as $t^2 + y^2 \to 0$

FIGURE 2.5.1 Graphing basic qualitative information for equations (1), (2), and (3).

Figure 2.5.1 suggests that for equation (1) the two isoclines of zero slope are also solutions to the DE, and that for equation (2) the isoclines of slope 1 are also solutions. You should confirm that algebraically in the equations.

We are almost ready to sketch solutions to equations (1), (2), and (3). But first think about *existence* and *uniqueness*. (See Sec. 1.5.) Does Picard's Theorem hold for these equations? Yes, it does, except for equation (3) at the origin. (We will look more closely at that point in just a minute.) For all other points we are guaranteed both existence and uniqueness, which means that solutions will not cross each other. Keeping this in mind, you can start at any point and sketch a solution that will follow the slope marks, without crossing any other solutions, as shown in Fig. 2.5.2.

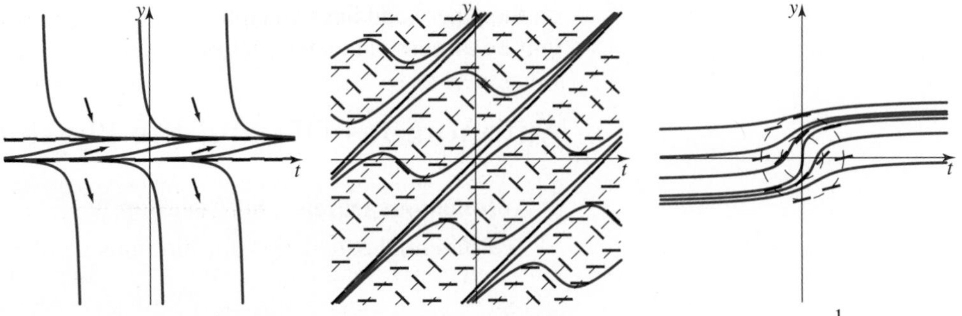

FIGURE 2.5.2 Sketching solutions for equations (1), (2), and (3).

An open-ended graphical DE solver on a calculator or computer can quickly give a more detailed picture from the numerical approximate solutions for many initial conditions. If the picture satisfies the appropriate qualitative properties shown in Fig. 2.5.1, you can reasonably expect that these numerical approximations are fairly accurate (usually indicating a small enough step size). Compare the computer-generated drawings of Fig. 2.5.3 with the sketches of Fig. 2.5.2, and confirm that the qualitative features are in agreement.

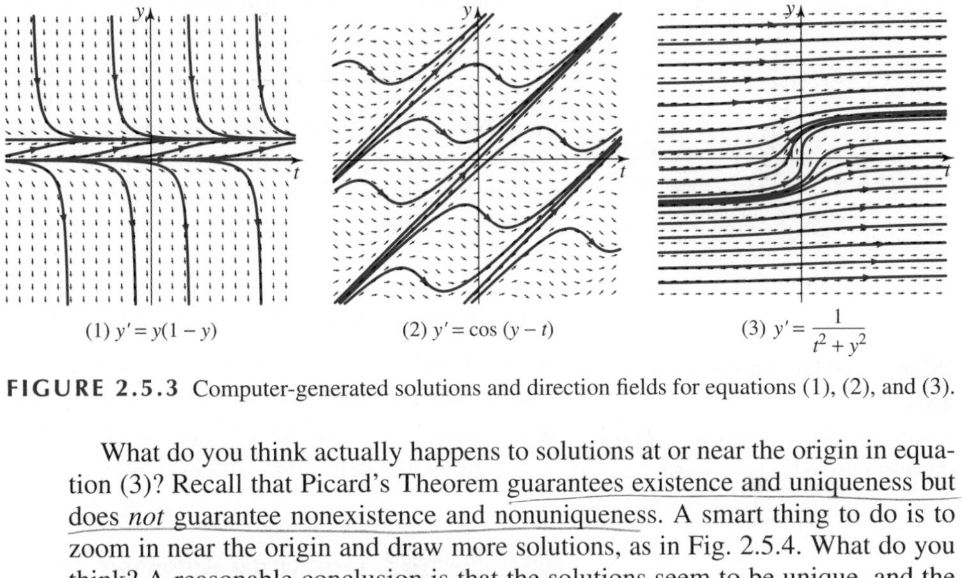

FIGURE 2.5.3 Computer-generated solutions and direction fields for equations (1), (2), and (3).

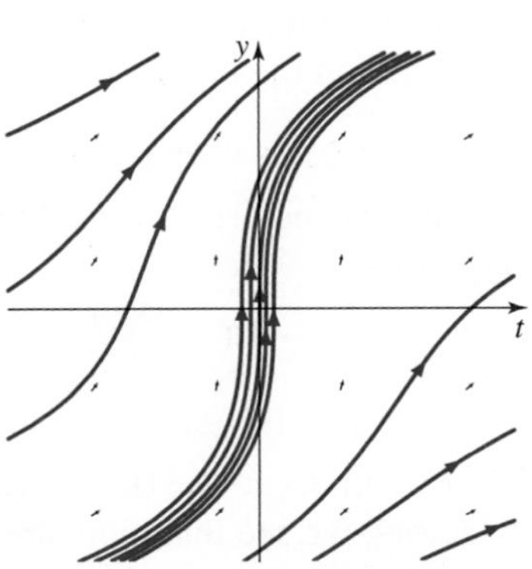

FIGURE 2.5.4 Adding detail by zooming in on the graph for equation (3).

What do you think actually happens to solutions at or near the origin in equation (3)? Recall that Picard's Theorem guarantees existence and uniqueness but does *not* guarantee nonexistence and nonuniqueness. A smart thing to do is to zoom in near the origin and draw more solutions, as in Fig. 2.5.4. What do you think? A reasonable conclusion is that the solutions seem to be unique, and the "failure" of the existence theorem seems to mean that at the origin (and only at the origin) the slope of the solution curve is vertical (thus exhibiting that a "nonexisting" slope can simply be vertical).

Equilibria and Stability

Recall from Secs. 1.2 and 1.3 that an *equilibrium* for a first-order DE is a value of c for which $y = c$ is a (constant) solution. Inspect equations (1), (2), and (3) and graphs of their solutions in Fig. 2.5.2 or Fig. 2.5.3. Which equations have equilibria? How many?

A nonlinear DE may have more than one equilibrium or none at all. Equations (2) and (3) have no equilibria. Although they have isoclines of zero slope or solutions that approach zero slope, neither has any constant solution. Equation (1) on the other hand has two equilibria. Can you identify the *stability* of each? This will be discussed in detail below.

Autonomous DEs and the Phase Line

Autonomous Differential Equation

A differential equation is **autonomous** if

$$dy/dt = f(y);$$

that is, if the independent variable t does not explicitly appear on the right-hand side of the equation.

For an *autonomous* DE, at any value of y the slopes dy/dt of solutions $y(t)$ do not depend on t.[1] This fact implies that on a ty graph all isoclines are horizontal lines and all solutions for any given y-value are horizontal translations. These features are visible for equation (1) in Figs. 2.5.2 and 2.5.3.

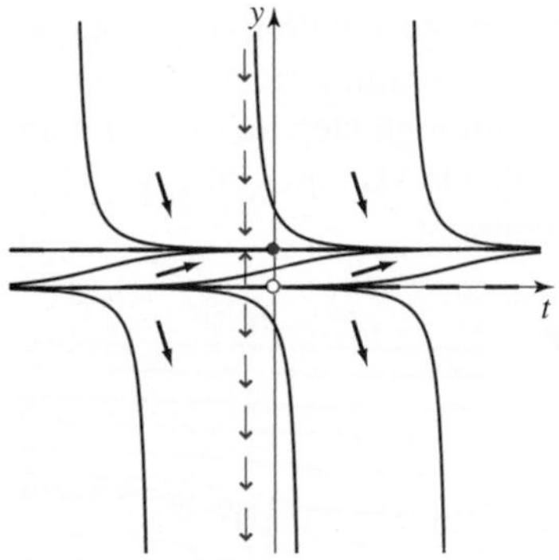

FIGURE 2.5.5 Phase line (in color) for equation (1), $y' = y(1 - y)$.

The equilibrium and stability features of an autonomous DE can be summarized in a single dimension along the y-axis by arrows pointing up or down to show whether slopes at the given y values are positive or negative, respectively. This graph is called the **phase line** and is shown in color for equation (1) on Fig. 2.5.5.

If the phase-line arrows point toward the equilibrium point on both sides, solutions approach it from both directions and it is **stable**; this is called a **sink** and is denoted on the diagram by a filled circle. An open circle represents an **unstable** equilibrium because solutions move away from it on both sides; this is called a **source**. The split circle indicates a **semistable** equilibrium point, called a **node**: solutions approach from one side but flow away on the other.

With the phase line, sketching solutions for an autonomous DE becomes very simple:

Logistic Phase Line

See the phase line in action—and see the relationship between y and y'.

- Locate the constant solutions at the equilibria.
- Observe the sign of y' between equilibria to know whether the slopes of the solutions are positive or negative, which gives the stability information.
- Recall that when the $y' = f(y)$ function meets the criteria for uniqueness of solutions, the solutions will never meet or actually cross in the ty graph.

EXAMPLE 1 **First-Order Autonomy** Let us discuss the nature of the solutions of two first-order autonomous differential equations:

(a) $y' = y(1 - y)(2 - y)$ (b) $y' = (y - 1)(y - 3)(y - 5)^2$

We analyze these DEs as follows.

(a) Set $y(1 - y)(2 - y) = 0$ to obtain equilibrium solutions at $y = 0$, $y = 1$, and $y = 2$. By looking at the signs of the factors, we see that there are

[1] **Autonomous** equations, also called **time-invariant** or **stationary** equations by engineers, were studied in Sec. 1.2, Problem 61 and Sec. 1.3, Problem 45. Equations in which t enters the right-hand side explicitly are called **time-varying** or **nonautonomous** equations.

positive slopes for $y > 2$ and $0 < y < 1$ and negative slopes for $1 < y < 2$ and $y < 0$. Plotting this information on the y-axis lets us see that the equilibrium at 1 is stable, while those at 0 and 2 are unstable. Typical solution curves are now sketched in Fig. 2.5.6(a).

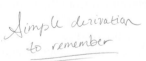

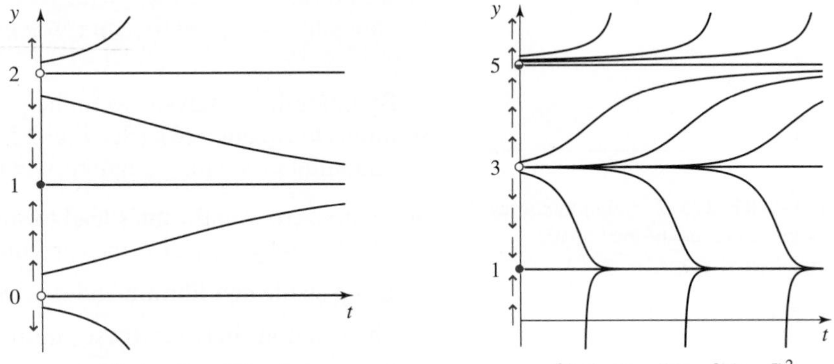

(a) $y' = y(1 - y)(2 - y)$ (b) $y' = (y - 1)(y - 3)(y - 5)^2$

FIGURE 2.5.6 Phase-line analyses for Example 1.

(b) By arguments similar to those for (a), we can obtain phase-line and solution graphs, as shown in Fig. 2.5.6(b). A semistable equilibrium occurs at $y = 5$ because the sign of y' does not change as y passes through 5. ■

From Linearity to Nonlinearity in the Real World

The unrestricted growth equation

$$\frac{dy}{dt} = ky, \quad k > 0, \tag{4}$$

which assumes that the rate of growth of a population is always proportional to its size, is linear and predicts exponential growth. But while exponential growth may occur *in the initial stages*, it cannot continue indefinitely. For long-range prediction we need models that take into account the interaction of the population with its environment. Population growth levels off as a result of limited food supplies, increased disease, crowding, and other factors. To build a model that takes such factors into consideration, we replace the constant growth rate k in equation (4) with a **variable growth rate** $k(y)$ that depends on the population size, giving the more general model

$$\frac{dy}{dt} = k(y)\, y. \tag{5}$$

Simple derivation to remember

For most populations, the growth rate $k(y)$ decreases with increasing y, so we make the simple choice of a decreasing linear function $k(y) = r - ay$, $a > 0$, $r > 0$. Substituting this function into (5), we obtain the **logistic** (or Verhulst[2]) **equation** $dy/dt = (r - ay)y$. Letting $L = r/a$, we obtain a conventional form of the equation, and the significance of the terms will soon be explained.

[2]Although the exponential growth equation $y' = ky$ goes back to 1798 and Thomas Malthus, it was the Belgian mathematician Pierre Verhulst (1804–1839) who argued in 1838 that the rate of growth in the Malthus equation could not remain constant indefinitely, but should diminish according to the law $k(y) = a - by$, resulting in the logistic equation we find useful today.

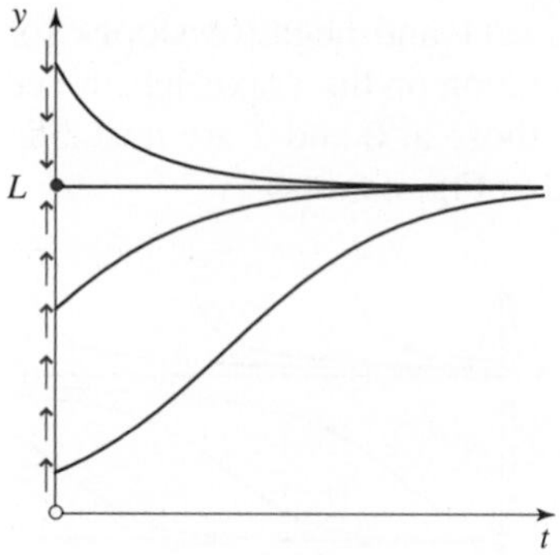

FIGURE 2.5.7 Solution curves (logistic curves) of the logistic equation $y' = ry(1 - y/L)$.

Logistic Equation

$$\frac{dy}{dt} = r\left(1 - \frac{y}{L}\right)y, \tag{6}$$

where the positive parameter r is called the **initial** (or **intrinsic**) **growth rate** and L is called the **carrying capacity**.

By phase-line analysis, as in the previous subsection, we produce a sketch of solutions to equation (6). (See Fig. 2.5.7.) Since the logistic equation is modeling a population problem, we ignore $y < 0$. We observe the following:

- *All* nonzero initial values lead to solutions that approach L asymptotically, which is why we call L the carrying capacity.
- L is a stable equilibrium and 0 is an unstable equilibrium.
- The solutions between these equilibria have an S-shape characteristic of the logistic curve.
- There is an inflection point between 0 and L, which is investigated in Problem 13.

Logistic Growth

Experiment with changing parameters and initial conditions.

Everything discovered so far about solutions of the logistic equation is the result of qualitative analysis of the DE (as summarized by the phase line). We will verify these properties by deriving a quantitative solution.

Analytic Solution of the Logistic Equation

The logistic equation (6) is separable,

$$\frac{dy}{(1 - \frac{y}{L})y} = r\,dt,$$

and can be solved using the *partial fraction decomposition*[3]

The partial fraction step is easy to check: combine the fractions on the right by obtaining a common denominator.

$$\frac{1}{y(1 - \frac{y}{L})} = \frac{1}{y} + \frac{\frac{1}{L}}{1 - \frac{y}{L}}.$$

Applying this to the separated form of equation (6) gives

$$\left(\frac{1}{y} + \frac{\frac{1}{L}}{1 - \frac{y}{L}}\right)dy = r\,dt,$$

and we integrate this to obtain

$$\ln|y| - \ln\left|1 - \frac{y}{L}\right| = rt + c, \tag{7}$$

where the constant of integration c is to be determined from the initial condition $y(0) = y_0$. From our qualitative analysis we know that if $0 < y_0 < L$, then $0 < y(t) < L$ for all future time; this means that $0 < y/L < 1$. Thus, both y and $1 - y/L$ are positive and we can drop the absolute values in equation (7). Therefore,

$$\ln\left(\frac{y}{1 - \frac{y}{L}}\right) = rt + c,$$

[3]Readers who are rusty on partial fraction decomposition can review the material in Appendix PF.

so if we write $C = e^c$, the general solution of (6) is given implicitly by

$$\frac{y}{1 - \frac{y}{L}} = Ce^{rt}. \tag{8}$$

Using the initial condition $y(0) = y_0$, so that $y = y_0$ when $t = 0$, equation (8) becomes

$$\frac{y_0}{1 - \frac{y_0}{L}} = C.$$

Substituting this value for C into (8) and solving for y we obtain solution (10) below.

Initial-Value Problem for the Logistic Equation
The solution for $t \geq 0$ of the logistic IVP

$$\frac{dy}{dt} = r\left(1 - \frac{y}{L}\right)y, \quad y(0) = y_0, \tag{9}$$

is given by

$$y(t) = \frac{L}{1 + \left(\frac{L}{y_0} - 1\right)e^{-rt}}, \tag{10}$$

where $r > 0$ is the **intrinsic growth rate** and $L > 0$ is the **carrying capacity**.

If $y_0 > L$, the same solution results; see Problem 13(a).

Table 2.5.1 U.S. population (in millions)

Year	Population
1900	76.1
1910	92.0
1920	105.7
1930	122.8
1940	131.7
1950	151.1
1960	179.3
1970	203.3
1980	226.5
1990	249.1
2000	271.3

EXAMPLE 2 **U.S. Population** Using the Bureau of Census population data in Table 2.5.1 and the logistic model, we will determine

(a) a theoretical maximum U.S. population,
(b) a predicted population for the year 2030, and
(c) an estimate for the population in 1790.

We let $t = 0$ represent the year 1900 and $t = 1$ the year 2000. Therefore, $t = 0.5$ is the year 1950, while $t = 1.3$ stands for 2030. Since the IVP (9), which is our model, contains two parameters, r and L, we need to use data from two points in addition to the initial condition at 1900, so we will use 1950 and 2000. The population is given by equation (10), together with the data

$$y(0) = 76.1, \quad y(0.5) = 151.1, \quad y(1) = 271.3.$$

With $y_0 = 76.1$ and $t = 0.5$, $y(0.5) = 151.1$, so equation (10) becomes

$$151.1 = \frac{L}{1 + \left(\frac{L}{76.1} - 1\right)e^{-\frac{r}{2}}}. \tag{11}$$

Now we use the second data point in equation (10); that is, we now let $y_0 = 76.1$ and $t = 1$, so $y(1) = 271.3$; this gives

$$271.3 = \frac{L}{1 + \left(\frac{L}{76.1} - 1\right)e^{-r}}. \tag{12}$$

While it is challenging to solve the system of equations (11) and (12) for the values of r and L, computer packages such as *MathCad* can do so efficiently. The result is $r \approx 1.6$, $L \approx 774$. Our completed logistic model for U.S. population is therefore

$$y(t) \approx \frac{774}{1+\left(\frac{774}{76.1}-1\right)e^{-1.6t}} = \frac{774}{1+9.17e^{-1.6t}}. \tag{13}$$

Now we can use equation (13) to answer the questions that we posed.

(a) The theoretical maximum population for the U.S. is $L = 774$ million.

(b) The predicted population for the year 2030 is $y(1.3) = 360.7$ million.

(c) The backward projection for 1790 involves using (13) with a negative t-value: $y(-1.1) = 14.3$ million. The actual population for the United States in 1790 was 4 million. (Can you explain why this discrepancy might occur?)

■

We need to realize that the accuracy of the predictions from the logistic model depends, among other things, on whether the parameters r and L remain constant. The basic logistic model does not take into account noncrowding influences on population, such as harvesting, immigration, wars, and technological advances. Problem 16 investigates additional effects of harvesting.

Populations with Minimum Thresholds

For many species of plants and animals, there is a critical population level known as the **threshold level** T; if the population falls below this level, the population will tend to zero—the species becomes extinct.[4] This happened to the *passenger pigeon*. As the result of indiscriminate slaughter, this bird became extinct in 1914. Due to its nesting and breeding habits, once the population fell below the threshold level, it could not recover. The simplest model of this phenomenon is the **threshold equation**.

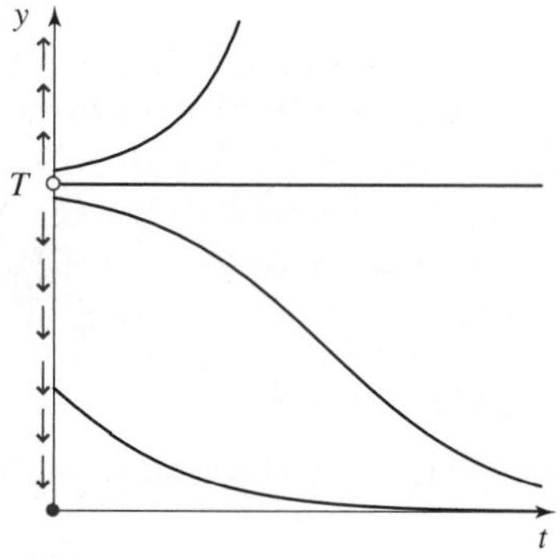

FIGURE 2.5.8 Analysis of the threshold equation

$y' = -r(1 - y/T)y.$

Threshold Equation

The **threshold equation** is nothing more than the logistic equation with a minus sign, the threshold level T replacing the carrying capacity L:

$$\frac{dy}{dt} = -r\left(1-\frac{y}{T}\right)y.$$

We have sketched the ty solutions graph for this equation in Fig. 2.5.8. The threshold equation has equilibrium points 0 and T but, in contrast to the logistic equation, 0 is stable and T is unstable. For initial values less than T, solutions tend to zero. Problem 34 investigates details of Fig. 2.5.8.

The quantitative solution is easy to obtain from that of the logistic equation by the change of variable $t = -\tau$. (See Problem 33.) The result is shown in the following box; compare with the typical solution curves sketched in Fig. 2.5.8.

[4] Epidemics exhibit threshold behavior, too. If the number of infected individuals exceeds a threshold level, the disease "takes off"; below that level, it dies out.

Initial-Value Problem for the Threshold Equation
The solution for $t \geq 0$ of the threshold IVP

$$\frac{dy}{dt} = -r\left(1 - \frac{y}{T}\right)y, \quad y(0) = y_0, \tag{14}$$

is given by

$$y(t) = \frac{T}{1 + \left(\frac{T}{y_0} - 1\right)e^{rt}}, \tag{15}$$

where $r > 0$ is the **initial growth rate** and $T > 0$ the **threshold level.**

Logistic Model in Another Context

A remarkable application of the logistic model was due to the geologist **M. King Hubbert** (1903–1989), who achieved fame due to his 1956 prediction of the decline of U.S. oil production in the 1970s.[5] He realized, by careful examination of the logistic curve, that when the *rate* dy/dt of oil yield began to drop, the inflection point on the logistic curve had been passed. This observation meant that the oil reserves were not nearly as high as others had predicted.

A reasonable explanation is the following, for the amount of oil $y(t)$ extracted from a particular reserve. Both in the beginning (when oil is first recovered) and in the end (when no oil remains), $dy/dt = 0$. In between these times dy/dt is positive and must reach a maximum (Fig. 2.5.9). There will be no values of y below zero or above the total in the reserve. You can see that once oil extraction hits the inflection point in the "middle," dy/dt begins to drop, so the reserve is already half-depleted. See Problem 29 to explore the possibility that the inflection point might not occur exactly when this reserve is half depleted, and to analyze the **Hubbert peak** in the graph of y' versus t that is commonly used in oil industry analysis.

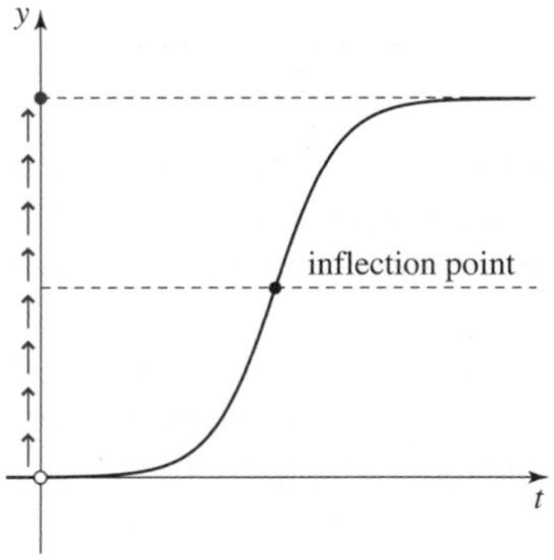

FIGURE 2.5.9 Hubbert's idealized explanation of oil reserve depletion.

Bifurcation

As you might imagine, the behavior of solutions to a differential equation depends on the values of parameters within the model, which in turn may correspond to real-world quantities one can control. Changes in the value of a parameter may cause the nature of the solutions of a DE to undergo dramatic qualitative change, such as suddenly having a different number and/or type of equilibrium solutions. A value of a parameter where such a change occurs is called a **bifurcation point**.

Nonlinear differential equations provide fertile ground for such sudden and drastic behavior changes. Let us begin by exploring a very simple example.

EXAMPLE 3 **Simple Bifurcation** We can see immediately that the nonlinear differential equation

$$y' = y(a - y)$$

[5]Hubbert's predictions have been holding up well in subsequent oil industry studies. In 1974 he predicted that current global oil production would peak in the late 1990s. Several internet sites on the World Wide Web provide discussion (sometimes heated) regarding corroboration of these tendencies, analyzing current trends and the reliability of available data. See also Colin J. Campbell and Jean H. Laherrère, "The End of Cheap Oil," *Scientific American*, March 1998, or Foster Morrison, *The Art of Modeling Dynamic Systems* (NY: Wiley-Interscience, 1991).

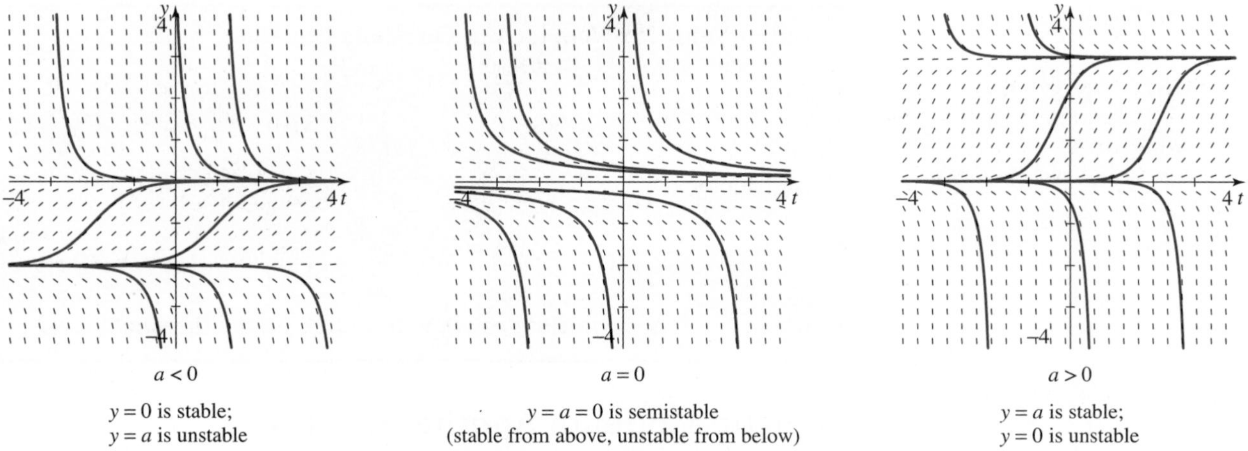

y
4
-4
4 t
-4
a < 0
y = 0 is stable;
y = a is unstable
a = 0
y = a = 0 is semistable
(stable from above, unstable from below)
a > 0
y = a is stable;
y = 0 is unstable

FIGURE 2.5.10 Before, at, and after bifurcation, for $y' = y(a - y)$.

has equilibria at $y = 0$ and at $y = a$. What is less obvious without a bit of thought is that the character of these equilibria changes depending on the values of a. Think about the sign of the slope, or sketch the direction fields and a few solutions (Fig. 2.5.10), to see the following:

As a decreases from positive to negative values, bifurcation occurs at $a = 0$, which is the bifurcation value of the parameter. Even the tiniest change between positive and negative values for a creates a serious effect on the solutions and their interpretation. Figure 2.5.11 is a **bifurcation diagram** that sums up the location and stability of equilibria for every value of a.

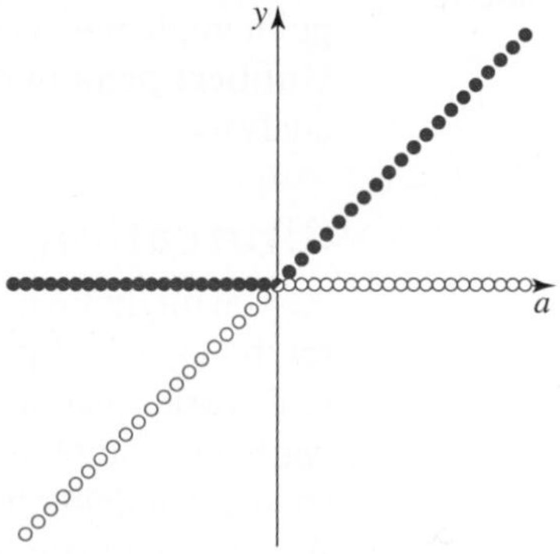

FIGURE 2.5.11 Bifurcation diagram for Example 3, showing for each value of a the equilibrium values for y—solid circles represent stability, open circles represent instability.

■

More elaborate examples of bifurcations are explored in Problems 35–38, and later with systems of nonlinear differential equations.

Summary

We have recognized the greater need for qualitative techniques to handle nonlinear DEs and applied them to a variety of equations. For autonomous first-order DEs the phase line provides a useful tool. Our major examples were the nonlinear autonomous growth and decay models based on the logistic and threshold differential equations.

2.5 Problems

Equilibria *In Problems 1–6, use direction fields and isoclines to make a qualitative sketch of the solutions, determine the equilibrium values (when they exist), and classify them as stable or unstable. Consider the parameters a and b to be positive in each case.*

1. $y' = ay + by^2$

2. $y' = ay - by^2$

3. $y' = -ay + by^2$

4. $y' = -ay - by^2$

5. $y' = e^y - 1$

6. $y' = y - \sqrt{y}$

Nonautonomous Sketching *Use the same directions for Problems 7–9 as for Problems 1–6. Describe in each case what differences are caused by the equations being nonautonomous.*

7. $y' = y(y - t)$ **8.** $y' = (y - t)^2$ **9.** $y' = \sin yt$

Inflection Points *For many DEs, the easiest way to pinpoint inflection points is not from the solution but from the DE itself. Find y'' by differentiating y', remembering to use the chain rule wherever y occurs. Then substitute for y' from the DE and set $y'' = 0$. Solve for y to find the inflection points (sometimes in terms of t). Use this technique to find inflection points for the solutions to the DEs in Problems 10–12.*

10. $y' = r\left(1 - \frac{y}{L}\right)y$ the logistic equation, Fig. 2.5.7

11. $y' = -r\left(1 - \frac{y}{T}\right)y$ the threshold equation, Fig. 2.5.8

12. $y' = \cos(y - t)$ equation (2), Fig. 2.5.3

13. Logistic Equation For the logistic IVP, we derived a solution formula (10) from the implicit solution (7) for the case of $0 < y_0 < L$.

(a) Show that the same solution formula (10) results from (7) if $y_0 > L$.

(b) Show that if $y_0 = L$, the solution must be *derived* from the DE (6), but that coincidentally the formula (10) also gives the correct answer.

(c) For each of the three cases $0 < y_0 < L$, $y_0 = L$, and $y_0 > L$, show how the solution formula (10) predicts the qualitative behavior of the solutions, shown in Fig. 2.5.8.

(d) Show that for $0 < y_0 < L$, the inflection point of the logistic curve (10) occurs at $y(t) = L/2$, where t has the value

$$t^* = \frac{1}{r}\ln\left(\frac{L}{y_0} - 1\right).$$

What is the rate of change at t^*?

14. Fitting the Logistic Law A population grows according to the logistic law with a limiting population of 5×10^9 individuals. The initial population of 0.2×10^9 begins growing by doubling every hour. What will the population be after 4 hours? HINT: To calculate the growth rate, assume that growth is initially exponential rather than logistic.

15. Culture Growth Suppose that we start at time $t_0 = 0$ with a sample of 1,000 cells. One day later we see that the population has doubled, and some time later we notice that the population has stabilized at 100,000. Assume a logistic growth model.

(a) What is the population after 5 days?

(b) How long does it take the population to reach 50,000 cells?

16. Logistic Model with Harvesting If the growth of a population follows the logistic model but is subject to "harvesting" (such as hunting or fishing), the model becomes

$$\frac{dy}{dt} = r\left(1 - \frac{y}{L}\right)y - h(t), \quad y(0) = y_0,$$

where $h(t)$ is the rate of harvesting.

(a) Show that when the harvest rate h is constant, a *maximum sustainable harvest* h_{max} is $rL/4$, which occurs when the population is half the carrying capacity. HINT: Set $y' = 0$ and use the quadratic formula on $r(1 - y/L)y - h = 0$ to find h_{max}.

(b) For constant h, create the phase line and a sketch of solutions. Show how the graphical information relates to the computation in (a), and tell what it implies with respect to policy decisions regarding hunting or fishing licensing. Compare with the case of *no* hunting.

Logistic with Harvest
Experiment with the harvesting rate, and watch out for extinction!

17. Campus Rumor A certain piece of dubious information about the cancellation of final exams began to spread one day on a college campus with a population of 80,000 students. Assume that initially one thousand students heard the rumor on the radio. Within a day 10,000 students had heard the rumor. Assume that the increase of the number x (in thousands) who had heard the rumor is proportional to the number of people who have heard the rumor and the number of people who have not heard the rumor. Determine $x(t)$.

18. Water Rumor A rumor about dihydrogen monoxide in the drinking water began to spread one day in a city of population 200,000. After one week, 1,000 people had been alarmed by the news. Assume that the rate of increase of the number N of people who have heard the rumor is proportional to the product of those who have heard the rumor and those who have not heard the rumor. Assume that $N(0) = 1$.

(a) Find an expression for $\dfrac{dN}{dt}$.

(b) Is this a logistic model? Explain.

(c) What are the equilibrium solutions, if any?

(d) Solve for $N(t)$, to find how long it takes for half the population to take notice.

(e) Suppose that at some point the problem is discussed in the newspaper, and some intelligent souls explain that there is no cause for alarm. Create a scenario and set up an IVP to model the spread of a counter-rumor.

19. Semistable Equilibrium Illustrate the **semistability** of the equilibrium point of

$$\frac{dy}{dt} = (1-y)^2, \quad y \geq 0,$$

by making a phase line and sketching typical solution curves.

20. Gompertz Equation The *Gompertz* equation

$$\frac{dy}{dt} = y(a - b\ln y),$$

where a and b are parameters, is used in actuarial studies, and to model growth of objects as diverse as tumors and organizations.[6]

(a) Show that the solution to the Gompertz equation is $y(t) = e^{a/b}e^{ce^{-bt}}$. HINT: Let $z = \ln y$, and use the chain rule to get a linear equation for dz/dt.

(b) Solve the IVP for this equation with $y(0) = y_0$.

(c) Describe the limiting value for $y(t)$ as $t \to \infty$. Assume that $a > 0$, and consider the cases $b > 0$ and $b < 0$.

21. Fitting the Gompertz Law In an experiment with bacteria, an initial population of about one (thousand) doubled in two hours, but at both 24 hours and 28 hours after the experiment began, there were only about 10 (thousand).

(a) Model this phenomenon using the Gompertz equation. HINT: Look first for the long-term level.

(b) Model the same phenomenon using the logistic equation; compare with (a).

Autonomous Analysis *For the first-order autonomous equations in Problems 22–27, complete the following.*

(a) *Sketch qualitative solution graphs.*

(b) *Highlight the equilibrium points of the equation, and draw phase-line arrows along the y-axis indicating the increasing or decreasing behavior of the solution $y = y(t)$. Classify their stability behavior.*

22. $y' = y^2$

23. $y' = -y(1-y)$

24. $y' = -y\left(1 - \dfrac{y}{L}\right)\left(1 - \dfrac{y}{M}\right)$

25. $y' = y - \sqrt{y}$

26. $y' = k(1-y)^2, k > 0$

27. $y' = y^2(4 - y^2)$

28. Stefan's Law Again According to **Stefan's Law of Radiation** (previously examined in Sec. 1.3, Problem 55 and Sec. 1.4, Problem 11), the rate of change of the radiation energy of a body at absolute temperature T is given by $dT/dt = k(M^4 - T^4)$, where $k > 0$ and M is the ambient or surrounding absolute temperature. Sketch typical solutions $T = T(t)$ for various initial temperatures $T_0 = T(0)$.

29. Hubbert Peak Refer to the final subsection "Logistic Model in Another Context" and Fig. 2.5.9 to explore the application of the logistic equation to the oil industry, as follows.

(a) The phrase "Hubbert peak" refers to a graph not shown, the graph of y' versus t. Sketch this missing graph from the information in the ty picture in Fig. 2.5.9. Your result should be bell-shaped, but not exactly in the same way as the Gaussian bell curve of Sec. 1.3, Example 5. Describe the differences.

(b) Show how much the pictures and arguments might differ if the inflection point occurs lower in the ty graph of Fig. 2.5.12. Explain why. Sketch a typical ty graph. Then sketch a typical ty' graph from your ty graph, as in part (a).

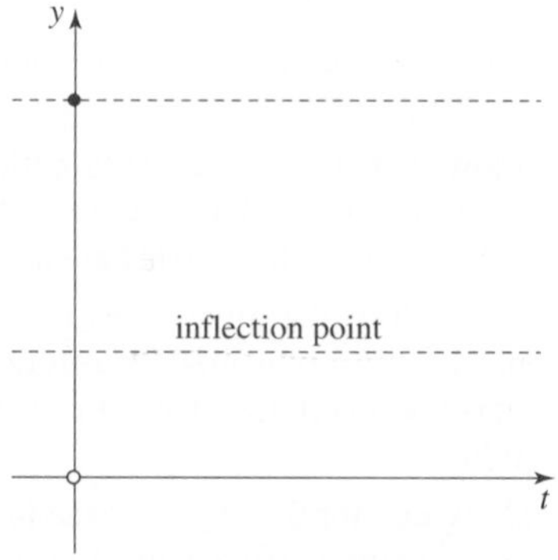

FIGURE 2.5.12 Changes to oil reserve depletion analysis for asymmetric case.

(c) Use your results from parts (a) and (b) to discuss to what extent the logistic equation can represent oil recovery phenomena in general.

[6]This equation dates back to 1825 when Benjamin Gompertz (1779–1865), an English mathematician, applied calculus to mortality rates.

30. Useful Transformation Solve the logistic IVP

$$y' = ky(1 - y), \quad y(0) = y_0,$$

by means of the change of variable $z = y/(1 - y)$. Solve the resulting DE in $z = z(t)$ and then resubstitute to obtain $y(t)$. (Your result should be equation (10) with $L = 1$.)

31. Chemical Reactions Two chemicals A and B react to form the substance X according to the law $dx/dt = k(100 - x)(50 - x)$, where k is a positive constant and x is the amount of substance X. Describe the amount of substance X, given the initial conditions in (a), (b), and (c). Sketch the direction field, equilibrium solutions, phase line, and solution starting from the initial value. Discuss the relative merits of each choice of initial conditions. Might a DE student question the validity of this model? Why?

(a) $x(0) = 0$ (b) $x(0) = 75$ (c) $x(0) = 150$

32. General Chemical Reaction of Two Substances When two chemicals A and B react to produce substance X, the amount x of substance X is described by the DE $dx/dt = k(a - x)^m(b - x)^n$, $a < b$, where a and b represent the amounts of substances A and B, and m and n are positive integers.

(a) Describe the nature of the solutions of this equation for odd or even values of m and n by analyzing graphs of x' versus x, thus determining the values of x for which the slope x' is positive or negative. On this xx' graph you can draw the phase line along the horizontal x-axis.

(b) Explain how different values of m and n affect the classification of equilibria. Which of the four odd/even exponent combinations gives a reasonable model if the manufacturing goal is to produce substance x?

33. Solving the Threshold Equation Make the change of variable $t = -\tau$ in the threshold IVP (14) and verify that this results in the IVP for the logistic equation. Use the solution of the logistic problem to verify the solution (15).

34. Limiting Values for Threshold Equation

(a) Show that if $y_0 < T$ the solution $y(t)$ of (14) tends to zero as $t \to \infty$.

(b) Show that if $y_0 > T$ the solution $y(t)$ of (14) "blows up" at time

$$t^* = \frac{1}{r}\ln\left(\frac{y_0}{y_0 - T}\right).$$

35. Pitchfork Bifurcation For the differential equation

$$\frac{dy}{dt} = \alpha y - y^3,$$

show that 0 is a bifurcation point of the parameter α as follows.

(a) Show that if $\alpha \le 0$ there is only one equilibrium point at 0 and it is stable.

(b) Show that if $\alpha > 0$ there are three equilibrium points: 0, which is unstable, and $\pm\sqrt{\alpha}$, which are stable.

(c) Then draw a *bifurcation diagram* for this equation. That is, plot the equilibrium points (as solid dots for stable equilibria and open dots for unstable equilibria) as a function of α, as in Fig. 2.5.11 for Example 3. Figure 2.5.13 shows values already plotted for $\alpha = -2$ and $\alpha = +2$; when you fill it in for other values of α, you should have a graph that looks like a pitchfork. Consequently, $\alpha = 0$ is called a **pitchfork bifurcation**; when the pitchfork branches at $\alpha = 0$, the equilibrium at $y = 0$ loses its stability.

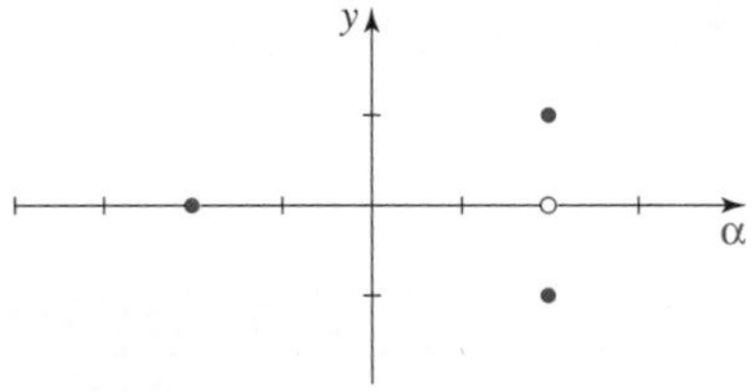

FIGURE 2.5.13 A start on the bifurcation diagram for $y' = \alpha y - y^3$ for Problem 35.

Pitchfork Bifurcation
Explore this equation (**Supercritical**) and its close relative $dy/dt = \alpha y + y^3$ (**Subcritical**).

36. Another Bifurcation Study the relationship between the values of the parameter b in the differential equation $dy/dt = y^2 + by + 1$ and the equilibrium points of the equation and their stability.

(a) Show that for $|b| > 2$ there are two equilibrium points; for $|b| = 2$, one; and for $|b| < 2$, none.

(b) Determine the bifurcation points for b (see Problem 35)—the b-values at which the solutions undergo qualitative change.

(c) Sketch solutions of the differential equation for different b-values (e.g., $b = -3, -2, -1, 0, 1, 2, 3$) in order to observe the change that takes place at the bifurcation points.

(d) Determine which of the equilibrium points are stable.

(e) Draw the bifurcation diagram for this equation; that is, plot the equilibrium points of this equation as a function of the parameter values for $-\infty < b < \infty$. For this equation, the bifurcation does *not* fall into the pitchfork class.

Saddle-Node Bifurcation
Explore this type of bifurcation for the equation $dy/dt = y^2 + r$.

Computer Lab: Bifurcation *In Problems 37 and 38 we study the effect of parameters on the solutions of differential equations. For each equation, do the following.*

(a) *Determine values of k where the number and/or nature of equilibrium points changes.*

(b) *Draw direction fields and sample solutions to the DE for different values of k.*

37. $y' = ky^2 + y + 1$

38. $y' = y^2 + y + k$

Computer Lab: Growth Equations *Four growth equations used by population theorists are given in Problems 39–42. Plot solutions for different values of their parameters, and try to determine their significance.*

39. $y' = r\left(1 - \frac{y}{L}\right)y$ (logistic equation)

40. $y' = -r\left(1 - \frac{y}{T}\right)y$ (threshold equation)

41. $y' = r\left(1 - \frac{\ln y}{L}\right)y$ (Gompertz equation)

42. $y' = re^{-\beta t}y$ (equation for decaying exponential rate)

43. Suggested Journal Entry Discuss the relative merits of the various growth equations that we have discussed in this chapter. Can you devise additional ones that might be better models in certain situations? To what uses might such models be put? Discuss the meaning of L in the logistic equation in terms of long-term behavior of populations. Does L play the same role in the Gompertz equation (when expressed as in Problem 41)?

2.6 Systems of Differential Equations: A First Look

SYNOPSIS: Surprisingly easily we can extend concepts and techniques to introduce autonomous first-order systems of differential equations and the vector fields they define in the phase plane. From the qualitative analysis of the phase plane, we can learn about equilibrium points and their stability, and obtain information about long-term behavior of solutions, such as boundedness and periodicity. Population models provide illuminating examples.

What Are Systems of Differential Equations?

Population studies involving two or more interacting species lead to systems of two or more "interlocking" or **coupled** differential equations. Similar situations arise in mechanical and electrical systems that have two or more interrelated components. We will encounter shortly an ecological example leading to the following system:

$$\begin{aligned} dx/dt &= 2x - xy, \\ dy/dt &= -3y + 0.5xy. \end{aligned} \tag{1}$$

We are considering *two* variables, x and y, that both depend on t. It is their *rate* equations that show their interrelation.

A simpler system, where the components are **decoupled**, is

$$\begin{aligned} dx/dt &= 2x, \\ dy/dt &= -3y. \end{aligned} \tag{2}$$

Analytic Definition of a Solution of a DE System

A **solution** of a system of two differential equations is a pair of functions $x(t)$ and $y(t)$ that simultaneously satisfies both equations.

In the decoupled system (2), each equation can be solved separately, and since these are linear, we even have an explicit solution:

$$\begin{aligned} x(t) &= c_1 e^{2t}, \\ y(t) &= c_2 e^{-3t}, \end{aligned} \tag{3}$$

where c_1 and c_2 are arbitrary constants.

System (1) is far more complicated to solve, because there is more than one dependent variable in each DE *and* the equations are nonlinear due to the xy terms.[1] However, we will be able to learn a lot about the behaviors of solutions from qualitative analysis, and you will see that calculators and computers can as easily make pictures for systems of two differential equations as for single DEs. In this chapter we will only discuss systems with two equations. Later, we use the tools of linear algebra to handle larger systems.

Autonomous First-Order Systems in Two Variables

Systems (1) and (2) are special cases of the two-dimensional first-order system

$$\begin{aligned} dx/dt &= P(x, y), \\ dy/dt &= Q(x, y), \end{aligned} \tag{4}$$

Parametric to Cartesian; Phase-Plane Drawing

Get a feel for how $x(t)$ and $y(t)$ combine to create an xy phase-plane trajectory and, in reverse, how real-time drawing in xy produces tx and ty graphs.

where P and Q depend *explicitly* on x and y and only *implicitly* on the underlying independent variable, time t. Such systems are called **autonomous**, as are single differential equations with time-invariant right-hand sides.

The functions $x(t)$ and $y(t)$ that satisfy the system (4) represent a parametric curve $(x(t), y(t))$, which we will graph in the xy-plane. Given a starting point $(x(0), y(0))$, the **initial condition**, we can visualize the solution as a curve that *has the correct* **tangent vector** *at each point*. See Fig. 2.6.1. We know from calculus that the slope of the tangent vector is $dy/dx = (dy/dt)/(dx/dt)$, for $dx/dt \neq 0$.

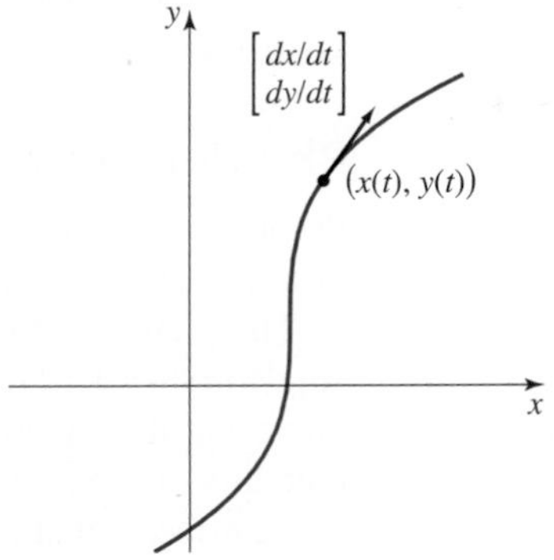

FIGURE 2.6.1 Notation for a parametric curve defined when x and y both depend on a parameter t.

[1] In general, if a system is composed of coupled equations it is not easy (and often impossible) to find analytic solutions. The major exception is a *linear* system with *constant coefficients*, which is the subject of Chapter 6.

Vector Fields

See how a vector field is constructed and how trajectories move through it.

Phase Plane for a DE System in Two Variables

For the system (4) of two differential equations,

- We call the xy-plane the **phase plane**.
- The collection of tangent vectors defined by the DE is called a **vector field**.
- The parametric curve defined by a solution $(x(t), y(t))$ is called a **trajectory**.

EXAMPLE 1 **Phase-Plane Trajectories** Consider the decoupled linear system (2) from a geometric point of view in the xy-plane. From the solution functions (3), we can describe the tangent vector to a solution of (2) at the point (x, y) as

$$\begin{bmatrix} dx/dt \\ dy/dt \end{bmatrix} = \begin{bmatrix} 2x \\ -3y \end{bmatrix}.$$

The vector field and some phase-plane trajectories of the solutions are shown in Fig. 2.6.2.

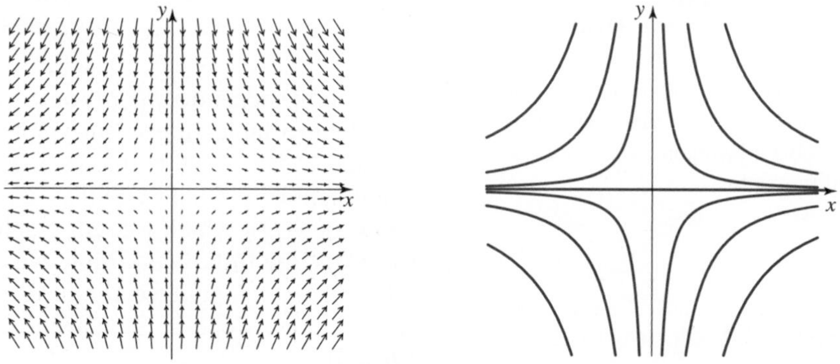

(a) Vector field (b) Solution curves

FIGURE 2.6.2 Autonomous system $dx/dt = 2x$, $dy/dt = -3y$, from system (2). ■

Phase Portraits for Autonomous Systems

Points (x, y) of the phase plane are called **states** of the system, and the collection of trajectories for various initial conditions is the **phase portrait** of the system.

The pictures in Fig. 2.6.2 are reminiscent of direction fields and solutions to single first-order DEs, as introduced in Sec. 1.2. In direction fields for $y' = f(t)$, however, all tangent vectors pointed to the right, because the t is increasing in that direction. Phase-plane trajectories for parametric equations are less restricted. Since t does not appear in the phase plane, looking at a completed solution curve, as in Fig. 2.6.2(b), gives no idea of the speed with which a moving point would trace it, or where you would be for a given value of t.

Equilibria and Stability

An **equilibrium point** for a two-dimensional system is an (x, y) point, where

$$dx/dt = 0 \quad \text{and} \quad dy/dt = 0$$

simultaneously. If a state is at equilibrium, it does not change.

- A **stable** equilibrium *attracts* (or at least keeps close) nearby solutions.
- An **unstable** equilibrium *repels* nearby solutions in at least one direction.

In the decoupled linear system (2) that gives the vector field shown in Fig. 2.6.2, the equilibrium at the origin is unstable because nearby trajectories are sent away in the horizontal direction.

Sketching Phase Planes

Computers can show vector fields and trajectories for an xy DE system as easily as for a single variable DE $y' = f(t, y)$, but the process becomes overly tedious by hand. An important hand-graphing tool for phase-plane analysis is an adaptation of *isoclines*, as discussed in Sec. 1.2.

Nullclines

- A v nullcline is an isocline of *vertical* slopes, where $dx/dt = 0$.
- An h nullcline is an isocline of *horizontal* slopes, where $dy/dt = 0$.

Equilibria occur at the points where a v nullcline intersects an h nullcline.

Fig. 2.6.2 confirms the location of the equilibrium at the intersection of the nullclines, which in this case are the axes.

The sign of dx/dt determines whether phase-plane trajectories are moving left or right, and the sign of dy/dt tells whether trajectories are moving up or down. These facts are summarized in Table 2.6.1, and you can confirm these behaviors on the solution curves in Fig. 2.6.2. You can also predict the proper combinations of left versus right, up versus down, for each quadrant.

Table 2.6.1 Directions for phase-plane trajectories

	–	0	+
$\frac{dx}{dt}$	←	|	→
$\frac{dy}{dt}$	↓	—	↑

Just calculating this general information of left/right, up/down is sufficient to predict the general character of solutions to a system of differential equations. (See Table 2.6.1.) You can thus avoid tedious detail in a hand drawing when you do not have access to a graphical DE solver, as you will see in Example 2.

We are now almost ready to rough-sketch some possible solutions. But first we must state an important principle: as with unique solutions to first-order DEs, trajectories may appear to merge (especially at an equilibrium), but they will not cross. The reason for this fact is that the existence and uniqueness theorem for autonomous *systems* of DEs applies to the *phase plane* and does *not* apply to the tx or ty graphs.

Sufficient conditions for existence of unique solutions to an initial-value problem for a system of DEs as given by equation (4) are that $P(x, y)$, $Q(x, y)$ and all four partial derivatives ($\partial P/\partial x$, $\partial P/\partial y$, $\partial Q/\partial x$, $\partial Q/\partial y$) are continuous. This results from adapting Picard's Theorem from Sec. 1.5.

Existence and Uniqueness:

Picard's Theorem can be extended to linear DE systems (Sec. 6.1) and to nonlinear DE systems (Sec. 7.1, Problems 39–40.)

Guiding Principle for Autonomous Systems

When existence and uniqueness conditions hold for an *autonomous* system, *phase-plane trajectories do not cross.*

We summarize the qualitative analysis for autonomous systems in two variables as follows. Further discussion will be given in Sec. 7.1, Problems 39–40.

Quick Sketching Outline for Phase Portraits

Nullclines and Equilibria:

- Where $dx/dt = 0$, slopes are vertical.
- Where $dy/dt = 0$, slopes are horizontal.
- Where $dx/dt = 0$ and $dy/dt = 0$: equilibria.

Left/Right Directions:

- Where dx/dt is positive, arrows point right.
- Where dx/dt is negative, arrows point left.

Up/Down Directions:

- Where dy/dt is positive, arrows point up.
- Where dy/dt is negative, arrows point down.

Check Uniqueness: Phase-plane trajectories do not cross where uniqueness holds.

EXAMPLE 2 **Rough-Sketching** Consider the system

$$\begin{aligned} dx/dt &= y - \cos x, \\ dy/dt &= x - y, \end{aligned} \tag{5}$$

and try to sketch the behavior of solutions in the xy phase plane.

From system (5), we see that vertical slopes occur when $y = \cos x$ and horizontal slopes occur when $x = y$. These nullclines intersect when $x = \cos x$. Without further ado, you can sketch this information and get a preliminary idea of what will happen to solutions. (See Fig. 2.6.3(a).)

From the signs of dx/dt and dy/dt in (5), you can add direction arrows, as shown in Fig. 2.6.3(b). For example, dx/dt is positive for $y > \cos x$, so arrows point right (rather than left) above the cosine curve that represents the nullcline of vertical slopes.

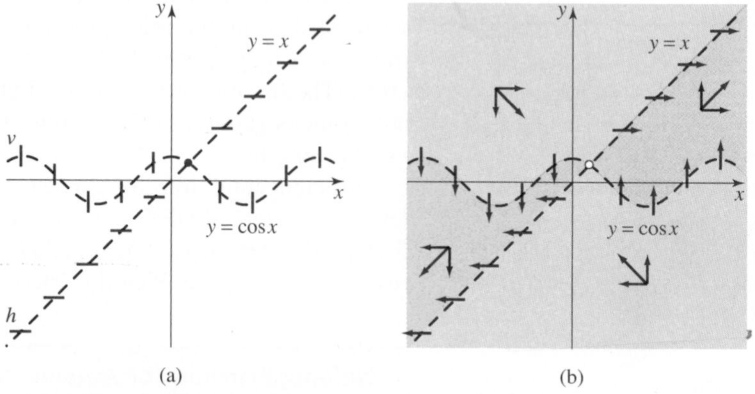

FIGURE 2.6.3 Graphs for Example 2, equation (5). (a) Nullclines and slope marks. (b) Directions are added to slope marks on isoclines and in regions between isoclines. Resultant general directions in these regions are shown in blue.

See if you can now rough-sketch possible solutions on Fig. 2.6.3(b). The trajectories should never cross, although they can appear to merge.

Without worrying about analytic solutions (which usually are not possible for nonlinear systems, and $\cos x$ makes this system nonlinear), or even hand-sketching the basic information in Fig. 2.6.3(b), you can ask a graphical DE solver to draw solutions—just as easily as for a single first-order DE. Figure 2.6.4 shows such a set of solutions. How does it compare with your rough sketch?

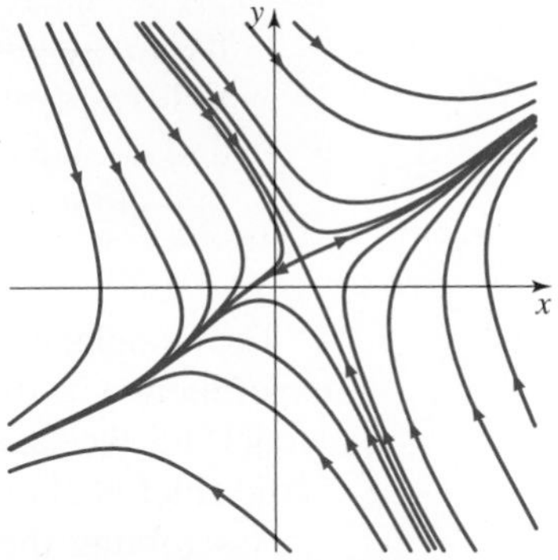

FIGURE 2.6.4 Actual trajectories for equation (5) of Example 2.

Add the nullclines to Fig. 2.6.4 and convince yourself that the actual solutions agree with Figs. 2.6.3(a) and 2.6.3(b). If your first guess at drawing solutions on Fig. 2.6.3(b) was not qualitatively the same as Fig. 2.6.4, try to track down what went wrong; this is how you build up experience and expertise. ■

Hudson Bay Data

Observed data on the populations of hares and lynxes over ninety years corroborates this model.

Lotka-Volterra

Here the prey are herbivores. You can see cycles of overpopulation and underpopulation in each species, and how they interact dynamically to create the phase portrait.

The Lotka-Volterra Predator-Prey Model

Let's consider a simplified ecological system in which two species occupy the same environment. One species, the **predator**, feeds on the other, the **prey**, while the prey feeds on something else readily available.[2] One example consists of foxes and rabbits in a woodland, where the foxes (predators) eat the rabbits (prey), and the rabbits in turn eat natural vegetation. Other examples are sharks (predators) and food fish (prey), ladybugs (predators) and aphids (prey), bass (predators) and sunfish (prey), and beetles (predators) and scale insects (prey). See Problems 10 and 13–15.

We illustrate the basic predator–prey model for the fox and rabbit system. Let F denote the population of foxes at time t and R the population of rabbits.[3] We must begin with some information about the behavior of foxes and rabbits. In particular, we must know how each population varies in the absence of the other, as well as how they interact—how each species is affected by the presence of the other. We will take the following statements as our starting point:

[2]American Alfred J. Lotka (1880–1949) was both a biological physicist and a mathematical demographer. He published in 1924 the first book in mathematical biology (*Elements of Physical Biology*, reprinted in 1956 by Dover as *Elements of Mathematical Biology*), in which he formulated these predator–prey equations. Lotka described an ecosystem in thermodynamic terms, as an energy transforming machine, and initiated the study of ecology.

The same predator–prey model was developed independently in 1926 by Italian mathematician Vito Volterra (1860–1940), who turned his attention to mathematical biology after World War I.

[3]Instead of the number of individuals, we can let F stand for the population in hundreds, population in thousands, or even the density in individuals per square mile; R can have similar units.

Predator-Prey Assumptions (Foxes and Rabbits)

- In the absence of foxes, the rabbit population follows the Malthusian growth law: $dR/dt = a_R R$, $a_R > 0$.
- In the absence of rabbits, the fox population will die off according to the law $dF/dt = -a_F F$, $a_F > 0$.
- When both foxes and rabbits are present, the number of interactions is proportional to the *product* of the population sizes, with the effect that the fox population increases while the rabbit population decreases (due to foxes eating rabbits). Thus, for positive proportionality constants c_R and c_F, the rate of change in the fox population is $+c_F RF$ (increase), that of the rabbit population $-c_R RF$ (decrease).

The modeler may not know the values of the parameters, a_R, a_F, c_R, c_F, but may experiment with different values, trying to make the behavior of the mathematical model "fit" the data observed in the field. When such a determination of parameters is the chief goal of the modeler's activity, it is called **system identification**.

Assembling the information about the rates of change due to separate and interactive assumptions, we obtain the following two-dimensional system.

Lotka-Volterra Equations for Predator-Prey Model

$$\begin{aligned} dR/dt &= \ \ a_R R - c_R RF, \\ dF/dt &= -a_F F + c_F RF. \end{aligned} \tag{6}$$

EXAMPLE 3 **Predator-Prey** We will suppose that as the result of suitable experimental work we have determined values for the parameters: $a_R = 2$, $a_F = 3$, $c_R = 1$, $c_F = 0.5$. System (6) then becomes

$$\begin{aligned} dR/dt &= \ \ 2R - \ \ \ RF, \\ dF/dt &= -3F + 0.5RF. \end{aligned} \tag{7}$$

The first step in our qualitative study of this system is to determine the constant or equilibrium solutions: that is, the states (R, F) for which dR/dt and dF/dt are both zero. Thus, we need to solve the system of two algebraic equations obtained from (7) when $dR/dt = dF/dt = 0$:

$$\begin{aligned} 2R - \ \ \ RF &= 0, \\ -3F + 0.5RF &= 0. \end{aligned}$$

Both equations factor and we obtain the solution points (0, 0) and (6, 2). The nonzero equilibrium point (6, 2) describes the situation when the populations are ecologically balanced. In this case, there are just enough rabbits to support the foxes and just enough foxes to keep the rabbits in check. The equilibrium point at (0, 0) corresponds to extinction for both species. These are plotted on the phase plane in Fig. 2.6.5(a). (Since R and F are population variables, we only use the first quadrant.)

We get additional help in our analysis by drawing the **nullclines**: the curves along which either dR/dt or dF/dt is zero. When $dR/dt = 0$, we obtain two nullclines of vertical slopes: $F = 2$ and $R = 0$. When $dF/dt = 0$, there

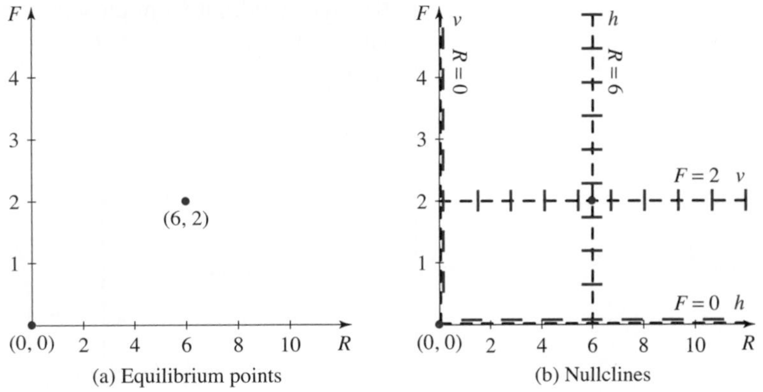

FIGURE 2.6.5 Phase-plane diagrams for system (7) of Example 3.

are two nullclines of horizontal slopes: $R = 6$ and $F = 0$. The nullclines are plotted in Fig. 2.6.5(b). The equilibria indeed occur at the points where a v nullcline of vertical slopes intersects an h nullcline of horizontal slopes, and *nothing* special happens when h meets h or v meets v.

The nullclines separate the first quadrant into four distinct regions that can be characterized as follows:

Region A	$dR/dt > 0, dF/dt > 0$
	Solution curves move up and to the right.
Region B	$dR/dt > 0, dF/dt < 0$
	Solution curves move down and to the right.
Region C	$dR/dt < 0, dF/dt > 0$
	Solution curves move up and to the left.
Region D	$dR/dt < 0, dF/dt < 0$
	Solution curves move down and to the left.

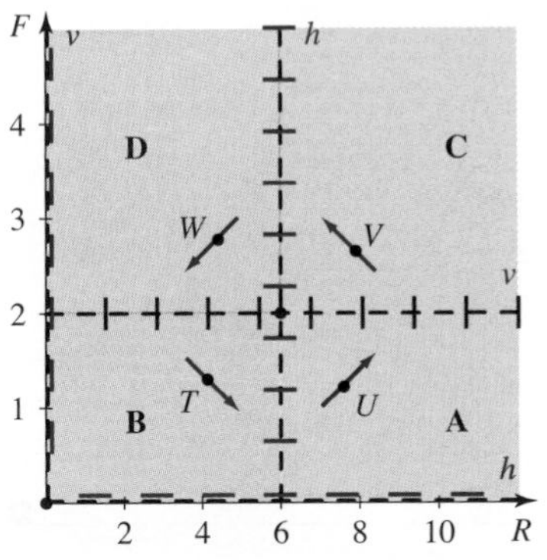

FIGURE 2.6.6 Phase-plane trends for system (7) of Example 3.

(Check any entry in this table by picking a point in the region and substituting its coordinates into system (7) to determine the signs of the derivatives.) As we can see from Fig. 2.6.6, this information indicates that solution curves wind around the equilibrium point (6, 2) in a counterclockwise direction.

For further insight into what we can learn from Fig. 2.6.6, suppose the system is at some point W. This could mean the foxes are starving after a hard winter, and their numbers are shrinking because rabbits are scarce. But as the state of the system is forced down to point T in region B, new rabbit litters are being born and R again increases. This pushes the state into region A, where at the point U we now see both species on the increase. As the state moves upward into region C, however, the greed of the growing fox population causes the rabbit population to decrease, typified by point V. This then drives the state into region D, in which the foxes begin to suffer because they have reduced the rabbit population too far, bringing us back toward point W, as shown in Fig. 2.6.7.

We can deduce that the equilibrium at (0, 0) is **unstable** because nearby solutions that start along the R axis move away. The stability of the equilibrium at (6, 2) is less clear—the analysis so far cannot tell whether nearby orbits spiral out, spiral in, or form a closed loop. It turns out in this case that all orbits inside the first quadrant form closed loops,[4] so the equilibrium at (6, 2) is **stable**, in an unusual manner—nearby orbits do not leave the vicinity, even in backward time.

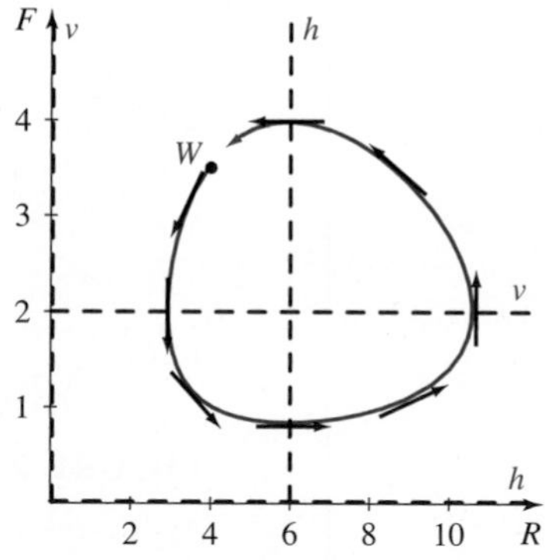

FIGURE 2.6.7 Phase-plane solution curve for system (7) of Example 3.

Numerical methods can be used to approximate a solution of the nonlinear system (7); a phase-plane trajectory is shown in Fig. 2.6.7.[5] It is impossible to tell from this trajectory, however, the rate of movement in time of the point representing the state of the system. It would be helpful to plot graphs of the population functions $R(t)$ and $F(t)$, as in the computer-generated graphs of Fig. 2.6.8. These are often called **time series** or *component solution functions*.

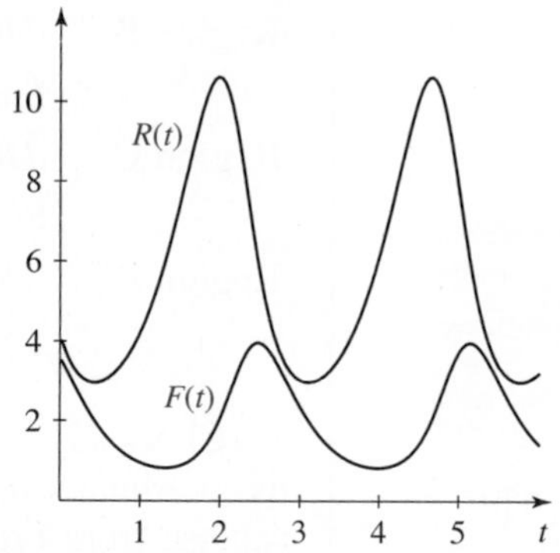

FIGURE 2.6.8 Component solution graphs for system (7) of Example 3.

---

[4]Martin Braun, in *Differential Equations and their Applications* (NY: Springer-Verlag, 1995), provides a nice proof of closed-loop orbits by making a new DE,

$$\frac{dF}{dR} = \frac{dF/dt}{dR/dt},$$

that does not depend explicitly on t. This new DE is solved explicitly by separation of the variables R and F. (The result gives the closed orbits in the RF plane that are hinted at in Fig. 2.6.8, but does not yield explicit formulas for $F(t)$ or $R(t)$. The numerical solutions shown in Fig. 2.6.8 show that $F(t)$ and $R(t)$ are periodic functions, but they are too "warped" to be normal sine or cosine functions.)

[5]Numerical methods for systems of differential equations are discussed in Chapter 7. They are a straightforward extension of those methods already discussed in Sec. 1.4 for a single first-order DE.

A model related to predator-prey involves two species in which the first depends for its survival on the second. It is therefore to the advantage of the first (the **parasite**) to keep the second (the **host**) alive and well. An example is an insect species that depends on certain plants for food; another is the case of a disease caused by a parasitic organism. (See Problem 16.)

The Competition Model

Another important population model describes systems in which two or more species compete for common resources. The species may or may not prey on each other. Several species of fish, for example, may compete for the same food supply but not feed on each other. Lions and hyenas, on the other hand, not only compete for a common food supply but will also kill their rivals, given the chance. Competition models are not limited to biology and ecology. Countries compete for trade, corporations for customers, and political parties for voters.

Competitive Exclusion

Herbivores en garde! Another species may be dining on your grassland. Is coexistence possible?

To illustrate the qualitative analysis of a system of differential equations modeling such a situation, we will consider the competition of sheep and rabbits for the limited grass resources on a certain range. (We will keep our model simple by ignoring such factors as predators and seasonal changes.) We let $R(t)$ and $S(t)$ denote the populations of rabbits and sheep, respectively, and list our assumptions about their independent and interactive characteristics.

Competition Assumptions (Rabbits and Sheep)

- Each species, in the absence of the other, will grow to carrying capacity according to the logistic law:

$$\frac{dR}{dt} = R(a_R - b_R R) = a_R R\left(1 - \frac{R}{L_R}\right),$$

where $L_R = a_R/b_R$ is the carrying capacity for rabbits, and

$$\frac{dS}{dt} = S(a_S - b_S S) = a_S S\left(1 - \frac{S}{L_S}\right),$$

where $L_S = a_S/b_S$ is the carrying capacity for sheep.

- When grazing together, each species has a negative effect on the other. (A sheep may nudge rabbits aside; too many rabbits discourage a sheep from grazing.) This introduces another term into each of the above equations. For positive constants c_R and c_S, the interactive contributions to the rates of change of the populations are $-c_R RS$ for rabbits and $-c_S RS$ for sheep.

Combining the growth and decay factors from these assumptions, we obtain the two-dimensional system of differential equations for the competition model:

Competition Model

$$\begin{aligned} dR/dt &= R(a_R - b_R R - c_R S), \\ dS/dt &= S(a_S - b_S S - c_S R). \end{aligned} \tag{8}$$

The values of the six parameters depend on the species under study.[6]

[6]Generally, $c_R > c_S$; this means that sheep bug rabbits more than rabbits bug sheep.

EXAMPLE 4 **Competition** We will assume that as the result of suitable experimentation, values have been determined that reduce system (8) to the following:

$$\begin{aligned} dR/dt &= R(3 - R - 2S), \\ dS/dt &= S(2 - S - R). \end{aligned} \tag{9}$$

We begin our quantitative analysis by solving the algebraic system

$$\begin{aligned} R(3 - R - 2S) &= 0, \\ S(2 - S - R) &= 0 \end{aligned}$$

to obtain four equilibrium points (0, 0), (0, 2), (3, 0), and (1, 1). These points, plotted in Fig. 2.6.9(a), have the following practical interpretations:

$(R, S) = (0, 2)$: the sheep have driven the rabbits to extinction;
$(R, S) = (3, 0)$: the rabbits have driven the sheep to extinction;
$(R, S) = (0, 0)$: both species have become extinct;
$(R, S) = (1, 1)$: the species are in balance at constant levels.

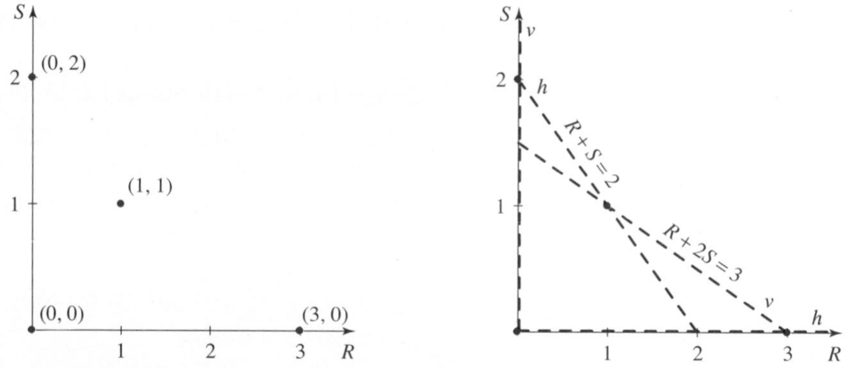

(a) Equilibrium points (b) Nullclines

FIGURE 2.6.9 Phase-plane diagrams for system (9).

Competitive Exclusion

Move the nullclines and see the behavior of competing populations respond.

To learn more about the stability of these equilibrium points, we determine the nullclines. When $dR/dt = 0$, the first equation in system (9) tells us that $3 - R - 2S = 0$ or $R = 0$; the R-nullclines are the S-axis and the line $R + 2S = 3$. Similarly, the S-nullclines are the R-axis and the line $R + S = 2$. The nullclines are plotted in Fig. 2.6.9(b). They separate the first quadrant into four regions as follows:

Region A $dR/dt > 0, dS/dt > 0$
Solution curves move up and to the right.

Region B $dR/dt > 0, dS/dt < 0$
Solution curves move down and to the right.

Region C $dR/dt < 0, dS/dt > 0$
Solution curves move up and to the left.

Region D $dR/dt < 0, dS/dt < 0$
Solution curves move down and to the left.

Along the R-nullcline, on which $dR/dt = 0$, tangent vectors are vertical; they are horizontal along the S-nullcline. Thus, we can sketch typical tangent vectors, as in Fig. 2.6.10. We can also add arrowheads to show the direction of motion of all horizontal and vertical slope marks, and draw conclusions about the equilibrium points: (0, 0) and (1, 1) are unstable because at least some nearby solutions move away; (0, 2) and (3, 0) are asymptotically stable because all nearby arrows point toward them.

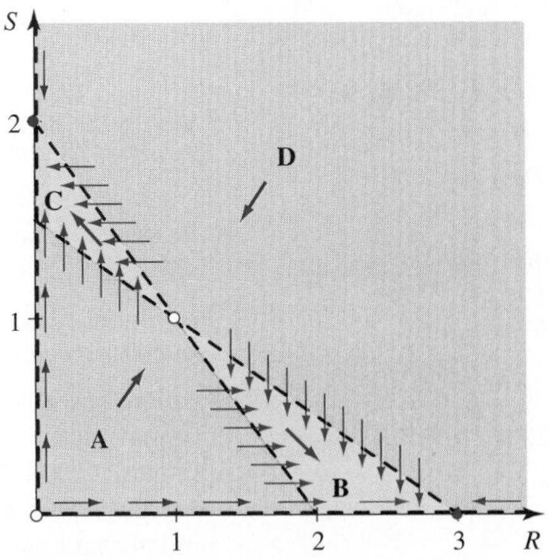

FIGURE 2.6.10 Rough vector field for system (9) showing nullclines and direction tendencies between nullclines.

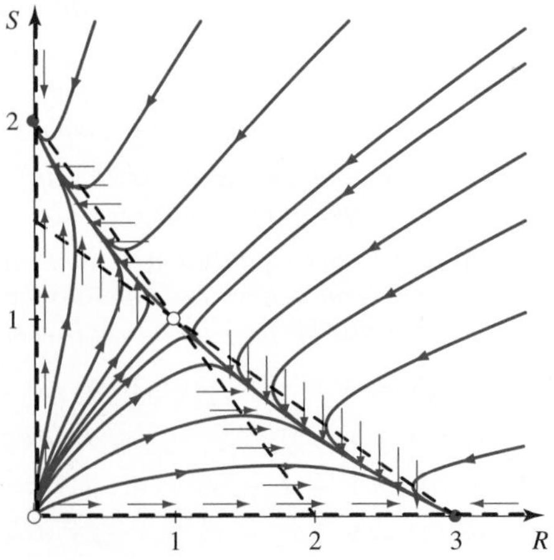

FIGURE 2.6.11 Solution curves for system (9).

Arrows on Nullclines:

The direction of the arrows on nullcline slope marks is determined by the sign of one derivative when the other is zero.

The numerically generated solution curves in Fig. 2.6.11 confirm our interpretations. Either the sheep will drive the rabbits to extinction or vice versa. Regardless of initial conditions, all solution curves eventually reach the stable equilibria at (0, 2) or (3, 0). In the real world, the instability of (1, 1) does not allow the system to remain there. The observation that two species competing for the same food cannot coexist is known to biologists as the **principle of competitive exclusion.**[7] ■

Example 4 gave us only one of four possible interactions of the nullclines in a competition model. The other arrangements predict different outcomes, determined by the locations and stability of the equilibrium points. See Problems 22–25, and Problem 32 with IDE.

Summary

We have introduced two-dimensional systems of differential equations, and have explored two important systems arising from problems in population biology: the predator–prey and competition models. Using phase-plane analysis, including vector fields, equilibrium points, and nullclines, we have examined the geometry of solution curves and issues of stability and instability. We hinted at other factors that might be included in a more detailed analysis (to be pursued in Chapters 6 and 7), and at such questions as the practical determination of parameters and the numerical solutions of systems.

[7]The interested reader may follow up these ideas in the following references: J. Maynard Smith, *Mathematical Ideas in Biology* (Cambridge: Cambridge University Press, 1968); L. Edelstein-Keshet, *Mathematical Models in Biology* (NY: Random House/Birkhauser, 1988); and D. Kaplan and L. Glass, *Understanding Nonlinear Dynamics* (NY: Springer-Verlag, 1995).

2.6 Problems

Predicting System Behavior *Consider the systems in Problems 1–8.*

(a) *Determine and plot the equilibrium points and nullclines for the systems.*

(b) *Show the direction of the vector field between the nullclines, as illustrated in Example 2 and Fig. 2.6.4.*

(c) *Sketch some solution curves starting near, but not on, the equilibrium point(s).*

(d) *Label each equilibrium as stable or unstable depending on the behavior of solutions that start nearby, and describe the long-term behavior of all the solutions.*

1. $dx/dt = y$
$dy/dt = x - 3y$

2. $dx/dt = 1 - x - y$
$dy/dt = x - y^2$

3. $dx/dt = 1 - x - y$
$dy/dt = 1 - x^2 - y^2$

4. $dx/dt = x + y$
$dy/dt = 2x + 2y$

5. $dx/dt = 4 - x - y$
$dy/dt = 3 - x^2 - y^2$

6. $dx/dt = y$
$dy/dt = 5x + 3y$

7. $dx/dt = 1 - x - y$
$dy/dt = x - |y|$

8. $dx/dt = x + 2y$
$dy/dt = x$

9. Creating a Predator-Prey Model Suppose that in the absence of foxes, a rabbit population increases by 15% per year; in the absence of rabbits, a fox population decreases by 25% per year; and in equilibrium, there are 1,000 foxes and 8,000 rabbits.

(a) Explain why the Volterra-Lotka equations are

$$dR/dt = 0.15R - 0.00015RF,$$
$$dF/dt = -0.25F + 0.00003125RF.$$

(b) Suppose we introduce an element of "harvesting," detrimental to the rate of growth of both prey and predator, so that the equations are modified as follows:

$$dR/dt = 0.15R - 0.00015RF - 0.1R,$$
$$dF/dt = -0.25F + 0.00003125RF - 0.1F.$$

Determine which population is most affected by this harvesting strategy, by calculating the new equilibrium point.

Lotka-Volterra with Harvest
A similar example allows you to explore possibilities.

10. Sharks and Sardines with Fishing We apply the classical predator–prey model of Volterra: $dx/dt = ax - bxy$, $dy/dt = -cx + dxy$, where x denotes the population of sardines (prey) and y the population of sharks (predators). We now subtract a term from each equation that accounts for the depletion of both species due to external fishing. If we fish each species at the same rate, then the Volterra model becomes

$$dx/dt = ax - bxy - fx,$$
$$dy/dt = -cy + dxy - fy,$$

where the constant $f \geq 0$ denotes the "fishing" effort.

(a) Find the equilibrium point of the system under fishing, and sketch a phase portrait for $f = 0.5$.

(b) Describe how the position of this fishing equilibrium has moved relative to the equilibrium point with no fishing (i.e., $f = 0$).

(c) When is it best to fish for sardines? For sharks? Just use common sense.

(d) Explain how this model describes the often unwanted consequences of spraying insecticide when a natural predator (good guys) controls an insect population (bad guys), but the insecticide kills both the natural predator and the insects?

Analyzing Competition Models *Consider the competition models for rabbits R and sheep S described in Problems 11 and 12. What are the equilibria, what do each signify, and which are stable?*

11. $dR/dt = R(1200 - 2R - 3S),$
$dS/dt = S(500 - R - S).$

12. $dR/dt = R(1200 - 3R - 2S),$
$dS/dt = S(500 - R - S).$

Finding the Model *The scenarios in Problems 13–16 describe the interaction of two and three different species of plants or animals. In each case, set up a system of differential equations that might be used to model the situation. It is not necessary at this stage to solve the systems.*

13. We have a population of rabbits (x) and foxes (y). In the absence of foxes, the rabbits obey the logistic population law. The foxes eat the rabbits, but will die from starvation if rabbits are not present. However, in this environmental system, hunters shoot rabbits but not foxes.

14. On Komodo Island we have three species: Komodo dragons (x), deer (y), and a variety of plants (z). The dragons eat the deer, the deer eat the plants, and the plants compete among themselves.

15. We have three species: violets, ants, and rodents. The violets produce seeds with density x. The violets in the absence of other species obey the logistic population law. Some of the violet seeds are eaten by the ants, whose density is y. The ants in the absence of other species also obey the logistic population law. And finally, rodents will die out unless they have violet seeds to eat. NOTE: Density is defined as population per unit area.

16. Host-Parasite Models Develop appropriate models of the two host-parasite scenarios suggested. HINT: You can adapt the predator-prey model.

(a) Consider a species of insect that depends for survival on a plant on which it feeds. It is to the advantage of the insect to keep the hosts alive. Write an appropriate system of differential equations for the parasite population $P(t)$ and the host population $H(t)$.

(b) How might your model adapt to the study of diseases caused by parasitic organisms?

Competition *Analyze the models for competition between two species given in Problems 17–20, using the following outline.* NOTE: *We require x and y to be nonnegative because this is a population model.*

(a) *Find and plot the equilibrium points and nullclines. Determine the directions of the vector fields between the nullclines.*

(b) *Decide whether the equilibrium points are unstable (repelling at least some nearby solutions) or stable (repelling no nearby solutions).*

(c) *Sketch portions of solution curves near equilibrium points; then complete the phase portrait of the system.*

(d) *Decide whether the two species described by the model can coexist. What conditions are required for coexistence when it is possible?*

17. $dx/dt = x(4 - 2x - y)$
$dy/dt = y(4 - x - 2y)$

18. $dx/dt = x(1 - x - y)$
$dy/dt = y(2 - x - y)$

19. $dx/dt = x(4 - x - 2y)$
$dy/dt = y(1 - 2x - y)$

20. $dx/dt = x(2 - x - 2y)$
$dy/dt = y(2 - 2x - y)$

21. Simpler Competition Consider a situation of two populations competing according to the simpler model

$$dx/dt = x(a - by),$$
$$dy/dt = y(c - dx),$$

where a, b, c, and d are positive constants. Find and sketch the equilibrium points and the nullclines, and determine directions of the vector field between the nullclines. Then determine the stability of the equilibrium points, and sketch the phase portrait of the system.

Nullcline Patterns *For the competition model*

$$dx/dt = x(a - bx - cy),$$
$$dy/dt = y(d - ex - fy),$$

where the parameters a, b, c, d, e, and f are all positive, the diagrams in Problems 22–25 show four possible positions of the nullclines and equilibrium points. In each case:

(a) *Draw arrows in each region between the nullclines to show directions of the vector field.*

(b) *Determine if each equilibrium point is stable or unstable.*

(c) *Draw the solution curves in a neighborhood of each equilibrium point.*

(d) *Sketch the phase portrait of the system.*

(e) *Draw a conclusion about the long-term fate of the species involved.*

22.

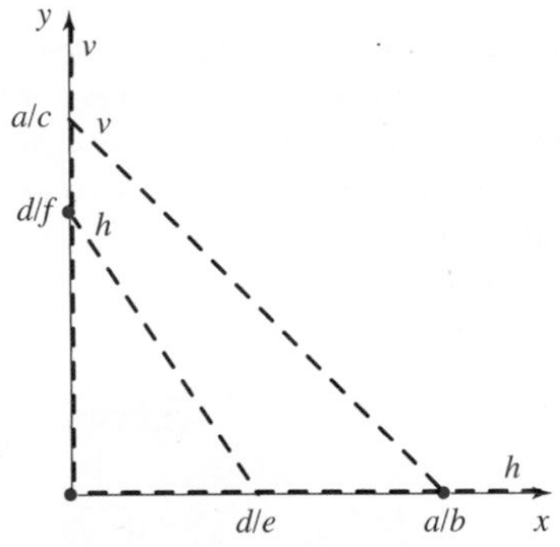

23.

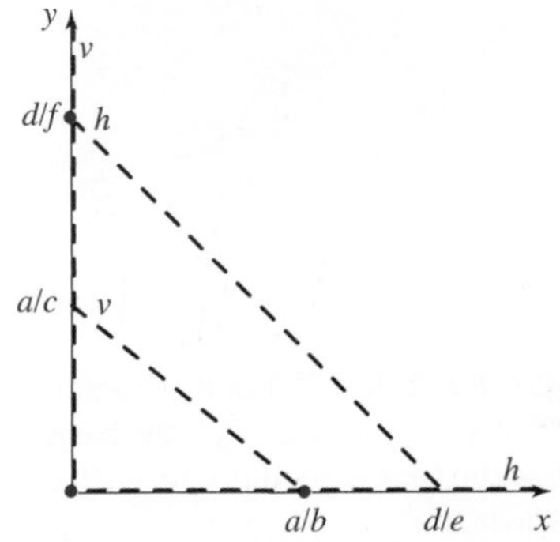

24.

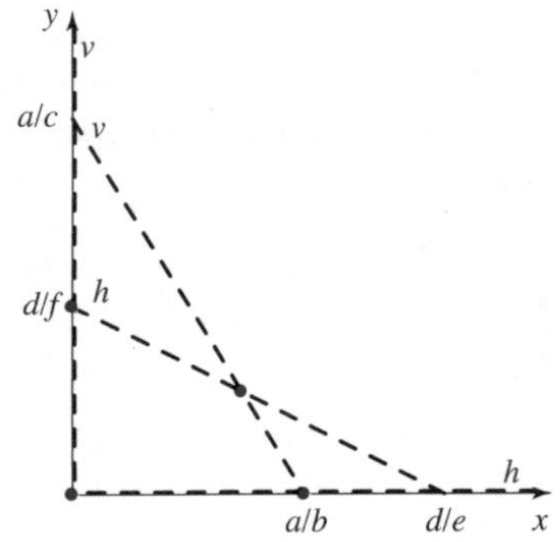

25.

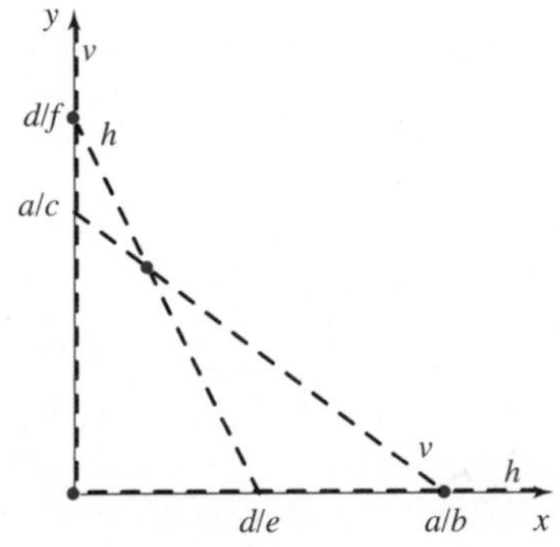

26. Unfair Competition Another model for competition arises when, in the absence of a second species, the first grows logistically and the second exponentially:

$$dx/dt = ax(1 - bx) - cxy,$$
$$dy/dt = dy - exy.$$

Show that for this model coexistence is impossible.

Basins of Attraction *Problems 27–30 each specify one of the competition scenarios in the previous set of problems. For each stable equilibrium point in the given model, find and color the set of points in the plane whose associated solution curves eventually approach that equilibrium point. An example is given in Fig. 2.6.12.*

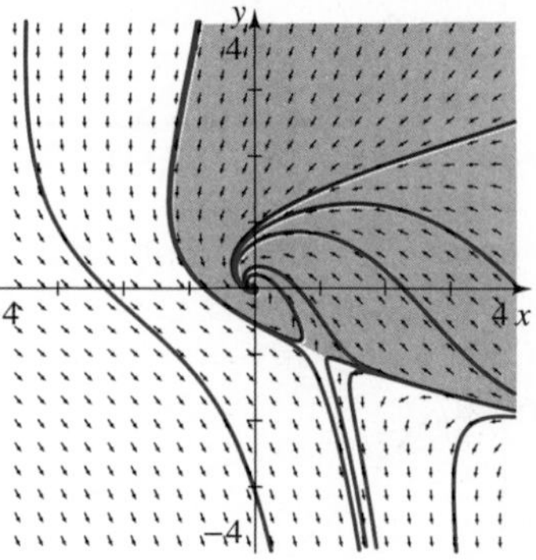

FIGURE 2.6.12 For the system $x' = -x - y$, $y' = x - y^2$, the basin of attraction for the equilibrium at (0, 0) is shaded.

Such a set of points is called the ***basin of attraction*** *for that equilibrium. Relate the coloring of these basins to your description of the long-term fate of the species given in the corresponding problem.*

27. Problem 2 **28.** Problem 3 **29.** Problem 18

30. Problem 21, with $a = b = c = d = 1$

31. Computer Lab: Parameter Investigation

Lotka-Volterra;
Lotka-Volterra with Harvest

Use these IDE tools to do the following:

(a) Analyze the effect of each of the parameters a_R, a_F, c_R, and c_F on the system (6).

(b) Analyze the effect of harvesting either species at a constant rate. Explain the different outcomes for harvesting predators versus harvesting prey.

32. Computer Lab: Competition Outcomes

Competitive Exclusion

Use this IDE tool to draw phase-plane trajectories for the competition model (8). Consider the four possible relations of the nullclines, and hand-sketch a phase portrait for each.

(a) Describe the different outcomes in terms of the equilibrium points and their stability.

(b) Describe whatever new insights you achieved with this interactive graphics exploration.

(c) For each stable equilibrium point in each of your four phase portraits from part (a), find and color its **basin of attraction**. (See Problems 27–30.) Relate these basins to your descriptions in part (a).

33. Suggested Journal Entry What additional significant ecological factors have been overlooked in our simplified fox-and-rabbit model? How might these be incorporated into the equations? Discuss similar issues for the competition model.

CHAPTER

3 Linear Algebra

The foundation of all modern mathematics, from computer animation to wavelets, is based upon the theory of matrices and vector spaces.

—Mark Parker, Carroll College

3.1 Matrices: Sums and Products

SYNOPSIS: We introduce arrays of numbers called matrices, which can be used to represent and manipulate large amounts of data efficiently. When considered as a generalization of ordinary numbers, these arrays can be added, subtracted, multiplied, and (in a sense) divided.

Introduction

Many readers will already have met matrices in precalculus, multivariable calculus, or discrete mathematics. By treating *arrays* of numbers as discrete entities with simpler names, we can state and solve a variety of mathematical and practical problems efficiently. For example, the system of algebraic equations

$$\begin{aligned} 7x + 4y &= 2, \\ -2x + y &= -7 \end{aligned}$$

and the system of differential equations

$$\begin{aligned} dx/dt &= 7x + 4y + t, \\ dy/dt &= -2x + y - e^t \end{aligned}$$

can be written compactly as $\mathbf{A}\vec{\mathbf{x}} = \vec{\mathbf{b}}$ and $\vec{\mathbf{x}}' = \mathbf{A}\vec{\mathbf{x}} + \vec{\mathbf{f}}$, respectively, where

$$\mathbf{A} = \begin{bmatrix} 7 & 4 \\ -2 & 1 \end{bmatrix}, \quad \vec{\mathbf{x}} = \begin{bmatrix} x \\ y \end{bmatrix}, \quad \vec{\mathbf{x}}' = \begin{bmatrix} dx/dt \\ dy/dt \end{bmatrix},$$

$$\vec{\mathbf{b}} = \begin{bmatrix} 2 \\ -7 \end{bmatrix}, \quad \text{and} \quad \vec{\mathbf{f}} = \begin{bmatrix} t \\ -e^t \end{bmatrix}.$$

In this section and the next we will introduce this notation and define the operations for vectors and matrices, which are the language of linear algebra. We will then be able to use them to solve efficiently linear systems of equations, both algebraic and differential, in any number of variables.

Basic Terminology

Matrix

A **matrix** is a rectangular array of **elements** or **entries** (numbers or functions) arranged in **rows** (horizontal) and **columns** (vertical):

$$\mathbf{A} = \begin{bmatrix} a_{11} & a_{12} & \cdots & a_{1n} \\ a_{21} & a_{22} & \cdots & a_{2n} \\ \vdots & \vdots & \ddots & \vdots \\ a_{m1} & a_{m2} & \cdots & a_{mn} \end{bmatrix}. \tag{1}$$

Square Matrix:

$$\underbrace{\begin{bmatrix} a_{11} & a_{12} & a_{13} \\ a_{21} & a_{22} & a_{23} \\ a_{31} & a_{32} & a_{33} \end{bmatrix}}_{\text{a } 3 \times 3 \text{ matrix}}$$

Diagonal elements are in color.

We say that a matrix **A**, having m rows and n columns, as shown in (1), is of **order** $m \times n$ (read as "m by n"). For an $n \times n$ **square matrix**, we may abbreviate and simply say order n. We will usually denote matrices by boldface capital letters, using the corresponding lowercase letter with subscripts (row first, then column) for the entries. If the order of the matrix is clear from the context, we sometimes write the shorthand $\mathbf{A} = [a_{ij}]$.

The matrix entries a_{ij} in an $n \times n$ matrix for which $i = j$ are called **diagonal elements**; all of them make up the (main) **diagonal** of a matrix.

An $m \times 1$ matrix is called a **column vector** and a $1 \times n$ matrix is called a **row vector**. An $m \times n$ matrix consists of n column vectors

$$\begin{bmatrix} a_{11} \\ a_{21} \\ \vdots \\ a_{m1} \end{bmatrix} \quad \begin{bmatrix} a_{12} \\ a_{22} \\ \vdots \\ a_{m2} \end{bmatrix} \quad \cdots \quad \begin{bmatrix} a_{1n} \\ a_{2n} \\ \vdots \\ a_{mn} \end{bmatrix}$$

and m row vectors

$$\begin{matrix} [a_{11} & a_{12} & \cdots & a_{1n}] \\ [a_{21} & a_{22} & \cdots & a_{2n}] \\ & \vdots & & \\ [a_{m1} & a_{m2} & \cdots & a_{mn}]. \end{matrix}$$

Entries in row vectors and column vectors are sometimes identified with single subscripts, as in the following example.

EXAMPLE 1 **Matrices and Order** Consider the following matrices:

$$\mathbf{A} = \begin{bmatrix} 1 & 3 \\ 2 & 4 \end{bmatrix}, \quad \mathbf{B} = \begin{bmatrix} 1 & 2 & -1 \\ 3 & 0 & 5 \end{bmatrix}, \quad \mathbf{C} = \begin{bmatrix} 3 & 4 & 5 \end{bmatrix}, \quad \mathbf{D} = \begin{bmatrix} 9 \\ 7 \end{bmatrix}.$$

(a) **A** is a 2×2 matrix with entries $a_{11} = 1, a_{12} = 3, a_{21} = 2, a_{22} = 4$.

(b) **B** is 2×3, $b_{11} = 1, b_{12} = 2, b_{13} = -1, b_{21} = 3, b_{22} = 0, b_{23} = 5$.

(c) **C** is a 1×3 matrix or **row vector**; $c_{11} = c_1 = 3, c_{12} = c_2 = 4, c_{13} = c_3 = 5$.

(d) **D** is a 2×1 matrix or **column vector**; $d_{11} = d_1 = 9, d_{21} = d_2 = 7$. ■

Equal Matrices

Two matrices of the same order are **equal** if their corresponding entries are equal. If matrices $\mathbf{A} = [a_{ij}]$ and $\mathbf{B} = [b_{ij}]$ are both $m \times n$, then

$$\mathbf{A} = \mathbf{B} \quad \text{if and only if} \quad a_{ij} = b_{ij}, \quad 1 \le i \le m,\ 1 \le j \le n. \tag{2}$$

EXAMPLE 2 Equality Among Matrices

(a) $\begin{bmatrix} 1 & 1 \\ 4 & 3 \end{bmatrix} = \begin{bmatrix} 3-2 & |-1| \\ 2^2 & \sqrt{9} \end{bmatrix}$ because corresponding elements are equal.

(b) Since the "(2, 1)-elements" are unequal, $\begin{bmatrix} 1 & 3 & 5 \\ 2 & 4 & 6 \end{bmatrix} \neq \begin{bmatrix} 1 & 3 & 5 \\ 6 & 4 & 6 \end{bmatrix}$.

(c) $[1 \quad 1 \quad 1] \neq \begin{bmatrix} 1 & 1 \\ 1 & 1 \end{bmatrix}$ regardless of entries: they are of different order. ■

Zero Matrices:

$$\mathbf{0}_{23} = \begin{bmatrix} 0 & 0 & 0 \\ 0 & 0 & 0 \end{bmatrix}; \mathbf{0}_{31} = \begin{bmatrix} 0 \\ 0 \\ 0 \end{bmatrix}$$

Diagonal Matrices:

$$\mathbf{A} = \begin{bmatrix} 2 & 0 & 0 \\ 0 & 4 & 0 \\ 0 & 0 & 1 \end{bmatrix}; \mathbf{B} = \begin{bmatrix} 0 & 0 \\ 0 & 0 \end{bmatrix}$$

B is also a zero matrix.

Identity Matrix:

$$\mathbf{I}_3 = \begin{bmatrix} 1 & 0 & 0 \\ 0 & 1 & 0 \\ 0 & 0 & 1 \end{bmatrix}$$

Special Matrices

- The $m \times n$ **zero matrix**, denoted $\mathbf{0}_{mn}$ (or just **0** if the order is clear from the context), has all its entries equal to 0.
- A **diagonal matrix** is a square matrix for which $a_{ij} = 0$ for all $i \neq j$. In general,

$$\mathbf{D} = \begin{bmatrix} a_{11} & 0 & \cdots & 0 \\ 0 & a_{22} & \ddots & \vdots \\ \vdots & \ddots & \ddots & 0 \\ 0 & \cdots & 0 & a_{nn} \end{bmatrix}.$$

  The diagonal elements $a_{11}, a_{22}, \ldots, a_{nn}$ may be zero or nonzero.
- The $n \times n$ **identity matrix**, denoted $\mathbf{I}_n$ (or just **I** if the order is clear from the context), is a diagonal matrix with all diagonal elements equal to 1. In general,

$$\mathbf{I}_n = \begin{bmatrix} 1 & 0 & \cdots & 0 \\ 0 & 1 & \ddots & \vdots \\ \vdots & \ddots & \ddots & 0 \\ 0 & \cdots & 0 & 1 \end{bmatrix}.$$

Matrix Arithmetic

Some operations of arithmetic, like addition and subtraction, carry over to matrices in a natural way. Others, like multiplication, turn out rather unexpectedly, as we shall see gradually.

Matrix Addition

Two matrices of the same order are added (or subtracted) by adding (or subtracting) corresponding entries and recording the results in a matrix of the same size. Using matrix notation, if $\mathbf{A} = [a_{ij}]$ and $\mathbf{B} = [b_{ij}]$ are both $m \times n$,

$$\begin{aligned} \mathbf{A} + \mathbf{B} &= [a_{ij}] + [b_{ij}] = [a_{ij} + b_{ij}], \\ \mathbf{A} - \mathbf{B} &= [a_{ij}] - [b_{ij}] = [a_{ij} - b_{ij}]. \end{aligned} \tag{3}$$

EXAMPLE 3 It All Adds Up

(a) Given two 2 × 3 matrices $\mathbf{A} = \begin{bmatrix} 1 & 2 \\ 3 & 2 \\ -1 & 0 \end{bmatrix}$ and $\mathbf{B} = \begin{bmatrix} 2 & 0 \\ -2 & 7 \\ 3 & 1/2 \end{bmatrix}$,

$$\mathbf{A}+\mathbf{B} = \begin{bmatrix} 1+2 & 2+0 \\ 3+(-2) & 2+7 \\ -1+3 & 0+1/2 \end{bmatrix} = \begin{bmatrix} 3 & 2 \\ 1 & 9 \\ 2 & 1/2 \end{bmatrix}$$

and

$$\mathbf{A}-\mathbf{B} = \begin{bmatrix} 1-2 & 2-0 \\ 3-(-2) & 2-7 \\ -1-3 & 0-1/2 \end{bmatrix} = \begin{bmatrix} -1 & 2 \\ 5 & -5 \\ -4 & -1/2 \end{bmatrix}.$$

(b) Given two 3 × 1 matrices $\mathbf{C} = [2 \quad 4 \quad 0]$ and $\mathbf{D} = [0 \quad 2 \quad -1]$,

$$\mathbf{C}+\mathbf{D} = [2+0 \quad 4+2 \quad 0+(-1)] = [2 \quad 6 \quad -1]$$

and

$$\mathbf{C}-\mathbf{D} = [2-0 \quad 4-2 \quad 0-(-1)] = [2 \quad 2 \quad 1].$$

■

Scalar Multiplication:

$$k\begin{bmatrix} a & b \\ c & d \end{bmatrix} = \begin{bmatrix} ka & kb \\ kc & kd \end{bmatrix},$$

where k is a real (or complex) number.

Multiplication by a Scalar

To find the product of a matrix and a *number*, real or complex, multiply each entry of the matrix by that number. This is called **multiplication by a scalar**. Using matrix notation, if $\mathbf{A} = [a_{ij}]$, then

$$c\mathbf{A} = [ca_{ij}] = [a_{ij}c] = \mathbf{A}c. \tag{4}$$

EXAMPLE 4 Scalar Times Matrix Consider the following matrices:

$$\mathbf{A} = \begin{bmatrix} 1 & 0 \\ 0 & 1 \end{bmatrix}, \quad \mathbf{B} = \begin{bmatrix} -9 \\ 0 \\ 6 \end{bmatrix}, \quad \mathbf{C} = [2 \quad -3 \quad 8], \quad \text{and} \quad \mathbf{D} = \begin{bmatrix} 2+i \\ 3-i \end{bmatrix}.$$

(a) $3\mathbf{A} = \begin{bmatrix} 3\cdot 1 & 3\cdot 0 \\ 3\cdot 0 & 3\cdot 1 \end{bmatrix} = \begin{bmatrix} 3 & 0 \\ 0 & 3 \end{bmatrix}.$

(b) $-\frac{2}{3}\mathbf{B} = \begin{bmatrix} -2/3\cdot -9 \\ -2/3\cdot 0 \\ -2/3\cdot 6 \end{bmatrix} = \begin{bmatrix} 6 \\ 0 \\ -4 \end{bmatrix}.$

(c) $\pi\mathbf{C} = [\pi\cdot 2 \quad \pi\cdot -3 \quad \pi\cdot 8] = [2\pi \quad -3\pi \quad 8\pi].$

(d) $i\mathbf{D} = \begin{bmatrix} i(2+i) \\ i(3-i) \end{bmatrix} = \begin{bmatrix} 2i+i^2 \\ 3i-i^2 \end{bmatrix} = \begin{bmatrix} -1+2i \\ 1+3i \end{bmatrix}.$

■

Properties of Matrix Addition and Scalar Multiplication

Suppose **A**, **B**, and **C** are $m \times n$ matrices and c and k are scalars. Then the following properties hold.

- $\mathbf{A}+\mathbf{B}$ is an $m \times n$ matrix. *(Closure)*
- $\mathbf{A}+\mathbf{B} = \mathbf{B}+\mathbf{A}$. *(Commutativity)*
- $\mathbf{A}+(\mathbf{B}+\mathbf{C}) = (\mathbf{A}+\mathbf{B})+\mathbf{C}$. *(Associativity)*

- $c(k\mathbf{A}) = (ck)\mathbf{A}$. *(Associativity)*
- $\mathbf{A} + \mathbf{0} = \mathbf{A}$. *(Zero Element)*
- $\mathbf{A} + (-\mathbf{A}) = \mathbf{0}$, where $-\mathbf{A}$ denotes $(-1)\mathbf{A}$. *(Inverse Element)*
- $c(\mathbf{A} + \mathbf{B}) = c\mathbf{A} + c\mathbf{B}$. *(Distributivity)*
- $(c + k)\mathbf{A} = c\mathbf{A} + k\mathbf{A}$. *(Distributivity)*

Each of these properties can be proved using corresponding properties of the entries, which are scalars. (See Problems 35–38.)

EXAMPLE 5 **Sample Proof** To prove the distributive property

$$c(\mathbf{A} + \mathbf{B}) = c\mathbf{A} + c\mathbf{B},$$

we suppose that a_{ij} and b_{ij} are the ijth elements of $\mathbf{A}$ and $\mathbf{B}$, respectively, and c is any scalar. Then, by the distributive property for real (complex) numbers, the ijth element of $c(\mathbf{A} + \mathbf{B})$ is $c(a_{ij} + b_{ij}) = ca_{ij} + cb_{ij}$, which is the ijth element of the sum $c\mathbf{A} + c\mathbf{B}$. ■

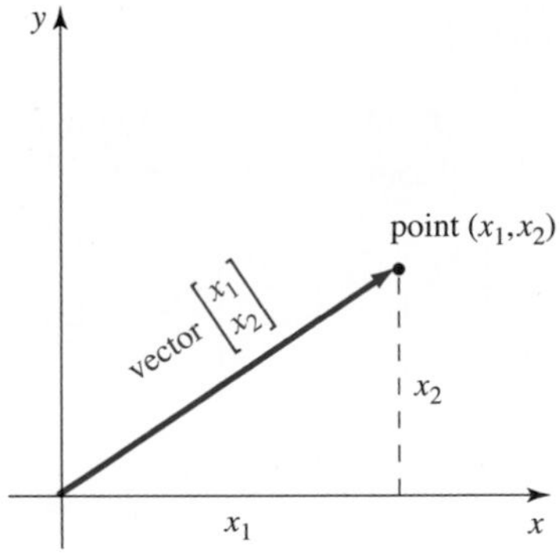

FIGURE 3.1.1 A vector in $\mathbb{R}^2$.

Vectors as Special Matrices

Multivariate calculus makes us familiar with two- and three-dimensional geometric vectors. In the coordinate plane $\mathbb{R}^2$, we write a vector $\vec{\mathbf{x}}$ as a column vector

$$\vec{\mathbf{x}} = \begin{bmatrix} x_1 \\ x_2 \end{bmatrix}$$

or as a point (x_1, x_2) in the plane. The scalars x_1 and x_2 are called first and second **coordinates** of $\vec{\mathbf{x}}$, respectively. We can interpret $\vec{\mathbf{x}}$ geometrically as a **position vector**; that is, as an arrow from the origin to the point (x_1, x_2). (See Fig. 3.1.1.)

It can often be more convenient, especially for large n, to express a vector in $\mathbb{R}^n$ using, *bracketed* row notation (with commas) rather than as a column vector; for example, as shown in margin. The use of brackets rather than parentheses distinguishes this notation from point coordinate notation $(x_1, x_2, \ldots, x_n)$.

Row Notation:

$$\vec{\mathbf{x}} = [\,x_1, x_2, \ldots, x_n\,]$$

is equivalent to

$$\vec{\mathbf{x}} = \begin{bmatrix} x_1 \\ x_2 \\ \vdots \\ x_n \end{bmatrix}.$$

The algebraic rules for adding vectors and multiplying scalars are special cases of the corresponding rules for matrices. We can visualize addition of vectors by means of the parallelogram law (see Problems 83–85) and multiplication by a scalar c as changing a vector's length. (See Fig. 3.1.2.) If $c < 0$, direction is reversed.

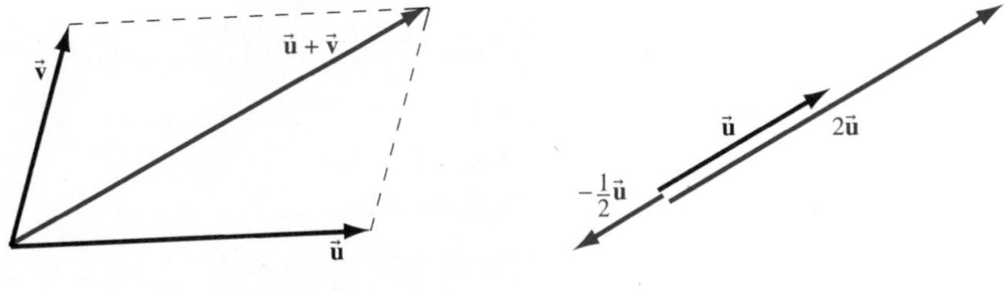

(a) Parallelogram sum (b) Scalar multiple

FIGURE 3.1.2 Vector operations in $\mathbb{R}^2$.

Moving on to three dimensions, we work with ordered triples of real numbers. We can visualize these triples geometrically as three-dimensional position vectors, from the origin to a point in three-space, and work with them in similar fashion to the two-dimensional case.

From our work with matrices of any size earlier in this section, we can consider equally well the collection of four- or five-element vectors. Our ability to visualize may stop with dimension three, but the algebra of these objects presents no difficulties.

$\mathbb{R}^n$ is the set of all vectors with n real number coordinates—that is, n-dimensional space. We define operations on vectors in n-dimensional space as follows:

Vector Addition and Scalar Multiplication

Let

$$\vec{\mathbf{x}} = \begin{bmatrix} x_1 \\ x_2 \\ \vdots \\ x_n \end{bmatrix} \quad \text{and} \quad \vec{\mathbf{y}} = \begin{bmatrix} y_1 \\ y_2 \\ \vdots \\ y_n \end{bmatrix}$$

be vectors in $\mathbb{R}^n$ and c be any scalar. Then $\vec{\mathbf{x}} + \vec{\mathbf{y}}$ and $c\vec{\mathbf{x}}$ are defined, respectively, as

$$\begin{bmatrix} x_1 \\ x_2 \\ \vdots \\ x_n \end{bmatrix} + \begin{bmatrix} y_1 \\ y_2 \\ \vdots \\ y_n \end{bmatrix} \equiv \begin{bmatrix} x_1 + y_1 \\ x_2 + y_2 \\ \vdots \\ x_n + y_n \end{bmatrix} \quad \text{and} \quad c \begin{bmatrix} x_1 \\ x_2 \\ \vdots \\ x_n \end{bmatrix} \equiv \begin{bmatrix} cx_1 \\ cx_2 \\ \vdots \\ cx_n \end{bmatrix}.$$

The properties of vector addition and scalar multiplication follow from corresponding properties for matrices. They are listed because of their importance. In Section 3.6, we will use them to characterize a *vector space*.

Zero Vector in $\mathbb{R}^n$:

$$\vec{\mathbf{0}} = \begin{bmatrix} 0 \\ 0 \\ \vdots \\ 0 \end{bmatrix}$$

Properties of Vector Addition and Scalar Multiplication

Suppose that $\vec{\mathbf{u}}$, $\vec{\mathbf{v}}$, and $\vec{\mathbf{w}}$ are vectors in $\mathbb{R}^n$ and c and k are scalars. Then the following properties hold.

- $\vec{\mathbf{u}} + \vec{\mathbf{v}}$ is a vector in $\mathbb{R}^n$. (*Closure*)
- $\vec{\mathbf{u}} + \vec{\mathbf{v}} = \vec{\mathbf{v}} + \vec{\mathbf{u}}$. (*Commutativity*)
- $\vec{\mathbf{u}} + (\vec{\mathbf{v}} + \vec{\mathbf{w}}) = (\vec{\mathbf{u}} + \vec{\mathbf{v}}) + \vec{\mathbf{w}}$. (*Associativity*)
- $c(k\vec{\mathbf{u}}) = (ck)\vec{\mathbf{u}}$. (*Associativity*)
- $\vec{\mathbf{u}} + \vec{\mathbf{0}} = \vec{\mathbf{u}}$. (*Zero Element*)
- $\vec{\mathbf{u}} + (-\vec{\mathbf{u}}) = \vec{\mathbf{0}}$. (*Inverse Element*)
- $c(\vec{\mathbf{u}} + \vec{\mathbf{v}}) = c\vec{\mathbf{u}} + c\vec{\mathbf{v}}$. (*Distributivity*)
- $(c + k)\vec{\mathbf{u}} = c\vec{\mathbf{u}} + k\vec{\mathbf{u}}$. (*Distributivity*)

The Scalar Product of Two Vectors

Before we define multiplication of a matrix by a matrix, we must look at the simplest case: the scalar product of a vector times a vector. Observe that the result is always a scalar.

Scalar Product vs. Vector Product: The *scalar* product is always a *scalar*:

vector · **vector** = *scalar*,

as opposed to the *vector* product from multivariate calculus and physics:

vector × **vector** = **vector**.

Scalar Product

The **scalar product** of a row vector $\vec{\mathbf{x}}$ and a column vector $\vec{\mathbf{y}}$ of equal length n is the result of adding the products of the corresponding entries as follows:

$$\vec{\mathbf{x}} \cdot \vec{\mathbf{y}} = [x_1 \quad x_2 \quad \cdots \quad x_n] \cdot \begin{bmatrix} y_1 \\ y_2 \\ \vdots \\ y_n \end{bmatrix}$$

$$\equiv x_1 y_1 + x_2 y_2 + \cdots + x_n y_n = \sum_{k=1}^{n} x_k y_k.$$

The scalar product is always a scalar.

EXAMPLE 6 **A Scalar Product**

$$\begin{bmatrix} 1 & 2 & 0 & 1 \end{bmatrix} \cdot \begin{bmatrix} 4 \\ 1 \\ 1 \\ 2 \end{bmatrix} = (1)(4) + (2)(1) + (0)(1) + (1)(2) = 8.$$

■

The scalar product is closely related to the familiar *dot* product of calculus or physics between two vectors of the same dimension, usually expressed as

[row vector] · [row vector].

The properties of scalar products will be investigated in Problems 88–91.

What could be the practical use of such a "strange" operation between two vectors? In fact, there are many uses.

For instance, every time a consumer places a purchase order (e.g., at McDonald's or amazon.com) he specifies quantities x_i of each item, which comprise a vector $\vec{\mathbf{x}}$. The corresponding unit prices y_i comprise another vector $\vec{\mathbf{y}}$, and the total cost on his invoice is a scalar product $\vec{\mathbf{x}} \cdot \vec{\mathbf{y}}$.

The scalar product also provides a useful calculation tool for the geometric concepts that follow.

Orthogonality

Two vectors in $\mathbb{R}^n$ are called **orthogonal** when their scalar product is zero.

In two and three dimensions, nonzero orthogonal vectors are geometrically **perpendicular** in the usual sense. The $\vec{\mathbf{0}}$ vector in $\mathbb{R}^n$ is orthogonal to every vector in $\mathbb{R}^n$.

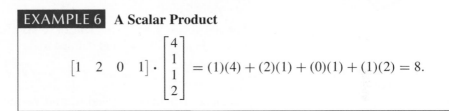

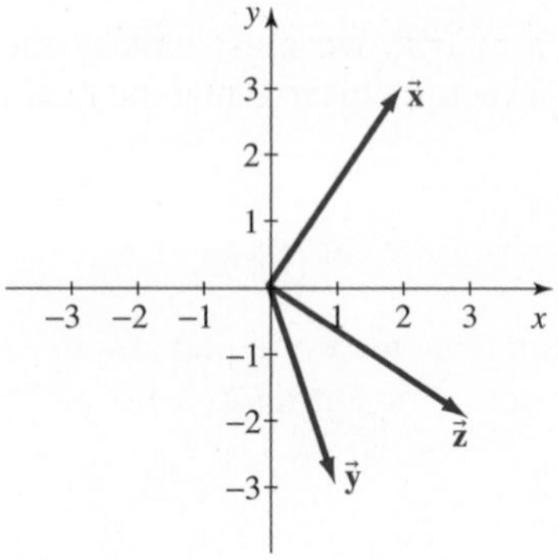

FIGURE 3.1.3 Which vectors are orthogonal?

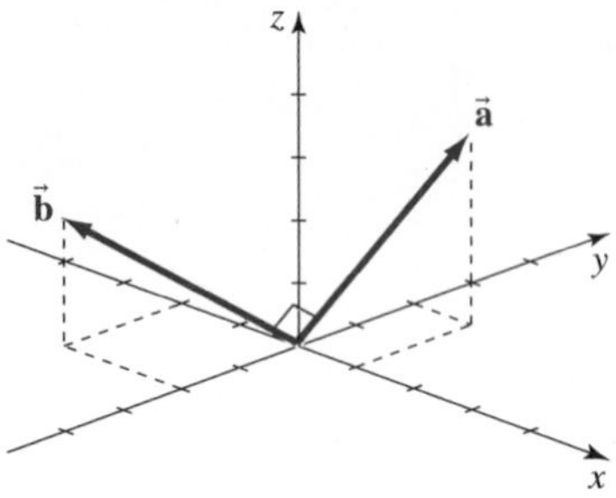

FIGURE 3.1.4 Orthogonal vectors in 3-space.

EXAMPLE 7 Orthogonal or Not

(a) If $\vec{\mathbf{x}} = [2, 3]$ and $\vec{\mathbf{y}} = [1, -3]$,

$$\vec{\mathbf{x}} \cdot \vec{\mathbf{y}} = (2)(1) + (3)(-3) = -7;$$

therefore $\vec{\mathbf{x}}$ and $\vec{\mathbf{y}}$ are not orthogonal.
But if $\vec{\mathbf{z}} = [3, -2]$,

$$\vec{\mathbf{x}} \cdot \vec{\mathbf{z}} = (2)(3) + (3)(-2) = 0,$$

so $\vec{\mathbf{x}}$ and $\vec{\mathbf{z}}$ are orthogonal. (See Fig. 3.1.3.)

(b) For $\vec{\mathbf{a}} = [1, 2, 3]$ and $\vec{\mathbf{b}} = [-2, -2, 2]$,

$$\vec{\mathbf{a}} \cdot \vec{\mathbf{b}} = (1)(-2) + (2)(-2) + (3)(2) = 0;$$

these vectors are orthogonal. (See Fig. 3.1.4.)

(c) The four-dimensional vectors $\vec{\mathbf{x}} = [1, 0, 1, 4]$ and $\vec{\mathbf{y}} = [0, 1, -4, 1]$ are orthogonal because $\vec{\mathbf{x}} \cdot \vec{\mathbf{y}} = (1)(0) + (0)(1) + (1)(-4) + (4)(1) = 0$, although we cannot *see* these vectors as perpendicular. ■

Absolute Value

For any vector $\vec{\mathbf{v}} \in \mathbb{R}^n$, the **length** or **absolute value** of $\vec{\mathbf{v}}$ is a nonnegative scalar, denoted by $\|\vec{\mathbf{v}}\|$ and defined to be

$$\|\vec{\mathbf{v}}\| \equiv \sqrt{\vec{\mathbf{v}} \cdot \vec{\mathbf{v}}}.$$

Vectors of length one are called **unit vectors**.

EXAMPLE 8 Lengths of Vectors Let $\vec{\mathbf{x}} = [1, 2, 0, 5]$. Then

$$\|\vec{\mathbf{x}}\| = \sqrt{1^2 + 2^2 + 0^2 + 5^2} = \sqrt{30}.$$

■

With the scalar product operation in our toolkit, we at last are ready to tackle multiplication of a matrix by a matrix, which is the key to the linear algebra shorthand for systems of equatons.

The Product of Two Matrices

Now we are ready to define the product of two matrices, provided that they are of "compatible orders." We can multiply **A** and **B** only when *the number of columns of* **A** (the left-hand factor) *equals the number of rows of* **B** (the right-hand factor).

Matrix Product

The **matrix product** of an $m \times r$ matrix **A** and an $r \times n$ matrix **B** is denoted

$$\mathbf{C} = \mathbf{AB},$$

where the ijth entry of the new matrix $\mathbf{C}$ is the scalar product of the ith row vector of $\mathbf{A}$ and the jth column vector of $\mathbf{B}$:

$$c_{ij} \equiv [a_{i1} \quad a_{i2} \quad \cdots \quad a_{ir}] \cdot \begin{bmatrix} b_{1j} \\ b_{2j} \\ \vdots \\ b_{rj} \end{bmatrix}$$

$$= a_{i1}b_{1j} + a_{i2}b_{2j} + \cdots + a_{ir}b_{rj} = \sum_{k=1}^{r} a_{ik}b_{kj}.$$

The product matrix $\mathbf{C}$ has order $m \times n$.

Matrix Machine

Observe what happens to a vector

$$\begin{bmatrix} a_1 \\ a_2 \end{bmatrix}$$

when it is premultiplied by a 2×2 matrix.

Illustration of the Matrix Product:

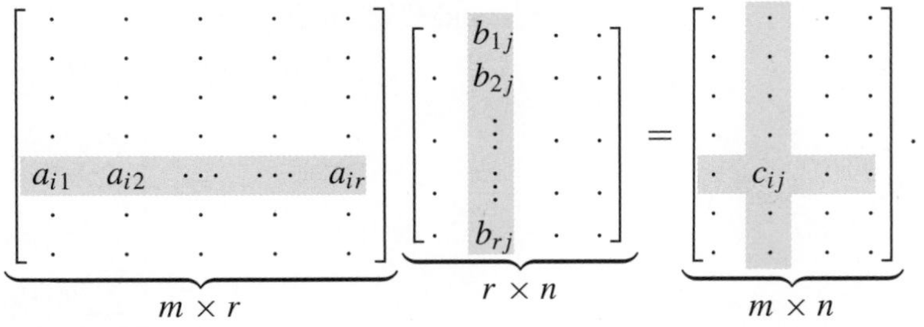

EXAMPLE 9 Product Practice

(a) For the matrices

$$\mathbf{A} = \begin{bmatrix} 1 & -1 & 3 \\ 0 & 4 & 2 \end{bmatrix} \quad \text{and} \quad \mathbf{B} = \begin{bmatrix} 3 & 1 \\ 2 & -4 \\ -1 & 0 \end{bmatrix},$$

here are the element-by-element calculations of the product $\mathbf{AB}$:

$$\mathbf{AB} = \begin{bmatrix} (1)(3) + (-1)(2) + (3)(-1) & (1)(1) + (-1)(-4) + (3)(0) \\ (0)(3) + (4)(2) + (2)(-1) & (0)(1) + (4)(-4) + (2)(0) \end{bmatrix}$$

$$= \begin{bmatrix} 3 - 2 - 3 & 1 + 4 + 0 \\ 0 + 8 - 2 & 0 - 16 + 0 \end{bmatrix} = \begin{bmatrix} -2 & 5 \\ 6 & -16 \end{bmatrix}. \tag{5}$$

(b) For the matrices

$$\mathbf{C} = \begin{bmatrix} 1 & 0 \\ 2 & 0 \\ 1 & 3 \end{bmatrix} \quad \text{and} \quad \mathbf{D} = \begin{bmatrix} 1 & 0 \\ 2 & 1 \end{bmatrix},$$

the matrix product is

$$\mathbf{CD} = \begin{bmatrix} 1 + 0 & 0 + 0 \\ 2 + 0 & 0 + 0 \\ 1 + 6 & 0 + 3 \end{bmatrix} = \begin{bmatrix} 1 & 0 \\ 2 & 0 \\ 7 & 3 \end{bmatrix}.$$

■

Matrix multiplication obeys some, but not all, of the familiar rules of algebra. Assuming that the matrices have compatible sizes, the associative and distributive rules hold.

Properties of Matrix Multiplication

- $(\mathbf{AB})\mathbf{C} = \mathbf{A}(\mathbf{BC})$. *(Associativity)*
- $\mathbf{A}(\mathbf{B} + \mathbf{C}) = \mathbf{AB} + \mathbf{AC}$. *(Distributivity)*
- $(\mathbf{B} + \mathbf{C})\mathbf{A} = \mathbf{BA} + \mathbf{CA}$. *(Distributivity)*
- In general, $\mathbf{AB} \neq \mathbf{BA}$, except in special cases. *(Noncommutativity)*

Identity matrices behave rather like the number 1, and zero matrices behave rather like the number 0. For an $m \times n$ matrix $\mathbf{A}$,

- $\mathbf{AI}_n = \mathbf{A}$, $\quad \mathbf{I}_m\mathbf{A} = \mathbf{A}$.
- $\mathbf{A0}_{np} = \mathbf{0}_{mp}$, $\quad \mathbf{0}_{qm}\mathbf{A} = \mathbf{0}_{qn}$, for any p and q.

EXAMPLE 10 Sample Proof of Distributivity For $\mathbf{A}$ of order $m \times p$, and $\mathbf{B}$ and $\mathbf{C}$ of order $p \times n$,

$$\begin{aligned}
\mathbf{A}(\mathbf{B} + \mathbf{C}) &= [a_{ij}][b_{ij} + c_{ij}] \\
&= \left[\sum_{k=1}^{p}(a_{ik})(b_{kj} + c_{kj})\right] \\
&= \left[\sum_{k=1}^{p} a_{ik}b_{kj} + a_{ik}c_{kj}\right] \\
&= \left[\sum_{k=1}^{p} a_{ik}b_{kj}\right] + \left[\sum_{k=1}^{p} a_{ik}c_{kj}\right] = \mathbf{AB} + \mathbf{AC}.
\end{aligned}$$

■

EXAMPLE 11 Commuters Beware

(a) For the matrices $\mathbf{A}$ and $\mathbf{B}$ of Example 9, we find that

$$\mathbf{BA} = \begin{bmatrix} 3 & 1 \\ 2 & -4 \\ -1 & 0 \end{bmatrix}\begin{bmatrix} 1 & -1 & 3 \\ 0 & 4 & 2 \end{bmatrix} = \begin{bmatrix} 3 & 1 & 11 \\ 2 & -18 & -2 \\ -1 & 1 & -3 \end{bmatrix}. \qquad (6)$$

Comparing equation (6) with equation (5), we see that the products $\mathbf{AB}$ and $\mathbf{BA}$ are indeed different (and even of different order).

(b) Furthermore, if we look at the product of matrices

$$\mathbf{P} = [1 \quad 0 \quad 2] \quad \text{and} \quad \mathbf{Q} = \begin{bmatrix} 1 & 0 \\ 1 & -1 \\ 0 & 2 \end{bmatrix},$$

we see that

$$\mathbf{PQ} = [1 + 0 + 0 \quad 0 + 0 + 4] = [1 \quad 4],$$

but the product $\mathbf{QP}$ is not defined because $\mathbf{Q}$ and $\mathbf{P}$ are not of compatible orders (i.e., $\mathbf{Q}$ is 3×2 and $\mathbf{P}$ is 1×3).

■

Commutativity of Matrices:
It is rarely the case that $\mathbf{AB} = \mathbf{BA}$. In fact, sometimes $\mathbf{BA}$ is not defined even though $\mathbf{AB}$ is. Diagonal matrices commute. Are there others?

Matrix and Transpose:
If $\mathbf{A} = \begin{bmatrix} 1 & 2 \\ 3 & 4 \\ 5 & 6 \end{bmatrix}$,

$\mathbf{A}^{\mathrm{T}} = \begin{bmatrix} 1 & 3 & 5 \\ 2 & 4 & 6 \end{bmatrix}$.

The Matrix Transpose

If we flip a matrix "diagonally" so that its rows become columns and its columns become rows, we get a new matrix called the **transpose** of the original matrix. We write $\mathbf{A}^{\mathrm{T}}$ for the transpose of $\mathbf{A}$; $[\,a_{ij}\,]^{\mathrm{T}} = [\,a_{ji}\,]$.

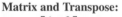

EXAMPLE 12 **Flip for Transposes** For matrices **A**, **B**, **C**, and **D**,

$$\mathbf{A} = \begin{bmatrix} 1 & 3 \\ 2 & 4 \end{bmatrix}, \quad \mathbf{B} = \begin{bmatrix} 1 & 2 & -1 \\ 3 & 0 & 5 \end{bmatrix}, \quad \mathbf{C} = [3 \quad 4 \quad 5], \quad \mathbf{D} = \begin{bmatrix} 9 \\ 7 \end{bmatrix},$$

the transposes are

$$\mathbf{A}^{\mathrm{T}} = \begin{bmatrix} 1 & 2 \\ 3 & 4 \end{bmatrix}, \quad \mathbf{B}^{\mathrm{T}} = \begin{bmatrix} 1 & 3 \\ 2 & 0 \\ -1 & 5 \end{bmatrix}, \quad \mathbf{C}^{\mathrm{T}} = \begin{bmatrix} 3 \\ 4 \\ 5 \end{bmatrix}, \quad \mathbf{D}^{\mathrm{T}} = [9 \quad 7].$$

■

Properties of Transposes

Assume **A** and **B** to be matrices of compatible orders.

- $(\mathbf{A}^{\mathrm{T}})^{\mathrm{T}} = \mathbf{A}$.
- $(\mathbf{A} + \mathbf{B})^{\mathrm{T}} = \mathbf{A}^{\mathrm{T}} + \mathbf{B}^{\mathrm{T}}$.
- $(k\mathbf{A})^{\mathrm{T}} = k\mathbf{A}^{\mathrm{T}}$, for any scalar k.
- $(\mathbf{AB})^{\mathrm{T}} = \mathbf{B}^{\mathrm{T}}\mathbf{A}^{\mathrm{T}}$.

The reader is asked to prove these properties in Problems 39–42.

Symmetric Matrix:

$$\underbrace{\begin{bmatrix} 1 & -2 & 3 \\ -2 & 0 & 4 \\ 3 & 4 & -1 \end{bmatrix}}_{\mathbf{A} = \mathbf{A}^{\mathrm{T}}}$$

If a matrix **A** is the same as its transpose $\mathbf{A}^{\mathrm{T}}$, we call **A** a **symmetric** matrix. A symmetric matrix must be square. The entries are reflected about the main diagonal.

Matrices with Function Entries

Matrices with entries that are functions rather than constants are important in this book because they will help us deal with differential equations and their solutions. In particular, we will be interested in such expressions as

$$\vec{\mathbf{x}}(t) = \begin{bmatrix} x_1(t) \\ x_2(t) \\ \vdots \\ x_n(t) \end{bmatrix} \quad \text{and} \quad \mathbf{A}(t) = \begin{bmatrix} a_{11}(t) & a_{12}(t) & \cdots & a_{1n}(t) \\ a_{21}(t) & a_{22}(t) & \cdots & a_{2n}(t) \\ \vdots & \vdots & \ddots & \vdots \\ a_{n1}(t) & a_{n2}(t) & \cdots & a_{nn}(t) \end{bmatrix}. \tag{7}$$

We say that the matrices $\vec{\mathbf{x}}(t)$ and $\mathbf{A}(t)$ are *continuous*, *piecewise continuous*, or *differentiable* provided that every entry has the required property. (The matrix is only as good as its *least* well-behaved element.)

The derivative of a matrix of functions is the matrix of the derivatives of the entries. The derivatives of the matrices in (7) are:

$$\vec{\mathbf{x}}'(t) = \frac{d\vec{\mathbf{x}}}{dt} = \begin{bmatrix} x_1'(t) \\ x_2'(t) \\ \vdots \\ x_n'(t) \end{bmatrix} = \left[x_i'(t)\right] \quad \text{and}$$

$$\mathbf{A}'(t) = \frac{d\mathbf{A}}{dt} = \begin{bmatrix} a_{11}'(t) & a_{12}'(t) & \cdots & a_{1n}'(t) \\ a_{21}'(t) & a_{22}'(t) & \cdots & a_{2n}'(t) \\ \vdots & \vdots & \ddots & \vdots \\ a_{n1}'(t) & a_{n2}'(t) & \cdots & a_{nn}'(t) \end{bmatrix} = \left[a_{ij}'(t)\right].$$

EXAMPLE 13 **Differentiating Arrays** For the expressions

$$\vec{\mathbf{x}} = \begin{bmatrix} x \\ y \end{bmatrix}, \quad \vec{\mathbf{g}}(t) = \begin{bmatrix} \ln t \\ -t^3 \\ \cos 2t \end{bmatrix}, \quad \text{and} \quad \mathbf{A}(t) = \begin{bmatrix} e^t & t^2 \\ \sin t & 2t \end{bmatrix},$$

we can calculate the derivatives to be, respectively,

$$\vec{\mathbf{x}}' = \begin{bmatrix} x' \\ y' \end{bmatrix}, \quad \vec{\mathbf{g}}'(t) = \begin{bmatrix} 1/t \\ -3t^2 \\ -2\sin 2t \end{bmatrix}, \quad \text{and} \quad \mathbf{A}'(t) = \begin{bmatrix} e^t & 2t \\ \cos t & 2 \end{bmatrix}.$$

■

Differentiation rules for combinations of matrices are pretty much like the ones learned in calculus for scalar functions. The tricky one is the product rule, where it is now essential to keep the products *in the same order* because of the noncommutativity of matrix multiplication.

Matrix Differentiation Rules

For differentiable matrices $\mathbf{A}(t)$ and $\mathbf{B}(t)$ and scalar constant c,

- $(\mathbf{A}(t) + \mathbf{B}(t))' = \mathbf{A}'(t) + \mathbf{B}'(t)$.
- $(c\mathbf{A}(t))' = c\mathbf{A}'(t)$.
- $(\mathbf{A}(t)\mathbf{B}(t))' = \mathbf{A}(t)\mathbf{B}'(t) + \mathbf{A}'(t)\mathbf{B}(t)$.

EXAMPLE 14 **Differentiating a Product of Matrices** For the matrices

$$\mathbf{A}(t) = \begin{bmatrix} \sin t & \cos t \\ 1 & 0 \end{bmatrix} \quad \text{and} \quad \mathbf{B}(t) = \begin{bmatrix} t^2 & t \\ 2t & 3 \end{bmatrix},$$

find the derivative of their product **AB**.

$$\begin{aligned}
(\mathbf{AB})'(t) &= \mathbf{A}(t)\mathbf{B}'(t) + \mathbf{A}'(t)\mathbf{B}(t) \\
&= \begin{bmatrix} \sin t & \cos t \\ 1 & 0 \end{bmatrix}\begin{bmatrix} 2t & 1 \\ 2 & 0 \end{bmatrix} + \begin{bmatrix} \cos t & -\sin t \\ 0 & 0 \end{bmatrix}\begin{bmatrix} t^2 & t \\ 2t & 3 \end{bmatrix} \\
&= \begin{bmatrix} (\sin t)2t + (\cos t)2 & \sin t \\ 2t & 1 \end{bmatrix} \\
&\quad + \begin{bmatrix} (\cos t)t^2 - (\sin t)2t & (\cos t)t - (\sin t)3 \\ 0 & 0 \end{bmatrix} \\
&= \begin{bmatrix} (2 + t^2)\cos t & t\cos t - 2\sin t \\ 2t & 1 \end{bmatrix}.
\end{aligned}$$

The same result can be obtained by multiplying **A** and **B**, *then* differentiating.

■

Historical Note

The term "matrix" was first mentioned in mathematical literature in an 1850 paper by English mathematician **James Joseph Sylvester** (1814–1897). The nontechnical meaning of the word is "a place within which something is produced or developed." For Sylvester, a matrix was an arrangement of quantities from which one could "produce" a number called the *determinant*, a quantity used in linear algebra to discuss systems of equations. (See Sec. 3.4.) The rules by which matrices are added and multiplied were developed by another English mathematician, **Arthur Cayley** (1821–1895), in connection with the study of *linear transformations* (to be studied in some detail in Chapter 5).

Cayley and Sylvester were giants of mathematics in the Victorian era, and lifetime collaborators.[1] They became close friends despite sharply contrasting personalities (Cayley was steady and serene, Sylvester turbulent), and each inspired the other to some of his most important work. Although he spent most of his career in London, Sylvester held a chair at Johns Hopkins University from 1876 to 1883. He helped elevate the status of mathematical research in North America, and founded in 1878 its first mathematical research journal, the *American Journal of Mathematics* (still in existence today).

Summary

We have defined matrices as arrays of numbers or functions, and learned their basic arithmetic. We have introduced the zero matrix, the identity matrix, diagonal matrices, symmetric matrices, and the transposed matrix. Important special matrices are (row or column) vectors. By computing the dot product of two vectors, we can tell whether they are orthogonal (a generalization of perpendicularity).

3.1 Problems

Do They Compute? *Calculate the quantities required in Problems 1–16, where*

$$\mathbf{A} = \begin{bmatrix} -1 & 0 & 3 \\ 2 & 1 & 2 \\ -1 & 0 & 1 \end{bmatrix}, \quad \mathbf{B} = \begin{bmatrix} 1 & 3 & 0 \\ 0 & 1 & 0 \\ 0 & 0 & 1 \end{bmatrix},$$

$$\mathbf{C} = \begin{bmatrix} 1 & 0 \\ 2 & 1 \\ 1 & 3 \end{bmatrix}, \quad \text{and} \quad \mathbf{D} = \begin{bmatrix} 3 & -1 & 0 \\ 2 & 1 & 2 \end{bmatrix},$$

or explain why they are undefined.

1. $2\mathbf{A}$ **2.** $\mathbf{A} + 2\mathbf{B}$ **3.** $2\mathbf{C} - \mathbf{D}$

4. $\mathbf{AB}$ **5.** $\mathbf{BA}$ **6.** $\mathbf{CD}$

7. $\mathbf{DC}$ **8.** $(\mathbf{DC})^{\mathrm{T}}$ **9.** $\mathbf{C}^{\mathrm{T}}\mathbf{D}$

10. $\mathbf{D}^{\mathrm{T}}\mathbf{C}$ **11.** $\mathbf{A}^2 = \mathbf{AA}$ **12.** $\mathbf{AD}$

13. $\mathbf{A} - \mathbf{I}_3$ **14.** $4\mathbf{B} - 3\mathbf{I}_3$ **15.** $\mathbf{C} - \mathbf{I}_3$

16. $\mathbf{AC}$

More Multiplication Practice *In Problems 17–22, compute the indicated products or explain why it is not possible.*

17. $[1 \quad 0 \quad -2] \begin{bmatrix} a & b \\ c & d \\ e & f \end{bmatrix}$ **18.** $\begin{bmatrix} a & b \\ c & d \end{bmatrix} \begin{bmatrix} d & -b \\ -c & a \end{bmatrix}$

19. $\begin{bmatrix} 2 & 0 \\ 1 & 1 \end{bmatrix} \begin{bmatrix} 1/2 & 0 \\ -1/2 & 1 \end{bmatrix}$ **20.** $[0 \quad 1 \quad 0] \begin{bmatrix} a & b & c \\ d & e & f \\ g & h & k \end{bmatrix}$

21. $[0 \quad 1] \begin{bmatrix} a & b & c \\ d & e & f \end{bmatrix} [1 \quad 1 \quad 0]$

22. $[1 \quad 1 \quad 0] \begin{bmatrix} a & b \\ c & d \\ e & f \end{bmatrix} \begin{bmatrix} 1 \\ 1 \end{bmatrix}$

23. Rows and Columns in Products Analyze the order of the product matrices or the factor matrices indicated in parts (a)–(c).

(a) If $\mathbf{AB}$ is a 6×5 matrix, how many columns does $\mathbf{B}$ have?

(b) If $\mathbf{AB}$ is a 4×7 matrix, how many rows does $\mathbf{A}$ have?

(c) If matrix $\mathbf{A}$ is a 2×6 matrix and $\mathbf{AB}$ is a 2×4 matrix, what order is matrix $\mathbf{B}$?

Which Rules Work for Matrix Multiplication? *For each of Problems 24–27, prove the statement in general or give a counterexample. In each problem,* $\mathbf{A}$, $\mathbf{B}$, *and* $\mathbf{I}$ *denote* $n \times n$ *matrices, and* $\vec{\mathbf{x}}$, $\vec{\mathbf{y}}$, *and* $\vec{\mathbf{z}}$ *denote column vectors in* $\mathbb{R}^n$. HINT: *For the proofs, use the properties of matrix multiplication and do not break the matrices down into elements.*

[1]Both Cayley and Sylvester worked as lawyers for a number of years to earn enough money to pursue mathematics!

24. $(\mathbf{A}+\mathbf{B})(\mathbf{A}-\mathbf{B})=\mathbf{A}^2-\mathbf{B}^2$

25. $(\mathbf{A}+\mathbf{B})^2=\mathbf{A}^2+2\mathbf{AB}+\mathbf{B}^2$

26. $(\mathbf{I}+\mathbf{A})^2=\mathbf{I}+2\mathbf{A}+\mathbf{A}^2$

27. $(\mathbf{A}+\mathbf{B})^2=\mathbf{A}^2+\mathbf{AB}+\mathbf{BA}+\mathbf{B}^2$

Find the Matrix *Find the nonzero matrices* **A**, **B**, *and* **C** *in Problems 28–30. If no such matrix exists, show why.*

28. $\mathbf{A}\begin{bmatrix}1 & 2\\ 3 & 4\end{bmatrix}=\begin{bmatrix}0 & 0\\ 0 & 0\end{bmatrix}$ **29.** $\begin{bmatrix}1 & 2 & 3\\ 0 & 1 & 0\end{bmatrix}\mathbf{B}=\mathbf{I}_2$

30. $\begin{bmatrix}1 & 2\\ 4 & 1\end{bmatrix}\mathbf{C}=\begin{bmatrix}2 & 0\\ 1 & 4\end{bmatrix}$

Commuters *In Problems 31–33, find all the* 2×2 *matrices that commute with the given matrix.*

31. $\begin{bmatrix}a & 0\\ 0 & a\end{bmatrix}$, where $a\in\mathbb{R}$

32. $\begin{bmatrix}1 & k\\ k & 1\end{bmatrix}$, where $k\in\mathbb{R}$, $k\neq 0$ **33.** $\begin{bmatrix}0 & 1\\ 1 & 0\end{bmatrix}$

34. Products with Transposes Use matrices

$$\mathbf{A}=\begin{bmatrix}1\\ 4\end{bmatrix}\quad\text{and}\quad\mathbf{B}=\begin{bmatrix}1\\ -1\end{bmatrix}$$

to find the indicated products for parts (a)–(d).

(a) $\mathbf{A}^{\mathrm{T}}\mathbf{B}$ (b) $\mathbf{AB}^{\mathrm{T}}$

(c) $\mathbf{B}^{\mathrm{T}}\mathbf{A}$ (d) $\mathbf{BA}^{\mathrm{T}}$

Reckoning *In Problems 35–38, prove the statements for* $m\times n$ *matrices* **A** *and* **B** *and scalars* c *and* d.

35. $\mathbf{A}-\mathbf{B}=\mathbf{A}+(-1)\mathbf{B}$ **36.** $\mathbf{A}+\mathbf{B}=\mathbf{B}+\mathbf{A}$

37. $(c+d)\mathbf{A}=c\mathbf{A}+d\mathbf{A}$ **38.** $c(\mathbf{A}+\mathbf{B})=c\mathbf{A}+c\mathbf{B}$

Properties of the Transpose *In Problems 39–42, either prove the properties in general using the fact that* $[a_{ij}]^{\mathrm{T}}=[a_{ji}]$, *or demonstrate the properties for general* 3×3 *matrices.*

39. $(\mathbf{A}^{\mathrm{T}})^{\mathrm{T}}=\mathbf{A}$ **40.** $(\mathbf{A}+\mathbf{B})^{\mathrm{T}}=\mathbf{A}^{\mathrm{T}}+\mathbf{B}^{\mathrm{T}}$

41. $(k\mathbf{A})^{\mathrm{T}}=k\mathbf{A}^{\mathrm{T}}$, for any scalar k

42. $(\mathbf{AB})^{\mathrm{T}}=\mathbf{B}^{\mathrm{T}}\mathbf{A}^{\mathrm{T}}$

43. Transposes and Symmetry Prove that if **A** is symmetric then so is $\mathbf{A}^{\mathrm{T}}$.

44. Symmetry and Products Give an example to show that the product of symmetric matrices is not necessarily symmetric.

45. Constructing Symmetry Show that for any $n\times n$ matrix **A**, the matrix $\mathbf{A}+\mathbf{A}^{\mathrm{T}}$ is always symmetric.

46. More Symmetry Demonstrate with an arbitrary 3×2 matrix **A** that $\mathbf{A}^{\mathrm{T}}\mathbf{A}$ and $\mathbf{AA}^{\mathrm{T}}$ are always symmetric. (In this case, they are not of the same order.)

Trace of a Matrix *Using the following definition, prove the properties of the trace in Problems 47–50.*

Trace

The **trace** of an $n\times n$ matrix $\mathbf{A}=[a_{ij}]$, denoted $\mathrm{Tr}\,\mathbf{A}$, is the sum of the diagonal elements:

$$\mathrm{Tr}\,\mathbf{A}=a_{11}+a_{22}+\cdots+a_{nn}=\sum_{k=1}^{n}a_{kk}.$$

47. $\mathrm{Tr}(\mathbf{A}+\mathbf{B})=\mathrm{Tr}\,\mathbf{A}+\mathrm{Tr}\,\mathbf{B}$ **48.** $\mathrm{Tr}(c\mathbf{A})=c\,\mathrm{Tr}\,\mathbf{A}$

49. $\mathrm{Tr}(\mathbf{A}^{\mathrm{T}})=\mathrm{Tr}\,\mathbf{A}$ **50.** $\mathrm{Tr}(\mathbf{AB})=\mathrm{Tr}(\mathbf{BA})$

Matrices Can Be Complex *Complex numbers can serve as entries in a matrix just as well as real numbers. Compute the expressions in Problems 51–58, where*

$$\mathbf{A}=\begin{bmatrix}1+i & 2i\\ 2 & 2-3i\end{bmatrix}\quad\textit{and}\quad\mathbf{B}=\begin{bmatrix}1 & -i\\ 2i & 1+i\end{bmatrix}.$$

51. $\mathbf{A}+2\mathbf{B}$ **52.** $\mathbf{AB}$ **53.** $\mathbf{BA}$ **54.** $\mathbf{A}^2$

55. $i\mathbf{A}$ **56.** $\mathbf{A}-2i\mathbf{B}$ **57.** $\mathbf{B}^{\mathrm{T}}$ **58.** $\mathrm{Tr}\,\mathbf{B}$

59. Real and Imaginary Components Any matrix **M** with complex number entries can be written $\mathbf{M}=\mathbf{R}+i\mathbf{S}$, where **R** and **S** are matrices of the same order as **M** but have real number entries. Obtain such decompositions for the matrices

$$\mathbf{A}=\begin{bmatrix}1+i & 2i\\ 2 & 2-3i\end{bmatrix}\quad\text{and}\quad\mathbf{B}=\begin{bmatrix}1 & -i\\ 2i & 1+i\end{bmatrix}.$$

60. Square Roots of Zero Are there any 2×2 matrices **A**, with elements not all zero, satisfying $\mathbf{A}^2=\mathbf{0}$? If so, give an example. If not, explain.

61. Zero Divisors If a and b are real or complex numbers such that $ab=0$, then either $a=0$ or $b=0$. Does this property hold for matrices? That is, if **A** and **B** are $n\times n$ matrices such that $\mathbf{AB}=\mathbf{0}$, is it true that we must have $\mathbf{A}=\mathbf{0}$ or $\mathbf{B}=\mathbf{0}$? Prove the result or find a counterexample. (Please do not do both.)

62. Does Cancellation Work? Suppose that $\mathbf{AB}=\mathbf{AC}$ for matrices **A**, **B**, and **C**. Is it true that **B** must equal **C**? Prove the result or find a counterexample.

63. Taking Matrices Apart Let **A** be an $n\times n$ matrix whose jth column is the column vector ($n\times 1$ matrix) $\mathbf{A}_j$; we can write this as $[\mathbf{A}_1\mid\mathbf{A}_2\mid\cdots\mid\mathbf{A}_n]$ (an example of a **partitioned matrix**). Let $\vec{\mathbf{x}}$ be the column vector $[x_1\ \ x_2\ \ \cdots\ \ x_n]^{\mathrm{T}}$.

(a) For $\mathbf{A} = \begin{bmatrix} 1 & 5 & 2 \\ -1 & 0 & 3 \\ 2 & 4 & 7 \end{bmatrix}$ and $\vec{\mathbf{x}} = \begin{bmatrix} 2 \\ 4 \\ 3 \end{bmatrix}$, verify that

$$\mathbf{A}\vec{\mathbf{x}} = x_1\mathbf{A}_1 + x_2\mathbf{A}_2 + \cdots + x_n\mathbf{A}_n. \quad (8)$$

(b) Show that in general the product of a matrix and a vector can be expressed as a *linear combination of the columns of the matrix*; that is, show that (8) is a true statement independent of the particular matrices in part (a).

Diagonal Matrices *For Problems 64 and 65, suppose that* **A** *and* **B** *are* $n \times n$ *diagonal matrices.*

64. Show that **AB** is diagonal. **65.** Show that **AB** = **BA**.

66. Upper Triangular Matrices A square $n \times n$ matrix $\mathbf{A} = [a_{ij}]$ is called **upper triangular** if $a_{ij} = 0$ for $i > j$. (All entries below the main diagonal are zero.)

(a) Give three examples of upper triangular matrices of different orders.

(b) For the 3×3 case, prove that if **A** and **B** are both upper triangular, then **AB** is also upper triangular.

(c) Prove part (b) for the general case (arbitrary n).

67. Hard Puzzle The **square root** of a matrix **A** is a matrix **R** such that **RR** = **A**. Show that the matrix

$$\mathbf{A} = \begin{bmatrix} 0 & 1 \\ 0 & 0 \end{bmatrix}$$

has *no* square root, while the matrix

$$\mathbf{B} = \begin{bmatrix} 1 & 0 \\ 0 & 1 \end{bmatrix}$$

has an *infinite number* of square roots.

Orthogonality *In Problems 68–71, find the real values of k for which the given vectors are orthogonal. If there are no such values, show why.*

68. $\begin{bmatrix} 1 \\ k \\ 0 \end{bmatrix}, \begin{bmatrix} 1 \\ 2 \\ 3 \end{bmatrix}$ **69.** $\begin{bmatrix} k \\ 2 \\ k \end{bmatrix}, \begin{bmatrix} 1 \\ 0 \\ 4 \end{bmatrix}$

70. $\begin{bmatrix} k \\ 0 \\ k^2 \end{bmatrix}, \begin{bmatrix} 1 \\ 2 \\ 3 \end{bmatrix}$ **71.** $\begin{bmatrix} 1 \\ 2 \\ k^2 \end{bmatrix}, \begin{bmatrix} -1 \\ 1 \\ -1 \end{bmatrix}$

Orthogonality and Subsets *In Problems 72–75, find the subset of* $\mathbb{R}^3$ *that is orthogonal to the given vectors. Sketch the subsets in* $\mathbb{R}^3$.

72. $\begin{bmatrix} 1 \\ 0 \\ 1 \end{bmatrix}$ **73.** $\begin{bmatrix} 1 \\ 0 \\ 1 \end{bmatrix}, \begin{bmatrix} 2 \\ 1 \\ 0 \end{bmatrix}$

74. $\begin{bmatrix} 1 \\ 0 \\ 1 \end{bmatrix}, \begin{bmatrix} 2 \\ 1 \\ 0 \end{bmatrix}, \begin{bmatrix} 3 \\ 4 \\ 5 \end{bmatrix}$ **75.** $\begin{bmatrix} 1 \\ 0 \\ 1 \end{bmatrix}, \begin{bmatrix} 2 \\ 1 \\ 0 \end{bmatrix}, \begin{bmatrix} 0 \\ -1 \\ -2 \end{bmatrix}$

Dot Products *Calculate the dot products of the vectors in Problems 76–81. Tell which pairs are orthogonal.*

76. $[2, 1] \cdot [-1, 2]$ **77.** $[-3, 0] \cdot [2, 1]$

78. $[2, 1, 2] \cdot [3, -1, 0]$ **79.** $[1, 0, -1] \cdot [1, 1, 1]$

80. $[5, 7, 5, 1] \cdot [-2, 4, -3, -3]$

81. $[7, 5, 1, 5] \cdot [4, -3, 2, 3]$

82. Lengths Show that the distance between the heads of any two vectors $\vec{\mathbf{u}}$ and $\vec{\mathbf{v}}$, as shown in Fig. 3.1.5, has length $\|\vec{\mathbf{u}} - \vec{\mathbf{v}}\|$.

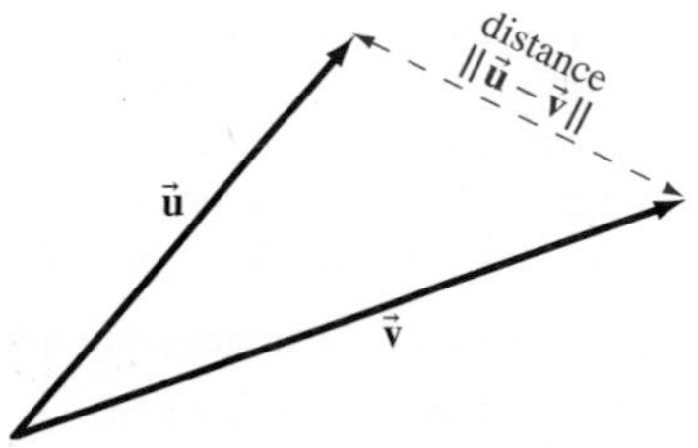

FIGURE 3.1.5 Diagram for Problem 82.

Geometric Vector Operations *For*

$$\mathbf{A} = \begin{bmatrix} 1 \\ 2 \end{bmatrix}, \quad \mathbf{B} = \begin{bmatrix} -3 \\ 1 \end{bmatrix}, \quad and \quad \mathbf{C} = \begin{bmatrix} -3 \\ -2 \end{bmatrix},$$

sketch geometrically the vectors described in Problems 83–85.

83. $\mathbf{A} + \mathbf{C}$ **84.** $\frac{1}{2}\mathbf{A} + \mathbf{B}$ **85.** $\mathbf{A} - 2\mathbf{B}$

Triangles *For the following pairs of vectors* $\vec{\mathbf{u}}$ *and* $\vec{\mathbf{v}}$ *given in Problems 86 and 87, show by an appropriate scalar product whether the triangle formed as in Fig. 3.1.5 is a right triangle. Sketch the triangle and identify the right angle. Confirm the Pythagorean Theorem.*

86. $[3, 2], [2, 3]$ **87.** $[2, -1, 2], [-1, 0, 1]$

Properties of Scalar Products *For Problems 88–91, consider the general scalar product on vectors* $\vec{\mathbf{a}}$, $\vec{\mathbf{b}}$, *and* $\vec{\mathbf{c}}$ *of the same dimension. Either prove the statement to be true or explain why it cannot be true. Take k to be a nonzero scalar.*

88. $\vec{\mathbf{a}} \cdot \vec{\mathbf{b}} \stackrel{?}{=} \vec{\mathbf{b}} \cdot \vec{\mathbf{a}}$ **89.** $\vec{\mathbf{a}} \cdot (\vec{\mathbf{b}} \cdot \vec{\mathbf{c}}) \stackrel{?}{=} (\vec{\mathbf{a}} \cdot \vec{\mathbf{b}}) \cdot \vec{\mathbf{c}}$

90. $k(\vec{\mathbf{a}} \cdot \vec{\mathbf{b}}) \stackrel{?}{=} (k\vec{\mathbf{a}}) \cdot \vec{\mathbf{b}}$ **91.** $\vec{\mathbf{a}} \cdot (\vec{\mathbf{b}} + \vec{\mathbf{c}}) \stackrel{?}{=} \vec{\mathbf{a}} \cdot \vec{\mathbf{b}} + \vec{\mathbf{a}} \cdot \vec{\mathbf{c}}$

92. Directed Graphs A **directed graph** is a finite set of points, called **nodes**, and an associated set of paths or **arcs**, each connecting two nodes in a given direction. (See Fig. 3.1.6.) Think of the arcs as strings or wires that can pass over or under each other; no actual "contact" takes place except at the nodes.

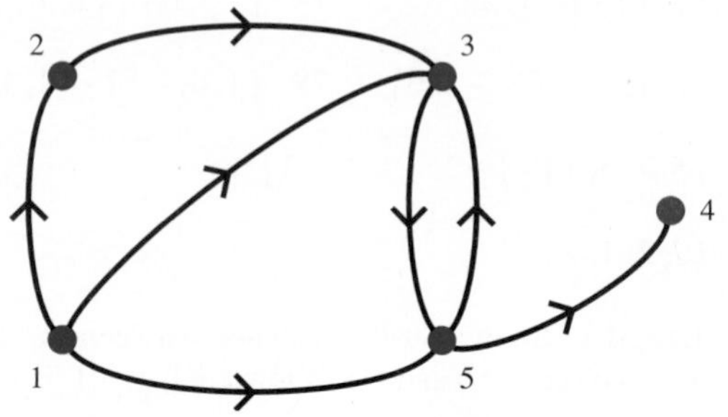

FIGURE 3.1.6 Directed graph (Problem 92).

Two nodes, i and j, are **adjacent** if there is an arc from i to j. (The arc from node 1 to node 3 is distinct from the arc from node 3 to node 1; the graph may include one, both, or neither.) If the graph has n nodes, its **adjacency matrix** is the $n \times n$ matrix $\mathbf{A} = [a_{ij}]$ defined by

$$a_{ij} = \begin{cases} 1 & \text{if there is an arc from node } i \text{ to node } j, \\ 0 & \text{if there is no such arc.} \end{cases}$$

(a) Write out the adjacency matrix for the directed graph in Fig. 3.1.6.

(b) Calculate the square of the adjacency matrix from part (a). What is the interpretation for the graph of an entry in this matrix? HINT: Two "consecutive" arcs, one from node i to node j, another from node j to node k, together form a "path" of length 2 from node i to node k.

93. Tournament Play The directed graph in Fig. 3.1.7 is called a **tournament graph** because every node is connected to every other node exactly once. The nodes represent players, and an arc from node i to node j stands for the fact that player i has beaten player j. Compute the adjacency matrix $\mathbf{T}$ of this tournament graph and rank the players by direct and indirect dominance.

HINT: Look at the meaning of $\mathbf{T}^2$ and $\mathbf{T} + \mathbf{T}^2$.

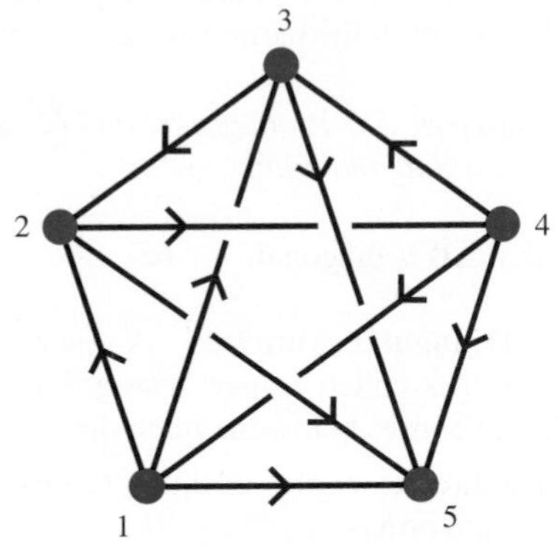

FIGURE 3.1.7 Tournament graph (Problem 93).

94. Suggested Journal Entry Contrast and compare matrix arithmetic with the arithmetic of real numbers. Respond to the following statements:

(a) Matrix arithmetic is built up from ordinary arithmetic because the elements of a matrix are real (or complex) numbers.

(b) Matrix arithmetic is a more general structure than ordinary arithmetic because all of real number arithmetic can be viewed as the special case for 1×1 matrices.

3.2 Systems of Linear Equations

SYNOPSIS: Systems of linear algebraic equations may represent too much, too little, or just the right amount of information to determine values of the variables constituting solutions. Using Gauss-Jordan reduction we can determine whether the system has many solutions, a unique solution, or none at all.

Introductory Example

We want to solve the following system of two linear equations in three variables:

$$\begin{aligned} 3x - 3y + 2z &= 6, \\ 3x + 6y - 2z &= 18. \end{aligned} \tag{1}$$

We suspect that two conditions do not provide enough information to find unique values for the three variables. Does this mean that there are many solutions?

What is a solution of (1)? It is a triple of numbers (x, y, z) that simultaneously satisfies each equation in system (1); each such solution represents a point in

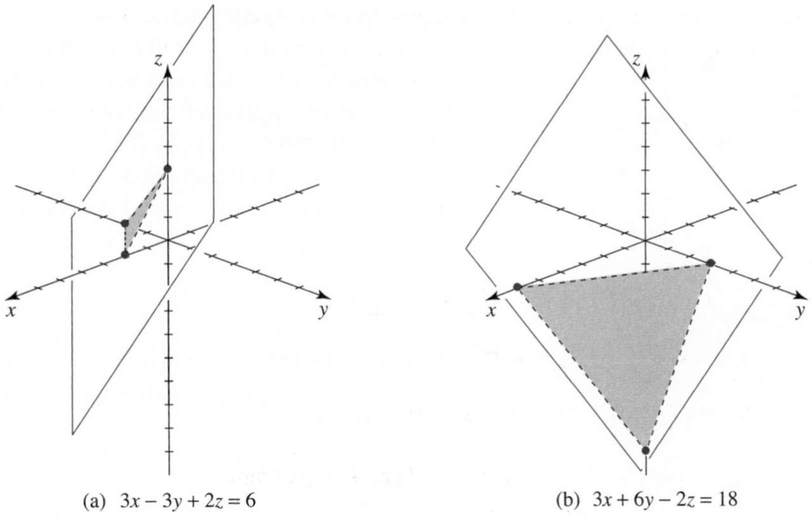

FIGURE 3.2.1 Planes represented by equations in system (1).

three-dimensional space. Since each equation in (1) represents a plane (Fig. 3.2.1), we need to know if these planes intersect, giving an entire line of solution points, or are parallel, resulting in no solutions. A third possibility is that both equations represent the same plane. We can eliminate that possibility by observing that the equations in (1) are not multiples of each other.

- If we replace the second equation of (1) by the result of adding -1 times the first equation to it, the new system of equations,

$$\begin{aligned} 3x - 3y + 2z &= 6, \\ 9y - 4z &= 12, \end{aligned} \tag{2}$$

  is equivalent to the original one: any solution of (1) must satisfy (2); any solution of (2) will also satisfy (1).

- Multiplying the first equation by $1/3$ and the second by $1/9$ also gives an equivalent system:

$$\begin{aligned} x - y + \frac{2}{3}z &= 2, \\ y - \frac{4}{9}z &= \frac{4}{3}. \end{aligned} \tag{3}$$

- Finally, we replace the first equation in (3) with the sum of the two equations, getting the system

$$\begin{aligned} x \quad + \frac{2}{9}z &= \frac{10}{3}, \\ y - \frac{4}{9}z &= \frac{4}{3}. \end{aligned} \tag{4}$$

We can get a solution for any choice of z just by using these formulas to compute the corresponding x- and y-values. To emphasize this, let t be the value chosen for z. Then we will have

$$x = -\frac{2}{9}t + \frac{10}{3}, \quad y = \frac{4}{9}t + \frac{4}{3}, \quad z = t. \tag{5}$$

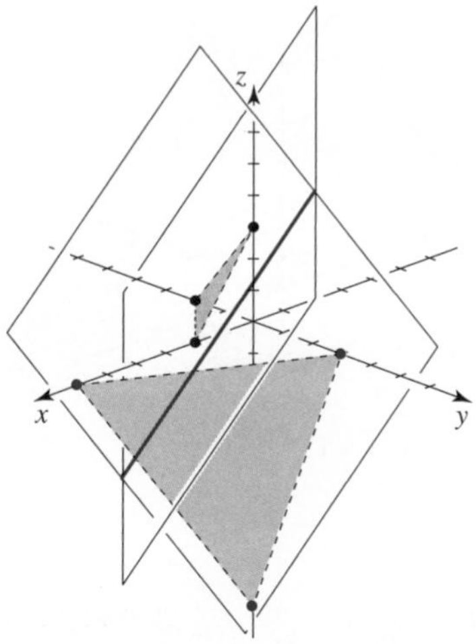

FIGURE 3.2.2 The solution to system (1) is the line of intersection of two planes.

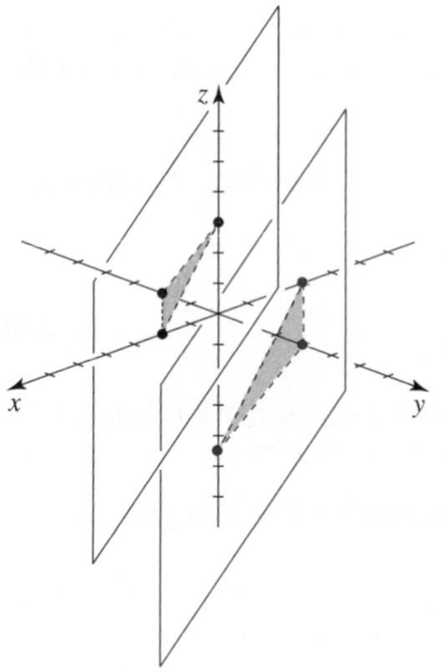

FIGURE 3.2.3 There is *no* solution to system (6) because the planes are parallel.

Equations (5) are parametric equations of the line of intersection of the two planes from the original system (1).[1] (See Fig. 3.2.2.) The original system was **underdetermined**; there was not enough information to determine a unique solution point. But our approach led to an infinite (one-parameter) family of solutions using parametric equations (5).

Had the given system (1) been just a *little* different, however, with a change in two coefficients in the second equation,

$$\begin{aligned} 3x - 3y + 2z &= 6, \\ -6x + 6y - 4z &= 18, \end{aligned} \tag{6}$$

the story would be *quite* different. Replace the second equation in (6) by the sum of that equation and two times the first one. The equivalent system that results is

$$\begin{aligned} 3x - 3y + 2z &= 6, \\ 0 &= 30, \end{aligned}$$

and the contradictory second equation tells us that there are no solutions at all. The two planes represented by the equations in (6) are parallel.[2] (See Fig. 3.2.3.)

These examples, systems (1) and (6), suggest the main questions of this section for linear systems of equations.

1. How can we tell whether a system has solutions?
2. If a system has solutions, how many are there?
3. How do we calculate solutions?

(Compare these questions with those asked about differential equations just before the summary of Sec. 1.5.)

Linear Systems and Matrices

An $m \times n$ **system of linear equations** is a set of m equations in n variables $x_1, x_2, \ldots, x_n$ of the form

$$\begin{aligned} a_{11}x_1 + a_{12}x_2 + \cdots + a_{1n}x_n &= b_1, \\ a_{21}x_1 + a_{22}x_2 + \cdots + a_{2n}x_n &= b_2, \\ &\vdots \\ a_{m1}x_1 + a_{m2}x_2 + \cdots + a_{mn}x_n &= b_m, \end{aligned} \tag{7}$$

where the a_{ij} and b_i are constants and the x_i are the unknown variables.[3] If the b_i are all equal to zero, system (7) is **homogeneous**. A **solution** is a point in $\mathbb{R}^n$ whose coordinates satisfy the system of equations (7).

---

[1]From the equations, we see that the line passes through the point $(10/3, 4/3, 0)$, and a direction vector is given by $[-2/9, 4/9, 1]$.

[2]The vector $[3, -3, 2]$ is perpendicular to both planes.

[3]As in Sec. 2.1 (where we defined linear differential equations), the label "linear" applies to the way the *dependent variables* x_i are treated. The *coefficients* a_{ij}, on the other hand, can be functions (even nonlinear functions) of t. But most of the linear systems of equations that we will consider throughout this text will be those with *constant* coefficients a_{ij}. Some exceptions occur in Chapters 2, 4, and 6, where Euler-Cauchy equations, variation of parameters, and qualitative approaches include the possibility of variable coefficients $a_{ij}(t)$.

Real Number Notation:

$\mathbb{R}^n$ is the set of all ordered n-tuples of real numbers:

$$\mathbb{R}^n = \{(x_1, x_2, \ldots, x_n) \mid x_i \in \mathbb{R}\}.$$

In vector notation, we write

$$\mathbb{R}^n = \left\{ \begin{bmatrix} x_1 \\ x_2 \\ \vdots \\ x_n \end{bmatrix} \middle| \; x_i \in \mathbb{R} \right\}.$$

If we rewrite system (7) as

$$\underbrace{\begin{bmatrix} a_{11} & a_{12} & \cdots & a_{1n} \\ a_{21} & a_{22} & \cdots & a_{2n} \\ \vdots & \vdots & \ddots & \vdots \\ a_{m1} & a_{m2} & \cdots & a_{mn} \end{bmatrix}}_{\mathbf{A}} \underbrace{\begin{bmatrix} x_1 \\ x_2 \\ \vdots \\ x_n \end{bmatrix}}_{\vec{\mathbf{x}}} = \underbrace{\begin{bmatrix} b_1 \\ b_2 \\ \vdots \\ b_m \end{bmatrix}}_{\vec{\mathbf{b}}},$$

we see the compact matrix-vector form

$$\mathbf{A}\vec{\mathbf{x}} = \vec{\mathbf{b}}. \tag{8}$$

System (8) is homogeneous if and only if $\vec{\mathbf{b}}$ is the zero vector.

EXAMPLE 1 Linear System Notation

(a) The system (1) in the introductory example is a 2×3 system. It has the form (8), where

$$\mathbf{A} = \begin{bmatrix} 3 & -3 & 2 \\ 3 & 6 & -2 \end{bmatrix}, \quad \vec{\mathbf{x}} = \begin{bmatrix} x \\ y \\ z \end{bmatrix}, \quad \text{and} \quad \vec{\mathbf{b}} = \begin{bmatrix} 6 \\ 18 \end{bmatrix}.$$

(b) The 3×2 system

$$\begin{aligned} x + y &= 1, \\ x - y &= 0, \\ x + 3y &= 2 \end{aligned} \tag{9}$$

has the form (8), with

$$\mathbf{A} = \begin{bmatrix} 1 & 1 \\ 1 & -1 \\ 1 & 3 \end{bmatrix}, \quad \vec{\mathbf{x}} = \begin{bmatrix} x \\ y \end{bmatrix}, \quad \text{and} \quad \vec{\mathbf{b}} = \begin{bmatrix} 1 \\ 0 \\ 2 \end{bmatrix}.$$

■

Now we ask: What is the meaning of **solutions** to a linear system? For small systems, we gain some insight from geometry, as we did in the introductory examples.

EXAMPLE 2 Three Planes Let us think about a 3×3 system, so that each equation has the form $ax + by + cz = d$. Since this equation represents a plane in three-dimensional space, the nature of the solutions to our system depends on the ways in which three planes in space can intersect one another. Figure 3.2.4 illustrates that the system may have no solutions, many solutions, or a unique solution. (Problems 5–9 explore similar considerations for a 2×2 system.) ■

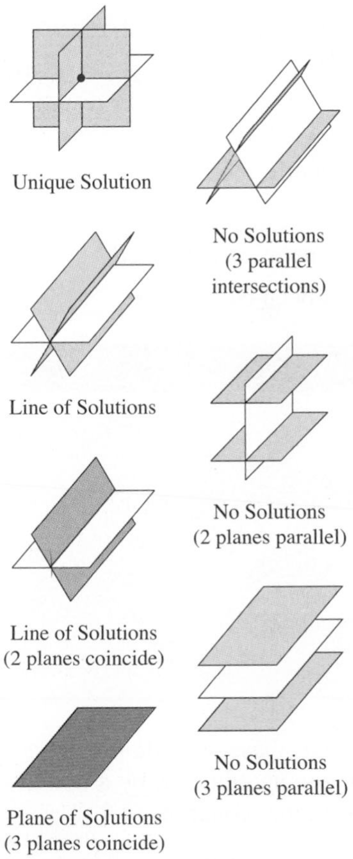

FIGURE 3.2.4 Intersections of three planes (Example 2).

An algebraic approach gives additional insight, as we have seen in the solutions of systems (1) and (6). We have a tendency to suppose that if there are more variables than equations the system will be underdetermined, while more equations than variables makes us expect an overdetermined system. Equal numbers of equations and variables, it seems, ought to be *just right* (so that the solution is uniquely determined). While these informal impressions are sometimes correct, they are often misleading.

System (1) followed the expectation: two equations in three variables, for which the underdetermined system had a one-parameter family of solutions. But system (6), which was also 2×3, had no solutions at all.

System (9) is instructive, too. It has three equations and only two variables, so our "rule of thumb" says it should be overdetermined. It turns out that this is not so. The system has a unique solution $x = 1/2$, $y = 1/2$.

These examples make it clear that we need a more systematic approach. In working with system (9), for example, we need to be able to discover that the third equation equals 2 times the first equation plus -1 times the second and contains no additional information; it is *redundant*. The tool we need is **Gauss-Jordan reduction**, which will enable us to transform systems into their simplest standard form; its basic steps are the following operations.

Elementary Row Operations

We are going to describe three operations for altering a row of a matrix. We will apply these to a matrix associated with a linear system, and the operations will correspond to algebraic manipulation of the equations of the system. Moreover, these manipulations will change the original system into an **equivalent system**—that is, a system that has the same set of solutions. It turns out that there is a strategy for applying these operations that leads to a standard form from which we can easily analyze solutions.

The matrix to which the operations will be applied is called the **augmented matrix** of the system $\mathbf{A}\vec{\mathbf{x}} = \vec{\mathbf{b}}$. It is formed by appending the entries of the column vector $\vec{\mathbf{b}}$ (right-hand side of the equation) to those of the coefficient matrix $\mathbf{A}$, creating a matrix that is now of order $m \times (n+1)$.[4] The augmented matrix of system (7) is

$$[\mathbf{A} \mid \vec{\mathbf{b}}] = \left[\begin{array}{cccc|c} a_{11} & a_{12} & \cdots & a_{1n} & b_1 \\ a_{21} & a_{22} & \cdots & a_{2n} & b_2 \\ \vdots & \vdots & \ddots & \vdots & \vdots \\ a_{m1} & a_{m2} & \cdots & a_{mn} & b_m \end{array}\right]. \tag{10}$$

Each elementary row operation described subsequently produces an equivalent system. In shorthand notation:

- R_i denotes the ith row of the matrix *before* the operation is applied;
- R_i^* stands for the ith row *after* the operation has been carried out.

Elementary Row Operations

- Interchange row i and row j:
$$R_i \leftrightarrow R_j \quad (\text{or } R_i^* = R_j,\ R_j^* = R_i).$$
- Multiply row i by a constant $c \neq 0$:
$$R_i^* = cR_i.$$
- Add c times row j to row i (leaving row j unchanged):
$$R_i^* = R_i + cR_j.$$

Applied to the equations corresponding to the rows, these operations represent *interchanging* the order of the equations, *multiplying* an equation by a nonzero constant, or *adding a multiple* of one equation to another equation.

---

[4]The optional vertical line between the entries of $\mathbf{A}$ and those of $\vec{\mathbf{b}}$ emphasizes the way the matrix is constructed.

By applying these elementary row operations to the augmented matrix of a linear system, we will reduce it to an equivalent system in which the solution is transparent.

This process can be done by many *different* sequences of row operations. We will begin with a straightforward example and illustrate a strategy.

EXAMPLE 3 **Solving an Algebraic System by Row Operations** Consider the system

$$\begin{aligned} x + y + z &= 3, \\ 2x - 3y - z &= -8, \\ -x + 2y + 2z &= 3. \end{aligned}$$

We set up the augmented matrix and proceed to choose row operations that will isolate the solutions.

A Solution Strategy:

Construct the augmented matrix $[\mathbf{A} \mid \vec{\mathbf{b}}]$.

$$\left[\begin{array}{rrr|r} 1 & 1 & 1 & 3 \\ 2 & -3 & -1 & -8 \\ -1 & 2 & 2 & 3 \end{array}\right]$$

Work from left to right to get zeros *below* the main diagonal of **A**.

$$\rightarrow \left[\begin{array}{rrr|r} 1 & 1 & 1 & 3 \\ 0 & 1 & 3 & -2 \\ 0 & 3 & 3 & 6 \end{array}\right] \begin{array}{l} \\ R_2^* = R_2 + 2R_3 \\ R_3^* = R_1 + R_3 \end{array}$$

$$\rightarrow \left[\begin{array}{rrr|r} 1 & 1 & 1 & 3 \\ 0 & 1 & 3 & -2 \\ 0 & 0 & -6 & 12 \end{array}\right] \begin{array}{l} \\ \\ R_3^* = R_3 - 3R_2 \end{array}$$

Multiply by scalars where necessary to make diagonal entries of **A** equal to 1.

$$\rightarrow \left[\begin{array}{rrr|r} 1 & 1 & 1 & 3 \\ 0 & 1 & 3 & -2 \\ 0 & 0 & 1 & -2 \end{array}\right] \begin{array}{l} \\ \\ R_3^* = -\frac{1}{6}R_3 \end{array}$$

Work from left to right to get zeros *above* the main diagonal of **A**.

$$\rightarrow \left[\begin{array}{rrr|r} 1 & 0 & -2 & 5 \\ 0 & 1 & 3 & -2 \\ 0 & 0 & 1 & -2 \end{array}\right] \begin{array}{l} R_1^* = R_1 - R_2 \\ \\ \\ \end{array}$$

$$\rightarrow \left[\begin{array}{rrr|r} 1 & 0 & 0 & 1 \\ 0 & 1 & 0 & 4 \\ 0 & 0 & 1 & -2 \end{array}\right] \begin{array}{l} R_1^* = R_1 + 2R_3 \\ R_2^* = R_2 - 3R_3 \\ \\ \end{array}$$

The final augmented matrix gives the solution:

$$x = 1, \quad y = 4, \quad z = -2. \qquad \blacksquare$$

Uniqueness of the RREF:

For any given matrix **A**, any sequence of legitimate row operations will result in the same RREF. However, many different matrices can have the same RREF.

For a hand calculation like Example 3, we could easily have chosen some different row operations, even for the same overall strategy, and, for different systems of algebraic equations, different row operations will be easier. Obviously, this process can become tedious and prone to arithmetic errors in the hands of mortals, but the idea is perfectly suited to computers. Using the key *ideas*, we proceed to formalize the computational process and make it sufficiently robust to handle all the special cases that may be encountered.

The form of the final augmented matrix of Example 3 is called **reduced row echelon form** (usually abbreviated as RREF). This form will be the goal of all such computations throughout this text.

Reduced Row Echelon Form (RREF)

A matrix is in reduced row echelon form if the following conditions are satisfied.

(i) Zero rows are at the bottom.

(ii) The leftmost nonzero entry of each nonzero row equals 1. This entry is called its **pivot** or **leading 1**.

(iii) Each pivot is further to the right than the pivot in the row above it.

(iv) Each pivot is the only nonzero entry in its column.

A column of any matrix **A** is called a **pivot column** if it corresponds to a column with a leading 1 in its RREF. The number of pivots is the same as the number of pivot columns.

RREF will produce an equivalent system from which the solution can be read immediately.

In Example 3 we saw

$$[\mathbf{A} \mid \vec{\mathbf{b}}] = \left[\begin{array}{rrr|r} 1 & 1 & 1 & 3 \\ 2 & -3 & -1 & -8 \\ -1 & 2 & 2 & 3 \end{array}\right]$$

had RREF

$$\left[\begin{array}{rrr|r} 1 & 0 & 0 & 1 \\ 0 & 1 & 0 & 4 \\ 0 & 0 & 1 & -2 \end{array}\right].$$

The pivot columns for both matrices (and all those between) are the first three columns, accented by shading. (Once we have found the pivot columns in the RREF, we can shade the pivot columns of **A** as well.)

A less complete process results in **row echelon form**, which differs from *reduced* row echelon form by weakening condition (iv) so that entries above the pivot are allowed to be nonzero. The system in Example 3 was in *row echelon form* after the third row operation:

$$\left[\begin{array}{rrr|r} 1 & 1 & 1 & 3 \\ 0 & 1 & 3 & -2 \\ 0 & 0 & 1 & -2 \end{array}\right].$$

EXAMPLE 4 **Reduced Row Echelon Form** Some of the following matrices are already in RREF form, while others fail to satisfy all four conditions listed above. (The matrix in (d) is in row echelon form, but not in *reduced* row echelon form.)

(a) $\begin{bmatrix} 1 & 0 & 0 \\ 0 & 1 & 0 \\ 0 & 0 & 1 \end{bmatrix}$ RREF with 3 pivot columns, as indicated

(b) $\begin{bmatrix} 1 & 0 & 2 \\ 0 & 1 & 0 \\ 0 & 0 & 0 \end{bmatrix}$ RREF with 2 pivot columns, as indicated

(c) $\begin{bmatrix} 1 & 2 & 0 & 3 & 2 \\ 0 & 0 & 2 & 2 & 2 \\ 0 & 0 & 0 & 0 & 0 \end{bmatrix}$ Rule (ii) violated in (2,3)-position

(d) $\begin{bmatrix} 1 & 2 \\ 0 & 1 \\ 0 & 0 \end{bmatrix}$ Rule (iv) violated in (1,2)-position

(e) $\begin{bmatrix} 1 & 0 & 0 & 1 \\ 0 & 1 & 0 & 2 \\ 0 & 0 & 0 & 0 \\ 0 & 0 & 0 & 0 \end{bmatrix}$ RREF with 2 pivot columns

(f) $\begin{bmatrix} 1 & 0 & 0 & 0 & 0 \\ 0 & 1 & 0 & 0 & 0 \\ 0 & 0 & 1 & 0 & 0 \\ 0 & 0 & 1 & 1 & 0 \end{bmatrix}$ Rules (iii) and (iv) violated, row 4

■

Gauss-Jordan Reduction

We now need a specific scheme or strategy that can be applied to a given matrix to force it to transform into RREF. The procedure called **Gauss-Jordan reduction**[5] will always work, and the result will always be the same: the RREF is unique.

The **Gauss-Jordan Algorithm** tells us how to use the RREF to solve algebraic systems of linear equations.

Gauss-Jordan Reduction Algorithm

The following procedure will solve the linear system $\mathbf{A}\vec{\mathbf{x}} = \vec{\mathbf{b}}$.

Step 1. Form the augmented matrix $[\mathbf{A} \mid \vec{\mathbf{b}}]$.

Step 2. Transform the augmented matrix to reduced row echelon form (RREF) using the elementary row operations.

Step 3. The linear system that corresponds to the matrix in reduced row echelon form, which was obtained in Step 2, has exactly the same solutions as the given linear system. For each nonzero row of the matrix in RREF solve for the unknown that corresponds to the leading 1 in the row. The rows consisting of all zeros can be ignored, because the corresponding equation is satisfied for any values of the variables.

Existence and Uniqueness of Solutions from the RREF

When an augmented matrix $[\mathbf{A} \mid \vec{\mathbf{b}}]$ is in RREF, we can inspect it for answers to our initial questions about the existence and uniqueness of solutions to the linear system $\mathbf{A}\vec{\mathbf{x}} = \vec{\mathbf{b}}$.

Are There Any Solutions?

An *inconsistent* system has no solutions, as in Example 5. *Consistent* systems are given in Example 3 and the introductory example, system (1).

Existence of Solutions to a Linear System

If the RREF of a linear system $\mathbf{A}\vec{\mathbf{x}} = \vec{\mathbf{b}}$ contains a row of the form

$$[0 \ \cdots \ 0 \mid k], \quad \text{where } k \text{ is nonzero,}$$

then the system has *no* solutions. (Such a row would require that we have $0x_1 + 0x_2 + \cdots + 0x_n = k \neq 0$, which is impossible.)

- If a system has no solutions it is called **inconsistent**.
- If a system has one or more solutions it is called **consistent**.

EXAMPLE 5 **Inconsistent System** We now consider the system

$$\begin{aligned} x + \ \ y + z &= 1, \\ x + 2y + z &= 4, \\ x + \ \ y + z &= 2, \end{aligned}$$

which has the augmented matrix

$$\left[\begin{array}{ccc|c} 1 & 1 & 1 & 1 \\ 1 & 2 & 1 & 4 \\ 1 & 1 & 1 & 2 \end{array}\right].$$

[5] See the historical note at the end of this section.

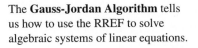

The RREF for an *inconsistent system* has at least one row

$$[0 \cdots 0 \mid k],$$

where k is nonzero.

Its reduced row echelon form is the matrix

$$\left[\begin{array}{ccc|c} 1 & 0 & 1 & 0 \\ 0 & 1 & 0 & 0 \\ 0 & 0 & 0 & 1 \end{array}\right].$$

The third equation of this equivalent system is $0 = 1$, a contradiction that tells us the system has no solutions; it is *inconsistent*. ■

If There Is a Solution, Is It the Only One?

Example 3 gave a *unique* solution, whereas system (1) of the introductory example is *underdetermined.*

Uniqueness of Solutions to a Linear System

A linear system $\mathbf{A}\vec{\mathbf{x}} = \vec{\mathbf{b}}$ must be *consistent* to have *any* solutions.

- If every column in the RREF is a pivot column, then there is only one solution, a **unique solution**.
- If one or more columns in the RREF is a nonpivot column, then there are infinitely many solutions. The system is **underdetermined.**

We have shown how we can obtain solutions to a linear system from the RREF by solving for each variable x_i that corresponds to a pivot column.

- The variables that correspond to a pivot column are called **basic** or **leading variables**. If we must solve them in terms of the remaining variables, which correspond to the nonpivot columns, then there are infinitely many solutions.
- The variables corresponding to the nonpivot columns are called **free variables**. They act as parameters and can be chosen to be any real number. Each choice corresponds to a distinct solution.

We appeal to the linear algebra principles from Section 3.1 to continue this discussion.

Superposition, Nonhomogeneous Principle, and RREF

For any $m \times n$ matrix $\mathbf{A}$, the function $L : \mathbb{R}^n \to \mathbb{R}^m$ defined by $L(\vec{\mathbf{x}}) = \mathbf{A}\vec{\mathbf{x}}$ is a linear operator. This fact follows directly from the Properties of Matrix Multiplication in Section 3.1; that is, for any $\vec{\mathbf{x}}, \vec{\mathbf{y}} \in \mathbb{R}^n$ and $c \in \mathbb{R}$,

$$L(\vec{\mathbf{x}} + \vec{\mathbf{y}}) = \mathbf{A}(\vec{\mathbf{x}} + \vec{\mathbf{y}}) = \mathbf{A}\vec{\mathbf{x}} + \mathbf{A}\vec{\mathbf{y}} = L(\vec{\mathbf{x}}) + L(\vec{\mathbf{y}}), \qquad \textit{(distributive property)}$$

$$cL(\vec{\mathbf{x}}) = c(\mathbf{A}\vec{\mathbf{x}}) = \mathbf{A}(c\vec{\mathbf{x}}) = L(c\vec{\mathbf{x}}). \qquad \textit{(multiplication by a scalar)}$$

Therefore the Superposition Principle and the Nonhomogeneous Principle apply to the operator L.

These principles allow us to write solutions of a nonhomogenous linear system $\mathbf{A}\vec{\mathbf{x}} = \vec{\mathbf{b}}$ *as*

$$\vec{\mathbf{x}} = \vec{\mathbf{x}}_h + \vec{\mathbf{x}}_p,$$

where $\vec{\mathbf{x}}_h$ *represents vectors in the set of all solutions of the associated homogeneous equation* $\mathbf{A}\vec{\mathbf{x}} = \vec{\mathbf{0}}$, *and* $\vec{\mathbf{x}}_p$ *is a particular solution to the original nonhomogeneous equation.* We use this format for solutions of consistent, underdetermined systems.

We can use the RREF of the augmented matrix $[\mathbf{A} \mid \vec{\mathbf{b}}]$ to find $\vec{\mathbf{x}}_p$, and then use the same RREF, with $\vec{\mathbf{b}}$ replaced by $\vec{\mathbf{0}}$, to find $\vec{\mathbf{x}}_h$.

EXAMPLE 6 **Revisiting Our Introductory Example** Writing linear system (1) in augmented matrix form, we find its RREF in standard fashion:

$$\left[\begin{array}{rrr|r} 3 & -3 & 2 & 6 \\ 3 & 6 & -2 & 18 \end{array}\right] \rightarrow \left[\begin{array}{rrr|r} 3 & -3 & 2 & 6 \\ 0 & 9 & -4 & 12 \end{array}\right] \quad R_2^* = R_2 + (-1)R_1$$

$$\rightarrow \left[\begin{array}{rrr|r} 1 & -1 & 2/3 & 2 \\ 0 & 1 & -4/9 & 4/3 \end{array}\right] \quad \begin{array}{l} R_1^* = (1/3)R_1 \\ R_2^* = (1/9)R_2 \end{array}$$

$$\rightarrow \left[\begin{array}{rrr|r} 1 & 0 & 2/9 & 10/3 \\ 0 & 1 & -4/9 & 4/3 \end{array}\right] \quad R_1^* = R_1 + R_2$$

which gives the equivalent linear system, matching (4), as

$$\begin{aligned} x \quad + \frac{2}{9}z &= \frac{10}{3}, \\ y - \frac{4}{9}z &= \frac{4}{3}. \end{aligned}$$

We can see that x and y are basic variables and z is a free variable, which gives us immediately, if we set $z = 0$,

$$\vec{\mathbf{x}}_p = \begin{bmatrix} 10/3 \\ 4/3 \\ 0 \end{bmatrix}.$$

The RREF for the associated homogeneous system can be obtained by replacing the rightmost column of the RREF with the zero vector:

$$\left[\begin{array}{rrr|r} 1 & 0 & 2/9 & 0 \\ 0 & 1 & -4/9 & 0 \end{array}\right].$$

The Superposition Principle gives

$$\vec{\mathbf{x}}_h = t \begin{bmatrix} 2/9 \\ -4/9 \\ 1 \end{bmatrix}, \quad -\infty < t < \infty.$$

The Nonhomogeneous Principle gives us the result that every solution to (1) can be written as $\vec{\mathbf{x}} = \vec{\mathbf{x}}_h + \vec{\mathbf{x}}_p$, which we can write simply as

$$\vec{\mathbf{x}} = t \begin{bmatrix} 2/9 \\ -4/9 \\ 1 \end{bmatrix} + \begin{bmatrix} 10/3 \\ 4/3 \\ 0 \end{bmatrix}, \quad -\infty < t < \infty.$$

Compare with our previous solution (5). The solution set is:

$$\left\{\vec{\mathbf{x}} \in \mathbb{R}^3 \mid \mathbf{A}\vec{\mathbf{x}} = \vec{\mathbf{b}}\right\} = \left\{ \begin{bmatrix} 10/3 \\ 4/3 \\ 0 \end{bmatrix} + t \begin{bmatrix} 2/9 \\ -4/9 \\ 1 \end{bmatrix} \middle| \, t \in \mathbb{R} \right\}.$$

■

EXAMPLE 7 **Inspecting the RREF** Suppose a system $\mathbf{A}\vec{\mathbf{x}} = \vec{\mathbf{b}}$ has an augmented matrix $[\mathbf{A} \mid \vec{\mathbf{b}}]$ in RREF of the form

$$\left[\begin{array}{cccccc|c} 1 & 0 & 0 & 0 & 1 & 0 & 4 \\ 0 & 1 & 0 & 0 & 0 & 0 & 3 \\ 0 & 0 & 1 & 0 & 0 & 2 & 2 \\ 0 & 0 & 0 & 1 & 2 & 3 & 0 \\ 0 & 0 & 0 & 0 & 0 & 0 & 0 \end{array}\right].$$

We know immediately that this is another underdetermined system, because there are more variables than equations, so we look to the RREF for details.

By inspection we can see that there are four pivot columns and two nonpivot columns, so that x_1, x_2, x_3, and x_4 are basic variables, and x_5 and x_6 are free variables. The solution is given by the corresponding equations

$$\begin{aligned} x_1 \quad + x_5 \quad &= 4, \\ x_2 \quad &= 3, \\ x_3 \quad + 2x_6 &= 2, \\ x_4 + 2x_5 + 3x_6 &= 0, \end{aligned} \qquad \text{or} \qquad \begin{aligned} x_1 &= -x_5 + 4, \\ x_2 &= 3, \\ x_3 &= -2x_6 + 2, \\ x_4 &= -2x_5 - 3x_6, \end{aligned}$$

where x_5 and x_6 can be any real numbers. It is standard practice to replace the variables x_5 and x_6 by r and s, respectively, to emphasize the parametric nature of the solution. We thus obtain the family of solutions

$$x_1 = -r + 4, \quad x_2 = 3, \quad x_3 = -2s + 2, \quad x_4 = -2r - 3s,$$

where r and s can be any real numbers. We can summarize as follows:

$$\vec{\mathbf{x}} = \vec{\mathbf{x}}_h + \vec{\mathbf{x}}_p = r\begin{bmatrix} -1 \\ 0 \\ 0 \\ -2 \\ 1 \\ 0 \end{bmatrix} + s\begin{bmatrix} 0 \\ 0 \\ -2 \\ -3 \\ 0 \\ 1 \end{bmatrix} + \begin{bmatrix} 4 \\ 3 \\ 2 \\ 0 \\ 0 \\ 0 \end{bmatrix}, \quad \text{for any } r, s \in \mathbb{R}.$$

Geometrically, this result represents a two-dimensional hyperplane in a six-dimensional space. The solution set is:

$$\left\{\vec{\mathbf{x}} \in \mathbb{R}^6 \mid \mathbf{A}\vec{\mathbf{x}} = \vec{\mathbf{b}}\right\} = \left\{ \begin{bmatrix} 4 \\ 3 \\ 2 \\ 0 \\ 0 \\ 0 \end{bmatrix} + r\begin{bmatrix} -1 \\ 0 \\ 0 \\ -2 \\ 1 \\ 0 \end{bmatrix} + s\begin{bmatrix} 0 \\ 0 \\ -2 \\ -3 \\ 0 \\ 1 \end{bmatrix} \,\middle|\, r, s \in \mathbb{R} \right\}.$$

■

Check the Calculation:
In matrix-vector form, we have found

$$\underbrace{\begin{bmatrix} 1 & 1 & 1 \\ 2 & -3 & -1 \\ -1 & 2 & 2 \end{bmatrix}}_{\mathbf{A}} \underbrace{\begin{bmatrix} 1 \\ 4 \\ -2 \end{bmatrix}}_{\vec{\mathbf{x}}} = \underbrace{\begin{bmatrix} 3 \\ 8 \\ -3 \end{bmatrix}}_{\vec{\mathbf{b}}}.$$

EXAMPLE 8 **Unique Solution** We can apply the Superposition and Nonhomogeneous Principles to Example 3, where for $\mathbf{A}\vec{\mathbf{x}} = \vec{\mathbf{b}}$ we have

$$[\mathbf{A} \mid \vec{\mathbf{b}}] = \left[\begin{array}{ccc|c} 1 & 1 & 1 & 3 \\ 2 & -3 & -1 & -8 \\ -1 & 2 & 2 & 3 \end{array}\right] \quad \text{and its RREF} \quad \left[\begin{array}{ccc|c} 1 & 0 & 0 & 1 \\ 0 & 1 & 0 & 4 \\ 0 & 0 & 1 & -2 \end{array}\right].$$

This gives a particular solution

$$\vec{\mathbf{x}}_p = \begin{bmatrix} 1 \\ 4 \\ -2 \end{bmatrix},$$

as found in Example 3, but is there also an $\vec{\mathbf{x}}_h$?

The associated homogeneous equation $\mathbf{A}\vec{\mathbf{x}} = \vec{\mathbf{0}}$ has RREF

$$\left[\begin{array}{ccc|c} 1 & 0 & 0 & 0 \\ 0 & 1 & 0 & 0 \\ 0 & 0 & 1 & 0 \end{array}\right],$$

which does give another unique solution, simply the zero vector. Consequently,

$$\vec{\mathbf{x}} = \vec{\mathbf{x}}_h + \vec{\mathbf{x}}_p = \vec{\mathbf{0}} + \begin{bmatrix} 1 \\ 4 \\ -2 \end{bmatrix} = \begin{bmatrix} 1 \\ 4 \\ -2 \end{bmatrix}.$$

Thus we see that $\vec{\mathbf{x}} = \vec{\mathbf{x}}_p$ indeed solves $\mathbf{A}\vec{\mathbf{x}} = \vec{\mathbf{b}}$ uniquely and completely. ■

The solution set for Example 8 is

$$\{\vec{\mathbf{x}} \in \mathbb{R}^3 \mid \mathbf{A}\vec{\mathbf{x}} = \vec{\mathbf{b}}\} = \left\{\begin{bmatrix} 1 \\ 4 \\ -2 \end{bmatrix}\right\}.$$

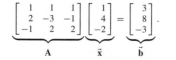

EXAMPLE 9 **Geometric Interpretation** Let us show how the Superposition and Nonhomogeneous Principles can be interpreted geometrically as well as algebraically. We will consider the nonhomogeneous linear algebraic equation

$$x + y = 1, \tag{11}$$

with

$$\vec{\mathbf{x}} = \begin{bmatrix} x \\ y \end{bmatrix}.$$

We know that this underdetermined but consistent system has infinitely many solutions; an example is shown in Fig. 3.2.5.

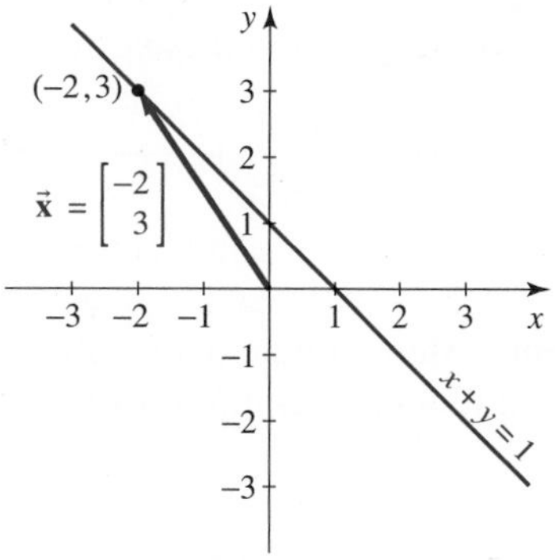

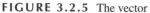

FIGURE 3.2.5 The vector $\begin{bmatrix} -2 \\ 3 \end{bmatrix}$ represents *one* solution to equation (11).

1. All solutions of the corresponding homogeneous equation $x + y = 0$ lie on a line through the origin and have coordinates $(c, -c)$; in vector form,

$$\vec{\mathbf{x}}_h = c\begin{bmatrix} 1 \\ -1 \end{bmatrix},$$

where c is an arbitrary constant.

2. Now pick an arbitrary solution of nonhomogeneous equation (11), say,

$$\vec{\mathbf{x}}_p = \begin{bmatrix} 0 \\ 1 \end{bmatrix}.$$

3. Adding the homogeneous solutions $\vec{\mathbf{x}}_h$ to this particular solution $\vec{\mathbf{x}}_p$ gives the general solution of equation (11):

$$\vec{\mathbf{x}} = \begin{bmatrix} x \\ y \end{bmatrix} = c\begin{bmatrix} 1 \\ -1 \end{bmatrix} + \begin{bmatrix} 0 \\ 1 \end{bmatrix}.$$

This is illustrated in Fig. 3.2.6 that follows.

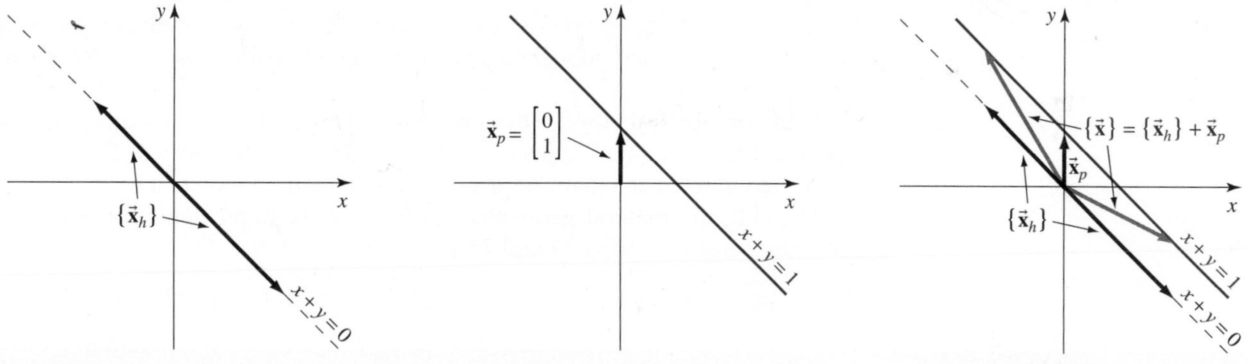

(A) Solutions of corresponding homogeneous equation $x + y = 0$ are vectors $\vec{\mathbf{x}}_h = c\begin{bmatrix} 1 \\ -1 \end{bmatrix}$ along this line.

(B) *Any* particular solution of the nonhomogeneous equation is chosen; we take $\vec{\mathbf{x}}_p = \begin{bmatrix} 0 \\ 1 \end{bmatrix}$.

(C) Adding the particular solution $\vec{\mathbf{x}}_p$ to *each* solution $\vec{\mathbf{x}}_h$ of the homogeneous system gives the collection of all solutions of the original equation—in other words, points on the line represented by position vectors $\vec{\mathbf{x}}$.

FIGURE 3.2.6 Decomposing two sample solutions of the nonhomogeneous linear algebraic equation, $x + y = 1$, for Example 9.

In other words, *the general solution of nonhomogeneous algebraic equation (11) is the general solution of the homogeneous equation plus a particular solution.*

Choosing a different particular solution such as (2, −1) would still give the same general solution, but the constant c would have to be adjusted. ■

Rank of a Matrix

As we have seen in Examples 6–9, in a consistent system nonunique solutions arise when one or more variables fail to correspond to a pivot column. Hence it is useful to focus on the number of pivot columns.

An augmented matrix $[\mathbf{A} \mid \vec{\mathbf{b}}]$ in RREF has exactly the same number of nonzero rows as pivot columns, so the rank is also the number of nonzero rows.

Rank

The **rank** r of a matrix equals the number of pivot columns in the RREF.

- If r is equal to the number of variables, there is a unique solution.
- If r is less than the number of variables, solutions are not unique.

The concept of rank is common in the language of linear algebra. We will encounter it again in Sections 3.6 and 5.2.

Historical Note

Systems of linear equations can be found in ancient Babylonian and Chinese texts dating back more than 2,000 years. Problem 73 restates such a system in today's mathematical notation.

Karl Friedrich Gauss (1777–1855) was a German mathematician and scientist, sometimes called the "prince of mathematicians" and often ranked as one of the three greatest mathematicians of all time (along with Newton and Archimedes). He gave a proof of the fundamental theorem of algebra in his doctoral dissertation, and published a groundbreaking work on number theory at the age of 24.

Wilhelm Jordan (1842–1899) was a German engineer whose contribution to solving linear equations appeared in his *Handbook of Geodesy* in 1888.

Although hand calculation for Gauss-Jordan elimination can be tedious, once understood, the method gives an algorithm easily implemented in a computer program. (See Problems 74 and 75.)

Summary

We have solved linear systems of equations by forming the augmented matrix of the system and finding its reduced row echelon form (RREF) using Gauss-Jordan reduction. This procedure leads to a simplified equivalent system whose solutions can be analyzed readily. The result is either a unique solution, a family of solutions with one or more parameters, or no solutions at all (inconsistent system). We have found that the Superposition and Nonhomogeneous Principles are an integral part of the process.

3.2 Problems

Matrix-Vector Form *Write each system in Problems 1–4 in matrix-vector form. Then write the augmented matrix of the system.*

1. $\begin{aligned} x + 2y &= 1 \\ 2x - y &= 0 \\ 3x + 2y &= 1 \end{aligned}$

2. $\begin{aligned} i_1 + 2i_2 + i_3 + 3i_4 &= 2 \\ i_1 - 3i_2 + 3i_3 &= 1 \end{aligned}$

3. $\begin{aligned} r + 2s + t &= 1 \\ r - 3s + 3t &= 1 \\ 4s - 5t &= 3 \end{aligned}$

4. $x_1 + 2x_2 - 3x_3 = 0$

Solutions in $\mathbb{R}^2$ *In Problems 5–9, match the systems of two linear equations in two variables with the following cases:*

(A) *intersecting lines*
(B) *parallel lines*
(C) *a single line*

5. $\begin{aligned} x + y &= 5 \\ 2x - y &= 4 \end{aligned}$

6. $\begin{aligned} 3x + 2y &= 6 \\ 9x + 6y &= 6 \end{aligned}$

7. $\begin{aligned} x - 2y &= 6 \\ -3x + 6y &= -18 \end{aligned}$

8. $\begin{aligned} x - 2y &= 6 \\ -3x + 6y &= 12 \end{aligned}$

9. $\begin{aligned} 2x - 5y &= 10 \\ x + y &= 2 \end{aligned}$

10. A Special Solution Set in $\mathbb{R}^3$ Write a system of three equations in three unknowns for which the solutions form a single plane in $\mathbb{R}^3$. Determine the parametric equation of the plane.

Reduced Row Echelon Form *Determine whether each of the matrices given in Problems 11–19 is in RREF or not. If not, explain which condition or conditions fail. Then use elementary row operations to obtain the RREF.*

11. $\begin{bmatrix} 1 & 0 \\ 0 & 0 \end{bmatrix}$

12. $\begin{bmatrix} 1 & 0 & 1 \\ 0 & 0 & 1 \end{bmatrix}$

13. $\begin{bmatrix} 1 & 3 & 1 \\ 0 & 0 & 2 \end{bmatrix}$

14. $\begin{bmatrix} 0 & 1 & 2 \\ 0 & 1 & 2 \\ 0 & 2 & 3 \end{bmatrix}$

15. $\begin{bmatrix} 1 & 4 & 3 \\ 0 & 0 & 0 \\ 0 & 0 & 0 \end{bmatrix}$

16. $\begin{bmatrix} 1 & 0 & 1 \\ 0 & 1 & 0 \\ 0 & 0 & 1 \end{bmatrix}$

17. $\begin{bmatrix} 1 & 2 & 3 & 1 \\ 0 & 1 & 3 & 2 \\ 0 & 0 & 0 & 1 \\ 0 & 0 & 0 & 0 \end{bmatrix}$

18. $\begin{bmatrix} 1 & 3 & 2 & 1 \\ 0 & 0 & 0 & 0 \\ 0 & 0 & 0 & 0 \\ 0 & 0 & 0 & 0 \end{bmatrix}$

19. $\begin{bmatrix} 1 & 3 & 0 & -1 & 0 \\ 0 & 0 & 1 & -2 & 0 \\ 0 & 0 & 0 & 0 & 1 \end{bmatrix}$

Gauss-Jordan Reduction *Use elementary row operations to reduce each matrix in Problems 20–23 to row echelon form and then to RREF. Then circle the pivot columns in the original matrix.*

20. $\begin{bmatrix} 1 & 3 & 8 & 0 \\ 0 & 1 & 2 & 1 \\ 0 & 1 & 2 & 4 \end{bmatrix}$

21. $\begin{bmatrix} 0 & 0 & 2 & 2 & -2 \\ 2 & 2 & 6 & 14 & 4 \end{bmatrix}$

22. $\begin{bmatrix} 1 & 0 & 0 \\ 2 & 4 & 6 \\ 5 & 8 & 12 \\ 0 & 8 & 12 \end{bmatrix}$

23. $\begin{bmatrix} 1 & 2 & 3 & 1 \\ 3 & 7 & 10 & 4 \\ 2 & 4 & 6 & 2 \end{bmatrix}$

Solving Systems *Use Gauss-Jordan reduction to transform the augmented matrix of each system in Problems 24–36 to RREF. Use it to discuss the solutions of the system (i.e., no solutions, a unique solution, or infinitely many solutions).*

24. $\begin{aligned} x + y &= 4 \\ x - y &= 0 \end{aligned}$

25. $\begin{aligned} y &= 2x \\ y &= x + 3 \end{aligned}$

26. $\begin{aligned} x + y + z &= 0 \\ y + z &= 1 \end{aligned}$

27. $\begin{aligned} 2x + 4y - 2z &= 0 \\ 5x + 3y &= 0 \end{aligned}$

28. $\begin{aligned} x - y - 2z &= 1 \\ 2x + 3y + z &= 2 \\ 5x + 4y + 2z &= 4 \end{aligned}$

29. $\begin{aligned} x_1 + 4x_2 - 5x_3 &= 0 \\ 2x_1 - x_2 + 8x_3 &= 9 \end{aligned}$

30. $\begin{aligned} x + z &= 2 \\ 2x - 3y + 5z &= 4 \\ 3x + 2y - z &= 4 \end{aligned}$

31. $\begin{aligned} x - y + z &= 0 \\ x + y &= 0 \\ x + 2y - z &= 0 \end{aligned}$

32. $\begin{aligned} x_1 + x_2 + 2x_3 &= 0 \\ 2x_1 - x_2 + x_3 &= 0 \\ 4x_1 + x_2 + 5x_3 &= 0 \end{aligned}$

33. $\begin{aligned} x_1 + x_2 + 2x_3 &= 1 \\ 2x_1 - x_2 + x_3 &= 2 \\ 4x_1 + x_2 + 5x_3 &= 4 \end{aligned}$

34. $\begin{aligned} x + 2y + z &= 2 \\ 2x - 4y - 3z &= 0 \\ -x + 6y - 4z &= 2 \\ x - y &= 4 \end{aligned}$

35. $\begin{aligned} x + 2y + z &= 2 \\ x - y &= 4 \\ 2x - y + 2z &= 0 \\ 3y + z &= -2 \end{aligned}$

36. $\begin{aligned} x_1 + 2x_3 - 4x_4 &= 1 \\ x_2 + x_3 - 3x_4 &= 2 \end{aligned}$

Using the Nonhomogeneous Principle *Determine the solution set $\mathbb{W}$ for the associated homogeneous systems in Problems 37–49. Then write the solutions to the systems in the original problems in the form $\vec{\mathbf{x}} = \vec{\mathbf{x}}_p + \vec{\mathbf{x}}_h$, where $\vec{\mathbf{x}}_h \in \mathbb{W}$.*

37. Problem 24 **38.** Problem 25 **39.** Problem 26

40. Problem 27 **41.** Problem 28 **42.** Problem 29

43. Problem 30 **44.** Problem 31 **45.** Problem 32

46. Problem 33 **47.** Problem 34 **48.** Problem 35

49. Problem 36

50. The RREF Example Consider the system

$$\begin{aligned} x_1 \quad + 2x_3 \quad + x_5 + 4x_6 &= 8,\\ 2x_2 \quad - 2x_4 - 4x_5 - 6x_6 &= 6,\\ x_3 \quad + 2x_6 &= 2,\\ 3x_1 \quad + x_4 + 5x_5 + 3x_6 &= 12,\\ - 2x_2 &= -6. \end{aligned}$$

Show that the RREF of its augmented matrix $[\mathbf{A} \mid \vec{\mathbf{b}}]$ is given in Example 6.

51. More Equations Than Variables Consider the system

$$\begin{aligned} 3x_1 + 5x_2 \quad &= 1,\\ 3x_1 + 7x_2 + 3x_3 &= 8,\\ 5x_2 \quad &= -5,\\ 2x_2 + 3x_3 &= 7,\\ x_1 + 4x_2 + x_3 &= 1. \end{aligned}$$

Find the RREF of the augmented matrix. Determine whether or not the system is consistent. If it is, find the solution or solutions, and give the subset $\mathbb{W}$ for the associated homogeneous system.

52. Consistency Can the system $\mathbf{A}\vec{\mathbf{x}} = \vec{\mathbf{0}}$ be inconsistent? (Assume that $\mathbf{A}$, $\vec{\mathbf{x}}$, and $\vec{\mathbf{0}}$ have the correct orders for $\mathbf{A}\vec{\mathbf{x}} = \vec{\mathbf{0}}$.) Explain why or why not.

Homogeneous Systems *In Problems 53–55, determine all the solutions of* $\mathbf{A}\vec{\mathbf{x}} = \vec{\mathbf{0}}$, *where the matrix shown is the RREF of the augmented matrix* $[\mathbf{A} \mid \vec{\mathbf{b}}]$.

53. $\left[\begin{array}{cccc|c} 1 & -2 & 0 & 5 & 0\\ 0 & 0 & 1 & 2 & 0\\ 0 & 0 & 0 & 0 & 0 \end{array}\right]$ **54.** $\left[\begin{array}{ccc|c} 1 & 0 & 2 & 0\\ 0 & 1 & 0 & 0 \end{array}\right]$

55. $\left[\begin{array}{cccc|c} 1 & -4 & 3 & 0 & 0 \end{array}\right]$

Making Systems Inconsistent *For each of the m × n matrices in Problems 56–60, determine the rank r of the given matrix* **A**. *If* $r < m$, *construct a vector* $\vec{\mathbf{b}}$ *so that the system* $\mathbf{A}\vec{\mathbf{x}} = \vec{\mathbf{b}}$ *is inconsistent.*

56. $\begin{bmatrix} 1 & 0 & 3\\ 0 & 2 & 4\\ 1 & 0 & 5 \end{bmatrix}$ **57.** $\begin{bmatrix} 4 & 5\\ 1 & 6\\ 3 & 1 \end{bmatrix}$ **58.** $\begin{bmatrix} 1 & 2 & -1\\ 1 & 0 & -3\\ 0 & -1 & 2 \end{bmatrix}$

59. $\begin{bmatrix} 1 & 1 & 2\\ 2 & -1 & 1\\ 4 & 1 & 5 \end{bmatrix}$ **60.** $\begin{bmatrix} 1 & -1 & 1\\ 1 & 1 & 0\\ 1 & 2 & -1 \end{bmatrix}$

Seeking Consistency *In Problems 61–65, determine the values of k, if any, that would make the augmented matrices shown those of consistent systems. If there is no such k, explain.*

61. $\left[\begin{array}{cc|c} 1 & 2 & 3\\ 2 & k & 0 \end{array}\right]$ **62.** $\left[\begin{array}{ccc|c} 1 & 2 & 1 & 2\\ 3 & 4 & 1 & k \end{array}\right]$

63. $\left[\begin{array}{cc|c} 1 & k & 0\\ k & 1 & 2 \end{array}\right]$ **64.** $\left[\begin{array}{cc|c} 2 & k & 0\\ 1 & 1 & 1\\ 3 & 3 & 0 \end{array}\right]$

65. $\left[\begin{array}{cccc|c} 1 & 0 & 0 & 1 & 2\\ 0 & 2 & 4 & 0 & 6\\ 1 & -1 & -2 & 1 & -1\\ 2 & 2 & 4 & 2 & k \end{array}\right]$

66. Not Enough Equations A linear system $\mathbf{A}\vec{\mathbf{x}} = \vec{\mathbf{b}}$ with fewer equations than unknowns has either infinitely many solutions or no solutions. Examine the augmented matrices $[\mathbf{A} \mid \vec{\mathbf{b}}]$ and decide which case applies to each matrix.

(a) $\left[\begin{array}{cccc|c} 2 & 1 & 0 & 0 & 3\\ 1 & -1 & 1 & 1 & 3\\ 2 & -3 & 4 & 4 & 9 \end{array}\right]$

(b) $\left[\begin{array}{cccc|c} 2 & 1 & 0 & 0 & 3\\ 1 & -1 & 1 & 1 & 3\\ 1 & 2 & -1 & -1 & -6 \end{array}\right]$

67. Not Enough Variables A linear system $\mathbf{A}\vec{\mathbf{x}} = \vec{\mathbf{b}}$ with fewer unknowns than equations can have infinitely many solutions, no solutions, or a unique solution. Construct the RREFs for the augmented matrices $[\mathbf{A} \mid \vec{\mathbf{b}}]$ that illustrate the three possible cases for a system of four equations in two unknowns.

68. True/False Questions If true, give an explanation. If false, give a counterexample.

(a) Different matrices cannot have the same RREF. True or false?

(b) If the rank of an $m \times n$ matrix is n, then the system $\mathbf{A}\vec{\mathbf{x}} = \vec{\mathbf{b}}$ has exactly one solution. True or false?

(c) If $\mathbf{A}$ is an $n \times n$ matrix and $\vec{\mathbf{b}}$ is a vector in $\mathbb{R}^n$ such that $\mathbf{A}\vec{\mathbf{x}} = \vec{\mathbf{b}}$ is inconsistent, then so is $\mathbf{A}\vec{\mathbf{x}} = \vec{\mathbf{c}}$ for any other nonzero vector $\vec{\mathbf{c}}$ in $\mathbb{R}^n$. True or false?

69. Equivalence of Systems When one system of equations is obtained from another by a sequence of elementary row operations on their augmented matrices, the systems are equivalent (have the same set of solutions) because the transformation can be reversed (hence a solution of the first satisfies the second and vice versa). Explain why this is true by giving an elementary row operation that reverses the effect of each basic one.

70. Homogeneous versus Nonhomogeneous The systems of Problems 32 and 33 differ only in their right-hand sides. Compare their solutions. Explain how this parallels the solution of homogeneous and nonhomogeneous linear first-order differential equations.

71. Solutions in Tandem A student who was asked to solve two systems,

$$\begin{aligned} 2x - y + z &= 6,\\ x + 2y \quad &= -3,\\ 3x + y - z &= -1 \end{aligned} \quad \text{and} \quad \begin{aligned} 2x - y + z &= -4,\\ x + 2y \quad &= 5,\\ 3x + y - z &= -1, \end{aligned}$$

noticed that they differed only in their right-hand sides. She formed the matrix

$$\underbrace{\begin{bmatrix} 2 & -1 & 1 \\ 1 & 2 & 0 \\ 3 & 1 & -1 \end{bmatrix}}_{\mathbf{A}} \underbrace{\begin{matrix} 6 \\ -3 \\ -1 \end{matrix}}_{\vec{\mathbf{b}}} \underbrace{\begin{matrix} -4 \\ 5 \\ -1 \end{matrix}}_{\vec{\mathbf{g}}}$$

and obtained its RREF using Gauss-Jordan elimination:

$$\left[\begin{array}{ccc|c|c} 1 & 0 & 0 & 1 & -1 \\ 0 & 1 & 0 & -2 & 3 \\ 0 & 0 & 1 & 2 & 1 \end{array}\right].$$

She concluded that the solutions of the two systems were given, respectively, by the vectors

$$\vec{\mathbf{x}}_b = \begin{bmatrix} 1 \\ -2 \\ 2 \end{bmatrix} \quad \text{and} \quad \vec{\mathbf{x}}_g = \begin{bmatrix} -1 \\ 3 \\ 1 \end{bmatrix}.$$

Explain why this is correct.

72. Tandem with a Twist

(a) Use the method of the previous problem to solve the systems

$$\begin{aligned} x + y &= 3, \\ 2y + z &= 2 \end{aligned} \quad \text{and} \quad \begin{aligned} x + y &= 5, \\ 2y + z &= 4. \end{aligned}$$

(b) Explain how the calculation of part (a) can be used to solve the matrix equation

$$\begin{bmatrix} 1 & 1 & 0 \\ 0 & 2 & 1 \end{bmatrix} \mathbf{X} = \begin{bmatrix} 3 & 5 \\ 2 & 4 \end{bmatrix}$$

for the unknown 3×2 matrix $\mathbf{X}$.

73. Two-Thousand-Year-Old Problem Find the area of two fields, given that one field yields 2/3 of a bushel of wheat per square yard and the other yields 1/2 a bushel per square yard. The total area of the two fields is 1,800 square yards and the total yield is 1,100 bushels. (This is a typical Babylonian problem, as mentioned in the historical note, with modernized units of measure.)

Computerizing *List (in appropriate order) the operations you would need to use to instruct a computer to solve* $\mathbf{A}\vec{\mathbf{x}} = \vec{\mathbf{b}}$ *by Gauss-Jordan elimination:*

74. in the 2×2 case. **75.** in the 3×3 case.

HINT: The strategy used in Example 3 is a good start for Problems 74 and 75, but it needs to be refined to meet all the formal rules for RREF and to carry out appropriate solutions in all the RREFs that might result. That is, explain how to deal with the coefficients in a given system to carry out the Gauss-Jordan steps. You need not worry about writing in an actual programming language, but you should list a set of steps to do the job. This is called "pseudocode."

76. Electrical Circuits The multiloop circuit shown in Fig. 3.2.7 has four junctions J_1, J_2, J_3, and J_4 and six branches carrying currents I_1, I_2, I_3, I_4, I_5, and I_6. **Kirchoff's Current Law** says *the sum of the currents at each junction must be zero.*[6]

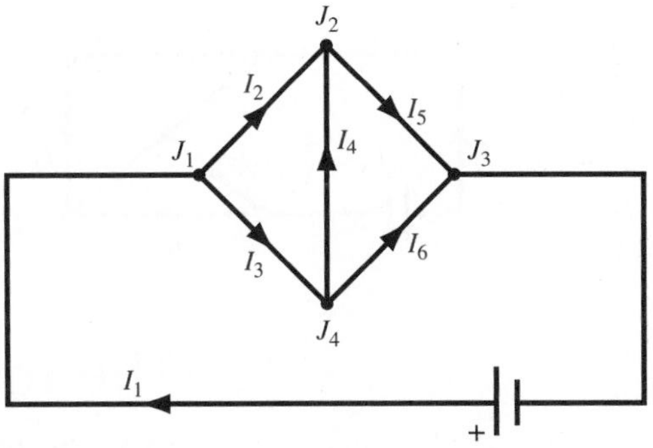

FIGURE 3.2.7 Multiloop circuit (Problem 76).

(a) Verify that Kirchoff's Current Law gives the following system of equations:

$$\begin{aligned} I_1 - I_2 - I_3 \qquad\qquad\qquad &= 0, \\ I_2 \qquad + I_4 - I_5 \qquad &= 0, \\ I_3 - I_4 \qquad - I_6 &= 0, \\ -I_1 \qquad\qquad + I_5 + I_6 &= 0. \end{aligned}$$

(b) Write the augmented matrix for the system of part (a) and transform it to RREF. How many parameters are needed to describe the set of solutions?

More Circuit Analysis *For each circuit in Problems 77–80, use Kirchoff's Current Law (Problem 76) to write a system of equations that must be satisfied by the currents.*

77.

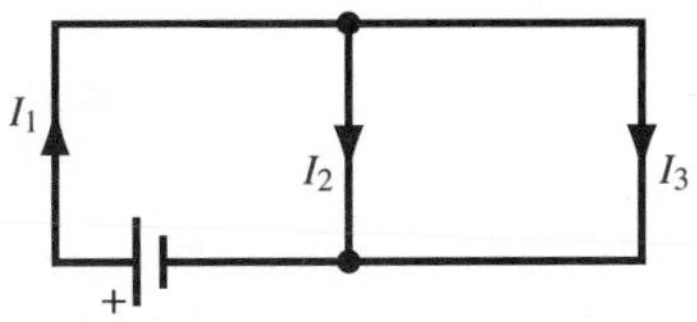

78.

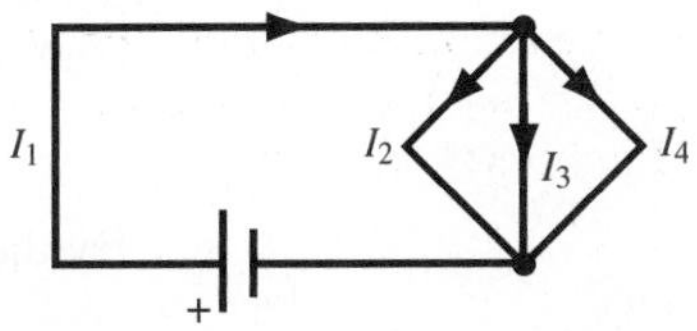

[6]Gustav Robert Kirchoff (1824–1887) was a German physicist who used topology to make important contributions to circuit theory. Kirchoff's Current Law together with his voltage law (See Sec. 3.4, Problem 43) are the foundation of circuit analysis.

79.

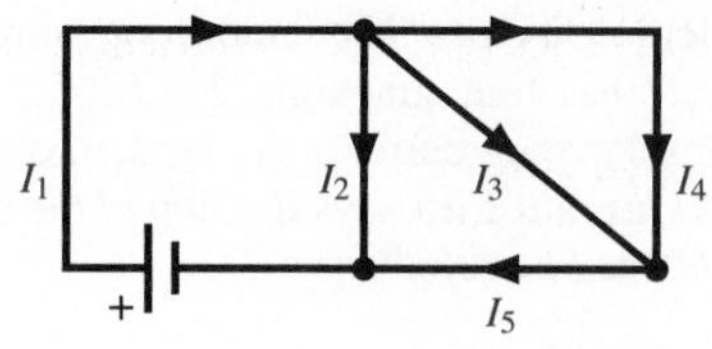

80.

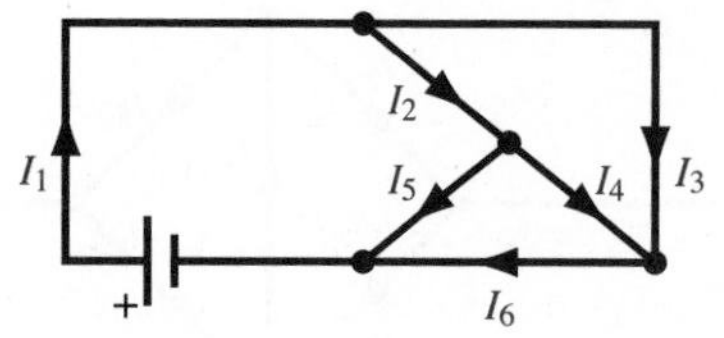

81. Suggested Journal Entry I Use your knowledge of the geometry of lines to give a complete discussion of all possible types of solutions for the 3 × 2 system

$$\begin{aligned} a_{11}x + a_{12}y &= b_1, \\ a_{21}x + a_{22}y &= b_2, \\ a_{31}x + a_{32}y &= b_3. \end{aligned}$$

82. Suggested Journal Entry II Formulate answers to the three questions posed at the end of the introductory example. (Pay special attention to Examples 5, 7, and 8.)

3.3 The Inverse of a Matrix

SYNOPSIS: We define the inverse of a square matrix and discover that not all square matrices have inverses. If the inverse of a square matrix exists, we can find it using Gauss-Jordan elimination. If the matrix of coefficients of a system of n linear equations in n unknowns has an inverse, we can use it to find the unique solution for the system.

Introductory Example

Suppose we are asked to solve a system of two linear equations in two variables,

$$\begin{aligned} x + y &= 1, \\ 4x + 5y &= 6. \end{aligned} \tag{1}$$

The system can be written compactly in the form

$$\mathbf{A}\vec{\mathbf{x}} = \vec{\mathbf{b}}, \tag{2}$$

where

$$\mathbf{A} = \begin{bmatrix} 1 & 1 \\ 4 & 5 \end{bmatrix}, \quad \vec{\mathbf{x}} = \begin{bmatrix} x \\ y \end{bmatrix}, \quad \text{and} \quad \vec{\mathbf{b}} = \begin{bmatrix} 1 \\ 6 \end{bmatrix}.$$

Suppose also that in the course of practicing our matrix multiplication, we have run across a matrix **W** with the surprising property that **WA** = **I**. In fact,

$$\mathbf{W} = \begin{bmatrix} 5 & -1 \\ -4 & 1 \end{bmatrix},$$

and

$$\begin{bmatrix} 5 & -1 \\ -4 & 1 \end{bmatrix} \begin{bmatrix} 1 & 1 \\ 4 & 5 \end{bmatrix} = \begin{bmatrix} 1 & 0 \\ 0 & 1 \end{bmatrix}.$$

Knowing a matrix with this property helps to solve equation (2). After multiplying each side of the equation *on the left* (remember that the product is not commutative), we get

$$\mathbf{W}(\mathbf{A}\vec{\mathbf{x}}) = \mathbf{W}\vec{\mathbf{b}}.$$

By the associative property (see Sec. 3.1), this is equivalent to

$$(\mathbf{WA})\vec{\mathbf{x}} = \mathbf{W}\vec{\mathbf{b}},$$

and since we calculated that **WA** = **I**, the equation becomes

$$\mathbf{I}\vec{\mathbf{x}} = \mathbf{W}\vec{\mathbf{b}}.$$

Because the identity matrix behaves like the number 1 (see Sec. 3.1), the left side simplifies to just $\vec{\mathbf{x}}$, and we have

$$\vec{\mathbf{x}} = \mathbf{W}\vec{\mathbf{b}}. \tag{3}$$

Resubstituting for $\vec{\mathbf{x}}$, $\mathbf{W}$, and $\vec{\mathbf{b}}$ gives

$$\begin{bmatrix} x \\ y \end{bmatrix} = \begin{bmatrix} 5 & -1 \\ -4 & 1 \end{bmatrix} \begin{bmatrix} 1 \\ 6 \end{bmatrix} = \begin{bmatrix} -1 \\ 2 \end{bmatrix}.$$

Therefore, by the definition of equality of matrices, $x = -1$ and $y = 2$.

But the best is yet to come. Suppose that we also need to solve 25 more systems exactly like system (1) except for the numbers on the right-hand sides. In other words, only the vector $\vec{\mathbf{b}}$ is different. This means that we can use the formula (3) for every one of these, by simply substituting the new vector $\vec{\mathbf{b}}$ each time. The matrix $\mathbf{W}$, with its special property, streamlines our calculations.

Inverse of a Matrix

Inverse of a Square Matrix:

Only square matrices can have inverses; however, not all square matrices *do* have inverses.

While there is no operation for matrices directly analogous to division of real or complex numbers, we have a useful substitute using the inverse of a matrix, which behaves rather like the reciprocal of a number. For certain *square* matrices we can find an associated matrix with the property that the product of the two is the identity matrix. (Compare this with the fact that for the number 1.25, we find that 0.8 is its reciprocal: $(1.25)(0.8) = 1$.) This was exactly what we found for the matrix $\mathbf{A}$ in the preceding example: matrix $\mathbf{W}$ had the property that $\mathbf{WA} = \mathbf{I}$, where $\mathbf{I}$ is the matrix that behaves a lot like the number 1. Furthermore, we can check that $\mathbf{AW} = \mathbf{I}$ also. $\mathbf{W}$ is the inverse of $\mathbf{A}$, denoted $\mathbf{A}^{-1}$.

The Inverse of a Matrix

If there exists, for an $n \times n$ matrix $\mathbf{A}$, another matrix $\mathbf{A}^{-1}$ of the same order such that

$$\mathbf{A}^{-1}\mathbf{A} = \mathbf{A}\mathbf{A}^{-1} = \mathbf{I}, \tag{4}$$

then $\mathbf{A}^{-1}$ is called the **inverse** of matrix $\mathbf{A}$, and $\mathbf{A}$ is said to be **invertible**.

The inverse matrix, if it exists, is unique. (See Problem 19.) Furthermore, since a matrix and its inverse commute, checking that *either* $\mathbf{A}^{-1}\mathbf{A} = \mathbf{I}$ *or* $\mathbf{A}\mathbf{A}^{-1} = \mathbf{I}$ is sufficient. Although not all matrices have inverses, we will learn several criteria for determining when a matrix is invertible. One such condition involves the determinant, defined in Sec. 3.4. Others will be presented in this chapter and in Chapter 5. (For a quick way to find the inverse of a 2×2 matrix, see Problem 15.)

EXAMPLE 1 **Invertible Matrices** We verified in our introductory example that the matrices $\mathbf{A}$ and $\mathbf{W}$ were inverses. (Actually we only calculated $\mathbf{WA}$, but $\mathbf{AW} = \mathbf{I}$ as well.) Both matrices are invertible, and we can write $\mathbf{A}^{-1}$ instead of $\mathbf{W}$. Another invertible matrix is

$$\mathbf{B} = \begin{bmatrix} 1 & -3 & 0 \\ 2 & -5 & 0 \\ 0 & -1 & 1 \end{bmatrix}, \quad \text{for which} \quad \mathbf{B}^{-1} = \begin{bmatrix} -5 & 3 & 0 \\ -2 & 1 & 0 \\ -2 & 1 & 1 \end{bmatrix}.$$

We check that

$$\begin{bmatrix} 1 & -3 & 0 \\ 2 & -5 & 0 \\ 0 & -1 & 1 \end{bmatrix} \begin{bmatrix} -5 & 3 & 0 \\ -2 & 1 & 0 \\ -2 & 1 & 1 \end{bmatrix} = \begin{bmatrix} 1 & 0 & 0 \\ 0 & 1 & 0 \\ 0 & 0 & 1 \end{bmatrix},$$

so $\mathbf{BB}^{-1} = \mathbf{I}$. The reader could verify that $\mathbf{B}^{-1}\mathbf{B} = \mathbf{I}$ as well. ■

Alternate Terminology:

A *noninvertible* matrix is sometimes called **singular**; an *invertible* matrix is then called **nonsingular**.

Some Properties of Invertible Matrices

- If $\mathbf{A}$ is invertible, then so is $\mathbf{A}^{-1}$, and
$$\left(\mathbf{A}^{-1}\right)^{-1} = \mathbf{A}. \tag{5}$$
- If $\mathbf{A}$ and $\mathbf{B}$ are invertible matrices of the same order, then their product $\mathbf{AB}$ is invertible. In fact,
$$(\mathbf{AB})^{-1} = \mathbf{B}^{-1}\mathbf{A}^{-1}. \tag{6}$$
- If $\mathbf{A}$ is invertible, then so is $\mathbf{A}^{\mathrm{T}}$, and
$$\left(\mathbf{A}^{\mathrm{T}}\right)^{-1} = \left(\mathbf{A}^{-1}\right)^{\mathrm{T}}. \tag{7}$$

EXAMPLE 2 **Inverse of a Matrix Product** We will use properties of matrix multiplication to verify formula (6) for the inverse of a product. In effect, due to the uniqueness of the inverse of a matrix, *if a matrix acts as inverse, it is the inverse*.

Since equation (6) claims that the matrix inverse to $\mathbf{AB}$ is $\mathbf{B}^{-1}\mathbf{A}^{-1}$, we shall verify the claim by multiplying the two together and applying the associative property of matrix multiplication:

$$\begin{aligned}
\left(\mathbf{B}^{-1}\mathbf{A}^{-1}\right)(\mathbf{AB}) &= \mathbf{B}^{-1}\left(\mathbf{A}^{-1}\mathbf{A}\right)\mathbf{B} && (\textit{associativity})\\
&= \mathbf{B}^{-1}(\mathbf{I})\mathbf{B} && (\textit{inverse})\\
&= \mathbf{B}^{-1}\mathbf{B} && (\textit{identity})\\
&= \mathbf{I}. && (\textit{inverse})
\end{aligned}$$

■

Inverse Matrix by Gauss-Jordan

In Sec. 3.2 we saw how to solve $\mathbf{A}\vec{\mathbf{x}} = \vec{\mathbf{b}}$ by reducing an augmented matrix $[\mathbf{A} \mid \vec{\mathbf{b}}]$ to reduced row echelon form (RREF); and, as explained there in Problem 71, it is possible to solve two similar systems in tandem. That is, for two systems

$$\mathbf{A}\vec{\mathbf{x}} = \vec{\mathbf{b}} \quad \text{and} \quad \mathbf{A}\vec{\mathbf{x}} = \vec{\mathbf{g}}$$

with the same coefficient matrix $\mathbf{A}$, we can obtain both solutions at once by reducing a twice-augmented matrix $[\,\mathbf{A} \mid \vec{\mathbf{b}} \mid \vec{\mathbf{g}}\,]$ to RREF, and taking as solutions of the two systems the next-to-last and last columns of the reduced matrix.

Furthermore, those two resulting column vector solutions $\vec{\mathbf{x}}_b$ and $\vec{\mathbf{x}}_g$ together form a *solution matrix* for the unknown matrix $\mathbf{X}$ in the combined matrix equation

$$\mathbf{A}\underbrace{\left[\vec{\mathbf{x}}_b \mid \vec{\mathbf{x}}_g\right]}_{\mathbf{X}} = \left[\vec{\mathbf{b}} \mid \vec{\mathbf{g}}\right].$$

We can put these ideas together to find the inverse of matrix $\mathbf{A}$.

Suppose that $\vec{\mathbf{x}}_1, \vec{\mathbf{x}}_2, \ldots, \vec{\mathbf{x}}_n$ are the columns of $\mathbf{A}^{-1}$. Then, because $\mathbf{AA}^{-1} = \mathbf{I}$, solving for $\mathbf{A}^{-1}$ is equivalent to solving the equations

$$\mathbf{A}\vec{\mathbf{x}}_1 = \vec{\mathbf{e}}_1, \quad \mathbf{A}\vec{\mathbf{x}}_2 = \vec{\mathbf{e}}_2, \ldots, \quad \mathbf{A}\vec{\mathbf{x}}_n = \vec{\mathbf{e}}_n,$$

where $\vec{\mathbf{e}}_1, \vec{\mathbf{e}}_2, \ldots, \vec{\mathbf{e}}_n$ are the columns of $\mathbf{I}$ (they are called **unit vectors**); that is,

$$\mathbf{A}\underbrace{\left[\vec{\mathbf{x}}_1 \mid \vec{\mathbf{x}}_2 \mid \cdots \mid \vec{\mathbf{x}}_n\right]}_{\mathbf{X} = \mathbf{A}^{-1}} = \underbrace{\left[\vec{\mathbf{e}}_1 \mid \vec{\mathbf{e}}_2 \mid \cdots \mid \vec{\mathbf{e}}_n\right]}_{\mathbf{I}}.$$

To Summarize:

$$\begin{aligned}
\mathbf{AX} &= \mathbf{I},\\
\mathbf{A}^{-1}\mathbf{AX} &= \mathbf{A}^{-1}\mathbf{I},\\
\mathbf{X} &= \mathbf{A}^{-1}.
\end{aligned}$$

To solve this, we need to find the RREFs of the augmented matrices

$$[\mathbf{A} \mid \vec{\mathbf{e}}_1], [\mathbf{A} \mid \vec{\mathbf{e}}_2], \ldots, [\mathbf{A} \mid \vec{\mathbf{e}}_n].$$

If we combine these into $[\mathbf{A} \mid \vec{\mathbf{e}}_1 \mid \vec{\mathbf{e}}_2 \mid \cdots \mid \vec{\mathbf{e}}_n]$, the $\vec{\mathbf{x}}_i$ can be found all at once.

If the RREF of a square matrix **A** is the identity matrix **I**, **A** is said to be **row equivalent to I**. This property holds if and only if **A** is invertible.

Setup for Finding the Inverse Matrix:

$$\left[\begin{array}{ccc|ccc} a_{11} & a_{12} & a_{13} & 1 & 0 & 0 \\ a_{21} & a_{22} & a_{23} & 0 & 1 & 0 \\ a_{31} & a_{32} & a_{33} & 0 & 0 & 1 \end{array}\right].$$

Inverse Matrix by RREF

For an $n \times n$ matrix **A**, the following procedure produces $\mathbf{A}^{-1}$, or shows that **A** is not invertible.

Step 1. Form the $n \times 2n$ matrix $\mathbf{M} = [\mathbf{A} \mid \mathbf{I}]$.

Step 2. Transform **M** into its RREF, **R**.

Step 3. If the first n columns of **R** form the identity matrix, then the last n columns form $\mathbf{A}^{-1}$. If the first n columns of **R** do *not* form the identity matrix, then **A** is *not* invertible.

If we put **A** and **I** side by side and reduce to RREF, the left half becomes the identity matrix and the right half turns into the inverse. In other words,

$$[\mathbf{A} \mid \mathbf{I}] \text{ becomes } [\mathbf{I} \mid \mathbf{A}^{-1}].$$

EXAMPLE 3 Inverses by RREF

(a) To find the inverse of $\mathbf{A} = \begin{bmatrix} 1 & 1 \\ 4 & 1 \end{bmatrix}$, form matrix

$$\mathbf{M}_\mathbf{A} = \left[\begin{array}{cc|cc} 1 & 1 & 1 & 0 \\ 4 & 1 & 0 & 1 \end{array}\right]$$

and reduce it to its RREF

$$\mathbf{R}_\mathbf{A} = \left[\begin{array}{cc|cc} 1 & 0 & -1/3 & 1/3 \\ 0 & 1 & 4/3 & -1/3 \end{array}\right].$$

A detailed calculation of $\mathbf{R}_\mathbf{A}$ could be given by the following row operations:

$$\left[\begin{array}{cc|cc} 1 & 1 & 1 & 0 \\ 4 & 1 & 0 & 1 \end{array}\right] \to \left[\begin{array}{cc|cc} 1 & 1 & 1 & 0 \\ 0 & -3 & -4 & 1 \end{array}\right] \quad R_2^* = R_2 - 4R_1$$

$$\to \left[\begin{array}{cc|cc} 1 & 1 & 1 & 0 \\ 0 & 1 & 4/3 & -1/3 \end{array}\right] \quad R_2^* = (-1/3)R_2$$

$$\to \left[\begin{array}{cc|cc} 1 & 0 & -1/3 & 1/3 \\ 0 & 1 & 4/3 & -1/3 \end{array}\right] \quad R_1^* = R_1 - R_2.$$

The left half of $\mathbf{R}_\mathbf{A}$ is the identity matrix of order 2, and we conclude that

$$\mathbf{A}^{-1} = \begin{bmatrix} -1/3 & 1/3 \\ 4/3 & -1/3 \end{bmatrix}.$$

(As a check on the Gauss-Jordan reduction, it is wise to verify either that $\mathbf{A}\mathbf{A}^{-1} = \mathbf{I}$ or that $\mathbf{A}^{-1}\mathbf{A} = \mathbf{I}$.)

(b) In attempting to determine the inverse of $\mathbf{C} = \begin{bmatrix} 3 & 0 & 3 \\ -1 & 2 & 1 \\ 1 & 1 & 2 \end{bmatrix}$, we row-reduce

$$\mathbf{M_C} = \left[\begin{array}{ccc|ccc} 3 & 0 & 3 & 1 & 0 & 0 \\ -1 & 2 & 1 & 0 & 1 & 0 \\ 1 & 1 & 2 & 0 & 0 & 1 \end{array}\right],$$

obtaining

$$\mathbf{R_C} = \left[\begin{array}{ccc|ccc} 1 & 0 & 1 & 0 & -1/3 & 2/3 \\ 0 & 1 & 1 & 0 & 1/3 & 1/3 \\ 0 & 0 & 0 & 1 & 1 & -2 \end{array}\right].$$

Since the left half of this reduced matrix is not the identity, **C** is *not* invertible.

(c) We calculate the inverse of the matrix $\mathbf{H} = \begin{bmatrix} 1 & 1 & 1 \\ 0 & 2 & 1 \\ 1 & 0 & 1 \end{bmatrix}$ by forming the matrix

$$\mathbf{M_H} = \left[\begin{array}{ccc|ccc} 1 & 1 & 1 & 1 & 0 & 0 \\ 0 & 2 & 1 & 0 & 1 & 0 \\ 1 & 0 & 1 & 0 & 0 & 1 \end{array}\right]$$

and calculating its RREF

$$\mathbf{R_H} = \left[\begin{array}{ccc|ccc} 1 & 0 & 0 & 2 & -1 & -1 \\ 0 & 1 & 0 & 1 & 0 & -1 \\ 0 & 0 & 1 & -2 & 1 & 2 \end{array}\right].$$

The left half of $\mathbf{R_H}$ is **I** and the right half gives

$$\mathbf{H}^{-1} = \begin{bmatrix} 2 & -1 & -1 \\ 1 & 0 & -1 \\ -2 & 1 & 2 \end{bmatrix}.$$

■

Invertible Matrices and Solutions to Linear Systems

The introductory example illustrated the advantage of knowing $\mathbf{A}^{-1}$ when trying to find the solutions to the matrix-vector equation

$$\mathbf{A}\vec{\mathbf{x}} = \vec{\mathbf{b}},$$

where **A** is a square $n \times n$ matrix and $\vec{\mathbf{b}}$ is any vector in $\mathbb{R}^n$. Given $\mathbf{A}^{-1}$, solving $\mathbf{A}\vec{\mathbf{x}} = \vec{\mathbf{b}}$ for $\vec{\mathbf{x}}$ is a simple process. We multiply both sides of the equation by $\mathbf{A}^{-1}$ to obtain the solution for $\vec{\mathbf{x}}$:

$$\begin{aligned} \mathbf{A}^{-1}(\mathbf{A}\vec{\mathbf{x}}) &= \mathbf{A}^{-1}\vec{\mathbf{b}}, \\ (\mathbf{A}^{-1}\mathbf{A})\vec{\mathbf{x}} &= \mathbf{A}^{-1}\vec{\mathbf{b}}, \\ \mathbf{I}\vec{\mathbf{x}} &= \mathbf{A}^{-1}\vec{\mathbf{b}}, \\ \vec{\mathbf{x}} &= \mathbf{A}^{-1}\vec{\mathbf{b}}. \end{aligned}$$

For an *invertible* matrix **A**, the solution $\vec{\mathbf{x}} = \mathbf{A}^{-1}\vec{\mathbf{b}}$ is unique by the following argument: If $\vec{\mathbf{y}}$ is also a solution, then it must satisfy $\mathbf{A}\vec{\mathbf{y}} = \vec{\mathbf{b}}$ and, by the same process as before, $\vec{\mathbf{y}} = \mathbf{A}^{-1}\vec{\mathbf{b}} = \vec{\mathbf{x}}$. ■

What happens if **A** is *not* invertible? If **A** is not invertible, then **A** is not row equivalent to the identity matrix, so there must be at least one column in the RREF of **A** that is not a pivot column. Because **A** is a square matrix, the RREF of the augmented matrix $[\mathbf{A} \mid \vec{\mathbf{b}}]$ must have n rows. Consequently, at least one row must be of the form $[0 \ \cdots \ 0 \mid k]$. If any such row has nonzero k, there are no solutions. If solutions do exist, any such row must have $k = 0$, which implies there must be infinitely many solutions.

Invertibility and Solutions
The matrix-vector equation $\mathbf{A}\vec{\mathbf{x}} = \vec{\mathbf{b}}$, where $\mathbf{A}$ is an $n \times n$ matrix, has

- a unique solution $\vec{\mathbf{x}} = \mathbf{A}^{-1}\vec{\mathbf{b}}$ if and only if $\mathbf{A}$ is invertible;
- either no solutions or infinitely many solutions if $\mathbf{A}$ is not invertible.

For the *homogeneous* equation $\mathbf{A}\vec{\mathbf{x}} = \vec{\mathbf{0}}$, there is always one solution, $\vec{\mathbf{x}} = \vec{\mathbf{0}}$, which is called the **trivial solution**. So we can see that if $\mathbf{A}$ is invertible the trivial solution is the only solution. And if $\mathbf{A}$ is not invertible, as shown in the preceding, there are infinitely many solutions.

The following list gives many of the characteristics of invertible matrices. Most follow directly from the fact that an $n \times n$ matrix is invertible if and only if it is row equivalent to $\mathbf{I}_n$.

Invertible Matrix Characterization

Let $\mathbf{A}$ be an $n \times n$ matrix. The following statements are equivalent:

- $\mathbf{A}$ is an invertible matrix.
- $\mathbf{A}^{\mathrm{T}}$ is an invertible matrix.
- $\mathbf{A}$ is row equivalent to $\mathbf{I}_n$.
- $\mathbf{A}$ has n pivot columns.
- The equation $\mathbf{A}\vec{\mathbf{x}} = \vec{\mathbf{0}}$ has only the trivial solution, $\vec{\mathbf{x}} = \vec{\mathbf{0}}$.
- The equation $\mathbf{A}\vec{\mathbf{x}} = \vec{\mathbf{b}}$ has a unique solution for every $\vec{\mathbf{b}}$ in $\mathbb{R}^n$.

There are many other ways to characterize invertible matrices. We will be adding to this list as we encounter new concepts.

The linear algebra that we have discussed so far will be very useful in solving bigger problems. Let us look at an example from differential equations.

Vector of Initial Conditions:

$$\vec{\mathbf{b}} = \begin{bmatrix} y(0) \\ y'(0) \\ y''(0) \end{bmatrix} = \begin{bmatrix} b_1 \\ b_2 \\ b_3 \end{bmatrix}.$$

EXAMPLE 4 **A Third-Order Initial-Value Problem** An engineering consultant finds that she must solve the following third-order initial-value problem:

$$y''' - 2y'' - y' + 2y = 0, \qquad y(0) = b_1, \quad y'(0) = b_2, \quad y''(0) = b_3. \quad (8)$$

She must solve this IVP for several different sets of initial conditions today, and expects to be asked to repeat the task for still other initial values tomorrow and next week.

The general solution is

$$y(t) = c_1 e^{2t} + c_2 e^{t} + c_3 e^{-t}, \quad (9)$$

where c_1, c_2, and c_3 are arbitrary constants. This solution can be found fairly quickly by methods of Chapter 4 (and can be verified by differentiation and

Vector of Coefficients of the DE Solution:

$$\vec{\mathbf{c}} = \begin{bmatrix} c_1 \\ c_2 \\ c_3 \end{bmatrix}.$$

substitution into the DE). The constants c_1, c_2, c_3 must be determined to satisfy the initial conditions. First she calculates the first and second derivatives of (9):

$$y'(t) = 2c_1e^{2t} + c_2e^t - c_3e^{-t},$$
$$y''(t) = 4c_1e^{2t} + c_2e^t + c_3e^{-t}.$$

Then the initial conditions become

$$\begin{aligned} y(0) &= c_1 + c_2 + c_3 = b_1, \\ y'(0) &= 2c_1 + c_2 - c_3 = b_2, \\ y''(0) &= 4c_1 + c_2 + c_3 = b_3. \end{aligned} \tag{10}$$

To solve the IVP (8), then, she must solve the 3×3 system of linear algebraic equations (10) for c_1, c_2, and c_3. In matrix-vector form, system (10) is

$$\underbrace{\begin{bmatrix} 1 & 1 & 1 \\ 2 & 1 & -1 \\ 4 & 1 & 1 \end{bmatrix}}_{\mathbf{A}} \underbrace{\begin{bmatrix} c_1 \\ c_2 \\ c_3 \end{bmatrix}}_{\vec{\mathbf{c}}} = \underbrace{\begin{bmatrix} b_1 \\ b_2 \\ b_3 \end{bmatrix}}_{\vec{\mathbf{b}}}. \tag{11}$$

Using linear algebra skills, the consultant is able to compute the *inverse* of the coefficient matrix, which is

$$\mathbf{A}^{-1} = \begin{bmatrix} -1/3 & 0 & 1/3 \\ 1 & 1/2 & -1/2 \\ 1/3 & -1/2 & 1/6 \end{bmatrix}.$$

Left-multiplying each side of equation (11) by $\mathbf{A}^{-1}$ reduces the left side to the product of the identity matrix and the vector of the c_i, so the result is

$$\underbrace{\begin{bmatrix} c_1 \\ c_2 \\ c_3 \end{bmatrix}}_{\vec{\mathbf{c}}} = \underbrace{\begin{bmatrix} -1/3 & 0 & 1/3 \\ 1 & 1/2 & -1/2 \\ 1/3 & -1/2 & 1/6 \end{bmatrix}}_{\mathbf{A}^{-1}} \underbrace{\begin{bmatrix} b_1 \\ b_2 \\ b_3 \end{bmatrix}}_{\vec{\mathbf{b}}}.$$

This formula will enable her to quickly compute the three coefficients in the solution for any set of initial conditions required. The key to using this strategy, of course, is coming up with that inverse matrix. ■

Application: Leontief Input/Output Model

In 1973 the Nobel Prize in economics was awarded to Professor Wassily Leontief of Harvard University for his development of **input-output analysis**, a body of knowledge valuable for studying interdependent industries, such as manufacturing, agriculture, energy, and so on.[1]

EXAMPLE 5 **Manufacturing Economics** Consider an economy consisting of two companies, A and B, called **interdependent industries**. (Normally, economists study dozens of interrelated companies or industries.) Suppose that

- for each \$1 worth of the product company A produces, it requires \$0.30 of its own product and \$0.50 worth of the product B produces; and

[1]Wassily Leontief (1906–1999) was born in Russia but emigrated to the United States in 1931. In addition to winning the Nobel Prize for his input-output analysis, Leontief also made contributions to *linear programming*, an important technique for solving linear systems with constraints.

- for every \$1 worth of product B produces, it requires \$0.40 worth of the product A produces and \$0.30 worth of its own product.

This type of situation is not unusual. If A were an electric power company and B a truck manufacturer, then both companies need electricity for operation and production and both need trucks for transportation of materials and services. We can put this **internal consumption** information in a **technological matrix** (see Table 3.3.1)

Flow of Goods in Table 3.3.1:

Goods flow from the row companies to the column companies.

Table 3.3.1 Technological matrix for two companies

		Input	
		Company A	**Company B**
Output	**Company A**	\$0.30	\$0.40
	Company B	\$0.50	\$0.30

or, simply,

$$\mathbf{T} = \begin{bmatrix} 0.3 & 0.4 \\ 0.5 & 0.3 \end{bmatrix},$$

where each entry represents the dollar value required to produce a dollar's worth of the column value.

Now, suppose that there is an **external demand** by other companies or consumers for \$150 of A's goods and \$100 of B's goods. The flow of goods (in dollar value) between A and B and the outside is shown in Fig. 3.3.1.

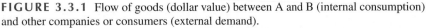

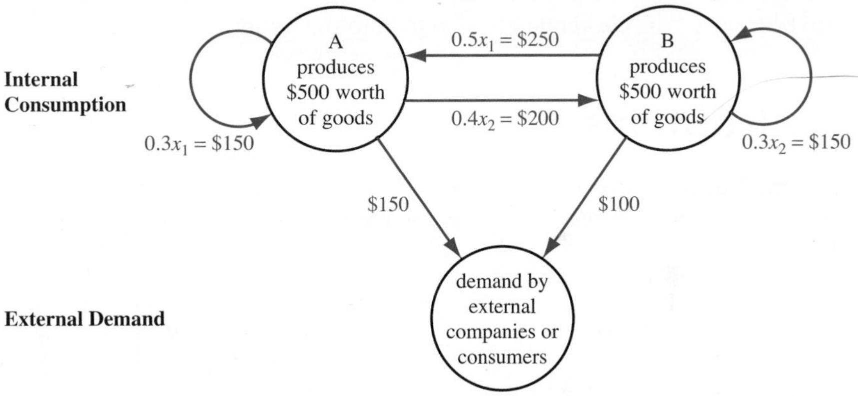

FIGURE 3.3.1 Flow of goods (dollar value) between A and B (internal consumption) and other companies or consumers (external demand).

The question is: How much should A and B produce to meet this external demand, keeping in mind that each must produce some for itself and the other interdependent company? To answer this question we denote as follows:

$$x_1 = \text{dollar value A should produce,}$$
$$x_2 = \text{dollar value B should produce,}$$

and use the basic fact of input-output analysis:

TOTAL OUTPUT = EXTERNAL DEMAND + INTERNAL CONSUMPTION

to write

$$x_1 = \$150 + 0.3x_1 + 0.4x_2,$$
$$x_2 = \$100 + 0.5x_1 + 0.3x_2.$$

In matrix form this is $\vec{\mathbf{x}} = \vec{\mathbf{d}} + \mathbf{T}\vec{\mathbf{x}}$, where

$$\vec{\mathbf{x}} = \begin{bmatrix} x_1 \\ x_2 \end{bmatrix} \quad \text{and} \quad \vec{\mathbf{d}} = \begin{bmatrix} 150 \\ 100 \end{bmatrix}.$$

Rewriting this system, we get

$$(\mathbf{I} - \mathbf{T})\vec{\mathbf{x}} = \vec{\mathbf{d}},$$

which has the solution

$$\vec{\mathbf{x}} = (\mathbf{I} - \mathbf{T})^{-1}\vec{\mathbf{d}} = \begin{bmatrix} 0.7 & -0.4 \\ -0.5 & 0.7 \end{bmatrix}^{-1} \begin{bmatrix} 150 \\ 100 \end{bmatrix}$$

$$= \frac{1}{29}\begin{bmatrix} 70 & 40 \\ 50 & 70 \end{bmatrix} \begin{bmatrix} 150 \\ 100 \end{bmatrix} = \begin{bmatrix} 500 \\ 500 \end{bmatrix}.$$

In other words, if each company produces \$500 worth of their respective products, then their respective external demands of \$150 and \$100 will be met. Each company must produce considerably more than its external demand, owing to the high internal consumption of products. ■

Summary

We found that certain square matrices have inverses. We used an adaptation of the Gauss-Jordan process to find the inverse of a square matrix if it exists or to show that an inverse does not exist. We also found that the action of an inverse of a matrix in matrix arithmetic is analogous to the action of a reciprocal in real number arithmetic.

If a matrix $\mathbf{A}$ of coefficients has an inverse $\mathbf{A}^{-1}$, we found the unique solution of the matrix equation $\mathbf{A}\vec{\mathbf{x}} = \vec{\mathbf{b}}$ to be $\vec{\mathbf{x}} = \mathbf{A}^{-1}\vec{\mathbf{b}}$. We applied this method to the Leontief model from economics.

3.3 Problems

Checking Inverses *In each of Problems 1–4, verify by multiplication that the given pair of matrices are inverses of one another.*

1. $\begin{bmatrix} 5 & 3 \\ 2 & 1 \end{bmatrix}, \begin{bmatrix} -1 & 3 \\ 2 & -5 \end{bmatrix}$

2. $\begin{bmatrix} 2 & -4 \\ 2 & 0 \end{bmatrix}, \begin{bmatrix} 0 & 1/2 \\ -1/4 & 1/4 \end{bmatrix}$

3. $\begin{bmatrix} 1 & 0 & 1 \\ 1 & 1 & -2 \\ 0 & 1 & 1 \end{bmatrix}, \begin{bmatrix} 3/4 & 1/4 & -1/4 \\ -1/4 & 1/4 & 3/4 \\ 1/4 & -1/4 & 1/4 \end{bmatrix}$

4. $\begin{bmatrix} -28 & -13 & 3 \\ 2 & 1 & 0 \\ -7 & -3 & 1 \end{bmatrix}, \begin{bmatrix} 1 & 4 & -3 \\ -2 & -7 & 6 \\ 1 & 7 & -2 \end{bmatrix}$

Matrix Inverses *Use row reduction to calculate the inverse of each matrix in Problems 5–14. Consider $k \neq 0$.*

5. $\begin{bmatrix} 2 & 0 \\ 1 & 1 \end{bmatrix}$

6. $\begin{bmatrix} 1 & 3 \\ 2 & 5 \end{bmatrix}$

7. $\begin{bmatrix} 0 & 1 & 1 \\ 5 & 1 & -1 \\ 3 & -3 & -3 \end{bmatrix}$

8. $\begin{bmatrix} 0 & 0 & 1 \\ 0 & 1 & 0 \\ 1 & 0 & 0 \end{bmatrix}$

9. $\begin{bmatrix} k & 0 & 0 \\ 0 & 1 & 0 \\ 0 & 0 & 1 \end{bmatrix}$

10. $\begin{bmatrix} 1 & 0 & 1 \\ 1 & -1 & 0 \\ 0 & 2 & 1 \end{bmatrix}$

11. $\begin{bmatrix} 1 & 0 & 0 & 0 \\ 0 & 1 & k & 0 \\ 0 & 0 & 1 & 0 \\ 0 & 0 & 0 & 1 \end{bmatrix}$

12. $\begin{bmatrix} 1 & 0 & 1 & 1 \\ 0 & 0 & 1 & 0 \\ 1 & 1 & 1 & 0 \\ 1 & 0 & 0 & 2 \end{bmatrix}$

13. $\begin{bmatrix} 1 & 0 & 0 & 0 \\ 0 & -1 & 0 & 0 \\ 0 & 1 & -2 & 0 \\ 1 & -1 & 3 & 3 \end{bmatrix}$

14. $\begin{bmatrix} 0 & 1 & 2 & 1 \\ 4 & 0 & 1 & 2 \\ 0 & 1 & 0 & 0 \\ 0 & 2 & 0 & 1 \end{bmatrix}$

15. **Inverse of the 2 x 2 Matrix** Verify that the inverse of a square matrix $\mathbf{A}$ of order two is given by the following

handy formula, if $ad - bc \neq 0$. That is, show that $\mathbf{AA}^{-1} = \mathbf{I}$.

$$\begin{bmatrix} a & b \\ c & d \end{bmatrix}^{-1} = \frac{1}{ad - bc}\begin{bmatrix} d & -b \\ -c & a \end{bmatrix}. \quad (12)$$

16. **Brute Force** Compute the inverse of matrix

$$\mathbf{A} = \begin{bmatrix} 1 & 3 \\ 1 & 2 \end{bmatrix}$$

by setting

$$\mathbf{A}^{-1} = \begin{bmatrix} a & b \\ c & d \end{bmatrix}$$

and solving a system of equations for a, b, c, and d.

Finding Counterexamples *For Problems 17 and 18, answer the given question by finding a proof or a counterexample in* $\mathbb{M}_{22}$.

17. Is it true that the sum of invertible $n \times n$ matrices is invertible?

18. Is it true that the only $n \times n$ matrices for which $\mathbf{A}^2 = \mathbf{A}$ are the zero matrix and the identity matrix?

19. **Unique Inverse** Show that if a matrix has an inverse it is unique.

20. **Invertible Matrix Method** Use the inverse matrix found in Problem 6 to solve the system

$$\begin{aligned} x_1 + 3x_2 &= -4, \\ 2x_1 + 5x_2 &= 10. \end{aligned}$$

21. **Solution by Invertible Matrix** Use the inverse matrix determined in Problem 7 to solve the system

$$\begin{aligned} y + z &= 5, \\ 5x + y - z &= 2, \\ 3x - 3y - 3z &= 0. \end{aligned}$$

More Solutions by Invertible Matrices *Use the inverse of the coefficient matrix to solve the systems in Problems 22 and 23.*

22. $$\begin{aligned} x - y + z &= 4 \\ x + y &= 1 \\ x + 2y - z &= 0 \end{aligned}$$

23. $$\begin{aligned} 4x + 3y - 2z &= 0 \\ 5x + 6y + 3z &= 10 \\ 3x + 5y + 2z &= 2 \end{aligned}$$

24. **Noninvertible 2 x 2 Matrices** Prove that for

$$\mathbf{A} = \begin{bmatrix} a & b \\ c & d \end{bmatrix}$$

A is not invertible if $ad = bc$.

Matrix Algebra with Inverses *In Problems 25–28, assume that* **A** *and* **B** *are invertible matrices of the same order.*

25. Simplify $(\mathbf{AB}^{-1})^{-1}$.

26. Simplify $\mathbf{B}(\mathbf{A}^2\mathbf{B}^2)^{-1}$.

27. If $\mathbf{A}(\mathbf{BA})^{-1}\vec{\mathbf{x}} = \vec{\mathbf{b}}$, solve for $\vec{\mathbf{x}}$.

28. Simplify $\left(\mathbf{A}^{-1}(\mathbf{BA}^{-1})^{-1}\mathbf{BA}^{-1}\right)^{-1}$.

29. **Question of Invertibility** What condition is required in order to solve for $\vec{\mathbf{x}}$ when $(\mathbf{A} + \mathbf{B})\vec{\mathbf{x}} = \vec{\mathbf{b}}$?

30. **Cancellation Works** Prove using matrix algebra that if **A**, **B**, and **C** are matrices such that $\mathbf{AB} = \mathbf{AC}$, and **A** is invertible, then $\mathbf{B} = \mathbf{C}$.

31. **An Inverse** Prove that if **A** is invertible and **B** is another square matrix such that $\mathbf{AB} = \mathbf{I}$, then $\mathbf{B} = \mathbf{A}^{-1}$.

Making Invertible Matrices *Choose a constant k so that the matrices given in Problems 32 and 33 are invertible. If no such k exists, say so and explain your reasoning.*

32. $\begin{bmatrix} 1 & 0 & k \\ 0 & 1 & 0 \\ 0 & 0 & 1 \end{bmatrix}$

33. $\begin{bmatrix} 1 & 0 & k \\ 0 & 1 & 0 \\ k & 0 & 1 \end{bmatrix}$

34. **Products and Noninvertibility** Let **A** and **B** be $n \times n$ matrices.
    (a) Show that if $\mathbf{BA} = \mathbf{I}_n$, then $\mathbf{AB} = \mathbf{I}_n$.
    (b) Show that if **AB** is invertible, then **A** must be invertible.

35. **Invertibility of Diagonal Matrices** Show that a diagonal matrix **A** is invertible if and only if all diagonal elements are nonzero. Give the form of $\mathbf{A}^{-1}$.

36. **Invertibility of Triangular Matrices** Show that an upper triangular matrix is invertible if and only if all diagonal elements are nonzero.

37. **Inconsistency** If the matrix-vector equation $\mathbf{A}\vec{\mathbf{x}} = \vec{\mathbf{b}}$ is inconsistent for some $\vec{\mathbf{b}}$ in $\mathbb{R}^n$, what can you determine about matrix **A**?

38. **Inverse of an Inverse** Prove the following property (stated in this section): "If **A** is invertible, then so is $\mathbf{A}^{-1}$, and $(\mathbf{A}^{-1})^{-1} = \mathbf{A}$."

39. **Inverse of a Transpose** Prove the following property (stated in this section): "If **A** is invertible, then so is $\mathbf{A}^{\mathrm{T}}$, and $(\mathbf{A}^{\mathrm{T}})^{-1} = (\mathbf{A}^{-1})^{\mathrm{T}}$."

40. **Elementary Matrices** If we perform a single row operation on an identity matrix, we obtain an **elementary matrix** $\mathbf{E}_{\mathrm{Int}}$, $\mathbf{E}_{\mathrm{Repl}}$, or $\mathbf{E}_{\mathrm{Scale}}$. Find the elementary matrices for each of the following row operations on $\mathbf{I}_3$.
    (a) Interchange rows 1 and 2 ($\mathbf{E}_{\mathrm{Int}}$).
    (b) Add k times row 1 to row 3 ($\mathbf{E}_{\mathrm{Repl}}$).
    (c) Multiply k times row 2 ($\mathbf{E}_{\mathrm{Scale}}$).

41. **Invertibility of Elementary Matrices** Explain why all elementary matrices must be invertible. Demonstrate this

property by finding the inverses of $\mathbf{E}_{\text{Int}}$, $\mathbf{E}_{\text{Repl}}$, and $\mathbf{E}_{\text{Scale}}$ in Problem 40.

Similar Matrices *Prove the statements in Problems 42–45 given the following definition.*

Similar Matrices

A matrix **B** is defined to be **similar** to matrix **A** (denoted by $\mathbf{B} \sim \mathbf{A}$) if there is an invertible matrix **P** such that $\mathbf{B} = \mathbf{P}^{-1}\mathbf{A}\mathbf{P}$.

42. Matrix **B** is similar to itself; that is, $\mathbf{B} \sim \mathbf{B}$.

43. If $\mathbf{B} \sim \mathbf{A}$, then $\mathbf{A} \sim \mathbf{B}$.

44. If $\mathbf{A} \sim \mathbf{B}$ and $\mathbf{B} \sim \mathbf{C}$, then $\mathbf{A} \sim \mathbf{C}$.

45. If $\mathbf{B} = \mathbf{P}^{-1}\mathbf{A}\mathbf{P}$ for some invertible **P**, then $\mathbf{B}^n = \mathbf{P}^{-1}\mathbf{A}^n\mathbf{P}$ for any positive integer n.

46. True/False Questions If true, explain. If false, give a 2×2 counterexample.

(a) A diagonal matrix is invertible if and only if its diagonal elements are nonzero. True or false?

(b) An upper triangular matrix is invertible if and only if its diagonal elements are nonzero. True or false?

(c) If **A** and **B** are $n \times n$ matrices such that **A** is invertible, then $\mathbf{A}\mathbf{B}\mathbf{A}^{-1} = \mathbf{B}$. True or false?

Leontief Model *Find the total outputs for each input-output matrix in Problems 47–50, with demands as given.*

47. $\mathbf{T} = \begin{bmatrix} 0.5 & 0 \\ 0 & 0.5 \end{bmatrix}, \quad \vec{\mathbf{d}} = \begin{bmatrix} 10 \\ 10 \end{bmatrix}$

48. $\mathbf{T} = \begin{bmatrix} 0 & 0.1 \\ 0.2 & 0 \end{bmatrix}, \quad \vec{\mathbf{d}} = \begin{bmatrix} 10 \\ 10 \end{bmatrix}$

49. $\mathbf{T} = \begin{bmatrix} 0.2 & 0.5 \\ 0.5 & 0.2 \end{bmatrix}, \quad \vec{\mathbf{d}} = \begin{bmatrix} 10 \\ 10 \end{bmatrix}$

50. $\mathbf{T} = \begin{bmatrix} 0.5 & 0.2 \\ 0.1 & 0.3 \end{bmatrix}, \quad \vec{\mathbf{d}} = \begin{bmatrix} 50 \\ 50 \end{bmatrix}$

51. How Much Is Left Over? In Example 5, suppose that A produces \$150 worth of its product and B produces \$250 worth of its product. What is the dollar value of both products available for external consumption?

52. Israeli Economy In 1966, Leontief used his input-output model to analyze the Israeli economy by dividing it into three segments: agriculture (A), manufacturing (M), and energy (E), as shown in the following technological matrix.

		Input		
		A	**M**	**E**
Output	**A**	\$0.30	\$0.00	\$0.00
	M	\$0.10	\$0.20	\$0.20
	E	\$0.05	\$0.01	\$0.02

The export demands on the Israeli economy (in thousands of Israeli pounds) are listed as follows.

Agriculture	\$140,000
Manufacturing	\$20,000
Energy	\$2,000

(a) Find $\mathbf{I} - \mathbf{T}$, where **T** is the technological matrix.

(b) Use computer software to find $(\mathbf{I} - \mathbf{T})^{-1}$.

(c) Find the total output for each sector required to meet both internal and external demand.

53. Suggested Journal Entry Discuss the similarities and dissimilarities between matrices and real numbers. Consider such issues as invertibility and cancellation laws. Can you think of comparisons going beyond the features so far presented in the text? To what extent can this comparison be adapted to the complex numbers instead?

3.4 Determinants and Cramer's Rule

SYNOPSIS: We introduce the determinant of a square matrix and use it to find a useful characterization of invertible matrices. If the matrix of coefficients of a system of n linear equations in n unknowns is invertible, we can use its determinant in Cramer's Rule to find the unique solution. We use matrices in the method of least squares to find a line of best fit for a given set of data.

Determinant of a Matrix

The determinant of a square matrix **A** is a *number* $|\mathbf{A}|$ associated with that matrix. This is different from the transpose of a matrix and the inverse of a matrix—those are matrices, but the determinant is a *scalar*.

Determinant of a 2 × 2 Matrix

The **determinant of a two-by-two matrix** is the product of the diagonal elements minus the product of the off-diagonal elements:

$$|\mathbf{A}| = \begin{vmatrix} a_{11} & a_{12} \\ a_{21} & a_{22} \end{vmatrix} = a_{11}a_{22} - a_{12}a_{21}. \tag{1}$$

Minors of a Matrix:

Cross out row 3 and column 2

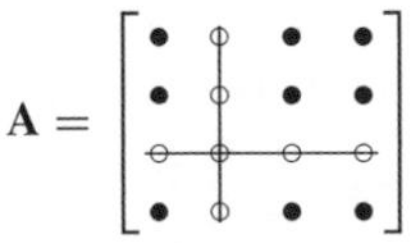

to obtain the minor of a_{32},

$$\mathbf{M}_{32} = \begin{bmatrix} \bullet & \bullet & \bullet \\ \bullet & \bullet & \bullet \\ \bullet & \bullet & \bullet \end{bmatrix}.$$

Given a formula for the 2 × 2 case, we develop a recursive procedure for calculating the determinant of an $n \times n$ matrix in terms of determinants of related matrices of order $(n-1) \times (n-1)$. For example, the determinant of a 4 × 4 matrix is expressed in terms of those of certain 3 × 3 matrices. Each of these, in turn, is calculated from the values of some 2 × 2 determinants. For these we use the basic formula (1). This recursive procedure, while apparently complicated, makes it possible to deal with arbitrarily large matrices in a systematic way.

Minors and Cofactors of a Matrix

Every element a_{ij} of an $n \times n$ matrix **A** has an associated minor and cofactor.

- The **minor** $\mathbf{M}_{ij}$ of a_{ij} is an $(n-1) \times (n-1)$ matrix obtained by deleting the ith row and the jth column of **A**.
- The **cofactor** of a_{ij} is the scalar

$$C_{ij} = (-1)^{i+j}|\mathbf{M}_{ij}|. \tag{2}$$

Signs of Matrix Cofactors:

The cofactor of an element of a matrix attaches alternate signs to the determinants of the minors, according to a checkerboard pattern:

$$\begin{bmatrix} + & - & + & - \\ - & + & - & + \\ + & - & + & - \\ - & + & - & + \end{bmatrix}.$$

For a_{32} the sign would be negative.

We can express the determinant of an $n \times n$ matrix in terms of the cofactors of any row or column.[1] (It does not matter which one because they all give the same answer.)[2]

Determinant of an $n \times n$ Matrix A

Choose any row or column and expand by the appropriate cofactor formula, using either **expansion by the ith row**:

$$|\mathbf{A}| = \sum_{j=1}^{n} a_{ij}C_{ij} = \sum_{j=1}^{n} a_{ij}(-1)^{i+j}|\mathbf{M}_{ij}| \tag{3}$$

or, equivalently, **expansion by the jth column**:

$$|\mathbf{A}| = \sum_{i=1}^{n} a_{ij}C_{ij} = \sum_{i=1}^{n} a_{ij}(-1)^{i+j}|\mathbf{M}_{ij}|. \tag{4}$$

Repeat this process, obtaining smaller matrices at each step. The definition is completed with the 2 × 2 case from equation (1).

---

[1]The subject of determinants was developed by many mathematicians in the late eighteenth and early nineteenth centuries. Despite its complicated definition, the determinant is a useful number that we will keep meeting in subsequent sections. The most complete work on the subject was written by French mathematician Augustin-Louis Cauchy (1789–1857); he first coined the word "determinant" in 1812. He also developed the method of evaluating determinants using expansion by minors along a row or down a column, usually known as Laplace's method.

[2]For a proof of the fact that the cofactors for any row or column can be used in calculating a determinant by Laplace expansion, see Otto Bretscher, *Linear Algebra* (Prentice Hall, 1997), 234.

EXAMPLE 1 **Computing a Determinant** We compute the determinant of matrix

$$\mathbf{A} = \begin{bmatrix} 3 & 1 & -1 \\ 2 & 1 & 3 \\ 0 & 1 & 2 \end{bmatrix},$$

using expansion by the first column. By equation (4),

$$|\mathbf{A}| = \sum_{i=1}^{3} a_{i1}C_{i1} = a_{11}C_{11} + a_{21}C_{21} + a_{31}C_{31}. \tag{5}$$

The minors associated with the entries of the first column are

$$\mathbf{M}_{11} = \begin{bmatrix} 1 & 3 \\ 1 & 2 \end{bmatrix}, \quad \mathbf{M}_{21} = \begin{bmatrix} 1 & -1 \\ 1 & 2 \end{bmatrix}, \quad \text{and} \quad \mathbf{M}_{31} = \begin{bmatrix} 1 & -1 \\ 1 & 3 \end{bmatrix},$$

so, by equations (2) and (5),

$$\begin{aligned} |\mathbf{A}| &= 3(-1)^{1+1}\begin{vmatrix} 1 & 3 \\ 1 & 2 \end{vmatrix} + 2(-1)^{2+1}\begin{vmatrix} 1 & -1 \\ 1 & 2 \end{vmatrix} + 0(-1)^{3+1}\begin{vmatrix} 1 & -1 \\ 1 & 3 \end{vmatrix} \\ &= 3\begin{vmatrix} 1 & 3 \\ 1 & 2 \end{vmatrix} - 2\begin{vmatrix} 1 & -1 \\ 1 & 2 \end{vmatrix} + 0\begin{vmatrix} 1 & -1 \\ 1 & 3 \end{vmatrix}. \end{aligned}$$

We complete the evaluation using equation (1) for the 2×2 determinants:

$$\begin{aligned} |\mathbf{A}| &= 3[(1)(2) - (3)(1)] - 2[(1)(2) - (-1)(1)] + 0[(1)(3) - (-1)(1)] \\ &= 3(-1) - 2(3) + 0(4) = -9. \end{aligned}$$

To confirm that the result $|\mathbf{A}| = -9$ was not dependent on the choice of column, try calculating $|\mathbf{A}|$ as an expansion by the third row. ■

EXAMPLE 2 **Another Determinant** The strategy for shortening the calculation of a determinant is to choose a row or column for expansion having the most zero entries. In computing $|\mathbf{A}|$ for the matrix

$$\mathbf{A} = \begin{bmatrix} 1 & 2 & 3 & 1 \\ 5 & 0 & 1 & -2 \\ 4 & 0 & 1 & 0 \\ 2 & 0 & 3 & 1 \end{bmatrix},$$

we choose the second column of $\mathbf{A}$ and calculate

$$|\mathbf{A}| = \begin{vmatrix} 1 & 2 & 3 & 1 \\ 5 & 0 & 1 & -2 \\ 4 & 0 & 1 & 0 \\ 2 & 0 & 3 & 1 \end{vmatrix} = 2(-1)^{1+2}\begin{vmatrix} 5 & 1 & -2 \\ 4 & 1 & 0 \\ 2 & 3 & 1 \end{vmatrix}.$$

In the 3×3 determinant from $\mathbf{M}_{12}$ we choose to expand by the second row:

$$\begin{aligned} |\mathbf{A}| = -2\begin{vmatrix} 5 & 1 & -2 \\ 4 & 1 & 0 \\ 2 & 3 & 1 \end{vmatrix} &= -2\left(4(-1)^{2+1}\begin{vmatrix} 1 & -2 \\ 3 & 1 \end{vmatrix} + 1(-1)^{2+2}\begin{vmatrix} 5 & -2 \\ 2 & 1 \end{vmatrix}\right) \\ &= 8(1+6) - 2(5+4) = 38. \end{aligned}$$

■

Row Operations and Determinants

Let **A** be a square matrix.

- If two rows of **A** are interchanged to produce matrix **B**, then
$$|\mathbf{B}| = -|\mathbf{A}|.$$
- If one row of **A** is multiplied by a constant k and then added to another row to produce matrix **B**, then
$$|\mathbf{B}| = |\mathbf{A}|.$$
- If one row of **A** is multiplied by k to produce matrix **B**, then
$$|\mathbf{B}| = k|\mathbf{A}|.$$

why?

Inverse Shortcut:

The shortcut for finding the inverse for a 2×2 matrix
$$\mathbf{A} = \begin{bmatrix} a & b \\ c & d \end{bmatrix},$$
found in Sec. 3.3, Problem 14, can be written more simply as
$$\mathbf{A}^{-1} = \frac{1}{|\mathbf{A}|}\begin{bmatrix} d & -b \\ -c & a \end{bmatrix}.$$

Proofs for these statements are outlined in Problems 37 and 38.

We perform *exactly* these operations to calculate the RREF for a square matrix **A**. If we do p row interchanges and scale the pivot elements to 1 by means of q multiplications by constants $k_q, k_{q-1}, \ldots, k_1$, and do any number of operations in which we add a multiple of one row to another to get the RREF **R**, we see that
$$|\mathbf{R}| = (-1)^p k_q k_{q-1} \cdots k_1 |\mathbf{A}|,$$
so $|\mathbf{A}| \neq 0$ if and only if $|\mathbf{R}| \neq 0$ if and only if the RREF of **A** is the identity matrix if and only if **A** is invertible. We have a new characterization for an invertible matrix.

Invertible Matrix Characterization (Using Determinants)

Let **A** be an $n \times n$ matrix. The following statements are equivalent:

- **A** is an invertible matrix.
- $|\mathbf{A}| \neq 0$.

If $|\mathbf{A}| = 0$, **A** is called **singular**, otherwise **A** is nonsingular.

The determinant of a product can be calculated simply by means of the following property. The proof is outlined in Problem 38.

Determinants of Sums:

For most matrices **A** and **B**
$$|\mathbf{A} + \mathbf{B}| \neq |\mathbf{A}| + |\mathbf{B}|.$$
Try some and see.

Determinants of Products of Matrices

For $n \times n$ matrices **A** and **B**, the determinant of **AB** is given by
$$|\mathbf{AB}| = |\mathbf{A}||\mathbf{B}|.$$

EXAMPLE 3 **Trying Out the Product** Let
$$\mathbf{A} = \begin{bmatrix} 1 & 4 & 0 \\ 1 & 0 & -2 \\ 0 & 1 & 0 \end{bmatrix} \quad \text{and} \quad \mathbf{B} = \begin{bmatrix} 1 & 2 & 0 \\ 0 & 1 & 1 \\ 1 & 3 & 0 \end{bmatrix}.$$
First find the determinant of the product,
$$|\mathbf{AB}| = \begin{vmatrix} 1 & 6 & 4 \\ -1 & -4 & 0 \\ 0 & 1 & 1 \end{vmatrix} = 1\begin{vmatrix} -4 & 0 \\ 1 & 1 \end{vmatrix} + 1\begin{vmatrix} 6 & 4 \\ 1 & 1 \end{vmatrix} = -4 + 2 = -2,$$

and then calculate $|\mathbf{A}|$ and $|\mathbf{B}|$,

$$|\mathbf{A}| = \begin{vmatrix} 1 & 4 & 0 \\ 1 & 0 & -2 \\ 0 & 1 & 0 \end{vmatrix} = -1 \begin{vmatrix} 1 & 0 \\ 1 & -2 \end{vmatrix} = -1(-2) = 2$$

and

$$|\mathbf{B}| = \begin{vmatrix} 1 & 2 & 0 \\ 0 & 1 & 1 \\ 1 & 3 & 0 \end{vmatrix} = -1 \begin{vmatrix} 1 & 2 \\ 1 & 3 \end{vmatrix} = -1(1) = -1.$$

Multiplying the determinants for **A** and **B**,

$$|\mathbf{A}|\,|\mathbf{B}| = (2)(-1) = -2,$$

we can see that $|\mathbf{AB}| = |\mathbf{A}|\,|\mathbf{B}| = -2$. ■

The interesting fact to remember about determinants is that they are scalars, so the usual rules for numbers apply. For instance, if $|\mathbf{A}|\,|\mathbf{B}| = 0$, then either $|\mathbf{A}|$ or $|\mathbf{B}|$ (or both) must equal zero just like any numbers. (Recall, however, that we can have matrices **A** and **B** such that $\mathbf{AB} = \mathbf{0}$ but neither $\mathbf{A} = \mathbf{0}$ nor $\mathbf{B} = \mathbf{0}$.)

We will list some more properties of determinants that either have been proved or will be explored in the problems.

Upper Triangular Matrix:

$$\begin{bmatrix} a_{11} & a_{12} & \cdots & a_{1n} \\ 0 & a_{22} & \cdots & a_{2n} \\ \vdots & \ddots & \ddots & \vdots \\ 0 & \cdots & 0 & a_{nn} \end{bmatrix}$$

Lower Triangular Matrix:

$$\begin{bmatrix} a_{11} & 0 & \cdots & 0 \\ a_{21} & a_{22} & \ddots & \vdots \\ \vdots & \vdots & \ddots & 0 \\ a_{n1} & a_{n2} & \cdots & a_{nn} \end{bmatrix}$$

Other Properties of Determinants

- $|\mathbf{A}^{\mathrm{T}}| = |\mathbf{A}|$.
- If $|\mathbf{A}| \neq 0$, then $|\mathbf{A}^{-1}| = \dfrac{1}{|\mathbf{A}|}$.
- If **A** is an upper triangular or lower triangular matrix, the determinant is the product of the diagonal elements,
$$|\mathbf{A}| = \prod_{i=1}^{m} a_{ii}.$$
- If one row or column of **A** consists entirely of zeros, then $|\mathbf{A}| = 0$.
- If two rows or two columns of **A** are equal, then $|\mathbf{A}| = 0$.

Cramer's Rule

A method for solving $n \times n$ systems of equations that have unique solutions using determinants, called **Cramer's Rule**, provides insight into the relationship between these two topics.[3] It is not as efficient as row reduction for numerical computation because the number of operations grows very rapidly as n increases, but it is often convenient for $n = 2$ or $n = 3$. Furthermore, Cramer's Rule is particularly useful in systems of linear differential equations where the coefficients are functions instead of constants.

[3]Gabriel Cramer (1704–1752) played a significant role in communicating mathematical ideas. He did not originate the rule bearing his name, but he developed notation that made it easier to state and apply.

Cramer's Rule

For the $n \times n$ matrix $\mathbf{A}$ having $|\mathbf{A}| \neq 0$, denote by $\mathbf{A}_i$ the matrix obtained from $\mathbf{A}$ by replacing its ith column with the column vector $\vec{\mathbf{b}}$. Then the ith component of the solution of the system $\mathbf{A}\vec{\mathbf{x}} = \vec{\mathbf{b}}$ is given by

$$x_i = \frac{|\mathbf{A}_i|}{|\mathbf{A}|}. \tag{6}$$

According to Cramer's Rule

$$\mathbf{A} = \begin{bmatrix} a & b \\ c & d \end{bmatrix},$$

$$\mathbf{A}_1 = \begin{bmatrix} e & b \\ f & d \end{bmatrix},$$

$$\mathbf{A}_2 = \begin{bmatrix} a & e \\ c & f \end{bmatrix}.$$

Proof We prove the theorem for the 2×2 case[4] by writing

$$\begin{aligned} ax_1 + bx_2 &= e, \\ cx_1 + dx_2 &= f. \end{aligned}$$

Multiplying the first equation by d, the second equation by $-b$, and adding the two equations, we arrive at

$$(ad - bc)x_1 = ed - bf,$$

$$x_1 = \frac{ed - bf}{ad - bc} = \frac{\begin{vmatrix} e & b \\ f & d \end{vmatrix}}{\begin{vmatrix} a & b \\ c & d \end{vmatrix}} = \frac{|\mathbf{A}_1|}{|\mathbf{A}|},$$

provided that $|\mathbf{A}| = ad - bc \neq 0$. By a similar argument, we can solve for x_2, getting

$$x_2 = \frac{af - ce}{ad - bc} = \frac{\begin{vmatrix} a & e \\ c & f \end{vmatrix}}{\begin{vmatrix} a & b \\ c & d \end{vmatrix}} = \frac{|\mathbf{A}_2|}{|\mathbf{A}|}.$$

□

EXAMPLE 4 Solutions by Cramer

(a) To solve the system of equations

$$\begin{aligned} x_1 + 2x_2 &= 5, \\ 2x_1 + 3x_2 &= 8 \end{aligned}$$

by Cramer's Rule, we first construct

$$\mathbf{A} = \begin{bmatrix} 1 & 2 \\ 2 & 3 \end{bmatrix}, \quad \mathbf{A}_1 = \begin{bmatrix} 5 & 2 \\ 8 & 3 \end{bmatrix}, \quad \text{and} \quad \mathbf{A}_2 = \begin{bmatrix} 1 & 5 \\ 2 & 8 \end{bmatrix}.$$

Then, by equation (6),

$$x_1 = \frac{|\mathbf{A}_1|}{|\mathbf{A}|} = \frac{-1}{-1} = 1 \quad \text{and} \quad x_2 = \frac{|\mathbf{A}_2|}{|\mathbf{A}|} = \frac{-2}{-1} = 2.$$

(b) Applying Cramer's Rule to solve the 3×3 system

$$\begin{aligned} 3x + 4y + z &= 4, \\ x \phantom{{}+4y} + 2z &= 3, \\ 2y + 4z &= 4, \end{aligned}$$

[4] This proof can be generalized to $n \times n$ systems. E.g., see Howard Anton and Chris Rorres, *Linear Algebra*, 7th ed. (NY: John Wiley and Sons, 1994).

we may write down the determinants of the various matrices directly:

$$x = \frac{\begin{vmatrix} 4 & 4 & 1 \\ 3 & 0 & 2 \\ 4 & 2 & 4 \end{vmatrix}}{\begin{vmatrix} 3 & 4 & 1 \\ 1 & 0 & 2 \\ 0 & 2 & 4 \end{vmatrix}} = \frac{-26}{-26} = 1, \quad y = \frac{\begin{vmatrix} 3 & 4 & 1 \\ 1 & 3 & 2 \\ 0 & 4 & 4 \end{vmatrix}}{\begin{vmatrix} 3 & 4 & 1 \\ 1 & 0 & 2 \\ 0 & 2 & 4 \end{vmatrix}} = \frac{0}{-26} = 0,$$

$$z = \frac{\begin{vmatrix} 3 & 4 & 4 \\ 1 & 0 & 3 \\ 0 & 2 & 4 \end{vmatrix}}{\begin{vmatrix} 3 & 4 & 1 \\ 1 & 0 & 2 \\ 0 & 2 & 4 \end{vmatrix}} = \frac{-26}{-26} = 1.$$

(c) The system of two equations in two variables,

$$x_1 + 2x_2 = 4,$$
$$2x_1 + \lambda x_2 = 3,$$

containing a parameter λ, can be solved by Cramer's Rule, showing clearly how the solutions depend on λ. We first identify

$$\mathbf{A} = \begin{bmatrix} 1 & 2 \\ 2 & \lambda \end{bmatrix}, \quad \mathbf{A}_1 = \begin{bmatrix} 4 & 2 \\ 3 & \lambda \end{bmatrix}, \quad \text{and} \quad \mathbf{A}_2 = \begin{bmatrix} 1 & 4 \\ 2 & 3 \end{bmatrix}.$$

Then, using equation (6), we find

$$x_1 = \frac{|\mathbf{A}_1|}{|\mathbf{A}|} = \frac{4\lambda - 6}{\lambda - 4} \quad \text{and} \quad x_2 = \frac{|\mathbf{A}_2|}{|\mathbf{A}|} = \frac{-5}{\lambda - 4}.$$

The system has no solution if $\lambda = 4$. ■

Method of Least Squares

A standard problem in the application of mathematics to the real world is to "fit" a curve to a set of experimental data. For some data and some curves this can be done exactly; in many cases an approximation is the best that we can expect.

A general strategy for finding the line $y = mx + k$ that best describes the trend of n data points is to determine k and m to minimize the quantity

$$F(k, m) = \sum_{i=1}^{n} [y_i - (k + mx_i)]^2, \tag{7}$$

the sum of the squares of the vertical distances between the data points and the line. To find the values of k and m that minimize (7), we must solve the 2×2 system

$$\frac{\partial F}{\partial k} = 0, \quad \frac{\partial F}{\partial m} = 0. \tag{8}$$

After differentiating and simplifying (see Problem 44), (8) becomes the following system:

Least Squares Method:

The best-fit straight line for n data points (x_i, y_i), $i = 1, 2, \ldots, n$, has intercept k and slope m as determined by the system

$$\begin{bmatrix} \sum_{i=1}^{n} 1 & \sum_{i=1}^{n} x_i \\ \sum_{i=1}^{n} x_i & \sum_{i=1}^{n} x_i^2 \end{bmatrix} \begin{bmatrix} k \\ m \end{bmatrix} = \begin{bmatrix} \sum_{i=1}^{n} y_i \\ \sum_{i=1}^{n} x_i y_i \end{bmatrix}. \tag{9}$$

Matrix-Vector Form of Example 5:

$$\begin{bmatrix} 1 & x_1 \\ 1 & x_2 \\ 1 & x_3 \\ 1 & x_4 \end{bmatrix} \begin{bmatrix} k \\ m \end{bmatrix} = \begin{bmatrix} y_1 \\ y_2 \\ y_3 \\ y_4 \end{bmatrix}.$$

EXAMPLE 5 **Least Squares Fit of College Grades** Consider the data in Table 3.4.1, listing high school and college grade point averages for four students. For each i, x_i denotes the high school GPA of the ith student, and y_i denotes the student's college GPA.

Table 3.4.1 Student GPAs

i	x_i	y_i
1	1.7	1.1
2	2.3	3.1
3	3.1	2.3
4	4.0	3.8

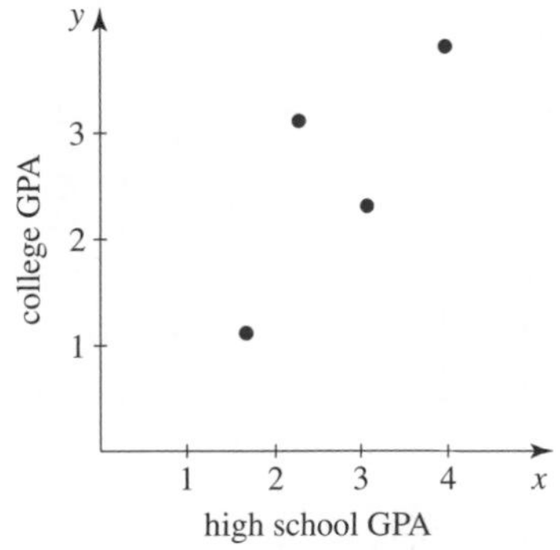

FIGURE 3.4.1 Scatter plot for student GPAs.

If we assume a linear relation, that is, that y is a linear function of x, we want to find values of the parameters k and m such that the line $y = k + mx$ contains the data points. The scatter plot of data points in Fig. 3.4.1 makes it clear that this is not possible: The points are obviously not collinear. This corresponds to the fact that the system of equations in k and m for the points to lie on this line is overdetermined:

$$\begin{aligned} k + 1.7m &= 1.1, \\ k + 2.3m &= 3.1, \\ k + 3.1m &= 2.3, \\ k + 4.0m &= 3.8 \end{aligned} \qquad \text{or, in general,} \qquad \begin{aligned} k + mx_1 &= y_1, \\ k + mx_2 &= y_2, \\ k + mx_3 &= y_3, \\ k + mx_4 &= y_4. \end{aligned}$$

Thus, for the data points of Table 3.4.1, the least squares formula (9) reduces to

$$\begin{bmatrix} 4 & 11.1 \\ 11.1 & 33.79 \end{bmatrix} \begin{bmatrix} k \\ m \end{bmatrix} = \begin{bmatrix} 10.3 \\ 31.33 \end{bmatrix},$$

and this system has the unique solution $[k, m] = [0.023, 0.92]$. The least squares line is $y = 0.023 + 0.92x$. It is superimposed on the data points in Fig. 3.4.2. ■

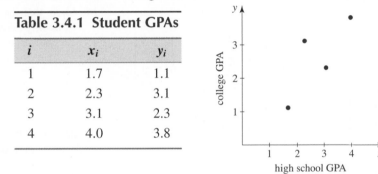

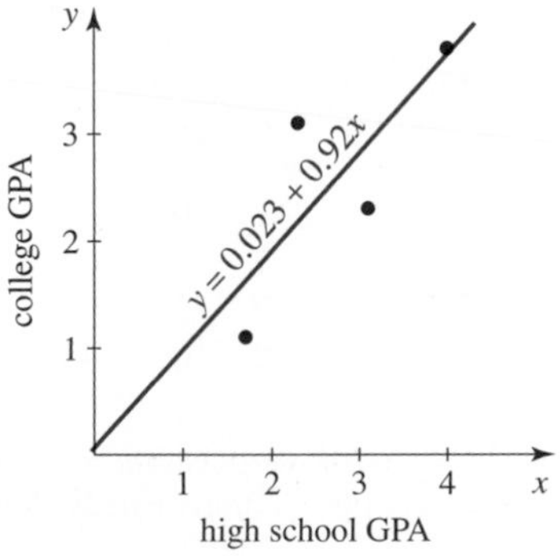

FIGURE 3.4.2 Least squares line for student GPAs (Example 5).

Obtaining the "least squares" system of equations (9) for a specific set of data can be simplified using matrix notation and properties of the transpose. This approach is outlined in Problem 45.

Summary

The determinant of a square matrix reveals whether or not the matrix is invertible. Cramer's Rule gives the explicit solution in terms of determinants for $n \times n$ systems that have unique solutions. The matrix formulation can help in determining the least squares line for sets of data.

3.4 Problems

Calculating Determinants *Use cofactor expansion and/or row reduction to evaluate the determinant of each matrix in Problems 1–7. (Choose your row or column carefully.)*

1. $\begin{bmatrix} 0 & 7 & 9 \\ 2 & 1 & -1 \\ 5 & 6 & 2 \end{bmatrix}$ **2.** $\begin{bmatrix} 1 & 2 & 3 \\ 0 & 1 & 0 \\ 1 & 0 & -3 \end{bmatrix}$

3. $\begin{bmatrix} 1 & 3 & 0 & -2 \\ 0 & 1 & -1 & 5 \\ -1 & -2 & 1 & 7 \\ 1 & 1 & 0 & -6 \end{bmatrix}$ **4.** $\begin{bmatrix} 1 & -4 & 2 & -2 \\ 4 & 7 & -3 & 5 \\ 3 & 0 & 8 & 0 \\ -5 & -1 & 6 & 9 \end{bmatrix}$

5. $\begin{bmatrix} 1 & 1 & 1 \\ 2 & 2 & 2 \\ 3 & 3 & 3 \end{bmatrix}$ **6.** $\begin{bmatrix} 0 & 0 & 1 \\ 0 & 2 & 1 \\ 3 & 1 & 1 \end{bmatrix}$

7. $\begin{bmatrix} 1 & 2 & 2 & 4 \\ -2 & 2 & -2 & 2 \\ 2 & 1 & -1 & -2 \\ -1 & -4 & 4 & 2 \end{bmatrix}$

Find the Properties *State which one of the row operations for determinants is illustrated in each of Problems 8–10.*

8. $\begin{vmatrix} 2 & 3 \\ -1 & 4 \end{vmatrix} = \begin{vmatrix} 2 & 3 \\ -3 & 1 \end{vmatrix}$ **9.** $\begin{vmatrix} 6 & 1 \\ 3 & -3 \end{vmatrix} = 3\begin{vmatrix} 6 & 1 \\ 1 & -1 \end{vmatrix}$

10. $\begin{vmatrix} 1 & -1 \\ 2 & 3 \end{vmatrix} = -\begin{vmatrix} 2 & 3 \\ 1 & -1 \end{vmatrix}$

Basketweave for 3 × 3 *There is a shortcut for finding the determinant of a* 3×3 *matrix. For example, let*

$$\mathbf{A} = \begin{bmatrix} 1 & 2 & 4 \\ 1 & 2 & 0 \\ 2 & 1 & 5 \end{bmatrix},$$

and repeat the first and second columns to the right of the original three columns:

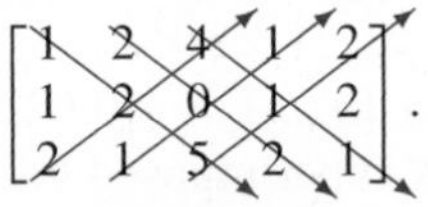

Now, add the products along the downward arrows and subtract the products along the upward arrows, getting

$$(10 + 0 + 4) - (16 + 0 + 10) = -12.$$

11. Check the result for $|\mathbf{A}|$ using the cofactor expansion.

12. Use the basketweave method to find the determinant for the matrix in Problem 1.

13. Use the basketweave method to find the determinant for the matrix in Problem 2.

14. Show that the basketweave method does not generalize to finding determinants for higher-order matrices. Try a generalized basketweave method on

$$\begin{vmatrix} 0 & 1 & 1 & 0 \\ 1 & 1 & 0 & 1 \\ 0 & 0 & 0 & 1 \\ 0 & 1 & 1 & 0 \end{vmatrix}$$

and show that it does not match the correct answer, which can be obtained by the cofactor method.

15. Triangular Determinants Show that the determinant of an upper triangular matrix is the product of its diagonal elements. The diagonal matrix is a special case. The statement for a lower triangular matrix can be proved in a similar fashion. HINT: For each cofactor expansion, use the first column.

Think Diagonal *Use the ideas of Problem 15 to evaluate the determinants in Problems 16–19.*

16. $\begin{vmatrix} -3 & 4 & 0 \\ 0 & 7 & 6 \\ 0 & 0 & 5 \end{vmatrix}$ **17.** $\begin{vmatrix} 4 & 0 & 0 \\ 0 & -3 & 0 \\ 0 & 0 & 1/2 \end{vmatrix}$

18. $\begin{vmatrix} 1 & 0 & 0 & 0 \\ -3 & 4 & 0 & 0 \\ 0 & 5 & -1 & 0 \\ 11 & 0 & -2 & 2 \end{vmatrix}$ **19.** $\begin{vmatrix} 6 & 22 & 0 & -3 \\ 0 & -1 & 0 & 4 \\ 0 & 0 & 13 & 0 \\ 0 & 0 & 0 & 4 \end{vmatrix}$

Invertibility *In Problems 20–22, what choices of k and m would make the matrices invertible?* HINT: *Check the determinant.*

20. $\begin{bmatrix} 1 & 0 & k \\ 0 & k & 1 \\ k & 0 & 4 \end{bmatrix}$ **21.** $\begin{bmatrix} 1 & k \\ k & -k \end{bmatrix}$ **22.** $\begin{bmatrix} 1 & 0 & m \\ 0 & 1 & 0 \\ k & 0 & 1 \end{bmatrix}$

Invertibility Test *In Problems 23–26, use the determinants of the matrices to test for the invertibility of the matrices.*

23. $\begin{bmatrix} 0 & -1 & 0 \\ 4 & 0 & 2 \\ 0 & -1 & 0 \end{bmatrix}$ **24.** $\begin{bmatrix} 1 & 3 \\ -1 & 2 \end{bmatrix}$

25. Problem 3 **26.** Problem 4

Product Verification *For the given* 2×2 *matrices* $\mathbf{A}$ *and* $\mathbf{B}$, *show directly (by finding the product* $\mathbf{AB}$*) that* $|\mathbf{AB}| = |\mathbf{A}|\,|\mathbf{B}|$.

27. $\mathbf{A} = \begin{bmatrix} 1 & 2 \\ 3 & 4 \end{bmatrix}$, $\mathbf{B} = \begin{bmatrix} 1 & 0 \\ 1 & 1 \end{bmatrix}$

28. $\mathbf{A} = \begin{bmatrix} 0 & 1 & 0 \\ 1 & 0 & 0 \\ 1 & 2 & 2 \end{bmatrix}$, $\mathbf{B} = \begin{bmatrix} 1 & 2 & 3 \\ -1 & 2 & 0 \\ 0 & 1 & -1 \end{bmatrix}$

29. Determinant of an Inverse Prove for invertible matrix $\mathbf{A}$ that

$$\left|\mathbf{A}^{-1}\right| = \frac{1}{|\mathbf{A}|}.$$

30. Do Determinants Commute? Let $\mathbf{A}$ and $\mathbf{B}$ be any two $n \times n$ matrices. Explain why $|\mathbf{AB}| = |\mathbf{BA}|$.

31. Determinant of Similar Matrices Matrix $\mathbf{A}$ is said to be **similar** to matrix $\mathbf{B}$ if there is an invertible matrix $\mathbf{P}$ such that $\mathbf{A} = \mathbf{P}^{-1}\mathbf{BP}$. (This will be discussed in Sec. 5.4 on diagonalization.) Show that similar matrices $\mathbf{A}$ and $\mathbf{B}$ have the same determinant.

32. Determinant of $\mathbf{A}^n$ The notation $\mathbf{A}^n$ denotes $\underbrace{\mathbf{AA}\cdots\mathbf{A}}_{n \text{ factors}}$.

(a) If $|\mathbf{A}^n| = 0$ for some integer n, then $\mathbf{A}$ must be noninvertible. Show why this result is true.

(b) If $|\mathbf{A}^n| \neq 0$ for some integer n, then $\mathbf{A}$ must be invertible. Verify this result for $n = 4$.

33. Determinants of Sums Give an example of square matrices $\mathbf{A}$ and $\mathbf{B}$ for which $|\mathbf{A}+\mathbf{B}| \neq |\mathbf{A}| + |\mathbf{B}|$.

34. Determinants of Sums Again Give an example of nonzero square matrices $\mathbf{A}$ and $\mathbf{B}$ such that $|\mathbf{A}+\mathbf{B}| = |\mathbf{A}| + |\mathbf{B}|$.

35. Scalar Multiplication Determine $|k\mathbf{A}|$ in terms of k and $|\mathbf{A}|$.

36. Inversion by Determinants Let $\mathbf{A}$ be a square matrix and let $\tilde{\mathbf{A}}$ be its **cofactor matrix**, the matrix obtained from $\mathbf{A}$ by replacing each of its elements by its cofactor. Then if $|\mathbf{A}| \neq 0$,

$$\mathbf{A}^{-1} = \frac{1}{|\mathbf{A}|}\left(\tilde{\mathbf{A}}^{\mathrm{T}}\right).$$

Use this formula to compute the inverse of the matrix

$$\mathbf{A} = \begin{bmatrix} 1 & 0 & 2 \\ 2 & 2 & 3 \\ 1 & 1 & 1 \end{bmatrix}.$$

Check your result by computing the inverse using row reduction.

37. Determinants of Elementary Matrices Find the determinants for each of the elementary matrices (Sec. 3.3, Problems 40 and 41) formed by the elementary row operations on the 3×3 identity matrix $\mathbf{I}$, described as follows.

(a) Interchange two rows of $\mathbf{I}$ to get matrix $\mathbf{E}_{\text{Int}}$. Find $|\mathbf{E}_{\text{Int}}|$.

(b) Replace a row by the sum of a multiple of another row and the original row to obtain matrix $\mathbf{E}_{\text{Repl}}$. Find $|\mathbf{E}_{\text{Repl}}|$.

(c) Scale a row by multiplying by a nonzero scalar k to obtain $\mathbf{E}_{\text{Scale}}$. Find $|\mathbf{E}_{\text{Scale}}|$.

The conclusions all extend to $n \times n$ matrices, and in combination with Problem 38 will prove the rules for the effects of row operations on determinants.

38. Determinant of a Product Complete the proof that the determinant of a product is the product of the determinants, using the results of the previous problem and, from Sec. 3.3, Problems 40 and 41.

(a) Show that if $\mathbf{A}$ is not invertible, then $|\mathbf{AB}| = |\mathbf{A}|\,|\mathbf{B}|$. HINT: By Problem 34 in Sec. 3.3, if $\mathbf{A}$ is not invertible, then neither is $\mathbf{AB}$.

(b) When $\mathbf{A}$ is invertible, you should use the fact that

$$\mathbf{AB} = (\mathbf{E}_p\mathbf{E}_{p-1}\cdots\mathbf{E}_1\mathbf{I})\mathbf{B},$$

where each $\mathbf{E}_j$ represents an elementary matrix for a row operation, to show that $|\mathbf{AB}| = |\mathbf{A}|\,|\mathbf{B}|$. HINT: First show that $|\mathbf{AB}| = (-1)^s k_1 k_2 \cdots k_s |\mathbf{B}|$ for some integer s and constants $k_1, k_2, \ldots, k_s$.

Cramer's Rule *Solve each of the systems in Problems 39–42 by employing Cramer's Rule. In Problem 40, λ is a parameter.*

39. $\begin{aligned} x + 2y &= 2 \\ 2x + 5y &= 0 \end{aligned}$

40. $\begin{aligned} x + y &= \lambda \\ x + 2y &= 1 \end{aligned}$

41. $\begin{aligned} x + y + 3z &= 5 \\ 2y + 5z &= 7 \\ x + 2z &= 3 \end{aligned}$

42. $\begin{aligned} x_1 + 2x_2 - x_3 &= 6 \\ 3x_1 + 8x_2 + 9x_3 &= 10 \\ 2x_1 - x_2 + 2x_3 &= -2 \end{aligned}$

43. The Wheatstone Bridge The **Wheatstone bridge** is a device used to measure an unknown resistance R_x by comparing it with known resistances R_1, R_2, and R_3.[5] The circuit diagram is shown in Fig. 3.4.3. The known resistances are adjusted so that no current from the cell with

[5] Sir Charles Wheatstone (1802–1875) was an English physicist with interests in acoustics and electricity. He invented the concertina (a small accordion) and the telegraph (before Morse). Although (as he acknowledged) he did not invent the Wheatstone bridge, it was his use of it that brought it to the attention of others.

voltage E_0 passes through the wire BD, in which there is an ammeter with resistance R_g.

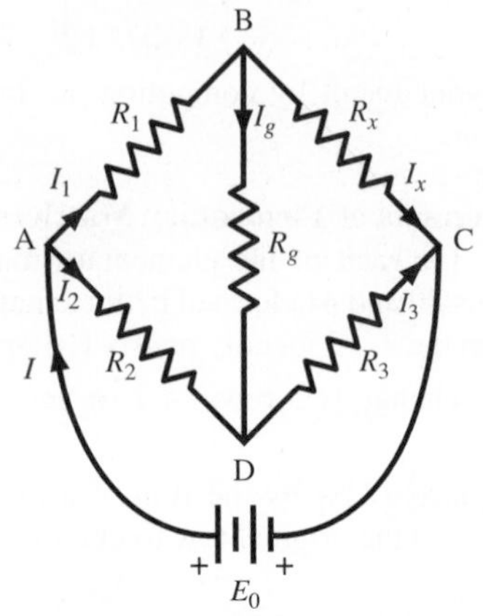

FIGURE 3.4.3 Wheatstone bridge (Problem 43).

(a) Use **Kirchoff's Current Law**[6] that *the algebraic sum of the currents at each junction must be zero* to show that the unknown currents I, I_1, I_2, I_3, and I_x satisfy the linear system

$$I_1 + I_2 = I, \quad I_x + I_3 = I,$$
$$I_g + I_x = I_1, \quad I_2 + I_g = I_3.$$

(b) **Kirchoff's Voltage Law** states that *the algebraic sum of all voltage drops around a closed circuit is zero.*[7] (Voltage equals resistance times current.) Use this law to show that

$$R_x I_x - R_3 I_3 - R_g I_g = 0,$$
$$R_1 I_1 + R_g I_g - R_2 I_2 = 0,$$
$$R_1 I_1 + R_x I_x = E_0.$$

(c) Show that the bridge is **balanced** (that is, $I_g = 0$) when $R_1 R_3 = R_2 R_x$.

44. Least Squares Derivation Derive equations (8) in the text for the parameters k and m in the least squares equation $y = k + mx$ by simplifying and solving the equations $\partial F/\partial k = 0$, $\partial F/\partial m = 0$, where F is as given in equation (7).

45. Alternative Derivation of Least Squares Equations Let

$$\mathbf{A} = \begin{bmatrix} 1 & x_1 \\ 1 & x_2 \\ 1 & x_3 \\ 1 & x_4 \end{bmatrix}, \quad \vec{\mathbf{x}} = \begin{bmatrix} k \\ m \end{bmatrix}, \quad \text{and} \quad \vec{\mathbf{b}} = \begin{bmatrix} y_1 \\ y_2 \\ y_3 \\ y_4 \end{bmatrix}.$$

(a) Show that equation (9) has matrix-vector form $\mathbf{A}\vec{\mathbf{x}} = \vec{\mathbf{b}}$.

(b) Show that premultiplying each side of the equation in part (a) by $\mathbf{A}^{\mathrm{T}}$ leads to the least squares equations (8) for $n = 4$.

46. Least Squares Calculation Obtain the least squares approximation $k + mx$ for the data that follows. Plot the points and the line.

x	y
0	1
1	1
2	3
3	3

Computer or Calculator *For the data sets in Problems 47 and 48, set up the system of equations to find a least squares fit. (A spreadsheet is a very fast way to compute the sums, and to make a scatter plot.) Then solve the system to find the actual least squares line $y = k + mx$. Finally, plot the points and the line.*

47.

x	y
1.6	1.7
3.2	5.3
6.9	5.1
8.4	6.5
9.1	8.0

48.

x	y
0.91	1.35
1.07	1.96
2.56	3.13
4.11	5.72
5.34	7.08
6.25	8.14

49. Least Squares in Another Dimension A chemist wishes to estimate the yield Y in a certain process that depends on the temperature T and the pressure P. A linear model for the process is assumed:

$$Y = a + b_1 T + b_2 P,$$

where a, b_1, and b_2 are unknown parameters. Assume that the observations are as follows.

Y	T	P
Y_1	T_1	P_1
Y_2	T_2	P_2
$\vdots$	$\vdots$	$\vdots$
Y_n	T_n	P_n

Derive equations for the coefficients of the least squares plane, generalizing the technique of the section for the least squares line.

[6] See Sec. 3.2, Problem 76.

[7] See "Modeling Electric Circuits" in Sec. 4.1.

50. Least Squares System Solution The overdetermined system of equations $\mathbf{A}\vec{\mathbf{x}} = \vec{\mathbf{b}}$ given by

$$\begin{bmatrix} 1 & 1 \\ 0 & 1 \\ -1 & 1 \end{bmatrix} \begin{bmatrix} x \\ y \end{bmatrix} = \begin{bmatrix} 1 \\ 2 \\ 1 \end{bmatrix} \tag{10}$$

has no solution. If you premultiply each side of the equation by $\mathbf{A}^{\mathrm{T}}$, however, you will obtain a system with a unique solution $\vec{\mathbf{x}} = [x_1, x_2]$ representing a **point** that *minimizes the sum of the squares of the differences between the left- and right-hand sides* of equation (10). It is called the **least squares solution** of $\mathbf{A}\vec{\mathbf{x}} = \vec{\mathbf{b}}$.

(a) Find the linear system $\mathbf{A}^{\mathrm{T}}\mathbf{A}\vec{\mathbf{x}} = \mathbf{A}^{\mathrm{T}}\mathbf{b}$.

(b) Solve the linear system found in (a) for the least squares solution.

(c) Plot the three lines determined by (10) and the point obtained in (b).

51. Suggested Journal Entry Suppose that a system of two equations in two variables consists of one linear and one nonlinear equation. How many solutions are possible? Give examples of various situations that may arise.

3.5 Vector Spaces and Subspaces

SYNOPSIS: We introduce the concept of vector space, fundamental to linear algebra, and the closely related idea of a subspace. We find familiar examples in the two- and three-dimensional Euclidean spaces, and other examples from algebra, calculus, and differential equations.

Introduction

Almost everyone likes vector spaces! Pure mathematicians like them because their theory is so elegant. Applied mathematicians like them because they provide a helpful setting for differential equations. Scientists and engineers like them because they provide visual models for complex concepts. Computer scientists seem to like them because they provide the matrices and systems of equations so ideal for number crunching. If there is anyone with doubts about vector spaces, it is Mother Nature. Vector spaces are incurably linear, while so many natural phenomena are nonlinear. But even the nonlinearities of the real world can often be approximated by linear structures.

Since the time of René Descartes, points in the plane have been designated by numerical coordinates (x_1, x_2), and the description of points in space by triples (x_1, x_2, x_3), came soon afterward.[1] William Rowan Hamilton broke out of the limitations of 3-dimensional space with his 4-dimensional "quaternions"; this research led to the notion of what we now call *vectors*.[2] The word "vector" came from the Latin word *vectus*, meaning "to carry."

In the decades around 1900, workers in both pure and applied mathematics kept turning up sets of objects that were quite different from points in the plane or in space but that obeyed many of the same rules. Identifying and studying the

[1]French philosopher and mathematician René Descartes (1596–1650) devoted his life to science and wrote many important works. His then-startling idea of Cartesian coordinates was added just as an appendix to his 1637 book on vortexes and the structure of the solar system, but it brought together algebra and geometry to the great enhancement of both. It was also Descartes who began our convention of using letters at the end of the alphabet for variables and reserving those at the beginning for constants, our use of exponent notation, and the $\sqrt{\ }$ symbol for square root.

[2]Sir William Rowan Hamilton (1805–1865), Ireland's greatest mathematician, was a genius who, by the age of 12, had not only mastered the languages of Europe, but had learned to read Greek, Latin, Sanskrit, Hebrew, Chinese, Persian, Arabic, Malay, Hindu, and several other languages. At this point he became interested in the challenge of mathematics and by the age of 17 was acknowledged by the Astronomer Royal of Ireland as already a mathematician (after Hamilton had found an error in Laplace's work on celestial mechanics).

generic structure that all these examples had in common proved very efficient, and the theory of **vector spaces** or **linear spaces** was born.

Vector Spaces

In studying the definition of a vector space, it can be helpful to think of the familiar two-dimensional "arrows" in the coordinate plane. But the definition we are going to give is *abstract*. The objects we work with are called vectors, but they may actually be functions, or points in five-dimensional space, or matrices of a particular size. All that matters is that they satisfy the requirements of the vector space definition. While geometric experience will help in grasping those requirements on a first acquaintance, it takes imagination to realize that vector spaces can be much more general.

We are assuming that vector spaces are *real* vector spaces in that the scalars are real numbers. We can extend this definition to complex vector spaces by allowing scalars to be complex numbers.

Important Vector Spaces:

The most important vector spaces for this course are $\mathbb{R}^n$, the solution spaces for linear algebraic systems, the DE solution spaces, and $\mathbb{M}_{mn}$.

Vector Space

A **vector space** $\mathbb{V}$ is a nonempty collection of objects called **vectors** for which are defined the operations

- *vector addition*, denoted $\vec{\mathbf{x}} + \vec{\mathbf{y}}$, and
- *scalar multiplication* (multiplication by a real constant), denoted $c\vec{\mathbf{x}}$,

that satisfy the following properties for all $\vec{\mathbf{x}}, \vec{\mathbf{y}}, \vec{\mathbf{z}} \in \mathbb{V}$ and $c, d \in \mathbb{R}$.

Closure Properties:

1. $\vec{\mathbf{x}} + \vec{\mathbf{y}} \in \mathbb{V}$.

2. $c\vec{\mathbf{x}} \in \mathbb{V}$.

Addition Properties:

3. There is a **zero vector** $\vec{\mathbf{0}}$ in $\mathbb{V}$ such that $\vec{\mathbf{x}} + \vec{\mathbf{0}} = \vec{\mathbf{x}}$. *(Additive Identity)*

4. For every vector $\vec{\mathbf{x}} \in \mathbb{V}$, there is a vector $-\vec{\mathbf{x}}$ in $\mathbb{V}$ (its **negative**) such that $\vec{\mathbf{x}} + (-\vec{\mathbf{x}}) = \vec{\mathbf{0}}$. *(Additive Inverse)*

5. $(\vec{\mathbf{x}} + \vec{\mathbf{y}}) + \vec{\mathbf{z}} = \vec{\mathbf{x}} + (\vec{\mathbf{y}} + \vec{\mathbf{z}})$. *(Associativity)*

6. $\vec{\mathbf{x}} + \vec{\mathbf{y}} = \vec{\mathbf{y}} + \vec{\mathbf{x}}$. *(Commutativity)*

Scalar Multiplication Properties:

7. $1\vec{\mathbf{x}} = \vec{\mathbf{x}}$. *(Scalar Multiplicative Identity)*

8. $c(\vec{\mathbf{x}} + \vec{\mathbf{y}}) = c\vec{\mathbf{x}} + c\vec{\mathbf{y}}$. *(First Distributive Property)*

9. $(c + d)\vec{\mathbf{x}} = c\vec{\mathbf{x}} + d\vec{\mathbf{x}}$. *(Second Distributive Property)*

10. $c(d\vec{\mathbf{x}}) = (cd)\vec{\mathbf{x}}$. *(Associativity)*

Do these properties look familiar? We hope so. (See Section 3.1.)

As we present examples of vector spaces, be alert for the definitions of the operations. Frequently, we refer to a "standard definition" when the "vectors" of our example are familiar objects, and they are added or multiplied in a familiar way. But sometimes we will find it instructive to define *new* operations on *old* objects in order to build a vector space with unusual but useful properties. It is important to keep your mind flexible and your imagination active.

Several useful properties of vector spaces are consequences of the basic requirements, such as the uniqueness of the zero vector and of the additive inverse of a vector, and the facts that $0\vec{\mathbf{x}} = \vec{\mathbf{0}}$ *and* $-\vec{\mathbf{x}} = (-1)\vec{\mathbf{x}}$ for all vectors $\vec{\mathbf{x}}$ in the space. The two closure requirements can be checked at once by verifying the following property:

Closure Under Linear Combination

0. $c\vec{\mathbf{x}} + d\vec{\mathbf{y}} \in \mathbb{V}$ whenever $\vec{\mathbf{x}}, \vec{\mathbf{y}} \in \mathbb{V}$ and $c, d \in \mathbb{R}$.

EXAMPLE 1 **The Vector Space $\mathbb{R}^n$** The familiar n-dimensional coordinate space $\mathbb{R}^n$ is a vector space. We can designate the vectors in $\mathbb{R}^n$ as points, such as $(x_1, x_2, \ldots, x_n)$, or row or column vectors with the same n entries or coordinates. In Sec. 3.1, the ten requirements were given as a special case for $\mathbb{M}_{mn}$. ■

EXAMPLE 2 **The Vector Space $\mathbb{M}_{mn}$** Let $\mathbb{M}_{mn}$ denote the collection of all $m \times n$ matrices with real entries. Looking back at the properties for addition and scalar multiplication of matrices in Sec. 3.1, we can see that the ten requirements for a vector space are met. In fact, the properties of $\mathbb{M}_{mn}$ and $\mathbb{R}^n$ and the operations of addition and scalar multiplication provided motivation for the general concept of vector spaces.

Here we explicitly check a few vector space properties in detail, just as a sampling of the entire list of properties that must be confirmed.

Real Number Properties Used in Checking the Properties for Example 2:

0. For any matrices $\mathbf{A} = [a_{ij}], \mathbf{B} = [b_{ij}] \in \mathbb{M}_{mn}$, and scalars $c, d \in \mathbb{R}$,

$$c\mathbf{A} + d\mathbf{B} = c[a_{ij}] + d[b_{ij}] = [ca_{ij} + db_{ij}] = [k_{ij}] \in \mathbb{M}_{mn},$$

where k_{ij} is a real number, for each element.

The sum of two real numbers is a real number.

5. For any matrices $\mathbf{A} = [a_{ij}], \mathbf{B} = [b_{ij}], \mathbf{C} = [c_{ij}] \in \mathbb{M}_{mn}$,

$$\begin{aligned}(\mathbf{A} + \mathbf{B}) + \mathbf{C} &= [a_{ij} + b_{ij}] + [c_{ij}] \\ &= [(a_{ij} + b_{ij}) + c_{ij}] \\ &= [a_{ij} + (b_{ij} + c_{ij})] \\ &= [a_{ij}] + [b_{ij} + c_{ij}] = \mathbf{A} + (\mathbf{B} + \mathbf{C}).\end{aligned}$$

For any real numbers a, b, and c, $(a + b) + c = a + (b + c)$.

10. For any scalars c and d and any matrix $\mathbf{A} = [a_{ij}] \in \mathbb{M}_{mn}$,

$$c(d\mathbf{A}) = c[da_{ij}] = [c(da_{ij})] = [(cd)a_{ij}] = (cd)\mathbf{A}.$$

For any real numbers a, c, and d, $c(da) = (cd)a$.

Therefore, $\mathbb{M}_{mn}$, with the operations of matrix addition and scalar multiplication, is a vector space. ■

Vector Function Spaces

Function spaces are important vector spaces in differential equations (as well as in many other branches of mathematical analysis). The "vectors" are functions defined on an interval I; they are added and multiplied in the usual way:

$$(f + g)(t) = f(t) + g(t) \quad \text{and} \quad (cf)(t) = cf(t), \qquad \text{for all } t \in I. \tag{1}$$

In Examples 3–6 we consider some candidates for function spaces.

EXAMPLE 3 DE Solution Space The set $\mathbb{V}$ of all solutions of the first-order linear homogeneous differential equation

$$y' + p(t)y = 0, \quad \text{defined on some interval } I, \tag{2}$$

where $p(t)$ is a continuous function on I, is a vector space. (For such a function space, the definitions (1) are considered to be "standard" and are not restated.)

To verify that these functions under these operations form a vector space, we need to check the ten requirements, but such properties as commutativity and associativity are well-known facts about functions from precalculus and calculus. The crucial requirement is the closure condition. So let us suppose that a and b are scalars and that $u(t)$ and $v(t)$ are solutions of (1), and check for closure under linear combination. We need to verify that $au(t) + bv(t)$ is also a solution of (2). We substitute and simplify:

$$\begin{aligned}(au + bv)' + p(t)(au + bv) &= au' + bv' + p(t)(au) + p(t)(bv)\\ &= a[u' + p(t)u] + b[v' + p(t)v]\\ &= a \cdot 0 + b \cdot 0 = 0.\end{aligned}$$

Since the calculation works when $a = b = 0$, the zero function is a solution, and the negative of a solution is again a solution because we can take $a = -1$ and $b = 0$. Verification of the remaining properties is left as an exercise. (See Problem 26.) ■

EXAMPLE 4 Second-Order DE Solution Space The set $\mathbb{V}$ of all solutions of the second-order linear homogeneous differential equation

$$y'' + p(t)y' + q(t)y = 0, \quad \text{defined on some interval } I, \tag{3}$$

where $p(t)$ and $q(t)$ are continuous functions on I, is a real vector space.

The homogeneous equation clearly has the zero function for a solution. To make the crucial check for closure properties, we will verify the property of closure under linear combination. So again we assume that a and b are scalars and that $u(t)$ and $v(t)$ are any two solutions of (3). Then

$$\begin{aligned}&(au + bv)'' + p(t)(au + bv)' + q(t)(au + bv)\\ &\quad = a[u'' + p(t)u' + q(t)u] + b[v'' + p(t)v' + q(t)v]\\ &\quad = a \cdot 0 + b \cdot 0 = 0.\end{aligned}$$

Verification of the other conditions is again left to the reader (see Problem 27), thus showing that the solutions $u(t)$ and $v(t)$ indeed form a vector space. ■

Solutions to *linear* and *homogeneous* DEs form a vector space. For example, consider a nonlinear DE, such as $y' + y^2 = 0$; which properties fail? It may be less obvious that it is also necessary for the DEs to be homogeneous or unforced, but the next example shows what can happen when that is not the case.

EXAMPLE 5 Nonhomogeneous Differential Equation The set of solutions of the first-order linear but nonhomogeneous differential equation

$$y' + 2ty = 1$$

is *not* a vector space: the zero function does not belong. ■

EXAMPLE 6 **Polynomial Space $\mathbb{P}_3$** Consider the space of all polynomials in t of degree ≤ 3. (The domain of these polynomial functions is $(-\infty, \infty)$.) A vector in $\mathbb{P}_3$ is given by

$$p(t) = a_0 + a_1 t + a_2 t^2 + a_3 t^3,$$

where a_1, a_2, and a_3 are real numbers. The sum of polynomials of degree ≤ 3 is always a polynomial of degree ≤ 3, as is the product of a scalar and a polynomial in $\mathbb{P}_3$. The other properties also follow from algebra. ■

In this book, the independent variable t is always *real*—we do not get into the complicated aspects of functions of a *complex* variable.

For future reference, we will list several important vector spaces and their usual notations. In each case the operations of addition and scalar multiplication are those from algebra.

Prominent Vector Spaces

- $\mathbb{R}^2$: The space of all ordered pairs (or 2-vectors)
- $\mathbb{R}^3$: The space of all ordered triples (or 3-vectors)
- $\mathbb{R}^n$: The space of all ordered n-tuples (or n-vectors)
- $\mathbb{P}$: The space of all polynomials
- $\mathbb{P}_n$: The space of all polynomials of degree less than or equal to n
- $\mathbb{M}_{mn}$: The space of all $m \times n$ matrices
- $C(I)$: The space of all continuous functions on the interval I (I may be an open or closed interval, finite or infinite)
- $C^n(I)$: The space of all functions on interval I (as above) having n continuous derivatives; C^n with no I specified is understood to mean $C^n(-\infty, \infty)$
- $\mathbb{C}^n$: The space of all ordered n-tuples of *complex* numbers $(a + bi, c + di, \ldots)$

Vector Subspaces

Often a vector space $\mathbb{V}$ has a subset $\mathbb{W}$ that is itself a vector space—a set of vectors that satisfy the ten conditions. Most of the addition and scalar multiplication properties are "inherited" from $\mathbb{V}$. If $\vec{\mathbf{x}}$ and $\vec{\mathbf{y}}$ are in $\mathbb{W}$, for example, then they are also in $\mathbb{V}$; then, so long as $\mathbb{W}$ is closed under addition, it is true that $\vec{\mathbf{x}} + \vec{\mathbf{y}} = \vec{\mathbf{y}} + \vec{\mathbf{x}}$ for vectors in $\mathbb{W}$. Checking whether a *subset* is a *subspace* is really a matter of checking for closure. We summarize this in the following theorem.

The Zero-Space Check:
The zero-space $\{\vec{\mathbf{0}}\}$ is always a subspace of any vector space. If $\vec{\mathbf{0}}$ is *not* in $\mathbb{W}$, then $\mathbb{W}$ is *empty* and is *not* a subspace.

Vector Subspace Theorem

A *nonempty* subset $\mathbb{W}$ of a vector space $\mathbb{V}$ is a **subspace** of $\mathbb{V}$ if it is closed under addition and scalar multiplication:

(i) If $\vec{\mathbf{u}}, \vec{\mathbf{v}} \in \mathbb{W}$, then $\vec{\mathbf{u}} + \vec{\mathbf{v}} \in \mathbb{W}$.

(ii) If $\vec{\mathbf{u}} \in \mathbb{W}$ and $c \in \mathbb{R}$, then $c\vec{\mathbf{u}} \in \mathbb{W}$.

As stated earlier, it is often efficient to verify both closure properties at once by verifying closure under "linear combinations":

$$\text{If } \vec{\mathbf{u}}, \vec{\mathbf{v}} \in \mathbb{W} \text{ and } a, b \in \mathbb{R}, \text{ then } a\vec{\mathbf{u}} + b\vec{\mathbf{v}} \in \mathbb{W}. \tag{4}$$

The theorem assumes that $\mathbb{W}$ is not empty, so there is at least a vector $\vec{\mathbf{u}}$ in $\mathbb{W}$. Consequently, $0\vec{\mathbf{u}} = \vec{\mathbf{0}}$ is in $\mathbb{W}$, so the zero vector *is* in $\mathbb{W}$. This fact also means that if the zero vector does not belong to a subset of $\mathbb{V}$, that subset cannot be a subspace; this process quickly screens out many candidates.

EXAMPLE 7 **Some Subsets of $\mathbb{R}^2$** Which of the subsets in Fig. 3.5.1 are also subspaces?

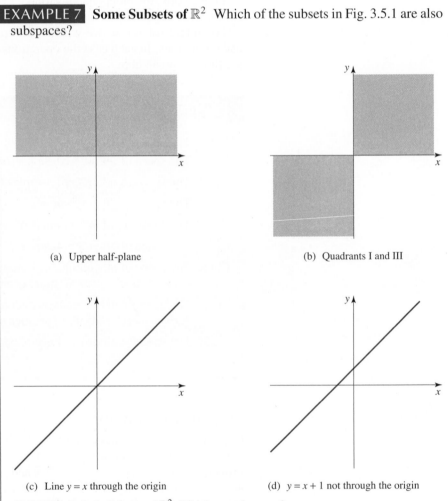

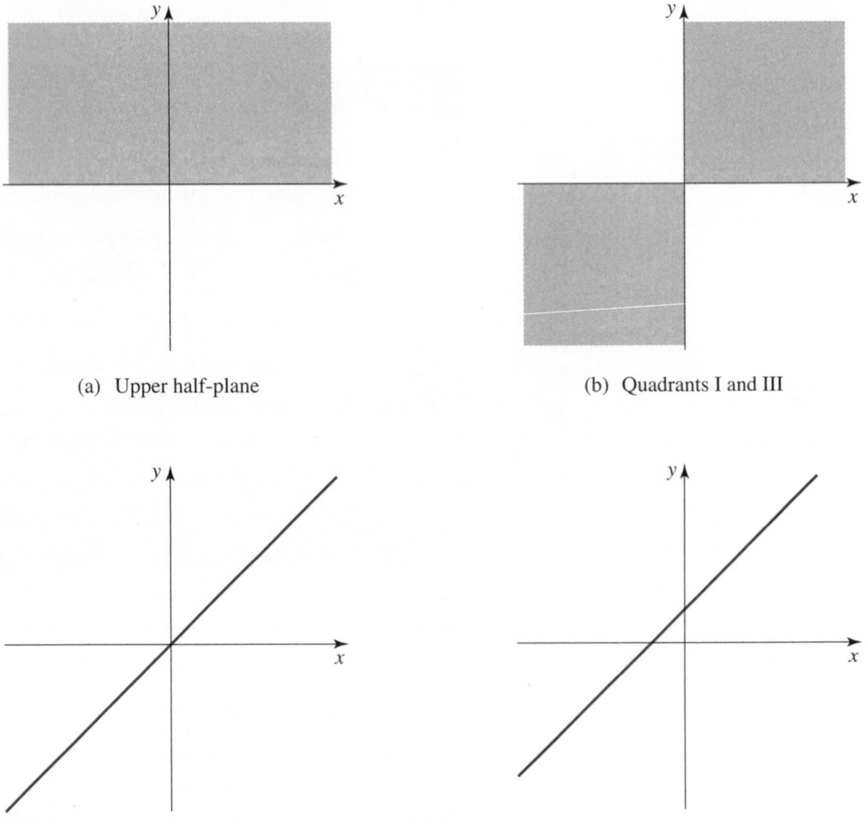

(a) Upper half-plane

(b) Quadrants I and III

(c) Line $y = x$ through the origin

(d) $y = x + 1$ not through the origin

FIGURE 3.5.1 Subsets of $\mathbb{R}^2$: Which are subspaces?

(a) The **upper half-plane**, given by the set $\{(x, y) \mid y \geq 0\}$, is *not* a subspace of $\mathbb{R}^2$ because it is not closed under scalar multiplication. The vector $(0, 1)$ is in the upper half-plane but its multiple, $(-1)(0, 1) = (0, -1)$, is not.

(b) The **quadrants I and III**, given by the set $\{(x, y) \mid xy \geq 0\}$, is *not* a subspace of $\mathbb{R}^2$ because it is not closed under addition: $(2, 1)$ and $(-1, -2)$ belong to the set but their sum, $(2, 1) + (-1, -2) = (1, -1)$, does not.

(c) The **line through the origin**, given by the set $\{(x, y) \mid x = y\}$, is a subspace of $\mathbb{R}^2$. If (s, s) and (t, t) are members of the set, then

$$a(s, s) + b(t, t) = (as, as) + (bt, bt) = (as + bt, as + bt)$$

belongs to the set as well.

(d) The **line not through the origin**, given by the set $\{(x, y) \mid y = x + 1\}$, is *not* a subspace of $\mathbb{R}^2$ because it does not contain $(0, 0)$. ■

In fact, the only subspaces of $\mathbb{R}^2$ are the following, as shown in Fig. 3.5.2:

- the zero subspace $\{(0, 0)\}$ (the zero vector alone);
- lines passing through the origin; and
- $\mathbb{R}^2$ itself.

Since the set consisting of the zero vector alone and the set $\mathbb{V}$ itself are always subspaces, we call them the **trivial subspaces**. This does not mean that they are unimportant, just that they are automatic. Thus we can state that *the only* ***nontrivial*** *subspaces of* $\mathbb{R}^2$ *are the lines through the origin*. It is important to remember that lines *not* through the origin are *not* subspaces.

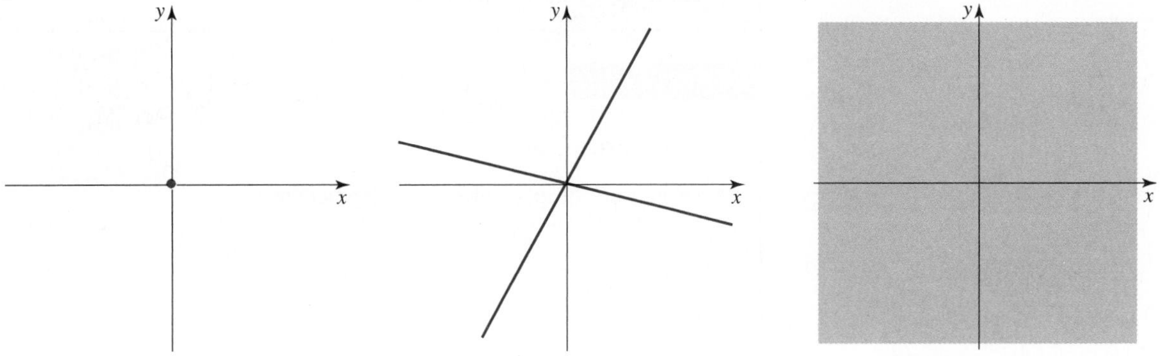

(a) Zero vector (b) Lines through the origin (c) Plane ($\mathbb{R}^2$ itself)

FIGURE 3.5.2 Subspaces of $\mathbb{R}^2$.

The subspaces of $\mathbb{R}^3$ are easy to classify as well. There are the two *trivial* ones, the **zero** subspace (the zero vector alone) and $\mathbb{R}^3$ itself. The others are lines that contain the origin and planes that contain the origin, as shown in Fig. 3.5.3. Lines and planes that do not pass through the origin do *not* constitute subspaces.

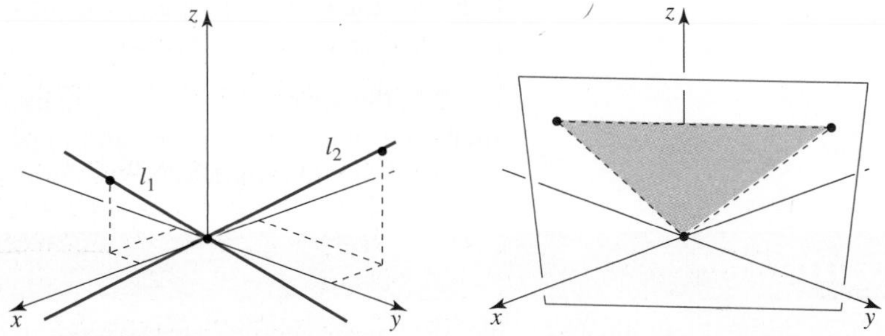

(a) Lines through the origin (b) Planes through the origin

FIGURE 3.5.3 Nontrivial subspaces of $\mathbb{R}^3$.

$\{x \in \mathbb{R}^n \mid \mathbf{A}\vec{\mathbf{x}} = \vec{\mathbf{0}}\}$ is a subspace of $\mathbb{R}^n$.

EXAMPLE 8 Solution Space for Homogeneous Linear Algebraic Systems The set of solutions of a homogeneous linear algebraic system $\mathbf{A}\vec{\mathbf{x}} = \vec{\mathbf{0}}$ is a subspace of $\mathbb{R}^n$, if $\mathbf{A}$ is an $m \times n$ matrix, $\vec{\mathbf{x}} \in \mathbb{R}^n$, and $\vec{\mathbf{0}}$ is the zero vector in $\mathbb{R}^m$. The fact that the set $\{\vec{\mathbf{x}} \in \mathbb{R}^n \mid \mathbf{A}\vec{\mathbf{x}} = \vec{\mathbf{0}}\}$ is closed under vector addition and scalar multiplication follows directly from the Superposition Principle. ■

$\{f \in \mathcal{C}[0, 1] \mid f(0) = 0\}$ is a subspace of $\mathcal{C}[0, 1]$.

EXAMPLE 9 A Restricted Function Space We will show that the set of all functions continuous on the closed interval $[0, 1]$, such that $f(0) = 0$, constitutes a vector space.

The set is contained in the prominent vector space $\mathcal{C}[0, 1]$. (It is customary to omit the extra set of parentheses when the interval is denoted using parentheses or brackets.) That makes our job simpler, because we can use the subspace theorem and simply check closure. Suppose that f and g are in the set. That means they are continuous on the interval and that $f(0) = 0$, $g(0) = 0$. Then, for any scalars a and b, $af + bg$ is again continuous on $[0, 1]$, by theorems from calculus. Also,

$$(af + bg)(0) = af(0) + bg(0) = a(0) + b(0) = 0.$$

By the Vector Subspace Theorem, the set is a subspace of $\mathcal{C}[0, 1]$ and therefore a vector space in its own right. ■

$\{\mathbf{A} \in \mathbb{M}_{22} \mid \operatorname{Tr}\mathbf{A} = 0\} = \left\{\begin{bmatrix} a & b \\ c & -a \end{bmatrix} \,\middle|\, a, b, c \in \mathbb{R}\right\}$ is a subspace of $\mathbb{M}_{22}$.

EXAMPLE 10 Zero-Trace 2 × 2 Matrices Let $\mathbb{W}$ be the set of all 2×2 matrices with zero trace. We use the Vector Subspace Theorem to prove that $\mathbb{W}$ is a subspace of $\mathbb{M}_{22}$, as follows.

- $\mathbb{W}$ is a nonempty subset of $\mathbb{M}_{22}$ because
$$\begin{bmatrix} 0 & 0 \\ 0 & 0 \end{bmatrix} \in \mathbb{W}.$$
- If
$$\begin{bmatrix} u_1 & u_2 \\ u_3 & u_4 \end{bmatrix} \in \mathbb{W} \quad \text{and} \quad \begin{bmatrix} v_1 & v_2 \\ v_3 & v_4 \end{bmatrix} \in \mathbb{W},$$
and r and s are any scalars, then
$$r\begin{bmatrix} u_1 & u_2 \\ u_3 & u_4 \end{bmatrix} + s\begin{bmatrix} v_1 & v_2 \\ v_3 & v_4 \end{bmatrix} = \begin{bmatrix} ru_1 + sv_1 & ru_2 + sv_2 \\ ru_3 + sv_3 & ru_4 + sv_4 \end{bmatrix},$$
and the resulting trace is
$$(ru_1 + sv_1) + (ru_4 + sv_4) = r(u_1 + u_4) + s(v_1 + v_4) = r0 + s0 = 0.$$

Hence the closure property for subspaces is satisfied. ■

$\mathbb{M}_{nn}$, the space of all $n \times n$ square matrices, has many interesting subspaces, including diagonal matrices and upper (or lower) triangular matrices. Some of these will be investigated in Problems 16–18.

Summary

We have defined and illustrated the concept of vector space, including the ten vector space requirements, and have learned how to identify subspaces. Important examples of vector spaces include those whose elements or "vectors" are geometric vectors, matrices, functions, and solutions of homogeneous differential equations.

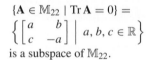

3.5 Problems

They Do Not All Look Like Vectors *For each vector space in Problems 1–10, give several examples of "vectors" in the space, including the zero vector and the negative of a typical vector.*

1. $\mathbb{R}^2$
2. $\mathbb{R}^3$
3. $\mathbb{R}^4$
4. $\mathbb{M}_{31}$
5. $\mathbb{M}_{23}$
6. $\mathbb{M}_{33}$
7. $\mathbb{P}_1$
8. $\mathbb{P}_2$
9. $\mathcal{C}(-\infty, \infty)$
10. $\mathcal{C}^2[0, 1]$

Are They Vector Spaces? *In each of Problems 11–23, decide whether or not the given set constitutes a vector space. Assume "standard" definitions of the operations.*

11. The set of vectors in the first quadrant of the plane
12. The set of vectors in the first octant of (x, y, z) space
13. The set of all pairs of real numbers (x, y) such that $x \geq y$
14. The set of all polynomials of degree two
15. The set of all polynomials of even degree
16. The set of all diagonal 2×2 matrices
17. The set of all 2×2 matrices with determinant equal to zero
18. The set of all invertible 2×2 matrices
19. The set of all 3×3 upper triangular matrices
20. The set of all continuous functions f defined on the interval $[0, 1]$ such that $f(0) = 1$
21. The set of all continuous functions f defined on the interval $[0, 1]$ such that $f(t) > 0$
22. The set of all differentiable functions on $(-\infty, \infty)$
23. The set of all functions integrable on the interval $[0, 1]$
24. **A Familiar Vector Space** Show that the set $\mathbb{R}$ of real numbers is itself a vector space. (We could write the name of this vector space as $\mathbb{R}^1$.)
25. **Not a Vector Space** Show that the set $\mathbb{Z}$ of integers (standard operations) is not a vector space by identifying at least one vector space property that fails.
26. **DE Solution Space** Verify or justify with results from algebra or calculus the vector space properties 3, 4, and 7–10 of a vector space for the solution space of Example 3.
27. **Another Solution Space** Verify or justify with results from algebra or calculus the vector space properties 3, 4, and 7–10 of a vector space for the solution space of Example 4.
28. **The Space $\mathcal{C}(-\infty, \infty)$** Verify or justify with results from algebra or calculus the fact that the continuous functions on the real line form a vector space.

Vector Space Properties *Show that the properties in Problems 29–32 hold in any vector space.*

29. *Unique Zero*: The zero element in a vector space is unique. HINT: Start with two zero elements and show that they must be equal.
30. *Unique Negative*: The negative of a vector is unique.
31. *Zero as Multiplier*: For any vector $\vec{\mathbf{v}}$, $0\vec{\mathbf{v}} = \vec{\mathbf{0}}$.
32. *Negatives as Multiples*: For any vector $\vec{\mathbf{v}}$, $-\vec{\mathbf{v}} = (-1)\vec{\mathbf{v}}$.
33. **A Vector Space Equation** Suppose that, for $c \in \mathbb{R}$ and $\vec{\mathbf{v}}$ in vector space $\mathbb{V}$, $c\vec{\mathbf{v}} = \vec{\mathbf{0}}$. Then show that either $c = 0$ or $\vec{\mathbf{v}} = \vec{\mathbf{0}}$.

Nonstandard Definitions *In Problems 34–36 we explore the possibility of defining a new vector space whose vectors are the familiar pairs (x, y) of real numbers but for which different operations are defined. (We are* not *in $\mathbb{R}^2$ any more, Toto!) For the operations given in each problem, decide whether the vector space requirements hold.*

34. $(x_1, y_1) + (x_2, y_2) \equiv (x_1 + x_2, 0)$,
    $c(x, y) \equiv (cx, y)$
35. $(x_1, y_1) + (x_2, y_2) \equiv (0, x_2)$,
    $c(x, y) \equiv (cx, cy)$
36. $(x_1, y_1) + (x_2, y_2) \equiv (x_1 + x_2, y_1 + y_2)$,
    $c(x, y) \equiv (\sqrt{c}\,x, \sqrt{c}\,y)$

Sifting Subsets for Subspaces *In each of Problems 37–46, decide whether the given* subset $\mathbb{W}$ *of the vector space* $\mathbb{V}$ *is or is not a* subspace *of* $\mathbb{V}$. *If not, identify at least one requirement that is not satisfied.*

37. $\mathbb{V} = \mathbb{R}^2$, $\quad \mathbb{W} = \{(x, y) \mid y = 0\}$
38. $\mathbb{V} = \mathbb{R}^2$, $\quad \mathbb{W} = \{(x, y) \mid x^2 + y^2 = 1\}$
39. $\mathbb{V} = \mathbb{R}^3$, $\quad \mathbb{W} = \{(x_1, x_2, x_3) \mid x_3 = 0\}$
40. $\mathbb{V} = \mathbb{P}_2$, $\quad \mathbb{W} = \{p(t) \mid \deg(p) = 2\}$

41. $\mathbb{V} = \mathbb{P}_3$, $\mathbb{W} = \{p(t) \mid p(0) = 0\}$

42. $\mathbb{V} = C[0, 1]$, $\mathbb{W} = \{f(t) \mid f(0) = 0\}$

43. $\mathbb{V} = C[0, 1]$, $\mathbb{W} = \{f(t) \mid f(0) = f(1) = 0\}$

44. $\mathbb{V} = C[a, b]$, $\mathbb{W} = \left\{ f(t) \,\middle|\, \int_a^b f(t)\,dt = 0 \right\}$

45. $\mathbb{V} = C^2[0, 1]$, $\mathbb{W} = \{f(t) \mid f'' + f = 0\}$

46. $\mathbb{V} = C^2[0, 1]$, $\mathbb{W} = \{f(t) \mid f'' + f = 1\}$

47. $\mathbb{V} = \mathbb{R}^n$,
$\mathbb{W} = \left\{ \vec{\mathbf{x}} \in \mathbb{R}^n \mid \mathbf{A}\vec{\mathbf{x}} = \vec{\mathbf{b}},\ \text{where } \mathbf{A} \in \mathbb{M}_{mn}, \vec{\mathbf{b}} \neq \vec{\mathbf{0}} \right\}$

48. $\mathbb{V} = \mathbb{R}^n$, $\mathbb{W} = \left\{ \vec{\mathbf{x}} \in \mathbb{R}^n \mid \mathbf{A}\vec{\mathbf{x}} = \vec{\mathbf{0}},\ \text{where } \mathbf{A} \in \mathbb{M}_{mn} \right\}$

49. Hyperplanes as Subspaces The subset $\mathbb{W}$ of $\mathbb{R}^4$ defined by

$$\mathbb{W} = \{(x, y, z, w) \mid ax + by + cz + dw = 0\},$$

where a, b, c, and d are real numbers not all zero, is a **hyperplane** through the origin. Show that $\mathbb{W}$ is a subspace of $\mathbb{R}^4$.

Are They Subspaces of $\mathbb{R}^n$? *Determine whether or not the subsets of $\mathbb{R}^n$ given in Problems 50–52 are subspaces. If not, show what required properties they fail to have.*

50. Is $\{(a, b, a - b, a + b) \mid a, b \in \mathbb{R}\}$ a subspace of $\mathbb{R}^4$?

51. Is $\{(a, 0, b, 1, c) \mid a, b, c \in \mathbb{R}\}$ a subspace of $\mathbb{R}^5$?

52. Is $\left\{(a, b, a^2, b^2) \mid a, b \in \mathbb{R}\right\}$ a subspace of $\mathbb{R}^4$?

Differentiable Subspaces *Let $\mathbb{D} = \mathbb{D}(-\infty, \infty)$ be the vector space of functions differentiable on the real line. Which of the following* subsets *of $\mathbb{D}$ are* subspaces *of $\mathbb{D}$?*

53. $\{f(t) \mid f' = 0\}$

54. $\{f(t) \mid f' = 1\}$

55. $\{f(t) \mid f' = f\}$

56. $\{f(t) \mid f' = f^2\}$

Property Failures *Find a subset of $\mathbb{R}^2$ fitting each description.*

57. Closed under vector addition but not under scalar multiplication

58. Closed under scalar multiplication but not under vector addition

59. Not closed under either vector addition or scalar multiplication

Solution Spaces of Homogeneous Linear Algebraic Systems *Solve the linear systems given in Problems 60–62 and determine their solution spaces.*

60.
$$\begin{aligned} x_1 - x_2 \phantom{{}+x_3} + 4x_4 + 2x_5 - x_6 &= 0 \\ 2x_1 - 2x_2 + x_3 + 2x_4 + 4x_5 - x_6 &= 0 \end{aligned}$$

61.
$$\begin{aligned} 2x_1 - 2x_2 + 4x_3 - 2x_4 &= 0 \\ 2x_1 + x_2 + 7x_3 + 4x_4 &= 0 \\ x_1 - 4x_2 - x_3 + 7x_4 &= 0 \\ 4x_1 - 12x_2 \phantom{{}-x_3} - 20x_4 &= 0 \end{aligned}$$

62.
$$\begin{aligned} 3x_1 \phantom{{}+3x_2} + 6x_3 + 3x_4 + 9x_5 &= 0 \\ x_1 + 3x_2 - 4x_3 - 8x_4 + 3x_5 &= 0 \\ x_1 - 6x_2 + 14x_3 + 19x_4 + 3x_5 &= 0 \end{aligned}$$

Nonlinear Differential Equations *Show that the solution sets of the following nonlinear differential equations are not vector spaces.*

63. $y' = y^2$

64. $y'' + \sin y = 0$

65. $y'' + \dfrac{1}{y} = 0$

DE Solution Spaces *Recall that the "general solution" of a DE means a family or set of solution functions $\{y \mid y$ satisfies the DE$\}$, where each y is a function on the interval determined by the domain of the DE. In Problems 66–69, does the general solution of each of the following DEs form a vector space or not? Explain.*

66. $y' + 2y = e^t$

67. $y' + y^2 = 0$

68. $y'' + ty = 0$

69. $y'' + (1 + \sin t)y = 0$

70. Line of Solutions If $\vec{\mathbf{p}}$ and $\vec{\mathbf{h}}$ are vectors in vector space $\mathbb{V}$, with $\vec{\mathbf{h}} \neq \vec{\mathbf{0}}$, then the *line through $\vec{\mathbf{p}}$ in the direction $\vec{\mathbf{h}}$* is defined to be the set $\{\vec{\mathbf{x}} \in \mathbb{V} \mid \vec{\mathbf{x}} = \vec{\mathbf{p}} + t\vec{\mathbf{h}},\ t \in \mathbb{R}\}$.

(a) Find the line in $\mathbb{R}^2$ through $\vec{\mathbf{p}} = (0, 1)$ in the direction $\vec{\mathbf{h}} = (2, 3)$.

(b) Find the line in $\mathbb{R}^3$ through $\vec{\mathbf{p}} = (2, 1, 3)$ in the direction $\vec{\mathbf{h}} = (2, -3, 0)$.

(c) Show that solutions of $y' + y = 0$ are a subspace of $\mathbb{V} = C^1(-\infty, \infty)$, and that every vector in this subspace is a multiple of e^{-t}.

(d) Show that the solutions of $y' + y = t$ form a line in $\mathbb{V}$ through $t - 1$ in the direction e^{-t}.

(e) Relate parts (c) and (d) to what you learned about solutions to homogeneous and nonhomogeneous differential equations in Sec. 2.1.

Orthogonal Complements *Prove the properties stated in Problems 71–73 using the following definition, illustrated by Fig. 3.5.4. Assume that $\mathbb{V}$ is a subspace of $\mathbb{R}^n$.*

Orthogonal Complement

Let $\mathbb{V}$ be a subspace of $\mathbb{R}^n$. A vector $\vec{\mathbf{u}}$ is *orthogonal* to subspace $\mathbb{V}$ provided that $\vec{\mathbf{u}}$ is orthogonal to every vector in $\mathbb{V}$. The set of all vectors in $\mathbb{R}^n$ that are orthogonal to $\mathbb{V}$ is called the **orthogonal complement** of $\mathbb{V}$, denoted

$$\mathbb{V}^{\perp} = \left\{ \vec{\mathbf{u}} \in \mathbb{R}^n \mid \vec{\mathbf{u}} \cdot \vec{\mathbf{v}} = 0 \text{ for every } \vec{\mathbf{v}} \in \mathbb{V} \right\}.$$

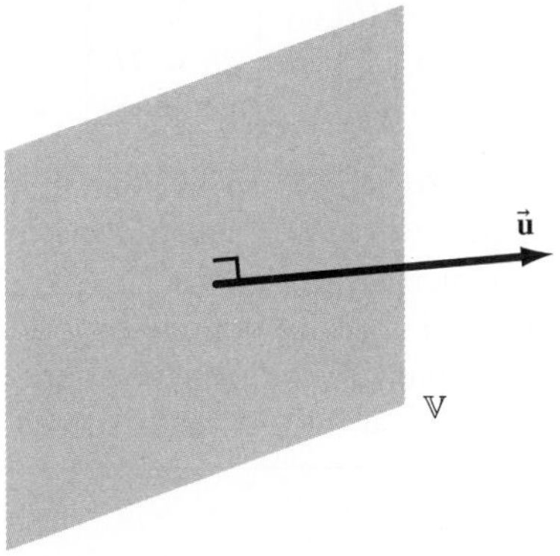

FIGURE 3.5.4 An orthogonal complement $\vec{\mathbf{u}}$ to a plane $\mathbb{V}$

71. $\mathbb{V}^{\perp}$ is a subspace of $\mathbb{R}^n$.

72. $\mathbb{V} \cap \mathbb{V}^{\perp} = \{\vec{\mathbf{0}}\}$

73. Suggested Journal Entry By Problem 70, the straight line in $\mathbb{R}^4$ through $\vec{\mathbf{p}} = (p_1, p_2, p_3, p_4)$ in the direction $\vec{\mathbf{h}} = (h_1, h_2, h_3, h_4)$ has parametric equations

$$x_1 = p_1 + th_1, \quad x_2 = p_2 + th_2,$$
$$x_3 = p_3 + th_3, \quad x_4 = p_4 + th_4,$$

where t is a real parameter. From Problem 49, a hyperplane in $\mathbb{R}^4$ is the solution set of the linear equation

$$a_1x_1 + a_2x_2 + a_3x_3 + a_4x_4 = a_0,$$

passing through the origin if and only if $a_0 = 0$. But you are now in the fourth dimension, with no pictures to guide you. How do you know a line is straight? How can you tell if a hyperplane is flat? How could you define and test these concepts?

3.6 Basis and Dimension

SYNOPSIS: We study the structure of a vector space by investigating linear independence, spanning sets, the basis of a vector space, and dimension. Earlier work on matrices and linear systems provides the necessary tools.

Spanning Sets

Given one or more vectors in a vector space, we can create more vectors by forming linear combinations: Multiply the vectors by scalars and add these together. This is the *only* way to make more vectors, because the vector space has only these two basic operations. Given vectors $\vec{\mathbf{v}}_1, \vec{\mathbf{v}}_2, \ldots, \vec{\mathbf{v}}_n$, for example, we can form a new vector

$$\vec{\mathbf{v}} = c_1\vec{\mathbf{v}}_1 + c_2\vec{\mathbf{v}}_2 + \cdots + c_n\vec{\mathbf{v}}_n,$$

where $c_1, c_2, \ldots, c_n$ are scalars, called a **linear combination** of the vectors. The set of *all* vectors that we can make from a given set of vectors in this way is called its span.

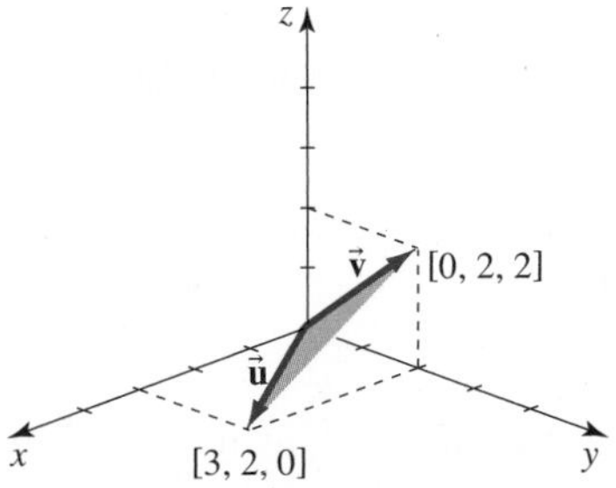

FIGURE 3.6.1 The span of two vectors in $\mathbb{R}^3$ is the entire plane of which the shaded area is a portion.

Span

The **span** of a set $\{\vec{\mathbf{v}}_1, \vec{\mathbf{v}}_2, \ldots, \vec{\mathbf{v}}_n\}$ of vectors in a vector space $\mathbb{V}$, denoted $\text{Span}\{\vec{\mathbf{v}}_1, \vec{\mathbf{v}}_2, \ldots, \vec{\mathbf{v}}_n\}$, is the set of all linear combinations of these vectors.

If $\text{Span}\{\vec{\mathbf{v}}_1, \vec{\mathbf{v}}_2, \ldots, \vec{\mathbf{v}}_n\} = \mathbb{V}$, we say that the set spans the vector space. The span of the vectors $\vec{\mathbf{u}}$ and $\vec{\mathbf{v}}$ in Fig. 3.6.1 is the plane they determine, because of the geometric interpretation of sums and scalar multiples. (Compare with Fig. 3.1.2.)

EXAMPLE 1 Picturing a Span Looking a little more carefully at the situation in Fig. 3.6.1, in which the two vectors are

$$\vec{\mathbf{u}} = \begin{bmatrix} 3 \\ 2 \\ 0 \end{bmatrix} \quad \text{and} \quad \vec{\mathbf{v}} = \begin{bmatrix} 0 \\ 2 \\ 2 \end{bmatrix},$$

we can write a typical vector of their span in the form

$$\begin{bmatrix} x \\ y \\ z \end{bmatrix} = a\vec{\mathbf{u}} + b\vec{\mathbf{v}} = a\begin{bmatrix} 3 \\ 2 \\ 0 \end{bmatrix} + b\begin{bmatrix} 0 \\ 2 \\ 2 \end{bmatrix} = \begin{bmatrix} 3a \\ 2a+2b \\ 2b \end{bmatrix}.$$

Equating components gives $x = 3a$, $y = 2a + 2b$, $z = 2b$. The first and third of these equations yield $a = (1/3)x$ and $b = (1/2)z$; substituting these in the y-equation produces $y = (2/3)x + z$. This is equivalent to $2x - 3y + 3z = 0$, which we recognize to be the equation of a plane containing the origin. A portion of this plane, which represents the span, is shaded in the figure. Although Span $\{\vec{\mathbf{u}}, \vec{\mathbf{v}}\}$ is a plane, it is not $\mathbb{R}^2$, because $\vec{\mathbf{u}}$ and $\vec{\mathbf{v}}$ are vectors from $\mathbb{R}^3$. ■

EXAMPLE 2 When Does an Additional Vector Change the Span? An additional vector in the set changes the span of the set only if it is not a linear combination of the original vectors in the set.

(a) Consider adding vector

$$\vec{\mathbf{w}} = \begin{bmatrix} -3 \\ 2 \\ 2 \end{bmatrix}$$

to the set in Example 1. Span $\{\vec{\mathbf{u}}, \vec{\mathbf{v}}, \vec{\mathbf{w}}\}$ = Span $\{\vec{\mathbf{u}}, \vec{\mathbf{v}}\}$ because

$$\vec{\mathbf{w}} = -1\vec{\mathbf{u}} + 2\vec{\mathbf{v}} \in \text{Span}\{\vec{\mathbf{u}}, \vec{\mathbf{v}}\},$$

which means that the vector $\vec{\mathbf{w}}$ remains in the plane spanned by $\vec{\mathbf{u}}$ and $\vec{\mathbf{v}}$.

(b) Now consider adding the vector

$$\vec{\mathbf{x}} = \begin{bmatrix} 1 \\ 1 \\ 0 \end{bmatrix}$$

to the set in Example 1. Is $\vec{\mathbf{x}} \in$ Span $\{\vec{\mathbf{u}}, \vec{\mathbf{v}}\}$? That is, can we write

$$\begin{bmatrix} 1 \\ 1 \\ 0 \end{bmatrix} = c_1\begin{bmatrix} 3 \\ 2 \\ 0 \end{bmatrix} + c_2\begin{bmatrix} 0 \\ 2 \\ 2 \end{bmatrix}?$$

Equating coefficients gives us an inconsistent set of linear equations

$$\begin{aligned} 1 &= 3c_1, \\ 1 &= 2c_1 + 2c_2, \\ 0 &= \phantom{2c_1 + {}} 2c_2. \end{aligned}$$

Consequently, $\vec{\mathbf{x}} \notin$ Span $\{\vec{\mathbf{u}}, \vec{\mathbf{v}}\}$.

(c) What is Span $\{\vec{\mathbf{u}}, \vec{\mathbf{v}}, \vec{\mathbf{x}}\}$? We can show that it is all of $\mathbb{R}^3$, as follows. Let

$$\vec{\mathbf{y}} = \begin{bmatrix} a \\ b \\ c \end{bmatrix}$$

be an arbitrary element of $\mathbb{R}^3$. Set

$$c_1 \begin{bmatrix} 3 \\ 2 \\ 0 \end{bmatrix} + c_2 \begin{bmatrix} 0 \\ 2 \\ 2 \end{bmatrix} + c_3 \begin{bmatrix} 1 \\ 1 \\ 0 \end{bmatrix} = \begin{bmatrix} a \\ b \\ c \end{bmatrix}$$

$$\begin{bmatrix} 3 & 0 & 1 \\ 2 & 2 & 1 \\ 0 & 2 & 0 \end{bmatrix} \begin{bmatrix} c_1 \\ c_2 \\ c_3 \end{bmatrix} = \begin{bmatrix} a \\ b \\ c \end{bmatrix}.$$

We see that for any a, b, c this system has a unique solution, because the determinant of the matrix of coefficients is nonzero. Thus we have confirmed that Span $\{\vec{\mathbf{u}}, \vec{\mathbf{v}}, \vec{\mathbf{x}}\} = \mathbb{R}^3$. ■

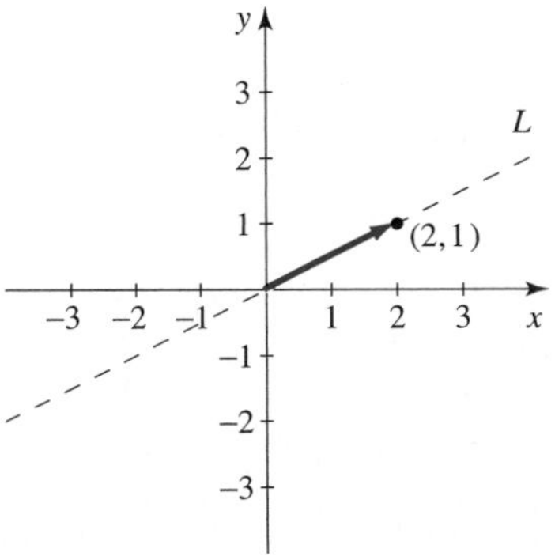

FIGURE 3.6.2 Span $\left\{\begin{bmatrix} 2 \\ 1 \end{bmatrix}\right\}$ is the line L (Example 3).

EXAMPLE 3 **Expressing Span as a Set of Vectors** We can see that

$$\text{Span}\left\{\begin{bmatrix} 2 \\ 1 \end{bmatrix}\right\} = \left\{ c \begin{bmatrix} 2 \\ 1 \end{bmatrix} \;\middle|\; c \in \mathbb{R} \right\}$$

is the set of all vectors on the line L, and that

$$\text{Span}\left\{\begin{bmatrix} 2 \\ 1 \end{bmatrix}\right\} = \text{Span}\left\{\begin{bmatrix} 2 \\ 1 \end{bmatrix}, \begin{bmatrix} 4 \\ 2 \end{bmatrix}, \begin{bmatrix} -2 \\ -1 \end{bmatrix}\right\},$$

because adding vectors along the same line does not change the span. See Fig. 3.6.2. ■

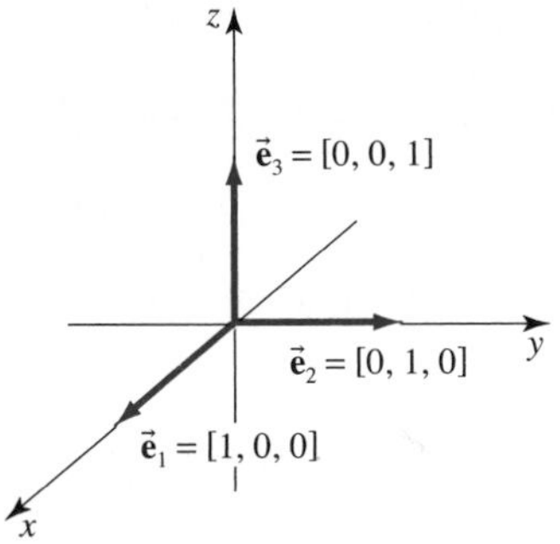

FIGURE 3.6.3
Span $\{\vec{\mathbf{e}}_1, \vec{\mathbf{e}}_2, \vec{\mathbf{e}}_3\}$ is all of $\mathbb{R}^3$ (Example 4).

EXAMPLE 4 **Spanning $\mathbb{R}^3$** If we form the span of the three vectors

$$\vec{\mathbf{e}}_1 = \begin{bmatrix} 1 \\ 0 \\ 0 \end{bmatrix}, \quad \vec{\mathbf{e}}_2 = \begin{bmatrix} 0 \\ 1 \\ 0 \end{bmatrix}, \quad \text{and} \quad \vec{\mathbf{e}}_3 = \begin{bmatrix} 0 \\ 0 \\ 1 \end{bmatrix}, \tag{1}$$

we obtain

$$\text{Span}\{\vec{\mathbf{e}}_1, \vec{\mathbf{e}}_2, \vec{\mathbf{e}}_3\} = \left\{ c_1 \begin{bmatrix} 1 \\ 0 \\ 0 \end{bmatrix} + c_2 \begin{bmatrix} 0 \\ 1 \\ 0 \end{bmatrix} + c_3 \begin{bmatrix} 0 \\ 0 \\ 1 \end{bmatrix} \;\middle|\; c_1, c_2, c_3 \in \mathbb{R} \right\};$$

that is (see Fig. 3.6.3),

$$\text{Span}\{\vec{\mathbf{e}}_1, \vec{\mathbf{e}}_2, \vec{\mathbf{e}}_3\} = \left\{ \begin{bmatrix} c_1 \\ c_2 \\ c_3 \end{bmatrix} \;\middle|\; c_1, c_2, c_3 \in \mathbb{R} \right\} = \mathbb{R}^3.$$ ■

The vectors $\vec{\mathbf{e}}_1$, $\vec{\mathbf{e}}_2$, and $\vec{\mathbf{e}}_3$ in equation (1) are the columns of identity matrix $\mathbf{I}_3$. They are called the **standard basis vectors** of $\mathbb{R}^3$. (We shall return to the meaning of "basis" later in the section.)

Suppose that we want to determine Span $\{\vec{\mathbf{e}}_1, \vec{\mathbf{e}}_2, \vec{\mathbf{e}}_3, \vec{\mathbf{u}}\}$, where

$$\vec{\mathbf{u}} = \begin{bmatrix} a \\ b \\ c \end{bmatrix}.$$

Span $\{\vec{\mathbf{e}}_1, \vec{\mathbf{e}}_2, \vec{\mathbf{e}}_3, \vec{\mathbf{u}}\}$ remains $\mathbb{R}^3$ because $\vec{\mathbf{u}} = a\vec{\mathbf{e}}_1 + b\vec{\mathbf{e}}_2 + c\vec{\mathbf{e}}_3 \in \text{Span}\,\{\vec{\mathbf{e}}_1, \vec{\mathbf{e}}_2, \vec{\mathbf{e}}_3\}$.

Spanning Sets in $\mathbb{R}^n$
A vector $\vec{\mathbf{b}}$ in $\mathbb{R}^n$ is in **Span$\{\vec{\mathbf{v}}_1, \vec{\mathbf{v}}_2, \ldots, \vec{\mathbf{v}}_n\}$**, where $\vec{\mathbf{v}}_1, \vec{\mathbf{v}}_2, \ldots, \vec{\mathbf{v}}_n$ are vectors in $\mathbb{R}^n$, provided that there is at least one solution of the matrix-vector equation $\mathbf{A}\vec{\mathbf{x}} = \vec{\mathbf{b}}$, where $\mathbf{A}$ is the matrix whose column vectors are $\vec{\mathbf{v}}_1, \vec{\mathbf{v}}_2, \ldots, \vec{\mathbf{v}}_n$.

It was no accident that the spans we found in Examples 1 and 4 were both vector spaces. This is an important fact.

If a bunch of vectors are in a vector space, the span of those vectors is a vector subspace.

Span Theorem
For a set of vectors $\{\vec{\mathbf{v}}_1, \vec{\mathbf{v}}_2, \ldots, \vec{\mathbf{v}}_n\}$ in vector space $\mathbb{V}$, Span$\{\vec{\mathbf{v}}_1, \vec{\mathbf{v}}_2, \ldots, \vec{\mathbf{v}}_n\}$ is a subspace of $\mathbb{V}$.

Proof The proof is a consequence of the subspace theorem of Sec. 3.5. Let $\vec{\mathbf{u}}$ and $\vec{\mathbf{w}}$ be two vectors in the span so that, for scalars r_i and s_i,

$$\vec{\mathbf{u}} = r_1\vec{\mathbf{v}}_1 + r_2\vec{\mathbf{v}}_2 + \cdots + r_n\vec{\mathbf{v}}_n \quad \text{and} \quad \vec{\mathbf{w}} = s_1\vec{\mathbf{v}}_1 + s_2\vec{\mathbf{v}}_2 + \cdots + s_n\vec{\mathbf{v}}_n.$$

Then, for any real numbers a and b,

$$\begin{aligned} a\vec{\mathbf{u}} + b\vec{\mathbf{v}} &= a(r_1\vec{\mathbf{v}}_1 + r_2\vec{\mathbf{v}}_2 + \cdots + r_n\vec{\mathbf{v}}_n) + b(s_1\vec{\mathbf{v}}_1 + s_2\vec{\mathbf{v}}_2 + \cdots + s_n\vec{\mathbf{v}}_n) \\ &= (ar_1 + bs_1)\vec{\mathbf{v}}_1 + (ar_2 + bs_2)\vec{\mathbf{v}}_2 + \cdots + (ar_n + bs_n)\vec{\mathbf{v}}_n. \end{aligned}$$

Therefore, $a\vec{\mathbf{u}} + b\vec{\mathbf{v}}$ is in the span, which is therefore closed under addition and scalar multiplication and is a subspace. □

Now we can specify vector subspaces in terms of the span of selected vectors from a vector space. For instance, a useful subspace is formed by the columns of a matrix.

$\text{Col}\underbrace{\left[\quad\right]}_{m\times n}$ is a subspace of $\mathbb{R}^m$.

Column Space
For any $m \times n$ matrix **A**, the **column space**, denoted Col **A**, is the span of the column vectors of **A**, and is a subspace of $\mathbb{R}^m$.

Look at the number m of elements in any column vector of an $m \times n$ matrix to see that Col **A** is a subspace of $\mathbb{R}^m$.

EXAMPLE 5 **Column Space** For the matrix

$$\mathbf{B} = \begin{bmatrix} 1 & 3 & 0 & 1 & -2 \\ 2 & 4 & 1 & 1 & 5 \end{bmatrix},$$

Col **B** is a subspace of $\mathbb{R}^2$. The span of the *five* column vectors is a subspace of *two*-dimensional $\mathbb{R}^2$ because each column vector is a vector in $\mathbb{R}^2$. ■

Linear Independence of Vectors

The two vectors (a, b) and (ka, kb), which lie on the same line through the origin, do not span the plane $\mathbb{R}^2$, because every linear combination of these vectors lies on the same line. To span the plane we need two vectors in different directions. The vectors (1, 2) and (2, 1), for example, will do the trick. We call them *linearly independent*.

Linear Independence
A set $\{\vec{\mathbf{v}}_1, \vec{\mathbf{v}}_2, \ldots, \vec{\mathbf{v}}_n\}$ of *vectors* in vector space $\mathbb{V}$ is **linearly independent** if no vector of the set can be written as a linear combination of the others. Otherwise it is **linearly dependent**.

If a set of vectors is linearly *dependent*, then, it must be possible to write at least one of them as a linear combination of the others. Thus a convenient criterion for linear *independence* is to show that if a linear combination of the $\vec{\mathbf{v}}_i$ is the zero vector, then all the coefficients must be zero:

$$c_1\vec{\mathbf{v}}_1 + c_2\vec{\mathbf{v}}_2 + \cdots + c_n\vec{\mathbf{v}}_n = \vec{\mathbf{0}} \quad \text{implies} \quad c_1 = c_2 = \cdots = c_n = 0. \tag{2}$$

Otherwise, if some $c_j \neq 0$, we could just divide by c_j and solve for $\vec{\mathbf{v}}_j$ in terms of the other vectors.

In the case of two vectors, determination of linear independence is particularly easy. Two vectors are linearly independent exactly when one is not a scalar multiple of the other. Thus (2, 3) and (2, 1) are linearly independent, while (2, 3) and (4, 6) are linearly dependent. NOTE: This rule does *not* extend to more than two vectors. As we will show in Examples 7 and 8, three or more vectors can be linearly dependent without being multiples of each other.

EXAMPLE 6 **Establishing Independence** To show that vectors [1, 1, 1], [1, 2, −1], and [1, 3, 2] are linearly independent in $\mathbb{R}^3$, we use criterion (2). Suppose that

$$c_1\begin{bmatrix}1\\1\\1\end{bmatrix} + c_2\begin{bmatrix}1\\2\\-1\end{bmatrix} + c_3\begin{bmatrix}1\\3\\2\end{bmatrix} = \vec{\mathbf{0}}.$$

This equation is equivalent to the following system of three equations in three unknowns c_1, c_2, and c_3, which can be written as the matrix-vector equation

$$\mathbf{A}\begin{bmatrix}c_1\\c_2\\c_3\end{bmatrix} = \begin{bmatrix}1 & 1 & 1\\1 & 2 & 3\\1 & -1 & 2\end{bmatrix}\begin{bmatrix}c_1\\c_2\\c_3\end{bmatrix} = \begin{bmatrix}0\\0\\0\end{bmatrix}. \tag{3}$$

Since the system is homogeneous, (0, 0, 0) is a solution. To invoke condition (2), we need to know that this is the unique solution of (3). This result follows either from computing the determinant of $\mathbf{A}$ to see that it is not zero (it turns out that $|\mathbf{A}| = 5$), or by reducing $\mathbf{A}$ to its RREF, which is $\mathbf{I}_3$. ■

Testing for Linear Independence

To test for linear independence of a set of n vectors $\vec{\mathbf{v}}_i$ in $\mathbb{R}^n$, we consider the system (2) in matrix-vector form:

$$\underbrace{\begin{bmatrix}| & | & & |\\ \vec{\mathbf{v}}_1 & \vec{\mathbf{v}}_2 & \cdots & \vec{\mathbf{v}}_n\\ | & | & & |\end{bmatrix}}_{\mathbf{A}}\underbrace{\begin{bmatrix}c_1\\c_2\\\vdots\\c_n\end{bmatrix}}_{\vec{\mathbf{x}}} = \vec{\mathbf{0}}.$$

The column vectors of $\mathbf{A}$ are linearly independent if and only if the solution $\vec{\mathbf{x}} = \vec{\mathbf{0}}$ is unique, which means that $c_i = 0$ for all i. Recall from Sections 3.3 and 3.4 that any one of the following satisfies this condition for a unique solution:

- $\mathbf{A}$ is invertible.
- $\mathbf{A}$ has n pivot columns.
- $|\mathbf{A}| \neq 0$.

EXAMPLE 7 **Adding a Vector** Let us look at the vectors [1, 1, 1], [1, 2, −1], [1, 3, 2], and [5, −1, 0] in $\mathbb{R}^3$, which includes the set from Example 6 with one more vector. The matrix-vector form of the corresponding homogeneous system is

$$\mathbf{A}\begin{bmatrix} c_1 \\ c_2 \\ c_3 \\ c_4 \end{bmatrix} = \underbrace{\begin{bmatrix} 1 & 1 & 1 & 5 \\ 1 & 2 & 3 & -1 \\ 1 & -1 & 2 & 0 \end{bmatrix}}_{3\times 4} \underbrace{\begin{bmatrix} c_1 \\ c_2 \\ c_3 \\ c_4 \end{bmatrix}}_{4\times 1} = \underbrace{\begin{bmatrix} 0 \\ 0 \\ 0 \end{bmatrix}}_{3\times 1}. \tag{4}$$

Because there are more columns than rows, we know that there will be at most three pivot columns in the RREF of $[\mathbf{A} \mid \vec{\mathbf{b}}]$ and at least one free variable. Consequently, the trivial solution to (4) is not unique, so the vectors are not linearly independent. ■

EXAMPLE 8 **Investigating Dependence** Let us find out whether the three vectors

$$\begin{bmatrix} 1 \\ 1 \\ 1 \end{bmatrix}, \quad \begin{bmatrix} -1 \\ 0 \\ 1 \end{bmatrix}, \quad \text{and} \quad \begin{bmatrix} -2 \\ 1 \\ 4 \end{bmatrix}$$

are independent or dependent. We begin as in Example 6, testing by the independence criterion (2). As in Example 6, we obtain the homogeneous linear system

$$\begin{bmatrix} 1 & -1 & -2 \\ 1 & 0 & 1 \\ 1 & 1 & 4 \end{bmatrix} \begin{bmatrix} c_1 \\ c_2 \\ c_3 \end{bmatrix} = \begin{bmatrix} 0 \\ 0 \\ 0 \end{bmatrix}. \tag{5}$$

The coefficient matrix has determinant zero, so system (5) has solutions other than the zero solution and the vectors are *dependent*.

But we can actually learn more about this set of vectors from the solution of (5). Its augmented matrix,

$$\left[\begin{array}{ccc|c} 1 & -1 & -2 & 0 \\ 1 & 0 & 1 & 0 \\ 1 & 1 & 4 & 0 \end{array}\right]$$

has the RREF

$$\left[\begin{array}{ccc|c} 1 & 0 & 1 & 0 \\ 0 & 1 & 3 & 0 \\ 0 & 0 & 0 & 0 \end{array}\right].$$

We can solve for the basic variables c_1 and c_2 in terms of the free variable c_3, to obtain

$$c_1 = -c_3, \quad c_2 = -3c_3.$$

If we choose $c_3 = -1$, then $c_1 = 1$ and $c_2 = 3$. Therefore,

$$(1)\begin{bmatrix} 1 \\ 1 \\ 1 \end{bmatrix} + (3)\begin{bmatrix} -1 \\ 0 \\ 1 \end{bmatrix} + (-1)\begin{bmatrix} -2 \\ 1 \\ 4 \end{bmatrix} = \vec{\mathbf{0}}.$$

We see, in fact, that each vector can be written as a linear combination of the other two. ■

Linear Independence of Vector Functions

The concept of linear independence carries over to function spaces (linear spaces whose "vectors" are functions). This is a useful concept in determining the size of the solution space for linear differential equations and systems. We restate the definition of linear dependence and independence in this important case, noting a vital additional condition:

"for *all* values of t in the interval I on which the functions are defined."

Linear Independence of Vector Functions

A set of vector functions $\{\vec{\mathbf{v}}_1(t), \vec{\mathbf{v}}_2(t), \ldots, \vec{\mathbf{v}}_n(t)\}$ in a vector space $\mathbb{V}$ is **linearly independent** on an interval I if, for *all* t in I, the only solution of

$$c_1\vec{\mathbf{v}}_1(t) + c_2\vec{\mathbf{v}}_2(t) + \cdots + c_n\vec{\mathbf{v}}_n(t) \equiv \vec{\mathbf{0}}$$

for scalars $c_1, c_2, \ldots, c_n$ is $c_i = 0$ for all i.

If for any value t_0 of t there is any solution with $c_i \neq 0$, the vector functions $\vec{\mathbf{v}}_1(t), \vec{\mathbf{v}}_2(t), \ldots, \vec{\mathbf{v}}_n(t)$ are **linearly dependent**.

EXAMPLE 9 **Independent Vector Functions** The vectors

$$\vec{\mathbf{v}}_1(t) = \begin{bmatrix} e^t \\ 0 \\ 2e^t \end{bmatrix}, \quad \vec{\mathbf{v}}_2(t) = \begin{bmatrix} e^{-t} \\ 3e^{-t} \\ 0 \end{bmatrix}, \quad \text{and} \quad \vec{\mathbf{v}}_3(t) = \begin{bmatrix} e^{2t} \\ e^{2t} \\ e^{2t} \end{bmatrix}$$

are linearly independent vector functions on $(-\infty, \infty)$. We shall show that the only constants c_1, c_2, c_3 for which

$$c_1\vec{\mathbf{v}}_1(t) + c_2\vec{\mathbf{v}}_2(t) + c_3\vec{\mathbf{v}}_3(t) = \vec{\mathbf{0}} \tag{6}$$

are $c_1 = c_2 = c_3 = 0$. Because (6) holds for all t, it must hold for $t = 0$. Therefore, we must have

$$c_1 \begin{bmatrix} 1 \\ 0 \\ 2 \end{bmatrix} + c_2 \begin{bmatrix} 1 \\ 3 \\ 0 \end{bmatrix} + c_3 \begin{bmatrix} 1 \\ 1 \\ 1 \end{bmatrix} = \begin{bmatrix} 0 \\ 0 \\ 0 \end{bmatrix}.$$

In matrix form, this system is

$$\mathbf{A} \begin{bmatrix} c_1 \\ c_2 \\ c_3 \end{bmatrix} = \begin{bmatrix} 1 & 1 & 1 \\ 0 & 3 & 1 \\ 2 & 0 & 1 \end{bmatrix} \begin{bmatrix} c_1 \\ c_2 \\ c_3 \end{bmatrix} = \begin{bmatrix} 0 \\ 0 \\ 0 \end{bmatrix}.$$

Its unique solution is $c_1 = c_2 = c_3 = 0$ because $|\mathbf{A}| = -1 \neq 0$, so the vectors are linearly independent. ■

Linear Independence of Functions

One way to check a set of *functions* $\{f_1(t), f_2(t), \ldots, f_n(t)\}$ in a vector space $\mathbb{W}$ for linear independence is to consider them as one-dimensional vectors,

$$\vec{\mathbf{v}}_i(t) = f_n(t).$$

EXAMPLE 10 **Independent Functions** Check the following functions for linear independence:

$$\vec{\mathbf{v}}_1(t) = e^t, \quad \vec{\mathbf{v}}_2(t) = 5e^{-t}, \quad \text{and} \quad \vec{\mathbf{v}}_3(t) = e^{3t}.$$

We investigate whether there are three nonzero scalars c_1, c_2, c_3 such that

$$c_1\vec{\mathbf{v}}_1(t) + c_2\vec{\mathbf{v}}_2(t) + c_3\vec{\mathbf{v}}_3(t) = \vec{\mathbf{0}}.$$

To evaluate three unknowns c_i, we need three equations. Since the preceding equation must hold for *all* t, we can choose any three values of t:

$$\begin{aligned}
&\text{for } t = 0: && c_1 + 5c_2 + c_3 = 0,\\
&\text{for } t = 1: && ec_1 + \frac{5}{e}c_2 + e^3 c_3 = 0,\\
&\text{for } t = -1: && \frac{1}{e}c_1 + 5ec_2 + \frac{1}{e^3}c_3 = 0.
\end{aligned}$$

The matrix of coefficients of this system has determinant

$$5(1/e^4 - e^4 + 2e^2 - 2/e^2) \neq 0,$$

so the system will have the unique solution $c_1 = c_2 = c_3 = 0$, implying that these functions are linearly *independent*. ■

As shown in Example 10, considering n functions as one-dimensional vectors requires us to choose n values of t in order to construct an n-dimensional system to solve for $c_1, c_2, \ldots, c_n$.

An alternative test for the linear independence of functions $f_1(t), f_2(t), \ldots, f_n(t)$ defined on a common interval I is based on the determinant test for invertible matrices (Sec. 3.4), provided that the functions can be differentiated as many times as needed. See Problem 32 to prove why for two functions. For testing three functions for linear independence, we must be able to differentiate them twice; for n functions, we must be able to differentiate them $n-1$ times.

Suppose that $f_1, f_2, \ldots, f_n$ are functions of t on some interval I, such that they can be differentiated $n-1$ times on I. We can set up the following n equations in n unknown constants, $c_1, c_2, \ldots, c_n$, by successive differentiation:

$$\begin{aligned}
c_1 f_1(t) + \quad c_2 f_2(t) + \cdots + \quad c_n f_n(t) &= 0,\\
c_1 f_1'(t) + \quad c_2 f_2'(t) + \cdots + \quad c_n f_n'(t) &= 0,\\
&\vdots\\
c_1 f_1^{(n-1)}(t) + c_2 f_2^{(n-1)}(t) + \cdots + c_n f_n^{(n-1)}(t) &= 0,
\end{aligned} \tag{7}$$

for every t in I. Here 0 denotes the function that is identically 0 for all t in I. We know that the only solution is the trivial one ($c_1 = c_2 = \cdots = c_n = 0$) if the determinant of the matrix coefficients of the c_i is not the *zero function*. This determinant is called *the Wronskian* and is denoted by $W[f_1, f_2, \ldots, f_n](t)$ or simply $W(t)$ when the functions are clear from the context.[1]

Wronskian of Functions $f_1, f_2, \ldots, f_n$ on I

$$W[f_1, f_2, \ldots, f_n](t) \equiv \begin{vmatrix} f_1(t) & f_2(t) & \cdots & f_n(t)\\ f_1'(t) & f_2'(t) & \cdots & f_n'(t)\\ \vdots & \vdots & \ddots & \vdots\\ f_1^{(n-1)}(t) & f_2^{(n-1)}(t) & \cdots & f_n^{(n-1)}(t) \end{vmatrix},$$

defined on I.

why?

[1]The Wronskian is named for Polish mathematician Jósef Maria Hoëné Wronski (1778–1853), who worked on the philosophical foundations of mathematics but also designed an innovative transportation vehicle and instruments for longitude. We will return to the Wronskian in Chapters 4 and 6.

NOTE: $W[f_1, f_2, \ldots, f_n]$ is defined on I provided that $f_1, f_2, \ldots, f_n$ have $n - 1$ derivatives.

The Wronskian and Trig Functions: Sometimes the linear dependence or independence of trigonometric functions can be checked more easily by using the definition and identities. The Wronskian can become complicated.

The Wronskain is for checking the linear independence of functions

The Wronskian and Linear Independence Theorem
If $W[f_1, f_2, \ldots, f_n](t) \neq 0$ for all t on the interval I, where $f_1, f_2, \ldots, f_n$ are defined, then $\{f_1, f_2, \ldots, f_n\}$ is a linearly independent set of functions on I.[2]

EXAMPLE 11 **Wronskian Check** Let $\{t^2 + 1, t^2 - 1, 2t + 5\}$ be a set of polynomials in $\mathbb{P}_3(t)$. We use the Wronskian to check for linear independence:

$$W(t) = \begin{vmatrix} t^2+1 & t^2-1 & 2t+5 \\ 2t & 2t & 2 \\ 2 & 2 & 0 \end{vmatrix}$$

$$= 2\begin{vmatrix} t^2-1 & 2t+5 \\ 2t & 2 \end{vmatrix} - 2\begin{vmatrix} t^2+1 & 2t+5 \\ 2t & 2 \end{vmatrix}$$

$$= 2\left[2t^2 - 2 - (4t^2 + 10t)\right] - 2\left[2t^2 + 2 - (4t^2 + 10t)\right] = -8 \neq 0.$$

Therefore, $\{t^2 + 1, t^2 - 1, 2t + 5\}$ is a linearly independent set of functions on $(-\infty, \infty)$. ■

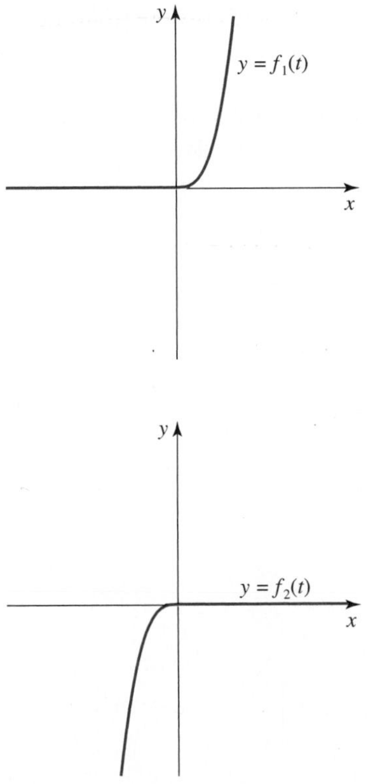

FIGURE 3.6.4 The functions of Example 12.

EXAMPLE 12 **Is The Converse True?** Suppose that we have the converse, that the Wronskian $W(f_1, f_2, \ldots, f_n) = 0$ over an entire interval I, where $f_1, f_2, \ldots, f_n$ are defined on I. Does this imply that $\{f_1, f_2, \ldots, f_n\}$ is linearly dependent?

The answer, somewhat surprisingly, is no.

Consider the following two functions, shown in Fig. 3.6.4, which provide a counterexample:

$$f_1(t) = \begin{cases} t^3 & \text{if } t \geq 0, \\ 0 & \text{if } t < 0, \end{cases} \quad \text{and} \quad f_2(t) = \begin{cases} 0 & \text{if } t \geq 0, \\ t^3 & \text{if } t < 0. \end{cases}$$

Then

$$f_1'(t) = \begin{cases} 3t^2 & \text{if } t \geq 0, \\ 0 & \text{if } t < 0, \end{cases} \quad \text{and} \quad f_2'(t) = \begin{cases} 0 & \text{if } t \geq 0, \\ 3t^2 & \text{if } t < 0, \end{cases}$$

and

$$W(f_1, f_2) = \begin{vmatrix} f_1 & f_2 \\ f_1' & f_2' \end{vmatrix} = f_1 f_2' - f_1' f_2$$

$$= \begin{cases} 0 & \text{if } t \geq 0, \\ 0 & \text{if } t < 0 \end{cases} - \begin{cases} 0 & \text{if } t \geq 0, \\ 0 & \text{if } t < 0 \end{cases} = 0.$$

However, f_1 can never be a scalar multiple of f_2, so they are linearly independent, not linearly dependent. ■

[2] If $\{f_1, f_2, \ldots, f_n\}$ is linearly dependent on I, then $W[f_1, f_2, \ldots, f_n](t) \equiv 0$ on I. So to show independence we only need to find one $t_0 \in I$ such that $W[f_1, f_2, \ldots, f_n](t_0) \neq 0$.

Under additional hypotheses, the converse question of Example 12 can have a different answer. In Section 4.2, we will use the Existence and Uniqueness Theorem to show that in the case where $f_1, f_2, \ldots, f_n$ are solutions of a linear homogeneous differential equation of order $\geq n$, then a zero Wronskian *does* imply linear dependence of the $\{f_1, f_2, \ldots, f_n\}$. MORAL: Be cautious.

Basis of a Vector Space

If we are looking for a set of vectors from which to build or generate a vector space $\mathbb{V}$, there are two opposing tendencies. A small set may not have enough vectors to generate everything; that is, to span the space. A large set may be linearly dependent and thus have more than we need. The ideal set is one that is *large enough* to span but *small enough* to be linearly independent. The set should, in other words, be as small as possible but no smaller. Such a set is called a **basis**: a *linearly independent spanning set*.

Basis of a Vector Space

The set $\{\vec{\mathbf{v}}_1, \vec{\mathbf{v}}_2, \ldots, \vec{\mathbf{v}}_n\}$ is a **basis** for vector space $\mathbb{V}$ provided that

(i) $\{\vec{\mathbf{v}}_1, \vec{\mathbf{v}}_2, \ldots, \vec{\mathbf{v}}_n\}$ is linearly independent; and

(ii) $\text{Span}\{\vec{\mathbf{v}}_1, \vec{\mathbf{v}}_2, \ldots, \vec{\mathbf{v}}_n\} = \mathbb{V}$.

EXAMPLE 13 **Standard Basis for $\mathbb{R}^3$** We saw in Example 4 that $\mathbb{R}^3$ is spanned by the set $\{\vec{\mathbf{e}}_1, \vec{\mathbf{e}}_2, \vec{\mathbf{e}}_3\}$, which we called **standard basis vectors**. To justify this term we need to show that these vectors really are linearly independent. The matrix they form, $\mathbf{I}_3$, is invertible, so the system

$$c_1 \begin{bmatrix} 1 \\ 0 \\ 0 \end{bmatrix} + c_2 \begin{bmatrix} 0 \\ 1 \\ 0 \end{bmatrix} + c_3 \begin{bmatrix} 0 \\ 0 \\ 1 \end{bmatrix} = \vec{\mathbf{0}}$$

has the unique solution $c_1 = c_2 = c_3 = 0$. ■

Standard Basis for $\mathbb{R}^n$

The **standard basis** for $\mathbb{R}^n$ is

$$\{\vec{\mathbf{e}}_1, \vec{\mathbf{e}}_2, \ldots, \vec{\mathbf{e}}_n\},$$

where

$$\vec{\mathbf{e}}_1 = \begin{bmatrix} 1 \\ 0 \\ 0 \\ \vdots \\ 0 \end{bmatrix}, \vec{\mathbf{e}}_2 = \begin{bmatrix} 0 \\ 1 \\ 0 \\ \vdots \\ 0 \end{bmatrix}, \ldots, \vec{\mathbf{e}}_n = \begin{bmatrix} 0 \\ 0 \\ 0 \\ \vdots \\ 1 \end{bmatrix}$$

are the column vectors of the identity matrix $\mathbf{I}_n$.

We can see that any invertible $n \times n$ matrix will have n linearly independent column vectors that form a basis for $\mathbb{R}^n$.

EXAMPLE 14 **Basis for a Hyperplane** To find a basis for the hyperplane in $\mathbb{R}^4$ that is the solution set of the equation

$$2x_1 + 3x_2 - 4x_3 - x_4 = 0, \tag{8}$$

we use the fact that a typical vector in the hyperplane is defined by choosing x_1, x_2, and x_3 arbitrarily and then determining x_4 from equation (8). This typical 4-vector is thus

$$\begin{bmatrix} x_1 \\ x_2 \\ x_3 \\ x_4 \end{bmatrix} = \begin{bmatrix} x_1 \\ x_2 \\ x_3 \\ 2x_1 + 3x_2 - 4x_3 \end{bmatrix} = x_1 \begin{bmatrix} 1 \\ 0 \\ 0 \\ 2 \end{bmatrix} + x_2 \begin{bmatrix} 0 \\ 1 \\ 0 \\ 3 \end{bmatrix} + x_3 \begin{bmatrix} 0 \\ 0 \\ 1 \\ -4 \end{bmatrix}. \tag{9}$$

Because x_1, x_2, and x_3 were arbitrary, equation (9) shows that the hyperplane is spanned by

$$\begin{bmatrix} 1 \\ 0 \\ 0 \\ 2 \end{bmatrix}, \quad \begin{bmatrix} 0 \\ 1 \\ 0 \\ 3 \end{bmatrix}, \quad \text{and} \quad \begin{bmatrix} 0 \\ 0 \\ 1 \\ -4 \end{bmatrix}.$$

To determine if these vectors are linearly independent, we form a matrix with these column vectors,

$$\mathbf{A} = \begin{bmatrix} 1 & 0 & 0 \\ 0 & 1 & 0 \\ 0 & 0 & 1 \\ 2 & 3 & -4 \end{bmatrix}, \quad \text{which has RREF} \quad \begin{bmatrix} 1 & 0 & 0 \\ 0 & 1 & 0 \\ 0 & 0 & 1 \\ 0 & 0 & 0 \end{bmatrix}.$$

Linear Independence of Columns: The columns of the matrix **A** corresponding to the columns of its RREF with leading 1s are linearly independent.

The RREF shows that these three vectors are linearly independent and therefore a basis for the hyperplane. Thus it seems that this subspace of $\mathbb{R}^4$ is what we should like to call three-dimensional. This leads to our final topic. ■

Dimension of a Vector Space

A vector space can have different bases. The **standard basis** for $\mathbb{R}^2$, for example, is $\{\vec{\mathbf{e}}_1, \vec{\mathbf{e}}_2\}$, for

$$\vec{\mathbf{e}}_1 = \begin{bmatrix} 1 \\ 0 \end{bmatrix} \quad \text{and} \quad \vec{\mathbf{e}}_2 = \begin{bmatrix} 0 \\ 1 \end{bmatrix},$$

the columns of $\mathbf{I}_2$, but another basis for $\mathbb{R}^2$ is given by

$$\left\{ \begin{bmatrix} 2 \\ 1 \end{bmatrix}, \begin{bmatrix} 1 \\ 2 \end{bmatrix} \right\}.$$

(The reader is asked to verify this in Problem 44.)

But, as we will prove in Appendix LT, the *number of vectors in a basis is always the same for a particular vector space*, and this fact allows us to define the **dimension** of a (finite-dimensional) vector space as the number of vectors in any basis. A vector space so large that no finite set of vectors spans it is called *infinite-dimensional*. Many of the function spaces that we study are infinite-dimensional. In studying particular differential equations, however, we are often able to limit our attention to finite-dimensional subspaces.

EXAMPLE 15 **Two-Dimensional Plane in $\mathbb{R}^4$** The solution of the system

$$\begin{aligned} x_1 + 2x_2 - x_3 + x_4 &= 0, \\ x_1 + 3x_2 + x_3 + 2x_4 &= 0 \end{aligned} \tag{10}$$

is a subspace in $\mathbb{R}^4$, the intersection of two 3-dimensional hyperplanes. What is its dimension? To answer this we rewrite (10) in reduced row echelon form:

$$\begin{aligned} x_1 \quad - 5x_3 - x_4 &= 0, \\ x_2 + 2x_3 + x_4 &= 0. \end{aligned} \tag{11}$$

The two free variables (those whose columns contain no pivot) tell us at once that the solution set of (10) is a two-parameter family. To exhibit a basis explicitly, we use equations (11) to write a typical vector of the subspace:

$$\begin{bmatrix} x_1 \\ x_2 \\ x_3 \\ x_4 \end{bmatrix} = \begin{bmatrix} 5r + s \\ -2r - s \\ r \\ s \end{bmatrix} = r \begin{bmatrix} 5 \\ -2 \\ 1 \\ 0 \end{bmatrix} + s \begin{bmatrix} 1 \\ -1 \\ 0 \\ 1 \end{bmatrix},$$

where $x_3 = r$ and $x_4 = s$ are arbitrary real numbers. The two vectors on the right are linearly independent and hence form a basis for the solutions. ■

The Dimension of the Column Space of a Matrix

In Sec. 3.2 we defined a pivot column of matrix **A** as a column that corresponds to a column in the RREF of **A** with a leading 1. We can see now that the pivot columns in a matrix are linearly independent and span Col **A**.

Properites of the Column Space of a Matrix

- The pivot columns of a matrix **A** form a basis for Col **A**.
- The dimension of the column space, called the **rank** of **A**, is the number of pivot columns in **A**,

$$\text{rank}\,\mathbf{A} = \dim(\text{Col}\,\mathbf{A}).$$

Thus we have another way of expressing the rank of a matrix, as previously defined in Section 3.2.

EXAMPLE 16 **Column Space Dimension** Let

$$\mathbf{A} = \begin{bmatrix} 1 & 0 & 3 & 5 & 7 \\ 0 & 2 & 4 & 6 & 8 \end{bmatrix}.$$

Its RREF is

$$\begin{bmatrix} 1 & 0 & 3 & 5 & 7 \\ 0 & 1 & 2 & 3 & 4 \end{bmatrix}.$$

The pivot column vectors of **A** are $\begin{bmatrix} 1 \\ 0 \end{bmatrix}$ and $\begin{bmatrix} 0 \\ 2 \end{bmatrix}$, and rank $\mathbf{A} = 2$, the dimension of the column space of **A**. ■

We can now add some new characterizations to our list for invertible matrices. Because an invertible $n \times n$ matrix **A** has the identity matrix **I** as its RREF, every column of **A** is a pivot column and rank $\mathbf{A} = n$.

Invertible Matrix Characterization (Basis for Col A)

Let $\mathbf{A}$ be an $n \times n$ matrix. The following statements are equivalent:

- $\mathbf{A}$ is invertible.
- The column vectors of $\mathbf{A}$ are linearly independent.
- Every column of $\mathbf{A}$ is a pivot column.
- The column vectors of $\mathbf{A}$ form a basis for Col $\mathbf{A}$.
- rank $\mathbf{A} = n$.

EXAMPLE 17 **Vector Space $\mathbb{P}_2$** It is easy to see that the vector space of polynomials of degree two or less is spanned by $\{1, t, t^2\}$, because a typical element has the form $a_1 + a_2 t + a_3 t^2$. If this set is linearly independent, it is a basis, and we can conclude that the dimension of $\mathbb{P}_2$ is 3. So, suppose that

$$c_1 + c_2 t + c_3 t^2 = 0 \tag{12}$$

holds for all t. Then in particular it must hold for the t-values -1, 0, and 1. Substituting these into (12) gives the system

$$\begin{aligned} c_1 - c_2 + c_3 &= 0, \\ c_1 \qquad\quad &= 0, \\ c_1 + c_2 + c_3 &= 0, \end{aligned}$$

and the augmented matrix $[\mathbf{A} \mid \vec{\mathbf{0}}]$ of this system has RREF

$$\left[\begin{array}{ccc|c} 1 & 0 & 0 & 0 \\ 0 & 1 & 0 & 0 \\ 0 & 0 & 1 & 0 \end{array}\right].$$

Therefore, $c_1 = c_2 = c_3 = 0$, so $\{1, t, t^2\}$ is a linearly independent set that forms a basis for $\mathbb{P}_2$. A shorthand notation for our conclusion is $\dim \mathbb{P}_2 = 3$. ■

EXAMPLE 18 **Infinite-Dimensional Space** The space $\mathbb{P}$ of all polynomials is infinite-dimensional. If we wrote down any finite set of polynomials as a possible basis, it would contain a polynomial of highest degree; suppose that highest degree is k. Then there would be no way to generate from this basis the polynomial t^{k+1}. Therefore, no finite set could ever span $\mathbb{P}$. We conclude that $\dim \mathbb{P} = \infty$. ■

There are a multitude of infinite-dimensional vector spaces. $C(I)$, $C^n(I)$, and $\mathbb{P}$ are examples.

EXAMPLE 19 **Dimension $\mathbb{M}_{23}$** A standard basis for $\mathbb{M}_{23}$ is the set

$$\left\{ \begin{bmatrix} 1 & 0 & 0 \\ 0 & 0 & 0 \end{bmatrix}, \begin{bmatrix} 0 & 1 & 0 \\ 0 & 0 & 0 \end{bmatrix}, \begin{bmatrix} 0 & 0 & 1 \\ 0 & 0 & 0 \end{bmatrix}, \begin{bmatrix} 0 & 0 & 0 \\ 1 & 0 & 0 \end{bmatrix}, \begin{bmatrix} 0 & 0 & 0 \\ 0 & 1 & 0 \end{bmatrix}, \begin{bmatrix} 0 & 0 & 0 \\ 0 & 0 & 1 \end{bmatrix} \right\}.$$

Therefore, $\dim \mathbb{M}_{23} = 6$. In fact, it is easy to see that we need a basis vector for each element of the matrix. We can generalize this observation to show that $\dim \mathbb{M}_{mn} = mn$. ■

EXAMPLE 20 **Dimension of Zero-Trace Matrices** Let us look at the subspace $\mathbb{W}$ of $\mathbb{M}_{22}$ in which every element has trace equal to zero:

$$\mathbb{W} = \left\{ \begin{bmatrix} a & b \\ c & d \end{bmatrix} \middle| \; a + d = 0 \right\}.$$

A basis for $\mathbb{W}$ is the set

$$\left\{ \begin{bmatrix} 1 & 0 \\ 0 & -1 \end{bmatrix}, \begin{bmatrix} 0 & 1 \\ 0 & 0 \end{bmatrix}, \begin{bmatrix} 0 & 0 \\ 1 & 0 \end{bmatrix} \right\},$$

which has three elements. Therefore, $\dim \mathbb{W} = 3$. (See Problem 75.) ■

EXAMPLE 21 **Dimension of a Solution Space** For all real t, e^t and e^{-t} are vectors in the solution space $\mathbb{S}$ for $x'' - x = 0$. We can check this fact by direct substitution. In the next chapter we will show that $\{e^t, e^{-t}\}$ is a basis for the solution space, so that $\dim \mathbb{S} = 2$. ■

Abstract Interpretation of Nonhomogeneous Linear Equations

Linear *algebraic* equations have solutions in terms of subspaces. In Sec. 3.2, Example 9, we noted that the solution set of $x + y = 0$ is a line through the origin, a one-dimensional subspace of $\mathbb{R}^2$. The solution set of $x + y = 1$, however, is not a subspace, because it is a line not containing the origin. (See Fig. 3.2.6.) But it *is* a line through $(0, 1)$ (a particular solution) in the direction of $(-1, 1)$ (a homogeneous solution), and is in fact given by

$$\underbrace{(x, y)}_{\vec{\mathbf{x}}} = \underbrace{(0, 1)}_{\vec{\mathbf{x}}_p} + \underbrace{c(-1, 1)}_{\vec{\mathbf{x}}_h}. \tag{13}$$

(Compare this with the language of Problem 70 in Sec. 3.5.)

First-order linear *differential* equations have a similar solution structure, as discussed in Sec. 2.1, Examples 7 and 8. Corresponding to the nonhomogeneous equation

$$y' + ay = b \tag{14}$$

is the homogeneous equation

$$y' + ay = 0 \tag{15}$$

having the family of solutions $\{ce^{-at}\}$, a one-dimensional subspace of $\mathcal{C}^1(\mathbb{R})$. The family of solutions of (14) is not a subspace because it does not include the zero function, but its solutions have a form similar to (13). Because $y = b/a$ is a particular solution of (14), the general solution of (14) is given by

$$y = \underbrace{\frac{b}{a}}_{\vec{\mathbf{y}}_p} + \underbrace{ce^{-at}}_{\vec{\mathbf{y}}_h}. \tag{16}$$

The Abstraction of Algebra

What is abstract for one generation of mathematicians and scientists becomes commonplace to the next. At one time the complex number $i = \sqrt{-1}$ was considered "imaginary" and rejected by no less a mathematical figure than René Descartes. Most of his contemporaries agreed, but a few persisted in carrying forward the development of these numbers. Today they are as concrete as the "real" numbers and indispensable in mathematics and science. Yesterday's outlandish idea becomes the working tool of today.

The idea of an abstract "space" whose elements are four-tuples, or matrices, or functions, or objects even less geometric in character, may at first strike us as unusual because geometric terms like *space*, *line*, *dimension*, *orthogonal*, and *intersection* have been transplanted into a new setting. In the long run, however, the abstraction of linear space structure allows us to cut through the clutter of particular circumstances to see an underlying pattern that can be studied once and applied widely. The more obviously applicable topics in differential equations often benefit from the efficiencies offered by the tools of linear algebra.

Summary

We have studied properties of subsets of vector spaces that span the vector space and that may be linearly independent or dependent. A minimal or independent spanning set forms a basis for a vector space and, if finite, provides a measure of its size or dimension. Matrices and row reduction are basic tools in these investigations.

3.6 Problems

The Spin on Spans *Determine whether the vectors in the set S span the vector space* $\mathbb{V}$.

1. $\mathbb{V} = \mathbb{R}^2$; $S = \{[0, 0], [1, 1]\}$

2. $\mathbb{V} = \mathbb{R}^3$; $S = \{[1, 0, 0], [0, 1, 0], [2, 3, 1]\}$

3. $\mathbb{V} = \mathbb{R}^3$; $S = \{[1, 0, -1], [2, 0, 4], [-5, 0, 2], [0, 0, 1]\}$

4. $\mathbb{V} = \mathbb{P}_2$; $S = \{1,\ t+1,\ t^2 - 2t + 3\}$

5. $\mathbb{V} = \mathbb{P}_2$; $S = \{t+1,\ t^2+1,\ t^2 - t\}$

6. $\mathbb{V} = \mathbb{M}_{22}$;
$$S = \left\{\begin{bmatrix} 1 & 1 \\ 0 & 0 \end{bmatrix}, \begin{bmatrix} 0 & 0 \\ 1 & 1 \end{bmatrix}, \begin{bmatrix} 1 & 0 \\ 1 & 0 \end{bmatrix}, \begin{bmatrix} 0 & 1 \\ 0 & 1 \end{bmatrix}\right\}$$

Independence Day *Decide whether the set S is a linearly independent subset of the given vector space* $\mathbb{V}$.

7. $\mathbb{V} = \mathbb{R}^2$; $S = \{[1, -1], [-1, 1]\}$

8. $\mathbb{V} = \mathbb{R}^2$; $S = \{[1, 1], [1, -1]\}$

9. $\mathbb{V} = \mathbb{R}^3$; $S = \{[1, 0, 0], [1, 1, 0], [1, 1, 1]\}$

10. $\mathbb{V} = \mathbb{R}^3$; $S = \{[2, -1, 4], [4, -2, 8]\}$

11. $\mathbb{V} = \mathbb{R}^3$; $S = \{[1, 1, 8], [-3, 4, 2], [7, -1, 3]\}$

12. $\mathbb{V} = \mathbb{P}_1$; $S = \{1, t\}$

13. $\mathbb{V} = \mathbb{P}_1$; $S = \{1+t, 1-t\}$

14. $\mathbb{V} = \mathbb{P}_2$; $S = \{t, 1-t\}$

15. $\mathbb{V} = \mathbb{P}_2$; $S = \{1+t, 1-t, t^2\}$

16. $\mathbb{V} = \mathbb{P}_2$; $S = \{t+3, t^2-1, 2t^2-t-5\}$

17. $\mathbb{V} = \mathbb{D}_{22}$ (the diagonal 2×2 matrices);
$$S = \left\{\begin{bmatrix} 1 & 0 \\ 0 & 0 \end{bmatrix}, \begin{bmatrix} 0 & 0 \\ 0 & 1 \end{bmatrix}\right\}$$

18. $\mathbb{V} = \mathbb{D}_{22}$; $S = \left\{\begin{bmatrix} 1 & 0 \\ 0 & 1 \end{bmatrix}, \begin{bmatrix} 1 & 0 \\ 0 & -1 \end{bmatrix}\right\}$

Function Space Dependence *Determine whether each subset S of* $C(\mathbb{R})$ *in Problems 19–25 is or is not linearly independent.*

19. $S = \{e^t, e^{-t}\}$

20. $S = \{e^t, te^t, t^2e^t\}$

21. $S = \{\sin t, \sin 2t, \sin 3t\}$

22. $S = \{1, \sin^2 t, \cos^2 t\}$

23. $S = \{1, t-1, (t-1)^2\}$

24. $S = \{e^t, e^{-t}, \cosh t\}$

25. $S = \{\sin^2 t, 4, \cos 2t\}$

Independence Testing *Determine whether or not the set of functions given in each of Problems 26–29 is linearly independent on* $(-\infty, \infty)$.

26. $\left\{ \begin{bmatrix} e^t \\ e^t \end{bmatrix}, \begin{bmatrix} 2e^{2t} \\ e^t \end{bmatrix} \right\}$

27. $\left\{ \begin{bmatrix} \sin t \\ \cos t \end{bmatrix}, \begin{bmatrix} \cos t \\ -\sin t \end{bmatrix} \right\}$

28. $\left\{ \begin{bmatrix} e^t \\ 2e^t \\ e^t \end{bmatrix}, \begin{bmatrix} e^{-t} \\ 2e^{-t} \\ e^t \end{bmatrix}, \begin{bmatrix} e^{2t} \\ 3e^{2t} \\ e^{2t} \end{bmatrix} \right\}$

29. $\left\{ \begin{bmatrix} e^{-t} \\ -4e^{-t} \\ e^{-t} \end{bmatrix}, \begin{bmatrix} e^{-t} \\ 0 \\ -e^{-t} \end{bmatrix}, \begin{bmatrix} 2e^{8t} \\ e^{8t} \\ 2e^{8t} \end{bmatrix} \right\}$

30. Twins? Is there any difference between the vector space spanned by the set $\{\cos t, \sin t\}$ and the vector space spanned by the set $S = \{\cos t + \sin t, \cos t - \sin t\}$?

31. A Questionable Basis Is the set

$$\left\{ \begin{bmatrix} 1 \\ 1 \\ 0 \end{bmatrix}, \begin{bmatrix} 0 \\ 1 \\ 1 \end{bmatrix}, \begin{bmatrix} 2 \\ 1 \\ -1 \end{bmatrix} \right\}$$

a basis for $\mathbb{R}^3$? If not, change one of the vectors to form a basis.

32. Wronskian Suppose that f and g are differentiable functions on the unit interval I. The **Wronskian** of f and g is given by

$$\begin{aligned} W[f, g](t) &\equiv \begin{vmatrix} f(t) & g(t) \\ f'(t) & g'(t) \end{vmatrix} \\ &= f(t)g'(t) - f'(t)g(t). \end{aligned} \tag{17}$$

Show that if $W[f, g](t) \neq 0$ for all t in I, then f and g are linearly independent. HINT: If $c_1 f(t) + c_2 g(t) \equiv 0$, then $c_1 f'(t) + c_2 g'(t) \equiv 0$; use Cramer's Rule to solve for c_1 and c_2.

33. Zero Wronskian Does Not Imply Linear Dependence We saw that a nonzero Wronskian implies linear independence but that the converse is *not* true. (See Example 12.) Another example follows:

(a) Show that the Wronskian is zero for the functions

$$f(t) = t^2 \quad \text{and} \quad g(t) = t\,|t| = \begin{cases} t^2 & \text{if } t \geq 0, \\ -t^2 & \text{if } t < 0. \end{cases}$$

(b) Show that the functions f and g are *not* linearly dependent but independent on $(-\infty, \infty)$.

34. Linearly Independent Exponentials Use the Wronskian to show that $\{e^{at}, e^{bt}\}$ is a linearly independent set if and only if $a \neq b$.

35. Looking Ahead Use the Wronskian to show that $\{e^t, te^t\}$ is linearly independent on $\mathbb{R}$.

36. Revisiting Linear Independence Use the Wronskian to show that the set $\{e^t, 5e^{-t}, e^{3t}\}$ is linearly independent. (See Example 7.)

Independence Checking *Use the Wronskian to check the subsets of* $C(\mathbb{R})$ *in Problems 37–42 for linear independence.*

37. $\{5, \cos t, \sin t\}$

38. $\{e^t, e^{-t}, 1\}$

39. $\{2 + t, 2 - t, t^2\}$

40. $\{3t^2 - 4, 2t, t^2 - 1\}$

41. $\{\cosh t, \sinh t\}$

42. $\{e^t \cos t, e^t \sin t\}$

Getting on Base in $\mathbb{R}^2$ *Determine whether the set given in each of Problems 43–48 is a basis for* $\mathbb{R}^2$. *Justify your answers.*

43. $\{[1, 1]\}$

44. $\{[1, 2], [2, 1]\}$

45. $\{[-1, -1], [1, 1]\}$

46. $\{[1, 0], [1, 1]\}$

47. $\{[1, 0], [0, 1], [1, 1]\}$

48. $\{[0, 0], [1, 1], [2, 2], [-1, -1]\}$

The Base for the Space *Determine whether or not the set* S *in each of Problems 49–56 is a basis for the specified vector space* $\mathbb{V}$.

49. $\mathbb{V} = \mathbb{R}^3$; $S = \{[1, 0, 0], [0, 1, 0]\}$

50. $\mathbb{V} = \mathbb{R}^3$; $S = \{[1, 0, 1], [1, 1, 0], [0, 1, 1]\}$

51. $\mathbb{V} = \mathbb{R}^3$; $S = \{[1, 0, 0], [0, 1, 0], [0, 1, 1], [1, 1, 0]\}$

52. $\mathbb{V} = \mathbb{P}_2$; $S = \{t^2 + 3t + 1, t^2 - 2t + 4\}$

53. $\mathbb{V} = \mathbb{P}_3$; $S = \{t^2, t + 3, t^3 + 4, t - 1, t^2 - 5t + 1\}$

54. $\mathbb{V} = \mathbb{P}_4$; $S = \{t^4, t + 3, t^3 + 4, t - 1, t^2 - 5t + 1\}$

55. $\mathbb{V} = \mathbb{M}_{22}$;

$$S = \left\{ \begin{bmatrix} 1 & 0 \\ 0 & 0 \end{bmatrix}, \begin{bmatrix} 0 & 1 \\ 0 & 0 \end{bmatrix}, \begin{bmatrix} 0 & 0 \\ 1 & 0 \end{bmatrix}, \begin{bmatrix} 1 & 1 \\ 1 & 1 \end{bmatrix} \right\}$$

56. $\mathbb{V} = \mathbb{M}_{23}$;

$$S = \left\{ \begin{bmatrix} 1 & 0 & 1 \\ 0 & 0 & 0 \end{bmatrix}, \begin{bmatrix} 1 & 1 & 0 \\ 0 & 0 & 0 \end{bmatrix}, \begin{bmatrix} 0 & 0 & 0 \\ 1 & 0 & 1 \end{bmatrix}, \begin{bmatrix} 0 & 0 & 0 \\ 1 & 1 & 0 \end{bmatrix}, \begin{bmatrix} 0 & 0 & 0 \\ 1 & 1 & 1 \end{bmatrix} \right\}$$

Sizing Them Up *For each of Problems 57 and 58, determine the dimension and find a basis for the subspace* $\mathbb{W}$ *of the vector space* $\mathbb{V}$.

57. $\mathbb{V} = \mathbb{R}^3$; $\mathbb{W} = \{[x_1, x_2, x_3] \mid x_1 + x_2 + x_3 = 0\}$

58. $\mathbb{V} = \mathbb{R}^4$; $\mathbb{W} = \{[x_1, x_2, x_3, x_4] \mid x_1 + x_3 = 0,\ x_2 = x_4\}$

Polynomial Dimensions *Find the dimension of the subspace of $\mathbb{P}_3$ spanned by the subset in Problems 59–61.*

59. $\{t, t-1\}$

60. $\{t, t-1, t^2+1\}$

61. $\{t^2, t^2-t-1, t+1\}$

62. Solution Basis Determine a basis for the solution set (a subspace of $\mathbb{R}^3$) of the system

$$\begin{aligned} x + y - z &= 0,\\ y - 5z &= 0. \end{aligned}$$

Solution Spaces for Linear Algebraic Systems *In Problems 63 and 64, determine bases and dimension for the solution spaces for the homogeneous systems (as given in Section 3.5, Problems 61 and 62).*

63.
$$\begin{aligned} 2x_1 - 2x_2 + 4x_3 - 2x_4 &= 0\\ 2x_1 + x_2 + 7x_3 + 4x_4 &= 0\\ x_1 - 4x_2 - x_3 + 7x_4 &= 0\\ 4x_1 - 12x_2 \qquad - 20x_4 &= 0 \end{aligned}$$

64.
$$\begin{aligned} 3x_1 \qquad + 6x_3 + 3x_4 + 9x_5 &= 0\\ x_1 + 3x_2 - 4x_3 - 8x_4 + 3x_5 &= 0\\ x_1 - 6x_2 + 14x_3 + 19x_4 + 3x_5 &= 0 \end{aligned}$$

DE Solution Spaces *For each of the differential equations given in Problems 65–70, consider their solution sets.*

(a) *Does the solution set form a subspace of the specified larger space? If not, explain.*

(b) *If the solution set is a subspace, find a basis and determine the dimension.*

65. $d^n y/dt^n = 0, \quad C^n(\mathbb{R})$

66. $y' - 2y = 0, \quad C^1(\mathbb{R})$

67. $y' - 2ty = 0, \quad C^1\left(-\frac{\pi}{2}, \frac{\pi}{2}\right)$

68. $y' + (\tan t)y = 0, \quad C^1(\mathbb{R})$

69. $y' + y^2 = 0, \quad C^1(\mathbb{R})$

70. $y' + (\cos t)y = 0, \quad C^1(\mathbb{R})$

Bases for Subspaces of $\mathbb{R}^n$ *In Problems 71–73, find a basis and the dimension of the given subspaces of $\mathbb{R}^n$.*

71. $\{[a, 0, b, a-b+c] \mid a, b, c \in \mathbb{R}\}$ as a subspace of $\mathbb{R}^4$

72. $\{[a, a-b, 2a+3b] \mid a, b, \in \mathbb{R}\}$ as a subspace of $\mathbb{R}^3$

73. $\{[x+y+z, x+y, 4z, 0] \mid x, y, z \in \mathbb{R}\}$ as a subspace of $\mathbb{R}^4$

74. Two-by-Two Basis Show that

$$\left\{\begin{bmatrix}1 & 0\\ 0 & 0\end{bmatrix}, \begin{bmatrix}0 & 1\\ 1 & 0\end{bmatrix}, \begin{bmatrix}0 & 0\\ 1 & 1\end{bmatrix}\right\}$$

is a linearly independent set in $\mathbb{M}_{22}$. Add another vector to the set to make it a basis for $\mathbb{M}_{22}$.

75. Basis for Zero Trace Matrices Show that

$$\left\{\begin{bmatrix}1 & 0\\ 0 & -1\end{bmatrix}, \begin{bmatrix}0 & 1\\ 0 & 0\end{bmatrix}, \begin{bmatrix}0 & 0\\ 1 & 0\end{bmatrix}\right\}$$

is a basis for the subspace of all 2×2 matrices in $\mathbb{M}_{22}$ with zero trace.

76. Hyperplane Basis Find a basis for the following hyperplane in $\mathbb{R}^4$: $x + 3y - 2z + 6w = 0$.

77. Symmetric Matrices Find the dimension and exhibit a basis for the subspace of all symmetric matrices in $\mathbb{M}_{22}$.

Making New Bases from Old *In Problems 78–80, a vector space and a basis will be given. Construct a different basis from the given basis. Make them substantially different, so that at least two vectors are not multiples of vectors in the original basis.*

78. $\{\vec{\mathbf{i}}, \vec{\mathbf{j}}, \vec{\mathbf{k}}\}$ in $\mathbb{R}^3$

79. $\left\{\begin{bmatrix}1 & 0\\ 0 & 0\end{bmatrix}, \begin{bmatrix}0 & 0\\ 0 & 1\end{bmatrix}\right\}$ in $\mathbb{D}$, the space of all diagonal matrices

80. $\{\sin t, \cos t\}$ in $\{y \in C''(\mathbb{R}) \mid y'' - y = 0\}$

81. Basis for $\mathbb{P}_2$ Do the vectors $\{t^2+t+1,\ t+1,\ 1\}$ form a basis for $\mathbb{P}_2$? If so, represent the polynomial $3t^2 + 2t + 1$ in terms of this basis.

82. True/False Questions If false, give a counterexample or a brief explanation.

(a) A solution set of a homogeneous system of linear algebraic equations, given by

$$\mathbb{W} = \left\{ s\begin{bmatrix}3\\0\\0\\-1\end{bmatrix} + t\begin{bmatrix}0\\2\\1\\1\end{bmatrix} \,\middle|\, s, t \in \mathbb{R} \right\},$$

is a subspace of $\mathbb{R}^4$. True or false?

(b) The dimension of $\mathbb{W}$ is 4. True or false?

(c) A basis for $\mathbb{W}$ is $\{[3, 0], [0, 2], [0, 1], [-1, 1]\}$. True or false?

83. Essay Question[3] Consider the homogeneous system of linear equations

$$\begin{aligned} x_1 + x_2 + x_3 - x_4 &= 0, \\ 3x_1 + 3x_2 + x_3 - 5x_4 &= 0, \\ 4x_1 + 4x_2 + 2x_3 - 6x_4 &= 0. \end{aligned}$$

and its solution set

$$\mathbb{W} = \left\{ s\begin{bmatrix} -1 \\ 1 \\ 0 \\ 0 \end{bmatrix} + t\begin{bmatrix} 2 \\ 0 \\ -1 \\ 1 \end{bmatrix} \middle| \; s, t \in \mathbb{R} \right\}.$$

Write a paragraph or two describing the solution set $\mathbb{W}$ of the system of equations. Write a clear exposition that includes appropriate use of the words *linear combination, span, subspace, vector space, linearly independent, basis,* and *dimension* in describing the solution set.

84. Convergent Sequence Space[4] Discuss the set $\mathbb{V}$ of all convergent sequences as a vector space, where the addition of "vectors" and multiplication by a scalar are defined by

$$\{a_n\} + \{b_n\} = \{a_n + b_n\} \quad \text{and} \quad c\{a_n\} = \{ca_n\}.$$

Describe the zero element and additive inverse elements for this vector space. Make a conjecture about the dimension of $\mathbb{V}$ and try to back it up. Give at least one example of a non-trivial subspace and discuss its dimension.

Cosets in $\mathbb{R}^3$ *Using the following definition, in Problems 85 and 86, find the $\mathbb{W}$-cosets for the given vectors $\vec{\mathbf{v}}$ and give a graphical description of each.*

> **Cosets**
>
> If $\mathbb{W}$ is a subspace of vector space $\mathbb{R}^n$ that includes the origin, and $\vec{\mathbf{v}}$ is a vector in $\mathbb{R}^n$, then the $\mathbb{W}$**-coset** of $\vec{\mathbf{v}}$, denoted $\vec{\mathbf{v}} + \mathbb{W}$, is the set defined by
>
> $$\vec{\mathbf{v}} + \mathbb{W} = \{\vec{\mathbf{v}} + \vec{\mathbf{w}} \mid \vec{\mathbf{w}} \text{ is in } \mathbb{W}\}.$$

Thus, in $\mathbb{R}^3$, if $\mathbb{W}$ is a plane passing through the origin, then the coset $\vec{\mathbf{v}} + \mathbb{W}$ is a plane parallel to $\mathbb{W}$ passing through $\vec{\mathbf{v}}$.

85. $\mathbb{W} = \{[x_1, x_2, x_3] \mid x_1 + x_2 + x_3 = 0\}$; $\vec{\mathbf{v}} = (0, 0, 1)$

86. $\mathbb{W} = \{[x_1, x_2, x_3] \mid x_3 = 0\}$; $\vec{\mathbf{v}} = (1, 1, 1)$

87. More Cosets As in Problems 85 and 86, describe the nature of a coset if the subspace $\mathbb{W}$ is a *line* through the origin. Illustrate for

$$\mathbb{W} = \{[x_1, x_2, x_3] \mid x_1 = t, x_2 = 3t, x_3 = 2t\} \quad \text{and} \quad \vec{\mathbf{v}} = (1, -2, 1).$$

88. Line in Function Space Interpret the general solution of the differential equation $y' + 2y = e^{-2t}$ as a line in a suitable function space.

89. Mutual Orthogonality Prove that nonzero mutually orthogonal vectors in a vector space $\mathbb{V}$ are automatically linearly independent.

90. Suggested Journal Entry I Discuss the distinction between *finite* and *finite-dimensional* for vector spaces. Can a vector space contain only a finite number of vectors? Explain.

91. Suggested Journal Entry II Does the set of *even* functions (where $f(t) = f(-t)$ on the interval $(-\infty, \infty)$) constitute a vector space? What about the set of polynomials of even degree? What is the relationship between these two families? How *big* are they?

92. Suggested Journal Entry III Look at Examples 9 and 10 and the paragraph about span and basis. Then consider the case of two vector functions with three components each. What size system would you need to determine whether the vectors were linearly independent? Consider the same question for the case of three vector functions of two components each.

---

[3] Thanks to Prof. J. Girolo, California Polytechnic State University.

[4] Courtesy of J. E. Hall, Westminster College.

CHAPTER

4 Higher-Order Linear Differential Equations

Nature is exceedingly simple and conformable to herself. Whatever reasoning holds for greater motions, should hold for lesser ones as well. The former depend upon the greater attractive forces of larger bodies, and I suspect that the latter depend upon the lesser forces, as yet unobserved, of insensible particles. For, from the forces of gravity, of magnetism and of electricity it is manifest that there are various kinds of natural forces, and that there may be still more kinds is not to be rashly denied. It is very well known that greater bodies act mutually upon each other by those forces, and I do not clearly see why lesser ones should not act on one another by similar forces.

—Sir Isaac Newton

4.1 The Harmonic Oscillator

SYNOPSIS: We introduce a central model in physics and engineering, the harmonic oscillator, the prototypic second-order linear differential equation. We derive initial-value problems describing mechanical vibrations and the behavior of electrical circuits, damped and undamped, and illustrate trajectories in the phase plane.

One of the most important differential equations is the linear second-order homogeneous equation

$$m\ddot{x} + b\dot{x} + kx = 0,$$

having constant coefficients m, b, and k with $m > 0$. It is the model for a class of phenomena collectively referred to as **damped harmonic oscillators**, which includes mass-spring systems, small oscillations of a pendulum, the motion of a charged particle in an electric field, and the alternating current in an *LRC*-circuit. Formulating and solving problems related to this equation are also stepping stones to understanding more complex models and other oscillatory systems, in biology, ecology, and

Newton's Dot Notation:

Scientists and engineers who work with many variables use Newton's dot notation for derivatives when the independent variable is time t:

$$\dot{x} = \frac{dx}{dt}, \qquad \ddot{x} = \frac{d^2x}{dt^2}.$$

meteorology, for example. The harmonic oscillator equation can be derived in a number of ways. We will do so first for a mass-spring system and again later using an *LRC*-circuit.

The Mass-Spring System

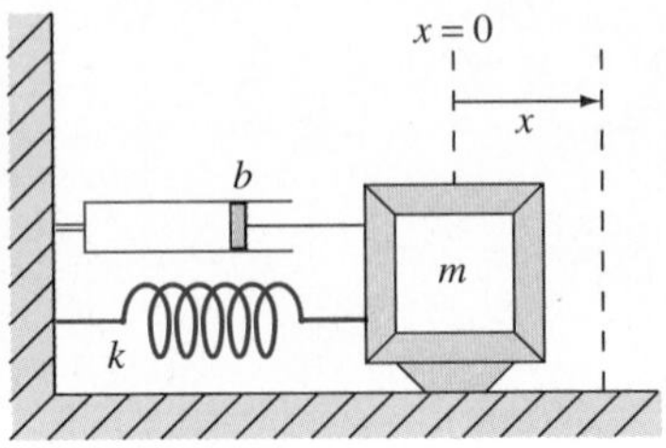

FIGURE 4.1.1 Mass-spring system.

Simple Harmonic Oscillator

Experiment with a variety of initial conditions and damping constants.

Consider an object of mass m on a tabletop. The object is attached to a spring that is in turn attached to the wall. (See Fig. 4.1.1.) The motion of the object is one-dimensional, from left to right (restrained by a track or groove), and we measure its displacement x from its rest position ($x = 0$) either to the left ($x < 0$) or to the right ($x > 0$). (The device pictured above the spring, called a *dashpot*, represents friction due to resistance of moving against a different medium, but we could substitute other damping or dissipative effects in the system, such as friction of the object and the table, or internal friction of the spring.)

We will model the motion using **Newton's Second Law of Motion**, $F = m\ddot{x}$, in which F stands for the sum of all the forces acting on the object. These forces are of three kinds.

1. **Restoring Force:** We assume that when the spring is compressed it tries to expand and when it is stretched it tries to contract. The basic principle of *linear* springs[1] is that the restorative force of the spring is proportional to the amount of stretching or compression:

$$F_{\text{restoring}} = -kx,$$

where k is the positive constant of proportionality and the negative sign indicates that the force opposes (points in the direction opposite to) the stretching or compression. This property is called **Hooke's Law**.[2] The **restoring constant** k measures the strength of the spring. The tiny springs in mechanical watches have small values of k while the springs in a car have large k values.

2. **Damping Force:** We will assume that the damping is due to friction between object and table and that it is proportional to the velocity,[3] acting in the direction opposite to the motion:

$$F_{\text{damping}} = -b\dot{x};$$

the **damping constant** $b > 0$ is small for a slippery surface like ice, large for a rough one like sandpaper.

3. **External Force:** We allow for external "driving" forces to affect the motion, such as wind, magnetic fields (if the object is iron), or shaking of the entire apparatus to the right or left:

$$F_{\text{external}} = f(t),$$

where $f(t)$ is the sum of all such forces. When $f(t) > 0$, the force acts to move the object to the right; when $f(t) < 0$, the force acts to the left.

---

[1]Linear springs exert a restorative force $-kx$ (a linear function); nonlinear models are sometimes used instead. Linear springs are classified as *hard* ($k > 2$) or *soft* ($k < 2$).

[2]Robert Hooke (1635–1703), an Englishman, was not only a physicist and mathematician but was noted as a craftsman who made significant improvements in astronomical instruments and watches.

[3]The assumption of proportionality to the velocity, reasonable for small velocities, may be replaced for larger velocities by proportionality to the square (or some other function) of the velocity. The damping constant may also depend on physical properties such as the Reynolds number, which measures the viscosity of the surrounding medium.

Applying Newton's Second Law, we equate the product of mass and acceleration to the sum of all the forces:

$$\text{mass} \times \text{acceleration} = F_{\text{restoring}} + F_{\text{damping}} + F_{\text{external}};$$
$$m\ddot{x} = -kx - b\dot{x} + f(t).$$

and so we have the equation for a simple harmonic oscillator.

Damping and Forcing of a Simple Harmonic Oscillator:

- $b = 0$ for undamped motion
- $b > 0$ for damped motion
- $f(t) \equiv 0$ for an unforced or free oscillator
- $f(t) \not\equiv 0$ for a forced or driven oscillator

Simple Harmonic Oscillator

The simple harmonic oscillator equation is

$$m\ddot{x} + b\dot{x} + kx = f(t), \tag{1}$$

a second-order nonhomogeneous linear differential equation with constant coefficients $m > 0$, $k > 0$, and $b \geq 0$.

- When $b = 0$, the motion is called **undamped**; otherwise it is **damped**.
- If $f(t) \equiv 0$, the equation is homogeneous,

$$m\ddot{x} + b\dot{x} + kx = 0, \tag{2}$$

and the motion is called **unforced**, **undriven**, or **free**; otherwise the motion is **forced** or **driven**.

We will turn to forced systems in Secs. 4.4–4.6; until then we will be analyzing homogeneous equations.

EXAMPLE 1 **Constructing the DE** A mass of 1 kilogram, resting on a tabletop, is attached to a spring and wall as in Fig. 4.1.1:

$$m = 1 \text{ kilogram}.$$

We discover by experimentation that it takes a force of 1 newton to push or pull the object 0.25 meters from its equilibrium position:

$$k = \frac{1 \text{ newton}}{0.25 \text{ meter}} = 4\,\frac{\text{newton}}{\text{meter}}.$$

We also measure the damping force of the object sliding on the table to be 0.50 newtons when the velocity is 0.25 meters per second:

$$b = \frac{0.5 \text{ newton}}{0.25 \text{ meter/sec}} = 2\,\frac{\text{newton sec}}{\text{meter}}.$$

The object is pulled to the right until the spring is stretched 0.50 meters and then released, without imparting any initial velocity:

$$x(0) = 0.50 \text{ meter} \quad \text{and} \quad \dot{x}(0) = 0 \text{ meter/sec}.$$

We can now formulate the IVP that describes the subsequent motion of the object, assuming that no external forces act on it. The differential equation of this unforced vibration is (2), and we have determined the initial conditions and parameters m, b, and k from the given information. The complete IVP is therefore

$$\ddot{x} + 2\dot{x} + 4x = 0, \qquad x(0) = 0.50, \quad \dot{x}(0) = 0.$$

■

Mass and Spring

Watch the action with a vertical mass-spring model, linked to evolving solution graphs.

A **second-order IVP** requires *two* initial conditions,

$$x(t_0) = x_0 \quad \text{and} \quad \dot{x}(t_0) = v_0,$$

which must be specified at the same point t_0 in the domain of the differential equation.

EXAMPLE 2 **Vertical Mass-Spring** The harmonic oscillator equation can also be used to describe the vertical motion of an object hanging from a spring attached to the ceiling. (The object must be pulled downward "perfectly," so there is no movement from side to side.) In this case, x measures the displacement up or down from the **equilibrium position**: the position of the object when the system is at rest with the downward gravitational force just balanced with the restoring force of the spring. (See Fig. 4.1.2.) ■

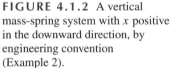

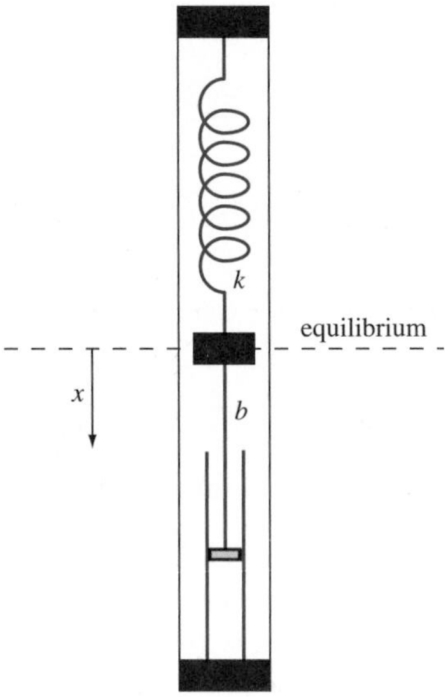

FIGURE 4.1.2 A vertical mass-spring system with x positive in the downward direction, by engineering convention (Example 2).

Importance of Units

It is important to use consistent units in any calculation.[4] In the mks system, length is measured in meters, weight or force in newtons, and mass in kilograms. A kilogram is that mass to which a force of one newton will give an acceleration of one meter per second every second. The energy unit is the joule (a newton-meter), which will be used in Sec. 4.6. The two (metric) systems, mks and cgs, as well as the more antiquated English (or engineering) system, are summarized in Table 4.1.1. In any (consistent) system of units, weight and mass are related by $w = mg$, where g is the acceleration due to gravity in the appropriate units.

Table 4.1.1 Units of measure

Quantity	mks	cgs	English
Force	newton (nt)	dyne	pound (lb)
Mass	kilogram (kg)	gram (gm)	slug (lb sec^2/ft)
Length	meter (m)	centimeter (cm)	foot (ft)
Value of g	9.8 m/sec^2	980.665 cm/sec^2	32 ft/sec^2
Energy	joule	erg	foot-pound (ft-lb)

Solution of the Undamped Unforced Oscillator

We will solve the damped and forced oscillator equations later in the chapter by various techniques. But we can get a quick solution of the undamped unforced case,

$$m\ddot{x} + kx = 0, \tag{3}$$

[4] As obvious as this fact may be, confusion of units has caused more than one disaster. For instance, in 1999 a Mars probe crashed on the surface of Mars when a Jet Propulsion Laboratory (JPL) subcontractor supplied critical computer data measured in the English system, whereas JPL was expecting data expressed in the metric system.

just by educated guessing. We know that $(\sin t)'' = -\sin t$ and $(\cos t)'' = -\cos t$, so it is not surprising that solutions of (3) involve sines and cosines. In fact, comparing

$$(\sin \omega_0 t)'' = -\omega_0^2 \sin \omega_0 t \qquad \text{and} \qquad \ddot{x} = -\frac{k}{m}x,$$

we see that a solution of (3) is given by

$$x(t) = \sin \omega_0 t, \qquad \text{where } \omega_0 = \sqrt{\frac{k}{m}}.$$

Another solution is given by $x(t) = \cos \omega_0 t$.

The Superposition Principle of Sec. 2.1 guarantees that every linear combination of solutions to a homogeneous linear DE is a solution.

Solution to the Undamped Unforced Oscillator

For the undamped unforced oscillator (3), solutions are

$$x(t) = c_1 \cos \omega_0 t + c_2 \sin \omega_0 t, \qquad \omega_0 = \sqrt{\frac{k}{m}}, \tag{4}$$

where c_1 and c_2 are arbitrary constants determined by initial conditions.

Trigonometric Identities:

$\cos(A - B) = \cos A \cos B + \sin A \sin B$

$\cos\theta = \cos(-\theta)$

Furthermore, the Solution Space Theorem, which will be presented in Sec. 4.2, guarantees that *every* solution will be given by (4).

We see that equation (4) represents a periodic motion. A sample solution is shown in Fig. 4.1.3. Using appropriate trigonometric identities, we can write (4) in an alternate polar form to show that it always represents a single pure sinusoidal motion (Problem 14).

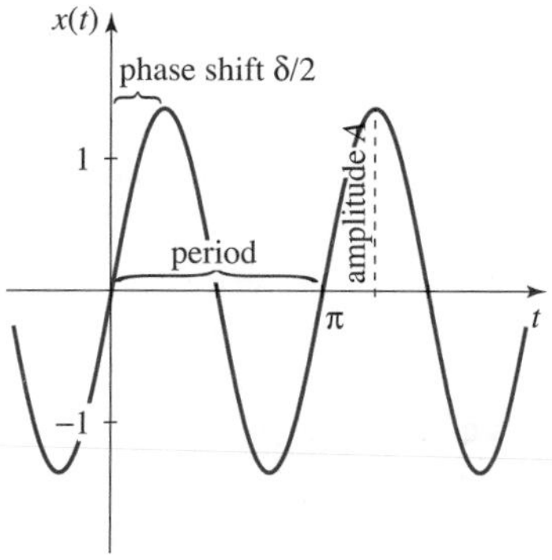

FIGURE 4.1.3 One solution to $\ddot{x} + 4x = 0$, where $\omega_0 = 2$, $\delta = \pi/2$, and phase shift $\delta/\omega_0 = \pi/4$.

Alternate Solution to the Undamped Unforced Oscillator

Solutions to the undamped unforced oscillator (3) may also be expressed as a family of sinusoidal oscillations given by

$$x(t) = A\cos(\omega_0 t - \delta). \tag{5}$$

- **Amplitude** A and **phase angle** δ (measured in radians) are arbitrary (but meaningful) constants that can be determined by initial conditions.
- The motion has **circular frequency** $\omega_0 = \sqrt{k/m}$, measured in radians per second or oscillations per 2π seconds, and **natural frequency** $f_0 = \omega_0/2\pi$, measured in oscillations per second.
- The **period** T of the oscillation (measured in seconds) is given by

$$T = \frac{1}{f_0} = \frac{2\pi}{\omega_0} = 2\pi\sqrt{\frac{m}{k}}. \tag{6}$$

- The solution (5) is a horizontal translation of $A\cos(\omega_0 t)$ with **phase shift** δ/ω_0.

It is a straightforward matter to convert solution forms (4) and (5) from one to the other. (See Problem 14 and the following summary.)

Conversion of Solutions to the Undamped Unforced Oscillator

The translation between the two forms (4) and (5) of the solution to the undamped unforced oscillator is given by

$$A = \sqrt{c_1^2 + c_2^2}, \quad \tan\delta = c_2/c_1, \tag{7}$$

and

$$c_1 = A\cos\delta, \quad c_2 = A\sin\delta. \tag{8}$$

See Problems 15–22 for practice converting solutions from one form to the other.

Simple Harmonic Oscillator

Bring this example to life! Set the initial condition by clicking in the $x\dot{x}$-plane, and look at the action.

EXAMPLE 3 **An Undamped IVP** To solve the initial-value problem

$$\ddot{x} + x = 0, \qquad x(0) = 0, \quad \dot{x}(0) = 1,$$

we identify $m = 1$ and $k = 1$ so that $\omega_0 = 1$.

(a) The general solution is

$$x(t) = c_1\cos t + c_2\sin t, \tag{9}$$

and differentiating gives

$$\dot{x}(t) = -c_1\sin t + c_2\cos t. \tag{10}$$

Substituting $t = 0$ into (9) and (10), we use the initial conditions to find $x(0) = c_1 = 0$ and $\dot{x}(0) = c_2 = 1$. Hence, the solution of the initial-value problem is

$$x(t) = \sin t.$$

(b) We could also write the solution of the differential equation in the alternate form

$$x(t) = A\cos(t - \delta).$$

Substituting the initial conditions yields

$$x(0) = A\cos\delta = 0 \quad \text{and} \quad \dot{x}(0) = A\sin\delta = 1,$$

giving $A = 1$ and $\delta = \pi/2$. Hence, we have

$$x(t) = A\cos(t - \delta) = \cos(t - \pi/2) = \sin t,$$

which is the same solution obtained in (a), although it did not seem so at first. ■

Phase Plane Description

Parametric to Cartesian

See how a point moving on a circle in a phase plane, $x\dot{x}$, is projected to the component curves, tx and $t\dot{x}$.

The solution of the undamped oscillator equation as an explicit function of time provides one way of understanding the model. We get other insights from the **phase plane** description of the relationship between the position x and the velocity $\dot{x}$.[5] We know from (5) that solutions are of the form $x(t) = A\cos(\omega_0 t - \delta)$. Differentiation gives

$$\dot{x} = \frac{dx}{dt} = -\omega_0 A\sin(\omega_0 t - \delta). \tag{11}$$

[5]The phase plane idea was introduced previously in Sec. 2.6. Here $\dot{x}$ plays the role that y did there.

Equations (5) and (11) for $x(t)$ and $\dot{x}(t)$ can be treated as parametric equations in t that make a simple graph in the $x\dot{x}$ phase plane. (See Fig. 4.1.4.) Solution graphs of $x(t)$ and $\dot{x}(t)$ are often called **time series** or **component graphs**.

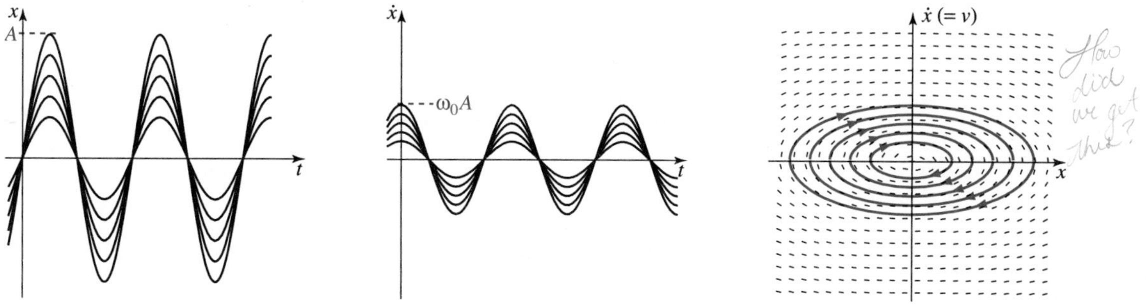

FIGURE 4.1.4 Some solution graphs for $x(t)$, $\dot{x}(t)$, and an $x\dot{x}$ phase portrait for the undamped oscillator $\ddot{x} + 0.25x = 0$.

We can learn a lot from these graphs and their relationships. For example, what do they tell us about ω_0? Where on the $x\dot{x}$ phase plane does each of the three linked trajectories begin? We can infer, from the comparative amplitudes of the time series graphs, that $0 < \omega_0 < 1$, and that $t = 0$ where each phase plane trajectory crosses the positive $\dot{x}$ axis. The time series graphs always show $x(0)$ and $\dot{x}(0)$ values on their vertical axes, so we know that on the phase plane graph $t = 0$ when $x = 0$ and $\dot{x}$ is positive. Problem 31 will direct some exploration of this use of phase plane portraits.

Phase Plane Drawing

Try your hand at drawing a shape in the phase plane and see the component tx and ty graphs unfold in real time.

Phase Portraits

For any autonomous second-order differential equation

$$\ddot{x} = F(x, \dot{x}),$$

the **phase plane** is the two-dimensional graph with x and $\dot{x}$ axes.

The phase plane has a **vector field** specified by the DE, which at any point in the phase plane gives a direction vector with

$$\text{horizontal component} \qquad dx/dt = \dot{x},$$
$$\text{vertical component} \qquad d\dot{x}/dt = \ddot{x}.$$

A **trajectory** is a path formed parametrically by the DE solutions $x(t)$ and $\dot{x}(t)$ as they follow the vector field. A graph showing phase plane trajectories is called a **phase portrait**.

A big advantage of phase portraits is that they can be graphed directly from the DE without having to solve it. The process is exactly the same as with direction fields for first-order equations. The details become transparent if we rewrite a second-order equation as a system of first-order equations by introducing a second variable y:

$$y = dx/dt = \dot{x}, \qquad \textit{(horizontal component)}$$
$$\dot{y} = d\dot{x}/dt = \ddot{x}. \qquad \textit{(vertical component)}$$

Conversion of a Second-Order Linear DE to a System

The second-order DE

$$m\ddot{x} + b\dot{x} + kx = f(t)$$

is equivalent to the system of first-order equations

$$\begin{aligned} \dot{x} &= y, \\ \dot{y} &= \ddot{x} = f(t) - (k/m)x - (b/m)y. \end{aligned} \tag{12}$$

Computers will draw vector fields and phase portraits for second-order DEs; however, many open-ended DE solvers require entering the equivalent systems form, so it is important to become comfortable with the conversion. (See Problems 60–64.) We will look at this process in more detail in Sec. 4.7.

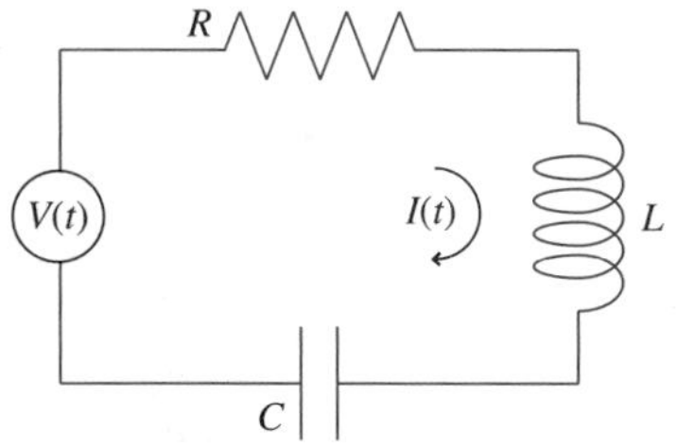

FIGURE 4.1.5 A series *LRC*-circuit; the arrow for $I(t)$ indicates the positive direction for the current.

Modeling Electrical Circuits

The current I in a wire, measured in *amperes* (amps), is a *flow* of charges, negative in the direction of the flow of electrons (negative charges); that is, the current I is the rate of change of the charge Q:

$$I(t) = \dot{Q}(t).$$

For a simple closed electrical series circuit like the one in Fig. 4.1.5, **Kirchoff's Voltage Law** tells us that *the input voltage equals the sum of the voltage drops around the circuit.* These voltage drops are of three kinds.

Resistance:

$V_R = RI$

1. **Drop across a Resistor:** The voltage drop across a resistor is proportional to the current $I(t)$ passing through the resistor (**Ohm's Law**[6]):

$$V_R(t) = RI(t), \tag{13}$$

where the constant of proportionality R is the **resistance** of the resistor, measured in *ohms*. A resistor is often a carbon device with resistance of 100 or 200 ohms, while the resistance of a copper wire is generally negligible.

Inductance:

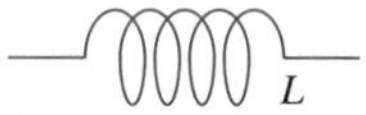

$V_L = L\dot{I}$

2. **Drop across an Inductor:** According to **Faraday's Law**,[7] *the voltage drop across an inductor is proportional to the time rate of change of the current passing through it*:

$$V_L(t) = L\dot{I}(t), \tag{14}$$

where the constant of proportionality L is the **inductance**, measured in *henries*. Inductors are generally coils of wire and are drawn as such.

---

[6]German physicist Georg Simon Ohm (1789–1854) discovered the law that carries his name. Although this work was to be of great influence in developing the theory of electrical circuits, its importance was not recognized by his colleagues for more than a decade.

[7]Michael Faraday (1791–1867), English chemist and physicist, united electricity and chemistry with his strong work on electrolysis, the process that liberates an element by passing electric current through a molten compound that contains the element. Equation (14) describes the electrical view; Faraday's other law of electrolysis gives the chemical view: the mass liberated by a given quantity of electricity is proportional to the atomic weight of the element liberated and is inversely proportional to its valence.

Capacitance:

$V_C = \frac{1}{C}Q$

3. **Drop across a Capacitor:** The voltage drop across a capacitor is proportional to the charge $Q(t)$ on the capacitor. The proportionality constant is written as $1/C$, where C is the **capacitance** of the capacitor, measured in *farads*:

$$V_C(t) = \frac{1}{C}Q(t).$$

A capacitor usually consists of two parallel plates separated by a gap through which no current flows; $Q(t)$ is the charge on one plate relative to the other. Although no current crosses the gap, the (alternating) current surges back and forth from plate to plate through the rest of the circuit. Since $I(t) = \dot{Q}(t)$, this voltage drop can be written

Current and Charge:
$I(t) = \dot{Q}(t)$

$$V_C(t) = \frac{1}{C}\int I(t)\,dt. \tag{15}$$

We can now apply Kirchoff's Voltage Law to the circuit of Fig. 4.1.5, where the voltage source $V(t)$ is a battery or electric generator:

Voltage Source:

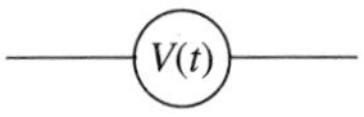

$$RI + L\dot{I} + \frac{1}{C}\int I(t)\,dt = V(t). \tag{16}$$

The left side of equation (16) is the sum of the voltage drops given by equations (13), (14), and (15) for the elements of the circuit. The result, containing both a derivative and an integral, is called an **integro-differential equation**. The simplest version of this equation for the series circuit equation uses again the fact that $I(t) = \dot{Q}(t)$.

Series Circuit Equation (Charge)

$$L\ddot{Q} + R\dot{Q} + \frac{1}{C}Q = V(t). \tag{17}$$

If we assume that there is no voltage source, so that $V(t) \equiv 0$,

$$L\ddot{Q} + R\dot{Q} + \frac{1}{C}Q = 0. \tag{18}$$

Series Circuits

Set values of R, L, and C for a circuit. Click an initial value on the phase plane and see the electrifying results.

The series circuit equation is just the harmonic oscillator equation in disguise. If an initial charge $Q(0) = Q_0$ and an initial current $I(0) = \dot{Q}(0) = I_0$ are given, we have an initial-value problem. The solution $Q(t)$ of this IVP and its derivative $\dot{Q}(t) = I(t)$ give the capacitor charge and circuit current for subsequent times.

We can derive a differential equation for the current I by differentiating (17) with respect to t.

Series Circuit Equation (Current)

$$L\ddot{I} + R\dot{I} + \frac{1}{C}I = \dot{V}(t). \tag{19}$$

For $\dot{V}(t) \equiv 0$, we get the homogeneous equation

$$L\ddot{I} + R\dot{I} + \frac{1}{C}I = 0. \tag{20}$$

Table 4.1.2 Electrical units

Quantity	Units
Voltage source $V(t)$	volt
Resistance R	ohm
Inductance L	henry
Capacitance C	farad
Charge $Q(t)$	coulomb
Current $I(t)$	ampere

Appropriate initial conditions for (19) or (20) would be $I(0) = I_0$, $\dot{I}(0) = \dot{I}_0$. The various circuit elements and their units are summarized in Table 4.1.2.

The Mechanical-Electrical Analog

We have noticed that equations (17) and (19) are just new instances of the harmonic oscillator equation (1). The mass on the table surges back and forth in the same mathematical pattern as the electrical current surging back and forth through the circuit of Fig. 4.1.5 (called the "*LRC*-circuit" because it contains an inductor, a resistor, and a capacitor). The analogy enables us to apply methods of electrical engineering to problems in mechanics and vice versa. The basis of simulation of mechanical systems by analog computers is this correspondence between mechanical and electrical elements. Resistance, for example, plays the part of the friction term in the mass-spring system. Table 4.1.3 summarizes these correlations.

Table 4.1.3 Mechanical-electrical analog

Mechanical System $m\ddot{x} + b\dot{x} + kx = f(t)$		**Electrical System** $L\ddot{Q} + R\dot{Q} + (1/C)Q = V(t)$	
Displacement	x	Q	Charge
Velocity	$\dot{x}$	$\dot{Q} = I$	Current
Mass	m	L	Inductance
Damping constant	b	R	Resistance
Spring constant	k	$1/C$	1/Capacitance
External force	$f(t)$	$V(t)$	Voltage source

The capacitor stores charge and hence stores potential energy as does a compressed or stretched spring. The inductor produces a "back-voltage" as the current increases through it, which tends to retard the charge, adding inertia to the system as does the mass in the mechanical system. The power of the analogy can be seen in the following example of a circuit without resistance.

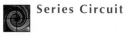

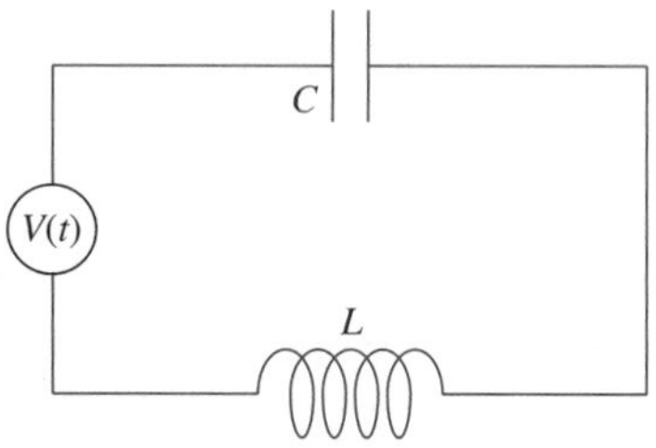

FIGURE 4.1.6 Comparison circuit (Example 4).

EXAMPLE 4 Comparison Circuit Consider a circuit composed of a capacitor with capacitance C and an inductor with inductance L hooked in series (Fig. 4.1.6), and suppose that at time $t = 0$ a charge Q_0 is put on the capacitor. The IVP is

$$L\ddot{Q} + \frac{1}{C}Q = 0, \qquad Q(0) = Q_0, \quad \dot{Q}(0) = 0.$$

By equation (9) we can see that the solution is

$$Q(t) = c_1 \cos \omega_0 t + c_2 \sin \omega_0 t, \qquad \omega_0 = \sqrt{\frac{1}{LC}},$$

where c_1 and c_2 can be determined from the initial conditions. The important point here is that we have an oscillating circuit. Charge moves from the capacitor through the coil, causing a change in current, which in turn causes a back-voltage, which causes the capacitor to reacquire its charge and so forth, forever. The lack of resistance acts like a lack of damping in a mechanical system. ■

Series Circuit

Set $R = 0$ and $V = 0$. Find conditions that make the circuit oscillate.

Summary

Formulations of the differential equations for the mass-spring system from mechanics and the *LRC*-circuit from electrical theory lead to the same linear second-order model called the harmonic oscillator. Solutions in the undamped case are sinusoidal vibrations.

4.1 Problems

The Undamped Oscillator *For Problems 1–8, find the simple harmonic motion described by the initial-value problem. See also Problems 23–30 and 32–39.*

1. $\ddot{x} + x = 0, \quad x(0) = 1, \quad \dot{x}(0) = 0$

2. $\ddot{x} + x = 0, \quad x(0) = 1, \quad \dot{x}(0) = 1$

3. $\ddot{x} + 9x = 0, \quad x(0) = 1, \quad \dot{x}(0) = 1$

4. $\ddot{x} + 4x = 0, \quad x(0) = 1, \quad \dot{x}(0) = -2$

5. $\ddot{x} + 16x = 0, \quad x(0) = -1, \quad \dot{x}(0) = 0$

6. $\ddot{x} + 16x = 0, \quad x(0) = 0, \quad \dot{x}(0) = 4$

7. $\ddot{x} + 16\pi^2 x = 0, \quad x(0) = 0, \quad \dot{x}(0) = \pi$

8. $4\ddot{x} + \pi^2 x = 0, \quad x(0) = 1, \quad \dot{x}(0) = \pi$

Graphing by Calculator *For the combinations of sine and cosine functions in Problems 9–13, do the following.*

(a) *Use a graphing calculator or computer to sketch the graph of each function.*

(b) *From your graphs, estimate the amplitude, period, and phase shift δ/ω_0 of the resulting oscillation.*

(c) *Write each function in the form $A\cos(\omega_0 t - \delta)$.*

9. $x(t) = \cos t + \sin t$

10. $x(t) = 2\cos t + \sin t$

11. $x(t) = 5\cos 3t + \sin 3t$

12. $x(t) = \cos 3t + 5\sin 3t$

13. $x(t) = -\cos 5t + 2\sin 5t$

14. Alternate Forms for Sinusoidal Oscillations To derive the conversion formulas (7) and (8), use the identity

$$\cos(\alpha - \beta) = \cos\alpha\cos\beta + \sin\alpha\sin\beta$$

from trigonometry to show that the family of sinusoidal oscillations $A\cos(\omega_0 t - \delta)$ can be written in the form

$$c_1\cos\omega_0 t + c_2\sin\omega_0 t,$$

where $c_1 = A\cos\delta$ and $c_2 = A\sin\delta$.

Single-Wave Forms of Simple Harmonic Motion *Rewrite Problems 15–18 in the form $A\cos(\omega_0 t - \delta)$ using the conversion equations (7).*

15. $\cos t + \sin t$

16. $\cos t - \sin t$

17. $-\cos t + \sin t$

18. $-\cos t - \sin t$

Component Form of Simple Harmonic Motion *Rewrite Problems 19–22 in the form $c_1\cos\omega_0 t + c_2\sin\omega_0 t$ using the conversion equations (8).*

19. $2\cos(2t - \pi)$

20. $\cos\left(t + \frac{\pi}{3}\right)$

21. $3\cos\left(t - \frac{\pi}{4}\right)$

22. $\cos\left(3t - \frac{\pi}{6}\right)$

Interpreting Oscillator Solutions *For Problems 23–30, determine the amplitude, phase angle, and period of the motion. (These are the equations of Problems 1–8 and 32–39.)*

23. $\ddot{x} + x = 0, \quad x(0) = 1, \quad \dot{x}(0) = 0$

24. $\ddot{x} + x = 0, \quad x(0) = 1, \quad \dot{x}(0) = 1$

25. $\ddot{x} + 9x = 0, \quad x(0) = 1, \quad \dot{x}(0) = 1$

26. $\ddot{x} + 4x = 0, \quad x(0) = 1, \quad \dot{x}(0) = -2$

27. $\ddot{x} + 16x = 0, \quad x(0) = -1, \quad \dot{x}(0) = 0$

28. $\ddot{x} + 16x = 0, \quad x(0) = 0, \quad \dot{x}(0) = 4$

29. $\ddot{x} + 16\pi^2 x = 0, \quad x(0) = 0, \quad \dot{x}(0) = \pi$

30. $4\ddot{x} + \pi^2 x = 0, \quad x(0) = 1, \quad \dot{x}(0) = \pi$

31. Relating Graphs For the oscillator DE $\ddot{x} + 0.25x = 0$, Fig. 4.1.4 shown previously linked solution graphs and phase portrait. Parts (a), (b), (c), and (d) relate to that figure.

(a) Mark on the phase portrait the starting points (where $t = 0$) for the trajectories shown.

(b) Write explicit solutions for $x(t)$ and $\dot{x}(t)$. What is the value of ω_0?

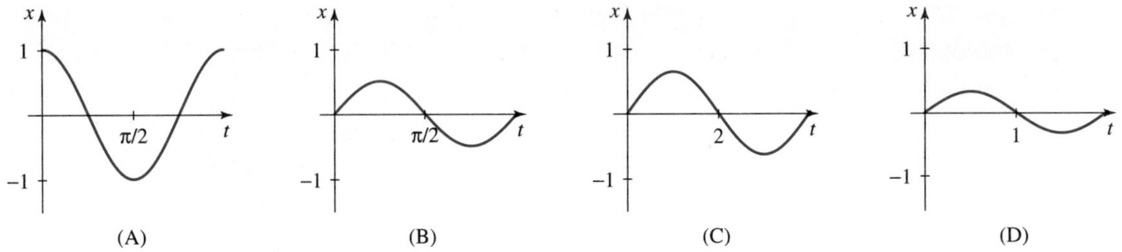

FIGURE 4.1.7 Graphs of the solutions that match the IVPs in Problems 40–43.

31. continued

(c) Label the t axis in the solutions graphs of Fig. 4.1.4 with the appropriate values for t. HINT: Consider where the solution graphs cross the axis.

(d) Given A as the amplitude of the solution with the largest oscillation, state the amplitudes of the other solutions shown.

Phase Portraits *For Problems 32–39, find $\dot{x}(t)$ and then sketch the trajectory for the IVP in the $x\dot{x}$ phase plane, with arrows showing the direction of motion. (These are the equations of Problems 1–8 and 23–30.) Explain how and why your phase portraits differ from each other and from Fig. 4.1.4.*

32. $\ddot{x} + x = 0, \quad x(0) = 1, \quad \dot{x}(0) = 0$

33. $\ddot{x} + x = 0, \quad x(0) = 1, \quad \dot{x}(0) = 1$

34. $\ddot{x} + 9x = 0, \quad x(0) = 1, \quad \dot{x}(0) = 1$

35. $\ddot{x} + 4x = 0, \quad x(0) = 1, \quad \dot{x}(0) = -2$

36. $\ddot{x} + 16x = 0, \quad x(0) = -1, \quad \dot{x}(0) = 0$

37. $\ddot{x} + 16x = 0, \quad x(0) = 0, \quad \dot{x}(0) = 4$

38. $\ddot{x} + 16\pi^2 x = 0, \quad x(0) = 0, \quad \dot{x}(0) = \pi$

39. $4\ddot{x} + \pi^2 x = 0, \quad x(0) = 1, \quad \dot{x}(0) = \pi$

Matching Problems *Match the IVPs in Problems 40–43 to the graphs in Fig. 4.1.7.*

40. $\ddot{x} + 4x = 0, \quad x(0) = 0, \quad \dot{x}(0) = 1$

41. $\ddot{x} + 4x = 0, \quad x(0) = 1, \quad \dot{x}(0) = 0$

42. $\ddot{x} + \pi^2 x = 0, \quad x(0) = 0, \quad \dot{x}(0) = 1$

43. $4\ddot{x} + \pi^2 x = 0, \quad x(0) = 0, \quad \dot{x}(0) = 1$

44. Changing Frequencies Consider the undamped harmonic oscillator defined by $\ddot{x} + \omega_0^2 x = 0$ with initial conditions $x(0) = 4$ and $\dot{x}(0) = 0$.

(a) For $\omega_0 = 0.5$, 1, and 2, the corresponding trajectories are plotted in Fig. 4.1.8(a) on the same tx-plane. In Fig. 4.1.8(b) the corresponding trajectories are plotted on the $x\dot{x}$ phase plane. Make a trace of both graphs and label each curve with the appropriate value of ω_0.

(b) If ω_0 is increased, we can see that the frequency of the oscillations in Fig. 4.1.8(a) increases (as expected, since ω_0 is the natural (circular) frequency). Describe what happens in the phase plane (Fig. 4.1.8(b)) if ω_0 is increased. How is $\dot{x}$ affected if ω_0 is increased?

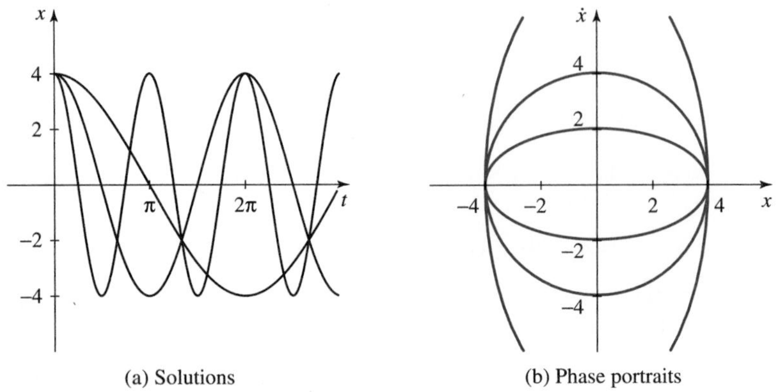

FIGURE 4.1.8 Graphs of $\ddot{x} + \omega_0^2 x = 0$ (Problem 44).

45. Detective Work Suppose that you have received the following two graphs without their equations. Show how you can infer the equations from graphical information.

(a) The graph in Fig. 4.1.9(a) represents $A \cos(t - \delta)$. Determine A and δ from the graph and write the equation of the curve in the form of equation (5).

(b) The graph in Fig. 4.1.9(b) represents $c_1 \cos t + c_2 \sin t$. Determine c_1 and c_2 from the graph and write the equation of the curve in the form of (4). HINT: Find A and δ first.

(a) Graph of $A \cos(t - \delta)$

(b) Graph of $c_1 \cos t + c_2 \sin t$

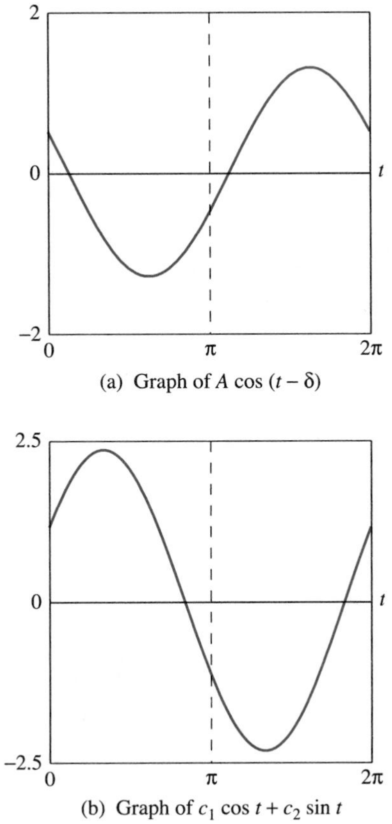

FIGURE 4.1.9 Simple harmonic motions (Problem 45).

46. Pulling a Weight An object of mass 2 kg, resting on a frictionless table, is attached to the wall by a spring as in Fig. 4.1.1. A force of 8 nt is applied to the mass, stretching the spring and moving the mass 0.5 m from its equilibrium position. The object is then released.

(a) Find the resulting motion of the object as a function of time.

(b) Determine the amplitude, period, and frequency of the motion.

(c) At what time does the mass first pass through the equilibrium position? What is its velocity at that time?

47. Finding the Differential Equation A mass of 500 gm is suspended from the ceiling by a frictionless spring. The mass stretches the spring 50 cm in coming to its equilibrium position, where the mass acting down is balanced exactly by the restoring force acting up. The object is then pulled down an additional 10 cm and released.

(a) Formulate the initial-value problem that describes the object's motion, setting x equal to the downward displacement from equilibrium.

(b) Solve for the motion of the object.

(c) Find the amplitude, phase angle, frequency, and period of the motion.

 Mass-Spring
Watch the motion of the mass linked with the graphs of the position and velocity displayed in a time graph and a phase plane.

48. Initial-Value Problems A 16-lb object is attached to the ceiling by a frictionless spring and stretches the spring 6 in. before coming to its equilibrium position. Formulate the initial-value problem describing the motion of the object under each of the following sets of conditions. Set x equal to the downward displacement from equilibrium.

(a) The object is pulled down 4 in. below its equilibrium position and released with an upward velocity of 4 ft/sec.

(b) The object is pushed up 2 in. and released with a downward velocity of 1 ft/sec.

49. One More Weight A 12-lb object attached to the ceiling by a frictionless spring stretches the spring 6 in. as it comes to its equilibrium. Find and solve the equation of motion if the object is initially pushed up 4 in. from its equilibrium and given an upward velocity of 2 ft/sec.

50. Comparing Harmonic Motions An object on a table attached to spring and wall as in Fig. 4.1.1 is pulled to the right, stretching the spring, and released. The same object is then pulled twice as far and released. What is the relationship between the two simple harmonic motions? Will the period of the second be twice that of the first? What about the amplitudes and frequencies?

Testing Your Intuition *Knowing (from Example 1) what you now do about the damped harmonic oscillator equation $m\ddot{x} + b\dot{x} + kx = 0$ and the meaning of the parameters m, b, and k, consider Problems 51–56. How would you expect the solution of each equation to behave? Can you imagine a physical system being modeled by the equation? What would you expect for its long-term behavior?*

51. $\ddot{x} + x + x^3 = 0$ **52.** $\ddot{x} + x - x^3 = 0$

$\omega = \sqrt{\frac{k}{m}}$ $\ddot{x} + x = x^3$

53. $\ddot{x} - x = 0$

54. $\ddot{x} + \frac{1}{t}\dot{x} + x = 0$

55. $\ddot{x} + (x^2 - 1)\dot{x} + x = 0$

56. $\ddot{x} + tx = 0$

57. *LR*-Circuit Consider the series LR-circuit shown in Fig. 4.1.10, in which a constant input voltage V_0 has been supplied until $t = 0$, when it is shut off.

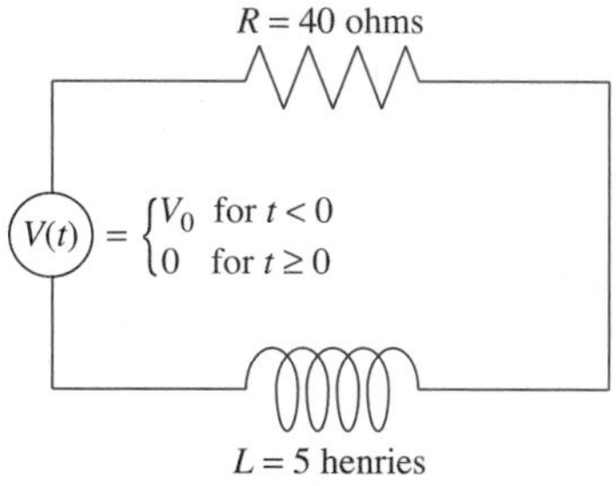

FIGURE 4.1.10 An *LR*-circuit (Problem 57).

(a) Before carrying out the mathematical analysis, describe what you think will happen to the circuit.

(b) For $t > 0$, use Kirchoff's voltage law to determine the sum of the voltage drops around the circuit. Set it equal to zero to obtain a first-order differential equation involving R, $\dot{I}$, L, and I. What are the initial conditions?

(c) Solve the DE in (b) for current I. Does your answer agree with (a)? Explain.

(d) Use the values $R = 40$ ohms, $L = 5$ henries, and $V_0 = 10$ volts to obtain an explicit solution.

58. *LC*-Circuit Consider the series LC-circuit shown in Fig. 4.1.11, in which, at $t = 0$, the current is 5 amps and there is no charge on the capacitor. Voltage V_0 is turned off at $t = 0$.

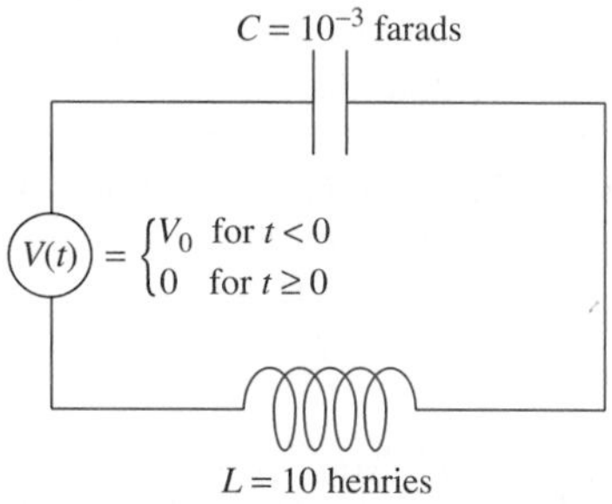

FIGURE 4.1.11 An *LC*-circuit (Problem 58).

(a) Before carrying out the mathematical analysis, describe what you think will happen to the charge on the capacitor.

(b) For $t > 0$, use Kirchoff's voltage law to determine the sum of the voltage drops around the circuit. Set it equal to zero to obtain a second-order differential equation involving L, $\dot{Q}$, Q, and C. What are the initial conditions?

(c) Solve the IVP in (b) for the charge Q on the capacitor. Does your result agree with part (a)?

(d) Obtain an explicit solution for $L = 10$ henries and $C = 10^{-3}$ farads.

59. A Pendulum Experiment A pendulum of length L is suspended from the ceiling so it can swing freely; θ denotes the angular displacement, in radians, from the vertical, as shown in Fig. 4.1.12. The motion is described by the *pendulum equation*,

$$\ddot{\theta} + \frac{g}{L}\sin\theta = 0.$$

Determine the period for small oscillations by using the approximation $\sin\theta \approx \theta$ (the linear pendulum). What is the relationship between the period of the pendulum and g, the acceleration due to gravity? If the sun is 400,000 times more massive than the earth, how much faster would the pendulum oscillate on the sun (provided it did not melt) than on the earth?

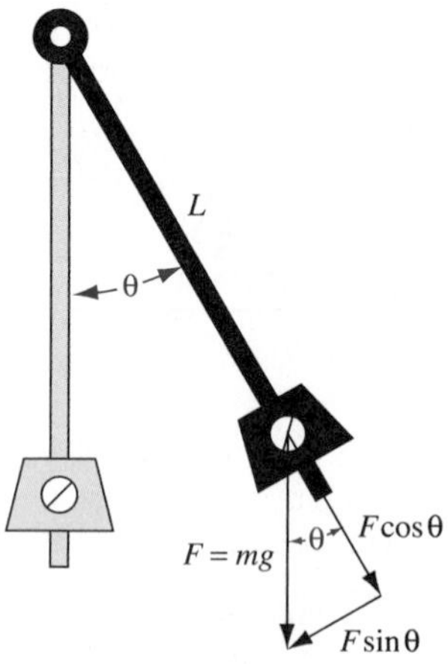

FIGURE 4.1.12 Simple pendulum (Problem 59).

Pendulums
This tool allows you to compare the motions of the linear, nonlinear, and forced pendulums with predictable and/or chaotic results.

Changing into Systems *For Problems 60–64, consider the second-order nonhomogeneous DEs. Write them as a system of first-order DEs as in (18).*

60. $4\ddot{x} - 2\dot{x} + 3x = 17 - \cos t$

61. $L\ddot{q} + R\dot{q} + \frac{1}{C}q = V(t)$

62. $5\ddot{q} + 15\dot{q} + \frac{1}{10}q = 5\cos 3t$

63. $t^2\ddot{x} + 4t\dot{x} + x = t\sin 2t$

64. $4\ddot{x} + 16x = 4\sin t$

65. Circular Motion A particle moves around the circle $x^2 + y^2 = r^2$ with a constant angular velocity of ω_0 radians per unit time. Show that the projection of the particle on the x axis satisfies the equation $\ddot{x} + \omega_0 x = 0$.

66. Another Harmonic Motion The mass-spring-pulley system shown in Fig. 4.1.13 satisfies the differential equation

$$\ddot{x} + \left(\frac{kR^2}{mR^2 + I}\right)x = 0,$$

where x is the displacement from equilibrium of the object of mass m. In this equation, R and I are, respectively, the radius and moment of inertia of the pulley, and k is the spring constant. Determine the frequency of the motion.

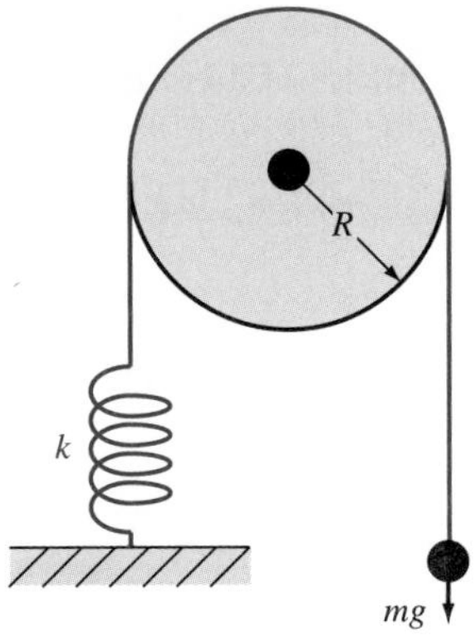

FIGURE 4.1.13 Mass-spring-pulley system (Problem 66).

67. Motion of a Buoy A cylindrical buoy with diameter 18 in. floats in water with its axis vertical, as shown in Fig. 4.1.14. When depressed slightly and released, its period of vibration is found to be 2.7 sec. Find the weight of the cylinder. HINT: **Archimedes' Principle** says that an object submerged in water is buoyed up by a force equal to the weight of the water displaced, where weight is the product of volume and density. The density of water is 62.5 lb/ft^3.

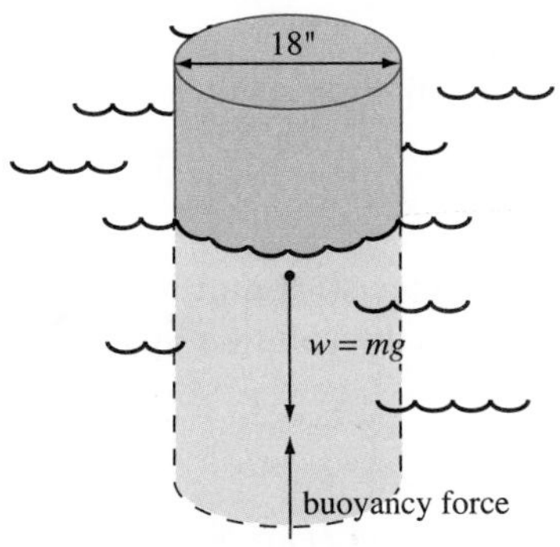

FIGURE 4.1.14 Motion of a buoy (Problem 67).

68. Los Angeles to Tokyo It can be shown that the force on an object inside a spherical homogeneous mass is directed towards the center of the sphere with a magnitude proportional to the distance from the center of the sphere. Using this principle, a train starting at rest and traveling in a vacuum without friction on a straight line tunnel from Los Angeles to Tokyo experiences a force in its direction of motion equal to $-kr\cos\theta$, where

- r is the distance of the train from the center of the earth,
- x is the distance of the train from the center of the tunnel,
- θ is the angle between r and x,
- $2d$ is the length of the tunnel between L.A. and Tokyo,
- R is the radius of the earth (4,000 miles),

as shown in Fig. 4.1.15.

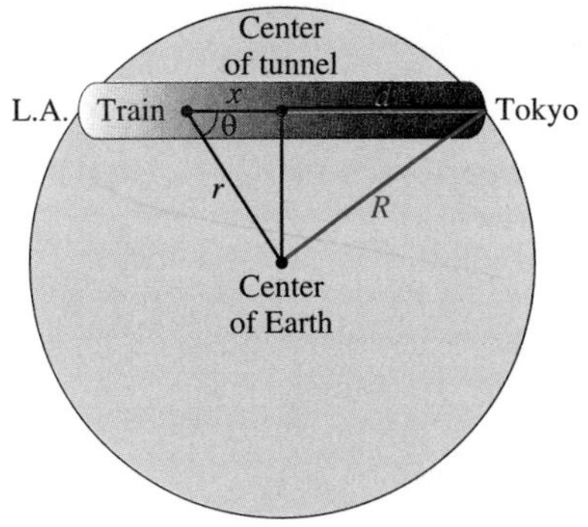

FIGURE 4.1.15 Tunnel from Los Angeles to Tokyo (Problem 68).

(a) Show that the train position x can be modeled by the initial-value problem

$$m\ddot{x} + kx = 0, \quad x(0) = d, \quad \dot{x}(0) = 0,$$

where x is the distance of the train to the center of the earth and R is the radius of the earth

(b) How long does it take the train to go from Los Angeles to Tokyo?

(c) Show that if a train starts at any point on earth and goes to another point on earth in this science fiction scenario, the time will be the *same* as calculated in part (a)!

69. Factoring Out Friction The damped oscillator equation (2) can be solved by a change of variable that "factors out the damping." Specifically, let $x(t) = e^{-(b/2m)t} X(t)$.

(a) Show that $X(t)$ satisfies

$$m\ddot{X} + \left(k - \frac{b^2}{4m}\right) X = 0. \tag{21}$$

(b) Assuming that $k - b^2/4m > 0$, solve equation (21) for $X(t)$; then show that the solution of equation (2) is

$$x(t) = Ae^{-(b/2m)t} \cos(\omega_0 t - \delta), \tag{22}$$

where $\omega_0 = \sqrt{4mk - b^2}/2m$.

70. Suggested Journal Entry With the help of equation (22) from Problem 69, describe the different short- and long-term behaviors of solutions of the mass-spring system in the damped and undamped cases. Illustrate with sketches.

4.2 Real Characteristic Roots

SYNOPSIS: For the linear second-order homogeneous differential equation with constant coefficients, we obtain a two-dimensional vector space of explicit solutions. When the roots of the characteristic equation are real, examples include overdamped and critically damped cases of the harmonic oscillator, with applications to mass-spring systems and LRC-circuits. We generalize our results at the end of this section.

We begin by solving a very straightforward set of DEs: linear homogeneous DEs with constant coefficients. After we have gained insight from this experience, we will examine more general linear DEs and prove the existence of bases for the solution spaces of the DEs.

Solving Constant Coefficient Second-Order Linear DEs

In the special case of a linear second-order homogeneous equation with constant coefficients, the custom is to write the DE in the form

$$ay'' + by' + cy = 0, \tag{1}$$

where a, b, and c are real constants and $a \neq 0$. The first-order linear examples in Sec. 2.3, which can be written $y' - ry = 0$, suggest that we try an exponential solution.[1] If we let $y = e^{rt}$ as a trial solution, then $y' = re^{rt}$ and $y'' = r^2 e^{rt}$, so (1) becomes

$$ar^2 e^{rt} + bre^{rt} + ce^{rt} = e^{rt}(ar^2 + br + c) = 0.$$

Because e^{rt} is never zero, (1) will be satisfied precisely when

Characteristic Equation

$$ar^2 + br + c = 0. \tag{2}$$

This quadratic equation, called the **characteristic equation** of the DE, is the key to finding solutions that form a basis for the solution space. By the quadratic

[1]Exponential solutions were first tried by Leonhard Euler (1707–1783), one of the greatest mathematicians of all time and surely the most prolific. He contributed to nearly every branch of the subject, and his amazing productivity continued even after he became blind in 1768.

formula, solutions to (2) occur when

$$r = \frac{-b \pm \sqrt{b^2 - 4ac}}{2a}.$$

We recall that, depending on the value of the *discriminant*

Discriminant

$$\Delta = b^2 - 4ac,$$

there are three possibilities for solutions of the quadratic:

1. two distinct real roots or zeros, (Section 4.2)
2. one real root (a "double root"), (Section 4.2)
3. a pair of conjugate complex roots. (Section 4.3)

Solutions to the characteristic equation are called **characteristic roots** or **eigenvalues** of the equation. (The term *eigenvalue* is from linear algebra, as we shall see in Sec. 5.3.) We will consider the implications for solutions of the DE when the characteristic roots are real in this section, and when they are complex in Sec. 4.3.

Case 1: Real Unequal Characteristic Roots ($\Delta > 0$)

The characteristic equation has real roots when its *discriminant* $\Delta = b^2 - 4ac$ is greater than or equal to zero. If $\Delta > 0$, the quadratic formula gives two distinct characteristic roots r_1 and r_2, and $e^{r_1 t}$ and $e^{r_2 t}$ are two independent solutions.

Solution of $ay'' + by' + cy = 0$ with Distinct Real Characteristic Roots
For $\Delta = b^2 - 4ac > 0$, the characteristic roots of the DE are

$$r_1 = \frac{-b + \sqrt{\Delta}}{2a} \quad \text{and} \quad r_2 = \frac{-b - \sqrt{\Delta}}{2a}. \tag{3}$$

The functions $e^{r_1 t}$ and $e^{r_2 t}$ are linearly independent solutions, and the general solution is given by

$$y(t) = c_1 e^{r_1 t} + c_2 e^{r_2 t}, \tag{4}$$

where c_1 and c_2 are arbitrary constants determined by initial conditions. The set $\{e^{r_1 t}, e^{r_2 t}\}$ forms a basis for the solution space.

For assurance that (4) gives *all* the solutions, see "Theoretical Considerations," later in this section.

EXAMPLE 1 **Real and Unequal Roots** To find the general solution of

$$y'' + 5y' + 6y = 0, \tag{5}$$

we write its characteristic equation

$$r^2 + 5r + 6 = 0,$$

which can be solved by factoring: $(r + 2)(r + 3) = 0$. The characteristic roots are $r_1 = -2$ and $r_2 = -3$. Two linearly independent solutions are given by e^{-2t} and e^{-3t}, and the general solution is

$$y(t) = c_1 e^{-2t} + c_2 e^{-3t}.$$

The set $\{e^{-2t}, e^{-3t}\}$ is a basis for the solution space $\mathbb{S}$, and $\dim \mathbb{S} = 2$.

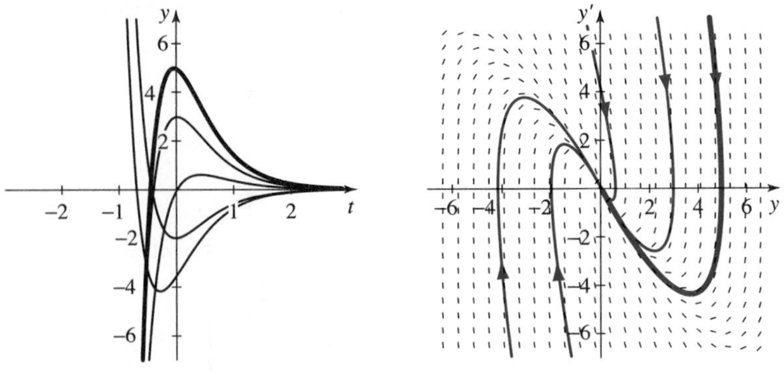

(a) Time series (b) Phase portrait

FIGURE 4.2.1 Some solutions for $y'' + 5y' + 6y = 0$ of Example 1.

In Fig. 4.2.1 the time series and phase portrait show the long-term behavior of the system. All solutions tend to 0 and remain there. The origin is an equilibrium. These behaviors are confirmed by the algebraic solution because each term approaches 0 as t increases. ■

Two initial values, $y(t_0)$ and $y'(t_0)$, are required to obtain the particular solution that passes through the point (t_0, y_0) with slope $y'(t_0)$.

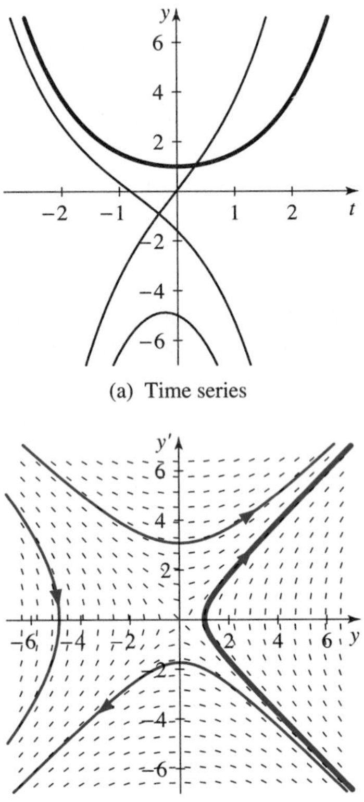

(a) Time series

(b) Phase portrait

FIGURE 4.2.2 Some solutions for $y'' - y = 0$ of Example 2.

EXAMPLE 2 **Flying Away** For the second-order DE

$$y'' - y = 0,$$

our solution follows a now-familiar pattern. We obtain the general solution by solving the characteristic equation $r^2 - 1 = 0$ for characteristic roots $r_1 = 1$ and $r_2 = -1$, and concluding that

$$y(t) = c_1 e^t + c_2 e^{-t}.$$

The set $\{e^t, e^{-t}\}$ is a basis for the solution space $\mathbb{S}$, and $\dim \mathbb{S} = 2$. Differentiating,

$$y'(t) = c_1 e^t - c_2 e^{-t}.$$

Figure 4.2.2 shows a solution graph and phase portrait, which demonstrate long-term behavior quite different from Example 1. In this case, solutions fly away from the equilibrium at the origin (after first heading towards it). Algebraically, this is the result of e^t increasing and e^{-t} decreasing as t gets larger.

For an IVP with initial conditions $y(0) = 1$ and $y'(0) = 0$, we can find the values of c_1 and c_2 for a particular solution. From $y(0) = c_1 + c_2 = 1$ and $y'(0) = c_1 - c_2 = 0$, it follows that $c_1 = c_2 = 1/2$. The solution to the IVP is

$$y(t) = \frac{1}{2}\left(e^t + e^{-t}\right) = \cosh t,$$

highlighted in Fig. 4.2.2. ■

Case 2: Real Repeated Characteristic Root ($\Delta = 0$)

When the discriminant of the characteristic equation (2) is zero, the equation has the **double root** $r = -\frac{b}{2a}$, giving only one solution function $e^{-(b/2a)t}$ of the exponential type.

To find a second independent solution of the DE (1), we use a device that resembles variation of parameters, as follows.[2]

- We have solutions of the form $ce^{-(b/2a)t}$, and we change the constant multiplier into a function $v(t)$.
- Substituting $v(t)e^{-(b/2a)t}$ into (1) leads eventually (Problem 48) to the condition $v'' = 0$, so v is a linear function of t.
- Picking $v(t) = t$, our candidate for a second independent solution is $te^{-(b/2a)t}$.

We can show that $e^{-(b/2a)t}$ and $te^{-(b/2a)t}$ are linearly independent (Problem 49), so we summarize the case as follows.

Solution of $ay'' + by' + cy = 0$ with Repeated Real Characteristic Root
For $\Delta = b^2 - 4ac = 0$, the characteristic root of the DE is

$$r = -\frac{b}{2a}. \tag{6}$$

The functions e^{rt} and te^{rt} are linearly independent solutions, and the general solution is given by

$$y(t) = c_1e^{rt} + c_2te^{rt}, \tag{7}$$

where c_1 and c_2 are arbitrary constants determined by initial conditions. The set $\{e^{rt}, te^{rt}\}$ is a basis for the solution space.

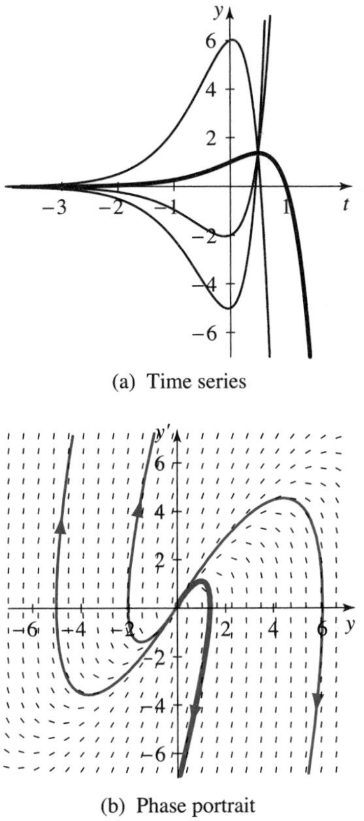

FIGURE 4.2.3 Some solutions of $y'' - 4y' + 4y = 0$ of Example 3.

EXAMPLE 3 **Repeated Root** To find a family of solutions for

$$y'' - 4y' + 4y = 0,$$

we solve the characteristic equation

$$r^2 - 4r + 4 = (r-2)^2 = 0,$$

which has double root $r = 2$. Two linearly independent solutions are given by e^{2t} and te^{2t}. The set $\{e^{2t}, te^{2t}\}$ is a basis for the solution space, and the general solution is

$$y(t) = c_1e^{2t} + c_2te^{2t}.$$

To solve the initial-value problem with $y(0) = 1$ and $y'(0) = 1$, we compute the derivative,

$$y' = 2c_1e^{2t} + c_2e^{2t} + 2c_2te^{2t},$$

and then substitute the initial conditions,

$$y(0) = c_1 = 1 \quad \text{and} \quad y'(0) = 2c_1 + c_2 = 1,$$

to find $c_1 = 1$ and $c_2 = 1 - 2c_1 = -1$. The solution to the IVP is

$$y(t) = e^{2t} - te^{2t},$$

highlighted in Fig. 4.2.3. With positive r, all solutions *leave* 0 faster and faster as t increases. Both terms of the algebraic solution increase without bound. ■

[2]See Sec. 2.2, equations (3) and (4). This device of replacing a constant by a function can be used to find a second solution in many situations where one solution is known, and is called *reduction of order*; see Problems 75–79.

Overdamped and Critically Damped Mass-Spring Systems

Let us return to the motion of an unforced mass-spring system, which obeys the harmonic oscillator equation of Sec. 4.1,

$$m\ddot{x} + b\dot{x} + kx = 0, \tag{8}$$

where $m > 0$ is the mass, $k > 0$ is the spring constant, and $b \geq 0$ is the damping constant. (See the previous section, especially Fig. 4.1.1.) The behavior of the system depends on the sign of the discriminant, which for (8) is

$$\Delta = b^2 - 4mk.$$

Mass-Spring

Observe the motion of overdamped and critically damped systems, linked to evolving solution graphs.

Overdamped Mass-Spring System

The motion of a mass-spring system (8) is called **overdamped** when we have $\Delta = b^2 - 4mk > 0$. Both characteristic roots are negative (see Problem 55), and solutions

$$x(t) = c_1 e^{r_1 t} + c_2 e^{r_2 t} \tag{9}$$

tend to zero without oscillation, crossing the t axis at most once.

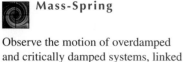

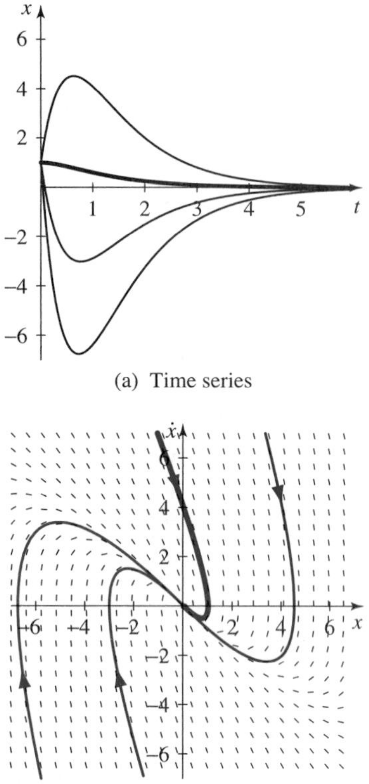

(a) Time series

(b) Phase portrait

FIGURE 4.2.4 Overdamped motion of $\ddot{x} + 3\dot{x} + 2x = 0$ of Example 4.

EXAMPLE 4 **Overdamping** We will solve for the motion of a mass-spring system like that of Fig. 4.1.1, in which the mass is 1 slug and the damping and spring constants are respectively $b = 3$ and $k = 2$. The object is initially pulled one foot to the right and released with no initial velocity.

The differential equation of this motion is

$$\ddot{x} + 3\dot{x} + 2x = 0$$

with characteristic equation $r^2 + 3r + 2 = (r+1)(r+2) = 0$. The discriminant is $\Delta = (3)^2 - 4(1)(2) = 1 > 0$, so the system is overdamped. The distinct real roots are negative, $r_1 = -1$ and $r_2 = -2$, and the general solution is

$$x(t) = c_1 e^{-t} + c_2 e^{-2t}.$$

The set $\{e^{-t}, e^{-2t}\}$ is a basis for the solution space.

Some solutions and their phase-plane trajectories are shown in Fig. 4.2.4, all starting at $x(0) = 1$ but with different initial velocities $\dot{x}(0)$. The equation of motion of any of these is derived by computing

$$\dot{x}(t) = -c_1 e^{-t} - 2c_2 e^{-2t}$$

and substituting initial conditions. For example, for $x(0) = 1$ and $\dot{x}(0) = 0$, we obtain $c_1 = 2$ and $c_2 = -1$, so the equation of motion for this IVP, highlighted in Fig. 4.2.4, is

$$x(t) = 2e^{-t} - e^{-2t}.$$

■

Critically Damped Mass-Spring System

The motion of a mass-spring system is called **critically damped** when $\Delta = b^2 - 4mk = 0$. The single characteristic root is negative,

$$r = -\frac{b}{2m}.$$

Solutions to the critically damped system, given by

$$x(t) = c_1 e^{rt} + c_2 t e^{rt}, \tag{10}$$

tend to zero, crossing the t axis at most once.

Critical Damping

Watch how solutions to a damped oscillator DE change as you move the b slider across the critical value. You can also see how the characteristic roots must become imaginary to produce oscillatory underdamping (Sec. 4.3).

EXAMPLE 5 **Critical Damping** Consider the mass-spring system (8) for which $m = 1$ kg and the damping and spring constants are $b = 6$ and $k = 9$, respectively. Assume that the object was pulled one foot to the right and given no initial velocity.

The discriminant is $\Delta = (6)^2 - 4(1)(9) = 0$, and so there is exactly one root $r = -3$. The set $\{e^{-3t}, te^{-3t}\}$ is a basis for the solution space. The solution is

$$x(t) = c_1 e^{-3t} + c_2 t e^{-3t},$$

from which we can compute

$$\dot{x}(t) = -3c_1 e^{-3t} - 3c_2 t e^{-3t} + c_2 e^{-3t}.$$

Substituting the initial conditions, $x(0) = 1$ and $\dot{x}(0) = 0$, which were implicit in the verbal description, we find that $c_1 = 1$ and $c_2 = 3$. The final solution to the IVP is

$$x(t) = (t + 3)e^{-3t},$$

as highlighted in Fig. 4.2.5.

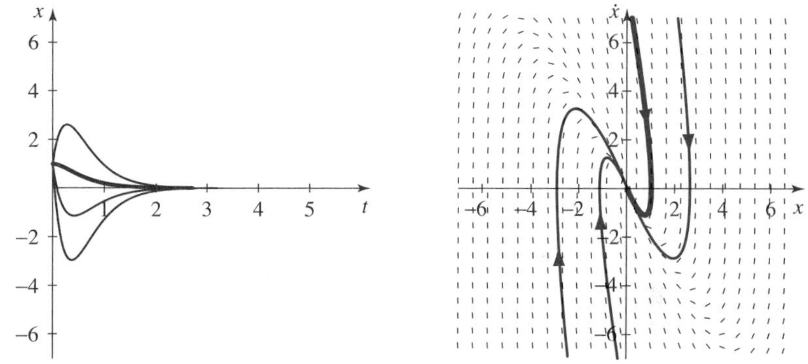

(a) Time series (b) Phase portrait

FIGURE 4.2.5 Critically damped motion of $\ddot{x} + 6\dot{x} + 9x = 0$ of Example 5. ■

The figures accompanying Examples 4 and 5 use the same initial conditions, demonstrating that critical damping (Fig. 4.2.5) gives solutions that approach the axis more quickly than overdamping (Fig. 4.2.4). Since the solutions shown all start with the same initial position, it is easy to see that the graphs of overdamped or critically damped motions may cross the positive t axis once or not at all, depending on the initial velocity, which appears as *slope* on the tx graphs.

The third and most common type of damping, called *underdamping*, will appear in Sec. 4.3 because it requires the characteristic roots to be imaginary. At the end of Sec. 4.3, we will give a summary of all three cases for damped oscillators.

LRC-Circuits

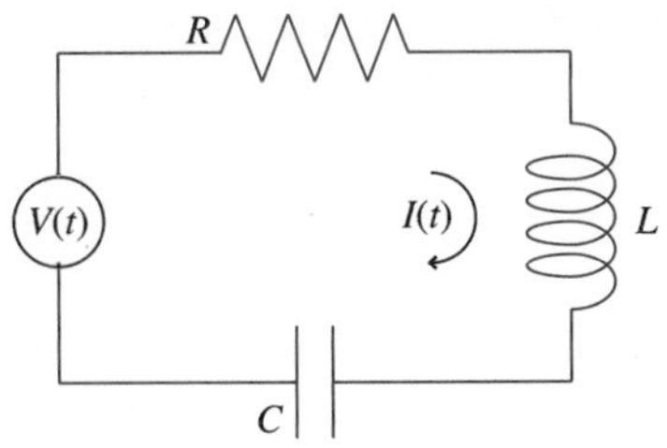

FIGURE 4.2.6 An *LRC*-circuit.

We learned in the previous section that the *LRC*-circuit (Fig. 4.2.6), in which a resistor (resistance R ohms), an inductor (inductance L henries), and a capacitor (capacitance C farads) are connected in series to a voltage source of $V(t)$ volts, is modeled by the initial-value problem

$$L\ddot{Q} + R\dot{Q} + \frac{1}{C}Q = V(t), \qquad Q(0) = Q_0, \quad \dot{Q}(0) = I(0) = I_0, \tag{11}$$

where $Q(t)$ is the charge in coulombs across the capacitor, $I(t)$ is the current in amps, and Q_0 and I_0 are the initial charge and current, respectively. When the charge $Q(t)$ has been obtained from (11), the current can be found from $\dot{Q}(t) = I(t)$.

EXAMPLE 6 **LRC-Circuit with No Input Voltage** Suppose a capacitor has been charged, and at $t = 0$ the voltage source is removed. We wish to find the current in an *LRC*-circuit with $R = 30$ ohms, $L = 1$ henry, $C = 1/200$ farads, and no voltage input. Initially, there is no current in the circuit and a charge across the capacitor of 10 coulombs, but now the charge begins to dissipate and create a current.

Using the formulation in (11), our initial-value problem is

$$\ddot{Q} + 30\dot{Q} + 200Q = 0, \qquad Q(0) = 10, \quad \dot{Q}(0) = 0.$$

The DE has characteristic equation $r^2 + 30r + 200 = 0$ with roots $r_1 = -10$ and $r_2 = -20$, so the general solution is

$$Q(t) = c_1e^{-10t} + c_2e^{-20t},$$

and its derivative is

$$\dot{Q}(t) = -10c_1e^{-10t} - 20c_2e^{-20t}.$$

The set $\{e^{-10t}, e^{-20t}\}$ is a basis for the solution space. Substituting the initial conditions gives

$$Q(0) = c_1 + c_2 = 10 \quad \text{and} \quad \dot{Q}(0) = -10c_1 - 20c_2 = 0.$$

Hence $c_1 = 20$ and $c_2 = -10$, and the charge is given by

$$Q(t) = 20e^{-10t} - 10e^{-20t}.$$

Differentiating gives the current:

$$I(t) = -200e^{-10t} + 200e^{-20t}.$$

Recall that the nonzero resistance acts as a damping factor. In the absence of an input voltage, both the charge across the capacitor and the current tend rapidly to zero. (See Fig. 4.2.7. While charge dissipates monotonically, the current first surges (negatively) before dying out.) Compare with Example 3 of Sec. 4.1. ■

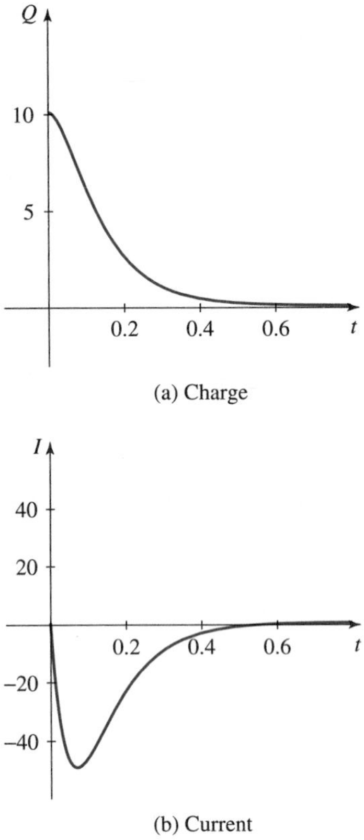

FIGURE 4.2.7 Charge and current solutions to IVP of Example 6.

Theoretical Considerations

The examples we have studied hint at what is true in general. For the linear homogeneous second-order equation

$$y'' + p(t)y' + q(t)y = 0, \tag{12}$$

with p and q continuous functions of t on some interval I, we will have a two-dimensional vector space of solutions; and, when the coefficients are constant, solutions will be built up from exponentials in some way. First, we state a basic existence and uniqueness result, which parallels Picard's Existence and Uniqueness Theorem as stated in Sec. 1.5. If we rewrite (12) as

$$y'' = f(t, y, y') = -p(t)y' - q(t)y = 0$$

then the function f and the partial derivatives $f_y = -q$ and $f_{y'} = -p$ are continuous because p and q are continuous. We can see that the Existence and Uniqueness Theorem given next is an extension of Picard's Theorem.

Restriction:

The Existence and Uniqueness Theorem requires *two* given conditions and both must be *initial* conditions. Problems 74–77 in Sec. 4.3 will show a reason for this restriction.

Existence and Uniqueness Theorem (Second-Order Version)
Let $p(t)$ and $q(t)$ be continuous on the open interval (a, b) containing t_0. For *any* A and B in $\mathbb{R}$, there exists a unique solution $y(t)$ defined on (a, b) to the initial-value problem

$$y'' + p(t)y' + q(t)y = 0, \quad y(t_0) = A,\ y'(t_0) = B.$$

We can use the Existence and Uniqueness Theorem to find that a basis exists for the general second-order equation (12).

Solution Space Theorem (Second-Order Version)
The solution space $\mathbb{S}$ for a second-order homogeneous differential equation has dimension 2.

Proof We need to show that for the linear homogeneous second-order equation

$$y'' + p(t)y' + q(t)y = 0,$$

with p and q continuous functions of t on some interval I, we have a two-dimensional vector space of solutions.

Step 1. The Existence and Uniqueness Theorem assures us that we can find a unique solution to (12) for *any* initial conditions. In particular, there exist two unique solutions $u(t)$ and $v(t)$ with respective initial conditions

$$\begin{aligned} u(t_0) &= 1, \\ u'(t_0) &= 0, \end{aligned} \quad \text{and} \quad \begin{aligned} v(t_0) &= 0, \\ v'(t_0) &= 1. \end{aligned}$$

Step 2. We check the linear independence of $u(t)$ and $v(t)$. Suppose that, for all $t \in (a, b)$,

$$c_1 u(t) + c_2 v(t) = 0.$$

Then from calculus we also have

$$c_1 u'(t) + c_2 v'(t) = 0.$$

In particular, for $t = t_0$:

$$\begin{bmatrix} u(t_0) & v(t_0) \\ u'(t_0) & v'(t_0) \end{bmatrix} \begin{bmatrix} c_1 \\ c_2 \end{bmatrix} = \begin{bmatrix} 1 & 0 \\ 0 & 1 \end{bmatrix} \begin{bmatrix} c_1 \\ c_2 \end{bmatrix} = \begin{bmatrix} 0 \\ 0 \end{bmatrix}.$$

This yields a unique solution $c_1 = c_2 = 0$. Alternatively, we can see that the Wronskian determinant $W = 1 \neq 0$.

Therefore, $\{u(t), v(t)\}$ is a linearly independent set on (a, b).

Step 3. To show that $u(t)$ and $v(t)$ *span* the solution space, we consider any $w(t)$ in the solution space. We will determine what conditions will be required to write $w(t)$ as a linear combination of $u(t)$ and $v(t)$. We must find real numbers k_1 and k_2 such that, for all $t \in (a, b)$,

$$w(t) = k_1 u(t) + k_2 v(t)$$

and

$$w'(t) = k_1 u'(t) + k_2 v'(t).$$

In particular, these equations must hold at $t = t_0$ from Step 1, so that

$$\begin{bmatrix} u(t_0) & v(t_0) \\ u'(t_0) & v'(t_0) \end{bmatrix} \begin{bmatrix} k_1 \\ k_2 \end{bmatrix} = \begin{bmatrix} 1 & 0 \\ 0 & 1 \end{bmatrix} \begin{bmatrix} k_1 \\ k_2 \end{bmatrix} = \begin{bmatrix} w(t_0) \\ w'(t_0) \end{bmatrix}.$$

From the fact that the determinant of the coefficient matrix is not zero, we are guaranteed a unique solution for k_i: $k_1 = w(t_0)$, $k_2 = w'(t_0)$. Thus $w(t)$ *can* be expressed in terms of $u(t)$ and $v(t)$, and uniqueness furthermore assures us this is the *only* solution.

Consequently, the solution space $\mathbb{S}$ *is* spanned by two functions $u(t)$ and $v(t)$. We have also shown that the solution space for a second-order linear homogeneous DE is two-dimensional. In other words, $\{u(t), v(t)\}$ is a basis for the solution space $\mathbb{S}$ and $\dim \mathbb{S} = 2$. □

We summarize what we have learned as follows.

Solutions of Homogeneous Linear DEs (Second-Order Version)
For any linear second-order homogeneous DE on (a, b),

$$y'' + p(t)y' + q(t)y = 0,$$

for which p and q are continuous on (a, b), *any* two linearly independent solutions $\{y_1, y_2\}$ form a basis of the solution space $\mathbb{S}$, and *every* solution y on (a, b) can be written as

$$y(t) = c_1 y_1(t) + c_2 y_2(t)$$

for some real numbers c_1 and c_2.

Generalizing to nth-Order DEs

For nth-order linear differential equations, the theorems all generalize as follows. An nth-order IVP requires n initial values.

Existence and Uniqueness Theorem
Let $p_1(t), p_2(t), \ldots, p_n(t)$ be continuous functions on the open interval (a, b) containing t_0. For *any* initial values $A_0, A_1, \ldots, A_{n-1} \in \mathbb{R}$, there exists a unique solution $y(t)$ defined on (a, b) to the initial-value problem

$$\begin{gathered} y^{(n)}(t) + p_1(t)y^{(n-1)}(t) + p_2(t)y^{(n-2)}(t) + \cdots + p_n(t)y(t) = 0, \\ y(t_0) = A_0,\ y'(t_0) = A_1, \ldots, y^{(n-1)}(t) = A_{n-1}. \end{gathered} \qquad (13)$$

As in the second-order case, this theorem enables us to guarantee the existence of a basis with n solutions for the solution space $\mathbb{S}$ for (13).

Solution Space Theorem
The solutions to a homogeneous linear differential equation of order n form an n-dimensional vector space.

The Solution Space Theorem is a fundamental idea in the study of linear DEs that allows us to predict at a glance the maximum number of linearly independent solutions to any homogeneous linear DE. The trick is to come up with a basis. If the equation has *constant coefficients*, we can do this explicitly. For nonconstant coefficients there are many known special cases but no general rules.

Solutions of Homogeneous Linear DEs (nth-Order Version)
For any linear nth-order homogeneous DE on (a, b),

$$y^{(n)}(t) + p_1(t)y^{(n-1)}(t) + p_2(t)y^{(n-2)}(t) + \cdots + p_n(t)y(t) = 0,$$

for which $p_1(t), p_2(t), \ldots, p_n(t)$ are continuous on (a, b), any n linearly independent solutions $\{y_1, y_2, \ldots, y_n\}$ form a basis of the solution space $\mathbb{S}$ and *every* solution y on (a, b) can be written as

$$y(t) = c_1y_1(t) + c_2y_2(t) + \cdots + c_ny_n(t)$$

for some real numbers $c_1, c_2, \ldots, c_n$.

In the first part of this section, we discovered how to find a basis for the special case of a second-order linear DE with constant coefficients. We start with the homogeneous case in this section and Sec. 4.3. Then in Secs. 4.4–4.6 we will find particular solutions for the nonhomogeneous case, to arrive at solutions to

$$y'' + p(t)y' + q(t)y = f(t)$$

in the expected form

$$y(t) = y_h(t) + y_p(t).$$

Linear Independence of Solutions and the Wronskian

The Solution Space Theorem provides us with the number of solutions in a basis for an nth-order linear homogeneous DE, namely n.

remember

- If we start with m solutions for the nth-order case, then if $m > n$, the solutions cannot be linearly independent.
- If $m = n$, we must test for linear independence (using the methods of Sec. 3.6).
- If $m < n$, the set of solutions does not span the space.

A Wronskian $W[y_1, y_2, \ldots, y_n]$ (see Sec. 3.5) on an interval (a, b) conveys more information in the test for linear independence when the functions $y_1, y_2, \ldots, y_n$ are solutions to the same nth-order linear homogeneous DE defined on (a, b) than it does when the functions are arbitrary functions in $\mathcal{C}^n(a, b)$.

This is because we have an Existence and Uniqueness Theorem for solutions to the DE.

The Wronskian Test for Linear Independence of DE Solutions

Suppose $\{y_1, y_2, \ldots, y_n\}$ is a set of solutions on (a, b) of an nth-order linear homogeneous DE,

$$L(y) = a_n(t)\frac{d^n y}{dt^n} + a_{n-1}\frac{d^{n-1}}{dt^{n-1}} + \cdots + a_1(t)\frac{dy}{dt} + a_0(t)y \equiv 0.$$

(i) If $W[y_1, y_2, \ldots, y_n] \neq 0$ at any point t on (a, b), the set $\{y_1, y_2, \ldots, y_n\}$ is linearly independent.

(ii) If $W[y_1, y_2, \ldots, y_n] \equiv 0$ on (a, b), the set $\{y_1, y_2, \ldots, y_n\}$ is linearly dependent.

The Wronskian test works in "both directions" only for n solutions to an nth-order linear homogeneous DE.

If the determinant of the Wronskin is nonzero, you get that the solution is linearly independent.

This may or may not mean that it is a basis of the solution space depending on whether it has the right number of dimensions.

Proof We proved (i) in Sec. 3.6. For (ii), suppose that $\{y_1, y_2, \ldots, y_n\}$ are solutions of $L(y) \equiv 0$ on (a, b). Let W denote $W[y_1, y_2, \ldots, y_n]$. Then we assume that $W(d) = 0$ for some $d \in (a, b)$, so we have

$$W(d) = \begin{vmatrix} y_1(d) & y_2(d) & \cdots & y_n(d) \\ y_1'(d) & y_2'(d) & \cdots & y_n'(d) \\ \vdots & \vdots & \ddots & \vdots \\ y_1^{(n-1)}(d) & y_2^{(n-1)}(d) & \cdots & y_n^{(n-1)}(d) \end{vmatrix} = 0$$

is the determinant for the matrix of coefficients for the linear system

$$c_1 y_1(d) + \quad c_2 y_2(d) + \cdots + \quad c_n y_n(d) = 0,$$

$$c_1 y_1'(d) + \quad c_2 y_2'(d) + \cdots + \quad c_n y_n'(d) = 0,$$

$$\vdots$$

$$c_1 y_1^{(n-1)}(d) + c_2 y_2^{(n-1)}(d) + \cdots + c_n y_n^{(n-1)}(d) = 0,$$

where $c_1, c_2, \ldots, c_n$ are unknowns. From Sec. 3.2, we see that $W(d) = 0$ implies that there exist an infinite number of solutions for the system. We pick one nonzero solution $\vec{c} = [c_1, c_2, \ldots, c_n]$. Consequently,

$$y = c_1 y_1 + c_2 y_2 + \cdots + c_n y_n$$

is a solution of $L(y) \equiv 0$ to which we can apply the Existence and Uniqueness Theorem. We select initial conditions at $t = d$ so that

$$y(d) = y'(d) = \cdots = y^{(n-1)}(d) = 0.$$

However, $y_0(t) \equiv 0$ on (a, b) is already a solution that satisfies these initial conditions. Because there can be only one such solution (uniqueness), we know that $y_0(t) = y(t)$ on (a, b) and $y \equiv 0 = c_1 y_1 + c_2 y_2 + \cdots + c_n y_n$, so that $\{y_1, y_2, \ldots, y_n\}$ is a linearly dependent set of solutions. ◻

EXAMPLE 7 Wronskians Everywhere

(a) Consider the set of solutions $A = \{2, t$ $(-\infty, \infty)$.

$$W = \begin{vmatrix} 2 & t-1 & t^2 & t^3+t \\ 0 & 1 & 2t & 3t^2+ \\ 0 & 0 & 2 & 6t \\ 0 & 0 & 0 & 6 \end{vmatrix}$$

$$= 2\begin{vmatrix} 1 & 2t & 3t^2+1 \\ 0 & 2 & 6t \\ 0 & 0 & 6 \end{vmatrix} = 2\begin{vmatrix} 2 & 6t \\ 0 & 6 \end{vmatrix} = 24 \neq 0.$$

A is a linearly independent set and hence a basis for the solution space.

(b) Consider the solutions $B = \{t, t+1, t^2-1, t^2\}$ to $\dfrac{d^4y}{dt^4} = 0$ on $(-\infty, \infty)$.

$$W = \begin{vmatrix} t & t+1 & t^2-1 & t^2 \\ 1 & 1 & 2t & 2t \\ 0 & 0 & 2 & 2 \\ 0 & 0 & 0 & 0 \end{vmatrix} = t\begin{vmatrix} 1 & 2t & 2t \\ 0 & 2 & 2 \\ 0 & 0 & 0 \end{vmatrix} = 0.$$

B is a linearly dependent set. (For example, the first function can be written as a linear combination of the other three: $t = (t+1) + (t^2-1) - (t^2)$.)

(c) Consider the set of solutions $C = \{1, t^2, t^3\}$ to $\dfrac{d^4y}{dt^4} = 0$ on $(-\infty, \infty)$.

$$W = \begin{vmatrix} 1 & t^2 & t^3 \\ 0 & 2t & 3t^2 \\ 0 & 2 & 6t \end{vmatrix} = \begin{vmatrix} 2t & 3t^2 \\ 2 & 6t \end{vmatrix} = 6t^2 = 0 \textit{ only} \text{ at } t = 0.$$

What happened? W is not identically 0, so we know $\{1, t^2, t^3\}$ is a linearly independent set. But the strong conclusion of the Wronskian test for solutions that occurred in (a) did not occur here because we had only three solutions of a fourth-order DE. ■

Summary

We have solved the linear second-order homogeneous differential equation with constant coefficients in the cases where the characteristic roots are real and either distinct or repeated. The result is a two-dimensional vector space of solutions. We solved the harmonic oscillator equation explicitly and then applied it to LRC-circuits and to overdamped and critically damped mass-spring systems. We determined the existence of a basis for the solution space for any linear homogeneous DE.

...ms

...eristic Roots *Determine the general solutions ...ential equations in Problems 1–14.*

1. ... $= 0$
2. $y'' - y' = 0$
3. $y'' - 9y = 0$
4. $y'' - y = 0$
5. $y'' - 3y' + 2y = 0$
6. $y'' - y' - 2y = 0$
7. $y'' + 2y' + y = 0$
8. $4y'' - 4y' + y = 0$
9. $2y'' - 3y' + y = 0$
10. $y'' - 6y' + 9y = 0$
11. $y'' - 8y' + 16y = 0$
12. $y'' - y' - 6y = 0$
13. $y'' + 2y' - y = 0$
14. $9y'' + 6y' + y = 0$

Initial Values Specified *For Problems 15–22, solve the initial-value problem.*

15. $y'' - 25y = 0,\quad y(0) = 1,\quad y'(0) = 0$
16. $y'' + y' - 2y = 0,\quad y(0) = 1,\quad y'(0) = 0$
17. $y'' + 2y' + y = 0,\quad y(0) = 0,\quad y'(0) = 1$
18. $y'' - 9y = 0,\quad y(0) = -1,\quad y'(0) = 0$
19. $y'' - 6y' + 9y = 0,\quad y(0) = 0,\quad y'(0) = -1$
20. $y'' + y' - 6y = 0,\quad y(0) = 1,\quad y'(0) = 1$
21. $y'' - y' = 0,\quad y(0) = 2,\quad y'(0) = -1$
22. $y'' - 4y' - 12y = 0,\quad y(0) = 1,\quad y'(0) = -1$

Bases and Solution Spaces *For each of the differential equations in Problems 23–26, give a basis and a solution space in terms of the basis.*

23. $y'' - 4y' = 0$
24. $y'' - 10y' + 25y = 0$
25. $5y'' - 10y' - 15y = 0$
26. $y'' + 2\sqrt{2}y' + 2y = 0$

Other Bases *Use the Solution Space Theorem to show that the sets given in Problems 27 and 28 are each a basis for the DE.*

27. $y' - 4y = 0;\quad \{e^{2t}, e^{-2t}\}, \{\cosh 2t, \sinh 2t\}, \{e^{2t}, \cosh 2t\}$
28. $y'' = 0;\quad \{1, t\}, \{t+1, t-1\}, \{2t, 3t-1\}$

The Wronskian Test *Use the Wronskian Test in Problems 29–31 to determine if the set of solutions is a basis for the given DE.*

29. $y^{(4)} = 0,\quad \{t+1, t-1, t^2+t, t^3\}$
30. $y''' - 10y'' - 15y' = 0,\quad \{te^{-5t}, e^{5t}, 2e^{5t} - 1\}$
31. $y^{(4)} = 0,\quad \{t+1, t^2+2t, t^2-2\}$
32. **Sorting Graphs** For the DE $\ddot{x} + 5\dot{x} + 6x = 0$ of Example 1, Fig. 4.2.8 adds to Fig. 4.2.1 the linked solution graph for $\dot{x}(t)$. Label the phase-plane trajectories from left to right as A, B, C, D, E. Then attach the same labels to the appropriate linked solutions $x(t)$ and $\dot{x}(t)$.

Relating Graphs *Problems 33–35 give linked solution graphs and a phase portrait for a single particular solution to $\ddot{x} + 5\dot{x} + 6x = 0$ from Example 1.*

Problems 36–39 give linked solution graphs and a phase portrait for a single particular solution to $\ddot{x} - \dot{x} - 6x = 0$. (See Problem 12.)

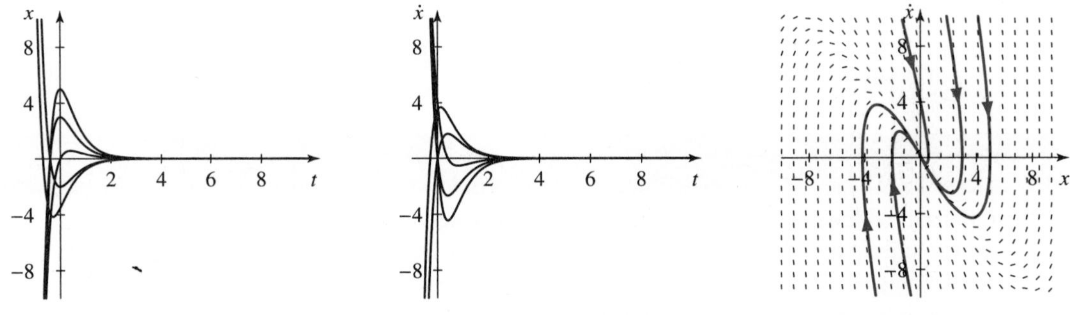

FIGURE 4.2.8 Graphs to sort for Problem 32, $\ddot{x} + 5\dot{x} + 6x = 0$.

For each Problem, relate the given graphs as follows:

(a) *Mark on the phase portrait the starting points (where $t = 0$) for the trajectories shown, and add arrows for the directions of each trajectory. Write down the initial conditions $x(0)$, $\dot{x}(0)$ for the phase-plane trajectory.*

(b) *Write the explicit solutions for $x(t)$ and $\dot{x}(t)$, then use your initial condition to solve the IVP.*

(c) *Describe how the graph for the solution $x(t)$ relates to its explicit formula from part (b).*

(d) *Describe how the graph for the solution $\dot{x}(t)$ relates to its explicit formula from part (b).*

For $\ddot{x} + 5\dot{x} + 6x = 0$:

33. Fig. 4.2.9 **34.** Fig. 4.2.10 **35.** Fig. 4.2.11

For $\ddot{x} - \dot{x} - 6x = 0$:

36. Fig. 4.2.12 **37.** Fig. 4.2.13

38. Fig. 4.2.14 **39.** Fig. 4.2.15

1) candy
2) reserve room

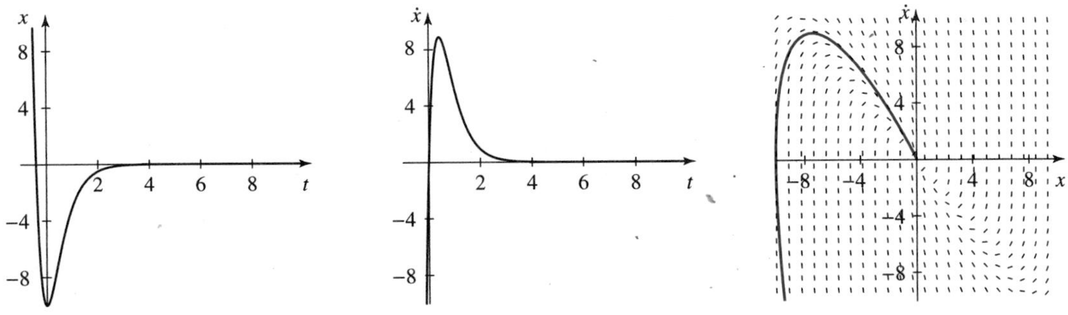

FIGURE 4.2.9 Graphs to relate for Problem 33, $\ddot{x} + 5\dot{x} + 6x = 0$.

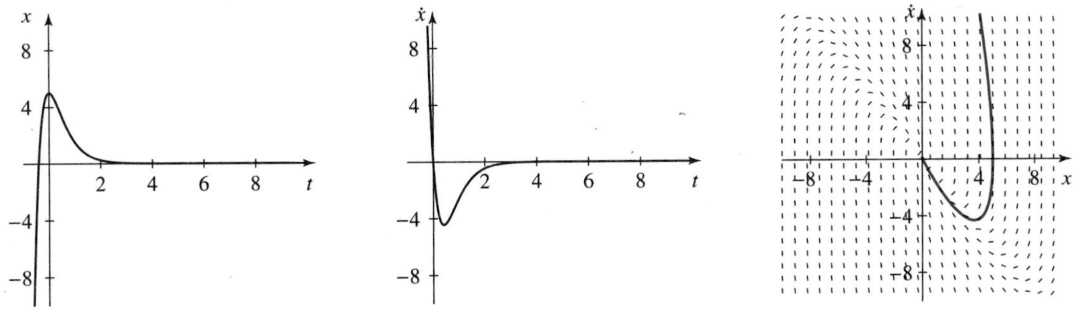

FIGURE 4.2.10 Graphs to relate for Problem 34, $\ddot{x} + 5\dot{x} + 6x = 0$.

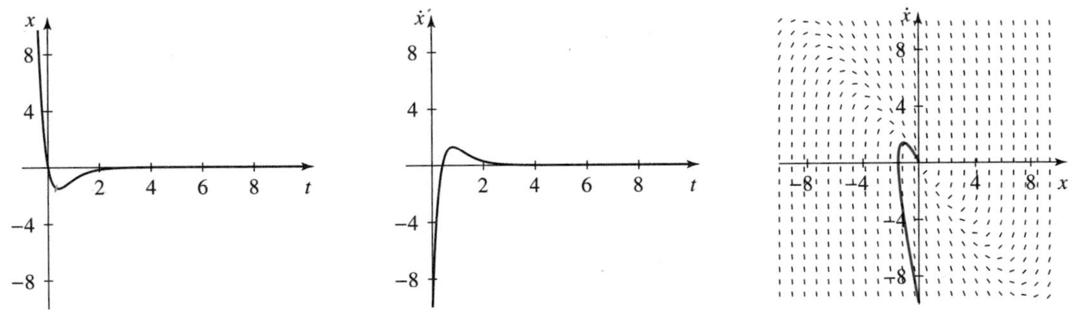

FIGURE 4.2.11 Graphs to relate for Problem 35, $\ddot{x} + 5\dot{x} + 6x = 0$.

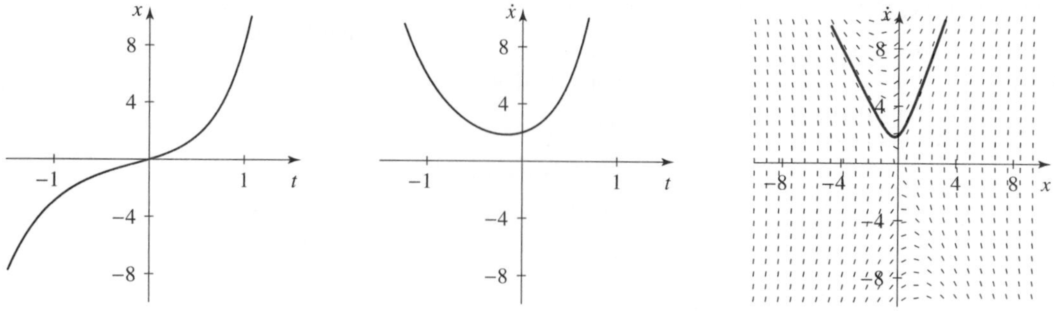

FIGURE 4.2.12 Graphs to relate for Problem 36, $\ddot{x} - \dot{x} - 6x = 0$.

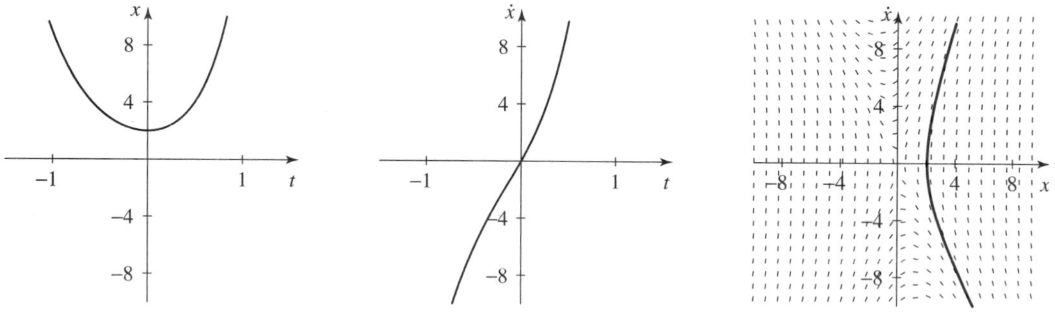

FIGURE 4.2.13 Graphs to relate for Problem 37, $\ddot{x} - \dot{x} - 6x = 0$.

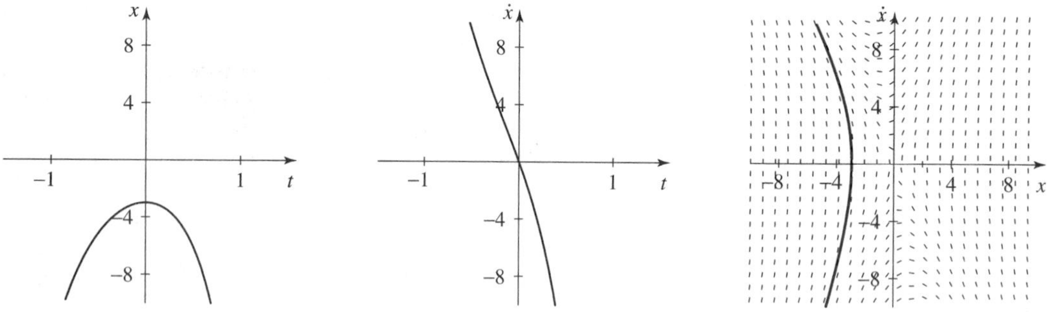

FIGURE 4.2.14 Graphs to relate for Problem 38, $\ddot{x} - \dot{x} - 6x = 0$.

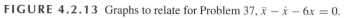

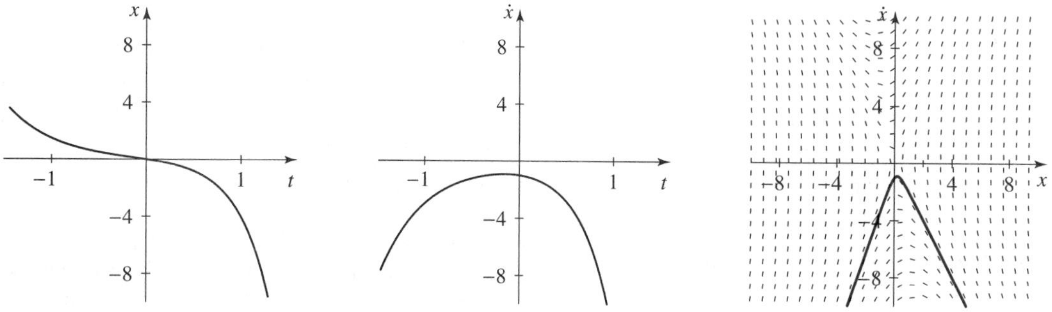

FIGURE 4.2.15 Graphs to relate for Problem 39, $\ddot{x} - \dot{x} - 6x = 0$.

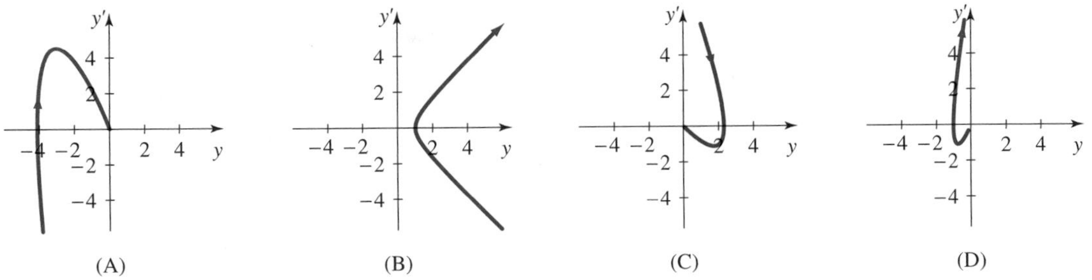

FIGURE 4.2.16 Phase portraits to match to Problems 40–43.

Phase Portraits *Match the ty solution graphs in Problems 40–43 to the corresponding yy′ phase-plane trajectory graph shown in Fig. 4.2.16.* HINT: *Keep in mind that the y-axis is vertical in one graph and horizontal in the other. You may also want to think about what the ty′ graph would look like.*

40.

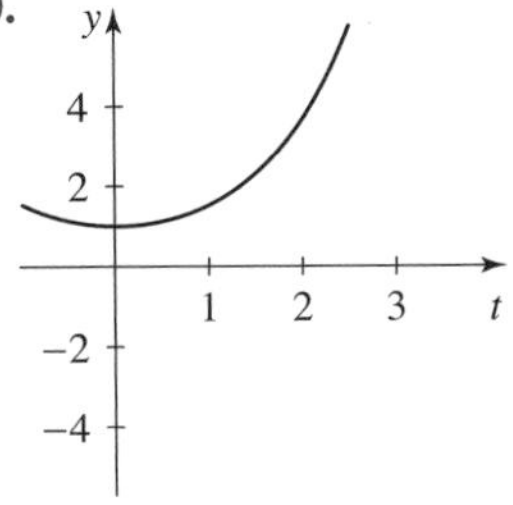

41.

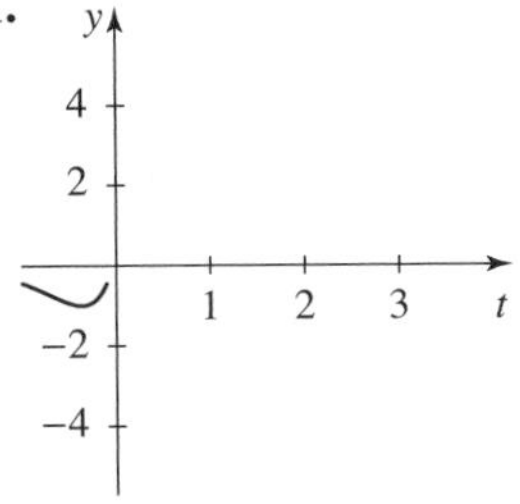

42.

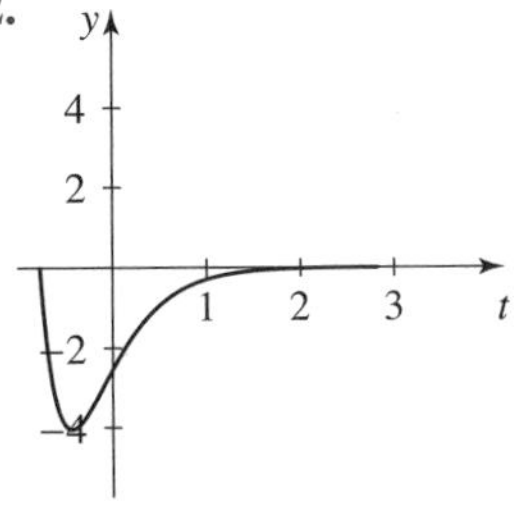

43.

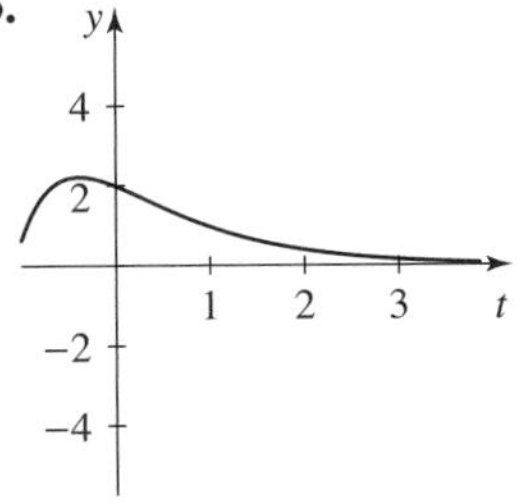

44. Independent Solutions Show that if r_1 and r_2 are distinct real characteristic roots of equation (1), then the solutions $e^{r_1 t}$ and $e^{r_2 t}$ are linearly independent.

45. Second Solution Verify that if the discriminant of equation (1) as given by $\Delta = b^2 - 4ac$ is zero, so that $b^2 = 4ac$ and the characteristic root is $r = -\frac{b}{2a}$, then substituting $y = v(t)e^{-(b/2a)t}$ into (1) leads to the condition $v''(t) = 0$.

46. Independence Again In the "repeated roots" case of equation (1), where $\Delta = b^2 - 4ac = 0$ and $r = -\frac{b}{2a}$, show that the solutions $e^{-(b/2a)t}$ and $te^{-(b/2a)t}$ are linearly independent.

47. Repeated Roots, Long-Term Behavior Show that in the "repeated roots" case of equation (1), the solution, which is given by $x(t) = c_1 e^{-(b/2a)t} + c_2 t e^{-(b/2a)t}$, for $\frac{b}{2a} > 0$, tends toward zero as t becomes large. HINT: You may need l'Hôpital's Rule: If, as x approaches a, both $f(x)$ and $g(x)$ approach zero, then the $\lim_{x\to a} f(x)/g(x)$ is indeterminate. But Marquis l'Hôpital (1661–1704) came to the rescue by publishing a result of Johann Bernoulli (1667–1748), that we can find the limit of the quotient by

$$\lim_{x\to a} \frac{f(x)}{g(x)} = \lim_{x\to a} \frac{f'(x)}{g'(x)},$$

providing, of course, that both derivatives exist in a neighborhood of a and approach nonzero limits.

48. Negative Roots Verify that in the overdamped mass-spring system, for which $\Delta = b^2 - 4mk > 0$, both characteristic roots are negative.

49. Circuits and Springs

(a) What conditions on the resistance R, the capacitance C and the inductance L in equation (11) correspond to overdamping and critical damping in the mass-spring system?

(b) Show that these conditions are directly analogous to $b > \sqrt{4mk}$ for overdamping and $b = \sqrt{4mk}$ for critical damping for the mass-spring system. Use Table 4.1.3.

50. A Test of Your Intuition We have two curves. The first starts at $y(0) = 1$ and its rate of increase equals its height; that is, it satisfies $y' = y$. The second curve also starts at $y(0) = 1$ with the same slope, and its second derivative, measuring upward curvature, equals its height; that is, it satisfies $y'' = y$. Which curve lies above the other? Make an educated guess before resolving the question analytically.

51. An Overdamped Spring The solution of the differential equation for an overdamped vibration has the form $x(t) = c_1 e^{r_1 t} + c_2 e^{r_2 t}$, with both c_1 and c_2 nonzero.

(a) Show that $x(t)$ is zero at most once.

(b) Show that $\dot{x}(t)$ is zero at most once.

52. A Critically Damped Spring The solution of the differential equation for a critically damped vibration has the form $x(t) = (c_1 + c_2 t)e^{rt}$, with both c_1 and c_2 nonzero.

(a) Show that $x(t)$ is zero at most once.

(b) Show that $\dot{x}(t)$ is zero at most once.

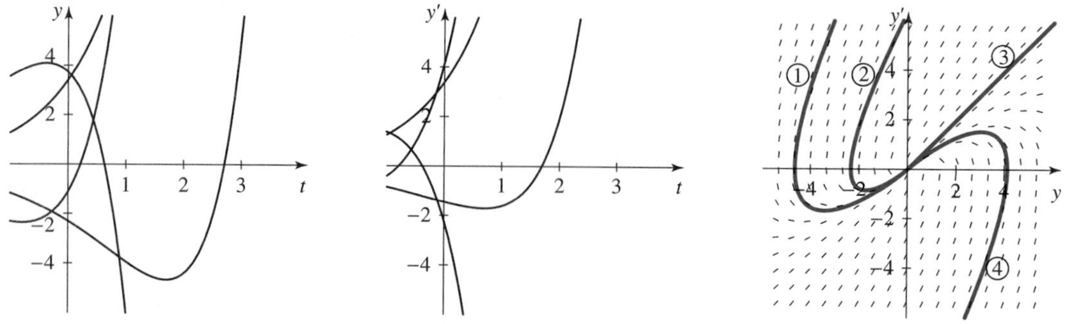

FIGURE 4.2.17 Graphs to link for Problem 53.

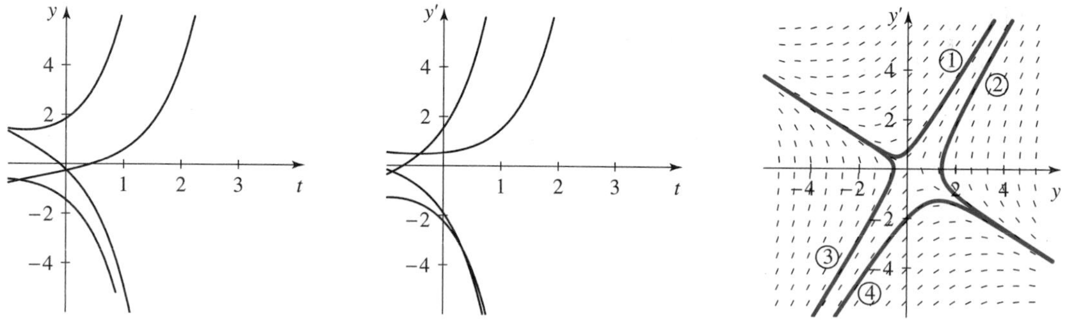

FIGURE 4.2.18 Graphs to link for Problem 54.

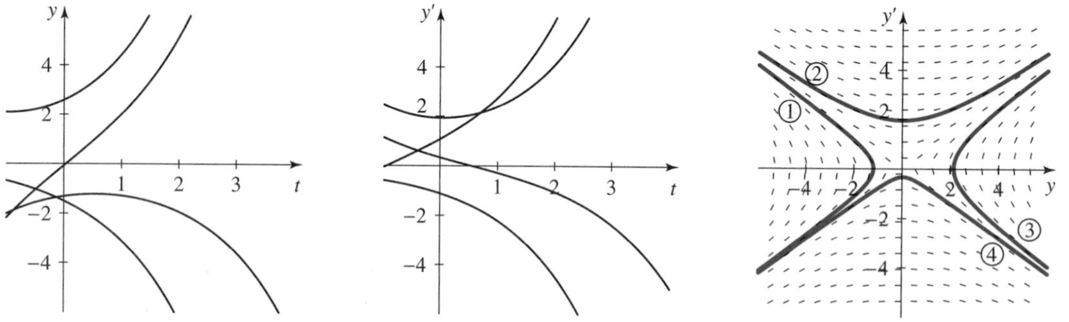

FIGURE 4.2.19 Graphs to link for Problem 55.

Linking Graphs *For the sets of ty, ty', and yy' graphs in Problems 53–55, match the corresponding trajectories. They are numbered on the phase portrait, so you can use those same numbers to identify the curves in the component solution graphs. On each phase-plane trajectory, mark the point where $t = 0$ and add arrowheads to show the direction of motion as t gets larger.*

53. Fig. 4.2.17 **54.** Fig. 4.2.18 **55.** Fig. 4.2.19

56. Damped Vibration A small object of mass 1 slug rests on a frictionless table and is attached, via a spring, to the wall. The damping constant is $b = 2$ lb sec/ft and the spring constant is $k = 1$ lb/ft. At time $t = 0$, the object is pulled 3 in. to the right and released. Show that the mass does not overshoot the equilibrium position at $x = 0$.

57. Surge Functions The function $x(t) = Ate^{-rt}$ can be used to model events for which there is a surge and die-off;

for example, the sales of a "hot" toy or the incidence of a highly infectious disease. This function can be obtained as the solution of a mass-spring system, $m\ddot{x} + b\dot{x} + kx = 0$. Assume $m = 1$. Find b and k and initial conditions $x(0)$ and $\dot{x}(0)$ in terms of parameters A and r that would yield the solution shown in Fig. 4.2.20.

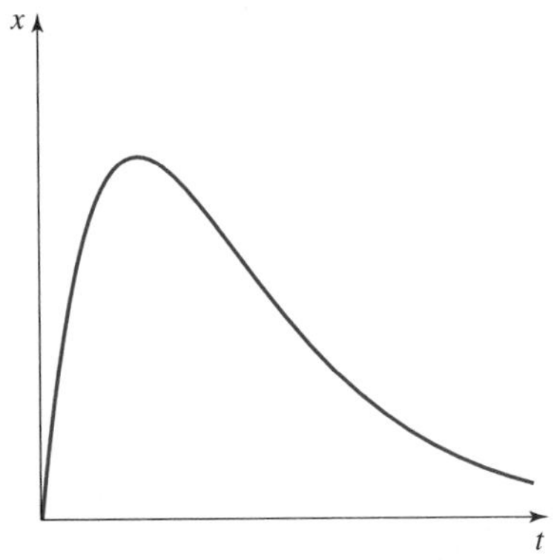

FIGURE 4.2.20 A particular solution to the IVP of Problem 57.

58. ***LRC*-Circuit I** A series *LRC*-circuit in a power grid has no input voltage, a resistor of 101 ohms, an inductor of 2 henries and a capacitor of 0.02 farads. Initially, the charge on the capacitor is 99 coulombs, and there is no current. (See Fig. 4.2.6.)
   (a) Determine the IVP for the charge across the capacitor.
   (b) Solve the IVP in (a) for the charge across the capacitor for $t > 0$.
   (c) Determine the current in the circuit for $t > 0$.
   (d) What are the long-term values of the charge and current?

Series Circuits
This tool provides help in visualizing the resulting current and long-term behavior in such circuits.

59. ***LRC*-Circuit II** A series *LRC*-circuit with no input voltage has a resistor of 15 ohms, an inductor of 1 henry, and a capacitor of 0.02 farads. Initially, the charge on the capacitor is 5 coulombs, and there is no current.
   (a) Determine the IVP for the charge across the capacitor.
   (b) Solve the IVP in (a) for the charge across the capacitor for $t > 0$.
   (c) Determine the current in the circuit for $t > 0$.
   (d) What are the long-term values of the charge and current?

60. **The Euler-Cauchy Equation** A well-known linear second-order equation with *variable* coefficients is the **Euler-Cauchy Equation**[3]

$$at^2y'' + bty' + cy = 0, \quad t > 0, \tag{14}$$

where $a, b, c \in \mathbb{R}$ and $a \neq 0$. Show by substituting $y = t^r$ that solutions of this form are obtained when r is a solution of the **Euler-Cauchy characteristic equation**

$$ar(r-1) + br + c = 0. \tag{15}$$

Then verify that if r_1 and r_2 are distinct solutions of (15), the general solution of (14) is given by

$$y(t) = c_1t^{r_1} + c_2t^{r_2}, \quad t > 0,$$

for arbitrary $c_1, c_2 \in \mathbb{R}$.

Euler-Cauchy Equations with Distinct Roots *Obtain, for $t > 0$, the general solution of the Euler-Cauchy equations in Problems 61–64.*

61. $t^2y'' + 2ty' - 12y = 0$

62. $4t^2y'' + 8ty' - 3y = 0$

63. $t^2y'' + 4ty' + 2y = 0$

64. $2t^2y'' + 3ty' - y = 0$

65. **Repeated Euler-Cauchy Roots** Verify that if the characteristic equation (15) for the Euler-Cauchy equation (14) has a repeated real root r, a second solution is given by $t^r \ln t$ and that t^r and $t^r \ln t$ are linearly independent.

Solutions for Repeated Euler-Cauchy Roots *Obtain, for $t > 0$, the general solution of the Euler-Cauchy equations in Problems 66–69.*

66. $t^2y'' + 5ty' + 4y = 0$

67. $t^2y'' - 3ty' + 4y = 0$

68. $9t^2y'' + 3ty' + y = 0$

69. $4t^2y'' + 8ty' + y = 0$

Computer: Phase-Plane Trajectories *Each of the functions in Problems 70–74 is the solution of a linear second-order differential equation with constant coefficients. In each case, do the following:*

   (a) *Determine the DE.*
   (b) *Calculate the derivative y' and the initial condition $y(0)$, $y'(0)$.*
   (c) *Plot the trajectory $[y(t), y'(t)]$ on the vector field in the yy'-plane.*

[3]The Euler-Cauchy equation can be recognized in standard form because the power of t is the same as the order of the derivative in each term (e.g., t^2y'').

70. $y(t) = 2e^{-t} + e^{-3t}$

71. $y(t) = e^{-t} + e^{-8t}$

72. $y(t) = e^{t} + e^{-t}$

73. $y(t) = e^{-t} + te^{-t}$

74. $y(t) = 3 + 2e^{2t}$

75. Reduction of Order[4] For a solution y_1 of

$$y'' + p(x)y' + q(x)y = 0 \qquad (16)$$

on interval I, such that y_1 is not the zero function on I, use the following steps to find the conditions on a function v of x such that

$$y_2 = vy_1$$

is a solution to equation (16) that is linearly independent from y_1 on I.

(a) Determine y_2' and y_2'' and substitute them into equation (16). Regroup and use the fact that y_1 is a solution of (16) to obtain

$$y_1 v'' + (2y_1' + py_1)v' = 0.$$

(b) Set $v' = w$. Solve the resulting first-order DE to obtain

$$v = \pm \int \frac{e^{-\int p(x)dx}}{y_1^2} dx \qquad (17)$$

so that

$$y_2 = y_1 \int \frac{e^{-\int p(x)dx}}{y_1^2} dx.$$

(c) Establish the fact that $\{y_1, y_2\}$ is a linearly independent set by showing that v cannot be a constant function on I. HINT: Show that v' cannot be identically zero on I.

Reduction of Order: Second Solution *Use the steps or the formula for y_2 developed in Problem 75 to find a second linearly independent solution to the second-order differential equations of Problems 76–79 for which y_1 is a known solution.* HINT: *Put the DE in standard form before using the formula.*

76. $y'' - 6y' + 9y = 0, \qquad y_1 = e^{3t}$

77. $y'' - 4y' + 4y = 0, \qquad y_1 = e^{2t}$

78. $t^2y'' - ty' + y = 0, \qquad y_1 = t$

79. $(t^2 + 1)y'' - 2ty' + 2y = 0, \quad y_1 = t$

Classical Equations *The equations in Problems 80–82 are some of the most famous differential equations in physics.*[5] *Use d'Alembert's reduction of order method described in Problem 75 along with the given solution y_1 to find a second solution $y_2(t)$.* HINT: *Be prepared for integrals that you cannot evaluate! Those answers should be left in terms of unevaluated integrals.*

80. $y'' - 2ty' + 4y = 0,$ **(Hermite's equation)**
$y_1(t) = 1 - 2t^2$

81. $(1 - t^2)y'' - ty' + y = 0,$ **(Chebyshev's equation)**
$y_1(t) = t$

Chebyshev's Equation
Graphical solutions give more insight than mere formulas.

82. $ty'' + (1 - t)y' + y = 0,$ **(Laguerre's equation)**
$y_1(t) = t - 1$

83. Lagrange's Adjoint Equation The integrating factor method, which was an effective method for solving first-order differential equations, is not a viable approach for solving second-order equations. To see what happens, even for the simplest equation, consider the differential equation

$$y'' + 3y' + 2y = f(t). \qquad (18)$$

Lagrange sought a function $\mu(t)$ such that if one multiplied the left-hand side of (18) by $\mu(t)$, one would get

$$\mu(t)[y'' + y' + y] = \frac{d}{dt}[\mu(t)y + g(t)y], \qquad (19)$$

where $g(t)$ is to be determined. In this way, the given differential equation would be converted to

$$\frac{d}{dt}[\mu(t)y' + g(t)y] = \mu(t)f(t),$$

which could be integrated, giving the first-order equation

$$\mu(t)y' + g(t)y = \int \mu(t)f(t)\,dt + c,$$

which could then be solved by first-order methods.

(a) Differentiate the right-hand side of (19) and set the coefficients of y, y', and y'' equal to each other to find $g(t)$.

---

[4]The reduction of order method of solving second-order DEs is of long standing, and is attributed to French mathematician Jean le Rond d'Alembert (1717–1783).

[5]These classical equations of physics were named for Charles Hermite (1822–1901) and Edmond Nicolas Laguerre (1834–1886), also Frenchmen, and for Russian mathematician Pafnuti Lvovich Chebyshev (1821–1894).

(b) Show that the integrating factor $\mu(t)$ satisfies the second-order homogeneous equation

$$\mu'' - \mu' + \mu = 0,$$

called the **adjoint equation** of (18). In other words, althought it is *possible* to find an "integrating factor" for second-order differential equations, to find it one must solve a new second-order equation for the integrating factor $\mu(t)$, which might be every bit as hard as the original equation. (In Sections 4.4 and 4.5, we will develop other methods.)

(c) Show that the adjoint equation of the general second-order linear equation

$$y'' + p(t)y' + q(t)y = f(t)$$

is the homogeneous equation

$$\mu'' - p(t)\mu' + [q(t) - p'(t)]\mu = 0.$$

84. Suggested Journal Entry The theory of linear second-order differential equations with constant coefficients depends on the nature of the solutions of a quadratic equation. Give other examples from precalculus or calculus where you have met a similar classification based on the sign of the discriminant of a quadratic.

4.3 Complex Characteristic Roots

SYNOPSIS: We complete the description of the two-dimensional solution space for the linear second-order homogeneous differential equation with constant coefficients for the case where the roots of the characteristic equation are complex numbers with imaginary terms. These solutions exhibit a variety of long-term behaviors, including periodic motions and damped oscillations.

Real and Complex Solutions

In solving the linear second-order homogeneous differential equation with constant coefficients,

$$ay'' + by' + cy = 0, \tag{1}$$

in the case of complex characteristic roots, we will encounter (nonreal) complex-valued solutions. It turns out, however, that the real and imaginary parts of these objects are also solutions and are, in fact, the actual real solutions that we want. Of course, we can always just verify directly, by substitution, that the real parts or the imaginary parts satisfy (1). But there is a general principle that can be checked too. If $u(t) + iv(t)$ is a solution of (1), then $u(t)$ and $v(t)$ are individual solutions as well, because

$$a(u + iv)'' + b(u + iv)' + c(u + iv) = (au'' + bu' + cu) + i(av'' + bv' + cv),$$

and a complex number is zero if and only if both its real and imaginary parts are zero.

In the previous section, we studied solutions of equation (1) for the cases in which the *discriminant* $\Delta = b^2 - 4ac$ is positive or zero. We now complete the discussion with Case 3, supposing that $\Delta < 0$.

Case 3: Complex Characteristic Roots ($\Delta < 0$)

When the discriminant $\Delta = b^2 - 4ac$ is negative, the characteristic equation $ar^2 + br + c = 0$ has the complex conjugate solutions

$$r_1 = -\frac{b}{2a} + i\frac{\sqrt{-\Delta}}{2a} = \alpha + i\beta \quad \text{and} \quad r_2 = -\frac{b}{2a} - i\frac{\sqrt{-\Delta}}{2a} = \alpha - i\beta. \tag{2}$$

The general solution can be written

$$y = k_1 e^{(\alpha + i\beta)t} + k_2 e^{(\alpha - i\beta)t}, \tag{3}$$

but $\{e^{(\alpha+i\beta)t}, e^{(\alpha-i\beta)t}\}$ is not a handy basis for interpretation. We shall use **Euler's formula,**[1]

$$e^{i\theta} = \cos\theta + i\sin\theta, \tag{4}$$

to convert (3) to a more meaningful form:

$$\begin{aligned} y &= k_1 e^{(\alpha+i\beta)t} + k_2 e^{(\alpha-i\beta)t} \\ &= k_1 e^{\alpha t}(\cos\beta t + i\sin\beta t) + k_2 e^{\alpha t}[\cos(-\beta t) + i\sin(-\beta t)] \\ &= e^{\alpha t}\,(k_1\cos\beta t + ik_1\sin\beta t + k_2\cos\beta t - ik_2\sin\beta t) \\ &= e^{\alpha t}[(k_1+k_2)\cos\beta t + i(k_1-k_2)\sin\beta t] \\ &= e^{\alpha t}(c_1\cos\beta t + c_2\sin\beta t). \end{aligned}$$

The new basis $\{e^{\alpha t}\cos\beta t, e^{\alpha t}\sin\beta t\}$ can be interpreted as oscillations, and we now have *real* solutions,

$$y_1(t) = e^{\alpha t}\cos\beta t \quad \text{and} \quad y_2(t) = e^{\alpha t}\sin\beta t,$$

which can be verified by direct substitution into (1). Moreover, y_1 and y_2 are linearly independent (see Problem 36), so they provide us with the general solution. In order to have a general *real* solution, the coefficients

$$c_1 = k_1 + k_2 \quad \text{and} \quad c_2 = i(k_1 - k_2)$$

must always be real numbers, even though k_1 and k_2 are assumed to be complex. (See Problem 37.)

Solution of $ay'' + by' + cy = 0$ with Complex Characteristic Roots
For $\Delta = b^2 - 4ac < 0$, the characteristic roots of the DE are

$$r_1, r_2 = \alpha \pm i\beta, \qquad \alpha = -\frac{b}{2a}, \quad \beta = \frac{\sqrt{-\Delta}}{2a}. \tag{5}$$

The functions $e^{\alpha t}\cos\beta t$ and $e^{\alpha t}\sin\beta t$ are linearly independent solutions, and the general solution is given by

$$y(t) = e^{\alpha t}(c_1\cos\beta t + c_2\sin\beta t), \tag{6}$$

where c_1 and c_2 are arbitrary constants determined by initial conditions. The set $\{e^{\alpha t}\cos\beta t, e^{\alpha t}\sin\beta t\}$ forms a basis for the solution space.

EXAMPLE 1 **Characteristic Roots May Be Complex** The characteristic equation of

$$y'' - 4y' + 13y = 0 \tag{7}$$

is

$$r^2 - 4r + 13 = 0,$$

[1] Recall that Euler's formula is a calculus fact, derived from Taylor series expansions of each of the functions. (See Problem 29.)

which has complex conjugate characteristic roots $r_1, r_2 = 2 \pm 3i$. The set $\{e^{2t}\cos 3t, e^{2t}\sin 3t\}$ forms a basis for the solution space, and the general solution of (7) is

$$y(t) = c_1 e^{2t}\cos 3t + c_2 e^{2t}\sin 3t.$$

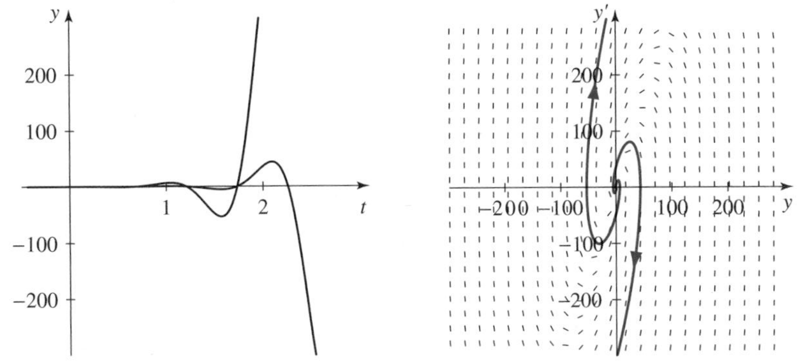

(a) Time series (b) Phase portrait

FIGURE 4.3.1 Two solutions of $y'' - 4y' + 13y = 0$ of Example 1.

(See Fig. 4.3.1.) Because e^{2t} grows ever larger as t increases, the solutions oscillate with ever-increasing amplitude (too big to see on this scale). ■

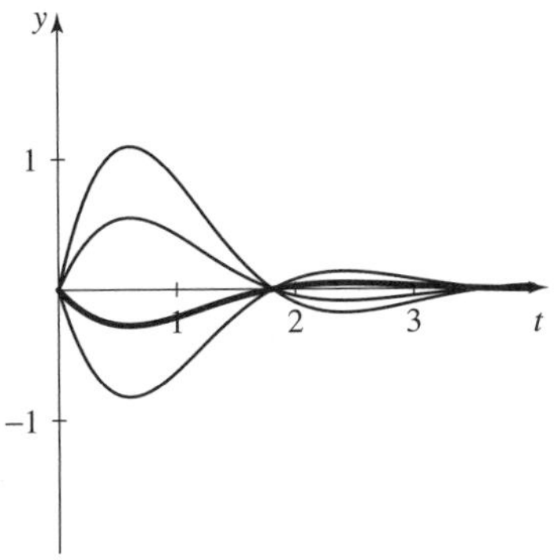

(a) Time series

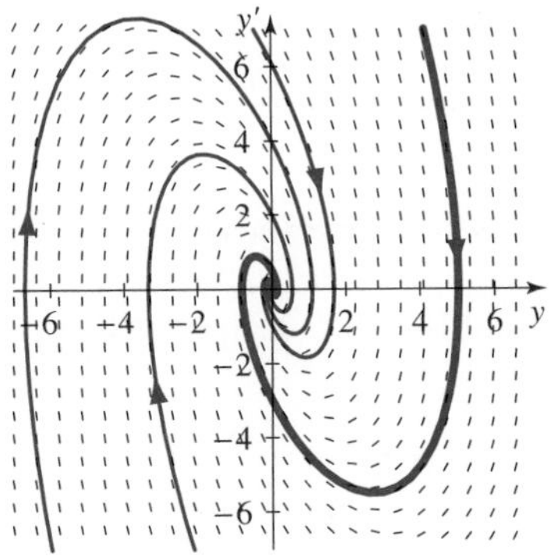

(b) Phase portrait

FIGURE 4.3.2 Some solutions to $y'' + 2y' + 4y = 0$ of Example 2.

EXAMPLE 2 **Initial-Value Problem** To find the general solution of

$$y'' + 2y' + 4y = 0, \qquad (8)$$

we write its characteristic equation $r^2 + 2r + 4 = 0$, which has roots *remember how to find characteristic equation.*

$$r_1, r_2 = \frac{-2 \pm \sqrt{4-16}}{2} = -1 \pm i\sqrt{3}.$$

The general solution is

$$y(t) = c_1 e^{-t}\cos\sqrt{3}\,t + c_2 e^{-t}\sin\sqrt{3}\,t,$$

and $\{e^{-t}\cos\sqrt{3}\,t, e^{-t}\sin\sqrt{3}\,t\}$ forms a basis for the solution space.

To find a particular solution to (8) for the initial conditions $y(0) = 0$ and $y'(0) = -1$, we differentiate,

$$y'(t) = -c_1 e^{-t}\cos\sqrt{3}\,t - \sqrt{3}c_1 e^{-t}\sin\sqrt{3}\,t$$
$$-c_2 e^{-t}\sin\sqrt{3}\,t + \sqrt{3}c_2 e^{-t}\cos\sqrt{3}\,t$$

and substitute initial conditions to determine that $c_1 = 0$ and $c_2 = -1/\sqrt{3}$. The particular solution to the IVP is

$$y(t) = -\frac{1}{\sqrt{3}}e^{-t}\sin\sqrt{3}\,t,$$

highlighted in Fig. 4.3.2. The fact that e^{-t} decreases toward zero means that trajectories spiral *in* toward the origin.

EXAMPLE 3 **Undamped Harmonic Oscillator** The differential equation

$$y'' + y = 0 \tag{9}$$

has characteristic equation $r^2 + 1 = 0$ with roots $\pm i$, so $\alpha = 0$ and $\beta = 1$. The set $\{\cos t, \sin t\}$ forms a basis for the solution space, and the general solution of (9) is

$$y(t) = c_1 \cos t + c_2 \sin t,$$

where c_1 and c_2 are arbitrary constants. Solutions are sinusoidal oscillations, as shown in Fig. 4.3.3, which is just what we would expect from a vibrating spring with no friction.

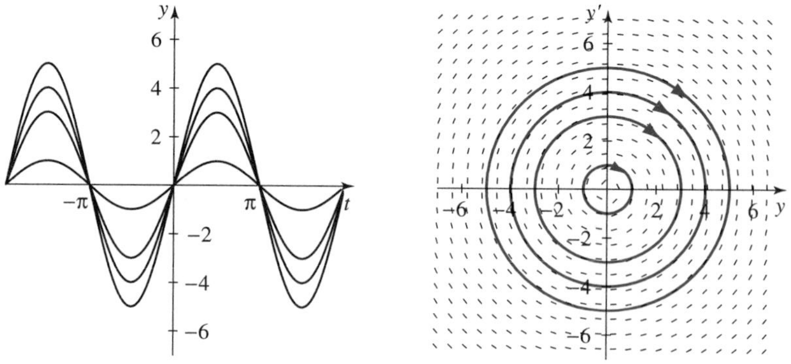

(a) Time series (b) Phase portrait

FIGURE 4.3.3 Some solutions to $y'' + y = 0$ of Example 3. ■

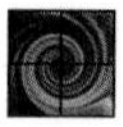

Mass and Spring

See the phase plane linked with the time graph.

Damped Systems with Complex Eigenvalues

Let us return to the motion of a damped mass-spring system

$$m\ddot{x} + b\dot{x} + kx = 0, \qquad m, b, k > 0, \tag{10}$$

first studied in Sec. 4.1. Recall from Sec. 4.2 that the motion is called *overdamped* when $\Delta > 0$ and *critically damped* when $\Delta = 0$. When the discriminant $\Delta = b^2 - 4mk < 0$, the characteristic roots are complex and we have the *third* type of damping, the most commonly encountered in modeling.

Underdamped Mass-Spring System

The motion of a mass-spring system (10) is called **underdamped** when $\Delta = b^2 - 4mk < 0$. Solutions are given by

$$x(t) = e^{-\frac{b}{2m}t}(c_1 \cos \omega_d t + c_2 \sin \omega_d t), \qquad \omega_d = \frac{\sqrt{4mk - b^2}}{2m}, \tag{11}$$

A sample solution is shown in Fig. 4.3.4. Using trigonometric identities, we rewrite (11) in alternate polar form and review the meanings of the various coefficients and parameters.

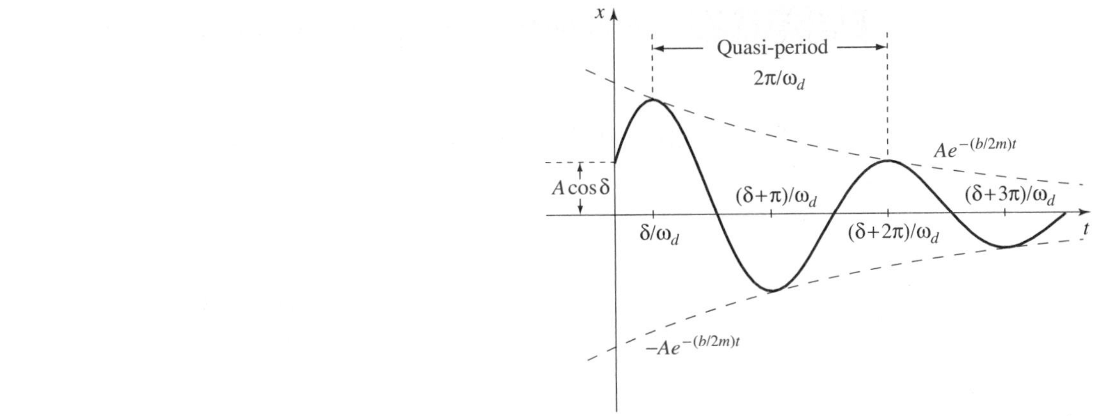

FIGURE 4.3.4 Underdamped oscillatory motion.

Alternate Solution to the Underdamped Unforced Oscillator

Solutions to the underdamped unforced oscillator may also be expressed as a family of sinusoidal oscillators with amplitude decreasing over time, given by

$$x(t) = A(t)\cos(\omega_d t - \delta), \quad \text{where } \omega_d = \frac{\sqrt{4mk - b^2}}{2m}, \tag{12}$$

where A and δ are arbitrary constants determined by initial conditions. The oscillator has

- **time-varying amplitude** $A(t) = Ae^{-(b/2m)t}$;
- **phase angle** δ (measured in radians); and
- **phase shift** $\varphi = \delta/\omega_d$.

The motion is not strictly periodic, oscillating with

- **circular quasi-frequency** $\omega_d = \dfrac{\sqrt{4mk - b^2}}{2m}$ (radians/sec);
- **natural quasi-frequency** $f_d = \dfrac{\omega_d}{2\pi}$ (in hertz); and
- **quasi-period** $T_d = \dfrac{1}{f_d} = \dfrac{2\pi}{\omega_d} = \dfrac{4\pi m}{\sqrt{4mk - b^2}}$ (measured in seconds).

The solution oscillates between two exponential curves $x(t) = \pm Ae^{-(b/2m)t}$, and the time required for the damped amplitude of the oscillation to decay from A to A/e is given by

- **time constant** $\tau = 2m/b$.

As with the undamped unforced oscillator (Sec. 4.1), it is a simple matter to convert solutions from (11) to (12), using $A = \sqrt{c_1^2 + c_2^2}$ and $\tan\delta = c_2/c_1$, or from (12) to (11), using $c_1 = A\cos\delta$ and $c_2 = A\sin\delta$.

EXAMPLE 4 **Underdamped Mass-Spring** A mass-spring system with damping,

$$\ddot{x} + \dot{x} + x = 0,$$

has characteristic equation $r^2 + r + 1 = 0$, whose roots are

$$r_1, r_2 = -\frac{1}{2} \pm i\frac{\sqrt{3}}{2}.$$

The general solution of the underdamped motion is given by

$$x(t) = e^{-t/2}\left[c_1 \cos\left(\frac{\sqrt{3}}{2}t\right) + c_2 \sin\left(\frac{\sqrt{3}}{2}t\right)\right],$$

and the set $\{e^{-t/2}\cos\frac{\sqrt{3}}{2}t, e^{-t/2}\sin\frac{\sqrt{3}}{2}t\}$ forms a basis for the solution space. If we substitute initial conditions $x(0) = 1$ and $\dot{x}(0) = 0$ into the general solution and its derivative,

$$\dot{x}(t) = e^{-t/2}\left[\left(-\frac{1}{2}c_1 + \frac{\sqrt{3}}{2}c_2\right)\cos\frac{\sqrt{3}}{2}t - \left(\frac{1}{2}c_2 + \frac{\sqrt{3}}{2}c_1\right)\sin\frac{\sqrt{3}}{2}t\right],$$

we obtain $c_1 = 1$, $c_2 = 1/\sqrt{3}$, and thus the particular solution is

$$x(t) = e^{-t/2}\left[\cos\left(\frac{\sqrt{3}}{2}t\right) + \frac{1}{\sqrt{3}}\sin\left(\frac{\sqrt{3}}{2}t\right)\right].$$

In alternate polar form this particular solution becomes

$$x(t) = \frac{2}{\sqrt{3}}e^{-t/2}\cos\left(\frac{\sqrt{3}}{2}t - \frac{\pi}{6}\right),$$

because

$$A = \sqrt{1^2 + \left(\frac{1}{\sqrt{3}}\right)^2} = \frac{2}{\sqrt{3}} \quad \text{and} \quad \delta = \tan^{-1}\left(\frac{\frac{1}{\sqrt{3}}}{1}\right) = \frac{\pi}{6}.$$

The oscillation, shown in Fig. 4.3.5, has

- time-varying amplitude $A(t) = Ae^{\alpha t} = \frac{2}{\sqrt{3}}e^{-\frac{t}{2}}$;
- circular quasi-frequency $\omega_d = \frac{\sqrt{3}}{2}$;
- natural quasi-frequency $f_d = \frac{\omega_d}{2\pi} = \frac{\frac{\sqrt{3}}{2}}{2\pi} = \frac{\sqrt{3}}{4\pi}$ hertz;
- quasi-period $T_d = \frac{2\pi}{\omega_d} = \frac{4\pi}{\sqrt{3}}$ seconds; and
- phase shift $\varphi = \frac{\delta}{\omega_d} = \frac{\frac{\pi}{6}}{\frac{\sqrt{3}}{2}} = \frac{\pi}{3\sqrt{3}}$ rad/sec. ■

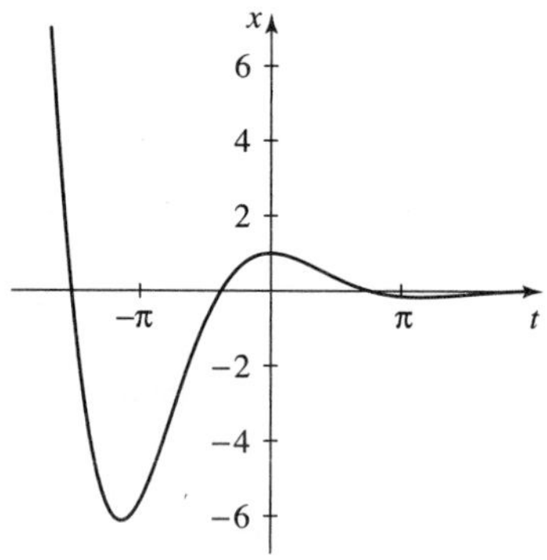

(a) Time series

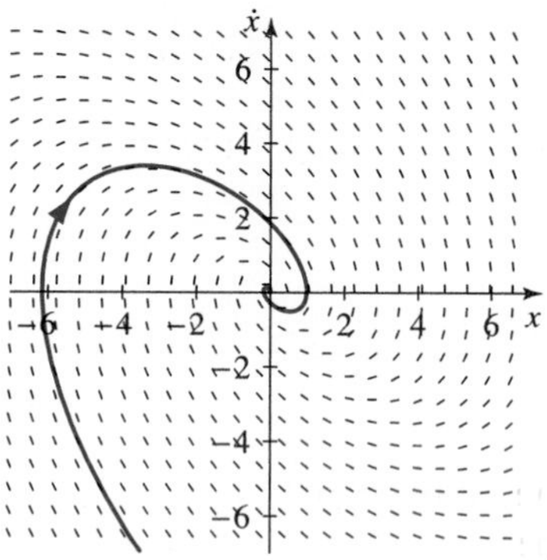

(b) Phase portrait

FIGURE 4.3.5 Solution to $\ddot{x} + \dot{x} + x = 0$ of Example 4 with initial conditions $x(0) = 1$, $\dot{x}(0) = 0$.

The Guitar String: A Qualitative Analysis

To demonstrate more clearly the solution of an undamped harmonic oscillator, such as the one in Example 3, we will switch from a mass-spring system to the vibrations of a guitar string. The same differential equation applies, but the guitar model has an additional attraction: we can *hear* it. We model it as a mass-spring system with the mass attached to two supports with two springs. (See Fig. 4.3.6.)

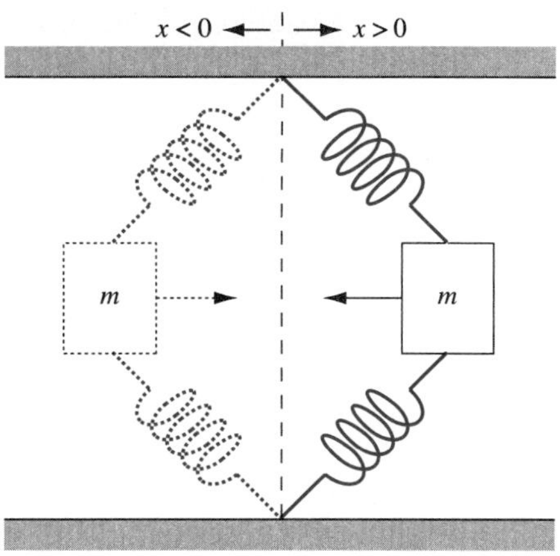

FIGURE 4.3.6 Guitar string as mass between two springs.

We assume that the vibrations of the string are in a plane. Displacement from rest is denoted by x, where $x > 0$ denotes displacement to the right of equilibrium and $x < 0$ denotes displacement to the left. The motion of the string is described by the equation

$$\ddot{x} + \omega_0^2 x = 0, \tag{13}$$

Frequency Units:
$f_0 = \dfrac{\omega_0}{2\pi}$ cycles/sec,
ω_0 is in radians/sec.

where ω_0 is the *circular frequency* (radians per second) at which the string vibrates, and its value depends on the tension and length of the string. But in music we speak of frequency f_0 in terms of *cycles* per second, so we will use the fact that $f_0 = \omega_0/(2\pi)$. For middle C, for example, the string vibrates at the natural frequency of 512 vibrations per second. Because there is no damping, the sound will last forever.

We know how to solve equation (13), which has no damping. In terms of f_0, the solution is

$$x = A\cos(2\pi f_0 t - \delta) \tag{14}$$

and

$$\dot{x} = -2\pi f_0 A \sin(2\pi f_0 t - \delta). \tag{15}$$

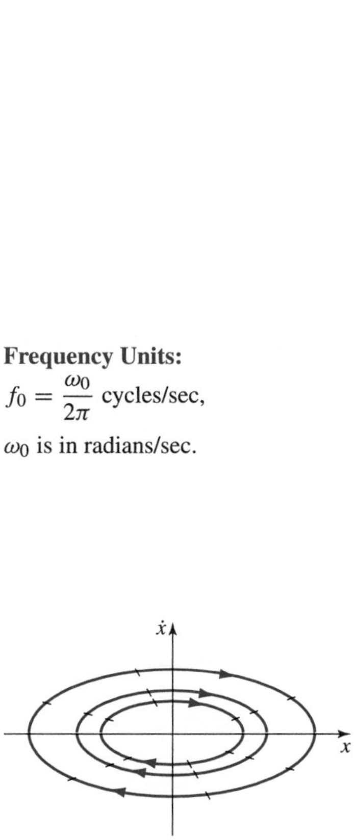

FIGURE 4.3.7 Phase-plane trajectories for the guitar string, with tickmarks at equal time intervals.

This gives a family of ellipses as phase-plane trajectories. The analytic solution (14) and (15) shows that the point describing the position and velocity of the "string" moves with the same frequency on all the ellipses, illustrated by the tick marks in Fig. 4.3.7. Soft notes have the same pitch as loud notes, a characteristic of a good guitar.

Each trajectory represents the solution with a given noise level. The trajectory that passes through the point $(x_0, 0)$ is the trajectory that corresponds to plucking the string by the amount $x(0) = x_0$. Plucking means no initial velocity, so the initial conditions are $x(0) = x_0$ and $\dot{x}(0) = 0$. By contrast, for a piano, the initial conditions for the *struck* string are $x(0) = 0$ and $\dot{x}(0) = v_0$.

Initial Conditions:
Plucking: $x(0) = x_0$, $\dot{x}(0) = 0$;
Striking: $x(0) = 0$, $\dot{x}(0) = v_0$.

Now think about what will happen to the guitar string if we add some damping.

EXAMPLE 5 **Underdamped Guitar String** To solve the DE

$$\ddot{x} + 2\dot{x} + 26x = 0,$$

we write the characteristic equation $r^2 + 2r + 26 = 0$ and determine its conjugate complex roots $-1 \pm 5i$. Hence $\alpha = -1$ and $\beta = 5$, and two independent solutions are given by $e^{-t}\cos 5t$ and $e^{-t}\sin 5t$. The general solution, for arbitrary c_1 and c_2, is given by

$$x(t) = c_1 e^{-t}\cos 5t + c_2 e^{-t}\sin 5t.$$

If we set $c_1 = c_2 = 1$, we obtain

$$x(t) = e^{-t}\cos 5t + e^{-t}\sin 5t \quad \text{and} \quad \dot{x}(t) = 4e^{-t}\cos 5t - 6e^{-t}\sin 5t.$$

Plotting $\dot{x}(t)$ versus $x(t)$ in the phase plane, we see the damped oscillation spiral toward its equilibrium point at the origin as the sound fades. (See Fig. 4.3.8.) ■

Summarizing Solutions for Real and Complex Characteristic Roots

The solutions of equation (1), discussed separately in three cases in the previous sections of this chapter, are summarized as follows.

Solutions to the Second-Order Linear DE with Constant Coefficients

The differential equation

$$ay'' + by' + cy = 0$$

has the characteristic equation

$$ar^2 + br + c = 0.$$

The quadratic formula gives rise to three different general solutions y_h for the DE, depending on the value of the discriminant $\Delta = b^2 - 4ac$.

Case 1 $\Delta > 0$	Real unequal roots: $r_1, r_2 = \dfrac{-b \pm \sqrt{b^2 - 4ac}}{2a}$	Overdamped motion: $y_h = c_1 e^{r_1 t} + c_2 e^{r_2 t}$
Case 2 $\Delta = 0$	Real repeated root: $r = -\dfrac{b}{2a}$	Critically damped motion: $y_h = c_1 e^{rt} + c_2 t e^{rt}$
Case 3 $\Delta < 0$	Complex conjugate roots: $r_1, r_2 = \alpha \pm \beta i$, $\alpha = -\dfrac{b}{2a}$, $\beta = \dfrac{\sqrt{4ac - b^2}}{2a}$	Underdamped motion: $y_h = e^{\alpha t}(c_1 \cos \beta t + c_2 \sin \beta t)$

Extensions to Higher-Order DEs

The methods of this section and the previous one generalize easily to higher-order differential equations. A homogeneous linear DE of order n, with constant coefficients,

$$a_n \frac{d^n y}{dt^n} + a_{n-1}\frac{d^{n-1}y}{dt^{n-1}} + a_{n-2}\frac{d^{n-2}y}{dt^{n-2}} + \cdots + a_1 \frac{dy}{dt} + a_0 y = 0 \tag{16}$$

has characteristic equation

$$a_n r^n + a_{n-1} r^{n-1} + a_{n-2} r^{n-2} + \cdots + a_1 r + a_0 = 0. \tag{17}$$

The Fundamental Theorem of Algebra guarantees that any polynomial (17) with real coefficients can be factored into linear and irreducible quadratic factors. (But it does not show us *how* to do it.) The methods of Sections 4.2 and 4.3 are applicable for each distinct linear or quadratic factor.

EXAMPLE 6 **Fourth-Order Equation** Consider the differential equation

$$\frac{d^4y}{dt^4} - 16y = 0.$$

Its characteristic equation,

$$r^4 - 16 = (r^2 - 4)(r^2 + 4) = (r + 2)(r - 2)(r^2 + 4) = 0,$$

has real unequal roots $r_1, r_2 = \pm 2$ and complex conjugate roots $r_3, r_4 = \pm 2i$.

The basis for the solution space is $\{e^{2t}, e^{-2t}, \cos 2t, \sin 2t\}$, and the general solution is

$$y = c_1e^{2t} + c_2e^{-2t} + c_3 \cos 2t + c_4 \sin 2t.$$

■

Although we know that the polynomial in (17) can be factored into linear and irreducible quadratic factors, the job may not be an easy one.

Factoring Characteristic Equations

If the coefficients in (17) are integers, we can select a rational factor q of a_0/a_n and substitute $r = q$ into the characteristic equation

$$f(r) = a_nr^n + \cdots + a_1r + a_0 = 0$$

to see if $f(q) = 0$. If so, $r - q$ is a factor of $f(r)$.

First look for a small integer q that divides a_0/a_n. Then try rational numbers.

EXAMPLE 7 **Factoring Characteristic Equations** Consider

$$y''' + y'' - 5y' + 3y = 0$$

and its characteristic equation

$$f(r) = r^3 + r^2 - 5r + 3 = 0.$$

We check factors of 3, namely ± 1 and ± 3, and find that $f(1) = 0$, so $r - 1$ is a factor of $f(r)$. Using long division,

$$f(r) = (r - 1)(r^2 + 2r - 3) = (r - 1)^2(r + 3) = 0,$$

so we have a real repeated root $r = 1$ and a single root $r_3 = -3$, and the general solution is

$$y = c_1e^t + c_2te^t + c_3e^{-3t}.$$

■

For each repeat of a root r in the characteristic equation of a higher-order DE, we need another power of t in the multiplier of e^{rt} to get another independent solution.

EXAMPLE 8 **Going On and On** Consider the fifth-order equation

$$\frac{d^5y}{dt^5} + 3\frac{d^4y}{dt^4} + 3\frac{d^3y}{dt^3} + \frac{d^2y}{dt^2} = 0.$$

The characteristic equation is

$$r^5 + 3r^4 + 3r^3 + r^2 = 0,$$

which factors into $(r+1)^3r^2 = 0$ with a triple root $r = -1$ and a double root $r = 0$. The solution is

$$y = \underbrace{(c_1 + c_2t + c_3t^2)e^{-t}}_{\text{for triple root}} + \underbrace{c_4 + c_5t}_{\text{for double root}}.$$

■

EXAMPLE 9 **Repeated Complex Roots** Consider

$$y^{(4)} + 8y'' + 16y = 0.$$

The characteristic equation is $r^4 + 8r^2 + 16 = 0$, which can be factored:

$$(r^2+4)^2 = 0,$$

yielding repeated complex conjugate roots

$$r = \pm 2i, \pm 2i.$$

The solution is

$$y = (c_1 + c_2t)\cos 2t + (c_3 + c_4t)\sin 2t.$$

■

Summary

We have completed the solution of the second-order linear homogeneous differential equation with constant coefficients for all cases of the characteristic roots (or eigenvalues): real and distinct, real and equal, or complex conjugates. The general solution of the equation generates a two-dimensional vector space in all three cases. The results are applied to overdamped, critically damped, and underdamped vibrations for the damped harmonic oscillator.

4.3 Problems

Solutions in General *For Problems 1–10, determine the general solution and give the basis $B = \{y_1, y_2\}$ for the solution space.*

1. $y'' + 9y = 0$

2. $y'' + y' + y = 0$

3. $y'' - 4y' + 5y = 0$

4. $y'' + 2y' + 8y = 0$

5. $y'' + 2y' + 4y = 0$

6. $y'' - 4y' + 7y = 0$

7. $y'' - 10y' + 26y = 0$

8. $3y'' + 4y' + 9y = 0$

9. $y'' - y' + y = 0$

10. $y'' + y' + 2y = 0$

Initial-Value Problems *Solve the IVPs in Problems 11–16.*

11. $y'' + 4y = 0, \quad y(0) = 1, \quad y'(0) = -1$

12. $y'' - 4y' + 13y = 0, \quad y(0) = 1, \quad y'(0) = 0$

13. $y'' + 2y' + 2y = 0, \quad y(0) = 1, \quad y'(0) = 0$

14. $y'' - y' + y = 0, \quad y(0) = 0, \quad y'(0) = 1$

15. $y'' - 4y' + 7y = 0, \quad y(0) = 0, \quad y'(0) = -1$

16. $y'' + 2y' + 5y = 0, \quad y(0) = 1, \quad y'(0) = -1$

Working Backwards *Write the standard form (equation (16) with leading coefficient $a_n = 1$) of the nth order linear homogeneous differential equation with real coefficients whose roots are given in Problems 17–20.*

17. 3rd order, $r = 1, 1, 1$

18. 3rd order, two of the roots are $r = 4, 1 - i$

19. 3rd order, two of the roots are $r = -2 + i, 2 + i$

20. 4th order, three of the roots are $2, -2, 4 + i$

Matching Problem *For Problems 21–28, determine which graph of the particular solution shown in Fig. 4.3.9 matches each differential equation.*

21. $y'' - y' = 0$

22. $y'' + y' = 0$

23. $y'' + 3y' + 2y = 0$

24. $y'' - 5y' + 6y = 0$

25. $y'' + y' + y = 0$

26. $y'' + y = 0$

27. $y'' + 4y' + 4y = 0$

28. $y'' - y' + y = 0$

29. Euler's Formula You can use the following process to justify Euler's formula

$$e^{i\theta} = \cos\theta + i\sin\theta.$$

(a) Write out explicitly the first dozen or so terms of the Maclaurin series (the Taylor expansion about the origin) given by

$$e^x = \sum_{n=0}^{\infty} \frac{x^n}{n!}.$$

(b) The series is valid for both real and complex numbers. Replace x by $i\theta$ and write the expression for $e^{i\theta}$.

(c) Simplify the results by using the periodicity of powers of i:

$$\begin{aligned} i^0 &= i^4 = i^8 = \cdots = 1, \\ i^1 &= i^5 = i^9 = \cdots = i, \\ i^2 &= i^6 = i^{10} = \cdots = -1, \\ i^3 &= i^7 = i^{11} = \cdots = -i. \end{aligned}$$

(d) Collect the real and imaginary terms.

(e) Obtain Euler's formula by recognizing the two Maclaurin series that appear in part (d).

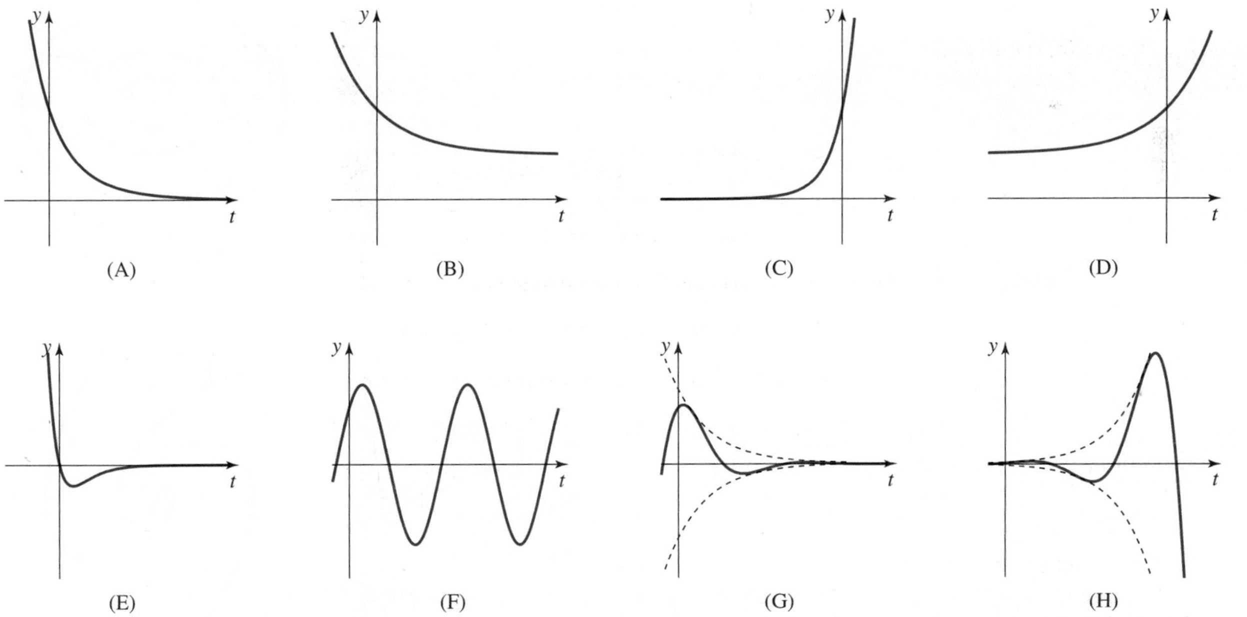

FIGURE 4.3.9 Particular solutions that match the differential equations in Problems 21–28. The dotted curves in (G) and (H) give the envelopes for the solutions.

Long-Term Behavior of Solutions *Suppose that r_1 and r_2 are the characteristic roots for $ay'' + by' + cy = 0$, so the solution is $y(t) = c_1e^{r_1t} + c_2e^{r_2t}$. For Problems 30–35, discuss the long-term solution behaviors for the given r_1, r_2 combinations. Assume $\beta \neq 0$.*

30. $r_1 < 0, r_2 < 0$ **31.** $r_1 < 0, r_2 = 0$ **32.** $r = \alpha \pm \beta i$

33. $r_1 = 0, r_2 = 0$ **34.** $r_1 > 0, r_2 < 0$ **35.** $r = \pm\beta i$

36. Linear Independence Verify that $e^{\alpha t}\cos\beta t$ and $e^{\alpha t}\sin\beta t$ are linearly independent on any interval.

37. Real Coefficients Suppose the roots of the characteristic equation for (1) are complex conjugates $\alpha \pm i\beta$, which gives rise to the general solution $y = k_1e^{(\alpha+i\beta)t} + k_2e^{(\alpha-i\beta)t}$, where k_1 and k_2 are any constants (even complex). Show that in order for the solution $y(t)$ to be real, k_1 and k_2 must be complex conjugates.

38. Solving $d^n y/dt^n = 0$

(a) Solve the equation

$$d^4y/dt^4 = 0 \tag{18}$$

by successive integration, getting $d^3y/dt^3 = k_3$ and $d^2y/dt^2 = k_3t + k_2$ to obtain $y(t)$.

(b) Determine the characteristic equation for (18) and use its roots to find the general solution. Compare this solution with the solution you found in (a).

(c) Generalize the process in (a) to solve $d^ny/dt^n = 0$.

Higher-Order DEs *Find the solutions for the following higher-order equations. Remember that for each repetition of a root, a term with an additional factor of t must be included. If a factorization of the characteristic equation $f(t) = 0$ is not obvious, look for a small integer q that satisfies $f(q) = 0$. Divide the characteristic equation by $r - q$.*

39. $y^{(5)} - 4y^{(4)} + 4y''' = 0$

40. $y''' + 4y'' - 7y' - 10y = 0$

41. $y^{(5)} - y' = 0$

42. $y''' - 4y'' + 5y' - 2y = 0$

43. $y''' + 6y'' + 12y' + 8y = 0$

44. $y^{(4)} - y = 0$

Linking Graphs *For the sets of ty, ty', and yy' graphs in Problems 45 and 46, match the corresponding trajectories. They are numbered on the phase portrait, so you can use those same numbers to identify the curves in the component solution graphs. On each phase-plane trajectory, mark the point where $t = 0$ and add arrowheads to show the direction of motion as t gets larger.*

45. Fig. 4.3.10

46. Fig. 4.3.11

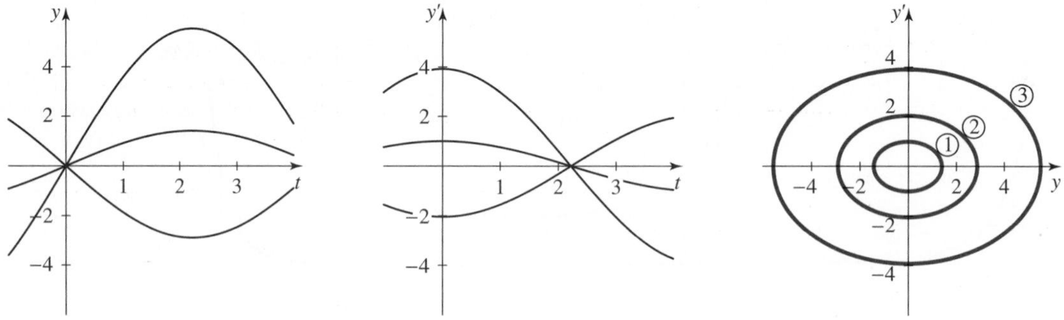

FIGURE 4.3.10 Graphs to be linked for Problem 45.

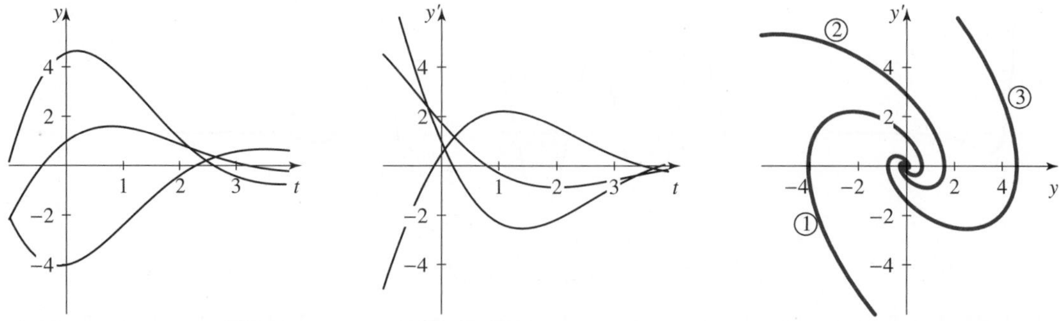

FIGURE 4.3.11 Graphs to be linked for Problem 46.

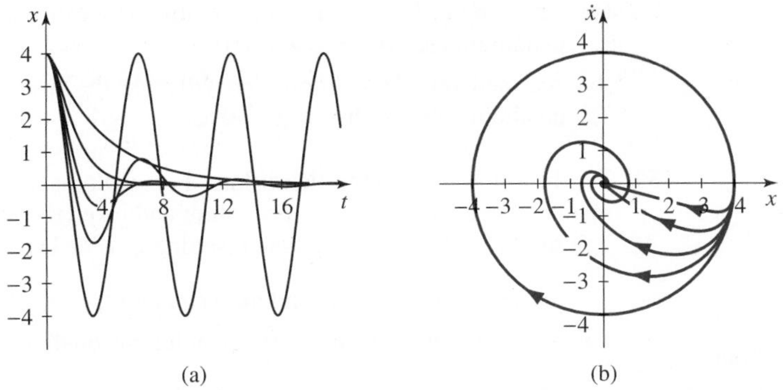

FIGURE 4.3.12 Solutions and phase-plane trajectories for Problem 47.

47. Changing the Damping Consider the mass-spring system

$$\ddot{x} + b\dot{x} + x = 0, \qquad x(0) = 4, \quad \dot{x}(0) = 0.$$

For damping coefficient $b = 0, 0.5, 1, 2, 4$, the corresponding solutions are plotted together in Fig. 4.3.12(a). Their phase-plane trajectories are plotted in Fig. 4.3.12(b). Make a trace of both graphs and label each curve with the appropriate value of b.

48. Changing the Spring Consider the mass-spring system

$$\ddot{x} + \dot{x} + kx = 0, \qquad x(0) = 4, \quad \dot{x}(0) = 0.$$

(a) For spring constant $k = 0.25, 0.5, 1, 2, 4$, the corresponding solutions are plotted together in Fig. 4.3.13(a). Their phase-plane trajectories are plotted in Fig. 4.3.13(b). Make a trace of both graphs and label each curve with the appropriate value of k.

(b) From your observations of the graphs in Fig. 4.3.13, do the oscillations increase in frequency and amplitude as the spring constant is increased (i.e., as the spring becomes "stiffer")? Explain.

49. Changing the Mass A mass-spring system has a mass m attached in standard fashion with a damping factor $b = 0$ and a spring constant $k = 16$.

(a) Discuss how the value of m affects the motion.

(b) How would the frequency of the motion be affected if the mass were doubled?

(c) Discuss how much damping would be required for the critical damping if the mass were increased.

50. Finding the Maximum

(a) For the mass-spring system for which $m = 1$, $b = 2$, $k = 3$, and $x(0) = 1$, $\dot{x}(0) = 0$, find the maximum displacement attained. (HINT: Differentiate the solution $x(t)$ and set $\dot{x}(t) = 0$ to find the critical point.)

(b) Do the same thing for $m = 1$, $b = 2$, $k = 10$, and $x(0) = 0$, $\dot{x}(0) = 2$ (underdamped).

(c) Do the same thing for $m = 1$, $b = 4$, $k = 4$, and $x(0) = 0$, $\dot{x}(0) = 2$ (critically damped).

Oscillating Euler-Cauchy *Euler-Cauchy equations were introduced in Sec. 4.2 Problem 60; Problems 51–54 consider Euler-Cauchy equations with nonreal characteristic roots. The solutions then have the final form*

$$y(t) = t^{\alpha}[c_1 \cos(\beta \ln t) + c_2 \sin(\beta \ln t)]. \tag{19}$$

51. Verify the solution (19). HINT: Use the relation

$$t^{\alpha \pm \beta i} = e^{(\alpha \pm \beta i)\ln t}.$$

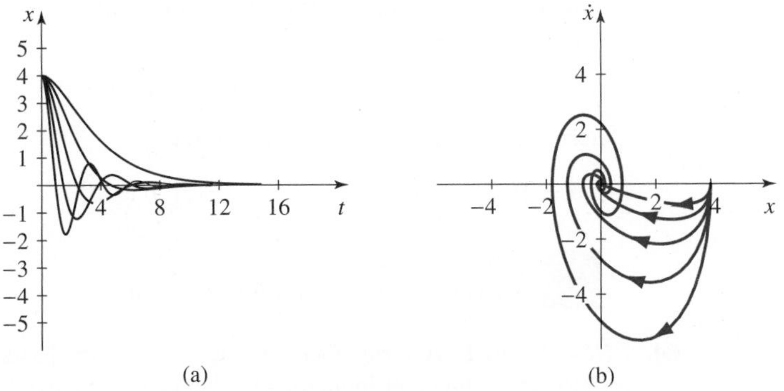

FIGURE 4.3.13 Solutions and phase-plane trajectories for Problem 48.

52. Solve $t^2y'' + 2ty' + y = 0$.

53. Solve $t^2y'' + 3ty' + 5y = 0$.

54. Solve $t^2y'' + 17ty' + 16y = 0$.

55. Third-Order Euler-Cauchy Use the substitution $y = t^r$ for $t > 0$ to obtain the characteristic equation for the following third-order Euler-Cauchy equation:

$$at^3y''' + bt^2y'' + cty' + dy = 0 \quad \text{for } t > 0.$$

Third-Order Euler-Cauchy Problems *Use the results from Problem 55 to solve the specific Euler-Cauchy equations of Problems 56 and 57.*

56. $t^3y''' + t^2y'' - 2ty' + 2y = 0$

57. $t^3y''' + 3t^2y'' + 5ty' = 0$

58. Inverted Pendulum Since the general solution of the linearized pendulum equation $\ddot{x} + (g/L)x = 0$ is the class of sinusoidal oscillations, for small displacements, the pendulum oscillates back and forth about its equilibrium point. The equation $\ddot{x} - (g/L)x = 0$ describes the *inverted pendulum*. (See Fig. 4.3.14(b).)

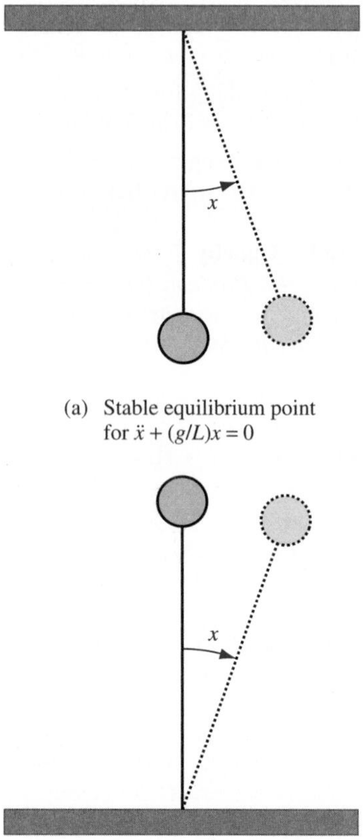

(a) Stable equilibrium point for $\ddot{x} + (g/L)x = 0$

(b) Unstable equilibrium point for $\ddot{x} - (g/L)x = 0$

FIGURE 4.3.14 Pendulum and inverted pendulum for Problem 59.

(a) Choosing $g/L = 1$, what is the motion of the inverted pendulum for $x(0) = 0$ and $\dot{x}(0) = 1$?

(b) Are there initial conditions that will make the inverted pendulum approach $x = \dot{x} = 0$ as $t \to \infty$?

59. Pendulum and Inverted Pendulum For *small* displacements, where $\sin x \approx x$, the pendulum and inverted pendulum of Fig. 4.3.14 are modeled (setting $g/L = 1$) by

$$\ddot{x} + x = 0 \quad \text{(linearized pendulum equation)},$$
$$\ddot{x} - x = 0 \quad \text{(linearized inverted pendulum equation)}.$$

Let us examine these models in the language of linear algebra.

(a) Show that fundamental solutions of the pendulum equation are e^{it} and e^{-it}, while those of the inverted pendulum are e^t and e^{-t}.

(b) Show that fundamental solutions of the pendulum equation are $\cos t$ and $\sin t$, while those of the inverted pendulum equation are $\cosh t$ and $\sinh t$ (hyperbolic cosine and hyperbolic sine).

(c) Are the solutions to both equations real? Explain.

60. Finding the Damped Oscillation Determine the constants for the damped oscillation

$$x(t) = e^{-t}(c_1 \cos t + c_2 \sin t),$$

subject to the initial conditions $x(0) = 1$ and $\dot{x}(0) = 1$. Graph the solution.

61. Extremes of Damped Oscillations Show that the maxima and minima of

$$x(t) = e^{\alpha t}(c_1 \cos \omega t + c_2 \sin \omega t)$$

for $\alpha < 0$ occur at equidistant values of t, adjacent values differing by π/ω.

62. Underdamped Mass-Spring System Find and graph the motion of a damped mass-spring system with mass $m = 0.25$ slugs, spring constant $k = 4$ lb/ft, and damping constant $b = 1$ lb sec/ft. The mass is initially pulled to the right, stretching the spring by 1 ft, and then released.

63. Damped Mass-Spring System The motion of a mass-spring system obeys

$$\ddot{x} + b\dot{x} + 64x = 0, \qquad x(0) = 1, \qquad \dot{x}(0) = 0.$$

Determine $x(t)$ and sketch the motion for

(a) $b = 10$; (b) $b = 16$; (c) $b = 20$.

Series Circuit
Visualize the solutions for *LRC*-circuits.

64. *LRC*-Circuit I A series *LRC*-circuit (Fig. 4.3.15) has a resistor of 8 ohms, an inductor of 1 henry and a capacitor of 0.04 farads. The initial charge on the capacitor is

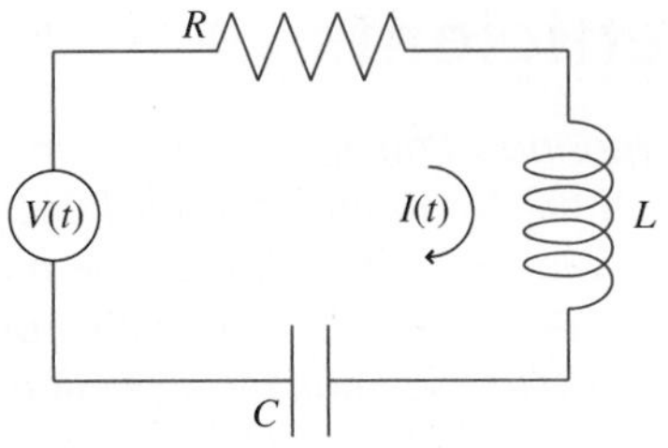

FIGURE 4.3.15 A series *LRC*-circuit; the arrow for $I(t)$ indicates the positive direction for the current.

1 coulomb, and there is initially no current in the circuit. Assume $V(t) = 0$ for $t > 0$.

(a) Formulate the IVP for the charge across the capacitor.

(b) Determine the charge across the capacitor for $t > 0$.

(c) Find the current in the circuit for $t > 0$.

(d) What are the long-term values of charge and current in the circuit?

65. *LRC*-Circuit II A series *LRC*-circuit has a resistor of 1 ohm, an inductor of 0.25 henries, and a capacitor of 0.25 farads. The initial charge on the capacitor is 1 coulomb, and there is initially no current in the circuit. Assume $V(t) = 0$ for $t > 0$.

(a) Formulate the IVP for the charge across the capacitor.

(b) Determine the charge across the capacitor for $t > 0$.

(c) Find the current in the circuit for $t > 0$.

(d) What are the long-term values of charge and current in the circuit?

66. Computer Lab: Damped Free Vibrations Improve your understanding of damped oscillations by working through Lab 9 of the IDE software package, skipping over parts 1.7 and 2.5 on Energy.

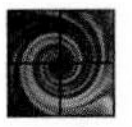

Linear Oscillators: Free Response

Lab 9 provides a simple visual and visceral introduction.

Effects of Nonconstant Coefficients *In our study of the damped mass-spring system with mass m, spring constant k, and damping constant b, we have used as our model the second-order linear differential equation $m\ddot{x} + b\dot{x} + kx = 0$ having constant coefficients. When coefficients change with time, the analytic solutions we have found for constant coefficients* do not work. *Explain why. Then, for Problems 67–73, consider some DEs in which m, b, and k change with time.*

(a) *Use your intuition and/or a computer to describe the motion of the system under these changing conditions.*

(b) *Use a computer to draw a solution $x(t)$ for $t > 0$, $x(0) = 2$, $\dot{x}(0) = 0$.* HINT: *When you need to avoid $t = 0$, try a trick like starting your plot at $t = 0.1$.*

(c) *Discuss what followed your intuition, what did not, and what further questions you might now ask.*

67. $\ddot{x} + \dfrac{1}{t}x = 0$

68. $\ddot{x} + \dfrac{1}{t}\dot{x} + x = 0$

69. $t\ddot{x} + x = 0$

70. $\ddot{x} + (x^2 - 1)\dot{x} + x = 0$

71. $\ddot{x} + (\sin t)\dot{x} + x = 0$

72. $\ddot{x} + \dfrac{1}{t}\dot{x} + tx = 0$

73. $\ddot{x} + (\sin 2t)x = 0$

Boundary-Value Problems *Two **boundary conditions** $y(a_1) = b_1$ and $y(a_2) = b_2$ can be used to specify the solution to what is now called a **boundary-value problem**, provided that the two conditions do not lead to a contradiction. Find all solutions of $y'' + y = 0$ satisfying the boundary conditions in Problems 74–77. If the given boundary condition leads to a contradiction, state this fact explicitly and show that it is so.*

74. $y(0) = 0,\ y(\pi/2) = 0$

75. $y(0) = 0,\ y(\pi/2) = 1$

76. $y(0) = 1,\ y(\pi) = 1$

77. $y(0) = 0,\ y(\pi/2) = 0$

Exact Second-Order Differential Equations *The differential equation*

$$y'' + p(t)y' + q(t)y = 0$$

*is called an **exact** second-order equation if it can be written in a form that can be integrated directly. An example is*

$$y'' + [g(t)y]' = 0,$$

where $g(t)$ is determined from $p(t)$ and $q(t)$. This DE can be integrated directly to get a first-order linear equation, which in turn can be integrated using the integrating factor method. For Problems 78–80, solve the given exact equations.

78. $y'' + \dfrac{1}{t}y' - \dfrac{1}{t^2}y = 0$ (HINT: $g(t) = 1/t$)

79. $y'' + \dfrac{2}{t}y' - \dfrac{2}{t^2}y = 0$

80. $(t^2 - 2t)y'' + 4(t-1)y' + 2y = 0$
(HINT: Let $g(t) = t^2 - 2t$ and show that the left-hand side equals $(gy)''$.)

81. Suggested Journal Entry The subsection entitled "Summarizing Solutions for Real and Complex Characteristic Roots" and Problems 30–35 on long-term behavior specify certain types of characteristic roots and tell how the solution evolves. Summarize these outcomes using the various behaviors as categories. That is, answer such questions as the following.

(a) When do solutions tend to zero as $t \to \infty$?

(b) When do solutions remain bounded but not tend to zero?

(c) When do solutions oscillate in an unbounded manner?

4.4 Undetermined Coefficients

SYNOPSIS: We extend the Superposition Principle to nonhomogeneous linear differential equations and apply it and the Nonhomogeneous Principle to the nonhomogeneous case. We introduce a widely useful scheme, the method of undetermined coefficients, for obtaining particular solutions of many nonhomogeneous equations. If common sense does not lead to a solution by inspection, undetermined coefficients may be the next-simplest alternative.

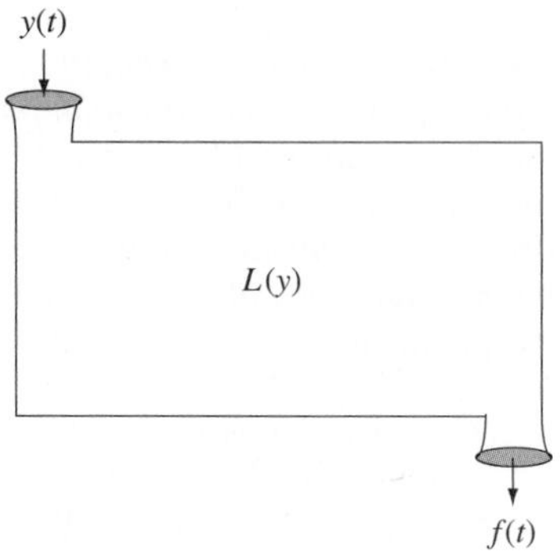

FIGURE 4.4.1 A linear operator L as a black box.

Combining Structure Principles

The fundamental structure given by the Superposition Principle and the Nonhomogeneous Principle, first studied in Sec. 2.1, extend far beyond the simple examples given to nonhomogeneous linear differential equations in general.

If L is a linear differential operator defined by

$$L(y) = a_n(t)y^{(n)} + a_{n-1}(t)y^{(n-1)} + \cdots + a_1(t)y' + a_0(t)y, \tag{1}$$

where all functions of t are assumed to be defined over some common interval I, then a general nonhomogeneous differential equation has the form $L(y) = f(t)$.

We can view the operator L as a "black box" to which we input a solution $y(t)$ that is operated on by L to obtain the forcing function $f(t)$ as output, as shown in Fig. 4.4.1.

We can restate the concept of superposition in terms of the linear differential operator, in a way that expands to *nonhomogeneous* linear DEs.

Superposition Principle for Nonhomogeneous Linear DEs

If $y_i(t)$ is a solution of $L(y) = f_i(t)$, for $i = 1, 2, \ldots, n$, and $c_1, c_2, \ldots, c_n$ are constants, then

$$y(t) = c_1y_1(t) + c_2y_2(t) + \cdots + c_ny_n(t)$$

is a solution of

$$L(y) = c_1f_1(t) + c_2f_2(t) + \cdots + c_nf_n(t).$$

At this point, we need to recall the Nonhomogeneous Principle.

Nonhomogeneous Principle for Linear DEs

The general solution of the nonhomogeneous linear differential equation $L(y) = f$ is

$$y = y_h + y_p,$$

where

- y_h is the general solution of $L(y) = 0$, and
- y_p is a particular solution of $L(y) = f$.

Combining these principles, as shown in Figs. 4.4.2 and 4.4.3, comes in handy for complicated problems. For instance, if we have solved $L(y) = f(t)$ for a particular $f(t)$, we can use that solution to jump-start solutions to other

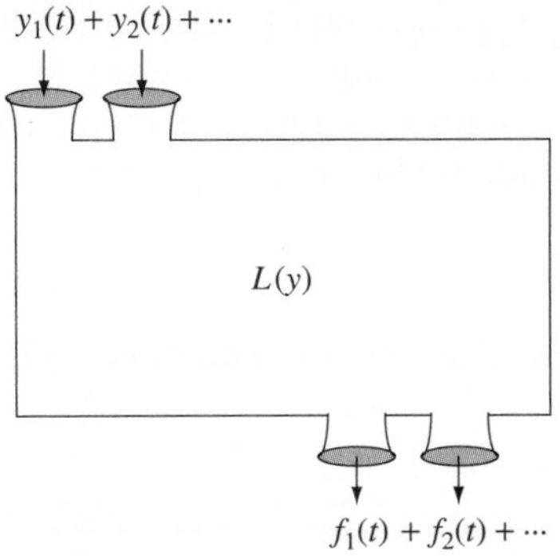

FIGURE 4.4.2 The Superposition Principle as a black box.

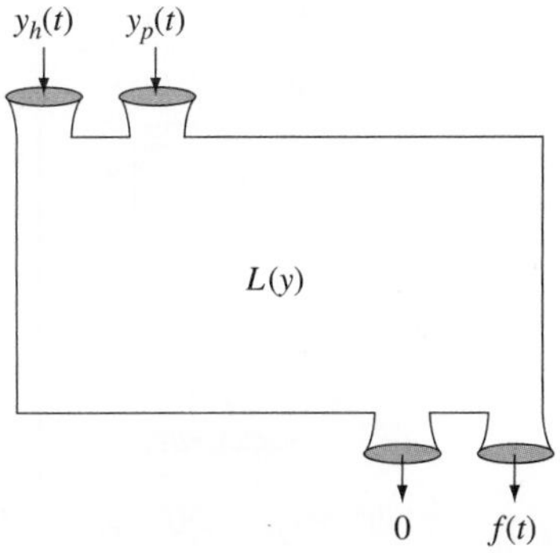

FIGURE 4.4.3 The Nonhomogeneous Principle as a black box.

DEs with the same linear operator and different but related forcing functions $f_1(t), f_2(t), \ldots, f_n(t)$.

Here we show some examples that focus on these principles. We have not yet explained *how* the various nonhomogeneous solutions are obtained. We will be able to find them later in this section by the method of undetermined coefficients.

EXAMPLE 1 Superposing Solutions Consider the nonhomogeneous equation

$$\underbrace{y'' - y' - 2y}_{L(y)} = \underbrace{2t + 1}_{f_1} \underbrace{- 2e^t}_{f_2}. \tag{2}$$

The brackets show how equation (2) can be broken down into two simpler nonhomogeneous problems. We can verify easily that the first, $L(y) = f_1$, has a particular solution $y_1 = -t$, and $L(y) = f_2$ has a solution $y_2 = e^t$. Then a particular solution of (2) is given by

$$y_p(t) = y_1 + y_2 = -t + e^t,$$

as shown in Fig. 4.4.4. The "ingredients," y_1 and y_2, create the desired outputs of L.

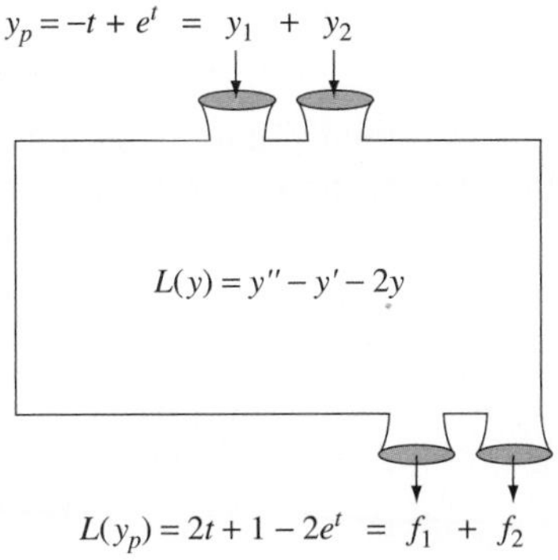

FIGURE 4.4.4 Example 1.

For the general solution of (2), we apply the Nonhomogeneous Principle. We know that the characteristic equation for (2) is $r^2 - r - 2 = (r-2)(r+1) = 0$. So

$$y_h(t) = c_1e^{2t} + c_2e^{-t},$$

and

$$y(t) = y_h + y_p = c_1e^{2t} + c_2e^{-t} - t + e^t. \quad ■$$

you can multiply solutions by eachother to get a different linear combination

EXAMPLE 2 Linear Combination of Forcing Terms Suppose that we want to keep the same operator but change the forcing function in Example 1, as follows:

$$\underbrace{y'' - y' - 2y}_{L(y)} = \underbrace{t + \frac{1}{2}}_{\frac{1}{2}f_1} + \underbrace{8e^t}_{-4f_2}, \tag{3}$$

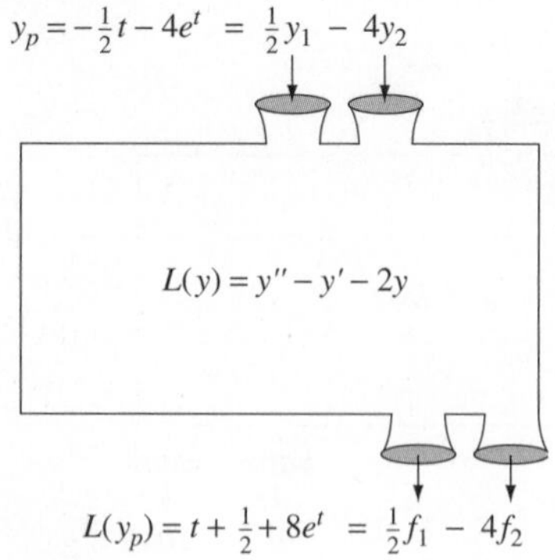

FIGURE 4.4.5 The particular solution of Example 2 as it relates to Example 1 with the same operator L.

where f_1 and f_2 are from Example 1. Since the forcing term is a linear combination of previous forcing terms (Example 1), the operator view of superposition allows us to "adjust the ingredients" y_1 and y_2 (found in Example 1) to get the desired output, as shown in Fig. 4.4.5. Therefore,

$$y_p = \frac{1}{2}y_1 - 4y_2 = -\frac{1}{2}t - 4e^t,$$

and the general solution of equation (3) is

$$y = c_1e^{2t} + c_2e^{-t} - \frac{1}{2}t - 4e^t.$$

■

For the case of a second-order DE with constant coefficients,

$$ay'' + by' + cy = f(t), \tag{4}$$

we have found y_h in Secs. 4.2 and 4.3, so now we simply seek y_p for $f(t) \neq 0$. The methods outlined in this section are the most direct and are widely applicable, but they are restricted to differential equations with *constant coefficients* and to *certain families* of forcing terms $f(t)$.

Solving by Inspection

Sometimes, especially after accumulating some experience, we can guess a solution just by "inspecting" the equation, without working through a series of calculations. The following observations show how to begin to recognize such situations.

EXAMPLE 3 **Constant Everything** If a linear differential equation has constant coefficients and the forcing term is constant, the particular solution can be seen by inspection.

(a) For the second-order equation

$$ay'' + by' + cy = d,$$

it is clear (at least in retrospect) that $y_p = d/c$ if $c \neq 0$.

(b) This idea works as well for the nth-order equation

$$a_ny^{(n)} + a_{n-1}y^{(n-1)} + \cdots + a_2y'' + a_1y' + a_0y = d,$$

where again $y_p = d/a_0$, provided that $a_0 \neq 0$.

(c) Inspection of the differential equation

$$y'' + y' = 1$$

leads at once to $y_p = t$.

■

Similar guesses "by inspection" are often possible for forcing functions $f(t)$ that are composed of simple functions.

EXAMPLE 4 **Particular Solutions by Inspection**

(a) The function $y_p = t$ is a particular solution of the differential equation

$$y'' + y = t.$$

(b) For the second-order differential equation

$$y'' - y = \sin t,$$

a particular solution may be seen by inspection to be $y_p = -\frac{1}{2}\sin t$. ∎

Nothing is so obvious as the solutions given in Examples 3 and 4, when pointed out by someone else. When one is on one's own, such solutions do not always jump out. The rest of this section will provide more constructive help.

Method of Undetermined Coefficients

The name of this method is unduly pessimistic, since after the method is applied, the coefficients *are* determined! Working from the undetermined to the determined is, after all, what problem solving is about.

There are a few limitations to the method: it works only for linear DEs with constant coefficients and certain types of forcing terms.

1. Must be linear DE
2. Certain forcing terms

> **Forcing Terms for the Method of Undetermined Coefficients**
>
> *know*
> - Polynomials in t.
> - Exponentials e^{at}.
> - Sinusoidal functions of the form $\cos kt$ and $\sin kt$.
> - Any finite products or sums of these families of functions.

Nevertheless, undetermined coefficients is probably the most frequently used method for finding particular solutions.

Don't forget to add homogeneous solution.

The method of undetermined coefficients is based on the fact that certain important sets of functions are closed under operators like L, where

$$L(y) = ay'' + by' + cy. \tag{5}$$

Consider the vector space $\mathbb{P}_n$ of polynomials of degree n or less that we studied in Sec. 3.5. If $y \in \mathbb{P}_n$, then $L(y) \in \mathbb{P}_n$, since we see from (5) that we have simply differentiated y, multiplied by constants, or added such objects together. This suggests that the nonhomogeneous differential equation

$$L(y) = f$$

might be expected to have a particular solution in $\mathbb{P}_n$ if $f \in \mathbb{P}_n$. We write down such a polynomial with *undetermined* coefficients, substitute it into the differential equation, and proceed to *determine* those coefficients!

EXAMPLE 5 **Along Came Poly** Let us find a particular solution for the problem

$$y'' - y' - 2y = 3t^2 - 1. \tag{6}$$

Our preceding observation suggests looking for our $y_p \in \mathbb{P}_2$, so we let

$$y_p = At^2 + Bt + C.$$

Then we calculate

$$y_p' = 2At + B \quad \text{and} \quad y_p'' = 2A,$$

and substitute these into (6):

$$2A - (2At + B) - 2(At^2 + Bt + C) = 3t^2 - 1.$$

By expanding and then collecting terms for each power of t on the left-hand side, we can express $L(y_p)$ as a member of $\mathbb{P}_2$ and compare it to the polynomial on the right-hand side:

$$(-2A)t^2 + (-2A - 2B)t + (2A - B - 2C) = 3t^2 - 1.$$

If these expressions represent the same element of $\mathbb{P}_2$, corresponding coefficients must be the same, namely

$$-2A = 3, \quad -2A - 2B = 0, \quad \text{and} \quad 2A - B - 2C = -1.$$

Solving from left to right gives first $A = -3/2$, then $B = -A = 3/2$; from $-3 - 3/2 - 2C = -1$ we get $C = -7/4$. Therefore,

$$y_p = -\frac{3}{2}t^2 + \frac{3}{2}t - \frac{7}{4}.$$

Since the homogeneous equation corresponding to (6) has characteristic equation $r^2 - r - 2 = (r - 2)(r + 1) = 0$, the general solution of (6) is

$$y = c_1e^{2t} + c_2e^{-t} - \frac{3}{2}t^2 + \frac{3}{2}t - \frac{7}{4}.$$

■

More Clannish Functions

what does this actually mean?

Another family of functions that is closed under differentiation, addition, and multiplication by constants (hence under the operator L of equation (5)) is the set of functions Ae^{kt}, where k is fixed and A is undetermined. Let us see how this works in an example.

EXAMPLE 6 **The Exp Family** We will look for a particular solution for

$$y'' - y' - 2y = 2e^{-3t} \tag{7}$$

in the form

$$y_p = Ae^{-3t}.$$

We calculate

$$y_p' = -3Ae^{-3t} \quad \text{and} \quad y_p'' = 9Ae^{-3t},$$

then substitute into equation (7):

$$9Ae^{-3t} + 3Ae^{-3t} - 2Ae^{-3t} = 2e^{-3t}.$$

Simplifying, $10Ae^{-3t} = 2e^{-3t}$, so $10A = 2$ and $A = 1/5$. The particular solution for (7) is

$$y_p = 1/5\, e^{-3t}.$$

■

A third family of functions with ingrown behavior consists of expressions of the form $A\cos kt + B\sin kt$, where k is fixed while A and B are undetermined. Differentiating, adding, and multiplying such expressions by a constant always leads to another such expression.

EXAMPLE 7 **The Trig Family** To find a particu

$$y'' - y' - 2y = 2\cos$$

we make the educated guess

$$y_p = A\cos 3t + B\,s$$

We cannot just use $y_p = A\cos 3t$, because d
us with "sin $3t$" terms as well. If we compute

$$y_p' = -3A\sin 3t + 3B\cos 3t \quad \text{and} \quad y_p'' = -9A\cos 3t$$

then substitute into (8), after simplification we will have

$$(-11A - 3B)\cos 3t + (3A - 11B)\sin 3t = 2\cos 3t. \tag{9}$$

Since the coefficient of $\sin 3t$ on the right-hand side of (9) is zero, we have two equations to determine A and B:

$$-11A - 3B = 2 \quad \text{and} \quad 3A - 11B = 0.$$

?

The solution of this system is $A = -11/65$ and $B = -3/65$, so

$$y_p = -\frac{11}{65}\cos 3t - \frac{3}{65}\sin 3t.$$

■

Mixing Families

If two families are mixed, we create a guess by combining both. This doesn't always work because sometimes you incorporate the general solution into your current solution such that the solutions cease to be LI.

The plot thickens when we mix these three types, but the combinations do not lead us any further than the trio of families discussed so far. If the forcing term is $3t^2e^t$, for example, we take our particular solution to have the form

$$y_p = (At^2 + Bt + C)e^t.$$

For a forcing term like $e^t \sin 2t$ we will use

$$y_p = e^t(A\cos 2t + B\sin 2t),$$

while a combination like $t\sin t$ as the driving function requires

$$y_p = (At + B)\cos t + (Ct + D)\sin t.$$

EXAMPLE 8 **Cross-Breeding** Following the preceding suggestion, we will look for a particular solution of the differential equation

$$y'' - y' - 2y = t^2e^t \tag{10}$$

$r^2 - r - 2 = 0$

in the form

$$y_p = (At^2 + Bt + C)e^t.$$

We get this big, long equation and then, because we are just looking for one particular solution, we set the values so that it works.

Differentiating, we obtain

$$y_p' = e^t[At^2 + (2A + B)t + (B + C)]$$

and

$$y_p'' = e^t[At^2 + (4A + B)t + (2A + 2B + C)].$$

Substituting into (10) and simplifying gives

$$e^t[t^2(-2A) + t(2A - 2B) + (2A + B - 2C)] = t^2e^t.$$

We interpret the right-hand side of this equation as $e^t[1 \cdot t^2 + 0 \cdot t + 0]$ and equate the coefficients of like terms:

$$-2A = 1, \quad 2A - 2B = 0, \quad 2A + B - 2C = 0.$$

Then $A = -1/2$, $B = -1/2$, and $C = -3/4$, giving us

$$y_p = e^t \left[-\frac{1}{2}t^2 - \frac{1}{2}t - \frac{3}{4} \right].$$

The general solution of (10) by the nonhomogeneous principle is

$$y = c_1 e^{2t} + c_2 e^{-t} + e^t \left[-\frac{1}{2}t^2 - \frac{1}{2}t - \frac{3}{4} \right]. \quad ■$$

Initial values are only required once you begin with homogeneous solution.

Serpent in Paradise

There is a difficulty in the procedure (as we have outlined it so far) if a term in our proposed y_p duplicates a term in the solution of the homogeneous equation. Suppose, for example, that we are solving

$$y'' - y' - 2y = 5e^{2t}. \tag{11}$$

1. Find homogenous using CE
2. Find particular using undetermined coefficients.

We already found in Example 5 that $y_h = c_1 e^{2t} + c_2 e^{-t}$. why

If we now try

$$y_p = Ae^{2t},$$

then $L(y_p) = 0$, because this is a solution of the homogeneous equation with $c_1 = A$ and $c_2 = 0$. More explicitly,

Review all of this!

$$y_p' = 2Ae^{2t} \quad \text{and} \quad y_p'' = 4Ae^{2t},$$

and substituting into (11) gives

$$y_p'' - y_p' - 2y_p = 4Ae^{2t} - 2Ae^{2t} - 2Ae^{2t} = 5e^{2t},$$

that is, $0 = 5e^{2t}$, which is contradictory.

The way out of this difficulty is similar to the strategy we employed for repeated characteristic roots in Sec. 4.2: we need to multiply by t. If we use

This solution ensures Linear Independence while still being solvable.

$$y_p = Ate^{2t}, \quad \text{why?}$$

a particular solution will emerge:

$$y_p' = e^{2t}(2At + A) \quad \text{and} \quad y_p'' = e^{2t}(4At + 4A).$$

Substituting into (11) now gives $3Ae^{2t} = 5e^{2t}$, so $A = 5/3$ and

$$y_p = \frac{5}{3}te^{2t}.$$

The general solution of (11) is

$$y = c_1 e^{2t} + c_2 e^{-t} + \frac{5}{3}te^{2t}.$$

EXAMPLE 9 **Serpent Bites Twice** In trying to solve the nonh

$$y'' - 2y' + y = 3e^t,$$

a student uses $y_p = Ae^t$, but then $y_p' = y_p'' = Ae^t$, and (12) reduces to 0 =
"Aha! I need to multiply by t," reasons the student. So our scholar now tak
$y_p = Ate^t$, which gives

$$y_p' = Ae^t + Ate^t \quad \text{and} \quad y_p'' = 2Ae^t + Ate^t,$$

and again (12) reduces to $0 = 3e^t$. "But you said to multiply by t," complains the student.

Well, so we did; but in *this* case it is necessary to multiply by t^2. Using

$$y_p = At^2e^t$$

gives

$$y_p' = 2Ate^t + At^2e^t \quad \text{and} \quad y_p'' = 2Ae^t + 4Ate^t + At^2e^t,$$

and now (12) reduces to $2Ae^t = 3e^t$, from which $A = 3/2$ and

$$y_p = \frac{3}{2}t^2e^t.$$

Why do we need t^2 instead of t? Let us look at the homogeneous equation $y'' - 2y' + y = 0$ with characteristic equation $r^2 - 2r + 1 = (r-1)^2 = 0$. The repeated characteristic root gives us

$$y_h = c_1e^t + c_2te^t,$$

so the first two attempts duplicated homogeneous solution terms. Only with the multiplier t^2 can we avoid this. ■

r² − 2r + 1 = ∅
(r − 1)(r − 1)
r = 1
eᵗ + teᵗ
for CE:
y'' − 2y' + y = 3eᵗ
yp = At²eᵗ
yp' = 2Ateᵗ + At²eᵗ
yp'' = 2Aeᵗ + 2Ateᵗ + 2Aeᵗ + At²eᵗ
= 4Aeᵗ + 2Ateᵗ + At²eᵗ
Solve for an A that works.
Why can we get away w/multiplying by t?

Always check the homogeneous solution first. What we must conclude from these examples is that we cannot proceed with the method of undetermined coefficients until we know what the homogeneous solution looks like. Our proposed particular solution y_p must take into account the form of y_h.

Since this method depends on making a wise prediction for y_p before solving for its coefficients, some authors refer to it as the **method of judicious guessing**. Part of being judicious is to check the homogeneous solution first. Table 4.4.1 is an attempt to reduce this guessing to reliable prediction.

Predicted Forms of y_p for the Method of Undetermined Coefficients

For a second-order linear DE

$$ay'' + by' + cy = f(t),$$

the method of undetermined coefficients uses the form of $f(t)$ to predict the form of $y_p(t)$, as shown in Table 4.4.1, on the next page.

Table 4.4.1 Predicting forms of particular solutions

Forcing Function $f(t)$	$\Rightarrow$	Particular Solution $y_p(t)$
(i) k		A_0
(ii) $P_n(t)$		$A_n(t)$
(iii) Ce^{kt}		A_0e^{kt}
(iv) $C\cos\omega t + D\sin\omega t$		$A_0\cos\omega t + B_0\sin\omega t$
(v) $P_n(t)e^{kt}$		$A_n(t)e^{kt}$
(vi) $P_n(t)\cos\omega t + Q_n(t)\sin\omega t$		$A_n(t)\cos\omega t + B_n(t)\sin\omega t$
(vii) $Ce^{kt}\cos\omega t + De^{kt}\sin\omega t$		$A_0e^{kt}\cos\omega t + B_0e^{kt}\sin\omega t$
(viii) $P_n(t)e^{kt}\cos\omega t + Q_n(t)e^{kt}\sin\omega t$		$A_n(t)e^{kt}\cos\omega t + B_n(t)e^{kt}\sin\omega t$

From a DE, 1) get CE., 2) find Roots; 3) find homogenous solution 4) find a particular solution which is independent of the homogenous solution.

- $P_n(t), Q_n(t), A_n(t), B_n(t) \in \mathbb{P}_n$ (hence $A_0, B_0 \in \mathbb{P}_0 = \mathbb{R}$), and k, ω, C, and D are real constants.
- In (iv) and (vi)–(viii), both terms must be included in y_p, even if only one of the terms is present in $f(t)$.

If any term or terms of y_p are found in y_h (i.e., if such terms are solutions of $ay'' + by' + cy = 0$), multiply the expression for y_p by t (or, if necessary, by t^2) to eliminate the duplication.

EXAMPLE 10 **Judicious Guessing** We will use our heads (or Table 4.4.1) to determine the correct form for the particular solution of

$$y'' + 2y' - 3y = f(t) \tag{13}$$

for each given forcing function $f(t)$. First, we need to solve the homogeneous companion equation $y'' + 2y' - 3y = 0$, which has the characteristic equation $r^2 + 2r - 3 = 0$. So

$$\{e^t, e^{-3t}\}$$

is a fundamental set of solutions to the associated homogeneous equation.

To find the particular solution y_p for the various forcing functions in (a)–(e), we can make judicious guesses using Table 4.4.1.

(a) $f(t) = t^2 + t - 3 \quad \Rightarrow y_p(t) = A_2t^2 + A_1t + A_0$

(b) $f(t) = e^{-t} \quad \Rightarrow y_p(t) = A_0e^{-t}$

(c) $f(t) = te^t \quad \Rightarrow y_p(t) = t(A_1t + A_0)e^t$

(d) $f(t) = 2t\cos 3t + \sin 3t$

$\quad \Rightarrow y_p(t) = (A_1t + A_0)\cos 3t + (B_1t + B_0)\sin 3t$

(e) $f(t) = te^{-2t}\sin t \quad \Rightarrow y_p(t) = [(A_1t+A_0)\cos t+(B_1t+B_0)\sin t]e^{-2t}$

We then determine the coefficients and write the general solution in each case as

$$y = c_1e^t + c_2e^{-3t} + y_p(t).$$

■

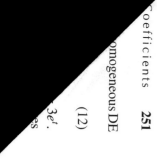

Summary

By suitably predicting the form of the particular solution of a nonhomogeneous problem whose forcing term is built from polynomials, exponentials, sines, and cosines, and which involves undetermined coefficients, we can determine those coefficients and calculate the particular solution explicitly. Correct prediction depends on knowing the general solution of the corresponding homogeneous problem. The same method can be extended to higher-order forced linear DEs with constant coefficients.

4.4 Problems

Inspection First *In Problems 1–8, determine a particular solution by inspecting the nonhomogeneous differential equations. If a particular solution is not obvious to you, use the method of undetermined coefficients.*

1. $y'' - y = t$
2. $y'' + y' = 2$
3. $y'' = 2$
4. $ty'' + y' = 4t$
5. $y'' - 2y' + 2y = 4$
6. $y'' - y = -2\cos t$
7. $y'' - y' + y = e^t$
8. $y''' + y' + y = 2t + 2$

Educated Prediction *You are given the nonhomogeneous differential equation*

$$y'' + 2y' + 5y = f(t).$$

Predict the form of y_p for the $f(t)$ given in Problems 9–12, remembering to use y_h as you set up y_p for undetermined coefficients. (You need not evaluate the coefficients.)

9. $f(t) = 2t^3 - 3t$
10. $f(t) = te^t$
11. $f(t) = 2\sin t$
12. $f(t) = 3e^{-t}\sin t$

Guess Again *Now you are given the nonhomogeneous differential equation*

$$y'' - 6y' + 9y = f(t).$$

Predict the form of y_p for the $f(t)$ given in Problems 13–16, remembering again to consider y_h. (You need not evaluate the coefficients.)

13. $f(t) = t\cos 2t$
14. $f(t) = te^{3t}$
15. $f(t) = e^{-t} + \sin t$
16. $f(t) = t^4 - t^2 + 1$

Determining the Undetermined *In Problems 17–40, obtain the general solution of the DE. If you cannot find y_p by inspection, use the method of undetermined coefficients.*

17. $y' = 1$
18. $y' + y = 1$
19. $y' + y = t$
20. $y'' = 1$
21. $y'' + 4y' = 1$
22. $y'' + 4y = 1$
23. $y'' + 4y' = t$
24. $y'' + y' - 2y = 3 - 6t$
25. $y'' + y = e^t + 3$
26. $y'' - y' - 2y = 6e^t$
27. $y'' + y' = 6\sin 2t$
28. $y'' + 4y' + 5y = 2e^t$
29. $y'' + 4y' + 4y = te^{-t}$
30. $y'' - y = t\sin t$
31. $y'' + y = 12\cos^2 t$
32. $y'' - y = 8te^t$
33. $y'' - 4y' + 4y = te^{2t}$
34. $y'' - 4y' + 3y = 20\cos t$
35. $y'' - 3y' + 2y = e^t \sin t$
36. $y'' + 3y' = \sin t + 2\cos t$
37. $y''' - 4y'' = 6t$
38. $y''' - 3y'' + 3y' - y = e^t$
39. $y^{(4)} - y = 10$
40. $y''' = y''$

Initial-Value Problems *Determine the solutions of the IVPs of Problems 41–52.*

41. $y'' + y' - 2y = 3 - 6t, \quad y(0) = -1, \quad y'(0) = 0$
42. $y'' + 4y' + 4y = te^{-t}, \quad y(0) = -1, \quad y'(0) = 1$
43. $y'' + 4y = t, \quad y(0) = 1, \quad y'(0) = -1$
44. $y'' + 2y' + y = 6\cos t, \quad y(0) = 1, \quad y'(0) = -1$
45. $4y'' + y = \cos 2t, \quad y(0) = 1, \quad y'(0) = 0$
46. $y'' + 9y = \cos 3t, \quad y(0) = 1, \quad y'(0) = -1$
47. $y'' - 3y' + 2y = 4e^{-t}, \quad y(0) = 1, \quad y'(0) = 0$
48. $y'' - 4y' + 3y = e^{-t} + t, \quad y(0) = 0, \quad y'(0) = 0$

49. $y'' - y' - 2y = 4\cos 2t, \quad y(0) = 0, \quad y'(0) = 0$

50. $y''' - 4y'' + 3y' = t^2,$
$y(0) = 1,\ y'(0) = 0,\ y''(0) = 0$

51. $y^{(4)} - y = e^{2t}, \quad y(0) = y'(0) = y''(0) = y'''(0) = 0$

52. $y^{(4)} = e^t, \quad y(0) = 1,\ y'(0) = y''(0) = y'''(0) = 0$

Trial Solutions *Use the method of undetermined coefficients to set up the particular solutions y_p in terms of A, B, C, …, but do* not *solve for the coefficients. (Remember that you need to find y_h first to make allowances for duplication.)*

53. $4y'' + y = t - \cos\left(\frac{t}{2}\right)$

54. $y''' - y'' = t^2 + e^t$

55. $y'' - 5y' + 6y = \cos t - te^t$

56. $y^{(4)} - y = te^t + \sin t$

57. Judicious Superposition

(a) Solve the homogeneous differential equation

$$y'' - y' - 6y = 0.$$

(b) Use undetermined coefficients to solve these nonhomogeneous problems:

(i) $y'' - y' - 6y = e^t$

(ii) $y'' - y' - 6y = e^{-t}$

(c) Recall that

$$\cosh t = \frac{1}{2}(e^t + e^{-t}) \quad \text{and} \quad \sinh t = \frac{1}{2}(e^t - e^{-t}).$$

Use this, the result from (b), and the Superposition Principle to solve this problem:

$$y'' - y' - 6y = \cosh t.$$

58. Wholesale Superposition Solve the equation $y' + y = e^t$ by substituting the power series for e^t,

$$e^t = 1 + t + \frac{t^2}{2!} + \frac{t^3}{3!} + \frac{t^4}{4!} + \cdots,$$

and solving the nonhomogeneous equation $y' + y = t^n/n!$ for values $n = 0, 1, 2, \ldots$. Then use the Superposition Principle to verify that the solution of the original equation is $\frac{1}{2}e^t$.

Discontinuous Forcing Functions *Solve the IVPs of Problems 59 and 60, in which the forcing function is discontinuous. You can do this by solving each problem on the indicated intervals and making certain that the solutions "match up"* smoothly *at the interval boundary (i.e., both y and y' must be continuous there).*

59. $y'' + y' = \begin{cases} 2 & \text{if } 0 \le t, \\ 1 & \text{if } t > 4, \end{cases} \quad y(0) = y'(0) = 0$

60. $y'' + 16y = \begin{cases} \cos t & \text{if } 0 \le t \le \pi, \\ 0 & \text{if } t > \pi, \end{cases} \quad y(0) = 1,\ y'(0) = 0$

Solutions of Differential Equations Using Complex Functions *There is a nice way to solve linear nonhomogeneous equations with constant coefficients whose right-hand sides consist of sine or cosine functions:*

$$ay'' + by' + cy = \begin{cases} R\cos \omega t, \\ R\sin \omega t. \end{cases} \tag{14}$$

The idea is to solve a modified equation in which we replace the sine or cosine term with the complex exponential $e^{i\omega t}$,

$$ay'' + by' + cy = Re^{i\omega t}, \tag{15}$$

where the complex constant $i = \sqrt{-1}$ is treated as any real constant. The important fact is that the real solution of (15) is the solution of (14) with a cosine on the right-hand side, and the imaginary solution of (15) is the solution of (14) with a sine on the right-hand side. Use this idea to solve Problems 61–63.

61. $y'' - 2y' + y = 2\sin t$

62. $y'' + 25y = 6\sin t$

63. $y'' + 25y = 20\sin 5t$

64. Complex Exponents Solve the differential equation

$$y'' - 3y' + 2y = 3e^{2it},$$

and verify that the real and complex parts of the solution are the solutions obtained when the right-hand side of the equation is replaced by $3\cos 2t$ and $3\sin 2t$, respectively.

65. Suggested Journal Entry Discuss whether you have seen any possible forcing functions in your physics or engineering class that would not be covered by the method of undetermined coefficients. Give one or two examples of forcing functions to which the method would be applicable.

4.5 Variation of Parameters

SYNOPSIS: The method of variation of parameters provides an alternative approach to determining the particular solution of a nonhomogeneous problem. It is not restricted to the case of constant coefficients and includes a broader class of forcing functions.

Introduction

In Sec. 2.2 we found a particular solution of the first-order nonhomogeneous equation $y' + p(t)y = f(t)$ using variation of parameters.[1] We now show how this method can be extended to finding a particular solution of the second-order nonhomogeneous equation

$$y'' + p(t)y' + q(t)y = f(t), \tag{1}$$

where $p(t)$, $q(t)$, and $f(t)$ are continuous functions. In addition to allowing variable coefficients $p(t)$ and $q(t)$, the nonhomogeneous term $f(t)$ is not restricted the way it was for the method of undetermined coefficients.

To apply **variation of parameters**, we first find two linearly independent solutions $y_1(t)$ and $y_2(t)$ of the corresponding homogeneous equation

$$y'' + p(t)y' + q(t)y = 0, \tag{2}$$

thus having the general solution[2]

$$y_h(t) = c_1y_1(t) + c_2y_2(t), \tag{3}$$

where c_1 and c_2 are arbitrary constants.

Variation of parameters might be thought of as a perturbation method, whereby we find the solution of the perturbed (nonhomogeneous) equation (1) by perturbing the solution (3) of the homogeneous equation. We do this by replacing the constants c_1 and c_2 with functions $v_1(t)$ and $v_2(t)$. We seek a particular solution of the form

$$y_p(t) = v_1(t)y_1(t) + v_2(t)y_2(t), \tag{4}$$

where $v_1(t)$ and $v_2(t)$ are unknown functions determined so that y_p satisfies the nonhomogeneous equation (1).

We find v_1 and v_2 by substituting (4) into the nonhomogeneous equation (1). We wish to obtain two equations in two unknowns, but we have only the condition $L(y_p) = f$. We must choose an auxiliary equation. Differentiating (4) by the product rule, we get

$$y_p' = v_1y_1' + v_2y_2' + v_1'y_1 + v_2'y_2. \tag{5}$$

Before calculating y_p'', we choose an auxiliary condition: that v_1 and v_2 satisfy

$$v_1'y_1 + v_2'y_2 = 0, \tag{6}$$

which reduces (5) to

$$y_p' = v_1y_1' + v_2y_2'. \tag{7}$$

Auxiliary Condition:

When $L(y) = f$ is a second-order DE, variation of parameters uses *two* conditions, that $v_1'y_1 + v_2'y_2 = 0$, as well as that $L(y_p) = f$.

[1] Variation of parameters is generally associated with finding particular solutions of second-order equations (or higher), although we used it in Sec. 2.2 to find particular solutions of first-order equations. First-order equations are more often solved by the integrating factor method, which finds both the homogeneous and particular solutions simultaneously, so there is no need for a special method for finding particular solutions.

[2] Generally there is no easy method for finding the general solution of a homogeneous equation with *variable* coefficients. Sometimes solutions can be found by inspection, transformations, reduction of order methods, series methods, and so on.

Differentiating again, from (7), we obtain

$$y_p'' = v_1 y_1'' + v_2 y_2'' + v_1' y_1' + v_2' y_2'. \tag{8}$$

Of course, we want y to satisfy the differential equation,

$$L(y) = y'' + py' + qy = f, \tag{9}$$

so we substitute (4), (7), and (8) into (9):

$$v_1 y_1'' + v_2 y_2'' + v_1' y_1' + v_2' y_2' + p(v_1 y_1' + v_2 y_2') + q(v_1 y_1 + v_2 y_2) = f.$$

Collecting terms, we have

$$v_1(y_1'' + py_1' + qy_1) + v_2(y_2'' + py_2' + qy_2) + (v_1' y_1' + v_2' y_2') = f. \tag{10}$$

But since $L(y_1) = 0$ and $L(y_2) = 0$, equation (10) reduces to

$$v_1' y_1' + v_2' y_2' = f. \tag{11}$$

So we are looking for functions v_1 and v_2 that satisfy both auxiliary conditions (6) and (11), guaranteeing that we have a solution of the DE. That is, we need v_1 and v_2 such that

$$\begin{aligned} y_1 v_1' + y_2 v_2' &= 0, \\ y_1' v_1' + y_2' v_2' &= f. \end{aligned} \tag{12}$$

This is a linear system in the "unknowns" v_1' and v_2'. After solving (12) for these, we will integrate to get v_1 and v_2, then substitute them into (4) to get our particular solution.

In practice, depending on the specific forms of y_1, y_2, and f, it may be most efficient to solve (12) by addition or substitution. But it is helpful to know that we *can* use Cramer's Rule from Sec. 3.4, obtaining

Cramer's Rule allows us to generalize easily to higher orders. (See Problem 17.)

$$v_1' = \frac{\begin{vmatrix} 0 & y_2 \\ f & y_2' \end{vmatrix}}{\begin{vmatrix} y_1 & y_2 \\ y_1' & y_2' \end{vmatrix}} \quad \text{and} \quad v_2' = \frac{\begin{vmatrix} y_1 & 0 \\ y_1' & f \end{vmatrix}}{\begin{vmatrix} y_1 & y_2 \\ y_1' & y_2' \end{vmatrix}}.$$

The denominator is the Wronskian $W(y_1, y_2) = y_1 y_2' - y_1' y_2$, and it is not zero because y_1 and y_2 are linearly independent. Thus we have

$$v_1' = -\frac{y_2 f}{W(y_1, y_2)} \quad \text{and} \quad v_2' = \frac{y_1 f}{W(y_1, y_2)}.$$

Integrate to solve for v_1 and v_2.

EXAMPLE 1 **Not in the Family** We will solve the nonhomogeneous DE

$$y'' + y = \sec t, \quad |t| < \frac{\pi}{2}.$$

The characteristic equation of $y'' + y = 0$ is $r^2 + 1 = 0$, so we have $r = \pm i$ and we can take $y_1 = \cos t$ and $y_2 = \sin t$. The Wronskian is

$$W(y_1, y_2) = \begin{vmatrix} \cos t & \sin t \\ -\sin t & \cos t \end{vmatrix} = \cos^2 t + \sin^2 t = 1.$$

Then

$$v_1' = -y_2 f = -\sin t \sec t = -\frac{\sin t}{\cos t} \quad \text{and} \quad v_2' = y_1 f = \cos t \sec t = 1.$$

Integrating, $v_1 = \ln(\cos t)$ and $v_2 = t$. (NOTE: $\ln|\cos t| = \ln(\cos t)$ because $|t| < \pi/2$.) Thus, by equation (4),

$$y_p = (\cos t)\ln(\cos t) + t \sin t,$$

and the general solution is given by

$$y = c_1 \cos t + c_2 \sin t + (\cos t)\ln(\cos t) + t \sin t.$$

■

We normally set to zero the constants of integration in calculating v_1 and v_2 from v_1' and v_2' because our goal is only *one particular solution*. If we replaced v_1 and v_2 by $v_1 + C_1$ and $v_2 + C_2$ in equation (4), we would obtain the general solution all at once:

$$\begin{aligned} y &= (v_1 + C_1)y_1 + (v_2 + C_2)y_2 \\ &= C_1 y_1 + C_2 y_2 + (v_1 y_1 + v_2 y_2) = y_h + y_p. \end{aligned}$$

EXAMPLE 2 **Sinusoidal Forcing** Find a particular solution for

$$y'' + y = 4 \sin t.$$

As in the previous example, $y_1 = \cos t$, $y_2 = \sin t$, and $W(y_1, y_2) = 1$. Thus $v_1' = -y_2 f = -4\sin^2 t$ and $v_2' = y_1 f = 4 \sin t \cos t$. Using double-angle formulas, we have

$$v_1' = -2(1 - \cos 2t) \quad \text{and} \quad v_2' = 2 \sin 2t.$$

Integrating,

$$v_1 = -2t + \sin 2t = -2t + 2\sin t \cos t \quad \text{and} \quad v_2 = -\cos 2t = 1 - 2\cos^2 t.$$

Therefore,

$$y_p = (-2t + 2\sin t \cos t)(\cos t) + (1 - 2\cos^2 t)(\sin t) = -2t \cos t + \sin t.$$

■

In Example 2 the "duplication" of functions in y_2 and f caused no difficulty, nor did it require any special procedure. This fact alerts us to an important difference between variation of parameters and the method of undetermined coefficients from Sec. 4.4.

The strategy in the method of undetermined coefficients is to *avoid* duplication with solutions of the homogeneous partner: we find those homogeneous solutions so we can steer away from them. The psychology of variation of parameters is just the reverse: we find those same homogeneous solutions and *use* them to construct the particular solution. The independent solutions of the homogeneous equation are stumbling blocks for undetermined coefficients, building blocks for variation of parameters.

Method of Variation of Parameters for Determining a Particular Solution y_p for $L(y) = y'' + p(t)y' + q(t)y = f(t)$

Step 1. Determine two linearly independent solutions y_1 and y_2 of the corresponding homogeneous equation $L(y) = 0$.

Step 2. Solve for v_1' and v_2' the system

$$\begin{aligned} y_1 v_1' + y_2 v_2' &= 0, \\ y_1' v_1' + y_2' v_2' &= f, \end{aligned}$$

or determine v_1' and v_2' from Cramer's Rule,

$$v_1' = \frac{\begin{vmatrix} 0 & y_2 \\ f & y_2' \end{vmatrix}}{\begin{vmatrix} y_1 & y_2 \\ y_1' & y_2' \end{vmatrix}} = \frac{-y_2 f}{W(y_1, y_2)}, \quad v_2' = \frac{\begin{vmatrix} y_1 & 0 \\ y_1' & f \end{vmatrix}}{\begin{vmatrix} y_1 & y_2 \\ y_1' & y_2' \end{vmatrix}} = \frac{y_1 f}{W(y_1, y_2)}, \tag{13}$$

where $W(y_1, y_2) = y_1 y_2' - y_1' y_2$ is the Wronskian.

Step 3. Integrate the results of Step 2 to find v_1 and v_2.

Step 4. Compute $y_p = v_1 y_1 + v_2 y_2$.

NOTE: The derivation is for an operator L having coefficient 1 for y'', so equations having a leading coefficient different from 1 must be divided by that coefficient in order to determine the standard form for $f(t)$.

This method extends to higher-dimensional systems. (See Problems 17–21.)

The Advantages of the Method

Variation of parameters supplements the method of undetermined coefficients in two significant ways. First, we can handle forcing terms for which undetermined coefficients does not apply, even if the coefficients for the DE are constant. Examples 1 and 2 illustrate this situation. The second advantage is that the method applies when the coefficients of y and y' are functions of t, provided that we can work out the homogeneous solution. (See Example 4.)

EXAMPLE 3 **An Unusual Forcing Term** To solve the nonhomogeneous problem

$$y'' - 2y' + y = \frac{e^t}{1+t^2},$$

we first find that the corresponding homogeneous equation has repeated characteristic root 1 and so we take $y_1 = e^t$ and $y_2 = te^t$. Then

$$W(y_1, y_2) = \begin{vmatrix} e^t & te^t \\ e^t & (t+1)e^t \end{vmatrix} = (t+1)e^{2t} - te^{2t} = e^{2t}.$$

Using equations (13), then,

$$v_1' = -\frac{y_2 f}{W} = -\frac{t}{1+t^2} \quad \text{and} \quad v_2' = \frac{y_1 f}{W} = \frac{1}{1+t^2}.$$

Integrating, we find

$$v_1 = -\frac{1}{2}\int \frac{2t\,dt}{1+t^2} = -\frac{1}{2}\ln(1+t^2) \quad \text{and} \quad v_2 = \tan^{-1} t.$$

Therefore,

$$y_p = v_1 y_1 + v_2 y_2 = -\frac{1}{2}e^t \ln(1+t^2) + te^t \tan^{-1} t,$$

and the general solution is

$$y = e^t\left[c_1 + c_2 t - \frac{1}{2}\ln(1+t^2) + t\tan^{-1} t\right].$$

■

EXAMPLE 4 **Variable Coefficients** In order to solve

$$t^2y'' - 2ty' + 2y = t\ln t, \quad t > 0, \tag{14}$$

we will first verify[3] that $y_1 = t$ and $y_2 = t^2$ are linearly independent solutions of the homogeneous equation

$$t^2y'' - 2ty' + 2y = 0. \tag{15}$$

For $y_1 = t$, $y_1' = 1$, and $y_1'' = 0$, equation (15) is satisfied. When we substitute $y_2 = t^2$, $y_2' = 2t$, and $y_2'' = 2$ into (15), the result is the same. These solutions are linearly independent because the Wronskian

$$W(y_1, y_2) = \begin{vmatrix} t & t^2 \\ 1 & 2t \end{vmatrix} = t^2,$$

which is nonzero on the domain $t > 0$. Equation (14) must be divided by t^2 so that L will have the form of (2):

$$y'' - \frac{2}{t}y' + \frac{2}{t^2}y = \frac{\ln t}{t}. \tag{16}$$

From (16) we identify $f(t) = (\ln t)/t$, so equations (13) give

$$v_1' = -\frac{y_2 f}{W} = -\frac{\ln t}{t} \quad \text{and} \quad v_2' = \frac{y_1 f}{W} = \frac{\ln t}{t^2}.$$

Integrating (using integration by parts for v_2) yields

$$v_1 = -\frac{1}{2}(\ln t)^2 \quad \text{and} \quad v_2 = -\frac{\ln t}{t} - \frac{1}{t}.$$

Therefore, from (4) we have

$$y_p = -\frac{t}{2}(\ln t)^2 - t\ln t - t;$$

the solution of (14) is

$$y = c_1 t + c_2 t^2 - \frac{t}{2}(\ln t)^2 - t\ln t - t.$$

■

Historical Note

As we first noted in Sec. 2.2 for first-order DEs, but have now extended to higher-dimensional systems, the two great eighteenth-century mathematicians **Leonhard Euler** (1707–1783) and **Joseph Louis Lagrange** (1736–1813) are credited with developing these techniques.

Euler solved the homogeneous second-order equation with constant coefficients $ay'' + by' + cy = 0$, and went on to solve related nonhomogeneous equations using the *method of undetermined coefficients*. It was Lagrange who found solutions to the nonhomogeneous equation using *variation of parameters*.

---

[3]The homogeneous solutions of equation (14) can be obtained by the Euler-Cauchy Method outlined in Problem 60 in Sec. 4.2.

Summary

Variation of parameters is a powerful method for obtaining particular solutions of nonhomogeneous problems, allowing variable coefficients and more general forcing functions. Its use in the variable coefficient case, however, is limited by the difficulty of obtaining quantitative formulas for the solution of the corresponding homogeneous problem.

4.5 Problems

Straight Stuff *Use variation of parameters to obtain a particular solution for the DEs in Problems 1–12; then determine the general solution.* NOTE: *In the event* $v_i = \int v_i' dt$ *is not integrable, use a dummy variable s and represent* v_i *as* $\int_{t_0}^{t} v_i'(s)ds$. *(Assume appropriate domains for t.)*

1. $y'' + y' = 4t$
2. $y'' - y' = e^{-t}$
3. $y'' - 2y' + y = \frac{1}{t}e^t$
4. $y'' + y = \csc t$
5. $y'' + y = \sec t \tan t$
6. $y'' - 2y' + 2y = e^t \sin t$
7. $y'' - 3y' + 2y = \frac{1}{1 + e^{-t}}$
8. $y'' + 2y' + y = e^{-t} \ln t$
9. $y'' + 4y = \tan 2t$
10. $y'' + 5y' + 6y = \cos e^t$
11. $y'' + y = \sec^2 t$
12. $y'' - y = \frac{e^t}{t}$

Variable Coefficients *Verify that the given* y_1 *and* y_2 *is a fundamental set of solutions for the homogeneous equation corresponding to the nonhomogeneous DE in Problems 13–16. Calculate a particular solution using variation of parameters and determine the general solution. (Assume appropriate domains for t.)*

13. $t^2y'' - 2ty' + 2y = t^3 \sin t$, $\quad y_1(t) = t,\ y_2(t) = t^2$
14. $t^2y'' + ty' - 4y = t^2(1 + t^2)$, $\quad y_1(t) = t^2,\ y_2(t) = t^{-2}$
15. $(1 - t)y'' + ty' - y = 2(t - 1)^2e^{-t}$, $\quad y_1(t) = t,\ y_2(t) = e^t$
16. $y'' + \frac{1}{t}y' + \left(1 - \frac{1}{4t^2}\right)y = t^{-1/2}$, $\quad y_1(t) = t^{-1/2} \sin t,\ y_2(t) = t^{-1/2} \cos t$
17. **Third-Order Theory** Consider the third-order nonhomogeneous equation

$$L(y) = y''' + p(t)y'' + q(t)y' + r(t)y = f(t).$$

Suppose that y_1, y_2, and y_3 form a fundamental set of solutions for $L(y) = 0$, so that

$$y_h = c_1y_1 + c_2y_2 + c_3y_3.$$

Imitate the development for the second-order equation to obtain the particular solution

$$y_p = v_1y_1 + v_2y_2 + v_3y_3,$$

where v_1', v_2', and v_3' must satisfy the system of equations

$$y_1v_1' + y_2v_2' + y_3v_3' = 0,$$
$$y_1'v_1' + y_2'v_2' + y_3'v_3' = 0,$$
$$y_1''v_1' + y_2''v_2' + y_3''v_3' = f.$$

Recall that, by Cramer's Rule (Sec. 3.4),

$$v_1' = \frac{W_1}{W},\quad v_2' = \frac{W_2}{W},\quad v_3' = \frac{W_3}{W},$$

where W is the Wronskian, and W_j is the Wronskian with the jth column replaced by

$$\begin{bmatrix} 0 \\ 0 \\ f \end{bmatrix}.$$

Third-Order DEs *Apply the method of Problem 17 to solve the third-order equations in Problems 18–20.*

18. $y''' - 2y'' - y' + 2y = e^t$
19. $y''' + y' = \sec t$
20. $y''' + 9y' = \tan 3t$

21. Method Choice Consider the third-order DE

$$y''' - y' = f(t),$$

and the relevance of $f(t)$ to the ease of solution. Solve each of the cases (a), (b), and (c) by whatever method you like. When is variation of parameters the best choice? When will you have to settle for an integral expression for $y(t)$ rather than an explicit formula?

(a) $f(t) = 2e^{-t}$ (b) $f(t) = \sin^2 t$ (c) $f(t) = \tan t$

22. Green's Function Representation Use variation of parameters to show that a particular solution of

$$y'' + y = f(t)$$

can be written in the form

$$y_p(t) = \int_0^t \sin(t - s) f(s) ds.$$

The function $\sin(t - s)$ is the *Green's function* for the differential equation.[4] HINT: Write equations (13) in integral form as

$$v_1(t) = \int_0^t -\frac{y_2(s) f(s)}{W(s)} ds$$

and

$$v_2(t) = \int_0^t \frac{y_1(s) f(s)}{W(s)} ds,$$

where $W(s)$ is the Wronskian, and substitute into equation (4).

23. Green Variation Use a calculation similar to that of Problem 22 to write a particular solution for

$$y'' - y = f(t)$$

using a Green's function:

$$y_p(t) = \int_0^t \sinh(t - s) f(s)\, ds.$$

24. Green's Follow-Up If you have studied differentiation under the integral sign in multivariable calculus, verify directly that the integral representations in Problems 22 and 23 satisfy the nonhomogeneous differential equation.

25. Suggested Journal Entry Discuss the further generalization of the method of variation of parameters, outlined in Problem 17 for order three, to orders four and higher. Comment on when this is a worthwhile avenue to pursue, and tell what you would try if it is not.

26. Suggested Journal Entry II Some professors prefer to teach only variation of parameters, because it works in all cases (provided that certain integrals exist). Others teach only the method of undetermined coefficients because it works for many applications. What do you think is the best course of action, and why?

4.6 Forced Oscillations

SYNOPSIS: We look specifically at forced vibrations of a harmonic oscillator, noting such phenomena as beats, resonance, and stability. Then we take a first look at the complications that arise in a forced damped oscillator.

Introduction

At this point we have the techniques to solve the DE for the mass-spring system (Fig. 4.6.1) for a variety of common forcing functions. Sinusoidal forcing functions (that fit the criteria of Sec. 4.4) have many applications in mechanical systems and elsewhere, such as in the analogous *LRC*-circuits.

[4]George Green (1793–1841) was a gifted son of an English baker. Despite only a year of formal schooling from age 8 to 9, Green learned a great deal of mathematics quite on his own, working on the top floor of the mill. In 1828 he published, through the Nottingham Subscription Library, "An Essay on the Application of Mathematical Analysis to the Theories of Electricity and Magnetism," concerning potential functions connecting volume and surface integrals. Although most readers could not understand such a specialized topic, one gentleman, Sir Edward Bromhead, recognized Green's talent. Through Bromhead's encouragement Green left the mill at age 40 to enter Cambridge as an undergraduate (in the same class as James Sylvester, acknowledged in Sec. 3.1). Green continued working in mathematics, but only after his death did others finally recognize the immense importance of his contributions.

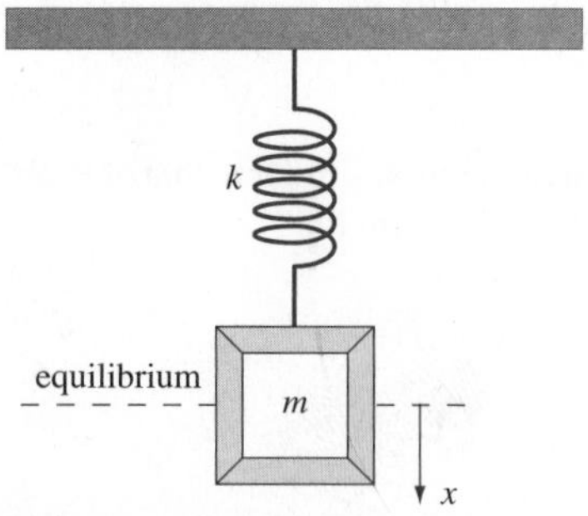

FIGURE 4.6.1 Mass-spring system, to which we can add external forcing.

Consider the following initial-value problem that models the now familiar mass-spring system (Sec. 4.1) with forcing:

$$m\ddot{x} + b\dot{x} + kx = F_0 \cos \omega_f t, \qquad x(0) = x_0, \quad \dot{x}(0) = v_0. \tag{1}$$

This DE has constant coefficients and one of the forcing functions that allows us to use the method of undetermined coefficients from Sec. 4.4.

Once we have found the values of the parameters m, b, k, F_0, and ω_f, the problem is of a type we can solve, with solutions of the form

$$x(t) = x_h + x_p.$$

EXAMPLE 1 **No Damping** Let us examine the case when $m = 1$ kg, $b = 0$, and $k = 1$ nt/m, and look at the IVP

$$\ddot{x} + x = 2\cos 3t, \qquad x(0) = 0, \quad \dot{x}(0) = 0. \tag{2}$$

We already know from Sec. 4.1 that

$$x_h = c_1 \cos t + c_2 \sin t.$$

Using the method of undetermined coefficients, guess that

$$x_p = A\cos 3t + B\sin 3t.$$

By differentiating and substituting back into (2), we obtain

$$x_p = -\frac{1}{4}\cos 3t$$

and

$$x = c_1 \cos t + c_2 \sin t - \frac{1}{4}\cos 3t,$$

so $x(0) = c_1 - \frac{1}{4} = 0$ and $\dot{x}(0) = c_2 = 0$. Consequently, the solution to the IVP (2) is

$$x = \frac{1}{4}\cos t - \frac{1}{4}\cos 3t.$$

■

General Solution of the Undamped System

We can always solve the DE of equation (1) for specific values of the parameters using the techniques of Sec. 4.4. But there are two justifications for solving (1) in general: it helps us understand the roles of the various parameters; and it changes the problem from solving the DE into a problem requiring merely substitution of parameters. It is rather tedious, however, so we will not present all the steps. We restate the DE (1) without damping and with forcing frequency ω_f and forcing amplitude F_0:

$$m\ddot{x} + kx = F_0 \cos \omega_f t. \tag{3}$$

Then, from Sec. 4.1, we know that

$$x_h = c_1 \cos \omega_0 t + c_2 \sin \omega_0 t, \qquad \text{where } \omega_0 = \sqrt{\frac{k}{m}}.$$

We must solve for x_p in two separate cases, depending on whether or not ω_0 is the same as ω_f.

CASE 1 Unmatched Frequencies ($\omega_f \neq \omega_0$) By the methods of Sec. 4.4,

$$x_p = A\cos\omega_f t + B\sin\omega_f t,$$

and we find that

$$A = \frac{F_0}{m\left(\omega_0^2 - \omega_f^2\right)} \quad \text{and} \quad B = 0.$$

($B = 0$ because $\dot{x}$ does not appear in equation (3).) Therefore, the complete solution of (3) is

$$x(t) = c_1\cos\omega_0 t + c_2\sin\omega_0 t + \frac{F_0}{m\left(\omega_0^2 - \omega_f^2\right)}\cos\omega_f t, \qquad \omega_f \neq \omega_0 \tag{4}$$

where c_1 and c_2 would be determined from initial conditions.

Alternatively, we can rewrite (4) as

$$x(t) = C\cos(\omega_0 t - \delta) + \frac{F_0}{m\left(\omega_0^2 - \omega_f^2\right)}\cos\omega_f t \tag{5}$$

and use initial conditions to evaluate C and δ. (Recall from Sec. 4.3 that $C = \sqrt{c_1^2 + c_2^2}$ and $\tan\delta = c_2/c_1$.)

Equation (5) makes it clear that when $\omega_0 \neq \omega_f$, the complete solution requires adding together *two sinusoidal functions of different frequencies*. Because both functions are periodic, their amplitudes will add in a periodic manner, so that sometimes they reinforce each other and sometimes they diminish each other. The regular periodic pattern produced in that fashion results in the phenomenon called **beats**, which are visible when ω_0 and ω_f are close but not equal. (See Fig. 4.6.2.) Further analysis is given in the following subsection, after Case 2.

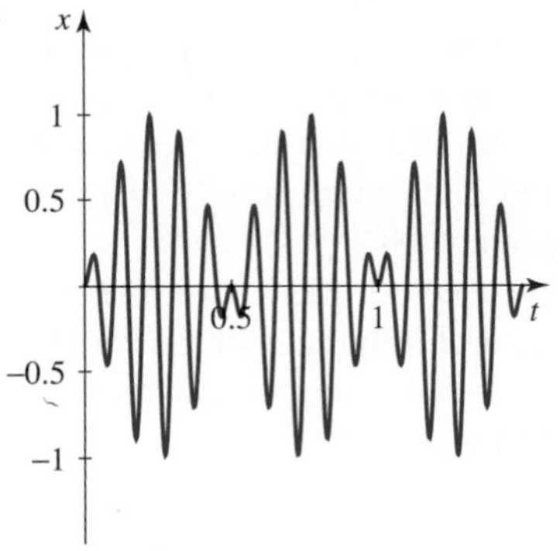

FIGURE 4.6.2 Beats: ω_f near but not equal to ω_0.

CASE 2 Resonance ($\omega_f = \omega_0$) In this case the forcing function is a solution to the homogeneous DE, so (by Sec. 4.4) we must include t as a factor of x_p,

$$x_p = At\cos\omega_0 t + Bt\sin\omega_0 t.$$

After several steps, we find that

$$A = 0 \quad \text{and} \quad B = \frac{F_0}{2m\omega_0},$$

so

$$x(t) = c_1\cos\omega_0 t + c_2\sin\omega_0 t + \frac{F_0}{2m\omega_0}t\sin\omega_0 t. \tag{6}$$

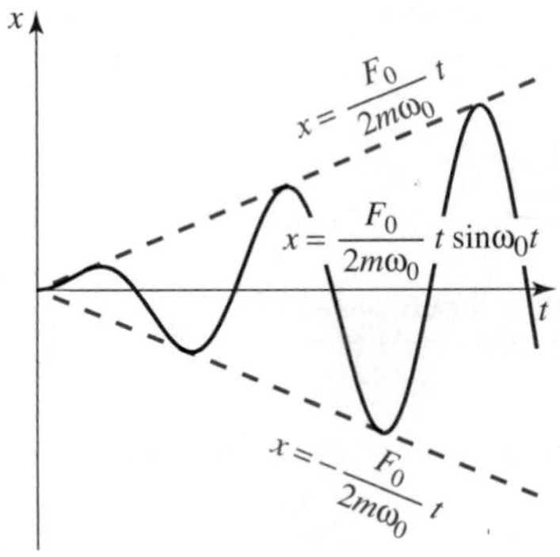

FIGURE 4.6.3 Pure resonance: $\omega_f = \omega_0$. The dashed lines form an *envelope* for the increasing amplitude.

The amplitude of the sine function in equation (6) is a linear function that grows with time t, as shown in Fig. 4.6.3. This phenomenon is called **pure resonance**. The forcing function has a frequency that exactly matches the natural frequency of the system, so the oscillations always reinforce and grow in amplitude until the system can no longer sustain them.

EXAMPLE 2 Forcing Frequency = Natural Frequency If we change the frequency of the forcing function in Example 1 to match the natural frequency,

$$\ddot{x} + x = 2\cos t, \qquad x(0) = 0, \quad \dot{x}(0) = 0, \tag{7}$$

then from (6) we can calculate

$$\frac{F_0}{2m\omega_0} = \frac{2}{2(1)(1)} = 1$$

to obtain

$$x_p = t \sin t \qquad \textit{(pure resonance)}$$

and

$$x = c_1 \cos t + c_2 \sin t + t \sin t.$$

From the initial conditions $x(0) = 0$ and $\dot{x}(0) = 0$, we obtain $c_1 = 0$ and $c_2 = 0$, so the solution to the IVP (7) is

$$x = t \sin t.$$

■

Damped Forced Vibrations

Remove the damping by setting $b = 0$. Then set $\omega\ (= \omega_f)$ at or near the natural frequency $\omega_0 = \sqrt{k/m}$ to see pure resonance or beats, respectively.

The phenomenon of resonance is common in our experience, as well as in myth and legend. A diver on a diving board increases the amplitude of the board's vibration by jumping with the board's natural frequency. Resonance occurs when microphones feed back the output of a sound system. Resonance magnifies the rattles of old cars at critical speeds. Ships roll and pitch more wildly when the waves' frequency matches the ship's natural frequency.

Soldiers crossing a bridge are ordered to break step to avoid creating a resonance between the cadence of the march and the natural frequency of the bridge.[1] And it is reputed that the walls of Jericho fell because the trumpet sounds set up waves that resonated with the natural frequency of those walls!

Analysis of Beats

As we saw in Fig. 4.6.2, the case when $\omega_f \neq \omega_0$ but ω_f is close to ω_0 in magnitude can give rise to a discernible periodic pattern that has a frequency, the beat frequency, lower than ω_0 or ω_f. Let us analyze this case in a little more depth.

If the system is initially at rest, so that $x(0) = \dot{x}(0) = 0$, solution (4) becomes

$$x(t) = -\frac{F_0}{m(\omega_0^2 - \omega_f^2)} \cos \omega_0 t + \frac{F_0}{m(\omega_0^2 - \omega_f^2)} \cos \omega_f t. \tag{8}$$

Using the trigonometric identity

$$\cos u - \cos v = -2 \sin\left(\frac{u - v}{2}\right) \sin\left(\frac{u + v}{2}\right),$$

equation (8) can be written

$$x(t) = \frac{2F_0}{m(\omega_0^2 - \omega_f^2)} \sin \frac{(\omega_0 - \omega_f)t}{2} \sin \frac{(\omega_0 + \omega_f)t}{2}. \tag{9}$$

When the difference between ω_f and ω_0 is small, $\omega_0 - \omega_f$ is much smaller than $\omega_0 + \omega_f$, and the factor

$$\sin \frac{(\omega_0 - \omega_f)t}{2}$$

[1] The collapse of the Tacoma Narrows Bridge has long been thought to have been caused by wind-driven resonant vibrations. However, recent analysis points to chaotic effects as well.

oscillates at a much slower rate than the factor

$$\sin \frac{(\omega_0 + \omega_f)t}{2}.$$

This means that equation (9) describes a rapidly oscillating function with frequency $(\omega_0 + \omega_f)/2$, oscillating inside the more slowly oscillating function with frequency $(\omega_0 - \omega_f)/2$. The functions

$$\pm \frac{2F_0}{m(\omega_0^2 - \omega_f^2)} \sin \frac{(\omega_0 - \omega_f)t}{2}$$

form an envelope for the solution (9), and the expression

$$\frac{2F_0}{m(\omega_0^2 - \omega_f^2)} \sin \frac{(\omega_0 - \omega_f)t}{2}$$

is called a **sinusoidal amplitude** for the more rapidly oscillating function in equation (9).

Solutions to the Undamped Forced Oscillator, $\omega_f \neq \omega_0$,

$$m\ddot{x} + kx = F_0 \cos \omega_f t:$$

The general solution is

$$x(t) = c_1 \cos \omega_0 t + c_2 \sin \omega_0 t + \frac{F_0}{m(\omega_0^2 - \omega_f^2)} \cos \omega_f t, \qquad \omega_0 = \sqrt{\frac{k}{m}},$$

where c_1 and c_2 are determined by initial conditions.

If the system starts from rest ($x(0) = 0$ and $\dot{x}(0) = 0$), the solution can be written

$$x(t) = \underbrace{\frac{2F_0}{m(\omega_0^2 - \omega_f^2)} \sin \frac{(\omega_0 - \omega_f)t}{2}}_{\text{sinusoidal amplitude}} \underbrace{\sin \frac{(\omega_0 + \omega_f)t}{2}}_{\text{more rapid oscillation within beats}}.$$

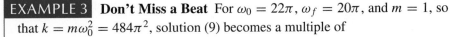

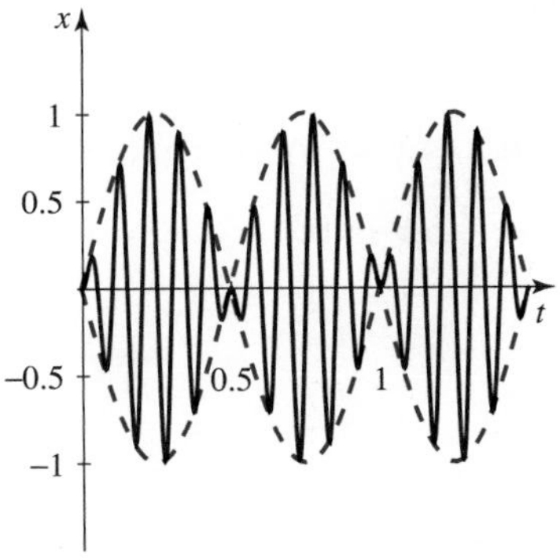

FIGURE 4.6.4 Beats, with their envelope (dashed), for Example 3, in the case where the sinusoidal amplitude of the envelope is 1.

EXAMPLE 3 Don't Miss a Beat For $\omega_0 = 22\pi$, $\omega_f = 20\pi$, and $m = 1$, so that $k = m\omega_0^2 = 484\pi^2$, solution (9) becomes a multiple of

$$\sin \pi t \sin 21\pi t \tag{10}$$

because

$$\frac{\omega_0 - \omega_f}{2} = \pi \quad \text{and} \quad \frac{\omega_0 + \omega_f}{2} = 21\pi.$$

The envelope curves defined by the sinusoidal amplitude are multiples of

$$x = \pm \sin \pi t. \tag{11}$$

The curves are graphed in Fig. 4.6.4, where (10) is solid and (11) is dotted. ■

The phenomenon of beats occurs in acoustics when two tuning forks vibrate with approximately the same frequency: one hears a periodic rising and falling in the noise level. The piano tuner uses this to check whether a particular note is in tune. If beats are detected, the note is not quite in tune; when the beats disappear, the note has been correctly tuned.

The Damped Forced Mass-Spring System

When we add damping to the mass-spring system, we return to equation (1) to model it:

$$m\ddot{x} + b\dot{x} + kx = F_0 \cos \omega_f t. \tag{12}$$

As usual we want to find the solution to the homogeneous DE and add it to a particular solution to get the general solution

$$x(t) = x_h + x_p.$$

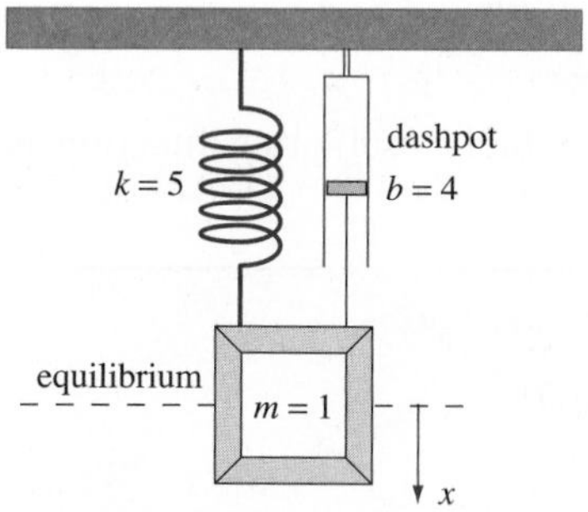

FIGURE 4.6.5 Mass-spring system with damping for Example 4.

EXAMPLE 4 **Adding Damping** Consider the IVP

$$\ddot{x} + 4\dot{x} + 5x = 10 \cos 3t, \qquad x(0) = \dot{x}(0) = 0. \tag{13}$$

The physical setup is shown in Fig. 4.6.5. Find the solutions $x(t)$.

Find x_h: For the associated homogeneous equation, the characteristic roots are $-2 \pm i$, so

$$x_h = e^{-2t}(c_1 \cos t + c_2 \sin t).$$

Find x_p: Use the method of undetermined coefficients to find a particular solution:

$$\begin{aligned} x_p &= A \cos 3t + B \sin 3t, \\ \dot{x}_p &= -3A \sin 3t + 3B \cos 3t, \\ \ddot{x}_p &= -9A \cos 3t - 9B \sin 3t. \end{aligned}$$

Substituting into DE (13) gives

$$\begin{aligned} &(-9A \cos 3t - 9B \sin 3t) \\ +4(&\ 3B \cos 3t - 3A \sin 3t) \\ +5(&\ \ A \cos 3t + \ B \sin 3t) = 10 \cos 3t, \end{aligned}$$

where we evaluate A and B as follows:

$$\begin{aligned} \text{coefficient of } \cos 3t\text{: } -9A + 12B + 5A &= 10, \\ \text{coefficient of } \sin 3t\text{: } -9B - 12A + 5B &= 0. \end{aligned}$$

Hence, we obtain

$$A = -\frac{1}{4} \quad \text{and} \quad B = \frac{3}{4},$$

so

$$x_p = -\frac{1}{4} \cos 3t + \frac{3}{4} \sin 3t.$$

The general solution for the DE in (13) is

$$x = e^{-2t}(c_1 \cos t + c_2 \sin t) - \frac{1}{4} \cos 3t + \frac{3}{4} \sin 3t,$$

so

$$\begin{aligned} \dot{x} = &-2e^{-2t}(c_1 \cos t + c_2 \sin t) + e^{-2t}(-c_1 \sin t + c_2 \cos t) \\ &+ \frac{3}{4} \sin 3t + \frac{9}{4} \cos 3t, \end{aligned}$$

and we can evaluate c_1 and c_2 from the initial conditions:

$$x(0) = 0 \quad \Rightarrow c_1 - \frac{1}{4} = 0 \quad \Rightarrow c_1 = \frac{1}{4};$$

$$\dot{x}(0) = 0 \quad \Rightarrow c_2 + \frac{7}{4} = 0 \quad \Rightarrow c_2 = -\frac{7}{4}.$$

Thus, for the IVP (13),

$$x(t) = \underbrace{e^{-2t}\left(\frac{1}{4}\cos t - \frac{7}{4}\sin t\right)}_{x_h \equiv \text{ transient}} \underbrace{- \frac{1}{4}\cos 3t + \frac{3}{4}\sin 3t}_{x_p \equiv \text{ steady-state periodic}}.$$

For a forced and damped oscillator, the homogeneous solution x_h is a **transient** solution, because for $b > 0$ it tends toward zero as time increases. The particular solution x_p may be a constant or periodic **steady-state** solution.

We can write these solutions in alternate forms to see explicitly the amplitude and phase angle for x_h and x_p:

$$x_h \approx \underbrace{1.768}_{\sqrt{\left(\frac{1}{4}\right)^2 + \left(-\frac{7}{4}\right)^2}} e^{-2t} \cos(t + \underbrace{1.4289}_{-\tan^{-1}(-7)})$$

and

$$x_p \approx \underbrace{0.791}_{\sqrt{\left(\frac{1}{4}\right)^2 + \left(\frac{3}{4}\right)^2}} \cos(3t + \underbrace{1.249}_{-\tan^{-1}(-3)}).$$

Figure 4.6.6 shows the solution to IVP (13) as the sum of x_h and x_p.

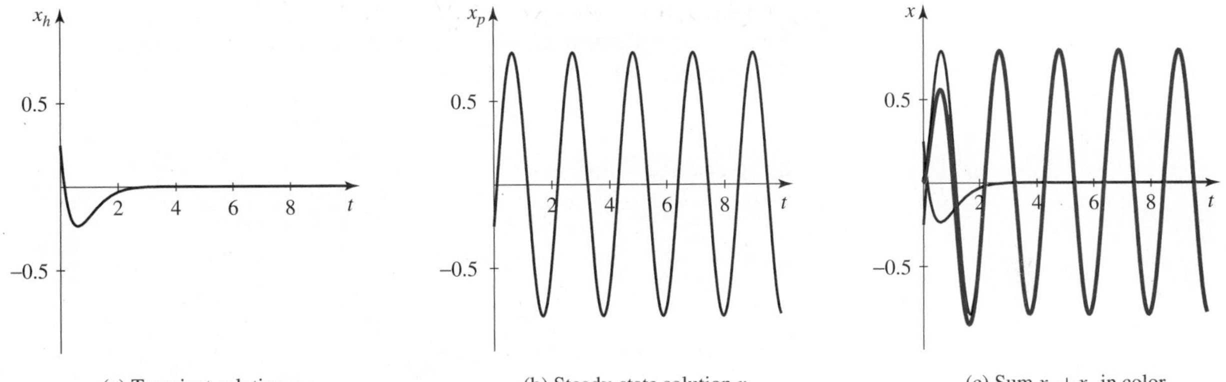

(a) Transient solution x_h (b) Steady-state solution x_p (c) Sum $x_h + x_p$ in color

FIGURE 4.6.6 Solution (in color) to the IVP of Example 4.

■

General Solution of the Damped Forced Mass-Spring System

Example 4 shows that there are many steps to solving an IVP for a nonhomogeneous linear second-order DE (12). We can summarize the general process.

Let us first look at the solutions to the associated homogeneous DE in this context. In Secs. 4.2 and 4.3, we learned that solving the characteristic equation

$$mr^2 + br + k = 0$$

by means of the quadratic formula gives rise to three distinct situations.

Damped Forced Vibrations

See the motion of the damped oscillator and watch the graphs of the transient and periodic steady-state solutions unfold.

Homogeneous Solutions x_h of a Damped Mass-Spring System
The damped mass-spring system

$$m\ddot{x} + b\dot{x} + kx = 0$$

has three different homogeneous solutions x_h, depending on the value of the discriminant $\Delta = b^2 - 4mk$.

Case 1 $\Delta > 0$	Overdamped motion: $x_h = c_1 e^{r_1 t} + c_2 e^{r_2 t}$	Real unequal roots: $r_1, r_2 = \dfrac{-b \pm \sqrt{b^2 - 4mk}}{2m}$
Case 2 $\Delta = 0$	Critically damped motion: $x_h = c_1 e^{rt} + c_2 t e^{rt}$	Real repeated root: $r = -\dfrac{b}{2m}$
Case 3 $\Delta < 0$	Underdamped motion: $x_h = e^{\alpha t}(c_1 \cos \beta t + c_2 \sin \beta t)$	Complex conjugate roots: $r_1, r_2 = \alpha \pm \beta i$ $\alpha = -\dfrac{b}{2m}, \ \beta = \dfrac{\sqrt{4mk - b^2}}{2m}$

To find a *particular* solution of the general damped forced mass-spring system (12), we use the method of undetermined coefficients (Sec. 4.4). We will confirm the details in Problem 34.

Particular Solution x_p of a Damped Mass-Spring System
The damped mass-spring system

$$m\ddot{x} + b\dot{x} + kx = F_0 \cos \omega_f t$$

has particular solution

$$x_p = A \cos \omega_f t + B \sin \omega_f t \tag{14}$$

with

Write these down!

$$A = \frac{m(\omega_0^2 - \omega_f^2) F_0}{m^2 (\omega_0^2 - \omega_f^2)^2 + (b\omega_f)^2}, \qquad B = \frac{b\omega_f F_0}{m^2 (\omega_0^2 - \omega_f^2)^2 + (b\omega_f)^2}, \tag{15}$$

and natural (circular frequency) $\omega_0 = \sqrt{k/m}$, which does not depend on damping b.

Alternatively, we can write x_p as a single periodic function

$$x_p = C \cos(\omega_f t - \delta), \tag{16}$$

where $C = \sqrt{A^2 + B^2}$ and $\tan \delta = B/A$, and the quadrant for δ is determined by the signs of B and A.

Thus, a somewhat shorter alternative to the multistep calculation of the particular solution x_p (as in Example 4) is to use formulas (15) for A and B and get x_p more directly from either (14) or (16). This only seems useful if we deal with enough problems of this type to be able to remember the formulas in (15).

Some Important Facts About Forced Damped Oscillators

Vibrations; Amplitude Response

Watch the effect of the input forcing frequency on the amplitude of the steady-state periodic solution.

A particular solution to (12) in the form (16) comes out to be the following (Problem 34 again):

$$x_p = \frac{F_0}{\sqrt{m^2\left(\omega_0^2 - \omega_f^2\right)^2 + (b\omega_f)^2}}\cos(\omega_f t - \delta) \tag{17}$$

with

$$\tan\delta = \frac{b\omega_f}{m\left(\omega_0^2 - \omega_f^2\right)}. \tag{18}$$

Notice from (17) that the response to the forcing term $F_0 \cos \omega_f t$ is also oscillatory with the same frequency ω_f but with a **phase lag** given by δ/ω_f and amplitude determined by the **amplitude factor**

$$A(\omega_f) = \frac{F_0}{\sqrt{m^2\left(\omega_0^2 - \omega_f^2\right)^2 + (b\omega_f)^2}}. \tag{19}$$

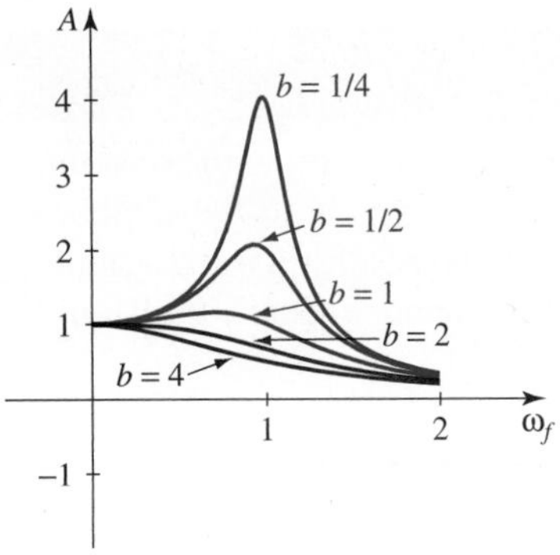

FIGURE 4.6.7 Amplitude response curves for forced damped oscillator with $k = m = 1$ and $\omega_0 = 1$.

The amplitude increases as ω_f approaches ω_0, and decreases when the difference in frequencies becomes large. When $A(\omega_f)$ is graphed as a function of the input frequency ω_f, the result is called the **amplitude response curve** (or **frequency response curve**). Several amplitude response curves are plotted in Fig. 4.6.7 for $k = m = 1$, $\omega_0 = 1$, and various values of the damping constant b. The maximum of the amplitude response curve occurs where we have "practical resonance."

To determine the behavior of the amplitude factor $A(\omega_f)$ given by (19) as a function of the input frequency ω_f, we can use differential calculus. (See Problem 36.) From the derivative,

$$A'(\omega_f) = \frac{-F_0}{\left[\sqrt{m^2\left(\omega_0^2 - \omega_f^2\right)^2 + (b\omega_f)^2}\right]^3}\,\omega_f\left(b^2 - 2mk + 2m^2\omega_f^2\right), \tag{20}$$

we can show the following.

Graphical Properties of the Amplitude Response Curve

The condition for an extremum, $A'(\omega_f) = 0$, of the amplitude response curve (19) is met only when

$$\omega_f = 0 \quad \text{or} \quad \omega_f = \sqrt{\frac{k}{m} - \frac{b^2}{2m^2}}. \tag{21}$$

(i) When $b^2 \geq 2mk$, we have $A'(\omega_f) = 0$ only when $\omega_f = 0$. This gives a *maximum* amplitude because, from (20), $A'(\omega_f) < 0$ for all $\omega_f > 0$. Substituting $\omega_f = 0$ into $A(\omega_f)$ gives the maximum amplitude,

$$A_{\max} = A(0) = \frac{F_0}{k}.$$

(ii) When $b^2 < 2mk$, we have $A'(\omega_f) = 0$ when $\omega_f = 0$, which now gives a relative *minimum*, or when $\omega_f = \sqrt{\frac{k}{m} - \frac{b^2}{2m^2}}$, which gives the *maximum* of $A(\omega_f)$.

Substituting the nonzero critical value into $A(\omega_f)$ gives the maximum amplitude,

$$A_{\max} = \frac{F_0}{b\sqrt{\frac{k}{m} - \frac{b^2}{4m^2}}}.$$

Pendulums

Choose *forced* pendulum to see a chaotic phase portrait arise before your eyes.

In Fig. 4.6.7 the black curves exhibit property (i) and the colored curves illustrate property (ii).

Under certain initial conditions, forced damped oscillators can also lead to *chaotic* motion.[2] We postpone discussion of this aspect to Sec. 7.5.

Summary

We modeled the motion of a forced mass-spring system using a second-order differential equation with constant coefficients. We included a typical real-world nonhomogeneous forcing term that is amenable to the method of undetermined coefficients, then combined the techniques of Secs. 4.2–4.4 to solve the resulting initial-value problem for both the damped and undamped cases. We introduced the concepts of resonance, beats, transient, and steady-state solutions in order to interpret the results.

4.6 Problems

Mass-Spring Problems *Find the position function $x(t)$ for each of the forced mass-spring systems in Problems 1–6. Find the amplitude and phase shift for x_{ss}, the steady-state solution.*

1. $m = 1, \quad b = 2, \quad k = 1, \quad F = 6\cos t$

2. $m = 1, \quad b = 2, \quad k = 3, \quad F = \cos 3t$

3. $m = 2, \quad b = 0, \quad k = 3, \quad F = 4\cos 8t$

4. $m = 2, \quad b = 2, \quad k = \frac{1}{2}, \quad F = \frac{5}{2}\cos t$

5. $m = 1, \quad b = 2, \quad k = 2, \quad F = 2\cos t$

6. $m = 1, \quad b = 4, \quad k = 5, \quad F = 2\cos 2t$

7. Pushing Up An 8-lb weight stretches a spring 4/3 ft to equilibrium. Then the weight starts from rest 2 ft above the equilibrium position. The damping force is $2.5\dot{x}$. An external force of $2\cos 2t$ is applied at $t = 0$. Find the position function $x(t)$ for $t > 0$.

8. Pulling Down A 16-lb weight stretches a spring 1 ft to equilibrium. Then the weight starts from rest 1 ft below the equilibrium position. The damping force is $6\dot{x}$. An external force of $4\cos 4t$ is applied at $t = 0$. Find the position function $x(t)$ for $t > 0$.

9. Mass-Spring Again A mass of 100 kg is attached to a long spring suspended from the ceiling. When the mass comes to rest at equilibrium, the spring has been stretched 20 cm. The mass is then pulled down 40 cm below the equilibrium point and released. Ignore any damping or external forces.

(a) Verify that $k = 4900$ nt/m.

(b) Solve for the motion of the mass.

(c) Find the amplitude and period of the motion.

(d) Now add damping to the system with damping coefficient given by $b = 500$ nt sec/m. Is the system underdamped, critically damped, or overdamped?

(e) Solve the damped system with the same initial conditions.

---

[2]See J. H. Hubbard, "What It Means to Understand a Differential Equation," *College Mathematics Journal* **25** no. 5 (1994), 372–384.

10. **Adding Forcing** Suppose the mass-spring system of Problem 9 (i.e., $m = 100$ kg, $k = 4900$ nt/m, $b = 500$ nt sec/m) is forced by an oscillatory function $f(t) = 100\cos\omega_f t$.

 (a) What value of ω_f will give the largest amplitude for the steady-state solution?

 (b) Find the steady-state solution when $\omega_f = 7$.

 (c) Now consider the system with no damping, $b = 0$, and $\omega_f = 7$. What is the *form* of the particular solution to the system? Do not solve for the constants.

11. **Electric Analog** Using a 4-ohm resistor, construct an *LRC*-circuit that is the analog of the mechanical system in Problem 10, in the sense that the two systems are governed by the same differential equation. That is, what values for L, C, and $V(t)$ will give a multiple of the following?

$$100\ddot{x} + 500\dot{x} + 4900x = 100\cos\omega_f t$$

12. **Damped Forced Motion I** Find the steady-state motion of a mass that vibrates according to the law

$$\ddot{x} + 8\dot{x} + 36x = 72\cos 6t.$$

13. **Damped Forced Motion II** A 32-lb weight is attached to a spring suspended from the ceiling, stretching the spring by 1.6 ft before coming to rest. At time $t = 0$ an external force of $f(t) = 20\cos 2t$ is applied to the system. Assume that the mass is acted upon by a damping force of $4\dot{x}$, where $\dot{x}$ is the instantaneous velocity in feet per second. Find the displacement of the weight with $x(0) = 0$ and $\dot{x}(0) = 0$.

14. **Calculating Charge** Consider the series circuit shown in Fig. 4.6.8, for which the inductance is 4 henries and the capacitance is 0.01 farads. There is negligible resistance. The input voltage is $10\cos 4t$. At time $t = 0$, the current and the charge on the capacitor are both zero. Determine the charge Q as a function of t.

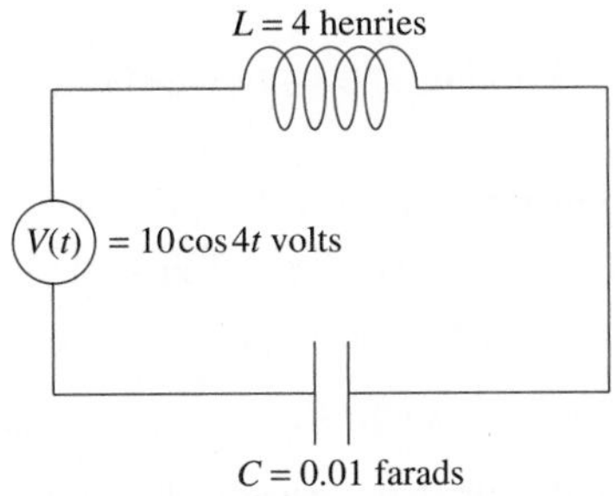

FIGURE 4.6.8 Circuit for Problem 14.

15. **Charge and Current** A resistor of 12 ohms is connected in series with an inductor of one henry, a capacitor of 0.01 farads, and a voltage source supplying $12\cos 10t$. At $t = 0$, the charge on the capacitor is zero and the current in the circuit is also zero.

 (a) Determine $Q(t)$, the charge on the capacitor as a function of time for $t > 0$.

 (b) Determine $I(t)$, the current in the circuit as a function of time for $t > 0$.

True/False Questions *For Problems 16 and 17, give a justification for your answer of true or false.*

16. True or false? For a forced damped mass-spring system with sinusoidal forcing, the frequency of the steady-state solution is the same as that of the forcing function.

17. True or false? For a forced damped mass-spring system with sinusoidal forcing, the amplitude of the steady-state solution is the same as that of the forcing function.

18. **Beats** Express $\cos 3t - \cos t$ in the form $A\sin\alpha t\sin\beta t$, and sketch its graph.

19. **The Beat Goes On** Express $\sin 3t - \sin t$ in the form $A\sin\alpha t\cos\beta t$, and sketch its graph.

Steady State *For Problems 20–22, find the steady-state solution having the form $x_{ss} = C\cos(\omega t - \delta)$, for the damped system.*

20. $\ddot{x} + 4\dot{x} + 4x = \cos t$

21. $\ddot{x} + 2\dot{x} + 2x = 2\cos t$

22. $\ddot{x} + \dot{x} + x = 4\cos 3t$

Resonance *A mass of one slug is hanging at rest on a spring whose constant is 12 lb/ft. At time $t = 0$, an external force of $f(t) = 16\cos\omega t$ lb is applied to the system.*

23. What is the frequency of the forcing function that is in resonance with the system?

24. Find the equation of motion of the mass with resonance.

25. **Ed's Buoy**[3] Ed is sitting on the dock and observes a cylindrical buoy bobbing vertically in calm water. He observes that the period of oscillation is 5 sec and that 4 ft of the buoy is above water when it reaches its maximum height and 2 ft above water when it is at its minimum height. An old seaman tells Ed that the buoy weighs 2000 lbs.

 (a) How will this buoy behave in rough waters, with the waves 6 ft from crest to trough, and with a period of 7 sec if you neglect damping?

 (b) Will the buoy ever be submerged?

[3]This problem is based on a problem taken from Robert E. Gaskell, *Engineering Mathematics* (Dryden Press, 1958).

26. General Solution of the Damped Forced System Consider the damped forced mass-spring equation

$$m\ddot{x} + b\dot{x} + kx = F_0 \cos \omega_f t.$$

(a) Verify, using the method of undetermined coefficients, that equations (14) and (15) give the particular solution.

(b) Using (15), verify equations (17) and (18), and show that in the damped case the transient solution will go to zero, so the particular solution will be the long-term or steady-state response of the system.

Phase Portrait Recognition *For Problems 27–30, match the phase-plane diagrams shown in Fig. 4.6.9 to the appropriate differential equations.*

27. $\ddot{x} + 0.3\dot{x} + x = \cos t$

28. $\ddot{x} + x = 0$

29. $\ddot{x} + x = \cos t$

30. $\ddot{x} + 0.3\dot{x} + x = 0$

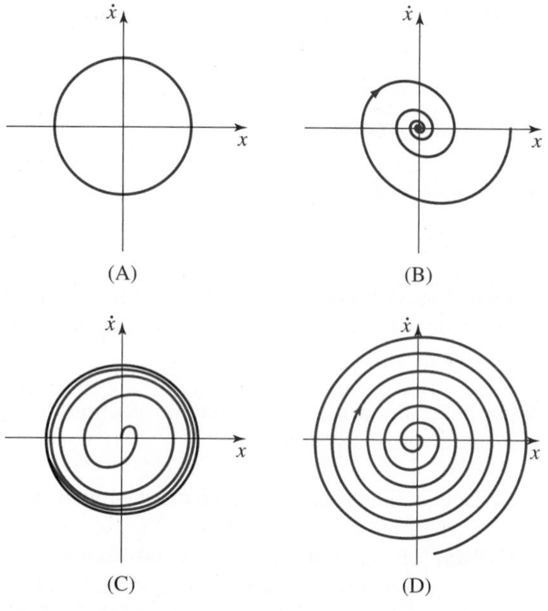

FIGURE 4.6.9 Phase portraits to match to Problems 27–30.

31. Matching 3D Graphs Another way of showing the interaction of the independent variable t and the dependent variables x and $\dot{x}$, for a mass-spring system modeled by $m\ddot{x} + b\dot{x} + kx = F(t)$ with initial conditions $x(0) = x_0$ and $\dot{x}(0) = v_0$, is by means of an $x\dot{x}t$ graph. Associate the properties (a)–(g) with the graphs shown in Fig. 4.6.10. Some properties may be associated with more than one graph, and vice versa.

(a) Pure resonance

(b) Beats

(c) Forced damped motion with a sinusoidal forcing function

(d) $x(0) > 0$; $\dot{x}(0) = 0$

(e) $x(0) = 0$; $\dot{x}(0) = 0$

(f) Steady-state periodic motion

(g) Unforced damped motion

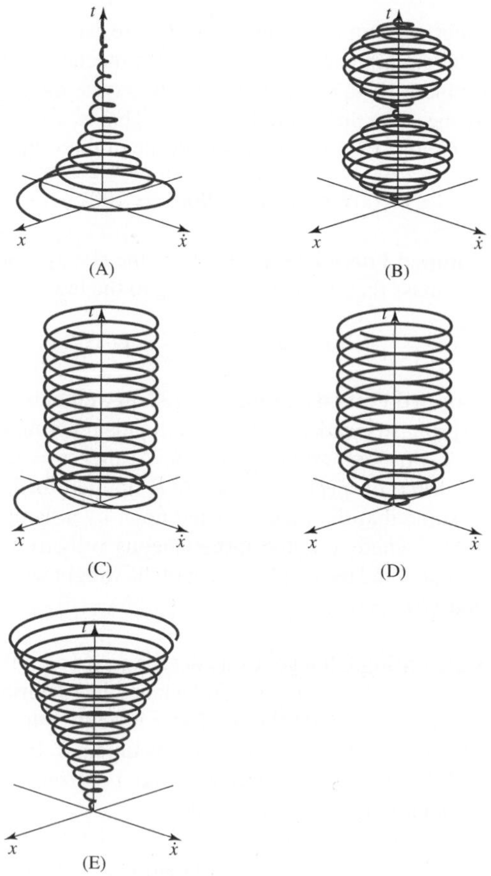

FIGURE 4.6.10 Trajectories in $tx y$-space for Problem 31.

32. Mass-Spring Analysis I Suppose that

$$x(t) = 4\cos 4t - 3\sin 4t + 5t\sin 4t$$

is the solution of a mass-spring system

$$m\ddot{x} + b\dot{x} + kx = F(t), \quad x(0) = x_0, \dot{x}(0) = v_0.$$

Assume that the homogeneous solution is not identically zero.

(a) Determine the part of the solution associated with the homogeneous DE.

(b) Calculate the amplitude of the oscillation of the homogeneous solution.

(c) Determine the amplitude of the particular solution.

(d) Which part of the solution will be unchanged if the initial conditions are changed?

(e) If the mass is 1 kg, what is the spring constant?

(f) Describe the motion of the mass according to the solution.

33. Electrical Version Suppose that

$$Q(t) = 4\cos 4t - 5\sin 4t + 6t\cos 4t$$

is the solution of a given *LRC* system

$$L\ddot{Q} + R\dot{Q} + \frac{1}{C}Q = V(t), \quad Q(0) = Q_0, \dot{Q}(0) = I_0.$$

Assume that the homogeneous solution is not identically zero.

(a) Determine the part of the solution that is the transient solution.

(b) Calculate the amplitude of the oscillation described by the transient solution.

(c) State which part of the solution is the steady-state solution.

(d) Which part of the solution will be unchanged if the initial conditions are changed?

(e) If the inductance is 1 henry, what is the capacitance?

(f) Describe what happens to the charge on the capacitor according to the solution.

34. Mass-Spring Analysis II Suppose that

$$x(t) = 3e^{-2t}\cos t - 2e^{-2t}\sin t + \sqrt{2}\cos(5t - \delta)$$

is the solution of a mass-spring system

$$m\ddot{x} + b\dot{x} + kx = F_0\cos\omega_f t, \quad x(0) = x_0, \dot{x}(0) = v_0.$$

Assume that the homogeneous solution is not identically zero.

(a) What part of the solution is the transient solution?

(b) If the mass is 1 kg, what is the damping constant b?

(c) Is the system underdamped, critically damped, or overdamped?

(d) What is the time-varying amplitude of the transient solution?

(e) What part of the solution is the steady-state solution?

(f) Find the angular frequency ω_f of the forcing function. Find the forcing amplitude F_0.

35. Perfect Aim Neglecting air friction, the planar motion $(x(t), y(t))$ of an object in a gravitational field is governed by the equations

$$\ddot{x} = 0 \quad \text{and} \quad \ddot{y} = -g,$$

where g is acceleration due to gravity. (See Fig. 4.6.11.) Suppose that you fire a dart gun, located at the origin, directly at a target located at (x_0, y_0). Suppose also that the dart has initial speed v_0, and at the exact moment the dart is fired the target object starts to fall.

(a) Find the vertical distance the dart and target as a function of time.

(b) Show that the dart will always hit the target.

(c) At what height will the dart hit the target?

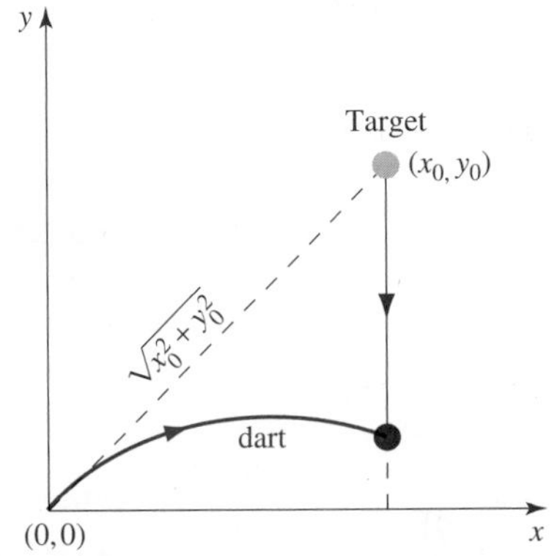

FIGURE 4.6.11 Paths of target and dart for Problem 35.

36. Extrema of the Amplitude Response Verify the extremal properties of the amplitude response function $A(\omega_f)$ in Fig. 4.6.7.

37. Suggested Journal Entry Sometimes resonance in a physical system is desirable, and sometimes it is undesirable. Discuss whether the resulting resonance would be helpful or destructive in the following systems (a)–(e). Then give at least one additional example on each side.

(a) Soldiers marching on a bridge with the same frequency as the natural frequency of the bridge

(b) A person rocking a car stuck in the snow with the same frequency as the natural frequency of the stuck car

(c) A child pumping a swing

(d) Vibrations caused by air passing over an airplane wing having the same frequency as the natural flutter of the wing

(e) Acoustic vibrations having the same frequency as the natural vibrations of a wine glass (the "Memorex experiment")

4.7 Conservation and Conversion

SYNOPSIS: We look at two types of physical systems: conservative systems, in which the available energy remains constant, and nonconservative systems, in which the available energy declines with time. Systems of first-order differential equations provide a unifying framework. We outline methods for converting second- and higher-order differential equations and systems into first-order systems.

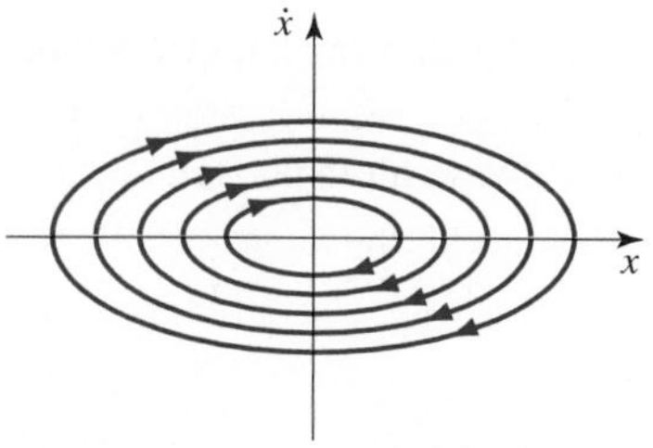

FIGURE 4.7.1 Phase portrait for an undamped harmonic oscillator, which conserves energy.

Energy of the Harmonic Oscillator

If we take the undamped unforced oscillator equation

$$m\ddot{x} + kx = 0 \tag{1}$$

and multiply both sides by $\dot{x}$, we obtain $m\dot{x}\ddot{x} + kx\dot{x} = 0$, which can be written in the form

$$\frac{d}{dt}\left[\frac{1}{2}m\dot{x}^2 + \frac{1}{2}kx^2\right] = 0, \tag{2}$$

as you can easily verify by differentiating. Since the derivative of the function in brackets in equation (2) is zero, the function must equal a constant, so we write

$$\underbrace{\frac{1}{2}m\dot{x}^2}_{\text{kinetic energy}} + \underbrace{\frac{1}{2}kx^2}_{\text{potential energy}} = E, \tag{3}$$

where

- E (the constant of integration) is the **total energy** of the system,
- $\frac{1}{2}m\dot{x}^2$ is the **kinetic energy** of the moving object, and
- $\frac{1}{2}kx^2$ is the **potential energy** of the spring.

Simple Harmonic Oscillator
Damped Vibrations: Energy
In both tools, the mass-spring model adds an energy graph. Compare results for damping coefficient $b = 0$ and $b > 0$.

Equation (3) states that *their sum remains constant, even though the terms themselves change with evolving time*. That is, equation (3) represents a **conservation of energy** principle for equation (1). We also say that the system is conservative.

If the object is initially at rest but is given an initial velocity $\dot{x}(0) = v_0$, the total energy can be evaluated as $E = \frac{1}{2}mv_0^2$. For each energy value E, the point (or vector) $(x(t), \dot{x}(t))$ describes an ellipse in the $x\dot{x}$-plane or phase plane with equation (3). (See Fig. 4.7.1.)

EXAMPLE 1 **Harmonic Oscillator** The total energy of the undamped mass-spring system described by the IVP

$$\ddot{x} + 4x = 0, \qquad x(0) = 0.5 \text{ m}, \qquad \dot{x}(0) = -3 \text{ m/sec}$$

is constant over time and is given by

$$E = \frac{1}{2}m\dot{x}^2 + \frac{1}{2}kx^2 = \frac{1}{2}(1)(-3)^2 + \frac{1}{2}(4)(0.5)^2 = 5 \text{ joules.}$$

■

Total Energy of the Damped Mass-Spring System

When we add damping to an unforced mass-spring system, the differential equation becomes $m\ddot{x} + b\dot{x} + kx = 0$, $x(0) = x_0$, $\dot{x}(0) = v_0$, and a phase portrait looks like Fig. 4.7.2. Here the sum of the kinetic and potential energies, given by

$$E(t) = \frac{1}{2}m[\dot{x}(t)]^2 + \frac{1}{2}k[x(t)]^2,$$

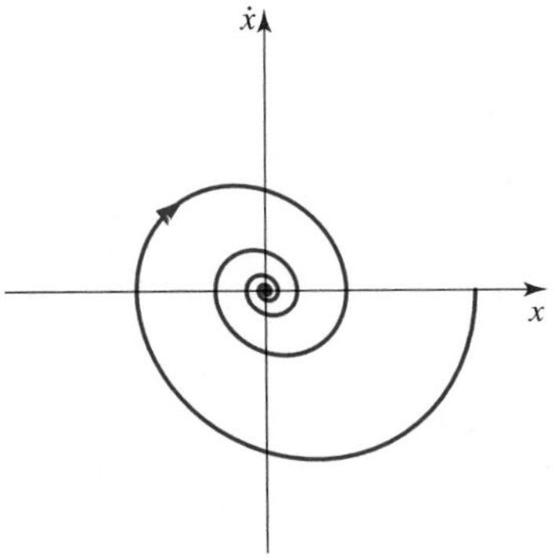

FIGURE 4.7.2 Phase portrait for a damped mass-spring system, which will lose energy.

is not constant but decreases over time due to heat loss. The amount of energy lost after time t is given by $E(0) - E(t)$. A mass-spring system with *damping* is called a **nonconservative system**.

EXAMPLE 2 **Heat Loss** The mass-spring system in Example 1 is now given damping and is described by the IVP

$$\ddot{x} + \dot{x} + 4x = 0, \qquad x(0) = 0.5 \text{ m}, \quad \dot{x}(0) = -3 \text{ m/sec}.$$

Suppose we observe the system after $t = 2$ sec and find $x(2) = -0.1$ m and $\dot{x}(2) = 2$ m/sec. The amount of energy of this system lost to heat during those two seconds is

$$E(0) - E(2) = \left[\frac{1}{2}(1)(-3)^2 + \frac{1}{2}(4)(0.5)^2\right] - \left[\frac{1}{2}(1)(2)^2 + \frac{1}{2}(4)(-0.1)^2\right]$$
$$= 2.98 \text{ joules}.$$

■

We summarize our energy information for oscillating systems as follows.

Energy of an Unforced Mass-Spring System
The total energy of the autonomous system

$$m\ddot{x} + b\dot{x} + kx = 0$$

is composed of three parts:

$$E = \underbrace{\frac{1}{2}m\dot{x}^2}_{\text{kinetic energy}} + \underbrace{\frac{1}{2}kx^2}_{\text{potential energy}} + \underbrace{\text{heat}}_{\text{loss created when damping exists}}$$

If heat loss is zero, the system is conservative.

Energy in General

A more general conservative physical system is represented by the autonomous differential equation

$$m\ddot{x} + F(x) = 0, \tag{4}$$

where $-F(x)$ is the restoring force, as in Sec. 4.1. The kinetic energy is still $\frac{1}{2}m\dot{x}^2$, and the potential energy $V(x)$ is given by the following equation.[1]

Conservative System:

In a conservative system, potential energy $V(x)$ must be a function of x alone.

$$V(x) = \int F(x)\,dx; \tag{5}$$

that is,

$$\frac{dV}{dx} = F(x). \tag{6}$$

For the case $F(x) = kx$ in equation (1), this gives

$$V(x) = \int kx\,dx = \frac{1}{2}kx^2.$$

[1]Physicists tend to define potential energy as the *negative* integral of the *restoring force*, which in our notation (4) is $-F(x)$. The result, $V(x) = -\int -F(x)dx$, is the same as (5).

(We choose the integration constant to be zero for convenience.) The quantity conserved is the total energy E, a function of position x and velocity $\dot{x}$:

$$E(x, \dot{x}) = \frac{1}{2}m\dot{x}^2 + V(x). \tag{7}$$

In physics a quantity like E is sometimes called a **constant of the motion** (or a **first integral of the differential equation**), and we say that (4) describes a **conservative system.**[2]

We can get useful insight into system (4) by graphing (7) as a surface in three-dimensional space; the energy $E(x, \dot{x})$ is the vertical coordinate above or below the $x\dot{x}$-plane. The resulting *energy surface* helps us understand the motion. Contour lines or level curves (curves in the $x\dot{x}$-plane along which E has a constant value) are the phase-plane trajectories of the motion. Extrema of the E-surface define the equilibrium points of the motion. For example, a local maximum corresponds to an unstable equilibrium, while a local minimum corresponds to a stable equilibrium.

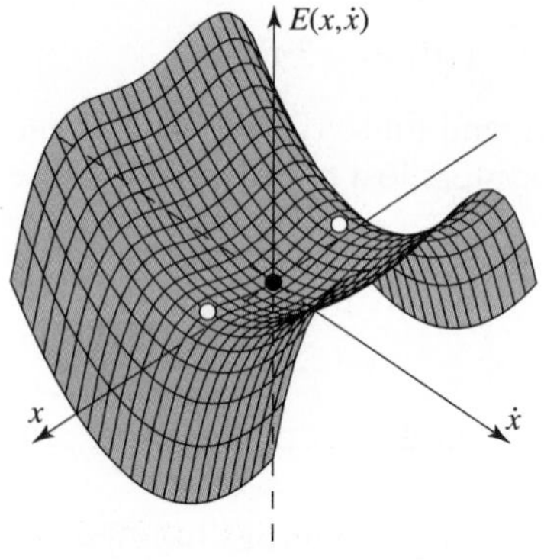

FIGURE 4.7.3 An energy surface $E(x, \dot{x})$.

Figure 4.7.3 shows such a surface. The local minimum at (0, 0) represents a stable equilibrium point, while saddle points at (−1, 0) and (1, 0) are unstable equilibrium points of the motion.

Energy of Conservative Systems
The DE for a conservative system must be *autonomous* in the form

$$m\ddot{x} + \frac{dV}{dx} = 0.$$

Then the phase-plane trajectories are *level curves* (constants) of the energy surface

$$E(x, \dot{x}) = \frac{1}{2}m\dot{x}^2 + V(x).$$

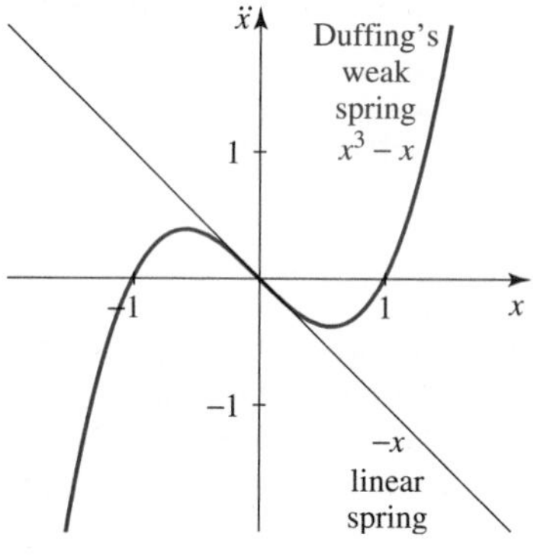

FIGURE 4.7.4 Restoring forces for Duffing's weak spring, in color, and the linear spring.

EXAMPLE 3 **Duffing's Weak Spring** Duffing's weak spring equation[3] is the second-order nonlinear differential equation

$$\ddot{x} + x - x^3 = 0. \tag{8}$$

Comparing it with (4), we see that the restoring force $-F(x) = x^3 - x$ (and $m = 1$). This "weak spring" is compared to the linear spring $-x$ in Fig. 4.7.4. It is weak because it only acts as a restoring force (opposing the displacement) for $-1 < x < 1$.

Let us investigate the motion (8) by determining the following:

(a) the kinetic and potential energies of the motion;

(b) the phase portrait from the contours of the energy function; and

(c) the location and nature of the equilibrium points.

[2]The concept of a conservative system has more general formulations in several branches of physics, where multivariable problems make heavy use of the tools of linear algebra. More information on conservative systems can be found in Grégoire Nicolis and Ilya Prigogine, *Exploring Complexity* (San Francisco: Freeman, 1989).

[3]Duffing's oscillators will be discussed in more detail in Sec. 7.5.

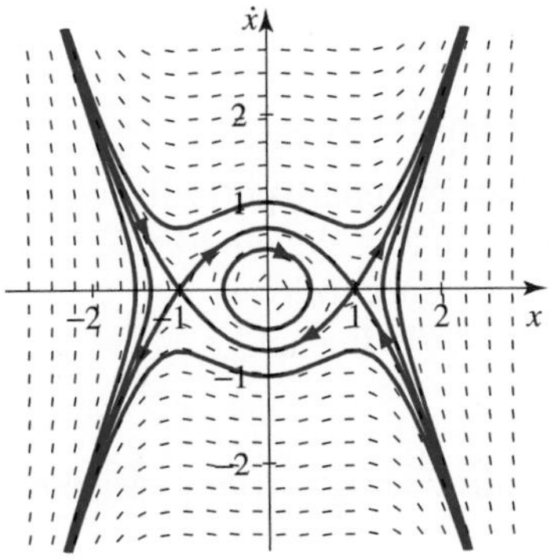

FIGURE 4.7.5 Level curves for the energy function of Duffing's weak spring.

We see that the kinetic energy KE $= \frac{1}{2}\dot{x}^2$ (by comparing (8) with Newton's law we see that $m = 1$), and the potential energy is given by

$$V(x) = \int F(x)dx = \int (x - x^3)\,dx = \frac{x^2}{2} - \frac{x^4}{4}.$$

Using a computer we can generate level curves for

$$E(x, \dot{x}) = \frac{1}{2}\dot{x}^2 + \frac{1}{2}x^2 - \frac{1}{4}x^4, \tag{9}$$

as shown in Fig. 4.7.5.

The equilibrium points occur at the solutions of the system

$$\frac{\partial E}{\partial \dot{x}} = 0, \quad \frac{\partial E}{\partial x} = 0,$$

that is, $\dot{x} = 0$ and $x - x^3 = 0$. Therefore, the equilibrium points are $(0, 0)$, $(1, 0)$, and $(-1, 0)$. There is a local minimum at $(0, 0)$; this point is a *stable* equilibrium for the system. The surface has saddle points at $(-1, 0)$ and $(1, 0)$; these are points of *unstable* equilibrium for the spring. Compare with the actual energy surface for this example, which is shown in Fig. 4.7.3 for $-2 \le x \le 2$, $-2 \le \dot{x} \le 2$. ■

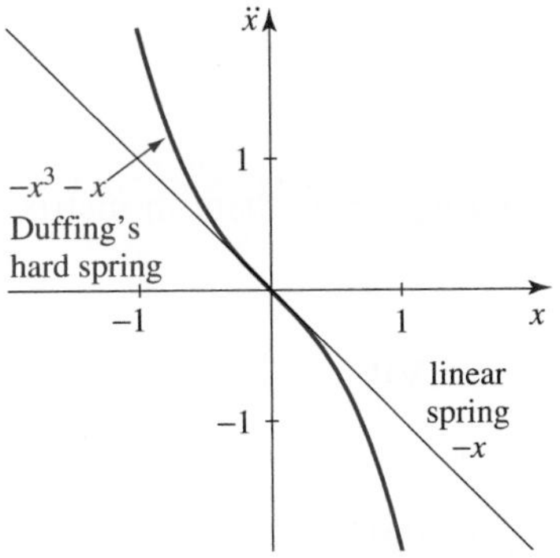

FIGURE 4.7.6 Restoring forces for Duffing's hard spring, in color, and the linear spring.

EXAMPLE 4 Duffing's Hard Spring Duffing's hard spring equation is the second-order nonlinear differential equation

$$\ddot{x} + x + x^3 = 0. \tag{10}$$

Now we have $-F(x) = -x^3 - x$, a "hard spring"; it is compared to the restoring force for the linear spring, $-x$, in Fig. 4.7.6. It resists deformation more vigorously than the linear spring for *all* deflections.

Investigating the same three questions as in the previous example, we find the following: kinetic energy KE $= \frac{1}{2}\dot{x}^2$, while the potential energy is given by

$$V(x) = \int F(x)dx = \int (x + x^3)\,dx = \frac{x^2}{2} + \frac{x^4}{4}.$$

Hence, the total energy is given by

$$E(x, \dot{x}) = \frac{1}{2}\dot{x}^2 + \frac{1}{2}x^2 + \frac{1}{4}x^4; \tag{11}$$

the contours are plotted in Fig. 4.7.7.

For system (10) there is only one equilibrium point, $(0, 0)$, and it is stable. ■

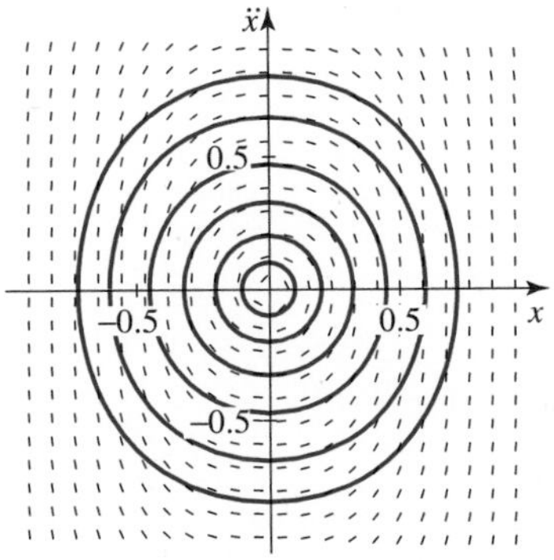

FIGURE 4.7.7 Level curves for the energy function of Duffing's hard spring.

Phase Portrait Clues

The examples of this section illustrate the fact that *phase-plane trajectories for conservative second-order DEs are in fact level curves of the total energy function*; the phase portrait gives a contour plot of $E(x, \dot{x})$. (See Figures 4.7.1, 4.7.5, and 4.7.7.)

For a nonconservative system, energy is not constant, and the trajectories have no such property. A phase portrait for a nonconservative system usually *shows* that it cannot be a contour plot for a surface. See Fig. 4.7.2 and imagine other trajectories (supposedly at different levels) spiraling between the existing one but all meeting at the center.

Problems 14–19 give more examples of both conservative and nonconservative systems.

Converting Second-Order DEs to Systems

Most graphic differential equation solvers that produce our phase portraits require the user to convert a second-order differential equation to a system of two first-order DEs, as shown in Sec. 4.1. A more general procedure follows.

Any second-order differential equation (not necessarily linear),

$$y'' = f(t, y, y'), \tag{12}$$

can be written as a system of two first-order DEs by introducing two new variables,

$$x_1 = y \quad \text{and} \quad x_2 = y'.$$

Then it follows that $x_1' = y'$ and $x_2' = y''$, and we get the system

$$\begin{aligned} x_1' &= x_2, \\ x_2' &= f(t, x_1, x_2). \end{aligned} \tag{13}$$

In vector form,

$$\vec{\mathbf{x}}' = \vec{\mathbf{f}}(t, \vec{\mathbf{x}}), \tag{14}$$

where

$$\vec{\mathbf{x}} = \begin{bmatrix} x_1 \\ x_2 \end{bmatrix} \quad \text{and} \quad \vec{\mathbf{f}}(t, \vec{\mathbf{x}}) = \begin{bmatrix} x_2 \\ f(t, x_1, x_2) \end{bmatrix}.$$

(We have already met 2×2 systems in Sec. 2.6, and we wrote them in matrix-vector form in Secs. 3.1 and 4.1.)

EXAMPLE 5 Conversion The harmonic oscillator IVP

$$y'' + y = 0, \qquad y(0) = 0, \quad y'(0) = 1$$

becomes, under the substitution $x_1 = y$, $x_2 = y'$, the system

$$\begin{aligned} x_1' &= x_2, \\ x_2' &= -x_1 \end{aligned}$$

with initial conditions $x_1(0) = 0$ and $x_2(0) = 1$. In matrix form, with

$$\vec{\mathbf{x}} = \begin{bmatrix} x_1 \\ x_2 \end{bmatrix} \quad \text{and} \quad \mathbf{A} = \begin{bmatrix} 0 & 1 \\ -1 & 0 \end{bmatrix},$$

this can be written

$$\vec{\mathbf{x}}' = \mathbf{A}\vec{\mathbf{x}}, \tag{15}$$

where the initial condition takes the form

$$\vec{\mathbf{x}}(0) = \begin{bmatrix} 0 \\ 1 \end{bmatrix}.$$

The unique solution of the system (15) is the vector

$$\vec{\mathbf{x}}(t) = \begin{bmatrix} x_1 \\ x_2 \end{bmatrix} = \begin{bmatrix} \sin t \\ \cos t \end{bmatrix}.$$

In the original problem, we have found both the position, which is given by $y(t) = x_1(t) = \sin t$, and the velocity $y'(t) = x_2(t) = \cos t$. ■

EXAMPLE 6 **Streamlined Substitution** We may convert the second-order DE

$$y'' + 3y' + \sin y = 0 \tag{16}$$

to a system with somewhat simpler notation (though the subscripted variables of Example 5 have an advantage when using linear algebra tools), proceeding as follows. Let $v = y'$. Then $v' = y''$ and (16) becomes the equation

$$v' + 3v + \sin y = 0.$$

Hence,

$$\begin{aligned} y' &= v, \\ v' &= -3v - \sin y. \end{aligned}$$

■

Generalizing to nth-order Differential Equations

Restricting our attention to **linear** nth-order differential equations with constant coefficients, we will convert the DE

$$y^{(n)} + a_{n-1}y^{(n-1)} + \cdots + a_2 y'' + a_1 y' + a_0 y = f(t)$$

to an $n \times n$ system by letting

$$x_1 = y,\ x_2 = y',\ x_3 = y'', \ldots,\ x_n = y^{(n-1)}. \tag{17}$$

If we differentiate the equations in (17) we have $x_1' = y' = x_2$, $x_2' = y'' = x_3$, and, eventually,

$$\begin{aligned} x_n' = y^{(n)} &= -a_{n-1}y^{(n-1)} - a_{n-2}y^{(n-2)} - \cdots - a_2 y'' - a_1 y' - a_0 y + f(t) \\ &= -a_{n-1}x_n - a_{n-2}x_{n-1} - \cdots - a_2 x_3 - a_1 x_2 - a_0 x_1 + f(t). \end{aligned}$$

In matrix-vector form, then,

$$\begin{bmatrix} x_1' \\ x_2' \\ x_3' \\ \vdots \\ x_{n-1}' \\ x_n' \end{bmatrix} = \underbrace{\begin{bmatrix} 0 & 1 & 0 & 0 & \cdots & 0 \\ 0 & 0 & 1 & 0 & \cdots & 0 \\ 0 & 0 & 0 & 1 & \cdots & 0 \\ \vdots & \vdots & \vdots & \vdots & \ddots & \vdots \\ 0 & 0 & 0 & 0 & \cdots & 1 \\ -a_0 & -a_1 & -a_2 & -a_3 & \cdots & -a_{n-1} \end{bmatrix}}_{\text{companion matrix of coefficients}} \begin{bmatrix} x_1 \\ x_2 \\ x_3 \\ \vdots \\ x_{n-1} \\ x_n \end{bmatrix} + \begin{bmatrix} 0 \\ 0 \\ 0 \\ \vdots \\ 0 \\ f(t) \end{bmatrix}.$$

This very special form of the **matrix of coefficients**, with zeros on the main diagonal (except in the lower right-hand corner), ones just above the main diagonal, and the negatives of the original coefficients on the bottom row, is sometimes called a **companion matrix**.

EXAMPLE 7 **A Three-by-Three Conversion** To convert the third-order IVP

$$y''' + 3y'' + 5y' + 2y = e^{-t}, \quad y(0) = 1, \quad y'(0) = 3, \quad y''(0) = 2 \tag{18}$$

into a system, we let

$$x_1 = y, \quad x_2 = y', \quad \text{and} \quad x_3 = y''.$$

Then $x_1' = y' = x_2$, $x_2' = y'' = x_3$, and $x_3' = y''' = -3y'' - 5y' - 2y + e^{-t} = -3x_3 - 5x_2 - 2x_1 + e^{-t}$. Hence,

$$\begin{aligned} x_1' &= x_2, \\ x_2' &= x_3, \\ x_3' &= -2x_1 - 5x_2 - 3x_3 + e^{-t}, \end{aligned}$$

where $x_1(0) = 1$, $x_2(0) = 3$, and $x_3(0) = 2$. Letting

$$\vec{\mathbf{x}} = \begin{bmatrix} x_1 \\ x_2 \\ x_3 \end{bmatrix}, \quad \mathbf{A} = \begin{bmatrix} 0 & 1 & 0 \\ 0 & 0 & 1 \\ -2 & -5 & -3 \end{bmatrix}, \quad \text{and} \quad \vec{\mathbf{f}} = \begin{bmatrix} 0 \\ 0 \\ e^{-t} \end{bmatrix},$$

the IVP (18) becomes

$$\vec{\mathbf{x}}' = \mathbf{A}\vec{\mathbf{x}} + \vec{\mathbf{f}}, \quad \vec{\mathbf{x}}(0) = \begin{bmatrix} 1 \\ 3 \\ 2 \end{bmatrix}.$$

■

NOTE: The peculiar form of the matrix of coefficients means that while every nth-order linear equation can be written as a system, we can expect that not every system can be converted to an nth-order linear DE. An exception to this observation is that of two *linear* first-order equations. In general, these *can* be so converted to a second-order DE.

EXAMPLE 8 **Converting Back** We will solve the 2×2 system

$$\begin{aligned} x_1' &= -2x_1 + x_2, \\ x_2' &= x_1 - 2x_2, \end{aligned} \tag{19}$$

by converting it into a second-order equation.

If we solve the first equation of (19) for x_2, we obtain

$$x_2 = x_1' + 2x_1. \tag{20}$$

Substituting (20) into the second equation of (19) gives

$$(x_1' + 2x_1)' = x_1 - 2(x_1' + 2x_1).$$

Then $x_1'' + 2x_1' = x_1 - 2x_1' - 4x_1$; that is,

$$x_1'' + 4x_1' + 3x_1 = 0. \tag{21}$$

The characteristic equation of (21) is $r^2 + 4r + 3 = (r+1)(r+3) = 0$, so the general solution of (21) is

$$x_1(t) = c_1e^{-t} + c_2e^{-3t}. \tag{22}$$

From (22) we find that $x_1' = -c_1e^{-t} - 3c_2e^{-3t}$, and we can substitute into (20) to obtain $x_2 = (-c_1e^{-t} - 3c_2e^{-3t}) + 2(c_1e^{-t} + c_2e^{-3t})$, which simplifies to

$$x_2(t) = c_1e^{-t} - c_2e^{-3t}. \tag{23}$$

Thus (22) and (23) give the general solution to system (19), obtained through conversion from solving the second-order equation (21).

■

Summary

We have looked at energy conservation in an important type of one-dimensional physical system described by autonomous second-order differential equations. We found that conservative systems can be analyzed using the phase-plane energy surface.

We have also seen that nth-order differential equations can be converted into systems of first-order differential equations, providing a uniform framework for their study. The reverse conversion is not possible in general but can be carried out for 2×2 linear systems.

4.7 Problems

1. **Total Energy of a Mass-Spring** Assume the IVP

$$\ddot{x} + x = 0, \qquad x(0) = 1, \quad \dot{x}(0) = -4$$

models a mass-spring system. Find the total energy of this system.

2. **Nonconservative Mass-Spring System** Assume the IVP

$$\ddot{x} + 2\dot{x} + 26x = 0, \qquad x(0) = 1, \quad \dot{x}(0) = 4$$

models a mass-spring system.

(a) Find the solution $x(t)$ and its derivative $\dot{x}$, and evaluate $x(\pi/5)$ and $\dot{x}(\pi/5)$.

(b) Calculate the total energy $E(t)$ of the system when $t = \pi/5$.

(c) Calculate the energy loss in the system due to friction in the time interval from $t = 0$ to $t = \pi/5$.

3. **General Formula for Total Energy in an *LC*-Circuit** Use the mechanical-electrical analog from Table 4.1.3 to determine total energy in an LC-circuit (i.e., with no resistance) without input voltage. Assume the initial charge on the capacitor to be Q_0 and the initial current in the circuit to be I_0.

4. **Energy in an *LC*-Circuit** Consider an LC-circuit with $L = 4$ henries and $C = 1/16$ farads, an initial charge of 4 coulombs, an initial current of 1 amp, and no input voltage. Determine the total energy of this circuit.

5. **Energy Loss in an *LRC*-Circuit** Consider an LRC-circuit with no input voltage, $L = 1$ henry, $R = 1$ ohm, and $C = 4$ farads, where there is no initial charge on the capacitor but an initial current of 2 amps. What is the total energy in this circuit at time t? What is the energy loss in this circuit at time t? HINT: The total energy is given by $E(t) = \frac{1}{2}L\dot{Q}^2 + \frac{1}{2}\frac{1}{C}Q^2$. First solve the appropriate IVP to find Q, then differentiate to get $\dot{Q}^2$.

Questions of Energy *Answer the following for the conservative equations in Problems 6–13.*

(a) *Determine the kinetic, potential, and total energies of the system.*

(b) *Determine the equilibrium points of the system.*

(c) *Plot the potential energy $V(x)$ and use its extrema to classify the equilibrium points as stable or unstable.*

6. $\ddot{x} - x + x^3 = 0$

7. $\ddot{x} - x - x^3 = 0$

8. $\ddot{x} - x + x^2 = 0$

9. $\ddot{x} + x^2 = 0$

10. $\ddot{x} - e^x - 1 = 0$

11. $\ddot{x} + (x - 1)^2 = 0$

12. $\ddot{x} = \dfrac{1}{x^2}$

13. $\ddot{x} = (x - 1)(x - 2)$

Conservative or Nonconservative? *Decide if the differential equations in Problems 14–19 are conservative or nonconservative and describe their phase portraits.*

14. $\ddot{x} + x^2 = 0$

15. $\ddot{x} + kx = 0$

16. $\ddot{x} + \dot{x} + x^2 = 1$

17. $\ddot{\theta} + \sin\theta = 0$

18. $\ddot{\theta} + \sin\theta = 1$

19. $\ddot{\theta} + \dot{\theta} + \sin\theta = 1$

20. **Time-Reversible Systems** Some mechanical systems have the property of **time-reversal symmetry**: their behavior looks the same whether time runs forward or backward. For example, if you ran the film of a swinging frictionless pendulum backwards, you would not be able to tell. This situation is in contrast to watching the film of a person jumping from a diving board, which is *irreversible*, like most real-life processes. All conservative systems $m\ddot{x} + F(x) = 0$ are time-reversible: if we make the change of variable $t \to -t$, the second derivative, and hence the equation, is unchanged. In (a)–(d) we explore time-reversibility.

(a) Show that a conservative equation $m\ddot{x} + F(x) = 0$ remains invariant under the change of variable from forward time t to backward time $\tau = -t$.

(b) Compare the forward and backward behavior of the harmonic oscillator

$$\ddot{x} + x = 0, \qquad x(0) = 1, \quad \dot{x}(0) = 0.$$

(c) Is throwing a ball into the air, described by

$$\ddot{x} = -mg, \qquad x(0) = 0, \quad \dot{x}(0) = 100,$$

a time-reversible system?

(d) Which of the following physical systems is time-reversible?

(i) A ball rolling on a flat surface

(ii) A ball rolling down a hill

(iii) Water running in a stream

(iv) A person taking a walk

(v) Electrons moving in a circuit

(vi) The motion of atomic particles

21. **Computer Lab: Undamped Spring** Use Part 1.7 of IDE Lab 9 to visualize how the potential and kinetic energies of a mass-spring system change during an oscillation but maintain a constant total energy.

Linear Oscillations: Free Response
Lab 9 clarifies the energies.

22. **Computer Lab: Damped Vibrations** Suppose that a small amount of damping is added to a previously undamped mass-spring system. Assume that both systems (undamped and damped) have the same initial energy. Compare what happens to their total available energy over time. Use Part 2.5 of IDE Lab 9 to investigate this. What happens when the damping is increased? Write up your conclusions.

Conversion of Equations *Write each of the differential equations in Problems 23–30 as a system of first-order equations. If the equation is linear, write the system in the matrix form* $\vec{\mathbf{x}} = \mathbf{A}\vec{\mathbf{x}} + \vec{\mathbf{f}}$.

23. $\ddot{x} + \omega_0^2 x = f(t)$ (forced harmonic oscillator)

24. $\ddot{\theta} + \dfrac{g}{L}\sin\theta = 0$ (pendulum equation)

25. $ay'' + by' + cy = 0$ (equivalent to mass-spring)

26. $L\ddot{Q} + R\dot{Q} + \dfrac{1}{C}Q = 0$ (*LRC*-circuit equation)

27. $t^2\ddot{x} + t\dot{x} + (t^2 - n^2)x = 0$ (Bessel's equation)

28. $\ddot{x} + (1 + \sin\omega t)x = 0$ (Mathieu's equation)

29. $(1 - t^2)y'' - 2ty' + n(n+1)y = 0$ (Legendre's equation)

30. $\dfrac{d^4y}{dt^4} + 3\dfrac{d^3y}{dt^3} + 2\dfrac{d^2y}{dt^2} + \dfrac{dy}{dt} + 4y = 1$ (fourth-order equation)

Conversion of IVPs *For Problems 31–34, transform the given IVP into an IVP for a first-order system.*

31. $y'' - y' + 2y = \sin t, \quad y(0) = 1, \quad y'(0) = 1$

32. $y''' + ty'' + y = 1, \quad y(0) = 0, \quad y'(0) = 1, y''(0) = 2$

33. $y'' + 3y' + 2z = e^{-t}, \quad y(0) = 0, \quad y'(0) = 1,$
$z'' + y + 2z = 1, \quad z(0) = 1, \quad z'(0) = 0$

34. $y''' + y' + 2z = 1, \quad y(0) = 1, \quad y'(0) = 0, y''(0) = 1,$
$z' + y + 2z = \sin t, \quad z(0) = 1$

Conversion of Systems *Rewrite each system in Problems 35–37 as a system of first-order equations.*

35. $\ddot{x}_1 + x_1 + 2x_2 = e^{-t},$
$\ddot{x}_2 \qquad + 2x_2 = 0$

36. $y''' = f(t, y, y', y'', z, z')$
$z'' = g(t, y, y', y'', z, z')$

37. $\ddot{x}_1 = a_{11}x_1 + a_{12}x_2 + a_{13}x_3$
$\ddot{x}_2 = a_{21}x_1 + a_{22}x_2 + a_{23}x_3$
$\ddot{x}_3 = a_{31}x_1 + a_{32}x_2 + a_{33}x_3$

Solving Linear Systems *Transform each system in Problems 38–41 into a second-order differential equation. Solve the second-order equation and then obtain a solution for the system from it.*

38. $x_1' = x_2$
$x_2' = -2x_1 - 3x_2$

39. $x_1' = 3x_1 - 2x_2$
$x_2' = 2x_1 - 2x_2$

40. $x_1' = x_1 + x_2$
$x_2' = 4x_1 + x_2$

41. $x_1' = x_2 + t$
$x_2' = -2x_1 + 3x_2 + 5$

Solving IVPs for Systems *Using the solution of an appropriate second-order equation as in Problems 38–41, obtain solutions of the following IVPs for* 2 × 2 *systems.*

42. $x_1' = 6x_1 - 3x_2, \quad x_1(0) = 2,$
$x_2' = 2x_1 + x_2, \quad x_2(0) = 3$

43. $x_1' = 3x_1 + 4x_2, \quad x_1(0) = 1,$
$x_2' = 2x_1 + x_2, \quad x_2(0) = -1$

44. **Counterexample** Devise an example of a system of first-order differential equations that cannot be transformed into a single higher-order differential equation. Justify your answer.

45. Coupled Mass-Spring System The system of two second-order differential equations

$$\begin{aligned} m_1\ddot{x}_1 &= -k_1x_1 + k_2(x_2 - x_1), \\ m_2\ddot{x}_2 &= -k_2(x_2 - x_1), \end{aligned} \qquad (24)$$

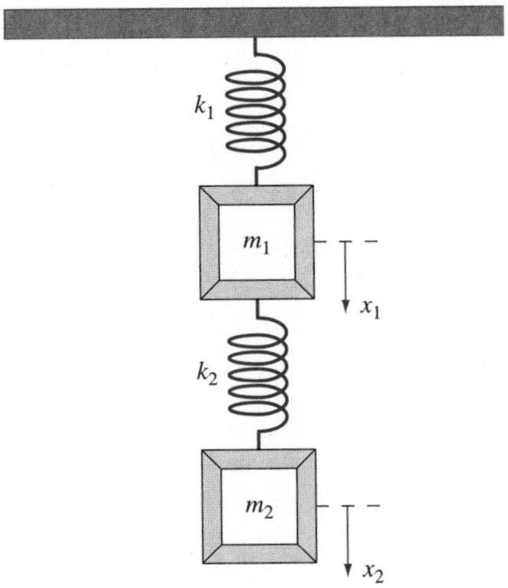

FIGURE 4.7.8 Mass-spring system (Problem 45).

describes the motion of the mass-spring system shown in Fig. 4.7.8. Rewrite this system as four first-order differential equations in the variables x_1, $v_1 = \dot{x}_1$, x_2, and $v_2 = \dot{x}_2$. Also write this system in matrix-vector form. Using your intuition about springs, can you find functions $x_1(t)$ and $x_2(t)$ that satisfy equation (24)? HINT: Set the masses and spring constants equal to 1 for simplicity.

46. Satellite Problem[4] The motion of a point mass in a force field that obeys an inverse-square law is governed by the differential equations

$$\ddot{r} = r(t)\dot{\theta}^2(t) - \frac{k}{r^2(t)} + u_1(t),$$

$$\ddot{\theta} = \frac{2\dot{\theta}(t)\dot{r}(t)}{r(t)} + \frac{1}{r(t)}u_2(t),$$

where k is a parameter, $u_1(t)$ applies a radial thrust, and $u_2(t)$ applies a tangential thrust. Convert this into a system of four first-order equations.

47. Two Inverted Pendulums[5] A cart of mass 1 unit has two inverted pendulums attached to its top; each pendulum has length 1 and a bob of mass m. An external force $u = u(t)$ is applied to the cart to stabilize the motion of the pendulums. (See Fig. 4.7.9.) For small $|\theta_1|$ and $|\theta_2|$, the motion of the pendulums is described by the system of two second-order nonhomogeneous linear differential equations

$$\ddot{\theta}_1 = (mg + 1)\theta_1 + mg\theta_2 - u(t),$$

$$\ddot{\theta}_2 = mg\theta_1 + (mg + 1)\theta_2 - u(t).$$

Convert this into a system of four first-order equations.

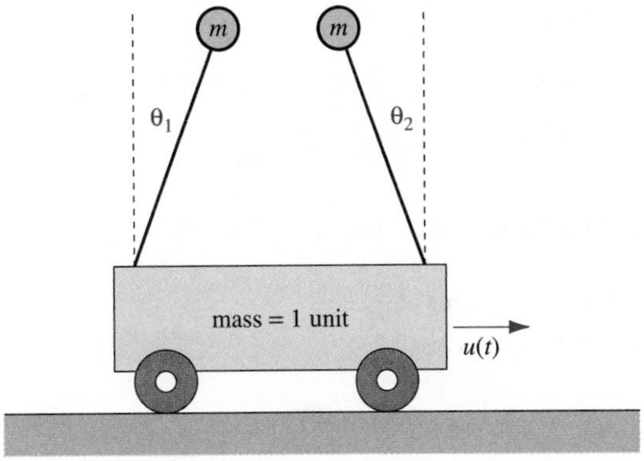

FIGURE 4.7.9 Two inverted pendulums (Problem 47).

48. Suggested Journal Entry How many things in nature can you think of that are time-reversible? Is the universe a time-reversible process? (If you know the answer to this question, contact the authors immediately!)

[4]This example is taken from the classic work on control by Roger Brockett, *Finite Dimensional Linear Systems* (NY: Wiley, 1970). For an introductory treatment of control theory, see Chapter 10.

[5]For a complete analysis of controlling two inverted pendulums, see Thomas Kailath, *Linear Systems* (NY: Prentice-Hall, 1980).

CHAPTER

5 Linear Transformations

In fact what is a mathematical creation? It does not consist in making new combinations with mathematical entities already known. Anyone could do that and the combinations so made would be infinite in number and most of them without interest. To create consists precisely in not making useless combinations but making those which are useful.

—*Henri Poincaré*

5.1 Linear Transformations

SYNOPSIS: We bring together the concepts of function from precalculus, matrix and vector space from linear algebra, derivative and integral from calculus, and differential operator from differential equations to focus on the unifying concept of linear transformation. Its importance lies in the fact that, in mapping one vector space to another, it preserves the linear structure—that is, vector addition and scalar multiplication.

Here we introduce a special kind of function for mapping one vector space to another, a mapping that is tailored specifically to preserve the vector space properties of scalar multiplication and vector addition. These functions are called *linear transformations*.

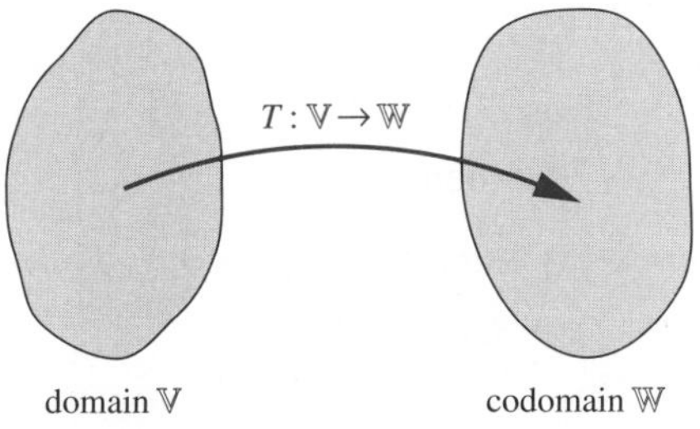

FIGURE 5.1.1 The domain $\mathbb{V}$ and codomain $\mathbb{W}$ of a linear transformation $T : \mathbb{V} \to \mathbb{W}$.

Linear Transformation

A **linear transformation** T on a vector space $\mathbb{V}$ to a vector space $\mathbb{W}$ is a function $T : \mathbb{V} \to \mathbb{W}$ that preserves *scalar multiplication* and *vector addition*. That is, for all $\vec{\mathbf{u}}, \vec{\mathbf{v}} \in \mathbb{V}$ and $c \in \mathbb{R}$,

$$T(c\vec{\mathbf{u}}) = cT(\vec{\mathbf{u}}); \tag{1}$$

$$T(\vec{\mathbf{u}} + \vec{\mathbf{v}}) = T(\vec{\mathbf{u}}) + T(\vec{\mathbf{v}}). \tag{2}$$

The vector space $\mathbb{V}$ is called the **domain** for T, and $\mathbb{W}$ is called the **codomain** (sometimes called the **target**). See Fig. 5.1.1.

The single condition

$$T(c\vec{\mathbf{u}} + d\vec{\mathbf{v}}) = cT(\vec{\mathbf{u}}) + dT(\vec{\mathbf{v}}) \tag{3}$$

is equivalent to (1) and (2) for vectors $\vec{\mathbf{u}}$, $\vec{\mathbf{v}}$ and scalars c, d.

A linear transformation always associates the zero vector of the domain with the zero vector of the codomain. Substituting $c = 0$ into equation (1) gives:

$$\begin{aligned} T(0 \cdot \vec{\mathbf{u}}) &= 0 \cdot T(\vec{\mathbf{u}}) \\ T(\vec{\mathbf{0}}) &= \vec{\mathbf{0}}. \end{aligned}$$

Alternatively, using (2) and setting $\vec{\mathbf{u}} = \vec{\mathbf{0}}$:

$$\begin{aligned} T(\vec{\mathbf{0}} + \vec{\mathbf{v}}) &= T(\vec{\mathbf{0}}) + T(\vec{\mathbf{v}}) \\ T(\vec{\mathbf{v}}) &= T(\vec{\mathbf{0}}) + T(\vec{\mathbf{v}}) \\ T(\vec{\mathbf{v}}) - T(\vec{\mathbf{v}}) &= T(\vec{\mathbf{0}}) \\ \vec{\mathbf{0}} &= T(\vec{\mathbf{0}}) \end{aligned}$$

Of course, the transformation may map nonzero vectors from the domain onto the zero vector of the codomain as well. We will have a lot more to say about the set of such vectors in Sec. 5.2.

Image of a Linear Transformation

The **image**, or **range**, of a linear transformation $T : \mathbb{V} \to \mathbb{W}$ is the set of vectors in $\mathbb{W}$ to which T maps the vectors in $\mathbb{V}$:

$$\operatorname{Im}(T) = \{\vec{\mathbf{w}} \in \mathbb{W} \mid \vec{\mathbf{w}} = T(\vec{\mathbf{v}}) \text{ for some } \vec{\mathbf{v}} \in \mathbb{V}\}.$$

The most obvious example of a linear transformation is the **identity map** id_n, which maps any vector $\vec{\mathbf{v}} \in \mathbb{R}^n$ to itself:

$$\mathrm{id}_n(\vec{\mathbf{v}}) = \vec{\mathbf{v}}.$$

Less trivial examples follow.

EXAMPLE 1 **Projection onto the xy-Plane** For the mapping $T : \mathbb{R}^3 \to \mathbb{R}^3$ defined by $T(x, y, z) = (x, y, 0)$, let us check conditions (1) and (2) for a linear transformation.

(a) Suppose that $(x, y, z) \in \mathbb{R}^3$ and $c \in \mathbb{R}$; then

$$T(c(x, y, z)) = T(cx, cy, cz) = (cx, cy, 0) = c(x, y, 0) = cT(x, y, z).$$

(b) Suppose that $(x_1, y_1, z_1), (x_2, y_2, z_2) \in \mathbb{R}^3$; then

$$\begin{aligned} T((x_1, y_1, z_1) + (x_2, y_2, z_2)) &= T(x_1 + x_2, y_1 + y_2, z_1 + z_2) \\ &= (x_1 + x_2, y_1 + y_2, 0) \\ &= (x_1, y_1, 0) + (x_2, y_2, 0) \\ &= T(x_1, y_1, z_1) + T(x_2, y_2, z_2). \end{aligned}$$

Therefore, T is linear. $\operatorname{Im}(T) = \mathbb{R}^2$, the xy-plane, a mere subset of $\mathbb{W}$. (See Fig. 5.1.2.) ■

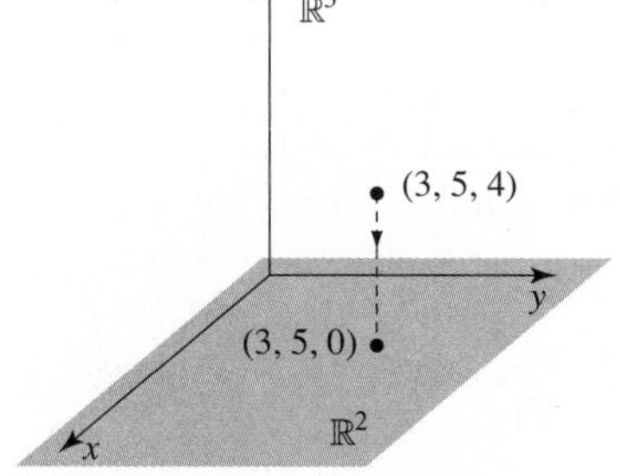

FIGURE 5.1.2 The transformation $T(x, y, z) = (x, y, 0)$ projects $\mathbb{R}^3$ onto a *plane* (Example 1).

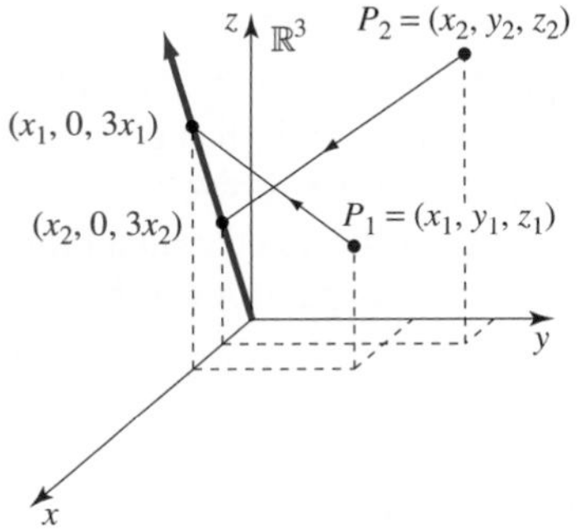

FIGURE 5.1.3 The transformation $T(x, y, z) = (x, 0, 3x)$ projects $\mathbb{R}^3$ onto a *line* (Example 2).

EXAMPLE 2 **Mapping $\mathbb{R}^3$ to a Line** We can confirm the single condition (3) to show that the mapping $T : \mathbb{R}^3 \to \mathbb{R}^3$ given by $T(x, y, z) = (x, 0, 3x)$ is a linear transformation.

Suppose that $(x_1, y_1, z_1), (x_2, y_2, z_2) \in \mathbb{R}^3$ and $c, d \in \mathbb{R}$; then

$$\begin{aligned} T(c(x_1, y_1, z_1) + d(x_2, y_2, z_2)) &= T(cx_1 + dx_2, cy_1 + dy_2, cz_1 + dz_2) \\ &= [cx_1 + dx_2, 0, 3(cx_1 + dx_2)] \\ &= c(x_1, 0, 3x_1) + d(x_2, 0, 3x_2) \\ &= cT(x_1, y_1, z_1) + dT(x_2, y_2, z_2). \end{aligned}$$

Therefore, T is linear. $\text{Im}(T) = \{(x, 0, 3x) \mid x \in \mathbb{R}\}$, a line in the xz coordinate plane in $\mathbb{R}^3$. (See Fig. 5.1.3.) ■

The geometric transformations of Examples 1 and 2 are called **projections**, because each $T : \mathbb{V} \to \mathbb{W}$ maps a vector space $\mathbb{V}$ to a lower-dimensional subspace of $\mathbb{W}$, which we have denoted $\text{Im}(T)$. In the next section we will see that the image of T is a subset of $\mathbb{W}$.

EXAMPLE 3 **Derivative Operator** Differentiation, the fundamental operation of beginning calculus, turns out to be a linear transformation. The **derivative operator** $D : \mathcal{C}^1[a, b] \to \mathcal{C}[a, b]$ is defined by

$$D(f) = f'.$$

Conditions (1) and (2) for linearity are satisfied by derivatives:

$$\begin{aligned} D(cf) &= cD(f), \\ D(f + g) &= D(f) + D(g), \end{aligned}$$

for any functions $f, g \in \mathcal{C}^1[a, b]$ and any constant c. ■

In similar fashion, one can confirm that $I : \mathcal{C}[a, b] \to \mathbb{R}$, the integration operator defined by

$$I(f) = \int_a^b f(t)\,dt,$$

is also a linear transformation. (See Problem 17.)

EXAMPLE 4 **Matrix Multiplication** If $\mathbf{A}$ is an $m \times n$ matrix and $\vec{\mathbf{x}}$ is a column n-vector, then $\mathbf{A}\vec{\mathbf{x}}$ can be considered a transformation $T : \mathbb{R}^n \to \mathbb{R}^m$, where $T(\vec{\mathbf{x}}) = \mathbf{A}\vec{\mathbf{x}}$. We leave it to the reader to confirm that the linearity properties are satisfied. ■

Geometry of Matrix Linear Transformations

When we consider the transformation T of Example 4, $T : \mathbb{R}^n \to \mathbb{R}^m$, which is defined by the $m \times n$ matrix $\mathbf{A}$, we see that the role of the matrix has changed. In Chapter 3 the emphasis was on combining matrices, and we showed ways in which they behaved like generalized numbers. Now these objects take on a more dynamic nature: they *do something*! For $\mathbf{A} = [a_{ij}]$, we can expand $\mathbf{A}\vec{\mathbf{x}} = \vec{\mathbf{b}}$:

The **identity map** id_n is defined by the square identity matrix $\mathbf{I}_n$.

$$\begin{bmatrix} a_{11} & a_{12} & \cdots & a_{1n} \\ a_{21} & a_{22} & \cdots & a_{2n} \\ \vdots & \vdots & \ddots & \vdots \\ a_{m1} & a_{m2} & \cdots & a_{mn} \end{bmatrix} \begin{bmatrix} x_1 \\ x_2 \\ \vdots \\ x_n \end{bmatrix} = \begin{bmatrix} b_1 \\ b_2 \\ \vdots \\ b_m \end{bmatrix}.$$

Multiplying by $\mathbf{A}$ transforms or changes any vector $\vec{\mathbf{x}} \in \mathbb{R}^n$ into another vector $\vec{\mathbf{b}} \in \mathbb{R}^m$.

For instance, if $m = n = 2$, the result of applying the transformation to all vectors in the plane is a transformed plane, but the origin always stays in place because $\mathbf{A}\vec{\mathbf{0}} = \vec{\mathbf{0}}$.

Shear:

A transformation that leaves coordinates *fixed* in one direction and *stretched* in another direction.

This term comes from shearing stress, which dislocates layers.

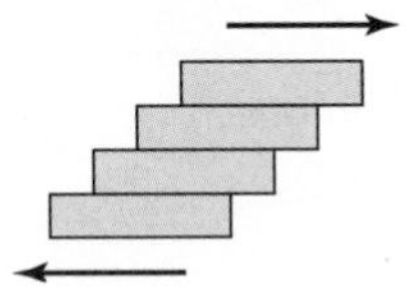

EXAMPLE 5 **Shear in the Plane** The matrix

$$\mathbf{A} = \begin{bmatrix} 1 & 1 \\ 0 & 1 \end{bmatrix}$$

defines a mapping that produces a **shear** of 1 unit in the x-direction. Let us explore the effect of the shear $\mathbf{A}$ on a square, parallel lines, and a unit circle.

(a)

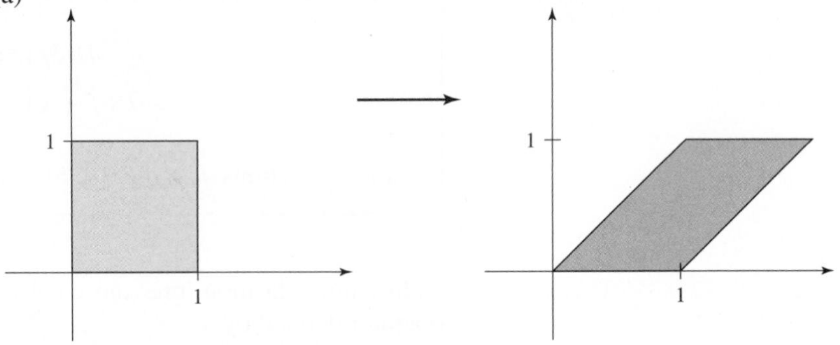

(b)

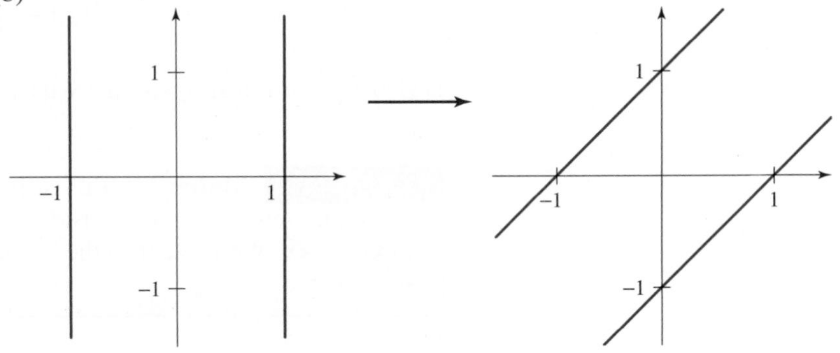

(c)

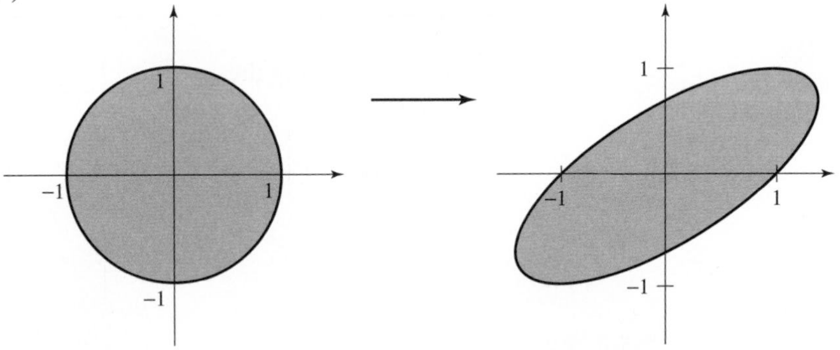

The shear **A** transformed the square into a parallelogram. The parallel lines remained parallel but in a new direction, and the circle became an ellipse. ■

EXAMPLE 6 **Rotation in the Plane** *Counterclockwise* rotation about the origin by an angle θ (Problem 81) is given by the matrix

$$\mathbf{R}_\theta = \begin{bmatrix} \cos\theta & -\sin\theta \\ \sin\theta & \cos\theta \end{bmatrix}.$$

For $\theta = \pi/6$, we can see what happens to any point (x, y) by calculating

$$\mathbf{R}_{\pi/6}\begin{bmatrix} x \\ y \end{bmatrix} = \begin{bmatrix} \cos\frac{\pi}{6} & -\sin\frac{\pi}{6} \\ \sin\frac{\pi}{6} & \cos\frac{\pi}{6} \end{bmatrix}\begin{bmatrix} x \\ y \end{bmatrix}$$

$$\approx \begin{bmatrix} 0.866 & -0.5 \\ 0.5 & 0.866 \end{bmatrix}\begin{bmatrix} x \\ y \end{bmatrix}$$

$$= \begin{bmatrix} 0.866x - 0.5y \\ 0.5x + 0.866y \end{bmatrix}.$$

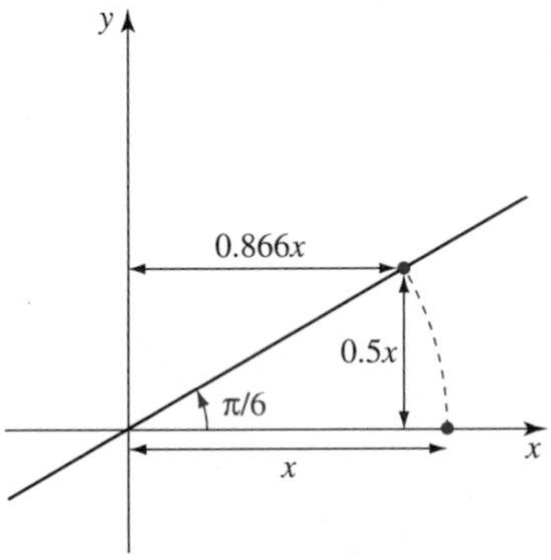

FIGURE 5.1.4 Results of a rotation transformation with matrix $\mathbf{R}_{\pi/6}$ (Example 6).

Points on the x-axis, with $y = 0$, are transformed to points with coordinates $(0.866x, 0.5x)$, as shown in Fig. 5.1.4. ■

We can represent a variety of familiar transformations using matrices. Several of these are shown in Table 5.1.1 on the next page. One can verify typical pairs of corresponding points under these mappings using the equation

$$\mathbf{A}\begin{bmatrix} x \\ y \end{bmatrix} = \begin{bmatrix} u \\ v \end{bmatrix},$$

with **A** the matrix in the first column, to compute the uv-coordinates for various xy-coordinates.

Table 5.1.1 Linear transformations of $\mathbb{R}^2$ onto $\mathbb{R}^2$

(a) Reflection about the x-axis:

$$\begin{bmatrix} 1 & 0 \\ 0 & -1 \end{bmatrix}.$$

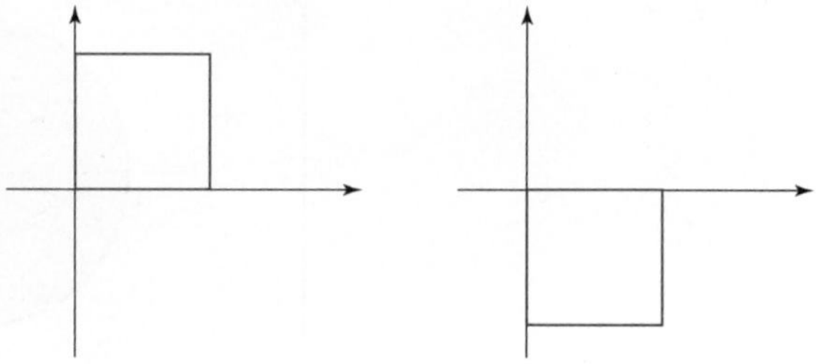

(b) Reflection about the y-axis:

$$\begin{bmatrix} -1 & 0 \\ 0 & 1 \end{bmatrix}.$$

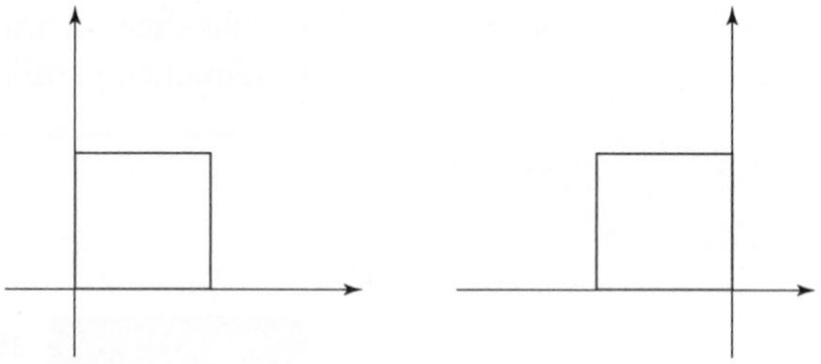

(c) Clockwise rotation of $\pi/4$ about the origin:

$$\begin{bmatrix} \cos\frac{\pi}{4} & \sin\frac{\pi}{4} \\ -\sin\frac{\pi}{4} & \cos\frac{\pi}{4} \end{bmatrix}.$$

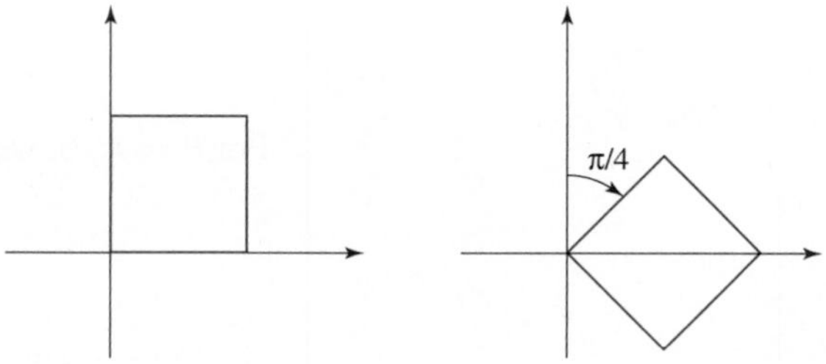

(d) Reflection about the line $y = x$:

$$\begin{bmatrix} 0 & 1 \\ 1 & 0 \end{bmatrix}.$$

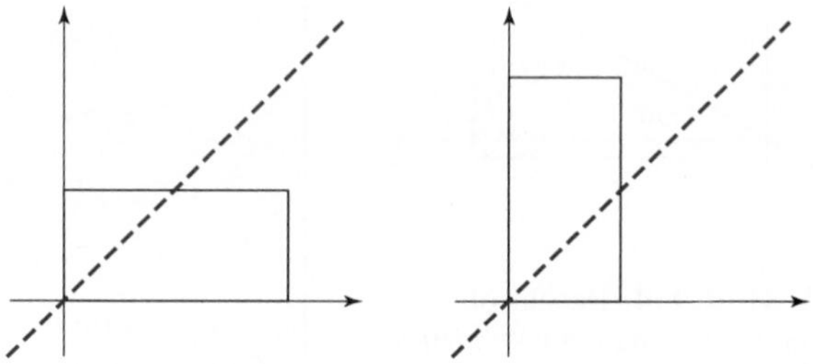

(e) Shear of 2 in the y-direction:

$$\begin{bmatrix} 1 & 0 \\ 2 & 1 \end{bmatrix}.$$

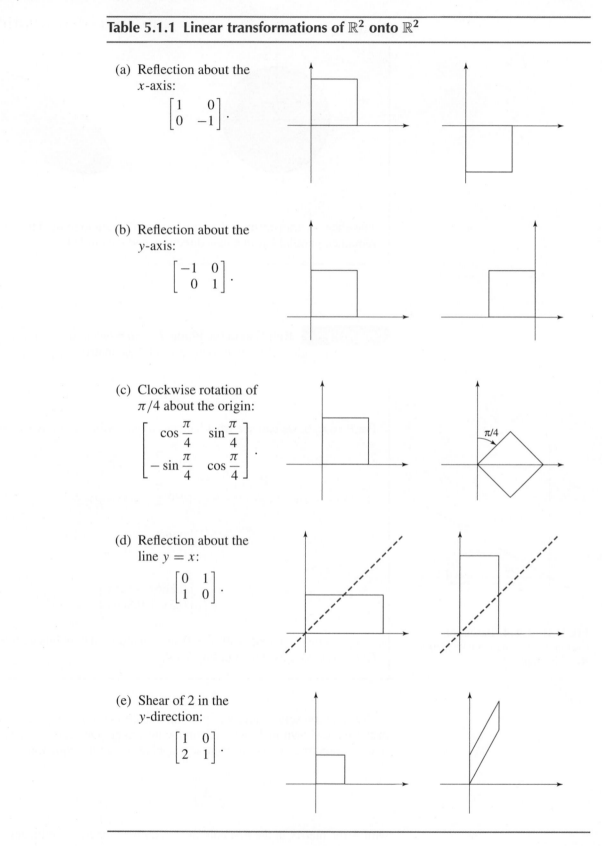

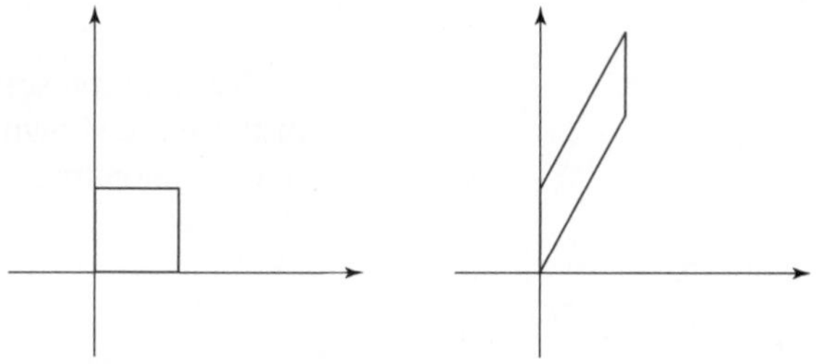

More Examples of Matrix Linear Transformations

EXAMPLE 7 **A Mapping from $\mathbb{R}^3$ to $\mathbb{R}^2$** The transformation $T : \mathbb{R}^3 \to \mathbb{R}^2$ defined by

$$T(\vec{\mathbf{v}}) = \mathbf{A}\vec{\mathbf{v}} = \begin{bmatrix} 1 & 1 & 2 \\ 2 & 3 & 5 \end{bmatrix} \vec{\mathbf{v}}$$

maps

$$\begin{bmatrix} v_1 \\ v_2 \\ v_3 \end{bmatrix} \quad \text{to} \quad \begin{bmatrix} 1 & 1 & 2 \\ 2 & 3 & 5 \end{bmatrix} \begin{bmatrix} v_1 \\ v_2 \\ v_3 \end{bmatrix} = \begin{bmatrix} v_1 + v_2 + 2v_3 \\ 2v_1 + 3v_2 + 5v_3 \end{bmatrix}.$$

What is the *image* of *this* transformation? A typical vector $\vec{\mathbf{u}}$ in the range is

$$\vec{\mathbf{u}} = \mathbf{A}\vec{\mathbf{v}} = v_1 \begin{bmatrix} 1 \\ 2 \end{bmatrix} + v_2 \begin{bmatrix} 1 \\ 3 \end{bmatrix} + v_3 \begin{bmatrix} 2 \\ 5 \end{bmatrix}.$$

It is easily verified that [1, 2] and [1, 3] are linearly independent in $\mathbb{R}^2$, so the **image**, which contains their span, must be exactly $\mathbb{R}^2$. The third vector [2, 5] is redundant; it has to be covered by the span of the first two vectors. ■

We now see that every $m \times n$ matrix $\mathbf{A}$ determines a linear transformation $T : \mathbb{R}^n \to \mathbb{R}^m$ defined by $T(\vec{\mathbf{v}}) = \mathbf{A}\vec{\mathbf{v}}$ for every $\vec{\mathbf{v}} \in \mathbb{R}^n$. Conversely, every linear transformation $T : \mathbb{R}^n \to \mathbb{R}^m$ determines a unique matrix $T(\vec{\mathbf{v}}) = \mathbf{A}\vec{\mathbf{v}}$, called the standard matrix of T, defined as follows:

Inverse Transformation: If the *inverse* matrix $\mathbf{A}^{-1}$ exists, it is associated with the *inverse* transformation $T^{-1} : \mathbb{R}^m \to \mathbb{R}^n$ defined by $T^{-1}(\vec{\mathbf{w}}) = \mathbf{A}^{-1}\vec{\mathbf{w}}$ for $\vec{\mathbf{w}} \in \mathbb{R}^n$.

The Standard Matrix for a Linear Transformation

Let $T : \mathbb{R}^n \to \mathbb{R}^m$ be a linear transformation. The **standard matrix** associated with T is defined by

$$\mathbf{A} = [\, T(\vec{\mathbf{e}}_1) \mid T(\vec{\mathbf{e}}_2) \mid \cdots \mid T(\vec{\mathbf{e}}_n) \,],$$

where the columns $T(\vec{\mathbf{e}}_j)$ are the images under T of the standard basis vectors $\vec{\mathbf{e}}_1, \vec{\mathbf{e}}_2, \ldots, \vec{\mathbf{e}}_n$.

Proof We can check that this matrix satisfies $T(\vec{\mathbf{v}}) = \mathbf{A}\vec{\mathbf{v}}$ by writing

$$\begin{aligned} T \begin{bmatrix} v_1 \\ v_2 \\ \vdots \\ v_n \end{bmatrix} &= T(v_1\vec{\mathbf{e}}_1 + v_2\vec{\mathbf{e}}_2 + \cdots + v_n\vec{\mathbf{e}}_n) \\ &= v_1 T(\vec{\mathbf{e}}_1) + v_2 T(\vec{\mathbf{e}}_2) + \cdots + v_n T(\vec{\mathbf{e}}_n) \\ &= [T(\vec{\mathbf{e}}_1) \mid T(\vec{\mathbf{e}}_2) \mid \cdots \mid T(\vec{\mathbf{e}}_n)] \begin{bmatrix} v_1 \\ v_2 \\ \vdots \\ v_n \end{bmatrix} = \mathbf{A} \begin{bmatrix} v_1 \\ v_2 \\ \vdots \\ v_n \end{bmatrix}. \end{aligned}$$

□

EXAMPLE 8 Finding Matrices for Transformations

(a) Find the standard matrix that will describe the transformation

$$T(x, y) = (x - y, x + y, 2x).$$

We seek a matrix **A** that will satisfy

$$\mathbf{A}\begin{bmatrix} x \\ y \end{bmatrix} = \begin{bmatrix} x - y \\ x + y \\ 2x \end{bmatrix}.$$

We can see that **A** must be 3×2, and for the multiplication to come out as planned, we must have

$$\mathbf{A}\left[T\begin{bmatrix} 1 \\ 0 \end{bmatrix} \middle| \; T\begin{bmatrix} 0 \\ 1 \end{bmatrix} \right] = \begin{bmatrix} 1 & -1 \\ 1 & 1 \\ 2 & 0 \end{bmatrix}.$$

(b) Let $D^2 : \mathbb{P}_3 \to \mathbb{P}_1$ be the second-derivative operator (on a subspace of the continuously twice-differentiable functions $\mathcal{C}^2$) so that for a typical cubic polynomial $ax^3 + bx^2 + cx + d$,

$$D^2(ax^3 + bx^2 + cx + d) = 6ax + 2b.$$

In matrix shorthand this becomes

$$\begin{bmatrix} \textit{matrix} \\ \textit{associated} \\ \textit{with } D^2 \end{bmatrix} \begin{bmatrix} a \\ b \\ c \\ d \end{bmatrix} = \begin{bmatrix} 6a \\ 2b \end{bmatrix}.$$

How do we determine the unknown matrix? We know that the unknown matrix must have 2 rows and 4 columns, so the required multiplications determine the matrix associated with D^2 accordingly:

$$\begin{bmatrix} 6 & 0 & 0 & 0 \\ 0 & 2 & 0 & 0 \end{bmatrix} \begin{bmatrix} a \\ b \\ c \\ d \end{bmatrix} = \begin{bmatrix} 6a \\ 2b \end{bmatrix}.$$

■

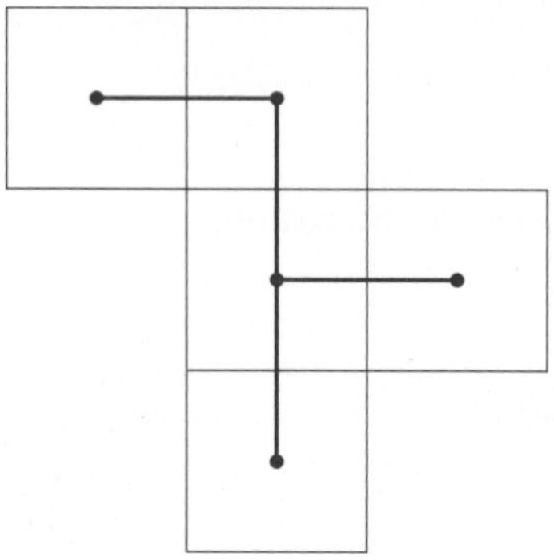

FIGURE 5.1.5 The r-pentomino.

In Appendix LT, we show how to find matrices associated with any linear transformation $T : \mathbb{V} \to \mathbb{W}$, where $\mathbb{V}$ and $\mathbb{W}$ are finite-dimensional.

Computer Graphics

Graphics programmers often use linear transformations to transform large collections of points. The "r"-shape in Fig. 5.1.5 (which looks somewhat like a lowercase letter "r") is a favorite of graphics programmers. Because of its lack of symmetry, it is useful for checking computer code for the handling of various linear transformations. These transformations, *which map line segments into line segments (or possibly a single point) while leaving an origin fixed*, include dilations and contractions, reflections, rotations, and shears. See Problems 54–66.

For the initial configuration of the r-shape (Fig. 5.1.5), we assume that the basis vectors are

$$\begin{bmatrix} 1 \\ 0 \end{bmatrix} \quad \text{and} \quad \begin{bmatrix} 0 \\ 1 \end{bmatrix},$$

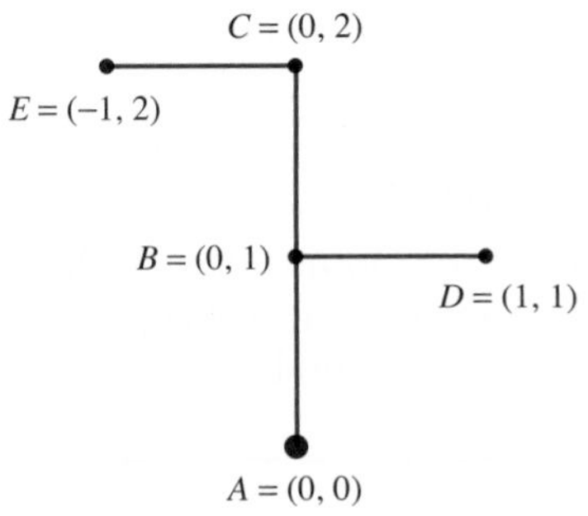

FIGURE 5.1.6 Coordinates for the r-shape; the origin has been set at A.

the standard basis (in standard order) for $\mathbb{R}^2$. Consequently, the matrix of any linear transformation must be consistent with operations on this basis.

Five points, and the lines that connect them, as shown in Fig. 5.1.6, determine the r-shape. We will work only with the points and set the origin at A.

We can make a matrix $\mathbf{X}$ to hold the information on the coordinates of these points, as shown in Table 5.1.2.

Table 5.1.2 Matrix for r-shape

Point	A	B	C	D	E
x-coordinate	0	0	0	1	−1
y-coordinate	0	1	2	1	2

Then, whatever the 2×2 transformation matrix $\mathbf{M}$, we can write the matrix for the transformed r-shape as the product $\mathbf{MX}$.

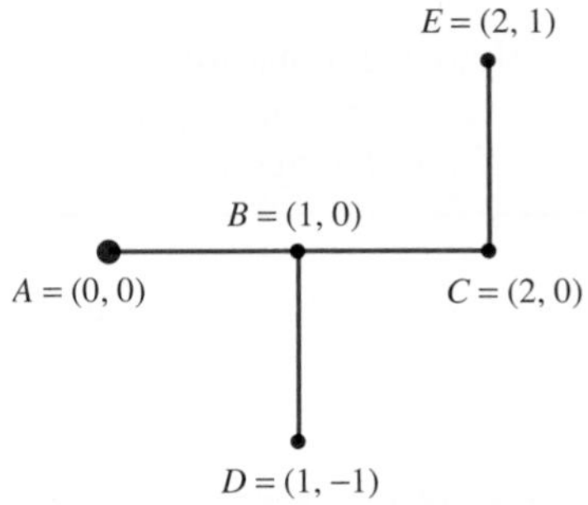

FIGURE 5.1.7 The clockwise rotation of Example 9.

EXAMPLE 9 **Clockwise Rotation** As shown in Example 6 and Problem 81, the matrix

$$\begin{bmatrix} \cos\theta & -\sin\theta \\ \sin\theta & \cos\theta \end{bmatrix}$$

rotates through an angle θ in the usual counterclockwise direction. So to rotate *clockwise* by 90°, we perform the indicated multiplication

$$\underbrace{\begin{bmatrix} 0 & 1 \\ -1 & 0 \end{bmatrix}}_{\mathbf{M}} \underbrace{\begin{bmatrix} 0 & 0 & 0 & 1 & -1 \\ 0 & 1 & 2 & 1 & 2 \end{bmatrix}}_{\vec{\mathbf{X}}} = \underbrace{\overset{\begin{matrix} A & B & C & D & E \end{matrix}}{\begin{bmatrix} 0 & 1 & 2 & 1 & 2 \\ 0 & 0 & 0 & -1 & 1 \end{bmatrix}}}_{\mathbf{M}\vec{\mathbf{X}}}$$

and observe the resulting rotated r-shape in Fig. 5.1.7. ■

Historical Note: The Power of Abstraction

The French mathematician **Jules Henri Poincaré** (1854–1912) is reputed to have said that, upon discovering a similarity between two different areas of mathematics, one should attempt to analyze it in depth.[1] Such parallels frequently make it possible to apply insights from one specialty to another, or even to develop a more general theory that includes both. Linearity has proved to be such a common property. In the early part of the nineteenth century, linearity properties in algebra and analysis developed independently. By the end of that century, German mathematician **David Hilbert** (1862–1943), Polish mathematician **Stefan Banach** (1892–1945), and many others contributed to a general theory of vector spaces and linear transformations that stimulated and enriched both matrix theory and differential equations, as well as other branches of pure and applied

[1] Poincaré, considered one of the greatest mathematical geniuses of all time, made many contributions to mathematics and other sciences. He made a point of always trying to develop his results from first principles; consequently, he was very good at explaining even complicated mathematics to others, and he was a popular science writer.

mathematics.[2] The blending of linear algebra and differential equations in this text is, in a sense, one of the fruits of their labor.

Some Common Linear Transformations

Table 5.1.3 lists several important examples of linear transformations. We have discussed the first four, and will discuss the last in Chapter 8. The reader is asked to verify the linearity of various linear transformations in the problems.

Table 5.1.3 Common linear transformations

I. $T : \mathbb{R}^n \to \mathbb{R}^m$	$T(\vec{\mathbf{x}}) = \mathbf{A}\vec{\mathbf{x}}$	Multiplication by an $m \times n$ matrix $\mathbf{A}$
II. $D^n : \mathcal{C}^n \to \mathcal{C}$	$D^n(f) = f^{(n)}$	nth-derivative operator
III. $I : \mathcal{C} \to \mathbb{R}$	$I(f) = \int_a^b f(t)\,dt$	Definite integral operator (fixed $[a, b]$)
IV. $L_n : \mathcal{C}^n \to \mathcal{C}$	$L_n(y) = y^{(n)} + a_1(t)y^{(n-1)} + \cdots + a_{n-1}(t)y' + a_n(t)y$	nth-order linear differential operator (continuous $a_1, a_2, \ldots, a_n$)
V. $\mathcal{L} : X \to Y$	$\mathcal{L}(f) = \int_0^\infty e^{-st} f(t)\,dt$	Laplace transform (for appropriate spaces X and Y). See Chapter 8.

Linear transformations $T : \mathbb{V} \to \mathbb{W}$ are sometimes called **linear operators**. If the domain and the codomain are the same (i.e., $T : \mathbb{V} \to \mathbb{V}$), then T is called a **linear operator on** $\mathbb{V}$.

Summary

Linear transformations are functions that map vector spaces into vector spaces, preserving vector addition and scalar multiplication. Examples include matrix multiplication operators, integration and differentiation operators, and the Laplace transform.

5.1 Problems

Checking Linearity *For the mapping defined in each of Problems 1–16, determine whether or not it is a linear transformation.*

1. $T : \mathbb{R}^2 \to \mathbb{R}$, $\quad T(x, y) = xy$
2. $T : \mathbb{R}^2 \to \mathbb{R}^2$, $\quad T(x, y) = (x + y, 2y)$
3. $T : \mathbb{R}^2 \to \mathbb{R}^2$, $\quad T(x, y) = (xy, 2y)$
4. $T : \mathbb{R}^2 \to \mathbb{R}^3$, $\quad T(x, y) = (x, 2, x + y)$
5. $T : \mathbb{R}^2 \to \mathbb{R}^3$, $\quad T(x, y) = (x, 0, 0)$
6. $T : \mathbb{R}^2 \to \mathbb{R}^4$, $\quad T(x, y) = (x, 1, y, 1)$

[2]Hilbert had immensely powerful insight that penetrated into the depths of a question and made unique connections with similar situations in far-flung mathematical fields. He made important contributions to many fields, including functional analysis, integral equations and quantum mechanics. His famous list of 23 open questions delivered at the Second International Congress of Mathematicians in Paris in 1899 showed the vitality of mathematics; many of these problems were solved in the twentieth century and brought forth new fields and new questions as a result.

Banach was clever at mathematics and attracted the supportive attention of key persons. He loved to work in cafés, either with others or in solitude; he produced fundamental results in topological vector spaces, and he also developed a systematic theory of functional analysis. Banach published many papers and textbooks, and began a journal and a set of monographs to help publish the work of others as well.

7. $T : \mathcal{C}[0, 1] \to \mathbb{R}$, $\quad T(f) = f(0)$

8. $T : \mathcal{C}[0, 1] \to \mathcal{C}[0, 1]$, $\quad T(f) = -f$

9. $T : \mathcal{C}^1[0, 1] \to \mathcal{C}[0, 1]$, $\quad T(f) = tf'(t)$

10. $T : \mathcal{C}^2[0, 1] \to \mathcal{C}[0, 1]$, $\quad T(f) = f'' + 2f' + 3f$

11. $T : \mathbb{P}_2 \to \mathbb{P}_1$, $\quad T(at^2 + bt + c) = 2at + b$

12. $T : \mathbb{P}_3 \to \mathbb{R}$, $\quad T(at^3 + bt^2 + ct + d) = a + b$

13. $T : \mathbb{M}_{22} \to \mathbb{M}_{22}$, $\quad T(\mathbf{A}) = \mathbf{A}^{\mathrm{T}}$

14. $T : \mathbb{M}_{22} \to \mathbb{R}$, $\quad T\begin{bmatrix} a & b \\ c & d \end{bmatrix} = \begin{vmatrix} a & b \\ c & d \end{vmatrix}$

15. $T : \mathbb{M}_{22} \to \mathbb{R}$, $\quad T(\mathbf{A}) = \mathrm{Tr}\,(\mathbf{A})$

16. $T : \mathbb{R}^n \to \mathbb{R}^m$, $T(\vec{\mathbf{x}}) = \mathbf{A}\vec{\mathbf{x}}$, where $\mathbf{A}$ is an $m \times n$ matrix

17. Integration Show that the integration operator $I : \mathcal{C}[a, b] \to \mathbb{R}$ defined by

$$I(f) = \int_a^b f(t)\,dt$$

is a linear transformation.

Linear Systems of DEs *Show that the systems of linear differential equations given in Problems 18 and 19 are linear transformations, where $x = x(t)$ and $y = y(t) \in \mathcal{C}^1(I)$.*

18. $T(x, y) = (x' - y, 2x + y')$

19. $T(x, y) = (x + y', y - 2x + y')$

Laying Linearity on the Line *Determine whether or not the mappings in Problems 20–25 are linear transformations from $\mathbb{R}$ to $\mathbb{R}$ (a and b are real constants).*

20. $T(x) = \sqrt{x}$ $\qquad$ **21.** $T(x) = ax + b$

22. $T(x) = \dfrac{1}{ax + b}$ $\qquad$ **23.** $T(x) = x^2$

24. $T(x) = \sin x$ $\qquad$ **25.** $T(x) = -\dfrac{3x}{2 + \pi}$

Geometry of a Linear Transformation *For Problems 26–28, let $T : \mathbb{R}^2 \to \mathbb{R}^2$ be the linear transformation given by $T(\vec{\mathbf{v}}) = \mathbf{A}\vec{\mathbf{v}}$, where*

$$\mathbf{A} = \begin{bmatrix} 1 & 2 \\ 0 & 1 \end{bmatrix}.$$

26. Verify that

$$T\begin{bmatrix} x \\ 0 \end{bmatrix} = \begin{bmatrix} x \\ 0 \end{bmatrix},$$

and explain why this means that the x-axis is mapped onto itself.

27. Verify that

$$T\begin{bmatrix} 0 \\ y \end{bmatrix} = \begin{bmatrix} 2y \\ y \end{bmatrix},$$

and explain why this means that the y-axis is mapped onto the line $y = x/2$.

28. Verify that

$$T\begin{bmatrix} x \\ y \end{bmatrix} = \begin{bmatrix} x \\ 0 \end{bmatrix} + \begin{bmatrix} 2y \\ y \end{bmatrix},$$

and use this fact to give a geometric interpretation of the mapping.

Geometric Interpretations in $\mathbb{R}^2$ *Construct a matrix representation for the transformations in Problems 29–31, and give a geometric interpretation of the mapping from $\mathbb{R}^2$ to $\mathbb{R}^2$. Make sketches to illustrate your conclusions.*

29. $T(x, y) = (x, -y)$ $\qquad$ **30.** $T(x, y) = (x, 0)$

31. $T(x, y) = (x, x)$

32. Composition of Linear Transformations

> **Composition Transformation**
>
> The **composition** $T \circ S : \mathbb{U} \to \mathbb{W}$ of two linear transformations $T : \mathbb{V} \to \mathbb{W}$ and $S : \mathbb{U} \to \mathbb{V}$ is defined by
>
> $$(T \circ S)(\vec{\mathbf{u}}) = T(S(\vec{\mathbf{u}})).$$

Show that the composition transformation is also a linear transformation.

Find the Standard Matrix *For each linear transformation $T : \mathbb{R}^n \to \mathbb{R}^m$ in Problems 33–40, determine the standard matrix* $\mathbf{A}$ *such that $T(\vec{\mathbf{v}}) = \mathbf{A}\vec{\mathbf{v}}$.*

33. $T(x, y) = x + 2y$ $\qquad$ **34.** $T(x, y) = (y, -x)$

35. $T(x, y) = (x + 2y, x - 2y)$

36. $T(x, y) = (x + 2y, x - 2y, y)$

37. $T(x, y, z) = (x + 2y, x - 2y, x + y - 2z)$

38. $T(v_1, v_2, v_3) = v_1 + v_3$

39. $T(v_1, v_2, v_3) = (v_1 + 2v_2, v_3, -v_1 + 4v_2 + 3v_3)$

40. $T(v_1, v_2, v_3) = (v_2, v_3, -v_1)$

Mapping and Images *For each linear transformation $T : \mathbb{R}^n \to \mathbb{R}^m$ given in Problems 41–48, compute the image under T of $\vec{\mathbf{u}}$, and find the vector(s), if any, that are mapped to $\vec{\mathbf{w}}$.*

41. $T(x, y) = (y, -x)$, $\qquad \vec{\mathbf{u}} = (0, 0)$, $\qquad \vec{\mathbf{w}} = (0, 0)$

42. $T(x, y) = (x + y, x)$, $\qquad \vec{\mathbf{u}} = (1, 0)$, $\qquad \vec{\mathbf{w}} = (3, 1)$

43. $T(x, y, z) = (x, y + z)$, $\vec{\mathbf{u}} = (0, 1, 2)$, $\vec{\mathbf{w}} = (1, 2)$

44. $T(u_1, u_2) = (u_1, u_1 + 2u_2)$, $\vec{\mathbf{u}} = (1, 2)$, $\vec{\mathbf{w}} = (1, 3)$

45. $T(u_1, u_2) = (u_1, u_1 + u_2, u_1 - u_2)$,
$\vec{\mathbf{u}} = (1, 1)$, $\vec{\mathbf{w}} = (1, 1, 0)$

46. $T(u_1, u_2) = (u_2, u_1, u_1 + u_2)$,
$\vec{\mathbf{u}} = (1, 2)$, $\vec{\mathbf{w}} = (2, 1, 3)$

47. $T(u_1, u_2, u_3) = (u_1 + u_3, u_2 - u_3)$,
$\vec{\mathbf{u}} = (1, 1, 1)$, $\vec{\mathbf{w}} = (0, 0)$

48. $T(u_1, u_2, u_3) = (u_1, u_2, u_1 + u_3)$,
$\vec{\mathbf{u}} = (1, 2, 1)$, $\vec{\mathbf{w}} = (0, 0, 1)$

Transforming Areas *For Problems 49–52, let* $T : \mathbb{R}^2 \to \mathbb{R}^2$ *be defined by* $T(\vec{\mathbf{v}}) = \mathbf{A}\vec{\mathbf{v}}$, *where*

$$\mathbf{A} = \begin{bmatrix} 1 & -1 \\ 2 & 1 \end{bmatrix}.$$

49. Determine the image under the map of the square having vertices (0, 0), (1, 0), (1, 1), and (0, 1). Calculate and compare the areas of the square and its image.

50. Repeat Problem 49 for the triangle with vertices (0, 0), (1, 1), and (−1, 1).

51. Repeat Problem 49 for the rectangle with vertices (0, 0), (1, 0), (1, 2), and (0, 2).

52. Calculate the determinant $|\mathbf{A}|$. Can you guess a connection with the results of Problems 49–51? What additional data might you collect?

53. Transforming Areas Again Repeat Problems 49–52 for the linear transformation defined by the matrix

$$\mathbf{B} = \begin{bmatrix} 2 & -1 \\ -4 & 3 \end{bmatrix}.$$

Do the results agree with any conclusion you drew from Problems 49–52? Can you argue, explain, or prove your conjecture?

54. Linear Transformations in the Plane

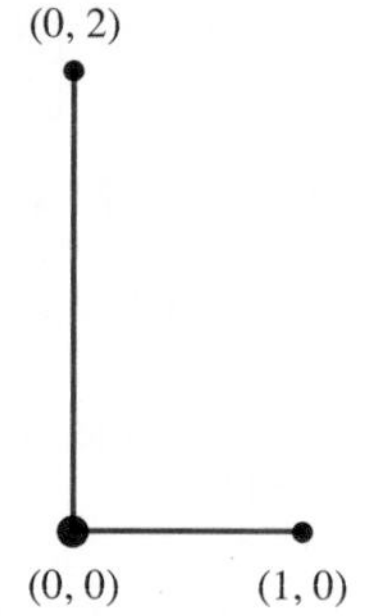

FIGURE 5.1.8 The L-shape used in Problem 54.

Images of the L-shape (Fig. 5.1.8) under various transformations of the plane are shown in Fig. 5.1.9. Each transformation is one of the following types (A)–(E):

(a) scaling (dilation or contraction);

(b) shear;

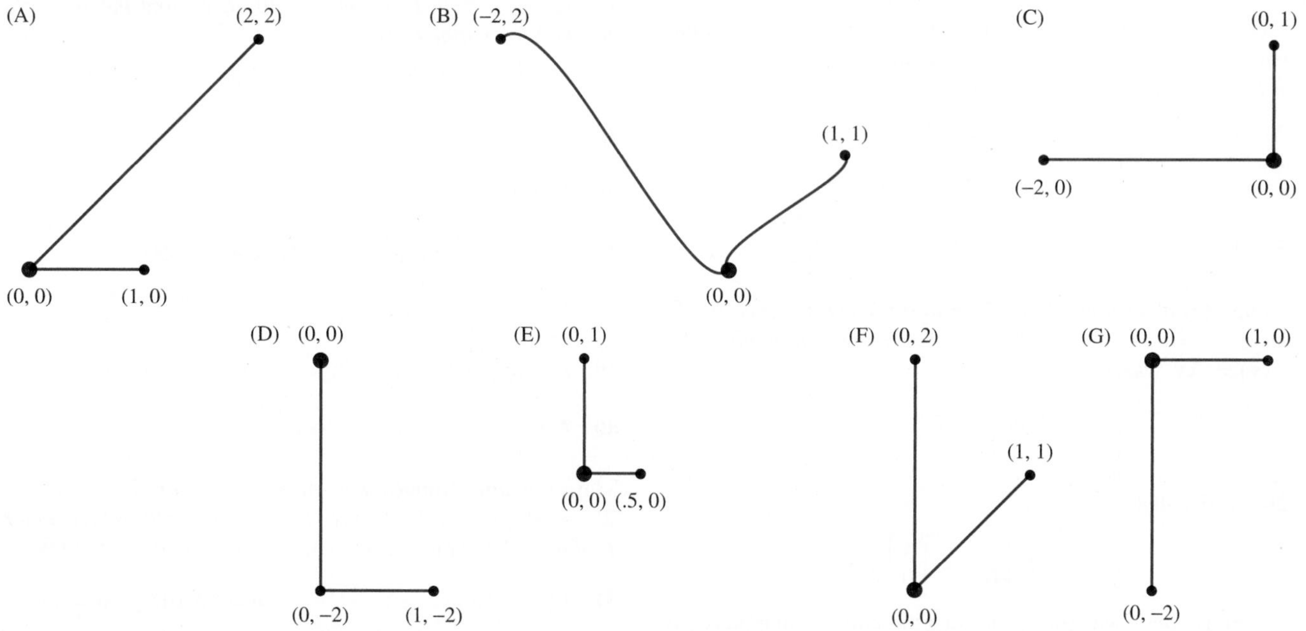

FIGURE 5.1.9 Linear transformations of the L-shape for Problems 54–59.

(c) rotation;
(d) reflection;
(e) nonlinear.

For each image, specify which type of transformation produced it. HINT: Consult Table 5.1.1.

Finding the Matrices *Each matrix in Problems 55–59 corresponds to one of the linear transformations in Problem 54. Match each matrix with the corresponding image from Fig. 5.1.9.* HINT: *Look at what happens to the unit vectors*

$$\vec{\mathbf{e}}_1 = \begin{bmatrix} 1 \\ 0 \end{bmatrix} \quad \text{and} \quad \vec{\mathbf{e}}_2 = \begin{bmatrix} 0 \\ 1 \end{bmatrix}.$$

55. $\mathbf{J} = \begin{bmatrix} 0 & -1 \\ 1 & 0 \end{bmatrix}$

56. $\mathbf{K} = \begin{bmatrix} 1 & 0 \\ 1 & 1 \end{bmatrix}$

57. $\mathbf{L} = \begin{bmatrix} 1 & 0 \\ 0 & -1 \end{bmatrix}$

58. $\mathbf{M} = \begin{bmatrix} .5 & 0 \\ 0 & .5 \end{bmatrix}$

59. $\mathbf{N} = \begin{bmatrix} 1 & 1 \\ 0 & 1 \end{bmatrix}$

60. Shear Transformation In Example 5, we looked at a shear transformation that produced a shear of one unit in the x-direction.

(a) What linear transformation matrix would perform a shear of one unit in the y-direction on the r-shape in Fig. 5.1.6? Which image in Fig. 5.1.10 corresponds to this transformation?

(b) Find the matrices for the other two shear transformations in Fig. 5.1.10.

61. Another Shear Transformation The matrix for a shear transformation of 2 units in the x-direction is

$$\mathbf{M} = \begin{bmatrix} 1 & 2 \\ 0 & 1 \end{bmatrix}.$$

Apply the transformation matrix $\mathbf{M}$ to the matrix shown in Table 5.1.2 for the r-shape in Fig. 5.1.6. Graph the transformed r-shape.

62. Clockwise Rotation In Example 6 we looked at a counterclockwise rotation about the origin. Write the matrix for a 30° *clockwise* rotation of the original r-shape. Graph the transformed r-shape.

63. Pinwheel The pinwheel in Fig. 5.1.11 is obtained from the r-shape (Fig. 5.1.6) by a shear transformation of -1 units in the y-direction followed by a succession of 30° rotations. (We are assuming that each successive transformation leaves a "print" so that the end result is the pinwheel shown.)

(a) Determine the matrix for the shear transformation and the number n of successive rotations required to complete the pinwheel.

(b) Is it true that $(\mathbf{R}_{30°})^n = \mathbf{I}$ for some n? Explain.

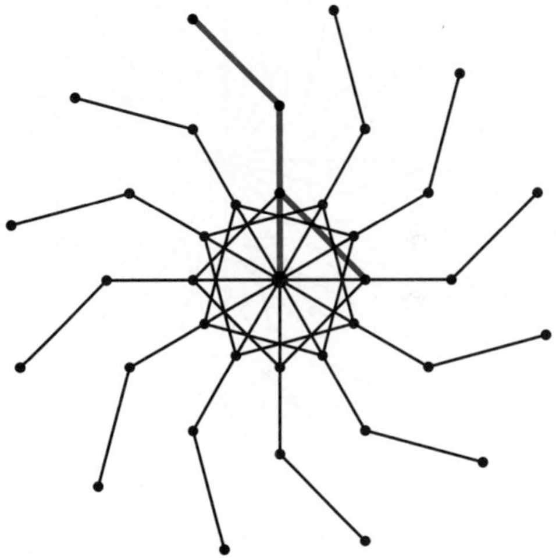

FIGURE 5.1.11 Pinwheel (scaled) for Problem 63.

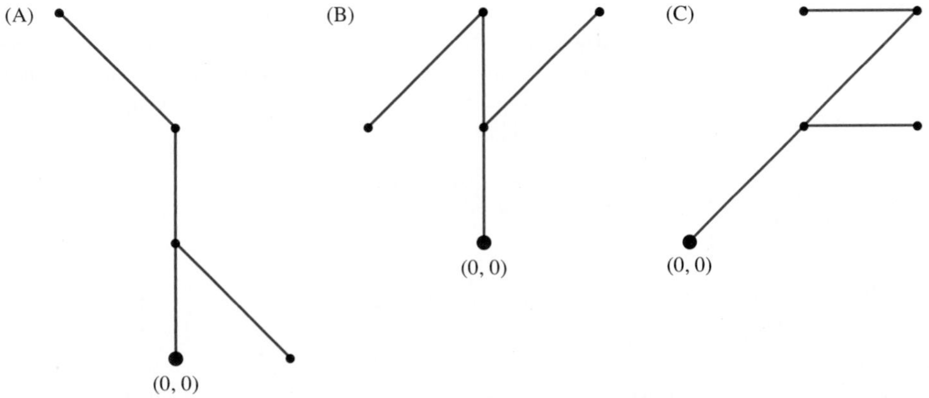

FIGURE 5.1.10 Shear transformations of the r-shape for Problem 60.

64. Flower Explain how the flower in Fig. 5.1.13 can be obtained from the F-shape in Fig. 5.1.12. Describe the succession of matrix transformations.

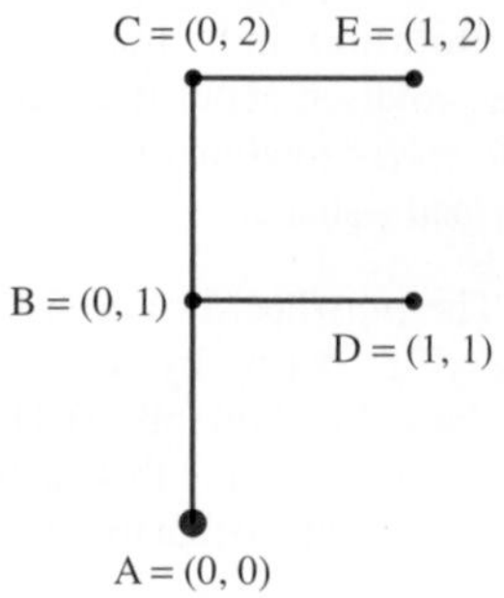

FIGURE 5.1.12 The F-shape used in Problem 64.

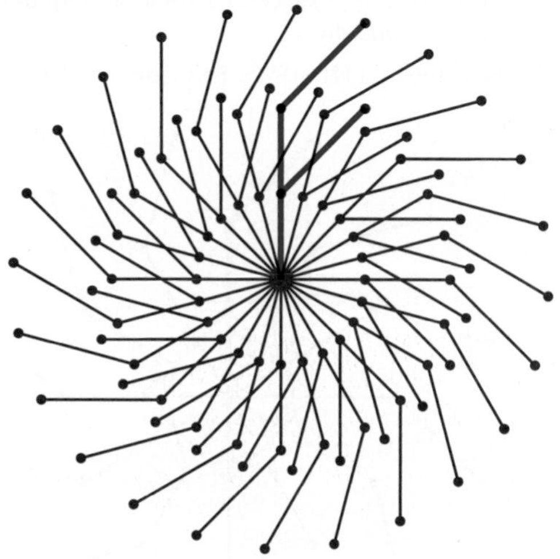

FIGURE 5.1.13 Flower (scaled) for Problem 64.

65. Successive Transformations Recall that linear transformations can be applied in succession by composition (defined in Problem 32). The corresponding process for linear transformation matrices is matrix multiplication. Consider a 1-unit shear in the y-direction followed by a counterclockwise rotation of 30°. Find the transformation matrix formed by the product of the two matrices. Sketch the transformed r-shape.

66. Reflections

(a) Reflect the r-shape (Fig. 5.1.6) about the x-axis and then about the y-axis. Find the transformation matrix

$$\mathbf{M} = \mathbf{R}_y\mathbf{R}_x,$$

where $\mathbf{R}_x$ is the matrix for reflection about the x-axis and $\mathbf{R}_y$ is the matrix for reflection about the y-axis, and sketch the transformed r-shape.

(b) What counterclockwise rotation is equivalent to these successive reflections? Illustrate with a sketch.

67. Derivative and Integral Transformations In the vector space $\mathcal{C}^\infty[a, b]$ of infinitely differentiable functions on the interval $[a, b]$, consider the derivative transformation D and the definite integral transformation I defined by

$$D(f)(x) = f'(x) \quad \text{and} \quad I(f)(x) = \int_a^x f(t)\,dt.$$

(a) Compute $(DI)(f) = D(I(f))$.

(b) Compute $(ID)(f) = I(D(f))$.

(c) Do these transformations **commute**? That is to say, is it true that $(DI)(f) = (ID)(f)$ for all vectors f in the space?

68. Anatomy of a Transformation The linear transformation $T : \mathbb{R}^2 \to \mathbb{R}^4$ is defined by $T(\vec{\mathbf{v}}) = \mathbf{A}\vec{\mathbf{v}}$, where

$$\mathbf{A} = \begin{bmatrix} 1 & -1 \\ 1 & 0 \\ 3 & 1 \\ 1 & 0 \end{bmatrix}.$$

(a) Determine the vectors in $\mathbb{R}^2$ that T maps to the zero vector in $\mathbb{R}^4$.

(b) Show that no vector in $\mathbb{R}^2$ is mapped to $[1, 1, 1, 1]$ in $\mathbb{R}^4$.

(c) Describe the subspace of $\mathbb{R}^4$ that is the image of T (that is, its range).

69. Anatomy of Another Transformation The linear transformation $T : \mathbb{R}^3 \to \mathbb{R}^2$ is defined by $T(\vec{\mathbf{v}}) = \mathbf{B}\vec{\mathbf{v}}$, where

$$\mathbf{B} = \begin{bmatrix} 1 & 1 & -1 \\ 2 & 2 & -3 \end{bmatrix}.$$

(a) Determine the vectors in $\mathbb{R}^3$ that T maps to the zero vector in $\mathbb{R}^2$.

(b) Find the vectors in $\mathbb{R}^3$ that T maps to $[1, 1]$ in $\mathbb{R}^2$.

(c) Describe the image (range) of the transformation T.

Functionals *Mappings from a vector space to the real numbers are sometimes called* ***functionals***.[3] *Determine whether the transformations in Problems 70–73 are* ***linear*** *functionals from* $\mathcal{C}[0, 1]$ *to* $\mathbb{R}$.

70. $T(f) = \dfrac{f(0) + f(1)}{2}$

71. $T(f) = \displaystyle\int_0^1 |f(t)|\,dt$

[3] Referring to a numerical-valued correspondence defined on a set of functions, the intended sense of *functional* was originally "function of a function."

72. $T(f) = -2\int_0^1 f(t)\,dt$

73. $T(f) = \sqrt{\int_0^1 f^2(t)\,dt}$

Further Linearity Checks *Verify that the mappings in Problems 74–76 are linear transformations.*

74. $L_1 : \mathcal{C}^1 \to \mathcal{C},\ L_1(y) = y' + p(t)y$ (p continuous)

75. $\mathcal{L} : X \to Y,\ \mathcal{L}(f) = \int_0^\infty e^{-st} f(t)\,dt$ (X and Y appropriate spaces)

76. $L : X \to \mathbb{R},\ L(a_n) = \lim_{n\to\infty} a_n$ (X the space of convergent real sequences)

Projections *Use the definition of projection, as stated here, for Problems 77–80.*

> **Projection**
>
> A linear transformation $T : \mathbb{V} \to \mathbb{W}$, where $\mathbb{W}$ is a subspace of $\mathbb{V}$, is called a **projection** provided that T, when restricted to $\mathbb{W}$, reduces to the identity mapping; that is, $T(\vec{\mathbf{w}}) = \vec{\mathbf{w}}$ for all vectors $\vec{\mathbf{w}}$ in the subspace $\mathbb{W}$.

77. Verify that the transformation in Example 1 is a projection. What is $\mathbb{W}$ in this case?

78. Verify that the transformation in Example 2 is a projection. What is $\mathbb{W}$ in this case?

79. Explain why the transformation $T : \mathbb{R}^3 \to \mathbb{R}^3$ given by $T(x, y, z) = (-x, 0, 3x)$ is not a projection. Identify subspace $\mathbb{W}$.

80. Is the linear transformation $T : \mathbb{R}^3 \to \mathbb{R}^3$ defined by $T(x, y, z) = (x + y, y, 0)$ a projection? Explain.

81. **Rotational Transformations** A mapping $T : \mathbb{R}^2 \to \mathbb{R}^2$ is given by $T(\vec{\mathbf{v}}) = \mathbf{A}\vec{\mathbf{v}}$, where

$$\mathbf{A} = \begin{bmatrix} \cos\theta & -\sin\theta \\ \sin\theta & \cos\theta \end{bmatrix}.$$

Show that T rotates every vector $\vec{\mathbf{v}} \in \mathbb{R}^2$ counterclockwise about the origin through angle θ. HINT: Express $\vec{\mathbf{v}}$ using polar coordinates,

$$\vec{\mathbf{v}} = \begin{bmatrix} r\cos\alpha \\ r\sin\alpha \end{bmatrix},$$

and use the identities for $\cos(\theta + \alpha)$ and $\sin(\theta + \alpha)$.

82. **Integral Transforms** If $K(s, t)$ is a continuous function of s and t on the square $0 \le s \le 1, 0 \le t \le 1$, and $f(t)$ is any continuous function of t for $0 \le t \le 1$, we can define the function F given by

$$F(s) = \int_0^1 K(s, t) f(t)\,dt.$$

Show that the mapping $T(f(t)) = F(s)$ is a linear transformation from $\mathcal{C}[0, 1]$ into itself.

Computer Lab: Matrix Machine *Use the matrix machine from IDE (Lab 15) to analyze each transformation from $\mathbb{R}^2$ to $\mathbb{R}^2$ in Problems 83–88, and answer the following questions.*

(a) *Which vectors are not moved by the transformation?*

(b) *Which nonzero vectors do not have their* direction *changed?*

(c) *Which vectors do not have their* magnitude *changed?*

(d) *Which vectors map onto the origin? This set is called the **nullspace** of the transformation.*

(e) *Which vectors, if any, map onto* $\begin{bmatrix} 1 \\ 0 \end{bmatrix}$?

(f) *Find the image of the transformation, and state whether it is all of $\mathbb{R}^2$ or a subset.*

Matrix Machine
Enter a matrix, then point/click to choose or change a vector; simultaneously you will see its transformation by the vector.

83. $\begin{bmatrix} 0 & 1 \\ -1 & 0 \end{bmatrix}$

84. $\begin{bmatrix} 1 & 1 \\ 1 & 1 \end{bmatrix}$

85. $\begin{bmatrix} 0 & 1 \\ 1 & 0 \end{bmatrix}$

86. $\begin{bmatrix} 1 & -2 \\ -2 & 3 \end{bmatrix}$

87. $\begin{bmatrix} 2 & 0 \\ 0 & 3 \end{bmatrix}$

88. $\begin{bmatrix} 1 & 2 \\ 1 & 0 \end{bmatrix}$

89. **Suggested Journal Entry** Give an intuitive description of the difference between a linear and a nonlinear transformation. Do you find that your impressions are more algebraic and computational or more geometric and pictorial?

5.2 Properties of Linear Transformations

SYNOPSIS: We continue our study of the linear transformation, learning how its kernel and image provide information about the nature of the mapping.

Introduction: Function Properties

We begin with a review of terms from calculus and precalculus.

> **Injectivity**
>
> A function $f : \mathbb{X} \to \mathbb{Y}$ is **one-to-one**, or **injective**, provided it is true that $f(u) = f(v)$ implies that $u = v$. That is, *different inputs give rise to different outputs*. See Fig. 5.2.1.

Picture This:

Which functions are injective (one-to-one) and which are surjective (onto)?

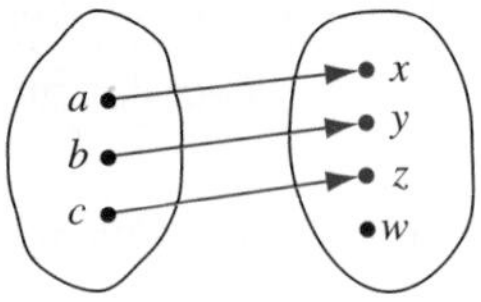

injective but not surjective

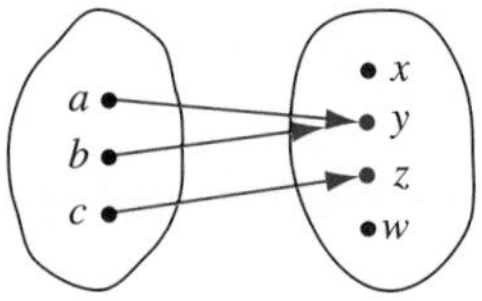

not injective, not surjective

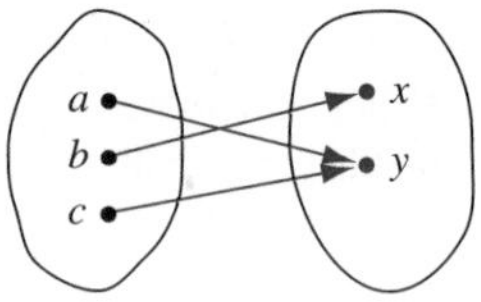

not injective but surjective

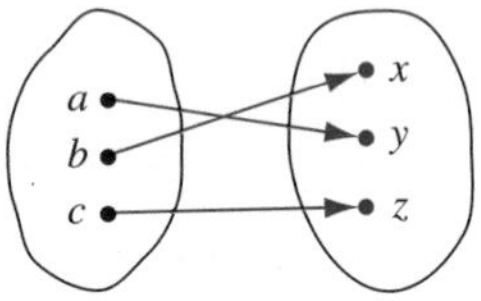

both injective and surjective

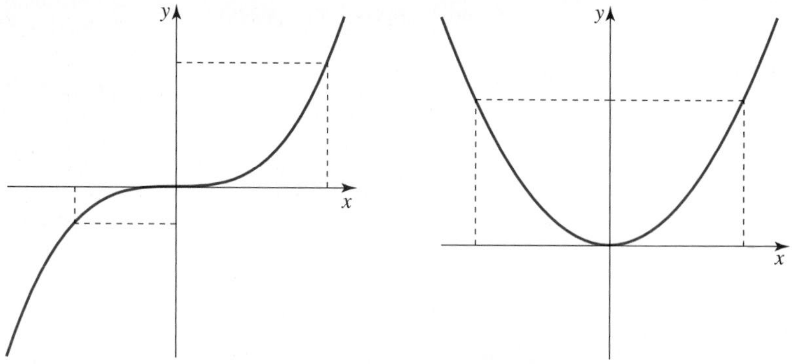

(a) $f(x) = x^3$ (b) $g(x) = x^2$

FIGURE 5.2.1 For $\mathbb{X} = \mathbb{Y} = \mathbb{R}$, $f(x)$ is injective but $g(x)$ is not.

> **Surjectivity**
>
> The set of *output* values of a function $f : \mathbb{X} \to \mathbb{Y}$ is a subset of the codomain $\mathbb{Y}$ and is called the *image* of the function. If the image is *all* of $\mathbb{Y}$, the function f is said to map **onto** $\mathbb{Y}$ or to be **surjective**. See Fig. 5.2.2.

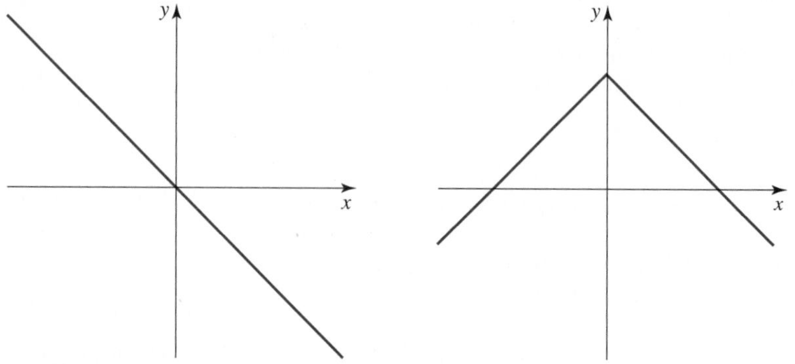

(a) $f(x) = -x$ (b) $g(x) = 1 - |x|$

FIGURE 5.2.2 $\mathbb{X} = \mathbb{Y} = \mathbb{R}$, $f(x)$ is surjective but $g(x)$ is not.

Because linear transformations are functions on vector spaces, these properties extend in a natural way. We shall consider them separately.

Image of a Linear Transformation

Recall (Sec. 5.1) that the **image**, or **range**, of a linear transformation $T : \mathbb{V} \to \mathbb{W}$ is the set of vectors in $\mathbb{W}$ to which T maps the vectors in $\mathbb{V}$. Linear transformations,

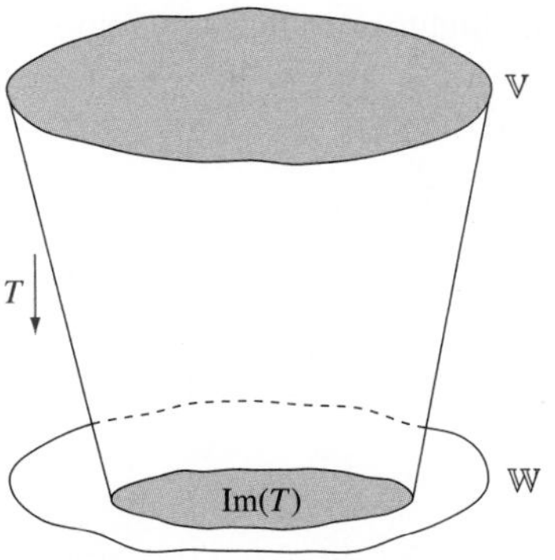

FIGURE 5.2.3 A linear transformation "spotlights" the image, a subspace of the codomain.

like the functions of calculus and precalculus, may have ranges that are subsets of their codomains or may actually map onto the entire codomain. If a transformation is not surjective, we want to look at the structure of the image, as shown in Fig. 5.2.3.

EXAMPLE 1 **The Image of a Projection** For the projection $T : \mathbb{R}^3 \to \mathbb{R}^3$ defined in Sec. 5.1, Example 1, as $T(x, y, z) = (x, y, 0)$, the image is $\mathbb{R}^2$. ■

EXAMPLE 2 **A Mapping from $\mathbb{R}^2$ to $\mathbb{R}^3$** The linear transformation $T : \mathbb{R}^2 \to \mathbb{R}^3$ defined by $T(\vec{\mathbf{v}}) = \mathbf{A}\vec{\mathbf{v}}$, where

$$\mathbf{A} = \begin{bmatrix} 1 & 1 \\ 1 & -1 \\ 2 & 1 \end{bmatrix},$$

maps a vector $\vec{\mathbf{v}}$ from domain $\mathbb{R}^2$ into a vector $\vec{\mathbf{u}}$ in codomain $\mathbb{R}^3$ as follows:

$$\mathbf{A}\vec{\mathbf{v}} = \begin{bmatrix} 1 & 1 \\ 1 & -1 \\ 2 & 1 \end{bmatrix} \begin{bmatrix} v_1 \\ v_2 \end{bmatrix} = \begin{bmatrix} v_1 + v_2 \\ v_1 - v_2 \\ 2v_1 + v_2 \end{bmatrix} = \vec{\mathbf{u}}. \tag{1}$$

What is the image of this transformation—that is, the subset of the codomain $\mathbb{R}^3$ that actually gets mapped onto by the transformation? From (1) and our knowledge of the matrix product, we know that

$$\mathbf{A}\begin{bmatrix} v_1 \\ v_2 \end{bmatrix} = v_1 \begin{bmatrix} 1 \\ 1 \\ 2 \end{bmatrix} + v_2 \begin{bmatrix} 1 \\ -1 \\ 1 \end{bmatrix} = \begin{bmatrix} u_1 \\ u_2 \\ u_3 \end{bmatrix}. \tag{2}$$

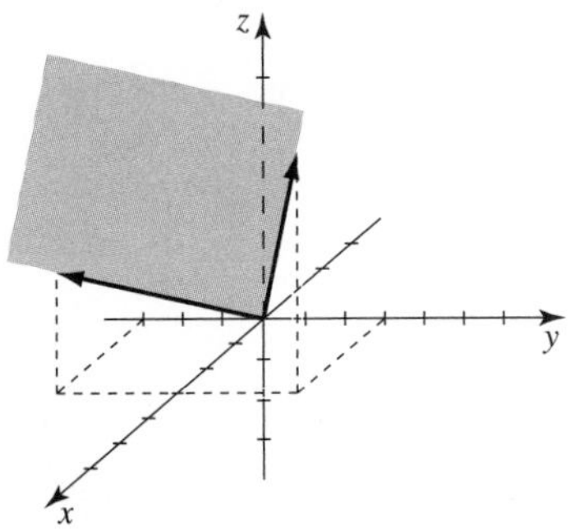

FIGURE 5.2.4 Two-dimensional image for the transformation of Example 2.

The image under the transformation T is Col **A**, the two-dimensional subspace of $\mathbb{R}^3$ spanned by the linearly independent vectors

$$\begin{bmatrix} 1 \\ 1 \\ 2 \end{bmatrix} \quad \text{and} \quad \begin{bmatrix} 1 \\ -1 \\ 1 \end{bmatrix}.$$

(Col **A** was defined in Sec. 3.6 as the span of the columns of **A**.) We can derive a scalar equation for this image, a plane through the origin, by writing (2) in the form

$$u_1 = v_1 + v_2, \quad u_2 = v_1 - v_2, \quad u_3 = 2v_1 + v_2$$

and eliminating v_1 and v_2 to obtain the equation of a plane[1] in $\mathbb{R}^3$

$$3u_1 + u_2 - 2u_3 = 0. \tag{3}$$

(See Fig. 5.2.4.) ■

In Examples 1 and 2, the transformation did not map onto its codomain but had for its image a subspace of that codomain. We sum up what we can learn from the image in the following theorem.

Image Theorem

Let $T : \mathbb{V} \to \mathbb{W}$ be a linear transformation from vector space $\mathbb{V}$ to vector space $\mathbb{W}$ with image $\text{Im}(T)$. Then

(i) $\text{Im}(T)$ is a subspace of $\mathbb{W}$;

(ii) T is surjective if and only if $\text{Im}(T) = \mathbb{W}$.

[1]We know from calculus that this plane (3) has normal vector $3\mathbf{i} + \mathbf{j} - 2\mathbf{k}$, the cross-product of the two given vectors $\mathbf{i} + \mathbf{j} + 2\mathbf{k}$ and $\mathbf{i} - \mathbf{j} + \mathbf{k}$ that lie in the plane.

Proof of the Image Theorem It is easy to see that the image is a subspace. Suppose that $\vec{\mathbf{w}}_1$ and $\vec{\mathbf{w}}_2$ are in the image, so there are vectors $\vec{\mathbf{v}}_1$ and $\vec{\mathbf{v}}_2$ in $\mathbb{V}$ such that $T(\vec{\mathbf{v}}_1) = \vec{\mathbf{w}}_1$ and $T(\vec{\mathbf{v}}_2) = \vec{\mathbf{w}}_2$. If c and d are scalars, then

$$T(c\vec{\mathbf{v}}_1 + d\vec{\mathbf{v}}_2) = cT(\vec{\mathbf{v}}_1) + dT(\vec{\mathbf{v}}_2) = c\vec{\mathbf{w}}_1 + d\vec{\mathbf{w}}_2.$$

We have found that there is a vector in $\mathbb{V}$ that maps to $c\vec{\mathbf{w}}_1 + d\vec{\mathbf{w}}_2$; therefore, $c\vec{\mathbf{w}}_1 + d\vec{\mathbf{w}}_2$ is in the image.

The second statement of the theorem is a restatement of the definition. □

Rank of a Linear Transformation

The dimension of the image of a linear transformation T is called its **rank**:

$$\text{rank}(T) \equiv \dim(\text{Im}(T)).$$

EXAMPLE 3 **Checking for the Image** For $T : \mathbb{R}^4 \to \mathbb{R}^2$ defined by

$$T(\vec{\mathbf{v}}) = \mathbf{A}\vec{\mathbf{v}} = \begin{bmatrix} 2 & -4 & 3 & 6 \\ -1 & 2 & -2 & -3 \end{bmatrix} \begin{bmatrix} v_1 \\ v_2 \\ v_3 \\ v_4 \end{bmatrix} = \vec{\mathbf{w}},$$

we can write

$$\vec{\mathbf{w}} = v_1 \begin{bmatrix} 2 \\ -1 \end{bmatrix} + v_2 \begin{bmatrix} -4 \\ 2 \end{bmatrix} + v_3 \begin{bmatrix} 3 \\ -2 \end{bmatrix} + v_4 \begin{bmatrix} 6 \\ -3 \end{bmatrix}.$$

It follows that

$$\text{Im}(T) = \text{Span}\left\{ \begin{bmatrix} 2 \\ -1 \end{bmatrix}, \begin{bmatrix} -4 \\ 2 \end{bmatrix}, \begin{bmatrix} 3 \\ -2 \end{bmatrix}, \begin{bmatrix} 6 \\ -3 \end{bmatrix} \right\},$$

a subset of $\mathbb{R}^2$ spanned by four vectors. To find its dimension, we determine the RREF of matrix $\mathbf{A}$, which is

$$\begin{bmatrix} 1 & -2 & 0 & 3 \\ 0 & 0 & 1 & 0 \end{bmatrix}.$$

The pivot columns of $\mathbf{A}$ are the column vectors that correspond to leading 1s in the RREF, so

$$\left\{ \begin{bmatrix} 2 \\ -1 \end{bmatrix}, \begin{bmatrix} 3 \\ -2 \end{bmatrix} \right\}$$

is a basis for $\text{Im}(T)$. The $\text{rank}(T) \equiv \dim(\text{Im}(T)) \equiv \dim(\text{Col }\mathbf{A}) = 2$.

In this case, T is surjective because $\text{Im}(T) = \mathbb{R}^2$. Therefore, the standard basis for $\mathbb{R}^2$ would also work, which would otherwise not automatically be the case. ■

Pivot columns:

For the basis for Col **A** must come from the original matrix **A**, not from the RREF.

Rank of a Matrix Multiplication Operator

For any linear transformation $T : \mathbb{R}^n \to \mathbb{R}^m$ defined by $T(\vec{\mathbf{x}}) = \mathbf{A}\vec{\mathbf{x}}$, where $\mathbf{A} \in \mathbb{M}_{mn}$ and $\vec{\mathbf{x}} \in \mathbb{V}$, the image of T is the column space of $\mathbf{A}$; that is, $\text{Im}(T) = \text{Col }\mathbf{A}$. The pivot columns of $\mathbf{A}$ form a basis for $\text{Im}(T)$. Consequently,

$$\begin{aligned} \text{rank}(T) &\equiv \dim(\text{Im}(T)) \equiv \dim(\text{Col }\mathbf{A}) \\ &= \text{the number of pivot columns in } \mathbf{A}. \end{aligned}$$

Kernel of a Linear Transformation

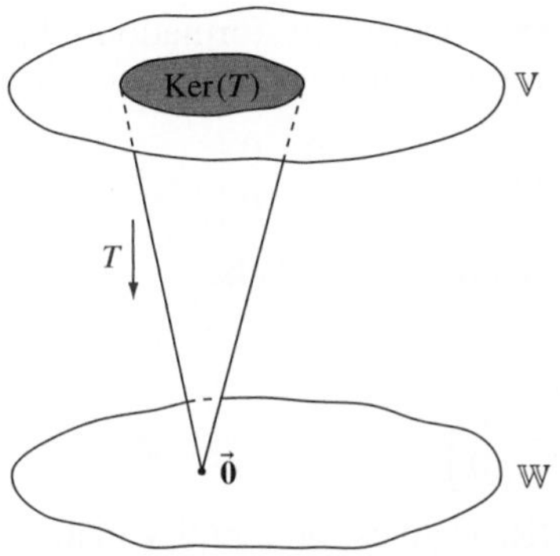

FIGURE 5.2.5 A linear transformation "squeezes" its kernel down to the zero vector.

Recall from the definition of a linear transformation that, for $T : \mathbb{V} \to \mathbb{W}$, the zero vector in the domain $\mathbb{V}$ maps to the zero vector in the codomain $\mathbb{W}$. It is possible to have other vectors of the domain map to zero as well. (See Fig. 5.2.5.)

Kernel of a Linear Transformation

The **kernel** (or **nullspace**) of a linear transformation $T : \mathbb{V} \to \mathbb{W}$, denoted $\text{Ker}(T)$, is the set of vectors in $\mathbb{V}$ mapped to the zero vector in $\mathbb{W}$:

$$\text{Ker}(T) = \{\vec{\mathbf{v}} \text{ in } \mathbb{V} \mid T(\vec{\mathbf{v}}) = \vec{\mathbf{0}}\}.$$

EXAMPLE 4 **Kernel of a Projection** Recall the projection $T : \mathbb{R}^3 \to \mathbb{R}^3$ (Example 1, Sec. 5.1) defined by $T(x, y, z) = (x, y, 0)$. The kernel of T is the z-axis:

$$\text{Ker}(T) = \{(0, 0, z) \mid z \in \mathbb{R}\}.$$

■

EXAMPLE 5 **Another Kernel** Let us revisit Example 7 of Sec. 5.1. The transformation $T : \mathbb{R}^3 \to \mathbb{R}^2$ defined by

$$T(\vec{\mathbf{v}}) = \mathbf{A}\vec{\mathbf{v}} = \begin{bmatrix} 1 & 1 & 2 \\ 2 & 3 & 5 \end{bmatrix} \vec{\mathbf{v}}$$

maps

$$\begin{bmatrix} v_1 \\ v_2 \\ v_3 \end{bmatrix} \quad \text{to} \quad \begin{bmatrix} 1 & 1 & 2 \\ 2 & 3 & 5 \end{bmatrix} \begin{bmatrix} v_1 \\ v_2 \\ v_3 \end{bmatrix} = \begin{bmatrix} v_1 + v_2 + 2v_3 \\ 2v_1 + 3v_2 + 5v_3 \end{bmatrix}.$$

What vectors does T map to $\vec{\mathbf{0}}$? That is, for what values of v_1, v_2, and v_3 do we have the following?

$$\begin{aligned} v_1 + v_2 + 2v_3 &= 0, \\ 2v_1 + 3v_2 + 5v_3 &= 0. \end{aligned}$$

The augmented matrix of this system,

$$\left[\begin{array}{ccc|c} 1 & 1 & 2 & 0 \\ 2 & 3 & 5 & 0 \end{array}\right],$$

has RREF

$$\left[\begin{array}{ccc|c} 1 & 0 & 1 & 0 \\ 0 & 1 & 1 & 0 \end{array}\right],$$

so v_3 is a free variable, and $v_1 = -v_3$ and $v_2 = -v_3$. Setting $v_3 = -s$, where s is a parameter, we find that the set of vectors mapped to zero is the one-dimensional subspace spanned by

$$\begin{bmatrix} 1 \\ 1 \\ -1 \end{bmatrix}.$$

(Parametric equations of this line through the origin in $\mathbb{R}^3$ are $v_1 = s$, $v_2 = s$, $v_3 = -s$.)

■

EXAMPLE 6 **Kernels of Different Sizes** Three linear transformations $T_{\mathbf{A}}$, $T_{\mathbf{B}}$, $T_{\mathbf{C}}$ from $\mathbb{R}^2$ to $\mathbb{R}^2$ are defined respectively by three matrices

$$\mathbf{A} = \begin{bmatrix} 1 & 1 \\ 4 & 1 \end{bmatrix}, \quad \mathbf{B} = \begin{bmatrix} 2 & 1 \\ 2 & 1 \end{bmatrix}, \quad \text{and} \quad \mathbf{C} = \begin{bmatrix} 0 & 0 \\ 0 & 0 \end{bmatrix}.$$

We solve the following systems to find and compare their kernels.

(a) $\mathbf{A}\vec{\mathbf{v}} = \vec{\mathbf{0}}$: We must solve

$$\begin{bmatrix} 1 & 1 \\ 4 & 1 \end{bmatrix} \begin{bmatrix} v_1 \\ v_2 \end{bmatrix} = \begin{bmatrix} 0 \\ 0 \end{bmatrix}.$$

Reducing this coefficient matrix to its RREF gives the identity matrix, so $v_1 = v_2 = 0$ and $\text{Ker}(T_{\mathbf{A}}) = \{\vec{\mathbf{0}}\}$.

(b) $\mathbf{B}\vec{\mathbf{v}} = \vec{\mathbf{0}}$: The RREF for matrix $\mathbf{B}$ is

$$\begin{bmatrix} 1 & 1/2 \\ 0 & 0 \end{bmatrix}.$$

Hence, $v_1 = -\frac{1}{2}v_2$, and replacing v_2 by the parameter $-2s$ gives the one-parameter family of solutions

$$\text{Ker}(T_{\mathbf{B}}) = \left\{ s \begin{bmatrix} 1 \\ -2 \end{bmatrix} \right\},$$

a one-dimensional subspace of the domain $\mathbb{R}^2$.

(c) $\mathbf{C}\vec{\mathbf{v}} = \vec{\mathbf{0}}$: It is clear from the matrix $\mathbf{C}$ of the third transformation that all vectors in the domain are mapped to zero and $\text{Ker}(T_{\mathbf{C}})$ is all of $\mathbb{R}^2$.

We have shown that it is possible for the kernel to be a subspace of dimension zero, one, or two of the domain $\mathbb{R}^2$. ■

These examples lead us to suspect (correctly) that kernels will turn out to be subspaces. Furthermore, the kernel tells whether or not the transformation is injective. (Consider the preceding examples and those of the previous section.) Let us emphasize these two points.

Kernel Theorem

Let $T : \mathbb{V} \to \mathbb{W}$ be a linear transformation from vector space $\mathbb{V}$ to vector space $\mathbb{W}$ with kernel $\text{Ker}(T)$. Then

(i) $\text{Ker}(T)$ is a subspace of $\mathbb{V}$;

(ii) T is injective if and only if $\text{Ker}(T) = \{\vec{\mathbf{0}}\}$.

Proof We have seen that the kernel always contains $\vec{\mathbf{0}}$, so it is nonempty. If vectors $\vec{\mathbf{u}}$ and $\vec{\mathbf{v}}$ are in $\text{Ker}(T)$, so that $T(\vec{\mathbf{u}}) = \vec{\mathbf{0}}$ and $T(\vec{\mathbf{v}}) = \vec{\mathbf{0}}$, and if c and d are scalars, then, by linearity,

$$T(c\vec{\mathbf{u}} + d\vec{\mathbf{v}}) = cT(\vec{\mathbf{u}}) + dT(\vec{\mathbf{v}}) = c\vec{\mathbf{0}} + d\vec{\mathbf{0}} = \vec{\mathbf{0}}.$$

This means that $c\vec{\mathbf{u}} + d\vec{\mathbf{v}}$ is in the kernel, and the kernel is a subspace by condition (4) of Sec. 3.5.

Suppose that T is injective, so that

$$T(\vec{\mathbf{u}}) = T(\vec{\mathbf{v}}) \quad \text{implies} \quad \vec{\mathbf{u}} = \vec{\mathbf{v}}.$$

If $\vec{\mathbf{w}}$ is in the kernel, then $T(\vec{\mathbf{w}}) = \vec{\mathbf{0}}$. But $T(\vec{\mathbf{0}}) = \vec{\mathbf{0}}$. Therefore,

$$T(\vec{\mathbf{w}}) = T(\vec{\mathbf{0}}), \quad \text{so} \quad \vec{\mathbf{w}} = \vec{\mathbf{0}}.$$

This means that the kernel contains only the zero vector.

On the other hand, let us assume that we know that $\text{Ker}(T) = \{\vec{\mathbf{0}}\}$, and want to prove that T is injective. If we know that $T(\vec{\mathbf{u}}) = T(\vec{\mathbf{v}})$, then by linearity we have

$$T(\vec{\mathbf{u}} - \vec{\mathbf{v}}) = T(\vec{\mathbf{u}}) - T(\vec{\mathbf{v}}) = \vec{\mathbf{0}},$$

and this says that $\vec{\mathbf{u}} - \vec{\mathbf{v}}$ is in the kernel. But the kernel contains only $\vec{\mathbf{0}}$, so $\vec{\mathbf{u}} - \vec{\mathbf{v}} = \vec{\mathbf{0}}$, that is, $\vec{\mathbf{u}} = \vec{\mathbf{v}}$. So $\text{Ker}(T) = \{\vec{\mathbf{0}}\}$ means that T is injective. □

The 2×2 geometric transformations in Table 5.1.1 illustrate kernels with dimension zero because only the point at the origin ends up at the origin; consequently, those examples are one-to-one.

EXAMPLE 7 **A Kernel in $\mathbb{R}^4$** Let us return to the transformation T of Example 3, defined by $T(\vec{\mathbf{v}}) = \mathbf{A}\vec{\mathbf{v}}$, where

$$\mathbf{A} = \begin{bmatrix} 2 & -4 & 3 & 6 \\ -1 & 2 & -2 & -3 \end{bmatrix}.$$

We determine the kernel by solving the homogeneous system $\mathbf{A}\vec{\mathbf{v}} = \vec{\mathbf{0}}$. The RREF for the augmented matrix is

$$\left[\begin{array}{cccc|c} 1 & -2 & 0 & 3 & 0 \\ 0 & 0 & 1 & 0 & 0 \end{array}\right],$$

so if $\vec{\mathbf{v}} = [v_1, v_2, v_3, v_4]$, the RREF tells us that $v_1 = 2v_2 - 3v_4$ and $v_3 = 0$. If we let $v_2 = r$ and $v_4 = s$, where r and s are parameters,

$$\vec{\mathbf{v}} = \begin{bmatrix} v_1 \\ v_2 \\ v_3 \\ v_4 \end{bmatrix} = \begin{bmatrix} 2r - 3s \\ r \\ 0 \\ s \end{bmatrix} = r \begin{bmatrix} 2 \\ 1 \\ 0 \\ 0 \end{bmatrix} + s \begin{bmatrix} -3 \\ 0 \\ 0 \\ 1 \end{bmatrix}$$

and

$$\text{Ker}(T) = \text{Span} \left\{ \begin{bmatrix} 2 \\ 1 \\ 0 \\ 0 \end{bmatrix}, \begin{bmatrix} -3 \\ 0 \\ 0 \\ 1 \end{bmatrix} \right\}.$$

The dimension of the kernel (sometimes called the **nullity**) of the transformation is 2. ■

EXAMPLE 8 **Kernel of a Differential Operator** A differential equation such as

$$y'' + y = f(t)$$

can be expressed in terms of the second-order linear differential operator $L_2 : \mathcal{C}^2 \to \mathcal{C}$, where $L_2(y) = y'' + y$. (See Table 5.1.3.) The Kernel Theorem tells us that the kernel of this transformation $L_2(y) = f$ is the set of solutions of the corresponding homogeneous equation $y'' + y = 0$, which is the two-dimensional subspace $\text{Ker}(L_2) = \text{Span}\{\sin t, \cos t\}$ of the function space $\mathcal{C}^2$, and $\dim(\text{Ker}(L_2)) = 2$. ■

EXAMPLE 9 Another Differential Operator Define the first-order linear differential operator $L_1 : \mathcal{C}^1 \to \mathcal{C}$ by

$$L_1(y) = y' + y.$$

Since the general solution of $y' + y = 0$ is $y = ce^{-t}$, $\text{Ker}(L_1) = \text{Span}\{e^{-t}\}$, a one-dimensional subspace of $\mathcal{C}^1$, so $\dim(\text{Ker}(L_1)) = 1$. ■

In Examples 8 and 9 we showed that the dimension of the kernel is exactly the number of linearly independent solutions one can expect from the order of a homogeneous differential equation.

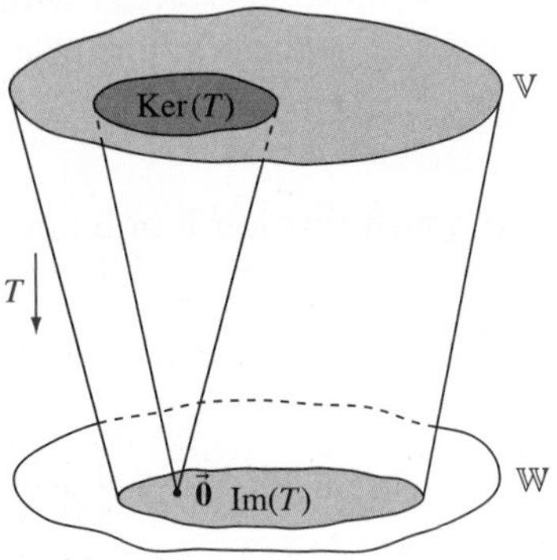

FIGURE 5.2.6 Kernel and image together.

Dimension Theorem

The information we get from Ker(T) and Im(T), shown in Fig. 5.2.6, combines for transformations on finite vector spaces, to give the following nice result.[2]

Dimension Theorem
Let $T : \mathbb{V} \to \mathbb{W}$ be a linear transformation from a finite vector space $\mathbb{V}$. Then

$$\dim(\text{Ker}(T)) + \dim(\text{Im}(T)) = \dim \mathbb{V}.$$

EXAMPLE 10 Illustrations of the Dimension Theorem

(a) In Example 1 of Sec. 5.1, $T : \mathbb{R}^3 \to \mathbb{R}^3$ defined by $T(x, y, z) = (x, y, 0)$ has $\dim(\text{Ker}(T)) = 1$ and $\dim(\text{Im}(T)) = 2$, so that

$$\dim(\text{Ker}(T)) + \dim(\text{Im}(T)) = \dim \mathbb{R}^3 = 3.$$

(b) In Example 7 we find that $T : \mathbb{R}^4 \to \mathbb{R}^2$ has $\dim \mathbb{R}^4 = 4$; hence we have $\dim(\text{Ker}(T)) = 2$. So

$$\dim(\text{Im}(T)) = 4 - 2 = 2,$$

which was exactly the result of Example 3.

(c) Let us look at $D^2 : \mathbb{P}_3(t) \to \mathbb{P}_1(t)$, where D^2 denotes the second-derivative operator, defined by

$$D^2(ax^3 + bx^2 + cx + d) = 6ax + 2b.$$

Then

$$\text{Ker}(D^2) = \{cx + d \mid c, d \in \mathbb{R}\} \quad \text{and} \quad \text{Im}(D^2) = \{6ax + 2b \mid a, b \in \mathbb{R}\}.$$

We have

$$\dim \mathbb{P}_3(t) = 4, \quad \dim(\text{Ker}(D^2)) = 2, \quad \text{and} \quad \dim(\text{Im}(D^2)) = 2,$$

which agrees with the Dimension Theorem.

[2]A proof for this theorem can be found in many higher-level texts in linear algebra, including E. M. Landesman and M. R. Hestenes, *Linear Algebra for Mathematics, Science and Engineering* (Prentice Hall, 1991), 353–355.

(d) Suppose that a 5×7 matrix $\mathbf{A}$ has four linearly independent column vectors. Let $T_{\mathbf{A}} : \mathbb{R}^7 \rightarrow \mathbb{R}^4$ be the associated linear transformation. Hence, we have the relation $\dim(\text{Col } \mathbf{A}) = \dim(\text{Im}(T)) = 4$, and $\dim(\text{Ker}(T)) = 7 - 4 = 3$.

■

See also Problems 61–64.

Solution of Nonhomogeneous Systems

One of the central ideas in differential equations comes directly from insights of linear algebra on the solutions of homogeneous and nonhomogeneous equations. We met this idea first in Sec. 2.1 for first-order linear DEs, and again in Sec. 4.4 for second-order linear DEs.

Nonhomogeneous Principle for Differential Equations
The general solution for a nonhomogeneous differential equation can be expressed in terms of a particular solution and the general solution of the corresponding homogeneous equation.

For example, we can see at once that the constant solution $y_p = 2$ satisfies the nonhomogeneous DE $y' + 2y = 4$, while the general solution of the corresponding homogenenous equation $y' + 2y = 0$ is given by $y_h = ce^{-2t}$. Then the general solution of $y' + 2y = 4$ is the sum: $y = y_h + y_p = ce^{-2t} + 2$.

We can expand this procedure to linear transformations in general, with the additional fact that the solution to the corresponding homogeneous equation is in the kernel of a transformation.

Nonhomogeneous Principle for Linear Transformations
Let $T : \mathbb{V} \rightarrow \mathbb{W}$ be a linear transformation from vector space $\mathbb{V}$ to vector space $\mathbb{W}$. Suppose that $\vec{\mathbf{v}}_p$ is *any* particular solution of the nonhomogenenous problem

$$T(\vec{\mathbf{v}}) = \vec{\mathbf{b}}. \tag{4}$$

Then the set S of all solutions of (4) is given by

$$S = \{\vec{\mathbf{v}}_p + \vec{\mathbf{v}}_h \mid \vec{\mathbf{v}}_h \in \text{Ker}(T)\}. \tag{5}$$

EXAMPLE 11 **Nonhomogeneous Algebraic Equations** The nonhomogeneous system

$$\begin{aligned} x_1 + x_2 + 3x_3 &= 4, \\ x_1 + 2x_2 + 5x_3 &= 6 \end{aligned}$$

can be described as $\mathbf{A}\vec{\mathbf{x}} = \vec{\mathbf{b}}$, where

$$\mathbf{A} = \begin{bmatrix} 1 & 1 & 3 \\ 1 & 2 & 5 \end{bmatrix} \quad \text{and} \quad \vec{\mathbf{b}} = \begin{bmatrix} 4 \\ 6 \end{bmatrix}.$$

The augmented matrix

$$\left[\begin{array}{ccc|c} 1 & 1 & 3 & 4 \\ 1 & 2 & 5 & 6 \end{array}\right] \quad \text{has RREF} \quad \left[\begin{array}{ccc|c} 1 & 0 & 1 & 2 \\ 0 & 1 & 2 & 2 \end{array}\right].$$

Replacing "free" variable x_3 (no pivot in the third column) by parameter s, we find that

$$x_1 = 2 - s, \quad x_2 = 2 - 2s, \quad x_3 = s. \tag{6}$$

It follows that

$$\vec{\mathbf{x}} = \begin{bmatrix} x_1 \\ x_2 \\ x_3 \end{bmatrix} = \begin{bmatrix} 2-s \\ 2-2s \\ s \end{bmatrix} = s\begin{bmatrix} -1 \\ -2 \\ 1 \end{bmatrix} + \begin{bmatrix} 2 \\ 2 \\ 0 \end{bmatrix} = \vec{\mathbf{x}}_h + \vec{\mathbf{x}}_p.$$

It is easy to check that

$$\vec{\mathbf{x}}_h = s\begin{bmatrix} -1 \\ -2 \\ 1 \end{bmatrix}$$

is the typical element of the one-dimensional kernel of the linear transformation $T(\vec{\mathbf{x}}) = \mathbf{A}\vec{\mathbf{x}}$ defined by $\mathbf{A}$, and that

$$\vec{\mathbf{x}}_p = \begin{bmatrix} 2 \\ 2 \\ 0 \end{bmatrix}$$

is a particular solution of $\mathbf{A}\vec{\mathbf{x}} = \vec{\mathbf{b}}$. The general solution of the nonhomogeneous system is therefore the set $S = \left\{\vec{\mathbf{x}}_p + \vec{\mathbf{x}}_h \mid \vec{\mathbf{x}}_h \in \text{Ker}(T)\right\}$.

Geometrically, the kernel is the line through the origin of $\mathbb{R}^3$ having parametric equations $x_1 = -s$, $x_2 = -2s$, and $x_3 = s$. The solution set S is the line parallel to the kernel through the point (2, 2, 0), another one-dimensional space, as shown in Fig. 5.2.7. (Its parametric equations are given by (6) above.) ■

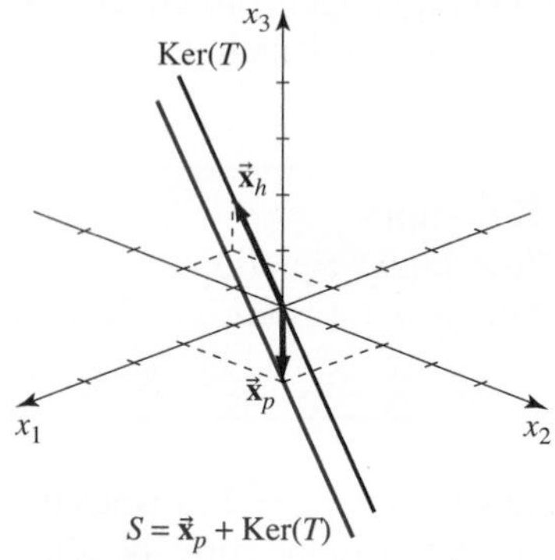

FIGURE 5.2.7 The solution set S of Example 11 is a translation of the kernel.

EXAMPLE 12 **Nonhomogeneous Differential Equation** To solve the second-order linear nonhomogenenous differential equation

$$y'' + y' - 2y = -4t, \tag{7}$$

we express it in terms of the operator $L_2 : C^2 \to C$ defined by

$$L_2(y) = y'' + y' - 2y.$$

The kernel of L_2 is the set of solutions of the homogeneous equation

$$y'' + y' - 2y = 0.$$

This DE has characteristic equation $r^2 + r - 2 = (r-1)(r+2) = 0$. Hence,

$$\text{Ker}(L_2) = \text{Span}\{e^t, e^{-2t}\} = \{c_1e^t + c_2e^{-2t} \mid c_1, c_2 \in \mathbb{R}\}.$$

One can verify easily that $L_2(2t+1) = -4t$, so $y_p = 2t+1$ is a particular solution of the nonhomogeneous problem $L_2(y) = -4t$. It follows that the solution set of (7) is

$$S = \{c_1e^t + c_2e^{-2t} + 2t + 1 \mid c_1, c_2 \in \mathbb{R}\}.$$

■

Summary

We have learned that the kernel of a linear transformation tells us when it is injective, and the image of a linear transformation tells us when it is surjective. The sum of the dimension of the kernel and the dimension of the image of a linear transformation is the dimension of its domain.

5.2 Problems

Finding Kernels *Find the kernel for the linear transformations in Problems 1–11. Describe the kernel.*

1. $T : \mathbb{R}^2 \to \mathbb{R}^2$, $T(x, y) = (-x, y)$

2. $T : \mathbb{R}^3 \to \mathbb{R}^2$, $T(x, y, z) = (2x + 3y - z, -x + 4y + 6z)$

3. $T : \mathbb{R}^3 \to \mathbb{R}^3$, $T(x, y, z) = (x, y, 0)$

4. $T : \mathbb{R}^3 \to \mathbb{R}^3$, $T(x, y, z) = (x - z, x - 2y, y - z)$

5. $D : C^1 \to C$, $D(f) = f'$

6. $D^2 : C^2 \to C$, $D^2(f) = f''$

7. $L_1 : C^1 \to C$, $L_1(y) = y' + p(t)y$

8. $L_n : C^n \to C$, $L_n(y) = y^{(n)} + a_1(t)y^{(n-1)} + \cdots + a_{n-1}(t)y' + a_n(t)y$

9. $T : \mathbb{M}_{23} \to \mathbb{M}_{32}$, $T(\mathbf{A}) = \mathbf{A}^{\mathrm{T}}$

10. $T : \mathbb{M}_{33} \to \mathbb{M}_{33}$, $T\begin{bmatrix} a & b & c \\ d & e & f \\ g & h & i \end{bmatrix} = \begin{bmatrix} a & 0 & 0 \\ 0 & e & 0 \\ 0 & 0 & i \end{bmatrix}$

11. $T : \mathbb{P}_2 \to \mathbb{P}_3$, $T(p) = \int_0^x p(t)\,dt$ for fixed x

Calculus Kernels *The transformations in Problems 12–15 should be familiar from calculus. Identify each transformation and give its kernel. (Problem 14 can have many correct answers.)*

12. $T : \mathbb{P}_2 \to \mathbb{P}_2$, $T(at^2 + bt + c) = 2at + b$

13. $T : \mathbb{P}_2 \to \mathbb{P}_2$, $T(at^2 + bt + c) = 2a$

14. $T : \mathbb{P}_2 \to \mathbb{P}_2$, $T(at^2 + bt + c) = 0$

15. $T : \mathbb{P}_3 \to \mathbb{P}_3$, $T(at^3 + bt^2 + ct + d) = 6at + 2b$

Superposition Principle *For Problems 16–20, suppose that $T : \mathbb{V} \to \mathbb{W}$ is a linear transformation from vector space $\mathbb{V}$ to vector space $\mathbb{W}$. Also suppose that $\vec{\mathbf{u}}_1$ is a solution of $T(\vec{\mathbf{u}}) = \vec{\mathbf{b}}_1$, and that $\vec{\mathbf{u}}_2$ is a solution of $T(\vec{\mathbf{u}}) = \vec{\mathbf{b}}_2$. Then $\vec{\mathbf{u}}_1 + \vec{\mathbf{u}}_2$ is a solution of $T(\vec{\mathbf{u}}) = \vec{\mathbf{b}}_1 + \vec{\mathbf{b}}_2$; this is called the* ***Superposition Principle****, as first introduced in Sec. 2.1.*

16. Use linearity to prove the Superposition Principle.

17. Show that $y = \cos t - \sin t$ is a solution of the nonhomogeneous linear equation $y'' - y' - 2y = 4\sin t - 2\cos t$.

18. Show that $y = t^2 - 2$ is a solution of

$$y'' - y' - 2y = 6 - 2t - 2t^2.$$

19. Use Problems 17 and 18 and the Superposition Principle to write the general solution of

$$y'' - y' - 2y = 4\sin t - 2\cos t + 6 - 2t - 2t^2.$$

20. Generalize the Superposition Principle to three or more terms.

Dissecting Transformations *In each of Problems 21–40, a transformation $T(\vec{\mathbf{v}}) = \mathbf{A}\vec{\mathbf{v}}$, $T : \mathbb{R}^n \to \mathbb{R}^m$, is given by a matrix $\mathbf{A}$. For each transformation, find the kernel, the image, and their dimensions. Determine whether the transformation is injective or surjective.*

21. $\begin{bmatrix} 0 & 0 \\ 0 & 0 \end{bmatrix}$

22. $\begin{bmatrix} 1 & 0 \\ 0 & -1 \end{bmatrix}$

23. $\begin{bmatrix} 1 & 0 \\ 0 & 0 \end{bmatrix}$

24. $\begin{bmatrix} 1 & 2 \\ 4 & 1 \end{bmatrix}$

25. $\begin{bmatrix} 1 & 2 \\ 2 & 4 \end{bmatrix}$

26. $\begin{bmatrix} 1 & 1 \\ 4 & 1 \end{bmatrix}$

27. $\begin{bmatrix} 1 & 1 & 1 \\ 1 & 2 & 1 \end{bmatrix}$

28. $\begin{bmatrix} 1 & 2 & 1 \\ 2 & 4 & 2 \end{bmatrix}$

29. $\begin{bmatrix} 1 & 2 & 1 \\ 2 & 1 & 2 \end{bmatrix}$

30. $\begin{bmatrix} 1 & 3 & 1 \\ 2 & 2 & 1 \end{bmatrix}$

31. $\begin{bmatrix} 1 & 1 \\ 1 & 2 \\ 1 & 1 \end{bmatrix}$

32. $\begin{bmatrix} 1 & 2 \\ 2 & 4 \\ 1 & 2 \end{bmatrix}$

33. $\begin{bmatrix} 0 & 0 \\ 0 & 0 \\ 0 & 0 \end{bmatrix}$

34. $\begin{bmatrix} 1 & 1 \\ 2 & 1 \\ 3 & 1 \end{bmatrix}$

35. $\begin{bmatrix} 1 & 2 & 1 \\ 0 & 1 & 1 \\ 0 & 0 & 1 \end{bmatrix}$

36. $\begin{bmatrix} 1 & 1 & 1 \\ 1 & 2 & 1 \\ 2 & 3 & 2 \end{bmatrix}$

37. $\begin{bmatrix} 1 & 2 & 1 \\ 2 & 4 & 1 \\ 1 & 1 & 1 \end{bmatrix}$

38. $\begin{bmatrix} 1 & 2 & 1 \\ 3 & 2 & 2 \\ 2 & 3 & 1 \end{bmatrix}$

39. $\begin{bmatrix} 1 & 2 & 0 \\ 0 & 1 & 1 \\ 0 & 0 & 1 \end{bmatrix}$

40. $\begin{bmatrix} 1 & 1 & 0 \\ 0 & 1 & 0 \\ 0 & 0 & 0 \end{bmatrix}$

Transformations and Linear Dependence

41. Let $T : \mathbb{R}^n \to \mathbb{R}^m$ be a linear transformation, and let $\{\vec{\mathbf{v}}_1, \vec{\mathbf{v}}_2, \vec{\mathbf{v}}_3\}$ be a linearly dependent set in $\mathbb{R}^n$. Prove that the set $\{T(\vec{\mathbf{v}}_1), T(\vec{\mathbf{v}}_2), T(\vec{\mathbf{v}}_3)\}$ is linearly dependent in $\mathbb{R}^m$.

42. Let $T : \mathbb{R}^n \to \mathbb{R}^m$ be a linear transformation, and let $\{\vec{\mathbf{v}}_1, \vec{\mathbf{v}}_2, \vec{\mathbf{v}}_3\}$ be a linearly independent set in $\mathbb{R}^n$. Give a counterexample to show that $\{T(\vec{\mathbf{v}}_1), T(\vec{\mathbf{v}}_2), T(\vec{\mathbf{v}}_3)\}$ need not be linearly independent in $\mathbb{R}^m$.

43. Let $T : \mathbb{R}^n \to \mathbb{R}^m$ be an injective linear transformation, and let $\{\vec{\mathbf{v}}_1, \vec{\mathbf{v}}_2, \vec{\mathbf{v}}_3\}$ be a linearly independent set in $\mathbb{R}^n$. Prove that $\{T(\vec{\mathbf{v}}_1), T(\vec{\mathbf{v}}_2), T(\vec{\mathbf{v}}_3)\}$ must be a linearly independent set in $\mathbb{R}^m$.

44. Prove that if a linear transformation T maps two linearly independent vectors onto a linearly dependent set, then the equation $T(\vec{\mathbf{x}}) = \vec{\mathbf{0}}$ has a nontrivial solution.

45. Consider the transformation $T : \mathbb{P}_2 \to \mathbb{R}^2$ defined by

$$T(p(t)) = \begin{bmatrix} p(0) \\ p(1) \end{bmatrix}.$$

For example, if $p(t) = t^2 - 6t + 4$, then

$$T(p(t)) = \begin{bmatrix} 4 \\ -1 \end{bmatrix}.$$

(a) Prove that T is a linear transformation.

(b) Find a basis for the kernel of T.

(c) Find a basis for the image of T.

Kernels and Images *Find the kernel and image of each linear transformation in Problems 46–51.*

46. $T : \mathbb{M}_{22} \to \mathbb{M}_{22}$, $\quad T(\mathbf{A}) = \mathbf{A}^{\mathrm{T}}$

47. $T : \mathbb{P}_3 \to \mathbb{P}_3$, $\quad T(p) = p'$

48. $T : \mathbb{M}_{22} \to \mathbb{M}_{22}$, $\quad T\begin{bmatrix} a & b \\ c & d \end{bmatrix} = \begin{bmatrix} a & b \\ b & c \end{bmatrix}$

49. $T : \mathbb{M}_{22} \to \mathbb{R}^2$, $\quad T\begin{bmatrix} a & b \\ c & d \end{bmatrix} = \begin{bmatrix} a+b \\ c+d \end{bmatrix}$

50. $T : \mathbb{R}^5 \to \mathbb{R}^5$, $\quad T(a, b, c, d, e) = (a, 0, c, 0, e)$

51. $T : \mathbb{R}^2 \to \mathbb{R}^3$, $\quad T(x, y) = (x + y, 0, x - y)$

Examples of Matrices *Give examples of matrices* **A** *in* $\mathbb{M}_{33}$ *such that* $T(\vec{\mathbf{x}}) = \mathbf{A}\vec{\mathbf{x}}$ *has the properties described in Problems 52–54.*

52. The Im(T) is the plane $2x - 3y + z = 0$.

53. The Im(T) is the line spanned by $\left\{\begin{bmatrix} 2 \\ 0 \\ 0 \end{bmatrix}\right\}$.

54. The Ker(T) is spanned by $\left\{\begin{bmatrix} 1 \\ 0 \\ 1 \end{bmatrix}, \begin{bmatrix} 0 \\ 1 \\ 2 \end{bmatrix}\right\}$.

True/False Questions *Answer Problems 55–60 true or false, and give a brief explanation or counterexample.*

55. If **A** is a square matrix, then Ker$(\mathbf{A}^2)$ = Ker(**A**). True or false?

56. If **A** is a square matrix, then Im$(\mathbf{A}^2)$ = Im(**A**). True or false?

57. If **A** is a square matrix, then Ker(**A**) = Ker(RREF). True or false?

58. If **A** is a square matrix, then Im(**A**) = Im(RREF). True or false?

59. If **A** and **B** are $n \times n$ matrices, then is it true or false that

$$\text{Ker}(\mathbf{A} + \mathbf{B}) = \text{Ker}(\mathbf{A}) + \text{Ker}(\mathbf{B})?$$

60. Im(**A**) for $\mathbf{A} = \begin{bmatrix} 1 & 1 \\ 1 & 1 \end{bmatrix}$ is a line in $\mathbb{R}^2$. True or false?

61. Detective Work A transformation $T : \mathbb{R}^4 \to \mathbb{R}^2$ is defined with matrix multiplication to be $T(\vec{\mathbf{v}}) = \mathbf{A}\vec{\mathbf{v}}$. It is known that the RREF of **A** is

$$\begin{bmatrix} 1 & -2 & 3 & 0 \\ 0 & 0 & 0 & 1 \end{bmatrix}.$$

Determine dim(Ker(T)) and dim(Im(T)). Is T one-to-one? Is it onto $\mathbb{R}^2$? Find bases for the kernel and image.

62. Detecting Dimensions Consider the transformation $T : \mathbb{R}^2 \to \mathbb{R}^4$ defined by $T(\vec{\mathbf{v}}) = \mathbf{B}\vec{\mathbf{v}}$. The RREF of **B** is

$$\begin{bmatrix} 1 & 0 \\ 0 & 1 \\ 0 & 0 \\ 0 & 0 \end{bmatrix}.$$

Determine dim(Ker(T)) and dim(Im(T)). Is T one-to-one? Is it onto $\mathbb{R}^4$?

63. Still Investigating For the transformation $T : \mathbb{R}^3 \to \mathbb{R}^4$ defined by $T(\vec{\mathbf{v}}) = \mathbf{A}\vec{\mathbf{v}}$, where **A** has RREF

$$\begin{bmatrix} 1 & 0 & 0 \\ 0 & 1 & 0 \\ 0 & 0 & 1 \\ 0 & 0 & 0 \end{bmatrix},$$

determine dim(Ker(T)) and dim(Im(T)). Is T one-to-one? Is it onto $\mathbb{R}^4$?

64. Dimension Theorem Again Consider transformation $T : \mathbb{R}^3 \to \mathbb{R}^3$ defined by $T(\vec{\mathbf{v}}) = \mathbf{C}\vec{\mathbf{v}}$, where **C** has RREF

$$\begin{bmatrix} 1 & -2 & 3 \\ 0 & 0 & 0 \\ 0 & 0 & 0 \end{bmatrix}.$$

Determine dim(Ker(T)) and dim(Im(T)) of transformation T, and decide whether it is injective and/or surjective.

65. The Inverse Transformation If $T : \mathbb{V} \to \mathbb{W}$ is an injective linear transformation, then we can define an inverse transformation $T^{-1} : \text{Im}(T) \to \mathbb{V}$ so that, for each $\vec{\mathbf{w}}$ in Im(T), $T^{-1}(\vec{\mathbf{w}}) = \vec{\mathbf{v}}$ if and only if $T(\vec{\mathbf{v}}) = \vec{\mathbf{w}}$. Show that T^{-1} is an injective and surjective linear transformation.

Review of Nonhomogeneous Algebraic Systems *Express the general solution for each system in Problems 66–71 as* *the sum of a particular solution and the solution of the corresponding homogeneous system.*

66. $x + y = 1$

67. $3x - y + z = -4$

68. $\begin{aligned} x + 2y &= 2 \\ 2x + y &= 2 \end{aligned}$

69. $\begin{aligned} x - 2y &= 5 \\ 2x + 4y &= -5 \end{aligned}$

70. $\begin{aligned} x + 2y - z &= 6 \\ 2x - y + 3z &= -3 \end{aligned}$

71. $\begin{aligned} x_1 + 3x_2 - 4x_3 &= 9 \\ -2x_1 + x_2 + 2x_3 &= -9 \\ -9x_1 + 15x_2 &= -3 \end{aligned}$

Review of Nonhomogeneous First-Order DEs *In each of Problems 72–77, express the general solution of the nonhomogeneous DE as the sum of a particular solution and the general solution of the corresponding homogenenous equation. The homogeneous equations are linear or separable; particular solutions (mostly constant) may be found by inspection.*

72. $y' - y = 3$

73. $y' + 2y = -1$

74. $y' + \frac{1}{t}y = \frac{1}{t}$

75. $y' + \frac{1}{t^2}y = \frac{2}{t^2}$

76. $y' + t^2 y = 3t^2$

77. $y' + ty = 1 + t^2$

Review of Nonhomogeneous Second-Order DEs *For each equation in Problems 78–81, express the general solution of the nonhomogeneous DE as the sum of a particular solution (each is a polynomial in t) and the general solution of the corresponding homogeneous DE.*

78. $y'' + y' - 2y = 2t - 3$

79. $y'' - 2y' + 2y = 4t - 6$

80. $y'' - 2y' + y = t - 3$

81. $y'' + y = 2t$

82. Suggested Journal Entry I The matrix of a linear transformation has been transformed to its reduced row echelon form. Discuss what information about the transformation you can obtain by knowing how many pivots there are and in which rows and columns they appear.

83. Suggested Journal Entry II The rows of an $m \times n$ matrix **A**, considered as n-vectors, span a subspace of $\mathbb{R}^n$ called the **row space** of **A**. Its columns span a subspace of $\mathbb{R}^m$ called the **column space** of **A**. If a linear transformation $T : \mathbb{R}^n \to \mathbb{R}^m$ is defined by $T(\vec{\mathbf{v}}) = \mathbf{A}\vec{\mathbf{v}}$, discuss the relationship to T of the row and column spaces of **A**.

5.3 Eigenvalues and Eigenvectors

SYNOPSIS: We study special vector directions (eigenvectors) and scalar multipliers (eigenvalues) associated with a square matrix or with a more general linear transformation. These eigenvectors and eigenvalues are useful both for understanding matrices (and the associated transformations) and for applying them to a variety of problems.

Matrix Machine

Construct a matrix and watch it transform vectors as fast as you move the mouse. A vector goes in with a click, and a transformed vector pops up.

Introductory Example

A linear transformation $T : \mathbb{R}^2 \to \mathbb{R}^2$ is defined by $T(\vec{\mathbf{u}}) = \mathbf{A}\vec{\mathbf{u}}$, where

$$\mathbf{A} = \begin{bmatrix} 1 & 2 \\ 2 & -2 \end{bmatrix}. \quad (1)$$

In general, T maps vector $\vec{\mathbf{u}}$ to a vector $T(\vec{\mathbf{u}})$ in a different direction. We have given examples of this in Fig. 5.3.1, showing $\vec{\mathbf{u}}$ and $T(\vec{\mathbf{u}})$ on the same diagram.

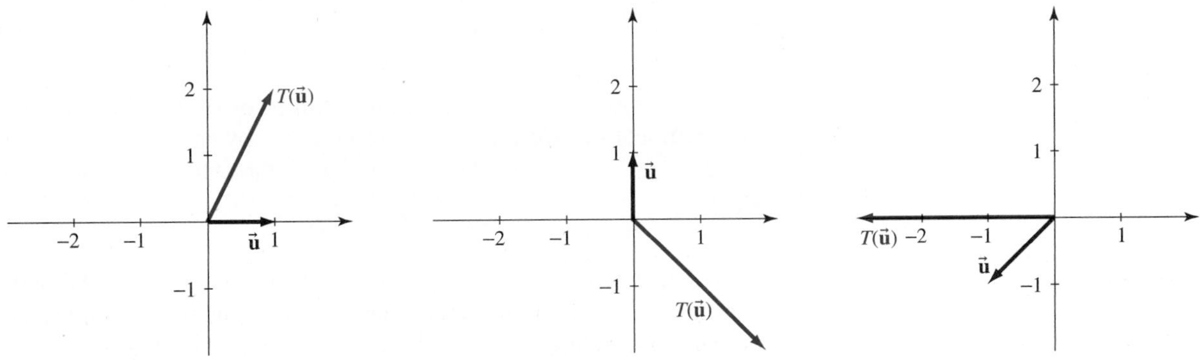

(a) $T\begin{bmatrix} 1 \\ 0 \end{bmatrix} = \begin{bmatrix} 1 \\ 2 \end{bmatrix}$ (b) $T\begin{bmatrix} 0 \\ 1 \end{bmatrix} = \begin{bmatrix} 2 \\ -2 \end{bmatrix}$ (c) $T\begin{bmatrix} -1 \\ -1 \end{bmatrix} = \begin{bmatrix} -3 \\ 0 \end{bmatrix}$

FIGURE 5.3.1 General vectors mapped by $T(\vec{\mathbf{u}}) = \mathbf{A}\vec{\mathbf{u}}$.

But something different happens for the special vectors in Fig. 5.3.2.

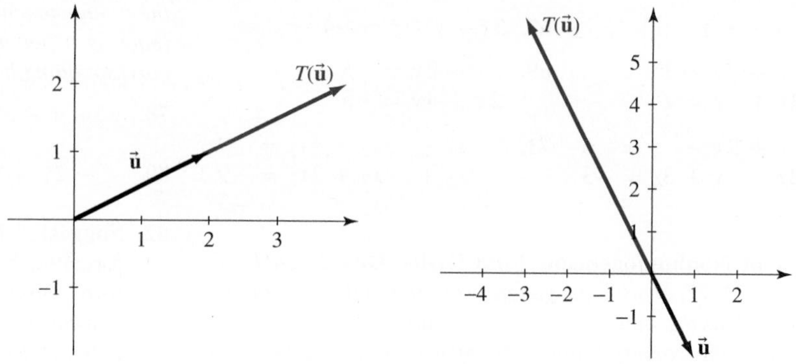

(a) $T\begin{bmatrix}2\\1\end{bmatrix} = \begin{bmatrix}4\\2\end{bmatrix} = 2\begin{bmatrix}2\\1\end{bmatrix}$ (b) $T\begin{bmatrix}1\\-2\end{bmatrix} = \begin{bmatrix}-3\\6\end{bmatrix} = -3\begin{bmatrix}1\\-2\end{bmatrix}$

FIGURE 5.3.2 Special vectors (in color), mapped by $T(\vec{\mathbf{u}}) = \mathbf{A}\vec{\mathbf{u}}$, that keep the same or opposite orientation.

Transforming vector [2, 1] by T gives a multiple of it, twice as long but in the same direction. For $[1, -2]$ the image is three times the length and in the opposite direction. In these special directions the transformation reduces to multiplication by a scalar. Such special directions (vectors) and corresponding multipliers are useful in understanding the mapping, and are important in applications as diverse as models of epidemics, behavior of economic variables, and the buckling of structural columns.

Vectors that are *not* rotated but simply stretched, shrunk, and/or reversed by a linear transformation are called "eigenvectors."

Eigenvalues and Eigenvectors

Given a square matrix **A** and the transformation it defines, the introductory example suggests that we look for vectors mapped onto multiples of themselves. (They point either in the same direction or in opposite directions.) We formulate this idea as follows.

Eigenvalue and Eigenvector

Let $T : \mathbb{V} \to \mathbb{V}$ be a linear transformation from vector space $\mathbb{V}$ into vector space $\mathbb{V}$. A scalar λ is an **eigenvalue** of T if there is a *nonzero* vector $\vec{\mathbf{v}} \in \mathbb{V}$ such that

$$T(\vec{\mathbf{v}}) = \lambda\vec{\mathbf{v}}.$$

Such a nonzero vector $\vec{\mathbf{v}}$ is called an **eigenvector** of T corresponding to λ.

If the linear transformation T is represented by an $n \times n$ matrix **A**, where $\mathbb{V} = \mathbb{R}^n$, and $T(\vec{\mathbf{v}}) = \mathbf{A}\vec{\mathbf{v}}$, then λ and $\vec{\mathbf{v}}$ are characterized by the equation

$$\mathbf{A}\vec{\mathbf{v}} = \lambda\vec{\mathbf{v}}. \qquad (2)$$

We will usually work with eigenvalues that are real numbers.[1] We will indicate later in this section how we can understand this equation when the eigenvalue is not real. Also, while the vector $\vec{\mathbf{v}} = \vec{\mathbf{0}}$ satisfies (2), we have excluded it as an

[1]The words *eigenvalue* and *eigenvector* are German-English hybrids ("eigen" means "belonging to" or "distinguished"). Eigenvalues are also called **proper values** or **characteristic values**.

Eigen-Engine

Find eigenvectors (with a click of the mouse) for 2×2 matrices.

eigenvector. Of course, if $\mathbf{A}\vec{\mathbf{v}} = \vec{\mathbf{0}}$, then $\vec{\mathbf{v}}$ is in the kernel of the transformation; the nonzero vectors of the kernel correspond to the eigenvalue $\lambda = 0$.

Computing Eigenvalues and Eigenvectors

If $\mathbf{I}$ is the identity matrix of the same size as $\mathbf{A}$, equation (2) may be written $\mathbf{A}\vec{\mathbf{v}} = \lambda\mathbf{I}\,\vec{\mathbf{v}}$, which is equivalent to $\mathbf{A}\vec{\mathbf{v}} - \lambda\mathbf{I}\,\vec{\mathbf{v}} = \vec{\mathbf{0}}$. Factoring the left-hand side gives

$$(\mathbf{A} - \lambda\mathbf{I})\vec{\mathbf{v}} = \vec{\mathbf{0}}. \tag{3}$$

While equation (3) always has the trivial solution $\vec{\mathbf{v}} = \vec{\mathbf{0}}$, we want eigenvectors that by definition are nonzero. But we know that nonzero solutions to (3) exist only if the coefficient matrix is singular; that is, when its determinant is zero. To find eigenvalues and eigenvectors, therefore, we must have

Characteristic Equation

$$|\mathbf{A} - \lambda\mathbf{I}| = 0,$$

called the **characteristic equation** of matrix $\mathbf{A}$. The polynomial in λ, denoted

$$p(\lambda) = |\mathbf{A} - \lambda\mathbf{I}|,$$

is called the **characteristic polynomial of A.**

A general procedure for finding eigenvalues and eigenvectors emerges.

Finding Eigenvalues and Eigenvectors for $n \times n$ Matrix A

Step 1. Write the characteristic equation,

$$|\mathbf{A} - \lambda\mathbf{I}| = 0. \tag{4}$$

Step 2. Solve the characteristic equation for the eigenvalues.

Step 3. For each eigenvalue λ_i, find the eigenvector(s) $\vec{\mathbf{v}}_i$ by solving the algebraic system

$$(\mathbf{A} - \lambda_i\mathbf{I})\vec{\mathbf{v}}_i = \vec{\mathbf{0}}. \tag{5}$$

Eigenvectors Are Not Unique:

An eigenvector is just a direction. Any nonzero multiple of $\vec{\mathbf{v}}_i$ serves just as well.

For large matrices with n greater than 2 or 3, these steps become cumbersome, but computer algebra systems (and some calculators) can come to the rescue, once the principles involved are understood.

EXAMPLE 1 **Confirming Our Experiment** The introductory example suggested that the matrix $\mathbf{A}$ given in equation (1) had eigenvalues 2 and -3, with corresponding eigenvectors

$$\begin{bmatrix} 2 \\ 1 \end{bmatrix} \quad \text{and} \quad \begin{bmatrix} 1 \\ -2 \end{bmatrix},$$

respectively. The captions of Fig. 5.3.2 show that the defining equation (2) for eigenvalues and eigenvectors is satisfied in each case. But what if we had to find these objects "starting from scratch"?

Step 1. For our matrix $\mathbf{A} = \begin{bmatrix} 1 & 2 \\ 2 & -2 \end{bmatrix}$, the characteristic equation (4) would be

$$\begin{bmatrix} 1-\lambda & 2 \\ 2 & -2-\lambda \end{bmatrix} = (1-\lambda)(-2-\lambda) - 4 = \lambda^2 + \lambda - 6 = 0. \quad (6)$$

Step 2. The characteristic equation (6) factors easily to $(\lambda - 2)(\lambda + 3) = 0$, so the eigenvalues $\lambda_1 = 2$ and $\lambda_2 = -3$ are readily apparent.

Step 3. To find the eigenvectors we now return to equation (5) and substitute our eigenvalues.

- For $\lambda_1 = 2$,

$$(\mathbf{A} - 2\mathbf{I})\vec{\mathbf{v}}_1 = \begin{bmatrix} 1-\lambda_1 & 2 \\ 2 & -2-\lambda_1 \end{bmatrix} = \begin{bmatrix} 1-2 & 2 \\ 2 & -2-2 \end{bmatrix}\vec{\mathbf{v}}_1 = \vec{\mathbf{0}}.$$

The augmented matrix for this homogeneous system is

$$\left[\begin{array}{cc|c} -1 & 2 & 0 \\ 2 & -4 & 0 \end{array}\right], \quad \text{which has RREF} \quad \left[\begin{array}{cc|c} 1 & -2 & 0 \\ 0 & 0 & 0 \end{array}\right],$$

giving us $v_1 = 2v_2$. If we let $v_2 = s$, then $v_1 = 2s$, and we find that

$$\vec{\mathbf{v}}_1 = \begin{bmatrix} 2s \\ s \end{bmatrix} = s\begin{bmatrix} 2 \\ 1 \end{bmatrix}.$$

We have found a whole family of eigenvectors belonging to the first eigenvalue: all nonzero multiples of

$$\vec{\mathbf{v}}_1 = \begin{bmatrix} 2 \\ 1 \end{bmatrix} \quad \text{for } \lambda_1 = 2.$$

- For $\lambda_2 = -3$, a similar calculation gives an augmented matrix

$$\left[\begin{array}{cc|c} 4 & 2 & 0 \\ 2 & 1 & 0 \end{array}\right], \quad \text{with RREF} \quad \left[\begin{array}{cc|c} 1 & 1/2 & 0 \\ 0 & 0 & 0 \end{array}\right],$$

so the eigenvectors belonging to the second eigenvalue are the nonzero multiples of

$$\vec{\mathbf{v}}_2 = \begin{bmatrix} 1 \\ -2 \end{bmatrix} \quad \text{for } \lambda_2 = -3.$$

■

EXAMPLE 2 **Characteristic Calculations** We seek the eigenvalues and eigenvectors for matrix

$$\mathbf{A} = \begin{bmatrix} 1 & 1 \\ 4 & 1 \end{bmatrix}.$$

Step 1. The characteristic equation is

$$p(\lambda) = |\mathbf{A} - \lambda\mathbf{I}| = \begin{vmatrix} 1-\lambda & 1 \\ 4 & 1-\lambda \end{vmatrix} = (1-\lambda)^2 - 4 = 0.$$

Step 2. This simplifies to

$$\lambda^2 - 2\lambda - 3 = (\lambda - 3)(\lambda + 1) = 0.$$

The eigenvalues are $\lambda_1 = 3$ and $\lambda_2 = -1$.

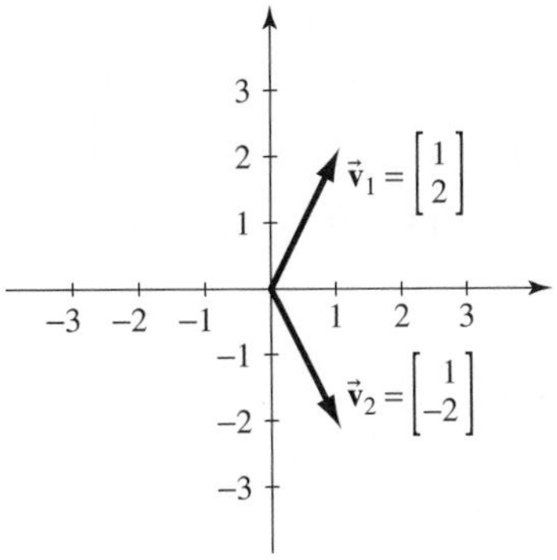

FIGURE 5.3.3 Eigenvectors for Example 2.

Step 3. We find the eigenvectors by solving equation (5).

- For $\lambda_1 = 3$, we must solve

$$(\mathbf{A} - 3\mathbf{I})\vec{\mathbf{v}}_1 = \begin{bmatrix} 1-3 & 1 \\ 4 & 1-3 \end{bmatrix} \begin{bmatrix} v_1 \\ v_2 \end{bmatrix} = \vec{\mathbf{0}}.$$

The augmented matrix for the system is

$$\left[\begin{array}{cc|c} -2 & 1 & 0 \\ 4 & -2 & 0 \end{array}\right], \quad \text{which has RREF} \quad \left[\begin{array}{cc|c} 1 & -1/2 & 0 \\ 0 & 0 & 0 \end{array}\right].$$

Therefore, $v_1 = \frac{1}{2} v_2$. Letting $v_2 = 2s$ yields $v_1 = s$, so for $s \neq 0$,

$$\vec{\mathbf{v}}_1 = s \begin{bmatrix} 1 \\ 2 \end{bmatrix} \quad \text{for } \lambda_1 = 3.$$

- For $\lambda_2 = -1$, we must solve

$$(\mathbf{A} - (-1)\mathbf{I})\vec{\mathbf{v}}_2 = \begin{bmatrix} 1-(-1) & 1 \\ 4 & 1-(-1) \end{bmatrix} \begin{bmatrix} v_1 \\ v_2 \end{bmatrix} = \vec{\mathbf{0}}.$$

The augmented matrix for this homogeneous system is

$$\left[\begin{array}{cc|c} 2 & 1 & 0 \\ 4 & 2 & 0 \end{array}\right], \quad \text{which has RREF} \quad \left[\begin{array}{cc|c} 1 & 1/2 & 0 \\ 0 & 0 & 0 \end{array}\right].$$

Now $v_1 = -\frac{1}{2} v_2$, so for $s \neq 0$,

$$\vec{\mathbf{v}}_2 = s \begin{bmatrix} 1 \\ -2 \end{bmatrix}, \quad \text{for } \lambda_2 = -1.$$

See Fig. 5.3.3. ■

EXAMPLE 3 **Eigenstuff in 3D** We want the eigenvalues and eigenvectors of

$$\mathbf{A} = \begin{bmatrix} 1 & 1 & -2 \\ -1 & 2 & 1 \\ 0 & 1 & -1 \end{bmatrix}.$$

Step 1. We form the characteristic equation

$$|\mathbf{A} - \lambda\mathbf{I}| = \begin{vmatrix} 1-\lambda & 1 & -2 \\ -1 & 2-\lambda & 1 \\ 0 & 1 & -1-\lambda \end{vmatrix} = 0.$$

Step 2. This simplifies to $\lambda^3 - 2\lambda^2 - \lambda + 2 = (\lambda - 2)(\lambda - 1)(\lambda + 1) = 0$, so the eigenvalues are 2, 1, and -1.

Step 3. For each eigenvalue, we find the eigenvector by solving $(\mathbf{A} - \lambda_i \mathbf{I})\vec{\mathbf{v}}_i = \vec{\mathbf{0}}$.

- For $\lambda_1 = 2$, the system is

$$\begin{bmatrix} -1 & 1 & -2 \\ -1 & 0 & 1 \\ 0 & 1 & -3 \end{bmatrix} \begin{bmatrix} v_1 \\ v_2 \\ v_3 \end{bmatrix} = \vec{\mathbf{0}}, \quad \text{with RREF} \quad \left[\begin{array}{ccc|c} 1 & 0 & -1 & 0 \\ 0 & 1 & -3 & 0 \\ 0 & 0 & 0 & 0 \end{array}\right].$$

Therefore, $v_1 = v_3$ and $v_2 = 3v_3$, so we replace the free variable v_3 by nonzero parameter s to get

$$\vec{\mathbf{v}}_1 = \begin{bmatrix} 1 \\ 3 \\ 1 \end{bmatrix} \quad \text{for } \lambda_1 = 2.$$

- For $\lambda_2 = 1$, we have

$$\begin{bmatrix} 0 & 1 & -2 \\ -1 & 1 & 1 \\ 0 & 1 & -2 \end{bmatrix} \begin{bmatrix} v_1 \\ v_2 \\ v_3 \end{bmatrix} = \vec{\mathbf{0}}, \quad \text{with RREF} \quad \left[\begin{array}{ccc|c} 1 & 0 & -3 & 0 \\ 0 & 1 & -2 & 0 \\ 0 & 0 & 0 & 0 \end{array}\right],$$

and

$$\vec{\mathbf{v}}_2 = \begin{bmatrix} 3 \\ 2 \\ 1 \end{bmatrix} \quad \text{for } \lambda_2 = 1.$$

- For $\lambda_3 = -1$, the system is

$$\begin{bmatrix} 2 & 1 & -2 \\ -1 & 3 & 1 \\ 0 & 1 & 0 \end{bmatrix} \begin{bmatrix} v_1 \\ v_2 \\ v_3 \end{bmatrix} = \vec{\mathbf{0}}, \quad \text{with RREF} \quad \left[\begin{array}{ccc|c} 1 & 0 & -1 & 0 \\ 0 & 1 & 0 & 0 \\ 0 & 0 & 0 & 0 \end{array}\right],$$

and

$$\vec{\mathbf{v}}_3 = \begin{bmatrix} 1 \\ 0 \\ 1 \end{bmatrix} \quad \text{for } \lambda_3 = -1.$$

■

Special Cases

Before we move on, let us take note of two special cases.

- **Triangular Matrices:** *The eigenvalues of an upper (or lower) triangular matrix appear on the main diagonal.* (See Problem 46.) Knowing this can save a lot of calculation!
- **2 x 2 Matrices:** For a 2×2 matrix $\mathbf{A} = \begin{bmatrix} a_{11} & a_{12} \\ a_{21} & a_{22} \end{bmatrix}$, we find that

$$\mathbf{A} - \lambda\mathbf{I} = \begin{bmatrix} a_{11} & a_{12} \\ a_{21} & a_{22} \end{bmatrix} - \begin{bmatrix} \lambda & 0 \\ 0 & \lambda \end{bmatrix} = \begin{bmatrix} a_{11} - \lambda & a_{12} \\ a_{21} & a_{22} - \lambda \end{bmatrix}$$

and

$$\begin{aligned} |\mathbf{A} - \lambda\mathbf{I}| &= (a_{11} - \lambda)(a_{22} - \lambda) - a_{12}a_{21} \\ &= \lambda^2 - (a_{11} + a_{22})\lambda + (a_{11}a_{22} - a_{12}a_{21}) = 0. \end{aligned}$$

Thus, the characteristic equation in the 2×2 case can be given in terms of the trace $(a_{11} + a_{22})$ and determinant $(a_{11}a_{22} - a_{12}a_{21})$ of $\mathbf{A}$, so we have

$$\lambda^2 - (\operatorname{Tr}\mathbf{A})\lambda + |\mathbf{A}| = 0. \tag{7}$$

Eigenspaces

Using linearity and the eigenvalue equation, it is easy to show (Problem 35) that the set of all eigenvectors belonging to an eigenvalue λ, together with the zero vector, form a subspace of $\mathbb{R}^n$; it is called the **eigenspace** $\mathbb{E}_\lambda$ of the eigenvalue.

Eigenspace Theorem for Linear Transformations

For each eigenvalue λ of a linear transformation $T : \mathbb{V} \to \mathbb{V}$, the eigenspace

$$\mathbb{E}_\lambda = \{\vec{\mathbf{v}} \in \mathbb{V} \mid T(\vec{\mathbf{v}}) = \lambda\vec{\mathbf{v}}\}$$

is a subspace of $\mathbb{V}$.

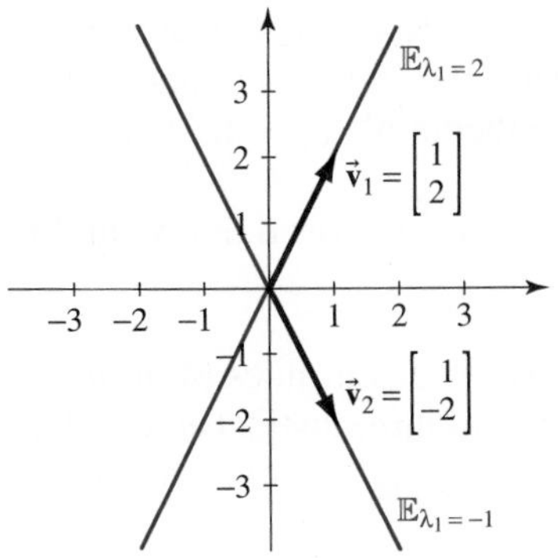

FIGURE 5.3.4 Eigenspaces for Example 2.

In our examples so far, the eigenspace for each λ has been a one-dimensional space, the nonzero multiples of a given eigenvector. (See Fig. 5.3.4.) Any point in such a one-dimensional subspace of $\mathbb{R}^2$ is transformed by $\mathbf{A}$ (simply by multiplication by λ) into another point in the eigenspace $\mathbb{E}_\lambda$.

Since the eigenspaces are subspaces of the domain on which the action of the linear transformation is so very simple, they help us break a linear problem down into several simpler ones. This is a pervasive theme in linear algebra; we can use the Superposition Principle to recombine the solutions of the simpler pieces. (Consider, for example, Problems 16–20 of Sec. 5.2.) Comparable methods for breaking down nonlinear problems have long been sought by mathematicians and scientists, but with much less success.[2]

EXAMPLE 4 **Eigenspaces** In Example 3, for the matrix

$$\mathbf{A} = \begin{bmatrix} 1 & 1 & -2 \\ -1 & 2 & 1 \\ 0 & 1 & -1 \end{bmatrix},$$

we have the following eigenspaces for the respective eigenvalues:

$$\mathbb{E}_{\lambda_1=2} = \text{Span}\left\{\begin{bmatrix} 1 \\ 3 \\ 1 \end{bmatrix}\right\}, \quad \mathbb{E}_{\lambda_2=1} = \text{Span}\left\{\begin{bmatrix} 3 \\ 2 \\ 1 \end{bmatrix}\right\},$$

and

$$\mathbb{E}_{\lambda_3=-1} = \text{Span}\left\{\begin{bmatrix} 1 \\ 0 \\ 1 \end{bmatrix}\right\}.$$

Each eigenvalue corresponds to a one-dimensional eigenspace, obtained in each case by adding the zero vector to the family of eigenvectors. ■

Distinct Eigenvalue Theorem

Let $\mathbf{A}$ be an $n \times n$ matrix. If $\lambda_1, \lambda_2, \ldots, \lambda_p$ are distinct eigenvalues with corresponding eigenvectors $\vec{\mathbf{v}}_1, \vec{\mathbf{v}}_2, \ldots, \vec{\mathbf{v}}_n$, then $\{\vec{\mathbf{v}}_1, \vec{\mathbf{v}}_2, \ldots, \vec{\mathbf{v}}_n\}$ is a set of linearly independent vectors.

Proof We begin with two distinct eigenvalues, $\lambda_1 \neq \lambda_2$, for matrix $\mathbf{A}$. If the associated eigenvectors $\vec{\mathbf{v}}_1$ and $\vec{\mathbf{v}}_2$ were linearly *dependent*, we would have (for some constant $c \neq 0$)

$$\vec{\mathbf{v}}_2 = c\vec{\mathbf{v}}_1. \tag{8}$$

If we multiply (8) by λ_2, we have

$$\lambda_2\vec{\mathbf{v}}_2 = c\lambda_2\vec{\mathbf{v}}_1. \tag{9}$$

If, on the other hand, we multiply (8) by $\mathbf{A}$, we have

$$\mathbf{A}\vec{\mathbf{v}}_2 = c\mathbf{A}\vec{\mathbf{v}}_1$$

or

$$\lambda_2\vec{\mathbf{v}}_2 = c\lambda_1\vec{\mathbf{v}}_1, \tag{10}$$

[2]More progress is being made today with nonlinear problems. An excellent source for the interested reader is Steven H. Strogatz, *Nonlinear Dynamics and Chaos* (Reading, MA: Addison-Wesley, 1994).

by the eigenvalue/eigenvector definition $\mathbf{A}\vec{\mathbf{v}}_i = \lambda_i \vec{\mathbf{v}}_i$. Recall also that the definition requires that $\vec{\mathbf{v}}_i \neq \vec{\mathbf{0}}$. Therefore, comparing the right-hand sides of (9) and (10) gives us $\lambda_1 = \lambda_2$, a *contradiction* to the theorem's hypothesis that the eigenvalues are distinct.

Thus, $\vec{\mathbf{v}}_1$ and $\vec{\mathbf{v}}_2$ *cannot* be linearly dependent; we have proved that $\vec{\mathbf{v}}_1$ and $\vec{\mathbf{v}}_2$ are linearly *independent*. □

The preceding proof extends to any $p \leq n$ distinct eigenvalues of an $n \times n$ matrix, as will be explored in Problem 36, for the case of three distinct eigenvalues and their corresponding eigenvectors.

Repeated Eigenvalues

EXAMPLE 5 **Multiple Eigenvalue** Let us determine the various "eigen-objects" for matrix

$$\mathbf{A} = \begin{bmatrix} -2 & 1 & 1 \\ 1 & -2 & 1 \\ 1 & 1 & -2 \end{bmatrix}.$$

Step 1. The characteristic equation $|\mathbf{A} - \lambda\mathbf{I}| = 0$ simplifies to $\lambda(\lambda + 3)^2 = 0$.

Step 2. The two solutions of this equation, $\lambda_1 = 0$ and $\lambda_2 = -3$, are the eigenvalues, but root -3 is a **double root** or root of **algebraic multiplicity** 2. (The eigenvalue 0 in this case is a **simple root** of algebraic multiplicity 1.) We continue our analysis much as before.

Step 3.

- For the eigenvalue $\lambda_1 = 0$, the system of equations to be solved is just $\mathbf{A}\vec{\mathbf{v}} = \vec{\mathbf{0}}$, and the eigenvectors we obtain, together with $\vec{\mathbf{0}}$, will be the kernel of the transformation defined by $\mathbf{A}$. The RREF of $\mathbf{A}$ is

$$\begin{bmatrix} 1 & 0 & -1 \\ 0 & 1 & -1 \\ 0 & 0 & 0 \end{bmatrix},$$

so $v_1 = v_3$ and $v_2 = v_3$, and

$$\vec{\mathbf{v}}_1 = s \begin{bmatrix} 1 \\ 1 \\ 1 \end{bmatrix} \quad \text{for } \lambda_1 = 0.$$

- For the double eigenvalue $\lambda_2 = -3$, the story is different. Now the equation $(\mathbf{A} - \lambda\mathbf{I})\vec{\mathbf{v}} = \vec{\mathbf{0}}$ takes the form

$$\begin{bmatrix} 1 & 1 & 1 \\ 1 & 1 & 1 \\ 1 & 1 & 1 \end{bmatrix} \begin{bmatrix} v_1 \\ v_2 \\ v_3 \end{bmatrix} = \vec{\mathbf{0}}, \quad \text{with RREF} \quad \left[\begin{array}{ccc|c} 1 & 1 & 1 & 0 \\ 0 & 0 & 0 & 0 \\ 0 & 0 & 0 & 0 \end{array}\right].$$

Hence, v_2 and v_3 are both free variables, while $v_1 = -v_2 - v_3$. If we let $v_2 = r$ and $v_3 = s$, then

$$\vec{\mathbf{v}} = \begin{bmatrix} -r - s \\ r \\ s \end{bmatrix} = r \begin{bmatrix} -1 \\ 1 \\ 0 \end{bmatrix} + s \begin{bmatrix} -1 \\ 0 \\ 1 \end{bmatrix}.$$

We have two linearly independent eigenvectors,

$$\vec{\mathbf{v}}_2 = \begin{bmatrix} -1 \\ 1 \\ 0 \end{bmatrix} \quad \text{and} \quad \vec{\mathbf{v}}_3 = \begin{bmatrix} -1 \\ 0 \\ 1 \end{bmatrix} \quad \text{for } \lambda_2 = -3 \text{ of multiplicity 2.}$$

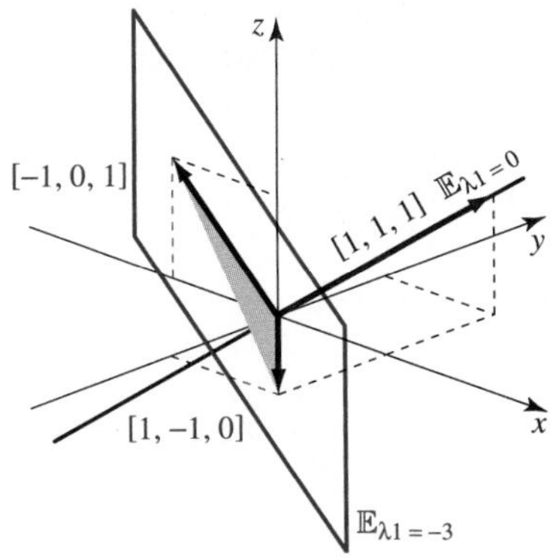

FIGURE 5.3.5 Example 5 eigenspaces: For $\lambda_1 = 0$, the eigenspace is a *line*. For $\lambda_2 = -3$, we find *two* linearly independent eigenvectors, which span a *plane*, a two-dimensional eigenspace. (For ease of visualization, we have drawn the negative of $\vec{\mathbf{v}}_2$.)

Their span,

$$\mathbb{E}_{\lambda_2=-3} = \text{Span}\left\{\begin{bmatrix}-1\\1\\0\end{bmatrix}, \begin{bmatrix}-1\\0\\1\end{bmatrix}\right\},$$

is the *two-dimensional eigenspace* belonging to double eigenvalue $\lambda_2 = -3$. *Any linear combination of these vectors is also an eigenvector* for -3. $\mathbb{E}_{\lambda_2=-3}$ is a *plane* in $\mathbb{R}^2$, spanned by the given vectors, or in fact by *any* two linearly independent eigenvectors in that plane. (See Fig. 5.3.5.) ■

Unfortunately, a double eigenvalue does not *always* have a two-dimensional eigenspace.

EXAMPLE 6 **Multiple Eigenvalue, Different Outcome** The matrix

$$\mathbf{A} = \begin{bmatrix}1&1&1\\0&1&1\\0&0&1\end{bmatrix}$$

has 1 as an eigenvalue of algebraic multiplicity 3. (Recall that the eigenvalues of upper triangular matrices appear on the main diagonal.) The system

$$(\mathbf{A}-\mathbf{I})\vec{\mathbf{v}} = \begin{bmatrix}0&1&1\\0&0&1\\0&0&0\end{bmatrix}\begin{bmatrix}v_1\\v_2\\v_3\end{bmatrix} = \vec{\mathbf{0}}$$

gives us $v_2 + v_3 = 0$ and $v_3 = 0$. If we let $v_1 = s$, we find only one solution,

$$\vec{\mathbf{v}}_1 = \begin{bmatrix}s\\0\\0\end{bmatrix} = s\begin{bmatrix}1\\0\\0\end{bmatrix} \quad \text{for } \lambda_1 = 1 \text{ of multiplicity 3.}$$

Consequently, the eigenspace belonging to $\lambda_1 = 1$ has dimension 1. ■

In Sec. 6.2, Examples 6 and 7, we will discuss how to handle the case (as in this last example) where the eigenvectors do not span the space of the transformation.

Nonreal Eigenvalues

If we proceed as in the previous example to find the eigenvalues for

$$\mathbf{A} = \begin{bmatrix}0&1\\-1&0\end{bmatrix}, \tag{11}$$

we form the characteristic equation

$$|\mathbf{A}-\lambda\mathbf{I}| = \begin{vmatrix}-\lambda&1\\-1&-\lambda\end{vmatrix} = \lambda^2+1 = 0,$$

whose only roots are the nonreal complex numbers $\lambda_1 = i$ and $\lambda_2 = -i$. What are we to make of these "eigenvalues"?

The situation is a little like our experience with solving quadratic equations, in which we are forced to consider nonreal solutions when the discriminant is negative. For some problems, "no real solutions" tells us what we need to know about what we are modeling: something cannot be done in the world of real-valued quantities. But we also learn that complex roots are useful in a broader context, such as obtaining solutions to the differential equations in Sec. 4.3, and later for DE systems in Sec. 6.3.

We can formally continue our analysis for the matrix in (11) by solving the equation $(\mathbf{A}-\lambda\mathbf{I})\vec{\mathbf{v}}=\vec{\mathbf{0}}$ for $\lambda_1=i$ and $\lambda_2=-i$. The resulting system for $\lambda_1=i$ is

$$\begin{bmatrix}-i & 1\\ -1 & -i\end{bmatrix}\begin{bmatrix}v_1\\ v_2\end{bmatrix}=\vec{\mathbf{0}},\quad \text{with RREF}\quad \left[\begin{array}{cc|c}1 & i & 0\\ 0 & 0 & 0\end{array}\right].$$

Therefore, v_2 is free, and if we let $v_2=s$, then $v_1=-is$. This gives

$$\vec{\mathbf{v}}_1=s\begin{bmatrix}-i\\ 1\end{bmatrix}\quad \text{for } \lambda_1=i.$$

Equivalent Eigenvectors:

With complex eigenvalues, equivalent eigenvectors can at first glance look very different. For example,

$$\vec{\mathbf{v}}_1=i\begin{bmatrix}-1\\ -i\end{bmatrix}=\begin{bmatrix}-i\\ 1\end{bmatrix}.$$

Alternatively, by shortcut (8) we find that

$$\vec{\mathbf{v}}_1=\begin{bmatrix}-1\\ -i\end{bmatrix}\quad \text{for } \lambda_1=i,$$

which is a multiple by $-i$ of $[-i,\,1]$ found previously, and

$$\vec{\mathbf{v}}_2=\begin{bmatrix}-1\\ i\end{bmatrix}\quad \text{for } \lambda_2=-i.$$

It is reasonable to ask what a complex eigenvector means. If we calculate

$$\mathbf{A}\vec{\mathbf{v}}_1=\mathbf{A}\begin{bmatrix}-1\\ -i\end{bmatrix}=\begin{bmatrix}0 & 1\\ -1 & 0\end{bmatrix}\begin{bmatrix}-1\\ -i\end{bmatrix}=\begin{bmatrix}-i\\ 1\end{bmatrix}=i\begin{bmatrix}-1\\ -i\end{bmatrix}=i\vec{\mathbf{v}}_1,$$

we do find that the eigenvector definition, equation (2), is satisfied. But we are in another world now: We're not in $\mathbb{R}^2$ any more! We will see that complex eigenvalues and eigenvectors arise in certain real transformations, but there will be no *real* eigenspace. Nevertheless, the transformations can have familiar geometric results.

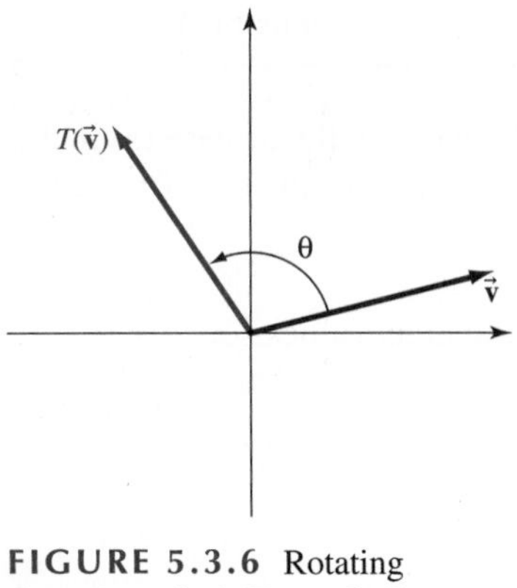

FIGURE 5.3.6 Rotating through angle θ (Example 7).

EXAMPLE 7 **Nonreal Eigenvalues** In Problem 81 of Sec. 5.1, it was shown that the transformation $T:\mathbb{R}^2\to\mathbb{R}^2$ defined by $T(\vec{\mathbf{v}})=\mathbf{A}\vec{\mathbf{v}}$, where

$$\mathbf{A}=\begin{bmatrix}\cos\theta & -\sin\theta\\ \sin\theta & \cos\theta\end{bmatrix},$$

rotates vectors in $\mathbb{R}^2$ through a counterclockwise angle of θ about the origin, as shown in Fig. 5.3.6. Now, for this matrix $\mathbf{A}$,

$$|\mathbf{A}-\lambda\mathbf{I}|=\begin{vmatrix}\cos\theta-\lambda & -\sin\theta\\ \sin\theta & \cos\theta-\lambda\end{vmatrix}=(\cos\theta-\lambda)^2+\sin^2\theta=0.$$

Therefore, $(\cos\theta-\lambda)^2=-\sin^2\theta$, and it follows that $\cos\theta-\lambda=\pm i\sin\theta$. Thus, we have for $\mathbf{A}$ the eigenvalues

$$\lambda_1,\lambda_2=\cos\theta\pm i\sin\theta,$$

and these are nonreal as long as $\theta\neq n\pi$.

The lack of real eigenvalues is consistent with the geometry of the transformation. If the plane is rotated about the origin by an angle other than a multiple of π, *no* vector can have the same (or opposite) direction as its image! ■

The biggest payoff from nonreal eigenvalues and eigenvectors will come in solving linear systems of differential equations, which we will explore in detail in Sec. 6.3.

A Larger Perspective

Some Properties of Eigenvalues

Let $\mathbf{A}$ be an $n \times n$ matrix.

(i) λ is an eigenvalue of $\mathbf{A}$ if and only if $|\mathbf{A} - \lambda\mathbf{I}| = 0$.

(ii) λ is an eigenvalue of $\mathbf{A}$ if and only if $(\mathbf{A} - \lambda\mathbf{I})\vec{\mathbf{v}} = \vec{\mathbf{0}}$ has a nontrivial solution.

(iii) $\mathbf{A}$ has a zero eigenvalue if and only if $|\mathbf{A}| = 0$.

(iv) $\mathbf{A}$ and $\mathbf{A}^{\mathrm{T}}$ have the same characteristic polynomials and the same eigenvalues.

(v) If λ is an eigenvalue of an invertible matrix $\mathbf{A}$, then $1/\lambda$ is an eigenvalue of $\mathbf{A}^{-1}$.

Properties (i) and (ii) follow from the definitions. Properties (iii), (iv), and (v) are covered in Problems 30, 41, and 31, respectively.

Although the properties listed are stated in terms of the eigenvalues of a matrix $\mathbf{A}$, we can just as easily consider the eigenvalues of a linear transformation $T : \mathbb{V} \to \mathbb{W}$. In fact, $\mathbf{A}$ could be replaced by T in every property. For each $m \times n$ matrix $\mathbf{A}$, there is an associated linear transformation $T : \mathbb{R}^n \to \mathbb{R}^m$ defined by $T(\vec{\mathbf{v}}) = \mathbf{A}\vec{\mathbf{v}}$. Furthermore, for each $T : \mathbb{V} \to \mathbb{W}$, where $\mathbb{V}$ is an n-dimensional vector space and $\mathbb{W}$ is m-dimensional, there is an associated $m \times n$ matrix. (See Appendix LT.)

In the more general vector space, $\mathcal{C}^1$, the eigenvectors of the derivative f' are solutions of $f' = \lambda f$. That is, these eigenvectors are *functions* $f(t) = ce^{\lambda t}$. The eigenvalues λ can be any real numbers.

EXAMPLE 8 **Eigenvalue of the Derivative Operator** Consider the linear transformation $D : \mathbb{P}_2 \to \mathbb{P}_2$ defined by the derivative $D(f) = f'$. Thus, for a typical vector in $\mathbb{P}_2$, say $ax^2 + bx + c$,

$$D(ax^2 + bx + c) = 2ax + b.$$

We can see that the only possible eigenvectors are the constant polynomials, in that they are mapped to zero. Consequently, $\lambda_1 = 0$ is the only eigenvalue, and its eigenspace is

$$\mathbb{E}_{\lambda_1=0} = \{c \mid c \in \mathbb{R}\} \subset \mathbb{P}_2.$$

■

A Bit of History

Eigenvalues were first introduced to mathematics in 1743 by Leonhard Euler, who showed that the nth-order linear homogeneous differential equation with constant coefficients has solutions of the form $y = e^{mt}$, where m satisfied a certain polynomial equation. It was, of course, the characteristic equation. Then he converted the DE to a system, as we learned to do in Sec. 4.4, and found that the same special values were associated with the matrix of this system. They were the eigenvalues. The examples that follow illustrate this historic connection.

EXAMPLE 9 **Characteristic Roots = Eigenvalues** The linear second-order equation

$$y'' - y' - 2y = 0 \tag{12}$$

has characteristic equation $r^2 - r - 2 = (r-2)(r+1) = 0$. We learned in Sec. 4.2 to use the solutions of this equation, the characteristic roots $r_1 = 2$ and $r_2 = -1$, to build the general solution

$$y = c_1 e^{2t} + c_2 e^{-t}$$

from the basic solutions e^{2t} and e^{-t}.

In Sec. 4.7 we learned to convert equation (12) into a system of two first-order equations by letting $x_1 = y$ and $x_2 = y'$. The resulting system,

$$\begin{aligned} x_1' &= x_2, \\ x_2' &= 2x_1 + x_2, \end{aligned} \tag{13}$$

has the matrix form $\vec{\mathbf{x}}' = \mathbf{A}\vec{\mathbf{x}}$, where

$$\vec{\mathbf{x}} = \begin{bmatrix} x_1 \\ x_2 \end{bmatrix} \quad \text{and} \quad \mathbf{A} = \begin{bmatrix} 0 & 1 \\ 2 & 1 \end{bmatrix}.$$

The characteristic equation $|\mathbf{A} - \lambda\mathbf{I}| = 0$ for this matrix $\mathbf{A}$ is $\lambda^2 - \lambda - 2 = 0$, the same as for the DE (12), so the characteristic roots of (12) are the eigenvalues of $\mathbf{A}$, the matrix for the corresponding system (13).

The correlation does not stop there, however. Corresponding to the solution $y = e^{2t}$ of (12) is the solution

$$\vec{\mathbf{x}}_1 = \begin{bmatrix} x_1 \\ x_2 \end{bmatrix} = \begin{bmatrix} y \\ y' \end{bmatrix} = \begin{bmatrix} e^{2t} \\ 2e^{2t} \end{bmatrix}$$

of system (13). We can calculate as follows for the eigenvalue $\lambda_1 = 2$:

$$\mathbf{A}\vec{\mathbf{x}}_1 = \begin{bmatrix} 0 & 1 \\ 2 & 1 \end{bmatrix} \begin{bmatrix} e^{2t} \\ 2e^{2t} \end{bmatrix} = \begin{bmatrix} 2e^{2t} \\ 4e^{2t} \end{bmatrix} = 2 \begin{bmatrix} e^{2t} \\ 2e^{2t} \end{bmatrix} = 2\vec{\mathbf{x}}_1 = \lambda_1 \vec{\mathbf{x}}_1,$$

so this solution

$$\vec{\mathbf{v}}_1 = \vec{\mathbf{x}}_1 = e^{2t} \begin{bmatrix} 1 \\ 2 \end{bmatrix} \quad \text{is an eigenvector for } \lambda_1 = 2.$$

Characteristic Correlations:

- The characteristic roots of a second-order linear DE are the eigenvalues of the corresponding system of first-order equations.
- The solutions of a second-order linear DE are combinations of the eigenvectors of the corresponding system of first-order equations.

A similar calculation shows that for the eigenvalue $\lambda_2 = -1$, we have $\mathbf{A}\vec{\mathbf{x}}_2 = (-1)\vec{\mathbf{x}}_2$, where $\vec{\mathbf{x}}_2$ is formed from the other basic solution $y = e^{-t}$ of (12). In this case,

$$\vec{\mathbf{v}}_2 = \vec{\mathbf{x}}_2 = \begin{bmatrix} e^{-t} \\ -e^{-t} \end{bmatrix} = e^{-t} \begin{bmatrix} 1 \\ -1 \end{bmatrix} \quad \text{is an eigenvector for } \lambda_2 = -1.$$

Hence, the solutions to (12) are combinations

$$\vec{\mathbf{x}} = \begin{bmatrix} x \\ y \end{bmatrix} = c_1\vec{\mathbf{x}}_1 + c_2\vec{\mathbf{x}}_2 = c_1\vec{\mathbf{v}}_1 + c_2\vec{\mathbf{v}}_2$$

of the eigenvectors of the linear transformation defined by $\vec{\mathbf{x}}' = \mathbf{A}\vec{\mathbf{x}}$.

Once again, as in Examples 1 and 2, the eigenspace for each eigenvalue is composed of multiples of a vector in $\mathbb{R}^2$. ■

Properties of Linear Homogeneous DEs with Distinct Eigenvalues

For the DE $\vec{\mathbf{x}}' = \mathbf{A}\vec{\mathbf{x}}$ with distinct eigenvalues, the following properties hold.

- The domain of the linear transformation is a vector space of vector functions.
- The solution set is also a vector space of vector functions.
- The eigenspace for each eigenvalue is a one-dimensional line in the direction of a vector in $\mathbb{R}^n$.

This connection between eigenvalues and solutions to DEs will be explored more carefully in Chapter 6.

EXAMPLE 10 **Complex Connections** Nonreal characteristic roots are obtained for the second-order equation

$$y'' + 2y' + 5y = 0; \qquad (14)$$

the characteristic equation is $r^2 + 2r + 5 = 0$, and its solutions are

$$r_1, r_2 = -1 \pm 2i.$$

Using the same substitution as in the previous example, we can convert (as in Sec. 4.7) equation (14) into the system $\vec{\mathbf{x}}' = \mathbf{A}\vec{\mathbf{x}}$, where

$$\mathbf{A} = \begin{bmatrix} 0 & 1 \\ -5 & -2 \end{bmatrix}.$$

The characteristic equation for $\mathbf{A}$ is

$$\lambda^2 + 2\lambda + 5 = 0, \qquad (15)$$

so the eigenvalues of $\mathbf{A}$ are the same as the characteristic roots of the DE (14),

$$\lambda = -1 \pm 2i.$$

To confirm the eigenvalue property (2), we use the solution to (14) from Sec. 4.3, $y = e^{\lambda t}$, where λ is one of the complex eigenvalues. Therefore, we have

$$\vec{\mathbf{x}} = \begin{bmatrix} y \\ y' \end{bmatrix} = \begin{bmatrix} e^{\lambda t} \\ \lambda e^{\lambda t} \end{bmatrix},$$

and can compute

$$\begin{aligned} \mathbf{A}\vec{\mathbf{x}} &= \begin{bmatrix} 0 & 1 \\ -5 & -2 \end{bmatrix} \begin{bmatrix} e^{\lambda t} \\ \lambda e^{\lambda t} \end{bmatrix} \\ &= \begin{bmatrix} \lambda e^{\lambda t} \\ (-5 - 2\lambda)e^{\lambda t} \end{bmatrix} = \begin{bmatrix} \lambda e^{\lambda t} \\ \lambda^2 e^{\lambda t} \end{bmatrix} \quad \text{(by equation (15))} \\ &= \lambda \begin{bmatrix} e^{\lambda t} \\ \lambda e^{\lambda t} \end{bmatrix} = \lambda \vec{\mathbf{x}}. \end{aligned}$$

Thus, the solution vector $\vec{\mathbf{x}}$ is indeed an eigenvector for λ. ■

Once again, a complex eigenvalue exhibits its defining behavior, but only if we operate in an expanded world of objects built from complex numbers rather than real numbers alone. The eigenvectors as well as the eigenvalues may consist of nonreal numbers.

Summary

We have learned to compute eigenvalues, eigenvectors, and eigenspaces for a square matrix, and have related these to the linear transformation defined by the matrix: *the transformation reduces to multiplication by a scalar on each eigenvector*. The characteristic roots of a linear second-order DE with constant coefficients turn out to be the eigenvalues of the matrix of the system to which it corresponds.

5.3 Problems

Computing Eigenstuff *For each matrix in Problems 1–16, compute its eigenvalues and eigenvector(s), and sketch the eigenspaces when the eigenvectors are real.*

1. $\begin{bmatrix} 2 & 0 \\ 0 & 1 \end{bmatrix}$ **2.** $\begin{bmatrix} 3 & 2 \\ 2 & 0 \end{bmatrix}$ **3.** $\begin{bmatrix} 1 & 2 \\ 1 & 2 \end{bmatrix}$

4. $\begin{bmatrix} 3 & 4 \\ -5 & -5 \end{bmatrix}$ **5.** $\begin{bmatrix} 1 & 3 \\ 1 & 3 \end{bmatrix}$ **6.** $\begin{bmatrix} 3 & 2 \\ -2 & -3 \end{bmatrix}$

7. $\begin{bmatrix} 1 & 1 \\ 1 & 1 \end{bmatrix}$ **8.** $\begin{bmatrix} 12 & -6 \\ 15 & -7 \end{bmatrix}$ **9.** $\begin{bmatrix} 1 & 4 \\ -4 & 11 \end{bmatrix}$

10. $\begin{bmatrix} 4 & 2 \\ -3 & 11 \end{bmatrix}$ **11.** $\begin{bmatrix} 3 & 5 \\ -1 & -1 \end{bmatrix}$ **12.** $\begin{bmatrix} 1 & 1 \\ 0 & 1 \end{bmatrix}$

13. $\begin{bmatrix} 2 & 4 \\ -1 & -2 \end{bmatrix}$ **14.** $\begin{bmatrix} 3 & 0 \\ 0 & 3 \end{bmatrix}$ **15.** $\begin{bmatrix} 2 & -1 \\ 1 & 4 \end{bmatrix}$

16. $\begin{bmatrix} 1 & 1 \\ -1 & -1 \end{bmatrix}$

17. Eigenvector Shortcut For a 2×2 matrix

$$\mathbf{A} = \begin{bmatrix} a & b \\ c & d \end{bmatrix}$$

with eigenvalue λ, show that if $b \neq 0$, then the corresponding eigenvector is

$$\vec{\mathbf{v}} = \begin{bmatrix} -b \\ a - \lambda \end{bmatrix}.$$

18. When Shortcut Fails The eigenvector shortcut of Problem 17 may *fail* when $b = 0$, forcing a return to the definition $\mathbf{A}\vec{\mathbf{v}} = \lambda\vec{\mathbf{v}}$ to find the eigenvector(s). For each eigenvalue in the following matrices, find the eigenvector(s). Discuss how/why the shortcut fails and why the definition succeeds.

(a) $\begin{bmatrix} 3 & 0 \\ 5 & 3 \end{bmatrix}$ (b) $\begin{bmatrix} 3 & 0 \\ 5 & 2 \end{bmatrix}$ (c) $\begin{bmatrix} 3 & 0 \\ 0 & 3 \end{bmatrix}$

More Eigenstuff *For each matrix in Problems 19–34, compute its eigenvalues, eigenvectors and the dimension of each eigenspace.*

19. $\begin{bmatrix} 2 & 0 & 0 \\ 1 & -1 & -2 \\ -1 & 0 & 1 \end{bmatrix}$ **20.** $\begin{bmatrix} 1 & 2 & -1 \\ 1 & 0 & 1 \\ 4 & -4 & 5 \end{bmatrix}$

21. $\begin{bmatrix} 1 & 2 & 2 \\ 2 & 0 & 3 \\ 2 & 3 & 0 \end{bmatrix}$ **22.** $\begin{bmatrix} 0 & 1 & -1 \\ 0 & -1 & 1 \\ 0 & 0 & 0 \end{bmatrix}$

23. $\begin{bmatrix} 0 & 1 & 1 \\ 1 & 0 & 1 \\ 1 & 1 & 0 \end{bmatrix}$ **24.** $\begin{bmatrix} 1 & 0 & 0 \\ -1 & 3 & 0 \\ 3 & 2 & -2 \end{bmatrix}$

25. $\begin{bmatrix} -1 & 0 & 1 \\ -1 & 3 & 0 \\ -4 & 13 & -1 \end{bmatrix}$ **26.** $\begin{bmatrix} 2 & 2 & 3 \\ 1 & 2 & 1 \\ 2 & -2 & 1 \end{bmatrix}$

27. $\begin{bmatrix} 1 & 0 & 0 \\ -4 & 3 & 0 \\ -4 & 2 & 1 \end{bmatrix}$ **28.** $\begin{bmatrix} 1 & 1 & 1 \\ 0 & 1 & 1 \\ 0 & 0 & 1 \end{bmatrix}$

29. $\begin{bmatrix} 3 & -2 & 0 \\ 1 & 0 & 0 \\ -1 & 1 & 3 \end{bmatrix}$ **30.** $\begin{bmatrix} 0 & 0 & 2 \\ -1 & 1 & 2 \\ -1 & 0 & 3 \end{bmatrix}$

31. $\begin{bmatrix} 2 & 1 & 8 & -1 \\ 0 & 4 & 0 & 0 \\ 0 & 0 & 6 & 0 \\ 0 & 0 & 0 & 4 \end{bmatrix}$ **32.** $\begin{bmatrix} 4 & 0 & 4 & 0 \\ 0 & 4 & 0 & 0 \\ 0 & 0 & 8 & 0 \\ -1 & -2 & 1 & 8 \end{bmatrix}$

33. $\begin{bmatrix} 2 & 0 & 1 & 2 \\ 0 & 2 & 0 & 0 \\ 0 & 0 & 6 & 0 \\ 0 & 0 & 1 & 4 \end{bmatrix}$ **34.** $\begin{bmatrix} 2 & 0 & 0 & 0 \\ 1 & -2 & 0 & 0 \\ 1 & 0 & 1 & 0 \\ 0 & 2 & 0 & 1 \end{bmatrix}$

35. Prove the Eigenspace Theorem Show that the set of eigenvectors belonging to a particular eigenvalue of an $n \times n$ matrix, together with the zero vector, is a subspace of $\mathbb{R}^n$. HINT: Use equation (2) and verify closure.

36. Distinct Eigenvalues Extended Extend the proof of the Distinct Eigenvalue Theorem for a 3×3 matrix $\mathbf{A}$ as follows: Show that if $\mathbf{A}$ has 3 distinct eigenvalues λ_1, λ_2, λ_3, then the corresponding eigenvectors $\vec{\mathbf{v}}_1$, $\vec{\mathbf{v}}_2$, $\vec{\mathbf{v}}_3$ are linearly independent. HINT: Use the fact that an eigenvector $\vec{\mathbf{v}}_i$ cannot be zero, and follow the steps shown in the proof for two distinct eigenvalues.

Invertible Matrices

37. Show that an invertible matrix cannot have a zero eigenvalue. In fact, you will have proved a characteristic of invertible matrices.

38. Suppose that λ is an eigenvalue of an invertible matrix $\mathbf{A}$. Show that $1/\lambda$ is an eigenvalue of $\mathbf{A}^{-1}$.

39. Give an example to illustrate Problem 38.

40. Similar Matrices

(a) Use the definition for similar matrices (i.e., $\mathbf{B} \sim \mathbf{A}$ if and only if $\mathbf{B} = \mathbf{P}^{-1}\mathbf{A}\mathbf{P}$ for some invertible matrix $\mathbf{P}$) to show that similar matrices have the same characteristic polynomials and eigenvalues.

(b) Show, using 2×2 matrices as examples, that the eigenvectors may be different for similar matrices.

41. Identity Eigenstuff What are the eigenvalues and eigenvectors of the following?

$$\mathbf{I}_2 = \begin{bmatrix} 1 & 0 \\ 0 & 1 \end{bmatrix}.$$

What about $\mathbf{I}_3$? What about $\mathbf{I}_n$?

42. Eigenvalues and Inversion If a matrix $\mathbf{A}$ has an inverse $\mathbf{A}^{-1}$, use equation (2) to show that $\mathbf{A}^{-1}$ has the same eigenvectors as $\mathbf{A}$. Determine a relationship between the eigenvalues of $\mathbf{A}$ and $\mathbf{A}^{-1}$. Illustrate with a suitable example.

Triangular Matrices *The eigenvalues of an upper triangular matrix and those of a lower triangular matrix appear on the main diagonal. Verify this fact for the matrices in Problems 43–45.*

43. $\begin{bmatrix} 1 & 1 \\ 0 & 1 \end{bmatrix}$ **44.** $\begin{bmatrix} 2 & 0 \\ -3 & -1 \end{bmatrix}$ **45.** $\begin{bmatrix} 1 & 0 & 3 \\ 0 & 4 & 1 \\ 0 & 0 & 2 \end{bmatrix}$

46. Use properties of determinants (Sec. 3.4) to demonstrate why the diagonal eigenvalue property holds in general for triangular matrices.

Eigenvalues of a Transpose *For Problems 47–49, let* $\mathbf{A}$ *be a square matrix and determine the following facts about its transpose.*

47. Show that $\mathbf{A}$ is invertible if and only if $\mathbf{A}^{\mathrm{T}}$ is invertible.

48. Show that $\mathbf{A}$ and $\mathbf{A}^{\mathrm{T}}$ have the same eigenvalues.

49. Give an example of matrices $\mathbf{A}$ and $\mathbf{A}^{\mathrm{T}}$ to show that the corresponding eigenvectors for a given λ are not the same.

50. Orthogonal Eigenvectors Let $\mathbf{A}$ be a **symmetric matrix** (that is, $\mathbf{A} = \mathbf{A}^{\mathrm{T}}$) with distinct eigenvalues λ_1 and λ_2. For such a matrix, if $\vec{\mathbf{v}}_1$ and $\vec{\mathbf{v}}_2$ are eigenvectors belonging to the distinct eigenvalues λ_1 and λ_2, respectively, then $\vec{\mathbf{v}}_1$ and $\vec{\mathbf{v}}_2$ are orthogonal.

(a) Illustrate this for

$$\mathbf{A} = \begin{bmatrix} 1 & 2 \\ 2 & 4 \end{bmatrix}.$$

(b) Prove fact for an $n \times n$ symmetric matrix. Use the fact that $\vec{\mathbf{v}}_1 \cdot \vec{\mathbf{v}}_2 = \vec{\mathbf{v}}_1^{\mathrm{T}}\vec{\mathbf{v}}_2$ (as a matrix product).

51. Another Eigenspace Find the eigenvalues, if any, and the corresponding eigenspaces for the linear transformation $T : \mathbb{P}_2 \to \mathbb{P}_2$ defined by $T(ax^2 + bx + c) = bx + c$.

52. Checking Up on Eigenvalues In a quadratic equation with leading coefficient 1, the negative of the coefficient of the linear term is the sum of the roots, and the constant term is the product of the roots.

(a) Prove these properties by expanding the factored quadratic

$$(x - \lambda_1)(x - \lambda_2) = 0.$$

(b) Compare this result to equation (5). Explain how to determine from a matrix, without solving the characteristic equation, the sum and product of its eigenvalues.

(c) Illustrate these results for the matrix

$$\begin{bmatrix} 3 & 2 \\ 2 & 0 \end{bmatrix}.$$

Looking for Matrices *For Problems 53–57, find all the* 2×2 *matrices with the desired properties.*

53. $\vec{\mathbf{v}} = \begin{bmatrix} 0 \\ 1 \end{bmatrix}$ is an eigenvector.

54. $\vec{\mathbf{v}} = \begin{bmatrix} 1 \\ 1 \end{bmatrix}$ is an eigenvector.

55. $\begin{bmatrix} 1 \\ 0 \end{bmatrix}$ and $\begin{bmatrix} -1 \\ 2 \end{bmatrix}$ are eigenvectors, with double eigenvalue $\lambda = 1$.

56. $\begin{bmatrix} 1 \\ 0 \end{bmatrix}$ and $\begin{bmatrix} -1 \\ 2 \end{bmatrix}$ are eigenvectors, with eigenvalues 1 and 2, respectively.

57. $\begin{bmatrix} 1 \\ -1 \end{bmatrix}$ and $\begin{bmatrix} 0 \\ 2 \end{bmatrix}$ are eigenvectors, with the same eigenvalue $\lambda = -1$.

Linear Transformations in the Plane *For Problems 58–62, find the eigenvalues, if any, and corresponding eigenvectors for the transformations in Table 5.1.1.*

58. Reflections about the x-axis.

59. Reflections about the y-axis.

60. Clockwise rotation of $\pi/4$ about the origin.

61. Reflection about the line $y = x$.

62. Shear of 2 in the y-direction.

Cayley-Hamilton *We have met these nineteenth-century mathematicians before: Cayley in Sec. 3.1 and Hamilton in Sec. 3.5. They proved the following theorem.*

> **Cayley-Hamilton Theorem**
> A matrix satisfies its own characteristic equation.

If $\lambda^2 + b\lambda + c = 0$ is the characteristic equation of the 2×2 matrix $\mathbf{A}$, *for example, then* $\mathbf{A}^2 + b\mathbf{A} + c\mathbf{I} = \mathbf{0}$. *Verify this for each matrix in Problems 63–66.*

63. $\begin{bmatrix} 1 & 1 \\ 4 & 1 \end{bmatrix}$

64. $\begin{bmatrix} 0 & 1 \\ -1 & 0 \end{bmatrix}$

65. $\begin{bmatrix} 1 & 1 & 0 \\ 0 & 1 & 1 \\ 0 & 0 & 1 \end{bmatrix}$

66. $\begin{bmatrix} 1 & 1 & 2 \\ 0 & 2 & 3 \\ 1 & 0 & 4 \end{bmatrix}$

Inverses by Cayley-Hamilton *For an invertible 3×3 matrix* $\mathbf{A}$, *we can write, using the Cayley-Hamilton Theorem,* $\mathbf{A}^3 + b\mathbf{A}^2 + c\mathbf{A} + d\mathbf{I} = \mathbf{0}$, *where b, c, and d are coefficients of the characteristic equation of* $\mathbf{A}$. *If we multiply through on the left by* $\mathbf{A}^{-1}$, *we get* $\mathbf{A}^2 + b\mathbf{A} + c\mathbf{I} + d\mathbf{A}^{-1} = \mathbf{0}$, *which can be solved for* $\mathbf{A}^{-1}$. *Use this method to calculate the inverses of Problems 67 and 68.*

67. $\begin{bmatrix} 2 & 0 & 0 \\ 1 & -1 & -3 \\ -1 & 0 & 1 \end{bmatrix}$

68. $\begin{bmatrix} 1 & 2 & -1 \\ 1 & 0 & 1 \\ 4 & -4 & 5 \end{bmatrix}$

69. Develop a Cayley-Hamilton formula for the inverse of a 2×2 matrix, and apply it to compute the inverses of the following.

(a) $\begin{bmatrix} 3 & 2 \\ -2 & -3 \end{bmatrix}$

(b) $\begin{bmatrix} 3 & 5 \\ -1 & -1 \end{bmatrix}$

70. Trace and Determinant as Parameters Express the eigenvalues of a 2×2 matrix in terms of its trace and its determinant.

71. Raising the Order Generalize the results of Problem 70 to the characteristic equation and eigenvalues of a 3×3 matrix. Then illustrate these results for the matrix

$$\begin{bmatrix} 1 & 2 & -1 \\ 1 & 0 & 1 \\ 4 & -4 & 5 \end{bmatrix}.$$

Eigenvalues and Conversion *Using the method of Sec. 4.7, convert each differential equation in Problems 72–75 to a system of first-order equations. Then verify that the characteristic roots of the DE are the same as the eigenvalues of the matrix of the converted linear system.*

72. $y'' - y' - 2y = 0$

73. $y'' - 2y' + 5y = 0$

74. $y''' + 2y'' - y' - 2y = 0$

75. $y''' - 2y'' - 5y' + 6y = 0$

Eigenfunction Boundary-Value Problems *For what values of the nonnegative constant λ in the equation $y'' + \lambda y = 0$ do there exist nonzero solutions satisfying the boundary conditions in Problems 76–78? The values of λ are called eigenvalues and the corresponding solutions are called eigenfunctions.*

76. $y(0) = 0$, $y(\pi) = 0$

77. $y'(0) = 0$, $y(\pi) = 0$

78. $y(-\pi) = y(\pi)$, $y'(-\pi) = y'(\pi)$

79. Computer Lab: Eigenvectors For each matrix (a)–(h), find the eigenvalues and eigenvectors. To make quick work of this, use computer software (e.g., IDE, Derive, Matlab, or other computer algebra systems). From your results, list conjectures (and illustrations) of what you might be able to predict for eigenvalues and eigenvectors from just looking at a 2×2 matrix (without calculations).

Eigen-Engine
For 2×2 matrices you can *see* the eigenvectors as well as their coordinate values.

(a) $\begin{bmatrix} 1 & 0 \\ 0 & 1 \end{bmatrix}$

(b) $\begin{bmatrix} 2 & 0 \\ 0 & 2 \end{bmatrix}$

(c) $\begin{bmatrix} 2 & 1 \\ 0 & 2 \end{bmatrix}$

(d) $\begin{bmatrix} 1 & 1 \\ 1 & 1 \end{bmatrix}$

(e) $\begin{bmatrix} 1 & 4 \\ 1 & 1 \end{bmatrix}$

(f) $\begin{bmatrix} 2 & 1 \\ -1 & 2 \end{bmatrix}$

(g) $\begin{bmatrix} 0 & 0 \\ 0 & 1 \end{bmatrix}$

(h) $\begin{bmatrix} 1 & 0 \\ 0 & 0 \end{bmatrix}$

80. Suggested Journal Entry Suppose that you had calculated the first few powers of a matrix $\mathbf{A}$. How could you use the Cayley-Hamilton Theorem (in the introduction to Problems 63–66) to compute higher powers of $\mathbf{A}$ without doing any further matrix multiplications? Could a similar scheme be used to find powers of $\mathbf{A}^{-1}$ for invertible $\mathbf{A}$?

5.4 Coordinates and Diagonalization

SYNOPSIS: We introduce coordinates relative to a basis and use matrices to change coordinates from one basis to another.

Also, we use eigenvectors to diagonalize a matrix; the result is a similar matrix with the eigenvalues on its diagonal and zeros elsewhere. We will see some special advantages of diagonalization in analyzing linear systems.

Introduction

Up to this point, we have used coordinates of a vector relative to the standard ordered basis for the vector space, usually $\mathbb{R}^n$. Now we are going to broaden the concept.[1]

From our discussion in Sec. 3.6, we know that if $\{\vec{\mathbf{b}}_1, \vec{\mathbf{b}}_2, \ldots, \vec{\mathbf{b}}_n\}$ is a basis for finite-dimensional vector space $\mathbb{V}$ and $\vec{\mathbf{v}}$ is any vector in $\mathbb{V}$, then

$$\vec{\mathbf{v}} = \beta_1\vec{\mathbf{b}}_1 + \beta_2\vec{\mathbf{b}}_2 + \cdots + \beta_n\vec{\mathbf{b}}_n,$$

because the basis is a spanning set.

These coordinates, the β_i, are *unique*. For, if we could have another set of coordinates, say

$$\vec{\mathbf{v}} = \delta_1\vec{\mathbf{b}}_1 + \delta_2\vec{\mathbf{b}}_2 + \cdots + \delta_n\vec{\mathbf{b}}_n,$$

then

$$\vec{\mathbf{v}} - \vec{\mathbf{v}} = (\beta_1 - \delta_1)\vec{\mathbf{b}}_1 + (\beta_2 - \delta_2)\vec{\mathbf{b}}_2 + \cdots + (\beta_n - \delta_n)\vec{\mathbf{b}}_n = \vec{\mathbf{0}}. \qquad (1)$$

Since the basis vectors are linearly independent, condition (1) implies that

$$\beta_1 - \delta_1 = 0, \beta_2 - \delta_2 = 0, \ldots, \beta_n - \delta_n = 0$$

and the δ_i were the same as the β_i after all.

We can now write a formal definition for coordinates.

Coordinates

Let $\vec{\mathbf{v}}$ be a vector in the finite-dimensional vector space $\mathbb{V}$, with basis $B = \{\vec{\mathbf{b}}_1, \vec{\mathbf{b}}_2, \ldots, \vec{\mathbf{b}}_n\}$ for $\mathbb{V}$. Then the **coordinates** of $\vec{\mathbf{v}}$ relative to B are the unique real numbers $\beta_1, \beta_2, \ldots, \beta_n$ such that

$$\vec{\mathbf{v}} = \beta_1\vec{\mathbf{b}}_1 + \beta_2\vec{\mathbf{b}}_2 + \cdots + \beta_n\vec{\mathbf{b}}_n. \qquad (2)$$

The **coordinate vector for $\vec{\mathbf{v}}$ relative to an ordered basis** B is the column vector

$$\vec{\mathbf{v}}_B = \begin{bmatrix} \beta_1 \\ \beta_2 \\ \vdots \\ \beta_n \end{bmatrix}_B$$

in $\mathbb{R}^n$, where $\beta_1, \beta_2, \ldots, \beta_n$ are the coordinates of $\vec{\mathbf{v}}$ relative to B.

---

[1] Appendix LT supplements this section by discussing additional concepts and theorems considered important in linear algebra.

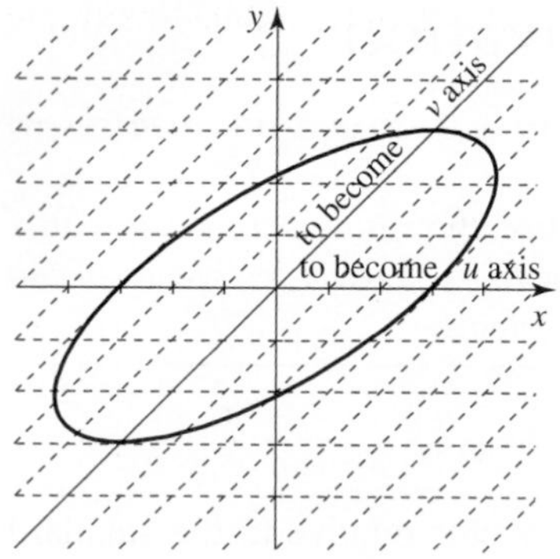

(a) Skewed ellipse, graph of $x^2 - 2xy + 2y^2 = 9$

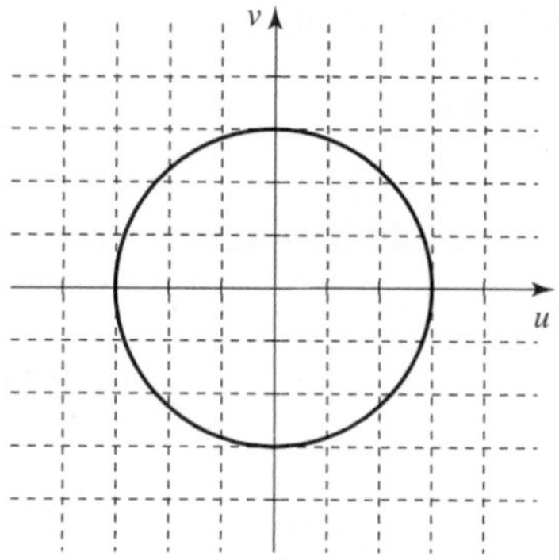

(b) After changing coordinates by $x = u + v$ and $y = v$

FIGURE 5.4.1 In new coordinates, the curve is simpler.

EXAMPLE 1 **A Motivating Example for Changing Bases** Changes of bases are sometimes encountered in earlier math courses. For example, a geometry student who has trouble wrapping his mind around the algebraic equation

$$x^2 - 2xy + 2y^2 = 9 \tag{3}$$

can use $x = u + v$ and $y = v$ to transform equation (3) into

$$(u + v)^2 + -2(u + v)v + 2v^2 = 9,$$

which simplifies to

$$u^2 + v^2 = 9 \tag{4}$$

and is immediately recognizable as a circle of radius 3. The right coordinate system makes all the difference! (See Fig. 5.4.1.)

Equation (3) describes a skewed ellipse in xy coordinates, but if we change to a new coordinate system with its origin at the center of the ellipse and axes parallel to the background grid lines drawn in Fig. 5.4.1(a), we get a circle of radius 3 on a uv coordinate system with axes orthogonal.

To get from the ellipse (3) to the circle (4), we used the linear transformation defined by the matrix

$$\begin{bmatrix} 1 & -1 \\ 0 & 1 \end{bmatrix},$$

which is the *inverse* of the matrix

$$\begin{bmatrix} 1 & 1 \\ 0 & 1 \end{bmatrix}$$

that produced the shear shown previously in Sec. 5.1, Example 5(c). ■

Changing Bases

Let us start with a straightforward example.

EXAMPLE 2 **Vectors in New Coordinates** The vector

$$\vec{\mathbf{u}} = \begin{bmatrix} 6 \\ 4 \end{bmatrix}$$

in $\mathbb{R}^2$ is expressed in terms of the standard basis vectors

$$\vec{\mathbf{e}}_1 = \begin{bmatrix} 1 \\ 0 \end{bmatrix} \quad \text{and} \quad \vec{\mathbf{e}}_2 = \begin{bmatrix} 0 \\ 1 \end{bmatrix},$$

which we studied in Sec. 3.6:

$$\vec{\mathbf{u}} = \begin{bmatrix} 6 \\ 4 \end{bmatrix} = 6\begin{bmatrix} 1 \\ 0 \end{bmatrix} + 4\begin{bmatrix} 0 \\ 1 \end{bmatrix} = 6\vec{\mathbf{e}}_1 + 4\vec{\mathbf{e}}_2.$$

The numbers 6 and 4 are the coordinates of $\vec{\mathbf{u}}$ relative to the *standard* basis $S = \{\vec{\mathbf{e}}_1, \vec{\mathbf{e}}_2\}$, as illustrated in Fig. 5.4.2.

But *any* pair of linearly independent vectors constitute a legitimate basis for $\mathbb{R}^2$. If we choose to use $B = \{\vec{\mathbf{b}}_1, \vec{\mathbf{b}}_2\}$,

$$\vec{\mathbf{b}}_1 = \begin{bmatrix} 1 \\ 1 \end{bmatrix} \quad \text{and} \quad \vec{\mathbf{b}}_2 = \begin{bmatrix} 1 \\ -1 \end{bmatrix},$$

as our basis, for example, then we must figure out how to express $\vec{\mathbf{u}}$ in terms of the new basis $B = \{\vec{\mathbf{b}}_1, \vec{\mathbf{b}}_2\}$, as shown in Fig. 5.4.3.

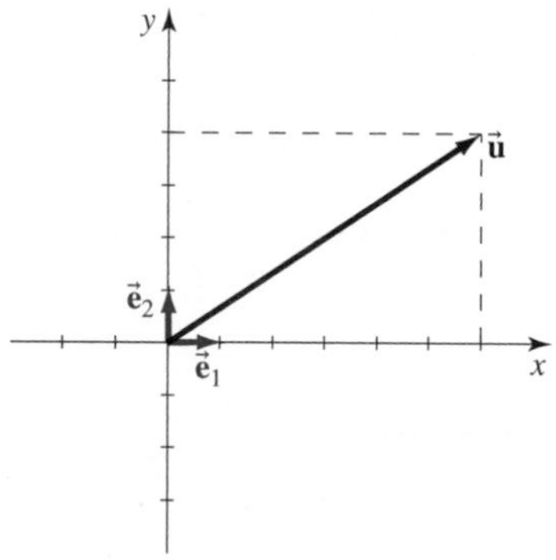

FIGURE 5.4.2 In standard coordinates, $\vec{\mathbf{u}} = 6\vec{\mathbf{e}}_1 + 4\vec{\mathbf{e}}_2$.

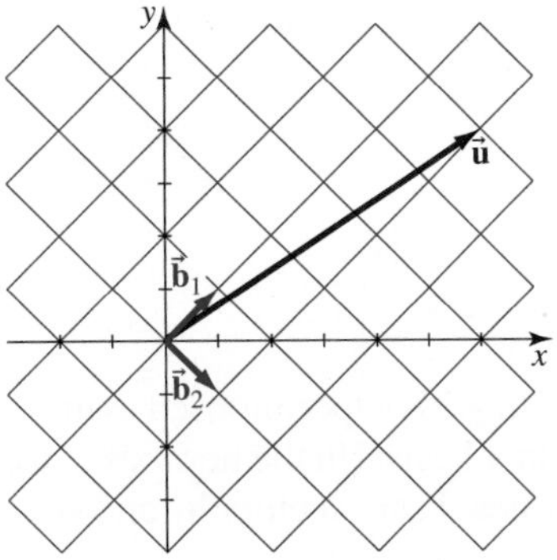

FIGURE 5.4.3 In nonstandard coordinates, $\vec{\mathbf{u}} = 5\vec{\mathbf{b}}_1 + \vec{\mathbf{b}}_2$.

The coordinate vector $\vec{\mathbf{u}}_B$ has coordinates β_1 and β_2 that we obtain by solving

$$\beta_1 \begin{bmatrix} 1 \\ 1 \end{bmatrix} + \beta_2 \begin{bmatrix} 1 \\ -1 \end{bmatrix} = \begin{bmatrix} 6 \\ 4 \end{bmatrix},$$

or, equivalently,

$$\underbrace{\begin{bmatrix} 1 & 1 \\ 1 & -1 \end{bmatrix}}_{\mathbf{M}_B} \underbrace{\begin{bmatrix} \beta_1 \\ \beta_2 \end{bmatrix}}_{\vec{\mathbf{u}}_B} = \underbrace{\begin{bmatrix} 6 \\ 4 \end{bmatrix}}_{\vec{\mathbf{u}}_S}.$$

To solve for the coordinate vector

$$\vec{\mathbf{u}}_B = \begin{bmatrix} \beta_1 \\ \beta_2 \end{bmatrix},$$

we multiply both sides by

$$\mathbf{M}_B^{-1} = \begin{bmatrix} 1/2 & 1/2 \\ 1/2 & -1/2 \end{bmatrix}$$

to obtain

$$\vec{\mathbf{u}}_B = \mathbf{M}_B^{-1}\vec{\mathbf{u}}_S = \begin{bmatrix} 1/2 & 1/2 \\ 1/2 & -1/2 \end{bmatrix} \begin{bmatrix} 6 \\ 4 \end{bmatrix} = \begin{bmatrix} 3+2 \\ 3-2 \end{bmatrix} = \begin{bmatrix} 5 \\ 1 \end{bmatrix}.$$

■

We knew $\mathbf{M}_B^{-1}$ would exist in Example 2 because a basis consists of linearly independent vectors, which are the column vectors of $\mathbf{M}_B$. We denote

$$\mathbf{M}_S = \mathbf{M}_B^{-1} = [\,\vec{\mathbf{b}}_1 \mid \vec{\mathbf{b}}_2\,]^{-1},$$

and call $\mathbf{M}_S$ the change of coordinate matrix from the standard basis to basis B.

Changing Bases in $\mathbb{R}^n$

Let $\vec{\mathbf{u}}_B$ be the coordinate vector relative to basis $B = \{\vec{\mathbf{b}}_1, \vec{\mathbf{b}}_2, \ldots, \vec{\mathbf{b}}_n\}$, and $\vec{\mathbf{u}}_S$ be the coordinate vector relative to the standard basis $S = \{\vec{\mathbf{e}}_1, \vec{\mathbf{e}}_2, \ldots, \vec{\mathbf{e}}_n\}$.

- To change coordinates from basis B to the standard basis:

$$\mathbf{M}_B\vec{\mathbf{u}}_B = \vec{\mathbf{u}}_S, \quad \text{where } \mathbf{M}_B = [\,\vec{\mathbf{b}}_1 \mid \vec{\mathbf{b}}_2 \mid \cdots \mid \vec{\mathbf{b}}_n\,].$$

- To change coordinates from the standard basis to basis B:

$$\mathbf{M}_S\vec{\mathbf{u}}_S = \vec{\mathbf{u}}_B, \quad \text{where } \mathbf{M}_S = \mathbf{M}_B^{-1} = [\,\vec{\mathbf{b}}_1 \mid \vec{\mathbf{b}}_2 \mid \cdots \mid \vec{\mathbf{b}}_n\,]^{-1}.$$

$\mathbf{M}_B$ is called the **change of coordinate matrix** from basis B to the standard basis, while $\mathbf{M}_S$ is the change of coordinate matrix from S to B.

See Appendix LT for a more general method for bases in an n-dimensional vector space $\mathbb{V}$.

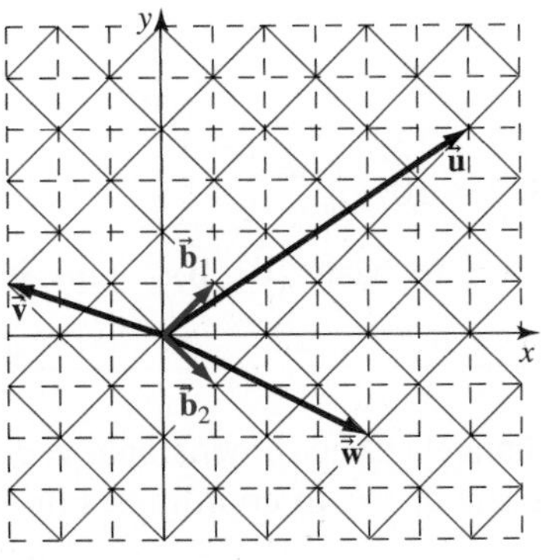

FIGURE 5.4.4 The vectors $\vec{\mathbf{u}}$, $\vec{\mathbf{v}}$, and $\vec{\mathbf{w}}$ of Example 3 with *both* grids superimposed (combining Fig. 5.4.2 and Fig. 5.4.3).

EXAMPLE 3 **There and Back Again** We return to Example 2 and continue to move between the standard basis S and the new basis B.

Given another vector

$$\vec{\mathbf{v}}_S = \begin{bmatrix} -3 \\ 1 \end{bmatrix},$$

we can use $\mathbf{M}_B^{-1}$ to find its "new" coordinates with respect to basis B:

$$\vec{\mathbf{v}}_B = \mathbf{M}_B^{-1}\vec{\mathbf{v}}_S = \begin{bmatrix} 1/2 & 1/2 \\ 1/2 & -1/2 \end{bmatrix} \begin{bmatrix} -3 \\ 1 \end{bmatrix} = \begin{bmatrix} -3/2 + 1/2 \\ -3/2 - 1/2 \end{bmatrix} = \begin{bmatrix} -1 \\ -2 \end{bmatrix}.$$

On the other hand, if we know the B-coordinates of a vector $\vec{\mathbf{w}}$,

$$\vec{\mathbf{w}}_B = \begin{bmatrix} 1 \\ 3 \end{bmatrix},$$

we can find its standard coordinates using $\mathbf{M}_B$:

$$\vec{\mathbf{w}}_S = \mathbf{M}_B \vec{\mathbf{w}}_B = \begin{bmatrix} 1 & 1 \\ 1 & -1 \end{bmatrix} \begin{bmatrix} 1 \\ 3 \end{bmatrix} = \begin{bmatrix} 4 \\ -2 \end{bmatrix}.$$

Vectors $\vec{\mathbf{u}}$, $\vec{\mathbf{v}}$, and $\vec{\mathbf{w}}$ are shown in Fig. 5.4.4, with grids superimposed to make it easy to read off their coordinates relative to either basis. (In the new basis, we walk along the diagonal grid in the direction of new basis vectors $\vec{\mathbf{b}}_1$ and $\vec{\mathbf{b}}_2$.) ■

The basis vectors represent the "axes" in the vector space, the one-dimensional subspaces (e.g., lines in $\mathbb{R}^n$) spanned by the individual basis vectors. For $\mathbb{R}^2$, they impose "grids" like those in Figs. 5.4.3 and 5.4.4, by which vectors can be located and measured.

EXAMPLE 4 **New Coordinates** Let us find the coordinates of the vector

$$\vec{\mathbf{u}}_S = \begin{bmatrix} 3 \\ -5 \end{bmatrix},$$

relative to another basis

$$B = \{\vec{\mathbf{b}}_1, \vec{\mathbf{b}}_2\} = \left\{ \begin{bmatrix} -2 \\ 1 \end{bmatrix}, \begin{bmatrix} 3 \\ 2 \end{bmatrix} \right\}.$$

We want β_1 and β_2 such that

$$\begin{bmatrix} 3 \\ -5 \end{bmatrix} = \beta_1 \begin{bmatrix} -2 \\ 1 \end{bmatrix} + \beta_2 \begin{bmatrix} 3 \\ 2 \end{bmatrix} = \begin{bmatrix} -2\beta_1 + 3\beta_2 \\ \beta_1 + 2\beta_2 \end{bmatrix}.$$

This is equivalent to

$$\begin{bmatrix} 3 \\ -5 \end{bmatrix} = \begin{bmatrix} -2 & 3 \\ 1 & 2 \end{bmatrix} \begin{bmatrix} \beta_1 \\ \beta_2 \end{bmatrix},$$

and we can solve this system by finding the matrix inverse to

$$\mathbf{M}_B = \begin{bmatrix} -2 & 3 \\ 1 & 2 \end{bmatrix},$$

namely,

$$\mathbf{M}_B^{-1} = \begin{bmatrix} -2/7 & 3/7 \\ 1/7 & 2/7 \end{bmatrix}.$$

Then,

$$\begin{bmatrix} \beta_1 \\ \beta_2 \end{bmatrix} = \begin{bmatrix} -2/7 & 3/7 \\ 1/7 & 2/7 \end{bmatrix} \begin{bmatrix} 3 \\ -5 \end{bmatrix} = \begin{bmatrix} -6/7 - 15/7 \\ 3/7 - 10/7 \end{bmatrix} = \begin{bmatrix} -3 \\ -1 \end{bmatrix}.$$

Thus, $(-3, -1)$ are the coordinates of $\vec{\mathbf{u}}$ relative to basis $\{\vec{\mathbf{b}}_1, \vec{\mathbf{b}}_2\}$. This is illustrated in Fig. 5.4.5. ■

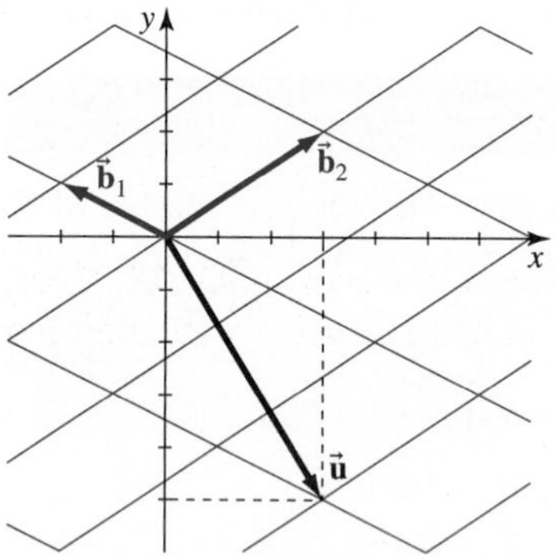

FIGURE 5.4.5 New coordinates for Example 4.

EXAMPLE 5 **Changing Bases in $\mathbb{R}^3$** Let us change a three-dimensional vector $\vec{\mathbf{u}}_S$, where $S = \{\vec{\mathbf{e}}_1, \vec{\mathbf{e}}_2, \vec{\mathbf{e}}_3\}$ is the standard basis of $\mathbb{R}^3$, to coordinates in a new basis

$$B = \left\{ \begin{bmatrix} 1 \\ 0 \\ 1 \end{bmatrix}, \begin{bmatrix} 1 \\ 2 \\ 0 \end{bmatrix}, \begin{bmatrix} 1 \\ 1 \\ 1 \end{bmatrix} \right\}.$$

Form the matrix

$$\mathbf{M}_B = \begin{bmatrix} 1 & 1 & 1 \\ 0 & 2 & 1 \\ 1 & 0 & 1 \end{bmatrix}.$$

Then any vector in the new basis takes the form $\vec{\mathbf{u}}_B = \mathbf{M}_B^{-1}\vec{\mathbf{u}}_S$, and from Sec. 3.3, Example 3(c), we know that

$$\mathbf{M}_B^{-1} = \begin{bmatrix} 2 & -1 & -1 \\ 1 & 0 & -1 \\ -2 & 1 & 2 \end{bmatrix}.$$

For instance, if $\vec{\mathbf{u}}_S = \begin{bmatrix} 1 \\ 2 \\ 3 \end{bmatrix}$, then $\vec{\mathbf{u}}_B = \mathbf{M}_B^{-1} \begin{bmatrix} 1 \\ 2 \\ 3 \end{bmatrix} = \begin{bmatrix} -3 \\ -2 \\ 6 \end{bmatrix}$. ■

Coordinates in Polynomial Spaces

The vector space $\mathbb{P}_2$ of real polynomials of degree two or less may be conveniently represented using basis $\{x^2, x, 1\}$. It is a **standard basis** for the space, although many other bases are possible.

Standard Basis for Polynomials of nth Degree:

$$S = \{x^n, x^{n-1}, \ldots, x^2, x, 1\}.$$

In this basis, polynomial $p(x) = 6x^2$ has coordinates (6, 0, 0), because

$$p(x) = 6x^2 = 6 \cdot x^2 + 0 \cdot x + 0 \cdot 1.$$

Similarly,

$$q(x) = x - 1 = 0 \cdot x^2 + 1 \cdot x + (-1) \cdot 1$$

has coordinates $(0, 1, -1)$, while

$$r(x) = 2x^2 + 3x$$

has coordinates (2, 3, 0).

Since the basis element $x^2 = 1 \cdot x^2 + 0 \cdot x + 0 \cdot 1$ has coordinates (1, 0, 0), while x and 1 correspond to (0, 1, 0) and (0, 0, 1), respectively, polynomial p may be represented by the coordinate vector in $\mathbb{R}^3$:

$$\vec{\mathbf{p}}_S = 6 \begin{bmatrix} 1 \\ 0 \\ 0 \end{bmatrix} + 0 \begin{bmatrix} 0 \\ 1 \\ 0 \end{bmatrix} + 0 \begin{bmatrix} 0 \\ 0 \\ 1 \end{bmatrix} = \begin{bmatrix} 6 \\ 0 \\ 0 \end{bmatrix}.$$

Similarly,

$$\vec{\mathbf{q}}_S = 0 \begin{bmatrix} 1 \\ 0 \\ 0 \end{bmatrix} + 1 \begin{bmatrix} 0 \\ 1 \\ 0 \end{bmatrix} + (-1) \begin{bmatrix} 0 \\ 0 \\ 1 \end{bmatrix} = \begin{bmatrix} 0 \\ 1 \\ -1 \end{bmatrix} \quad \text{and} \quad \vec{\mathbf{r}}_S = \begin{bmatrix} 2 \\ 3 \\ 0 \end{bmatrix}.$$

EXAMPLE 6 **Alternative Basis** Let us show that $\{x^2 + 1, x^2 - 1, x\}$ is also a basis for $\mathbb{P}_2$, and find the corresponding coordinates for the polynomials

$$p(x) = 6x^2, \quad q(x) = x - 1, \quad \text{and} \quad r(x) = 2x^2 + 3x.$$

Our proposed new basis vectors $\vec{\mathbf{b}}_1$, $\vec{\mathbf{b}}_2$, and $\vec{\mathbf{b}}_3$ are represented in the standard basis by

$$B = \left\{ \begin{bmatrix} 1 \\ 0 \\ 1 \end{bmatrix}, \begin{bmatrix} 1 \\ 0 \\ -1 \end{bmatrix}, \begin{bmatrix} 0 \\ 1 \\ 0 \end{bmatrix} \right\},$$

and, as in Sec. 3.6, we may demonstrate their independence by noting that the matrix

$$\mathbf{M}_B = \begin{bmatrix} 1 & 1 & 0 \\ 0 & 0 & 1 \\ 1 & -1 & 0 \end{bmatrix} \quad \text{has RREF} \quad \begin{bmatrix} 1 & 0 & 0 \\ 0 & 1 & 0 \\ 0 & 0 & 1 \end{bmatrix}.$$

But these independent vectors span a subspace of $\mathbb{R}^3$ of dimension 3, and $\mathbb{R}^3$ itself is 3-dimensional. Therefore, the vectors span $\mathbb{R}^3$ and form a basis. Consequently, the corresponding basis vectors in $\mathbb{P}_2$ form a basis for $\mathbb{P}_2$.

What are the coordinates relative to this new basis of $p(x) = 6x^2$? Since p is represented by the coordinate vector

$$\vec{\mathbf{p}}_S = \begin{bmatrix} 6 \\ 0 \\ 0 \end{bmatrix},$$

for the standard basis $\{x^2, x, 1\}$, we want to find coordinates β_1, β_2, and β_3 such that

$$\vec{\mathbf{p}}_S = \begin{bmatrix} 6 \\ 0 \\ 0 \end{bmatrix} = \beta_1 \begin{bmatrix} 1 \\ 0 \\ 1 \end{bmatrix} + \beta_2 \begin{bmatrix} 1 \\ 0 \\ -1 \end{bmatrix} + \beta_3 \begin{bmatrix} 0 \\ 1 \\ 0 \end{bmatrix};$$

that is,

$$\begin{bmatrix} 1 & 1 & 0 \\ 0 & 0 & 1 \\ 1 & -1 & 0 \end{bmatrix} \begin{bmatrix} \beta_1 \\ \beta_2 \\ \beta_3 \end{bmatrix} = \begin{bmatrix} 6 \\ 0 \\ 0 \end{bmatrix}.$$

We can solve this system directly by reducing its augmented matrix to RREF, or we can calculate that the inverse matrix to

$$\mathbf{M}_B = \begin{bmatrix} 1 & 1 & 0 \\ 0 & 0 & 1 \\ 1 & -1 & 0 \end{bmatrix} \quad \text{is} \quad \mathbf{M}_B^{-1} = \begin{bmatrix} 1/2 & 0 & 1/2 \\ 1/2 & 0 & -1/2 \\ 0 & 1 & 0 \end{bmatrix}.$$

Then the new coordinates are given by

$$\begin{bmatrix} \beta_1 \\ \beta_2 \\ \beta_3 \end{bmatrix} = \mathbf{M}_B^{-1}\vec{\mathbf{p}}_S = \mathbf{M}_B^{-1} \begin{bmatrix} 6 \\ 0 \\ 0 \end{bmatrix} = \begin{bmatrix} 1/2 & 0 & 1/2 \\ 1/2 & 0 & -1/2 \\ 0 & 1 & 0 \end{bmatrix} \begin{bmatrix} 6 \\ 0 \\ 0 \end{bmatrix} = \begin{bmatrix} 3 \\ 3 \\ 0 \end{bmatrix} = \vec{\mathbf{p}}_B.$$

So, we have found that

$$p(x) = 6x^2 = 3 \cdot (x^2 + 1) + 3 \cdot (x^2 - 1) + 0 \cdot x.$$

Because we have found $\mathbf{M}_B^{-1}$, we can quickly convert $q(x)$ and $r(x)$ as well:

$$\mathbf{M}_B^{-1}\vec{\mathbf{q}}_S = \mathbf{M}_B^{-1} \begin{bmatrix} 0 \\ 1 \\ -1 \end{bmatrix} = \begin{bmatrix} 1/2 & 0 & 1/2 \\ 1/2 & 0 & -1/2 \\ 0 & 1 & 0 \end{bmatrix} \begin{bmatrix} 0 \\ 1 \\ -1 \end{bmatrix} = \begin{bmatrix} -1/2 \\ 1/2 \\ 1 \end{bmatrix} = \vec{\mathbf{q}}_B$$

and

$$\mathbf{M}_B^{-1}\vec{\mathbf{r}}_S = \mathbf{M}_B^{-1} \begin{bmatrix} 2 \\ 3 \\ 0 \end{bmatrix} = \begin{bmatrix} 1/2 & 0 & 1/2 \\ 1/2 & 0 & -1/2 \\ 0 & 1 & 0 \end{bmatrix} \begin{bmatrix} 2 \\ 3 \\ 0 \end{bmatrix} = \begin{bmatrix} 1 \\ 1 \\ 3 \end{bmatrix} = \vec{\mathbf{r}}_B.$$

Using the new coordinates, we can confirm that the polynomials $q(x)$ and $r(x)$ could be written alternatively as

$$q(x) = x - 1 = -\frac{1}{2} \cdot (x^2 + 1) + \frac{1}{2} \cdot (x^2 - 1) + 1 \cdot x$$

and

$$r(x) = 2x^2 + 3x = 1 \cdot (x^2 + 1) + 1 \cdot (x^2 - 1) + 3 \cdot x.$$

■

For an extended discussion of finding matrices for transformations in vector spaces of functions, see Appendix LT.

The Right Point of View for DEs

We turn our attention to differential equations with an illustrative example.

EXAMPLE 7 **Decoupling a DE** Suppose we would like to solve the system of linear differential equations

$$\begin{aligned} x_1' &= x_1 + x_2, \\ x_2' &= 4x_1 + x_2. \end{aligned} \tag{5}$$

Because each equation depends on both unknown functions, we need a new idea. We would like to make a change of basis so that our system is **decoupled**. That is, we seek a new basis in which we can solve each DE *separately* for each component. We want to make a change of variable $\vec{\mathbf{x}} = \mathbf{P}\vec{\mathbf{u}}$, much as we changed coordinates in the earlier examples. First, we rewrite system (5) in matrix form as $\vec{\mathbf{x}}' = \mathbf{A}\vec{\mathbf{x}}$, with

$$\mathbf{A} = \begin{bmatrix} 1 & 1 \\ 4 & 1 \end{bmatrix}.$$

Then, if we let $\vec{\mathbf{x}} = \mathbf{P}\vec{\mathbf{u}}$, the system

$$\vec{\mathbf{x}}' = \mathbf{A}\vec{\mathbf{x}}$$

is transformed to

$$\mathbf{P}\vec{\mathbf{u}}' = \mathbf{A}\mathbf{P}\vec{\mathbf{u}}$$

or

$$\vec{\mathbf{u}}' = (\mathbf{P}^{-1}\mathbf{A}\mathbf{P})\vec{\mathbf{u}}. \tag{6}$$

For the matrix $\mathbf{P}$, we choose the columns to be the eigenvectors of the matrix $\mathbf{A}$, which were calculated with the eigenvalues in Sec. 5.3, Example 2, to be

$$\lambda_1 = 3, \quad \vec{\mathbf{v}}_1 = \begin{bmatrix} 1 \\ 2 \end{bmatrix};$$

$$\lambda_2 = -1, \quad \vec{\mathbf{v}}_2 = \begin{bmatrix} 1 \\ -2 \end{bmatrix}.$$

Thus, we choose

$$\mathbf{P} = \begin{bmatrix} 1 & 1 \\ 2 & -2 \end{bmatrix} \quad \text{with its inverse} \quad \mathbf{P}^{-1} = \begin{bmatrix} 1/2 & 1/4 \\ 1/2 & -1/4 \end{bmatrix}.$$

Now, we apply (6) to $\mathbf{A}$ to get

$$\mathbf{P}^{-1}\mathbf{A}\mathbf{P} = \begin{bmatrix} 1/2 & 1/4 \\ 1/2 & -1/4 \end{bmatrix} \begin{bmatrix} 1 & 1 \\ 4 & 1 \end{bmatrix} \begin{bmatrix} 1 & 1 \\ 2 & -2 \end{bmatrix} = \begin{bmatrix} 3 & 0 \\ 0 & -1 \end{bmatrix},$$

hence the transformed system

$$\begin{bmatrix} u_1 \\ u_2 \end{bmatrix}' = \begin{bmatrix} 3 & 0 \\ 0 & -1 \end{bmatrix} \begin{bmatrix} u_1 \\ u_2 \end{bmatrix}.$$

In component form, this gives

$$\begin{aligned} u_1' &= 3u_1, \\ u_2' &= -u_2. \end{aligned} \tag{7}$$

System (7) is much easier to solve than (5) because the unknown functions are decoupled. We see immediately that $u_1 = c_1e^{3t}$ and $u_2 = c_2e^{-t}$. Linear algebra has changed a complicated linked system of DEs into a system we can solve in our heads! The price: translate between $\vec{\mathbf{u}}$ and $\vec{\mathbf{x}}$, using $\mathbf{P}$. To get the answer for (5), then, we use the fact that $\vec{\mathbf{x}} = \mathbf{P}\vec{\mathbf{u}}$. Thus,

$$\begin{bmatrix} x_1 \\ x_2 \end{bmatrix} = \begin{bmatrix} 1 & 1 \\ 2 & -2 \end{bmatrix} \begin{bmatrix} c_1e^{3t} \\ c_2e^{-t} \end{bmatrix} = \begin{bmatrix} c_1e^{3t} + c_2e^{-t} \\ 2c_1e^{3t} - 2c_2e^{-t} \end{bmatrix},$$

and the solution of (5) is given by

$$x_1 = c_1e^{3t} + c_2e^{-t} \quad \text{and} \quad x_2 = 2c_1e^{3t} - 2c_2e^{-t}$$

for arbitrary constants c_1 and c_2, as we could easily verify by differentiating and substituting.

We are not going to explain the trick completely now, but note a remarkable coincidence. The entries 3 and -1 in the diagonal matrix $\mathbf{P}^{-1}\mathbf{AP}$ are the eigenvalues of $\mathbf{A}$ (by Sec. 5.3, Example 2), listed in the same order as the eigenvectors we chose for $\mathbf{P}$. Problem 65 suggests further exploration of this example. ■

Diagonalizing a Matrix

Eigenvectors make good basis vectors. This observation is the point of the illustrative example just completed. In particular, eigenvectors provide a way to replace the original matrix with one that is diagonal. To see how this process works for a general $n \times n$ matrix $\mathbf{A}$, suppose that $\mathbf{A}$ has linearly independent eigenvectors $\vec{\mathbf{v}}_1, \vec{\mathbf{v}}_2, \ldots, \vec{\mathbf{v}}_n$, with corresponding eigenvalues $\lambda_1, \lambda_2, \ldots, \lambda_n$ not necessarily distinct. Suppose that

$$\mathbf{P} = \begin{bmatrix} | & | & & | \\ \vec{\mathbf{v}}_1 & \vec{\mathbf{v}}_2 & \cdots & \vec{\mathbf{v}}_n \\ | & | & & | \end{bmatrix} \quad \text{and} \quad \mathbf{D} = \begin{bmatrix} \lambda_1 & 0 & \cdots & 0 \\ 0 & \lambda_2 & \cdots & 0 \\ \vdots & \vdots & \ddots & \vdots \\ 0 & 0 & \cdots & \lambda_n \end{bmatrix}.$$

Discussion We will make a computation to find an expression for $\mathbf{D}$ in terms of $\mathbf{A}$ and $\mathbf{P}$. Of course, we know that $\mathbf{A}\vec{\mathbf{v}}_i = \lambda_i\vec{\mathbf{v}}_i$, for all $1 \le i \le n$.

Because of the way that matrices are multiplied, it is easy to see that for any matrix $\mathbf{M}$, $\mathbf{M}\vec{\mathbf{e}}_i$ is the ith column of $\mathbf{M}$, where $\{\vec{\mathbf{e}}_1, \vec{\mathbf{e}}_2, \ldots, \vec{\mathbf{e}}_n\}$ is the standard basis of $\mathbb{R}^n$. Then the ith column of $\mathbf{AP}$ is

$$(\mathbf{AP})\vec{\mathbf{e}}_i = \mathbf{A}(\mathbf{P}\vec{\mathbf{e}}_i) = \mathbf{A}\vec{\mathbf{v}}_i = \lambda_i\vec{\mathbf{v}}_i.$$

Also, the ith column of $\mathbf{PD}$ is

$$(\mathbf{PD})\vec{\mathbf{e}}_i = \mathbf{P}(\mathbf{D}\vec{\mathbf{e}}_i) = \mathbf{P}(\lambda_i\vec{\mathbf{e}}_i) = \lambda_i(\mathbf{P}\vec{\mathbf{e}}_i) = \lambda_i\vec{\mathbf{v}}_i.$$

Hence, $\mathbf{AP}$ and $\mathbf{PD}$ have ith columns equal for all i, so $\mathbf{AP} = \mathbf{PD}$. Because $\vec{\mathbf{v}}_1, \vec{\mathbf{v}}_2, \ldots, \vec{\mathbf{v}}_n$ are linearly independent, $\mathbf{P}$ is invertible, so

$$\mathbf{D} = \mathbf{P}^{-1}\mathbf{AP} = \begin{bmatrix} \lambda_1 & 0 & \cdots & 0 \\ 0 & \lambda_2 & \cdots & 0 \\ \vdots & \vdots & \ddots & \vdots \\ 0 & 0 & \cdots & \lambda_n \end{bmatrix}.$$

We say that $\mathbf{P}$ **diagonalizes** the matrix $\mathbf{A}$, and $\mathbf{A}$ is **diagonalizable**. □

A Question of Order:

The order of eigenvalues in $\mathbf{D}$, which is not unique, determines the order of the eigenvectors in $\mathbf{P}$. Furthermore, different basis vectors of the eigenspaces can be used.

We have just proved a theorem. The argument given above works for square matrices of any size, provided that we have enough linearly independent eigenvectors. (These eigenvectors might correspond to n distinct eigenvalues, or some eigenvalue might have multiplicity greater than one but have as many linearly independent eigenvectors as its multiplicity.)

Diagonalization Theorem

An $n \times n$ matrix $\mathbf{A}$ is diagonalizable

(i) if and only if it has n linearly independent (real) eigenvectors;

(ii) if and only if the sum of the dimensions of its eigenspaces is n.

The diagonalization process *excludes* matrices with repeated eigenvalues if they do not have sufficient eigenvectors (as in Sec. 5.3, Example 6).[2]

Once we have computed the eigenvectors of $\mathbf{A}$, we can diagonalize $\mathbf{A}$ as follows:

Diagonalization of a Matrix

For an $n \times n$ matrix $\mathbf{A}$ with n linearly independent eigenvectors:

Step 1. Construct an $n \times n$ diagonal matrix $\mathbf{D}$ of the eigenvalues λ_i, for $1 \le i \le n$. (NOTE: An eigenvalue with multiplicity m appears m times.)

Step 2. Construct another $n \times n$ matrix $\mathbf{P}$ with the eigenvectors $\vec{\mathbf{v}}_i$ as columns, *listed in the order corresponding to the eigenvalues* λ_i *in* $\mathbf{D}$.

The following equations are all true and equivalent:

$$\mathbf{AP} = \mathbf{PD} \tag{8}$$

$$\mathbf{A} = \mathbf{PDP}^{-1} \tag{9}$$

$$\mathbf{D} = \mathbf{P}^{-1}\mathbf{AP} \tag{10}$$

Equation (8) is handiest for checking calculations; equations (9) and (10) are handiest for proving theorems, as we will see in Sections 6.5 and 6.6.

EXAMPLE 8 **Three-by-Three Diagonalization** A 3×3 illustration comes from Sec. 5.3, Example 3, where

$$\mathbf{A} = \begin{bmatrix} 1 & 1 & -2 \\ -1 & 2 & 1 \\ 0 & 1 & -1 \end{bmatrix},$$

and we found eigenvalues $\lambda_1 = 2$, $\lambda_2 = 1$, and $\lambda_3 = -1$, with respective eigenvectors

$$\vec{\mathbf{v}}_1 = \begin{bmatrix} 1 \\ 3 \\ 1 \end{bmatrix}, \quad \vec{\mathbf{v}}_2 = \begin{bmatrix} 3 \\ 2 \\ 1 \end{bmatrix}, \quad \text{and} \quad \vec{\mathbf{v}}_3 = \begin{bmatrix} 1 \\ 0 \\ 1 \end{bmatrix}.$$

---

[2]We found in Sec. 5.3 that a multiple eigenvalue might have too few eigenvectors to form a basis of the eigenspace. See Problems 60–62 for an example of a procedure similar to diagonalization, which can be useful with such a double eigenvalue.

With

$$\mathbf{D} = \begin{bmatrix} 2 & 0 & 0 \\ 0 & 1 & 0 \\ 0 & 0 & -1 \end{bmatrix} \quad \text{and} \quad \mathbf{P} = \begin{bmatrix} 1 & 3 & 1 \\ 3 & 2 & 0 \\ 1 & 1 & 1 \end{bmatrix},$$

we can verify equation (8),

$$\mathbf{AP} = \mathbf{PD}.$$

■

EXAMPLE 9 **Diagonalizing with a Double Eigenvalue** The matrix

$$\mathbf{A} = \begin{bmatrix} -2 & 1 & 1 \\ 1 & -2 & 1 \\ 1 & 1 & -2 \end{bmatrix}$$

from Sec. 5.3, Example 6, turned out to have a single eigenvalue with an eigenvector, and a double eigenvalue with two linearly independent eigenvectors:

$$\lambda_1 = 0 \quad \text{with} \quad \vec{\mathbf{v}}_1 = \begin{bmatrix} 1 \\ 1 \\ 1 \end{bmatrix}$$

and

$$\lambda_2 = -3 \quad \text{with} \quad \vec{\mathbf{v}}_2 = \begin{bmatrix} -1 \\ 1 \\ 0 \end{bmatrix} \quad \text{and} \quad \vec{\mathbf{v}}_3 = \begin{bmatrix} -1 \\ 0 \\ 1 \end{bmatrix}.$$

Hence,

$$\mathbf{P} = \begin{bmatrix} 1 & -1 & -1 \\ 1 & 1 & 0 \\ 1 & 0 & 1 \end{bmatrix} \quad \text{and} \quad \mathbf{P}^{-1} = \begin{bmatrix} 1/3 & 1/3 & 1/3 \\ -1/3 & 2/3 & -1/3 \\ -1/3 & -1/3 & 2/3 \end{bmatrix},$$

and we can verify equation (10), that

$$\mathbf{P}^{-1}\mathbf{AP} = \begin{bmatrix} 0 & 0 & 0 \\ 0 & -3 & 0 \\ 0 & 0 & -3 \end{bmatrix}$$

gives the eigenvalues in appropriate order on the diagonal.

■

EXAMPLE 10 **Nondiagonalizable Matrix** The matrix

$$\mathbf{A} = \begin{bmatrix} 1 & 1 \\ 0 & 1 \end{bmatrix}$$

has double eigenvalue 1, but only one independent eigenvector of the form

$$c \begin{bmatrix} 1 \\ 0 \end{bmatrix}.$$

This matrix cannot be diagonalized, because it does not have enough linearly independent eigenvectors.

■

Linear systems of equations with nondiagonalizable matrices require additional techniques (a generalization of eigenvectors) in order to find solutions. This procedure will be discussed, for DEs, in Sec. 6.2, Example 7.

Similarity

A square matrix **B** is **similar** to matrix **A** (in shorthand, $\mathbf{B} \sim \mathbf{A}$) if there exists an invertible matrix **P** such that $\mathbf{B} = \mathbf{P}^{-1}\mathbf{A}\mathbf{P}$.

Diagonalizability means similarity to a diagonal matrix. Similar matrices share many properties. (See Problem 55.) They have the same eigenvectors and the same characteristic equation.

NOTE: The similarity of matrices is *not* the same as row equivalence.

Glances Backward and Forward

One might reasonably ask *why* we diagonalize a matrix to see the eigenvalues on the diagonal, when we have to *find* the eigenvalues and eigenvectors to do so—it sounds rather circular. However, a diagonal matrix can be a useful conceptual tool that has practical applications as well.

Factoring: For example, one nice result of the diagonalization process is that we can *factor* a diagonalizable matrix **A** into a product of three matrices, each with its own significance.

$$\text{If } \mathbf{P}^{-1}\mathbf{A}\mathbf{P} = \mathbf{D}, \text{ then } \mathbf{A} = \mathbf{P}\mathbf{D}\mathbf{P}^{-1}. \tag{11}$$

P lists the eigenvectors, **D** lists the eigenvalues and, as we will see in Sec. 6.5, $\mathbf{P}^{-1}$ provides the constants in an IVP.

Natural Coordinates: Many physicists view diagonalization as finding the *natural* coordinates for a physical process. *Diagonalization is a change of basis to a coordinate system with the eigenvectors as axes.*

Decoupling: In Example 7, which started off this differential equation discussion, we used eigenvectors to diagonalize the matrix of the system of differential equations, in order to **decouple** the two equations. This decoupling meant that we separated or segregated the unknown functions to make finding the solution easier. We will exploit this device further in Sec. 6.5.

Powers of Matrices: We can use (11) to calculate that, if **M** is a diagonalizable matrix, then, as we will explore in Problem 49(a),

$$\mathbf{M}^k = \mathbf{P}\mathbf{D}^k\mathbf{P}^{-1}. \tag{12}$$

For large k, (12) can provide a great shortcut for calculation, as we will see for DEs in Sec. 6.6.

Summary

Change of basis (and hence of coordinates) in a finite-dimensional vector space can be effected by multiplying by a suitable square matrix. A basis consisting of the eigenvectors of a given matrix can be used to find a diagonalization that is similar to the original matrix, a device useful in solving IVPs for differential equations or iterative equations.

5.4 Problems

Changing Coordinates I *In Problems 1–3, let $S = \{\vec{\mathbf{e}}_1, \vec{\mathbf{e}}_2\}$ be the standard basis, and*

$$B = \{\vec{\mathbf{b}}_1, \vec{\mathbf{b}}_2\} = \left\{\begin{bmatrix} 3 \\ -2 \end{bmatrix}, \begin{bmatrix} -4 \\ 3 \end{bmatrix}\right\}$$

be a new basis for $\mathbb{R}^2$.

1. Calculate the coordinate-change matrices $\mathbf{M}_B$ to go from B to S and $\mathbf{M}_B^{-1}$ as in Example 2.

2. Convert the vectors

$$\begin{bmatrix} 3 \\ 8 \end{bmatrix}, \quad \begin{bmatrix} 2 \\ -1 \end{bmatrix}, \quad \text{and} \quad \begin{bmatrix} 0 \\ 1 \end{bmatrix}$$

from the standard basis to the basis B.

3. Vectors

$$\begin{bmatrix} 3 \\ -1 \end{bmatrix}, \quad \begin{bmatrix} 2 \\ 2 \end{bmatrix}, \quad \text{and} \quad \begin{bmatrix} 1 \\ 0 \end{bmatrix}$$

are expressed in basis B. Find their standard basis representations.

Changing Coordinates II *In Problems 4–6, let $S = \{\vec{\mathbf{e}}_1, \vec{\mathbf{e}}_2\}$ be the standard basis, and*

$$B = \{\vec{\mathbf{b}}_1, \vec{\mathbf{b}}_2\} = \left\{\begin{bmatrix} 1 \\ -1 \end{bmatrix}, \begin{bmatrix} -1 \\ 2 \end{bmatrix}\right\}$$

be a new basis for $\mathbb{R}^2$.

4. Calculate the coordinate change matrices $\mathbf{M}_B$ to go from B to S and $\mathbf{M}_B^{-1}$ as in Example 2.

5. Convert the vectors

$$\begin{bmatrix} 1 \\ 3 \end{bmatrix}, \quad \begin{bmatrix} -1 \\ 1 \end{bmatrix}, \quad \text{and} \quad \begin{bmatrix} 4 \\ 5 \end{bmatrix}$$

from the standard basis to the basis B.

6. Vectors

$$\begin{bmatrix} 2 \\ 2 \end{bmatrix}, \quad \begin{bmatrix} 1 \\ -1 \end{bmatrix}, \quad \text{and} \quad \begin{bmatrix} 1 \\ 0 \end{bmatrix}$$

are expressed in basis B. Find their standard basis representations.

Changing Coordinates III *For Problems 7–9, the standard basis is $S = \{\vec{\mathbf{e}}_1, \vec{\mathbf{e}}_2, \vec{\mathbf{e}}_3\}$. Let*

$$B = \left\{\begin{bmatrix} 1 \\ 0 \\ 0 \end{bmatrix}, \begin{bmatrix} 1 \\ 1 \\ 0 \end{bmatrix}, \begin{bmatrix} 1 \\ 1 \\ 1 \end{bmatrix}\right\}$$

be a new basis for $\mathbb{R}^3$.

7. Calculate the coordinate change matrices $\mathbf{M}_B$ to go from B to S and $\mathbf{M}_B^{-1}$ as in Example 5.

8. Convert the vectors

$$\begin{bmatrix} 1 \\ 0 \\ 1 \end{bmatrix}, \quad \begin{bmatrix} 2 \\ 3 \\ 0 \end{bmatrix}, \quad \text{and} \quad \begin{bmatrix} 0 \\ 4 \\ 3 \end{bmatrix}$$

from the standard basis to the basis B.

9. Vectors

$$\begin{bmatrix} 1 \\ 0 \\ -1 \end{bmatrix}, \quad \begin{bmatrix} 1 \\ 1 \\ 3 \end{bmatrix}, \quad \text{and} \quad \begin{bmatrix} -2 \\ 1 \\ 1 \end{bmatrix}$$

are expressed in basis B. Find their standard basis representations.

Changing Coordinates IV *For Problems 10–12, the standard basis is $S = \{\vec{\mathbf{e}}_1, \vec{\mathbf{e}}_2, \vec{\mathbf{e}}_3\}$. Let*

$$B = \left\{\begin{bmatrix} 1 \\ 0 \\ 0 \end{bmatrix}, \begin{bmatrix} 0 \\ 0 \\ 1 \end{bmatrix}, \begin{bmatrix} 2 \\ 1 \\ -1 \end{bmatrix}\right\}$$

be a new basis for $\mathbb{R}^3$.

10. Calculate the coordinate change matrices $\mathbf{M}_B$ to go from B to S and $\mathbf{M}_B^{-1}$ as in Example 5.

11. Convert the vectors

$$\begin{bmatrix} 1 \\ -2 \\ 1 \end{bmatrix}, \quad \begin{bmatrix} -1 \\ 1 \\ 0 \end{bmatrix}, \quad \text{and} \quad \begin{bmatrix} 3 \\ 0 \\ 2 \end{bmatrix}$$

from the standard basis to basis B.

12. Vectors

$$\begin{bmatrix} -1 \\ -1 \\ -4 \end{bmatrix}, \quad \begin{bmatrix} 1 \\ -1 \\ 3 \end{bmatrix}, \quad \text{and} \quad \begin{bmatrix} 3 \\ 1 \\ 4 \end{bmatrix}$$

are expressed in basis B. Find their standard basis representations.

Polynomial Coordinates I *For Problems 13–15, we take $S = \{x^2, x, 1\}$ as the standard basis in $\mathbb{P}_2$ and introduce a new basis $N = \{2x^2 - x, x^2, x^2 + 1\}$.*

13. Compute coordinate change matrices $\mathbf{M}_N$ to go from N to S and $\mathbf{M}_N^{-1}$ as in Example 6.

14. Express in terms of the new basis N the standard basis polynomials

$$p(x) = x^2 + 2x + 3,$$
$$q(x) = x^2 - 2,$$
$$r(x) = 4x - 5.$$

15. Vector representations of three polynomials relative to the basis N are

$$\vec{\mathbf{u}}_N = \begin{bmatrix} 1 \\ 0 \\ 2 \end{bmatrix}, \quad \vec{\mathbf{v}}_N = \begin{bmatrix} -2 \\ 2 \\ 3 \end{bmatrix}, \quad \text{and} \quad \vec{\mathbf{w}}_N = \begin{bmatrix} -1 \\ -1 \\ 0 \end{bmatrix}.$$

Calculate standard representations of $u(x)$, $v(x)$, and $w(x)$.

Polynomial Coordinates II *For Problems 16–18, we take $S = \{x^3, x^2, x, 1\}$ as the standard basis in $\mathbb{P}_3$ and introduce a new basis $Q = \{x^3, x^3 + x, x^2, x^2 + 1\}$.*

16. Compute coordinate change matrices $\mathbf{M}_Q$ to go from Q to S and $\mathbf{M}_Q^{-1}$ as in Example 6.

17. Find the coordinate vectors of these standard basis polynomials in terms of the new basis Q:

$$p(x) = x^3 + 2x^2 + 3,$$
$$q(x) = x^2 - x - 2,$$
$$r(x) = x^3 + 1.$$

18. Vector representations of three polynomials relative to the basis Q are

$$\vec{\mathbf{u}}_Q = \begin{bmatrix} 1 \\ -1 \\ 0 \\ 2 \end{bmatrix}, \quad \vec{\mathbf{v}}_Q = \begin{bmatrix} -2 \\ 0 \\ -2 \\ 0 \end{bmatrix}, \quad \text{and} \quad \vec{\mathbf{w}}_Q = \begin{bmatrix} 3 \\ -1 \\ 4 \\ 2 \end{bmatrix}.$$

Calculate standard representations of $u(x)$, $v(x)$, and $w(x)$.

Matrix Representations for Polynomial Transformations *Using the standard basis $S = \{t^4, t^3, t^2, t, 1\}$ for $\mathbb{P}_4$, determine a matrix representing the transformation from $\mathbb{P}_4$ to $\mathbb{P}_4$ given in each of Problems 19–24. Then apply your matrix to the following polynomials:*

(a) $g(t) = t^4 - t^3 + t^2 - t + 1$

(b) $q(t) = t^4 + 2t^2 + 4$

(c) $r(t) = -4t^4 + 3t^3$

(d) $w(t) = t^4 - 8t^2 + 16$

19. $T(f(t)) = f''(t)$ **20.** $T(f(t)) = f(0)$

21. $T(f(t)) = f'''(t)$ **22.** $T(f(t)) = f(-t)$

23. $T(f(t)) = f'(t) - 2f(t)$ **24.** $T(f(t)) = f''(t) + f(t)$

Diagonalization *In Problems 25–48, determine whether each matrix $\mathbf{A}$ is diagonalizable. If it is, determine a matrix $\mathbf{P}$ that diagonalizes it and compute $\mathbf{P}^{-1}\mathbf{AP}$. You can obtain $\mathbf{P}^{-1}\mathbf{AP}$ directly from careful construction of a diagonal matrix with eigenvalues along the diagonal in the* ***proper order***.

25. $\begin{bmatrix} 3 & 2 \\ -2 & -3 \end{bmatrix}$ **26.** $\begin{bmatrix} 1 & -1 \\ 1 & 3 \end{bmatrix}$ **27.** $\begin{bmatrix} 1 & 2 \\ 2 & 1 \end{bmatrix}$

28. $\begin{bmatrix} 1 & 3 \\ 1 & 3 \end{bmatrix}$ **29.** $\begin{bmatrix} 3 & 1 \\ -1 & 5 \end{bmatrix}$ **30.** $\begin{bmatrix} 0 & -1 \\ 1 & 0 \end{bmatrix}$

31. $\begin{bmatrix} 12 & -6 \\ 15 & -7 \end{bmatrix}$ **32.** $\begin{bmatrix} 3 & 1/2 \\ 0 & 3 \end{bmatrix}$ **33.** $\begin{bmatrix} 4 & -2 \\ 1/2 & 2 \end{bmatrix}$

34. $\begin{bmatrix} 1 & 4 \\ -4 & 1 \end{bmatrix}$ **35.** $\begin{bmatrix} 1 & 0 & 1 \\ 0 & 1 & 0 \\ 1 & 0 & 1 \end{bmatrix}$ **36.** $\begin{bmatrix} 0 & 1 & -1 \\ 0 & -1 & 1 \\ 0 & 0 & 0 \end{bmatrix}$

37. $\begin{bmatrix} 4 & 2 & 0 \\ -1 & 1 & 0 \\ 0 & 1 & 2 \end{bmatrix}$ **38.** $\begin{bmatrix} 4 & 1 & -1 \\ 2 & 5 & -2 \\ 1 & 1 & 2 \end{bmatrix}$

39. $\begin{bmatrix} 3 & -1 & 1 \\ 7 & -5 & 1 \\ 6 & -6 & 2 \end{bmatrix}$ **40.** $\begin{bmatrix} 0 & 0 & 1 \\ 0 & 1 & 2 \\ 0 & 0 & 1 \end{bmatrix}$

41. $\begin{bmatrix} 1 & 1 & 1 \\ 0 & 0 & 1 \\ 0 & 0 & 1 \end{bmatrix}$ **42.** $\begin{bmatrix} 4 & 2 & 3 \\ 2 & 1 & 2 \\ -1 & 2 & 0 \end{bmatrix}$

43. $\begin{bmatrix} 1 & 0 & 0 \\ -4 & 3 & 0 \\ -4 & 2 & 1 \end{bmatrix}$ **44.** $\begin{bmatrix} 3 & -2 & 0 \\ 1 & 0 & 0 \\ -1 & 1 & 3 \end{bmatrix}$

45. $\begin{bmatrix} 0 & 0 & 2 \\ -1 & 1 & 2 \\ -1 & 0 & 3 \end{bmatrix}$ **46.** $\begin{bmatrix} 2 & 1 & 8 & -1 \\ 0 & 4 & 0 & 0 \\ 0 & 0 & 6 & 0 \\ 0 & 0 & 0 & 4 \end{bmatrix}$

47. $\begin{bmatrix} 4 & 0 & 4 & 0 \\ 0 & 4 & 0 & 0 \\ 0 & 0 & 8 & 0 \\ -1 & -2 & 1 & 8 \end{bmatrix}$ **48.** $\begin{bmatrix} 2 & 0 & 1 & 2 \\ 0 & 2 & 0 & 0 \\ 0 & 0 & 6 & 0 \\ 0 & 0 & 1 & 4 \end{bmatrix}$

49. Powers of a Matrix Suppose that $\mathbf{A}$ is a diagonalizable matrix that has been written in the form $\mathbf{A} = \mathbf{PDP}^{-1}$, where $\mathbf{D}$ is diagonal.

(a) Show that for positive integer k, $\mathbf{A}^k = \mathbf{PD}^k\mathbf{P}^{-1}$.

(b) Use the result of part (a) to compute $\mathbf{A}^{50}$ for

$$\mathbf{A} = \begin{bmatrix} 1 & 1 \\ 4 & 1 \end{bmatrix}.$$

(c) Show that if $\mathbf{D}$ is diagonal, then $\mathbf{D}^k$ is diagonal.

(d) Is the equation in part (a) true for $k = -1$? Could this be useful in finding the inverse of a matrix?

50. Determinants and Eigenvalues Let $\mathbf{A}$ be an $n \times n$ matrix such that its characteristic polynomial is

$$|\mathbf{A} - \lambda\mathbf{I}| = (\lambda - \lambda_1)(\lambda - \lambda_2)\cdots(\lambda - \lambda_n),$$

where the λ_i are distinct.

(a) Explain why

$$|\mathbf{A}| = \lambda_1\lambda_2\cdots\lambda_n.$$

(b) If the λ_i are not distinct but $\mathbf{A}$ is diagonalizable, would the same property hold? Explain.

Constructing Counterexamples *In Problems 51–53, construct the required examples.*

51. Construct a 2×2 matrix that is invertible but not diagonalizable.

52. Construct a 2×2 matrix that is diagonalizable but not invertible.

53. Construct a 2×2 matrix that is neither invertible nor diagonalizable.

54. Computer Lab: Diagonalization Use appropriate computer software to diagonalize (if possible) the following matrices.

(a) $\begin{bmatrix} 1 & 1 & 1 & 1 \\ 0 & 1 & 1 & 1 \\ 0 & 0 & 1 & 1 \\ 0 & 0 & 0 & 1 \end{bmatrix}$

(b) $\begin{bmatrix} -2 & 1 & 1 & 0 & 0 \\ 1 & -2 & 1 & 0 & 0 \\ 1 & 1 & -2 & 0 & 0 \\ 0 & 0 & 0 & 1 & 1 \\ 0 & 0 & 0 & 4 & 1 \end{bmatrix}$

(c) $\begin{bmatrix} 3 & 0 & 0 & 0 \\ 0 & 1 & 1 & 0 \\ 0 & 1 & 1 & 0 \\ 0 & 0 & 0 & 5 \end{bmatrix}$

55. Similar Matrices We have defined matrix $\mathbf{B}$ to be similar to matrix $\mathbf{A}$ (denoted by $\mathbf{B} \sim \mathbf{A}$) if there is an invertible matrix $\mathbf{P}$ such that $\mathbf{B} = \mathbf{P}^{-1}\mathbf{AP}$. Prove the following:

(a) Similar matrices have the same characteristic polynomial and the same eigenvalues.

(b) Similar matrices have the same determinant and the same trace.

(c) Show by example (using 2×2 matrices) that similar matrices can have different eigenvectors.

56. How Similar Are They? Let

$$\mathbf{A} = \begin{bmatrix} 4 & -2 \\ 1 & 1 \end{bmatrix} \quad \text{and} \quad \mathbf{B} = \begin{bmatrix} -3 & 10 \\ -3 & 8 \end{bmatrix}.$$

(a) Show that $\mathbf{A} \sim \mathbf{B}$.

(b) Verify that $\mathbf{A}$ and $\mathbf{B}$ share the properties discussed in parts (a) and (b) of Problem 55.

57. Computer Lab: Similarity Challenge Repeat Problem 55 for the more challenging case of two 3×3 matrices, using a computer algebra system.

$$\mathbf{A} = \begin{bmatrix} 1 & 2 & -3 \\ 2 & 0 & 1 \\ 1 & -3 & 1 \end{bmatrix} \quad \text{and} \quad \mathbf{B} = \begin{bmatrix} 1 & -19 & 58 \\ 1 & 12 & -27 \\ 5 & 15 & -11 \end{bmatrix}.$$

58. Orthogonal Matrices An orthogonal matrix $\mathbf{P}$ is a square matrix whose transpose equals its inverse:

$$\mathbf{P}^{\mathrm{T}} = \mathbf{P}^{-1}.$$

(a) Show that this is equivalent to the condition $\mathbf{PP}^{\mathrm{T}} = \mathbf{I}$.

(b) Use part (a) to show that the column vectors of an orthogonal matrix are orthogonal vectors. (See Sec. 3.1.)

59. Orthogonally Diagonalizable Matrices A matrix $\mathbf{A}$ is orthogonally diagonalizable if there is an orthogonal matrix $\mathbf{P}$ that diagonalizes it. Show that the matrix

$$\mathbf{A} = \begin{bmatrix} 4 & 2 \\ 2 & 7 \end{bmatrix}$$

is orthogonally diagonalizable. HINT: Symmetric matrices have orthogonal eigenvectors.

60. When Diagonalization Fails Prove that, for a 2×2 matrix $\mathbf{A}$ with a double eigenvalue but a single eigenvector $\vec{\mathbf{v}} = [v_1, v_2]$, $v_2 \neq 0$, the matrix

$$\mathbf{Q} = \begin{bmatrix} v_1 & 1 \\ v_2 & 0 \end{bmatrix} \quad \text{and its inverse} \quad \mathbf{Q}^{-1}$$

can provide a change of basis for $\mathbf{A}$, such that $\mathbf{Q}^{-1}\mathbf{AQ}$ is a triangular matrix (which will have the eigenvalues on the diagonal, as shown in Sec. 5.3, Problems 43–46).

Triangularizing *Apply the procedure of Problem 60 to triangularize the matrices of Problems 61–62.*

61. $\begin{bmatrix} 2 & -1 \\ 4 & 6 \end{bmatrix}$

62. $\begin{bmatrix} 1 & 1 \\ -1 & -1 \end{bmatrix}$

63. Suggested Journal Entry I Discuss how you might go about defining coordinates for a vector in an infinite-dimensional vector space, $C[0, 1]$, for example. Would it make a difference if you just wanted to be able to approximate vectors rather than obtaining exact representations?

64. Suggested Journal Entry II Explain and elaborate the assertion by Gilbert Strang of the need to emphasize that "Diagonalizability is concerned with the eigenvectors. Invertibility is concerned with the eigenvalues."[3]

---

[3]Gilbert Strang, *Linear Algebra and Its Applications*, 3rd edition (Harcourt Brace Jovanovich, 1988), 256.

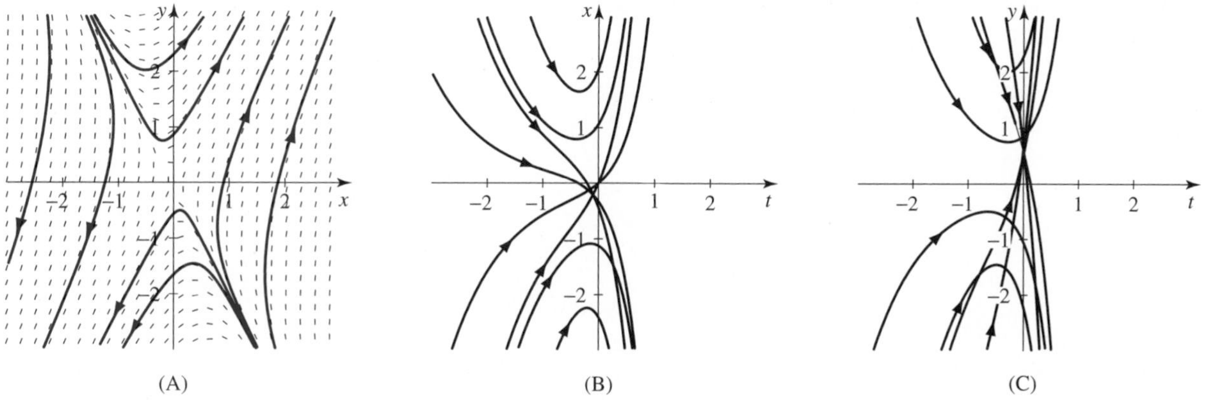

FIGURE 5.4.6 Phase portrait and solution graphs in xy coordinates for Problem 65.

65. Suggested Journal Entry III Figure 5.4.6(A) shows the phase portrait of Example 7,

$$\begin{bmatrix} x \\ y \end{bmatrix}' = \begin{bmatrix} 1 & 1 \\ 4 & 1 \end{bmatrix} \begin{bmatrix} x \\ y \end{bmatrix}.$$

(a) Add to the phase portrait the eigenvectors [1, 2] and [1, −2], which go with the eigenvalues 3 and −1, respectively.

(b) Why might we call the eigenvectors "nature's coordinate system," and the original xy coordinates "human laboratory coordinates"?

(c) Figure 5.4.6(B) and (C) show solution graphs for $x(t)$ and $y(t)$; what sort of solution graphs would you expect if new coordinate axes are aligned along the eigenvectors?

(d) Relate your discoveries in parts (b) and (c) to the following support statement for diagonalization: "One of the goals of all science is to find the simplest way to describe a physical system, and eigenvectors just happen to be the way to do it for linear systems of differential equations."

CHAPTER

6 Linear Systems of Differential Equations

In scientific thought we adopt the simplest theory which will explain the facts under consideration and enable us to predict new facts.

—J. B. S. Haldane

6.1 Theory of Linear DE Systems

SYNOPSIS: *We use the methods of linear algebra to describe the structure of solutions to systems of first-order linear differential equations. We apply this general framework to interpret our earlier results for second-order linear differential equations.*

Linear versus Nonlinear

Perhaps we should pause here for a reminder of the role of linearity in mathematical modeling. We have stated before that, on the whole, "real-world" systems are *not* linear. But a great many of them are *approximately* linear, especially in the vicinity of certain critical points that are of special interest in applications. A grasp of linear systems is a significant first step toward dealing with nonlinear systems, to be covered in Chapter 7.

Linear systems provide useful models for situations in which a change in any variable depends linearly on *all* other variables. Examples are blood flow through different organs of the body, cascades of tanks in chemical processing, linked mechanical components like springs and balances, or the components of an electrical power network.

An Overview of Linear Systems

The following themes and topics, some of which we have already begun to explore, will be the building blocks for this chapter.

- The solution space of an nth-order homogeneous linear differential equation has dimension n.
- Principles of nonhomogeneity and superposition allow us to assemble solutions from simpler parts.
- Linear differential equations of nth order can be converted to systems of n first-order differential equations.

- Numerical methods for first-order differential equations extend naturally to systems.
- The case of constant coefficients allows us to generate closed-form solutions.
- Diagonalization of the coefficient matrix in the constant coefficient case permits decoupling of the system.

Due to these facts, *linear systems are the most tractable.*

The two disciplines of linear algebra and differential equations merge most effectively (and visually) in this chapter. We begin with the algebraic aspects, which allow us to extend what we have learned from second-order DEs (Chapter 4) and systems of two first-order DEs (Sec. 2.6) to higher dimensions ($n > 2$).

The link between an nth-order linear homogeneous DE with constant coefficients and its associated linear system is due to Euler (as we recounted for $n = 2$ in Sec. 5.3, Examples 9 and 10). In this chapter, we further explore this historic interplay between linear algebra and DEs. Let us now develop the details.

Linear First-Order DE System

An n-dimensional **linear first-order DE system** on open interval I is one that can be written as a matrix-vector equation

$$\vec{\mathbf{x}}'(t) = \mathbf{A}(t)\vec{\mathbf{x}}(t) + \vec{\mathbf{f}}(t). \tag{1}$$

- $\mathbf{A}(t)$ is an $n \times n$ matrix of continuous functions on I.
- $\vec{\mathbf{f}}(t)$ is an $n \times 1$ vector of continuous functions on I.
- $\vec{\mathbf{x}}(t)$ is an $n \times 1$ **solution vector** of differentiable functions on I that satisfies (1).

If $\vec{\mathbf{f}}(t) \equiv \vec{\mathbf{0}}$, the system is **homogeneous**,

$$\vec{\mathbf{x}}'(t) = \mathbf{A}(t)\vec{\mathbf{x}}(t). \tag{2}$$

EXAMPLE 1 Solution Checking

(a) The homogeneous linear first-order system

$$\begin{aligned} x' &= 3x - 2y, \\ y' &= x, \\ z' &= -x + y + 3z \end{aligned} \tag{3}$$

can be written in matrix-vector form $\vec{\mathbf{x}}' = \mathbf{A}\vec{\mathbf{x}}$ as

$$\vec{\mathbf{x}}' = \begin{bmatrix} 3 & -2 & 0 \\ 1 & 0 & 0 \\ -1 & 1 & 3 \end{bmatrix} \vec{\mathbf{x}}, \quad \text{where } \vec{\mathbf{x}} = \begin{bmatrix} x \\ y \\ z \end{bmatrix}.$$

The vector

$$\vec{\mathbf{x}}_h = \begin{bmatrix} 2e^{2t} \\ e^{2t} \\ e^{2t} \end{bmatrix}$$

is a solution, which we confirm by showing that $\mathbf{A}\vec{\mathbf{x}}_h = \vec{\mathbf{x}}_h'$, as follows:

$$\mathbf{A}\vec{\mathbf{x}}_h = \begin{bmatrix} 3 & -2 & 0 \\ 1 & 0 & 0 \\ -1 & 1 & 3 \end{bmatrix} \begin{bmatrix} 2e^{2t} \\ e^{2t} \\ e^{2t} \end{bmatrix}$$

$$= \begin{bmatrix} 6e^{2t} - 2e^{2t} \\ 2e^{2t} \\ -2e^{2t} + e^{2t} + 3e^{2t} \end{bmatrix} = \begin{bmatrix} 4e^{2t} \\ 2e^{2t} \\ 2e^{2t} \end{bmatrix} = \begin{bmatrix} 2e^{2t} \\ e^{2t} \\ e^{2t} \end{bmatrix}' = \vec{\mathbf{x}}_h'.$$

(b) The nonhomogeneous linear first-order system

$$\begin{aligned} x' &= 3x - 2y \qquad + 2 - 2e^t, \\ y' &= x \qquad\qquad - e^t, \\ z' &= -x + y + 3z + e^t - 1 \end{aligned}$$

can be written in matrix-vector form $\vec{\mathbf{x}}' = \mathbf{A}\vec{\mathbf{x}} + \vec{\mathbf{f}}(t)$ as

$$\vec{\mathbf{x}}' = \begin{bmatrix} 3 & -2 & 0 \\ 1 & 0 & 0 \\ -1 & 1 & 3 \end{bmatrix} \vec{\mathbf{x}} + \begin{bmatrix} 2 - 2e^t \\ -e^t \\ e^t - 1 \end{bmatrix},$$

and has a particular solution

$$\vec{\mathbf{x}}_p = \begin{bmatrix} e^t \\ 1 \\ 0 \end{bmatrix},$$

which we confirm by showing that $\vec{\mathbf{x}}_p' - \mathbf{A}\vec{\mathbf{x}}_p = \vec{\mathbf{f}}(t)$, as follows:

$$\vec{\mathbf{x}}_p' - \mathbf{A}\vec{\mathbf{x}}_p = \begin{bmatrix} e^t \\ 1 \\ 0 \end{bmatrix}' - \begin{bmatrix} 3 & -2 & 0 \\ 1 & 0 & 0 \\ -1 & 1 & 3 \end{bmatrix} \begin{bmatrix} e^t \\ 1 \\ 0 \end{bmatrix}$$

$$= \begin{bmatrix} e^t - (3e^t - 2) \\ 0 - e^t \\ 0 - (-e^t + 1) \end{bmatrix} = \begin{bmatrix} 2 - 2e^t \\ -e^t \\ e^t - 1 \end{bmatrix} = \vec{\mathbf{f}}(t).$$

■

Now that we have some familiarity with manipulating a linear first-order DE system in matrix form, we can use linear algebra to lay the groundwork for actually finding solutions.

Applying Principles of Linear Algebra

The foundation for further analysis of linear systems begins with a definition and a theorem, similar to those in previous chapters, but appropriately adapted to systems.

Initial-Value Problem (IVP) for a Linear DE System

For a linear DE system, an **initial-value problem** is the combination of system (1) and an initial value vector:

$$\vec{\mathbf{x}}' = \mathbf{A}(t)\vec{\mathbf{x}} + \vec{\mathbf{f}}(t), \quad \vec{\mathbf{x}}(t_0) = \vec{\mathbf{x}}_0 = \begin{bmatrix} c_1 \\ c_2 \\ \vdots \\ c_n \end{bmatrix},$$

where $c_1, c_2, \ldots, c_n$ are real constants.

As we have seen before (in Sec 1.5 for first-order DEs and Sec. 4.2 for second-order DEs), *existence* of solutions is assured if the function elements of $\mathbf{A}(t)$ and $\vec{\mathbf{f}}(t)$ are continuous; *uniqueness* requires initial conditions and continuity of partial derivatives. In the case of a *linear* system, the partial derivatives $\partial f(t)/\partial x_i$ are automatically continuous because they consist simply of the coefficients of $\vec{\mathbf{x}}$, which are the elements of $\mathbf{A}(t)$.

Existence and Uniqueness Theorem for Linear DE Systems
Given an $n \times n$ matrix function $\mathbf{A}(t)$ and an $n \times 1$ vector function $\vec{\mathbf{f}}(t)$, both continuous on an open interval I containing t_0, and a constant n-vector $\vec{\mathbf{x}}_0$, there exists a unique vector function $\vec{\mathbf{x}}(t)$ such that

$$\vec{\mathbf{x}}' = \mathbf{A}(t)\vec{\mathbf{x}} + \vec{\mathbf{f}}(t) \quad \text{and} \quad \vec{\mathbf{x}}(t_0) = \vec{\mathbf{x}}_0.$$

The familiar storyline of Chapters 2–4 applies to finding solutions of linear DE systems. We know that solutions *exist* for

$$\vec{\mathbf{x}}' = \mathbf{A}(t)\vec{\mathbf{x}} + \vec{\mathbf{f}}(t). \tag{4}$$

To *find* them we first find the general solution $\vec{\mathbf{x}}_h$ of the associated homogeneous DE

$$\vec{\mathbf{x}}' = \mathbf{A}\vec{\mathbf{x}},$$

which we begin in the next section. Eventually, in Sec. 6.7, we will finish solving the nonhomogeneous IVP by finding a particular solution $\vec{\mathbf{x}}_p$ to (4). The complete solution of (4) will be the familiar construction

$$\vec{\mathbf{x}} = \vec{\mathbf{x}}_h + \vec{\mathbf{x}}_p.$$

Homogeneous Linear Systems

We state the Superposition Principle in the context of homogeneous linear systems.

The Superposition Principle for Homogeneous Linear DE Systems
Let $\vec{\mathbf{x}}_1, \vec{\mathbf{x}}_2, \ldots, \vec{\mathbf{x}}_n$ be solution vectors for the homogeneous equation

$$\vec{\mathbf{x}}' = \mathbf{A}(t)\vec{\mathbf{x}} \qquad \text{on } I. \tag{5}$$

Then, any linear combination of these solution vectors is also a solution vector for (5). That is,

$$\vec{\mathbf{x}} = c_1\vec{\mathbf{x}}_1 + c_2\vec{\mathbf{x}}_2 + \cdots + c_n\vec{\mathbf{x}}_n$$

is a solution on I for any real constants $c_1, c_2, \ldots, c_n$.

The proof is completely analogous to the proof in Sec. 2.1 for solution functions of a single first-order homogeneous linear DE.

EXAMPLE 2 **Here Comes That Superposition Again** It is easily verified that, for system (3) in Example 1,

$$\vec{\mathbf{x}}_1 = \begin{bmatrix} 0 \\ 0 \\ e^{3t} \end{bmatrix}, \quad \vec{\mathbf{x}}_2 = \begin{bmatrix} 2e^{2t} \\ e^{2t} \\ e^{2t} \end{bmatrix}, \quad \text{and} \quad \vec{\mathbf{x}}_3 = \begin{bmatrix} e^{t} \\ e^{t} \\ 0 \end{bmatrix}$$

are all solutions to $\vec{\mathbf{x}}' = \mathbf{A}\vec{\mathbf{x}}$. Thus, the Superposition Principle guarantees that

$$c_1 \begin{bmatrix} 0 \\ 0 \\ e^{3t} \end{bmatrix} + c_2 \begin{bmatrix} 2e^{2t} \\ e^{2t} \\ e^{2t} \end{bmatrix} + c_3 \begin{bmatrix} e^t \\ e^t \\ 0 \end{bmatrix}$$

is a solution for any constants c_1, c_2, and c_3. ■

To find *all* the solutions of a linear system of n first-order DEs, we must find enough linearly independent solutions to construct a *basis* for the *solution space*. How many such solutions are necessary and sufficient?

Solution Space Theorem for Homogeneous Linear DE Systems
If

$$\vec{\mathbf{x}}' = \mathbf{A}(t)\vec{\mathbf{x}},$$

where $\mathbf{A}$ is an $n \times n$ matrix, then the set of solutions $\vec{\mathbf{x}}(t)$ is a vector space of dimension n.

This theorem is analogous to the one stated in Chapter 4. Later in this section, we will prove a special case of this theorem, where all entries of $\mathbf{A}$ are constant.

According to the Solution Space Theorem, for an $n \times n$ linear system we seek *n linearly independent* solutions $\vec{\mathbf{x}}_1, \vec{\mathbf{x}}_2, \ldots, \vec{\mathbf{x}}_n$ to form a basis for the solution space. Then the Superposition Principle gives us the following.

Solution Theorem for Homogeneous Linear DE Systems
For n linearly independent solutions $\vec{\mathbf{x}}_1, \vec{\mathbf{x}}_2, \ldots, \vec{\mathbf{x}}_n$ of

$$\vec{\mathbf{x}}' = \mathbf{A}(t)\vec{\mathbf{x}},$$

the general solution is

$$\vec{\mathbf{x}}_h = c_1\vec{\mathbf{x}}_1 + c_2\vec{\mathbf{x}}_2 + \cdots + c_n\vec{\mathbf{x}}_n, \quad c_1, c_2, \ldots, c_n \in \mathbb{R}.$$

EXAMPLE 3 **Completing a General Solution** For system (3) of Examples 1 and 2 we have verified three solutions:

$$\vec{\mathbf{x}}_1 = \begin{bmatrix} 0 \\ 0 \\ e^{3t} \end{bmatrix}, \quad \vec{\mathbf{x}}_2 = \begin{bmatrix} 2e^{2t} \\ e^{2t} \\ e^{2t} \end{bmatrix}, \quad \text{and} \quad \vec{\mathbf{x}}_3 = \begin{bmatrix} e^t \\ e^t \\ 0 \end{bmatrix}.$$

To show that $\{\vec{\mathbf{x}}_1, \vec{\mathbf{x}}_2, \vec{\mathbf{x}}_3\}$ are linearly independent on $(-\infty, \infty)$, we proceed as follows:

Step 1. Choose one point, say $t_0 = 0$, in $(-\infty, \infty)$.

Step 2. Calculate $\vec{\mathbf{x}}_1(t_0), \vec{\mathbf{x}}_2(t_0), \vec{\mathbf{x}}_3(t_0)$ and construct the column space matrix. For $t_0 = 0$,

$$\mathbf{C} = \begin{bmatrix} 0 & 2 & 1 \\ 0 & 1 & 1 \\ 1 & 1 & 0 \end{bmatrix}.$$

Step 3. Test for linear independence by calculating the determinant $|\mathbf{C}|$. For $t_0 = 0$, $|\mathbf{C}| \neq 0$, so the vectors of the set $\{\vec{\mathbf{x}}_1, \vec{\mathbf{x}}_2, \vec{\mathbf{x}}_3\}$ are linearly independent.

Thus the general solution for $\vec{\mathbf{x}}' = \mathbf{A}\vec{\mathbf{x}}$ is

$$\vec{\mathbf{x}}_h = c_1\vec{\mathbf{x}}_1 + c_2\vec{\mathbf{x}}_2 + c_3\vec{\mathbf{x}}_3.$$

■

Alternate Solution Expressions

The language of linear algebra allows several ways of expressing solutions to linear systems of DEs.

EXAMPLE 4 **How to State Solutions** The complete general solution of system (3) found in Example 3 is

$$\vec{\mathbf{x}}_h = c_1 \begin{bmatrix} 0 \\ 0 \\ e^{3t} \end{bmatrix} + c_2 \begin{bmatrix} 2e^{2t} \\ e^{2t} \\ e^{2t} \end{bmatrix} + c_3 \begin{bmatrix} e^t \\ e^t \\ 0 \end{bmatrix}.$$

By making a column matrix of $\vec{\mathbf{x}}_1$, $\vec{\mathbf{x}}_2$, and $\vec{\mathbf{x}}_3$ and a column vector of the constants c_1, c_2, c_3, we can write another equivalent format:

$$\vec{\mathbf{x}}_h = \begin{bmatrix} 0 & 2e^{2t} & e^t \\ 0 & e^{2t} & e^t \\ e^{3t} & e^{2t} & 0 \end{bmatrix} \begin{bmatrix} c_1 \\ c_2 \\ c_3 \end{bmatrix}.$$

■

There is a special term for the last alternative in Example 4.

Fundamental Matrix

For a basis of n linearly independent solutions of $\vec{\mathbf{x}}' = \mathbf{A}\vec{\mathbf{x}}$, the matrix $\mathbf{X}(t)$ whose *columns* are the vector solutions $\vec{\mathbf{x}}_1, \vec{\mathbf{x}}_2, \ldots, \vec{\mathbf{x}}_n$ is called a **fundamental matrix** for the system.

Properties of a Fundamental Matrix X for $\vec{\mathbf{x}}' = \mathbf{A}\vec{\mathbf{x}}$:

(i) $|\mathbf{X}| \neq 0$,

(ii) $\mathbf{X}'(t) = \mathbf{A}\mathbf{X}(t)$.

Thus, for a solution to a 3-dimensional system $\vec{\mathbf{x}}' = \mathbf{A}\vec{\mathbf{x}}$, we can write

$$\vec{\mathbf{x}}_h = \underbrace{\begin{bmatrix} | & | & | \\ \vec{\mathbf{x}}_1, & \vec{\mathbf{x}}_2, & \vec{\mathbf{x}}_3 \\ | & | & | \end{bmatrix}}_{\mathbf{X}(t)} \begin{bmatrix} c_1 \\ c_2 \\ c_3 \end{bmatrix}, \quad c_1, c_2, c_3 \in \mathbb{R}. \tag{6}$$

It follows from properties of the matrix product that $\mathbf{X}'(t) = \mathbf{A}\mathbf{X}(t)$.

The fundamental matrix is not unique. A different set of n linearly independent solutions would produce a different $\mathbf{X}(t)$. The general solution formula (6) would still hold, but the c_i would be specific to any particular solution.

Homogeneous Linear Systems with Constant Coefficients

For the simplest n-dimensional linear DE system $\vec{\mathbf{x}}' = \mathbf{A}\vec{\mathbf{x}}$, where $\mathbf{A}$ is an $n \times n$ matrix of constants, the evolution of the Solution Space Theorem stated previously is as follows.

Step 1. *The solution space is a vector space.*

We know from examples and problems in Secs. 3.5 and 3.6 that the set of vector functions of the form

$$\vec{\mathbf{x}} = \begin{bmatrix} x_1(t) \\ x_2(t) \\ \vdots \\ x_n(t) \end{bmatrix},$$

having components in $\mathcal{C}^1$, is a vector space under the standard definitions of addition and multiplication by a scalar. If $\vec{\mathbf{u}}(t)$ and $\vec{\mathbf{v}}(t)$ are solutions of (5), so that

$$\vec{\mathbf{u}}' = \mathbf{A}\vec{\mathbf{u}} \quad \text{and} \quad \vec{\mathbf{v}}' = \mathbf{A}\vec{\mathbf{v}},$$

and if c_1 and c_2 are scalars, then

$$\begin{aligned}(c_1\vec{\mathbf{u}} + c_2\vec{\mathbf{v}})' &= c_1\vec{\mathbf{u}}' + c_2\vec{\mathbf{v}}' \\ &= c_1\mathbf{A}\vec{\mathbf{u}} + c_2\mathbf{A}\vec{\mathbf{v}} = \mathbf{A}(c_1\vec{\mathbf{u}}) + \mathbf{A}(c_2\vec{\mathbf{v}}) = \mathbf{A}(c_1\vec{\mathbf{u}} + c_2\vec{\mathbf{v}}).\end{aligned}$$

Thus, the solution space is a subspace and, hence, a *vector space.*

Step 2. *The Existence and Uniqueness Theorem gives a special set B of n solution vectors.*

Because the constant coefficients are continuous on $\mathbb{R}$, the Existence and Uniqueness Theorem tells us that we can find a solution

$$\vec{\mathbf{x}}_i(t) \quad \text{with} \quad \vec{\mathbf{x}}_i(0) = \vec{\mathbf{e}}_i$$

for $i = 1, 2, \ldots, n$, where $\vec{\mathbf{e}}_i$ is the ith standard basis vector in $\mathbb{R}^n$. That is,

$$\vec{\mathbf{x}}_1(0) = \begin{bmatrix}1\\0\\0\\\vdots\\0\end{bmatrix}, \quad \vec{\mathbf{x}}_2(0) = \begin{bmatrix}0\\1\\0\\\vdots\\0\end{bmatrix}, \quad \ldots, \quad \vec{\mathbf{x}}_n(0) = \begin{bmatrix}0\\0\\0\\\vdots\\1\end{bmatrix}.$$

This step creates a special set of solution vectors $B = \{\vec{\mathbf{x}}_1, \vec{\mathbf{x}}_2, \ldots, \vec{\mathbf{x}}_n\}$.

Step 3. *The vectors in B are linearly independent.*

To show this fact, we suppose that we have constants $a_1, a_2, \ldots, a_n$ such that

$$a_1\vec{\mathbf{x}}_1(t) + a_2\vec{\mathbf{x}}_2(t) + \cdots + a_n\vec{\mathbf{x}}_n(t) = \vec{\mathbf{0}}$$

or, equivalently,

$$\begin{bmatrix} | & | & & | \\ \vec{\mathbf{x}}_1(t), & \vec{\mathbf{x}}_2(t), & \cdots, & \vec{\mathbf{x}}_n(t) \\ | & | & & | \end{bmatrix} \begin{bmatrix}a_1\\a_2\\\vdots\\a_n\end{bmatrix} = \vec{\mathbf{0}}.$$

Then this result must be true for *all* t, including $t = 0$. But we have assumed that

$$\begin{bmatrix} | & | & & | \\ \vec{\mathbf{x}}_1(0), & \vec{\mathbf{x}}_2(0), & \cdots, & \vec{\mathbf{x}}_n(0) \\ | & | & & | \end{bmatrix} = \mathbf{I}_n.$$

Consequently,

$$\mathbf{I}_n \begin{bmatrix}a_1\\a_2\\\vdots\\a_n\end{bmatrix} = \mathbf{0} \quad \Rightarrow \quad \begin{bmatrix}a_1\\a_2\\\vdots\\a_n\end{bmatrix} = \vec{\mathbf{0}},$$

and $\vec{\mathbf{x}}_1, \vec{\mathbf{x}}_2, \ldots, \vec{\mathbf{x}}_n$ are linearly independent.

Step 4. *The set B spans the solution space.*

To show this, we need to show that an arbitrary solution $\vec{\mathbf{u}}(t)$ of (5) is a linear combination of the $\vec{\mathbf{x}}_i(t)$. Suppose that

$$\vec{\mathbf{u}}(0) = \vec{\mathbf{b}} = \begin{bmatrix} b_1 \\ b_2 \\ \vdots \\ b_n \end{bmatrix},$$

and let us form $\vec{\mathbf{v}}(t)$ as follows:

$$\vec{\mathbf{v}}(t) = b_1\vec{\mathbf{x}}_1(t) + b_2\vec{\mathbf{x}}_2(t) + \cdots + b_n\vec{\mathbf{x}}_n(t).$$

Then $\vec{\mathbf{v}}(t)$ is also a solution of (5) by superposition. But

$$\vec{\mathbf{v}}(0) = b_1\vec{\mathbf{x}}_1(0) + b_2\vec{\mathbf{x}}_2(0) + \cdots + b_n\vec{\mathbf{x}}_n(0) = b_1\vec{\mathbf{e}}_1 + b_2\vec{\mathbf{e}}_2 + \cdots + b_n\vec{\mathbf{e}}_n = \vec{\mathbf{b}}.$$

Hence, by the Existence and Uniqueness Theorem, $\vec{\mathbf{u}}$ and $\vec{\mathbf{v}}$ are the same solution and $\vec{\mathbf{u}} = b_1\vec{\mathbf{x}}_1 + b_2\vec{\mathbf{x}}_2 \cdots + b_n\vec{\mathbf{x}}_n$. Therefore,

$$\text{Span } B = \text{solution space.}$$

Graphical Views

For a system of DEs in two variables, we have *three* planar graphs: tx, ty, and xy. We have seen these before, in Secs. 2.6 and 4.1.

Parametric to Cartesian; Phase Plane Drawing

See how xy, tx, and xy graphs relate.

Graphs for Two-Dimensional DE Systems

- The tx and ty graphs showing the individual solution functions $x(t)$ and $y(t)$ are called **component graphs**, **solution graphs**, or **time series**.
- The xy graph is the **phase plane**. The **trajectories** in the phase plane are the parametric curves described by $x(t)$ and $y(t)$.

Trajectories on a phase plane create a **phase portrait**.

Let us examine and compare three oscillatory systems.

EXAMPLE 5 **Simple Harmonic Motion** The familiar equation

$$x'' + 0.1x = 0$$

can be written in system form as

$$\begin{aligned} x' &= y, \\ y' &= -0.1x \end{aligned} \quad \text{or} \quad \begin{bmatrix} x \\ y \end{bmatrix}' = \begin{bmatrix} 0 & 1 \\ -0.1 & 0 \end{bmatrix} \begin{bmatrix} x \\ y \end{bmatrix}.$$

Any version of these equations produces solutions of the form

$$\begin{aligned} x(t) &= c_1 \cos\sqrt{0.1}\,t + c_2 \sin\sqrt{0.1}\,t, \\ y(t) &= x'(t) = -\sqrt{0.1}c_1 \sin\sqrt{0.1}\,t + \sqrt{0.1}c_2 \cos\sqrt{0.1}\,t, \end{aligned}$$

as we have shown in Chapter 4. The graphs in Fig. 6.1.1 result.

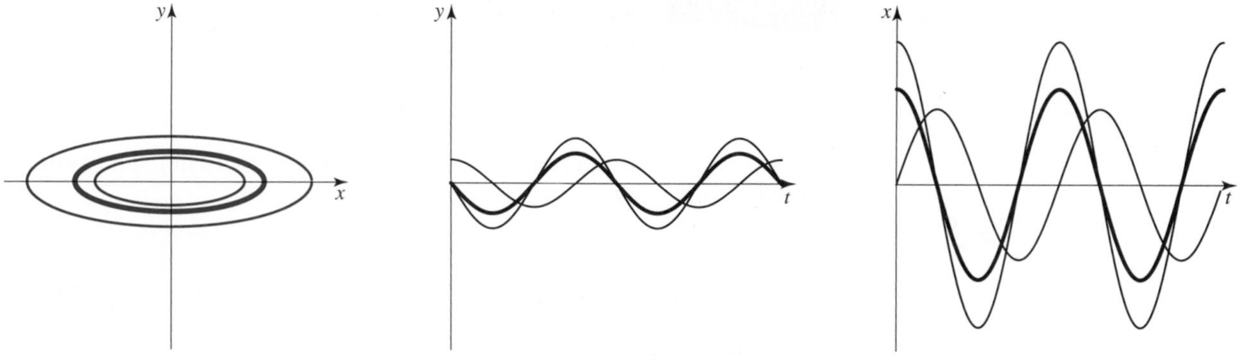

FIGURE 6.1.1 Undamped oscillator—phase portrait and component graphs for Example 5. ■

EXAMPLE 6 **Damped Harmonic Motion** The second-order DE

$$x'' + 0.05x' + 0.1x = 0$$

is equivalent to the system

$$\begin{aligned} x' &= y, \\ y' &= -0.1x - 0.05y \end{aligned} \quad \text{or} \quad \begin{bmatrix} x \\ y \end{bmatrix}' = \begin{bmatrix} 0 & 1 \\ -0.1 & -0.05 \end{bmatrix} \begin{bmatrix} x \\ y \end{bmatrix},$$

with solutions of the form

$$x(t) \approx e^{-0.025t}(c_1 \cos 0.32t + c_2 \sin 0.32t),$$

$$\begin{aligned} y(t) \approx\ & e^{-0.025t}(-0.32c_1 \sin 0.32t + 0.32c_2 \cos 0.32t) \\ & - 0.025e^{-0.025t}(c_1 \cos 0.32t + c_2 \sin 0.32t), \end{aligned}$$

as illustrated in Fig. 6.1.2.

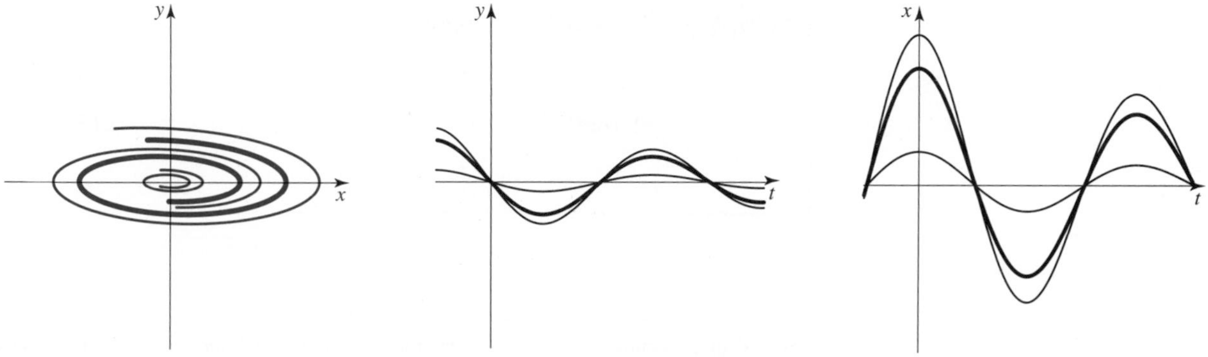

FIGURE 6.1.2 Damped oscillator—phase portrait and component graphs for Example 6. ■

EXAMPLE 7 Nonautonomous Complications Let's compare Example 5 to a nonautonomous version,

$$x'' + 0.1x = 0.5\cos t, \tag{7}$$

which represents a periodically forced harmonic oscillator. The system form of equation (7), is

$$\begin{aligned} x' &= y, \\ y' &= -0.1x + 0.5\cos t \end{aligned} \quad \text{or} \quad \begin{bmatrix} x \\ y \end{bmatrix}' = \begin{bmatrix} 0 & 1 \\ -0.1 & 0 \end{bmatrix} \begin{bmatrix} x \\ y \end{bmatrix} + \begin{bmatrix} 0 \\ 0.5\cos t \end{bmatrix}.$$

We will discuss (briefly in Sec. 6.4, Example 3, and more thoroughly in Chapter 8) how to find analytic formulas for the solutions to this equation, but with computers we can *draw* the solutions from a numerical calculation of the sort we presented in Sec. 1.4.[1] Euler's method proceeds in a fashion similar to that for a single equation, but because both x' and y' depend on both x and y, we need extra columns to calculate both the derivatives.[2] (See Appendix SS.)

The data for a single trajectory, after entering the quantities in boldface for initial conditions and step size, looks like Table 6.1.1, and the graphs that result are shown in Fig. 6.1.3.

Table 6.1.1 An Euler's method trajectory $x_{n+1} = x_n + hx'(t_n)$; $y_{n+1} = y_n + hy'(t_n)$ for Example 7 with step size $h = 0.1$ and $x(0) = 1$, $y(0) = 0$.

t_n	x_n	y_n	$x' = y$	$y' = -0.1x + 0.5\cos t$
0	**1**	**0**	0	0.4
0.1	1	0.04	0.04	0.3975
0.2	1.004	0.0798	0.0798	0.3896
0.3	1.0120	0.1187	0.1187	0.3765
0.4	1.0238	0.1564	0.1564	0.3581
0.5	1.0395	0.1922	0.1922	0.3348
0.6	1.0587	0.2257	0.2257	0.3068
0.7	1.0813	0.2563	0.2563	0.2743
0.8	1.1069	0.2838	0.2838	0.2377
0.9	1.1353	0.3075	0.3075	0.1973
1	1.1660	0.3273	0.3273	0.1535
⋮	⋮	⋮	⋮	⋮

[1] We will discuss numerical methods more thoroughly in Sec. 7.3; for nonlinear equations it is usually the only approach. See also Appendix SS.

[2] The Euler's method spreadsheet for all 2×2 linear systems looks like Table 6.1.1. You need only enter t, x, y, step size h, and the specific formulas for x' and y'. See Problem 12.

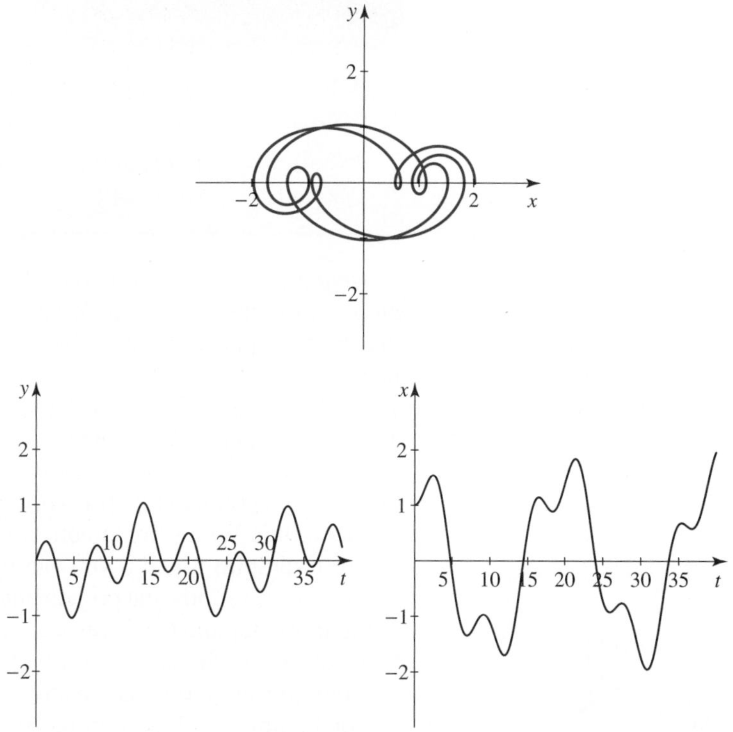

FIGURE 6.1.3 Forced undamped oscillator—phase portrait and component graphs for a *single* solution in Example 7. ■

The graphs shown in Examples 5–7 should be familiar. However, a two-dimensional system of DEs has another, important, *three*-dimensional graph, txy, shown in Fig. 6.1.4.

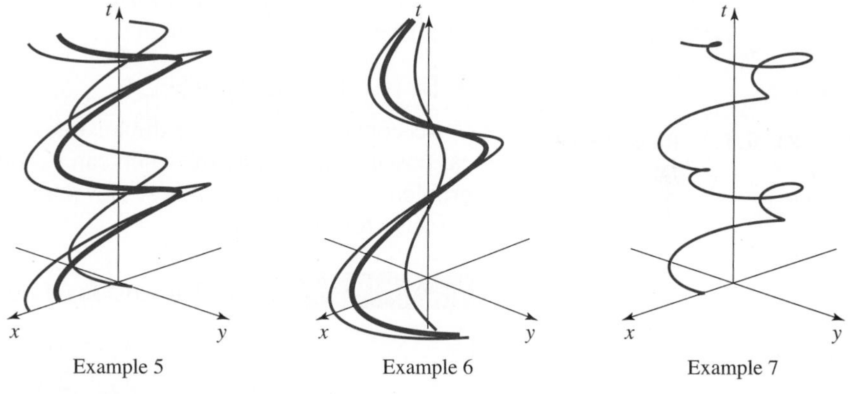

FIGURE 6.1.4 Three-dimensional txy views of solutions of two-dimensional systems.

Each of the planar graphs shows what you would see by looking down the "other" axis of the three-dimensional graph. For instance, if you look down the t-axis you see the phase portrait in the xy-plane.

A Graphical Look at Uniqueness

The Existence and Uniqueness Theorem gives specific information that can be interpreted as to whether trajectories intersect. That is, the issue of uniqueness for an n-dimensional system is equivalent to the question of whether two or more solutions can emanate from the same point in $(n+1)$-space.

Graphical Properties of Uniqueness in an n-Dimensional DE System

- For a linear system of differential equations in $\mathbb{R}^n$, solutions do not cross in $t, x_1, x_2, \ldots, x_n$-space (that is, $\mathbb{R}^n \times \mathbb{R}$, or $\mathbb{R}^{n+1}$).
- For an *autonomous* linear system in $\mathbb{R}^n$, trajectories *also* do not cross in $x_1, x_2, \ldots, x_n$-space ($\mathbb{R}^n$).

Hence, for a nonautonomous *equation, the uniqueness will be observed only in an* $(n+1)$*-dimensional graph.* The vector field in the phase space is not constant—the slopes in phase space evolve over time. This situation is a consequence of the fact that if t is explicit in the equations, there is no constant vector field in phase space. In Example 7 we added a nonconstant $f(t)$ to the equations of Example 5, and you see a tangled xy-phase portrait, even for a single solution.

Examples 5–7 illustrate these properties, despite the fact that every three-dimensional graph in Fig. 6.1.4 *appears* to show trajectories that cross themselves or each other. With some practice, you will be able to interpret them as follows.

For Examples 5 and 6 you can imagine that the txy graphs are composed of nonintersecting coils that project onto the nonintersecting phase-plane trajectories in Figures 6.1.1 and 6.1.2, respectively. Those phase portraits for Examples 5 and 6 illustrate the extra feature of nonintersection in the phase plane, which is true only for autonomous equations.

For Example 7, however, the single trajectory shown in Fig. 6.1.3 definitely intersects itself in the phase plane, so it may be less clear that the crossings we see in the txy view shown in Fig. 6.1.4 are not really crossings in 3D space. Nevertheless, the Existence and Uniqueness Theorem tells us that the trajectory does *not* cross in 3D space. We have rotated the txy graph in 3D space to give a more convincing view in Fig. 6.1.5.

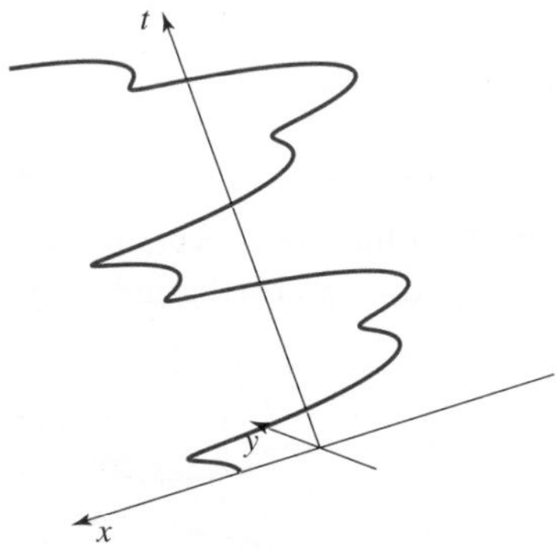

FIGURE 6.1.5 Rotated 3D view for the single trajectory of Example 7.

If you consider the t-axis as stretching the phase-plane view out in the t direction, trajectories will not be *able* to cross—points with the same (x, y) coordinates will have *different* t coordinates.

A Brief Look at $n > 2$

If we consider systems of dimension $n > 2$, there will be even more graphical possibilities, many of which can become too complicated to visualize. Each problem suggests its own best views. (See the pictures in Sec. 7.1 of the Lorenz attractor for an example.)

EXAMPLE 8 **Three Dimensions** In matrix-vector form, the 3×3 system

$$\begin{aligned} x_1' &= 3x_1 - x_2 + x_3, \\ x_2' &= 2x_1 \phantom{{}-x_2} + x_3, \\ x_3' &= x_1 - x_2 + 2x_3 \end{aligned} \tag{8}$$

becomes $\vec{\mathbf{x}}' = \mathbf{A}\vec{\mathbf{x}}$, where

$$\mathbf{A} = \begin{bmatrix} 3 & -1 & 1 \\ 2 & 0 & 1 \\ 1 & -1 & 2 \end{bmatrix}.$$

Three independent solutions are

$$\vec{\mathbf{u}}(t) = \begin{bmatrix} 0 \\ e^t \\ e^t \end{bmatrix}, \quad \vec{\mathbf{v}}(t) = \begin{bmatrix} e^{2t} \\ e^{2t} \\ 0 \end{bmatrix}, \quad \text{and} \quad \vec{\mathbf{w}}(t) = \begin{bmatrix} te^{2t} \\ te^{2t} \\ e^{2t} \end{bmatrix}.$$

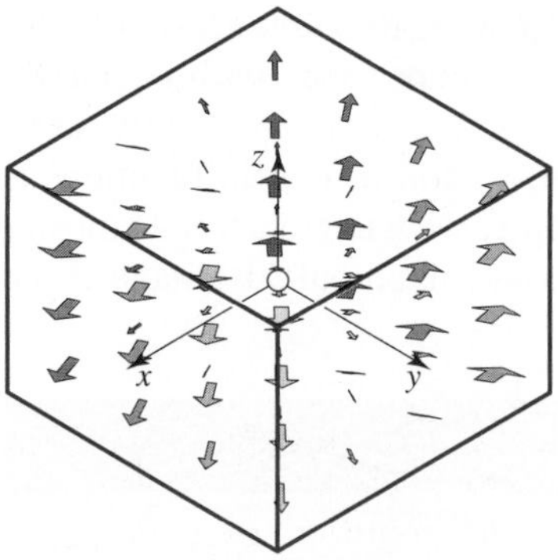

(a) Vector field

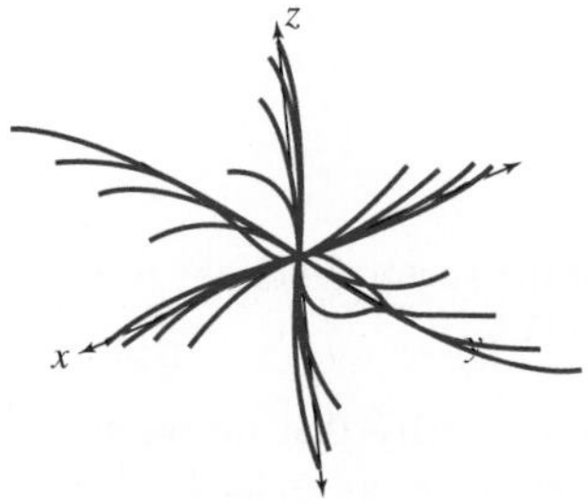

(b) Some trajectories

FIGURE 6.1.6 The three-dimensional system of Example 8.

(See Problem 11.) A fundamental matrix for the system is therefore

$$\mathbf{X}(t) = \begin{bmatrix} 0 & e^{2t} & te^{2t} \\ e^t & e^{2t} & te^{2t} \\ e^t & 0 & e^{2t} \end{bmatrix},$$

and the general solution of (8) is given by

$$\begin{aligned} x_1(t) &= \qquad\quad c_2e^{2t} + c_3te^{2t}, \\ x_2(t) &= c_1e^t + c_2e^{2t} + c_3te^{2t}, \\ x_3(t) &= c_1e^t \qquad\qquad + c_3e^{2t}. \end{aligned}$$

Some computer programs can draw a three-dimensional vector field, such as the one shown in Fig. 6.1.6(a). Although this representation is harder to read than a two-dimensional vector field, it nevertheless can depict useful information such as sources, sinks, and fixed points. Figure 6.1.6(b) shows some sample trajectories, all emanating from points near the origin. ■

For any system with $n > 2$, you will find that a judicious choice of two-dimensional views can be invaluable in understanding solution behaviors even of large systems. The following example shows what can be done with an intimidating linear system far beyond the other examples of this section.

EXAMPLE 9 **A Solution Must Exist** Consider the IVP

$$\vec{\mathbf{x}}' = \begin{bmatrix} t & 0 & 0 & \sqrt{t^2+1} \\ 0 & 1 & 2 & -1 \\ 1 & -1 & 3 & 0 \\ 0 & 1 & 1 & t \end{bmatrix}\vec{\mathbf{x}} + \begin{bmatrix} \cos t \\ \sin t \\ t^3 \\ e^{t^2} \end{bmatrix}, \quad \vec{\mathbf{x}}(-1) = \begin{bmatrix} -2 \\ 1 \\ 1 \\ -1.5 \end{bmatrix},$$

and suppose that you hope to see a solution on the t-interval $(-4, 4)$. Although the system is indeed linear, it has *four* dimensions, it is *nonhomogeneous*, and its coefficients are *variable*, not constant. Nevertheless, the Existence and Uniqueness Theorem applies and assures us that there *is* a unique solution $\vec{\mathbf{x}}(t)$ on the interval $(-4, 4)$. This unique solution may be impossible to find algebraically, but we are assured that it exists. Consequently, although the uniqueness requires five dimensions (four for $\vec{\mathbf{x}}$ and one for t), the theorem allows us to solve numerically with some confidence (using a *small* stepsize) and to draw (with appropriate software) various views in 2D, such as the component graphs: tx_1, tx_2, tx_3, and tx_4. (See Fig. 6.1.7.)

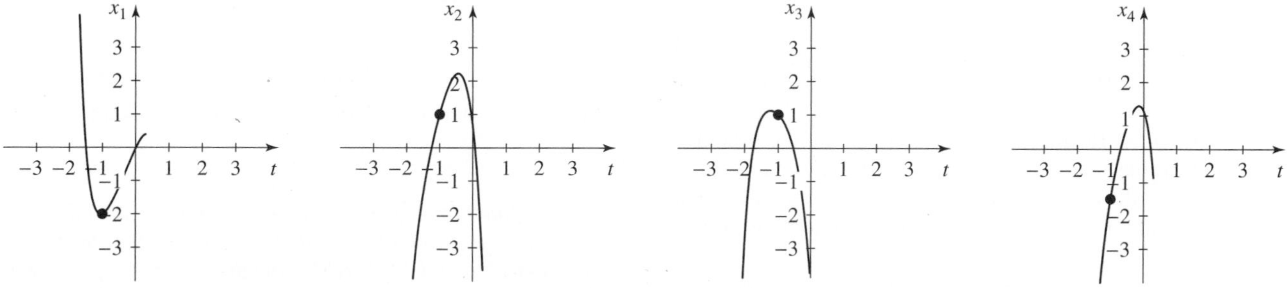

FIGURE 6.1.7 Component graphs for the IVP of Example 9. The initial condition is highlighted with a dot. We see that as t increases from its initial value of -1, x_1 is increasing and x_3 is decreasing, while x_2 and x_4 both rise, then fall to negative values. This sort of information tells a mathematical modeler what to expect, for instance, in a physical experiment to test the model. ■

Analyzing a system of differential equations by treating its solutions as "points" in a suitable vector space of functions builds on our experience with geometric vector spaces, and helps us to understand the structure more intuitively. (The study of vector spaces whose elements are scalar, vector, or matrix functions is called *functional analysis*.) For the rest of this chapter we will concentrate mostly on the 2×2 case. This restriction will allow us to take maximal advantage of the phase plane as a geometric tool.

Summary

Linear systems of differential equations generalize linear nth-order equations. The solution space of the homogeneous linear system of order n is an n-dimensional vector space. The general solution is then a linear combination of the functions forming a basis for this space, a fundamental set of solutions.

6.1 Problems

Breaking Out Systems *For Problems 1–4, rewrite each vector system as a system of first-order DEs.*

1. $\vec{\mathbf{x}}' = \begin{bmatrix} 1 & 2 \\ 4 & -1 \end{bmatrix}\vec{\mathbf{x}}$

2. $\vec{\mathbf{x}}' = \begin{bmatrix} 1 & 0 \\ 0 & -1 \end{bmatrix}\vec{\mathbf{x}} + \begin{bmatrix} 0 \\ 1 \end{bmatrix}$

3. $\vec{\mathbf{x}}' = \begin{bmatrix} 4 & 3 \\ -1 & -1 \end{bmatrix}\vec{\mathbf{x}} + e^{-t}\begin{bmatrix} 1 \\ 0 \end{bmatrix}$

4. $\vec{\mathbf{x}}' = \begin{bmatrix} 0 & 1 & 0 \\ 0 & 0 & 1 \\ -2 & 1 & 3 \end{bmatrix}\vec{\mathbf{x}} + \sin t\begin{bmatrix} 0 \\ 0 \\ 1 \end{bmatrix}$

Checking It Out *In Problems 5–8, verify that the given vector functions satisfy the system; then give a fundamental matrix* $\mathbf{X}(t)$ *and the general solution* $\vec{\mathbf{x}}(t)$.

5. $\vec{\mathbf{x}}' = \begin{bmatrix} 1 & 3 \\ 3 & 1 \end{bmatrix}\vec{\mathbf{x}}, \quad \vec{\mathbf{u}}(t) = \begin{bmatrix} e^{4t} \\ e^{4t} \end{bmatrix}, \quad \vec{\mathbf{v}}(t) = \begin{bmatrix} e^{-2t} \\ -e^{-2t} \end{bmatrix}$

6. $\vec{\mathbf{x}}' = \begin{bmatrix} 4 & -1 \\ 2 & 1 \end{bmatrix}\vec{\mathbf{x}}, \quad \vec{\mathbf{u}}(t) = \begin{bmatrix} e^{3t} \\ e^{3t} \end{bmatrix}, \quad \vec{\mathbf{v}}(t) = \begin{bmatrix} e^{2t} \\ 2e^{2t} \end{bmatrix}$

7. $\vec{\mathbf{x}}' = \begin{bmatrix} 1 & 1 \\ 4 & 1 \end{bmatrix}\vec{\mathbf{x}}, \quad \vec{\mathbf{u}}(t) = \begin{bmatrix} e^{-t} \\ -2e^{-t} \end{bmatrix}, \quad \vec{\mathbf{v}}(t) = \begin{bmatrix} e^{3t} \\ 2e^{3t} \end{bmatrix}$

8. $\vec{\mathbf{x}}' = \begin{bmatrix} 0 & 1 \\ -1 & 0 \end{bmatrix}\vec{\mathbf{x}}, \quad \vec{\mathbf{u}}(t) = \begin{bmatrix} \sin t \\ \cos t \end{bmatrix}, \quad \vec{\mathbf{v}}(t) = \begin{bmatrix} \cos t \\ -\sin t \end{bmatrix}$

9. Uniqueness in the Phase Plane Illustrate the graphical implications that trajectories *can* cross in tx and ty space, but not in the phase plane, by drawing phase-plane trajectories *and* component graphs for the autonomous system

$$\vec{\mathbf{x}}' = \begin{bmatrix} 0 & 1 \\ -1 & 0 \end{bmatrix}\vec{\mathbf{x}}.$$

Explain how different component solution graphs can have the same phase-plane trajectory, by marking starting points and directions on the phase-plane trajectory.

10. Verification Verify that the vector functions $\vec{\mathbf{u}}$, $\vec{\mathbf{v}}$, and $\vec{\mathbf{w}}$ in Example 9 satisfy the IVP.

11. Third-Order Verification Verify that the vector functions $\vec{\mathbf{u}}$, $\vec{\mathbf{v}}$, and $\vec{\mathbf{w}}$ of Example 8 are indeed solutions of system (8), and that they are linearly independent.

12. Euler's Method Numerics Set up an Euler's method spreadsheet for each of the following systems:
(a) Example 5 (b) Example 6 (c) Example 9

Choose an initial condition that corresponds to Fig. 6.1.1, 6.1.2, or 6.1.7 as appropriate, use a step size $h = 0.1$, and confirm that the solutions indeed set off in the directions shown in the figures.

NOTE: If you go far enough, you may see a gradual divergence from the pictured solutions, which were calculated by a fancier method (Runge-Kutta) that provides good accuracy with fewer steps than Euler's method. Describe any such observations.

Finding Trajectories *If the first-order equation*

$$\frac{dy}{dx} = \frac{cx + dy}{ax + by}$$

can be solved for y as a function of x, the graph of such a function is a phase-plane trajectory for the system

$$x' = ax + by,$$
$$y' = cx + dy.$$

Alternatively, one may obtain trajectories by solving

$$\frac{dx}{dy} = \frac{ax + by}{cx + dy}$$

for x as a function of y. Use one or both of these techniques to determine and sketch phase-plane trajectories for the systems of Problems 13 and 14. Predict how the speed of a point traveling along a trajectory will be affected by the position of the point.

13. $x' = x$
$y' = y$

14. $x' = y$
$y' = -x$

15. **Computer Check** Use a calculator or computer with an open-ended graphic DE solver to draw phase-plane trajectories for the systems in Problems 13 and 14. Discuss any differences that appear from your previous work.
NOTE: One way to do this problem is to use IDE.

Matrix Element Input
Just enter the matrix to see the vector field and trajectories.

16. **Computer Lab: Skew-Symmetric Systems**

Skew-Symmetric Matrix
If $\mathbf{A} = -\mathbf{A}^{\mathrm{T}}$, then matrix $\mathbf{A}$ is **skew-symmetric**.

For solutions of $\vec{\mathbf{x}}' = \mathbf{A}\vec{\mathbf{x}}$ with $\mathbf{A}$ skew-symmetric, the length of vector $\vec{\mathbf{x}}$ is constant. What does this mean in terms of phase-plane trajectories? Use IDE's **Matrix Element Input** tool, or an open-ended graphic DE solver, to verify this for the following systems, and explain the role of the parameter k.

(a) $\vec{\mathbf{x}}' = \begin{bmatrix} 0 & 1 \\ -1 & 0 \end{bmatrix}\vec{\mathbf{x}}$ (b) $\vec{\mathbf{x}}' = \begin{bmatrix} 0 & k \\ -k & 0 \end{bmatrix}\vec{\mathbf{x}},\ k > 0$

The Wronskian *Consider the following useful definition, an extension of that given in Sec. 3.6.*

Wronskian of Solutions

For an $n \times n$ linear system $\vec{\mathbf{x}}' = \mathbf{A}(t)\vec{\mathbf{x}}$, if $\vec{\mathbf{x}}_1, \vec{\mathbf{x}}_2, \ldots, \vec{\mathbf{x}}_n$ are solutions on a t-interval I and $\mathbf{X}(t)$ is the matrix whose columns are the $\vec{\mathbf{x}}_i$, then the determinant of $\mathbf{X}$ is called the **Wronskian** of these solutions, and we write

$$W \begin{bmatrix} | & | & & | \\ \vec{\mathbf{x}}_1(t), & \vec{\mathbf{x}}_2(t), & \cdots, & \vec{\mathbf{x}}_n(t) \\ | & | & & | \end{bmatrix} = |\mathbf{X}(t)|.$$

We have shown (at the end of Sec. 4.2) that the Wronskian of a set of linear DE solutions is either identically zero on I or never zero on I. Hence, if the Wronskian is not zero, the solutions are a fundamental set. In Problems 17–22, the given functions are solutions of a homogeneous system. Calculate their Wronskian in order to decide whether they form a fundamental set.

17. $\left\{ e^{2t}\begin{bmatrix} 2 \\ 1 \end{bmatrix}, e^{t}\begin{bmatrix} 1 \\ 0 \end{bmatrix} \right\}$

18. $\left\{ e^{3t}\begin{bmatrix} 1 \\ 1 \end{bmatrix}, e^{-t}\begin{bmatrix} 2 \\ -3 \end{bmatrix} \right\}$

19. $\left\{ e^{t}\begin{bmatrix} 2 \\ 1 \end{bmatrix}, e^{t}\begin{bmatrix} 1 \\ 0 \end{bmatrix} \right\}$

20. $\left\{ \begin{bmatrix} 3e^{4t} \\ e^{4t} \end{bmatrix}, \begin{bmatrix} e^{4t} \\ e^{4t} \end{bmatrix} \right\}$

21. $\left\{ \begin{bmatrix} e^{t}\cos t \\ -e^{t}\sin t \end{bmatrix}, \begin{bmatrix} e^{t}\sin t \\ e^{t}\cos t \end{bmatrix} \right\}$

22. $\left\{ \begin{bmatrix} \cos 3t \\ -\sin 3t \end{bmatrix}, \begin{bmatrix} \sin 3t \\ \cos 3t \end{bmatrix} \right\}$

23. **Suggested Journal Entry** After reviewing Secs. 2.6 and 4.1, rewrite "the story up to now" in your own words, and indicate how the results in the present section advance "the story."

24. **Suggested Journal Entry II** For an $n \times n$ homogeneous linear system, how might you address the issue of why the solution space can be of dimension less than n if the system is *algebraic*, but not if it is a system of DEs?
HINT: Contrast the results of Sec. 3.2 with the Solution Space Theorem of this section.

6.2 Linear Systems with Real Eigenvalues

SYNOPSIS: To construct explicit solutions of homogeneous linear systems with constant coefficients, we use the eigenvalues and eigenvectors of the matrix of coefficients. We study here the two-dimensional cases for which the eigenvalues are real, and examine their portraits in the phase plane.

New Building Blocks

We begin by building solutions for the 2×2 system

$$\vec{\mathbf{x}}' = \mathbf{A}\vec{\mathbf{x}}, \tag{1}$$

where $\mathbf{A}$ is a constant matrix, from what we have learned about solutions to

$$ay'' + by' + cy = 0. \tag{2}$$

Even though the system is more general, we know that there is an underlying connection. When the second-order equation is converted to an equivalent system, the eigenvalues of the system matrix are the characteristic roots of the second-order equation. How can we make use of this fact? If r_1 and r_2 are the characteristic roots for equation (2), the solutions are built, one way or another, from $e^{r_1 t}$ and $e^{r_2 t}$. We need to find the corresponding building blocks for system (1).

Because solutions of (1) must be vectors, we will try something of the form

$$\vec{\mathbf{x}} = e^{\lambda t}\vec{\mathbf{v}}. \tag{3}$$

Substituting (3) into (1) gives

$$\lambda e^{\lambda t}\vec{\mathbf{v}} = \mathbf{A}e^{\lambda t}\vec{\mathbf{v}},$$

which is equivalent to

$$e^{\lambda t}\mathbf{A}\vec{\mathbf{v}} - \lambda e^{\lambda t}\mathbf{I}\vec{\mathbf{v}} = \vec{\mathbf{0}}.$$

In factored form, we have

$$e^{\lambda t}(\mathbf{A} - \lambda\mathbf{I})\vec{\mathbf{v}} = \vec{\mathbf{0}}.$$

Because $e^{\lambda t}$ is never zero, we need to find λ and $\vec{\mathbf{v}}$ such that

Characteristic Equation for Eigenvalues

$$(\mathbf{A} - \lambda\mathbf{I})\vec{\mathbf{v}} = \vec{\mathbf{0}}. \tag{4}$$

But a scalar λ and a nonzero vector $\vec{\mathbf{v}}$ satisfying (4) are no more nor less than an *eigenvalue* and *eigenvector* of matrix $\mathbf{A}$. We have got it!

Solving Homogeneous Linear 2 x 2 DE Systems with Constant Coefficients

For a two-dimensional system of homogeneous linear differential equations $\vec{\mathbf{x}}' = \mathbf{A}\vec{\mathbf{x}}$, where $\mathbf{A}$ is a matrix of constants that has eigenvalues λ_1 and λ_2 with corresponding eigenvectors $\vec{\mathbf{v}}_1$ and $\vec{\mathbf{v}}_2$, we obtain two solutions:

$$e^{\lambda_1 t}\vec{\mathbf{v}}_1 \quad \text{and} \quad e^{\lambda_2 t}\vec{\mathbf{v}}_2.$$

If $\lambda_1 \neq \lambda_2$, these two solutions are *linearly independent* and form a basis for the solution space. Thus, the general solution, for arbitrary constants c_1 and c_2, is

$$\vec{\mathbf{x}}(t) = c_1 e^{\lambda_1 t}\vec{\mathbf{v}}_1 + c_2 e^{\lambda_2 t}\vec{\mathbf{v}}_2. \tag{5}$$

If $\lambda_1 = \lambda_2$, then there may be only one linearly independent eigenvector; additional tactics may be required to obtain a basis of two vectors for the solution space. (See the subsection Repeated Eigenvalues later in this section.)

Matrix Element Input

See how changing the matrix $\mathbf{A}$ affects the phase portraits for $\vec{\mathbf{x}}' = \mathbf{A}\vec{\mathbf{x}}$.

From our study of the eigenstuff of a matrix $\mathbf{A}$ in Chapter 5, we expect to find different results depending on the nature of the eigenvalues. Are they real or nonreal? If real, are they distinct (unequal)? This section deals with the cases involving *real* eigenvalues (first distinct, then repeated). The nonreal cases will be covered in Sec. 6.3.

Distinct Real Eigenvalues

If matrix $\mathbf{A}$ has two different real eigenvalues, $\lambda_1 \neq \lambda_2$, with corresponding eigenvectors $\vec{\mathbf{v}}_1$ and $\vec{\mathbf{v}}_2$, then $\vec{\mathbf{v}}_1$ and $\vec{\mathbf{v}}_2$ are linearly independent (see the Distinct Eigenvalue Theorem following Sec. 5.3, Example 4). We use this fact to show that $e^{\lambda_1 t}\vec{\mathbf{v}}_1$ and $e^{\lambda_2 t}\vec{\mathbf{v}}_2$ are also linearly independent. Indeed, if

$$c_1 e^{\lambda_1 t}\vec{\mathbf{v}}_1 + c_2 e^{\lambda_2 t}\vec{\mathbf{v}}_2 = \vec{\mathbf{0}},$$

then, for $t = 0$,

$$c_1\vec{\mathbf{v}}_1 + c_2\vec{\mathbf{v}}_2 = \vec{\mathbf{0}} \quad \text{and} \quad c_1 = c_2 = 0.$$

Hence, these solutions form a fundamental set.

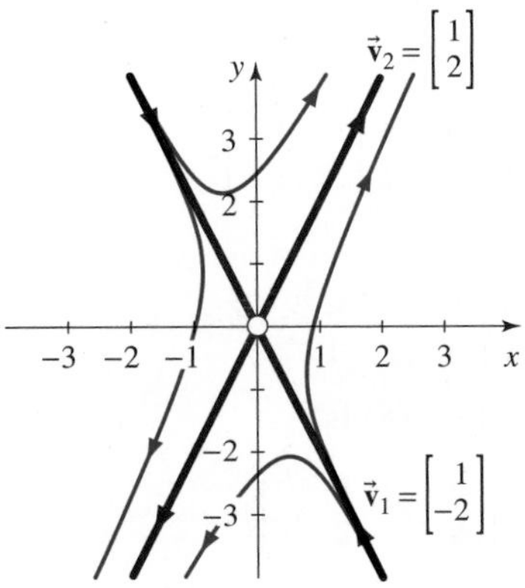

FIGURE 6.2.1 Phase portrait for Example 1, with eigenvalues $\lambda_1 = -1$, $\lambda_2 = 3$ and an *unstable* equilibrium at the origin.

EXAMPLE 1 **Opposite-Sign Eigenvalues** To find the general solution of

$$\vec{\mathbf{x}}' = \mathbf{A}\vec{\mathbf{x}} = \begin{bmatrix} 1 & 1 \\ 4 & 1 \end{bmatrix}\vec{\mathbf{x}}, \tag{6}$$

we recall (Sec. 5.3, Example 2) that $\mathbf{A}$ has eigenvalues

$$\lambda_1 = -1 \quad \text{and} \quad \lambda_2 = 3$$

with corresponding eigenvectors

$$\vec{\mathbf{v}}_1 = \begin{bmatrix} 1 \\ -2 \end{bmatrix} \quad \text{and} \quad \vec{\mathbf{v}}_2 = \begin{bmatrix} 1 \\ 2 \end{bmatrix}.$$

Thus, the general solution of system (6) is

$$\vec{\mathbf{x}} = c_1 e^{-t}\begin{bmatrix} 1 \\ -2 \end{bmatrix} + c_2 e^{3t}\begin{bmatrix} 1 \\ 2 \end{bmatrix}.$$

Some sample phase-plane trajectories are shown in Fig. 6.2.1, with the eigenvectors in bold. ■

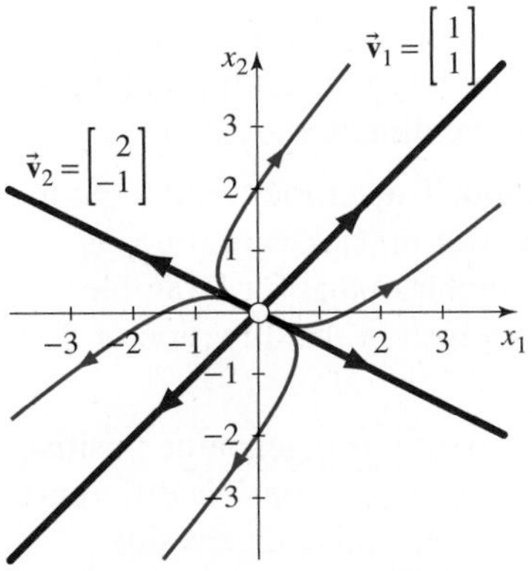

FIGURE 6.2.2 Phase portrait for Example 2, with positive eigenvalues $\lambda_1 = 4$, $\lambda_2 = 1$ and an unstable equilibrium at the origin.

EXAMPLE 2 **Positive Eigenvalues** The system

$$\vec{\mathbf{x}}' = \begin{bmatrix} 2 & 2 \\ 1 & 3 \end{bmatrix}\vec{\mathbf{x}} \tag{7}$$

is solved by first determining the eigenstuff. The characteristic equation is

$$\begin{vmatrix} 2-\lambda & 2 \\ 1 & 3-\lambda \end{vmatrix} = 0, \quad \text{or} \quad \lambda^2 - 5\lambda + 4 = 0.$$

By the methods of Sec. 5.3, the eigenvalues are

$$\lambda_1 = 4 \quad \text{and} \quad \lambda_2 = 1,$$

with corresponding eigenvectors

$$\vec{\mathbf{v}}_1 = \begin{bmatrix} 1 \\ 1 \end{bmatrix} \quad \text{and} \quad \vec{\mathbf{v}}_2 = \begin{bmatrix} 2 \\ -1 \end{bmatrix}.$$

The general solution of (7) is

$$\vec{\mathbf{x}} = c_1 e^{4t}\begin{bmatrix} 1 \\ 1 \end{bmatrix} + c_2 e^{t}\begin{bmatrix} 2 \\ -1 \end{bmatrix}.$$

As shown in Fig. 6.2.2, phase-plane trajectories move away from an unstable equilibrium at the origin. ■

EXAMPLE 3 **Negative Eigenvalues** The initial-value problem

$$\vec{\mathbf{x}}' = \begin{bmatrix} -2 & 1 \\ 1 & -2 \end{bmatrix}\vec{\mathbf{x}}, \quad \vec{\mathbf{x}}(0) = \begin{bmatrix} 3 \\ 1 \end{bmatrix} \tag{8}$$

has two negative eigenvalues,

$$\lambda_1 = -1 \quad \text{and} \quad \lambda_2 = -3,$$

with corresponding eigenvectors

$$\vec{\mathbf{v}}_1 = \begin{bmatrix} 1 \\ 1 \end{bmatrix} \quad \text{and} \quad \vec{\mathbf{v}}_2 = \begin{bmatrix} 1 \\ -1 \end{bmatrix}.$$

So the general solution to system (8) is

$$\vec{\mathbf{x}}(t) = c_1 e^{-t}\begin{bmatrix} 1 \\ 1 \end{bmatrix} + c_2 e^{-3t}\begin{bmatrix} 1 \\ -1 \end{bmatrix}.$$

Then, for the IVP,

$$\vec{\mathbf{x}}(0) = c_1\begin{bmatrix} 1 \\ 1 \end{bmatrix} + c_2\begin{bmatrix} 1 \\ -1 \end{bmatrix} = \begin{bmatrix} 1 & 1 \\ 1 & -1 \end{bmatrix}\begin{bmatrix} c_1 \\ c_2 \end{bmatrix} = \begin{bmatrix} 3 \\ 1 \end{bmatrix},$$

which can be written as an augmented matrix,

$$\left[\begin{array}{cc|c} 1 & 1 & 3 \\ 1 & -1 & 1 \end{array}\right], \quad \text{with RREF} \quad \left[\begin{array}{cc|c} 1 & 0 & 2 \\ 0 & 1 & 1 \end{array}\right],$$

to find $c_1 = 2$ and $c_2 = 1$. So the solution of the IVP (8) is

$$\vec{\mathbf{x}}(t) = 2e^{-t}\begin{bmatrix} 1 \\ 1 \end{bmatrix} + e^{-3t}\begin{bmatrix} 1 \\ -1 \end{bmatrix} = \begin{bmatrix} 2e^{-t} + e^{-3t} \\ 2e^{-t} - e^{-3t} \end{bmatrix}.$$

Fig. 6.2.3 shows some general solutions, which move toward a stable equilibrium at the origin. The grey curve is the solution to the IVP. ■

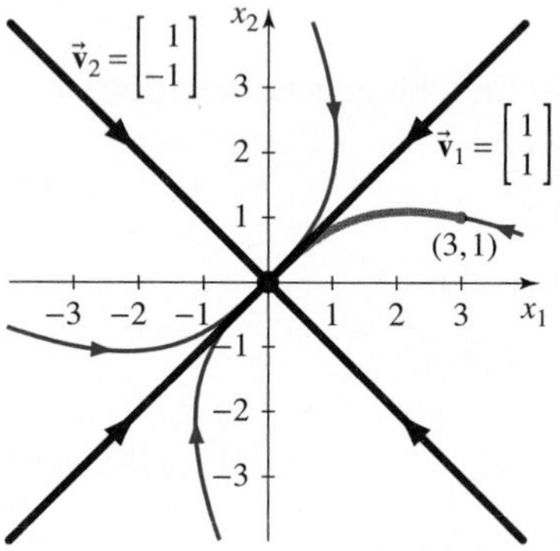

FIGURE 6.2.3 Phase portrait for Example 3, with negative eigenvalues $\lambda_1 = -1, \lambda_2 = -3$ and a *stable* equilibrium at the origin. The grey curve is the solution to the IVP.

Behavior of Solutions

The systems of Examples 1–3 exhibit different phase portraits because of differing combinations of signs of their eigenvalues. We recap the major possibilities of real characteristic roots shown in Sec. 4.2.

- Example 3 (Fig. 6.2.3) is a *stable* equilibrium. All trajectories tend to the origin as $t \to \infty$, because both eigenvalues are negative. Furthermore, because e^{-3t} decays faster than e^{-t}, solutions not lying on the lines along the eigenvectors approach the origin asymptotic to the direction of slower decay.
- Example 2 (Fig. 6.2.2) is an *unstable* equilibrium. Trajectories tend to ∞ as $t \to \infty$, because both eigenvalues are positive. But in backward time, as $t \to -\infty$, they tend to the origin in a pattern similar to that for Example 3. The main difference between these cases is a reversal of the direction of flow, outward for Example 2, inward for Example 3.
- Example 1 (Fig. 6.2.1) is a *saddle* equilibrium. With one eigenvalue positive and the other negative, the behavior along the "eigendirections" is different: inward for the negative eigenvalue, outward for the positive eigenvalue. Trajectories in the sectors between these directions flow inward along the direction associated with the negative eigenvalue, outward along the positive eigenvector's direction.

We will have more to say about such phase portraits in Sec. 6.4.

Sketching Phase Portraits for 2 × 2 Systems

Although we use many computer-generated phase portraits like Figs. 6.2.1–6.2.3, you will find it convenient to be able to sketch trajectories by hand. In the case of *linear* DE systems, we have (as you may already have observed) a very powerful tool for sketching phase portraits: *eigenvalues* and *eigenvectors*. Throughout this chapter, we will see that *all* trajectories are "guided" by these gifts from linear algebra. This material will be covered in detail in Sec. 6.4, but we can use the following principles to sketch a phase portrait.

Phase Plane Role of Real Eigenvectors and Eigenvalues

For an *autonomous* and *homogeneous* two-dimensional linear DE system:

- Trajectories move toward or away from the equilibrium according to the *sign of the eigenvalues* (negative or positive, respectively) associated with the eigenvectors.
- Along each *eigenvector* is a unique trajectory called a **separatrix** that separates trajectories curving one way from those curving another way.
- The *equilibrium occurs at the origin*, and the phase portrait is *symmetric about this point*.

Symmetry About the Origin:

For linear DE systems, every trajectory drawn gives another "for free."

For such a 2 × 2 system with distinct real eigenvalues, there are three possible combinations of signs for the eigenvalues—opposite, both positive, or both negative. Each possibility gives rise to a particular type of equilibrium, shown in Figs. 6.2.1–6.2.3 with Examples 1–3, and summarized at the end of this subsection in Fig. 6.2.5.

Adding the preceding properties to familiar principles gives a quick sketch of the phase portrait, as follows.

Sketching the Vector Field:

Drawing an entire vector field by hand would be tedious, but you can use the principle to calculate and *check the slope of a trajectory at any given point.*

- Draw the eigenvectors with arrows according to the sign of the eigenvalues.
- All trajectories must follow the *vector field* (Secs. 2.6 and 4.1).
- The Existence and Uniqueness Theorem (Sec. 6.1) tells us that *phase-plane trajectories cannot cross.* (Phase-plane trajectories may *appear* to meet at an equilibrium, but they actually never get there.)

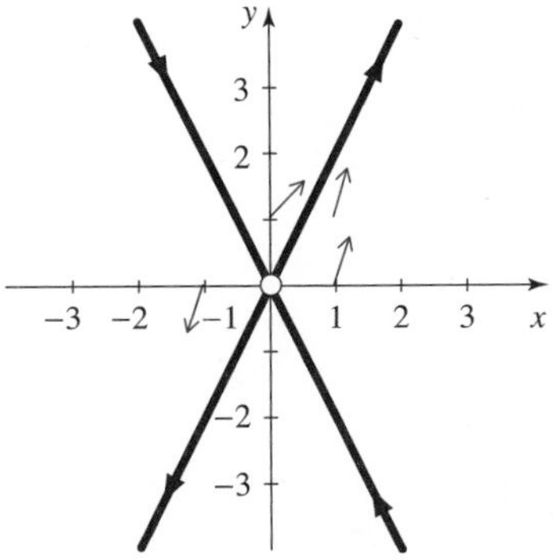

FIGURE 6.2.4 A few slope vectors for the system of Examples 1 and 2.

EXAMPLE 4 **Constructing a Vector Field** We write system (6) of Example 1 in component form,

$$x' = f(x, y) = x + y,$$
$$y' = g(x, y) = 4x + y,$$

and draw (in Fig. 6.2.4) a few vector elements with slopes calculated as follows:

At $(1, 0)$,	$x' = 1$,	$y' = 4$,	*right* 1, *up* 4.
At $(1, 1)$,	$x' = 2$,	$y' = 5$,	*right* 1, *up* 5.
At $(0, 1)$,	$x' = 1$,	$y' = 1$,	*right* 1, *up* 1.
At $(-1, 0)$,	$x' = -1$,	$y' = -4$,	*left* 1, *down* 4.
At $(0, 0)$,	$x' = 0$,	$y' = 0$,	*no motion*.

The system is in equilibrium at the origin. ■

Speed $\equiv \|\vec{\mathbf{x}}'(t)\|$

$$= \sqrt{\left(\frac{dx}{dt}\right)^2 + \left(\frac{dy}{dt}\right)^2}$$

in two dimensions.

Speed and Shape of Trajectories

- "Speed" along a trajectory in the direction of an eigenvector depends on the *magnitude* (absolute value) of the associated eigenvalue: "fast" for the eigenvalue with the largest magnitude, or "slow" for the eigenvalue with the smallest magnitude.
- Trajectories become parallel to the fast eigenvectors further away from the origin, and tangent to the slow eigenvectors—closer to the origin, in the cases of source or sink, further from the origin for a saddle.

"Fast" and "slow" directions for Examples 1–3 are marked (in Fig. 6.2.5.)

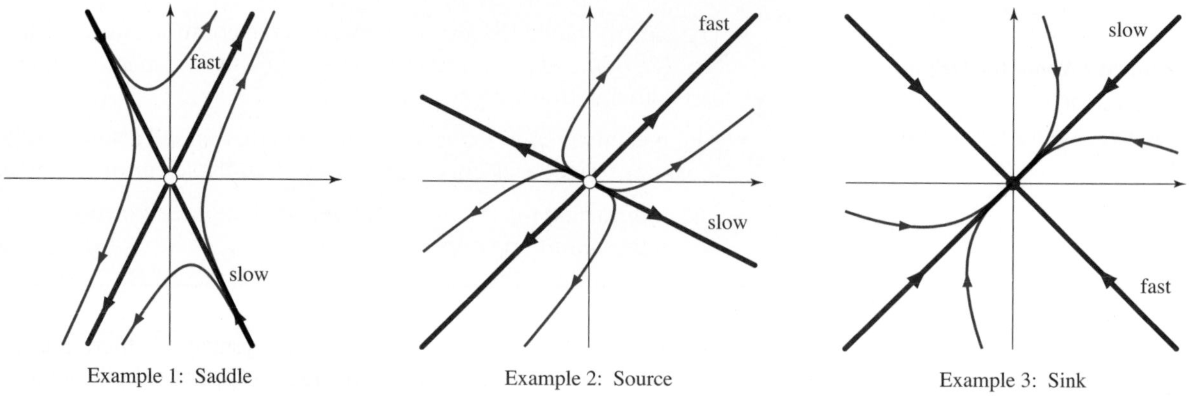

FIGURE 6.2.5 Real and distinct eigenvectors—the three possibilities. The eigenvector solutions are separatrices between trajectories curving one way and those curving another way. The fast eigenvectors are those for the eigenvalues with larger magnitude.

Repeated Real Eigenvalues

We learned in Chapter 5 that, if the characteristic equation has a double root, there may still be two linearly independent eigenvectors belonging to it, or there may be only one. If there are two, nothing new is needed, but if there is no second independent eigenvector, a solution of a different form must be found.

In the 2×2 case, the eigenspace belonging to a double eigenvalue can be *two*-dimensional only if the matrix is a multiple of the identity matrix. (See Problem 36.) We see the consequences in the following example.

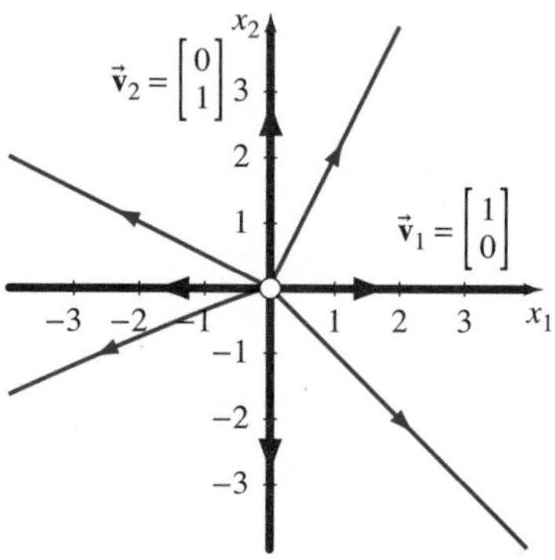

FIGURE 6.2.6 Phase portrait for Example 5, with an unstable equilibrium at the origin, a double eigenvalue $\lambda_1 = \lambda_2 = 3$, and two independent eigenvectors along the axes.

EXAMPLE 5 **Repeated Eigenvalues** For the system

$$\vec{\mathbf{x}}' = \mathbf{A}\vec{\mathbf{x}} = \begin{bmatrix} 3 & 0 \\ 0 & 3 \end{bmatrix}\vec{\mathbf{x}}, \tag{9}$$

the matrix $\mathbf{A} = 3\mathbf{I}$ has *double* eigenvalue

$$\lambda_1 = \lambda_2 = 3,$$

and *two* linearly independent eigenvectors are

$$\vec{\mathbf{v}}_1 = \begin{bmatrix} 1 \\ 0 \end{bmatrix} \quad \text{and} \quad \vec{\mathbf{v}}_2 = \begin{bmatrix} 0 \\ 1 \end{bmatrix}.$$

The general solution is

$$\vec{\mathbf{x}} = c_1e^{3t}\begin{bmatrix} 1 \\ 0 \end{bmatrix} + c_2e^{3t}\begin{bmatrix} 0 \\ 1 \end{bmatrix} = \begin{bmatrix} c_1e^{3t} \\ c_2e^{3t} \end{bmatrix}.$$

All solutions tend to infinity as $t \to \infty$, and to the origin as $t \to -\infty$, as in Fig. 6.2.6. (For a negative eigenvalue, the directions are reversed.) All trajectories lie along half-lines extending from the unstable equilibrium at the origin.

Because system (9) has component form

$$x_1' = 3x_1,$$
$$x_2' = 3x_2,$$

it is clear that $x_1 = c_1 e^{3t}$ and $x_2 = c_2 e^{3t}$. Hence, $x_2 = (c_2/c_1)x_1$. We will explore decoupled systems, of which this is an example, in Sec. 6.5. ■

EXAMPLE 6 **One Eigenvector Shy** To solve the system

$$\vec{\mathbf{x}}' = \mathbf{A}\vec{\mathbf{x}} = \begin{bmatrix} 2 & -1 \\ 4 & 6 \end{bmatrix}\vec{\mathbf{x}}, \tag{10}$$

we first calculate the eigenstuff for **A**, and find a *double* eigenvalue

$$\lambda_1 = \lambda_2 = 4$$

and a *single* independent eigenvector

$$\vec{\mathbf{v}} = \begin{bmatrix} 1 \\ -2 \end{bmatrix}.$$

Thus, *one* solution is

$$\vec{\mathbf{x}}_1 = e^{4t}\begin{bmatrix} 1 \\ -2 \end{bmatrix}, \tag{11}$$

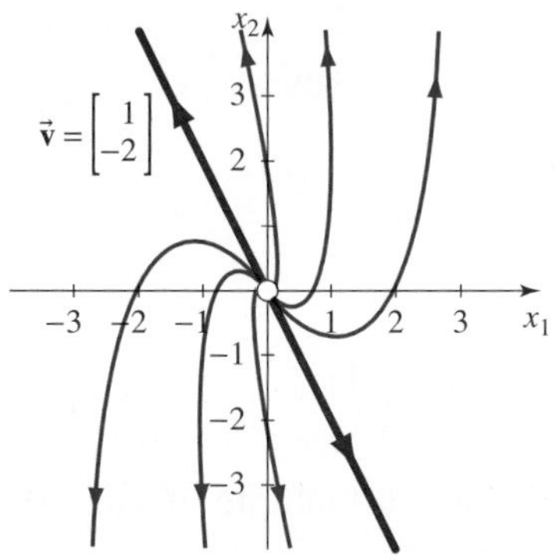

FIGURE 6.2.7 Phase portrait for Example 6, with an unstable equilibrium at the origin, a *double* eigenvalue $\lambda_1 = \lambda_2 = 4$, and a single eigenvector.

but a 2×2 system needs *another* solution to form the basis for the solution space. We can *see* this need if we hand-sketch a quick phase portrait (Problem 37) from nullclines and vectors at any selected points, or use a computer graphics DE solver, as in Fig. 6.2.7.

The trick we used in Chapter 4 to get a second solution for a second-order equation with double characteristic root will not work here. If we try

$$\vec{\mathbf{x}}_2 = te^{4t}\vec{\mathbf{v}} \tag{12}$$

and substitute it into (10), our solution evaporates, as we shall demonstrate.

Differentiating $\vec{\mathbf{x}}_2$ gives

$$\vec{\mathbf{x}}_2' = e^{4t}\vec{\mathbf{v}} + 4te^{4t}\vec{\mathbf{v}},$$

and we know that $\mathbf{A}\vec{\mathbf{v}} = \lambda\vec{\mathbf{v}}$, so

$$\mathbf{A}\vec{\mathbf{x}}_2 = \mathbf{A}te^{4t}\vec{\mathbf{v}} = \lambda te^{4t}\vec{\mathbf{v}} = 4te^{4t}\vec{\mathbf{v}}.$$

The DE $\vec{\mathbf{x}}_2' = \mathbf{A}\vec{\mathbf{x}}_2$ becomes

$$e^{4t}\vec{\mathbf{v}} + 4te^{4t}\vec{\mathbf{v}} = te^{4t}\mathbf{A}\vec{\mathbf{v}}, \tag{13}$$

which is only true if $\vec{\mathbf{v}} = \vec{\mathbf{0}}$, a contradiction! ■

A better guess at a second (linearly independent) solution to equation (10) of Example 6 must involve *another* vector. The next example will introduce a useful procedure to deal with a case of insufficient eigenvectors. You will see that the result, called a *generalized eigenvector*, is not really an eigenvector as defined in Chapter 5, but it does the job of completing the basis for the solution space.

EXAMPLE 7 **One Eigenvector Shy: To the Rescue** We still seek a second solution to equation (10) from Example 6. Because equation (13) contains terms in both e^{4t} and te^{4t}, it seems reasonable to add to our failed try in (12) another term that multiplies the pesky e^{4t} by a new vector $\vec{\mathbf{u}}$. So we try

$$\vec{\mathbf{x}}_2 = te^{4t}\vec{\mathbf{v}} + e^{4t}\vec{\mathbf{u}}.$$

Now, $\vec{\mathbf{x}}_2' = e^{4t}\vec{\mathbf{v}} + 4te^{4t}\vec{\mathbf{v}} + 4e^{4t}\vec{\mathbf{u}}$, and the DE (10) becomes

$$e^{4t}\vec{\mathbf{v}} + 4te^{4t}\vec{\mathbf{v}} + 4e^{4t}\vec{\mathbf{u}} = \mathbf{A}(te^{4t}\vec{\mathbf{v}} + e^{4t}\vec{\mathbf{u}}).$$

Equating coefficients of te^{4t} and e^{4t} gives $4\vec{\mathbf{v}} = \mathbf{A}\vec{\mathbf{v}}$ and $\vec{\mathbf{v}} + 4\vec{\mathbf{u}} = \mathbf{A}\vec{\mathbf{u}}$, or

$$(\mathbf{A} - 4\mathbf{I})\vec{\mathbf{v}} = \vec{\mathbf{0}} \quad \text{and} \quad (\mathbf{A} - 4\mathbf{I})\vec{\mathbf{u}} = \vec{\mathbf{v}}. \tag{14}$$

To solve equations (14) for $\vec{\mathbf{u}}$ and $\vec{\mathbf{v}}$, we notice that the first restates the fact that $\vec{\mathbf{v}}$ is the eigenvector belonging to $\lambda = 4$, so we still have

$$\vec{\mathbf{v}} = \begin{bmatrix} 1 \\ -2 \end{bmatrix}.$$

We can use this in the second equation in (14) to find $\vec{\mathbf{u}}$:

$$(\mathbf{A} - 4\mathbf{I})\vec{\mathbf{u}} = \begin{bmatrix} -2 & -1 \\ 4 & 2 \end{bmatrix}\begin{bmatrix} u_1 \\ u_2 \end{bmatrix} = \begin{bmatrix} 1 \\ -2 \end{bmatrix},$$

so $u_1 + \frac{1}{2}u_2 = -\frac{1}{2}$. If we let $u_1 = k$, then $u_2 = -2k - 1$, and

$$\vec{\mathbf{u}} = \begin{bmatrix} k \\ -2k - 1 \end{bmatrix} = k\begin{bmatrix} 1 \\ -2 \end{bmatrix} + \begin{bmatrix} 0 \\ -1 \end{bmatrix}.$$

Then,

$$\vec{\mathbf{x}}_2 = te^{4t}\begin{bmatrix} 1 \\ -2 \end{bmatrix} + ke^{4t}\begin{bmatrix} 1 \\ -2 \end{bmatrix} + e^{4t}\begin{bmatrix} 0 \\ -1 \end{bmatrix}. \tag{15}$$

We can drop the middle term of (15), which is just a multiple of our first solution (11), so our second solution is

$$\vec{\mathbf{x}}_2 = te^{4t}\begin{bmatrix} 1 \\ -2 \end{bmatrix} + e^{4t}\begin{bmatrix} 0 \\ -1 \end{bmatrix} = e^{4t}\begin{bmatrix} t \\ -2t - 1 \end{bmatrix}. \tag{16}$$

We leave it to the reader (Problem 44) to verify that solutions (11) and (16) are linearly independent. The general solution of system (10) is therefore

$$\vec{\mathbf{x}} = c_1\vec{\mathbf{x}}_1 + c_2\vec{\mathbf{x}}_2 = c_1e^{4t}\begin{bmatrix} 1 \\ -2 \end{bmatrix} + c_2e^{4t}\begin{bmatrix} t \\ -2t - 1 \end{bmatrix}.$$

Figure 6.2.7 (Example 6) shows typical phase-plane trajectories for system (10), which has an unstable equilibrium at the origin. The positive eigenvalue causes all solutions to tend to infinity along a trajectory that becomes parallel to the eigenvector as $t \to \infty$. As $t \to -\infty$, the solutions tend to the origin, asymptotic to the line $x_2 = -2x_1$ along the single eigenvector. The generalized eigenvector $\vec{\mathbf{u}}$ includes a variable t, so it *cannot* be drawn as a second stable vector on the phase portrait. ■

We can summarize the procedure worked out in Example 7 as follows.

Generalized Eigenvector and the Phase Portrait:

The generalized eigenvector helps to create a *basis* for the solution space; however, the generalized eigenvector does *not* have significance in the phase portrait.

Creating a Generalized Eigenvector for a System with Insufficient Eigenvectors

If a homogeneous linear 2×2 system of first-order DEs has repeated eigenvalue λ with only a single eigenvector, a second linearly independent solution can be created as follows:

Step 1. Find an eigenvector $\vec{\mathbf{v}}$ corresponding to λ.

Step 2. Find a nonzero vector $\vec{\mathbf{u}}$ so that

$$(\mathbf{A} - \lambda\mathbf{I})\vec{\mathbf{u}} = \vec{\mathbf{v}}.$$

Step 3. Then $\vec{\mathbf{x}}(t) = c_1 e^{\lambda t}\vec{\mathbf{v}} + c_2 e^{\lambda t}(t\vec{\mathbf{v}} + \vec{\mathbf{u}})$.

The vector $\vec{\mathbf{u}}$ is called a **generalized eigenvector** of $\mathbf{A}$ corresponding to λ.

Generalizing to Higher Dimensions

We have concentrated in this section on 2-dimensional homogeneous linear DE systems with constant coefficients. Most results can be generalized to an n-dimensional system as follows.

Solving n-Dimensional Homogeneous Linear DE Systems with Constant Coefficients

For an n-dimensional system of homogeneous linear differential equations $\vec{\mathbf{x}}' = \mathbf{A}\vec{\mathbf{x}}$, where $\mathbf{A}$ is a matrix of constants that has eigenvalues $\lambda_1, \lambda_2, \ldots, \lambda_n$ with corresponding eigenvectors $\vec{\mathbf{v}}_1, \vec{\mathbf{v}}_2, \ldots, \vec{\mathbf{v}}_n$, we obtain solutions:

$$e^{\lambda_1 t}\vec{\mathbf{v}}_1, e^{\lambda_2 t}\vec{\mathbf{v}}_2, \ldots, e^{\lambda_n t}\vec{\mathbf{v}}_n.$$

If $\lambda_i \neq \lambda_j$ for all $i \neq j$, these solutions are *linearly independent* and form a basis for the solution space. Thus, the general solution, for arbitrary constants $c_1, c_2, \ldots, c_n \in \mathbb{R}^n$, is

$$\vec{\mathbf{x}}(t) = c_1 e^{\lambda_1 t}\vec{\mathbf{v}}_1 + c_2 e^{\lambda_2 t}\vec{\mathbf{v}}_2 + \cdots + c_n e^{\lambda_n t}\vec{\mathbf{v}}_n.$$

The case of repeated eigenvalues ($\lambda_i = \lambda_j$ for some $i \neq j$) requires either independent eigenvectors or generalized eigenvectors.

EXAMPLE 8 **Repeated Eigenvalues in 3D** We wish to solve the system

$$\begin{aligned} x_1' &= 3x_1 + x_2 - x_3, \\ x_2' &= x_1 + 3x_2 - x_3, \\ x_3' &= 3x_1 + 3x_2 - x_3, \end{aligned} \quad \text{or} \quad \vec{\mathbf{x}}' = \mathbf{A}\vec{\mathbf{x}} = \begin{bmatrix} 3 & 1 & -1 \\ 1 & 3 & -1 \\ 3 & 3 & -1 \end{bmatrix}\vec{\mathbf{x}}. \tag{17}$$

The characteristic equation is $\lambda^3 - 5\lambda^2 + 8\lambda - 4 = 0$, giving the eigenvalues

$$\lambda_1 = 1 \quad \text{and} \quad \lambda_2 = \lambda_3 = 2$$

with corresponding eigenvectors

$$\vec{\mathbf{v}}_1 = \begin{bmatrix} 1 \\ 1 \\ 3 \end{bmatrix} \qquad \text{for } \lambda_1 = 1,$$

and

$$\vec{\mathbf{v}}_2 = \begin{bmatrix} 1 \\ -1 \\ 0 \end{bmatrix} \quad \text{and} \quad \vec{\mathbf{v}}_3 = \begin{bmatrix} 1 \\ 0 \\ 1 \end{bmatrix} \qquad \text{for } \lambda_2 = \lambda_3 = 2.$$

The double eigenvalue has *two* linearly independent eigenvectors. Hence, we have found a fundamental set of solutions to the DE system:

$$\vec{\mathbf{x}}_1 = e^t \begin{bmatrix} 1 \\ 1 \\ 3 \end{bmatrix}, \quad \vec{\mathbf{x}}_2 = e^{2t} \begin{bmatrix} 1 \\ -1 \\ 0 \end{bmatrix}, \quad \vec{\mathbf{x}}_3 = e^{2t} \begin{bmatrix} 1 \\ 0 \\ 1 \end{bmatrix}.$$

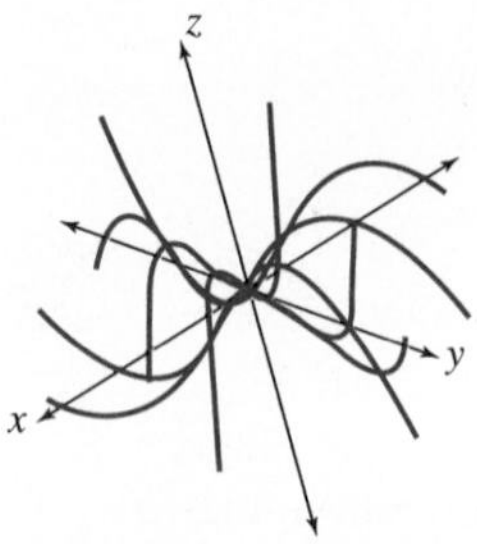

FIGURE 6.2.8
Some trajectories in 3-D for Example 8.

The general solution to the original system (17) is then

$$\vec{\mathbf{x}}(t) = \begin{bmatrix} x_1(t) \\ x_2(t) \\ x_3(t) \end{bmatrix} = c_1 e^t \begin{bmatrix} 1 \\ 1 \\ 3 \end{bmatrix} + c_2 e^{2t} \begin{bmatrix} 1 \\ -1 \\ 0 \end{bmatrix} + c_3 e^{2t} \begin{bmatrix} 1 \\ 0 \\ 1 \end{bmatrix}.$$

The trajectories, shown in Fig. 6.2.8, all emanate from an infinitesimal neighborhood of the origin. ■

The method for solving a system with insufficient eigenvectors extends to systems larger than 2×2. For an eigenvalue of multiplicity m with fewer than m eigenvectors, we can get from one eigenvector $\vec{\mathbf{v}}$ another solution of the form

$$\vec{\mathbf{x}}(t) = e^{\lambda t}(t\vec{\mathbf{v}} + \vec{\mathbf{u}}).$$

(See Problems 38 and 39 for extension of this process to $m > 2$.)

Applications to Multiple Compartment Models

We can extend the mixing problems we met in Sec. 2.4 to a system of several compartments, each with its own equation for

$$\text{RATE OF CHANGE} = \text{RATE IN} - \text{RATE OUT}. \tag{18}$$

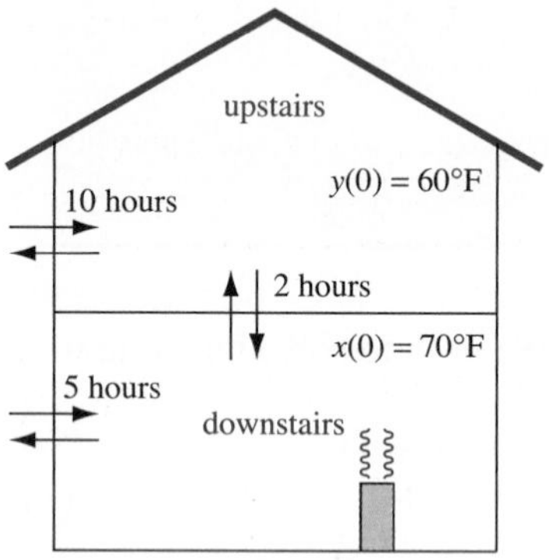

FIGURE 6.2.9 Flow of heat in a two-story house (Example 9).

EXAMPLE 9 **Heating Problem** Professor West lives in a two-story house, shown in Fig. 6.2.9. One winter night, with the outside temperature at 0°F, her furnace fails.

Suppose that the time constants in Professor West's house, which specify the rate of heat flow between the rooms, are 5 hours between downstairs and the outside, 10 hours between upstairs and the outside, and 2 hours between the two floors. If the temperature when the furnace fails is 70°F downstairs and 60°F upstairs, what are the future temperatures on each level of the house?

Letting $x(t)$ and $y(t)$ denote the temperature downstairs and upstairs, respectively, Newton's Law of Cooling states that the rate of change in the temperature of a room is proportional to the difference between the temperature of the room and the temperature of the surrounding medium. In this problem the rate at which the temperature downstairs changes depends on the addition of heat (now zero) from the furnace, along with heat gain (or loss) from upstairs and the outside. Also, the rate of change of temperature upstairs depends on the heat gain (or loss) from downstairs and the outside. Hence, the two unknowns $x(t)$ and $y(t)$ satisfy the initial-value problem

$$x' = -\frac{1}{5}[x(t) - 0] - \frac{1}{2}[x(t) - y(t)] = -\frac{7}{10}x(t) + \frac{1}{2}y(t),$$

$$y' = -\frac{1}{2}[y(t) - x(t)] - \frac{1}{10}[y(t) - 0] = \frac{1}{2}x(t) - \frac{3}{5}y(t),$$

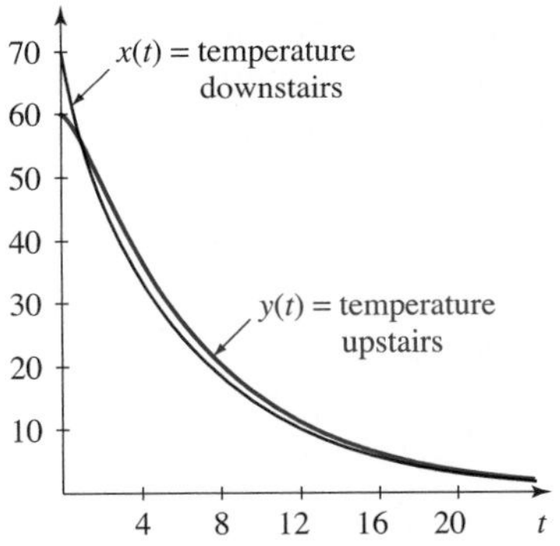

FIGURE 6.2.10 Temperature in the upstairs and downstairs over a 24-hour period (Example 9).

where $x(0) = 70$ and $y(0) = 60$. Writing the system in matrix form, we have

$$\begin{bmatrix} x' \\ y' \end{bmatrix} = \frac{1}{10}\begin{bmatrix} -7 & 5 \\ 5 & -6 \end{bmatrix}\begin{bmatrix} x \\ y \end{bmatrix}, \quad \begin{bmatrix} x(0) \\ y(0) \end{bmatrix} = \begin{bmatrix} 70 \\ 60 \end{bmatrix}.$$

Finding the eigenvalues of the coefficient matrix of this system, we find the general solution to be

$$\begin{bmatrix} x(t) \\ y(t) \end{bmatrix} \approx c_1 e^{-0.15t}\begin{bmatrix} 0.9 \\ 1 \end{bmatrix} + c_2 e^{-1.15t}\begin{bmatrix} -1.1 \\ 1 \end{bmatrix}.$$

Substituting the initial conditions $x(0) = 70$ and $y(0) = 60$ gives $c_1 = 68$ and $c_2 = -8$. The results are graphed in Fig. 6.2.10. After a couple of hours, the temperature upstairs is slightly higher, though by 24 hours both stories are approaching zero degrees. ■

EXAMPLE 10 **Two-Tank Mixing Problem** Consider the two tanks shown in Fig. 6.2.11. Each tank initially holds 100 gal of water in which 10 lb of salt has been dissolved. Fresh water flows into tank 1 at a rate of 3 gal/min, and the well-stirred mixture flows into tank 2 at a rate of 5 gal/min. The well-stirred mixture in tank 2 is simultaneously pumped back into tank 1 at a rate of 2 gal/min and out of tank 2 at a rate of 3 gal/min. Determine the initial-value problem that describes this system.

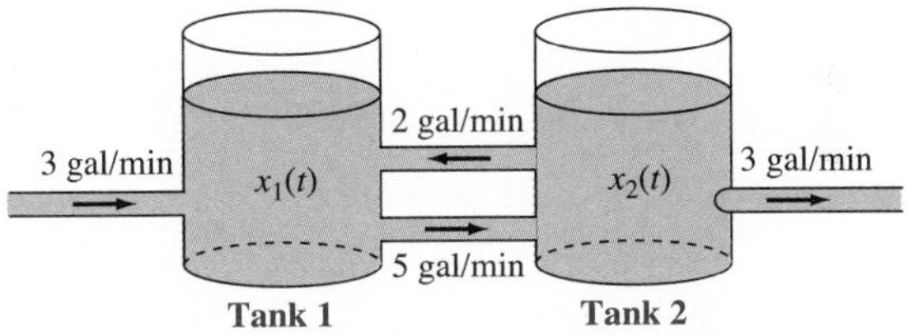

FIGURE 6.2.12 Two-tank arrangement for Example 10.

Let $x_1(t)$ and $x_2(t)$ be the amount of salt (in lb) in tank 1 and tank 2, respectively. We set up the IVP in lb/min, using equation (18):

$$x_1' = \underbrace{\left(0\frac{\text{lb}}{\text{gal}}\right)\left(3\frac{\text{gal}}{\text{min}}\right) + \left(\frac{x_2 \text{ lb}}{100 \text{ gal}}\right)\left(2\frac{\text{gal}}{\text{min}}\right)}_{\text{RATE IN}} - \underbrace{\left(\frac{x_1 \text{ lb}}{100 \text{ gal}}\right)\left(5\frac{\text{gal}}{\text{min}}\right)}_{\text{RATE OUT}},$$

$$x_2' = \underbrace{\left(\frac{x_1 \text{ lb}}{100 \text{ gal}}\right)\left(5\frac{\text{gal}}{\text{min}}\right)}_{\text{RATE IN}} - \underbrace{\left[\left(\frac{x_2 \text{ lb}}{100 \text{ gal}}\right)\left(2\frac{\text{gal}}{\text{min}}\right) + \left(\frac{x_2 \text{ lb}}{100 \text{ gal}}\right)\left(3\frac{\text{gal}}{\text{min}}\right)\right]}_{\text{RATE OUT}}.$$

With the addition of the initial conditions

$$x_1(0) = x_2(0) = 10,$$

this DE system simplifies to the following, in matrix-vector form:

$$\begin{bmatrix} x_1 \\ x_2 \end{bmatrix}' = \begin{bmatrix} -0.05 & 0.02 \\ 0.05 & -0.05 \end{bmatrix}\begin{bmatrix} x_1 \\ x_2 \end{bmatrix}, \quad \vec{\mathbf{x}}(0) = \begin{bmatrix} 10 \\ 10 \end{bmatrix}.$$

The pure water input makes this system homogeneous. In both tanks, "gallons in" equals "gallons out," which means that the volume stays constant at 100 gal and the matrix of coefficients has constant entries.

The solutions for $x_1(t)$ and $x_2(t)$ are graphed in Fig. 6.2.12. ■

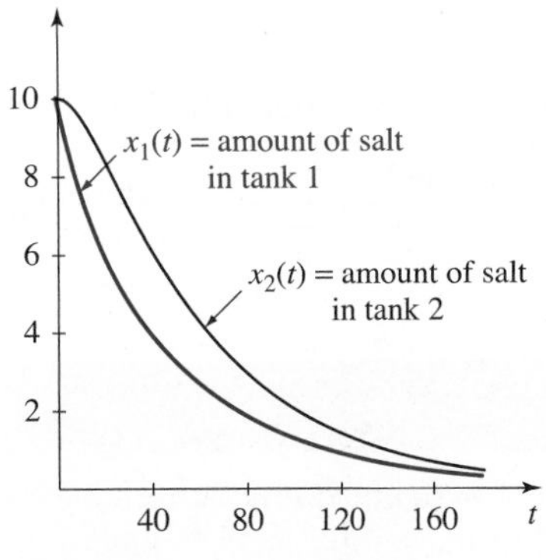

FIGURE 6.2.11 Graph of solutions for Example 10.

EXAMPLE 11 **An Electrical Network** A multiloop electrical network, such as is shown in Fig. 6.2.13, can be modeled by a system of differential equations, using Kirchoff's Laws of Currents and Voltages. (See Sec. 3.4, Problem 43.)

- By *Kirchoff's First Law* (currents at junctions):

$$I_1 = I_2 + I_3.$$

- By *Kirchoff's Second Law* (voltages around loops):

$$L_1 \frac{dI_1}{dt} + R_2 I_1 + R_1 I_3 = V(t),$$

$$L_2 \frac{dI_2}{dt} - R_1 I_3 = 0.$$

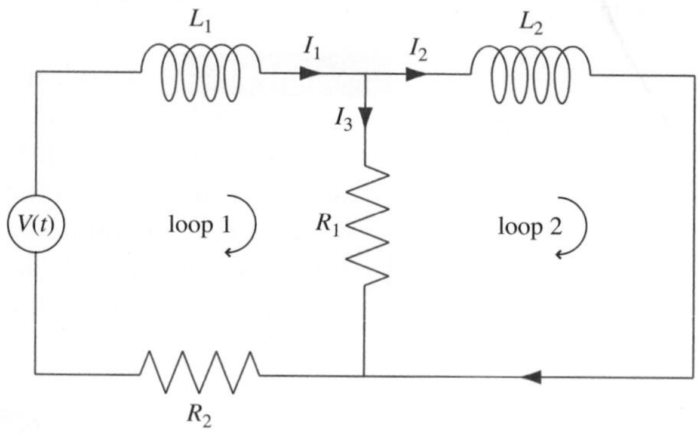

FIGURE 6.2.13 Multiloop electrical network for Example 11.

Substituting $I_3 = I_1 - I_2$ yields the nonhomogeneous linear DE system

$$\begin{bmatrix} I_1 \\ I_2 \end{bmatrix}' = \begin{bmatrix} -\dfrac{R_1 + R_2}{L_1} & \dfrac{R_1}{L_1} \\ \dfrac{R_1}{L_2} & -\dfrac{R_1}{L_2} \end{bmatrix} \begin{bmatrix} I_1 \\ I_2 \end{bmatrix} + \begin{bmatrix} \dfrac{V(t)}{L_1} \\ 0 \end{bmatrix}.$$

We can draw the following conclusions:

- If there is no input voltage (i.e., $V(t) \equiv 0$ for $t \geq 0$), then the system is homogeneous.
- If the eigenvalues are real, the currents grow or decay exponentially.
- If the eigenvalues are complex, the currents oscillate. ■

Summary

We have obtained explicit solutions for 2×2 and 3×3 homogeneous linear systems with constant coefficients when the eigenvalues of the coefficient matrix are real. Typical phase portraits illuminate long-term behavior of such solutions. We introduced several applications.

6.2 Problems

Sketching Second-Order DEs *In Problems 1–4, find the constant solution(s) $x(t) \equiv k$ for each second-order equation, and determine the behavior as follows:*

(a) *Rewrite the equation as a system of two first-order equations.*

(b) *Find the equilibrium solution(s) of the equivalent first-order system.*

(c) *Deduce the behavior of the trajectories about the fixed point as $t \to \infty$—for example, flying away from the fixed point, orbiting it, approaching it.*

(d) *Describe the physical behavior of solutions to the DE. Tell what it means for a mass–spring system.*

1. $x'' + x' + x = 0$

2. $x'' - x' + x = 0$

3. $x'' + x = 1$

4. $x'' + 2x' + x = 2$

Matching Game *Match each system of Problems 5–8 with one of the vector fields in Fig. 6.2.14.* Hint: *Use information from eigenvalues (Sec. 5.3) and nullclines (Sec. 2.5).*

5. $x' = x$, $y' = y$

6. $x' = -x$, $y' = -y$

7. $\vec{\mathbf{x}}' = \begin{bmatrix} 0 & 1 \\ 1 & -1 \end{bmatrix} \vec{\mathbf{x}}$

8. $\vec{\mathbf{x}}' = \begin{bmatrix} 1 & 1 \\ 1 & -1 \end{bmatrix} \vec{\mathbf{x}}$

Solutions in General *Find the general solutions for Problems 9–22. Sketch the eigenvectors and a few typical trajectories. (Show your method.)*

9. $\vec{\mathbf{x}}' = \begin{bmatrix} -4 & 2 \\ 2 & -1 \end{bmatrix} \vec{\mathbf{x}}$

10. $\vec{\mathbf{x}}' = \begin{bmatrix} 2 & 1 \\ -3 & 6 \end{bmatrix} \vec{\mathbf{x}}$

11. $\vec{\mathbf{x}}' = \begin{bmatrix} 1 & -1 \\ 2 & 4 \end{bmatrix} \vec{\mathbf{x}}$

12. $\vec{\mathbf{x}}' = \begin{bmatrix} 10 & -5 \\ 8 & -12 \end{bmatrix} \vec{\mathbf{x}}$

13. $\vec{\mathbf{x}}' = \begin{bmatrix} 5 & -1 \\ 3 & 1 \end{bmatrix} \vec{\mathbf{x}}$

14. $\vec{\mathbf{x}}' = \begin{bmatrix} 1 & 2 \\ 4 & 3 \end{bmatrix} \vec{\mathbf{x}}$

15. $\vec{\mathbf{x}}' = \begin{bmatrix} 1 & 0 \\ -2 & 2 \end{bmatrix} \vec{\mathbf{x}}$

16. $\vec{\mathbf{x}}' = \begin{bmatrix} 3 & 3 \\ -1 & -1 \end{bmatrix} \vec{\mathbf{x}}$

17. $\vec{\mathbf{x}}' = \begin{bmatrix} 3 & -2 \\ 2 & -2 \end{bmatrix} \vec{\mathbf{x}}$

18. $\vec{\mathbf{x}}' = \begin{bmatrix} 4 & 3 \\ -4 & -4 \end{bmatrix} \vec{\mathbf{x}}$

19. $\vec{\mathbf{x}}' = \begin{bmatrix} 1 & -2 \\ 3 & -4 \end{bmatrix} \vec{\mathbf{x}}$

20. $\vec{\mathbf{x}}' = \begin{bmatrix} 5 & -2 \\ -2 & 8 \end{bmatrix} \vec{\mathbf{x}}$

21. $\vec{\mathbf{x}}' = \begin{bmatrix} 4 & -3 \\ 8 & -6 \end{bmatrix} \vec{\mathbf{x}}$

22. $\vec{\mathbf{x}}' = \begin{bmatrix} 5 & 3 \\ -1 & 1 \end{bmatrix} \vec{\mathbf{x}}$

Repeated Eigenvalues *Find the general solutions for Problems 23 and 24. Sketch the eigenvectors and a few typical trajectories. (Show your method.)*

23. $\vec{\mathbf{x}}' = \begin{bmatrix} -1 & 1 \\ -4 & 3 \end{bmatrix} \vec{\mathbf{x}}$

24. $\vec{\mathbf{x}}' = \begin{bmatrix} 3 & 2 \\ -8 & -5 \end{bmatrix} \vec{\mathbf{x}}$

Solutions in Particular *Solve the IVPs in Problems 25–34. Sketch the trajectory.*

25. $\vec{\mathbf{x}}' = \begin{bmatrix} -2 & 1 \\ -5 & 4 \end{bmatrix} \vec{\mathbf{x}}, \quad \vec{\mathbf{x}}(0) = \begin{bmatrix} 1 \\ 3 \end{bmatrix}$

26. $\vec{\mathbf{x}}' = \begin{bmatrix} 1 & -3 \\ -2 & 2 \end{bmatrix} \vec{\mathbf{x}}, \quad \vec{\mathbf{x}}(0) = \begin{bmatrix} 1 \\ -1 \end{bmatrix}$

27. $\vec{\mathbf{x}}' = \begin{bmatrix} 2 & 0 \\ 0 & 3 \end{bmatrix} \vec{\mathbf{x}}, \quad \vec{\mathbf{x}}(0) = \begin{bmatrix} 5 \\ 4 \end{bmatrix}$

28. $\vec{\mathbf{x}}' = \begin{bmatrix} -2 & 4 \\ 1 & 1 \end{bmatrix} \vec{\mathbf{x}}, \quad \vec{\mathbf{x}}(0) = \begin{bmatrix} -1 \\ 1 \end{bmatrix}$

29. $\vec{\mathbf{x}}' = \begin{bmatrix} 1 & 1 \\ 1 & 1 \end{bmatrix} \vec{\mathbf{x}}, \quad \vec{\mathbf{x}}(0) = \begin{bmatrix} 2 \\ 3 \end{bmatrix}$

30. $\vec{\mathbf{x}}' = \begin{bmatrix} -3 & 2 \\ 1 & -2 \end{bmatrix} \vec{\mathbf{x}}, \quad \vec{\mathbf{x}}(0) = \begin{bmatrix} -1 \\ 6 \end{bmatrix}$

31. $\vec{\mathbf{x}}' = \begin{bmatrix} -2 & 1 \\ 4 & -2 \end{bmatrix} \vec{\mathbf{x}}, \quad \vec{\mathbf{x}}(0) = \begin{bmatrix} 2 \\ 4 \end{bmatrix}$

32. $\vec{\mathbf{x}}' = \begin{bmatrix} 1 & 12 \\ 3 & 1 \end{bmatrix} \vec{\mathbf{x}}, \quad \vec{\mathbf{x}}(0) = \begin{bmatrix} 0 \\ 1 \end{bmatrix}$

FIGURE 6.2.14 Vector fields to match to the systems in Problems 5–8.

33. $\vec{\mathbf{x}}' = \begin{bmatrix} 1 & -1 \\ 2 & 4 \end{bmatrix}\vec{\mathbf{x}}, \qquad \vec{\mathbf{x}}(0) = \begin{bmatrix} 1 \\ 0 \end{bmatrix}$

34. $\vec{\mathbf{x}}' = \begin{bmatrix} 1 & 2 \\ 2 & 1 \end{bmatrix}\vec{\mathbf{x}}, \qquad \vec{\mathbf{x}}(0) = \begin{bmatrix} 1 \\ 3 \end{bmatrix}$

35. Creating New Problems

(a) Find a 3×3 matrix with a double eigenvalue $\lambda_1 = \lambda_2$ that has only one eigenvector, and a separate eigenvalue $\lambda_3 \neq \lambda_1$ with another eigenvector.

(b) Find a 3×3 matrix with a triple eigenvalue and **two** linearly independent eigenvectors.

36. Repeated Eigenvalue Theory Suppose that

$$\vec{\mathbf{x}}' = \begin{bmatrix} a & b \\ c & d \end{bmatrix}\vec{\mathbf{x}}.$$

(a) Show that the system has a double eigenvalue if and only if the condition $(a-d)^2 + 4bc = 0$ is satisfied, and that the eigenvalue is $\frac{1}{2}(a+d)$.

(b) Show that if the condition in (a) holds and $a = d$, the eigenspace will be *two*-dimensional only if the matrix

$$\begin{bmatrix} a & b \\ c & d \end{bmatrix}$$

is diagonal.

(c) Show that if the condition in (a) holds and $a \neq d$, the eigenvectors belonging to $\frac{1}{2}(a+d)$ are linearly dependent; that is, scalar multiples of

$$\begin{bmatrix} 2b \\ d-a \end{bmatrix}.$$

(d) Show that the general solution of the system with double eigenvalue and $a \neq d$ is

$$c_1 e^{\lambda t}\begin{bmatrix} 2b \\ d-a \end{bmatrix} + c_2 e^{\lambda t}\left(t\begin{bmatrix} 2b \\ d-a \end{bmatrix} + \begin{bmatrix} 0 \\ 2 \end{bmatrix}\right),$$

where $\lambda = \frac{1}{2}(a+d)$.

37. Quick Sketch For equation (10) of Example 6, show the calculations for vectors at whatever points you choose. Confirm that your result has the same characteristics as Fig. 6.2.7. Then sketch some trajectories on your graph.

38. Generalized Eigenvectors Suppose that we wish to extend the method described for finding one generalized eigenvector to finding two (or more) generalized eigenvectors. Let's look at the case where λ has multiplicity 3 but has only one linearly independent eigenvector $\vec{\mathbf{v}}$. First, we find $\vec{\mathbf{u}}_1$ by the method described in this section. Then we find $\vec{\mathbf{u}}_2$ such that

$$(\mathbf{A} - \lambda\mathbf{I})\vec{\mathbf{u}}_2 = \vec{\mathbf{u}}_1, \quad \text{or} \quad (\mathbf{A} - \lambda\mathbf{I})^2\vec{\mathbf{u}}_2 = \vec{\mathbf{v}}$$

(We continue in this fashion to obtain $\vec{\mathbf{v}}, \vec{\mathbf{u}}_1, \vec{\mathbf{u}}_2, \ldots, \vec{\mathbf{u}}_r$, for $r < m$, where m is the multiplicity of λ and r is the number of "missing" eigenvectors for λ.)

(a) Show that

$$\vec{\mathbf{x}}_1 = e^{\lambda t}\vec{\mathbf{v}},$$
$$\vec{\mathbf{x}}_2 = (t\vec{\mathbf{v}} + \vec{\mathbf{u}}_1)e^{\lambda t},$$
$$\vec{\mathbf{x}}_3 = \left(\frac{1}{2}t^2\vec{\mathbf{v}} + t\vec{\mathbf{u}}_1 + \vec{\mathbf{u}}_2\right)e^{\lambda t}$$

are solutions of $\vec{\mathbf{x}}' = \mathbf{A}\vec{\mathbf{x}}$, given that $\mathbf{A}\vec{\mathbf{v}} = \lambda\vec{\mathbf{v}}$ and λ has multiplicity 3 and $r = 2$.

(b) Show that the vectors $\vec{\mathbf{x}}_1$, $\vec{\mathbf{x}}_2$, and $\vec{\mathbf{x}}_3$ are linearly independent.

(c) Solve $\vec{\mathbf{x}}' = \begin{bmatrix} 1 & 1 & 1 \\ 0 & 1 & 1 \\ 0 & 0 & 1 \end{bmatrix}\vec{\mathbf{x}}.$

39. One Independent Eigenvector Consider $\vec{\mathbf{x}}' = \mathbf{A}\vec{\mathbf{x}}$ for

$$\mathbf{A} = \begin{bmatrix} 0 & 0 & 1 \\ 1 & 0 & -3 \\ 0 & 1 & 3 \end{bmatrix}.$$

(a) Show that $\mathbf{A}$ has eigenvalue $\lambda = 1$ with multiplicity 3, and that all eigenvectors are scalar multiples of

$$\vec{\mathbf{v}} = \begin{bmatrix} 1 \\ -2 \\ 1 \end{bmatrix}.$$

(b) Use part (a) to find a solution of the system in the form

$$\vec{\mathbf{x}}_1 = e^t\vec{\mathbf{v}}.$$

(c) Find a second solution in the form

$$\vec{\mathbf{x}}_2 = te^t\vec{\mathbf{v}} + e^t\vec{\mathbf{u}},$$

where vector $\vec{\mathbf{u}}$ is to be determined. HINT: Find $\vec{\mathbf{u}}$ that satisfies $(\mathbf{A} - \mathbf{I})\vec{\mathbf{u}} = \vec{\mathbf{v}}$.

(d) Use the result of part (c) to find a third solution of the system of the form

$$\vec{\mathbf{x}}_3 = \frac{1}{2}t^2e^t\vec{\mathbf{v}} + te^t\vec{\mathbf{u}} + e^t\vec{\mathbf{w}}.$$

HINT: Find $\vec{\mathbf{w}}$ that satisfies $(\mathbf{A} - \mathbf{I})\vec{\mathbf{w}} = \vec{\mathbf{u}}$.

Solutions in Space *Find the general solutions for Problems 40 and 41.*

40. $\vec{\mathbf{x}}' = \begin{bmatrix} 3 & 2 & 2 \\ 1 & 4 & 1 \\ -2 & -4 & -1 \end{bmatrix}\vec{\mathbf{x}}$

41. $\vec{\mathbf{x}}' = \begin{bmatrix} -1 & 1 & 0 \\ 1 & 2 & 1 \\ 0 & 3 & -1 \end{bmatrix}\vec{\mathbf{x}}$

Spatial Particulars *Obtain solutions for the IVPs of Problems 42 and 43.*

42. $\vec{\mathbf{x}}' = \begin{bmatrix} 1 & -1 & 0 \\ 0 & -1 & 3 \\ -1 & 1 & 0 \end{bmatrix} \vec{\mathbf{x}}, \quad \vec{\mathbf{x}}(0) = \begin{bmatrix} 0 \\ 0 \\ 1 \end{bmatrix}$

43. $\vec{\mathbf{x}}' = \begin{bmatrix} 1 & 1 & 0 \\ 1 & 1 & 0 \\ 0 & 0 & -1 \end{bmatrix} \vec{\mathbf{x}}, \quad \vec{\mathbf{x}}(0) = \begin{bmatrix} 2 \\ 4 \\ 2 \end{bmatrix}$

44. Verification of Independence Show that the solutions obtained in Example 7 of this section are linearly independent.

45. Adjoint Systems The linear system

$$\vec{\mathbf{x}}' = \mathbf{A}\vec{\mathbf{x}} \tag{19}$$

has a "cousin" system

$$\vec{\mathbf{w}}' = -\mathbf{A}^{\mathrm{T}}\vec{\mathbf{w}}, \tag{20}$$

called its **adjoint**. (Taking the negative of the transpose twice returns the original matrix, so each system is the adjoint of the other.)

(a) Determine the system adjoint to $\vec{\mathbf{x}}' = \begin{bmatrix} 0 & 1 \\ 1 & 0 \end{bmatrix} \vec{\mathbf{x}}$.

(b) Establish that for the solutions $\vec{\mathbf{x}}$ and $\vec{\mathbf{w}}$ of adjoint systems (19) and (20) it is true that

$$\frac{d}{dt}\{\vec{\mathbf{w}}^{\mathrm{T}}\vec{\mathbf{x}}\} = \vec{\mathbf{w}}'^{\mathrm{T}}\vec{\mathbf{x}} + \vec{\mathbf{w}}^{\mathrm{T}}\vec{\mathbf{x}}' = 0,$$

so $\vec{\mathbf{w}}^{\mathrm{T}}\vec{\mathbf{x}} \equiv$ constant. HINT: $(\mathbf{AB})^{\mathrm{T}} = \mathbf{B}^{\mathrm{T}}\mathbf{A}^{\mathrm{T}}$.

(c) Solve the IVP consisting of the system of part (a) and

$$\vec{\mathbf{x}}(0) = \begin{bmatrix} 1 \\ 0 \end{bmatrix}.$$

(d) Solve the IVP consisting of the adjoint of the system of part (a) and the initial condition

$$\vec{\mathbf{w}}(0) = \begin{bmatrix} 0 \\ 1 \end{bmatrix}.$$

(e) For the initial conditions in parts (c) and (d), $\vec{\mathbf{w}}^{\mathrm{T}}(0)\vec{\mathbf{x}}(0) = 0$. What can you conclude about the paths $\vec{\mathbf{x}}(t)$ and $\vec{\mathbf{w}}(t)$ if they are plotted on the same set of axes?

46. Cauchy-Euler Systems The system $t\vec{\mathbf{x}}' = \mathbf{A}\vec{\mathbf{x}}$, where $\mathbf{A}$ is a constant matrix and $t > 0$, is called a **Cauchy-Euler system**.

(a) Show that the Cauchy-Euler system has a solution of the form $\vec{\mathbf{x}} = t^{\lambda}\vec{\mathbf{v}}$, where λ is an eigenvalue of $\mathbf{A}$ and $\vec{\mathbf{v}}$ is a corresponding eigenvector.

(b) Solve the Cauchy-Euler system

$$t\vec{\mathbf{x}}' = \begin{bmatrix} 3 & -2 \\ 2 & -2 \end{bmatrix} \vec{\mathbf{x}}, \quad t > 0.$$

Computer Lab: Predicting Phase Portraits *Check your intuition for each of the systems in Problems 47–50, following these steps.*

(a) *Sketch what you think the direction field looks like.*

(b) *Use an open-ended solver to draw the vector field.*

(c) *Solve the system analytically and compare with results in (a) and (b). Explain or reconcile any differences.*

47. $x' = x$, $y' = -y$

48. $x' = 0$, $y' = -y$

49. $x' = x + y$, $y' = x + y$

50. $x' = y$, $y' = x$

51. Radioactive Decay Chain The radioactive isotope of iodine, I-135, decays into the radioactive isotope Xe-135 of xenon; this in turn decays into another (stable) product. The half-lives of iodine and xenon are 6.7 hours and 9.2 hours, respectively.

(a) Write a system of differential equations describing the amounts of I-135 and Xe-135 present at any time.

(b) Obtain the general solution of the system found in part (a).

52. Multiple Compartment Mixing I Consider two large tanks, connected as shown in Fig. 6.2.15. Tank A is initially filled with 100 gal of water in which 25 lb of salt has been dissolved. Tank B is initially filled with 100 gal of pure water. Pure water is poured into tank A at the constant rate of 4 gal/min. The well-mixed solution from tank A is constantly being pumped to tank B at a rate of 6 gal/min, and the solution in tank B is constantly being pumped to tank A at the rate of 2 gal/min. The solution in tank B also exits the tank at the rate of 4 gal/min.

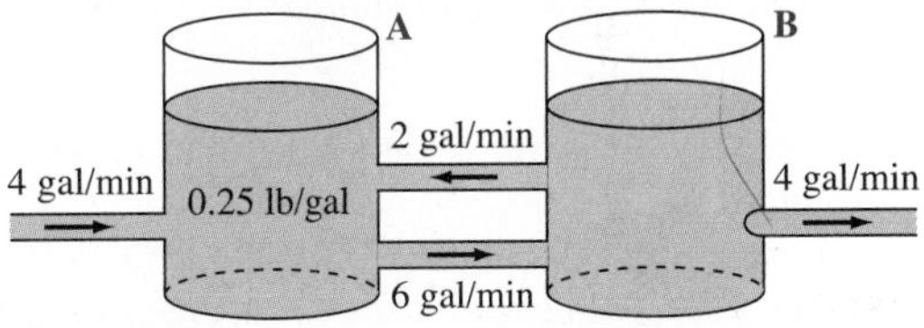

FIGURE 6.2.15 Two-tank arrangement for Problem 52.

(a) Find the amount of salt in each tank at any time.

(b) Draw graphs to show how the salt level in each tank changes with respect to time.

(c) Does the amount of salt in tank B ever exceed that in tank A?

(d) What is the long-term behavior in each tank?

53. Multiple Compartment Mixing II Repeat Problem 52, but change the initial volume in tank A to 150 gal.

54. Mixing and Homogeneity Why is the linear system that models the arrangement in Problem 52 homogeneous? Change the problem statement as simply as possible to keep the same matrix of coefficients **A** in the system for the new problem

$$\vec{\mathbf{x}}' = \mathbf{A}\vec{\mathbf{x}} + \vec{\mathbf{f}}(t), \quad \text{where} \quad \vec{\mathbf{f}}(t) = \begin{bmatrix} 2 \\ 0 \end{bmatrix}.$$

55. Aquatic Compartment Model A simple three-compartment model that describes nutrients in a food chain has been studied by M. R. Cullen.[1] (See Fig. 6.2.16.) For example, the constant $a_{31} = 0.04$ alongside the arrow connecting compartment 1 (phytoplankton) to compartment 3 (zooplankton) means that at any given time, nutrients pass from the phytoplankton compartment to the zooplankton compartment at the rate of $0.04x_1$ per hour. Find the linear system $\vec{\mathbf{x}}' = \mathbf{A}\vec{\mathbf{x}}$ that describes the amount of nutrients in each compartment.

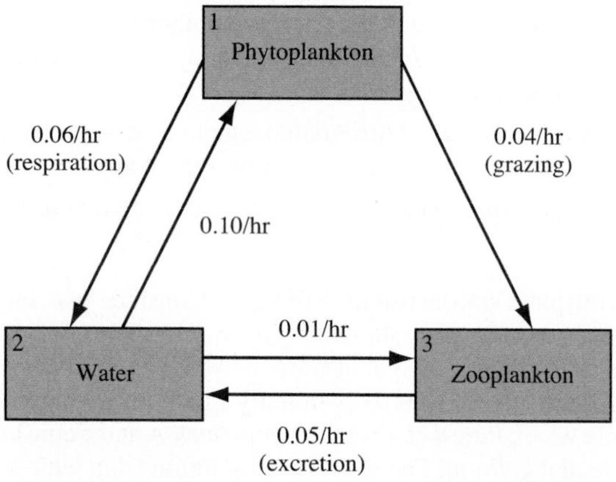

FIGURE 6.2.16 Aquatic compartment model for Problem 55.

Electrical Circuits *Use Kirchoff's Laws to determine a homogeneous linear* 2×2 *system that models the circuits in Problems 56 and 57. The input voltage* $V(t) = 0$ *for* $t \geq 0$.

56. Determine the general solutions for the currents I_1, I_2, and I_3 if $R_1 = R_2 = R_3 = 4$ ohms and $L_1 = L_2 = 2$ henries.

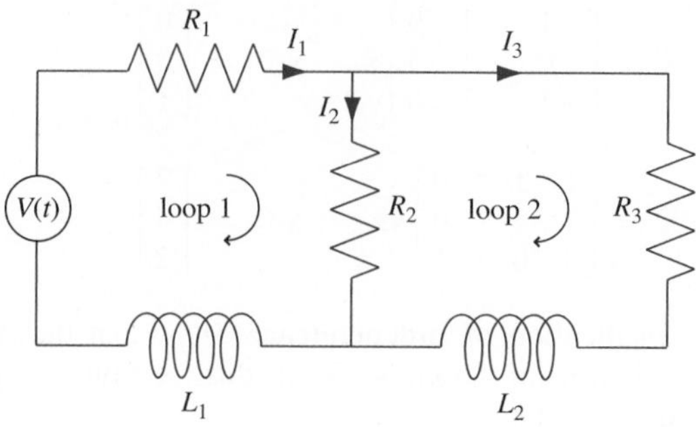

57. Find general solutions for the currents I_1, I_2, and I_3, if $R_1 = 4$ ohms, $R_3 = 6$ ohms, $L_1 = 1$ henry, $L_2 = 2$ henries.

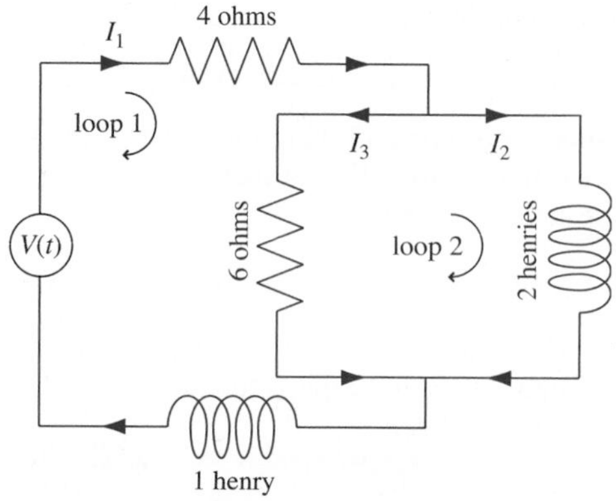

58. Suggested Journal Entry Suppose that **A** is a 3×3 matrix with three distinct eigenvalues. What kinds of long-term behavior (both as $t \to \infty$ and as $t \to -\infty$) are possible for solutions of the system $\vec{\mathbf{x}}' = \mathbf{A}\vec{\mathbf{x}}$, according to various possible combinations of signs of the eigenvalues? What kinds of three-dimensional geometry might be associated with these various cases?

6.3 Linear Systems with Nonreal Eigenvalues

SYNOPSIS: We construct explicit solutions for homogeneous linear systems with constant coefficients in cases for which the eigenvalues are nonreal. We examine their portraits in the phase plane for 2×2 *systems.*

Complex Building Blocks

In seeking solutions of the 2×2 linear system of differential equations

$$\vec{\mathbf{x}}' = \mathbf{A}\vec{\mathbf{x}} \tag{1}$$

in the previous section, we substituted $\vec{\mathbf{x}} = e^{\lambda t}\vec{\mathbf{v}}$ into the equation and found that we must have a scalar and nonzero vector $\vec{\mathbf{v}}$ such that

$$(\mathbf{A} - \lambda\mathbf{I})\vec{\mathbf{v}} = \vec{\mathbf{0}}, \tag{2}$$

[1] Adapted from M. R. Cullen, *Mathematics for the Biosciences* (PWS Publishers, 1983).

where eigenvalue λ is a solution of the characteristic equation

$$|\mathbf{A} - \lambda\mathbf{I}| = 0. \tag{3}$$

If the quadratic equation (3) has negative discriminant, the solutions are the complex conjugates

$$\lambda_1 = \alpha + i\beta \quad \text{and} \quad \lambda_2 = \alpha - i\beta,$$

where α and β are real and $\beta \neq 0$.

In Sec. 5.3 we discussed complex eigenvalues, but not the consequences on eigenvectors. The eigenvector $\vec{\mathbf{v}}_1$ belonging to λ_1, determined from equation (2), will in general have complex components.

Taking the complex conjugate of

Matrix Element Input

Try matrices with complex eigenvalues and observe the results in the phase plane.

$$(\mathbf{A} - \lambda_1\mathbf{I})\vec{\mathbf{v}}_1 = \vec{\mathbf{0}}$$

yields (by Appendix CN)

$$(\mathbf{A} - \overline{\lambda}_1\mathbf{I})\overline{\vec{\mathbf{v}}}_1 = (\mathbf{A} - \lambda_2\mathbf{I})\overline{\vec{\mathbf{v}}}_1 = \vec{\mathbf{0}},$$

so $\vec{\mathbf{v}}_2 = \overline{\vec{\mathbf{v}}_1}$.

We have shown the following.

Find One, Get One Free:

We need only find one eigenvector for one nonreal eigenvalue, because the other eigenvector is its complex conjugate, which corresponds to the other eigenvalue, also a complex conjugate.

Complex Eigenvalues and Eigenvectors

For a real matrix **A**, nonreal eigenvalues come in complex conjugate pairs,

$$\lambda_1, \lambda_2 = \alpha \pm i\beta,$$

with α, β real numbers and $\beta \neq 0$.

The corresponding eigenvectors are also complex conjugate pairs and can be written

$$\vec{\mathbf{v}}_1, \vec{\mathbf{v}}_2 = \vec{\mathbf{p}} \pm i\vec{\mathbf{q}},$$

where $\vec{\mathbf{p}}$ and $\vec{\mathbf{q}}$ are real vectors.

EXAMPLE 1 **Complex Eigenstuff for a Matrix** The characteristic equation of the matrix

$$\mathbf{A} = \begin{bmatrix} 6 & -1 \\ 5 & 4 \end{bmatrix} \tag{4}$$

is $(6 - \lambda)(4 - \lambda) + 5 = 0$, which simplifies to $\lambda^2 - 10\lambda + 29 = 0$. We obtain for **A** the two eigenvalues

$$\lambda_1, \lambda_2 = 5 \pm 2i$$

with eigenvectors

$$\vec{\mathbf{v}}_1 = \begin{bmatrix} 1 \\ 1 - 2i \end{bmatrix} \quad \text{and} \quad \vec{\mathbf{v}}_2 = \overline{\vec{\mathbf{v}}}_1 = \begin{bmatrix} 1 \\ 1 + 2i \end{bmatrix}.$$

Alternatively, we can write

$$\vec{\mathbf{v}}_1, \vec{\mathbf{v}}_2 = \underbrace{\begin{bmatrix} 1 \\ 1 \end{bmatrix}}_{\vec{\mathbf{p}}} \pm i \underbrace{\begin{bmatrix} 0 \\ -2 \end{bmatrix}}_{\vec{\mathbf{q}}}.$$

■

Solving the DE System

Let us return to solving system (1) in general for the case of nonreal eigenvalues. For $\lambda_1, \lambda_2 = \alpha \pm \beta i$ and the corresponding complex conjugate eigenvectors $\vec{\mathbf{v}}_1$ and $\vec{\mathbf{v}}_2$, we can write

$$\vec{\mathbf{x}} = c_1 e^{\lambda_1 t}\vec{\mathbf{v}}_1 + c_2 e^{\lambda_2 t}\vec{\mathbf{v}}_2.$$

Beware:

Recall that eigenvectors are unique only up to multiplication by a scalar. However a *nonreal* scalar can result in eigenvectors that *look* very different, for the same nonreal eigenvalue. Solutions to DEs by either form of the eigenvectors can be shown to be equivalent (but that may require a whole separate exercise).

However, to analyze the qualitative behavior of the trajectories when the eigenvalues and eigenvectors are nonreal, we write solutions in terms of the *real* vectors $\vec{\mathbf{p}}$ and $\vec{\mathbf{q}}$ (the real and imaginary parts of $\vec{\mathbf{v}}_1$, respectively).

For eigenvalue $\lambda_1 = \alpha + i\beta$ and corresponding eigenvector $\vec{\mathbf{v}}_1 = \vec{\mathbf{p}} + i\vec{\mathbf{q}}$, a solution will take the form

$$\vec{\mathbf{x}}(t) = e^{\lambda_1 t}\vec{\mathbf{v}}_1 = e^{(\alpha+i\beta)t}(\vec{\mathbf{p}} + i\vec{\mathbf{q}}). \tag{5}$$

As in Chapter 4, we will find that the *real* and *imaginary* parts of the complex solution (5) are *real* and *linearly independent* solutions of system (1).

Step 1. Suppose that

$$\vec{\mathbf{x}}(t) = \vec{\mathbf{x}}_{\mathrm{Re}}(t) + i\vec{\mathbf{x}}_{\mathrm{Im}}(t)$$

is a complex vector solution of (1), with $\vec{\mathbf{x}}_{\mathrm{Im}}(t) \neq \vec{\mathbf{0}}$. Then

$$\vec{\mathbf{x}}'(t) = \vec{\mathbf{x}}'_{\mathrm{Re}}(t) + i\vec{\mathbf{x}}'_{\mathrm{Im}}(t) = \mathbf{A}\vec{\mathbf{x}}_{\mathrm{Re}}(t) + i\mathbf{A}\vec{\mathbf{x}}_{\mathrm{Im}}(t) = \mathbf{A}\vec{\mathbf{x}}(t).$$

If we equate separately the real and imaginary parts of this equation, we find that

$$\vec{\mathbf{x}}'_{\mathrm{Re}}(t) = \mathbf{A}\vec{\mathbf{x}}_{\mathrm{Re}}(t) \quad \text{and} \quad \vec{\mathbf{x}}'_{\mathrm{Im}}(t) = \mathbf{A}\vec{\mathbf{x}}_{\mathrm{Im}}(t),$$

so $\vec{\mathbf{x}}_{\mathrm{Re}}(t)$ and $\vec{\mathbf{x}}_{\mathrm{Im}}(t)$ are *separate* and *real* solutions of (1).

Step 2. For the complex solution (5), we can determine the real and imaginary parts by using Euler's formula, $e^{i\theta} = \cos\theta + i\sin\theta$, to write

$$\begin{aligned} e^{\lambda_1 t}\vec{\mathbf{v}}_1 &= e^{\alpha t}(\cos\beta t + i\sin\beta t)(\vec{\mathbf{p}} + i\vec{\mathbf{q}}) \\ &= \underbrace{e^{\alpha t}(\cos\beta t\,\vec{\mathbf{p}} - \sin\beta t\,\vec{\mathbf{q}})}_{\vec{\mathbf{x}}_{\mathrm{Re}}(t)} + i\,\underbrace{e^{\alpha t}(\sin\beta t\,\vec{\mathbf{p}} + \cos\beta t\,\vec{\mathbf{q}})}_{\vec{\mathbf{x}}_{\mathrm{Im}}(t)}. \end{aligned}$$

Step 3. Since $\vec{\mathbf{x}}_{\mathrm{Re}}(t)$ and $\vec{\mathbf{x}}_{\mathrm{Im}}(t)$ are *linearly independent* solutions (Problem 37), and we need only two (by the Solution Space Theorem of Sec. 6.1), the general solution of (1) is given by

$$\vec{\mathbf{x}}(t) = c_1\vec{\mathbf{x}}_{\mathrm{Re}}(t) + c_2\vec{\mathbf{x}}_{\mathrm{Im}}(t)$$

for arbitrary constants c_1 and c_2. Any solutions derived from λ_2 and $\vec{\mathbf{v}}_2$ will be linear combinations of the solutions already determined from λ_1 and $\vec{\mathbf{v}}_1$.

At this point, we have developed a complete strategy for solving equation (1):

Solving a Two-Dimensional DE System $\vec{\mathbf{x}}' = \mathbf{A}\vec{\mathbf{x}}$ with Nonreal Eigenvalues $\lambda_1, \lambda_2 = \alpha \pm i\beta$

Step 1. For one eigenvalue λ_1, find its corresponding eigenvector $\vec{\mathbf{v}}_1$. The second eigenvalue λ_2 and its eigenvector $\vec{\mathbf{v}}_2$ are complex conjugates of the first. The eigenvectors are of the form $\vec{\mathbf{v}}_1, \vec{\mathbf{v}}_2 = \vec{\mathbf{p}} \pm i\vec{\mathbf{q}}$.

Step 2. Construct the linearly independent real ($\vec{\mathbf{x}}_{\text{Re}}$) and imaginary ($\vec{\mathbf{x}}_{\text{Im}}$) parts of the solutions as follows:

$$\begin{aligned} \vec{\mathbf{x}}_{\text{Re}} &= e^{\alpha t}(\cos\beta t\,\vec{\mathbf{p}} - \sin\beta t\,\vec{\mathbf{q}}), \\ \vec{\mathbf{x}}_{\text{Im}} &= e^{\alpha t}(\sin\beta t\,\vec{\mathbf{p}} + \cos\beta t\,\vec{\mathbf{q}}). \end{aligned} \tag{6}$$

Step 3. The general solution is

$$\vec{\mathbf{x}}(t) = c_1\vec{\mathbf{x}}_{\text{Re}}(t) + c_2\vec{\mathbf{x}}_{\text{Im}}(t). \tag{7}$$

Formulas (6) and (7) are more complicated than those for solutions of systems with real eigenvectors, because the linearly independent solutions $\vec{\mathbf{x}}_{\text{Re}}$ and $\vec{\mathbf{x}}_{\text{Im}}$ involve not one but *two* vectors each. Fortunately, they are the same two vectors in each case, so the general solution still involves only two vectors, $\vec{\mathbf{p}}$ and $\vec{\mathbf{q}}$.

We shall first give examples of the typical trajectory behaviors for systems with nonreal eigenvalues; a subsection on *interpreting* these trajectories and their formulas will follow.

EXAMPLE 2 **Complex Eigenstuff for a DE System** To solve the system

$$\vec{\mathbf{x}}' = \mathbf{A}\vec{\mathbf{x}} = \begin{bmatrix} 6 & -1 \\ 5 & 4 \end{bmatrix}\vec{\mathbf{x}}, \tag{8}$$

we recall from Example 1 that the eigenvalues of $\mathbf{A}$ are $\lambda_1, \lambda_2 = 5 \pm 2i$, and an eigenvector belonging to $\lambda_1 = 5 + 2i$ is

$$\vec{\mathbf{v}}_1 = \begin{bmatrix} 1 \\ 1 \end{bmatrix} + i\begin{bmatrix} 0 \\ -2 \end{bmatrix}.$$

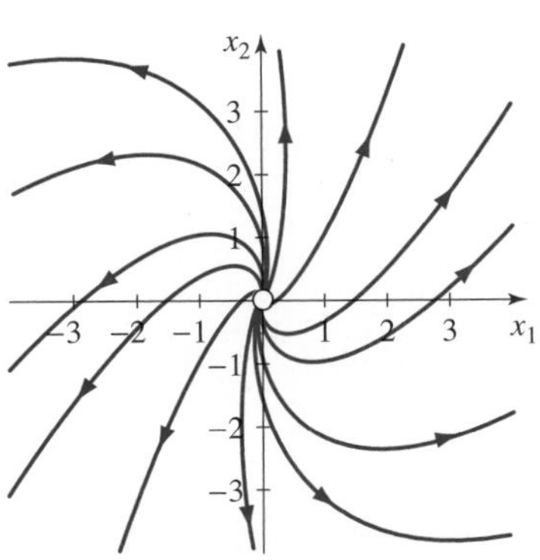

FIGURE 6.3.1 Phase-plane trajectories for Example 2, with eigenvalues $\lambda_1, \lambda_2 = 5 \pm 2i$ and an *unstable* equilibrium at the origin.

Hence, by (6), a fundamental set of solutions for system (8) is given by

$$\vec{\mathbf{x}}_{\text{Re}}(t) = e^{5t}\cos 2t\begin{bmatrix} 1 \\ 1 \end{bmatrix} - e^{5t}\sin 2t\begin{bmatrix} 0 \\ -2 \end{bmatrix} = e^{5t}\begin{bmatrix} \cos 2t \\ \cos 2t + 2\sin 2t \end{bmatrix},$$

$$\vec{\mathbf{x}}_{\text{Im}}(t) = e^{5t}\sin 2t\begin{bmatrix} 1 \\ 1 \end{bmatrix} + e^{5t}\cos 2t\begin{bmatrix} 0 \\ -2 \end{bmatrix} = e^{5t}\begin{bmatrix} \sin 2t \\ \sin 2t - 2\cos 2t \end{bmatrix}.$$

The general solution of system (8) is therefore

$$\vec{\mathbf{x}}(t) = e^{5t}\left(c_1\begin{bmatrix} \cos 2t \\ \cos 2t + 2\sin 2t \end{bmatrix} + c_2\begin{bmatrix} \sin 2t \\ \sin 2t - 2\cos 2t \end{bmatrix}\right), \tag{9}$$

where c_1 and c_2 are arbitrary real constants.

Because $\alpha > 0$, all trajectories spiral outward to infinity as $t \to \infty$. Some typical trajectories are shown in Fig. 6.3.1. (Backward trajectories, as $t \to -\infty$, spiral inward toward the origin.) ■

EXAMPLE 3 **Spirals Reverse** To solve the system

$$\vec{\mathbf{x}}' = \mathbf{A}\vec{\mathbf{x}} = \begin{bmatrix} 0 & 1 \\ -5 & -2 \end{bmatrix} \vec{\mathbf{x}}, \tag{10}$$

we solve the characteristic equation $|\mathbf{A} - \lambda\mathbf{I}| = 0$ and obtain the eigenvalues

$$\lambda_1 = -1 + 2i \quad \text{and} \quad \lambda_2 = -1 - 2i.$$

We therefore have $\alpha = -1$ and $\beta = 2$. To determine corresponding eigenvectors, we now write system (2) for λ_1,

$$\begin{bmatrix} 1 - 2i & 1 \\ -5 & -2 + 1 - 2i \end{bmatrix} \vec{\mathbf{v}} = \mathbf{0},$$

and find

$$\vec{\mathbf{v}}_1 = \begin{bmatrix} 1 \\ -1 + 2i \end{bmatrix} = \begin{bmatrix} 1 \\ -1 \end{bmatrix} + i \begin{bmatrix} 0 \\ 2 \end{bmatrix} = \vec{\mathbf{p}} + i\vec{\mathbf{q}}.$$

From equation (6), then, we obtain

$$\vec{\mathbf{x}}_{\text{Re}}(t) = e^{-t} \cos 2t \begin{bmatrix} 1 \\ -1 \end{bmatrix} - e^{-t} \sin 2t \begin{bmatrix} 0 \\ 2 \end{bmatrix} = \begin{bmatrix} e^{-t} \cos 2t \\ e^{-t}(-\cos 2t - 2 \sin 2t) \end{bmatrix},$$

$$\vec{\mathbf{x}}_{\text{Im}}(t) = e^{-t} \sin 2t \begin{bmatrix} 1 \\ -1 \end{bmatrix} + e^{-t} \cos 2t \begin{bmatrix} 0 \\ 2 \end{bmatrix} = \begin{bmatrix} e^{-t} \sin 2t \\ e^{-t}(-\sin 2t + 2 \cos 2t) \end{bmatrix}.$$

The general solution of system (10) is given, for arbitrary constants c_1 and c_2, by

$$\vec{\mathbf{x}}(t) = e^{-t} \left(c_1 \begin{bmatrix} \cos 2t \\ -\cos 2t - 2 \sin 2t \end{bmatrix} + c_2 \begin{bmatrix} \sin 2t \\ -\sin 2t + 2 \cos 2t \end{bmatrix} \right). \tag{11}$$

Because $\alpha < 0$, phase-plane trajectories (11) spiral toward the origin as $t \to \infty$ (and outward as $t \to -\infty$). Some typical trajectories are shown in Fig. 6.3.2.

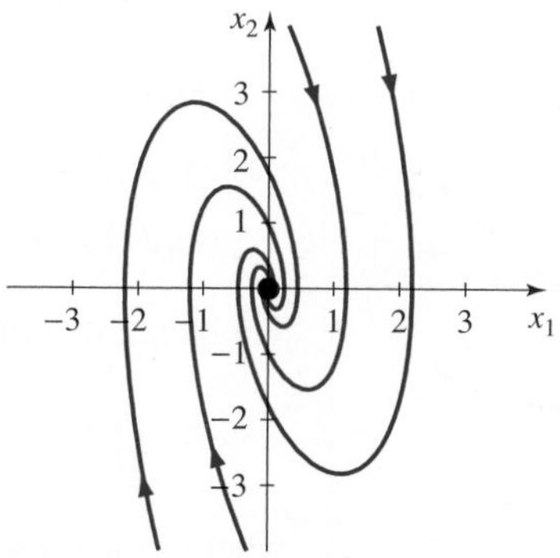

FIGURE 6.3.2 Phase-plane trajectories for Example 3, with eigenvalues $\lambda_1, \lambda_2 = -1 \pm 2i$ and a *stable* equilibrium at the origin.

■

EXAMPLE 4 **Purely Imaginary** For the system

$$\vec{\mathbf{x}}' = \mathbf{A}\vec{\mathbf{x}} = \begin{bmatrix} 4 & -5 \\ 5 & -4 \end{bmatrix}\vec{\mathbf{x}}, \tag{12}$$

the eigenvalues are purely imaginary (that is, $\alpha = 0$), because the characteristic equation of $\mathbf{A}$ is

$$|\mathbf{A} - \lambda\mathbf{I}| = \begin{bmatrix} 4-\lambda & -5 \\ 5 & -4-\lambda \end{bmatrix} = 0,$$

which simplifies to $\lambda^2 + 9 = 0$. Therefore, we have

$$\lambda_1 = 3i = 0 + 3i \quad \text{and} \quad \lambda_2 = \overline{\lambda}_1 = -3i = 0 - 3i$$

with eigenvectors

$$\vec{\mathbf{v}}_1, \vec{\mathbf{v}}_2 = \begin{bmatrix} 5 \\ 4 \mp 3i \end{bmatrix} = \begin{bmatrix} 5 \\ 4 \end{bmatrix} \pm i \begin{bmatrix} 0 \\ -3 \end{bmatrix} = \vec{\mathbf{p}} \pm i\vec{\mathbf{q}}.$$

By equation (6), we find

$$\vec{\mathbf{x}}_{\text{Re}} = \cos 3t \begin{bmatrix} 5 \\ 4 \end{bmatrix} - \sin 3t \begin{bmatrix} 0 \\ -3 \end{bmatrix} \quad \text{and} \quad \vec{\mathbf{x}}_{\text{Im}} = \sin 3t \begin{bmatrix} 5 \\ 4 \end{bmatrix} + \cos 3t \begin{bmatrix} 0 \\ -3 \end{bmatrix}.$$

The general solution of system (12) is, for arbitrary c_1 and c_2,

$$\vec{\mathbf{x}}(t) = c_1 \begin{bmatrix} 5\cos 3t \\ 4\cos 3t + 3\sin 3t \end{bmatrix} + c_2 \begin{bmatrix} 5\sin 3t \\ 4\sin 3t - 3\cos 3t \end{bmatrix}.$$

This time there is no exponential growth or decay factor, because $\alpha = 0$. Trajectories are closed curves that enclose the origin. Some typical curves are plotted in Fig. 6.3.3. Such solutions are **periodic**, repeating their motions after returning to the initial point from tracing the closed orbit. We can learn the direction of the arrows for the trajectories by plotting x_1' and x_2' at points of interest.

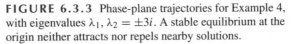

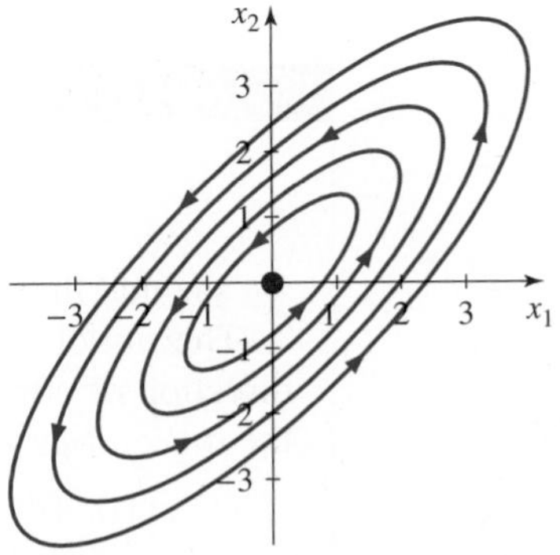

FIGURE 6.3.3 Phase-plane trajectories for Example 4, with eigenvalues $\lambda_1, \lambda_2 = \pm 3i$. A stable equilibrium at the origin neither attracts nor repels nearby solutions. ■

Behavior of Solutions

The systems of Examples 2–4 exhibit different phase portraits because they have different types of nonreal eigenvalues.

- Example 2 (Fig. 6.3.1) is an *unstable* equilibrium. Trajectories spiral outward from the origin, growing without bound, because $\alpha > 0$.
- Example 3 (Fig. 6.3.2) is an **asymptotically stable** equilibrium. Trajectories spiral toward the origin, decaying to zero, because $\alpha < 0$. Technically they never reach zero (because of uniqueness the origin is a separate, fixed-point solution), but they get ever closer.
- Example 4 (Fig. 6.3.3) is a *stable* equilibrium. Trajectories are closed loops and represent periodic motion. In contrast to Example 3, the equilibrium at the origin does not attract nearby solutions, but it does not repel them either, so we call it stable but not asymptotically stable. This happens whenever $\alpha = 0$.

Sketching Phase Portraits

If eigenvalues are nonreal, we know to expect spiral phase portraits. But because the eigenvectors are also nonreal, they do not appear in the phase plane. However, quick sketches can easily be made using *nullclines* (introduced in Sec. 2.6), which we summarize as follows.

Nullclines for a DE System

For a two-dimensional DE system

$$x' = f(x, y),$$
$$y' = g(x, y),$$

- the ***v*-nullcline** is the set of all points with *vertical* slope, which occur on the curve obtained by solving $x' = f(x, y) = 0$;
- the ***h*-nullcline** is the set of all points with *horizontal* slope, which occur on the curve obtained by solving $y' = g(x, y) = 0$.

When an h- and a v-nullcline intersect, an *equilibrium* or *fixed point* occurs.

Adding a few direction vectors to the nullcline sketch at key points is enough to show the general trajectory behavior for a linear DE system.

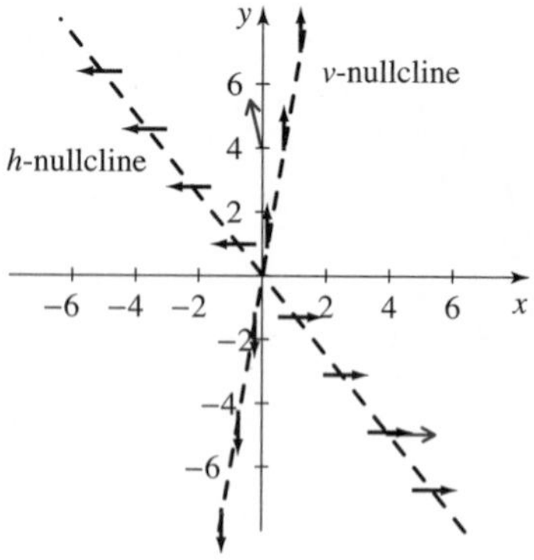

FIGURE 6.3.4 Nullclines and sample vectors (in color) for Example 5.

EXAMPLE 5 **Quick Sketch** Returning again to Example 2 with

$$\vec{\mathbf{x}}' = \begin{bmatrix} 6 & -1 \\ 5 & 4 \end{bmatrix} \vec{\mathbf{x}},$$

we see that the v-nullcline is $y = 6x$ and the h-nullcline is $y = -5x/4$.

The nullclines give us an indication of the flow of the trajectories. The direction of circulation can be determined by checking the tangent vectors to the trajectories at convenient points. For example, as shown in Fig. 6.3.4,

$$\text{at } (0, 4), \quad dx/dt = 6(0) - 1(4) = -4 \quad (\textit{to the left});$$
$$dy/dt = 5(0) + 4(4) = 16 \quad (\textit{upward}).$$

$$\text{at } (4, -5), \quad dx/dt = 6(4) - 1(-5) = 29 \quad (\textit{to the right});$$
$$dy/dt = 5(4) + 4(-5) = 0 \quad (\textit{no vertical motion}).$$

■

Interpreting the Solutions

If we take some liberties and rewrite the solution formulas (6) for $\vec{\mathbf{x}}_{\text{Re}}$ and $\vec{\mathbf{x}}_{\text{Im}}$ in a form that is easy to interpret and remember, we have the following.[1]

Real Solutions from Nonreal Eigenvalues

For $\vec{\mathbf{x}}' = \mathbf{A}\vec{\mathbf{x}}$ with nonreal eigenvalues $\lambda_1, \lambda_2 = \alpha \pm \beta i$ and complex eigenvectors $\vec{\mathbf{v}}_1, \vec{\mathbf{v}}_2 = \vec{\mathbf{p}} \pm \vec{\mathbf{q}}i$, arrange the components of the solution as

$$\begin{bmatrix} \vec{\mathbf{x}}_{\text{Re}} \\ \vec{\mathbf{x}}_{\text{Im}} \end{bmatrix} = \underbrace{e^{\alpha t}}_{\text{expansion}} \underbrace{\begin{bmatrix} \cos\beta t & -\sin\beta t \\ \sin\beta t & \cos\beta t \end{bmatrix}}_{\text{rotation}} \underbrace{\begin{bmatrix} \vec{\mathbf{p}} \\ \vec{\mathbf{q}} \end{bmatrix}}_{\text{tilt and shape}}. \qquad (13)$$

Each factor of equation (13) has a particular meaning.

(a) The first factor, $e^{\alpha t}$, determines *expansion* or *contraction*.

- If $\alpha > 0$, trajectories spiral outward from the origin, representing solutions that *grow without bound*.
- If $\alpha < 0$, trajectories spiral inward toward the origin, representing solutions that *decay to zero*.
- If $\alpha = 0$, trajectories are closed loops, representing *periodic solutions*.

(b) The second factor is the familiar *rotation matrix* (Sec. 5.1, Example 6). The angle of rotation, βt, is ever-increasing as t increases, so trajectories *spiral* around the origin, counterclockwise for $\beta > 0$.

(c) The third factor, containing $\vec{\mathbf{p}}$ and $\vec{\mathbf{q}}$, determines *tilt* and *shape* of the *elliptical trajectories* that would result if $\alpha = 0$.

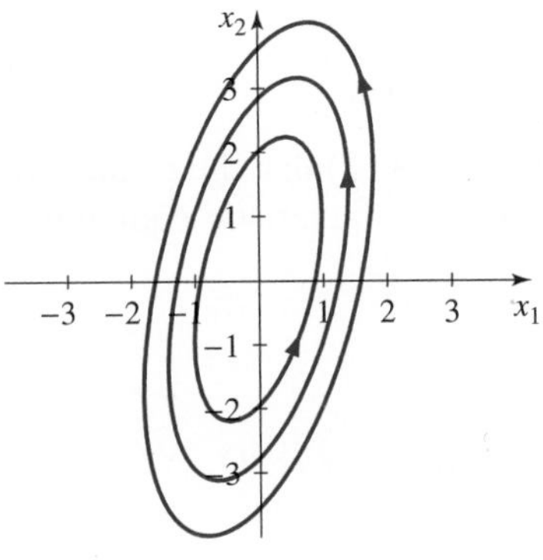

(a) Setting α to 0 to show only rotation and tilt/shape.

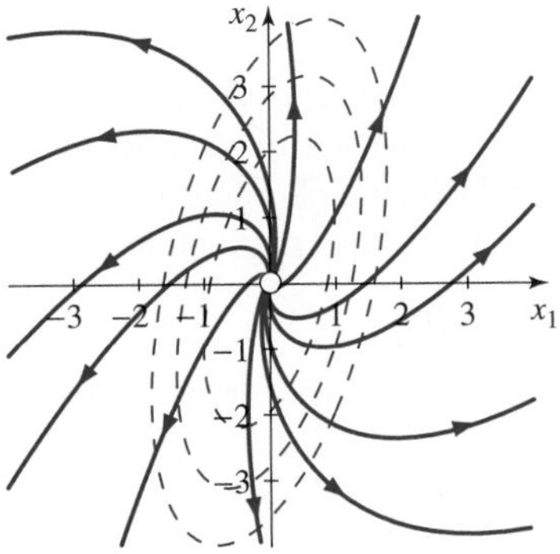

(b) Complete solution, including expansion factor $e^{\alpha t}$.

FIGURE 6.3.5 Effects of the factors in (13) applied to the general solution (9) of Example 2.

Thus, the *eigenvalues* $(\alpha \pm i\beta)$ control expansion and rotation while the *eigenvectors* $(\vec{\mathbf{p}} \pm i\vec{\mathbf{q}})$ determine shape and tilt of the spiraling trajectories. Problems 19–29 explore some details of these relationships.

EXAMPLE 6 **Interpretation** In Example 2, with general solution (9), we found $\alpha = 5$ and expanding trajectories with counterclockwise rotation and some asymmetry, due to the fact that $\vec{\mathbf{p}}$ and $\vec{\mathbf{q}}$ are not perpendicular.

If we plot the parametric equations (9) without the expansion factor, e^{5t}, each trajectory becomes an ellipse, tilted and stretched, as in Fig. 6.3.5(a).

Then in Fig. 6.3.5(b) we plot several trajectories of the complete solution equations (9) to show how the rotational and elliptical factors affect the expanding trajectories of Example 2. ■

---

[1]In equation (13), the elements of the first and last matrices are vectors rather than real numbers, but the familiar matrix multiplication rules can be applied on the right, treating $\vec{\mathbf{p}}$ and $\vec{\mathbf{q}}$ simply as elements to yield exactly the equations in (6).

Second-Order Equations versus Two-by-Two Systems

In Sec. 4.3 we solved the second-order differential equation

$$ay'' + by' + cy = 0 \tag{14}$$

in the case where the characteristic roots were the nonreal solutions $\alpha \pm i\beta$ of the characteristic equation $ar^2 + br + c = 0$. We found that $e^{\alpha t}\cos\beta t$ and $e^{\alpha t}\sin\beta t$ formed a fundamental set of solutions for this DE.

We then found, in Sec. 4.4, that an equation like (14) can be converted into a 2×2 system of first-order equations by setting $x_1 = y$ and $x_2 = y'$. In this case,

Romeo and Juliet

Try a lighthearted example of a 2×2 oscillating system.

$$\begin{aligned} x_1' &= x_2, \\ x_2' &= -\frac{c}{a}x_1 - \frac{b}{a}x_2. \end{aligned} \tag{15}$$

System (15) has matrix-vector form

$$\vec{\mathbf{x}}' = \mathbf{A}\vec{\mathbf{x}}, \quad \text{where} \quad \mathbf{A} = \begin{bmatrix} 0 & 1 \\ -c/a & -b/a \end{bmatrix}. \tag{16}$$

We have already mentioned (Sec. 5.3, just before Example 8) that the connection between (14) and (16) caught Euler's attention. The characteristic equation of matrix $\mathbf{A}$ is $a\lambda^2 + b\lambda + c = 0$. Solving (16) for

$$\vec{\mathbf{x}}(t) = \begin{bmatrix} x_1(t) \\ x_2(t) \end{bmatrix}$$

gives the solution $y(t) = x_1(t)$ of (14), and $y'(t)$ is given by $x_2(t)$. We confirm this fact in the following example.

EXAMPLE 7 **Making the Connection** Let's consider the DE

$$y'' + 2y' + 5y = 0. \tag{17}$$

We learned in Sec. 4.3 that this equation represents an *underdamped harmonic oscillator*. The discriminant, $\Delta = 2^2 - 4(5) = -16$, of its characteristic equation $r^2 + 2r + 5 = 0$, tells the story. We found that the characteristic roots are $r_1, r_2 = -1 \pm 2i$, and the general second-order solution is given by

$$y(t) = e^{-t}(c_1 \cos 2t + c_2 \sin 2t). \tag{18}$$

For the system approach to equation (17), the conversion procedure (15) gives

$$\vec{\mathbf{x}}' = \begin{bmatrix} 0 & 1 \\ -5 & -2 \end{bmatrix}\vec{\mathbf{x}},$$

which is the system (11) of Example 3, with general system solution (12)

$$\begin{bmatrix} x_1(t) \\ x_2(t) \end{bmatrix} = e^{-t}\left(c_1\begin{bmatrix} \cos 2t \\ -\cos 2t - 2\sin 2t \end{bmatrix} + c_2\begin{bmatrix} \sin 2t \\ -\sin 2t + 2\cos 2t \end{bmatrix}\right).$$

The first component $x_1(t)$ is in agreement with equation (18) for the solution of a second-order DE, and we can compute from (18) that

$$\begin{aligned} y' &= -e^{-t}(c_1\cos 2t + c_2 \sin 2t) + e^{-t}(-2c_1 \sin 2t + 2c_2 \cos 2t) \\ &= e^{-t}[(-c_1 + 2c_2)\cos 2t + (-2c_1 - c_2)\sin 2t] \end{aligned}$$

agrees with the second component $x_2(t)$. *The connection is complete!* ■

Summary

For the 2×2 homogeneous linear system with constant coefficients, nonreal eigenvalues yield real-valued solutions. The resulting families of trajectories are characterized by the real part α of the complex eigenvalues, spiraling inward if α is negative, spiraling outward if α is positive, and forming simple closed curves about the origin if α is zero, corresponding to periodic solutions. These results complete our generalization of the solutions of second-order linear equations with constant coefficients from Chapter 4.

6.3 Problems

Solutions in General *Find the general solutions for Problems 1–12, and handsketch the phase portraits from nullclines and/or vector fields. (You can check your results by using a CAS, a solver, or IDE Matrix Element Input.)*

Matrix Element Input
Enter the matrix to see the direction field; point and click to add trajectories.

1. $\vec{\mathbf{x}}' = \begin{bmatrix} 0 & 1 \\ -1 & 0 \end{bmatrix} \vec{\mathbf{x}}$

2. $\vec{\mathbf{x}}' = \begin{bmatrix} -1 & 2 \\ -1 & -3 \end{bmatrix} \vec{\mathbf{x}}$

3. $\vec{\mathbf{x}}' = \begin{bmatrix} 1 & 2 \\ -2 & 1 \end{bmatrix} \vec{\mathbf{x}}$

4. $\vec{\mathbf{x}}' = \begin{bmatrix} 6 & -1 \\ 5 & 2 \end{bmatrix} \vec{\mathbf{x}}$

5. $\vec{\mathbf{x}}' = \begin{bmatrix} 1 & 1 \\ -2 & -1 \end{bmatrix} \vec{\mathbf{x}}$

6. $\vec{\mathbf{x}}' = \begin{bmatrix} 2 & -4 \\ 2 & -2 \end{bmatrix} \vec{\mathbf{x}}$

7. $\vec{\mathbf{x}}' = \begin{bmatrix} 3 & -2 \\ 4 & -1 \end{bmatrix} \vec{\mathbf{x}}$

8. $\vec{\mathbf{x}}' = \begin{bmatrix} 2 & -5 \\ 1 & -2 \end{bmatrix} \vec{\mathbf{x}}$

9. $\vec{\mathbf{x}}' = \begin{bmatrix} 1 & -1 \\ 5 & -3 \end{bmatrix} \vec{\mathbf{x}}$

10. $\vec{\mathbf{x}}' = \begin{bmatrix} -2 & -3 \\ 3 & -2 \end{bmatrix} \vec{\mathbf{x}}$

11. $\vec{\mathbf{x}}' = \begin{bmatrix} -3 & -1 \\ 2 & -1 \end{bmatrix} \vec{\mathbf{x}}$

12. $\vec{\mathbf{x}}' = \begin{bmatrix} 2 & 4 \\ -2 & -2 \end{bmatrix} \vec{\mathbf{x}}$

Solutions in Particular *Solve the IVPs in Problems 13–16.*

13. $\vec{\mathbf{x}}' = \begin{bmatrix} 1 & -1 \\ 1 & 1 \end{bmatrix} \vec{\mathbf{x}}, \quad \vec{\mathbf{x}}(0) = \begin{bmatrix} -1 \\ 1 \end{bmatrix}$

14. $\vec{\mathbf{x}}' = \begin{bmatrix} 0 & -4 \\ 1 & 0 \end{bmatrix} \vec{\mathbf{x}}, \quad \vec{\mathbf{x}}(0) = \begin{bmatrix} 1 \\ 1 \end{bmatrix}$

15. $\vec{\mathbf{x}}' = \begin{bmatrix} -3 & 2 \\ -1 & -1 \end{bmatrix} \vec{\mathbf{x}}, \quad \vec{\mathbf{x}}(0) = \begin{bmatrix} 1 \\ 1 \end{bmatrix}$

16. $\vec{\mathbf{x}}' = \begin{bmatrix} 1 & -5 \\ 1 & -3 \end{bmatrix} \vec{\mathbf{x}}, \quad \vec{\mathbf{x}}(0) = \begin{bmatrix} 5 \\ 4 \end{bmatrix}$

17. Nonreal Conditions Suppose that $\mathbf{A}$ is a 2×2 matrix with nonreal eigenvalues $\lambda = \alpha \pm \beta i$.

(a) Show that one of the nondiagonal elements, but not both, must be negative in order for the eigenvalues to be nonreal.

(b) Show that eigenvalues are imaginary ($\alpha = 0$) if and only if $\operatorname{Tr} \mathbf{A} = 0$ and $|\mathbf{A}| > 0$.

18. Rotation Direction Show that for $\vec{\mathbf{x}}' = \mathbf{A}\vec{\mathbf{x}}$, with nonreal eigenvalues of

$$\mathbf{A} = \begin{bmatrix} a & b \\ c & d \end{bmatrix},$$

rotation along trajectories is determined as follows.

- if b is negative, rotation is counterclockwise;
- if c is negative, rotation is clockwise.

HINT: See Problem 17(a).

19. Complexities of Complex Eigenvectors Elliptical phase-plane trajectories occur in a 2×2 DE system with purely imaginary eigenvalues, which happens when $\operatorname{Tr} \mathbf{A} = 0$ and $|\mathbf{A}| > 0$. Thus, for $\lambda = \pm\beta i$,

$$\mathbf{A} = \begin{bmatrix} a & b \\ c & -a \end{bmatrix} \quad \text{and} \quad \beta^2 = |\mathbf{A}|.$$

(a) Show that

$$\vec{\mathbf{v}} = \begin{bmatrix} -b \\ a - \beta i \end{bmatrix} = \underbrace{\begin{bmatrix} -b \\ a \end{bmatrix}}_{\vec{\mathbf{p}}} + \underbrace{\begin{bmatrix} 0 \\ -\beta \end{bmatrix}}_{\vec{\mathbf{q}}} i$$

is an eigenvector for $\lambda = \beta i$.

(b) *Recall that any scalar multiple of $\vec{\mathbf{v}}$ is also an eigenvector. However, if the scalar multiple is also complex, the result is far from obviously "the same."* Multiplying the vector $\vec{\mathbf{v}}$ in (a) by the complex scalar $\frac{1}{b}(a + \beta i)$ gives a new vector

$$\vec{\mathbf{v}}^* = \frac{1}{b}(a + \beta i)\vec{\mathbf{v}}_1 = \begin{bmatrix} -a - \beta i \\ -c \end{bmatrix}.$$

Show that $\vec{\mathbf{v}}_1^*$ is also an eigenvector for $\lambda_1 = \beta i$.

(c) Explain how writing

$$\vec{\mathbf{v}}^* = \underbrace{\begin{bmatrix} -a \\ -c \end{bmatrix}}_{\vec{\mathbf{p}}^*} + \underbrace{\begin{bmatrix} -\beta \\ 0 \end{bmatrix}}_{\vec{\mathbf{q}}^*} i$$

gives real and imaginary parts, $\vec{\mathbf{p}}^*$ and $\vec{\mathbf{q}}^*$, for the complex eigenvector that are completely different from the real and imaginary parts, $\vec{\mathbf{p}}$ and $\vec{\mathbf{q}}$, for $\vec{\mathbf{v}}$.

20. Elliptical Shape and Tilt[2] For a 2×2 linear DE system

$$\vec{\mathbf{x}}' = \begin{bmatrix} a & b \\ c & -a \end{bmatrix} \vec{\mathbf{x}}$$

with purely imaginary eigenvalues $\lambda_1, \lambda_2 = \pm \beta i$ and complex eigenvectors of the particular form given in Problem 19(a),

$$\vec{\mathbf{v}}_1, \vec{\mathbf{v}}_2 = \underbrace{\begin{bmatrix} -b \\ a \end{bmatrix}}_{\vec{\mathbf{p}}} \pm \underbrace{\begin{bmatrix} 0 \\ -\beta \end{bmatrix}}_{\vec{\mathbf{q}}} i,$$

we can make a quick handsketch of the elliptical trajectories as follows.

(a) All the elliptical trajectories are concentric and similar, so we can get all the key information from just one. We have from equation (7) that

$$\vec{\mathbf{x}}(t) = c_1 \vec{\mathbf{x}}_{\text{Re}} + c_2 \vec{\mathbf{x}}_{\text{Im}}.$$

By choosing the particular solution where $c_1 = 1$, $c_2 = 0$, we reduce the solution to the single equation (6),

$$\vec{\mathbf{x}}_{\text{Re}}(t) = \cos \beta t \, \vec{\mathbf{p}} - \sin \beta t \, \vec{\mathbf{q}}.$$

Calculate $\vec{\mathbf{x}}(t) = \vec{\mathbf{x}}_{\text{Re}}(t)$ and $\vec{\mathbf{x}}'(t)$, then do parts (b), (c).

(b) Show that for $t = 0$,

- $\vec{\mathbf{x}}(0) = \vec{\mathbf{p}}$ gives an initial condition, with
- $\vec{\mathbf{x}}'(0) = -\beta \vec{\mathbf{q}}$ the initial velocity as a *vertical* tangent, pointing *upward.*

Show also that for $\beta t = \pi$,

- $\vec{\mathbf{x}}(\pi/\beta) = -\vec{\mathbf{p}}$ gives another point on the same elliptical trajectory, with
- $\vec{\mathbf{x}}'(\pi/\beta) = \beta \vec{\mathbf{q}}$ as velocity, pointing *downward.*

Then show that for $\beta t = \pi/2$,

- $\vec{\mathbf{x}}(\pi/2\beta) = -\vec{\mathbf{q}}$ is on the same ellipse, with
- $\vec{\mathbf{x}}'(\pi/2\beta) = -\beta \vec{\mathbf{p}}$ as velocity, anti-parallel to $\vec{\mathbf{p}}$,

and that for $\beta t = 3\pi/2$,

- $\vec{\mathbf{x}}(3\pi/2\beta) = \vec{\mathbf{q}}$ is also on the same ellipse, with
- $\vec{\mathbf{x}}'(\beta t = \pi) = \beta \vec{\mathbf{p}}$ as the velocity, parallel to $\vec{\mathbf{p}}$.

Hence, the four vectors $\pm\vec{\mathbf{p}}$ and $\pm\vec{\mathbf{q}}$, with their tangent vectors, define a parallelogram into which the ellipse must fit. An example is shown in Figure 6.3.6; other examples are given in Problems 21–24. (See the Caution note for Problems 26–29 for a discussion of the parameter βt.)

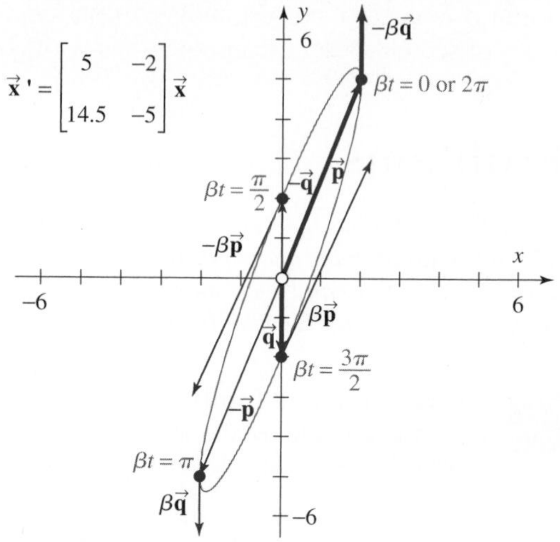

FIGURE 6.3.6 For a 2×2 linear DE $\vec{\mathbf{x}}' = \mathbf{A}\vec{\mathbf{x}}$ with nonreal eigenvalues, the real and imaginary parts of the eigenvectors $\pm\vec{\mathbf{p}}$, $\pm\vec{\mathbf{q}}$, together with the appropriate velocity vectors for their endpoints, determine the shape and tilt of an elliptical trajectory. (See Problem 20.) For this example, with $\beta = 2$, we have drawn the velocity vectors at half scale.

"Boxing" the Ellipse *For Problems 21–24*

(a) *Find and plot the real and imaginary parts of the eigenvectors,* $\pm\vec{\mathbf{p}}$, $\pm\vec{\mathbf{q}}$, *as found in Problem 19(a).*

(b) *Add the appropriate velocity vectors at the end of each (See Problem 20) and rough-sketch the elliptical trajectory that passes through these four points.*

(c) *Compare your sketch with a computer phase portrait from a graphic DE solver and explain any discrepancies.*

21. $\vec{\mathbf{x}}' = \begin{bmatrix} 4 & -5 \\ 5 & -4 \end{bmatrix} \vec{\mathbf{x}}$

22. $\vec{\mathbf{x}}' = \begin{bmatrix} -1 & -1 \\ 5 & 1 \end{bmatrix} \vec{\mathbf{x}}$

23. $\vec{\mathbf{x}}' = \begin{bmatrix} 1 & -1 \\ 2 & -1 \end{bmatrix} \vec{\mathbf{x}}$

24. $\vec{\mathbf{x}}' = \begin{bmatrix} -1 & 1 \\ -2 & 1 \end{bmatrix} \vec{\mathbf{x}}$

25. Tilt with Precision[3] For a system as described in Problem 20, we have the condition that at the ends of the major and minor axes of an elliptical trajectory, *the velocity vector must be orthogonal to the position vector.* That is, we must have

$$\vec{\mathbf{x}}(t) \cdot \vec{\mathbf{x}}'(t) = 0 \tag{19}$$

[2] Courtesy of Bjørn Felsager, Haslev Gymnasium, Denmark.

[3] Courtesy of Professor John Cantwell, St. Louis University.

Using your equations from Problem 20(a), show that condition (19) is satisfied when

$$\tan 2\beta t = \frac{2\vec{\mathbf{p}}\cdot\vec{\mathbf{q}}}{\|\vec{\mathbf{q}}\|^2 - \|\vec{\mathbf{p}}\|^2}. \qquad (20)$$

Problems 26–29 ask you to apply this principle to several examples.

Axes for Ellipses *For Problems 26–29*

(a) *Using equation (20) from Problem 25, solve for the two values of βt^* for which (19) is satisfied.*

(b) *Substitute (one at a time) each value of βt^* found in part (a) into the solutions found in Problem 20 (a) to calculate the endpoints of the major and minor axes of the elliptical trajectories for the specified linear system of DEs.*

(c) *Use a computer phase portrait from a graphic DE solver to confirm your results. Discuss any discrepancies.*

CAUTION: *This is a problem in coordinates, not angles. The quantity βt is a* parameter *in the basis described by $\vec{\mathbf{p}}$, $\vec{\mathbf{q}}$, and is definitely* not *an angle in the phase-plane. See, for example Figure 6.3.6, which makes this clear: the four points on the trajectory shown are separated by increments of $\beta t = \pi/2$, but the angles pictured are far from equal. Along the elliptical trajectory, the angular velocity in the xy-plane for angle theta with the x-axis is* not *constant.*

26. $\vec{\mathbf{x}}' = \begin{bmatrix} 4 & -5 \\ 5 & -4 \end{bmatrix}\vec{\mathbf{x}}$ **27.** $\vec{\mathbf{x}}' = \begin{bmatrix} -1 & -1 \\ 5 & 1 \end{bmatrix}\vec{\mathbf{x}}$

28. $\vec{\mathbf{x}}' = \begin{bmatrix} 1 & -1 \\ 2 & -1 \end{bmatrix}\vec{\mathbf{x}}$ **29.** $\vec{\mathbf{x}}' = \begin{bmatrix} -1 & 1 \\ -2 & 1 \end{bmatrix}\vec{\mathbf{x}}$

30. 3 x 3 System Solve the system

$$\vec{\mathbf{x}}' = \begin{bmatrix} -1 & 0 & 0 \\ 0 & 0 & 2 \\ 0 & -2 & 0 \end{bmatrix}\vec{\mathbf{x}} = \mathbf{A}\vec{\mathbf{x}}$$

as follows.

(a) Show that the eigenvalues of **A** are $\lambda_1 = -1$, $\lambda_2 = 2i$, and $\lambda_3 = -2i$.

(b) Use an eigenvector for λ_1 to obtain one solution,

$$\vec{\mathbf{x}}_1 = e^{-t}\begin{bmatrix} 1 \\ 0 \\ 0 \end{bmatrix}.$$

(c) Show that

$$\begin{bmatrix} 0 \\ 1 \\ i \end{bmatrix} = \begin{bmatrix} 0 \\ 1 \\ 0 \end{bmatrix} + i\begin{bmatrix} 0 \\ 0 \\ 1 \end{bmatrix}$$

is an eigenvector belonging to λ_2, and obtain two more solutions,

$$\vec{\mathbf{x}}_2 = \begin{bmatrix} 0 \\ \cos 2t \\ -\sin 2t \end{bmatrix} \quad\text{and}\quad \vec{\mathbf{x}}_3 = \begin{bmatrix} 0 \\ \sin 2t \\ \cos 2t \end{bmatrix}.$$

(d) Write the general solution and obtain its components in the form

$$x = c_1 e^{-t},$$
$$y = c_2\cos 2t + c_3\sin 2t,$$
$$z = c_3\cos 2t - c_2\sin 2t.$$

(e) Write the IVP solution for

$$\vec{\mathbf{x}}(0) = \begin{bmatrix} 1 \\ 0 \\ 1 \end{bmatrix}$$

in the form of part (d).

(f) Discuss the geometry of the solution curve in part (e) from its parametric equations. The curve is a **helix**.

Threefold Solutions *Solve the 3×3 systems given in Problems 31–34.*

31. $\vec{\mathbf{x}}' = \begin{bmatrix} 1 & 0 & -1 \\ 0 & 2 & 0 \\ 1 & 0 & 1 \end{bmatrix}\vec{\mathbf{x}}$ **32.** $\vec{\mathbf{x}}' = \begin{bmatrix} 0 & 1 & 0 \\ 0 & 0 & 1 \\ -1 & 0 & 0 \end{bmatrix}\vec{\mathbf{x}}$

33. $\vec{\mathbf{x}}' = \begin{bmatrix} 1 & 0 & 0 \\ 2 & 1 & -2 \\ 3 & 2 & 1 \end{bmatrix}\vec{\mathbf{x}}$ **34.** $\vec{\mathbf{x}}' = \begin{bmatrix} -3 & 1 & -2 \\ 0 & -1 & -1 \\ 2 & 0 & 0 \end{bmatrix}\vec{\mathbf{x}}$

Triple IVPs *Solve the initial-value problems given in Problems 35 and 36.*

35. $\vec{\mathbf{x}}' = \begin{bmatrix} 3 & 0 & -1 \\ 0 & -3 & -1 \\ 0 & 2 & -1 \end{bmatrix}\vec{\mathbf{x}}, \quad \vec{\mathbf{x}}(0) = \begin{bmatrix} -5 \\ 13 \\ -26 \end{bmatrix}$

36. $\vec{\mathbf{x}}' = \begin{bmatrix} 0 & 1 & 0 \\ -1 & 0 & -1 \\ 0 & 1 & 0 \end{bmatrix}\vec{\mathbf{x}}, \quad \vec{\mathbf{x}}(0) = \begin{bmatrix} 0 \\ 1 \\ 1 \end{bmatrix}$

37. Matter of Independence Show that the real and imaginary parts, $\vec{\mathbf{x}}_{\text{Re}}$ and $\vec{\mathbf{x}}_{\text{Im}}$, of the solutions in equation (7), as given in equations (6), are linearly independent.

38. Skew-Symmetric Systems Recall (Sec. 6.1, Problem 16) that matrix **A** is *skew-symmetric* if $\mathbf{A} = -\mathbf{A}^{\mathrm{T}}$. Solutions of $\vec{\mathbf{x}}' = \mathbf{A}\vec{\mathbf{x}}$, where **A** is skew-symmetric, have constant length for all t. Find explicit formulas for solutions to the system

$$\vec{\mathbf{x}}' = \begin{bmatrix} 0 & k \\ -k & 0 \end{bmatrix}\vec{\mathbf{x}}, \quad k \text{ real},$$

and verify that its length is constant, as was shown graphically in Sec. 6.1, Problem 16.

39. Coupled Mass-Spring System Suppose that two equal masses $m_1 = m_2 = m$ are attached to three springs, each having the same spring constant $k_1 = k_2 = k_3 = k$, where the two outside springs are attached to walls. The masses slide in a straight line on a frictionless surface. The system is set in motion by holding the left mass in its equilibrium position while at the same time pulling the right mass to the right of its equilibrium a distance d. (See Fig. 6.3.7.)

We denote by $x(t)$ and $y(t)$ the positions of the respective masses m_1 and m_2 from their respective equilibrium positions. Then $(y-x)$ is the stretch or compression of the middle spring. Since the only forces acting on the masses are the forces due to the connecting springs, Hooke's Law says that

- the force on m_1 due to the left spring $= -k_1 x$,
- the force on m_1 due to the middle spring $= -k_2(y-x)$,
- the force on m_2 due to the middle spring $= -k_2(y-x)$,
- the force on m_2 due to the right spring $= -k_3 y$.

Hence, we have the initial-value problem

$$\begin{aligned} m_1\ddot{x} &= -k_1 x - k_2(y-x),\ x(0)=0,\ \dot{x}(0)=0; \\ m_2\ddot{y} &= -k_2(y-x) - k_3 y,\ y(0)=2,\ \dot{y}(0)=0. \end{aligned} \tag{21}$$

If we let $x_1 = x$, $x_2 = \dot{x}$, $x_3 = y$, and $x_4 = \dot{y}$, we can write these two equations (20) as the 4×4 system of equations

$$\begin{bmatrix} \dot{x}_1 \\ \dot{x}_2 \\ \dot{x}_3 \\ \dot{x}_4 \end{bmatrix} = \begin{bmatrix} 0 & 1 & 0 & 0 \\ \dfrac{-k_1+k_2}{m_1} & 0 & \dfrac{-k_2}{m_1} & 0 \\ 0 & 0 & 0 & 1 \\ \dfrac{k_2}{m_2} & 0 & -\dfrac{k_2+k_3}{m_2} & 0 \end{bmatrix} \begin{bmatrix} x_1 \\ x_2 \\ x_3 \\ x_4 \end{bmatrix}.$$

Set $k_i = m_i = 1$, and show how these equations can be solved using eigenvalues and eigenvectors to find

$$x(t) = \cos t - \cos\sqrt{3}t \tag{22}$$
$$y(t) = \cos t + \cos\sqrt{3}t \tag{23}$$

as illustrated in Fig. 6.3.8.

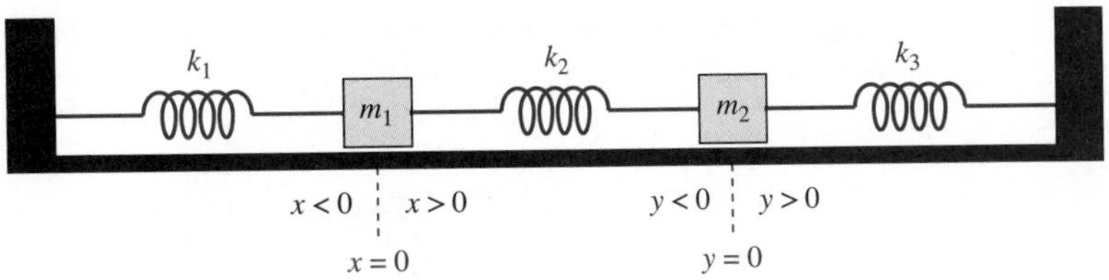

FIGURE 6.3.7 Mass-spring configuration for Problem 39.

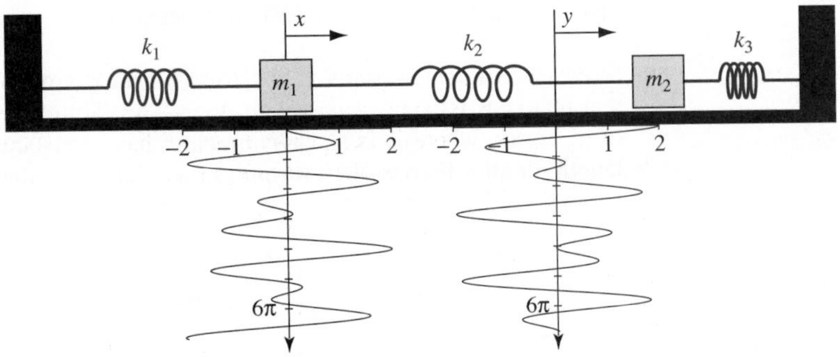

FIGURE 6.3.8 Solution graphs (21) and (22) with mass-spring configuration in Problem 39, if $m_1 = m_2 = k_1 = k_2 = k_3 = 1$, $x(0) = 0$, $y(0) = 2$.

Computer Lab: Time Series *For Problems 40 and 41, use a calculator or computer to plot the phase portrait for the given IVP. Then plot component graphs $x_1(t)$ and $x_2(t)$ for each IVP as functions of t. How do these graphs relate to the corresponding trajectory in the phase plane?*

40. $\vec{\mathbf{x}}' = \mathbf{A}\vec{\mathbf{x}} = \begin{bmatrix} 0 & 1 \\ -5 & -2 \end{bmatrix}\vec{\mathbf{x}}, \qquad \vec{\mathbf{x}}(0) = \begin{bmatrix} 2 \\ 2 \end{bmatrix}.$

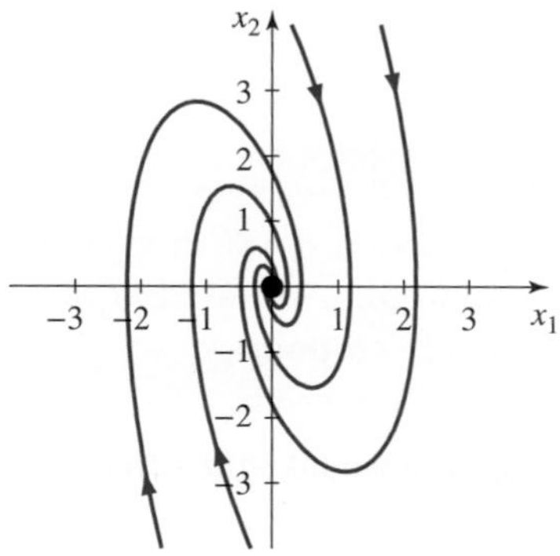

(See Example 3.)

41. $\vec{\mathbf{x}}' = \mathbf{A}\vec{\mathbf{x}} = \begin{bmatrix} 4 & -5 \\ 5 & -4 \end{bmatrix}\vec{\mathbf{x}}, \qquad \vec{\mathbf{x}}(0) = \begin{bmatrix} 2 \\ -2 \end{bmatrix}.$

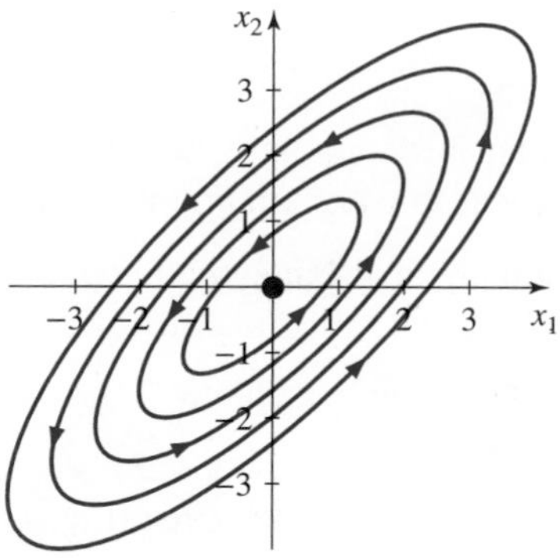

(See Example 4.)

42. Suggested Journal Entry The graphs in Fig. 6.3.9 are solution curves from three different autonomous linear systems of DEs,

$$\vec{\mathbf{x}}' = \mathbf{A}\vec{\mathbf{x}},$$

where

$$\vec{\mathbf{x}} = \begin{bmatrix} x \\ y \\ z \end{bmatrix}.$$

What can you deduce about the eigenvalues (and eigenvectors) that will apply to all three DEs? What can you deduce about the differences in z' between the three cases?

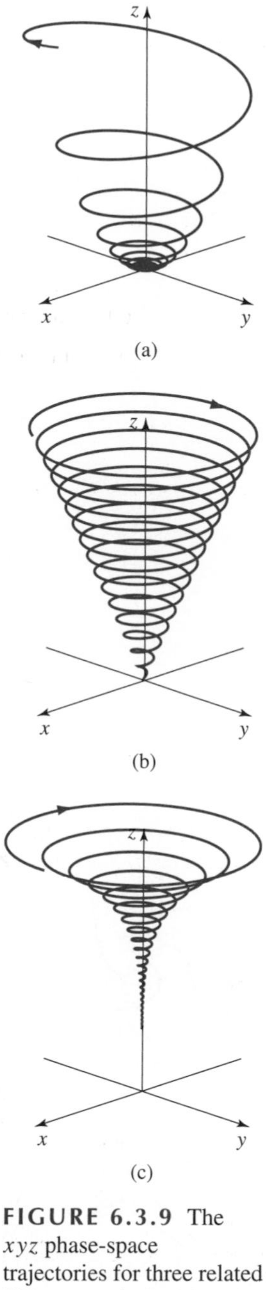

FIGURE 6.3.9 The xyz phase-space trajectories for three related DE systems in Problem 42.

6.4 Stability and Linear Classification

SYNOPSIS: We characterize the stability and instability of equilibrium solutions of linear systems of differential equations. We classify the stability of fixed points in the phase portraits that correspond to equilibrium solutions for homogeneous 2×2 linear systems with constant coefficients.

Stability of Equilibrium Solutions

We can now refine and summarize what we have learned about the behaviors of solutions to linear DE systems. To begin, we revisit the central concept of equilibrium.

> **Equilibrium Solution**
>
> A constant solution $\vec{\mathbf{x}} \equiv \vec{\mathbf{c}}$ of the autonomous system $\vec{\mathbf{x}}' = \vec{\mathbf{f}}(\vec{\mathbf{x}})$ (such that $\vec{\mathbf{f}}(\vec{\mathbf{c}}) = \vec{\mathbf{0}}$) is called an **equilibrium solution**. An equilibrium solution in the phase plane is simply a point, called a **fixed point**.

For convenience, throughout this section we take the origin $\vec{\mathbf{x}} \equiv \vec{\mathbf{0}}$ as our fixed point.

An equilibrium solution represents a constant steady state of the real-world system being modeled. What is important about such a solution is the nature of "nearby" solutions. Small disturbances in the physical system will disturb the conditions from $\vec{\mathbf{c}}$ to a slightly different vector $\vec{\mathbf{x}}_0$, and the solution starting at this new point may or may not return the system to its steady state.

Loosely speaking, $\vec{\mathbf{x}} \equiv \vec{\mathbf{c}}$ is stable if solutions that start close, stay close.[1] If $\vec{\mathbf{c}}$ is *unstable*, there are initial conditions $\vec{\mathbf{x}}_0$ arbitrarily close to $\vec{\mathbf{c}}$ such that the solutions starting there do *not* remain close to $\vec{\mathbf{c}}$.

> **Stability of Equilibrium Solutions**
>
> An equilibrium solution $\vec{\mathbf{x}} \equiv \vec{\mathbf{c}}$ of an autonomous system $\vec{\mathbf{x}}' = \vec{\mathbf{f}}(\vec{\mathbf{x}})$ is **stable** if solutions that start sufficiently near to $\vec{\mathbf{c}}$ remain bounded.
>
> - If nearby solutions not only remain close but actually tend to $\vec{\mathbf{c}}$ as a limit as $t \to \infty$, the equilibrium solution is called **asymptotically stable**.
> - If nearby solutions are neither attracted nor repelled, the equilibrium solution is called **neutrally stable**.
>
> An equilibrium solution that is *not* stable is called **unstable**.

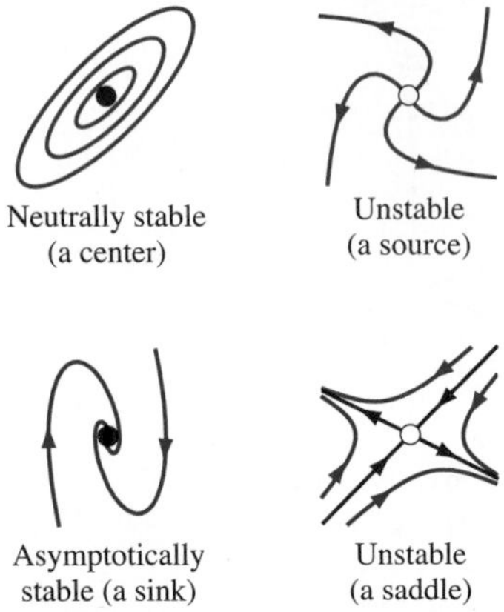

FIGURE 6.4.1 Various equilibrium patterns.

We have seen examples of these different types, as in Fig. 6.4.1, throughout this chapter. Now we will see how they can be organized to predict their behavior. We shall create a catalog that includes the various transition behaviors that occur at boundaries between types.

This section is devoted to the classification of equilibrium solutions of 2×2 homogeneous linear DE systems with constant coefficients. This restriction may seem like a very special case, but it provides the visual insights that serve as

---

[1]The isolated equilibrium solution $\vec{\mathbf{x}} \equiv \vec{\mathbf{c}}$ (*isolated* means that there is a disc around $\vec{\mathbf{c}}$ containing no other equilibrium solution) is stable if for each disc S about $\vec{\mathbf{c}}$ there exists a positive number b (depending on S) with the following property: for any initial condition $\vec{\mathbf{x}}_0$ within b units of $\vec{\mathbf{c}}$, the solution starting at $\vec{\mathbf{x}}_0$ remains in S for all $t \geq 0$.

a starting point for studying both the higher-dimensional linear systems of this chapter and the nonlinear systems of Chapter 7.

Let's start again with the linear system:

$$\begin{aligned} x' &= ax + by, \\ y' &= cx + dy, \end{aligned} \quad \text{or} \quad \vec{\mathbf{x}}' = \begin{bmatrix} a & b \\ c & d \end{bmatrix} \vec{\mathbf{x}} = \mathbf{A}\vec{\mathbf{x}}. \tag{1}$$

We set the characteristic polynomial of $\mathbf{A}$ equal to zero to find the eigenvalues:

$$|\mathbf{A} - \lambda \mathbf{I}| = \lambda^2 - \underbrace{(a+d)}_{\mathrm{Tr}\,\mathbf{A}}\lambda + \underbrace{(ad - bc)}_{|\mathbf{A}|} = 0.$$

By the quadratic formula, we solve for the eigenvalues in terms of trace and determinant of $\mathbf{A}$:

$$\lambda = \frac{\mathrm{Tr}\,\mathbf{A} \pm \sqrt{(\mathrm{Tr}\,\mathbf{A})^2 - 4|\mathbf{A}|}}{2}, \tag{2}$$

- The sign of the discriminant $\Delta = (\mathrm{Tr}\,\mathbf{A})^2 - 4|\mathbf{A}|$ determines whether we have two distinct real eigenvalues, one repeated real eigenvalue, or a complex conjugate pair of eigenvalues.
- The fact that we now have two parameters $\mathrm{Tr}\,\mathbf{A}$ and $|\mathbf{A}|$, instead of the original four matrix entries, means that we can construct a **parameter plane** graph of $|\mathbf{A}|$ with respect to $\mathrm{Tr}\,\mathbf{A}$, where the coordinates of points on the parameter plane determine the eigenvalues as given in (2). (See Fig. 6.4.2.)

Parameter Plane Animation

Take a tour through the parameter plane and watch the phase portraits shift and change.

Parameter Plane Input; Matrix Element Input

Pick a point in the trace-determinant plane and click on the phase plane to start a trajectory. Enter the matrix elements for $\mathbf{A}$ and click on the phase plane to see the trajectory for the solution of $\vec{\mathbf{x}}' = \mathbf{A}\vec{\mathbf{x}}$.

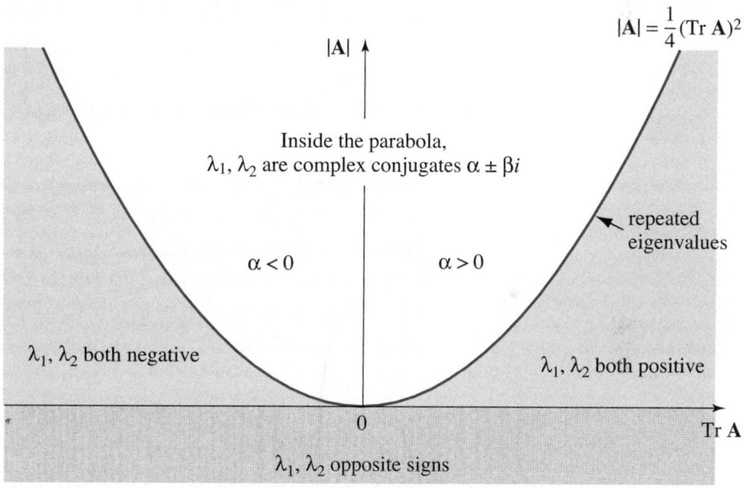

FIGURE 6.4.2 Trace-determinant plane for $\vec{\mathbf{x}}' = \mathbf{A}\vec{\mathbf{x}}$, eigenvalue view.

- The condition on the discriminant

$$\Delta = (\mathrm{Tr}\,\mathbf{A})^2 - 4|\mathbf{A}| = 0$$

gives us a parabola in the parameter plane.

Once we find the eigenvalues, what is the role of the corresponding eigenvectors? For eigenvalues λ_1 and λ_2 with corresponding eigenvectors $\vec{\mathbf{v}}_1$ and $\vec{\mathbf{v}}_2$, we obtain two solutions for (1):

$$e^{\lambda_1 t}\vec{\mathbf{v}}_1 \quad \text{and} \quad e^{\lambda_2 t}\vec{\mathbf{v}}_2.$$

If these solutions are linearly independent, then they form a basis of the solution space for (1). Of course, that is not always the case, because we can have repeated eigenvalues with a single eigenvector. (See Sec. 6.2, Examples 5 and 6.)

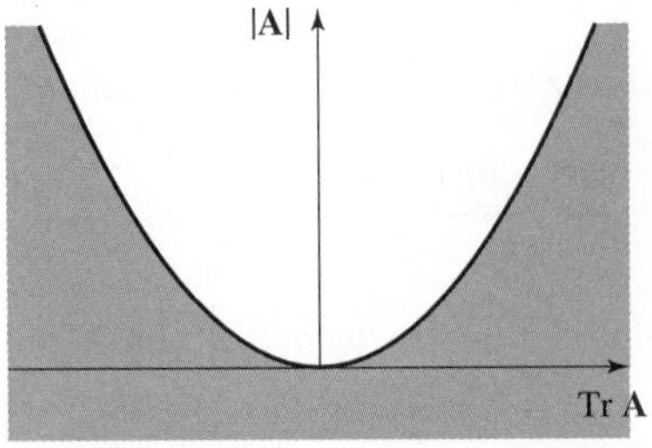

FIGURE 6.4.3 Real distinct eigenvalues occur *outside* the parabola in the trace-determinant plane.

Real Distinct Eigenvalues ($\Delta > 0$)

Node Behaviors

When $\Delta = (\mathrm{Tr}\,\mathbf{A})^2 - 4|\mathbf{A}| > 0$ (in the shaded area of Fig. 6.4.3), we have real eigenvalues $\lambda_1 \neq \lambda_2$ with corresponding linearly independent eigenvectors $\vec{\mathbf{v}}_1$ and $\vec{\mathbf{v}}_2$, and general solution

$$\vec{\mathbf{x}} = c_1 e^{\lambda_1 t}\vec{\mathbf{v}}_1 + c_2 e^{\lambda_2 t}\vec{\mathbf{v}}_2. \tag{3}$$

The *signs* of the eigenvalues direct the trajectory behavior in the phase portrait.

It is common to label the eigendirections *fast* and *slow,* depending on the *magnitude* of the *eigenvalues*. In all cases (see Sec. 6.2) trajectories run parallel to the fast eigenvector and tangent to the slow eigenvector.

Attracting Node ($\lambda_1 < \lambda_2 < 0$). When λ_1 and λ_2 are both negative, both terms of the solution (3) tend toward zero as $t \to \infty$, so the fixed point is asymptotically stable and is said to be an **attracting node** or a **node sink**. The term $e^{\lambda_1 t}$ tends toward zero faster than $e^{\lambda_2 t}$, so trajectories tend to approach the origin along a path tangent to $\vec{\mathbf{v}}_2$. We call $\vec{\mathbf{v}}_1$ the **fast eigendirection** and $\vec{\mathbf{v}}_2$ the **slow eigendirection**. (See Fig. 6.4.4 and Sec. 6.2, Example 3.)

Repelling Node ($0 < \lambda_1 < \lambda_2$). When both λ_1 and λ_2 are positive, solution (3) tends to become infinite as $t \to \infty$. The vectors $\vec{\mathbf{v}}_1$ and $\vec{\mathbf{v}}_2$ are the slow and fast eigendirections, respectively. The $e^{\lambda_2 t}$ term grows much faster than the $e^{\lambda_1 t}$ term, so trajectories tend to become parallel to (but not asymptotic to) the fast eigendirection $\vec{\mathbf{v}}_2$. The origin is called a **repelling node** or a **node source**. It is clearly an unstable fixed point. (See Fig. 6.4.5 and Sec. 6.2, Example 2.)

Saddle Point ($\lambda_1 < 0 < \lambda_2$). When eigenvalues λ_1 and λ_2 have different signs, solutions still have form (3), but the terms $e^{\lambda_1 t}$ and $e^{\lambda_2 t}$ behave quite differently as $t \to \infty$. The term $e^{\lambda_1 t}$ tends toward zero, and $e^{\lambda_2 t}$ tends toward infinity. In the phase plane, the eigenvector $\vec{\mathbf{v}}_1$ is pointed toward the origin and the eigenvector $\vec{\mathbf{v}}_2$ is pointed away from the origin. Trajectories actually slow down as they approach the origin along $\vec{\mathbf{v}}_1$ and speed up again as they leave along $\vec{\mathbf{v}}_2$. You can *see* this phenomenon as a graphics solver evolves a trajectory using a very small step size. (Remember that we can't see t in the phase portrait.) The trajectory in the phase plane will tend toward $\vec{\mathbf{v}}_2$ asymptotically. The unstable fixed point at the origin is called a **saddle point**. (See Fig. 6.4.6 and Sec. 6.2, Example 1.)

Borderline Cases. The cases in which one of the eigenvalues is zero, or there is a repeated eigenvalue are rare and will be handled separately in the subsection on borderline cases. First, we classify fixed points for the second major category of solutions, when the eigenvalues are not real numbers.

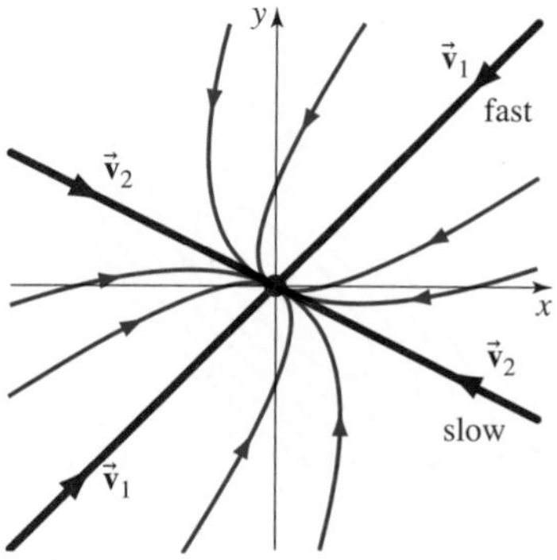

FIGURE 6.4.4 An attracting node (or node sink) is asymptotically stable; it occurs when $\lambda_1 < \lambda_2 < 0$.

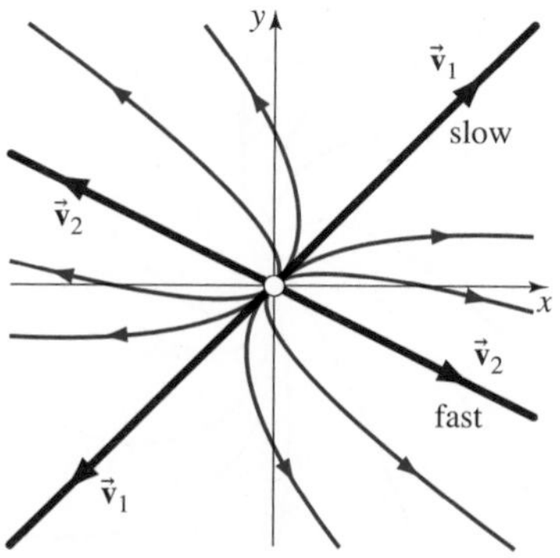

FIGURE 6.4.5 A repelling node (or node source) is unstable; it occurs when $0 < \lambda_1 < \lambda_2$.

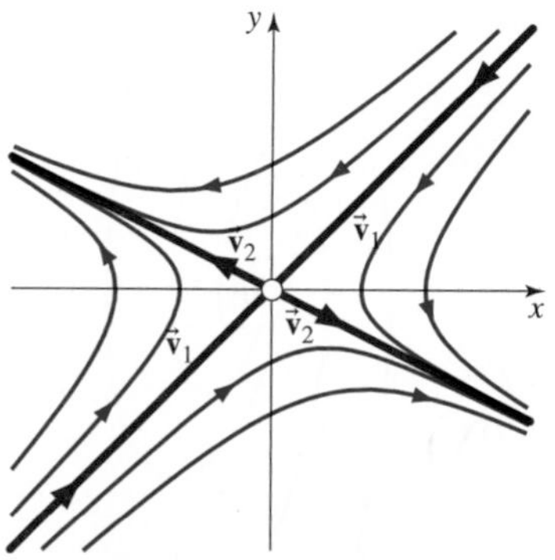

FIGURE 6.4.6 A saddle is also unstable; it occurs when $\lambda_1 < 0 < \lambda_2$. Here $|\lambda_1| > |\lambda_2|$.

Complex Conjugate Eigenvalues ($\Delta < 0$)

Spiraling Behaviors

When $\Delta = (\text{Tr}\,\mathbf{A})^2 - 4|\mathbf{A}| < 0$ (in the shaded area of Fig. 6.4.7), we get nonreal eigenvalues,

$$\lambda_1 = \alpha + \beta i \quad \text{and} \quad \lambda_2 = \alpha - \beta i,$$

where $\alpha = \text{Tr}\,\mathbf{A}/2$ and $\beta = \sqrt{-\Delta}$. Notice that α and β are real, and $\beta \neq 0$.

Recall from Sec. 6.3, equation (6), that the real solutions are given by

$$\begin{aligned} \vec{\mathbf{x}}_{\text{Re}} &= e^{\alpha t}(\cos \beta t\, \vec{\mathbf{p}} - \sin \beta t\, \vec{\mathbf{q}}), \\ \vec{\mathbf{x}}_{\text{Im}} &= e^{\alpha t}(\sin \beta t\, \vec{\mathbf{p}} + \cos \beta t\, \vec{\mathbf{q}}). \end{aligned} \tag{4}$$

For complex eigenvalues, stability behavior of solutions depends on the *sign* of α.

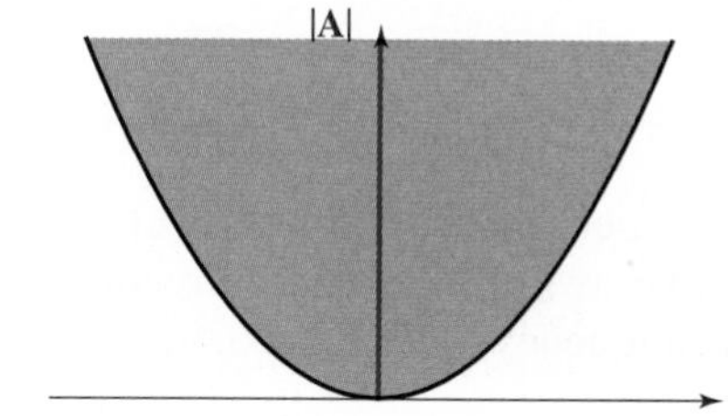

FIGURE 6.4.7 Complex conjugate eigenvalues occur *inside* the parabola in the trace-determinant plane.

Attracting Spiral ($\alpha < 0$). When α is negative, solutions decay to the origin, because in (4) the factor $e^{\alpha t} \to 0$ as $t \to \infty$. The trajectories are descending spirals toward the fixed point at the origin, which is asymptotically stable and is called an **attracting spiral** or **spiral sink**. (See Fig. 6.4.8 and Sec. 6.3, Example 3.)

Repelling Spiral ($\alpha > 0$). When α is positive, solutions grow, because in (4) the factor $e^{\alpha t} \to \infty$ as $t \to \infty$. The trajectories are spirals without bound away from the origin. The fixed point at the origin is called a **repelling spiral** or **spiral source** and is clearly unstable. (See Fig. 6.4.9 and Sec. 6.3, Example 2.)

Center ($\alpha = 0$). When α is zero, the eigenvalues are purely imaginary. System (4) reduces (Sec. 6.3, equation (6)) to

$$\begin{aligned} \vec{\mathbf{x}}_{\text{Re}} &= \cos \beta t\, \vec{\mathbf{p}} - \sin \beta t\, \vec{\mathbf{q}}, \\ \vec{\mathbf{x}}_{\text{Im}} &= \sin \beta t\, \vec{\mathbf{p}} + \cos \beta t\, \vec{\mathbf{q}}. \end{aligned}$$

The phase-plane trajectories are closed loops about the fixed point $\vec{\mathbf{x}} = \vec{\mathbf{0}}$, representing periodic motion. (See Fig. 6.4.10 and Sec. 6.3, Example 4.) The fixed point is called a **center**. A center is neither attracting nor repelling, so it is neutrally stable. This happens whenever $\text{Tr}\,\mathbf{A} = 0$, which in the trace-determinant plane occurs *along the positive vertical axis.*

The nodes and spirals are the usual cases, because eigenvalues in general are nonzero and unequal. Zero eigenvalues or repeated real eigenvalues are rarities that occur only on the borders of the various regions.

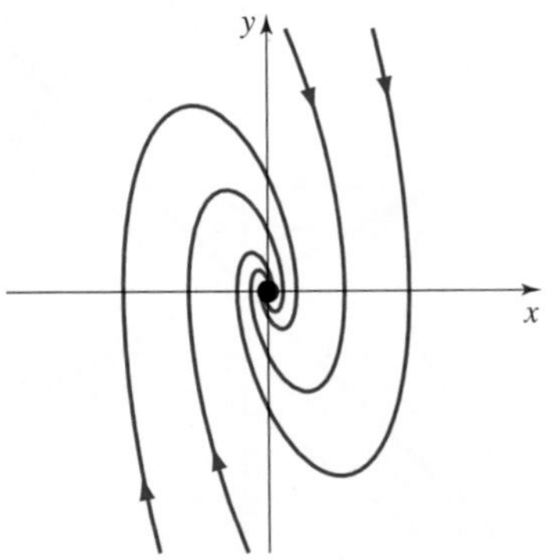

FIGURE 6.4.8 An attracting spiral (or spiral sink) is stable; it occurs when $\alpha < 0$.

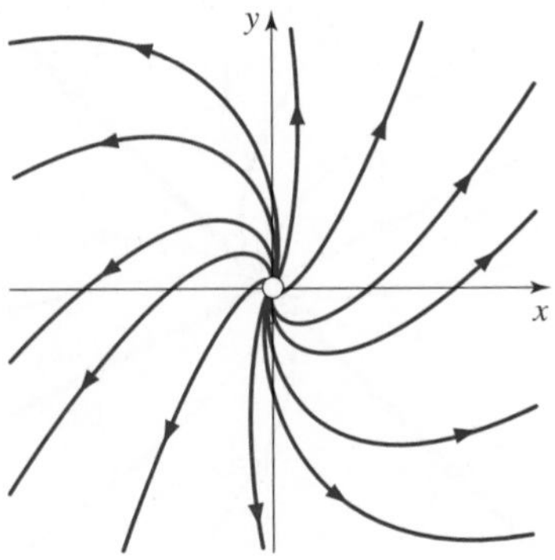

FIGURE 6.4.9 A repelling spiral (or spiral source) is unstable; it occurs when $\alpha > 0$.

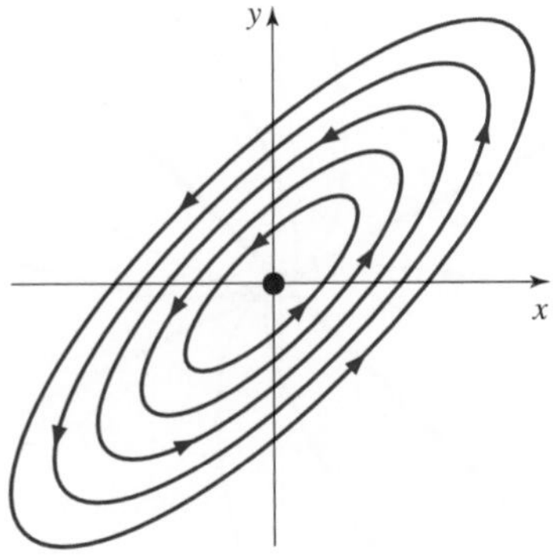

FIGURE 6.4.10 A center is *neutrally* stable; it occurs when $\alpha = 0$.

Borderline Case: Zero Eigenvalues ($|\mathbf{A}| = 0$)

When $|\mathbf{A}| = 0$ (see Fig. 6.4.11), at least one eigenvalue is zero. If *one* eigenvalue is zero, we get a row of nonisolated fixed points in the eigendirection associated with that eigenvalue, and the phase-plane trajectories are all straight lines in the direction of the other eigenvector.

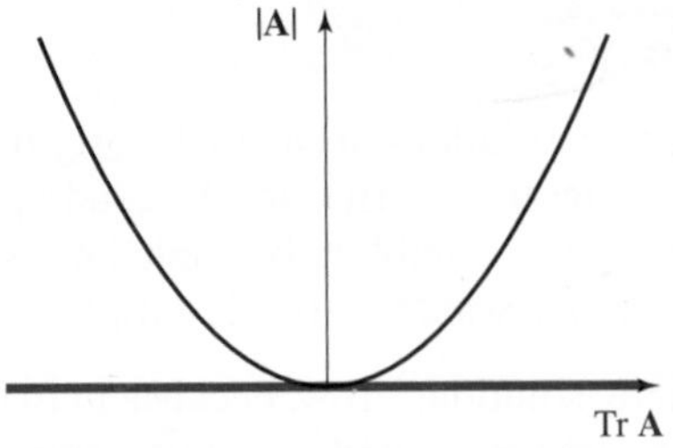

FIGURE 6.4.11 Zero eigenvalues occur on the horizontal axis in the trace-determinant plane.

EXAMPLE 1 **Single Zero Eigenvalue** The system

$$\vec{\mathbf{x}}' = \begin{bmatrix} 2 & 1 \\ 6 & 3 \end{bmatrix} \vec{\mathbf{x}}$$

has characteristic equation $\lambda^2 - 5\lambda = 0$. The eigenvalues are

$$\lambda_1 = 0 \quad \text{and} \quad \lambda_2 = 5$$

with corresponding eigenvectors

$$\vec{\mathbf{v}}_1 = \begin{bmatrix} 1 \\ -2 \end{bmatrix} \quad \text{and} \quad \vec{\mathbf{v}}_2 = \begin{bmatrix} 1 \\ 3 \end{bmatrix}.$$

Along $\vec{\mathbf{v}}_1$, we have $\vec{\mathbf{x}}' = \vec{\mathbf{0}}$, so there is a line of equilibrium points, unstable because $\lambda_2 = 5 > 0$. Other trajectories move away from $\vec{\mathbf{v}}_1$ in directions parallel to $\vec{\mathbf{v}}_2$. None of these can *cross* $\vec{\mathbf{v}}_1$, by uniqueness. (See Fig. 6.4.12.)

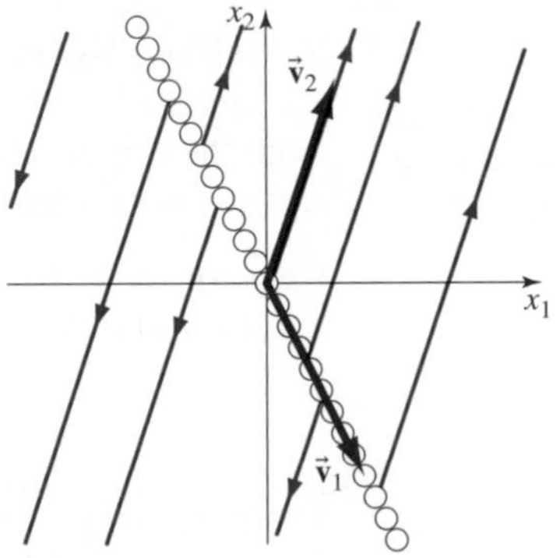

FIGURE 6.4.12 When $\lambda_1 = 0$, as in Example 1, a line of equilibria occur along $\vec{\mathbf{v}}_1$. The stability of the equilibria depends on the sign of λ_2, and all trajectories run to or from $\vec{\mathbf{v}}_1$ parallel to $\vec{\mathbf{v}}_2$. ■

If *two* eigenvalues are zero (a special case of repeated eigenvalues), there is only one eigenvector, along which we have a row of nonisolated fixed points.

Trajectories from any other point in the phase plane must also go parallel to the one eigenvector, in directions specified by the system.

EXAMPLE 2 **Double Zero Eigenvalue** The system

$$\vec{\mathbf{x}}' = \begin{bmatrix} 3 & 9 \\ -1 & -3 \end{bmatrix} \vec{\mathbf{x}}$$

has characteristic equation $\lambda^2 = 0$, so we have a repeated eigenvalue

$$\lambda = 0$$

with a single eigenvector

$$\vec{\mathbf{v}} = \begin{bmatrix} -3 \\ 1 \end{bmatrix}.$$

Along $\vec{\mathbf{v}}$, we have $\vec{\mathbf{x}}' = \vec{\mathbf{0}}$, so we again have a line of equilibrium points. At any other point in the phase plane, motion is parallel to $\vec{\mathbf{v}}$, with direction determined by the system. (See Fig. 6.4.13.) For example:

- At $(1, 0)$, $x' = 3$, $y' = -1$ indicates motion to the right and down.
- At $(-1, 0)$, $x' = -3$, $y' = 1$ indicates motion to the left and up.

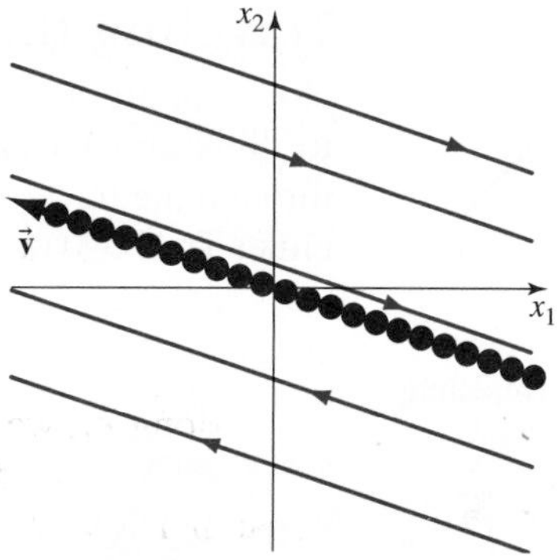

FIGURE 6.4.13 When $\lambda_1, \lambda_2 = 0$, as in Example 2, a line of *neutrally stable* equilibria occur along a single eigenvector $\vec{\mathbf{v}}$. Neither attracted nor repelled, all other trajectories are parallel to $\vec{\mathbf{v}}$.

■

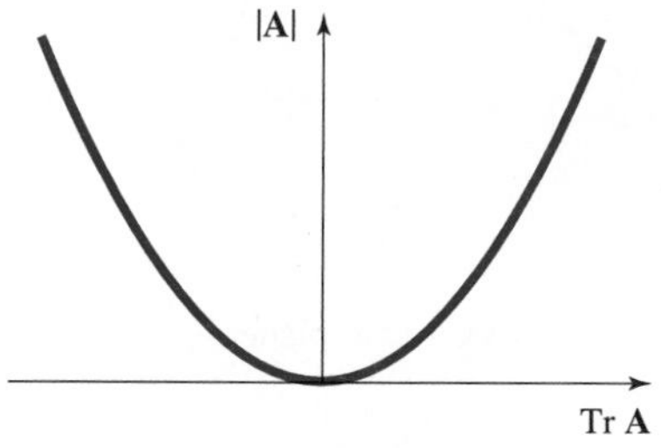

FIGURE 6.4.14 Real repeated eigenvalues occur *on* the parabola in the trace-determinant plane.

Borderline Case: Real Repeated Eigenvalues ($\Delta = 0$)

The points corresponding to real repeated eigenvalues are located on the parabola $\Delta = (\text{Tr}\,\mathbf{A})^2 - 4|\mathbf{A}| = 0$ (see Fig. 6.4.14), which is the border that separates the regions between spirals and nodes (i.e., regions of nonreal and real distinct eigenvalues, respectively). The resulting phase plane behavior depends on whether there are one or two linearly independent eigenvectors for λ. In other words, is the geometric multiplicity one or two? It can be either; we will consider the more common possibility first.

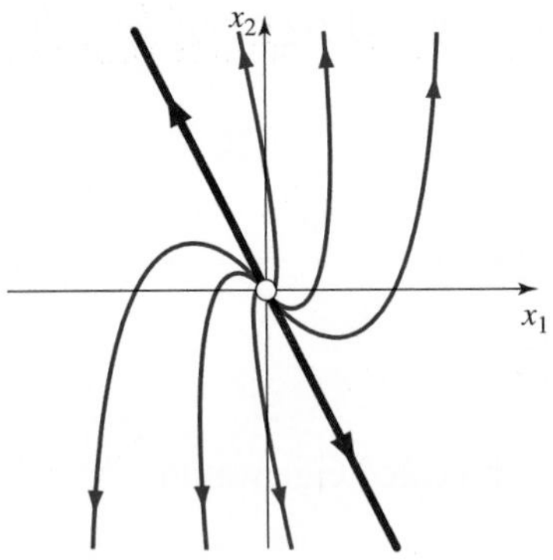

FIGURE 6.4.15 When a double eigenvalue $\lambda > 0$ has *one* independent eigenvector, a repelling *degenerate node* occurs, as in Sec. 6.2, Examples 5 and 6.

Degenerate Node (Fig. 6.4.15). When a repeated eigenvalue λ has *one* linearly independent eigenvector, the fixed point is called a **degenerate node**, attracting and asymptotically stable for negative λ, repelling and unstable for positive λ. The *sign* of the eigenvalue gives the stability. Because there is only one eigendirection, the following hold:

- If $\lambda > 0$, trajectories tend to infinity, parallel to $\vec{\mathbf{v}}$.
- If $\lambda < 0$, trajectories approach the origin parallel to $\vec{\mathbf{v}}$. (See Problem 4.)

If $\lambda = 0$, which occurs at the origin in the trace-determinant plane, a line of fixed points lie along the eigenvector. All other trajectories are parallel to the eigenvector. (See Example 2 and Fig. 6.4.13.)

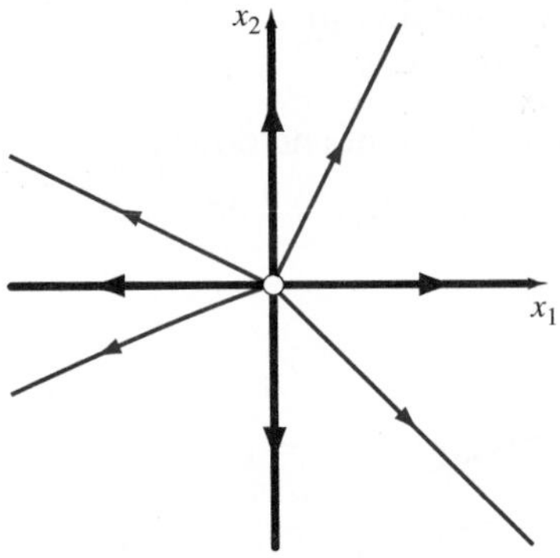

FIGURE 6.4.16 When a double eigenvalue $\lambda > 0$ has *two* independent eigenvectors, a repelling *star node* occurs, as in Sec. 6.2, Example 4.

Star Node (Fig. 6.4.16). When a repeated eigenvalue λ has *two* linearly independent eigenvectors, they span the plane. In consequence, *every* vector is an eigenvector for λ. Every trajectory is a straight line, approaching the origin if λ is negative and repelling out from the origin if λ is positive. The fixed point is called an **attracting** or **repelling star node** and is stable or unstable accordingly.

Traveling through the Parameter Plane

Now that all the major and borderline cases have been explained, we are ready to think about what will happen as equation parameters change and we move through the trace-determinant plane, shown again in Fig. 6.4.17. For $\vec{\mathbf{x}}' = \mathbf{A}\vec{\mathbf{x}}$, if either $\operatorname{Tr}\mathbf{A} > 0$ or $|\mathbf{A}| < 0$, the system is unstable.

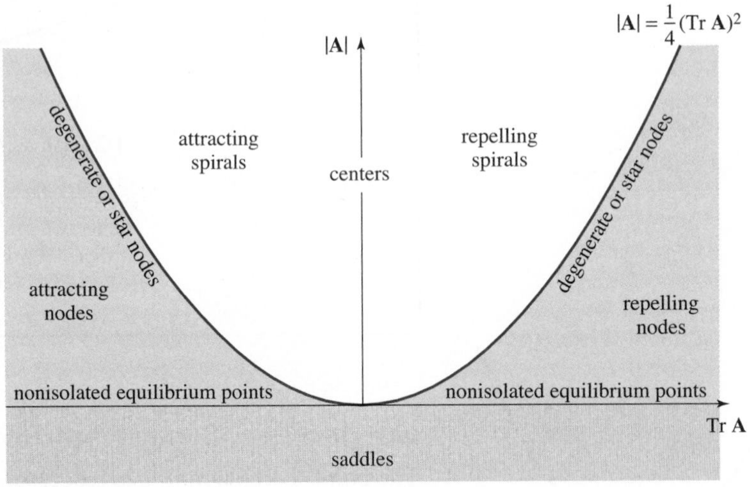

FIGURE 6.4.17 Trace-determinant plane for $\vec{\mathbf{x}}' = \mathbf{A}\vec{\mathbf{x}}$, phase-plane behavior view.

Four Animation Path

Head for the borders and slide along the four curves between regions in the parameter plane.

Stable behaviors are the exception rather than the rule. (See Fig. 6.4.18.)

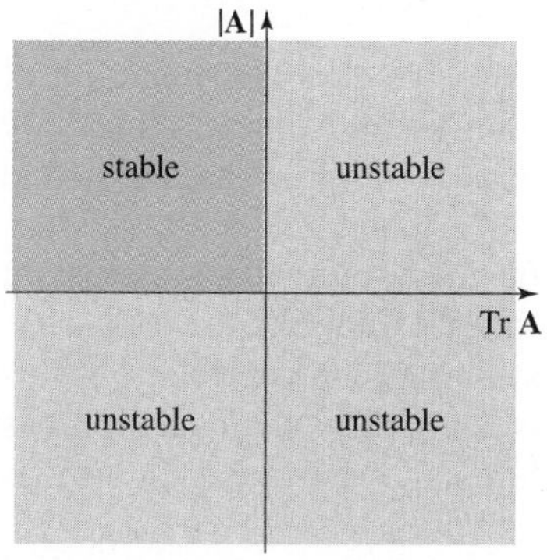

FIGURE 6.4.18 Only the second quadrant of the trace-determinant plane has stable solutions for a 2×2 DE system $\vec{\mathbf{x}}' = \mathbf{A}\vec{\mathbf{x}}$.

EXAMPLE 3 **To Damp or Not to Damp** Let's return to the unforced damped harmonic oscillator of Chapter 4:

$$m\ddot{x} + b\dot{x} + kx = 0, \quad \text{with } m, k > 0,\ b \geq 0. \tag{5}$$

Rewriting (5) as a linear system gives us

$$\begin{aligned} \dot{x} &= y, \\ \dot{y} &= -\frac{k}{m}x - \frac{b}{m}y, \end{aligned} \quad \text{or} \quad \dot{\vec{\mathbf{x}}} = \begin{bmatrix} 0 & 1 \\ -k/m & -b/m \end{bmatrix} \vec{\mathbf{x}},$$

with

$$\operatorname{Tr} \mathbf{A} = -\frac{b}{m} \leq 0 \quad \text{and} \quad |\mathbf{A}| = \frac{k}{m} > 0.$$

From (2)

$$\lambda_1, \lambda_2 = \frac{-\dfrac{b}{m} \pm \sqrt{\dfrac{b^2}{m^2} - 4\dfrac{k}{m}}}{2} = -\frac{b}{2m} \pm \frac{1}{2m}\sqrt{b^2 - 4mk}.$$

Suppose that we trace a path of points on an appropriate portion of the parameter plane in Fig. 6.4.19, giving us the regions that correspond to

- $b = 0$ (undamped),
- $0 < b < \sqrt{4mk}$ (underdamped),
- $b = \sqrt{4mk}$ (critically damped),
- $b > \sqrt{4mk}$ (overdamped).

The insets in the figure show the typical phase portrait behaviors you would see.

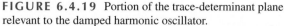

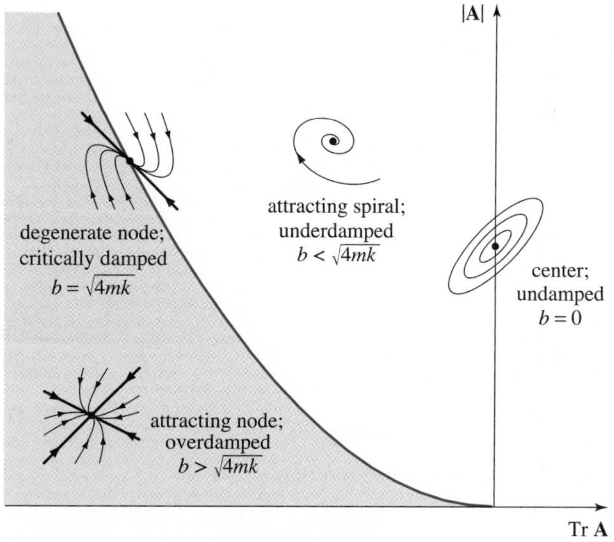

FIGURE 6.4.19 Portion of the trace-determinant plane relevant to the damped harmonic oscillator. ■

Apology for a Special Case

As we come to the end of this section, you may think we have lavished a lot of attention on a situation that is very special in that the system is linear, it has only one equilibrium point, and that equilibrium point is at the origin. There are good reasons for all of this.

We have chosen in this chapter to deal with linear systems having constant coefficients because we can obtain explicit quantitative solutions to work with. Such quantitative representations allow us to see properties and interrelationships that are harder to spot from purely numerical or even graphical results, helpful as they are. Even *more* special was our focus on 2-dimensional systems, but this is where we see nonambiguous pictures.

Having only one equilibrium point helps us to focus on the variety of typical configurations, one to a system. This result is a feature of linearity, as long as the system is unforced.

With this detailed anatomy of equilibrium points established, however, we will find that it can be transformed, with modest adjustments, to help analyze nonlinear systems with multiple equilibrium points, many of them not at the origin. Look well to your nodes and spirals—they will come back, often a little squeezed, stretched, or twisted, in broader contexts as we go on.

Summary

We defined stability, asymptotic stability, and instability for equilibrium solutions of autonomous systems $\vec{\mathbf{x}}' = \mathbf{f}(\vec{\mathbf{x}})$. Then, for 2×2 *linear homogeneous constant coefficient* systems $\vec{\mathbf{x}}' = \mathbf{A}\vec{\mathbf{x}}$, we studied the geometric configurations that characterize the equilibrium solutions, namely nodes (including degenerate nodes and star nodes), spiral points, centers, and saddle points. Finally, we classified these geometries according to their stability properties.

6.4 Problems

Classification Verification *For the systems in Problems 1–6, verify that the equilibrium point at the origin has the geometric character claimed, and determine its stability behavior.*

1. $\vec{\mathbf{x}}' = \begin{bmatrix} 1 & 1 \\ 4 & -2 \end{bmatrix}\vec{\mathbf{x}}$ (saddle point)

2. $\vec{\mathbf{x}}' = \begin{bmatrix} 0 & 1 \\ -1 & 0 \end{bmatrix}\vec{\mathbf{x}}$ (center)

3. $\vec{\mathbf{x}}' = \begin{bmatrix} -2 & 0 \\ 0 & -2 \end{bmatrix}\vec{\mathbf{x}}$ (star node)

4. $\vec{\mathbf{x}}' = \begin{bmatrix} -2 & 1 \\ 0 & -2 \end{bmatrix}\vec{\mathbf{x}}$ (degenerate node)

5. $\vec{\mathbf{x}}' = \begin{bmatrix} 2 & 1 \\ 3 & 4 \end{bmatrix}\vec{\mathbf{x}}$ (node)

6. $\vec{\mathbf{x}}' = \begin{bmatrix} 0 & 1 \\ -1 & -1 \end{bmatrix}\vec{\mathbf{x}}$ (spiral point)

7. Undamped Spring Convert the equation of the undamped mass-spring system,

$$\ddot{x} + \omega_0^2 x = 0,$$

into a system. Determine its equilibrium point or points, and classify the geometry and stability of each.

8. Damped Spring Convert the equation of the damped vibrating spring,

$$m\ddot{x} + b\dot{x} + kx = 0,$$

into a system (mass m, damping constant b, and spring constant k are all positive). Show that the origin is an equilibrium solution, and classify its geometry and stability as functions of m, b, and k.

9. One Zero Eigenvalue Suppose that $\lambda_1 = 0$ but $\lambda_2 \neq 0$, and show the following:

(a) There is a line of equilibrium points.

(b) Solutions starting off the equilibrium line tend toward the line if $\lambda_2 < 0$, away from it if $\lambda_2 > 0$.

10. Zero Eigenvalue Example Consider the system

$$x_1' = 0,$$
$$x_2' = -x_1 + x_2.$$

(a) Show that the eigenvalues are $\lambda_1 = 0$ and $\lambda_2 = 1$.

(b) Find the equilibrium points of the system.

(c) Obtain the general solution of the system.

(d) Show that the solution curves are straight lines.

Both Eigenvalues Zero *In Problems 11–14, you will investigate the nature of the solution of the* 2×2 *system* $\vec{\mathbf{x}}' = \mathbf{A}\vec{\mathbf{x}}$ *when both eigenvalues of* **A** *are zero. Find the solution, plot any fixed points on the phase portrait, and indicate the pertinent information in your sketch.*

11. $\vec{\mathbf{x}}' = \begin{bmatrix} 0 & 1 \\ 0 & 0 \end{bmatrix} \vec{\mathbf{x}}$

12. $\vec{\mathbf{x}}' = \begin{bmatrix} 2 & 1 \\ -4 & -2 \end{bmatrix} \vec{\mathbf{x}}$

13. $\vec{\mathbf{x}}' = \begin{bmatrix} 3 & -9 \\ 1 & -3 \end{bmatrix} \vec{\mathbf{x}}$

14. $\vec{\mathbf{x}}' = \begin{bmatrix} -4 & 2 \\ -8 & 4 \end{bmatrix} \vec{\mathbf{x}}$

15. Zero Again Consider the linear system

$$\vec{\mathbf{x}}' = \begin{bmatrix} 1 & -2 \\ 1 & -2 \end{bmatrix} \vec{\mathbf{x}}.$$

(a) Find the eigenvalues and eigenvectors.

(b) Draw typical solution curves in the phase plane.

16. All Zero Describe the phase portrait of the system

$$\vec{\mathbf{x}}' = \begin{bmatrix} 0 & 0 \\ 0 & 0 \end{bmatrix} \vec{\mathbf{x}}.$$

17. Stability[2] Classify geometry and stability properties of

$$\vec{\mathbf{x}}' = \begin{bmatrix} k & 0 \\ 0 & -1 \end{bmatrix} \vec{\mathbf{x}}$$

for the following values of parameter k.

(a) $k < -1$ (b) $k = -1$ (c) $-1 < k < 0$
(d) $k = 0$ (e) $k > 0$

18. Bifurcation Point Bifurcation points are values of a parameter of a system at which the behavior of the solutions change qualitatively. Determine the bifurcation points of the system

$$\vec{\mathbf{x}}' = \begin{bmatrix} 0 & 1 \\ -1 & k \end{bmatrix} \vec{\mathbf{x}}.$$

19. Interesting Relationships The system $\vec{\mathbf{x}}' = \mathbf{A}\vec{\mathbf{x}}$, where

$$\mathbf{A} = \begin{bmatrix} a & b \\ c & d \end{bmatrix},$$

has eigenvalues λ_1 and λ_2. Show the following:

(a) $\operatorname{Tr} \mathbf{A} = \lambda_1 + \lambda_2$ (b) $|\mathbf{A}| = \lambda_1 \lambda_2$

Interpreting the Trace-Determinant Graph *For* $\vec{\mathbf{x}}' = \mathbf{A}\vec{\mathbf{x}}$, *Fig. 6.4.17 represents possible combinations of values of trace* TrA *and determinant* $|\mathbf{A}|$. *In Problems 20–27, establish the facts about the equilibrium solution at* $\vec{\mathbf{x}} = \vec{\mathbf{0}}$.

20. If $|\mathbf{A}| > 0$ and $(\operatorname{Tr} \mathbf{A})^2 - 4|\mathbf{A}| > 0$, the origin is a node.

21. If $|\mathbf{A}| < 0$, the origin is a saddle point.

22. If $\operatorname{Tr} \mathbf{A} \neq 0$ and $(\operatorname{Tr} \mathbf{A})^2 - 4|\mathbf{A}| < 0$, the origin is a spiral point.

23. If $\operatorname{Tr} \mathbf{A} = 0$ and $|\mathbf{A}| > 0$, the origin is a center.

24. If $(\operatorname{Tr} \mathbf{A})^2 - 4|\mathbf{A}| = 0$ and $\operatorname{Tr} \mathbf{A} \neq 0$, the origin is a degenerate or star node.

25. If $\operatorname{Tr} \mathbf{A} > 0$ or $|\mathbf{A}| < 0$, the origin is unstable.

26. If $|\mathbf{A}| > 0$ and $\operatorname{Tr} \mathbf{A} = 0$, the origin is neutrally stable. This is the case of purely imaginary eigenvalues.

27. If $\operatorname{Tr} \mathbf{A} < 0$ and $|\mathbf{A}| > 0$, the origin is asymptotically stable.

28. Suggested Journal Entry What can you say about the relationship between diagonalization of matrix **A** and the geometry and stability of the equilibrium solution at the origin of the system? Develop your response using specific examples.

---

[2]Inspired by Steven H. Strogatz, *Nonlinear Dynamics and Chaos* (Reading: Addison-Wesley, 1994), an excellent treatment of dynamical systems at a more advanced level.

6.5 Decoupling a Linear DE System

SYNOPSIS: Diagonalizing the matrix of a 2×2 system of homogeneous linear differential equations with constant coefficients provides an alternative solution method to that of the previous sections for 2×2 systems. This method generalizes more readily to larger systems in a significant number of situations, including those with forcing terms, where its effect is to decouple the variables.

Diagonalization Is the Key

We looked at the notion of diagonalization in Sec. 5.4, in connection with the change of basis in a vector space. We return to it now because it complements our explicit solutions of systems developed earlier in this chapter. For many systems it is true that eigenvectors form a superior basis. They define directions in which the behavior of the system simplifies to dependence on one variable at a time. Not only is this useful in the process of *quantitative* solution, as we shall show, but it also points the way to helpful *qualitative* interpretations. For example, look again at how real eigenvectors relate to the phase portraits in Sec. 6.2. Diagonalization allows us to use the eigenvectors as *axes* in a transformed coordinate system, as you will see soon in Example 1, Fig. 6.5.1.

In Sec. 5.4, Example 7 showed how diagonalizing can simplify a system of differential equations. We now elaborate on this idea, for linear DE systems *with constant coefficients*, if the matrices are also *diagonalizable*. Let us review.

Diagonalization of a Matrix

A diagonalizable $n \times n$ matrix $\mathbf{A}$ has n eigenvalues and n linearly independent eigenvectors. We can construct:

- $\mathbf{D}$, a diagonal matrix whose diagonal elements are the eigenvalues of $\mathbf{A}$;
- $\mathbf{P}$, a matrix whose columns are the eigenvectors, listed in the order corresponding to the order of the eigenvalues in $\mathbf{D}$.

$\mathbf{P}$ is a change-of-basis matrix such that

$$\mathbf{A} = \mathbf{P}\mathbf{D}\mathbf{P}^{-1} \quad \text{and} \quad \mathbf{D} = \mathbf{P}^{-1}\mathbf{A}\mathbf{P}. \tag{1}$$

We say that $\mathbf{P}$ **diagonalizes** $\mathbf{A}$.

Checking Calculations:
For checking calculations,

$$\mathbf{A}\mathbf{P} = \mathbf{P}\mathbf{D}$$

is a useful variation of (1).

The equations in (1) are quite useful, for checking your calculations, or for proving many amazing results, such as the following.

Decoupling a Homogeneous Linear System

Matrix Element Input

Experiment with the difference in phase portraits for diagonal matrices as opposed to nondiagonal matrices.

Decoupling a Homogeneous Linear DE System

For a linear DE system

$$\vec{\mathbf{x}}' = \mathbf{A}\vec{\mathbf{x}} \tag{2}$$

with diagonalizable matrix $\mathbf{A}$, the change of variables

$$\vec{\mathbf{x}} = \mathbf{P}\vec{\mathbf{w}} \tag{3}$$

transforms system (2) into a **decoupled** system

$$\vec{\mathbf{w}}' = \mathbf{D}\vec{\mathbf{w}}, \tag{4}$$

where each component equation involves a single variable and can be easily solved to find $\vec{\mathbf{w}}$. The general solution $\vec{\mathbf{x}}$ to (2) follows from (3).

Real eigenvectors in $\vec{\mathbf{w}}$ always lie along the axes.

Proof If $\vec{\mathbf{x}} = \mathbf{P}\vec{\mathbf{w}}$, then $\vec{\mathbf{w}} = \mathbf{P}^{-1}\vec{\mathbf{x}}$ and

$$\vec{\mathbf{w}}' = \mathbf{P}^{-1}\vec{\mathbf{x}}' = \mathbf{P}^{-1}\mathbf{A}\vec{\mathbf{x}} = \mathbf{P}^{-1}(\mathbf{PDP}^{-1})\vec{\mathbf{x}} = \mathbf{D}\vec{\mathbf{w}}.$$

In system (4) the variables are no longer mixed together as they are in (2). Each component equation of (4) is a first-order linear DE in a single dependent variable, and you learned to solve those in Chapter 2. Once you have found each component solution w_i, you can find $\vec{\mathbf{x}} = \mathbf{P}\vec{\mathbf{w}}$ by simple matrix multiplication. □

EXAMPLE 1 **Decoupling** To apply diagonalization to the system

$$\vec{\mathbf{x}}' = \mathbf{A}\vec{\mathbf{x}} = \begin{bmatrix} 2 & 2 \\ 1 & 3 \end{bmatrix}\vec{\mathbf{x}}, \tag{5}$$

we determine that $\mathbf{A}$ has eigenvalues

$$\lambda_1 = 4 \quad \text{and} \quad \lambda_2 = 1,$$

with corresponding eigenvectors

$$\vec{\mathbf{v}}_1 = \begin{bmatrix} 1 \\ 1 \end{bmatrix} \quad \text{and} \quad \vec{\mathbf{v}}_2 = \begin{bmatrix} 2 \\ -1 \end{bmatrix}.$$

From the eigenvalues and eigenvectors, respectively, we construct

$$\mathbf{D} = \begin{bmatrix} 4 & 0 \\ 0 & 1 \end{bmatrix} \quad \text{and} \quad \mathbf{P} = \begin{bmatrix} 1 & 2 \\ 1 & -1 \end{bmatrix}.$$

We can replace the system corresponding to (5) with the transformed system corresponding to $\vec{\mathbf{w}}' = \mathbf{D}\vec{\mathbf{w}}$:

$$\begin{aligned} x_1' &= 2x_1 + 2x_2, \\ x_2' &= \ \ x_1 + 3x_2, \end{aligned} \quad \Rightarrow \quad \begin{aligned} w_1' &= 4w_1 \\ w_2' &= w_2. \end{aligned}$$

The solution to the $\vec{\mathbf{w}}$ system is immediate:

$$w_1(t) = c_1e^{4t} \quad \text{and} \quad w_2(t) = c_2e^t.$$

By (3), the general solution to (5) is

$$\vec{\mathbf{x}}(t) = \mathbf{P}\vec{\mathbf{w}} = \begin{bmatrix} 1 & 2 \\ 1 & -1 \end{bmatrix}\begin{bmatrix} c_1e^{4t} \\ c_2e^t \end{bmatrix} = \begin{bmatrix} c_1e^{4t} + 2c_2e^t \\ c_1e^{4t} - \ \ c_2e^t \end{bmatrix}. \tag{6}$$

This is the same result that we obtained in Sec. 6.2, Example 2, with reversed c_i because the eigenvalues were listed in reverse order.

A look at the phase portraits of the two systems in Example 1 shows what decoupling accomplishes. The eigenvectors in $\vec{\mathbf{x}}$-space have been rotated (individually) to align with the axes in $\vec{\mathbf{w}}$-space. (See Fig. 6.5.1.) ■

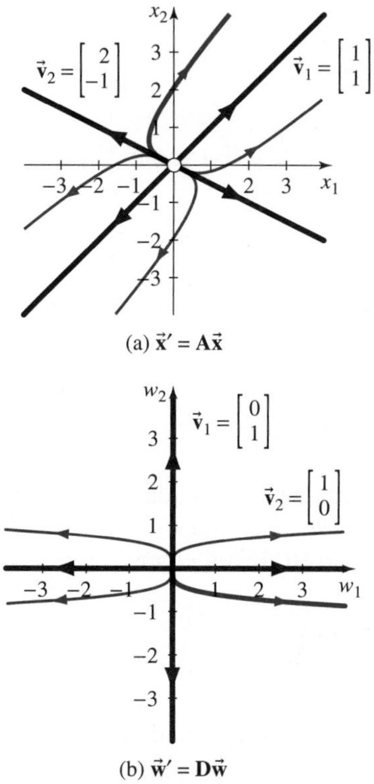

FIGURE 6.5.1 Phase portraits for a system before and after decoupling. The eigenvectors lie along the axes in the new coordinate system.

Example 1 illustrates that to get the general solution to a linear homogeneous DE system it is only necessary to find $\mathbf{D}$ for equation (4) and $\mathbf{P}$ for equation (3).

Computing $\mathbf{P}^{-1}$ allows us to (a) check our calculations and/or (b) incorporate an initial condition $\vec{\mathbf{x}}_0$. We can evaluate the c_i in the general solution using this fact: $\vec{\mathbf{x}}_0 = \mathbf{P}\vec{\mathbf{w}} = \mathbf{P}\vec{\mathbf{c}}$, so

$$\vec{\mathbf{c}} = \mathbf{P}^{-1}\vec{\mathbf{x}}_0.$$

EXAMPLE 2 **Exploiting $\mathbf{P}^{-1}$** Returning to system (5) of Example 1, we compute

$$\mathbf{P}^{-1} = \begin{bmatrix} 1/3 & 2/3 \\ 1/3 & -1/3 \end{bmatrix}.$$

(a) We can confirm our computation by multiplication, using either of the equations in (1):

$$\mathbf{D} = \mathbf{P}^{-1}\mathbf{A}\mathbf{P} = \begin{bmatrix} 1/3 & 2/3 \\ 1/3 & -1/3 \end{bmatrix}\begin{bmatrix} 2 & 2 \\ 1 & 3 \end{bmatrix}\begin{bmatrix} 1 & 2 \\ 1 & -1 \end{bmatrix} = \begin{bmatrix} 4 & 0 \\ 0 & 1 \end{bmatrix}.$$

(b) To find the solution that passes through (0, 2), we compute the c_i:

$$\vec{\mathbf{c}} = \mathbf{P}^{-1}\vec{\mathbf{x}}_0$$

$$\begin{bmatrix} c_1 \\ c_2 \end{bmatrix} = \begin{bmatrix} 1/3 & 2/3 \\ 1/3 & -1/3 \end{bmatrix}\begin{bmatrix} 0 \\ 2 \end{bmatrix} = \begin{bmatrix} 4/3 \\ -2/3 \end{bmatrix}.$$

Substitution into the general solution (6) gives

$$\vec{\mathbf{x}} = \mathbf{P}\vec{\mathbf{w}} = \begin{bmatrix} 1 & 2 \\ 1 & -1 \end{bmatrix}\begin{bmatrix} 4/3e^{4t} \\ -2/3e^{t} \end{bmatrix} = \begin{bmatrix} 4/3e^{4t} - 4/3e^{t} \\ 4/3e^{4t} + 2/3e^{t} \end{bmatrix}.$$

This particular solution is highlighted in Fig. 6.5.1. ■

EXAMPLE 3 **Detripling?** Let's consider the 3×3 system $\vec{\mathbf{x}}' = \mathbf{A}\vec{\mathbf{x}}$ having matrix

$$\mathbf{A} = \begin{bmatrix} -2 & 1 & 1 \\ 1 & -2 & 1 \\ 1 & 1 & -2 \end{bmatrix}.$$

From Sec. 5.3, Example 5, we know that the characteristic equation is given by $|\mathbf{A} - \lambda\mathbf{I}| = \lambda(\lambda + 3)^2 = 0$, with eigenvalues $\lambda_1 = 0$ and $\lambda_2 = \lambda_3 = -3$. Although there is a repeated eigenvalue, we found sufficient eigenvectors

$$\vec{\mathbf{v}}_1 = \begin{bmatrix} 1 \\ 1 \\ 1 \end{bmatrix}, \quad \vec{\mathbf{v}}_2 = \begin{bmatrix} -1 \\ 1 \\ 0 \end{bmatrix}, \quad \text{and} \quad \vec{\mathbf{v}}_3 = \begin{bmatrix} -1 \\ 0 \\ 1 \end{bmatrix}$$

to diagonalize $\mathbf{A}$. Thus, we can write from these eigenvalues and eigenvectors

$$\mathbf{D} = \begin{bmatrix} 0 & 0 & 0 \\ 0 & -3 & 0 \\ 0 & 0 & -3 \end{bmatrix} \quad \text{and} \quad \mathbf{P} = \begin{bmatrix} 1 & -1 & -1 \\ 1 & 1 & 0 \\ 1 & 0 & 1 \end{bmatrix}.$$

With $\vec{\mathbf{x}} = \mathbf{P}\vec{\mathbf{w}}$, we convert the system $\vec{\mathbf{x}}' = \mathbf{A}\vec{\mathbf{x}}$ into the simpler system $\vec{\mathbf{w}}' = \mathbf{D}\vec{\mathbf{w}}$:

$$\begin{aligned} x_1' &= -2x_1 + x_2 + x_3, \\ x_2' &= x_1 - 2x_2 + x_3, \\ x_3' &= x_1 + x_2 - 2x_3, \end{aligned} \quad \Rightarrow \quad \begin{aligned} w_1' &= 0, \\ w_2' &= -3w_2, \\ w_3' &= -3w_3. \end{aligned}$$

Solving $\vec{\mathbf{w}}' = \mathbf{D}\vec{\mathbf{w}}$ gives

$$\vec{\mathbf{w}}(t) = \begin{bmatrix} c_1 \\ c_2e^{-3t} \\ c_3e^{-3t} \end{bmatrix}.$$

Therefore, the DE system $\vec{\mathbf{x}}' = \mathbf{A}\vec{\mathbf{x}}$ has solution

$$\vec{\mathbf{x}} = \mathbf{P}\vec{\mathbf{w}} = \begin{bmatrix} 1 & -1 & -1 \\ 1 & 1 & 0 \\ 1 & 0 & 1 \end{bmatrix} \begin{bmatrix} c_1 \\ c_2e^{-3t} \\ c_3e^{-3t} \end{bmatrix} = \begin{bmatrix} c_1 - (c_2 + c_3)e^{-3t} \\ c_1 + c_2e^{-3t} \\ c_1 + c_3e^{-3t} \end{bmatrix}.$$

■

Decoupling Nonhomogeneous Systems

The major advantage of the diagonalization process with differential equations is that it can be extended to nonhomogeneous systems.

Decoupling a Nonhomogeneous Linear DE System

To decouple a linear system

$$\vec{\mathbf{x}}' = \mathbf{A}\vec{\mathbf{x}} + \mathbf{f}(t), \tag{7}$$

where $n \times n$ matrix $\mathbf{A}$ has n linearly independent eigenvectors, proceed as follows.

Step 1. Calculate the eigenvalues and find the corresponding n independent eigenvectors of $\mathbf{A}$.

Step 2. Form the diagonal matrix $\mathbf{D}$ whose diagonal elements are the eigenvalues and the matrix $\mathbf{P}$ whose columns are the n eigenvectors, listed in the same order as their corresponding eigenvalues. Then find $\mathbf{P}^{-1}$.

Step 3. Let

$$\vec{\mathbf{x}} = \mathbf{P}\vec{\mathbf{w}}, \tag{8}$$

and solve the decoupled system

$$\vec{\mathbf{w}}' = \mathbf{D}\vec{\mathbf{w}} + \mathbf{P}^{-1}\vec{\mathbf{f}}(t). \tag{9}$$

Step 4. Solve (7) using (8) and the solution to (9).

Proof

$$\begin{aligned} \vec{\mathbf{w}}' = (\mathbf{P}^{-1}\vec{\mathbf{x}})' &= \mathbf{P}^{-1}\vec{\mathbf{x}}' \\ &= \mathbf{P}^{-1}(\mathbf{A}\vec{\mathbf{x}} + \vec{\mathbf{f}}) \\ &= \mathbf{P}^{-1}\mathbf{A}\vec{\mathbf{x}} + \mathbf{P}^{-1}\vec{\mathbf{f}} \\ &= \mathbf{P}^{-1}(\mathbf{P}\mathbf{D}\mathbf{P}^{-1})\vec{\mathbf{x}} + \mathbf{P}^{-1}\vec{\mathbf{f}} \\ &= \mathbf{D}\vec{\mathbf{w}} + \mathbf{P}^{-1}\vec{\mathbf{f}}. \end{aligned}$$

The transformed system (9) is also nonhomogeneous, with forcing function $\mathbf{P}^{-1}\mathbf{f}$. The component equations are nonhomogeneous first-order linear equations of the type solved in Sec. 2.1. □

EXAMPLE 4 **Nonhomogeneous Decoupling** We will decouple

$$x_1' = -3x_1 + x_2,$$
$$x_2' = x_1 - 3x_2 + e^{-t},$$

a nonhomogeneous 2 × 2 system, which has matrix-vector form (7), where

$$\mathbf{A} = \begin{bmatrix} -3 & 1 \\ 1 & -3 \end{bmatrix} \quad \text{and} \quad \mathbf{f}(t) = \begin{bmatrix} 0 \\ e^{-t} \end{bmatrix}.$$

We can readily calculate that **A** has eigenvalues $\lambda_1 = -2$ and $\lambda_2 = -4$ with independent eigenvectors

$$\vec{\mathbf{v}}_1 = \begin{bmatrix} 1 \\ 1 \end{bmatrix} \quad \text{and} \quad \vec{\mathbf{v}}_2 = \begin{bmatrix} 1 \\ -1 \end{bmatrix},$$

respectively, and you can confirm that

$$\mathbf{D} = \begin{bmatrix} -2 & 0 \\ 0 & -4 \end{bmatrix}, \quad \mathbf{P} = \begin{bmatrix} 1 & 1 \\ 1 & -1 \end{bmatrix}, \quad \text{and} \quad \mathbf{P}^{-1} = \begin{bmatrix} 1/2 & 1/2 \\ 1/2 & -1/2 \end{bmatrix}.$$

The new forcing term is

$$\mathbf{P}^{-1}\mathbf{f} = \begin{bmatrix} 1/2 & 1/2 \\ 1/2 & -1/2 \end{bmatrix} \begin{bmatrix} 0 \\ e^{-t} \end{bmatrix} = \begin{bmatrix} \frac{1}{2}e^{-t} \\ -\frac{1}{2}e^{-t} \end{bmatrix},$$

Recycling:

Once we have **D**, **P**, and $\mathbf{P}^{-1}$, we can use them over and over with different forcing functions.

and our decoupled system is

$$\vec{\mathbf{w}}' = \mathbf{D}\vec{\mathbf{w}} + \mathbf{P}^{-1}\mathbf{f} = \begin{bmatrix} -2 & 0 \\ 0 & -4 \end{bmatrix} \begin{bmatrix} w_1 \\ w_2 \end{bmatrix} + \begin{bmatrix} \frac{1}{2}e^{-t} \\ -\frac{1}{2}e^{-t} \end{bmatrix},$$

or

$$\begin{aligned} w_1' &= -2w_1 + \tfrac{1}{2}e^{-t}, \\ w_2' &= -4w_2 - \tfrac{1}{2}e^{-t}. \end{aligned} \tag{10}$$

Solving the decoupled equations (10) separately, with methods of Sec. 2.1,

$$w_1 = c_1e^{-2t} + \tfrac{1}{2}e^{-t},$$
$$w_2 = c_2e^{-4t} - \tfrac{1}{6}e^{-t}.$$

Now we can use $\vec{\mathbf{x}} = \mathbf{P}\vec{\mathbf{w}}$ to obtain

$$\vec{\mathbf{x}} = \begin{bmatrix} 1 & 1 \\ 1 & -1 \end{bmatrix} \begin{bmatrix} w_1 \\ w_2 \end{bmatrix} = \begin{bmatrix} w_1 + w_2 \\ w_1 - w_2 \end{bmatrix} = \begin{bmatrix} c_1e^{-2t} + c_2e^{-4t} + \frac{1}{3}e^{-t} \\ c_1e^{-2t} - c_2e^{-4t} + \frac{2}{3}e^{-t} \end{bmatrix},$$

from which we can write

$$x_1(t) = c_1e^{-2t} + c_2e^{-4t} + \frac{1}{3}e^{-t} \quad \text{and} \quad x_2(t) = c_1e^{-2t} - c_2e^{-4t} + \frac{2}{3}e^{-t}.$$

■

The examples in this section have all used matrices with *real* eigenvalues. The process of decoupling is valid for the cases with *nonreal* eigenvalues, but the complex number calculations can become something of a tangle and cause more trouble than enlightenment. (See Problem 23.)

Summary

When a matrix can be diagonalized, the corresponding system of linear DEs can be decoupled, reducing its solution to a set of one-variable problems. This process extends to nonhomogeneous linear DEs, which is our main motivation for this section.

6.5 Problems

Decoupling Homogeneous Linear Systems *For Problems 1–10, construct appropriate diagonalizing matrices, decouple the linear systems, then solve the systems.*

1. $\vec{\mathbf{x}}' = \begin{bmatrix} -1 & -2 \\ -2 & 2 \end{bmatrix}\vec{\mathbf{x}}$

2. $\vec{\mathbf{x}}' = \begin{bmatrix} 0 & -1 \\ -3 & 2 \end{bmatrix}\vec{\mathbf{x}}$

3. $\vec{\mathbf{x}}' = \begin{bmatrix} 0 & -1 \\ -1 & 0 \end{bmatrix}\vec{\mathbf{x}}$

4. $\vec{\mathbf{x}}' = \begin{bmatrix} 2 & 3 \\ 1 & 4 \end{bmatrix}\vec{\mathbf{x}}$

5. $\vec{\mathbf{x}}' = \begin{bmatrix} 2 & -3 \\ 2 & -5 \end{bmatrix}\vec{\mathbf{x}}$

6. $\vec{\mathbf{x}}' = \begin{bmatrix} 0 & 1 \\ 1 & 0 \end{bmatrix}\vec{\mathbf{x}}$

7. $\vec{\mathbf{x}}' = \begin{bmatrix} 1 & 1 & 1 \\ 1 & 1 & 1 \\ 1 & 1 & 1 \end{bmatrix}\vec{\mathbf{x}}$

8. $\vec{\mathbf{x}}' = \begin{bmatrix} 0 & 0 & 0 \\ 0 & 1 & 0 \\ 1 & 0 & 1 \end{bmatrix}\vec{\mathbf{x}}$

9. $\vec{\mathbf{x}}' = \begin{bmatrix} 1 & 0 & 0 \\ -4 & 3 & 0 \\ -4 & 2 & 1 \end{bmatrix}\vec{\mathbf{x}}$

10. $\vec{\mathbf{x}}' = \begin{bmatrix} 3 & -2 & 0 \\ 1 & 0 & 0 \\ -1 & 1 & 3 \end{bmatrix}\vec{\mathbf{x}}$

Decoupling Nonhomogeneous Linear Systems *For Problems 11–20, construct appropriate diagonalizing matrices, decouple the linear systems, then solve the nonhomogeneous systems. (A computer algebra system is recommended for Problems 19 and 20.)*

11. $\vec{\mathbf{x}}' = \begin{bmatrix} 0 & 1 \\ 1 & 0 \end{bmatrix}\vec{\mathbf{x}} + \begin{bmatrix} 1 \\ 1 \end{bmatrix}$

12. $\vec{\mathbf{x}}' = \begin{bmatrix} -3 & 1 \\ 1 & -3 \end{bmatrix}\vec{\mathbf{x}} + \begin{bmatrix} \sin t \\ 0 \end{bmatrix}$

13. $\vec{\mathbf{x}}' = \begin{bmatrix} 1 & 1 \\ 1 & 1 \end{bmatrix}\vec{\mathbf{x}} + \begin{bmatrix} t \\ 1 \end{bmatrix}$

14. $\vec{\mathbf{x}}' = \begin{bmatrix} 5 & 4 \\ 1 & 2 \end{bmatrix}\vec{\mathbf{x}} + \begin{bmatrix} 5t \\ 0 \end{bmatrix}$

15. $\vec{\mathbf{x}}' = \begin{bmatrix} 1 & 4 \\ 2 & 3 \end{bmatrix}\vec{\mathbf{x}} + \begin{bmatrix} t \\ 2t \end{bmatrix}$

16. $\vec{\mathbf{x}}' = \begin{bmatrix} 1 & 4 \\ -4 & 11 \end{bmatrix}\vec{\mathbf{x}} + \begin{bmatrix} e^t \\ e^t \end{bmatrix}$

17. $\vec{\mathbf{x}}' = \begin{bmatrix} 1 & 0 & 0 \\ -4 & 3 & 0 \\ -4 & 2 & 1 \end{bmatrix}\vec{\mathbf{x}} + \begin{bmatrix} 1 \\ 0 \\ 1 \end{bmatrix}$

18. $\vec{\mathbf{x}}' = \begin{bmatrix} 3 & -2 & 0 \\ 1 & 0 & 0 \\ -1 & 1 & 3 \end{bmatrix}\vec{\mathbf{x}} + \begin{bmatrix} 4 \\ 6 \\ 1 \end{bmatrix}$

19. $\vec{\mathbf{x}}' = \begin{bmatrix} 4 & 1 & -1 \\ 2 & 5 & -2 \\ 1 & 1 & 2 \end{bmatrix}\vec{\mathbf{x}} + \begin{bmatrix} 1 \\ t \\ t^2 \end{bmatrix}$

20. $\vec{\mathbf{x}}' = \begin{bmatrix} 0 & 0 & 1 & 0 \\ 0 & 0 & 0 & 1 \\ 1 & 0 & 0 & 0 \\ 0 & 1 & 0 & 0 \end{bmatrix}\vec{\mathbf{x}} + \begin{bmatrix} t \\ 0 \\ -t \\ 1 \end{bmatrix}$

21. Working Backwards Find a matrix with eigenvalues $\lambda_1 = 1$, $\lambda_2 = -1$, and corresponding eigenvectors

$$\vec{\mathbf{v}}_1 = \begin{bmatrix} 1 \\ 1 \end{bmatrix} \quad \text{and} \quad \vec{\mathbf{v}}_2 = \begin{bmatrix} 1 \\ 2 \end{bmatrix}.$$

22. Jordan Form For

$$\vec{\mathbf{x}}' = \mathbf{A}\vec{\mathbf{x}} = \begin{bmatrix} 2 & 1 \\ -1 & 4 \end{bmatrix}\vec{\mathbf{x}}, \tag{11}$$

decoupling hits a snag: The matrix **A** has a multiple eigenvalue λ with only one independent eigenvector, so **A** cannot be diagonalized. It *can*, however, be put in the **Jordan form**

$$\mathbf{J} = \begin{bmatrix} \lambda & 1 \\ 0 & \lambda \end{bmatrix}$$

by a procedure a little like that for diagonalization.[1] Let $\mathbf{P} = [\vec{\mathbf{v}} \,|\, \vec{\mathbf{w}}]$ be the 2×2 matrix with columns $\vec{\mathbf{v}}$ and $\vec{\mathbf{w}}$, where $\vec{\mathbf{v}}$ is an eigenvector belonging to λ; that is, a solution of $(\mathbf{A} - \lambda\mathbf{I})\vec{\mathbf{v}} = \vec{\mathbf{0}}$, and $\vec{\mathbf{w}}$ is a solution of

$$(\mathbf{A} - \lambda\mathbf{I})\vec{\mathbf{w}} = \vec{\mathbf{v}}.$$

(In other words, $\vec{\mathbf{w}}$ is a generalized eigenvector of **A**.)

(a) Calculate λ, $\vec{\mathbf{w}}$, **P**, and $\mathbf{P}^{-1}$; verify that $\mathbf{P}^{-1}\mathbf{A}\mathbf{P} = \mathbf{J}$.

(b) Show how you can use a modified decoupling method to solve the linear DE system (11).

23. Complex Decoupling Find the solution for

$$\vec{\mathbf{x}}' = \begin{bmatrix} 0 & 1 \\ -1 & 0 \end{bmatrix}\vec{\mathbf{x}}$$

by the methods of Sec. 6.3 or 4.3, then find the solution by decoupling and confirm that you can get the same result.

24. Suggested Journal Entry In Example 1, we constructed phase portraits in the x_1x_2-plane and the w_1w_2-plane of the diagonalized version of the system. For other examples in Sec. 6.2, discuss how the phase portrait in the w_1w_2-plane would differ from the illustrated phase portrait in the x_1x_2-plane. Can you generalize this to higher-order systems with real eigenvalues?

---

[1] When a matrix can't be diagonalized, it can be converted into a matrix with "Jordan blocks" strung along the diagonal. Marie Ennemond Camille Jordan (1838–1922), a French mathematician, is the gentleman whose name is associated with this form of a matrix. He is *not* the same person we met in Sec. 3.2, who was associated with Gauss-Jordan elimination.

6.6 Matrix Exponential

SYNOPSIS: We define $e^{\mathbf{A}}$ and $e^{\mathbf{A}t}$ for an $n \times n$ matrix $\mathbf{A}$, and use them to show alternate ways of solving linear DE systems, both homogeneous and nonhomogeneous.

Constant Matrix Exponential

For any constant a, we have seen that the general solution of the first-order equation $y' = ay$ is $y = ce^{at}$, where c is an arbitrary constant. We now show that it is possible to define an $n \times n$ matrix $e^{\mathbf{A}t}$ so that

$$\vec{\mathbf{x}} = e^{\mathbf{A}t}\vec{\mathbf{c}} \tag{1}$$

is a solution of the linear system $\vec{\mathbf{x}}' = \mathbf{A}\vec{\mathbf{x}}$, where $\mathbf{A}$ is an $n \times n$ constant *matrix* and $\vec{\mathbf{c}}$ is an $n \times 1$ *vector* of arbitrary constants. In (1), the vector $\vec{\mathbf{c}}$ postmultiplies $e^{\mathbf{A}t}$ in order that the product $e^{\mathbf{A}t}\vec{\mathbf{c}}$ be an $n \times 1$ vector.

Our first task is to define $e^{\mathbf{A}}$ for a constant square matrix $\mathbf{A}$. One approach to defining the exponent of a matrix is motivated by the series expansion of the *scalar* exponential function,

$$e^x = 1 + x + \frac{x^2}{2!} + \frac{x^3}{3!} + \cdots + \frac{x^k}{k!} + \cdots, \tag{2}$$

which converges for all real or complex x. Replacing the number 1 in (2) by the identity matrix $\mathbf{I}$, and x by an $n \times n$ matrix $\mathbf{A}$, gives us the matrix exponential.

Constant Matrix Exponential

Given a constant $n \times n$ matrix $\mathbf{A}$,

$$e^{\mathbf{A}} = \mathbf{I} + \mathbf{A} + \frac{\mathbf{A}^2}{2!} + \frac{\mathbf{A}^3}{3!} + \cdots + \frac{\mathbf{A}^k}{k!} + \cdots. \tag{3}$$

The computation of the matrix exponential is not an easy matter, due to the fact that it is an infinite series, but in some cases, such as when $\mathbf{A}$ is a diagonal matrix, the computation is straightforward.

EXAMPLE 1 **Matrix Exponential of a Diagonal Matrix** The powers of a 2×2 diagonal matrix

$$\mathbf{A} = \begin{bmatrix} a & 0 \\ 0 & b \end{bmatrix} \quad \text{are} \quad \mathbf{A}^n = \begin{bmatrix} a^n & 0 \\ 0 & b^n \end{bmatrix},$$

so the matrix exponential is simply

$$\begin{aligned} e^{\mathbf{A}} &= \mathbf{I} + \mathbf{A} + \frac{\mathbf{A}^2}{2!} + \cdots \\ &= \begin{bmatrix} 1 & 0 \\ 0 & 1 \end{bmatrix} + \begin{bmatrix} a & 0 \\ 0 & b \end{bmatrix} + \begin{bmatrix} a^2/2 & 0 \\ 0 & b^2/2 \end{bmatrix} + \cdots \\ &= \begin{bmatrix} 1 + a + a^2/2 + \cdots & 0 \\ 0 & 1 + b + b^2/2 + \cdots \end{bmatrix} = \begin{bmatrix} e^a & 0 \\ 0 & e^b \end{bmatrix}. \end{aligned}$$

Hence, the matrix exponential of a 2×2 diagonal matrix can be obtained by simply raising the numbers on the diagonal to their exponentials. The $n \times n$

diagonal matrix follows along the same lines. The exponential matrix of

$$\mathbf{A} = \begin{bmatrix} a_1 & 0 & \cdots & 0 \\ 0 & a_2 & \cdots & 0 \\ \vdots & \vdots & \ddots & \vdots \\ 0 & 0 & \cdots & a_n \end{bmatrix} \quad \text{is simply} \quad e^{\mathbf{A}} = \begin{bmatrix} e^{a_1} & 0 & \cdots & 0 \\ 0 & e^{a_2} & \cdots & 0 \\ \vdots & \vdots & \ddots & \vdots \\ 0 & 0 & \cdots & e^{a_n} \end{bmatrix}.$$

■

Nilpotent matrices are another class of matrices for which the matrix exponential can easily be found.

Nilpotent Matrix

A square matrix $\mathbf{A}$ is called **nilpotent** if $\mathbf{A}^n = \mathbf{0}$ for some positive integer n.

In the case of a nilpotent matrix, series (3) terminates after a finite number of terms, so the matrix exponential $e^{\mathbf{A}}$ is a finite sum. One class of nilpotent matrices is the class of triangular matrices in which all entries on the main diagonal of the matrix are zero.

EXAMPLE 2 **Nilpotent Matrix** To find the matrix exponential $e^{\mathbf{A}}$ of the triangular matrix

$$\mathbf{A} = \begin{bmatrix} 0 & -1 & 2 \\ 0 & 0 & 1 \\ 0 & 0 & 0 \end{bmatrix},$$

we need only compute

$$\mathbf{A}^2 = \begin{bmatrix} 0 & 0 & -1 \\ 0 & 0 & 0 \\ 0 & 0 & 0 \end{bmatrix} \quad \text{and} \quad \mathbf{A}^3 = \begin{bmatrix} 0 & 0 & 0 \\ 0 & 0 & 0 \\ 0 & 0 & 0 \end{bmatrix}.$$

For $n \geq 3$, $\mathbf{A}^n = \mathbf{0}$, so

$$\begin{aligned} e^{\mathbf{A}} &= \mathbf{I} + \mathbf{A} + \frac{1}{2!}\mathbf{A}^2 \\ &= \begin{bmatrix} 1 & 0 & 0 \\ 0 & 1 & 0 \\ 0 & 0 & 1 \end{bmatrix} + \begin{bmatrix} 0 & -1 & 2 \\ 0 & 0 & 1 \\ 0 & 0 & 0 \end{bmatrix} + \frac{1}{2!}\begin{bmatrix} 0 & 0 & -1 \\ 0 & 0 & 0 \\ 0 & 0 & 0 \end{bmatrix} \\ &= \begin{bmatrix} 1 & -1 & 3/2 \\ 0 & 1 & 1 \\ 0 & 0 & 1 \end{bmatrix}. \end{aligned}$$

■

The matrix exponential $e^{\mathbf{A}}$ satisfies most of the properties of scalar exponentials.

The fact that the matrix exponential always has an inverse means that the columns of $e^{\mathbf{A}}$ are linearly independent. Property (ii) also states that finding the inverse of a matrix exponential is trivial, because one simply replaces the matrix $\mathbf{A}$ in the power series (3) with its negative, $-\mathbf{A}$. (Problems 16 and 17 ask the reader to verify properties (ii) and (iii).)

Properties of the Matrix Exponential $e^{\mathbf{A}}$

(i) $e^{\mathbf{0}} = \mathbf{I}_n$, where $\mathbf{0}$ is the $n \times n$ zero matrix.

(ii) $\left(e^{\mathbf{A}}\right)^{-1} = e^{-\mathbf{A}}$.

(iii) If $\mathbf{AB} = \mathbf{BA}$, then $e^{\mathbf{A}+\mathbf{B}} = e^{\mathbf{A}}e^{\mathbf{B}}$.

Matrix Exponential Function

If t is a scalar variable, then by replacing the constant matrix $\mathbf{A}$ with $t\mathbf{A}$ we arrive at the matrix exponential *function*.[1]

Matrix Exponential Function

Given an $n \times n$ constant matrix $\mathbf{A}$,

$$e^{\mathbf{A}t} = \mathbf{I} + t\mathbf{A} + \frac{t^2}{2!}\mathbf{A}^2 + \cdots + \frac{t^k}{k!}\mathbf{A}^k + \cdots. \tag{4}$$

One can show that series (4) converges to an $n \times n$ matrix for all t. (See Problem 20.)

EXAMPLE 3 Nilpotent Matrix Function Using the matrix

$$\mathbf{A} = \begin{bmatrix} 0 & -1 & 2 \\ 0 & 0 & 1 \\ 0 & 0 & 0 \end{bmatrix}$$

from Example 2, we calculate the matrix exponential *function* $e^{\mathbf{A}t}$ as follows:

$$\begin{aligned} e^{\mathbf{A}t} &= \mathbf{I} + t\mathbf{A} + \frac{t^2}{2!}\mathbf{A}^2 \\ &= \begin{bmatrix} 1 & 0 & 0 \\ 0 & 1 & 0 \\ 0 & 0 & 1 \end{bmatrix} + t\begin{bmatrix} 0 & -1 & 2 \\ 0 & 0 & 1 \\ 0 & 0 & 0 \end{bmatrix} + \frac{t^2}{2!}\begin{bmatrix} 0 & 0 & -1 \\ 0 & 0 & 0 \\ 0 & 0 & 0 \end{bmatrix} \\ &= \begin{bmatrix} 1 & -t & 2t - t^2/2 \\ 0 & 1 & t \\ 0 & 0 & 1 \end{bmatrix}. \end{aligned}$$

■

Differentiation of the Matrix Exponential Function

$$\frac{d}{dt}e^{\mathbf{A}t} = \mathbf{A}e^{\mathbf{A}t}. \tag{5}$$

Proof The derivative of $e^{\mathbf{A}t}$ follows along the same lines as the derivative of the scalar function $de^{ax}/dx = ae^{ax}$. To verify the vector case, we differentiate the power series (4) term by term, getting

$$\begin{aligned} \frac{d}{dt}e^{\mathbf{A}t} &= \frac{d}{dt}\left(\mathbf{I} + \mathbf{A}t + \frac{t^2}{2!}\mathbf{A}^2 + \frac{t^3}{3!}\mathbf{A}^3 + \cdots\right) \\ &= \mathbf{0} + \mathbf{A} + t\mathbf{A}^2 + \frac{t^2}{2!}\mathbf{A}^3 + \cdots \\ &= \mathbf{A}\left(\mathbf{I} + \mathbf{A}t + \frac{t^2}{2!}\mathbf{A}^2 + \cdots\right) = \mathbf{A}e^{\mathbf{A}t}. \end{aligned}$$

□

[1]Sometimes we write the matrix exponential involving t as $e^{t\mathbf{A}}$ and sometimes as $e^{\mathbf{A}t}$. While it is good mathematical form to put scalars in front of matrices ($t\mathbf{A}$), it is also general convention to put variables after constants ($\mathbf{A}t$).

Homogeneous Linear DE Systems

Matrix Exponential Solution of $\vec{\mathbf{x}}' = \mathbf{A}\vec{\mathbf{x}}$

The general solution of

$$\vec{\mathbf{x}}' = \mathbf{A}\vec{\mathbf{x}}, \tag{6}$$

where $\mathbf{A}$ is a constant $n \times n$ matrix, is given by

$$\vec{\mathbf{x}} = e^{\mathbf{A}t}\vec{\mathbf{c}}, \tag{7}$$

where $\vec{\mathbf{c}}$ is an $n \times 1$ vector of arbitrary constants.

If an initial condition, $\vec{\mathbf{x}}(0) = \vec{\mathbf{x}}_0$, is added to (6), then the solution to the resulting IVP is

$$\vec{\mathbf{x}} = e^{\mathbf{A}t}\vec{\mathbf{x}}_0. \tag{8}$$

Proof By direct substitution of (7) in (6), we see that

$$\vec{\mathbf{x}}' = \frac{d}{dt}e^{\mathbf{A}t}\vec{\mathbf{c}} = \mathbf{A}e^{\mathbf{A}t}\vec{\mathbf{c}} = \mathbf{A}\vec{\mathbf{x}}.$$

$e^{\mathbf{A}t} = \mathbf{X}(t)$

Hence, the matrix exponential $e^{\mathbf{A}t}$ is a **fundamental matrix** $\mathbf{X}(t)$ of the linear system $\vec{\mathbf{x}}' = \mathbf{A}\vec{\mathbf{x}}$.

Substituting an initial condition, $\vec{\mathbf{x}}(0) = \vec{\mathbf{x}}_0$, yields $\vec{\mathbf{c}} = \vec{\mathbf{x}}_0$; hence the solution of the initial-value problem is equation (8). □

EXAMPLE 4 **An Old Favorite** For the harmonic oscillator system of Sec 4.1,

$$x' = y,$$
$$y' = -x,$$

we write the system in matrix form $\vec{\mathbf{x}}' = \mathbf{A}\vec{\mathbf{x}}$, where

$$\mathbf{A} = \begin{bmatrix} 0 & 1 \\ -1 & 0 \end{bmatrix},$$

to find the matrix exponential solution. Computing powers of $\mathbf{A}$, we find that

$$\mathbf{A}^2 = -\mathbf{I}, \quad \mathbf{A}^3 = -\mathbf{A}, \quad \mathbf{A}^4 = \mathbf{I}, \quad \mathbf{A}^5 = \mathbf{A}, \quad \ldots,$$

so the matrix exponential $e^{\mathbf{A}t}$, or fundamental matrix, is

$$\begin{aligned} e^{\mathbf{A}t} &= \mathbf{I} + t\mathbf{A} - \frac{t^2}{2!}\mathbf{I} - \frac{t^3}{3!}\mathbf{A} + \frac{t^4}{4!}\mathbf{I} + \cdots \\ &= \begin{bmatrix} 1 & 0 \\ 0 & 1 \end{bmatrix} + t\begin{bmatrix} 0 & 1 \\ -1 & 0 \end{bmatrix} - \frac{t^2}{2!}\begin{bmatrix} 1 & 0 \\ 0 & 1 \end{bmatrix} \\ &\quad - \frac{t^3}{3!}\begin{bmatrix} 0 & 1 \\ -1 & 0 \end{bmatrix} + \frac{t^4}{4!}\begin{bmatrix} 1 & 0 \\ 0 & 1 \end{bmatrix} + \cdots \\ &= \begin{bmatrix} 1 - \frac{t^2}{2!} + \frac{t^4}{4!} - \cdots & t - \frac{t^3}{3!} + \frac{t^5}{5!} - \cdots \\ -t + \frac{t^3}{3!} - \frac{t^5}{5!} + \cdots & 1 - \frac{t^2}{2!} + \frac{t^4}{4!} - \cdots \end{bmatrix} = \begin{bmatrix} \cos t & \sin t \\ -\sin t & \cos t \end{bmatrix}. \end{aligned}$$

The general solution for the harmonic oscillator can be written as

$$\vec{\mathbf{x}}(t) = e^{\mathbf{A}t}\vec{\mathbf{c}} = \begin{bmatrix} \cos t & \sin t \\ -\sin t & \cos t \end{bmatrix}\begin{bmatrix} c_1 \\ c_2 \end{bmatrix}$$

$$= \begin{bmatrix} c_1 \cos t + c_2 \sin t \\ -c_1 \sin t + c_2 \cos t \end{bmatrix} = c_1 \begin{bmatrix} \cos t \\ -\sin t \end{bmatrix} + c_2 \begin{bmatrix} \sin t \\ \cos t \end{bmatrix}.$$

■

Alternate Interpretations of the Matrix Exponential

The matrix exponential $e^{\mathbf{A}t}$ can always be computed from the definition (4), which involves an infinite series, but this approach might only be easily applied in cases with inherent repetition. Fortunately, there are other ways to find $e^{\mathbf{A}t}$ that do not use the definition. Some ways follow in this section; then in Sec. 8.3 we will find the matrix exponential using the Laplace transform.

The solution of the initial-value problem

$$\vec{\mathbf{x}}' = \mathbf{A}\vec{\mathbf{x}}, \quad \vec{\mathbf{x}}(0) = \vec{\mathbf{x}}_0,$$

can be written very compactly in terms of a fundamental matrix $\mathbf{X}$, whose columns are linearly independent solutions $\vec{\mathbf{x}}_i$ of $\vec{\mathbf{x}}' = \mathbf{A}\vec{\mathbf{x}}$. (See Sec. 6.1.) We obtain

$$e^{\mathbf{A}t} = \mathbf{X}(t)\mathbf{X}^{-1}(0), \tag{9}$$

by the following argument.

Proof The general solution of $\vec{\mathbf{x}}' = \mathbf{A}\vec{\mathbf{x}}$ is

$$\vec{\mathbf{x}} = a_1\vec{\mathbf{x}}_1 + a_2\vec{\mathbf{x}}_2 + \cdots + a_n\vec{\mathbf{x}}_n,$$

or

$$\vec{\mathbf{x}} = \mathbf{X}(t)\vec{\mathbf{a}},$$

where $\vec{\mathbf{a}}$ is a constant vector with elements $a_1, a_2, \ldots, a_n$. For initial values $\vec{\mathbf{x}}(0) = \vec{\mathbf{x}}_0$, we have

$$\vec{\mathbf{x}}(0) = \mathbf{X}(0)\vec{\mathbf{a}} = \vec{\mathbf{x}}_0, \quad \text{or} \quad \vec{\mathbf{a}} = \mathbf{X}^{-1}(0)\vec{\mathbf{x}}_0.$$

Hence,

$$\vec{\mathbf{x}} = \mathbf{X}(t)\mathbf{X}^{-1}(0)\vec{\mathbf{x}}_0$$

is the unique solution of the initial-value problem. But the solution can also be expressed uniquely as the matrix exponential $\vec{\mathbf{x}} = e^{\mathbf{A}t}\vec{\mathbf{x}}_0$, so

$$e^{\mathbf{A}t}\vec{\mathbf{x}}_0 = \mathbf{X}(t)\mathbf{X}^{-1}(0)\vec{\mathbf{x}}_0.$$

Hence,

$$e^{\mathbf{A}t} = \mathbf{X}(t)\mathbf{X}^{-1}(0),$$

and the matrix exponential $e^{\mathbf{A}t}$ is a fundamental matrix for the initial-value problem $\vec{\mathbf{x}}' = \mathbf{A}\vec{\mathbf{x}}$, $\vec{\mathbf{x}}(0) = \vec{\mathbf{x}}_0$. □

EXAMPLE 5 **Back Door to the Matrix Exponential** We saw in Sec. 6.2, Example 1, that the linear system $\vec{\mathbf{x}}' = \mathbf{A}\vec{\mathbf{x}}$ with

$$\mathbf{A} = \begin{bmatrix} 1 & 1 \\ 4 & 1 \end{bmatrix}$$

has eigenvalues $\lambda_1 = 3$ and $\lambda_2 = -1$; it has a fundamental matrix

$$\mathbf{X}(t) = \begin{bmatrix} e^{3t} & e^{-t} \\ 2e^{3t} & -2e^{-t} \end{bmatrix}, \quad \text{so} \quad \mathbf{X}^{-1}(0) = \frac{1}{4}\begin{bmatrix} 2 & 1 \\ 2 & -1 \end{bmatrix}.$$

Using (9), we can calculate that

$$e^{\mathbf{A}t} = \mathbf{X}(t)\mathbf{X}^{-1}(0) = \frac{1}{4}\begin{bmatrix} 2e^{3t} + 2e^{-t} & e^{3t} - e^{-t} \\ 4e^{3t} - 4e^{-t} & 2e^{3t} + 2e^{-t} \end{bmatrix}.$$

■

Another useful interpretation of $e^{\mathbf{A}t}$ is as an $n \times n$ matrix whose ith column is the unique solution of the initial-value problem

$$\vec{\mathbf{x}}' = \mathbf{A}\vec{\mathbf{x}}, \quad \vec{\mathbf{x}}_i(0) = \vec{\mathbf{e}}_i,$$

where $\vec{\mathbf{e}}_i$ is an $n \times 1$ vector with a 1 in the ith position and zeros elsewhere.

Nonhomogeneous Linear DE Systems

You may wonder about the value of the matrix exponential $e^{\mathbf{A}t}$ if we can find the solution to $\vec{\mathbf{x}}' = \mathbf{A}\vec{\mathbf{x}}$ by finding the eigenvalues and eigenvectors of $\mathbf{A}$. One appeal is that it allows us to solve problems, such as the nonhomogeneous linear system

$$\vec{\mathbf{x}}' = \mathbf{A}\vec{\mathbf{x}} + \vec{\mathbf{f}}(t),$$

using notation that we used with the scalar nonhomogeneous linear equation

$$x' + p(t)y = f(t),$$

which we solved by the integrating factor in Sec. 2.2.

Let us rewrite the linear system in an analogous way as

$$\vec{\mathbf{x}}' - \mathbf{A}\vec{\mathbf{x}} = \vec{\mathbf{f}}(t),$$

where $\mathbf{A}$ is a constant matrix and $\vec{\mathbf{f}}(t)$ is an $n \times 1$ vector of functions. We multiply each side of the equation by the matrix exponential $e^{-\mathbf{A}t}$, the matrix equivalent of the scalar integrating factor $e^{\int p(t)dt}$, getting

$$e^{-\mathbf{A}t}\left(\vec{\mathbf{x}}' - \mathbf{A}\vec{\mathbf{x}}\right) = e^{-\mathbf{A}t}\vec{\mathbf{f}}(t). \tag{10}$$

By matrix differentiation, we find

$$\frac{d}{dt}\left[e^{-\mathbf{A}t}\vec{\mathbf{x}}(t)\right] = e^{-\mathbf{A}t}(-\mathbf{A})\vec{\mathbf{x}}(t) + e^{-\mathbf{A}t}\vec{\mathbf{x}}'(t) = e^{-\mathbf{A}t}(\vec{\mathbf{x}}' - \mathbf{A}\vec{\mathbf{x}}). \tag{11}$$

Hence, (10) can be written as

$$\frac{d}{dt}\left[e^{-\mathbf{A}t}\vec{\mathbf{x}}(t)\right] = e^{-\mathbf{A}t}\vec{\mathbf{f}}(t).$$

Integrating, we find

$$e^{-\mathbf{A}t}\vec{\mathbf{x}}(t) = \int_0^t e^{-\mathbf{A}s}\vec{\mathbf{f}}(s)ds + \vec{\mathbf{c}},$$

and using the property $e^{\mathbf{A}t}e^{-\mathbf{A}t} = \mathbf{I}$, we obtain the general solution

$$\vec{\mathbf{x}}(t) = e^{\mathbf{A}t}\vec{\mathbf{c}} + e^{\mathbf{A}t}\int_0^t e^{-\mathbf{A}s}\vec{\mathbf{f}}(s)ds.$$

This proves the following theorem.

Matrix Exponential Solution to Nonhomogeneous DE Systems
If $\mathbf{A}$ is a constant $n \times n$ matrix and $\vec{\mathbf{f}}(t)$ is an $n \times 1$ vector of functions, then the solution of the linear system

$$\vec{\mathbf{x}}' = \mathbf{A}\vec{\mathbf{x}} + \vec{\mathbf{f}}(t),$$

given in terms of the matrix exponential, is

$$\vec{\mathbf{x}}(t) = e^{\mathbf{A}t}\vec{\mathbf{c}} + e^{\mathbf{A}t}\int_0^t e^{-\mathbf{A}s}\vec{\mathbf{f}}(s)ds. \tag{12}$$

If initial conditions $\vec{\mathbf{x}}(0) = \vec{\mathbf{x}}_0$ are supplied, the unique solution is given by

$$\vec{\mathbf{x}}(t) = e^{\mathbf{A}t}\vec{\mathbf{x}}_0 + e^{\mathbf{A}t}\int_0^t e^{-\mathbf{A}s}\vec{\mathbf{f}}(s)ds. \tag{13}$$

Matrix Integration:
Integration of a matrix is done element by element, using the Fundamental Theorem of Calculus for Definite Integrals.

EXAMPLE 6 **Nonhomogeneous Linear System** To solve $\vec{\mathbf{x}}' = \mathbf{A}\vec{\mathbf{x}} + \vec{\mathbf{f}}(t)$, where

$$\mathbf{A} = \begin{bmatrix} 0 & 1 \\ -1 & 0 \end{bmatrix} \quad \text{and} \quad \vec{\mathbf{f}}(t) = \begin{bmatrix} t \\ 0 \end{bmatrix},$$

we recall from Example 4 that

$$e^{\mathbf{A}t} = \begin{bmatrix} \cos t & \sin t \\ -\sin t & \cos t \end{bmatrix}, \quad \text{so} \quad e^{-\mathbf{A}t} = \left(e^{\mathbf{A}t}\right)^{-1} = \begin{bmatrix} \cos t & -\sin t \\ \sin t & \cos t \end{bmatrix}.$$

For the solution, we must calculate

$$\begin{aligned} \int_0^t e^{-\mathbf{A}s}\vec{\mathbf{f}}(s)ds &= \int_0^t \begin{bmatrix} \cos s & -\sin s \\ \sin s & \cos s \end{bmatrix}\begin{bmatrix} s \\ 0 \end{bmatrix} ds \\ &= \int_0^t \begin{bmatrix} s\cos s \\ s\sin s \end{bmatrix} ds && \textit{(integration by parts)} \\ &= \begin{bmatrix} t\sin t + \cos t \\ -t\cos t + \sin t \end{bmatrix} - \begin{bmatrix} 1 \\ 0 \end{bmatrix}. && \textit{(evaluation of definite integral)} \end{aligned}$$

Hence, from (13), we have the solution

$$\begin{aligned} \vec{\mathbf{x}}(t) &= \vec{\mathbf{x}}_h + \vec{\mathbf{x}}_p \\ &= e^{\mathbf{A}t}\vec{\mathbf{c}} + e^{\mathbf{A}t}\int_0^t e^{-\mathbf{A}s}\vec{\mathbf{f}}(s)\,ds \\ &= \begin{bmatrix} \cos t & \sin t \\ -\sin t & \cos t \end{bmatrix}\begin{bmatrix} c_1 \\ c_2 \end{bmatrix} \\ &\quad + \begin{bmatrix} \cos t & \sin t \\ -\sin t & \cos t \end{bmatrix}\left(\begin{bmatrix} t\sin t + \cos t \\ -t\cos t + \sin t \end{bmatrix} - \begin{bmatrix} 1 \\ 0 \end{bmatrix}\right) \\ &= \begin{bmatrix} \cos t & \sin t \\ -\sin t & \cos t \end{bmatrix}\begin{bmatrix} c_1 \\ c_2 \end{bmatrix} + \begin{bmatrix} 1 - \cos t \\ \sin t - t \end{bmatrix}. \end{aligned}$$

■

Connecting the Matrix Exponential and Eigenfunctions

We have seen that one method of solution of $\vec{\mathbf{x}}' = \mathbf{A}\vec{\mathbf{x}}$ leads to the matrix exponential $\vec{\mathbf{x}} = e^{\mathbf{A}t}\vec{\mathbf{c}}$, while in Secs. 6.2 and 6.3 we found the solution in terms of eigenvalues and eigenfunctions. To obtain a connection between these approaches, we revisit Sec. 6.5.

Recall that if $\mathbf{A}$ is a *diagonalizable* $n \times n$ matrix, then

$$\mathbf{A} = \mathbf{P}\mathbf{D}\mathbf{P}^{-1},$$

where $\mathbf{D}$ is a diagonal matrix of all n eigenvalues of $\mathbf{A}$, and $\mathbf{P}$ is the matrix having the corresponding n linearly independent eigenvectors as columns.

Also recall that for the IVP system

$$\vec{\mathbf{x}}' = \mathbf{A}\vec{\mathbf{x}}, \quad \vec{\mathbf{x}}(0) = \vec{\mathbf{x}}_0,$$

the change of variables $\vec{\mathbf{x}} = \mathbf{P}\vec{\mathbf{w}}$ leads to an easily solved *decoupled* system

$$\vec{\mathbf{w}}' = \mathbf{D}\vec{\mathbf{w}} = \begin{bmatrix} \lambda_1 & 0 & \cdots & 0 \\ 0 & \lambda_2 & \ddots & \vdots \\ \vdots & \ddots & \ddots & 0 \\ 0 & \cdots & 0 & \lambda_n \end{bmatrix} \vec{\mathbf{w}}$$

with solution

$$\vec{\mathbf{w}}(t) = \begin{bmatrix} e^{\lambda_1 t} & 0 & \cdots & 0 \\ 0 & e^{\lambda_2 t} & \ddots & \vdots \\ \vdots & \ddots & \ddots & 0 \\ 0 & \cdots & 0 & e^{\lambda_n t} \end{bmatrix} \vec{\mathbf{w}}_0,$$

so that

$$\vec{\mathbf{x}}(t) = \mathbf{P}\vec{\mathbf{w}} = \mathbf{P} \begin{bmatrix} e^{\lambda_1 t} & 0 & \cdots & 0 \\ 0 & e^{\lambda_2 t} & \ddots & \vdots \\ \vdots & \ddots & \ddots & 0 \\ 0 & \cdots & 0 & e^{\lambda_n t} \end{bmatrix} \mathbf{P}^{-1}\vec{\mathbf{x}}_0. \tag{14}$$

With the matrix exponential of this section we solved the same IVP by equation (8) to get

$$\vec{\mathbf{x}} = e^{\mathbf{A}t}\vec{\mathbf{x}}_0,$$

and we recognize the matrix in (14) as $e^{\mathbf{D}t}$, so (14) becomes

$$\vec{\mathbf{x}} = \mathbf{P}e^{\mathbf{D}t}\mathbf{P}^{-1}\vec{\mathbf{x}}_0.$$

Comparing these last two equations gives yet another way to calculate the matrix exponential, simply in terms of its eigenvalues and eigenvectors.

Matrix Exponential from Eigenfunctions

For a diagonalizable matrix $\mathbf{A}$,

$$e^{\mathbf{A}t} = \mathbf{P}e^{\mathbf{D}t}\mathbf{P}^{-1}, \tag{15}$$

where $\mathbf{D}$ is a diagonal matrix of all eigenvalues of $\mathbf{A}$, and $\mathbf{P}$ is the matrix having the corresponding eigenvectors as columns.

In Problems 21–23, you can show explicitly that equations (15) and (9) give the same results for $e^{\mathbf{A}t}$, whenever both are applicable.

Summary

We defined the matrix exponential $e^{\mathbf{A}t}$ so that the homogeneous linear DE system with constant matrix $\mathbf{A}$,

$$\vec{\mathbf{x}}' = \mathbf{A}\vec{\mathbf{x}},$$

has a solution that can be written as $\vec{\mathbf{x}} = e^{\mathbf{A}t}\vec{\mathbf{c}}$. We saw that this idea is analogous to the solution of first-order linear DEs, and extended the technique to IVPs and nonhomogeneous systems. In the process, we found several different ways to *compute* the matrix exponential.

6.6 Problems

Matrix Exponential Functions *Find the matrix exponential* $e^{\mathbf{A}t}$ *of the matrices given in Problems 1–6.*

1. $\mathbf{A} = \begin{bmatrix} 1 & 0 \\ 0 & -1 \end{bmatrix}$ **2.** $\mathbf{A} = \begin{bmatrix} -i & 0 \\ 0 & i \end{bmatrix}$

3. $\mathbf{A} = \begin{bmatrix} 1 & 0 \\ 1 & 0 \end{bmatrix}$ **4.** $\mathbf{A} = \begin{bmatrix} 0 & 1 \\ 0 & 0 \end{bmatrix}$

5. $\mathbf{A} = \begin{bmatrix} 1 & 0 & 0 \\ 0 & 2 & 0 \\ 0 & 0 & 3 \end{bmatrix}$ **6.** $\mathbf{A} = \begin{bmatrix} 0 & 1 & 1 \\ 0 & 0 & 1 \\ 0 & 0 & 0 \end{bmatrix}$

DE Solutions Using Matrix Exponentials *Using matrix exponentials, find the general solutions of the linear systems given in Problems 7–14.*

7. $x' = x$, $y' = y$ **8.** $x' = y$, $y' = x$

9. $x' = x + y$, $y' = x$ **10.** $x' = y + z$, $y' = z$, $z' = 0$

11. $\vec{\mathbf{x}}' = \begin{bmatrix} -1 & 0 \\ 0 & 2 \end{bmatrix}\begin{bmatrix} x_1 \\ x_2 \end{bmatrix} + \begin{bmatrix} 1 \\ 0 \end{bmatrix}$

12. $\vec{\mathbf{x}}' = \begin{bmatrix} 2 & 0 \\ 0 & 3 \end{bmatrix}\begin{bmatrix} x_1 \\ x_2 \end{bmatrix} + \begin{bmatrix} 0 \\ 6 \end{bmatrix}$

13. $\vec{\mathbf{x}}' = \begin{bmatrix} 0 & 1 \\ 1 & 0 \end{bmatrix}\begin{bmatrix} x_1 \\ x_2 \end{bmatrix} + \begin{bmatrix} 1 \\ 1 \end{bmatrix}$

14. $\vec{\mathbf{x}}' = \begin{bmatrix} 0 & 1 \\ -1 & 0 \end{bmatrix}\begin{bmatrix} x_1 \\ x_2 \end{bmatrix} + \begin{bmatrix} 1 \\ 0 \end{bmatrix}, \quad \begin{bmatrix} x_1(0) \\ x_2(0) \end{bmatrix} = \begin{bmatrix} 1 \\ 1 \end{bmatrix}$

15. Products of Matrix Exponentials Suppose that

$$\mathbf{A} = \begin{bmatrix} 0 & -1 \\ 0 & 0 \end{bmatrix} \quad \text{and} \quad \mathbf{B} = \begin{bmatrix} 0 & 0 \\ 1 & 0 \end{bmatrix}.$$

(a) Find $e^{\mathbf{A}t}$ and $e^{\mathbf{B}t}$.

(b) Find $e^{(\mathbf{A}+\mathbf{B})t}$.

(c) Does $e^{(\mathbf{A}+\mathbf{B})t} = e^{\mathbf{A}t}e^{\mathbf{B}t}$?

Properties of Matrix Exponentials *Verify the properties of the matrix exponentials in Problems 16 and 17.*

16. If $e^{\mathbf{A}}$ is the matrix exponential for a square constant matrix $\mathbf{A}$, then its inverse is given by $\left(e^{\mathbf{A}}\right)^{-1} = e^{-\mathbf{A}}$.

17. If $\mathbf{AB} = \mathbf{BA}$, then $e^{\mathbf{A}+\mathbf{B}} = e^{\mathbf{A}}e^{\mathbf{B}}$.

18. Nilpotent Example Suppose that

$$\mathbf{A} = \begin{bmatrix} 1 & 1 & -1 \\ 1 & 0 & -1 \\ 1 & 1 & -1 \end{bmatrix}.$$

(a) Show that $\mathbf{A}$ is nilpotent; that is, that there exists an integer n such that $\mathbf{A}^n = \mathbf{0}$.

(b) Solve the linear system $\vec{\mathbf{x}}' = \mathbf{A}\vec{\mathbf{x}}$.

19. An Exponential Pattern Suppose

$$\mathbf{A} = \begin{bmatrix} 0 & 1 \\ 1 & 0 \end{bmatrix}.$$

(a) Show that $\mathbf{A}^{2n} = \mathbf{I}$ and $\mathbf{A}^{2n+1} = \mathbf{A}$, for some positive integer n.

(b) Use the results from part (a) to show that

$$e^{\mathbf{A}t} = \begin{bmatrix} \cosh t & \sinh t \\ \sinh t & \cosh t \end{bmatrix}.$$

(c) Find the general solution of $\vec{\mathbf{x}}' = \mathbf{A}\vec{\mathbf{x}}$.

20. Nilpotent Criterion Show that a matrix is nilpotent if and only if its eigenvalues are zero.

Fundamental Matrices *Verify that* $e^{\mathbf{A}t} = \mathbf{X}(t)\mathbf{X}^{-1}(0)$ *and* $e^{\mathbf{A}t} = \mathbf{P}e^{\mathbf{D}t}\mathbf{P}^{-1}$ *give the same result for the matrix exponential of the matrices in Problems 21–23.*

21. $\mathbf{A} = \begin{bmatrix} 1 & 2 \\ 0 & 1 \end{bmatrix}$ **22.** $\mathbf{A} = \begin{bmatrix} 1 & 1 & 1 \\ 0 & 2 & 1 \\ 0 & 0 & 3 \end{bmatrix}$

23. $\begin{bmatrix} 1 & 1 \\ 4 & 1 \end{bmatrix}$ (Example 5)

Computer Lab *CAS commands for finding the matrix exponential $e^{\mathbf{A}t}$ are*

- *Maple with (linalg):* *exponential (A*t);*
- *Mathematica:* *MatrixExp(At);*
- *Matlab:* *syms t, expm(A*t).*

Find the matrix exponential of the matrices given in Problems 24–25.

24. $\mathbf{A} = \begin{bmatrix} 0 & 0 & 0 & 1 \\ 0 & 0 & 1 & 0 \\ 0 & 1 & 0 & 0 \\ 1 & 0 & 0 & 0 \end{bmatrix}$ **25.** $\mathbf{A} = \begin{bmatrix} 0 & 0 & 0 & 1 \\ 0 & 0 & -1 & 0 \\ 0 & 1 & 0 & 0 \\ -1 & 0 & 0 & 0 \end{bmatrix}$

Computer DE Solutions *Use a CAS and matrix exponentials to solve the linear system $\vec{\mathbf{x}}' = \mathbf{A}\vec{\mathbf{x}}$ for the matrices given in Problems 26–30. If initial conditions $\vec{\mathbf{x}}(0) = \vec{\mathbf{x}}_0$ are given, find the unique solution.*

26. $\mathbf{A} = \begin{bmatrix} 3 & -2 \\ 2 & -2 \end{bmatrix}$

27. $\mathbf{A} = \begin{bmatrix} 1 & 5 \\ -2 & -1 \end{bmatrix}, \quad \vec{\mathbf{x}}_0 = \begin{bmatrix} 1 \\ 1 \end{bmatrix}$

28. $\mathbf{A} = \begin{bmatrix} 1 & 1 & 1 \\ 2 & 1 & -1 \\ -8 & -5 & -3 \end{bmatrix}$

29. $\mathbf{A} = \begin{bmatrix} 3 & 1 & 0 \\ 0 & 3 & 1 \\ 0 & 0 & 3 \end{bmatrix}, \quad \vec{\mathbf{x}}(0) = \begin{bmatrix} 1 \\ 0 \\ 0 \end{bmatrix}$

30. $\mathbf{A} = \begin{bmatrix} 6 & 3 & -2 \\ -4 & -1 & 2 \\ 13 & 9 & -3 \end{bmatrix}, \quad \vec{\mathbf{x}}(0) = \begin{bmatrix} 1 \\ 0 \\ 0 \end{bmatrix}$

31. Suggested Journal Entry We have now solved linear DE systems, with constant coefficients, by eigenvalues and eigenvectors, by decoupling, and by matrix exponentials. List the restrictions and advantages of each method. Discuss, on the basis of your experience, which methods you prefer in which situations (e.g., general solutions versus IVPs; homogeneous versus nonhomogeneous DEs; real versus nonreal eigenvalues).

6.7 Nonhomogeneous Linear Systems

SYNOPSIS: We see how the common techniques of undetermined coefficients and variation of parameters can be extended to find a particular solution of the nonhomogeneous linear system $\vec{\mathbf{x}}' = \mathbf{A}(t)\vec{\mathbf{x}} + \vec{\mathbf{f}}(t)$ with forcing term $\vec{\mathbf{f}}(t)$.

Introduction

Most of this chapter has been focused on solving *homogeneous* linear DE systems, although the techniques of *decoupling* (Sec. 6.5) and the *matrix exponential* (Sec. 6.6) extended to nonhomogeneous systems. In this section, we return to the basic ideas of linear nonhomogenous equations in general.

We have seen in Sec. 2.1 that the general solution of a linear nonhomogeneous differential equation can be expressed as the sum of its homogeneous solutions and any particular solution:

$$x = x_h + x_p.$$

The same principle holds for nonhomogeneous linear systems of differential equations:

$$\vec{\mathbf{x}} = \vec{\mathbf{x}}_h + \vec{\mathbf{x}}_p.$$

We now show how the two methods given in Chapter 4—*undetermined coefficients* and *variation of parameters*—can be used to find particular solutions to nonhomogeneous systems.

Undetermined Coefficients

If $\mathbf{A}$ is a constant matrix, the method of undetermined coefficients, studied in Sec. 4.4, can be extended to finding a particular solution of a linear system of nonhomogeneous equations

$$\vec{\mathbf{x}}' = \mathbf{A}\vec{\mathbf{x}} + \vec{\mathbf{f}}(t). \tag{1}$$

The simplest case is *constant* forcing. To find a particular solution of the nonhomogeneous system

$$\vec{\mathbf{x}}' = \mathbf{A}\vec{\mathbf{x}} + \vec{\mathbf{b}}, \tag{2}$$

where $\vec{\mathbf{b}}$ is a constant vector, we try a constant $\vec{\mathbf{x}}_p$, which gives $\vec{\mathbf{x}}'_p = 0$. Substituting $\vec{\mathbf{x}}_p$ and $\vec{\mathbf{x}}'_p = \vec{\mathbf{0}}$ into (2) gives

$$\mathbf{A}\vec{\mathbf{x}}_p + \vec{\mathbf{b}} = \vec{\mathbf{0}}, \quad \text{or} \quad \vec{\mathbf{x}}_p = -\mathbf{A}^{-1}\vec{\mathbf{b}},$$

so the general solution to a system (2), with constant forcing, takes the form

$$\vec{\mathbf{x}} = \vec{\mathbf{x}}_h + \vec{\mathbf{x}}_p = \vec{\mathbf{x}}_h - \mathbf{A}^{-1}\vec{\mathbf{b}}.$$

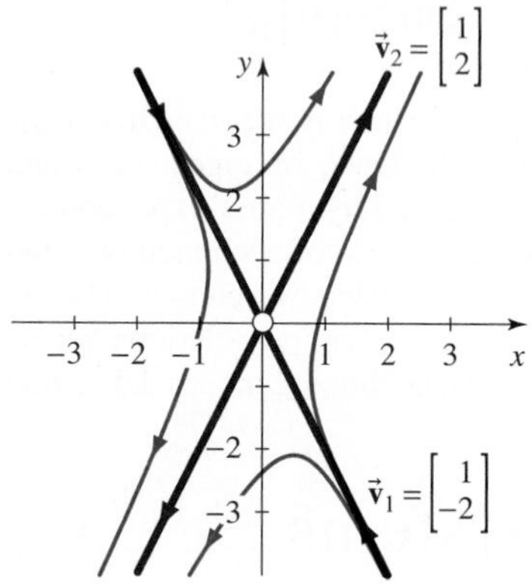

(a) Homogeneous system

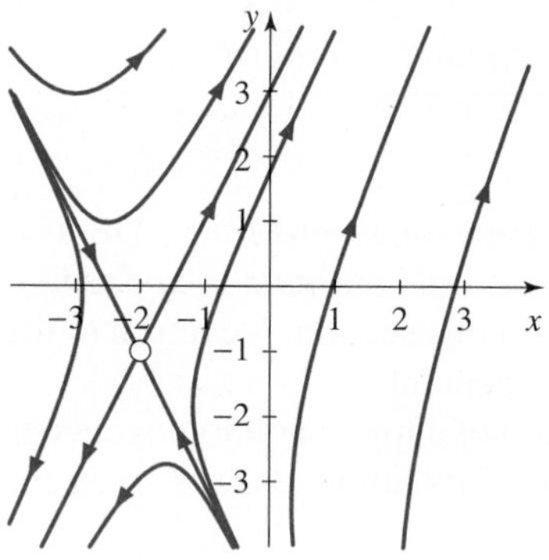

(b) Nonhomogeneous system

FIGURE 6.7.1 Phase portraits for (a) homogeneous and (b) nonhomogeneous solutions to Example 1 are simply a translation.

EXAMPLE 1 **Constant Forcing** We wish to find the general solution of

$$\vec{\mathbf{x}}' = \mathbf{A}\vec{\mathbf{x}} + \vec{\mathbf{b}} = \begin{bmatrix} 1 & 1 \\ 4 & 1 \end{bmatrix}\begin{bmatrix} x_1 \\ x_2 \end{bmatrix} + \begin{bmatrix} 3 \\ 9 \end{bmatrix}, \tag{3}$$

In Sec. 6.2, Example 1, we determined the solution of the corresponding homogeneous system $\vec{\mathbf{x}}' = \mathbf{A}\vec{\mathbf{x}}$ to be

$$\vec{\mathbf{x}}_h = c_1 e^{3t}\begin{bmatrix} 1 \\ 2 \end{bmatrix} + c_2 e^{-t}\begin{bmatrix} 1 \\ -2 \end{bmatrix}.$$

To find a particular solution, we find the inverse of $\mathbf{A}$:

$$\mathbf{A} = \begin{bmatrix} 1 & 1 \\ 4 & 1 \end{bmatrix} \quad \Rightarrow \quad \mathbf{A}^{-1} = \frac{1}{3}\begin{bmatrix} -1 & 1 \\ 4 & -1 \end{bmatrix},$$

and calculate

$$\vec{\mathbf{x}}_p = -\mathbf{A}^{-1}\vec{\mathbf{b}} = -\frac{1}{3}\begin{bmatrix} -1 & 1 \\ 4 & -1 \end{bmatrix}\begin{bmatrix} 3 \\ 9 \end{bmatrix} = \begin{bmatrix} -2 \\ -1 \end{bmatrix}.$$

The general solution of system (3) is

$$\vec{\mathbf{x}}(t) = \underbrace{c_1 e^{3t}\begin{bmatrix} 1 \\ 2 \end{bmatrix} + c_2 e^{-t}\begin{bmatrix} 1 \\ -2 \end{bmatrix}}_{\vec{\mathbf{x}}_h} + \underbrace{\begin{bmatrix} -2 \\ -1 \end{bmatrix}}_{\vec{\mathbf{x}}_p}.$$

Figure 6.7.1 shows that the phase portrait for (3) is simply a translation of the homogeneous system, because the forcing is a constant vector. ■

More generally, the method of undetermined coefficients applies to (1) whenever $\mathbf{A}$ is a matrix of constant coefficients and the forcing vector $\vec{\mathbf{f}}(t)$ is restricted to the same families of functions described in Sec. 4.4:

- **(i)** polynomials in t,
- **(ii)** e^{at},
- **(iii)** $\cos kt$, $\sin kt$, and
- **(iv)** finite sums and products of the above functions.

The idea is to choose a particular *form* for a particular solution depending on $\vec{\mathbf{f}}(t)$, substitute it into the differential equation, and determine the coefficients that satisfy the equation. This particular solution is then added to the homogeneous solutions to give the general solution.

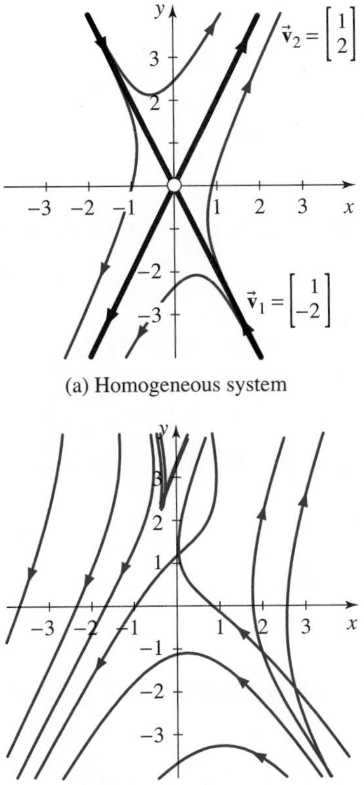

(a) Homogeneous system

(b) Nonhomogeneous system

FIGURE 6.7.2 Phase portraits for (a) homogeneous and (b) nonhomogeneous solutions to Example 2 are less clearly related, because forcing is not constant, and there is no equilibrium.

EXAMPLE 2 Undetermined Coefficients for a Nonhomogeneous System

We need only one particular solution $\mathbf{x}_p(t)$ to solve the nonhomogeneous 2×2 system

$$\vec{\mathbf{x}}' = \mathbf{A}\vec{\mathbf{x}} + \vec{\mathbf{f}}(t) = \begin{bmatrix} 1 & 1 \\ 4 & 1 \end{bmatrix} \vec{\mathbf{x}} + \begin{bmatrix} t-2 \\ 4t-1 \end{bmatrix}. \tag{4}$$

Because the elements of $\vec{\mathbf{f}}(t)$ are polynomials in t, we predict that

$$\vec{\mathbf{x}}_p = \begin{bmatrix} at+b \\ ct+d \end{bmatrix}, \quad \text{and} \quad \vec{\mathbf{x}}_p' = \begin{bmatrix} a \\ c \end{bmatrix}.$$

We can substitute $\vec{\mathbf{x}}_p$ into (4):

$$\begin{aligned} \begin{bmatrix} a \\ c \end{bmatrix} &= \begin{bmatrix} 1 & 1 \\ 4 & 1 \end{bmatrix} \begin{bmatrix} at+b \\ ct+d \end{bmatrix} + \begin{bmatrix} t-2 \\ 4t-1 \end{bmatrix} \\ &= \begin{bmatrix} at+b+ct+d \\ 4at+4b+ct+d \end{bmatrix} + \begin{bmatrix} t-2 \\ 4t-1 \end{bmatrix} \\ &= \begin{bmatrix} at+b+ct+d+t-2 \\ 4at+4b+ct+d+4t-1 \end{bmatrix} \\ &= t\begin{bmatrix} a+c+1 \\ 4a+c+4 \end{bmatrix} + \begin{bmatrix} b+d-2 \\ 4b+d-1 \end{bmatrix}. \end{aligned}$$

Equating corresponding coefficients leads to a system of four equations in a, b, c, and d:

$$\begin{aligned} a &= b+d-2, & a+c+1 &= 0, \\ c &= 4b+d-1, & 4a+c+4 &= 0. \end{aligned}$$

with augmented matrix

$$\left[\begin{array}{cccc|c} 1 & -1 & 0 & -1 & -2 \\ 1 & 0 & 1 & 0 & -1 \\ 0 & -4 & 1 & -1 & -1 \\ 4 & 0 & 1 & 0 & -4 \end{array}\right] \quad \text{and RREF} \quad \left[\begin{array}{cccc|c} 1 & 0 & 0 & 0 & -1 \\ 0 & 1 & 0 & 0 & 0 \\ 0 & 0 & 1 & 0 & 0 \\ 0 & 0 & 0 & 1 & 1 \end{array}\right].$$

Hence, $a = -1$, $b = 0$, $c = 0$, and $d = 1$. With these values, we obtain

$$\vec{\mathbf{x}}_p = \begin{bmatrix} -t \\ 1 \end{bmatrix}.$$

By adding $\vec{\mathbf{x}}_p$ to the homogeneous solution $\mathbf{x}_h(t)$ found in Sec. 6.2, Example 1, we obtain the general solution

$$\vec{\mathbf{x}}(t) = c_1 e^{3t} \begin{bmatrix} 1 \\ 2 \end{bmatrix} + c_2 e^{-t} \begin{bmatrix} 1 \\ -2 \end{bmatrix} + \begin{bmatrix} -t \\ 1 \end{bmatrix}.$$

Figure 6.7.2 shows how the nonconstant forcing term distorts the homogeneous system and produces trajectories that cross, because the vector field is no longer constant. ■

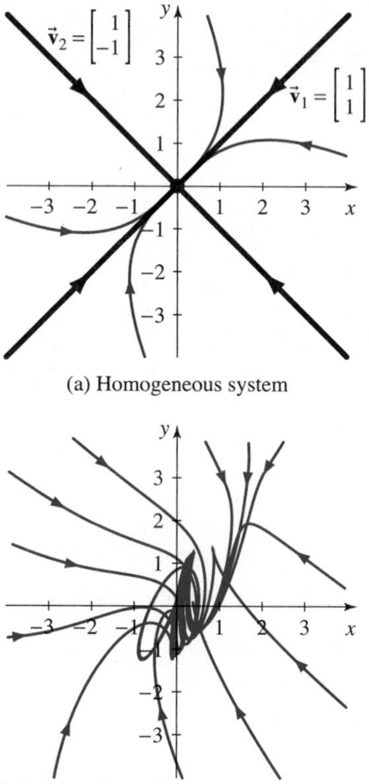

FIGURE 6.7.3 Phase portraits for (a) homogeneous and (b) nonhomogeneous solutions to Example 3, again distorted by nonconstant forcing to have no equilibrium.

EXAMPLE 3 **Systematic Guess** To find a particular solution for the nonhomogeneous 2×2 system

$$\vec{\mathbf{x}}' = \mathbf{A}\vec{\mathbf{x}} + \vec{\mathbf{f}} = \begin{bmatrix} -2 & 1 \\ 1 & -2 \end{bmatrix} \vec{\mathbf{x}} + \begin{bmatrix} \sin 2t \\ 3\cos 2t \end{bmatrix}, \tag{5}$$

we make the judicious guess

$$\vec{\mathbf{x}}_p = \begin{bmatrix} A\cos 2t + B\sin 2t \\ C\cos 2t + D\sin 2t \end{bmatrix} \quad \text{so that} \quad \vec{\mathbf{x}}'_p = \begin{bmatrix} 2B\cos 2t - 2A\sin 2t \\ 2D\cos 2t - 2C\sin 2t \end{bmatrix}.$$

Substituting $\vec{\mathbf{x}}_p$ into (5) and collecting terms yields

$$\cos t \begin{bmatrix} 2B \\ 2D \end{bmatrix} + \sin t \begin{bmatrix} -2A \\ -2C \end{bmatrix} = \cos t \begin{bmatrix} -2A + C \\ A - 2C + 3 \end{bmatrix} + \sin t \begin{bmatrix} -2B + D + 1 \\ B - 2D \end{bmatrix}$$

and a system of four algebraic equations in A, B, C, and D:

$$\begin{aligned} 2B &= -2A + C, & -2A &= -2B + D + 1, \\ 2D &= A - 2C + 3, & -2C &= B - 2D. \end{aligned}$$

This system has the augmented matrix

$$\left[\begin{array}{cccc|c} 2 & 2 & -1 & 0 & 0 \\ 1 & 0 & -2 & -2 & -3 \\ 2 & -2 & 0 & 1 & -1 \\ 0 & 1 & 2 & -2 & 0 \end{array}\right] \quad \text{with RREF} \quad \left[\begin{array}{cccc|c} 1 & 0 & 0 & 0 & -21/65 \\ 0 & 1 & 0 & 0 & 38/65 \\ 0 & 0 & 1 & 0 & 34/65 \\ 0 & 0 & 0 & 1 & 53/65 \end{array}\right],$$

so our particular solution is

$$\vec{\mathbf{x}}_p = \frac{1}{65} \begin{bmatrix} -21\cos 2t + 38\sin 2t \\ 34\cos 2t + 53\sin 2t \end{bmatrix}.$$

We calculated $\vec{\mathbf{x}}_h$ in Sec. 6.2, Example 3, so we can write the general solution of system (5) as

$$\vec{\mathbf{x}}(t) = c_1 e^{-t} \begin{bmatrix} 1 \\ 1 \end{bmatrix} + c_2 e^{-3t} \begin{bmatrix} 1 \\ -1 \end{bmatrix} + \frac{1}{65} \begin{bmatrix} -21\cos 2t + 38\sin 2t \\ 34\cos 2t + 53\sin 2t \end{bmatrix}.$$

Figure 6.7.3 shows the complicated distortions produced by sinusoidal forcing terms. ■

EXAMPLE 4 **Two-Tank Mixing Problem** Consider the two tanks shown in Fig. 6.7.4, where initially each tank contains 100 gal of fresh water. A salt solution with concentration 1 lb/gal is pumped into Tank 1 at the rate of 3 gal/min, and the solution in Tank 1 is pumped to Tank 2 at a rate of 4 gal/min. The

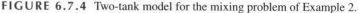

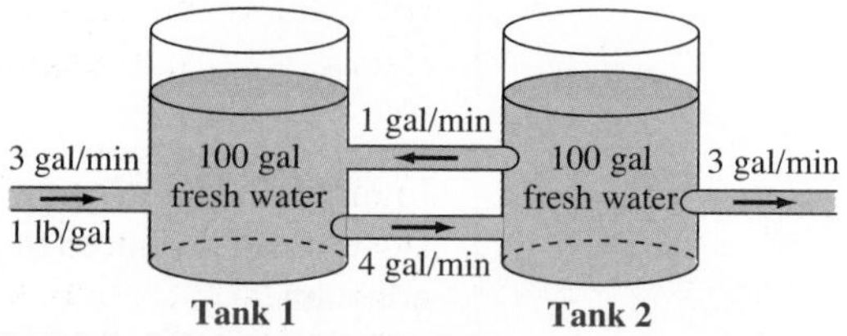

FIGURE 6.7.4 Two-tank model for the mixing problem of Example 2.

solution in Tank 2 is pumped back into Tank 1 at a rate of 1 gal/min and also to the outside at the rate of 3 gal/min. How much salt is in each tank at any time? What is the steady state of the solution?

Calling $x_1(t)$ and $x_2(t)$ the amount of salt (in lb) in Tank 1 and Tank 2, respectively, the rate of change (in lb/min) of salt in each tank will be

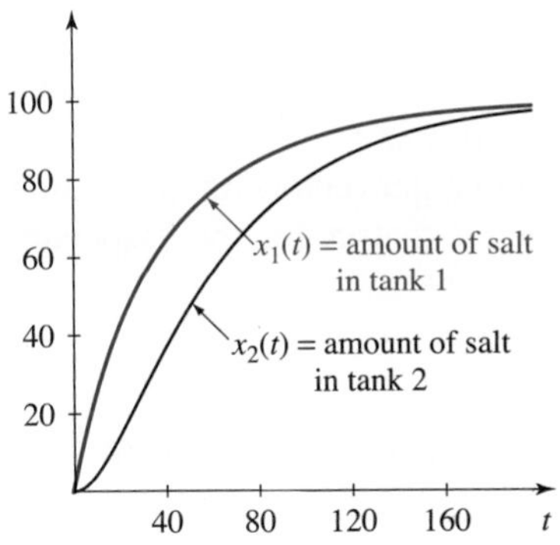

FIGURE 6.7.5 Graph of solutions for Example 4.

$$x_1' = \underbrace{\left(1\frac{\text{lb}}{\text{gal}}\right)\left(3\frac{\text{gal}}{\text{min}}\right) + \left(\frac{x_2\text{ lb}}{100\text{ gal}}\right)\left(1\frac{\text{gal}}{\text{min}}\right)}_{\text{RATE IN}} - \underbrace{\left(\frac{x_1\text{ lb}}{100\text{ gal}}\right)\left(4\frac{\text{gal}}{\text{min}}\right)}_{\text{RATE OUT}},$$

$$x_2' = \underbrace{\left(\frac{x_1\text{ lb}}{100\text{ gal}}\right)\left(4\frac{\text{gal}}{\text{min}}\right)}_{\text{RATE IN}} - \underbrace{\left(\frac{x_2\text{ lb}}{100\text{ gal}}\right)\left(4\frac{\text{gal}}{\text{min}}\right)}_{\text{RATE OUT}},$$

which reduces to

$$\begin{aligned} x_1' &= -0.04x_1 + 0.01x_2 + 3, \\ x_2' &= 0.04x_1 - 0.04x_2. \end{aligned}$$

The initial-value problem can be written in matrix-vector form as a nonhomogeneous system of DEs,

$$\vec{\mathbf{x}}' = \mathbf{A}\vec{\mathbf{x}} + \vec{\mathbf{f}}(t) = \begin{bmatrix} x_1' \\ x_2' \end{bmatrix} = \begin{bmatrix} -0.04 & 0.01 \\ 0.04 & -0.04 \end{bmatrix}\vec{\mathbf{x}} + \begin{bmatrix} 3 \\ 0 \end{bmatrix}, \quad \vec{\mathbf{x}}(0) = \begin{bmatrix} 0 \\ 0 \end{bmatrix},$$

because at $t = 0$ there is no salt in either tank. The eigenvalues and eigenvectors of $\mathbf{A}$ can easily be found to be

$$\lambda_1 = -0.02, \quad \vec{\mathbf{v}}_1 = \begin{bmatrix} 1 \\ 2 \end{bmatrix} \quad \text{and} \quad \lambda_2 = -0.06, \quad \vec{\mathbf{v}}_1 = \begin{bmatrix} 1 \\ -2 \end{bmatrix},$$

which yields the homogeneous solution:

$$\vec{\mathbf{x}}_h = c_1e^{-0.02t}\begin{bmatrix} 1 \\ 2 \end{bmatrix} + c_2e^{-0.06t}\begin{bmatrix} 1 \\ -2 \end{bmatrix}. \tag{6}$$

Because the forcing term is a constant vector, the particular solution is

$$\begin{aligned} \vec{\mathbf{x}}_p = -\mathbf{A}^{-1}\vec{\mathbf{b}} &= -\begin{bmatrix} -0.04 & 0.01 \\ 0.04 & -0.04 \end{bmatrix}^{-1}\begin{bmatrix} 3 \\ 0 \end{bmatrix} \\ &= \frac{25}{3}\begin{bmatrix} 4 & 1 \\ 4 & 4 \end{bmatrix}\begin{bmatrix} 3 \\ 0 \end{bmatrix} = \begin{bmatrix} 100 \\ 100 \end{bmatrix}. \end{aligned}$$

Hence the general solution is

$$\vec{\mathbf{x}}(t) = c_1e^{-0.02t}\begin{bmatrix} 1 \\ 2 \end{bmatrix} + c_2e^{-0.06t}\begin{bmatrix} 1 \\ -2 \end{bmatrix} + \begin{bmatrix} 100 \\ 100 \end{bmatrix}. \tag{7}$$

Substituting the initial conditions $x_1(0) = x_2(0) = 0$ into (7), we find $c_1 = -75$ and $c_2 = -25$, giving the solution

$$\begin{aligned} x_1 &= -75e^{-0.025t} - 25e^{-0.06t} + 100, \\ x_2 &= -150e^{-0.025t} + 50e^{-0.06t} + 100. \end{aligned}$$

These curves are drawn in Fig. 6.7.5. It takes about 2.5 hours before both tanks get within 10% of the steady-state value of 100 lb in each tank. ■

Variation of Parameters

The method of variation of parameters described in Sec. 4.5 can be adapted to the more general nonhomogeneous system

$$\vec{\mathbf{x}}' = \mathbf{A}(t)\vec{\mathbf{x}} + \vec{\mathbf{f}}(t), \tag{8}$$

where the elements of the matrix $\mathbf{A}(t)$ can be *functions of t*.

If $\mathbf{A}$ is an $n \times n$ matrix with n linearly independent eigenvectors, then we know from the basic theory (Sec. 6.1) that a fundamental matrix for the associated homogeneous system

$$\vec{\mathbf{x}}' = \mathbf{A}(t)\vec{\mathbf{x}} \tag{9}$$

is a matrix $\mathbf{X}(t)$ whose columns are linearly independent solutions of (9).[1] We also know that

$$\mathbf{X}'(t) = \mathbf{A}(t)\mathbf{X}(t), \tag{10}$$

and that the general solution of (9) is given by

$$\vec{\mathbf{x}}_h = \mathbf{X}(t)\vec{\mathbf{c}}, \tag{11}$$

where $\vec{\mathbf{c}}$ is an arbitrary constant vector. The idea is to replace $\vec{\mathbf{c}}$ in (11) by a vector function $\vec{\mathbf{v}}(t)$, and to determine $\vec{\mathbf{v}}$ so that

$$\vec{\mathbf{x}}_p = \mathbf{X}(t)\vec{\mathbf{v}}(t) \tag{12}$$

will be a solution of (8). That means that we require

$$\begin{aligned}
(\mathbf{X}\vec{\mathbf{v}})' &= \mathbf{A}\mathbf{X}\vec{\mathbf{v}} + \vec{\mathbf{f}},\\
\mathbf{X}'\vec{\mathbf{v}} + \mathbf{X}\vec{\mathbf{v}}' &= \mathbf{A}\mathbf{X}\vec{\mathbf{v}} + \vec{\mathbf{f}},\\
\mathbf{A}\mathbf{X}\vec{\mathbf{v}} + \mathbf{X}\vec{\mathbf{v}}' &= \mathbf{A}\mathbf{X}\vec{\mathbf{v}} + \vec{\mathbf{f}}, \quad \text{(using (10))}\\
\mathbf{X}\vec{\mathbf{v}}' &= \vec{\mathbf{f}},\\
\vec{\mathbf{v}}' &= \mathbf{X}^{-1}\vec{\mathbf{f}}.
\end{aligned}$$

$\mathbf{X}(t)$ has linearly independent column vectors, so $\mathbf{X}(t)$ is invertible.

Then,

$$\vec{\mathbf{v}} = \int \mathbf{X}^{-1}(t)\vec{\mathbf{f}}(t)\,dt + \vec{\mathbf{k}},$$

and for a particular solution we can choose $\vec{\mathbf{k}} = \vec{\mathbf{0}}$. Substituting into (12), we have

$$\vec{\mathbf{x}}_p = \mathbf{X}(t)\vec{\mathbf{v}} = \mathbf{X}(t)\int \mathbf{X}^{-1}(t)\vec{\mathbf{f}}(t)\,dt. \tag{13}$$

Hence, the general solution of (8) is

$$\vec{\mathbf{x}}(t) = \underbrace{\mathbf{X}(t)\vec{\mathbf{c}}}_{\vec{\mathbf{x}}_h} + \underbrace{\mathbf{X}(t)\int \mathbf{X}^{-1}(t)\vec{\mathbf{f}}(t)dt}_{\vec{\mathbf{x}}_p}, \tag{14}$$

and we can determine that $\vec{\mathbf{c}} = \mathbf{X}^{-1}(0)\vec{\mathbf{x}}_0$ for the initial conditions $\vec{\mathbf{x}}(0) = \vec{\mathbf{x}}_0$. This fact leads to the following theorem.

[1] A fundamental matrix exists only if there are n linearly independent solutions $\vec{\mathbf{x}}_i(t)$. If eigenvalues are repeated, it may be necessary to use generalized eigenvectors (Sec. 6.2, Example 6).

General Solution of $\vec{\mathbf{x}}' = \mathbf{A}(t)\vec{\mathbf{x}} + \vec{\mathbf{f}}(t)$

Let $\mathbf{A}(t)$ be an $n \times n$ matrix whose elements are continuous functions on the interval under consideration, and let $\vec{\mathbf{f}}(t)$ be an $n \times 1$ vector with continuous elements. If $\mathbf{X}(t)$ is a fundamental matrix for $\vec{\mathbf{x}}' = \mathbf{A}(t)\vec{\mathbf{x}}$, then the general solution of the nonhomogeneous linear system

$$\vec{\mathbf{x}}' = \mathbf{A}(t)\vec{\mathbf{x}} + \vec{\mathbf{f}}(t), \quad \vec{\mathbf{x}}(0) = \vec{\mathbf{x}}_0$$

is

$$\vec{\mathbf{x}}(t) = \mathbf{X}(t)\mathbf{X}^{-1}(0)\vec{\mathbf{x}}_0 + \mathbf{X}(t)\int_0^t \mathbf{X}^{-1}(s)\vec{\mathbf{f}}(s)ds. \tag{15}$$

EXAMPLE 5 **Playing the System** We can use variation of parameters to find a particular solution for system (1) of Example 1,

$$\vec{\mathbf{x}}' = \mathbf{A}\vec{\mathbf{x}} + \vec{\mathbf{f}} = \begin{bmatrix} 1 & 1 \\ 4 & 1 \end{bmatrix}\vec{\mathbf{x}} + \begin{bmatrix} t-2 \\ 4t-1 \end{bmatrix}. \tag{16}$$

We know that

$$\vec{\mathbf{x}}_h = c_1 e^{3t}\begin{bmatrix} 1 \\ 2 \end{bmatrix} + c_2 e^{-t}\begin{bmatrix} 1 \\ -2 \end{bmatrix},$$

so a fundamental matrix and its inverse are given by

$$\mathbf{X}(t) = \begin{bmatrix} e^{3t} & e^{-t} \\ 2e^{3t} & -2e^{-t} \end{bmatrix} \quad \text{and} \quad \mathbf{X}^{-1}(t) = \frac{1}{4}\begin{bmatrix} 2e^{-3t} & e^{-3t} \\ 2e^{t} & -e^{t} \end{bmatrix}.$$

Then,

$$\vec{\mathbf{v}}'(t) = \mathbf{X}^{-1}(t)\vec{\mathbf{f}}(t) = \frac{1}{4}\begin{bmatrix} 2e^{-3t} & e^{-3t} \\ 2e^{t} & -e^{t} \end{bmatrix}\begin{bmatrix} t-2 \\ 4t-1 \end{bmatrix} = \frac{1}{4}\begin{bmatrix} e^{-3t}(6t-5) \\ -e^{t}(2t+3) \end{bmatrix}.$$

Integration by parts of each component yields

$$\vec{\mathbf{v}}(t) = \frac{1}{4}\begin{bmatrix} e^{-3t}(1-2t) \\ -e^{t}(2t+1) \end{bmatrix}.$$

Then, by (13), a particular solution of (16) is given by

$$\vec{\mathbf{x}}_p = \mathbf{X}(t)\vec{\mathbf{v}}(t) = \frac{1}{4}\begin{bmatrix} e^{3t} & e^{-t} \\ 2e^{3t} & -2e^{-t} \end{bmatrix}\begin{bmatrix} e^{-3t}(1-2t) \\ -e^{t}(2t+1) \end{bmatrix} = \begin{bmatrix} -t \\ 1 \end{bmatrix},$$

exactly as we found in Example 3 by undetermined coefficients. ■

EXAMPLE 6 **Variation of Parameters** Use variation of parameters to solve an IVP with the same matrix **A** as Example 5 but different forcing functions,

$$\vec{\mathbf{x}}' = \begin{bmatrix} 1 & 1 \\ 4 & 1 \end{bmatrix}\begin{bmatrix} x_1 \\ x_2 \end{bmatrix} + e^{t}\begin{bmatrix} 2 \\ 0 \end{bmatrix}, \quad \vec{\mathbf{x}}(0) = \begin{bmatrix} 1 \\ 0 \end{bmatrix}. \tag{17}$$

Recall from Example 5 that a fundamental matrix is

$$\mathbf{X}(t) = \begin{bmatrix} e^{3t} & e^{-t} \\ 2e^{3t} & -2e^{-t} \end{bmatrix} \quad \text{with} \quad \mathbf{X}^{-1}(t) = \frac{1}{4}\begin{bmatrix} 2e^{-3t} & e^{-3t} \\ 2e^{t} & -e^{t} \end{bmatrix}.$$

Using formula (13) for a particular solution, we have

$$\begin{aligned}
\vec{\mathbf{x}}_p &= \mathbf{X}(t)\int_0^t \mathbf{X}^{-1}(s)\vec{\mathbf{f}}(s)ds \\
&= \begin{bmatrix} e^{3t} & e^{-t} \\ 2e^{3t} & -2e^{-t} \end{bmatrix} \int_0^t \frac{1}{4}\begin{bmatrix} 2e^{-3s} & e^{-3s} \\ 2e^{s} & -e^{s} \end{bmatrix}\begin{bmatrix} 2e^{3s} \\ 0 \end{bmatrix} ds \\
&= \frac{1}{4}\begin{bmatrix} e^{3t} & e^{-t} \\ 2e^{3t} & -2e^{-t} \end{bmatrix} \int_0^t \begin{bmatrix} 4 \\ 4e^{4s} \end{bmatrix} ds \\
&= \frac{1}{4}\begin{bmatrix} e^{3t} & e^{-t} \\ 2e^{3t} & -2e^{-t} \end{bmatrix} \begin{bmatrix} 4t \\ e^{4t}-1 \end{bmatrix} \\
&= \frac{1}{4}\begin{bmatrix} -e^{-t}+(4t+1)e^{3t} \\ 2e^{-t}+(8t-2)e^{3t} \end{bmatrix}.
\end{aligned}$$

Hence, the general solution of (17) is given by

$$\begin{aligned}
\vec{\mathbf{x}}(t) &= \mathbf{X}(t)\mathbf{X}^{-1}(0)\vec{\mathbf{x}}_0 + \mathbf{X}(t)\int_0^t \mathbf{X}^{-1}(s)\vec{\mathbf{f}}(s)ds \\
&= \begin{bmatrix} e^{3t} & e^{-t} \\ 2e^{3t} & -2e^{-t} \end{bmatrix} \frac{1}{4}\begin{bmatrix} 2 & 1 \\ 2 & -1 \end{bmatrix}\begin{bmatrix} 1 \\ 0 \end{bmatrix} + \frac{1}{4}\begin{bmatrix} -e^{-t}+(4t+1)e^{3t} \\ 2e^{-t}+(8t-2)e^{3t} \end{bmatrix} \\
&= \frac{1}{4}\begin{bmatrix} e^{-t}+(4t+3)e^{3t} \\ -2e^{-t}+(8t+2)e^{3t} \end{bmatrix}.
\end{aligned}$$

■

Summary

The Nonhomogeneous Theorem for Linear Systems can be applied to obtain the general solution of

$$\vec{\mathbf{x}}' = \mathbf{A}(t)\vec{\mathbf{x}} + \vec{\mathbf{f}}.$$

Various methods for finding a particular solution $\vec{\mathbf{x}}_p$ include undetermined coefficients and variation of parameters, in addition to decoupling (Sec. 6.5) and the matrix exponential (Sec. 6.6).

6.7 Problems

1. **Superposition for Systems** Given that

$$L(\vec{\mathbf{x}}) = \vec{\mathbf{x}}' - \begin{bmatrix} 1 & 2 \\ 0 & 1 \end{bmatrix}\vec{\mathbf{x}},$$

and that

$$\vec{\mathbf{x}}_1 = \begin{bmatrix} e^t \\ e^t \end{bmatrix} \text{ is a solution for } L(\vec{\mathbf{x}}_1) = \begin{bmatrix} -2e^t \\ 0 \end{bmatrix} = \vec{\mathbf{f}}_1$$

and

$$\vec{\mathbf{x}}_2 = \begin{bmatrix} 1 \\ -1 \end{bmatrix} \text{ is a solution for } L(\vec{\mathbf{x}}_2) = \begin{bmatrix} 1 \\ 1 \end{bmatrix} = \vec{\mathbf{f}}_2,$$

find a particular solution to

$$L(\vec{\mathbf{x}}) = \begin{bmatrix} e^t+2 \\ 2 \end{bmatrix}.$$

2. **Superposition for Systems Once More** Given that $L(\vec{\mathbf{x}}) = \vec{\mathbf{f}}$ is a 2×2 linear system of equations, and that

$$\vec{\mathbf{x}}_1 = \begin{bmatrix} t \\ 1 \end{bmatrix} \text{ is a solution for } L(\vec{\mathbf{x}}) = \begin{bmatrix} 1+t \\ -1-3t \end{bmatrix} = \vec{\mathbf{f}}_1$$

and

$$\vec{\mathbf{x}}_2 = \begin{bmatrix} 1 \\ 2 \end{bmatrix} \text{ is a solution for } L(\vec{\mathbf{x}}) = \begin{bmatrix} 1 \\ -5 \end{bmatrix} = \vec{\mathbf{f}}_2,$$

find a particular solution to

$$L(\vec{\mathbf{x}}) = \begin{bmatrix} 2t+5 \\ -6t-17 \end{bmatrix}.$$

3. Nonhomogeneous Illustration Illustrate the Superposition Principle for nonhomogeneous linear systems for

$$\vec{\mathbf{x}}' = \mathbf{A}\vec{\mathbf{x}} + \vec{\mathbf{f}} = \begin{bmatrix} 1 & 1 \\ 4 & 1 \end{bmatrix}\vec{\mathbf{x}} + \begin{bmatrix} t-2+e^t \\ 4t-1-4e^t \end{bmatrix}$$

and its particular solution

$$\vec{\mathbf{x}}_p = \begin{bmatrix} e^t - t \\ 1 - e^t \end{bmatrix}$$

by finding the general solution.

Systematic Prediction *Use the method of undetermined coefficients to solve the nonhomogeneous system in each of Problems 4–7.*

4. $\vec{\mathbf{x}}' = \begin{bmatrix} 1 & 4 \\ 1 & 1 \end{bmatrix}\vec{\mathbf{x}} + \begin{bmatrix} 3 \\ 0 \end{bmatrix}$ **5.** $\vec{\mathbf{x}}' = \begin{bmatrix} 1 & 4 \\ 1 & 1 \end{bmatrix}\vec{\mathbf{x}} + \begin{bmatrix} 0 \\ 9t \end{bmatrix}$

6. $\vec{\mathbf{x}}' = \begin{bmatrix} 1 & 4 \\ 1 & 1 \end{bmatrix}\vec{\mathbf{x}} + \begin{bmatrix} 0 \\ e^t \end{bmatrix}$

7. $\vec{\mathbf{x}}' = \begin{bmatrix} 1 & 4 \\ 1 & 1 \end{bmatrix}\vec{\mathbf{x}} + \begin{bmatrix} 0 \\ 10\sin t \end{bmatrix}$

8. System Superposition Prove the Superposition Principle for nonhomogeneous systems of n linear first-order equations. Show how it follows from the linearity of L.

Variation of Parameters *For Problems 9–15, use variation of parameters to obtain a particular solution for the nonhomogeneous system and then find the general solution.*

9. $\vec{\mathbf{x}}' = \begin{bmatrix} 1 & 1 \\ 4 & 1 \end{bmatrix}\vec{\mathbf{x}} + \begin{bmatrix} -3 \\ -9 \end{bmatrix}$

10. $\vec{\mathbf{x}}' = \begin{bmatrix} 1 & 1 \\ 4 & 1 \end{bmatrix}\vec{\mathbf{x}} + \begin{bmatrix} e^t \\ -4e^t \end{bmatrix}$

11. $\vec{\mathbf{x}}' = \begin{bmatrix} 0 & -1 \\ 3 & 4 \end{bmatrix}\vec{\mathbf{x}} + \begin{bmatrix} 3t \\ 9 \end{bmatrix}$

12. $\vec{\mathbf{x}}' = \begin{bmatrix} 1 & 1 \\ 4 & 1 \end{bmatrix}\vec{\mathbf{x}} + \begin{bmatrix} 2e^{3t} \\ 0 \end{bmatrix}$

13. $\vec{\mathbf{x}}' = \begin{bmatrix} 2 & 2 \\ 1 & 3 \end{bmatrix}\vec{\mathbf{x}} + \begin{bmatrix} 1 \\ -t \end{bmatrix}$

14. $\vec{\mathbf{x}}' = \begin{bmatrix} -4 & 2 \\ 2 & -1 \end{bmatrix}\vec{\mathbf{x}} + \begin{bmatrix} t^{-1} \\ 2t^{-1}+4 \end{bmatrix}, \quad t > 0$

15. $\vec{\mathbf{x}}' = \begin{bmatrix} 4 & -2 \\ 8 & -4 \end{bmatrix}\vec{\mathbf{x}} + \begin{bmatrix} t^{-3} \\ -t^{-2} \end{bmatrix}, \quad t > 0$

16. $\vec{\mathbf{x}}' = \begin{bmatrix} 0 & -1 \\ 1 & 0 \end{bmatrix}\vec{\mathbf{x}} + \begin{bmatrix} 0 \\ \tan t \end{bmatrix}$

17. Two-Tank Mixing Problem Two tanks, each with capacity 100 gal, are initially filled with fresh water. Brine containing 1 lb of salt per gallon flows into the first tank at a rate of 4 gal/min, and the dissolved mixture flows into the second tank at a rate of 6 gal/min. The resultant stirred mixture is simultaneously pumped back into the first tank at the rate of 2 gal/min and out of the second tank at the rate of 4 gal/min. (See Fig. 6.7.6.)

(a) Find the initial-value problem that describes the future amount of salt in the two tanks.

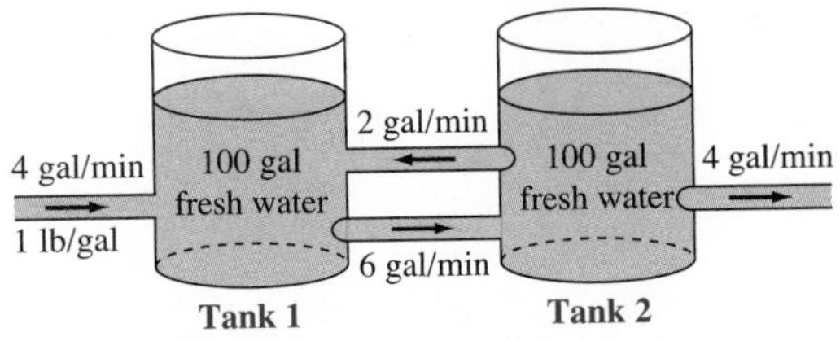

FIGURE 6.7.6 Two-tank model for Problem 17.

(b) The solution for the system (provided by Maple) is

$$\begin{bmatrix} x_1(t) \\ x_2(t) \end{bmatrix} = -157.47e^{-0.025t}\begin{bmatrix} 0.50 \\ 0.87 \end{bmatrix} + 42.53e^{-0.095t}\begin{bmatrix} -0.50 \\ 0.87 \end{bmatrix} + \begin{bmatrix} 100 \\ 100 \end{bmatrix}.$$

Show that the solution satisfies the initial conditions in part (a), and check that the eigenvalues of the matrix in (a) agree with those displayed in (b).

(c) Plot the height functions for both tanks on the same graph. Is there an equilibrium solution?

18. Two-Loop Circuit Find the currents I_1 and I_2 in the two-loop circuit in Fig. 6.7.7, when initially both currents are zero. (This is the same circuit as in Sec. 6.2, Problem 61.) What are the steady states of the currents? Plot $I_1(t)$ and $I_2(t)$ on the same graph.

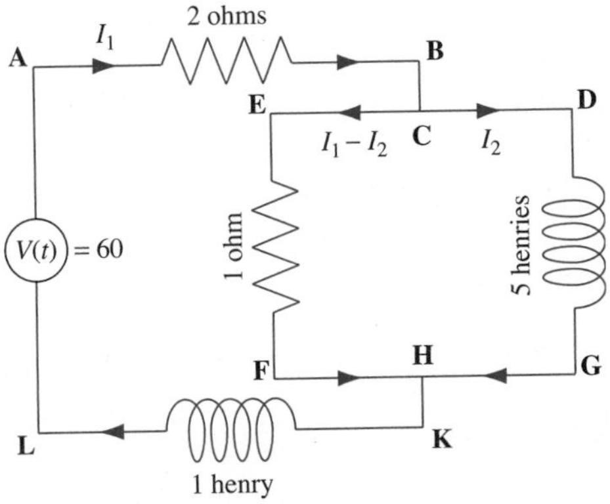

FIGURE 6.7.7 Two-loop circuit for Problem 18.

19. Multiple-Loop *RL* Circuit with AC Input Initially there is no current in the circuit in Figure 6.7.8. Find the currents I_1 and I_2 at future values of time when an AC voltage of $220 \sin t$ is applied to the circuit. Plot $I_1(t)$ and $I_2(t)$ on the same graph. Also plot their phase portrait, and discuss how the graphs are related.

20. Suggested Journal Entry We have used a variety of methods to find a particular solution of a nonhomogeneous system of first-order linear DEs: undetermined coefficients, variation of parameters, decoupling, and the matrix exponential. Compare the methods for applicability, ease of use, and so on.

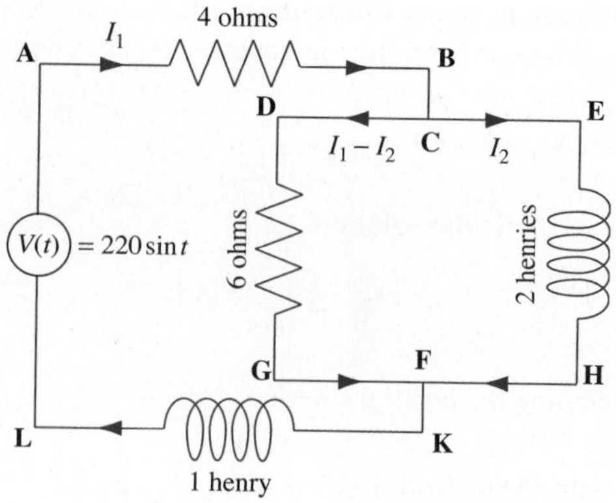

FIGURE 6.7.8 Two-loop circuit for Problem 19.

CHAPTER 7 Nonlinear Systems of Differential Equations

Not only in research, but also in the everyday world of politics and economics, we would all be better off if more people realized that simple nonlinear systems do not necessarily possess simple dynamical properties.[1]

—Robert May

7.1 Nonlinear Systems

SYNOPSIS: We extend our study of homogeneous systems to include nonlinearity. In the process, we broaden phase-portrait analysis to include multiple equilibrium points and limit cycles, and observe their effects on the geometry and stability of solutions.

Introduction

In this chapter we turn to *nonlinear* systems, where there will seldom be an analytic solution with formulas. Qualitative analysis, with its focus on equilibria (of which there can be more than one) and stability, becomes far more important.

Before computer graphics became readily available in the early 1980s, the subject of nonlinear systems was not easily accessible. But since then it has become almost trivial to create phase portraits of 2×2 systems, including nonlinear ones, as easily as for a single first-order DE in one variable. (See Sec. 2.6.) With a little experience, freshmen and nonmathematicians can readily analyze a nonlinear 2×2 system, instead of needing to wait for specialized graduate courses. This section aims to give some of that experience.

You will recognize many familiar themes in this new nonlinear setting:

- equilibria and stability,
- nullclines,
- existence and uniqueness for autonomous systems,
- *linear* systems that provide the foundations for analysis, and
- the linking of tx and ty graphs to the phase-plane trajectories.

[1] Reprinted by permission from *Nature* (**261:** 459–467) 1976, Macmillan Magazines Ltd.

There is also a new feature: limit cycles. We will use *qualitative analysis* to draw the pictures, then focus on *interpretation* of the pictures in terms of real-world models.

Autonomous 2 x 2 Systems

In this chapter we will study the autonomous system

$$\begin{aligned} x' &= dx/dt = f(x, y), \\ y' &= dy/dt = g(x, y), \end{aligned} \tag{1}$$

consisting of two differential equations in two dependent variables, with the functions f and g no longer restricted to being linear. With vector notation,

$$\mathbf{x} = \begin{bmatrix} x \\ y \end{bmatrix} \quad \text{and} \quad \mathbf{x}' = \frac{d\mathbf{x}}{dt} = \begin{bmatrix} f(x, y) \\ g(x, y) \end{bmatrix}. \tag{2}$$

No matrix is involved when the system is not linear.

When the components $x(t)$ and $y(t)$ of a solution of (1) are plotted parametrically in the xy-plane (the phase plane), the **solution curves** or **trajectories** represent their interaction. We will see a much greater variety of phase portraits than those of linear systems.

Compare the nonlinear phase portraits of Fig. 7.1.1 with the linear phase portraits classified in Sec. 6.4. What new aspects do you notice?

Vector Fields; Two-Dimensional Equations

Try a selection of nonlinear examples. Compare linear and nonlinear examples from menu.

- Nonlinear systems can have *more than one equilibrium, or none.*
- Phase portraits for *nonlinear* systems include some *local* patterns that look suspiciously like the patterns you studied for *linear* systems in Sec. 6.4.
- The way that *locally linear* phase portraits *fit together* is decidedly *nonlinear.*
- The phase portrait in Fig. 7.1.1(d) shows a new feature—the *limit cycle.* The limit cycle is a dark loop, attracting spiral trajectories from both inside and outside.

Each of these statements needs to be examined in some detail.

Qualitative Analysis

Parametric to Cartesian; Phase Plane Drawing

Review the connections between xy, tx, and ty graphs.

How do trajectories of the general autonomous 2×2 system (1) move about the phase plane? Are there principles prescribing what they can do and what they cannot do? Two such rules concern uniqueness (see Sec. 2.6) and continuity. We will deal primarily with 2×2 systems that satisfy uniqueness criteria. Some systems that do not, together with a theorem, are discussed in Problems 39–40 and in Sec. 7.5.

Properties of Phase-Plane Trajectories in a Nonlinear 2 × 2 System

(i) When uniqueness holds, phase-plane trajectories cannot cross.

(ii) When the given functions f and g are continuous, trajectories are continuous (no breaks) and smooth (no corners or cusps).

If you experiment with drawings of curves that satisfy these rules, you will find that, unless they are closed curves, they are always coming from somewhere and are always headed for somewhere as well. What is more, the "somewhere" can be a point, a closed curve, or "infinity." Curves in three or more dimensions can behave more wildly, but we will stick to two-dimensional portraits in this section.

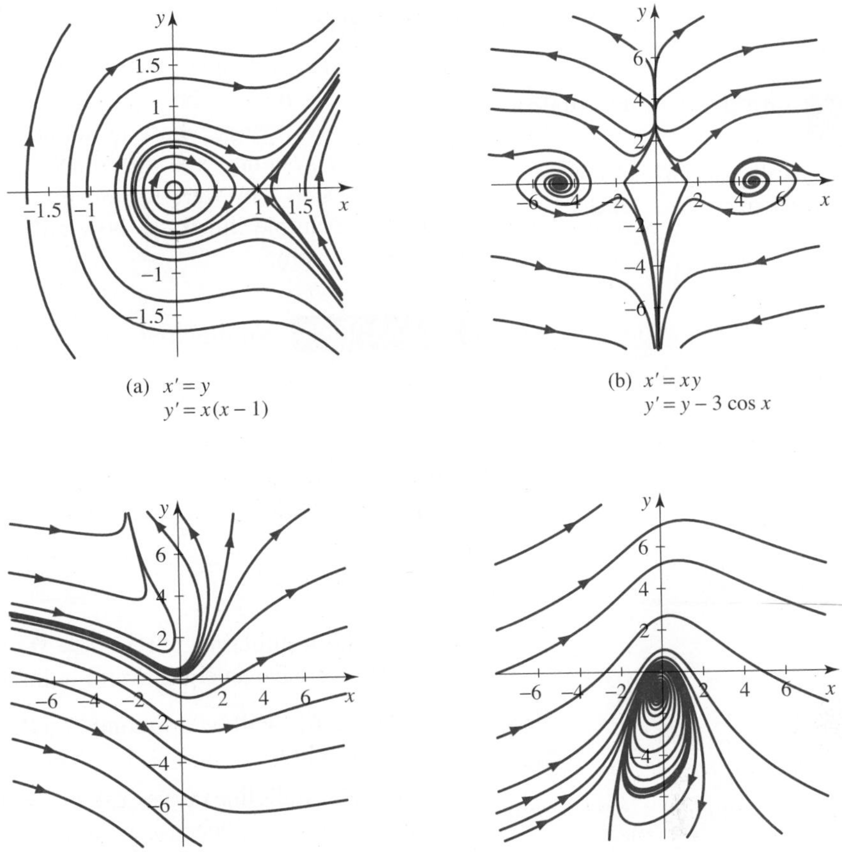

(a) $x' = y$
$y' = x(x-1)$

(b) $x' = xy$
$y' = y - 3\cos x$

(c) $x' = x^2 - y + 2$
$y' = y + 2x$

(d) $x' = 1 + y + x^2$
$y' = y/2 - 3x$

FIGURE 7.1.1 Typical phase portraits for 2×2 autonomous nonlinear systems.

Recall from Chapter 6 that the critical question to ask about an equilibrium solution is whether or not it is stable: Do nearby solutions stay close or wander away? The definitions of **stable**, **asymptotically stable**, **neutrally stable**, and **unstable** equilibria should be reviewed at this point. (Sec. 6.4.) You should be able to identify the stability of the equilibria in Fig. 7.1.1. (See Problems 10–13.)

Equilibria

For nonlinear systems, it is common for phase portraits to contain **equilibrium points** where the system is at rest. Unlike the linear systems studied in Chapter 6, nonlinear systems can have more than one equilibrium solution (which may occur at points other than the origin), or none at all. Equilibrium points can be found by determining where $x' = 0$ and $y' = 0$; that is, for system (1), by *simultaneously* solving the algebraic equations

$$x' = f(x, y) = 0,$$
$$y' = g(x, y) = 0,$$

either exactly or by numerical approximation. Try this out on the systems and phase portraits in Fig. 7.1.1.

Another method for locating equilibria is to use **nullclines**, introduced in Chapters 2 and 6, and now to be revisited. Graphically, *equilibria occur at the intersections of nullclines of horizontal slopes with nullclines of vertical slopes.*

Nullclines

Competitive Exclusion

See how the nullclines interact to create different phase portraits.

We saw in Secs. 2.6 and 6.3 how nullclines, curves where either $x' = 0$ or $y' = 0$, are a valuable tool in the qualitative analysis of systems of linear differential equations. We now see how nullclines can be used to help analyze solutions of nonlinear systems. The difference here is that nullclines are not necessarily straight lines, as they were for linear systems, but curves in the phase plane. The following example illustrates how nullclines can be used to analyze a difficult nonlinear system.

EXAMPLE 1 **Nonlinearity Twists the Phase Portrait**[2] Although the nonlinear system

$$\begin{aligned} x' &= x + e^{-y}, \\ y' &= -y, \end{aligned} \tag{3}$$

does not have a closed-form solution, the system can be analyzed using qualitative tools. We find the nullclines by solving

$$\begin{aligned} x' &= x + e^{-y} = 0, && (v\text{-nullcline}) \\ y' &= -y = 0. && (h\text{-nullcline}) \end{aligned}$$

Where the v-nullcline $x + e^{-y} = 0$ and the h-nullcline $y = 0$ intersect, at $(-1, 0)$, we have an equilibrium point. (See Fig. 7.1.2.)

- Trajectories will be horizontal on the x-axis (h-nullcline), with movement to the right for $x' = x + 1 > 0$ and to the left when $x' = x + 1 < 0$.
- On the v-nullcline curve, $x + e^{-y} = 0$, movement is down when $y > 0$ and up when $y < 0$, because $y' = -y$.

The nullclines partition the xy-plane into four distinct "quadrants," where x' and y' have different signs. Figure 7.1.2 shows the general directions of trajectories in each quadrant. From the arrows you can get a good sense of the way the solutions flow, especially if you zoom in on the equilibrium. We can carry this analysis further by drawing a direction field of the system superimposed with several solutions. (See Fig. 7.1.3.) We can conclude from Fig. 7.1.3 that the equilibrium point $(-1, 0)$ is unstable and looks like a warped saddle.

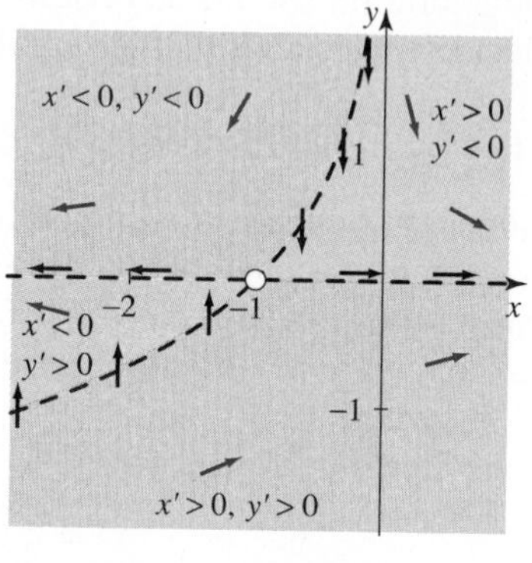

FIGURE 7.1.2 Nullcline information for Example 1.

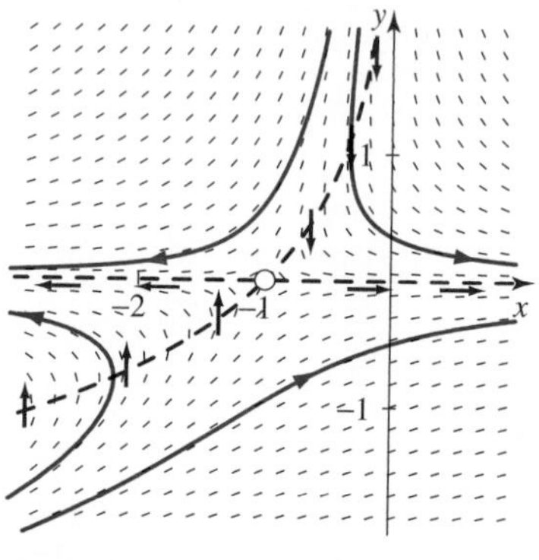

FIGURE 7.1.3 Vector field with trajectories for Example 1. ■

[2] Adapted from Steven H. Strogatz, *Nonlinear Dynamics and Chaos* (Reading, MA: Addison-Wesley, 1994).

Problems 10–13 ask you to sketch the nullclines on the graphs in Fig. 7.1.1 and identify the equilibria and their stabilities. You should make this practice a habit when you meet any 2×2 nonlinear DE system. It greatly clarifies the behavior of the system, and is the proper focus for describing that behavior.

EXAMPLE 2 **Extraterrestrial** The nonlinear system

$$x' = xy,$$
$$y' = 9 - x^2 - y^2,$$

has four equilibrium points, $(\pm 3, 0)$ and $(0, \pm 3)$, which are found by solving the simultaneous equations $x' = 0$, $y' = 0$. The v-nullclines (solving $x' = 0$) are the x and y axes, and the h-nullcline (solving $y' = 0$) is the circle $x^2 + y^2 = 9$. We draw these curves, superimposing the nullclines with horizontal and vertical arrows, in Fig. 7.1.4.

- The vertical arrows point up inside the circle $x^2 + y^2 = 9$, where $y' = 9 - x^2 - y^2 > 0$, and down outside the circle.
- The horizontal arrows point to the right in the first and third quadrants, because $x' = xy > 0$ there, and to the left in the second and fourth quadrants, because $x' = xy < 0$ there.

If we zoom in on the equilibria, the arrows on and between the nullclines let us deduce that $(0, \pm 3)$ are *unstable* equilibria, and that solutions flow around the other equilibria at $(\pm 3, 0)$. Adding a few trajectories around $(\pm 3, 0)$ shows that the equilibria on the x-axis are neutrally stable, because nearby trajectories are neither attracted nor repelled. Locally, the points $(\pm 3, 0)$ look like center points with periodic solutions circling around them. ■

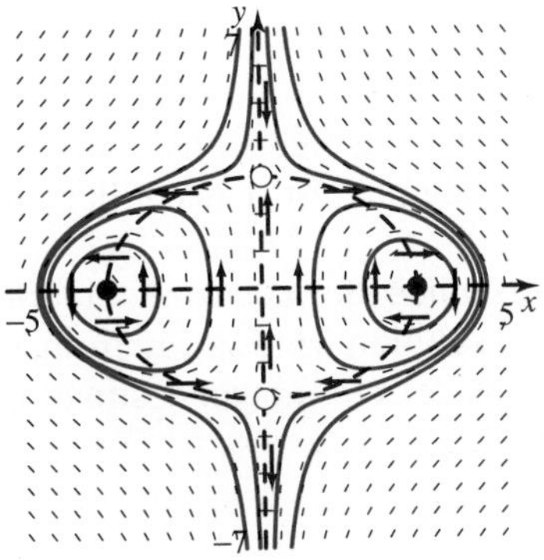

FIGURE 7.1.4 Phase portrait for Example 2. (The x- and y-axes have different scales, so the circle nullcline is distorted.)

Limit Cycles

A new geometric feature not encountered in our analysis of linear systems is the *limit cycle*.

> **Limit Cycle**
>
> A **limit cycle** is a closed curve (representing a periodic solution) to which other solutions tend by winding around more and more closely from either the inside or outside (in either forward or backward time).

Glider; Chemical Oscillator

These examples have limit cycles for certain parameter values.

In Chapter 6, closed orbits came only in families about a center. By contrast, *a limit cycle is isolated*: there is a strip surrounding it that contains no other closed orbit, such as the dark cycle of Fig. 7.1.1(d). (Notice that the closed orbits in Fig. 7.1.1(a) are *not* isolated.)

Limit cycles are not as easy to *find* as equilibria (which can be located algebraically),[3] but they will show up in computer phase portraits of nonlinear 2×2 systems, so you should learn to understand what they represent.

In Fig. 7.1.5 we see different ways in which nearby solutions may relate to a limit cycle. *Stable* behavior occurs when the solution wraps closer to the limit

[3]See the Poincaré-Bendixson theorem in J. H. Hubbard and B. H. West, *Differential Equations: A Dynamical Systems Approach, Part 2: Higher Dimensional Systems* (TAM 18, NY: Springer-Verlag, 1995), Chapter 8.

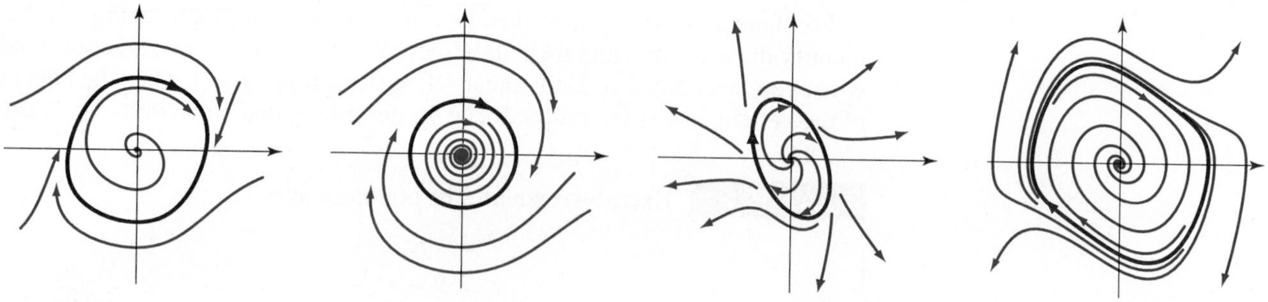

(a) Stable from both the outside and the inside

(b) Stable from the outside, unstable from the inside

(c) Stable from the inside, unstable from the outside

(d) Unstable from both the outside and the inside

FIGURE 7.1.5 Behavior of solutions near a limit cycle.

cycle as $t \to \infty$. In a physical system, a stable limit cycle is frequently a desired behavior, representing ongoing motion that neither comes to rest at an equilbrium nor flies off to infinity.

On the other hand, an *unstable* limit cycle occurs when the *backward* solution ($t \to -\infty$) tends to the limit cycle, and the forward solution winds away from it, either to an equilibrium point or another limit cycle inside or to another destination outside (equilibrium point, limit cycle, or "infinity").

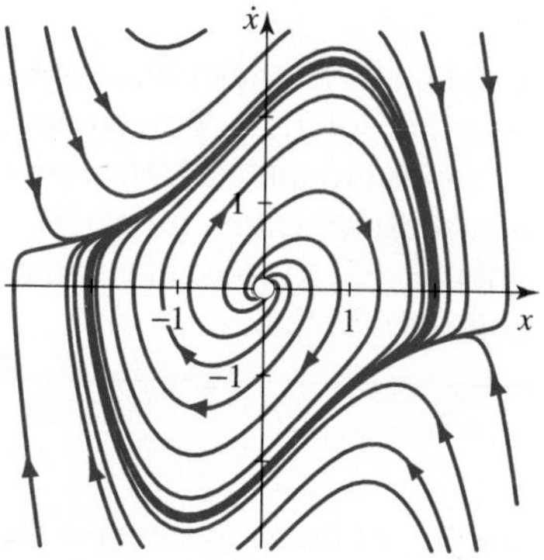

FIGURE 7.1.6 Phase portrait for the van der Pol system (4), with $\varepsilon = 1$. All solutions tend to the limit cycle as $t \to \infty$.

EXAMPLE 3 **van der Pol Equation** The nonlinear second-order DE

$$\ddot{x} - \varepsilon(1 - x^2)\dot{x} + x = 0,$$

called the van der Pol equation,[4] can be converted as usual to a system of first-order equations

$$\begin{aligned} \dot{x} &= y, \\ \dot{y} &= \varepsilon(1 - x^2)y - x. \end{aligned} \tag{4}$$

For certain values of ε, the phase portrait generates a limit cycle that attracts all trajectories as $t \to \infty$. See, for example, Fig. 7.1.6. ■

van der Pol Circuit

See how some choices of ε create a limit cycle in the phase portrait, and see the effects evolve in time series and a model electric circuit.

Component Solution Graphs

We may plot separately, in the tx- or ty-plane, the individual behavior of each component as a function of time; these plots are called **time series**. Frequently we will call on the computer to create phase portraits and time series, which it can do numerically to remarkable accuracy, as we shall study in Sec. 7.3. The important task is to see how these graphs are linked. Figure 7.1.7 shows the time series for a single trajectory in the phase plane. (Other examples of linked trajectories were given in Chapter 4.)

Integrable Solutions

As we remarked earlier, quantitative solutions can be obtained and plotted for linear systems of the type studied in Chapter 6, but this is less often the case for

[4]Balthazar van der Pol (1889–1959) was a Dutch physicist and engineer who in the 1920s developed mathematical models (still in use) for the internal voltages and currents of radios. For a slightly more general form of this equation and for its application to electrical circuits, see R. Borrelli and C. Coleman, *Differential Equations: A Modeling Perspective* (NY: Wiley, 1998).

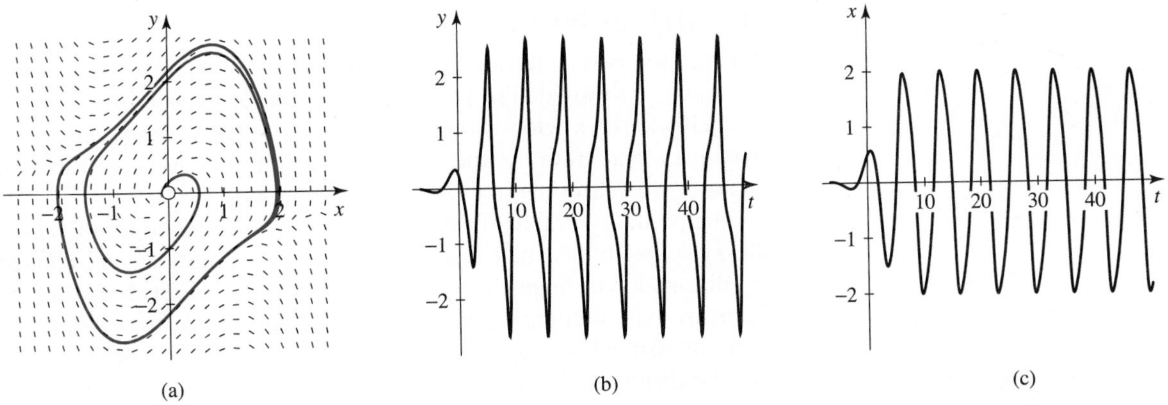

FIGURE 7.1.7 Phase portrait (a) and time series (b) and (c) for a single trajectory from Fig. 7.1.6. All three views show attraction to the limit cycle.

nonlinear problems. An exception is the situation in which the calculus identity

$$\frac{dy}{dx} = \frac{dy/dt}{dx/dt} = \frac{g(x, y)}{f(x, y)} \tag{5}$$

transforms (1) into a first-order equation in x and y that can be solved, explicitly or implicitly, to give a family of solution curves in the xy-plane. (Of course, this device may work for linear systems as well as nonlinear ones.)

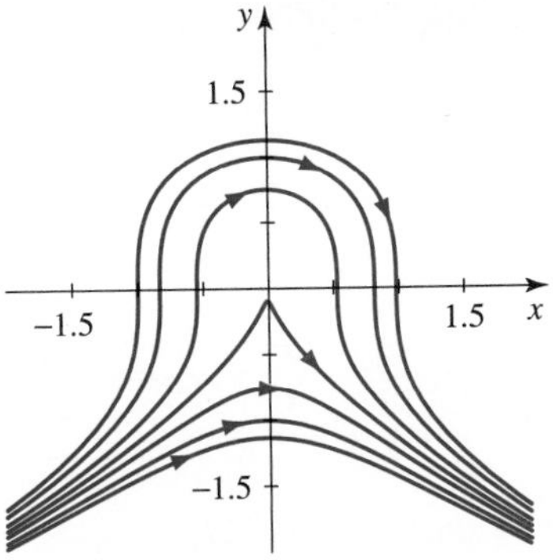

FIGURE 7.1.8 Phase portrait for Example 4 by direct integration.

EXAMPLE 4 Semicubical Parabolas For the nonlinear system

$$x' = y^2,$$
$$y' = -x,$$

identity (5) leads to the first-order equation

$$\frac{dy}{dx} = \frac{y'}{x'} = -\frac{x}{y^2}.$$

This separable equation has the family of implicit solution curves

$$3x^2 + 2y^3 = c,$$

several of which are shown in Fig. 7.1.8. ■

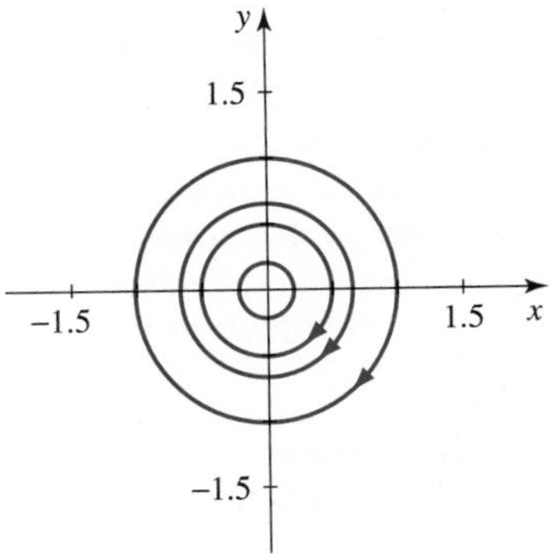

FIGURE 7.1.9 Phase portrait for Example 5 by direct integration.

EXAMPLE 5 Around in Circles The linear system

$$x' = y,$$
$$y' = -x,$$

leads to the separable DE

$$\frac{dy}{dx} = \frac{y'}{x'} = -\frac{x}{y},$$

using (5), and solution curves are a family $x^2 + y^2 = c$ of circles. (See Fig. 7.1.9.) Alternatively, you could have obtained the circles from the linear algebra methods of Chapter 6. ■

Historical Note

For the first two centuries after the invention of calculus, the study of differential equations provided some stunning successes in description and prediction of the behavior of a wide variety of physical systems, from electricity to thermodynamics, and from hydrodynamics to astronomy. Most of this work involved quantitative analysis of the differential equations.

The predictive value of differential equations ran into trouble, however, when the systems studied had unstable equilibrium points: Small changes in initial conditions led to huge changes in behavior. It was French mathematician Henri Poincaré who recognized this problem at the turn of the twentieth century. He pinpointed it in his essay, *Science and Method*, as the fault of the initial conditions, not the differential equations. His answer: Study the whole family of solutions that start near the equilibrium solution, in order to develop a more comprehensive description of its behavior. Thus the qualitative theory of differential equations was born.

When the sensitivity to initial conditions or parameters becomes *hypersensitivity*, we encounter the theory of **chaotic systems**. We will have more to say about these systems in Secs. 7.4, 7.5, and 9.3.

Summary

In extending phase-plane analysis to nonlinear systems, we encounter multiple equilibrium points and limit cycles. Analysis is facilitated by studying nullclines and the regions into which they separate the plane.

7.1 Problems

Review of Classifications *For each system in Problems 1–5, determine the dependent variables, given that t is the independent variable, and the parameters; determine whether the system is autonomous or nonautonomous, linear or nonlinear, and (if linear) homogeneous or nonhomogeneous.*

1. $x' = x + ty$
$y' = 2x + y + \gamma \sin t$

2. $u' = 3u + 4v$
$v' = -2u + \sin t$

3. $x_1' = \kappa x_2$
$x_2' = -\sin x_1$

4. $p' = q$
$q' = pq - \sin t$

5. $S' = -rSI$
$I' = rSI - \gamma I$
$R' = \gamma I$

Verification Review *In each of Problems 6–9, show that the system is satisfied by the given set of functions.*

6. $x' = x$ $\{x = e^t,\ y = e^t\}$
$y' = y$

7. $x' = y$ $\{x = \sin t,\ y = \cos t\}$
$y' = -x$

8. $x' = y + t$ $\left\{x = -\frac{3}{2}t + \frac{3}{4},\ y = -t - \frac{3}{2}\right\}$
$y' = -2x + 3y + 5$

9. $x' = x$ $\{x = 0,\ y = \sin 2t,\ z = \cos 2t\}$
$y' = 2z$
$z' = -2y$

A Habit to Acquire *For each of the nonlinear systems in Problems 10–13, make a graph of the nullclines with arrows on and between them showing the direction of solutions. Identify each equilibrium and label it stable or unstable. Interpret these sketches in terms of the systems' phase portraits, shown in Fig. 7.1.1(a)–(d), to clarify the behaviors of the trajectories. Then for each graph write a paragraph to describe the behaviors.*

10. $x' = y$
$y' = x(x - 1)$

11. $x' = xy$
$y' = y - 3\cos x$

12. $x' = x^2 - y + 2$
$y' = y + 2x$

13. $x' = 1 + y + x^2$
$y' = y/2 - 3x$

Phase Portraits from Nullclines *For the nonlinear systems in Problems 14–19, determine the equilibrium solutions, if any, and sketch the h- and v-nullclines, drawing appropriate arrows on and between them, to indicate the direction of the solution curves. If the system has equilibrium points, determine if they are stable or unstable. Add some typical solutions and write a description of their behaviors. Identify any limit cycles.*

14. $x' = xy$
$y' = y - x^2 + 1$

15. $x' = y - \ln|x|$
$y' = x - \ln|y|$

16. $x' = y + x(1 - x^2 - y^2)$
$y' = -x + y(1 - x^2 - y^2)$

17. $x' = 1 - x^2 - y^2$
$y' = x$

18. $x' = y - x^2 + 1$
$y' = y + x^2 - 1$

19. $x' = |x| - y - 1$
$y' = |x| + y - 1$

Equilibria for Second-Order DEs *For the differential equations in Problems 20–25, find and classify the constant solutions as follows:*

(a) *Rewrite the second-order equation as a system of two first-order equations.*
(b) *Draw the nullclines for the first-order system, labeled with appropriate arrows, and find the equilibria.*
(c) *Deduce whether the equilibrium points of the nonlinear system are stable, thereby determining the stability of the constant solutions of the second-order DE.*
(d) *Identify any periodic solutions and state whether they are limit cycles.*

20. $x'' + (x^2 - 1)x' + x = 0$

21. $\theta'' + (g/L)\sin\theta = 0$

22. $x'' - \dfrac{x}{x-1} = 0$

23. $\ddot{x} + \dot{x}^2 + x^2 = 0$

24. $x'' + |x|x' + x = 0$

25. $\ddot{x} + (\dot{x}^2 - 1)\dot{x} + x = 0$

26. Creative Challenge Create an interesting phase portrait by choosing nullclines that intersect at key points and arranging to make them stable (filled dots) or unstable (open dots).[5] HINT: Think about how ET was "designed" in Example 2.

Finding Equations of Trajectories *Use the identity*

$$\frac{dy}{dx} = \frac{y'}{x'} = \frac{g(x, y)}{f(x, y)}$$

to sketch and find equations for the phase-plane trajectories of the systems in Problems 27–30.

27. $x' = y$
$y' = x$

28. $x' = y$
$y' = -x$

29. $x' = y(x^2 + 1)$
$y' = 2xy^2$

30. $x' = 1$
$y' = x + y$

Nonlinear Systems from Applications *For each of the systems in Problems 31–33, find the equilibrium points and draw sample trajectories in the phase plane. Discuss the long-term behavior of solutions in terms of the equilibria.*

31. $\dot{x} = 2xy$
$\dot{y} = y^2 - x^2 - 1$
(Electric field between two charges)

32. $\dot{x} = 2xy$
$\dot{y} = y^2 - x^2$
(Dipole system)

33. $\dot{x} = y$
$\dot{y} = -x - \text{sgn } y,$
(Coulomb damping)

where

$$\text{sgn } y = \begin{cases} 1 & \text{if } y > 0, \\ 0 & \text{if } y = 0, \\ -1 & \text{if } y < 0. \end{cases}$$

34. Sequential Solution Determine the general solution (containing two arbitrary constants) for the system

$$x' = -2x,$$
$$y' = xy^2,$$

by solving the first equation and substituting the result into the second.

Polar Limit Cycles *For the polar coordinate systems in Problems 35–38, determine the limit cycles in the xy-phase plane and discuss their stability; find quantitative solutions when possible.* HINT: *With polar coordinates, if some constant value k for r causes $\dot{r}$ to be zero, then there is a circular limit cycle at $r = k$. Explain. What does $\dot{\theta} = 1$ mean? You should be able to sketch typical xy-phase-plane trajectories by hand.*

35. $\dot{r} = (1 - r)^2$
$\dot{\theta} = 1$

36. $\dot{r} = r(a - r)$
$\dot{\theta} = 1$

37. $\dot{r} = r(1 - r)(2 - r)$
$\dot{\theta} = 1$

38. $\dot{r} = r(1 - r)(2 - r)(3 - r)^2$
$\dot{\theta} = 1$

Testing Existence and Uniqueness *Picard's Existence and Uniqueness Theorem, given in Sec. 1.5, extends to higher-dimensional systems as follows. The linear equations of*

[5]Our colleagues Robert Borrelli and Courtney Coleman at Harvey Mudd College had wonderful success with an assignment for groups of students simply to create a phase portrait that looked like a *cat*. Various versions resulted, and their students learned a great deal about equilibria and stability in the process.

Chapters 4 and 6 satisfy it automatically, but for nonlinear systems you can run into trouble.

Existence and Uniqueness Theorem, Extended
For an n-dimensional system of first-order DEs:

$$\frac{dx_1}{dt} = f_1(t, x_1, x_2, \ldots, x_n),$$

$$\frac{dx_2}{dt} = f_2(t, x_1, x_2, \ldots, x_n),$$

$$\vdots$$

$$\frac{dx_n}{dt} = f_n(t, x_1, x_2, \ldots, x_n),$$

where all f_i are continuous on a t-interval I and on a region R in $\mathbb{R}^n$, where $a_i < x_i < b_i$, with any initial point $(t_0, \vec{\mathbf{x}}_0) \in I \times R$, there exists a positive number h such that the initial-value problem

$$\vec{\mathbf{x}}' = \vec{\mathbf{f}}(t, \vec{\mathbf{x}}), \quad \vec{\mathbf{x}}(t_0) = \vec{\mathbf{x}}_0,$$

has a solution $\vec{\mathbf{x}}(t)$ for t in the interval $(t_0 - h, t_0 + h)$.
If, furthermore, $\partial f_i / \partial x_j$ is also continuous in R for all i, j, then that solution is *unique*.

For each of the 2×2 *systems in Problems 39 and 40:*

(a) *Tell where (and why) you would expect difficulties with existence and/or uniqueness.*

(b) *Sketch a phase portrait that will illustrate what does (or does not) happen, and explain.*

39. $x' = 1 + x$, $y' = (1 + x)\sqrt{y}$

40. $x' = x/y$, $y' = x - y/x$

41. Hamiltonian for the Harmonic Oscillator Hamiltonian mechanics[6] is based on the Hamiltonian function $H(p, q)$, representing the total energy in terms of the **generalized coordinate** p and **generalized momentum** q. (Newtonian mechanics focuses on forces.) The **Hamiltonian system** is then defined by

$$\dot{q} = \frac{\partial H}{\partial p} \quad \text{and} \quad \dot{p} = -\frac{\partial H}{\partial q}.$$

For the undamped mass-spring system with mass m, spring constant k, and displacement x, we let $q = x$ and $p = m\dot{x}$ (the momentum).

(a) Show that the kinetic energy of the mass is $\dfrac{p^2}{2m}$.

(b) Show that the total energy is $H(p, q) = \dfrac{p^2}{2m} + \dfrac{kq^2}{2}$.

(c) Derive the corresponding Hamiltonian system.

Computer Lab: Phase-Plane Analysis *For the systems in Problems 42–47, use appropriate software to carry out the following investigation:*

(a) *Draw a vector field.*

(b) *Draw sample solution curves.*

(c) *Determine the equilibrium points.*

(d) *Determine the stability behavior of the equilibrium points.*

(e) *Discuss the long-term behavior of the system.*

(f) *Identify any periodic solutions and state whether they are limit cycles.*

42. $x' = x(x - y)$, $y' = y(1 - y)$

43. $x' = x - x^2$, $y' = -y$

44. $x' = 1 - |x|$, $y' = x - y$

45. $x' = x(2 - x - y)$, $y' = -y$

46. $x' = x + y - x^3$, $y' = -x$

47. $x' = \sin(xy)$, $y' = \cos(x + y)$

48. Computer Lab: Graphing in Two Dimensions Do IDE Lab 17 to help answer the following questions, which become even more important for nonlinear DEs than for linear DEs: What do second-order differential equations have in common with systems of two first-order equations? Why are phase planes and vector fields so important? How do they relate to $x(t)$ and $y(t)$ time series? What information can you squeeze out of the nullclines?

Graphing Two-Dimensional Equations
Lab 17 uses several tools to bring to interactive life the concepts discussed to this point for two-dimensional systems of DEs. Graphs appear instantly and can be manipulated. Part 4 is especially useful for building intuition.

49. Computer Lab: The Glider If you've ever played with a balsa-wood glider, you know that it flies in a wavy path if you throw it gently and does loop-the-loops if you throw it hard. Do IDE Lab 19 to see how this is all explained by nonlinear phase-plane analysis.[7]

---

[6]Named for William Rowan Hamilton. (See Sec. 3.5.)

[7]Model development and analysis by Steven Strogatz, Cornell University.

The Glider
Lab 19 provides all explanations, including the physics. Every phase plane trajectory evolves simultaneously with time series and an animation of the corresponding glider flight.

50. **Computer Lab: Nonlinear Oscillators** A child on a swing asks to be started as high as you can, with only a single initial push. Small-angle assumptions no longer hold, so this is a case of an unforced nonlinear oscillator. Work IDE Lab 20 to answer the question of whether a loop-the-loop is a possible outcome.

Pendulums
Try the nonlinear unforced options, both damped and undamped.

51. **Suggested Journal Entry** Discuss the distinction between quantitative and qualitative methods in the analysis of differential equations and systems. Contrast the advantages and limitations of each approach.

7.2 Linearization

SYNOPSIS: We will study the behavior of solutions of an autonomous nonlinear 2 × 2 system near an equilibrium point by analyzing a related linear system called the linearization. This merger of linear algebra and calculus makes it possible to classify the stability behavior of equilibria and limit cycles for many nonlinear systems.

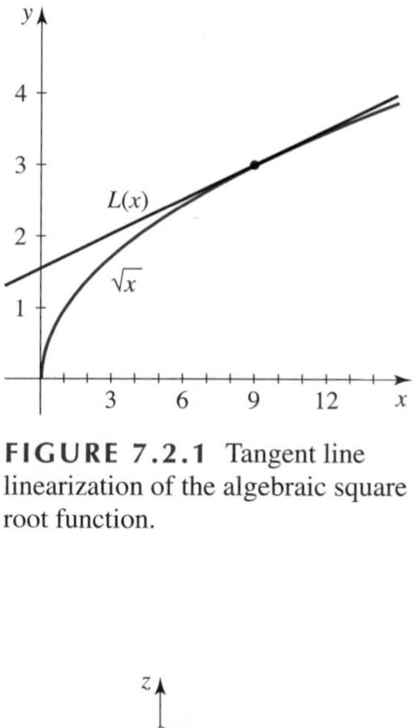

FIGURE 7.2.1 Tangent line linearization of the algebraic square root function.

Linearization of a Function

The student of single-variable calculus learns to approximate $\sqrt{10}$ by using the linearization

$$L(x) = 3 + \frac{1}{6}(x - 9) \quad \text{to the function} \quad f(x) = \sqrt{x}.$$

The result is that

$$\sqrt{10} = f(10) \approx L(10) = 3 + \frac{1}{6}(10 - 9) = 3\frac{1}{6}.$$

The linearization is just the tangent line to the graph of the square root function at $x = 9$, calculated from

$$L(x) = f(x_0) + (x - x_0)f'(x_0) \quad \text{for } x_0 = 9. \tag{1}$$

(See Fig. 7.2.1.)

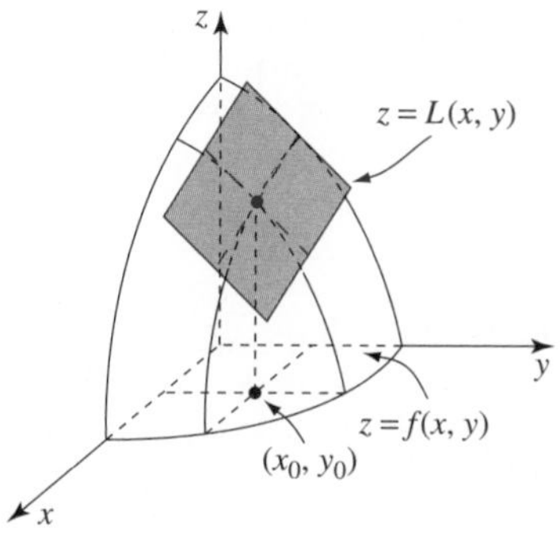

FIGURE 7.2.2 Tangent plane linearization for an algebraic function of two variables.

A similar calculation for the two-variable function $z = f(x, y)$ near the point (x_0, y_0) in the domain of f leads to

$$L(x, y) = f(x_0, y_0) + (x - x_0)f_x(x_0, y_0) + (y - y_0)f_y(x_0, y_0), \tag{2}$$

where f_x and f_y are continuous partial derivatives of f. In this case, $L(x, y)$ represents the tangent plane at (x_0, y_0) to the surface $z = f(x, y)$. (See Fig. 7.2.2.)

As a consequence of Taylor's Theorem for one and two variables, the errors in approximations (1) and (2) for smooth functions f are of the same order of magnitude as $(x - x_0)^2$ and $(y - y_0)^2$. Hence, behavior of the linearization is very similar to that of the original function in a suitable neighborhood of the point in question.

Informal Approach with DEs

Now we turn to the autonomous DE system

$$x' = f(x, y), \qquad y' = g(x, y),$$

where f and g are differentiable functions, and study the behavior of solutions near an equilibrium point (x_e, y_e), where we know that

$$f(x_e, y_e) = 0, \qquad g(x_e, y_e) = 0.$$

When (x_e, y_e) is not at the origin, we will *translate coordinates* to make (x_e, y_e) the *new* origin, then we shall replace f and g with their linearizations at that new origin. The result will be a linear system with a unique equilibrium point at the translated origin; this is the type of problem we studied in detail in Chapter 6. The stability behavior of the nonlinear system will usually be similar to that of the linearized system.

In some cases, the algebraic form of component functions f and g is sufficiently simple that the linearization can be obtained by "inspection." In this case, inspection means noting that for small values of the variables we can *ignore higher-order terms*.

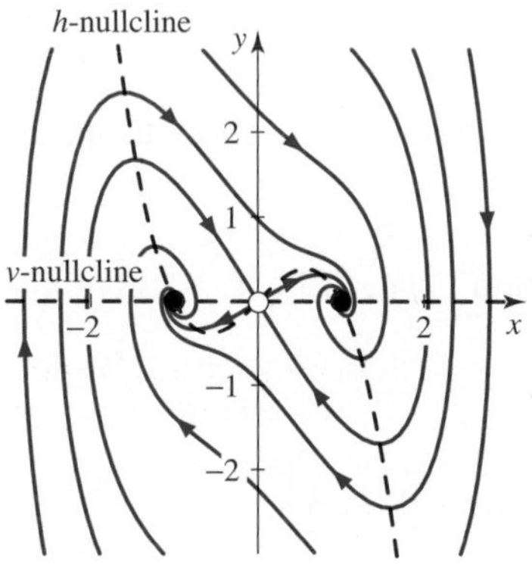

FIGURE 7.2.3 Phase portrait for nonlinear system (3), for Example 1.

EXAMPLE 1 **Linearization by Inspection** The phase portrait for

$$\begin{aligned} x' &= y, \\ y' &= -y + x - x^3, \end{aligned} \tag{3}$$

is shown in Fig. 7.2.3. The equilibrium solutions are found from the simultaneous solution of equations $y = 0$ and $-y + x - x^3 = 0$; they are $(0, 0)$, $(1, 0)$, and $(-1, 0)$. We shall analyze their stability as we linearize.

(a) Since x^3 is much smaller than x near $x = 0$, the linearization of (3) near $(0, 0)$ is just

$$\begin{aligned} x' &= \quad\; y, \\ y' &= x - y, \end{aligned} \quad \text{or} \quad \vec{\mathbf{x}}' = \mathbf{A}\vec{\mathbf{x}} = \begin{bmatrix} 0 & 1 \\ 1 & -1 \end{bmatrix} \vec{\mathbf{x}}. \tag{4}$$

A has eigenvalues $-1/2 \pm \sqrt{5}/2$ of opposite sign, so the origin is an unstable saddle point. The origin is also an unstable solution for system (3), but the nonlinear phase portrait (Fig 7.2.3, near the origin) is a distortion of the picture for the linearization (Fig. 7.2.4, center).

(b) To study the behavior of system (3) near the equilibrium solution $(1, 0)$, consider $(1, 0)$ as an "origin" using the transformation $u = x - 1$ and $v = y$; then $u' = x'$, $v' = y'$, and $v' = -v + (u + 1) - (u + 1)^3$. Therefore, we have

$$\begin{aligned} u' &= v, \\ v' &= -2u - v - 3u^2 - u^3. \end{aligned}$$

Dropping the higher-order terms gives the linear system

$$\begin{bmatrix} u \\ v \end{bmatrix}' = \begin{bmatrix} 0 & 1 \\ -2 & -1 \end{bmatrix} \begin{bmatrix} u \\ v \end{bmatrix}, \tag{5}$$

having eigenvalues $-1/2 \pm i\sqrt{7}/2$. Since the real part of this pair of complex conjugate eigenvalues is negative, the translated origin is an asymptotically stable attracting spiral point for the linearized system (5).

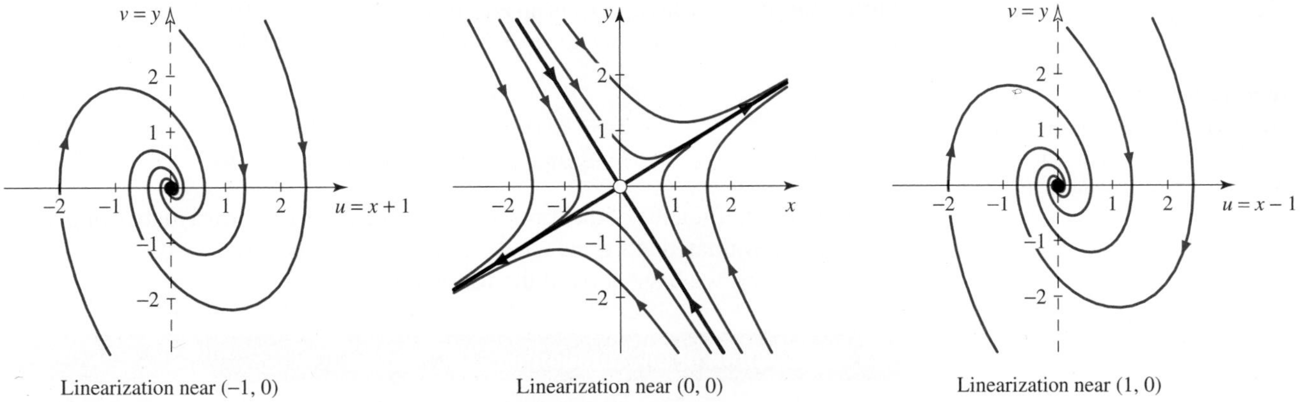

FIGURE 7.2.4 Linearized systems for (3), in Example 1. Translated axes are dashed, with empty arrow heads.

We thus conclude that $(1, 0)$ is an asymptotically stable solution for the nonlinear system (3). (See Fig. 7.2.4, right.)

(c) A similar analysis shows that the equilibrium point $(-1, 0)$ is also an asymptotically stable solution: in fact, the linearization at $(-1, 0)$ turns out to be the same system as that at $(1, 0)$, as shown in Fig. 7.2.4 on the left. You can see that this is expected from examining the two outer equilibria for the original nonlinear system in Fig. 7.2.3. ■

Formal Linearization

When linearizing the autonomous system

$$x' = f(x, y), \quad y' = g(x, y), \tag{6}$$

is not just a matter of dropping higher-order terms from a polynomial expression, we exploit the linearization (2) based on **Taylor's Theorem**. We start by translating an equilibrium solution to the origin with the transformation

$$(u, v) = (x - x_e, y - y_e).$$

Because x_e and y_e are constant, $u' = x'$ and $v' = y'$, and (6) becomes

$$\begin{aligned} u' &= f(u + x_e, v + y_e), \\ v' &= g(u + x_e, v + y_e). \end{aligned} \tag{7}$$

When $u = v = 0$, the right-hand sides of the component equations in (7) become $f(x_e, y_e)$ and $g(x_e, y_e)$, and these are zero because (x_e, y_e) is an equilibrium solution of (6). The Taylor expansions of f and g about (x_e, y_e) are

Taylor Series Expansions

$$\begin{aligned} f(x, y) &= f(x_e, y_e) + (x - x_e) f_x(x_e, y_e) + (y - y_e) f_y(x_e, y_e) + R_1(x, y), \\ g(x, y) &= g(x_e, y_e) + (x - x_e) g_x(x_e, y_e) + (y - y_e) g_y(x_e, y_e) + R_2(x, y), \end{aligned}$$

where we assume that the remainder terms R_1 and R_2 are second-order and small for (x, y) near (x_e, y_e),[1] in the sense that

$$\lim_{(x,y)\to(x_e,y_e)} \frac{R_1(x, y)}{\sqrt{x^2 + y^2}} = 0 \quad \text{and} \quad \lim_{(x,y)\to(x_e,y_e)} \frac{R_2(x, y)}{\sqrt{x^2 + y^2}} = 0. \tag{8}$$

[1]Recall from calculus that the Taylor remainder term for a linear approximation to $f(x, y)$ is

$$R(x, y) = (x - x_0)^2 f_{xx}(x_c, y_c) + 2(x - x_0)(y - y_0) f_{xy}(x_c, y_c) + (y - y_0)^2 f_{yy}(x_c, y_c),$$

where x_c is between x and x_0, and y_c is between y and y_0.

Furthermore, because (x_e, y_e) is an equilibrium solution of (6), $f(x_e, y_e) = 0$ and $g(x_e, y_e) = 0$. So, when we drop R_1 and R_2 from the Taylor series expansions, we have

Almost Linear:

Systems of differential equations $x' = f(x, y), y' = g(x, y)$, for which (9) holds, are called *almost linear* systems.

$$\begin{aligned} u' &= uf_x(x_e, y_e) + vf_y(x_e, y_e), \\ v' &= ug_x(x_e, y_e) + vg_y(x_e, y_e). \end{aligned} \tag{9}$$

In (9) the coefficients of the translated variables u and v form a matrix of partial derivatives, evaluated at (x_e, y_e), called the *Jacobian matrix*.[2]

In summary, we have derived the following:

Linearization of an Autonomous DE System

For $x' = f(x, y)$ and $y' = g(x, y)$, f and g twice-differentiable, the linearized system at an equilibrium point (x_e, y_e) translated by $u = x - x_e$ and $v = y - y_e$,

$$\begin{bmatrix} u \\ v \end{bmatrix}' = \mathbf{J}(x_e, y_e)\begin{bmatrix} u \\ v \end{bmatrix}, \quad \text{where } \mathbf{J}(x_e, y_e) = \begin{bmatrix} f_x(x_e, y_e) & f_y(x_e, y_e) \\ g_x(x_e, y_e) & g_y(x_e, y_e) \end{bmatrix}$$

is the **Jacobian matrix**. If **J** is nonsingular, the linearized system has a unique equilibrium point at $(u, v) = (0, 0)$, and the techniques of Sec. 6.4 can be used on **J** to classify its behavior.

In the end, the Jacobian matrix for the linearization (9) is calculated directly from the original system (6), and you need not make an explicit transformation of (6) in terms of u and v. We can say that the system (6) is **almost linear** at (x_e, y_e).

- When all the eigenvalues of the Jacobian matrix are negative or have negative real parts, the equilibrium solution of the original system is asymptotically stable.
- If any of the eigenvalues of the Jacobian matrix are positive or have positive real parts, the equilibrium solution is unstable.
- If the Jacobian matrix has real eigenvalues of opposite sign, the nonlinear equilibrium point will behave something like a saddle, though the solutions that approach the equilibrium point will not in general do so along straight lines (as you can observe in Example 2).
- The one case where linearization fails to predict nonlinear behavior is when the Jacobian matrix has purely imaginary eigenvalues. The linearization will have a center equilibrium, but the perturbation usually causes a spiral that can be either stable or unstable. (See Problems 11 and 12.)

NOTE: Linearization and the Jacobian only concern equilibria; they *cannot* find limit cycles, which are solely a nonlinear phenomenon.

[2]Prussian mathematician Carl Jacobi (1804–1851) entered university at age 12 and studied mathematics, classics, and philosophy, much on his own. In his university teaching he introduced the seminar method to keep students abreast of the latest mathematics. Jacobi's research in differential equations for dynamics and in determinants came together in the important matrix discussed here. However, it was Cauchy in 1815 who actually first introduced the "Jacobian."

EXAMPLE 2 **Analysis** The nonlinear system

$$\begin{aligned} x' &= y, \\ y' &= x(x-4), \end{aligned} \tag{10}$$

has equilibria at (0, 0) and (4, 0), and Jacobian matrix

$$\mathbf{J}(x_e, y_e) = \begin{bmatrix} 0 & 1 \\ 2x_e - 4 & 0 \end{bmatrix}.$$

At equilibrium point (0, 0), the Jacobian matrix for the linearized system is

$$\mathbf{J}(0,0) = \begin{bmatrix} 0 & 1 \\ -4 & 0 \end{bmatrix} \quad \text{with eigenvalues } \lambda_1, \lambda_2 = \pm 2i.$$

The equilibrium point is a center, which is neutrally stable.

At (4, 0),

$$\mathbf{J}(4,0) = \begin{bmatrix} 0 & 1 \\ 4 & 0 \end{bmatrix} \quad \text{with eigenvalues } \lambda_1, \lambda_2 = \pm 2.$$

This equilibrium point is a saddle, which is unstable.

Figure 7.2.5 shows the phase portraits for the nonlinear system (10) and the linearized systems at (0, 0) and (4, 0). As shown in Examples 1 and 2, the nonlinear phase portrait incorporates the portraits of the linearizations, with distortions to maintain uniqueness of solutions.

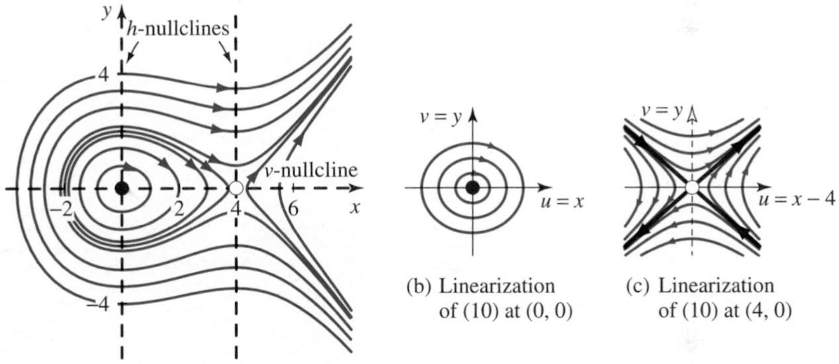

(b) Linearization of (10) at (0, 0)

(c) Linearization of (10) at (4, 0)

(a) Phase portrait for system (10)

FIGURE 7.2.5 Phase portrait (a) for a nonlinear system (10) with two equilibrium points, and the corresponding linearizations (b) and (c).

■

EXAMPLE 3 **Damped Pendulum** To analyze the stability of the equilibrium points of the (nonlinear) pendulum equation with damping,

$$\theta'' + \theta' + \sin\theta = 0, \tag{11}$$

we first convert it, using $x = \theta$ and $y = \theta'$, to the autonomous system

$$\begin{aligned} x' &= f(x,y) = y, \\ y' &= g(x,y) = -y - \sin x. \end{aligned} \tag{12}$$

The equilibrium solutions of (11) are $(n\pi, 0)$ for $n = 0, \pm 1, \pm 2, \ldots$.

Pendulums

See how the phase portraits change from linear to nonlinear, undamped to damped.

In IDE all the nonlinear pendulum tools have an infinite number of equilibria along the horizontal axis. The pattern of equilibria repeats with period 2π, so the phase plane is drawn only from $-\pi$ to π; a trajectory that goes off on the right comes back on the left.

Evaluating the Jacobian matrix

$$\mathbf{J}(x, y) = \begin{bmatrix} f_x & f_y \\ g_x & g_y \end{bmatrix} = \begin{bmatrix} 0 & 1 \\ -\cos x & -1 \end{bmatrix}$$

at (0, 0) gives

$$\mathbf{J}(0, 0) = \begin{bmatrix} 0 & 1 \\ -1 & -1 \end{bmatrix} \quad \text{with eigenvalues } \lambda_1, \lambda_2 = -1/2 \pm i\sqrt{3}/2.$$

The origin is an attracting spiral point for the linearized system and an asymptotically stable equilibrium for (12). From this fact, we can deduce that every trajectory at the top or bottom of the phase portrait is directed toward the horizontal axis.

To examine the equilibrium point at $(\pi, 0)$, evaluate the Jacobian matrix there to obtain

$$\mathbf{J}(\pi, 0) = \begin{bmatrix} 0 & 1 \\ 1 & -1 \end{bmatrix} \quad \text{with eigenvalues } \lambda_1, \lambda_2 = -1/2 \pm \sqrt{5}/2.$$

Because the eigenvalues are real and of opposite signs, the linearization at $(\pi, 0)$ has a saddle point. Hence, the pendulum has unstable "saddlelike" behavior at $(\pi, 0)$.

The persistent reader will be able to show further that the equilibrium solutions $(k\pi, 0)$ for odd k are unstable, while for even k they are asymptotically stable. (See Fig. 7.2.6.)

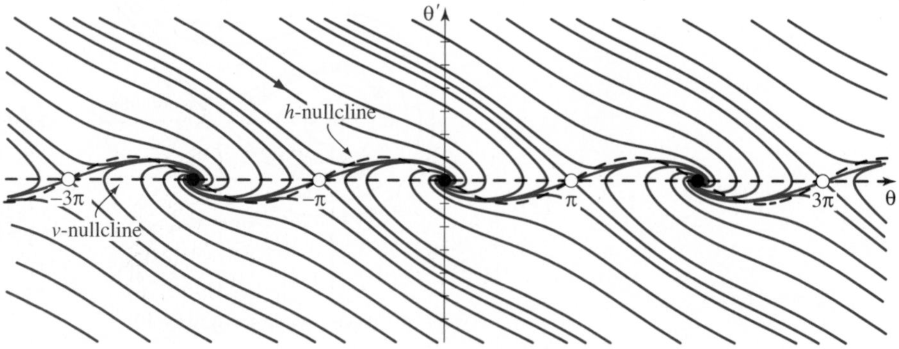

FIGURE 7.2.6 Phase portrait for the damped nonlinear pendulum (11). The equilibria are all on the θ-axis—saddles at odd multiples of π, attracting spirals at even multiples of π.

When Linearization Fails

In Section 6.4, we summarized the geometry and stability properties of the equilibrium at the origin for the 2×2 linear system $\vec{\mathbf{x}}' = \mathbf{A}\vec{\mathbf{x}}$, characterized according to the nature of the eigenvalues of $\mathbf{A}$. Much of this analysis carries over to nonlinear systems, *near the equilibrium.* The conspicuous exception is that of the *center equilibrium*, which is stable but not asymptotically stable. Small perturbations can tip such solutions either way, and no general prediction is possible. (Compare Problems 11 and 12.)

Table 7.2.1 Stabilities versus eigenvalues

	Eigenvalues	Linearized System		Nonlinear System	
		Geometry	Stability	Geometry	Stability
Real distinct roots	$\lambda_1 < \lambda_2 < 0$	Attracting node	Asymptotically stable	Attracting node	Asymptotically stable
	$0 < \lambda_2 < \lambda_1$	Repelling node	Unstable	Repelling node	Unstable
	$\lambda_1 < 0 < \lambda_2$	Saddle	Unstable	Saddle	Unstable
Real repeated roots	$\lambda_1 = \lambda_2 < 0$	**Attracting star or degenerate node**	Asymptotically stable	**Attracting node or spiral**	Asymptotically stable
	$\lambda_1 = \lambda_2 > 0$	**Repelling star or degenerate node**	Unstable	**Repelling node or spiral**	Unstable
Complex conjugate roots	$\alpha > 0$	Repelling spiral	Unstable	Repelling spiral	Unstable
	$\alpha < 0$	Attracting spiral	Asymptotically stable	Attracting spiral	Asymptotically stable
	$\alpha = 0$	**Center**	**Stable**	**Center or spiral**	**Uncertain**

Table 7.2.1 summarizes the relationships between the linear and nonlinear results.[3] The only differences are highlighted in boldface.

Stability of Nonlinear Systems

For the nonlinear system

$$\begin{aligned} x' &= f(x, y), \\ y' &= g(x, y), \end{aligned} \quad \text{with Jacobian} \quad \mathbf{J} = \begin{bmatrix} f_x(x_e, y_e) & f_y(x_e, y_e) \\ g_x(x_e, y_e) & g_y(x_e, y_e) \end{bmatrix},$$

let the eigenvalues of $\mathbf{J}$ at equilibrium solution (x_e, y_e) be λ_1 and λ_2 (real case) or $\alpha \pm i\beta$ (nonreal case). The geometry and stability characteristics about that equilibrium are related as shown in Table 7.2.1.

The linear and nonlinear systems differ at an equilibrium only when the linear system is on a border that involves nonreal eigenvalues (the parabola separating nodes from spirals, or the vertical half-axis that separates attracting spirals from

[3]Details of the proof of this classification can be found in J. H. Hubbard and B. H. West, *Systems of Ordinary Differential Equations* (NY: Springer-Verlag, 1991), Chapter 8.

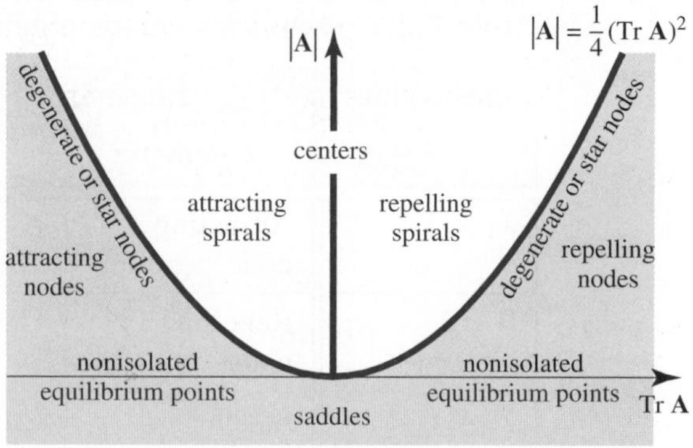

FIGURE 7.2.7 Equilibrium behaviors for linearized systems. Along the highlighted borders that involve nonreal eigenvalues, the nonlinear systems can have different equilibrium behaviors.

repelling spirals). Figure 7.2.7 repeats the key information from Sec. 6.4. On these borderline cases, the nonlinear perturbation can throw the stability to either side.

EXAMPLE 4 **An Uncertain Case** We will analyze the stability of the equilibrium solution at the origin of the nonlinear system

$$\begin{aligned} x' &= f(x, y) = y - x\sqrt{x^2 + y^2}, \\ y' &= g(x, y) = -x - y\sqrt{x^2 + y^2}. \end{aligned} \tag{13}$$

Calculating the Jacobian matrix at (0, 0), we obtain

$$\mathbf{J} = \begin{bmatrix} 0 & 1 \\ -1 & 0 \end{bmatrix} \quad \text{with eigenvalues } \lambda_1, \lambda_2 = \pm i.$$

The nonzero solutions for the linearization are periodic; the orbits are circles about the origin, which is a center point, as shown in Fig. 7.2.8(b).

However, as can be seen in Fig. 7.2.8(a), the origin is *not* a center point for nonlinear system (13), and solutions to the nonlinear system are not periodic.

To understand the discrepancy between the phase portraits for the nonlinear and linearized systems, interpret (13) as the sum of the vector field

$$\begin{bmatrix} y \\ -x \end{bmatrix},$$

always tangent to circles about the origin, and

$$\begin{bmatrix} -x\sqrt{x^2 + y^2} \\ -y\sqrt{x^2 + y^2} \end{bmatrix},$$

which points inward toward the origin. The sum tends inward, and solutions spiral inward; the origin is asymptotically stable. See Fig. 7.2.8(a).

It is illuminating to pause and extend this discussion to look at the closely related system

$$\begin{aligned} x' &= f(x, y) = y + x\sqrt{x^2 + y^2}, \\ y' &= g(x, y) = -x + y\sqrt{x^2 + y^2}. \end{aligned} \tag{14}$$

System (14) has the same linearization as system (13); by reasoning similar to that preceding, we can predict that the solutions to (14) will spiral outward, making the equilibrium at the origin unstable. ■

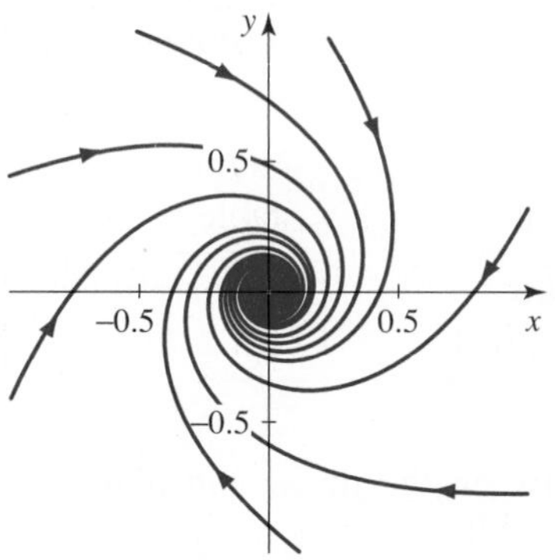

(a) Nonlinear system

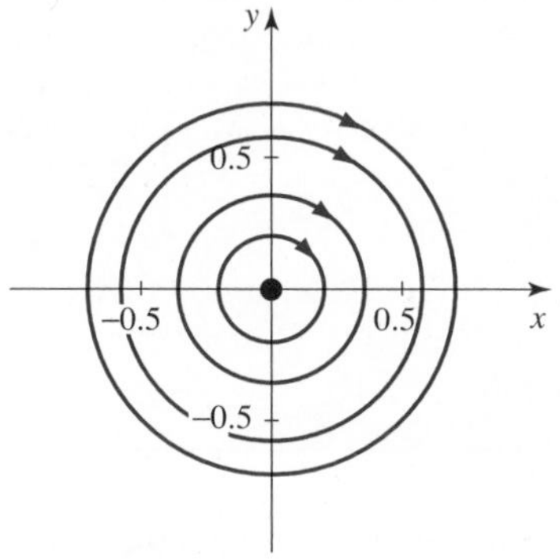

(b) Linearized system at (0,0)

FIGURE 7.2.8 A nonlinear phase portrait (a) for system (13) and the phase portrait (b) for its linearization.

Summary

An equilibrium solution of a nonlinear autonomous system can be analyzed by studying a closely related linear system called the linearization. In most cases, the classification of this linearization according to the scheme of Chapter 6 predicts the stability of the solution of the nonlinear system, and often its geometry as well.

7.2 Problems

Original Equilibrium *In Problems 1–6, show that each system has an equilibrium point at the origin. Compute the Jacobian, then discuss the type and stability of the equilibrium point. Find and describe other equilibria if they exist.*

1. $x' = -2x + 3y + xy$
$y' = -x + y - 2xy^2$

2. $x' = -y - x^3$
$y' = x - y^3$

3. $x' = x + y + 2xy$
$y' = -2x + y + y^3$

4. $x' = y$
$y' = -\sin x - y$

5. $x' = x + y^2$
$y' = x^2 + y^2$

6. $x' = \sin y$
$y' = -\sin x + y$

Unusual Equilibria *For each system in Problems 7–9, determine the type and stability of each real equilibrium point by calculating the Jacobian matrix at each equilibrium.*

7. $x' = 1 - xy$
$y' = x - y^3$

8. $x' = x - 3y + 2xy$
$y' = 4x - 6y - xy$

9. $x' = 4x - x^3 - xy^2$
$y' = 4y - x^2y - y^3$

10. Linearization Completion Complete the analysis started in Example 1 by providing the details of the linearization about the point $(-1, 0)$ for $x' = y$, $y' = -y + x - x^3$.

Uncertainty *Because a center equilibrium is stable but not asymptotically stable, nonlinear perturbation can have different outcomes, shown in Problems 11 and 12.*

11. Determine the stability of the equilibrium solutions of the **strong** spring $\ddot{x} + \dot{x} + x + x^3 = 0$.

12. Determine the stability of the equilibrium solutions of the **weak** spring $\ddot{x} + \dot{x} + x - x^3 = 0$.

13. Liénard Equation[4] A generalized damped mass-spring equation, the Liénard equation, is $\ddot{x} + p(x)\dot{x} + q(x) = 0$. If $q(0) = 0$, $\dot{q}(0) > 0$, and $p(0) > 0$, show that the origin is a stable equilibrium point.

14. Conservative Equation A second-order DE of the form $\ddot{x} + F(x) = 0$ is called a **conservative differential equation**. (See Sec. 4.7.) Find the equilibrium points of the conservative equation $\ddot{x} + x - x^2 - 2x^3 = 0$ and determine their type and stability.

15. Predator-Prey Equations In Sec. 2.6 we introduced the Lotka-Volterra predator–prey system

$$x' = (a - by)x,$$
$$y' = (cx - d)y,$$

and determined its equilibrium points $(0, 0)$ and $(d/c, a/b)$. Use the Jacobian matrix to analyze the stability around the equilibrium point $(d/c, a/b)$. Interpret the trajectories of this system as plotted in Fig. 2.6.7.

Lotka-Volterra
This tool lets you experiment on screen.

16. van der Pol's Equation Show that the zero solution of van der Pol's equation, $\ddot{x} - \varepsilon(1 - x^2)\dot{x} + x = 0$, is unstable for any positive value of parameter ε.

van der Pol
This tool lets you experiment on screen.

---

[4]Alfred Liénard (1869–1958) was a French mathematician and applied physicist.

Damped Mass-Spring Systems *The second-order linear DE $m\ddot{x} + b\dot{x} + kx = 0$ models vibrations of a mass m attached to a spring with spring constant k and damping constant b.*

For the nonlinear variations in Problems 17–20, use your intuition to decide whether the zero solution ($x = \dot{x} \equiv 0$) is stable or unstable. Check your intuition by transforming to a first-order system and linearizing.

17. $\ddot{x} + \dot{x}^3 + x = 0$

18. $\ddot{x} + \dot{x} - \dot{x}^3 + x = 0$

19. $\ddot{x} + \dot{x} + \dot{x}^3 + x = 0$

20. $\ddot{x} - \dot{x} + x = 0$

Liapunov Functions *An alternative approach to determining stability is the* ***direct method*** *of Liapunov.*[5] *Liapunov assumes the existence of a* ***positive-definite*** *energylike function $L(x, y)$ with continuous first partial derivatives.*[6] *His theorem states that if $(0, 0)$ is an isolated equilibrium solution of $x' = f(x, y)$ and $y' = g(x, y)$, and if*

$$\frac{dL}{dt} = \frac{\partial L}{\partial x}\frac{dx}{dt} + \frac{\partial L}{\partial y}\frac{dy}{dt} = L_x x' + L_y y'$$

(the derivative of L along the trajectory) is ***negative definite*** *on a neighborhood of the origin, then the origin is asymptotically stable.*

Use Liapunov's direct method to verify the asymptotic stability of the origin for each system in Problems 21 and 22, after checking that the given function L is a legitimate Liapunov function.

21. $x' = y - 2x^3$

$y' = -2x - 3y^5$

$L(x, y) = 2x^2 + y^2$

22. $x' = 2y - x^3$

$y' = -x^3 - y^5$

$L(x, y) = x^4 + 4y^2$

23. A Bifurcation Point If a nonlinear system depends on a parameter k (such as a damping constant, spring constant, or chemical concentration), a critical value k_0 where the qualitative behavior of the system changes is called a **bifurcation point**. Show that $k = 0$ is a bifurcation point for the system

$$x' = -x(y^2 + 1), \quad y' = y^2 + k, \qquad (15)$$

as follows. Illustrate each part with a phase portrait.

(a) Show that (15) has two equilibrium points for $k < 0$.

(b) Show that (15) has one equilibrium point for $k = 0$.

(c) Show that (15) has no equilibrium points for $k > 0$.

(d) Calculate the linearization about the equilibrium point for $k = 0$. Relate the phase portraits for (b) and (d).

2D Saddle-Node Bifurcation
This tool lets you experiment on screen with a similar example.

Computer Lab: Trajectories *Rewrite the second-order equation in each of Problems 24–27 as a first-order system with $x' = y$. Use appropriate software to sketch trajectories using the direction field for $dy/dx = y'/x'$. Compare with behaviors of the linearized systems (see Chapter 4), and explain what is different and why.*

24. $x'' + x \sin x = 0$

25. $x'' + x - 0.1(x^2 + 2x^3) = 0$

26. $x'' - (1 - x^2)x' + x = 0$

27. $x'' + x - 0.25x^2 = 0$

28. Computer Lab: Competition Work IDE Lab 22 to get a visceral feel for how changing parameters affects the location and character of the equilibria. This system was discussed in detail in Sec. 2.6.

Competitive Exclusion
Because changing parameters can change relative positions of the nullclines, very different scenarios can result.

29. Suggested Journal Entry I Consider the tangent line linearization $L(x)$ to the graph of a function $f(x)$ of one variable, and discuss its relative predictive value for the behavior of f in the cases $L'(x_0) > 0$, $L'(x_0) = 0$, and $L'(x_0) < 0$. Can you draw an analogy to the linearization of an autonomous system of DEs?

30. Suggested Journal Entry II Summarize the relationship between a nonlinear system and its linearization at an equilibrium point, both geometrically and in regard to stability.

---

[5] Aleksandr M. Liapunov (1857–1918) was a Russian mathematician whose **direct method** (or *second method*) was the conclusion of his doctoral dissertation (1892). He argued intuitively that an asymptotically stable equilibrium point of a physical system must correspond to a point of minimum potential energy.

[6] Function $L(x, y)$ is **positive definite** on domain D containing the origin if $L(0, 0) = 0$ and $L(x, y) > 0$ at all other points of D; it is **negative definite** on D if $L(0, 0) = 0$ and $L(x, y) < 0$ at all other points of D.

7.3 Numerical Solutions

SYNOPSIS: Since most nonlinear DE systems do not even have analytic solutions, we must rely seriously on numerical methods. We generalize Euler's method to systems of two or more first-order equations. Using this method, or its more accurate cousins, provides numerical and graphical approximations to trajectories and time-series graphs.

Euler's Method in Higher Dimensions

Numerical methods move from point to point along the locally determined direction. Euler's method, while not the most accurate, is the easiest to understand and therefore a good place to start.[1]

A typical 2×2 system of first-order differential equations is

$$\begin{aligned} x' &= f(t, x, y), \\ y' &= g(t, x, y). \end{aligned} \tag{1}$$

We consider the initial-value problem in which $(x, y) = (x_0, y_0)$ when $t = t_0$. The solution of this problem is represented parametrically in three-dimensional space by a curve (t, x, y). (See Fig. 7.3.1.) More frequently we will examine the two-dimensional projections of the 3D graphs.

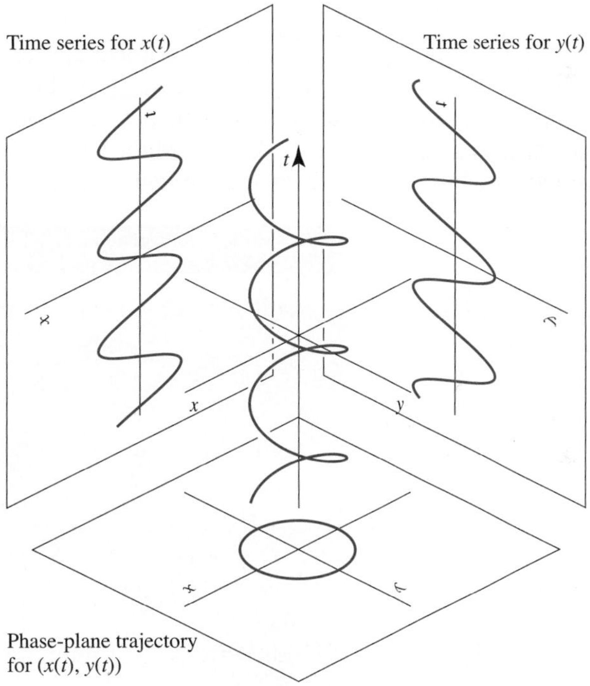

FIGURE 7.3.1 A typical three-dimensional trajectory and projections: phase portrait and time series.

Euler's method for a single equation consisted of piecing together line segments, beginning at the initial t-value and following the direction field for

[1] Fancier numerical approximation methods such as Runge-Kutta or Adams-Bashforth (Sec. 1.4) can be adapted in similar fashion.

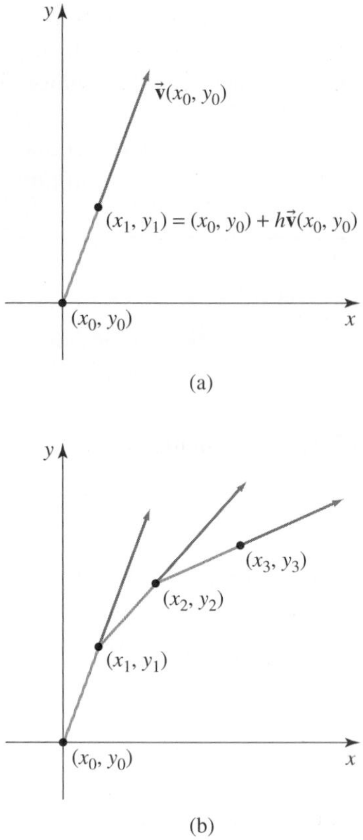

FIGURE 7.3.2 Piecing together Euler-approximate solutions for an approximate trajectory, in color.

subsequent equally-spaced t-intervals. This is pictured in Fig. 1.4.1, which you may wish to review.

In the case of a 2×2 system like (1), we wish to move in the phase plane from point (x_0, y_0) in the direction of the field, given by the vector

$$\vec{\mathbf{v}}(t_0) = \begin{bmatrix} f(t_0, x_0, y_0) \\ g(t_0, x_0, y_0) \end{bmatrix}.$$

We move along this vector as t increases from t_0 to $t_1 = t_0 + h$, where h is the step size, from

$$\begin{bmatrix} x_0 \\ y_0 \end{bmatrix} \quad \text{to} \quad \begin{bmatrix} x_1 \\ y_1 \end{bmatrix} = \begin{bmatrix} x_0 \\ y_0 \end{bmatrix} + h \begin{bmatrix} f(t_0, x_0, y_0) \\ g(t_0, x_0, y_0) \end{bmatrix}.$$

Arriving at (x_1, y_1), we move off in the new direction determined at that point,

$$\vec{\mathbf{v}}(t_1) = \begin{bmatrix} f(t_1, x_1, y_1) \\ g(t_1, x_1, y_1) \end{bmatrix},$$

to reach the next vertex of the piecewise-approximate solution:

$$\begin{bmatrix} x_2 \\ y_2 \end{bmatrix} = \begin{bmatrix} x_1 \\ y_1 \end{bmatrix} + h \begin{bmatrix} f(t_1, x_1, y_1) \\ g(t_1, x_1, y_1) \end{bmatrix}.$$

Continuing the process yields the sequence of vertices,

$$\begin{bmatrix} x_0 \\ y_0 \end{bmatrix}, \begin{bmatrix} x_1 \\ y_1 \end{bmatrix}, \ldots, \begin{bmatrix} x_i \\ y_i \end{bmatrix}, \ldots,$$

of the piecewise-linear approximation to the solution.

Figure 7.3.2 suggests how the first few segments of an Euler-approximate solution may be pieced together in the phase plane. While a solution point may move along a solution curve at a variable speed,[2] the segments of the piecewise approximation represent motions of roughly equal duration.

Euler's Method for a 2 × 2 DE System

For the initial-value problem

$$\vec{\mathbf{x}}' = \begin{bmatrix} x \\ y \end{bmatrix}' = \vec{\mathbf{f}}(t, \vec{\mathbf{x}}) = \begin{bmatrix} f(t, x, y) \\ g(t, x, y) \end{bmatrix}, \quad \vec{\mathbf{x}}(0) = \begin{bmatrix} x_0 \\ y_0 \end{bmatrix}, \tag{2}$$

use the formulas

$$t_{n+1} = t_n + h, \tag{3}$$

$$x_{n+1} = x_n + hf(t_n, x_n, y_n), \tag{4}$$

$$y_{n+1} = y_n + hg(t_n, x_n, y_n), \tag{5}$$

with step size h to compute iteratively the points

$$\begin{bmatrix} x_1 \\ y_1 \end{bmatrix}, \begin{bmatrix} x_2 \\ y_2 \end{bmatrix}, \ldots, \begin{bmatrix} x_K \\ y_K \end{bmatrix}.$$

The piecewise-linear path connecting $\vec{\mathbf{x}}(0)$ and these points is the Euler approximation to the phase-plane trajectory of the IVP for the interval $t_0 \le t \le t_K$.

[2]The solution point in the xy-plane moves with speed $\sqrt{\dot{x}^2 + \dot{y}^2}$.

While our attention in this chapter has been focused largely on autonomous (time-invariant) systems, no such restriction is necessary in applying Euler's method. Likewise the generalization to systems of three or more equations should be fairly obvious. (See Problems 13–15 and Appendix SS.)

EXAMPLE 1 **Euler at Work** We use Euler's method with step size $h = 0.1$ to approximate the solution for $0 \le t \le 1$ of the initial-value problem

$$\vec{\mathbf{x}}' = \begin{bmatrix} x \\ y \end{bmatrix}' = \begin{bmatrix} -2x \\ xy^2 \end{bmatrix}, \quad \vec{\mathbf{x}}(0) = \begin{bmatrix} 2 \\ 1 \end{bmatrix}. \tag{6}$$

In the notation of the boxed algorithm, $x_0 = 2$, $y_0 = 1$, $t_0 = 0$, $f(t, x, y) = -2x$, and $g(t, x, y) = xy^2$. (This example happens to be autonomous.)

Step size $h = 0.1$ will give t-values 0, 0.1, 0.2, ..., 1.0, using equation (3). For equations (4) and (5), we will have

$$x_{n+1} = x_n + 0.1(-2x_n),$$
$$y_{n+1} = y_n + 0.1\left(x_n y_n^2\right).$$

For example,

$$x_1 = x_0 + (0.1)(-2x_0) = 2 + (0.1)(-2 \cdot 2) = 1.6;$$
$$y_1 = y_0 + (0.1)\left(x_0 y_0^2\right) = 1 + (0.1)(2 \cdot 1^2) = 1.2.$$

Completing subsequent calculations of this type for index values to $n = 10$, we can compute the results in the first six columns of Table 7.3.1. It is the same process that we carried out in Sec. 1.4, but there are more columns to tabulate.

A spreadsheet program is an ideal tool for carrying out these computations. Enter the four bold numbers and set up the formulas to do the rest. See Appendix SS.

Table 7.3.1 Numerical computation for IVP (6)

n	t_n	x_n	y_n	$x'_n = f(x_n, y_n)$	$y'_n = g(x_n, y_n)$	Exact $x(t_n)$	Exact $y(t_n)$
0	**0**	**2**	**1**	−4	2	2	1
1	**0.1**	1.6	1.2	−3.2	2.304	1.6375	1.2214
2	0.2	1.28	1.4304	−2.56	2.6189	1.3406	1.4918
3	0.3	1.024	1.6923	−2.048	2.9326	1.0976	1.8221
4	0.4	0.8192	1.9856	−1.6384	3.2296	0.8987	2.2255
5	0.5	0.6554	2.3085	−1.3107	3.4926	0.7358	2.7183
6	0.6	0.5243	2.6578	−1.0486	3.7034	0.6023	3.3201
7	0.7	0.4194	3.0281	−0.8389	3.8460	0.4931	4.0552
8	0.8	0.3355	3.4127	−0.6711	3.9080	0.4038	4.9530
9	0.9	0.2684	3.8035	−0.5369	3.8834	0.3306	6.0496
10	1.0	0.2147	4.1918	−0.4295	3.7735	0.2707	7.3891

The initial-value problem (6) is based on the system studied earlier in Sec. 7.1, Problem 34; its quantitative solution is

$$\begin{aligned} x &= 2e^{-2t}, \\ y &= e^{2t}. \end{aligned} \tag{7}$$

The values of these functions calculated at $t_0, t_1, \ldots, t_{10}$ are listed in the final two columns of Table 7.3.1 for comparison with the Euler-approximate values. The curve and its piecewise approximation are plotted in Fig. 7.3.3.

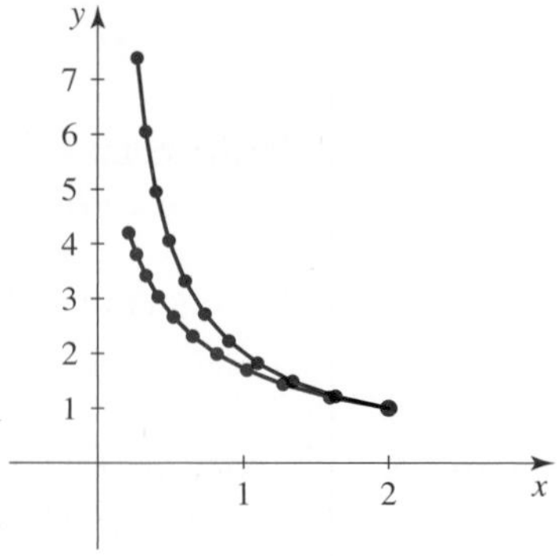

FIGURE 7.3.3 Solution trajectory (above) and Euler approximation (below, in color) for IVP (6), where the initial value is the rightmost point.

■

Euler's method becomes less accurate with successive steps. At each step the slope throughout the interval is approximated by the slope at one end; the resulting errors accumulate. More accurate approaches, like the **Runge-Kutta method** (Problems 16–21), which average the slope values, cut down on this *discretization error*, but increase the *roundoff error* because additional arithmetic operations are required. (Revisit the error graph in Fig. 1.4.3.)

Accurate numerical methods are of great utility in studying the behavior of nonlinear systems for which quantitative solutions are frequently unavailable. They form the basis of all "solver" programs for differential equations and systems.

EXAMPLE 2 **An Epidemic Model** We will use a standard model for the spread of an infectious disease to study a flu epidemic in a small liberal arts college community of 1,000 persons. This mutant flu strain causes severe intestinal distress for several days, after which the victim recovers and is immune. Let's measure time t in days, and let

$S(t)$ = "susceptibles"	(those who have not had the flu and can catch it);
$I(t)$ = "infecteds"	(people who are currently sick and can give the flu to others);
$R(t)$ = "recovereds"	(individuals who have had the flu but are now well and immune to it).

At any time,

$$S + I + R = 1000, \tag{8}$$

so any two of the variables determine the third. The so-called *SIR*-model is derived from three assumptions:

- The rate at which susceptibles become infected is proportional to the product (a measure of possible interactions) of these two populations: $S' = -\alpha IS$; the constant of proportionality α is called the **infection rate**.

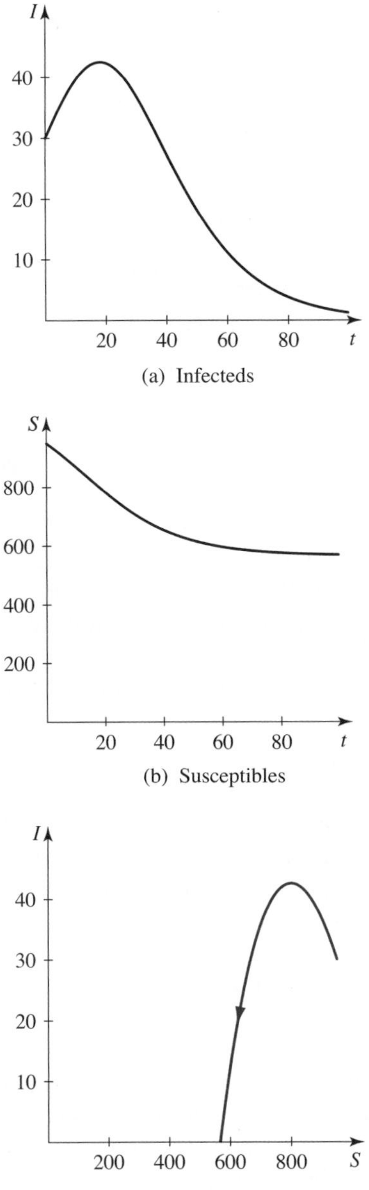

FIGURE 7.3.4 Populations of infected and susceptible individuals for the IVP of Example 2. The phase portrait (c) shows that the epidemic peaks and dies out, and at what levels, but graphs (a) and (b) are necessary to show the time dimension.

- The population of those recovered increases at a rate proportional to the number who are sick, $R' = \beta I$, where β is the **recovery rate**.
- The rate of change of the number infected is the difference between the rate at which healthy people get sick and the rate at which sick people get well: $I' = \alpha I S - \beta I$.

We therefore have the system of differential equations

$$\begin{aligned} S' &= -\alpha I S, \\ I' &= \alpha I S - \beta I, \\ R' &= \beta I, \end{aligned} \tag{9}$$

as a model for the epidemic. Because the first two equations in (9) are independent of R, we can solve the subsystem

$$\begin{aligned} S' &= -\alpha I S, \\ I' &= \alpha I S - \beta I, \end{aligned} \tag{10}$$

and obtain R by subtraction using equation (8).

For particular values of α and β, Euler's method can be used to generate predictions of how many will be sick, and provisions can be made to care for them. With $S(0) = 950$ and $I(0) = 30$, for example, and estimating $\alpha = 0.00025$ and $\beta = 0.2$, we find (using step size $h = 0.1$) that

$$\begin{aligned} S_{n+1} &= S_n + (0.1)(-0.00025 I_n S_n), \\ I_{n+1} &= I_n + (0.1)(0.00025 I_n S_n - 0.2 I_n). \end{aligned}$$

Computation using a spreadsheet (see Appendix SS) or a DE solver provides the graphs in Fig. 7.3.4. The number of sick individuals peaks at about 43 persons on or about the twentieth day. By day 100 the number of susceptibles has leveled off at about 560 and the epidemic seems to have run its course, but there are still plenty of susceptibles for another epidemic next year. ■

You may find it instructive to set up a spreadsheet (see Appendix SS) in such a way that the parameters α and β can be changed and the rest of the calculations automatically updated. Experimenting with these parameter values, one can observe their effect on the output.

You can do the same kind of experimentation with parameters more "directly" or "viscerally" with a graphic solver that allows you to see more than one solution at a time. That is, it *graphs* each solution as it is calculated numerically, and does it so quickly that you can choose many initial conditions on one graph. Such a picture is easily worth a thousand words (or many more numbers), as in the following.

EXAMPLE 3 **Finding Bifurcation Values** Recall from Sec. 2.5 that bifurcation means a drastic change in solution behavior as a parameter changes value. Consider

$$\begin{bmatrix} x \\ y \end{bmatrix}' = \begin{bmatrix} -1 & 1 \\ \alpha & \alpha \end{bmatrix} \begin{bmatrix} x \\ y \end{bmatrix}.$$

The value of $\alpha = 0$ is a bifurcation parameter, because $\alpha > 0$ gives a saddle equilibrium, while α slightly less than zero gives an attracting node. (See Fig. 7.3.5.) You can find other surprises in Problem 26.

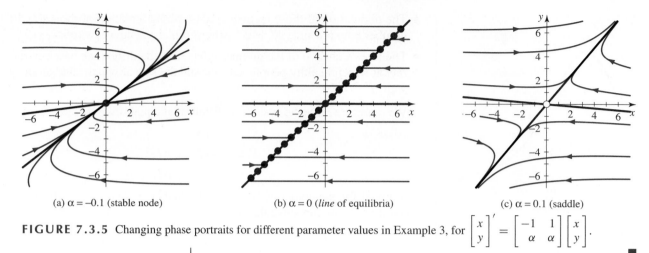

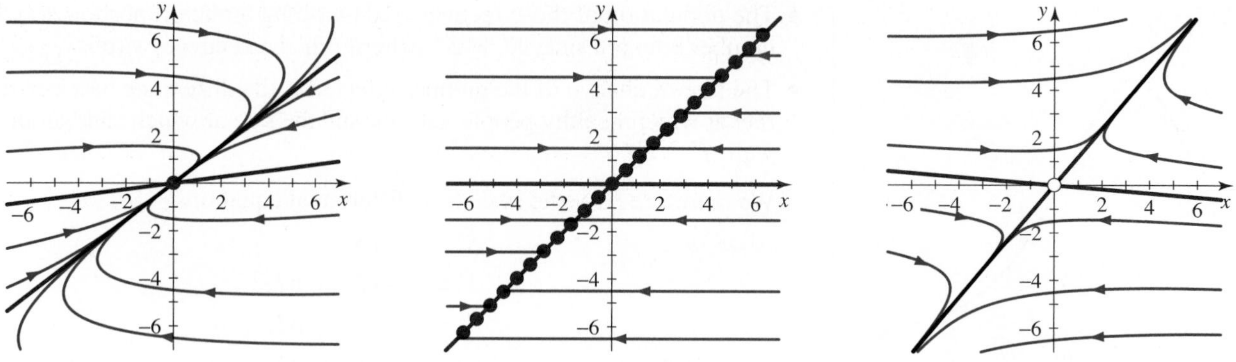

(a) $\alpha = -0.1$ (stable node) (b) $\alpha = 0$ (*line* of equilibria) (c) $\alpha = 0.1$ (saddle)

FIGURE 7.3.5 Changing phase portraits for different parameter values in Example 3, for $\begin{bmatrix} x \\ y \end{bmatrix}' = \begin{bmatrix} -1 & 1 \\ \alpha & \alpha \end{bmatrix}\begin{bmatrix} x \\ y \end{bmatrix}$.

On a spreadsheet approach you would see the bifurcation of Example 3 as a change from solutions approaching zero to solutions flying off to infinity. Finding the parameter value where this occurs can be tedious for more complicated systems.

For nonlinear systems there are even more possibilities for bifurcations. Limit cycles are often involved. (See Problems 27 and 28.)

Asking Questions

Chemical Oscillator

Try different values of b and see limit cycles appear (or disappear) in the phase portraits; see the difference in the color behavior of the mixed chemicals.

Predator-Prey; Lotka-Volterra

See discussion questions with Lab 21 for an example of how to seek changes in parameter values that achieve desired changes of real-world behaviors. The qualitative approach is the key.

Since it is usually not possible to find analytic solutions to nonlinear DE systems, the necessity of relying on numerical methods becomes paramount. We need more than rough approximations to solutions. Thus, numerical analysis has become a very important subject in mathematics.

Sophisticated numerical methods appeared in the nineteenth century, but Newton himself developed approaches to calculation from the earliest days of calculus. Recall Newton's method for solving algebraic equations using the derivative. Astronomers were especially zealous in developing computational skills to deal with masses of observational data. Their "science of comptometry" required unbelievable patience and produced some astounding results.

The age of the microchip, however, has opened a brave new world of numerical investigation, and contemporary hardware and software tools allow students who are just beginning the study of differential equations to develop exploratory skills and obtain startling insights into the behavior of systems. As your experience increases, you will learn how best to approach problems and what questions to ask.

What range of parameters should you study? What initial conditions should you consider? What are the equilibrium points, and are they stable? Is the model any good? Do your differential equations seem to have periodic solutions when the real system does not? Or does it? Does the motion of your system get faster as the amplitude gets larger? What does this mean about the physical system? Have you discovered something new? Do any solutions go to infinity, and what does that mean? Which variables should be plotted to get the clearest picture?

Embrace the opportunities offered by modern technology: Your power to analyze differential equations will surely benefit!

Summary

We extend Euler's method for approximating solutions of differential equations to systems of differential equations that are not necessarily linear or even autonomous. This method and its more accurate cousins (such as the Runge-Kutta method), together with such devices as the spreadsheet for implementation, provide us with important and flexible investigative tools whose real power was unknown before the age of the microchip.

7.3 Problems

Spreadsheet Calculation *Use a spreadsheet (or other comparable software) to make Euler-approximate solutions of each IVP in Problems 1–4, from $t = 0$ to $t = 1$, using different step sizes (e.g., $h = 0.1$ and $h = 0.05$), with initial condition $x(0) = 1$, $y(0) = 1$. Graph both approximate trajectories on the same xy plot, labeling each curve with its h value. Discuss where and how the step size makes a difference in the approximate solutions, and whatever you can infer about the exact solutions.*

1. $x' = y$, $y' = -x + x^3 - y$

2. $x' = y$, $y' = -x - x^3 - y$

3. $x' = y$, $y' = -x - y^3$

4. $x' = y$, $y' = -x - y - y^3$

Changing Views *Do the following for each system in Problems 5–10.*

(a) *Analyze behavior by making an xy phase portrait, plus a tx and/or ty graph associated with a typical trajectory.*

(b) *Describe the motion along trajectories in terms of*

$$\text{speed} = \sqrt{\dot{x}^2 + \dot{y}^2}.$$

5. $\dot{x} = y^2$, $\dot{y} = x^2$

6. $\dot{x} = x + y$, $\dot{y} = x + y$

7. $\dot{x} = y$, $\dot{y} = -x$

8. $\dot{x} = y$, $\dot{y} = -x + x^3$

9. $\dot{x} = y$, $\dot{y} = -x - x^3$

10. $\dot{x} = y$, $\dot{y} = -\sin x$

Changing Parameters *In Problems 11 and 12, experiment with several different values of ε to find changes in long-term behavior. Describe the effects on both phase portraits and equilibria.*

11. $\dot{x} = x + \varepsilon x(1 - x^2 - y^2)$
$\dot{y} = -x + \varepsilon y(1 - x^2 - y^2)$

12. $\dot{x} = y$
$\dot{y} = -x + \varepsilon(1 - x^2)y$

Euler for 3 x 3 Systems *An easy extension can be made to formulas (3)–(5) to adapt them to approximate solutions of the system*

$$x' = f(t, x, y, z),$$
$$y' = g(t, x, y, z),$$
$$z' = q(t, x, y, z).$$

If we begin at time $t = t_0$ with $x(t_0) = x_0$, $y(t_0) = y_0$, and $z(t_0) = z_0$, using step size h, then

$$t_{n+1} = t_n + h,$$
$$x_{n+1} = x_n + hf(t_n, x_n, y_n, z_n),$$
$$y_{n+1} = y_n + hg(t_n, x_n, y_n, z_n),$$
$$z_{n+1} = z_n + hq(t_n, x_n, y_n, z_n).$$

Use this method with $h = 0.1$ and $h = 0.05$ to estimate the solution at $t = 1$ for the IVP in each of Problems 13–15. Comment on the differences in results; in Problem 13, compare with the analytic solution.

13. $x' = x + y$
$y' = y + z$ $\quad x(0) = 1,\ y(0) = 1,\ z(0) = 1$
$z' = -y + 2z$

14. $x' = -x + xy$
$y' = y + xz$ $\quad x(0) = 1,\ y(0) = 1,\ z(0) = 2$
$z' = -y + yz$

15. $x' = x + y$
$y' = -x + tz$ $\quad x(0) = 2,\ y(0) = 1,\ z(0) = 1$
$z' = z + x^2$

16. Epidemic Use Euler's method with $h = 0.1$ to tabulate the values of S, I, and R in the epidemic model of Example 2 for 20 days, using $\alpha = 0.0001$ and $\beta = 0.15$.

17. Epidemic Formula From equations (10) and the fact that $dI/dS = I'/S'$, show that

$$I = I_0 + S_0 - S + \frac{\beta}{\alpha} \ln\left(\frac{S}{S_0}\right).$$

Using $\alpha = 0.00025$ and $\beta = 0.2$, plot trajectories in the SI-plane for various initial values I_0 and S_0, and interpret the results.

18. Bug Race Three bugs start at the point (0, 1) and race according to their respective differential equations:

Bug A:	Bug B:	Bug C:
$x' = y,$	$x' = y,$	$x' = y,$
$y' = -x;$	$y' = -x + x^3;$	$y' = -x - x^3.$

When the gun sounds, they all race around the origin aiming to arrive back at the starting point, but not necessarily at the same time. Which bug wins?

Runge-Kutta Method *A more accurate numerical solution scheme than Euler's is the* ***Runge-Kutta method****, quite widely used in actual practice. (See Problems 19–22 in Sec. 1.4.) Instead of using the slope at the beginning of an interval, a weighted average of slopes is used.*

Runge-Kutta Method for 2 x 2 DE Systems

The formulas for the solution of IVP

$$x' = f(t, x, y), \quad x(t_0) = x_0;$$
$$y' = g(t, x, y), \quad y(t_0) = y_0,$$

with step size h are as follows:

$$x_{n+1} = x_n + \frac{h}{6}(k_{n1} + 2k_{n2} + 2k_{n3} + k_{n4}),$$

$$y_{n+1} = y_n + \frac{h}{6}(m_{n1} + 2m_{n2} + 2m_{n3} + m_{n4}),$$

where

$$k_{n1} = f(t_n, x_n, y_n),$$
$$k_{n2} = f\left(t_n + \frac{h}{2}, x_n + \frac{h}{2}k_{n1}, y_n + \frac{h}{2}m_{n1}\right),$$
$$k_{n3} = f\left(t_n + \frac{h}{2}, x_n + \frac{h}{2}k_{n2}, y_n + \frac{h}{2}m_{n2}\right),$$
$$k_{n4} = f(t_n + h, x_n + hk_{n3}, y_n + hm_{n3});$$

and

$$m_{n1} = g(t_n, x_n, y_n),$$
$$m_{n2} = g\left(t_n + \frac{h}{2}, x_n + \frac{h}{2}k_{n1}, y_n + \frac{h}{2}m_{n1}\right),$$
$$m_{n3} = g\left(t_n + \frac{h}{2}, x_n + \frac{h}{2}k_{n2}, y_n + \frac{h}{2}m_{n2}\right),$$
$$m_{n4} = g(t_n + h, x_n + hk_{n3}, y_n + hm_{n3}).$$

For Problems 19–24, approximate solutions by the Runge-Kutta method, using appropriate computer software. Compare with the Euler's method results obtained in Problems 1–6, both numerically (for $x(1)$ and $y(1)$ when $x(0) = 1$, $y(0) = 1$) and graphically (with tx plots, where time is explicit), using the same step size for both approximations. Explain what information comes from tx plots that does not show up on xy plots.

19. IVP of Problem 1

20. IVP of Problem 2

21. IVP of Problem 3

22. IVP of Problem 4

23. IVP of Problem 5

24. IVP of Problem 6

25. Proper Step Size The solution of the IVP

$$\begin{aligned} x' &= y, \\ y' &= -x, \end{aligned} \qquad x(0) = 1, y(0) = 0,$$

is $x(t) = \sin t$, $y(t) = \cos t$, and the solution curve in the phase plane is the clockwise unit circle. Use a computer and Euler's method to see how small a step size is required to guarantee returning close to the starting point. Repeat with Runge-Kutta, and compare the results of these two methods.

26. Additional Bifurcations Return to Example 3, where we found a bifurcation value $\alpha_1 = 0$ for

$$\begin{bmatrix} x \\ y \end{bmatrix}' = \begin{bmatrix} -1 & 1 \\ \alpha & \alpha \end{bmatrix} \begin{bmatrix} x \\ y \end{bmatrix}.$$

(a) Try $\alpha = -2$. How has the equilibrium changed?

(b) Locate the second bifurcation value α_2 between -0.1 and -2 where the equilibrium behavior makes the change noted in (a). HINT: Comparison of your answer to (a) with Fig. 7.3.5(a) should suggest how to find an equation to solve.

(c) There is yet another bifurcation value α_3 occurring for some $\alpha < -2$, as you may have noticed with your calculations in (b). Find α_3 and explain, with eigenvectors, what happens when α passes through this value.

(d) Make a set of phase portraits to extend Fig. 7.3.5(a) past α_3, showing how the character of trajectories and equilibria changes around each of the bifurcation values. Add eigenvectors as appropriate. Write a verbal description of the progression of phase portraits.

27. Hopf Bifurcation[3] Work IDE Lab 25 Part 1 to investigate the bifurcation involving a limit cycle that creates the chemical oscillator.

Chemical Oscillator
Some values of the parameter cause colors to change back and forth as shown in Sec. 1.1, Example 12.

28. Saddle Node Bifurcation Work IDE Lab 25 Part 2 to study another common bifurcation.

2D Saddle-Node Bifurcation
Try another example where changing a parameter drastically changes the phase portrait.

29. Suggested Journal Entry For the *SIR* model, discuss how the parameters, or even the form of the equations, might be affected by such factors as vaccines or quarantines.

---

[3]Heinz Hopf (1894–1971) was attracted to mathematics in 1917 when, during a leave from German military service, he attended a university class on set theory. After World War I he studied mathematics more formally and made prominent contributions to algebraic topology and other fields; many of these were results obtained from Hopf's studies of vector fields.

7.4 Chaos, Strange Attractors, and Period Doubling

SYNOPSIS: We introduce the concept of chaotic behavior and strange attractors for solutions of differential equations, a subject where you have to take time to experiment and to think. There exist no quick formulas for answering the questions that arise.

Introducing Chaos

Linear differential equations describe a world of Platonic purity in which physical phenomena behave with clockwork regularity. And although the world outside the classroom is seldom linear, with turbulence, irregularity, and randomness to be expected, scientists long clung to the time-honored belief that such erratic phenomena were just incidental "noise" and not reflective of the true underlying mechanisms. The philosophy was to concentrate on problems that were linear or readily linearized.

In the latter half of the twentieth century, however, the scientific community began to realize that turbulent and unpredictable behavior reflects deeper relationships not previously appreciated. **Deterministic chaos** is not random; it evolves from precise (nonlinear) equations. **Chaotic motion**—motion that is erratic and seemingly random, highly sensitive to slight changes in conditions—emerges from deterministic systems and is not just the result of noise in the observations. Chaotic motions have been observed in ecology, economics, physics, chemistry, and engineering: convection of heat in a fluid, forced nonlinear pendulums, stirred chemical reactions, cardiac fluctuations, and stock market variations.

New examples of deterministic chaos appear regularly. Parallel to these natural phenomena has been the discovery, due in large part to computer technology, of similarly unpredictable behavior in nonlinear mathematical systems.

At the intersection of the theoretical and the practical was the experience in 1963 of MIT meteorologist Edward Lorenz, investigating a differential equations model of thermally induced air convection.[1] He used a computer to approximate the solution of the nonlinear system

Lorenz Equations

$$\begin{aligned} dx/dt &= 10(y - x), \\ dy/dt &= rx - y - xz, \\ dz/dt &= xy - (8/3)z. \end{aligned} \tag{1}$$

(The variables x, y, and z represent rotation speed and nonlinear temperature distributions within convection cells; the parameter r was set at 28.) To check his results, Lorenz ran his program again, but this time he specified the initial conditions to only three decimal places rather than the six places used for the first run. Returning to his office from fetching a cup of coffee, he was surprised to find the new results quite different from the previous ones.

Lorenz did some checking to verify that the problems had been correctly entered, and eventually realized that the result was due to very small changes in

[1]For this interesting story and more on Lorenz, see J. Gleick, *Chaos: Making a New Science* (NY Viking, 1987). Lorenz' original paper, "Deterministic Nonperiodic Flow," appeared in the *Journal of Atmospheric Sciences* **20** (1963), 130.

the initial conditions; he had entered

$$(-6.299, 11.677, 10.970) \quad \text{instead of} \quad (-6.298702, 11.676666, 10.970053).$$

System (1) is *hypersensitive* to the initial conditions: after only a short period of time, for $t > 6$, the solutions are quite different. (The graphs for one component are shown in Fig. 7.4.1.) Lorenz later told *Discover* magazine, "I knew then that if the real atmosphere behaved like this [mathematical model], long-range weather prediction was impossible."

Lorenz Equations: Discovery 1963

For one of the variables, watch the two solutions evolve and diverge.

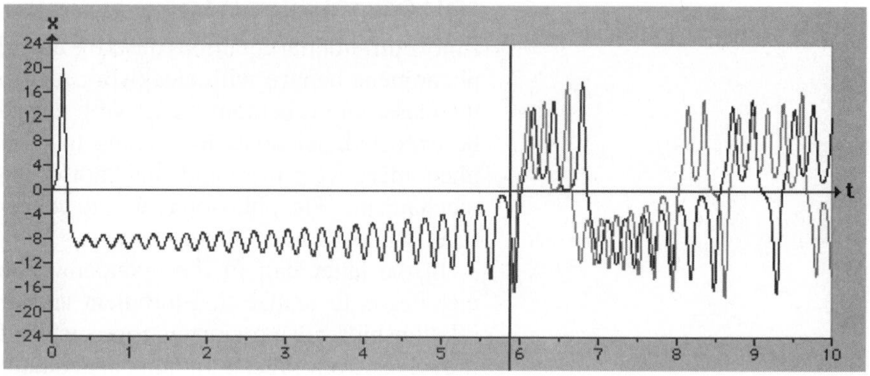

FIGURE 7.4.1 Sensitivity of solutions of the Lorenz system (1) to initial conditions.

$$dx/dt = 10(y - x),$$
$$dy/dt = rx - y - xz,$$
$$dz/dt = xy - (8/3)z.$$

Lest we give the impression that nothing was known about chaotic systems before Lorenz experienced his moment of enlightenment in 1963, we should be reminded that the eminent mathematician Poincaré was well aware at the beginning of the twentieth century that the solutions to certain differential equations in celestial mechanics were extremely sensitive to initial conditions, leading him to speculate that the motions of orbiting planets might not be completely predictable.

Characteristics of Deterministic Chaos

Extreme sensitivity to initial conditions has already been mentioned as a symptom of chaotic motion that is not random. Furthermore, the deterministic chaotic motion discovered by Lorenz is bounded. You can see in Example 1 that different initial conditions yield different time series and trajectories—but the region of phase space where they operate is clearly defined, and after some time the phase portraits look roughly the same! We say that the trajectories are following a **strange attractor** that limits the region where they will be found.[2]

Lorenz Equations: Sensitive Dependence

Click once to choose initial conditions that differ only by .001 in x. Watch the time series and 2D views evolve, and see the differences between the trajectories chaotically expand and shrink back. You can clear transients with a click to get a clearer view.

Since any numerical integration algorithm involves error, and a chaotic system is hypersensitive to changes in initial conditions, it might seem to be hopeless to try to approximate and graph such solutions. However, if we ignore the *transient* solution curve (whatever happens before the solution lands on the attractor), we may be able to learn something from the limit behavior of solutions. That is, the strange attractor (if there is one) may be seen more easily if the first hundred (or more) points are not plotted.

[2]Strange attractors were formally defined in 1971 by Belgian physicist David Ruelle and Dutch mathematician Floris Takens. The problem is that we don't know where the solution will be *within* the strange attractor at a given point in time.

EXAMPLE 1 **The Lorenz Attractor** In Fig. 7.4.2 we compare two different initial conditions for the Lorenz system (1) and invite you to experiment and create your own examples.

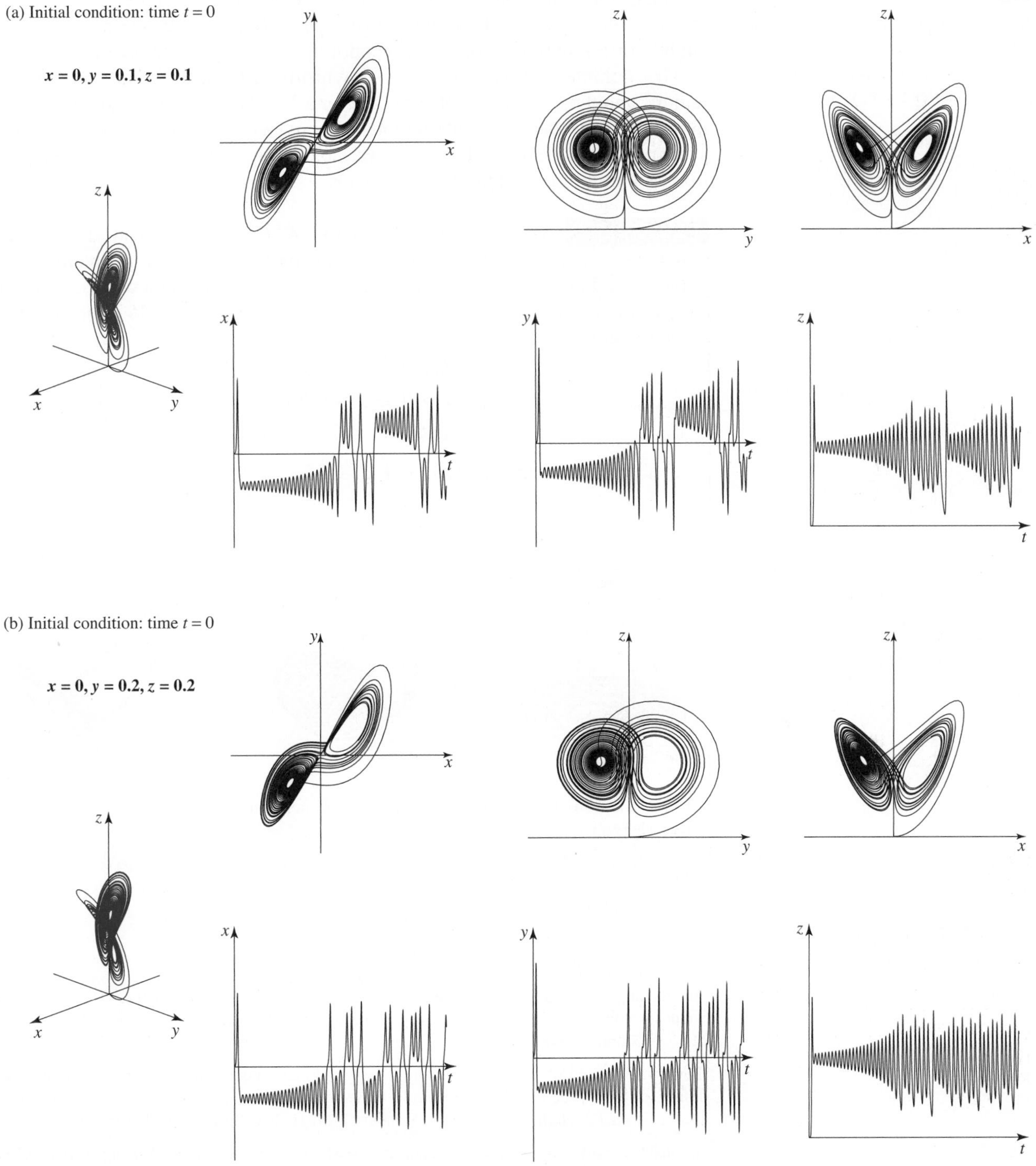

FIGURE 7.4.2 Phase portraits, projections, and time series for Lorenz equations, with two different initial conditions, for $r = 28$.

■

Mathematicians long wondered whether this intriguing attractor might not just be an artifact of the numerical computations necessary to display it; it was several decades after its discovery before a young mathematician in Sweden, Warwick Tucker, proved in his Ph.D. thesis that the Lorenz attractor is truly what it appears to be—the orbits are definitely chaotic.[3]

Lorenz Equations: Phase Plane $0 \le r \le 30$

See a simulation of the Lorenz model linked to evolving trajectories and solutions. This tool plots xy pictures *and* time series for x, y, and z.

The Lorenz equations satisfy existence and uniqueness requirements (as stated for systems in Sec. 7.1, Problems 39–40), so each trajectory determined by even slightly different initial conditions cannot intersect any of the others.

This nonintersection of trajectories whirling around differently in the same region cannot happen in the phase plane for a 2×2 autonomous system—another dimension is needed.[4] For autonomous systems (with no explicit t) like the Lorenz system, that means that at least three variables coupled by three differential equations is a minimum requirement for chaos.

EXAMPLE 2 **The Roessler Attractor** A German medical doctor, Otto Roessler, was led to study chaos phenomena from his work in chemistry and theoretical biology.[5] He tried to find the simplest system that would exhibit the sort of behaviors discovered in the Lorenz attractor, and in 1979 he published the system

Roessler Equations

$$\begin{aligned} dx/dt &= -y - z, \\ dy/dt &= x + 0.2y, \\ dz/dt &= -rz + xz + 0.2, \end{aligned} \tag{2}$$

with the phase portrait shown in Fig. 7.4.3.

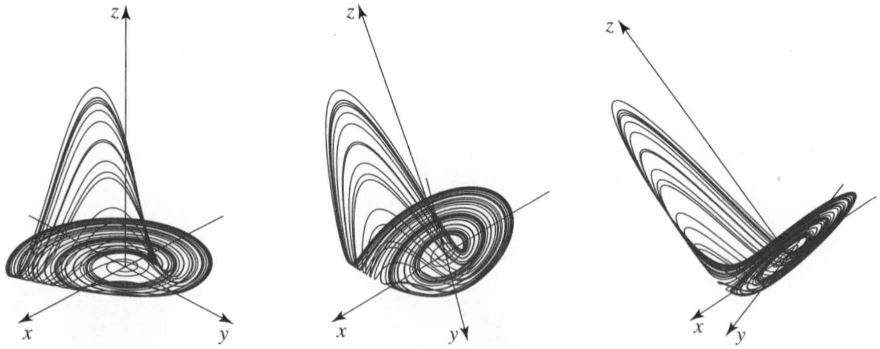

FIGURE 7.4.3 Various views of the three-dimensional strange attractor for the Roessler equations with $r = 5.7$.

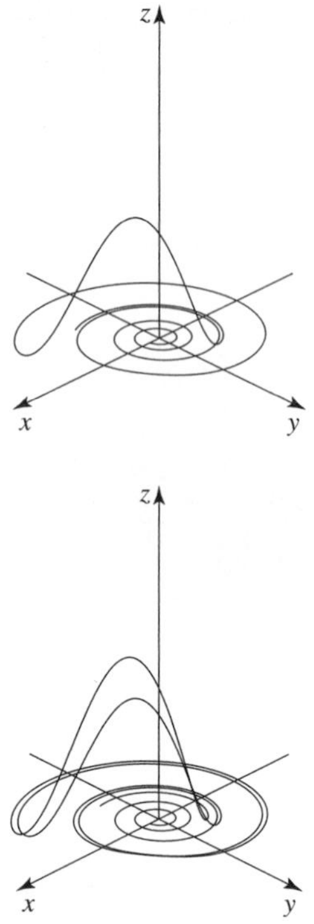

FIGURE 7.4.4 Earlier stages in plotting Roessler attractor in Fig. 7.4.3.

Roessler likened the way this attractor develops to pulling taffy—every once in a while stretching the figure in the z direction and folding it back down to the xy-plane. Figure 7.4.4 shows how the particular version of the attractor shown

[3]See the article by Ian Stewart in the August 31, 2000 issue of *Nature* (pp. 948–949) for a lovely perspective on the import of this discovery, or the article by M. Viana in *Mathematical Intelligencer* **22** no. 3 (Summer 2000), 6–9.

[4]For a justification of the need for more than two dimensions to create deterministic chaos and strange attractors, see J. H. Hubbard and B. H. West, *Differential Equations: A Dynamical Systems Approach, Part 2: Higher Dimensional Systems* (TAM 18, NY: Springer-Verlag 1995), Chapter 8.

[5]J. Gleick provides an engaging account of Roessler's efforts in *Chaos: Making a New Science* (Viking 1987), 141. An original source is O. E. Roessler, "Continuous Chaos—Four Prototype Equations," *Bifurcation Theory and Applications in Scientific Disciplines*, edited by O. Gurel and O. E. Roessler, *Annals of the New York Academy of Sciences* **316** (1979), 376.

in Fig. 7.4.3 evolved as the time interval was increased. Problems 6–8 provide a guide to further analysis of this strange attractor. ■

Period-Doubling Route to Chaos

For most systems, chaos appears only for certain parameter intervals. The Lorenz and Roessler equations are both examples of a common phenomenon called **period doubling**, in which a nice periodic cycle, forming a loop in phase space, exists over an interval of r, then suddenly splits into a doubly twisted loop, which soon splits again into a quadruply twisted loop. For each doubling, the time series suddenly shows for each low-high pair an extra set of lows and highs at new levels, as shown in Fig. 7.4.5. The number of relative highs is called the **period** of the cycle.

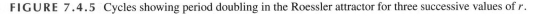

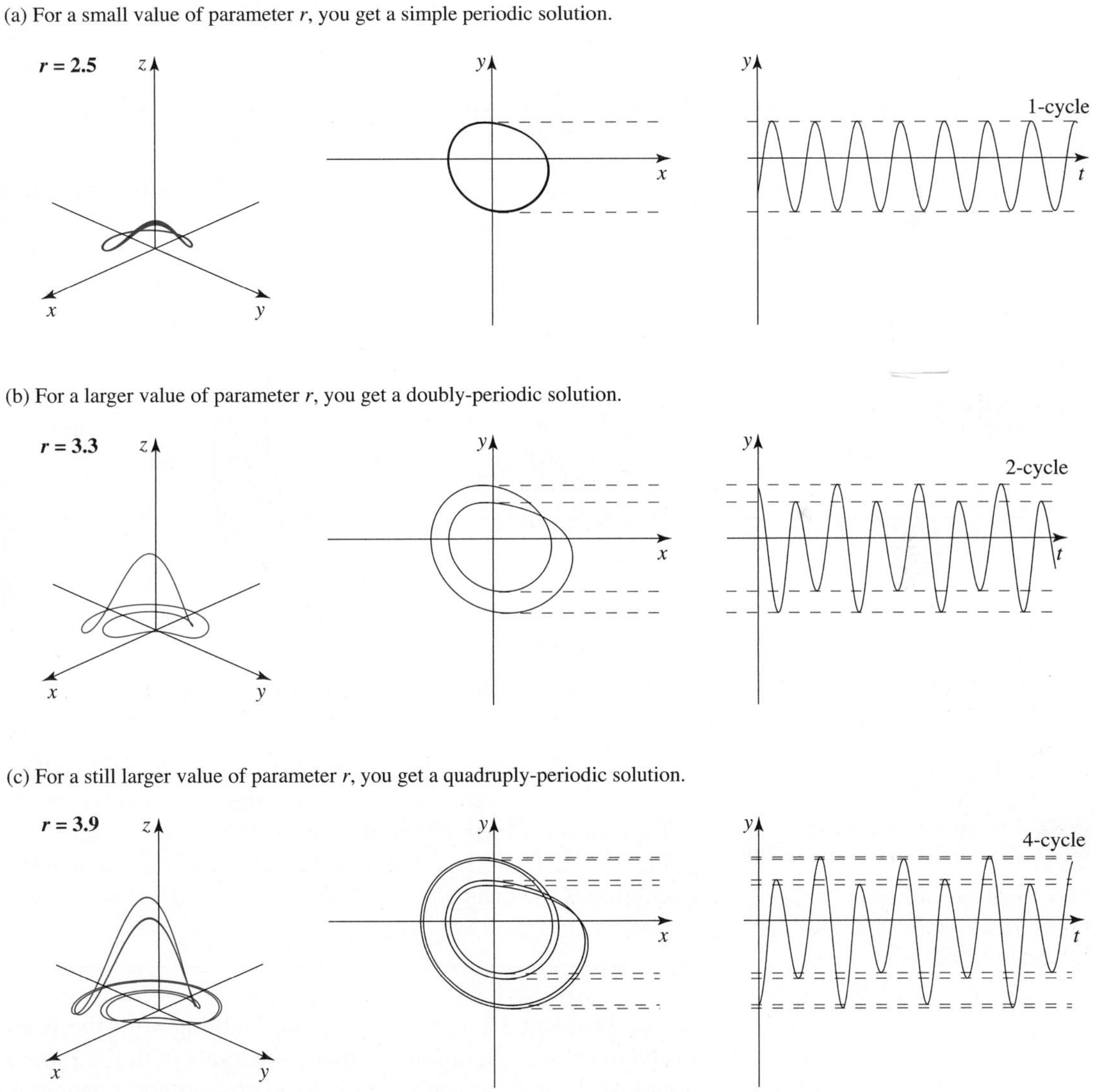

(a) For a small value of parameter r, you get a simple periodic solution.
r = 2.5
z
x
y
y
x
y
1-cycle
t
(b) For a larger value of parameter r, you get a doubly-periodic solution.
r = 3.3
z
x
y
y
x
y
2-cycle
t
(c) For a still larger value of parameter r, you get a quadruply-periodic solution.
r = 3.9
z
x
y
y
x
y
4-cycle
t

FIGURE 7.4.5 Cycles showing period doubling in the Roessler attractor for three successive values of r.

EXAMPLE 3 **Period Doubling with Lorenz** If we vary the parameter r in the Lorenz equations (1), we find that chaos does not occur at all for values much below the 28 that Lorenz was using in 1963. Many other values of r produce lovely closed orbits instead of chaos. And there are many intervals of chaos associated either above or below, with a period-doubling sequence of closed orbits. For instance, a more complicated closed orbit (a triple loop) appears in the Lorenz equations (1) with $r = 156$, as shown in Fig. 7.4.6.

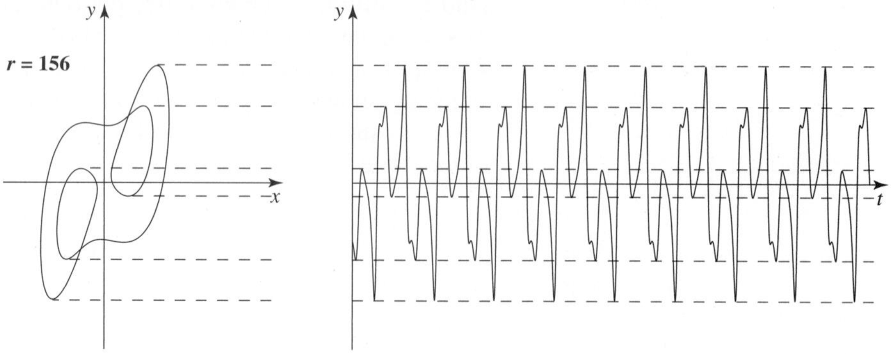

FIGURE 7.4.6 A period-three cycle in the Lorenz equations (1). The extra bumps in the ty graph corresponds to relative extrema; they do not represent additional loops or cycles.

For values of r that descend from 152 to 140, we see a typical progression of phase portraits from a closed triple loop to a *double* closed triple loop to a *quadruple* closed triple loop, and so on, doubling the number of loops over smaller and smaller r intervals, until they become a thick chaotic band, as shown in Fig. 7.4.7.

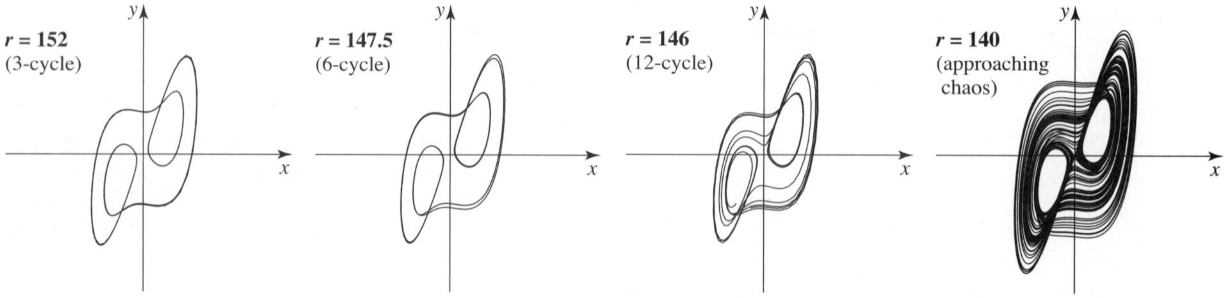

FIGURE 7.4.7 Period-doubling orbits for different values of r in the Lorenz equations (1).

Lorenz Equations: Phase Plane $0 < r < 300$

Vary r and look for cycles. A button clears transients at any time so you can get to a clean orbit.

It is difficult to see the cycles clearly in graphs as small as these. For a better view *and* the advantages of watching the cycles evolve, try the IDE tool **Lorenz Equations: Phase Plane $0 < r < 300$**. You can set r to the values used in Fig. 7.4.7. There is a button to "clear transients," which erases the screen but continues the orbit from the last point erased—this way you can see the cycle cleanly once you are on it. ■

See Problem 11 to explore period-doubling with the Roessler attractor, and Problem 9 for exploration of other r-intervals with the Lorenz attractor. Period-doubling is not the only route to chaos—other phenomena such as *quasi-periodicity* and *intermittency* can occur. Problem 10 explores this aspect.

Summary

Extreme sensitivity to initial conditions and chaotic behavior of solutions call our attention to ways in which solutions of nonlinear problems act badly. These behaviors can appear when a system of differential equations is more than two-dimensional, necessitating at least three variables in an autonomous system. The discovery of strange attractors, on the other hand, points to the existence of a *pattern* underlying such chaotic behavior: some order within chaos. One very common such pattern is the period-doubling of closed orbits as a parameter changes, until chaos results. Yet even such a universal pattern is hard to pin down in detail for any given example. You have to learn how to experiment.

7.4 Problems

1. **Equilibrium Analysis** For the Lorenz equations (1) with $r = 28$, find the equilibria and determine the stability of each. Explain how this information helps to explain the pattern of the strange attractor.

2. **Hypersensitivity** Using a graphical DE solver, demonstrate the extreme sensitivity to initial conditions of the Lorenz equations (1) with $r = 28$, with your own choice of initial conditions close together. How long does it take for the solutions to visibly diverge? Describe your results as a presentation for a nonbeliever.

3. **Long-Term Behavior** Experiment as in Problem 2, but create much longer time series. Do the phase portraits look even more similar? What about the time series, if you think about chopping off the beginnings for this set? Explain.

4. **Lorenz yz** For the Lorenz equations (1) with $r = 28$, solve numerically for a long period of time and plot the variables y and z parametrically in the yz-plane. Use the initial condition $(0, 1, 25)$. Explain the approximate symmetry that results.

5. **Linearized Lorenz Equations**
   (a) For small values of x, y, and z, show that the Lorenz system (1) can be approximated by the linearization
   $$\mathbf{x}' = \mathbf{A}\mathbf{x} = \begin{bmatrix} -10 & 10 & 0 \\ 28 & -1 & 0 \\ 0 & 0 & -8/3 \end{bmatrix}\mathbf{x}.$$
   (b) Solve this system with the same initial conditions as in Problem 2, and plot the variables y and z parametrically in the yz-plane. Compare with your results in Problem 2.
   (c) Discuss the extent to which linearization is useful in a chaotic system.

6. **Roessler Views** For the Roessler system (2), examine other graphical representations. You may find a graphical DE solver very helpful, but you may be able to deduce qualitative answers from the pictures in Example 2.
   (a) Sketch xy, yz, and xz graphs. Explain their appearances.
   (b) What can you expect the time series tx, ty, and tz to look like? After making a conjecture, make computer images and compare. Explain any discrepancies.

7. **Roessler Analysis** Consider the Roessler system (2).
   (a) Find the equilibria and explain the stability of each (as in Problem 1).
   (b) Make a hypersensitivity study (as in Problem 2).

8. **Sensitivity** Chaotic strange attractors are sensitive to more than just initial conditions. Experiment with the Roessler attractor using the initial conditions $x(0) = 1$, $y(0) = 0$, and $z(0) = 0$. Use a graphical DE solver to compare results in each of the following.
   (a) Compare the results of Euler's method with those of Runge-Kutta for the numerical calculations that create the graphs. Describe the differences.
   (b) Compare the results of changing the step size, with each of the methods. Do you think one or the other method is less subject to this sort of sensitivity?
   (c) If you have the opportunity, try the same experiments on different graphical DE solvers. That is, enter the same equations, the same method, the same step size, and the same initial condition but use different solvers (e.g., different computer systems). What is different and what is the same?
   (d) If there is so much sensitivity affecting the graphs, to what extent can we rely on our graphs in examining these strange attractors?

9. **Varying Lorenz** Work IDE Lab 27, Part 3, concerning the equilibrium values of the Lorenz system with different choices for the parameter r.

Lorenz Equations:
Parameter Grid $0 \le r \le 30$,
Phase Plane $0 \le r \le 30$,
Phase Plane $0 \le r \le 320$

The standard Lorenz example has $r = 28$, but there are many even more interesting behaviors when r takes different values. See also Problem 10.

10. Bifurcations in Lorenz System Work IDE Lab 27, Part 4 and make a bifurcation analysis for different values of r. You will be led to observe period doubling, noisy periodicity, and intermittency on the way to chaos.

11. Varying Roessler A more general Roessler system can be written

$$\begin{aligned} dx/dt &= -y - z, \\ dy/dt &= x + 0.2y, \\ dz/dt &= -rz + xz + 0.2, \end{aligned} \tag{3}$$

where we have seen that $r = 5.7$ produces a strange attractor. The graphs in Fig. 7.4.5 were made using $r = 2.5$, 3.3, and 3.9. If your graphical DE solver can draw really fine lines, do the following and describe your results.

(a) Try to find (or at least bracket) bifurcation values for r that signal the change from a single loop to a double loop and again from a double loop to a quadruple loop.

(b) Find more period doubling after $r = 3.9$—can you find an 8-cycle? a 16-cycle? (The bifurcation values for r that signal period-doubling changes are very hard to find, but cycle examples will tell you where to look if you need to.) You will see that these subsequent bifurcations are closer and closer together, and, consequently, more difficult to find and to illustrate. When orbits have split but are very close together, they can easily appear as a single orbit, but if you can examine a time series closely enough, or its numerical orbit, you may be able to see them separate.

(c) As r increases much past 3.9 you will see chaotic orbits as separate cycles merge into thick bands, but you will see gaps in these bands for a while. Then you will find a value of r that suddenly, in the midst of all this chaos, gives a lovely clear cycle of period 3. If you are very lucky, you may be able to see this bifurcate into a cycle of period 6, and so on.

12. Period Doubling Elsewhere Three-dimensional autonomous systems of differential equations are not the only places we can see a period-doubling route to chaos. *Discrete* dynamical systems can exhibit this phenomenon as well and, in fact, using fewer dimensions! An example is numerical solution by *Euler's method* of the first-order logistic equation from Sec. 2.4.[6]

(a) Sketch a direction field for $y' = y(1 - y)$, with some sample solutions. Locate the equilibria and label them stable or unstable. What do you expect for a numerical approximate solution starting at $y(0) = 0.5$?

(b) Make a separate Euler-approximate solution for each of the following step sizes: $h = 1.8$, 2.2, 2.4, 2.55, 2.57, and 2.6. Obtain both a graph and a numerical list of the last hundred iterations and identify what kind of cycles you see. HINT: In the case of the larger values of h, it can be difficult to separate values of y_n that are close together. Use at least four decimal places (preferably eight) in your list of iterates; the number of graphical levels will be clearer if you plot points only, without the connecting lines, for some 500 or 1,000 iterates.

(c) Try to locate (or bracket) one or two of the bifurcation values of h between the values at which the number of cycles changes. (This is harder to do for the higher h values.)

(d) Try some other values of h and report what happens. You might wish to revisit these results when you reach Sec. 9.3.

13. Suggested Journal Entry Discuss how the concepts of this section support the idea that in studying nonlinear differential equations, we discover *both* order within chaos and chaos within order.

7.5 Chaos in Forced Nonlinear Systems

SYNOPSIS: Forcing terms bring the possibility of deterministic chaos to nonlinear systems of differential equations; new concepts like Poincaré sections aid in analysis.

Forced Dissipative Chaos

Of particular interest in the study of chaos are the forced dissipative systems: nonlinear systems subject to frictional damping and driven by external forces. If a nonconservative system has no "input" forcing, its motion will die out, but driven by an external force, it can exhibit interesting behavior. Standard examples, which result in turbulence, are eddies in a stream or the airflow behind an airplane wing. (See Fig. 7.5.1.)

[6]Adapted by permission from R. Borrelli and C. Coleman, *Differential Equations: A Modeling Approach* (NY Wiley, 1998), 137–141.

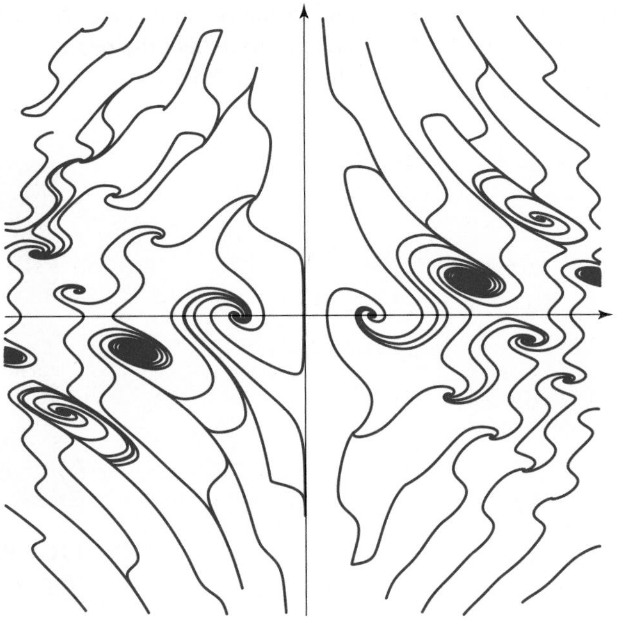

FIGURE 7.5.1 Eddies in a turbulent stream.

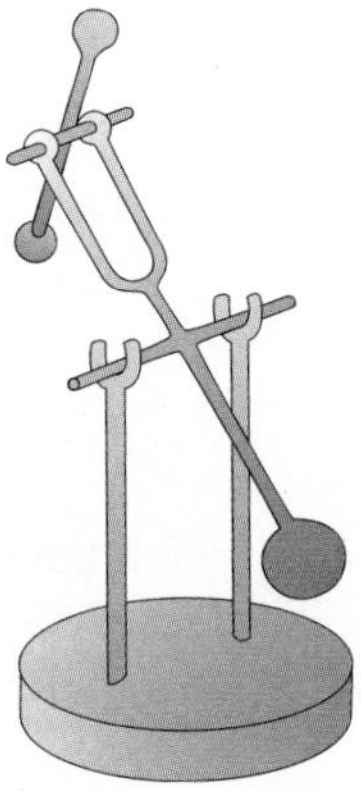

FIGURE 7.5.2 Chaos toy.

A system of linked pendulums, which can be purchased in many novelty shops, presents another example of a forced dissipative system whose motion is chaotic. The largest pendulum is driven by electrical impulses generated in the base of the toy, while the two smaller attached ones are driven by repelling magnets. (See Fig. 7.5.2.) When the large pendulum moves past a sensor in the base, a magnetic field generates a current, which exerts a torque on the pendulum. A slight disturbance will start the system oscillating; observing this motion will quickly convince you of the sensitivity of its long-term behavior to the initial conditions. The motion appears to be random or chaotic.[1]

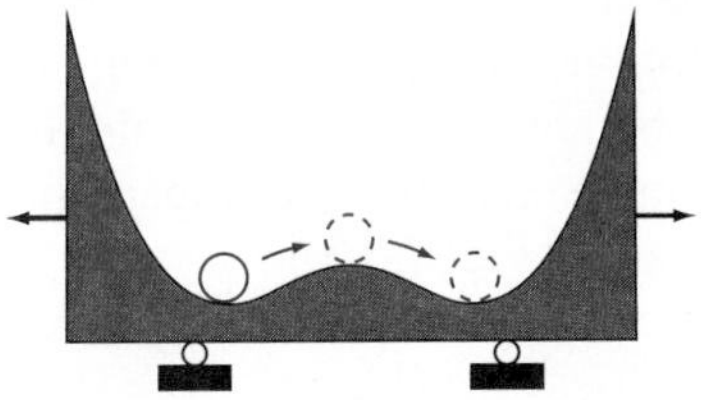

FIGURE 7.5.3 Ball rolling on an oscillating warped surface.

Another forced dissipative system resulting in chaotic motion consists of a ball moving on a warped surface with two depressions.[2] The surface vibrates periodically, while the ball jiggles from one depression to the other (Fig. 7.5.3). No matter how accurately the initial conditions are known, the system is so sensitive that it is impossible to predict its future.

Duffing Oscillator

It is interesting to examine a nonlinear vibration closely related to the linear mass-spring systems studied in Chapters 4 and 6. This particular model, studied by Francis Moon and colleagues at Cornell University, consists of a slender metal strip clamped at the top by a rigid framework, with two magnets near the free

---

[1]Prof. Alan Wolf and fellow physicists at the Cooper Union have analyzed this motion and verified that it is deterministically chaotic. See A. Wolf and T. Bessoir, "Diagnosing Chaos in the Space Circle," *Physica D* **50** (1991), 239–258. For a shorter method using Fourier analysis, see F. C. Moon, *Chaotic Vibrations* (NY: Wiley, 1987), 148–150.

[2]See F. C. Moon, *Chaotic Vibrations* (NY: Wiley, 1987), 75–76, 273–274. There are a number of physical systems that give rise to the same equations. A more detailed source on this example is N. B. Tufillaro and A. M. Albano (1986), "Chaotic Dynamics of a Bouncing Ball," *American Journal of Physics* **54** (10), 939–944.

end of the beam located to pull it in opposite directions.[3] Each magnet is strong enough to deflect the beam in its direction from the central equilibrium position. (See Fig. 7.5.4.)

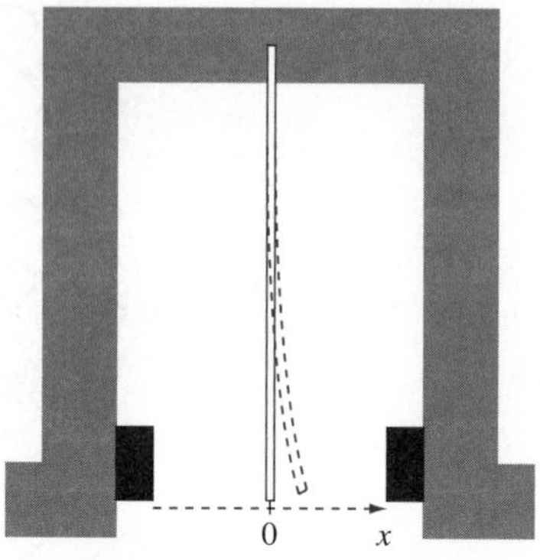

FIGURE 7.5.4 Magnet and steel strip system for Duffing oscillator.

There are three equilibrium points. The central position—in which the equal but opposite forces of the magnets are in balance—is unstable. The least deflection in either direction will give one magnet an advantage over the other, and the strip will bend to one of the two stable equilibrium points.

The entire apparatus is now shaken from left to right to left with a sinusoidally varying driving motor. The amplitude of this motion is large enough to shake the strip away from either stable equilibrium position. The idea is to determine the horizontal displacement $x(t)$ of the lower end of the strip as a function of the amplitude and frequency of the shaking.

This model is representative of many systems that are predictable until disturbed by driving forces. For "small" shaking, the behavior is predictable, with the solution remaining "trapped" in the influence of one magnet or the other. For larger shaking, however, the results are unpredictable, or chaotic, as the solution shifts erratically back and forth between the two stable equilibria. (Compare this with the rolling ball on an oscillating warped surface of Fig. 7.5.3.)

Duffing Oscillator

See how a model of this shaking physical system creates a chaotic phase-plane trajectory.

A mathematical model for the three-equilibrium situation described above is Duffing's equation[4]

$$m\ddot{x} + b\dot{x} - kx + x^3 = F_0 \cos \omega t, \tag{1}$$

a forced "nonlinear spring." Special cases have appeared in Sec. 7.2, Problems 11 and 12, and Sec. 7.3, Problems 8 and 9. To compare Duffing's equation with the linear forced harmonic oscillator,

$$m\ddot{x} + b\dot{x} + kx = F_0 \cos \omega t, \tag{2}$$

[3] See S. H. Strogatz, *Nonlinear Dynamics and Chaos* (Addison Wesley, 1994) or F. C. Moon, *Chaotic Vibrations* (NY: Wiley, 1987).

[4] German engineer Georg Duffing studied mechanical devices and their vibrations. He first published, in 1918, an investigation of the equation that bears his name. But J. Gleick tells us in *Chaos: Making a New Science* (Viking, 1987) that physicists in the 1970s were startled to learn from Joseph Ford of the Georgia Institute of Technology that *chaotic* behavior was associated with this model, long used as a standard example for nonchaotic behavior.

set $m = k = 1$ and compare the homogeneous companion equations of (2) and (1), respectively:

$$\ddot{x} = -b\dot{x} - x, \qquad \text{(linear)} \tag{3}$$

$$\ddot{x} = -b\dot{x} + x(1 - x^2). \qquad \text{(nonlinear)} \tag{4}$$

- In the linear case, equation (3), the term $-x$ on the right-hand side works as a restoring force, contributing a negative acceleration for positive displacement x and a positive acceleration for negative displacement.
- In the nonlinear case, the nonlinear term $+x(1 - x^2)$ in (4) has an opposite effect, repelling away from the origin, so long as x is small. But when $|x| > 1$, this term changes signs, restoring the motion inward (toward either 1 or -1). (See Fig. 7.5.5.)

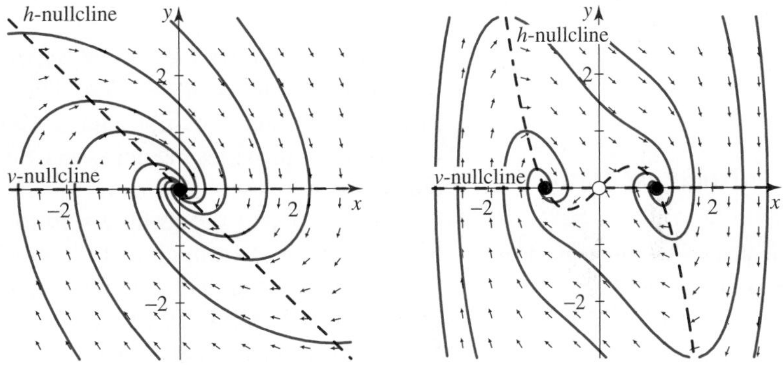

(a) Unforced linear spring:
$\dot{x} = y$
$\dot{y} = -y - x$
one equilibrium

(b) Unforced Duffing's spring:
$\dot{x} = y$
$\dot{y} = -y + x(1 - x^2)$
three equilibria

FIGURE 7.5.5 Comparison of linear and nonlinear springs: different nullclines of horizontal slopes give different equilibria.

When we add the forcing term, results are not so simple. For one thing, because of the explicit t in the equation, we can no longer draw a vector field. (The slopes are constantly changing!)

EXAMPLE 1 **Forced Duffing Oscillator** Returning to equation (1), with $m, k, \omega = 1$ and $b, F_0 = 0.25$,

$$\ddot{x} + 0.25\dot{x} - x + x^3 = 0.25 \cos t, \tag{5}$$

we solve numerically for three very close but different initial conditions, shown in Fig. 7.5.6.

- In the first case, the trajectory goes into a limit cycle around the leftmost equilibrium for the homogeneous case.
- In the second case, it is attracted to a limit cycle around the rightmost equilibrium.
- In the third case, the orbit switches back and forth in a chaotic way.

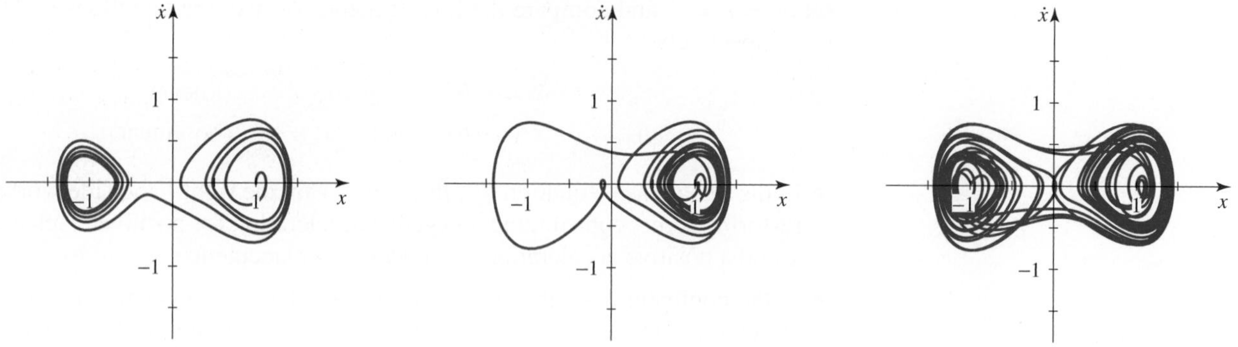

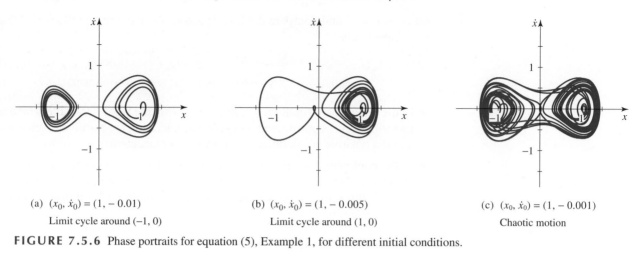

(a) $(x_0, \dot{x}_0) = (1, -0.01)$
Limit cycle around $(-1, 0)$

(b) $(x_0, \dot{x}_0) = (1, -0.005)$
Limit cycle around $(1, 0)$

(c) $(x_0, \dot{x}_0) = (1, -0.001)$
Chaotic motion

FIGURE 7.5.6 Phase portraits for equation (5), Example 1, for different initial conditions.

Example 1 shows the sort of *sensitivity to initial conditions* that we found in the chaotic systems of Sec. 7.4. Those systems were autonomous and required three dimensions to exhibit chaos. Here our system has only two variables, but the explicit t provides the necessary third dimension, and we indeed have another class of systems with deterministic chaotic behavior.

Duffing Oscillator

This interactive illustration provides a quick way to experiment.

Figure 7.5.6(c) shows how solutions of Duffing's equation (5) diverge, even though they start from almost identical initial positions. These solutions appear as random as the path of a fly buzzing around in your kitchen. However, a *strange attractor* is operating in this case, as in the Lorenz and Roessler equations of Sec. 7.4. If time continues to move forward and produces a longer trajectory, can you predict any region(s) that may never be visited by the buzzing fly?

Poincaré Sections and Strange Attractors

Period of the Poincaré Section:

In a forced oscillation, the natural period for a Poincaré section is the period of the forcing term.

A good way to analyze the behavior of a forced dissipative system like Duffing's equation is to forget about the *continuous* motion and examine the solution at a sequence of discrete times. We look at the solution under a strobe light, freezing it at particular instants. The question, of course, is when to flash the light. For equation (5), the light is flashed at multiples of the period of the driving term: We look at times t that are a multiple of 2π; that is, $t = 0, 2\pi, 4\pi, \ldots$, labeled $A, B, C, \ldots$ in Fig. 7.5.7. This discrete representation of the solution is called a **Poincaré section** of the solution.

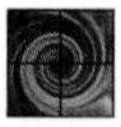

Forced Damped Pendulum: Poincaré Section

This tool demonstrates the Poincaré section for a different equation, but you will see that the results are similar. Watch an orbit evolve and see how the strobe light captures a point at periodic time intervals.

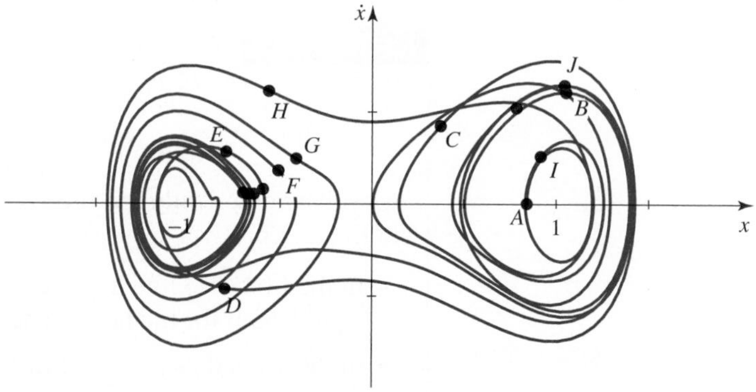

FIGURE 7.5.7 Poincaré section points (in color) for the solution of Duffing's equation (5) with initial condition $(1, -0.001)$.

As more points are added to the Poincaré section, suppressing the portions of the solution curve between these points, a fascinating picture emerges. (See Fig. 7.5.8.) Even more amazing is the fact that this same long-term pattern emerges regardless of what initial point is chosen! Despite the chaotic divergence for solution curves in Fig. 7.5.6, there is an underlying set of *limiting points* that characterize long-term behavior. This set of points is another strange attractor (or chaotic attractor) similar in principle to the Lorenz and Roessler attractors (Sec. 7.4).

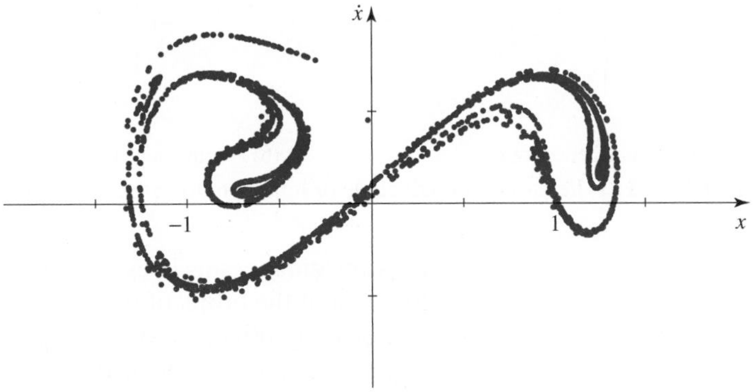

FIGURE 7.5.8 The strange attractor in the Poincaré section of Duffing's equation.

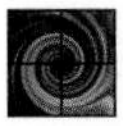

Forced Damped Pendulum: Poincaré Section

Watching how and where the stated points occur on the trajectory can illuminate the point about *different rates at different places*.

You might be curious about the fact that the strange attractor in the Poincaré section lacks the symmetry indicated in the phase portrait of solutions for Duffing's equation. This is a good time to remember that the Poincaré section is marking points at equal *time* intervals, and between each of two successive points in the Poincaré section lies an entire trajectory calculation, which is traversed at different rates in different places.

Unlike a solution curve, a strange attractor in the Poincaré section is not smooth. Nevertheless, its periodic sampling of the solution curve, which reminds us of recording periodic observations of physical or natural systems, reveals an inherent structure that would not otherwise be apparent. In a real sense, the Poincaré section helps us to bring order out of chaos.

How Chaos Comes About: Stretching and Folding in the Phase Plane

We can also use the Poincaré section to show *how* this chaos comes about. Consider an open disk C_0 of initial points surrounding the origin. One iteration over a time interval of 2π stretches this circle in one direction and shrinks it so much in another direction that it appears as just a curve segment C_1. (Imagine stretching out a little rubber-band circle until it appears one-dimensional.) At the same time, the iteration is bending the segment that now appears one-dimensional. (See Fig. 7.5.9(a).) Eventually you see, in further iterations C_2, C_3, C_4, and C_5, that the "segments" tend also to fold back and switch directions, becoming ever more complicated. (See Fig. 7.5.9(b) and (c).)

The *stretching/shrinking* behavior is the result of the *saddle* character of the equilibrium at the origin. The *folding* behavior is caused by the *nonlinearity* in the Duffing equation. These are the essential ingredients for the extreme sensitivity to initial conditions. For instance, you can see from the sequence illustrated in Fig. 7.5.9, on the next page, that two points close on the original circle can be stretched far apart in the first iteration, and probably even farther apart in the next; however, the folding in the third iteration may or may not bring the images of

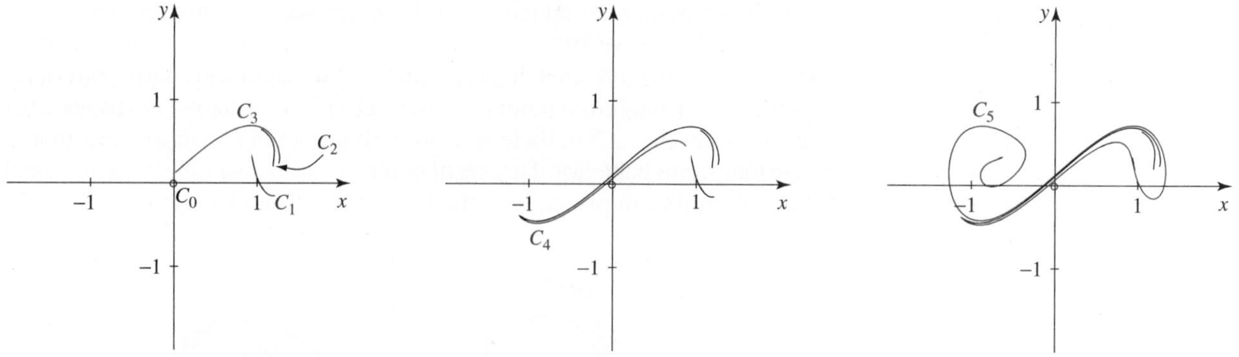

(a) First three iterates of C_0 (b) Fourth iterate of C_0, in color (c) Fifth iterate of C_0, in color

FIGURE 7.5.9 Iterates of a small circle of initial points in the Poincaré section of the forced Duffing equation.

those original points back to being close again. And this is only the beginning of the saga of the orbits of two very close initial conditions.

Forced nonlinear systems, especially when chaos results, need a great deal of exploration and further development of skills of description and asking questions. There are no easy formulas, but we have given some of the ideas that will help you explore such fascinating systems. Other ideas and other systems are introduced in the problems.

Historical Note

The strange attractor takes its name from the fact that in some vague limiting sense it "attracts" the points of the section (and because of its generally unusual appearance). Before the discovery of such sets of points (impossible until the age of the computer), the consensus among mathematicians was that there were but three types of bounded motion exhibited by differential equations:

- equilibrium points or motions converging to them;
- periodic orbits or motions converging to them; and
- almost periodic orbits or motions converging to them.

Strange attractors represent a completely different kind of set in the phase plane in which solutions "live." Since strange attractors were discovered, the solutions of many nonlinear differential equations have been discovered to "move" toward such sets. (See Problems 7–12.)

Mary Cartwright (1900–1998) and **John Littlewood** (1885–1977) were English mathematicians who spent some twenty years working on nonlinear circuit equations like van der Pol's (see Problem 15) under Britain's Department of Science and Industrial Research, which in the late 1930s recognized the importance of nonlinear DEs for radio engineers and scientists. Their collaboration helps set the stage for dynamical systems as a discipline.

Summary

Duffing's oscillator and other forced nonlinear systems show a marked contrast in behavior to related autonomous systems; trajectories exhibit extreme sensitivity to initial conditions and chaotic behavior. On the other hand, in these cases the study of Poincaré sections and their strange attractors gives us ways to find meaning and pattern underlying such chaotic behavior: order from chaos. As with the nonlinear chaotic systems of Sec. 7.4, exploration and qualitative analysis are necessary to begin to understand chaotic forced systems.

7.5 Problems

Please remember that these problems are more open-ended exploratory exercises than the plug-and-chug type you can find with linear systems. They are excellent for group projects, and may lead you to follow other questions that arise during your experimentation. A good habit to develop is the listing of such questions as you work.

1. **Damped Pendulum** We return to an old friend, the nonlinear equation for pendulum motion with damping and forcing,

$$\ddot{\theta} + b\dot{\theta} + k \sin\theta = F \cos\omega t, \tag{6}$$

and set the damping coefficient $b = 0.1$ and $k = 1$.

(a) For the unforced case ($F = 0$), find the equilibria and classify them; then sketch a phase portrait, highlighting the trajectory that starts with $\theta(0) = 3$ and $\dot{\theta}(0) = 1$.

(b) Add forcing by setting $F = 1$ and $\omega = 1$. Use a graphical DE solver to draw the phase-plane trajectory that starts with $\theta(0) = 3$ and $\dot{\theta}(0) = 1$. Explain why it is necessary to calculate this trajectory numerically.

HINT: You should use a small step size, and a good approximation method like Runge-Kutta, to stay as close as possible to the continuous trajectory specified by the equation. Carry your calculation far enough so that you can see the long-term behavior. You probably want to clear or ignore the first hundred or more points in order to see the long-term behavior without the distraction of the transient aspect, or you can plot $\theta(t)$ over a long enough time interval to see the steady state clearly emerge.

Pendulums
Although this tool gives a less exact answer to a slightly different problem, you will get a good feeling for the requested linear/nonlinear comparison if you watch how the trajectories evolve in the different cases provided. The graphs are linked to a model of the pendulum motion.

(c) Compare the phase portraits in (a) and (b). Describe and explain the differences in these trajectories, which start at the same initial conditions.

2. **Linearized Damped Pendulum** Linearize equation (6) from Problem 1 about the nonlinear equilibrium $(2\pi, 0)$ in the "center" of the trajectories, and repeat parts (a)–(c) for the linearized equation.

3. **Solution of the Linearized Pendulum** Solve analytically the unforced and forced linearized pendulum equations from Problem 2, and relate the solution equations to your graphs.

4. **Nonlinear versus Linear Pendulums** Discuss the differences between the nonlinear and linear models of Problems 1 and 2, first for the unforced cases and then for the forced cases.

5. **Chaos Exploration of the Forced Damped Pendulum** Chaos appears in the forced damped pendulum discussed in Problem 1. We will take a look in terms of *sensitivity to initial conditions* in the particular case

$$\ddot{\theta} + 0.1\dot{\theta} + \sin\theta = \cos t.$$

There are many directions in which to continue exploration of this chaotic regime, once you are convinced that whatever is going on is not easy to describe.[5]

(a) Plot a trajectory for $\theta(0) = 3$, $\dot{\theta}(0) = 1$, then a trajectory for $\theta(0) = -1$, $\dot{\theta}(0) = 1$. These trajectories settle on orbits that cycle around *different* phase-plane points. Explain in terms of equilibria in the unforced nonlinear system, and tell what these different settling levels mean in terms of the motion of the pendulum. HINT: The time series for $\theta(t)$ can be very helpful.

Forced Damped Pendulum
A model of the pendulum motion gives a good idea of how the phase portrait and time series graphs evolve. The problem allows a deeper exploration.

(b) To investigate the sensitivity to initial conditions that are close together, keep $\dot{\theta}(0) = 1$ and plot $\theta(t)$ as a time series for different values of $\theta(0)$, such as 1, 1.05, 1.1, 1.15, 1.2, 1.25, 1.3, 1.35, 1.4, 1.45, 1.5, over a long enough interval to feel they have settled to a steady state. Make a list of your initial conditions and the equilibria where their orbits settle, and describe the lack of pattern.

NOTE: Different graphing solvers will probably produce different results, because the sensitivity to initial conditions is that extreme.[6]

---

[5] See J. H. Hubbard, "What It Means to Understand a Differential Equation," *College Mathematics Journal* **25** no. 5 (Nov. 1994). This article discusses the situation posed in this problem, and, furthermore, shows how this chaos can be *controlled*! For an interactive hands-on approach to this particular equation, see the multimedia software package *ODE Architect* by the C-ODE-E Consortium (NY Wiley, 1998), Module 12, and Chapter 12 of the companion book, both titled *Chaos and Control.*

[6] See R. Borrelli and C. Coleman, "Computers, Lies and the Fishing Season," *College Mathematics Journal* (Nov. 1994).

6. **Period Doubling; Poincaré Sections in Forced Damped Pendulum** IDE Lab 26, Parts 1 and 2, examines both Poincaré sections and the period doubling that leads to chaos. Answer the questions in this lab.

Chaos in Forced Nonlinear Oscillators
Lab 26 uses Poincaré sections to see period doubling most easily (when you *clear* transient solutions).

NOTE: The phase plane in this IDE tool is shown only from $-\pi$ to π; larger values of $\theta(t)$ are mapped mod 2π, so what goes off the graph on the right comes back on the left. To see cycles clearly, you will want to "clear" transient solutions after a few cycles and let the orbit continue once it has reached the cycle.

7. **Double-Well Potential** A nonlinear second-order DE

$$\ddot{x} + \dot{x} - x + x^3 = F \cos \omega t$$

governs the motion of a particle under a potential[7]

$$V(x) = 0.25x^4 - 0.5x^2.$$

(a) For the unforced case ($F = 0$), find the equilibria and classify them; then sketch a phase portrait, highlighting the trajectory that starts with $\theta(0) = -1$ and $\dot{\theta}(0) = 1$.

(b) Add forcing by setting $F = 1$ and $\omega = 1$. Plot the phase-plane trajectory starting at $\theta(0) = -1$ and $\dot{\theta}(0) = 1$.

(c) Remove damping by deleting the $\dot{x}$ term, and repeat (a) and (b).

(d) Compare and describe your results. Which orbits appear to be periodic? Which appear to be chaotic? Why? Can you be sure?

8. **Forced Duffing Oscillator, A Route to Chaos** Use a graphical DE solver to compare the unforced ($F = 0$) Duffing equation with

$$\ddot{x} + 0.25\dot{x} - x + x^3 = F \cos t. \qquad (7)$$

(a) For initial conditions in the neighborhood of the origin (say a disk of radius less than 1) for $F = 0$, find and explain how to predict whether the orbit will end at the right or left equilibrium. Make a qualitatively based argument.

(b) Now change F slowly from 0. Some good values to try are $F = 0.001, 0.01, 0.1, 0.2, \ldots$. For what order of magnitude do you begin to see chaotic behavior? Describe what changes first.

9. **Period-Doubling Exploration, Forced Duffing Oscillator** Consider a more general Duffing oscillator:

$$\ddot{x} + b\dot{x} - kx + x^3 = F \cos \omega t,$$
$$0 \le b \le 0.5, \quad b_0 = 0.15;$$
$$-2 \le k \le 2\ , \quad k_0 = 1.00;$$
$$0 \le F \le 0.5, \quad F_0 = 0.30;$$
$$0 \le \omega \le 2\ , \quad \omega_0 = 1.00,$$

Duffing Oscillator
Vary the different parameters to quickly and easily explore this system.

A particularly useful feature of the IDE tools is the ability in chaotic examples to *clear transients*, that is, to erase the orbit and continue from where it had stopped. This ability leaves cycles clearly visible, which highlights the period-doubling phenomenon.

(a) Work IDE Lab 26, Part 3.

(b) Try some other values of (b, k, F, ω) and see if you can find another cycle. Sketch whatever cycles result and label each with its parameter values and period. You might try adjusting parameters to see more period doubling. List the parameter values and periods that result.

(c) Some known combinations[8] of (b, k, F, ω) that will give cycles require an F value outside the range of the IDE Duffing Oscillator tool, but you may be able to experiment with an open-ended graphical DE solver. Set $b = 0.2$, $k = 0$, $\omega = 1$, and try $F = 5$, 16.5, and 23.5. Sketch these cycles and label each with its parameter values and period. Use a ruler to mark peak levels, as was done in Sec. 7.4. Try adjusting parameters to see more period doubling. List the parameter values and periods that result.

Chemical Oscillators *Chemists have made a number of simplified models, even for implausible autocatalytic systems.*[9] *Problems 10 and 11 take a look at two of these and provide good opportunities for a class to pool their results. See also Sec. 7.3, Problem 27.*

10. Tomita and Kai added forcing to simplified chemical kinetic equations that represent the **Brusselator** system

[7] Forced double-well oscillator: P. J. Holmes, "A Nonlinear Oscillator with a Strange Attractor," *Philo. Trans. R. Soc. London A* **292** (1979), 419–448. See also F. C. Moon, *Chaotic Vibrations* (NY Wiley, 1987).

[8] See A. V. Holden, *Chaos* (Princeton University Press, 1986), 28–31.

[9] See A. V. Holden, *Chaos* (Princeton University Press, 1986), 25–33.

proposed by Prigogine and Lefever in 1968:[10]

$$\dot{x} = A + x^2 y - (B + 1)x + F \cos \omega t,$$
$$\dot{y} = Bx - x^2 y.$$

Start with $A = 0.4$, $B = 1.2$, $F = 0.05$, and $\omega = 0.6$, which should settle into a nice cycle. Raise the ω values toward 1.0, at the necessary increments to produce period doubling and eventual chaos. Use a computer to draw these solutions on tx and/or ty graphs, carrying t far enough to see about 10 relative peaks of a steady-state solution. Label each tx or ty graph with its parameter values and period, and use a ruler to mark peak levels (as was done in Sec. 7.4).

11. Tomita and Daido studied the **Glycolytic Oscillator** and added forcing to an earlier model published by Higgins in 1964:[11]

$$\dot{x} = -xy^2 + A + F \cos \omega t,$$
$$\dot{y} = xy^2 - y.$$

(a) Start with $A = 0.999$, $F = 0.42$, and $\omega = 2$, which should settle into a nice cycle. For $\omega = 1.75$, you should see a chaotic orbit; find a value of ω in between that shows period doubling. Sketch these cycles and label each with its parameter values and period. Use a ruler to mark peak levels, as was done in Sec. 7.4.

(b) Now vary ω in the opposite direction, to 3.5 and 4.5, to see examples of *quasiperiodic* orbits, another indication of lurking chaos. Sketch these "cycles" and label each with its parameter values and period, and discuss the differences caused by the different values of ω. Predict what will happen for another value of ω between 3 and 5, plot the trajectory, and comment on the results. Don't be surprised if it's not what you predicted. Such is the nature of chaos experiments!

van der Pol Oscillator

As a precursor to Problem 12, explore the famous circuit with varying resistance in the *unforced* case

$$\ddot{x} - \varepsilon(1 - x^2)\dot{x} + x = 0.$$

Adding *forcing* creates chaotic behavior, as outlined in Problem 12.

12. Forced van der Pol Equation A forced van der Pol equation

$$\ddot{x} - \varepsilon(1 - x^2)\dot{x} + x = F \cos \omega t \tag{8}$$

can also produce interesting experiments. For a start, try the following sets of parameter values.

(a) $\varepsilon = 0.10$, $F = 0.5$, $\omega = 1$

(b) $\varepsilon = 1$, $F = 0.5$, $\omega = 1$

(c) $\varepsilon = 1$, $F = 1$, $\omega = 0.3$

(d) $\varepsilon = 1$, $F = 1$, $\omega = 0.4$

For equation (8), in each case, plot a phase-plane trajectory and a tx time series carrying t far enough to see at least 10 relative peaks of steady-state oscillation. Add to your time-series graph the horizontal lines for the maximum and minimum amplitudes, and describe what happens between them. Is there a pattern to what happens between successive maxima?

Poincaré Sections for Periodic Functions *The functions $x(t)$ in Problems 13 and 14 do not lead to chaotic behavior like we saw in solutions for the Duffing oscillator, but they illuminate other behaviors.*

13. Find the points of an $x\dot{x}$ Poincaré section for the following functions and periods, starting at $t = 0$. In each case, tell **(i)** what difference a different t_0 might make, and **(ii)** what period would give a Poincaré section that is a single point.

(a) $x(t) = \sin t$, period π

(b) $x(t) = \sin t$, period $\pi/4$

(c) $x(t) = \sin 2t$, period π

(d) $x(t) = \sin 2t + \sin t$, period π

14. We add a twist to the previous problem by making one of the frequencies an irrational number. With computer assistance (e.g., a spreadsheet—see Appendix SS for plot procedure), find the $x\dot{x}$ Poincaré section for

$$x(t) = \sin 2\pi t + \sin t$$

by plotting the first 1,000 points of the sequence $\{(x(t), \dot{x}(t)) \mid t = 0, 2\pi, 4\pi, \ldots\}$. The resulting set of points lies on a *closed curve* in the $x\dot{x}$-plane. This tells us that $x(t)$ is *quasiperiodic* with two incommensurate frequencies present, and does not display chaotic behavior.

[10] For the Brusselator, see I. Prigogine and R. Lefever, "Symmetry Breaking Instabilities in Dissipative Systems," *Journal of Chemistry and Physics* (1968), 1695–1700; K. Tomita and T. Kai, "Stroboscopic Phase Portrait and Strange Attractors," *Physical Letters* **66A** (1978), 91–93.

[11] For the glycolytic oscillator, see J. Higgins, "A Chemical Mechanism for Oscillation of Glycolytic Intermediates in Yeast Cells," *Proceedings of the National Academy of Sciences* (USA) **51** (1964), 989–994; K. Tomita and H. Daido, "Possibility of Chaotic Behavior and Multi-basins in Forced Glycolytic Oscillator," *Physical Letters* **79A** (1980), 133–137.

Stagecoach Wheels and the Poincaré Section *The Poincaré section can often determine the natural frequency of a system. Consider the simple dynamical system consisting of the point moving clockwise on the unit circle with frequency ω_0, as in Fig. 7.5.10.*

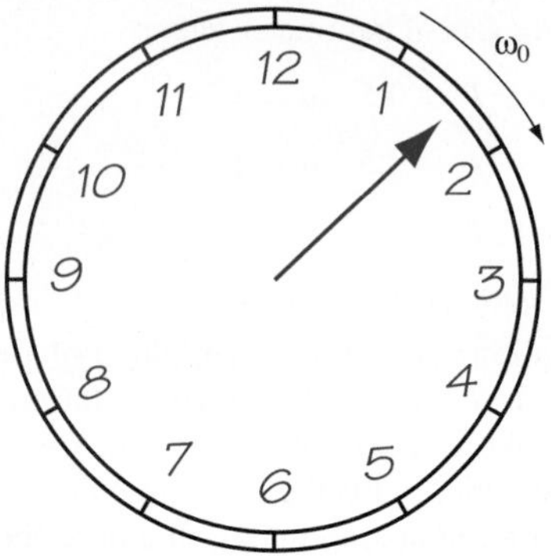

FIGURE 7.5.10 Clockface dynamical system.

Visualize this as the second hand of a clock. Instead of observing the clock continuously, view it under a strobe light flashing with frequency ω_s. When the frequency $\omega_s = \omega_0$, the strobe light flashes "in synch" with the motion, so the clock hand will appear motionless. If $\omega_s = (4/3)\omega_0$, the strobe light will flash four times while the second hand rotates three times; the strobed hand will appear to move counterclockwise in 15-second jumps, as shown in Fig. 7.5.11.[12] *Problems 15–18 investigate the relationships between the frequencies of clock and strobe.*

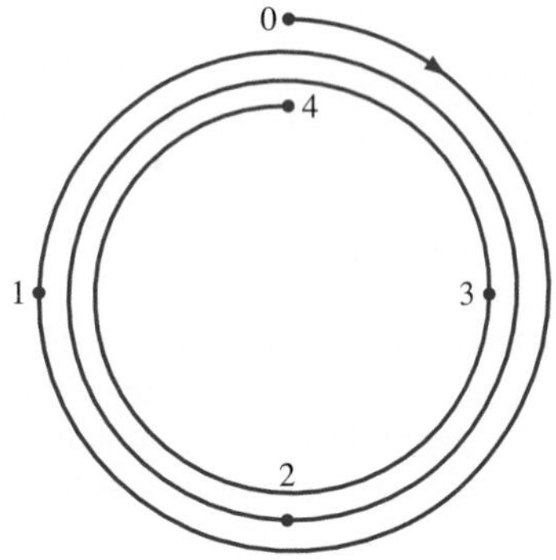

FIGURE 7.5.11 Schematic sketch of strobed motion (numbered dots) on clockhand rotations (spiral path) for

$$\omega_s = \frac{4}{3}\omega_0.$$

15. How will the hand of the clock seem to move when the relationship between natural frequency ω_0 and strobe frequency ω_s is as follows?

(a) $\omega_s = \frac{1}{2}\omega_0$ (b) $\omega_s = \frac{1}{4}\omega_0$

(c) $\omega_s = 2\omega_0$ (d) $\omega_s = 4\omega_0$

(e) $\omega_s = \frac{2}{3}\omega_0$ (f) $\omega_s = \frac{3}{4}\omega_0$

16. Try to generalize what you found in Problem 13 by determining the apparent motion ("strobed motion") of the hand of the clock for $\omega_s = (p/q)\omega_0$, where p/q is a rational number in lowest terms. HINT: Look at some specific fractions, but try not to look at Problem 17.

17. You should have found in answering the previous problem that the Poincaré section will consist of p equally spaced points on the circle. The order will be such that the strobe first catches the qth of these p points, starting at the top and moving clockwise. The strobe continues to skip $q-1$ clockwise positions with each flash. For the following values of p/q, determine the Poincaré section and tell whether the strobed motion appears to be clockwise, counterclockwise, or erratic:

(a) $\frac{5}{2}$ (b) $\frac{5}{3}$ (c) $\frac{8}{3}$ (d) $\frac{12}{11}$ (e) $\frac{100}{101}$

18. Consider the power of the Poincaré section as follows: by observing the strobed points, it is possible to determine the natural frequency of the system. For each set of conditions below, find ω_0 in terms of ω_s.

(a) The strobed motion has period 2 and moves in 30-second jumps.

(b) The strobed motion has period 4 and moves in 45-second jumps.

(c) The strobed motion has period 12 and moves in 25-second jumps.

19. Suggested Journal Entry Discuss the differences between a system that exhibits chaotic behavior and one whose behavior is purely random. Is there a difference?

[12]In old Western movies, stagecoach wheels often seemed to rotate backwards, which is explained by Fig. 7.5.11.

CHAPTER 8 Laplace Transforms

Such is the advantage of a well constructed language that its simplified notation often becomes the source of profound theories.

—Pierre Simon de Laplace

8.1 The Laplace Transform and Its Inverse

SYNOPSIS: We study the Laplace transform, a linear transformation between function spaces, as a bypass strategy for changing linear differential equations and systems into algebraic problems that are easier to solve.

A Bypass Strategy

The strategy of transforming a hard problem into an equivalent but more tractable one is common in mathematics (and other areas).[1] In this chapter, we discuss the solution strategy of the Laplace transform $\mathcal{L} : \mathbb{V} \to \mathbb{W}$, a linear transformation between vector spaces of functions. We will get to a formal definition after a bit of discussion, but Fig. 8.1.1 illustrates the general idea.

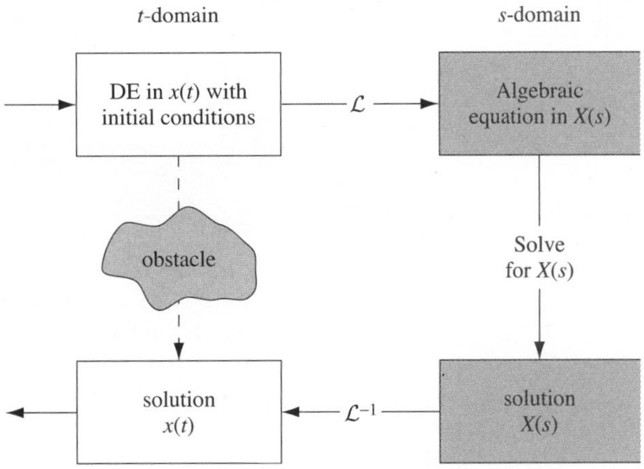

FIGURE 8.1.1 A bypass strategy using the Laplace transform.

Laplace: Solver

Try out this strategy with some easy forcing function.

[1]For more on this way of analyzing problem-solving strategies, see Z. A. Melzak, *Bypasses* (NY: John Wiley, 1983).

We concentrate in this section on the transform step for elementary functions. In Sec. 8.2, we will use the Laplace transform to reformulate differential equations as algebraic equations.[2] After solving the algebraic equations, we will transform from $\mathbb{W}$ back to $\mathbb{V}$ using the inverse transform $\mathcal{L}^{-1}$. In later sections we will show how the Laplace transform can be applied to nonelementary functions. Step functions and delta functions (Sec. 8.3) are particularly useful examples.

Definition and Properties

Laplace Transform

The **Laplace transform** $\mathcal{L}\{f(t)\}$ of a *suitable* function $f(t)$ defined on $[0, \infty)$ is the function $F(s)$ given by

$$\mathcal{L}\{f(t)\} = F(s) \equiv \int_0^\infty e^{-st} f(t)\,dt = \lim_{b\to\infty} \int_0^b e^{-st} f(t)\,dt, \qquad (1)$$

where s may be complex.[3]

If this limit fails to exist, then $f(t)$ is "unsuitable" for the Laplace transform. Functions $f(t)$ in the domain of $\mathcal{L}$ are those for which the integral exists. A visual interpretation is shown in Fig. 8.1.2 for $f(t) = \sin t$.

Laplace: Definition

Visualize the workings of the definition! A slider lets you set the parameter s and a rollover leads you past the intricacies of an improper integral.

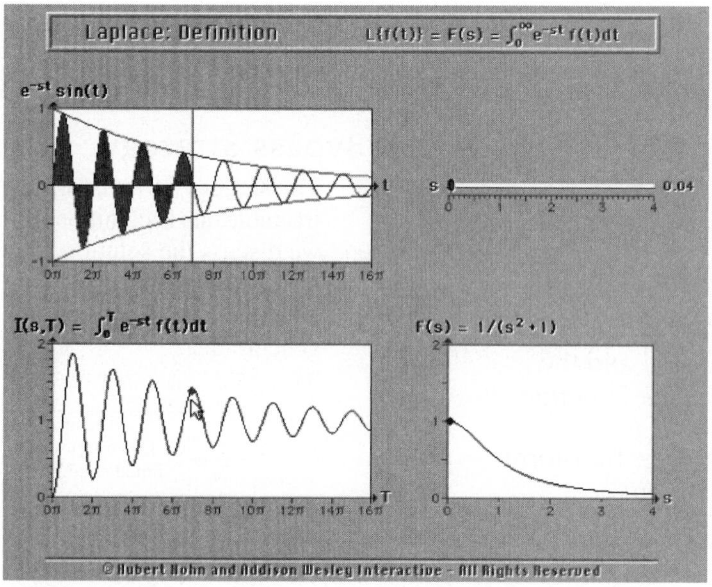

FIGURE 8.1.2 The Laplace transform is the limit of an integral that calculates the net area trapped between $e^{-st} f(t)$ and the horizontal axis. The peaks of the integral (lower left) can be seen to occur where $e^{-st} f(t)$ (upper left) crosses the axis. Because of the negative exponential factor and certain properties of $f(t)$, this limit exists (lower right).

---

[2]See the Historical Note at the end of Sec. 8.2.

[3]As in most textbooks at this level, we usually treat s as a real variable; however, Example 4 will illustrate the power of allowing s to take on complex values. In IDE Lab 14, the *Laplace: Vibrations and Poles* tool shows the need for the s-domain to be a region in the complex plane. The *poles* of a Laplace transform are the zeroes of its denominator, the values of s where $F(s)$ becomes infinite.

It follows from properties of integrals that $\mathcal{L}$ is a *linear* transformation. (See Problem 19.)

Linearity of the Laplace Transform
If $F(s) = \mathcal{L}\{f(t)\}$ and $G(s) = \mathcal{L}\{g(t)\}$, then

$$\mathcal{L}\{af(t) + bg(t)\} = aF(s) + bG(s), \quad \text{for } a, b \in \mathbb{R} \text{ (or } \mathbb{C}). \tag{2}$$

EXAMPLE 1 **Simplest Transform** For the constant function $f(t) \equiv 1$, for $t \geq 0$, we calculate

$$\mathcal{L}\{f(t)\} = \mathcal{L}\{1\} = \int_0^{\infty} e^{-st} \cdot 1\, dt = \lim_{b\to\infty} \int_0^b e^{-st}\, dt$$

$$= \lim_{b\to\infty} \left[-\frac{e^{-st}}{s}\right]_0^b = \lim_{b\to\infty} \left[-\frac{e^{-sb}}{s} + \frac{1}{s}\right].$$

If $s > 0$, then $e^{-sb} = 1/e^{sb} \to 0$ as $b \to \infty$, so

$$\mathcal{L}\{1\} = F(s) = \frac{1}{s}, \quad s > 0. \tag{3}$$

Thus we have our first "transform pair": $f(t) = 1$, $F(s) = 1/s$, for $s > 0$. (See Fig. 8.1.3.)

Laplace: Transformer

Find these graphs and transform pairs in an interactive lookup.

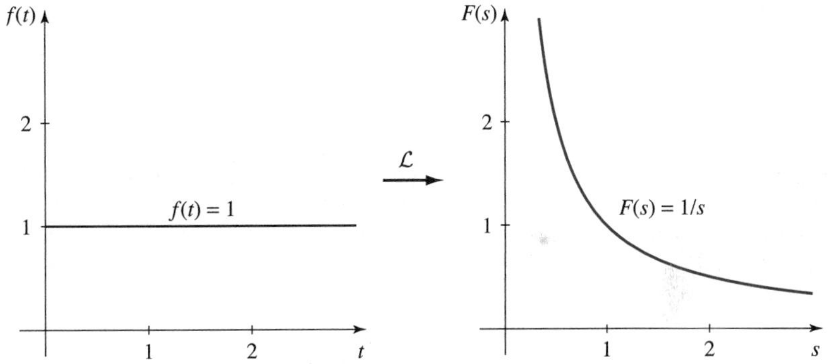

FIGURE 8.1.3 Transform pair $f(t) = 1$, $F(s) = 1/s$ for Example 1. ■

The Laplace transform always exists for an important class of functions $f(t)$:

piecewise continuous[4] *functions on* $[0, \infty)$ *of exponential order* α.

Exponential order means that $f(t)$ grows no faster than an exponential function. The function $f(t) = 1$, used in Example 1, and those $f(t)$ given in Examples 2–4 to follow are of exponential order. The function $x = e^{t^2}$ is *not* of exponential order because t^2 outgrows kt for any constant k.

[4] A function is *piecewise continuous* on an interval $[a, b]$ if there exist a finite number of points in the interval so that the function is continuous on each of the resulting open subintervals and has (finite) one-sided limits at each end of each subinterval. A function is *piecewise continuous* on $[0, \infty)$ if it is piecewise continuous on $[0, b)$ for each $0 < b < \infty$.

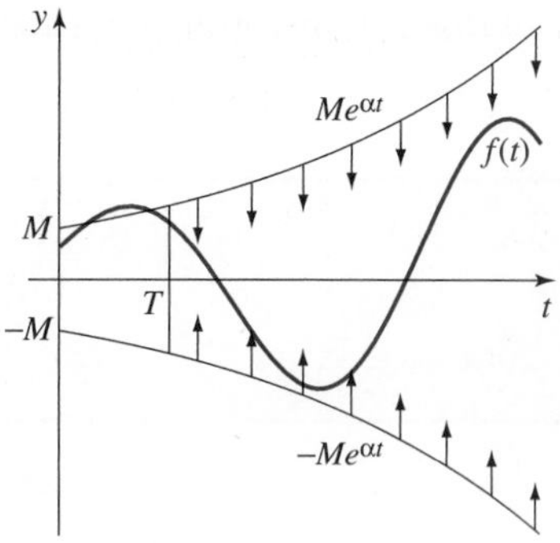

FIGURE 8.1.4 Function $f(t)$ of exponential order α, estimating a value of T.

Specifically, $f(t)$ is of exponential order α for real number α if there exist positive constants T and M such that

$$|f(t)| \leq Me^{\alpha t}, \quad t \geq T. \tag{4}$$

This property is illustrated in Fig. 8.1.4.

These ideas give us the following theorem, which says that f must be "dampable" by some exponential function in order for the Laplace transform to exist.[5]

> **Existence Theorem for Laplace Transforms**
> If $f(t)$ is piecewise continuous on $[0, \infty)$ and of exponential order α, then the Laplace transform $F(s) = \mathcal{L}\{f(t)\}$ exists for $s > \alpha$.

EXAMPLE 2 **Transform of the Exponential** For $f(t) = e^{at}$, a any real constant, we compute

$$\begin{aligned}\mathcal{L}\{f(t)\} = \mathcal{L}\{e^{at}\} &= \int_0^\infty e^{-st}e^{at}\,dt \\ &= \lim_{b\to\infty}\int_0^b e^{-(s-a)t}\,dt \\ &= \lim_{b\to\infty}\left[-\frac{e^{-(s-a)b}}{s-a} + \frac{1}{s-a}\right],\end{aligned}$$

so

$$\mathcal{L}\{e^{at}\} = F(s) = \frac{1}{s-a}, \quad s > a. \tag{5}$$

Thus we have our second transform pair: $f(t) = e^{at}$, $F(s) = 1/(s-a)$. (See Fig. 8.1.5.)

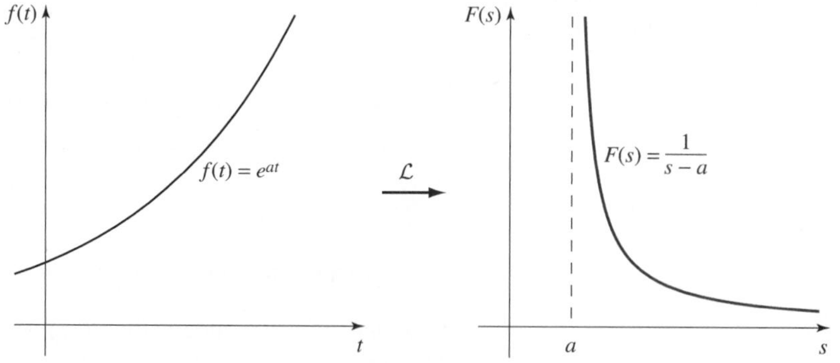

FIGURE 8.1.5 Transform pair $f(t) = e^{at}$, $F(s) = 1/(s-a)$ for Example 2. ■

Laplace: Transformer

You can find this graph and transform pair, as well as many others.

[5] For a proof of the Existence Theorem, see M. Braun, *Differential Equations and Their Applications*, 3rd ed. (NY: Springer-Verlag, 1983). An equivalent discussion is given in R. Borrelli and C. Coleman, *Differential Equations: A Modeling Perspective* (NY: Wiley, 1998), 301–302 and 305–306.

EXAMPLE 3 **Exploiting Linearity** Using (3) and (5) from Examples 1 and 2, together with the linearity property (2), we can calculate the Laplace transform of $f(t) = 5 - 3e^{-2t}$ as follows:

$$\mathcal{L}\{f(t)\} = \mathcal{L}\{5 - 3e^{-2t}\} = 5\mathcal{L}\{1\} - 3\mathcal{L}\{e^{-2t}\},$$

so

$$F(s) = 5\left(\frac{1}{s}\right) - 3\left[\frac{1}{s - (-2)}\right] = \frac{2s + 10}{s(s + 2)}.$$

■

EXAMPLE 4 **Laplace Transform of Sine and Cosine** The definition (1) of a Laplace transform can be used to determine $\mathcal{L}\{\cos kt\}$ and $\mathcal{L}\{\sin kt\}$ directly; however, the integrals require a double integration by parts. We can use linearity to simplify the process if we extend (2) to complex constants.

From Euler's formula and the linearity property we know that

$$\mathcal{L}\{e^{ikt}\} = \mathcal{L}\{\cos kt + i \sin kt\} = \mathcal{L}\{\cos kt\} + i\mathcal{L}\{\sin kt\}. \tag{6}$$

From Example 2, equation (5), we also know that, for $s > \text{Re}(ik) = 0$,

$$\begin{aligned} \mathcal{L}\{e^{ikt}\} &= \frac{1}{s - ik} \\ &= \frac{1}{s - ik} \cdot \frac{s + ik}{s + ik} \\ &= \frac{s + ik}{s^2 - (i^2k^2)} \end{aligned}$$

$$\mathcal{L}\{e^{ikt}\} = \frac{s}{s^2 + k^2} + i\frac{k}{s^2 + k^2}. \tag{7}$$

By equating the real and imaginary parts of (6) and (7), we obtain:

$$\mathcal{L}\{\cos kt\} = \frac{s}{s^2 + k^2} \quad \text{and} \quad \mathcal{L}\{\sin kt\} = \frac{k}{s^2 + k^2}, \quad s > 0.$$

■

Inverse Transform

Inverse Laplace Transform

A function $f(t)$ whose transform is $F(s)$ is called the **inverse Laplace transform** of F, and we write $f(t) = \mathcal{L}^{-1}\{F(s)\}$.

EXAMPLE 5 **Working Backward** Of what function $f(t)$ is

$$F(s) = \frac{2s - 14}{(s + 1)(s - 3)}$$

the Laplace transform? By the theory of partial fractions,[6]

$$F(s) = \frac{4}{s + 1} - \frac{2}{s - 3} = 4 \cdot \frac{1}{s + 1} - 2 \cdot \frac{1}{s - 3}.$$

---

[6]Example 5 uses the partial fraction decomposition of $(2s - 14)/(s + 1)(s - 3)$. If you aren't familiar with this idea or the procedure for obtaining the decomposition, see Appendix PF.

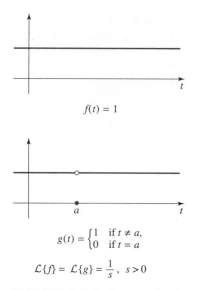

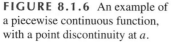

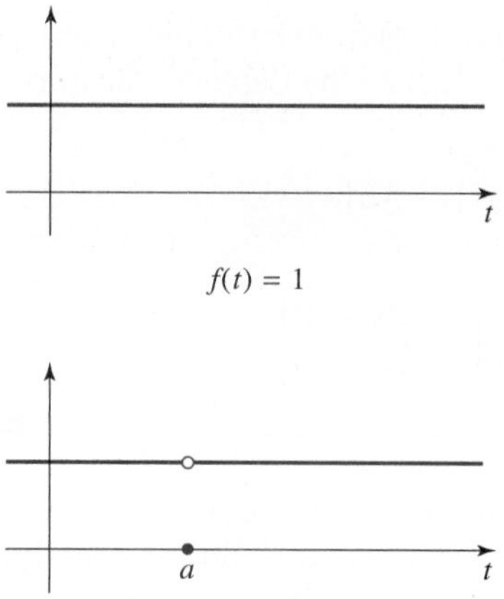

$f(t) = 1$

$g(t) = \begin{cases} 1 & \text{if } t \neq a, \\ 0 & \text{if } t = a \end{cases}$

$\mathcal{L}\{f\} = \mathcal{L}\{g\} = \frac{1}{s}, \ s > 0$

FIGURE 8.1.6 An example of a piecewise continuous function, with a point discontinuity at a.

Then, by linearity and (5),

$$F(s) = 4\mathcal{L}\{e^{-t}\} - 2\mathcal{L}\{e^{3t}\} = \mathcal{L}\{4e^{-t} - 2e^{3t}\},$$

so $f(t) = 4e^{-t} - 2e^{3t}$. ■

You may wonder how we know that there *is* a function $f(t)$ having such a transform $F(s)$, and whether there is only one such object. With certain exceptions involving how piecewise continuous functions are defined at their jump points, it is true that $\mathcal{L}$ is an injective or one-to-one transformation.[7]

Since Laplace transforms involve *integrals*, point discontinuities do not affect their values. For inverse Laplace transforms we simply choose the continuous version. (See Fig. 8.1.6.)

Turning the Tables

Practical application of the Laplace transform depends on having a substantial store of "transform pairs" (like those we calculated in Examples 1, 2, and 4) for transforming and then transforming back. Table 8.1.1 will be adequate for our needs; a longer version is inside the back cover, and more extensive collections are available in various reference books. Problems 21, 22, 24, and 31 provide verification of entries in the table that are not covered by the preceding examples.

Of course, a table of transform pairs is also a table of inverse transform pairs. In using Table 8.1.1 to implement the return trip "$\mathcal{L}^{-1}$," we often need a little

Laplace: Transformer

These transform pairs are available at a click of the cursor.

Table 8.1.1 Short table of Laplace transforms

$f(t)$	$F(s) = \mathcal{L}\{f(t)\}$	
(i) 1	$\frac{1}{s}$	$s > 0$
(ii) t^n	$\frac{n!}{s^{n+1}}$	$s > 0$, n a positive integer
(iii) e^{at}	$\frac{1}{s-a}$	$s > a$
(iv) $t^n e^{at}$	$\frac{n!}{(s-a)^{n+1}}$	$s > a$, n a positive integer
(v) $\sin bt$	$\frac{b}{s^2+b^2}$	$s > 0$
(vi) $\cos bt$	$\frac{s}{s^2+b^2}$	$s > 0$
(vii) $e^{at}\sin bt$	$\frac{b}{(s-a)^2+b^2}$	$s > a$
(viii) $e^{at}\cos bt$	$\frac{s-a}{(s-a)^2+b^2}$	$s > a$
(ix) $\sinh bt$	$\frac{b}{s^2-b^2}$	$s > \lvert b\rvert$
(x) $\cosh bt$	$\frac{s}{s^2-b^2}$	$s > \lvert b\rvert$

[7] It is a little more ticklish to characterize the range of $\mathcal{L}$ (that is, the domain of $\mathcal{L}^{-1}$), and for such matters the reader will need to consult more advanced texts. In particular, a formula for $\mathcal{L}^{-1}$ in terms of integrals can be written, but it involves functions of a complex variable, even for real results.

algebraic resourcefulness, as subsequent examples will illustrate, including completing the square, partial fractions, and either adding and subtracting the same quantity, or multiplying and dividing by the same quantity instead.

EXAMPLE 6 **Inverse Transform** We use partial fraction decomposition and Table 8.1.1 to find the inverse Laplace transform of

$$F(s) = \frac{2s^2 - s + 11}{(s-1)(s^2+5)} \quad \Rightarrow \quad F(s) = \frac{2}{s-1} - \frac{1}{s^2+5}.$$

We adapt the first term to entry (iii) and the second term to entry (v) with $b = \sqrt{5}$. By writing F in the form

$$F(s) = 2\underbrace{\left(\frac{1}{s-1}\right)}_{\mathcal{L}\{e^t\}} - \frac{1}{\sqrt{5}}\underbrace{\left[\frac{\sqrt{5}}{s^2+(\sqrt{5})^2}\right]}_{\mathcal{L}\{\sin(\sqrt{5}t)\}},$$

we obtain

$$f(t) = \mathcal{L}^{-1}\{F(s)\} = 2e^t - \frac{1}{\sqrt{5}}\sin(\sqrt{5}\,t).$$

■

EXAMPLE 7 **Playing Square** Imagine that

$$F(s) = \frac{s+1}{s^2+4s+13}$$

is the Laplace transform of a function $f(t)$. To determine $f(t)$, we need to make some preliminary rearrangements:

$$\begin{aligned} F(s) = \frac{s+1}{s^2+4s+4+9} &= \frac{(s+2)-1}{(s+2)^2+3^2} \\ &= \frac{s+2}{(s+2)^2+3^2} - \frac{1}{3}\left[\frac{3}{(s+2)^2+3^2}\right] \\ &= \mathcal{L}\{e^{-2t}\cos 3t\} - \frac{1}{3}\mathcal{L}\{e^{-2t}\sin 3t\}. \end{aligned}$$

Therefore, using Table 8.1.1, entries (vii) and (viii),

$$f(t) = e^{-2t}\cos 3t - \frac{1}{3}e^{-2t}\sin 3t. \tag{8}$$

Forcing functions $f(t)$ of type (8) are tedious to handle with undetermined coefficients or variation of parameters.

■

Summary

We defined a Laplace transform of a function $f(t)$ as an improper integral of $e^{-st}f(t)$ from 0 to ∞, and found that it was a linear operator. The Laplace transform is guaranteed to exist for piecewise continuous functions that are of exponential order. We determined the Laplace transforms for some elementary functions and constructed a table of transform pairs (the transform and its inverse). We will find this table to be very useful in the solution of linear initial-value problems.

8.1 Problems

Transforms from the Definition *Use the integral definition to calculate the Laplace transform of the function in each of Problems 1–10.*

1. $f(t) = 5$
2. $f(t) = t$
3. $f(t) = e^{2t}$
4. $f(t) = e^{-t}$
5. $f(t) = \sin 2t$
6. $f(t) = \cos 3t$
7. $f(t) = t^2$
8. $f(t) = \begin{cases} 1 & \text{if } 0 \le t < 4, \\ 0 & \text{if } t \ge 4 \end{cases}$
9. $f(t) = \begin{cases} t-1 & \text{if } 0 \le t < 2, \\ 1 & \text{if } t \ge 2 \end{cases}$
10. $f(t) = \begin{cases} 1-t^2 & \text{if } 0 \le t < 1, \\ 0 & \text{if } t \ge 1 \end{cases}$

Transforms with Tools *Use linearity and Table 8.1.1 to determine the Laplace transform of the function in each of Problems 11–18.*

11. $f(t) = a + bt + ct^2$
12. $f(t) = 1 + e^{-t}$
13. $f(t) = e^{2t} + e^{-2t}$
14. $f(t) = 3 + t + e^{-t}\sin 2t$
15. $f(t) = (t + 3t^2)e^{-t}$
16. $f(t) = t^3 e^{-3t} + 4e^{-t}\cos 3t$
17. $f(t) = 2e^{at} - e^{-at}$
18. $f(t) = te^{-3t} + 2\sin t$

19. **Linearity of the Laplace Transform** Prove that the Laplace transform $\mathcal{L}$ is a linear transform by showing that, for all f and g whose transforms exist,

$$\mathcal{L}\{f+g\} = \mathcal{L}\{f\} + \mathcal{L}\{g\},$$

$$\mathcal{L}\{cf\} = c\mathcal{L}\{f\},$$

where $c \in \mathbb{C}$ is an arbitrary constant.

20. **Is There a Product Rule?** We have seen that the Laplace transform of a sum is the sum of the transforms. Is there a product rule that says $\mathcal{L}\{fg\}$ is equal to $\mathcal{L}\{f\}\mathcal{L}\{g\}$? Show that the answer is no. HINT: Try some examples. Section 8.4 will provide later enlightenment.

21. **Laplace Transform of Damped Sine and Cosine Functions** To verify entries (vii) and (viii) of Table 8.1.1, consider the identity

$$e^{(a+ik)t} = e^{at}e^{ikt} = e^{at}(\cos kt + i\sin kt).$$

(a) Show that

$$\mathcal{L}\{e^{(a+ik)t}\} = \frac{1}{s-(a+ik)} = \frac{s-a}{(s-a)^2+k^2} + i\frac{k}{(s-a)^2+k^2}.$$

(b) Use the result from (a) and the fact that the Laplace transform is a linear transformation to show that

$$\mathcal{L}\{e^{at}\cos kt\} = \frac{s-a}{(s-a)^2+k^2}$$

and

$$\mathcal{L}\{e^{at}\sin kt\} = \frac{k}{(s-a)^2+k^2}.$$

(See Example 4 for a similar proof.)

22. **Laplace Transforms for Hyperbolic Functions** Use the linearity property of the Laplace transform and the fact that $\mathcal{L}\{e^{at}\} = 1/(s-a)$ for $s > a$ to verify entries (ix) and (x) of Table 8.1.1. HINT: Use the definitions of the hyperbolic trigonometric functions:

$$\sinh t \equiv \frac{e^t - e^{-t}}{2} \quad \text{and} \quad \cosh t \equiv \frac{e^t + e^{-t}}{2}.$$

23. **Using Hyperbolic Functions** Use the linearity property and Table 8.1.1 to determine:

(a) $\mathcal{L}\{\cosh^2 bt - \sinh^2 bt\}$ for $s > |b|$;

(b) $\mathcal{L}\{t^2\cosh bt\}$ for $s > |b|$.

24. **Power Rule** Verify entry (ii) of Table 8.1.1, as follows.

(a) Show that

$$\mathcal{L}\{t^n\} = \frac{n}{s}\mathcal{L}\{t^{n-1}\} \quad \text{for } n \ge 2$$

using integration by parts.

(b) Use part (a) to give a proof by induction that

$$\mathcal{L}\{t^n\} = \frac{n!}{s^{n+1}}, \quad n = 1, 2, \ldots.$$

25. **Multiplier Rule** Assume that f is a function such that

$$\frac{d}{ds}\int_0^\infty e^{-st}f(t)\,dt = \int_0^\infty \frac{d}{ds}(e^{-st})f(t)\,dt.$$

Use this assumption to show that if $\mathcal{L}\{f(t)\} = F(s)$, then

$$\mathcal{L}\{tf(t)\} = -\frac{dF(s)}{ds}.$$

Multiplier Applications *Use the multiplier rule of Problem 25 to compute the Laplace transform of the function given in Problems 26–30.*

26. te^{at} **27.** $t \sin 3t$ **28.** $t \cosh bt$

29. $3t \cos at$ **30.** $-2t \sinh 2t$

31. Exponential Shift Show that if $\mathcal{L}\{f(t)\} = F(s)$ for $s > 0$, then

$$\mathcal{L}\{e^{at} f(t)\} = F(s-a), \quad s > a.$$

Using the Shift *Use the exponential shift property of Problem 31 to calculate the Laplace transform of the functions given in Problems 32–37.*

32. $t^n e^{at}$ **33.** $e^t \sin 2t$

34. $e^{-t} \cos 3t$ **35.** $e^{2t} \cosh 3t$

36. $e^{-3t} \sinh t$ **37.** $te^{2t} \sin 3t$

38. Linearity of the Inverse Show that the inverse Laplace transform is linear; that is, for functions $F(s)$ and $G(s)$ in the range of $\mathcal{L}$ and real constants a and b,

$$\mathcal{L}^{-1}\{aF(s) + bG(s)\} = a\mathcal{L}^{-1}\{F(s)\} + b\mathcal{L}^{-1}\{G(s)\}.$$

39. Out of Order Show that the function $f(t) = e^{t^2}$ is not of exponential order α for any real α.

Inverse Transforms *Determine the inverse Laplace transform $f(t)$ of the function $F(s)$ in each of Problems 40–54.*

40. $\dfrac{1}{s^3}$ **41.** $\dfrac{2}{s} + \dfrac{3}{s-1} + \dfrac{7}{s^3}$

42. $\dfrac{5}{s^2+3}$ **43.** $\dfrac{3}{s-3} + \dfrac{4}{s+3}$

44. $\dfrac{1}{s^2+3s}$ **45.** $\dfrac{s+1}{s^2+2s+10}$

46. $\dfrac{1}{s^2+4s+4}$ **47.** $\dfrac{3s+5}{s^2-6s+25}$

48. $\dfrac{s+1}{s^2+s-2}$ **49.** $\dfrac{5}{s^2+s-6}$

50. $\dfrac{2s+4}{s^2-1}$ **51.** $\dfrac{7}{s^2+4s+7}$

52. $\dfrac{2s+16}{s^2+4s+13}$ **53.** $\dfrac{4}{s^2(s^2+4)}$

54. $\dfrac{3}{(s^2+1)(s^2+4)}$

55. Computer Exploration Work IDE Lab 14, Parts 1.1–1.3 to get a head start on Sec. 8.2.

Laplace: Definition;
Laplace: Transformer
Explore visually the process of Laplace transformation.

56. Suggested Journal Entry An interesting property of a Laplace transform is that it is "one-sided." What happens when a Laplace transform is taken of a function such as $\sin t$, for which the natural domain is $(-\infty, \infty)$? Do we just ignore the negative portion of the curve? What might be the practical justification for such a behavior?

8.2 Solving DEs and IVPs with Laplace Transforms

SYNOPSIS: We show how the Laplace transform converts a linear differential equation to an algebraic equation, including the initial conditions, that can be solved easily. Then the algebraic solution can be transformed back to give the DE solution.

We extend our catalog of Laplace transforms to include derivatives and multiples of $f(t)$. Our final example introduces Bessel functions.

Introduction

We have solved differential equations using a variety of techniques: the integrating factor method, the Euler-Lagrange two-part method, change of variables, trial-and-error methods, variation of parameters, and so on. We now show how the Laplace transform can be used to solve an initial-value problem,

$$ay'' + by' + cy = f(t), \qquad y(0) = y_0, \quad y'(0) = y'_0, \tag{1}$$

for linear differential equations with constant coefficients. The Laplace transform is a linear transform, so we can transform each term of the equation, getting

$$a\mathcal{L}\{y''\} + b\mathcal{L}\{y'\} + c\mathcal{L}\{y\} = \mathcal{L}\{f\}, \tag{2}$$

which will involve the initial conditions, as we shall explain.

Transform of the Derivative

The basis for applying Laplace transforms to problems in differential equations is the way that the transform relates to derivatives. Assuming that both $f(t)$ and $f'(t)$ have transforms and that $f(t)$ has exponential order α, we derive formulas for the transforms of derivatives, as follows. By the definition of the Laplace transform,

$$\mathcal{L}\{f'(t)\} = \int_0^\infty e^{-st} f'(t)\,dt = \lim_{b\to\infty} \int_0^b e^{-st} f'(t)\,dt. \tag{3}$$

Integration by Parts:

$$\int u\,dv = uv - \int v\,du$$

Integrating the last integral by parts gives

$$\int_0^b \underbrace{e^{-st}}_{u}\,\underbrace{f'(t)\,dt}_{dv} = \left[\underbrace{e^{-st}}_{u}\,\underbrace{f(t)}_{v}\right]_0^b - \int_0^b \underbrace{f(t)}_{v}\left[\underbrace{-se^{-st}dt}_{du}\right],$$

$$= e^{-sb} f(b) - f(0) + s\int_0^b e^{-st} f(t)\,dt. \tag{4}$$

Taking the limit as $b \to \infty$, (4) becomes

$$\mathcal{L}\{f'(t)\} = \lim_{b\to\infty} e^{-sb} f(b) - f(0) + s\mathcal{L}\{f(t)\}. \tag{5}$$

Differentiation in the t-domain looks a lot like multiplication in the s-domain. For example, when $f(0) = 0$, $\mathcal{L}\{f'\} = s\mathcal{L}\{f\}$.

Because f is of exponential order α, we know that $|f(b)| \le Me^{\alpha b}$. It can be shown that $\lim_{b\to\infty} e^{-sb} f(b) = 0$, and (5) becomes

$$\mathcal{L}\{f'(t)\} = s\mathcal{L}\{f(t)\} - f(0). \tag{6}$$

We can easily calculate $\mathcal{L}\{f''(t)\}$ using (6) by treating $f'(t)$ as the function and $f''(t)$ as its derivative. We get

$$\mathcal{L}\{f''(t)\} = s\mathcal{L}\{f'(t)\} - f'(0). \tag{7}$$

Substituting (6) into (7) gives

$$\begin{aligned}\mathcal{L}\{f''(t)\} &= s[s\mathcal{L}\{f(t)\} - f(0)] - f'(0)\\ &= s^2\mathcal{L}\{f(t)\} - sf(0) - f'(0).\end{aligned}$$

This calculation can be iterated to produce the following general result.

Laplace: Derivative

See the derivative theorem in action.

Derivative Theorem for Laplace Transforms

If $f, f', \dots, f^{(n-1)}$ are continuous on $[0, \infty)$ and $f^{(n)}$ is piecewise continuous on $[0, \infty)$, and if $f, f', \dots, f^{(n)}$ are of exponential order α, then for $s > \alpha$, $n = 1, 2, \dots,$

$$\mathcal{L}\{f^{(n)}\} = s^n\mathcal{L}\{f\} - s^{n-1}f(0) - s^{n-2}f'(0) - \cdots - f^{(n-1)}(0).$$

In particular,

$$\mathcal{L}\{f'\} = s\mathcal{L}\{f\} - f(0), \qquad (n = 1)$$

$$\mathcal{L}\{f''\} = s^2\mathcal{L}\{f\} - sf(0) - f'(0), \qquad (n = 2)$$

$$\mathcal{L}\{f'''\} = s^3\mathcal{L}\{f\} - s^2f(0) - sf'(0) - f''(0). \qquad (n = 3)$$

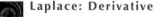

So Laplace transforms convert derivatives to algebraic formulas. In fact, the linearity property and the derivative theorem allow us to change linear DEs with initial conditions (in the t-domain) into algebraic equations (in the s-domain), as promised in Fig. 8.1.1. Furthermore, the initial conditions are built in!

Solving Linear DEs with the Laplace Transform

In the previous section, we outlined a general strategy for applying Laplace transforms in terms of the bypass. More specifically, we follow three main steps, as listed in the box below.

Finding General Solutions:

Although one generally associates the Laplace transform with finding the particular solution of an IVP, you can find the general solution of a DE by using arbitrary constants for the initial conditions, such as

$$y(0) = c_1, \quad y'(0) = c_2.$$

(See Problems 14 and 15.)

Laplace Transform Strategy

Step 1. Using the Laplace transform $\mathcal{L}$, transform the IVP with unknown function $y(t)$ into an algebraic problem with unknown function $Y(s)$.

Step 2. Solve the algebraic problem for $Y(s)$.

Step 3. Manipulating $Y(s)$ algebraically if necessary, use the inverse Laplace transform $\mathcal{L}^{-1}$ to transform $Y(s)$ into the IVP solution $y(t)$.

The use of the Laplace transform in solving initial-value problems is analogous to the way the logarithm was used to multiply large numbers in previous centuries, before calculators and computers were around. To multiply two numbers, one takes their logarithms, adds the logarithms, then finds the inverse logarithm of the sum. Let's see, without any more delay, how to use Laplace transforms to solve an initial-value problem.

EXAMPLE 1 **Initial-Value Problem** Suppose

$$y'' - 2y' - 3y = 0, \qquad y(0) = 2, \quad y'(0) = -10. \tag{8}$$

We will denote the Laplace transform of the solution function $y(t)$ by

$$Y(s) = \mathcal{L}\{y(t)\}.$$

By linearity (because $\mathcal{L}\{0\} = 0$),

$$\mathcal{L}\{y''\} - 2\mathcal{L}\{y'\} - 3\mathcal{L}\{y\} = 0. \tag{9}$$

Now, calculate the transforms of the derivatives using the initial conditions:

$$\begin{aligned}\mathcal{L}\{y''\} &= s^2\mathcal{L}\{y\} - sy(0) - y'(0)\\ &= s^2Y(s) - 2s + 10,\end{aligned} \tag{10}$$

and

$$\begin{aligned}\mathcal{L}\{y'\} &= s\mathcal{L}\{y\} - y(0)\\ &= sY(s) - 2.\end{aligned} \tag{11}$$

Substituting (10) and (11) into the transformed differential equation (9), we get

$$[s^2Y(s) - 2s + 10] - 2[sY(s) - 2] - 3Y(s) = 0, \tag{12}$$

an algebraic equation in $Y(s)$. We solve (12) for $Y(s)$ by elementary manipulations:

$$(s^2 - 2s - 3)Y(s) = 2s - 14,$$

$$Y(s) = \frac{2s - 14}{s^2 - 2s - 3} = \frac{2s - 14}{(s + 1)(s - 3)}.$$

Therefore, the *Laplace transform* of the solution to (8) is $Y(s)$, and the solution itself is

$$y(t) = \mathcal{L}^{-1}\{Y(s)\}.$$

In Sec. 8.1, Example 5, we found that $\mathcal{L}^{-1}\{Y(s)\} = 4e^{-t} - 2e^{3t}$, so the solution of the IVP is $y(t) = 4e^{-t} - 2e^{3t}$. ■

EXAMPLE 2 **Forced Motion** To solve the IVP

$$y'' + 3y' + 2y = e^{-3t}, \qquad y(0) = 0, \quad y'(0) = 1,$$

we apply the Laplace transform $\mathcal{L}$ to the DE, which gives

$$\mathcal{L}\{y''\} + 3\mathcal{L}\{y'\} + 2\mathcal{L}\{y\} = \mathcal{L}\{e^{-3t}\}.$$

By the derivative theorem,

$$s^2\mathcal{L}\{y\} - sy(0) - y'(0) + 3[s\mathcal{L}\{y\} - y(0)] + 2\mathcal{L}\{y\} = \mathcal{L}\{e^{-3t}\}.$$

Therefore,

$$s^2Y(s) - 1 + 3sY(s) + 2Y(s) = \frac{1}{s+3}.$$

Solving for $Y(s)$, we have

$$(s^2 + 3s + 2)Y(s) = 1 + \frac{1}{s+3} = \frac{s+4}{s+3},$$

$$Y(s) = \frac{s+4}{(s^2+3s+2)(s+3)}.$$

Using the partial fraction decomposition for the right-hand side yields

$$Y(s) = \frac{1/2}{s+3} - \frac{2}{s+2} + \frac{3/2}{s+1}.$$

By Table 8.1.1, entry (iii),

$$y(t) = \frac{1}{2}e^{-3t} - 2e^{-2t} + \frac{3}{2}e^{-t}$$

is the solution of the IVP. ■

EXAMPLE 3 **Sinusoidal Forcing** We apply Laplace transforms to both sides of

$$y'' + 4y = \sin t, \qquad y(0) = 0, \quad y'(0) = 1$$

to get

$$s^2\mathcal{L}\{y\} - sy(0) - y'(0) + 4\mathcal{L}\{y\} = \mathcal{L}\{\sin t\},$$

$$s^2Y(s) - 1 + 4Y(s) = \frac{1}{s^2+1},$$

$$(s^2+4)Y(s) = 1 + \frac{1}{s^2+1} = \frac{s^2+2}{s^2+1}.$$

Laplace Trade-off:

As you can observe, the trade-off for avoiding obstacles on the DE side (as depicted in Fig. 8.1.1) is the use of partial fractions on the algebraic side when finding inverse Laplace transforms.

Thus,

$$Y(s) = \frac{s^2+2}{(s^2+1)(s^2+4)} = \frac{1/3}{s^2+1} + \frac{2/3}{s^2+4},$$

and $y(t) = \frac{1}{3}\sin t + \frac{1}{3}\sin 2t$ is the solution of the IVP. ■

EXAMPLE 4 **Increase in Order** No additional theory is needed to solve the third-order IVP

$$y''' + y' = e^t, \qquad y(0) = 0, \quad y'(0) = 0, \quad y''(0) = 0.$$

Taking Laplace transforms and using the derivative theorem,

$$s^3\mathcal{L}\{y\} - s^2 \cdot 0 - s \cdot 0 - 0 + s\mathcal{L}\{y\} - 0 = \mathcal{L}\{e^t\},$$

$$(s^3+s)Y(s) = \frac{1}{s-1}.$$

Then,

$$Y(s) = \frac{1}{(s-1)(s^3+s)} = -\frac{1}{s} + \frac{1/2}{s-1} + \frac{(1/2)s - 1/2}{s^2+1}$$

$$= -\frac{1}{s} + \frac{1/2}{s-1} + \frac{(1/2)s}{s^2+1} - \frac{1/2}{s^2+1},$$

and

$$y(t) = -1 + \frac{1}{2}e^t + \frac{1}{2}\cos t - \frac{1}{2}\sin t.$$

■

Laplace Transform of $e^{at}f(t)$

Using the definition of the Laplace transform to find the Laplace transform is about as impractical as using the definition of the derivative to find derivatives. Calculus uses rules of operation, such as the product rule, chain rule, and so on, to find derivatives without resorting to the definition. We introduce two useful rules for the Laplace transform. The first is usually called the **translation property**, which says that multiplication by e^{at} in the t-domain equals translation in the s-domain.

Translation Property:

The Laplace transform of e^{at} times a function can be obtained by taking the transform of the function, then replacing s by $s-a$.

Translation Property for Multiplication by e^{at}

If the Laplace transform $F(s) = \mathcal{L}\{f(t)\}$ exists for $s > \alpha$, then

$$\mathcal{L}\{e^{at}f(t)\} = F(s-a), \qquad \text{for } s > a + \alpha.$$

Proof The translation property is derived from the definition of the Laplace transform as follows:

$$\mathcal{L}\{e^{at}f(t)\} = \int_0^\infty e^{-st}e^{at}f(t)\,dt = \int_0^\infty e^{-(s-a)t}f(t)\,dt = F(s-a). \quad \square$$

EXAMPLE 5 **Translation Property** To find $\mathcal{L}\{e^{at}\cos bt\}$, we first refer back to Table 8.1.1. We know that

$$\mathcal{L}\{\cos bt\} = F(s) = \frac{s}{s^2+b^2}.$$

Using the translation property, we have

$$\mathcal{L}\{e^{at}\cos bt\} = F(s-a) = \frac{s-a}{(s-a)^2+b^2}.$$

■

The inverse of the translation property, which states that

$$\mathcal{L}\{F(s-a)\} = e^{at} f(t), \tag{13}$$

is useful because many solutions of second-order equations are of this form—for example, a damped vibration $e^{-t}\sin 2t$. The following example illustrates this idea.

EXAMPLE 6 **Inverse Variation of the Translation Property**

(a) We can calculate

$$\mathcal{L}^{-1}\left\{\frac{1}{s^2+6s+10}\right\}$$

using the inverse of the translation property. Because $s^2+6s+10$ cannot be factored into real linear factors, we complete its square, getting $s^2+6s+10=(s+3)^2+1$. Hence, we have

$$\mathcal{L}^{-1}\left\{\frac{1}{s^2+6s+10}\right\} = \mathcal{L}^{-1}\left\{\frac{1}{(s+3)^2+1}\right\} = e^{-3t}\sin t.$$

(b) Similarly, to calculate

$$\mathcal{L}^{-1}\left\{\frac{3s-1}{s^2+2s+5}\right\},$$

we complete the square of the denominator, getting $s^2+2s+5=(s+1)^2+4$. We also break the fraction into two separate fractions:

$$\frac{3s-1}{s^2+2s+5} = \frac{3(s+1)}{(s+1)^2+4} - \frac{4}{(s+1)^2+4},$$

so that we can use the translation property to obtain

$$\begin{aligned}\mathcal{L}^{-1}\left\{\frac{3s-1}{s^2+2s+5}\right\} &= 3\mathcal{L}^{-1}\left\{\frac{s+1}{(s+1)^2+4}\right\} - 4\mathcal{L}^{-1}\left\{\frac{1}{(s+1)^2+4}\right\}\\ &= 3e^{-t}\cos 2t - 4e^{-t}\sin 2t\\ &= e^{-t}(3\cos 2t - 4\sin 2t)\end{aligned}$$

■

Laplace Transform of $t^n f(t)$

A useful rule for solving differential equations with coefficients that are powers of t is the "multiplication by t^n" rule. By its use the Laplace transform converts a differential equation whose coefficients are powers of t to a differential equation in the transform.

Multiplication by t^n Rule:
The Laplace transform of t^n times a function can be obtained by taking the nth derivative of the transform of the function, and then multiplying by -1 if n is odd.

Multiplication by t^n Rule for the Laplace Transform
If $f(t)$ is a piecewise continuous function on $[0, \infty)$ and of exponential order α, then for $s > \alpha$,

$$\mathcal{L}\{t^n f(t)\} = (-1)^n \frac{d^n F}{ds^n}(s), \quad \text{where } n \text{ is a positive integer.}$$

Proof We prove the result for $n = 1$, using the fact that the derivative and integral can be interchanged, giving

$$\begin{aligned} \frac{d}{ds}F(s) &= \frac{d}{ds}\int_0^\infty e^{-st} f(t)\,dt \\ &= \int_0^\infty \frac{d}{ds} e^{-st} f(t)\,dt \\ &= \int_0^\infty te^{-st} f(t)\,dt = -\mathcal{L}\{tf(t)\}. \end{aligned}$$

The result for arbitrary n can be obtained through repeated differentiation. □

EXAMPLE 7 **Multiplication by t** To find $\mathcal{L}\{t \cos bt\}$, we first recall from Table 8.1.1 that

$$\mathcal{L}\{\cos bt\} = \frac{s}{s^2 + b^2}.$$

Then we use the Multiplication by t^n Rule to get

$$\mathcal{L}\{t \cos bt\} = -\frac{d}{ds}\left(\frac{s}{s^2 + b^2}\right) = \frac{s^2 - b^2}{(s^2 + b^2)^2}.$$

■

Variable Coefficients

Linear differential equations with variable coefficients can sometimes be solved by use of the Laplace transform. Bessel's equation of order zero is one such example.

EXAMPLE 8 **Bessel's Equation** To solve Bessel's equation,

$$ty'' + y' + ty = 0, \tag{14}$$

with initial conditions $y(0) = 1$, $y'(0) = 0$, we first compute

$$\mathcal{L}\{y'\} = s\mathcal{L}\{y\} - y(0) = s\mathcal{L}\{y\} - 1,$$

$$\mathcal{L}\{y''\} = s^2\mathcal{L}\{y\} - sy(0) - y'(0) = s^2\mathcal{L}\{y\} - s.$$

Because y and y'' are multiplied by t, application of the Multiplication by t^n Rule gives

$$-\frac{d}{ds}(s^2 Y(s) - s) + (sY(s) - 1) - \frac{d}{ds}Y(s) = 0, \tag{15}$$

where $Y(s) \equiv \mathcal{L}\{y\}$. After differentiation and simplification, we arrive at the following first-order differential equation in $Y(s)$:

$$(s^2 + 1)Y'(s) + sY(s) = 0.$$

Separating variables, we have

$$\frac{Y'(s)}{Y(s)} = \frac{-s}{s^2+1},$$

whose general solution is

$$Y(s) = \frac{c}{\sqrt{s^2+1}}, \tag{16}$$

where c is an arbitrary constant. To find the inverse[1] of $Y(s)$, we use the binomial series for $1/\sqrt{s^2+1}$, getting

$$Y(s) = \frac{c}{\sqrt{s^2+1}} = \frac{c}{s}\left(1 - \frac{1}{2}\cdot\frac{1}{s^2} + \frac{1}{2!}\cdot\frac{3}{2}\cdot\frac{1}{s^4} + \cdots + \frac{1\cdot 3\cdot 5\cdots(2n-1)}{2^n n!}\cdot\frac{(-1)^n}{s^{2n}} + \cdots\right). \tag{17}$$

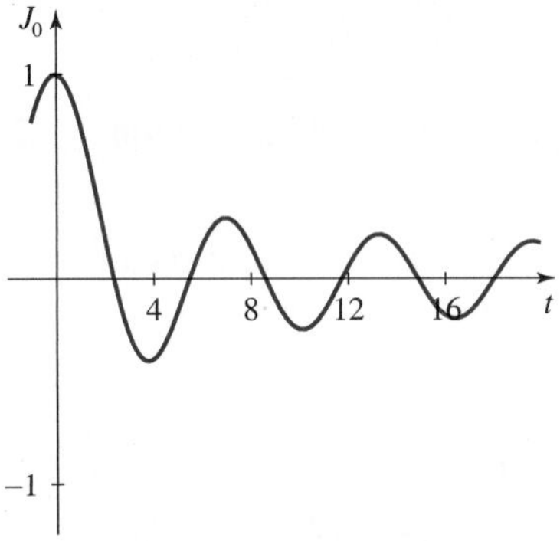

FIGURE 8.2.1 Example 8: Bessel's function of order zero.

Finally, taking the inverse of each term in this series, we arrive at the solution,[2] which is generally denoted $J_0(t)$:

$$y(t) \equiv J_0(t) = c\left(1 - \frac{t^2}{2^2} + \frac{t^4}{2^2 4^2} - \frac{t^6}{2^2 4^2 6^2} + \cdots\right). \tag{18}$$

If we apply the initial condition $y(0) = 1$, we see that the constant $c = 1$. This power series is called **Bessel's function of order zero**, and it is one of the most useful special functions. It arises in problems in partial differential equations that have cylindrical symmetry, most famously in analyzing vibrations of drumheads. The graph of $J_0(t)$ is shown in Fig. 8.2.1. It looks similar to the solution of a damped oscillator, but the axis crossings occur at nonperiodic t intervals. ■

Historical Note

The Laplace transform is named after French mathematician **Pierre Simon de Laplace** (1749–1827), who used the transform in his seminal treatise on probability, *Mécanique Céleste*. It was the English engineer **Oliver Heaviside** (1850–1925), however, who used the *operational calculus* in the nineteenth century to solve differential equations by techniques similar to the methods of this section. Heaviside was one of the first industrial mathematicians. While working for the Great Northern Telegraph Company, he made numerous contributions to the mathematical study of transmission lines and cables. The mathematical community would not accept his work because he failed to justify his techniques; he had an aversion to "mathematical rigor." His reply to the mathematicians who criticized his methods was, "Should I refuse my dinner because I do not understand the process of digestion?" Only in the twentieth century have Heaviside's uses of Laplace transforms been justified from a precise mathematical standpoint.

[1] At this point one would most likely seek out a table of Laplace transforms. See inside back cover.

[2] We assume that the inverse Laplace transform can be found by taking the inverse of each term in the infinite series.

Summary

By taking a detour from one function space to another using the Laplace transform, we are now able to transform initial-value problems for linear differential equations and systems into algebraic equations that can be solved without integration. The inverse Laplace transform returns us to the main road and the solution to the original problems. We have found several additional simplifying results for Laplace transforms:

$$\mathcal{L}\{f'(t)\}, \quad \mathcal{L}\{f^{(n)}(t)\}, \quad \mathcal{L}\{e^{at}f(t)\}, \quad \mathcal{L}\{t^n f(t)\}.$$

We then demonstrated the use of Laplace transforms in solving Bessel's equation of zero order, arriving at an important power series $J_0(t)$, the zero-order Bessel function.

8.2 Problems

First-Order Problems *Use Laplace transforms to solve the IVPs in Problems 1–4.*

1. $y' = 1, \quad y(0) = 1$

2. $y' - y = 0, \quad y(0) = 1$

3. $y' - y = e^t, \quad y(0) = 1$

4. $y' + y = e^{-t}, \quad y(0) = -1$

Transformations at Work *Solve the IVPs in Problems 5–12 using Laplace transforms.*

5. $y'' - 3y' + 2y = 0; \quad y(0) = 1, \quad y'(0) = 0$

6. $y'' + 2y' = 4; \quad y(0) = 1, \quad y'(0) = -4$

7. $y'' + 9y = 20e^{-t}; \quad y(0) = 0, \quad y'(0) = 1$

8. $y'' + 9y = \cos 3t; \quad y(0) = 1, \quad y'(0) = -1$

9. $y'' + 3y' + 2y = 6; \quad y(0) = 0, \quad y'(0) = 2$

10. $y'' + y' + y = 1; \quad y(0) = 0, \quad y'(0) = 0$

11. $y'' + y' + y = \sin t; \quad y(0) = 0, \quad y'(0) = 0$

12. $y'' + y' + y = e^{-t}; \quad y(0) = 0, \quad y'(0) = 1$

General Solutions *If you let $y(0) = A$ and $y'(0) = B$, the method of Laplace transforms applied to the DE*

$$ay'' + by' + cy = f(t)$$

will produce the general solution in terms of the arbitrary constants A and B. Apply this to the DEs in Problems 13 and 14, and compare to the results you would get from nontransform methods.

13. $y'' - y = t$

14. $y'' + 3y' + 2y = 0$

Raising the Stakes *Solve the IVPs in Problems 15 and 16 using Laplace transforms. Again, compare with nontransform methods.*

15. $y''' - y'' - y' + y = 6e^t;$
    $y(0) = 0, y'(0) = 0, y''(0) = 0$

16. $y^{(4)} - y = 0;$
    $y(0) = 1,$
    $y'(0) = 0, y''(0) = -1, y'''(0) = 0$

17. **Which Grows Faster?** Which curve grows faster: the curve that grows proportionally to its *value* or the curve that grows proportionally to its *accumulated area*? In other words, for the DEs

$$y' = ky$$
$$y' = k\int_0^t y(t)\,dt,$$

where $k > 0$ and $y(0) = 1$, which solution is larger? HINT: Use the *integral rule* for the Laplace transform,

$$\mathcal{L}\left\{\int_0^t f(t)dt\right\} = \frac{1}{s}F(s).$$

18. **Laplace Transform Using Power Series** Find the Laplace transform of $f(t) = e^t$ by the following steps.

    (a) Write the power series of e^t expanded around $t = 0$.

    (b) Find the Laplace transform of each term in the power series.

    (c) Identify the power series found in part (b) to get the transform of e^t.

Operator Methods *Symbolic methods are methods for solving differential equations in which the derivatives are replaced by* **symbolic operators** $D = d/dt$, $D^2 = d^2/dt^2, \ldots$. *Hence, the differential equation*

$$y'' + y' - 2y = 0$$

would be replaced by the algebraic *equation*

$$D^2 y + Dy - 2y = 0$$

or

$$(D^2 + D - 2)y = 0$$

or

$$(D - 1)(D + 2)y = 0.$$

As with the Laplace transform, one turns a differential equation into an algebraic equation. Problems 19–20 ask you to solve two second-order DEs using symbolic operators.

19. Solve $y'' + 3y' + 2y = 1$ by the following steps.

(a) Write the differential equation in operator form, and show that it can be written as

$$(D + 1)(D + 2)y = 1.$$

(b) Show that y can be found by solving two first-order equations:

$$(D + 1)v = 1,$$
$$(D + 2)y = v.$$

(c) Find y from part (b) by solving the first equation to get v, then substitute v into the second equation to obtain y. HINT: Keep in mind that $(D + 1)v = 1$ is simply $v' + v = 1$, and $(D + 2)y = v$ is $y' + 2y = v$.

20. Find a particular solution of

$$y'' + y = t^6$$

using symbolic operators. HINT: Rewrite the differential equation in operator form, solve for y by dividing by $D^2 + 1$, write the power series for

$$\frac{1}{D^2 + 1} = 1 - D^2 + D^4 - \cdots,$$

then let this power series "operate" on t^6. This gives a particular solution. Verify the answer.

21. Bessel Functions with IDE In general, Bessel's equation of order p is written

$$t^2 y'' + ty' + (t^2 - p^2)y = 0. \qquad (19)$$

Because it is singular at $t = 0$, equation (19) will not generally have a Maclaurin series solution. However, as you have seen in Example 8 for $p = 0$, there may be another kind of series solution $J_0(t)$, called the *Bessel function of the first kind*, which this problem will investigate. For further exploration of Bessel functions of the first kind, see IDE Lab 30, Part 4: *The Vibration of a Drumhead.*

(a) Show that (19) reduces to (14) when $p = 0$.

Bessel Functions 1st Kind
See a quick analysis of Bessel functions of various orders.

(b) Example 8 ended with the intriguing statement that, unlike the solution curve for a damped oscillator as in Chapter 4, the graph of the Bessel function $J_0(t)$ crosses the horizontal axis at nonconstant intervals of t. Use IDE (or other software) to find the first few zeroes of $J_0(t)$, and describe what happens to the intervals between axis crossings as t increases.

(c) Use IDE to explore $J_p(t)$ for a few other values of p. Describe differences in behavior from $J_0(t)$. MORAL: This is a good demonstration of why a single example as in part (b) does not tell the whole story. We have given only a beginner's taste of this subject.

22. Computer Exploration Work IDE Lab 14, Parts 1.4–1.7, to explore why linear DEs act like algebraic equations in Laplace transform space.

Laplace: Derivative;
Laplace: Transformer
These tools create more visual intuition and explanation.

23. Laplace and Vibration Work IDE Lab 14, parts 1.8–1.10, for more insight on damped oscillators.

Laplace: Transformer;
Laplace: Vibration and Poles
See exactly how initial values relate to the character of the eigenvalues in a damped oscillator.

24. Suggested Journal Entry Other transforms of particular interest to engineers and mathematicians are the Fourier transform and the Zeta transform. Look them up in an advanced engineering mathematics text (e.g., *Advanced Engineering Mathematics* by Zill and Cullen or *Introduction to Applied Mathematics* by Strang). Determine what kinds of functions they are usually applied to and what kind of problems they are used to solve. Do they exhibit the same kind of "one-sidedness" as the Laplace transform?

8.3 The Step Function and the Delta Function

SYNOPSIS: We introduce the step function and the delta function, two functions that play an important role in physical problems. We use these functions to model discontinuous forcing functions for a variety of physical situations.

Unit Step Function

Many physical systems are controlled by on/off switches, which are modeled by discontinuous functions. For example, the input voltage of a circuit can be turned on or off, or stepped up or down; mechanical systems can experience abrupt changes in forcing. Most of these systems can be represented using step functions.[1]

Step Functions

The **unit step function**

$$\text{step}(t) = \begin{cases} 0 & \text{if } t < 0, \\ 1 & \text{if } t \geq 0. \end{cases} \tag{1}$$

The **translated step function**

$$\text{step}(t-a) = \begin{cases} 0 & \text{if } t < a, \\ 1 & \text{if } t \geq a. \end{cases} \tag{2}$$

The *unit step function* represents flipping a switch from off to on at time $t = 0$, while the *translated step function* represents flipping the switch at time $t = a$. These functions are illustrated in Fig. 8.3.1.

Laplace Endpoints:
Laplace transforms involve integrals, so the endpoint values of the step function can be disregarded. This means that step function definitions may be written with $>$ rather than $\geq$, and the graphs could have open circles on both sides of the step.

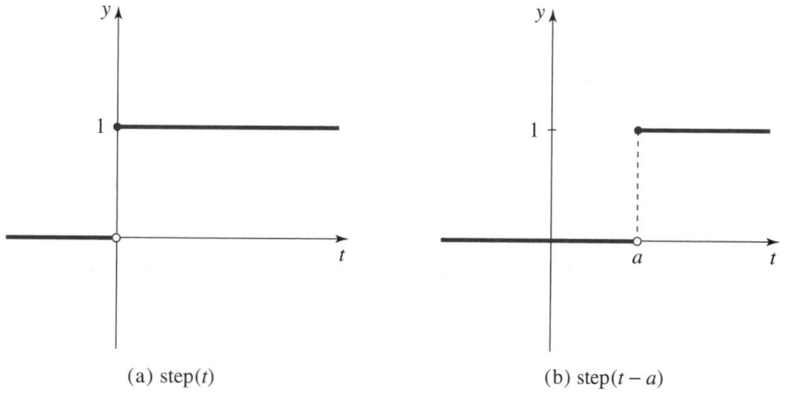

(a) step(t) (b) step($t-a$)

FIGURE 8.3.1 Step functions.

Laplace Transform of the Step Function

$$\mathcal{L}\{\text{step}(t-a)\} = \frac{e^{-as}}{s}. \tag{3}$$

[1]The unit step function is often called the Heaviside function, after Oliver Heaviside (1850–1925). See the Historical Note in Sec. 8.2.

Proof From the definition of the Laplace transform, you can confirm that

$$\begin{aligned}\mathcal{L}\{\text{step}(t-a)\} &= \int_0^\infty e^{-st}\,\text{step}(t-a)\,dt\\ &= \int_a^\infty e^{-st}\,dt\\ &= \lim_{b\to\infty}\left[-\frac{e^{-st}}{s}\right]_a^b\\ &= \lim_{b\to\infty} -\frac{1}{s}[e^{-sb}-e^{-sa}] = \frac{e^{-as}}{s}.\end{aligned}$$

□

EXAMPLE 1 **Piecewise Functions** The step function is useful in representing functions with a piecewise definition. The reader can verify that

$$f(t) = \begin{cases} 2 & \text{if } t < 3,\\ -4 & \text{if } 3 \le t < 4,\\ 1 & \text{if } t \ge 4\end{cases} = 2 - 6\,\text{step}(t-3) + 5\,\text{step}(t-4)$$

and that

$$\mathcal{L}\{f(t)\} = F(s) = \frac{2 - 6e^{-3s} + 5e^{-4s}}{s}.$$

Figure 8.3.2 shows the Laplace transform of piecewise $f(t)$ to be a smooth and continuous function $F(s)$. The term $2/s$ dominates.

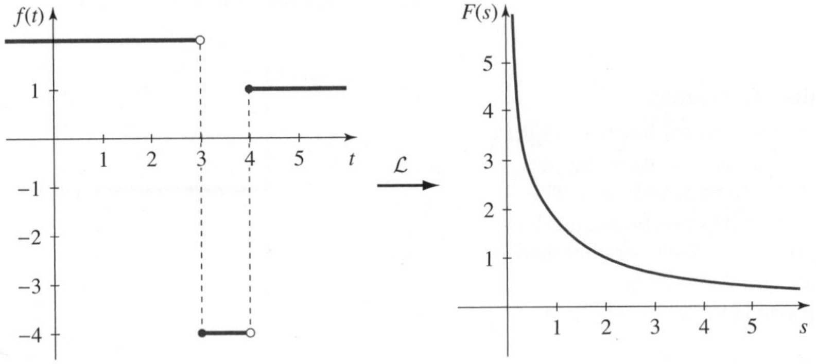

FIGURE 8.3.2 Example 1: Piecewise function $f(t)$ and its Laplace transform. ■

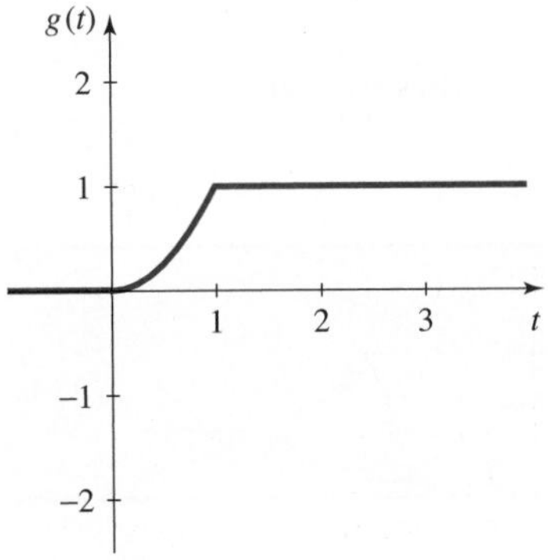

FIGURE 8.3.3 Piecewise function of Example 2.

EXAMPLE 2 **Another Piecewise Function** Using the unit step function and its translates, one can write the piecewise function

$$g(t) = \begin{cases} 0 & \text{if } t < 0,\\ t^2 & \text{if } 0 \le t < 1,\\ 1 & \text{if } t \ge 1\end{cases}$$

as a single formula

$$g(t) = t^2\,\text{step}(t) + (1-t^2)\,\text{step}(t-1),$$

illustrated in Fig. 8.3.3.

In this case,

$$\begin{aligned}\mathcal{L}\{g(t)\} &= \int_0^\infty t^2 e^{-st}\,\text{step}(t)\,dt + \int_0^\infty (1-t^2)e^{-st}\,\text{step}(t-1)\,dt\\ &= \int_0^\infty t^2 e^{-st}\,dt + \int_1^\infty e^{-st}\,dt - \int_1^\infty t^2 e^{-st}\,dt\\ &= \int_0^1 t^2 e^{-st}\,dt + \int_1^\infty e^{-st}\,dt\\ &= \frac{2}{s} - e^{-s}\left(\frac{1}{s} + \frac{2}{s^2} + \frac{2}{s^3}\right) + \frac{1}{s}e^{-s}\\ &= \frac{2}{s^3} - 2e^{-s}\left(\frac{1}{s^2} + \frac{1}{s^3}\right).\end{aligned}$$

■

Delayed Functions

In many applications we are interested in *delaying*, or shifting, a given function.

Laplace: Shift and Step

This is not a new dance step; it is only a look at the construction of a delayed function.

Delayed Function

For a given function $g(t)$, the **delayed function**

$$f(t) = \begin{cases} 0 & \text{if } t < c,\\ g(t-c) & \text{if } t \ge c\end{cases}$$

shifts $g(t)$ to the right c units from the origin, and replaces it by zero to the left of $t = c$. Using the unit step function, the delayed function can also be written as

$$f(t) = g(t-c)\,\text{step}(t-c). \tag{4}$$

EXAMPLE 3 **Imposing a Delay** When we shift the function $g(t) = \sqrt{t}$ by 3 units, we obtain the delayed function

$$f(t) = \begin{cases} 0 & \text{if } t < 3,\\ \sqrt{t-3} & \text{if } t \ge 3\end{cases} = \sqrt{t-3}\,\text{step}(t-3).$$

Figure 8.3.4 shows how the delayed function "wakes up" at $t = 3$.

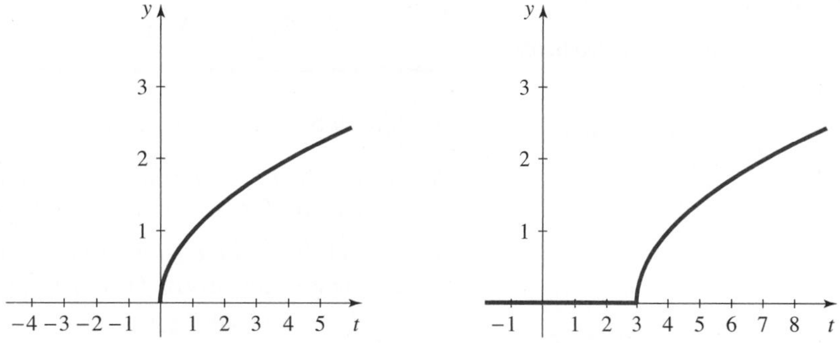

(a) Original function $g(t) = \sqrt{t}$ (b) Delayed function $f(t) = \sqrt{t-3}\,\text{step}(t-3)$

FIGURE 8.3.4 Example 3: Delaying a function $g(t)$ for $t = 3$ units.

■

Laplace Transforms for Delayed Functions

To apply the method of the previous section to differential equations with delayed forcing functions, we need to know the Laplace transform of such drivers. Let us suppose that we have a forcing function $f(t)$ delayed $t = c$ units. That is, the right-hand side of our differential equation is $f(t-c)\,\text{step}(t-c)$, and

$$\begin{aligned}\mathcal{L}\{f(t-c)\,\text{step}(t-c)\} &= \int_0^\infty e^{-st} f(t-c)\,\text{step}(t-c)\,dt \\ &= \lim_{b\to\infty}\int_0^b e^{-st} f(t-c)\,\text{step}(t-c)\,dt,\end{aligned} \tag{5}$$

where we can assume $b > c$ because b will tend to infinity. Furthermore, by definition, $\text{step}(t-c) = 0$ for $t < c$ and 1 for $t \geq c$, so

$$\int_0^b e^{-st} f(t-c)\,\text{step}(t-c)\,dt = \int_c^b e^{-st} f(t-c)\,dt. \tag{6}$$

Making the change of variable $w = t - c$, so that $dw = dt$, we have

$$\begin{aligned}\int_c^b e^{-st} f(t-c)\,dt &= \int_0^{b-c} e^{-s(w+c)} f(w)\,dw \\ &= e^{-cs}\int_0^{b-c} e^{-sw} f(w)\,dw.\end{aligned} \tag{7}$$

Combining (5), (6), and (7) yields

$$\begin{aligned}\mathcal{L}\{f(t-c)\,\text{step}(t-c)\} &= \lim_{b\to\infty}\int_0^b e^{-st} f(t-c)\,\text{step}(t-c)\,dt \\ &= \lim_{b\to\infty}\int_c^b e^{-st} f(t-c)\,dt \\ &= \lim_{b\to\infty} e^{-cs}\int_0^{b-c} e^{-sw} f(w)\,dw \\ &= e^{-cs}\int_0^\infty e^{-sw} f(w)\,dw \\ &= e^{-cs}F(s),\end{aligned}$$

where $F(s) = \mathcal{L}\{f(t)\}$.

We have proved the following result:

Laplace: Shifting Theorem

The delay theorem is illustrated here.

Delay Theorem (or Shifting Theorem)

$$\mathcal{L}\{f(t-c)\,\text{step}(t-c)\} = e^{-cs}\mathcal{L}\{f(t)\}, \quad c > 0.$$

In plain words the theorem[2] tells us the following.

1. To get the *Laplace transform* of the delayed function $f(t-c)\,\text{step}(t-c)$, transform $f(t)$ to $F(s)$ and multiply by e^{-cs}.
2. To get the *inverse transform* of an expression of the form $e^{-cs}F(s)$, find the inverse transform $f(t)$ of $F(s)$, and impose the delay by replacing $f(t)$ by $f(t-c)\,\text{step}(t-c)$.

[2]If $f(t)$ is of exponential order α, then the delay theorem (and its alternate form) need the restriction $s > \alpha$.

Because we commonly encounter an expression of the form $g(t)\,\text{step}(t-c)$, it is often useful to substitute $g(t)$ for $f(t-c)$ in the delay theorem, thus writing it a different way.

Alternate Form of the Delay Theorem

$$\mathcal{L}\{g(t)\,\text{step}(t-c)\} = e^{-cs}\mathcal{L}\{g(t+c)\}.$$

EXAMPLE 4 **Transforming a Delayed Function** To find the Laplace transform of the function

$$h(t) = t^2\,\text{step}(t-1),$$

we choose the alternate form of the delay theorem, with $c = 1$ and $g(t) = t^2$. Hence,

$$\begin{aligned}\mathcal{L}\{h(t)\} = \mathcal{L}\{t^2\,\text{step}(t-1)\} &= e^{-s}\mathcal{L}\{(t+1)^2\}\\ &= e^{-s}\mathcal{L}\{t^2+2t+1\}\\ &= e^{-s}\left(\frac{2}{s^3}+\frac{2}{s^2}+\frac{1}{s}\right).\end{aligned}$$

Compare this simplicity with a direct calculation that would require integration by parts of $\int t^2e^{-ts}dt$. ■

Q. When do we expect a delayed function in the solution?

A. Anytime e^{-cs} appears in the transform.

EXAMPLE 5 **Reverse Delay Play** To transform the function

$$F(s) = \frac{1-e^{-3s}}{s^2} = \frac{1}{s^2} - \frac{e^{-3s}}{s^2}$$

back to $f(t)$, we treat the second term as the transform of a delay. Thus,

$$\mathcal{L}^{-1}\{F(s)\} = t - \underbrace{(t-3)\,\text{step}(t-3)}_{\mathcal{L}^{-1}\{e^{-3s}/s^2\}}.$$

■

An old engineering trick uses a *filter* or *chopper function*, to pull out a *piece* of a function over an interval $[a, b)$.

Chopper Function

$$\text{step}(t-a) - \text{step}(t-b) = \begin{cases} 0 & \text{if } t < a,\\ 1 & \text{if } a \le t < b,\\ 0 & \text{if } t \ge b.\end{cases} \tag{8}$$

The chopper will cut out the part of the function between a and b for any particular function $f(t)$, as shown in Fig. 8.3.5 on the next page. The chopped

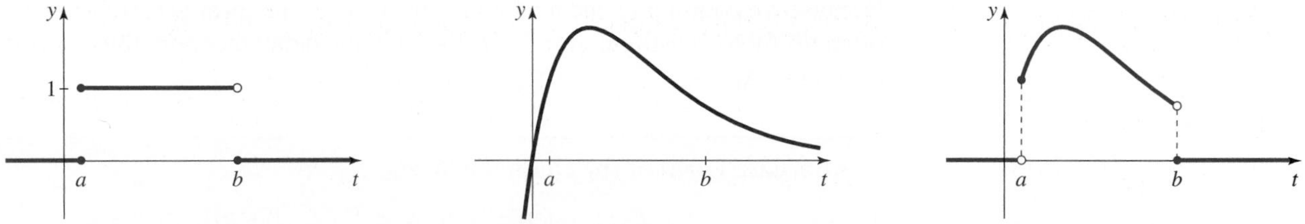

The chopper function
$y = \text{step}(t-a) - \text{step}(t-b)$

$y = f(t)$

$y = f(t)\left[\text{step}(t-a) - \text{step}(t-b)\right]$

FIGURE 8.3.5 Applying the chopper function (on the left) to $f(t)$ (in the middle) gives the chopped function shown on the right.

function (on the right of Fig. 8.3.5) is

$$f(t)[\text{step}(t-a) - \text{step}(t-b)],$$

which has, by the alternate form of the delay theorem, Laplace transform

$$\begin{aligned}&\mathcal{L}\{f(t)[\text{step}(t-a) - \text{step}(t-b)]\}\\ &\quad = e^{-as}\mathcal{L}\{f(t+a)\} - e^{-bs}\mathcal{L}\{f(t+b)\}.\end{aligned} \tag{9}$$

EXAMPLE 6 **The Alternate Way** We will use the chopper function and the alternate form of the delay theorem to find the Laplace transform of

$$f(t) = \begin{cases} 0 & \text{if } t < 1, \\ -\sin \pi t & \text{if } 1 \le t < 2, \\ 0 & \text{if } t \ge 2 \end{cases}$$

(shown in Fig. 8.3.6), which can be written by (8) as

$$f(t) = -\sin \pi t[\text{step}(t-1) - \text{step}(t-2)].$$

Finding the Laplace transform by (9), where $a = 1$ and $b = 2$,

$$\mathcal{L}\{f(t)\} = -e^{-s}\mathcal{L}\{-\sin \pi(t+1)\} + e^{-2s}\mathcal{L}\{\sin \pi(t+2)\}.$$

Using the trigonometric identity for $\sin(A+B)$ and simplifying, we have

$$\begin{aligned}\mathcal{L}\{f(t)\} &= -e^{-s}\mathcal{L}\{\sin \pi t \underbrace{\cos \pi}_{-1} + \cos \pi t \underbrace{\sin \pi}_{0}\}\\ &\quad + e^{-2s}\mathcal{L}\{\sin \pi t \underbrace{\cos 2\pi}_{+1} + \cos \pi t \underbrace{\sin 2\pi}_{0}\}\\ &= \mathcal{L}\{\sin \pi t\}(e^{-s} + e^{-2s})\\ &= \frac{\pi}{s^2 + \pi^2}(e^{-s} + e^{-2s}).\end{aligned}$$

■

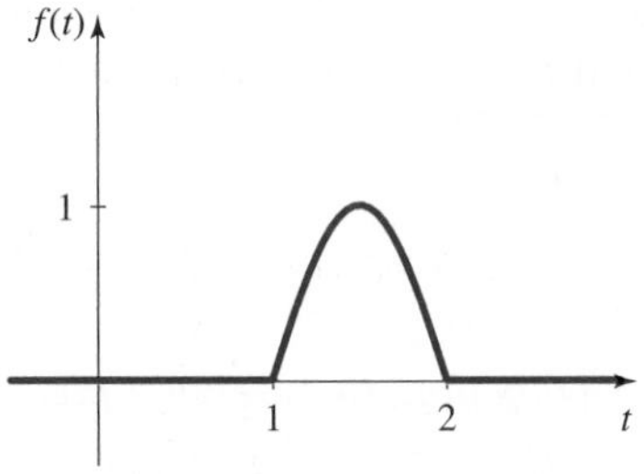

FIGURE 8.3.6 The function $f(t)$ of Example 6.

Laplace Transforms Can Expedite Solving DEs with Piecewise Forcing Terms

If a differential equation has a piecewise forcing term, solving with earlier methods is cumbersome: a separate solution must be calculated for each time interval, and the solutions "matched" by finding appropriate initial values on each new interval. The Laplace transform allows us to solve such problems more efficiently.

Laplace: Solver

See the process work for this same initial-value problem.

- First do the partial fraction decomposition on the rational function (i.e., without the exponential).
- Then use the delay theorem to find the inverse transform.

EXAMPLE 7 **On/Off Forcing Term** We will solve the IVP

$$x'' + x = f(t) = \begin{cases} 1 & \text{if } 0 \le t < \pi, \\ 0 & \text{if } t \ge \pi, \end{cases} \qquad x(0) = 0, \quad x'(0) = 0.$$

By rewriting the forcing term, we have

$$x'' + x = 1 - \text{step}(t - \pi) \qquad x(0) = 0, \quad x'(0) = 0.$$

Taking transforms and substituting the initial conditions yields

$$s^2 X(s) + X(s) = \mathcal{L}\{1 - \text{step}(t - \pi)\}. \tag{10}$$

By the delay theorem, $\mathcal{L}\{1 \cdot \text{step}(t - \pi)\} = e^{-\pi s} \cdot \frac{1}{s}$, so equation (10) becomes

$$s^2 X(s) + X(s) = \frac{1}{s} - \frac{e^{-\pi s}}{s};$$

therefore,

$$X(s) = \frac{1 - e^{-\pi s}}{s(s^2+1)} = \frac{1}{s(s^2+1)} - e^{-\pi s}\frac{1}{s(s^2+1)}$$

$$= \left(\frac{1}{s} - \frac{s}{s^2+1}\right) - e^{-\pi s}\left(\frac{1}{s} - \frac{s}{s^2+1}\right).$$

The delay theorem gives us

$$x(t) = \mathcal{L}^{-1}\{X(s)\} = (1 - \cos t) - [1 - \cos(t - \pi)]\,\text{step}(t - \pi).$$

In piecewise format,

$$x(t) = \begin{cases} 1 - \cos t & \text{if } 0 \le t < \pi, \\ 1 - \cos t - [1 - \cos(t - \pi)] & \text{if } t \ge \pi; \end{cases}$$

this simplifies, using the trigonometric identity for $\cos(t - \pi)$, to

$$x(t) = \begin{cases} 1 - \cos t & \text{if } 0 \le t < \pi, \\ -2\cos t & \text{if } t \ge \pi. \end{cases}$$

(We are a little informal about this description at the "switch" point, but you should check that it is well defined by continuity.) We have graphed the "input" $f(t)$ and the "output" $x(t)$ in Fig. 8.3.7.

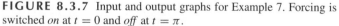

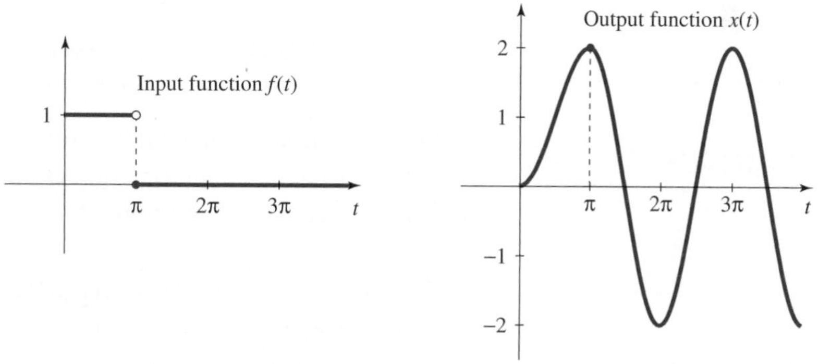

FIGURE 8.3.7 Input and output graphs for Example 7. Forcing is switched *on* at $t = 0$ and *off* at $t = \pi$.

■

The Delta Function

Physical systems frequently involve impulsive forces that act over very short periods of time, such as a bat hitting a baseball or subatomic particles colliding. To deal with these, physicist Paul Dirac invented a "functionlike" object we call the **Dirac delta function** or **unit impulse function** $\delta(t)$.[3]

Let us start with a special function

$$f_h(t) = \begin{cases} 0 & \text{if } t < 0, \\ 1/h & \text{if } 0 \le t < h, \\ 0 & \text{if } t \ge h, \end{cases}$$

such that $\int_{-\infty}^{\infty} f_h(t)\,dt = 1$. Typical graphs of $f_h(t)$ are shown in Fig. 8.3.8. Dirac then suggested that

$$\delta(t) \equiv \lim_{h \to 0} f_h(t).$$

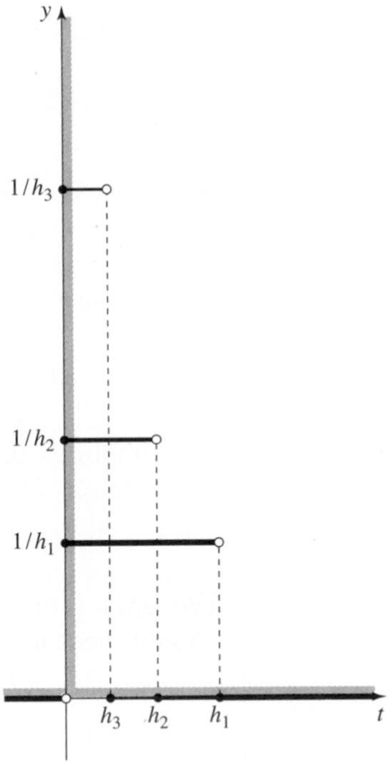

FIGURE 8.3.8 Towards the unit impulse function (color) as a limit. The *area* under $f_h(t)$ is exactly 1 for every h, where $0 < h < \infty$.

This definition is not valid in ordinary mathematics, because $\lim_{h \to 0}(1/h)$ does not exist, but Dirac's delta function simultaneously satisfies two conditions that no ordinary function could fulfill.[4]

---

[3]Paul A. M. Dirac (1902–1984), an English physicist, received the Nobel prize at age 31 for his work in quantum theory.

[4]Today, δ-functions are included in the class of generalized functions and have a solid place in modern mathematics. They were somewhat suspect when Dirac introduced them in the early 1930s. Nevertheless, it is so useful to have such a "function"—and the Laplace transform works as if the function did exist—that mathematicians took the time and effort to build an elaborate edifice (called distribution theory) that gives generalized functions a solid foundation.

Dirac Delta Function

The **Dirac delta function** or **unit impulse function** $\delta(t)$ is defined by *two* conditions:

$$\textbf{(i)}\ \ \delta(t) = \begin{cases} 0 & \text{if } t \neq 0, \\ \lim\limits_{h \to 0} \left(\dfrac{1}{h}\right) & \text{if } t = 0 \end{cases} \qquad \textbf{(ii)}\ \ \int_{-\infty}^{\infty} \delta(t)\, dt = 1.$$

One way to visualize the delta function is as a limit, in the sense of a sequence of functions that are zero everywhere except near the origin, where they have "spikes"; in the limit, the spikes become taller but thinner in order to satisfy condition (ii).

To find the Laplace transform of the unit impulse function, we will first calculate the transform of f_h, and then examine what happens as h tends to zero. Since f_h is zero for $t > h$,

Laplace: Delta Function

Get a feel for the limiting function.

$$\mathcal{L}\{f_h(t)\} = \int_0^{\infty} e^{-st} f_h(t)\, dt = \int_0^{h} e^{-st} f_h(t)\, dt$$

$$= \frac{1}{h} \int_0^{h} e^{-st}\, dt = \frac{1 - e^{-hs}}{hs}.$$

Using l'Hôpital's rule, we find that

$$\lim_{h \to 0} \mathcal{L}\{f_h(t)\} = 1,$$

and we define the Laplace transform of the unit impulse function as follows.

Laplace Transform of the Delta Function

$$\mathcal{L}\{\delta(t)\} = 1; \tag{11}$$

$$\mathcal{L}\{\delta(t-a)\} = e^{-as}. \tag{12}$$

We can find (12) by a similar calculation, or by writing a delayed δ-function as

$$\delta(t-a)\, \text{step}(t-a)$$

and applying the delay theorem.

Laplace Transforms Solve DEs with δ-Functions

IVPs with δ-functions are easier to solve than those with step functions.

Problems involving shifts and delays can be solved without Laplace transforms, but the transforms are a big help, as we saw above. For systems subject to impulses modeled by the delta function, Laplace transform methods are essential.

The delta function plays an important role in medical research, because a dose of medication administered by an injection can be entered into the drug input rate as a delta function.

EXAMPLE 8 **Drug Therapy** Suppose that a medical researcher administers to a patient 100 mg of a drug, and the same amount 24 hours later. Let us assume that the half-life of the drug in the patient's body is 24 hours. We can use our knowledge of first-order differential equations to find the future amount $y = y(t)$ of the drug present in the body. This is basically a mixing problem (Sec. 2.4), so we write

$$y' = \underbrace{100\delta(t) + 100\delta(t-1)}_{\text{rate in}} - \underbrace{ky}_{\text{rate out}}, \tag{13}$$

where we have measured time in days. The input is graphed in Fig. 8.3.9(a).

We can evaluate the decay constant k in equation (13) using the half-life information (as in Sec. 2.3). The decay equation $y' = -ky$ has the solution $y = y_0 e^{-kt}$. A half-life of 24 hours means that

$$\frac{1}{2} y_0 = y_0 e^{-k},$$

so $k = \ln 2 \approx 0.7$, and we can rewrite (13) as

$$y' + 0.7y = 100\delta(t) + 100\delta(t-1). \tag{14}$$

Taking the Laplace transform of (14), we get

$$s\mathcal{L}\{y\} + 0.7\mathcal{L}\{y\} = 100(1 + e^{-s}),$$

or

$$\begin{aligned} \mathcal{L}\{y\} &= 100\left(\frac{1+e^{-s}}{s+0.7}\right) \\ &= 100\left(\frac{1}{s+0.7}\right) + 100\left(\frac{e^{-s}}{s+0.7}\right) \\ &= 100\mathcal{L}(e^{-0.7t}) + 100e^{-s}\mathcal{L}(e^{-0.7t}). \end{aligned}$$

Hence, the future amount of the drug is

$$\begin{aligned} y(t) &= 100e^{-0.7t} + 100e^{-0.7(t-1)}\,\text{step}(t-1) \\ &= \begin{cases} 0 & \text{if } t < 0, \\ 100e^{-0.7t} & \text{if } 0 < t < 1, \\ 100(1+e^{0.7})e^{-0.7t} & \text{if } 1 < t. \end{cases} \end{aligned} \tag{15}$$

(See Fig. 8.3.9(b).) This is exactly the sort of problem in which endpoint information may not be known exactly.

You can easily sketch the solution graph in Fig. 8.3.9(b) by hand without using the solution formula (15) at all, just from the input and half-life information.

- The *jumps* in the solution at $t = 0$ and $t = 1$ are caused by the injection of the drug. An impulse of magnitude $A\delta(t - t_0)$ will give rise to a jump by amount A at $t = t_0$ in the solution $y(t)$ of the first-order equation.
- After every 24-hour interval, the drug remaining in the body is reduced by half.

■

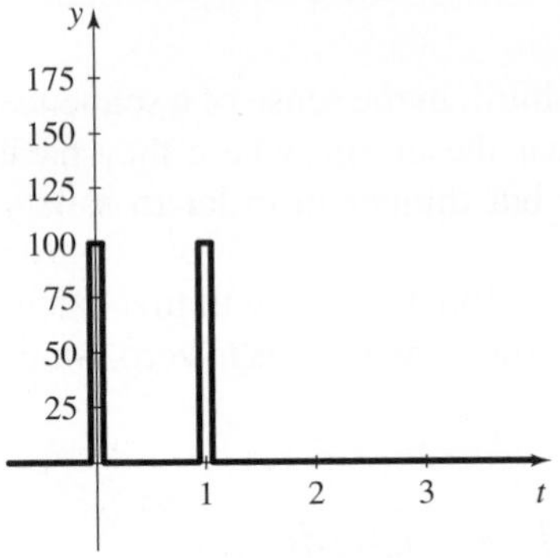

(a) Input rate function

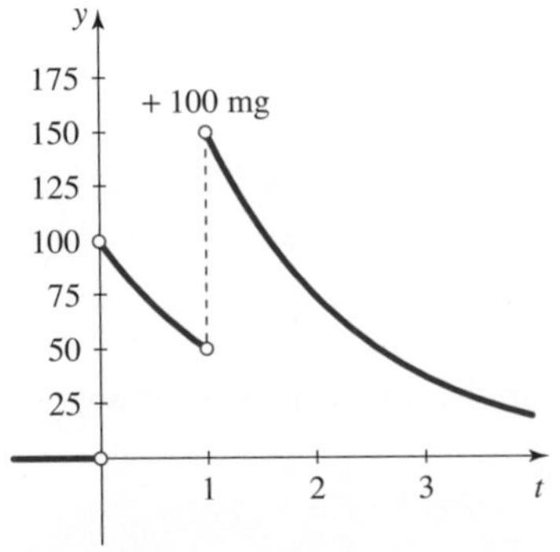

(b) The amount of drug in the body (in milligrams)

FIGURE 8.3.9 Input and output functions for Example 8.

Impulse Response Function

It is important to know how a linear system reacts to a sudden jarring or delta function input. This output is referred to as the **impulse response function**. Knowing how a linear system reacts to a delta function input allows us to find the output of the system to *any* input. (We will learn more about this when we study the convolution in Sec. 8.4.) The following examples find the impulse response to a few linear systems.

Laplace: Solver

A similar IVP with different initial conditions but the same DE is there for you to inspect.

Engineers typically measure displacement of a mass on a spring as *positive* in the *downward* direction.

EXAMPLE 9 **Impulse Response** A mass is attached to a spring and released from rest 1 unit below its equilibrium position. After the mass vibrates for π seconds, it is struck by a hammer in the downwards direction, exerting unit force on the mass. Assuming that the system is governed by the initial-value problem

$$\ddot{y} + y = \delta(t - \pi), \qquad y(0) = 1, \quad \dot{y}(0) = 0, \tag{16}$$

where $y(t)$ represents the *downward* displacement from equilibrium at time t, determine the subsequent motion of the mass.

Taking the Laplace transform of each side of equation (16), we obtain

$$s^2 Y(s) - s + Y(s) = e^{-\pi s}.$$

Solving for $Y(s)$ gives

$$Y(s) = \frac{s}{s^2+1} + \frac{e^{-\pi s}}{s^2+1} = \mathcal{L}\{\cos t\} + e^{-\pi s}\mathcal{L}\{\sin t\}.$$

Using the alternate form of the delay theorem, we find the impulse response

$$\begin{aligned} y(t) &= \cos t + \underbrace{\sin(t-\pi)}_{-\sin t}\operatorname{step}(t-\pi) \\ &= \begin{cases} \cos t & \text{if } 0 \le t < \pi, \\ \cos t - \sin t & \text{if } \pi \le t. \end{cases} \end{aligned} \tag{17}$$

The graph of $y(t)$ is shown in Fig. 8.3.10.

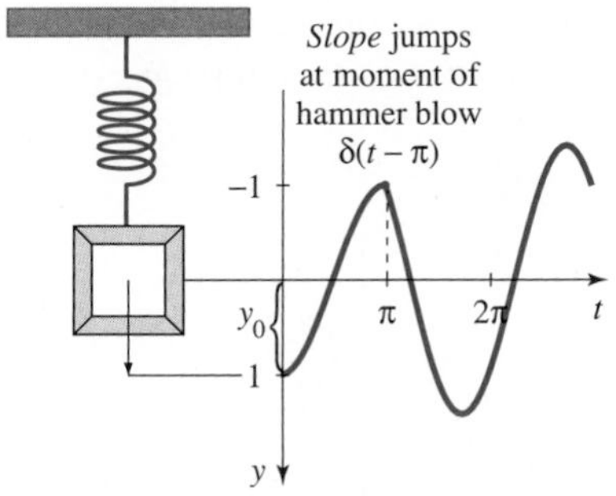

FIGURE 8.3.10 The impulse response for Example 9 shows a (downward) jump in the *slope* $\dot{y}$ at $t = \pi$.

The response to $\delta(t - t_0)$ for this *second*-order DE causes a jump of +1 in the *first derivative* of the solution, at $t = t_0$. ■

Laplace: Solver

Try a damped example with a δ-function.

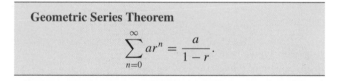

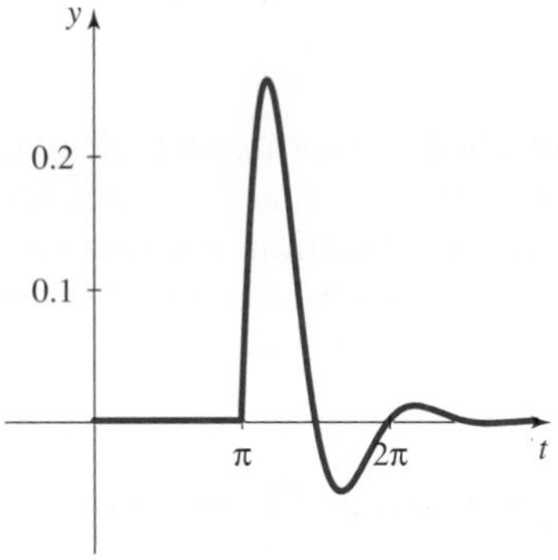

FIGURE 8.3.11 The impulse response for Example 10. The function $\frac{1}{2}\sin 2t$ wakes up and starts when $t = \pi$.

EXAMPLE 10 **Deltas and Damping** We find the impulse response of the system

$$x'' + 2x' + 5x = \delta(t - \pi), \qquad x(0) = 0, \quad x'(0) = 0.$$

The object is at rest at time zero, and remains so until the impulse occurs at time $t = \pi$. Taking transforms and simplifying, we have

$$(s^2 + 2s + 5)X(s) = e^{-\pi s};$$

therefore,

$$X(s) = e^{-\pi s} \cdot \frac{1}{(s+1)^2 + 2^2} = e^{-\pi s} \cdot \frac{1}{2} \cdot \frac{2}{(s+1)^2 + 2^2}.$$

Then, by the delay theorem and the transform table, we have

$$x(t) = \frac{1}{2}\,\text{step}(t - \pi)e^{-(t-\pi)} \sin 2(t - \pi).$$

Thus,

$$x(t) = \begin{cases} 0 & \text{if } 0 \le t < \pi, \\ \frac{1}{2}e^{-t+\pi} \sin 2t & \text{if } t \ge \pi. \end{cases}$$

The impulse response is graphed in Fig. 8.3.11. ■

Summary

The unit step function and delayed functions enlarge our repertoire for forcing functions in nonhomogeneous linear DEs, and the unit impulse function (which is not quite a function) allows us to deal with an input that causes a sudden change. Laplace transforms in these cases ease or enable solutions of DEs with such discontinuous forcing terms.

8.3 Problems

Stepping Out *Use step functions to write the function in each of Problems 1–4 as a single (unbranched) expression.*

1. $f(t) = \begin{cases} 0 & \text{if } t < 0, \\ a & \text{if } 0 \le t < 1, \\ b & \text{if } 1 \le t < 2, \\ c & \text{if } t \ge 2. \end{cases}$

2. $f(t) = \begin{cases} 1 & \text{if } t < 2, \\ e^t & \text{if } 2 \le t < 3, \\ 2 & \text{if } t \ge 3. \end{cases}$

3. $f(t) = \begin{cases} 1 & \text{if } t < 1, \\ 4t - t^2 & \text{if } 1 \le t < 4, \\ 1 & \text{if } t \ge 4. \end{cases}$

4. $f(t) = \begin{cases} 0 & \text{if } t < 2, \\ \sin \pi t & \text{if } 2 \le t < 4, \\ 0 & \text{if } t \ge 4. \end{cases}$

Geometric Series *A geometric series is a series in which each successive term is obtained by multiplying the preceding term by a constant ratio r. If $|r| < 1$, the following theorem provides the sum of the infinite series.*

Geometric Series Theorem

$$\sum_{n=0}^{\infty} ar^n = \frac{a}{1-r}.$$

We can use this result to write the Laplace transforms in "closed form" (i.e., without the "..." at the end of the expression) for functions in the next set of problems. In each of Problems 5–9, write the functions f as a sum of unit step functions, take their Laplace transforms, and then write $\mathcal{L}\{f\}$ in closed form using the Geometric Series Theorem.

5. Find the Laplace transform for the staircase function on $[0, \infty)$, defined by $f(t) = n$, $n \le t < n + 1$ for each whole number n.

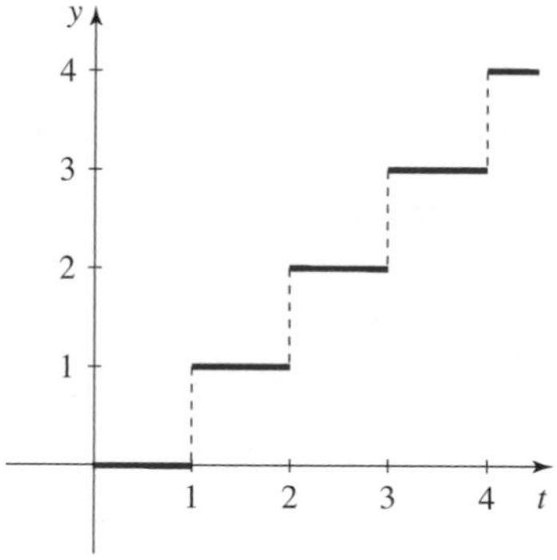

6. Find the Laplace transform for another staircase function, defined on $[0, \infty)$ by $f(t) = n + 1$, $n \le t < n + 1$ for each whole number n.

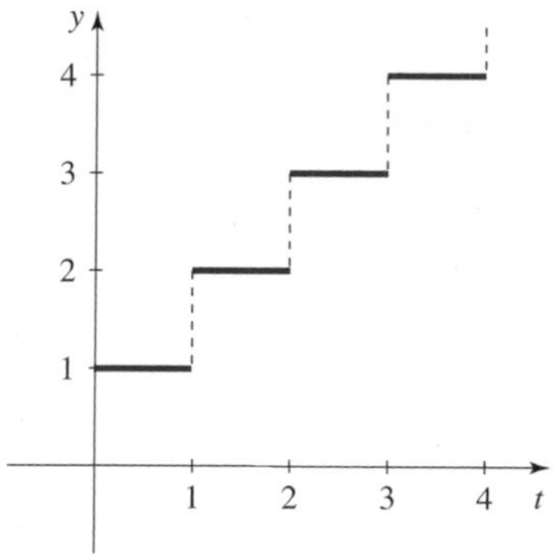

7. Find the Laplace transform for the descending (and diminishing) staircase function f, defined on $[0, \infty)$ by $f(t) = 1/(2^n)$, $n \le t < n + 1$ for each whole number n.

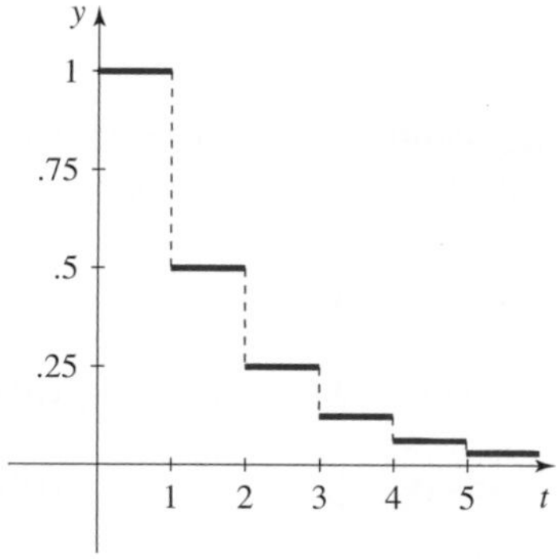

8. Find the Laplace transform for the square wave function f, as shown, defined on $[0, \infty)$.

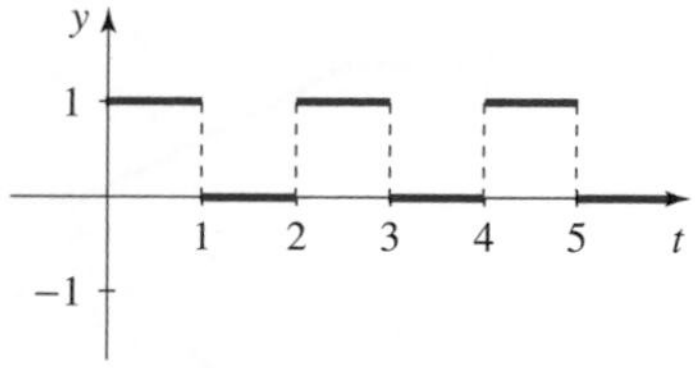

9. Find the Laplace transform for the square wave function f, as shown, defined on $[0, \infty)$.

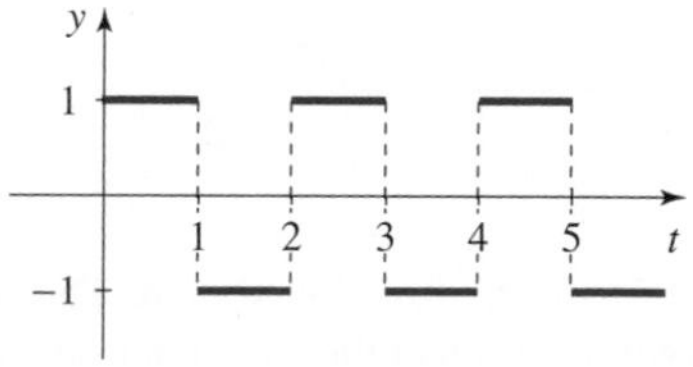

Piecewise Continuous Functions *In some applications we will need to use pieces of continuous functions and combine them to make a new piecewise continuous function. For the functions described in Problems 10–15, use the chopper function and the alternate form of the delay theorem to find the Laplace transforms.*

10. Find $\mathcal{L}\{f\}$ for the triangular function shown, defined by

$$f(t) = \begin{cases} 0 & \text{if } t < 0, \\ t & \text{if } 0 \le t < 1, \\ 2 - t & \text{if } 1 \le t < 2, \\ 0 & \text{if } t \ge 2. \end{cases}$$

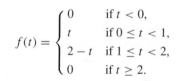

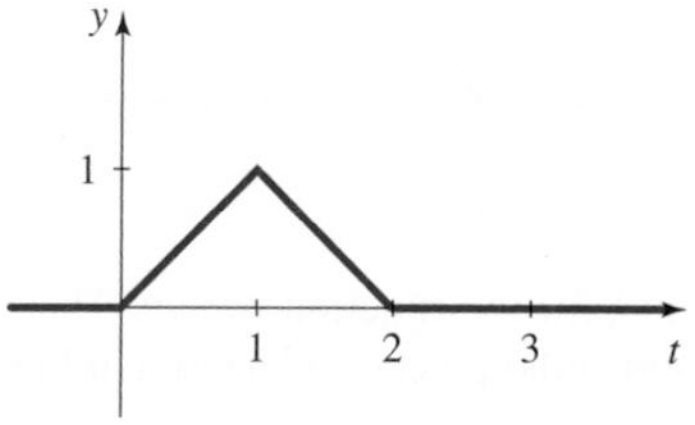

We can use the chopper function to write

$$f(t) = t[\text{step}(t) - \text{step}(t-1)] + (2-t)[\text{step}(t-1) - \text{step}(t-2)].$$

11. Find $\mathcal{L}\{f\}$ for the function shown. (Note that $f(t) = 0$ for $t > 3$.)

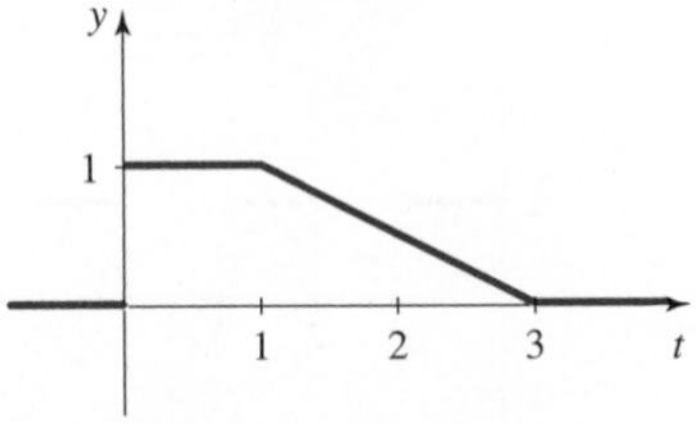

12. Find $\mathcal{L}\{f\}$ for the function shown. (Note that $f(t) = 2$ for $t > 3$.)

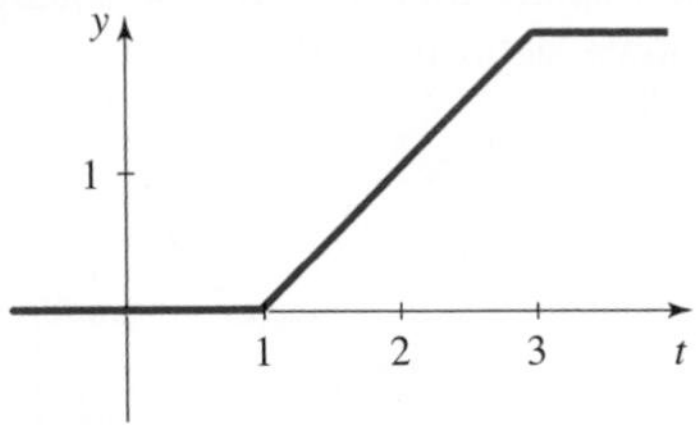

13. Find $\mathcal{L}\{f\}$ for the function shown. The function is made up of a single hump of the sine function and a straight line.

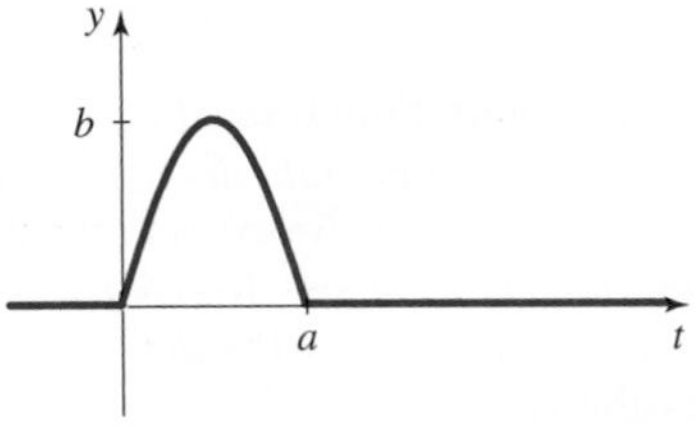

NOTE: A sine function of period P can be expressed as

$$f(t) = \sin\left(\frac{2\pi t}{P}\right).$$

14. Find $\mathcal{L}\{f\}$ for the function shown. The function is made up of a sinusoidal piece and straight lines.

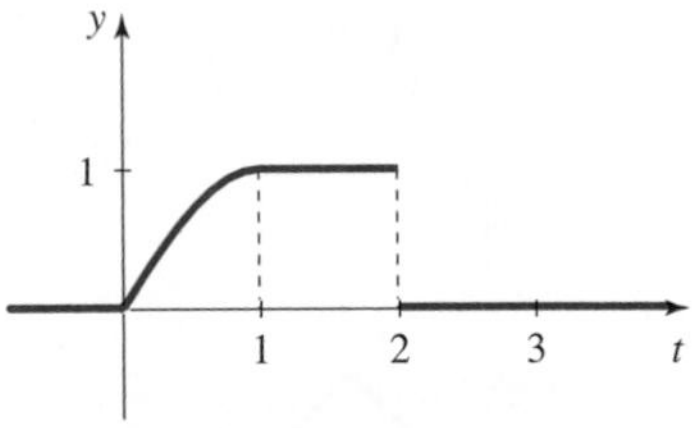

15. Find $\mathcal{L}\{f\}$ for the function shown. The function is made up of sinusoidal pieces and a straight line.

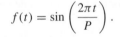

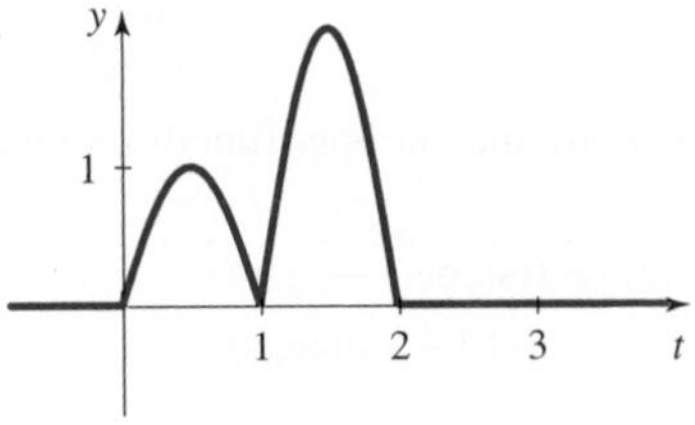

Transforming Delta *Determine the Laplace transform of the function in each of Problems 16–19.*

16. $\delta(t-1) + 2\delta(t-2) + 3\delta(t-3)$

17. $\delta(t) - 2\delta(t-\pi) + \delta(t-2\pi)$

18. $\delta(t) - \delta(t-1) + \delta(t-2) - \delta(t-3) + \delta(t-4) - \cdots$
$= \sum_{n=0}^{\infty}(-1)^n\delta(t-n)$

19. $\delta(t) + \delta(t-\pi) + \delta(t-2\pi) + \delta(t-3\pi) + \delta(t-4\pi) + \cdots$
$= \sum_{n=0}^{\infty}\delta(t-n\pi)$

Laplace Step by Step *Obtain Laplace transforms for the functions in Problems 20–26.*

20. $1 - \text{step}(t-1)$

21. $1 - 2\,\text{step}(t-1) + \text{step}(t-2)$

22. $(t-1)\,\text{step}(t-1)$

23. $\sin(t-\pi)\,\text{step}(t-\pi)$

24. $e^t\,\text{step}(t-3)$

25. $\text{step}(1-e^{-t})$

26. $t^2\text{step}(t-2)$

Inverse Transforms *Obtain the inverse Laplace transform for each function given in Problems 27–32.*

27. $\dfrac{e^{-s}}{s}$ **28.** $\dfrac{e^{-s}}{s^2}$ **29.** $\dfrac{e^{-2s}}{s-3}$

30. $\dfrac{e^{-4s}}{s+4}$ **31.** $\dfrac{e^{-s}}{s(s+1)}$

32. $\dfrac{e^{-s} - 2e^{-2s} + 2e^{-3s} - e^{-4s}}{s}$

Transforming Solutions *In each of Problems 33–36, solve the IVP and graph both the forcing function and the solution.*

33. $x' = 1 - \text{step}(t-1);$ $x(0) = 0$

34. $x' = 1 - 2\,\text{step}(t-1) + \text{step}(t-2);\ x(0) = 0$

35. $x'' + x = \text{step}(t-3);$ $x(0) = 0,\ x'(0) = 1$

36. $x'' + x = \text{step}(t-\pi) - \text{step}(t-2\pi);\ x(0) = 0,\ x'(0) = 1$

Periodic Formula *In Problems 8 and 9, we found that the Geometric Series Theorem can be used to write the Laplace transforms of square wave functions in closed form. These functions are examples of periodic functions with period P, where $f(t)$ is defined on $[0, P]$; then $f(t) = f(t - P)$ for each t in (P, ∞). By applying the Geometric Series Theorem to any periodic function defined on $[0, \infty)$ that is piecewise continuous and of exponential order, we can determine a formula for the Laplace transform of a periodic function:*

$$\mathcal{L}\{f\} = \frac{1}{1 - e^{-sP}} \int_0^P e^{-st} f(t)\, dt. \tag{18}$$

The integral is only over the first period of the function.

In Problems 37–45, we find the Laplace transforms of the periodic functions shown. HINT: *Look back at Problems 10–15.*

37. Consider the triangular wave with period $P = 2$ shown. Use the answer for Problem 10 to obtain the Laplace transform $\mathcal{L}\{f\}$, and then multiply it by $1/(1 - e^{-2s})$ to get the transform for the periodic function.

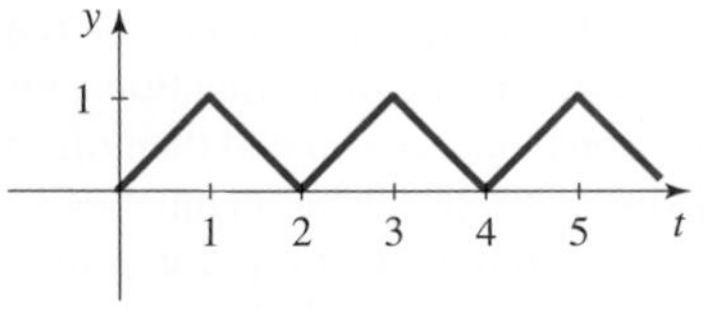

38. Find the Laplace transform for the new triangular wave shown. (See Problem 10.)

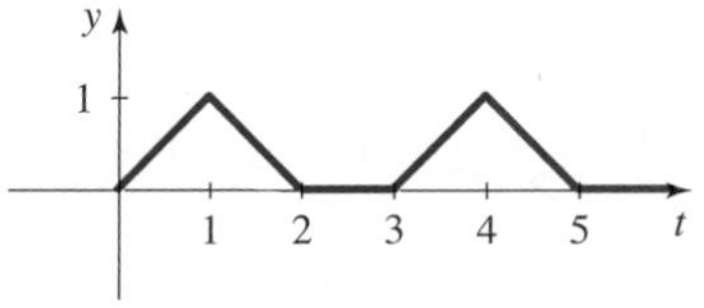

39. Find the Laplace transform for the sawtooth wave shown.

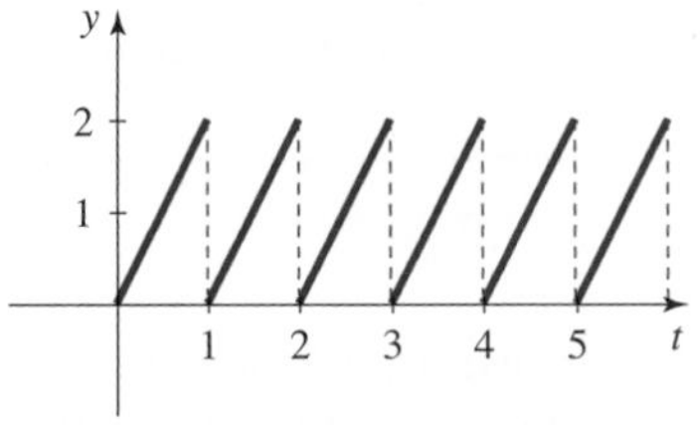

40. Find the Laplace transform for $|\sin t|$ (called by engineers the full-wave rectification of $\sin t$), as shown. (See Problem 13.)

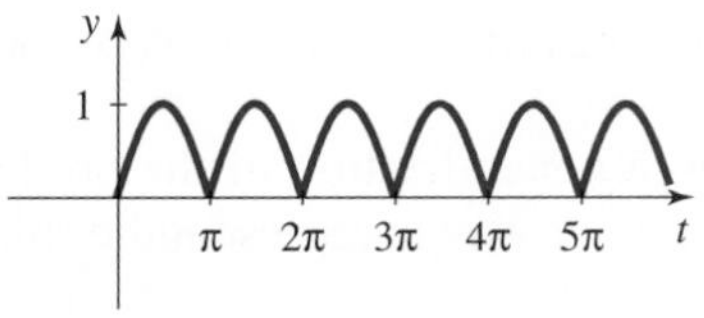

41. Find the Laplace transform for the half-wave rectification of $\sin t$ (ignoring the negative parts by making them zero), as shown. (See Problem 13.)

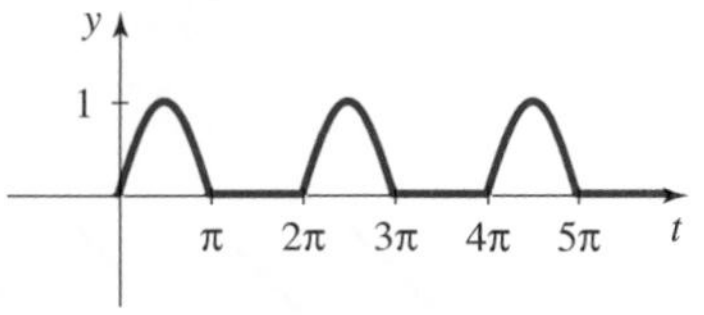

42. Find the Laplace transform for the periodic function shown. (See Problem 14.) On the interval $(0, 1)$, the curve is sinusoidal.

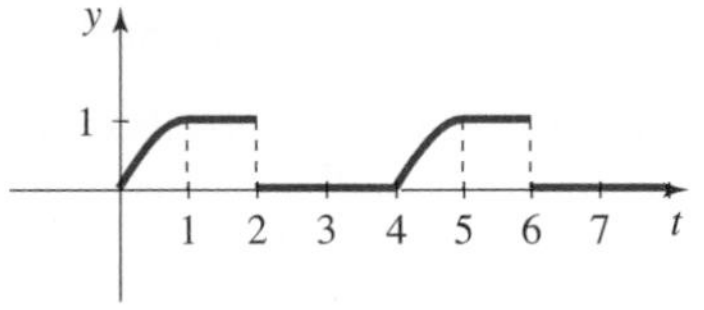

43. Find the Laplace transform for the periodic function shown. (See Problem 11.)

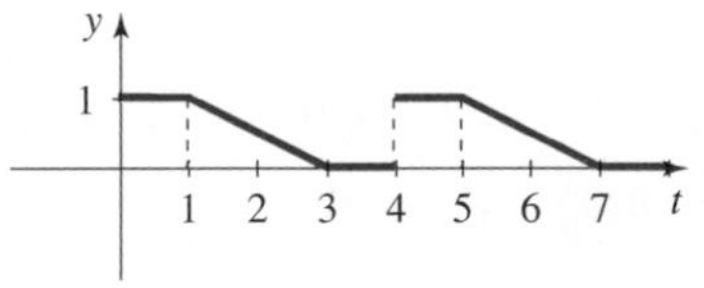

44. Find the Laplace transform of the periodic function shown. (See Problem 15.) The curved portions are pieces of sinusoidal functions.

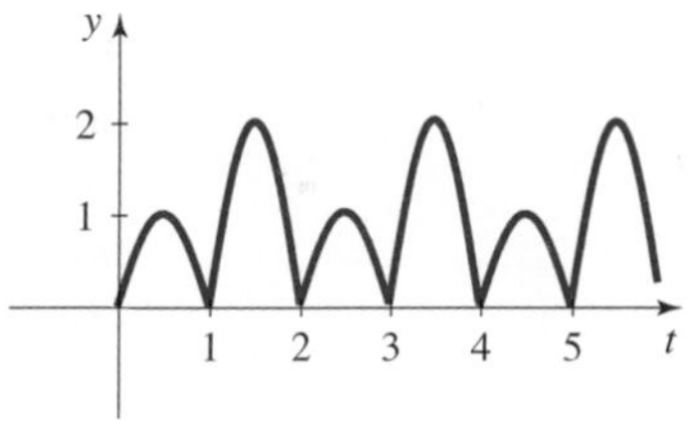

45. Find the Laplace transform for the square wave of Problem 8 by using the formula for periodic functions. Show that the result you obtain is equal to the result from Problem 8. HINT: Factor the denominator of the expression you obtained from the periodic formula.

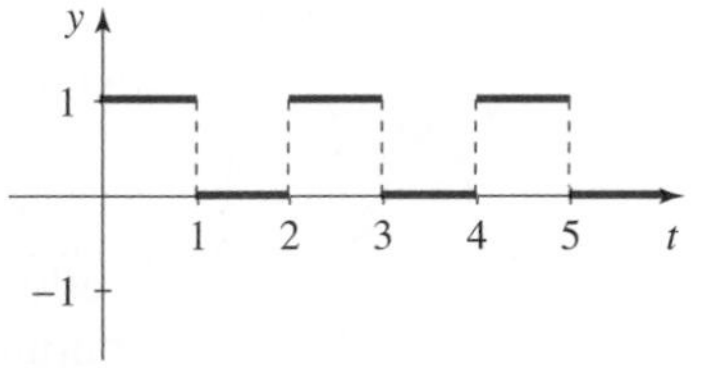

46. **Sawblade** Solve the initial-value problem $x'+x = f(t)$, $x(0) = 0$, where the forcing function

$$f(t) = t - \text{step}(t-1) - \text{step}(t-2) - \cdots$$

is the triangular sawtooth wave shown.

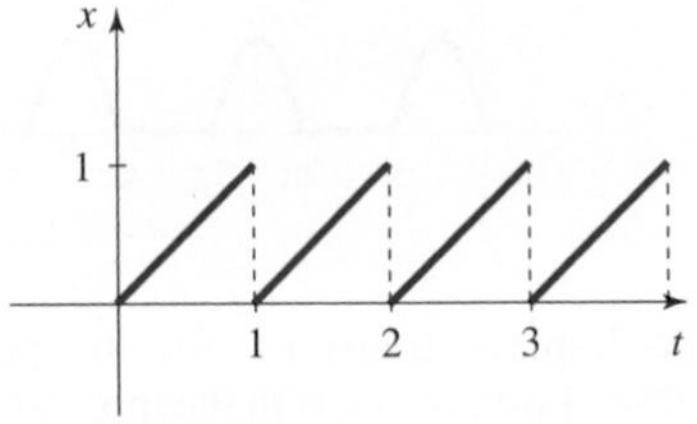

47. **Square Wave Input** Solve the initial-value problem $x'' + x = f(t)$, $x(0) = 0$, $x'(0) = 0$, where the rectangular wave shown is given by the forcing function

$$f(t) = 1 - 2\text{step}(t-1) + 2\text{step}(t-2) - \cdots.$$

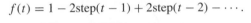

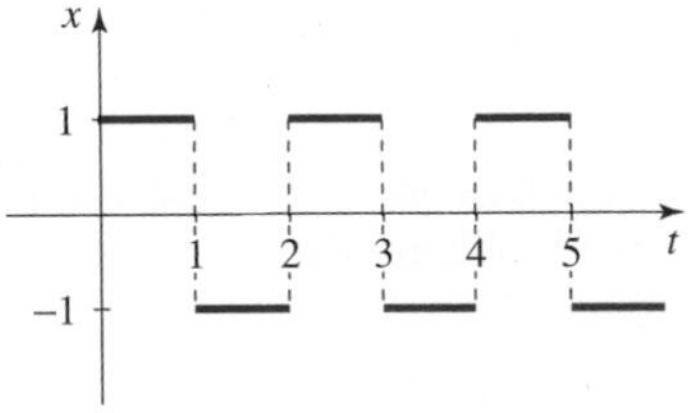

Solve on Impulse *Solve the IVP in each of Problems 48–52.*

48. $x' = \delta(t)$; $x(0) = 0$

49. $x' = \delta(t) - \delta(t-1)$; $x(0) = 0$

50. $x'' + x = \delta(t - 2\pi)$; $x(0) = 0$, $x'(0) = 0$

51. $x'' + x = -\delta(t - \pi) + \delta(t - 2\pi)$; $x(0) = 0$, $x'(0) = 1$

52. $x'' + x = \delta(t - 2\pi)$; $x(0) = 1$, $x'(0) = 0$

53. **Laplace with Forcing Functions** Work IDE Lab 14, Part 2, to visualize the effects of the theorems in this section.

Laplace: Translation;
Laplace: Solver;
Laplace: Shifting Theorem;
Laplace: Shift and Step;
Laplace: Delta Function
Explore how Laplace transforms interact with forced oscillators.

54. **Suggested Journal Entry** In the laboratory, it is impossible to generate force, voltage, or other physical quantities that are precisely represented by δ-functions, since the length of time the quantity would be required to "act" is zero. To approximate delta functions, engineers generate step functions that turn on and then off in a very short period of time. Discuss what techniques would be required to generate forcing functions for a mass-spring system, input voltages for a simple *LRC*-circuit, and injection of drugs into the bloodstream, which might be reresented by δ-functions.

8.4 The Convolution Integral and the Transfer Function

SYNOPSIS: The convolution integral and the transfer function play an important role in the description of linear systems and have many applications from home finance to quantum mechanics. For example, quantum systems subjected to radiation move to higher energy states—the quantum numbers of the energy states are convolutions.

Convolution

In Sec. 8.1, we found that the Laplace transform of the sum of two functions is equal to the sum of their transforms. The same is not true for the product, as we explored in Sec. 8.1, Problem 10. Consider, for example, that

$$\mathcal{L}\{t^n\} = \frac{n!}{s^{n+1}} \quad \text{and} \quad \mathcal{L}\{e^{at}\} = \frac{1}{s-a}, \quad \text{but} \quad \mathcal{L}\{t^n e^{at}\} = \frac{n!}{(s-a)^{n+1}}.$$

The result is not a product, but the ideas of the transforms of the factors are in some fashion merged.

The missing link is *convolution*, a generalization of the product. Defined as follows, the convolution provides another way to represent the solutions to input/output problems.

Laplace: Convolution Theorem

Here is an interactive version. See it to believe it.

Convolution

If $f(t)$ and $g(t)$ are piecewise continuous functions on $[0, \infty)$, the **convolution** of f and g is defined by

$$(f * g)(t) = \int_0^t f(t-w)g(w)\,dw. \tag{1}$$

The convolution is often denoted $f(t) * g(t)$.

The idea of convolution leads to a useful theorem relating to Laplace transforms.[1]

Convolution Theorem

If $f(t)$ and $g(t)$ are piecewise continuous functions on $[0, \infty)$ of exponential order α, then

$$\mathcal{L}\{(f * g)(t)\} = \mathcal{L}\{f(t)\} \cdot \mathcal{L}\{g(t)\} \equiv F(s) \cdot G(s), \quad s > \alpha. \tag{2}$$

The inverse form of (2) is

$$\mathcal{L}^{-1}\{F(s) \cdot G(s)\} = (f * g)(t), \quad s > \alpha. \tag{3}$$

We will use the convolution theorem to advantage whenever we want to find the inverse Laplace transform of a *product* of two functions $F(s)$ and $G(s)$ for which we know individually $\mathcal{L}^{-1}\{F(s)\}$ and $\mathcal{L}^{-1}\{G(s)\}$.

Useful Trigonometric Identities

$$\sin A \sin B = \frac{1}{2}[\cos(A-B) - \cos(A+B)]$$

$$\cos A \cos B = \frac{1}{2}[\cos(A+B) + \cos(A-B)]$$

$$\sin A \cos B = \frac{1}{2}[\sin(A+B) + \sin(A-B)]$$

EXAMPLE 1 **Resonance** Consider the case in which the frequency of the forcing function is the natural frequency of the system, such as the IVP

$$y'' + y = \sin t, \qquad y(0) = 1, \quad y'(0) = 0. \tag{4}$$

Using the methods of Sec. 8.2, we take the Laplace transform of both sides of the DE and solve for $Y(s) = \mathcal{L}\{y(t)\}$ to get

$$Y(s) = \frac{s}{s^2+1} + \frac{1}{(s^2+1)^2}. \tag{5}$$

The inverse Laplace transform of (5) is

$$y(t) = \cos t + \underbrace{\mathcal{L}^{-1}\left\{\frac{1}{(s^2+1)^2}\right\}}_{\sin t \,*\, \sin t},$$

[1] For a proof of the convolution theorem, see M. Braun, *Differential Equations and Their Applications*, 3rd ed. (NY: Springer-Verlag, 1983), or R. Borrelli and C. Coleman, *Differential Equations; A Modeling Perspective* (NY: Wiley, 1998).

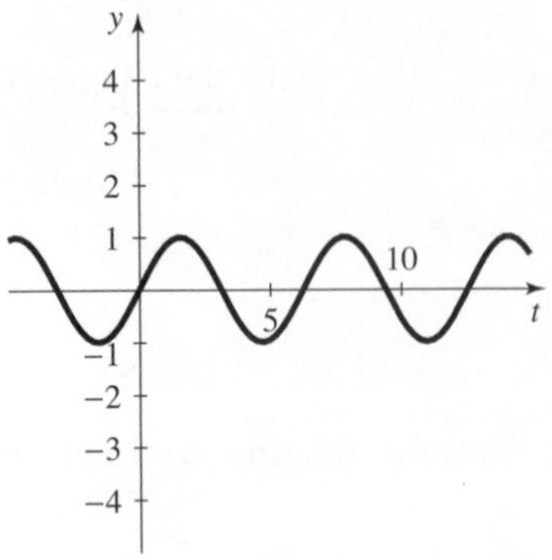

(a) Forcing function

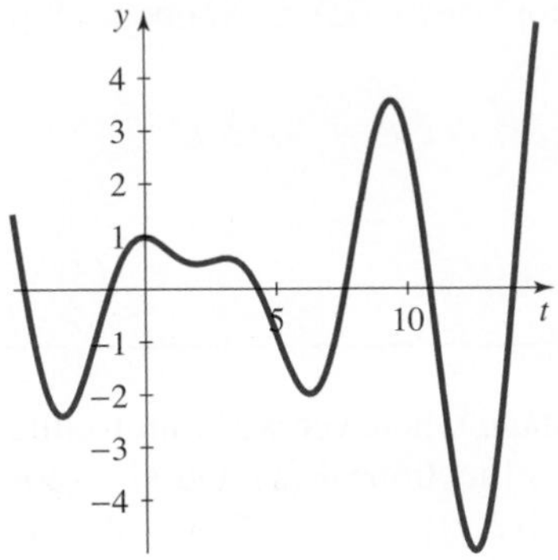

(b) Output function

FIGURE 8.4.1 Example 1: Graphs of the forcing function and the output function.

which can be completed using the convolution theorem as follows:

$$
\begin{aligned}
\sin t * \sin t &= \int_0^t \sin(t-w)\sin w\, dw \\
&= \frac{1}{2}\int_0^t [\cos(t-2w) - \cos t]\, dw \\
&= \frac{1}{2}\left[-\frac{\sin(t-2w)}{2}\right]_0^t - \frac{1}{2}t\cos t \\
&= \frac{1}{2}\left[\frac{\sin t}{2} + \frac{\sin t}{2}\right] - \frac{1}{2}t\cos t \\
&= \frac{\sin t - t\cos t}{2}.
\end{aligned}
$$

So the solution to the IVP (4) is

$$y(t) = \cos t + \frac{\sin t - t\cos t}{2}.$$

See Fig. 8.4.1 for graphs of the forcing function and the output function. ■

Properties of the Convolution

Many of the properties of convolution follow from the properties of integrals.

Convolution Properties

If $f(t)$ and $g(t)$ are piecewise continuous functions on $(0, \infty)$, then

- $f * g = g * f$, (commutative property)
- $f * (g * h) = (f * g) * h$, (associative property)
- $f * (g + h) = f * g + f * h$, (distributive property)
- $f * 0 = 0$. (zero multiplication)

EXAMPLE 2 **Sample Proof of Commutativity** We can verify the commutative property of the convolution as follows. By definition,

$$(f * g)(t) = \int_0^t f(t-w)g(w)\, dw. \tag{6}$$

Through a change of variable from t to $u = t - w$, equation (6) becomes

$$(f * g)(t) = \int_t^0 f(u)g(t-u)(-du) \tag{7}$$

$$= \int_0^t g(t-u)f(u)\, du, \tag{8}$$

and the definition of convolution leads to the conclusion that

$$(f * g)(t) = (g * f)(t).$$

■

We ask the reader to prove the remaining convolution properties in Problems 1–3.

The Transfer Function

In Sec. 2.2, we solved the first-order equation $y' + ay = f(t)$ using the integrating factor method. We found that the solution to this equation is

$$y(t) = \int_0^t e^{-a(t-w)} f(w)dw,$$

which we now recognize as a convolution, namely $y = e^{-at} * f(t)$. In fact, *all* solutions of linear differential equations with constant coefficients are convolutions.

Consider a physical system with two different forcing terms, $\delta(t)$ and $f(t)$, modeled by the same second-order DE:

$$ah'' + bh' + ch = \delta(t), \qquad h(0) = h'(0) = 0; \tag{9}$$

$$ay'' + by' + cy = f(t), \qquad y(0) = y'(0) = 0. \tag{10}$$

We say such systems are in **zero initial state** *because all initial conditions are zero*. We shall show how their solutions are related.

Laplace: Vibrations and Poles

Look at the mass-spring problem again from the standpoint of the transfer function.

NOTE: The *poles* of the transfer function are zeroes of the denominator of $H(s)$, which are also the roots of the characteristic equation for the DE.

Forcing with a Delta Function

The system

$$a_n h^{(n)} + a_{n-1} h^{(n-1)} + \cdots + a_1 h' + a_0 h = \delta(t),$$

$$h(0) = h'(0) = \cdots = h^{(n-1)}(0) = 0$$

has a solution $h(t)$, which is an **impulse response function**, as introduced in Sec. 8.3. Its Laplace transform,

$$H(s) = \mathcal{L}\{h(t)\} = \frac{1}{a_n s^n + a_{n-1} s^{n-1} + \cdots + a_1 s + a_0}, \tag{11}$$

is called the **transfer function** of the system.

Returning to equation (10), the IVP strategy of Sec. 8.2 gives

$$\mathcal{L}\{y(t)\} = \frac{1}{as^2 + bs + c}\mathcal{L}\{f(t)\},$$

which tells us that

$$\mathcal{L}\{y\} = \mathcal{L}\{h\}\mathcal{L}\{f\}.$$

Staying in the s-Domain:

Some engineers prefer to stay in the s-domain and use transfer functions to characterize the systems. (See Sec. 10.1, Problems 17–22.)

Consequently, by the convolution theorem, we have for (10) the solution

$$y(t) = h(t) * f(t) = \int_0^t h(t - w) f(w)\, dw. \tag{12}$$

Equation (12) states a very important result:

> *The output of a linear system is the convolution of the impulse response function and the input function of the system.*

In other words, once we know how a linear system reacts to a delta function, we know how it reacts to *any* input. We summarize this discussion as the following theorem.

Solution of a Linear IVP as a Convolution

Let $a_0, a_1, \ldots, a_n$ be real numbers and $h(t)$ the impulse response defined by

$$\begin{aligned} a_n h^{(n)} + a_{n-1} h^{(n-1)} + \cdots + a_1 h' + a_0 h &= \delta(t), \\ h(0) = h'(0) = \cdots = h^{(n-1)}(0) &= 0. \end{aligned} \tag{13}$$

If the convolution $h * f$ exists for some forcing function f on an interval $0 < t < T$, then the initial-value problem

$$\begin{aligned} a_n y^{(n)} + a_{n-1} y^{(n-1)} + \cdots + a_1 y' + a_0 y &= f(t), \\ y(0) = y'(0) = \cdots = y^{(n-1)}(0) &= 0 \end{aligned} \tag{14}$$

has the solution

$$y(t) = (h * f)(t) = \int_0^t h(t - w) f(w)\, dw, \quad t \in (0, T). \tag{15}$$

EXAMPLE 3 Second-Order IVPs in Zero Initial State We can use the impulse response $h(t)$ to write the solution of a second-order IVP as a convolution. To find the solution of

$$y'' + y = f(t), \qquad y(0) = 0, \quad y'(0) = 0, \tag{16}$$

we first find the impulse response by solving

$$h'' + h = \delta(t), \qquad h(0) = 0, \quad h'(0) = 0$$

to get $h(t) = \sin t$. Hence, the solution of (16) is the convolution

$$y(t) = \sin t * f(t) = \int_0^t \sin(t - w) f(w)\, dw.$$

■

EXAMPLE 4 Third-Order IVP in Zero Initial State Write the solution of the following IVP as a convolution:

$$y''' + 6y'' + 12y' + 8y = f(t), \quad y(0) = y'(0) = y''(0) = 0. \tag{17}$$

Taking the Laplace transform of

$$h''' + 6h'' + 12h' + 8h = \delta(t), \qquad h(0) = h'(0) = h''(0) = 0,$$

gives

$$\mathcal{L}\{h''' + 6h'' + 12h' + 8h\} = 1,$$

and the transfer function

$$H(s) = \frac{1}{s^3 + 6s^2 + 12s + 8}.$$

The solution of (17) has Laplace transform

$$Y(s) = H(s)F(s) = \frac{1}{s^3 + 6s^2 + 12s + 8} F(s) = \frac{1}{(s+2)^3} F(s).$$

Thus, by the convolution theorem,

$$y(t) = \frac{1}{2} t^2 e^{-2t} * f(t).$$

■

The use of unit impulse functions and the convolution integral does not usually save computation time, due to the fact that the integral must be calculated, although convolution integrals can often be evaluated by means of a computer algebra system. The major advantage of the transfer function is that $H(s)$ contains all the information from the initial-value problem except the forcing function, and $H(s)$ can be multiplied by $F(s)$ for *any* appropriate forcing function $f(t)$. Then, by taking the inverse Laplace transform of this product via the convolution, we obtain the solution.

EXAMPLE 5 **On/Off by Transfer** We return to Sec. 8.3, Example 7, the IVP

$$x'' + x = 1 - \text{step}(t - \pi), \qquad x(0) = 0, \quad x'(0) = 0.$$

Because the initial conditions are all zero, we can use equation (14), with coefficients $a = c = 1$ and $b = 0$, to find the transfer function

$$H(s) = \mathcal{L}\{h(t)\} = \frac{1}{s^2 + 1}$$

and its inverse Laplace transform $h(t) = \sin t$.

By equation (15), we know that

$$x(t) = h(t) * f(t), \quad \text{where } f(t) = 1 - \text{step}(t - \pi),$$

so the solution of the IVP is the convolution

$$x(t) = \sin t * [1 - \text{step}(t - \pi)],$$

which can be evaluated as follows:

$$\begin{aligned} x(t) &= \int_0^t \sin(t - w)[1 - \text{step}(w - \pi)]\, dw \\ &= \begin{cases} \displaystyle\int_0^t \sin(t - w)\, dw & \text{if } 0 \le t < \pi, \\ \displaystyle\int_0^\pi \sin(t - w)\, dw + \int_\pi^\infty 0\, dw & \text{if } t \ge \pi \end{cases} \\ &= \begin{cases} \displaystyle\int_0^t \sin t \cos w - \cos t \sin w\, dw & \text{if } 0 \le t < \pi, \\ \displaystyle\int_0^\pi \sin t \cos w - \cos t \sin w\, dw & \text{if } t \ge \pi \end{cases} \\ &= \begin{cases} 1 - \cos t & \text{if } 0 \le t < \pi, \\ -2\cos t & \text{if } t \ge \pi. \end{cases} \end{aligned}$$

For a graph of the solution, see Fig. 8.3.7. ■

Applications

The convolution integral does not seem particularly intuitive, until you look at a few examples to see how it arises in everyday life.

In general, a convolution

$$y(t) = h(t) * f(t) = \int_0^t h(t - w) f(w)\, dw \tag{18}$$

can be interpreted as the total amount of some "substance" (money, radioactive material, heat, energy, concentration ... anything at all) at time t, if the substance is added at a *rate* $f(t)$, and $h(t)$ is the amount that the substance changes (grows or decays) as the result of the dynamics of the system.

Moreover, the convolution integral is practical because it gives a formula for the output of a linear system with constant coefficients that can be applied to different inputs $f(t)$.

In the laboratory, the δ-function, or rather a step-function of very short duration, can be used to test an unknown system in zero initial state that is assumed to be linear. The resulting solution function $h(t)$, the transfer function, is approximated using a continuous approximation to obtain solutions for a variety of forcing terms.

EXAMPLE 6 **Savings** Suppose that we make continuous deposits into a savings account at a nonconstant rate of $R(t)$ dollars per unit time, and that the bank pays interest at an annual rate of 6%. How much money will be in the account at any future time t?

The answer is indeed a convolution, and the question is nontrivial because different deposits earn interest for different lengths of time. To solve this problem, we simplify it by subdividing the time interval into small subintervals and making periodic deposits at the endpoints of these intervals, as illustrated in Fig. 8.4.2.

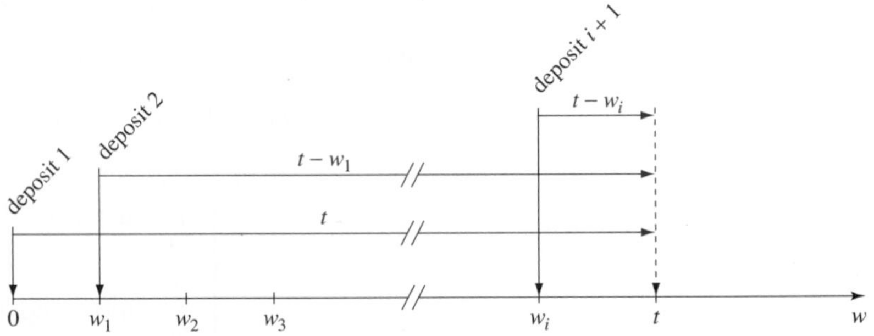

FIGURE 8.4.2 Example 6: Making deposits at equally spaced intervals.

Because we want to know the future value of the account at time t, it is best to denote the time scale by the new variable w, and to think of t as an arbitrary point on the w axis. Under this convention, if we deposit $R(w_i)$ dollars at time w_i, this deposit will grow in value, due to compound interest, to the value $R(w_i)e^{0.06(t-w_i)}$ at time t. The $t - w_i$ in the expression is due to the fact that the length of time the deposit collects interest is $t - w_i$. Adding up the deposits and the interest from the $n + 1$ deposits, the total amount in the bank will be

$$\text{total amount from finite payments} = \sum_{i=0}^{n} R(w_i)e^{0.06(t-w_i)}. \tag{19}$$

We now let the time intervals between deposits approach zero and arrive at the total amount of the account from continuous payments at rate $R(t)$, or the convolution

$$\text{total amount} = \int_0^t R(w)e^{0.06(t-w)}dw = e^{0.06t} * R(t), \tag{20}$$

where $e^{0.06t}$ is the impulse response to the delta function input of a \$1 deposit at time $t = 0$.

See Problem 37 to show that calculation of the convolution in equation (20) gives the same solution that we obtained in Sec. 2.3 equations (10) and (11), for the case where $R(t) = a$ and the constant a is the total amount deposited continuously in each year.

■

EXAMPLE 7 **Lake Pollution** Suppose that a pollutant is dumped into a lake at a rate $f(t)$, whereupon it immediately begins to degrade exponentially at a rate $h(t) = e^{-kt}$. What is the amount of pollutant in the lake at time t?

In the previous example, we made deposits into a bank account and then the money grew. Here, we add something that decreases instead of growing. Hence, the future amount $y(t)$ of pollutant in the lake at time t is given by

$$y(t) = h(t) * f(t) = \int_0^t e^{-k(t-w)} f(w)\, dw.$$

■

Summary

The Convolution Theorem gives us a new kind of product in the t-domain. The inverse form of the theorem relates products of transforms in the s-domain to the convolution of functions in the t-domain. The transfer function, defined as the transform of the impulse response function with initial conditions zero, can be seen to contain all the information about the original linear differential equation with constant coefficients.

8.4 Problems

Convolution Properties *Verify the convolution properties in Problems 1–3 for the functions to which the convolution theorem applies.*

1. $f * (g * h) = (f * g) * h$

2. $f * (g + h) = f * g + f * h$

3. $f * 0 = 0$

Calculating Convolutions *Find the convolutions in Problems 4–10 from the integral definition of a convolution.*

4. $1 * 1$ **5.** $1 * 1 * 1$ **6.** $1 * t$

7. $t * t$ **8.** $t * t * t * \cdots * t$ (k factors)

9. $e^t * e^{-t}$ **10.** $e^{at} * e^{-at}$

11. First-Order Convolution Equation Inasmuch as the convolution is sometimes called "generalized" multiplication, what is the solution of the equation for $a * t = b$, for $a, b \in \mathbb{R}$, $a \neq 0$? Is it equal to $t = b/a$, as it is for $at = b$?

Convoluted Solutions *In each of Problems 12–17, obtain a solution of the IVP in terms of a convolution.*

12. $x' = f(t)$; $x(0) = 0$

13. $x' = f(t)$; $x(0) = 1$

14. $x' + x = f(t)$; $x(0) = 0$

15. $x' + x = f(t)$; $x(0) = 1$

16. $x'' + x = f(t)$; $x(0) = 1$, $x'(0) = 0$

17. $x'' + 3x' + 2x = f(t)$; $x(0) = 0$, $x'(0) = 0$

Transfer and Impulse Response Functions *In each of Problems 18–21, determine the transfer function and impulse response function of the input/output system. Write the output as a convolution of the impulse response and input functions (take all initial conditions to be zero).*

18. $x' = f(t)$ **19.** $x' + ax = f(t)$

20. $x'' + x = f(t)$ **21.** $x'' + 4x' + 5x = f(t)$

Inverse of Convolution Theorem *For Problems 22–28, find the given inverse Laplace transform in terms of a convolution, and evaluate the convolution.*

22. $\mathcal{L}^{-1}\left\{\frac{1}{s^2}\right\}$ **23.** $\mathcal{L}^{-1}\left\{\frac{1}{s^3}\right\}$

24. $\mathcal{L}^{-1}\left\{\frac{1}{s(s+1)}\right\}$ **25.** $\mathcal{L}^{-1}\left\{\frac{4}{s^2(s-2)}\right\}$

26. $\mathcal{L}^{-1}\left\{\frac{1}{s^2(s^2+1)}\right\}$ **27.** $\mathcal{L}^{-1}\left\{\frac{1}{(s^2+1)^2}\right\}$

28. $\mathcal{L}^{-1}\left\{\dfrac{1}{(s^2+k^2)^2}\right\}$ (See Example 1.)

29. Nonzero Initial State Suppose that $h(t)$, the unit impulse response function, is the solution to

$$ah''+bh'+ch=\delta(t)$$

in zero initial state. Use $h(t)$ to find the Laplace transform of the solution of the IVP

$$ax''+bx'+cx=f(t),\quad x(0)=x_0,\, x'(0)=x_1.$$

Nonzero Practice *Solve the following problems using the method of Problem 29.*

30. $x''+x'+x=\delta(t-2),\qquad x(0)=1,\quad x'(0)=0$

31. $x''+4x=4\cos t,\qquad x(0)=1,\quad x'(0)=-1$

Fractional Calculus *Although calculus students learn how to integrate and differentiate functions once, twice, three times, and so on, there is a theory of* ***fractional calculus*** *that defines integrals and derivatives of any real (or even complex) order. Problems 32–34 provide an introduction to this subject.*

32. Using the following definition, find the one-half integrals in (a)–(c).

> **One-Half Integral**
>
> The **one-half integral** (or semi-integral) of a function $f(t)$ is defined as the convolution
>
> $$I_{1/2}(f)=\frac{1}{\sqrt{\pi}}(t^{-1/2}*f).$$

(a) $I_{1/2}(1)$ (b) $I_{1/2}(t)$ (c) $I_{1/2}(at^2+bt+c)$

33. Verify that two half-integrals equal the whole integral

$$I_{1/2}(I_{1/2}(f))(t)=\int_0^t f(w)\,dw.$$

HINT: $\mathcal{L}\{t^{-1/2}\}=\sqrt{\pi/s}$.

34. Using the definition of the one-half integral as given in Problem 32, we have the following definition.

> **One-Half Derivative**
>
> The **one-half derivative** (or semi-derivative) of a function $f(t)$ is defined by
>
> $$\frac{d^{1/2}}{dt^{1/2}}f(t)=\frac{d}{dt}I_{1/2}(f).$$

As with whole derivatives, fractional derivatives may or may not exist. Find the following one-half derivatives and compare them with first derivatives.

(a) $\dfrac{d^{1/2}}{dt^{1/2}}(1)$ (b) $\dfrac{d^{1/2}}{dt^{1/2}}(t)$ (c) $\dfrac{d^{1/2}}{dt^{1/2}}(at^2+bt+c)$

35. Trendy Savings A corporation uses a continuous model for commissions from the sales of a trendy model, using a surge function $f(t)=10^6te^{-t}$. These proceeds are deposited immediately into a savings account that accrues 8% interest, compounded continuously.

(a) Set up an IVP for the amount A in the savings account at time t, given that $A(0)=0$.

(b) Write the solution $A(t)$ as a convolution.

(c) By means of a computer algebra system (CAS), evaluate the convolution integral to determine the solution. On one graph, plot the surge function and the solution function.

36. Investment and Savings Upon graduation from college, John makes an investment that makes continuous deposits of $r(t)=10000e^{0.01t}$ dollars into a savings account. If the bank pays annual interest of 4%, compounded continuously, find and evaluate the convolution integral that approximates the value of his account. What is the value of John's savings after 20 years?

37. Consistency Check In Sec. 2.3, equations (10) and (11), we calculated that savings $A(t)$ from continuous deposits $R(t)=a$ would be

$$A(t)=A_0e^{rt}+\frac{a}{r}(e^{rt}-1),$$

where r is the interest rate and a is the amount deposited over one year. Substitute $R(t)=a$ into equation (20) of Example 6 and compute the convolution to show that the convolution integral of this section indeed gives the same result.

38. Lake Pollutant A pollutant is dumped into a lake at a rate $r(t)=2e^{0.05t}$ kg/day. Once in the lake, the pollutant degrades according to the law $e^{-0.10t}$. Write and evaluate the convolution integral that gives the amount $y(t)$ of pollutant in the lake at time t and evaluate this integral.

39. Radioactive Decay Chain A radioactive substance is created from another radioactive substance at the rate $r(t)=e^{0.001t}$ (grams/year). The substance created then decays at the rate $e^{-0.01t}$. Write the convolution integral that gives the amount $y(t)$ of the substance present at time t and evaluate this integral.

Volterra Integral Equation *An equation of the type*

$$y(t)=g(t)+\int_0^t k(t-w)y(w)\,dw, \tag{21}$$

where g and k are known functions and the unknown function y appears under the integral sign, is called a ***Volterra***

integral equation.[2] *Because the integral in this equation is the convolution $k * y$, it is possible to solve this equation using the Laplace transform. For Problems 40–44, solve the given Volterra integral equation.*

40. $y(t) = 1 - \int_0^t y(w)\,dw$

41. $y(t) = t - \int_0^t (t-w)y(w)\,dw$

42. $y(t) = t^3 + \int_0^t \sin(t-w)y(w)\,dw$

43. $y(t) = e^t\left[1 + \int_0^t e^{-w}y(w)\,dw\right]$

44. $y(t) = \cos t + \int_0^t \sin(t-w)y(w)\,dw$

45. General Solution of Volterra's Equation Show that the general solution of the Volterra integral equation (21) is given by

$$y(t) = \mathcal{L}^{-1}\left\{\frac{G(s)}{1-K(s)}\right\},$$

where $G(s)$ and $K(s)$ are the Laplace transforms of $g(t)$ and $k(t)$, respectively.

46. Looking for the Current Consider the series circuit, shown in Fig. 8.4.3, for which $L = 1$ henry, $R = 10$ ohms, and $C = 1/25$ farad.

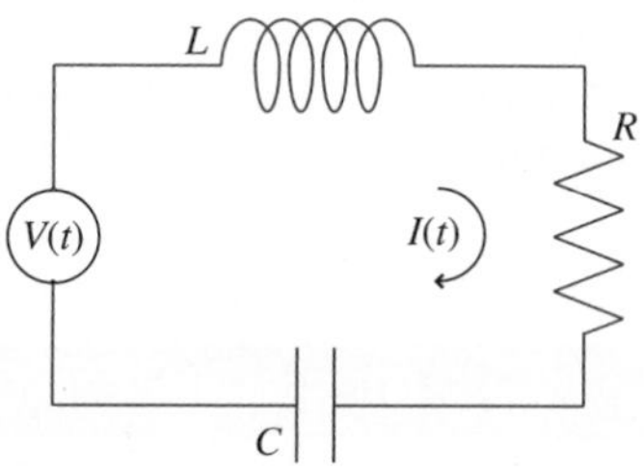

FIGURE 8.4.3 *LRC*-circuit.

(a) Write the integrodifferential equation for the current as a function of time for $V(t) = 12 - 24\,\text{step}(t-1)$.

(b) Assuming that $I(0) = 0$ and $I'(0) = 0$, use the transfer function to find the solution for the IVP, written as a convolution.

(c) Use a computer algebra system to evaluate the convolution integral.

Transfer Functions for Circuits *For the following two circuits, in zero initial state, determine the transfer function and indicate the solution for the unknown function of t as a convolution.*

47. Find $H(s)$ and $Q(t)$ for the circuit in Fig. 8.4.4.

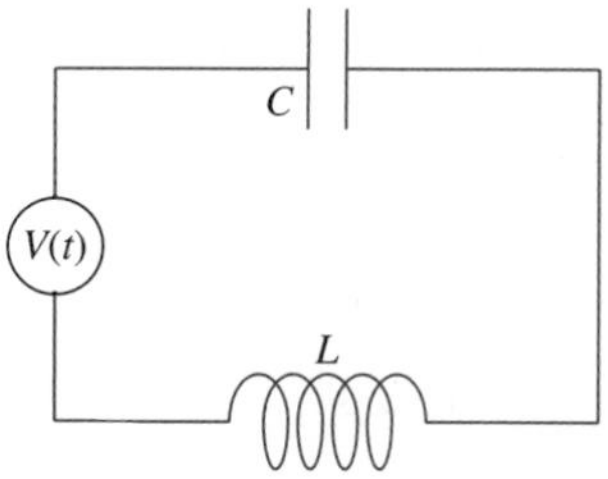

FIGURE 8.4.4 *LC*-Circuit for Problem 47.

48. Find $H(s)$ and $I(t)$ for the circuit in Fig. 8.4.5.

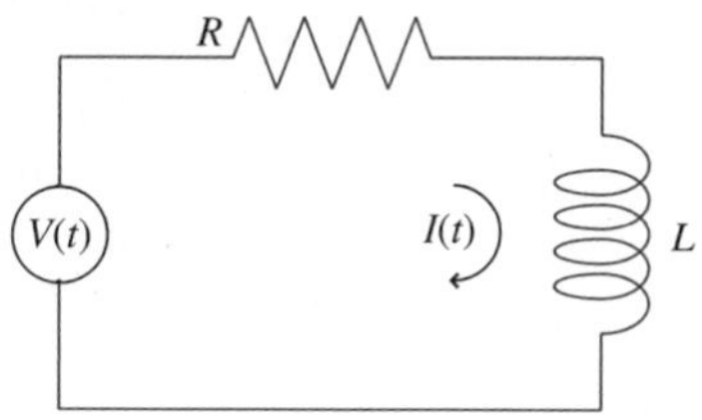

FIGURE 8.4.5 *LR*-Circuit for Problem 48.

49. Interesting Convolution Show that $y(t) = \sin t * \sin t$ is the solution of

$$y'' + y = \sin t, \qquad y(0) = y'(0) = 0.$$

50. Duhamel's Principle Consider the IVP

$$ay'' + by' + cy = f(t), \qquad y(0) = y'(0) = 0.$$

Duhamel's Principle[3] states that the solution of this initial-value problem can be expressed as

$$y(t) = \int_0^t f(t-w)z'(w)\,dw,$$

where $z(t)$ satisfies the same initial-value problem but with $f(t)$ replaced by 1. That is,

$$az'' + bz' + cz = 1, \qquad z(0) = z'(0) = 0.$$

Verify Duhamel's Principle.

[2]We met Italian mathematician Vito Volterra (1860–1940) in Sec. 2.6, concerning his mathematical description of predator-prey systems. However, long before he turned his attention to mathematical biology, his work (1884–1896) on integral equations such as (21) claimed more fame among mathematicians working with differential equations.

[3]Frenchman Jean-Marie Duhamel (1797–1872) had a long and productive career in mathematics and physics. He is known particularly for his mathematical theory of heat, from which arose Duhamel's Principle.

51. **Using Duhamel's Principle** Use Duhamel's Principle (stated in Problem 51) to find the solution of the initial-value problem

$$y'' - y = f(t), \qquad y(0) = y'(0) = 0.$$

52. **Interesting Integral Equation** Solve the integral equation

$$\int_0^t y(\omega)\, d\omega = y(t) * y(t)$$

and verify your results.

53. **Computer Exploration** Work IDE Lab 14, Part 3, to further explore the convolution process.

Laplace: Convolution Example; Laplace: Convolution Theorem; Laplace: Solver
Visualization of convolution fills in missing steps of solver tool.

54. **Suggested Journal Entry** Look for the appearance of the convolution integral in your field, whether it is engineering, mathematics, or science. Consider why this integral occurs in so many applications. What kind of processes are being described? What do they have in common?

8.5 Laplace Transform Solution of Linear Systems

SYNOPSIS: We will extend the method of Laplace transforms for linear initial-value problems to solving initial-value problems for linear systems of differential equations. The advantage of the method is that we can solve the transformed IVP by linear algebraic methods.

Introduction

We have seen that the Laplace transform is a powerful tool for solving a single linear differential equation with constant coefficients by transforming it into an algebraic equation, solving the algebraic equation, and transforming it back to obtain the solution of the differential equation. The Laplace transform can also be used to solve a *system* of linear differential equations with constant coefficients by taking the transform of each component of the system, and transforming it to a *system* of linear algebraic equations.

Laplace Transform Strategy for Linear DE Systems

The initial-value problem

$$\vec{\mathbf{x}}' = \mathbf{A}\vec{\mathbf{x}} + \vec{\mathbf{f}}(t), \qquad \vec{\mathbf{x}}(0) = \vec{\mathbf{x}}_0, \tag{1}$$

where $\mathbf{A}$ is an $n \times n$ matrix of real constants and $\vec{\mathbf{f}}(t)$ is an $n \times 1$ vector of functions of t, can be solved using the following strategy:

Step 1. Using the Laplace transform $\mathcal{L}$, transform the IVP (1) with unknown $\vec{\mathbf{x}}(t)$ into an algebraic problem with unknown $\vec{\mathbf{X}}(s)$.

$$s\vec{\mathbf{X}}(s) - \vec{\mathbf{x}}_0 = \mathbf{A}\vec{\mathbf{X}}(s) + \vec{\mathbf{F}}(s), \tag{2}$$

or equivalently,

$$(s\mathbf{I} - \mathbf{A})\vec{\mathbf{X}}(s) = \vec{\mathbf{x}}_0 + \vec{\mathbf{F}}(s), \tag{3}$$

where $\vec{\mathbf{X}}(s) = \mathcal{L}\{\vec{\mathbf{x}}(t)\}$ and $\vec{\mathbf{F}}(s) = \mathcal{L}\{\vec{\mathbf{f}}(t)\}$ are the vectors of transforms of entries of $\vec{\mathbf{x}}(t)$ and $\vec{\mathbf{f}}(t)$, respectively.

Step 2. Solve the algebraic system for $\vec{\mathbf{X}}(s)$:

$$\vec{\mathbf{X}}(s) = (s\mathbf{I} - \mathbf{A})^{-1}(\vec{\mathbf{x}}_0 + \vec{\mathbf{F}}(s)). \tag{4}$$

Step 3. Manipulating $\vec{\mathbf{X}}(s)$ algebraically if necessary, use the inverse Laplace transform $\mathcal{L}^{-1}$ to transform $\vec{\mathbf{X}}(s)$ into the IVP solution $\vec{\mathbf{x}}(t)$.

The Laplace solution to a system of linear DEs *can* be written

$$\begin{aligned}\vec{\mathbf{x}}(t) &= \mathcal{L}^{-1}\{(s\mathbf{I} - \mathbf{A})^{-1}(\vec{\mathbf{x}}_0 + \vec{\mathbf{F}}(s))\} \\ &= \underbrace{\mathcal{L}^{-1}\{(s\mathbf{I} - \mathbf{A})^{-1}\vec{\mathbf{x}}_0\}}_{\vec{\mathbf{x}}_h} + \underbrace{\mathcal{L}^{-1}\{(s\mathbf{I} - \mathbf{A})^{-1}\vec{\mathbf{F}}(s)\}}_{\vec{\mathbf{x}}_p},\end{aligned}$$

but problems are usually *solved* by retracing the steps.

EXAMPLE 1 **Homogeneous Linear System** We will solve the initial-value problem

$$\begin{aligned} x_1' &= x_1 + x_2, \\ x_2' &= 4x_1 + x_2, \end{aligned} \qquad x_1(0) = 1, \quad x_2(0) = 0 \tag{5}$$

in component form using Laplace transforms.

Step 1. Denoting $X_1(s) = \mathcal{L}\{x_1\}$ and $X_2(s) = \mathcal{L}\{x_2\}$, and using the derivative and linear properties of the transform, we find the component form of (2):

$s\vec{\mathbf{X}}(s) - \vec{\mathbf{x}}_0 = \mathbf{A}\vec{\mathbf{X}}(s)$

$$\begin{aligned} sX_1(s) - x_1(0) &= X_1(s) + X_2(s), \\ sX_2(s) - x_2(0) &= 4X_1(s) + X_2(s). \end{aligned} \tag{6}$$

Substituting into (6) the initial conditions $x_1(0) = 1$, $x_2(0) = 0$, we rewrite the equations in form (3):

$(s\mathbf{I} - \mathbf{A})\vec{\mathbf{X}}(s) = \vec{\mathbf{x}}_0$

$$\begin{aligned} (s-1)X_1(s) - \quad X_2(s) &= 1, \\ -4X_1(s) + (s-1)X_2(s) &= 0. \end{aligned}$$

Step 2. Solving this algebraic system yields

$\vec{\mathbf{X}}(s) = (s\mathbf{I} - \mathbf{A})^{-1}\vec{\mathbf{x}}_0$

$$\begin{aligned} X_1(s) &= \frac{s-1}{s^2 - 2s - 3} = \frac{1}{2(s-3)} + \frac{1}{2(s+1)}, \\ X_2(s) &= \frac{4}{s^2 - 2s - 3} = \frac{1}{s-3} - \frac{1}{s+1}. \end{aligned} \tag{7}$$

Step 3. The solution of (5) is the inverse transform of (7):

$$\begin{aligned} x_1(t) &= \frac{1}{2}(e^{3t} + e^{-t}), \\ x_2(t) &= e^{3t} - e^{-t}. \end{aligned} \tag{8}$$

This solution is the same as that for Sec. 6.2, Example 1, although the previous example did not include initial conditions. ■

EXAMPLE 2 Nonhomogeneous Linear System The Laplace transform strategy will also solve the nonhomogeneous initial-value problem

$$\begin{aligned} x_1' &= -2x_1 + x_2 + 1, \\ x_2' &= x_1 - 2x_2, \end{aligned} \qquad x_1(0) = 0, \quad x_2(0) = 1. \tag{9}$$

Step 1. The Laplace transform of (9) is

$s\vec{\mathbf{X}}(s) - \vec{\mathbf{x}}_0 = \mathbf{A}\vec{\mathbf{X}}(s) + \vec{\mathbf{F}}(s)$

$$\begin{aligned} sX_1(s) &= -2X_1(s) + X_2(s) + \frac{1}{s}, \\ sX_2(s) - 1 &= X_1(s) - 2X_2(s), \end{aligned} \tag{10}$$

or

$(s\mathbf{I} - \mathbf{A})\vec{\mathbf{X}}(s) = \vec{\mathbf{x}}_0 + \vec{\mathbf{F}}(s)$

$$\begin{aligned} (s+2)X_1(s) - X_2(s) &= \frac{1}{s} \\ -X_1(s) + (s+2)X_2(s) &= 1. \end{aligned} \tag{11}$$

Step 2. Solving (11) for $X_1(s)$ and $X_2(s)$, we get

$\vec{\mathbf{X}}(s) = (s\mathbf{I} - \mathbf{A})^{-1}(\vec{\mathbf{x}}_0 + \vec{\mathbf{F}}(s))$

$$\begin{aligned} X_1(s) &= \frac{2}{s(s+3)} = \frac{2}{3s} - \frac{2}{3(s+3)}, \\ X_2(s) &= \frac{s+1}{s(s+3)} = \frac{1}{3s} + \frac{2}{3(s+3)}. \end{aligned} \tag{12}$$

Step 3. The inverse Laplace transform of (12) reveals the solution:

$$\begin{aligned} x_1(t) &= \frac{2}{3}(1 - e^{-3t}), \\ x_2(t) &= \frac{1}{3}(1 + 2e^{-3t}). \end{aligned}$$

■

EXAMPLE 3 Lidocaine Metabolism The drug Lidocaine is commonly used in the treatment of *ventricular arrhythmias* (irregular heartbeat). The schematic model shown in Fig. 8.5.1 is a widely used model of Lidocaine kinetics. It shows a two-compartment model predicting the amount of Lidocaine in the blood (m_1) and in the tissue (m_2) resulting from a 2-mg injection of Lidocaine into the bloodstream.

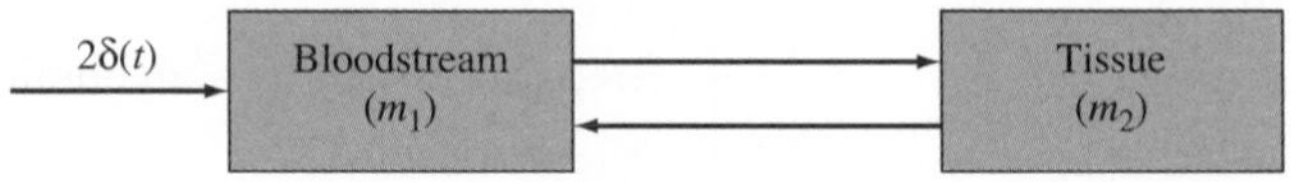

FIGURE 8.5.1 Example 3: Lidocaine two-compartment model.

We assume that 2 mg of Lidocaine is injected into the bloodstream, which then moves into the tissues of the heart. Since an injection of a drug is often represented mathematically by a delta function, the initial-value problem that describes the amount (in milligrams) of Lidocaine in the bloodstream (m_1) and in the tissue (m_2) is

$$\begin{aligned} \dot{m}_1 &= -m_1 + 2\delta(t), \\ \dot{m}_2 &= -m_1 - m_2, \end{aligned} \qquad m_1(0) = 0, \quad m_2(0) = 0, \tag{13}$$

where $\delta(t)$ is the delta function. Find the future amount of Lidocaine in the bloodstream and heart tissue.

One of the special features of the Laplace transform is its ability to work with impulse functions.

Denoting $M_1(s) = \mathcal{L}\{m_1\}$ and $M_2(s) = \mathcal{L}\{m_2\}$, and taking the Laplace transform of (14), we find

$$sM_1(s) = -M_1(s) + 2,$$
$$sM_2(s) = -M_1(s) - M_2(s).$$

Solving for $M_1(s)$ and $M_2(s)$, we get

$$M_1(s) = \frac{2}{s+1},$$
$$M_2(s) = \frac{2}{(s+1)^2}.$$

Hence,

$$m_1(t) = \mathcal{L}^{-1}\left\{\frac{2}{s+1}\right\} = 2e^{-t},$$
$$m_2(t) = \mathcal{L}^{-1}\left\{\frac{2}{(s+1)^2}\right\} = 2te^{-t}.$$

Graphs of these solutions are shown in Fig. 8.5.2. ■

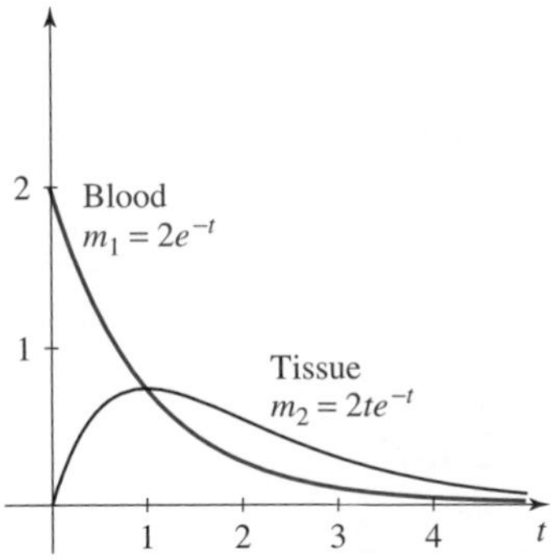

FIGURE 8.5.2 Example 3: metabolism of Lidocaine in the blood and tissue.

EXAMPLE 4 **Coupled Mass-Spring System** Two masses (m_1, m_2) are connected to each other and two walls by three springs (k_1, k_2, k_3), as shown in Fig. 8.5.3. The masses slide in a straight line on a frictionless surface. The system is set in motion by holding the left mass in its equilibrium position while at the same time pulling the right mass a distance p to the right of its equilibrium. What is the subsequent motion of the masses?

Motion of the Springs:

- When $x > 0$, the left spring (k_1) is stretched, so the force pulls m_1 to the left.
- When $y - x > 0$, the middle spring (k_2) is stretched, pulling m_1 to the right and m_2 to the left;
- When $y > 0$, the right spring (k_3) is compressed, pushing m_2 to the left.

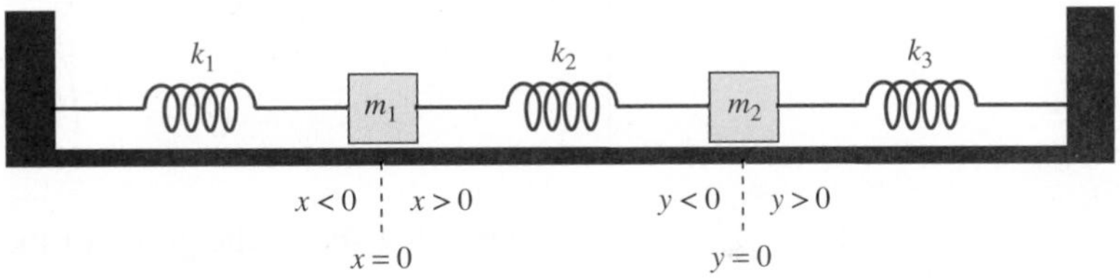

FIGURE 8.5.3 Example 4: Coupled vibrating system.

Denoting by $x(t)$ and $y(t)$ the positions of the masses m_1 and m_2 with respect to their equilibrium positions, Hooke's Law tells us the following about the forces acting on the masses due to the connecting springs:

- The force on m_1 due to the left spring is $-k_1x$.
- The force on m_1 due to the middle spring is $k_2(y - x)$.
- The force on m_2 due to the middle spring is $-k_2(y - x)$.
- The force on m_2 due to the right spring is $-k_3y$.

Hence, we have the following initial-value problem:

$$\begin{aligned} m_1\ddot{x} &= -k_1x + k_2(y - x), \qquad x(0) = 0, \quad \dot{x}(0) = 0, \\ m_2\ddot{y} &= -k_2(y - x) - k_3y), \qquad y(0) = p, \quad \dot{y}(0) = 0. \end{aligned} \tag{14}$$

IVP equation (14) for Example 4:

$$m_1\ddot{x} = -k_1x + k_2(y - x),$$
$$m_2\ddot{y} = -k_2(y - x) - k_3y;$$
$$x(0) = 0, \quad \dot{x}(0) = 0;$$
$$y(0) = p, \quad \dot{y}(0) = 0.$$

Taking the Laplace transform of the equations in (14), we get

$$\begin{aligned} m_1[s^2X(s) - sx(0) - \dot{x}(0)] &= -k_1X(s) + k_2[Y(s) - X(s)], \\ m_2[s^2Y(s) - sy(0) - \dot{y}(0)] &= -k_2[Y(s) - X(s)] - k_3Y(s). \end{aligned} \tag{15}$$

Letting $m_1 = m_2 = m$ and $k_1 = k_2 = k_3 = k$, and substituting the initial conditions, we get

$$\begin{aligned} (ms^2 + 2k)X(s) - kY(s) &= 0, \\ -kX(s) + \left(ms_2^2k\right)Y(s) &= msp. \end{aligned}$$

Solving for $X(s)$ and $Y(s)$, we find

$$\begin{aligned} X(s) &= \frac{kmps}{(ms^2 + k)(ms^2 + 3k)} \\ &= \frac{mps}{2(ms^2 + k)} - \frac{mps}{2(ms^2 + 3k)} \\ &= \frac{ps}{2(s^2 + k/m)} - \frac{ps}{2(s^2 + 3k/m)} \end{aligned}$$

and

$$\begin{aligned} Y(s) &= \frac{mps(ms^2 + 2k)}{(ms^2 + k)(ms^2 + 3k)} \\ &= \frac{mps}{2(ms^2 + k)} + \frac{mps}{2(ms^2 + 3k)} \\ &= \frac{ps}{2(s^2 + k/m)} + \frac{ps}{2(s^2 + 3k/m)}. \end{aligned}$$

Hence, the solution is

$$\begin{aligned} x(t) &= \frac{p}{2}\left(\cos\sqrt{\frac{k}{m}}t - \cos\sqrt{\frac{3k}{m}}t\right), \\ y(t) &= \frac{p}{2}\left(\cos\sqrt{\frac{k}{m}}t + \cos\sqrt{\frac{3k}{m}}t\right). \end{aligned} \tag{16}$$

Figure 8.5.4 shows the graph of the solution when $p = 2$ and $k = m = 1$.

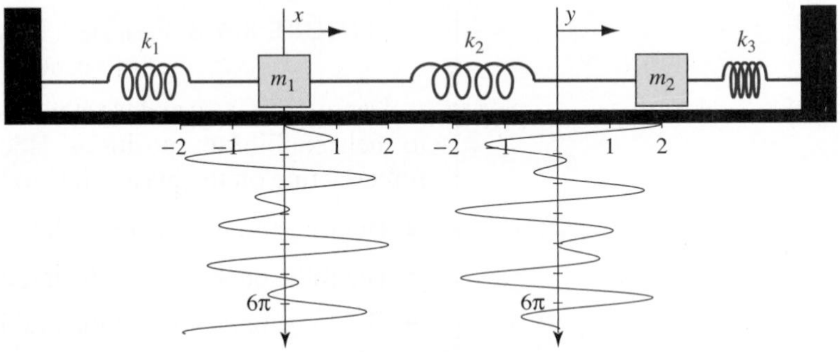

FIGURE 8.5.4 Example 4: Motion of vibrating masses for $p = 2$, $k = m = 1$.

Solving two second-order equations, we have reached the same solution that required four first-order equations in Sec. 6.3, Problem 41. ■

Matrix Exponential

Recall the Laplace transform for the exponential function,

$$\mathcal{L}\{e^{at}\} = \frac{1}{s-a}.$$

This would suggest that the Laplace transform for the matrix exponential, introduced in Sec. 6.6, might be

$$\mathcal{L}\{e^{\mathbf{A}t}\} = (s\mathbf{I} - \mathbf{A})^{-1}.$$

We can verify that this is true by taking the Laplace transform of the formal definition of the matrix exponential

Check the result by showing that

$$(s\mathbf{I} - \mathbf{A})\left(\frac{1}{s}\mathbf{I} + \frac{1}{s^2} + \cdots\right) = \mathbf{I}.$$

$$\begin{aligned}\mathcal{L}\{e^{\mathbf{A}t}\} &= \mathcal{L}\left\{\mathbf{I} + \mathbf{A}t + \mathbf{A}^2\frac{t^2}{2!} + \cdots\right\}\\ &= \frac{1}{s}\mathbf{I} + \frac{1}{s^2}\mathbf{A} + \frac{1}{s^3}\mathbf{A}^2 + \cdots\\ &= (s\mathbf{I} - \mathbf{A})^{-1}.\end{aligned}$$

Hence, yet another way to express the matrix exponential is as follows.

Matrix Exponential as the Inverse of a Laplace Transform

The matrix exponential function $e^{\mathbf{A}t}$ of a matrix $\mathbf{A}$ can be written as

$$e^{\mathbf{A}t} = \mathcal{L}^{-1}\{(s\mathbf{I} - \mathbf{A})^{-1}\}. \tag{17}$$

EXAMPLE 5 **Matrix Exponential from Laplace** We will use the Laplace transform to find the matrix exponential function $e^{\mathbf{A}t}$ of the matrix

$$\mathbf{A} = \begin{bmatrix} 6 & 1 \\ -4 & 2 \end{bmatrix}.$$

We first compute

$$s\mathbf{I} - \mathbf{A} = \begin{bmatrix} s-6 & -1 \\ 4 & s-2 \end{bmatrix}$$

and its determinant

$$|s\mathbf{I} - \mathbf{A}| = (s-6)(s-2) + 4 = s^2 - 8s + 16 = (s-4)^2,$$

then use them to calculate

$$(s\mathbf{I} - \mathbf{A})^{-1} = \frac{1}{(s-4)^2}\begin{bmatrix} s-2 & 1 \\ -4 & s-6 \end{bmatrix}.$$

Hence, using partial fractions for each entry, we have

$$\begin{aligned}e^{\mathbf{A}t} &= \mathcal{L}^{-1}\{(s\mathbf{I} - \mathbf{A})^{-1}\}\\ &= \mathcal{L}^{-1}\left\{\begin{bmatrix} \dfrac{1}{(s-4)} + \dfrac{2}{(s-4)^2} & \dfrac{1}{(s-4)^2} \\ \dfrac{-4}{(s-4)^2} & \dfrac{1}{(s-4)} - \dfrac{2}{(s-4)^2} \end{bmatrix}\right\}\\ &= \begin{bmatrix} e^{4t} + 2te^{4t} & te^{4t} \\ -4te^{4t} & e^{4t} - 2te^{4t} \end{bmatrix}\\ &= e^{4t}\begin{bmatrix} 1+2t & t \\ -4t & 1-2t \end{bmatrix}.\end{aligned}$$

■

Summary

We have seen that the Laplace transform can be extended from solving initial-value problems for single linear equations to solving initial-value problems for linear systems $\vec{\mathbf{x}}' = \mathbf{A}\vec{\mathbf{x}} + \vec{\mathbf{f}}(t)$, $\vec{\mathbf{x}}(0) = \vec{\mathbf{x}}_0$, where $\mathbf{A}$ is an $n \times n$ matrix of constants. The application of the Laplace transform yields an algebraic system in the s-domain that can be solved by standard methods in linear algebra to find

$$\vec{\mathbf{X}}(s) = (s\mathbf{I} - \mathbf{A})^{-1}(\vec{\mathbf{x}}_0 + \vec{\mathbf{F}}(s)).$$

Then the solution in the t-domain is

$$\vec{\mathbf{x}}(t) = \mathcal{L}^{-1}\{\vec{\mathbf{X}}(s)\}.$$

8.5 Problems

Laplace for Systems *For Problems 1–6, solve the given initial-value problem using Laplace transforms.*

1. $\begin{aligned}\dot{x} &= y,\\ \dot{y} &= -x,\end{aligned} \quad x(0) = 0, \quad y(0) = 1$

2. $\begin{aligned}\dot{x} &= x - y,\\ \dot{y} &= 2x + 4y,\end{aligned} \quad x(0) = -1, \quad y(0) = 1$

3. $\begin{aligned}\dot{x} &= y,\\ \dot{y} &= -2x + 3y + 12e^{4t},\end{aligned} \quad x(0) = 1, \quad y(0) = 1$

4. $\begin{aligned}\dot{x} &= y,\\ \dot{y} &= -x + 2\cos t,\end{aligned} \quad x(0) = 0, \quad y(0) = 0$

5. $\begin{aligned}\dot{x} &= y + e^{3t},\\ \dot{y} &= -2x + 3y,\end{aligned} \quad x(0) = 0, \quad y(0) = 0$

6. $\begin{aligned}\dot{x} &= -y + t,\\ \dot{y} &= 3x + 4y - 2 - 4t,\end{aligned} \quad x(0) = 0, \quad y(0) = 0$

7. $\vec{\mathbf{x}}' = \begin{bmatrix} 1 & 4 \\ 1 & 1 \end{bmatrix}\vec{\mathbf{x}} + \begin{bmatrix} -10 \\ 1 \end{bmatrix}, \quad \vec{\mathbf{x}}(0) = \begin{bmatrix} 0 \\ 0 \end{bmatrix}$

8. $\vec{\mathbf{x}}' = \begin{bmatrix} 3 & -3 \\ 2 & -2 \end{bmatrix}\vec{\mathbf{x}} + \begin{bmatrix} 4 \\ -1 \end{bmatrix}, \quad \vec{\mathbf{x}}(0) = \begin{bmatrix} 1 \\ 0 \end{bmatrix}$

9. $\vec{\mathbf{x}}' = \begin{bmatrix} 2 & 1 \\ -3 & 6 \end{bmatrix}\vec{\mathbf{x}} + \begin{bmatrix} e^{5t} \\ e^{5t} \end{bmatrix}, \quad \vec{\mathbf{x}}(0) = \begin{bmatrix} 0 \\ 1 \end{bmatrix}$

10. $\vec{\mathbf{x}}' = \begin{bmatrix} 0 & -1 \\ 3 & 4 \end{bmatrix}\vec{\mathbf{x}} + \begin{bmatrix} t \\ -4t - 2 \end{bmatrix}, \quad \vec{\mathbf{x}}(0) = \begin{bmatrix} 1 \\ 1 \end{bmatrix}$

General Solutions of Linear Systems *It is possible to find the general solution of a system of linear equations by letting the initial conditions be arbitrary constants. For Problems 11–12, find the general solution of the given linear system.*

11. $\begin{aligned}\dot{x} &= x + y\\ \dot{y} &= 4x + y\end{aligned}$

12. $\begin{aligned}\dot{x} &= -x - 4y\\ \dot{y} &= x - y\end{aligned}$

13. **More Complicated Linear System** Using $\vec{\mathbf{x}} = \begin{bmatrix} x \\ y \end{bmatrix}$ and $\dot{\vec{\mathbf{x}}} = \begin{bmatrix} \dot{x} \\ \dot{y} \end{bmatrix}$, write the following linear system in matrix form. Then find the Laplace transform $\vec{\mathbf{X}}(s)$ and solve the system

$$\begin{aligned}\dot{x} + 4x + \dot{y} &= 0,\\ \dot{x} - 2x + y &= 0,\end{aligned} \quad x(0) = 0, \quad y(0) = 1.$$

HINT: The coefficients of $\vec{\mathbf{x}}$ and $\dot{\vec{\mathbf{x}}}$ are matrices.

14. **Higher-Order Systems** Use the Laplace transform to solve the linear system

$$\begin{aligned}\dot{x}_1 &= x_2 + t^2, & x_1(0) &= 1,\\ \dot{x}_2 &= x_3, & x_2(0) &= 1,\\ \dot{x}_3 &= x_4, & x_3(0) &= 1,\\ \dot{x}_4 &= x_1, & x_4(0) &= -1.\end{aligned}$$

15. **Finding General Solutions** Consider the linear system

$$\begin{aligned}\dot{x} &= a_{11}x + a_{12}y + f_1(t), & x(0) &= c_1,\\ \dot{y} &= a_{21}x + a_{22}y + f_2(t), & y(0) &= c_2,\end{aligned}$$

where a_{11}, a_{12}, a_{21}, a_{22}, c_1, and c_2 are constants. Use the Laplace transform to show that the solution can be written in the form

$$\begin{aligned}x(t) &= x_h(t) + x_p(t),\\ y(t) &= y_h(t) + y_p(t),\end{aligned}$$

where $x_h(t)$, $y_h(t)$ depend on c_1, c_2 and $x_p(t)$, $y_p(t)$ depend on $f_1(t)$, $f_2(t)$.

16. Drug Metabolism[1] A sustained-release drug is taken orally at prescribed intervals to maintain a constant input rate r. If the mass of the drug in the gastrointestinal tract and bloodstream are denoted x_1 and x_2, respectively, then the rates of change x_1' and x_2' can be shown to be

$$x_1' = -k_1 x_1 + r,$$
$$x_2' = k_1 x_1 - k_2 x_2.$$

The positive constants k_1 and k_2 are rate constants that vary from person to person. Taking the initial conditions to be $x_1(0) = 1$, $x_2(0) = 0$, use the Laplace transform to solve this initial-value problem for $k_1 = k_2 = k$ and $r = 0$.

Mass-Spring Variations *Use the results of Example 4 to find the solutions for Problems 17 and 18. Assume* $x(0) = 0$, $x'(0) = 1$, $y(0) = 0$, *and* $y'(0) = 0$.

17. $m_1 = m_2 = 2$, $k_1 = k_2 = k_3 = 1$

18. $m_1 = m_2 = 2$, $k_1 = 0$, $k_2 = 1$, $k_3 = 0$

19. A Three-Compartment Model A person ingests a semi-toxic chemical, which enters the bloodstream at the constant rate R, whereupon it distributes itself in the bloodstream, tissue, and bone. It is excreted in urine and sweat at rates u and s, respectively. The variables x_1, x_2, and x_3 define the concentrations of the chemical in the three areas. (See Fig. 8.5.5.) The equation for x_1 is given by

$$\frac{dx_1}{dt} = R - ux_1 - k_{12}x_1 + k_{21}x_2.$$

Find the equations for x_2 and x_3.

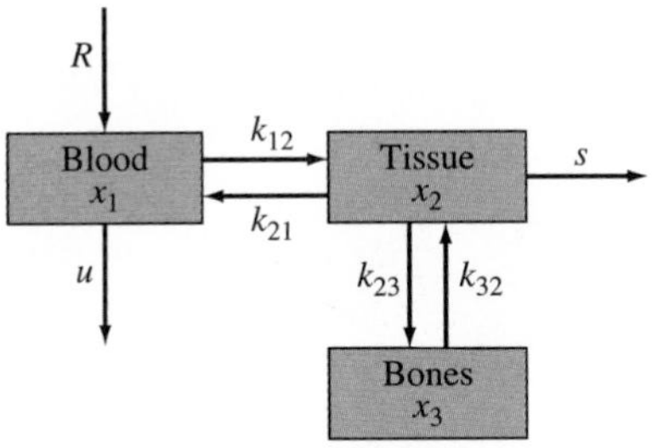

FIGURE 8.5.5 Problem 15: Three-compartment model

20. Vibration with a Free End Two springs and two masses vibrate on a frictionless surface, as illustrated in Fig. 8.5.6. The system is set in motion by holding the mass m_1 in its equilibrium position and pulling the mass m_2 to the right of its equilibrium position by 1 foot and then releasing both masses. Assume that $m_1 = 1$ slug, $m_2 = 2$ slugs, $k_1 = 4$ lb/ft, and $k_2 = 2$ lb/ft.

(a) Determine the equations of motion of the masses.

(b) Find the Laplace transform of the solution to the equations found in part (a).

(c) If you have access to a CAS, find solution formulas.

(d) Discuss the meaning of your results.

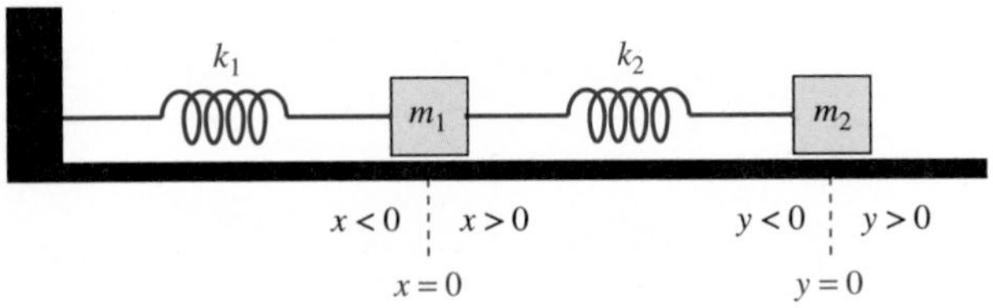

FIGURE 8.5.6 Problem 16: Mass-spring system with free end.

Comparing Laplace *In Problems 21–23, use Laplace transforms to rework the systems problems previously solved by the matrix exponential in Sec. 6.6, Problems 11–13, and compare the method of Laplace transforms with previous methods. In each case assume* $\vec{\mathbf{x}}(0) = \vec{\mathbf{0}}$.

21. $\vec{\mathbf{x}}' = \begin{bmatrix} -1 & 0 \\ 0 & 2 \end{bmatrix} \begin{bmatrix} x_1 \\ x_2 \end{bmatrix} + \begin{bmatrix} 1 \\ 0 \end{bmatrix}$

22. $\vec{\mathbf{x}}' = \begin{bmatrix} 2 & 0 \\ 0 & 3 \end{bmatrix} \begin{bmatrix} x_1 \\ x_2 \end{bmatrix} + \begin{bmatrix} 0 \\ 6 \end{bmatrix}$

23. $\vec{\mathbf{x}}' = \begin{bmatrix} 0 & 1 \\ 1 & 0 \end{bmatrix} \begin{bmatrix} x_1 \\ x_2 \end{bmatrix} + \begin{bmatrix} 1 \\ 1 \end{bmatrix}$

24. Suggested Journal Entry I Now that you are practiced with Laplace transforms in solving systems of first-order IVPs, compare the convenience of using Laplace transforms with the earlier methods in Chapter 6.

25. Suggested Journal Entry II Is it possible to use an impulse function or a step function as one or more of the elements in the forcing input to a mass-spring system, such as the coupled mass-spring system in Example 4? There are some nice examples of these kinds of inputs in biological systems. For instance, Example 3 considers a delta function that represents the injection of a drug. Each differential equation corresponds to a compartment (bloodstream, heart), and the forcing term tells us what compartment is getting the drug when. Discuss another system that might have inputs that can be represented by delta functions, and explain roughly how each DE might be set up.

[1] Adapted from N. H. McClamrock, *State Models of Dynamic Systems* (NY: Springer-Verlag, 1980).

CHAPTER

9 Discrete Dynamical Systems

Big whorls have little whorls
Which feed on their velocity,
And little whorls have lesser whorls
And so on to viscosity.

—Lewis F. Richardson

9.1 Iterative Equations

SYNOPSIS: *We introduce first- and second-order linear iterative equations (or difference equations), which are the discrete counterparts of differential equations for modeling quantities whose changes occur step by step rather than in a flow. We develop solutions (which are sequences) and sample some applications.*

How Many Rabbits?

The Italian merchant and mathematician Leonardo of Pisa proposed this problem at the beginning of the thirteenth century:

> *A pair of newly born rabbits, male and female, were placed in a hutch. In two months these rabbits began their breeding cycle and produced one pair of rabbits, one male and one female. The original rabbits and their offspring continued to breed in this manner; that is, the first pair of offspring appearing at the parental age of two months and then a new pair every month thereafter—always one male and one female. All rabbits survived the first year. What then is the total number of pairs of rabbits at the end of each month during the first year?*[1]

Initially there is one pair of rabbits ($y_0 = 1$), and at the end of the first month, there is *still* one pair ($y_1 = 1$).

- At the end of the second month, however, the first pair of rabbits has a pair of offspring and so there are *two* rabbit pairs ($y_2 = 2$).

[1] *Liber Abaci* (1202), Leonardo's book that brought Arabic numerals to the Western world, contains this problem.

- At the end of the third month, the first rabbit pair has another pair and so there are three rabbit pairs ($y_3 = 3$).
- At the end of the fourth month, both the first and the second rabbit pairs have litters and so there are *five* rabbit pairs ($y_4 = 5$).

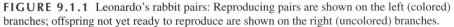

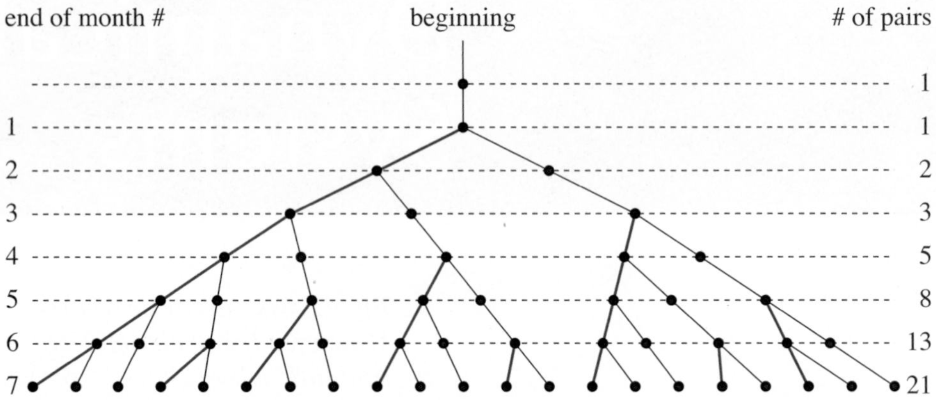

FIGURE 9.1.1 Leonardo's rabbit pairs: Reproducing pairs are shown on the left (colored) branches; offspring not yet ready to reproduce are shown on the right (uncolored) branches.

Continuing in this manner, the number of rabbit pairs at the end of each month will be the number of rabbit pairs the month before (assuming they all live to the next month) *plus* the number of rabbit pairs two months prior (since all pairs of rabbits can have offspring after two months). (See Fig. 9.1.1.) In other words,

$$\#\text{ pairs at } (n+2) \text{ months} = \#\text{ pairs at } (n+1) \text{ months} + \#\text{ pairs at } n \text{ months}$$

or

$$y_{n+2} = y_{n+1} + y_n, \quad n = 0, 1, 2, \ldots, \tag{1}$$

where $y_0 = 1$ and $y_1 = 1$. Equation (1) is an example of a **second-order iterative** (or **difference**) **equation**; together with the initial conditions $y_0 = 1$ and $y_1 = 1$ it forms a **discrete initial-value problem**.

If we wish to answer Leonardo of Pisa's question as to the number of rabbit pairs present at the end of each month of the first year, we must find the sequence:

$$\begin{aligned}
y_0 &= 1, \\
y_1 &= 1, \\
y_2 &= y_1 + y_0 = 1 + 1 = 2, \\
y_3 &= y_2 + y_1 = 2 + 1 = 3, \\
y_4 &= y_3 + y_2 = 3 + 2 = 5, \\
&\vdots \\
y_{12} &= y_{11} + y_{10} = 144 + 89 = 233.
\end{aligned}$$

So one rabbit pair generates 233 rabbit pairs in one year.

Leonardo's nickname was Fibonacci, so equation (1) is called the **Fibonacci sequence**. This particular sequence has become famous because of its many amazing interactions with mathematics and real-world phenomena.[2] Here are

[2]For many years a quarterly journal of the Fibonacci Association published nothing but mathematical curiosities associated with this sequence. A nice illustrative set of examples from nature, as well as mathematics, is contained in M. Boles and R. Newman, *Universal Patterns*, rev. ed. (Bedford, MA: Pythagorean Press, 1990).

a few examples: The number of clockwise and counterclockwise spirals in a sunflower, pine cone, or pineapple are always successive Fibonacci numbers. Branching patterns in nature are associated with Fibonacci numbers. Fibonacci sequences arise in music theory. Certain diagonal sums in Pascal's triangle of binomial coefficients follow the Fibonacci sequence. The ratios of successive Fibonacci numbers tend to the **golden mean**, $(1+\sqrt{5})/2 = 1.618\ldots.$

The Fibonacci sequence is one of the earliest recursively defined sequences to be studied. It is typical of models of processes in which change takes place in a discrete or step-wise fashion, rather than continuously (as in the earlier chapters of this book).

There are interesting parallels between iterative equations and differential equations, and our earlier work provides some useful clues and guesses about how to find solutions in this new situation. We will find that, as for differential equations, it is rare to find closed-form solutions for iterative equations unless they are (surprise!) *linear*.

Yet there are significant ways in which the two subjects differ. For example, derivatives are no longer the appropriate tool. Other methods must be devised. There are genuine surprises lurking in this topic, as you will see, especially in Secs. 9.2 and 9.3. We will begin in this section by analyzing the simplest cases, all linear, with *constant coefficients*.

First-Order Linear Iterative Equations

First-Order Linear Iterative Equation

The **first-order linear iterative equation** is an equation of the form

$$y_{n+1} = ay_n + b, \quad n = 0, 1, 2, \ldots. \tag{2}$$

We frequently assume that an initial value y_0 has been specified, and give the **solution sequence** y_n in terms of y_0. If y_0 is not specified, it appears in the solution in a role parallel to that of a constant of integration in the solution of a differential equation.

Let us emphasize that a solution of an iterative equation is a *sequence* y_n of real numbers, and therefore a function defined on the positive integers $\{1, 2, 3, \ldots\}$ or the nonnegative integers $\{0, 1, 2, 3, \ldots\}$; context will make clear which.

Equation (2), useful in finance and population modeling, can be solved using the formula for a series, the sum of a sequence. From (2) we have

$$\begin{aligned}
y_1 &= ay_0 + b, \\
y_2 &= ay_1 + b = a(ay_0 + b) + b \\
&= a^2 y_0 + ab + b \\
&= a^2 y_0 + b(a+1), \\
y_3 &= ay_2 + b = a[a^2 y_0 + b(a+1)] + b \\
&= a^3 y_0 + ab(a+1) + b \\
&= a^3 y_0 + b(a^2 + a + 1).
\end{aligned}$$

In general, then,

$$y_n = a^n y_0 + b(\underbrace{a^{n-1} + a^{n-2} + \cdots + a + 1}_{\text{geometric series}}). \tag{3}$$

The quantity in parentheses on the right-hand side of (3) is a geometric series with ratio a. If $a \neq 1$, its sum is $(a^n - 1)/(a - 1)$, so our solution is given by

$$y_n = a^n y_0 + b\left(\frac{a^n - 1}{a - 1}\right), \quad a \neq 1. \tag{4}$$

If $a = 1$, the parenthetical quantity on the right-hand side of (3) is equal to n, so in that case $y_n = (1^n)y_0 + nb = y_0 + nb$.

Solution of a First-Order Linear Iterative Equation

The general solution of

$$y_{n+1} = ay_n + b, \quad n = 0, 1, 2, \ldots, \tag{5}$$

is given by the sequence

$$y_n = \begin{cases} a^n y_0 + b\left(\dfrac{a^n - 1}{a - 1}\right) & \text{if } a \neq 1, \\ y_0 + nb & \text{if } a = 1. \end{cases} \tag{6}$$

Solutions are *geometric* series for $a \neq 1$, *arithmetic* series for $a = 1$.

EXAMPLE 1 **Compound Interest** Suppose that you deposit \$100 in a savings account paying interest at a rate $r = 0.08$ (that is, 8%), *compounded annually*.[3] At the end of the first year you will have \$108. At the end of the second year you earn \$8 more on your original \$100, plus 8% of the additional \$8 on deposit during the second year for a total of \$116.64. If y_n is the balance after n years at a rate of r,

$$y_{n+1} = (1 + r)y_n,$$

an equation of type (5) with $a = 1 + r$ and $b = 0$. Hence,

$$y_n = (1 + r)^n y_0.$$

The value of \$100 invested at 8% for 50 years (the **future value** of your hundred bucks) is $y_{50} = (1.08)^{50}(\$100) = \$4{,}690.16$. ■

EXAMPLE 2 **Money for the Future** The typical reader of this book, at age about 20, may be looking forward to retiring in about 50 years from a career as a differential equations professor. Suppose that you decide to start saving now by depositing \$500 annually into a long-term bond fund paying 12% interest compounded annually. What will you have at retirement?

Each year you credit your interest plus the annual deposit, so if y_n is the amount after n years, then $y_{n+1} = y_n + 0.12y_n + 500$. Hence we have the linear iterative equation

$$y_{n+1} = 1.12y_n + 500, \quad y_0 = 500;$$

its solution from (6) is

$$y_n = (1.12)^n(500) + 500\left(\frac{1.12^n - 1}{1.12 - 1}\right).$$

Upon retirement, you will have over a million dollars ($y_{50} \approx \$1{,}344{,}510$)! ■

[3]Most banks compound daily; 8% annually is (8/365)% per day.

Iterative equations are useful in studying biological populations that have natural rhythms (such as annual mating and reproductive cycles), or those that are monitored only on a periodic basis. The following example illustrates such a model.

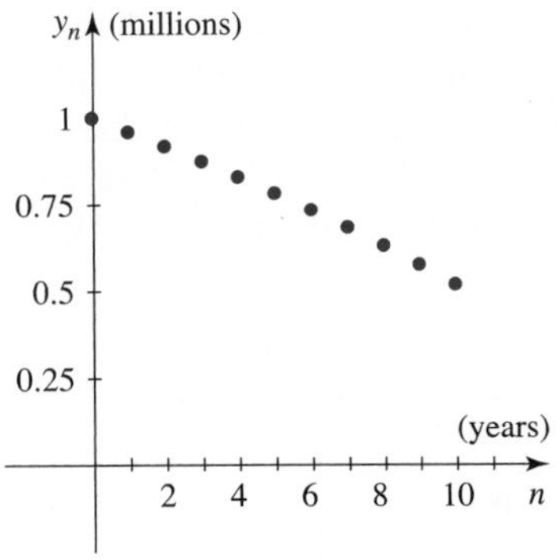

FIGURE 9.1.2 Controlled fishery.

EXAMPLE 3 **Controlled Fishery** A lake has a trout population of one million, which grows at an annual rate of 4% in the absence of fishing. Fishing regulations allow 80,000 fish to be caught per year. How should a fisheries biologist estimate the future size of the fish population?

Denoting by y_n the fish population after n years, the population is modeled by the linear iterative equation $y_{n+1} - y_n = 0.04y_n - 80{,}000$, where the difference $y_{n+1} - y_n$ is the change in the fish population from the end of year n to the end of year $n + 1$. This change results from the natural growth $0.04y_n$ less the harvest of 80,000 trout by fishermen. Since the equation is not yet in the standard form (5), we rewrite it as

$$y_{n+1} = 1.04y_n - 80{,}000, \qquad y_0 = 1{,}000{,}000.$$

The solution of this first-order equation is

$$y_n = 1{,}000{,}000(1.04)^n - 80{,}000\left(\frac{1.04^n - 1}{1.04 - 1}\right). \tag{7}$$

Equation (7) simplifies (see Problem 13) to

$$y_n = 1{,}000{,}000(2 - 1.04^n). \tag{8}$$

Figure 9.1.2 shows a plot of these predictions for the first ten years.

Further analysis is suggested in Problem 13. (Would you expect the pond to become fished out?) ■

Equilibria and Stability

This is a good time to start thinking about long-term behavior and questions of stability for solutions of iterative equations. Does the solution sequence *converge* or *diverge?* Does it reach an *equilibrium solution* or *fixed point?* Such questions can be attacked algebraically, graphically, or numerically. Keep these questions in mind as you read; some anti-intuitive surprises are in store.

Equilibria are the values y_e for which iteration will not change the value.

Equilibrium or Fixed Point

An **equilibrium** (or **fixed point**) is a value y_e for which $y_e = f(y_e)$. For a first-order linear iterative equation, an equilibrium occurs when $y_e = ay_e + b$, or

$$y_e = \frac{b}{1-a}.$$

A fixed point can either attract or repel. Think about how you might tell which.

The availability of computers to perform many iterations quickly allows us to investigate solutions of iterative equations directly, whether closed-form expressions are available or not. The spreadsheet is particularly useful for this, not only calculating a table of values of sequence $\{y_n\}$, but graphing y_n versus n (the **time series**). These graphs are often easier to interpret if the points are connected by line segments as they are plotted.

EXAMPLE 4 **Iterative Behaviors** We can iterate by hand, calculator, or spreadsheet to calculate the first twenty values (given $y_0 = 1$) of the solution sequences for three iterative equations:

$$\text{(a) } y_{n+1} = 0.8y_n + 2, \quad \text{(b) } y_{n+1} = -0.8y_n + 2, \quad \text{(c) } y_{n+1} = -1.04y_n.$$

A spreadsheet such as *Excel* is particularly handy, because once it is set up, you can automatically generate the values *and* graphs for any iterative equation.

- To obtain the list of y_n values for equation (a), we enter the initial index 0 in cell A1 and the initial value 1 in cell B1. The formula = A1 + 1 in cell A2 increments the index. Entering the formula = (0.8) * B1 + 2 into cell B2 calculates the next term of the sequence. The beginning of this process is shown in the spreadsheets below in Fig. 9.1.3. Simply drag the block < A2 : B2 > down to complete the tabulation for the first twenty terms.

Spreadsheet Formulas

	A	B
1	insert n_0	insert y_0
2	= A1 + 1	= (0.8) * B1 + 2

Spreadsheet Output

	A	B
1	0	1
2	1	2.8

FIGURE 9.1.3 Spreadsheet iteration for first-order iterative IVP.

- To obtain the time-series graphs, we select the n and y_n columns and ask for a Chart, where "First column contains X-values for XY-chart." Then we select the Line Graph format that joins the points.

Figure 9.1.4 shows partial lists of values and the time series for each of the three sequences of this example. Which view presents the data most quickly and forcefully? Within the figure, we describe the long-term behavior that you would expect in each case, which you can check against the lists of values for confirmation.

■

(a)

n	y_n
0	1
1	2.8
2	4.24
3	5.392
4	6.3136
5	7.05088
6	7.640704
7	8.1125632
8	8.49005056
9	8.79204045
10	9.03363236
11	9.22690589
12	9.38152471
13	9.50521977
14	9.60417581
15	9.68334065

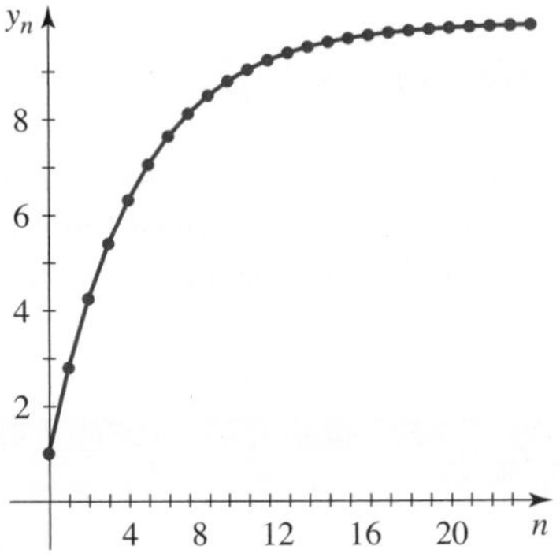

The equation

$$y_{n+1} = 0.8y_n + 2$$

has an **attracting** fixed point

$$\begin{aligned} y_e &= 0.8y_e + 2, \\ 0.2y_e &= 2, \\ y_e &= 10 \end{aligned}$$

because

$$y_n \to 10$$

(from below), as $n \to \infty$.

(b)

n	y_n
0	1
1	1.2
2	1.04
3	1.168
4	1.0656
5	1.14752
6	1.081984
7	1.1344128
8	1.09246976
9	1.12602419
10	1.09918065
11	1.12065548
12	1.10347561
13	1.11721951
14	1.10622439
15	1.11502049

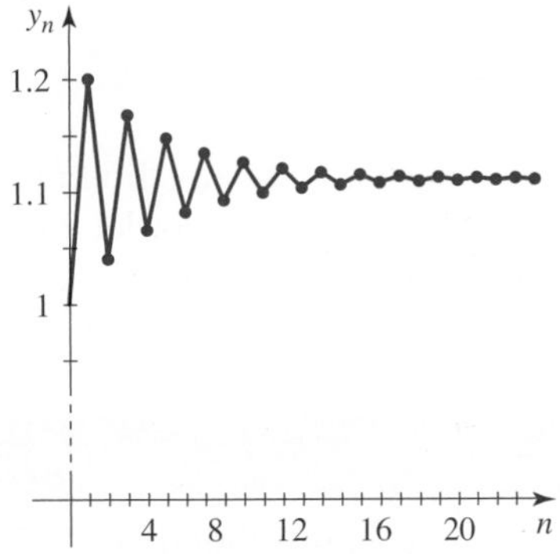

The equation

$$y_{n+1} = -0.8y_n + 2$$

also has an attracting fixed point

$$\begin{aligned} y_e &= -0.8y_e + 2, \\ 1.8y_e &= 2, \\ y_e &= \frac{2}{1.8} = 1.111\ldots \end{aligned}$$

because

$$y_n \to 1.111\ldots, \text{ as } n \to \infty.$$

In this case, iterates *oscillate* above and below the fixed point.

(c)

n	y_n
0	1
1	−1.04
2	1.0816
3	−1.124864
4	1.16985856
5	−1.2166529
6	1.26531902
7	−1.31593178
8	1.36856905
9	−1.42331181
10	1.48024428
11	−1.53945406
12	1.60103222
13	−1.66507351
14	1.73167645
15	−1.80094351

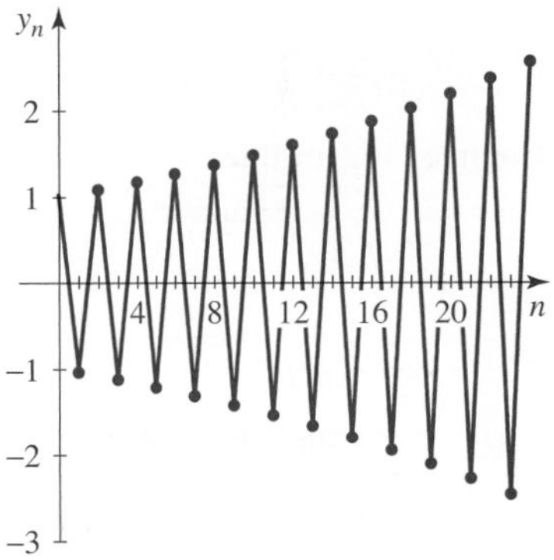

The equation

$$y_{n+1} = -1.04y_n$$

has a **repelling** fixed point:

$$\begin{aligned} y_e &= -1.04y_e, \\ y_e &= 0. \end{aligned}$$

Only a seed $y_0 = 0$ will stay at zero. For all other seeds y_0, the iterates move further and further away from zero, alternating positive and negative values.

FIGURE 9.1.4 Lists of values and time-series graphs for three solution sequences ($y_0 = 1$).

Second-Order Linear Iterative Equations

Second-Order Linear Homogeneous Iterative Equation

The **second-order linear homogeneous iterative equation** with constant coefficients is

$$ay_{n+2} + by_{n+1} + cy_n = 0, \quad a \neq 0. \tag{9}$$

Fixed Points:

Linear and homogeneous iterative systems of the form (9) have fixed points at $y_e = 0$. Think about whether these might be attracting or repelling.

When we studied the parallel second-order *differential* equation

$$ay'' + by' + cy = 0$$

in Chapter 4, we found basic solutions of the form $e^{\lambda t}$, where λ is an eigenvalue or characteristic root—that is, a solution of the characteristic equation

$$a\lambda^2 + b\lambda + c = 0. \tag{10}$$

Equation (10) is also the **characteristic equation** for iterative equation (9), but (as we will show in Sec. 9.2) the solution of (9) corresponding to eigenvalue λ now turns out to be

$$y_n = \lambda^n.$$

If we set $y_n = \lambda^n$ for any n, then $y_{n+1} = \lambda^{n+1}$ and $y_{n+2} = \lambda^{n+2}$, so equation (9) becomes

$$\lambda^n(a\lambda^2 + b\lambda + c) = 0,$$

and unless $\lambda = 0$, equation (10) must hold. The algebraic form of the solution to an iterative second-order equation depends on the nature of the eigenvalues in a now-familiar way, *but the geometric behavior of solutions is entirely different.* We will develop this further in the next section.

Complex Eigenvalues:

For $r = \sqrt{\alpha^2 + \beta^2}$ and $\theta = \tan^{-1}\left(\dfrac{\beta}{\alpha}\right)$,

$$\begin{aligned}\lambda &= \alpha + \beta i\\ &= r\cos\theta + i(r\sin\theta)\\ &= re^{i\theta}.\end{aligned}$$

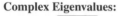

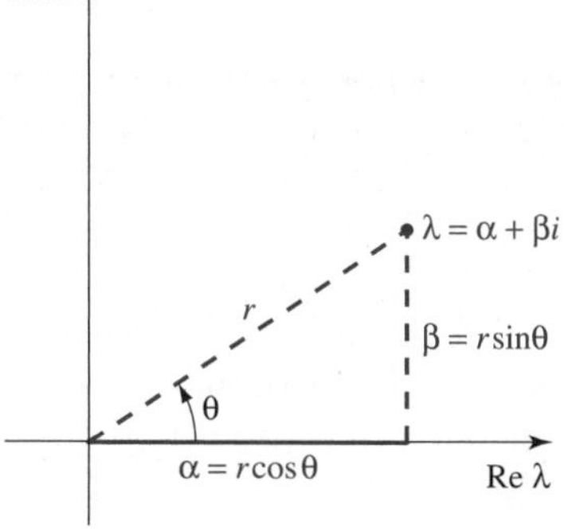

FIGURE 9.1.5 The polar form of a complex number (Appendix CN).

Solution of a Second-Order Linear Homogeneous Iterative Equation

The general solution of

$$ay_{n+2} + by_{n+1} + cy_n = 0, \quad a \neq 0, \quad n - 0, 1, 2, \ldots$$

is given by a sequence depending on the *eigenvalues* (solutions of the *characteristic equation* $a\lambda^2 + b\lambda + c = 0$) as follows:

- For real, distinct eigenvalues λ_1 and λ_2,
$$y_n = c_1\lambda_1^n + c_2\lambda_2^n.$$
- For repeated real eigenvalue λ,
$$y_n = c_1\lambda^n + c_2 n\lambda^n.$$
- For complex eigenvalues $\lambda_1, \lambda_2 = re^{\pm i\theta}$ (see Fig. 9.1.5),
$$y_n = r^n(c_1\cos n\theta + c_2\sin n\theta).$$

The *fixed point* occurs at $y_e = 0$.

Iterative equations can be useful in modeling problems related to the spread of disease, in which statistics are compiled at discrete time intervals (weekly, monthly, yearly, etc.). The next example illustrates a simple epidemic model.

EXAMPLE 5 **Spread of Disease** A country is experiencing a slowly spreading virus. The rate of increase of newly infected persons is 10% per year. We know that 20,000 were infected in the first year and 25,000 new cases were detected in the second year, and we want to model the number y_n (in thousands) of newly infected individuals in year n. We write

$$y_{n+2} - y_{n+1} = 1.10(y_{n+1} - y_n). \tag{11}$$

The difference on the left-hand side of (11) is the rate of increase "this year," while the parenthetical difference on the right-hand side of (11) is the rate of increase "last year" (since the rate is "per year," the rate and the actual difference are alike). Using the initial data provided, we have the second-order discrete initial-value problem

$$y_{n+2} - 2.1y_{n+1} + 1.1y_n = 0, \qquad y_0 = 20, \quad y_1 = 25. \tag{12}$$

The characteristic equation is $\lambda^2 - 2.1\lambda + 1.1 = 0$, having distinct real eigenvalues $\lambda_1 = 1$ and $\lambda_2 = 1.1$. The general solution for (12) is

$$y_n = c_1(1)^n + c_2(1.1)^n,$$

where c_1 and c_2 are determined by y_0 and y_1. When $n = 0$, $y_0 = c_1 + c_2 = 20$; when $n = 1$, $y_1 = c_1 + 1.1c_2 = 25$. Solving the system

$$\begin{aligned} c_1 + \quad c_2 &= 20, \\ c_1 + 1.1c_2 &= 25, \end{aligned}$$

we find $c_1 = -30$ and $c_2 = 50$, so that

$$y_n = 50(1.1)^n - 30, \quad n = 0, 1, 2, \ldots.$$

The data are graphed for twelve years in Fig. 9.1.6. You are asked to think about long-term behavior and interpretations in Problem 35.

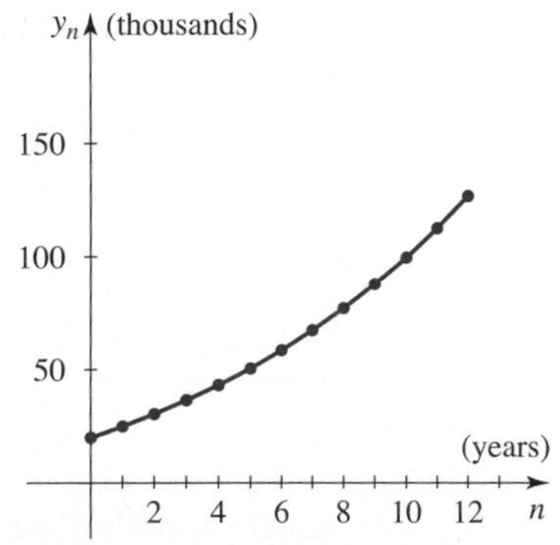

FIGURE 9.1.6 Epidemic model for Example 5 with a repelling fixed point at $y_e = 0$.

■

EXAMPLE 6 **Complex Eigenvalues** In solving the second-order iterative equation

$$2y_{n+2} - 2y_{n+1} + y_n = 0,$$

we solve the characteristic equation $2\lambda^2 - 2\lambda + 1 = 0$ and get the pair of complex conjugate eigenvalues

$$\lambda = \frac{1}{2} \pm \frac{1}{2}i.$$

In polar form, we have

$$\lambda = \left(1/\sqrt{2}\right)e^{\pm i\pi/4}.$$

Therefore, the solution sequence is

$$y_n = \left(\frac{1}{\sqrt{2}}\right)^n \left(c_1 \cos\frac{n\pi}{4} + c_2 \sin\frac{n\pi}{4}\right).$$

If we impose initial conditions $y_0 = 5$ and $y_1 = 0$, then we get two equations for c_1 and c_2:

$$y_0 = 5 = c_1, \qquad y_1 = 0 = \frac{1}{\sqrt{2}}\left[c_1\left(\frac{1}{\sqrt{2}}\right) + c_2\left(\frac{1}{\sqrt{2}}\right)\right].$$

Therefore, $c_1 = 5$ and $c_2 = -5$, giving the IVP solution

$$y_n = \frac{5}{\left(\sqrt{2}\right)^n}\left(\cos\frac{n\pi}{4} - \sin\frac{n\pi}{4}\right).$$

The resulting points for $n = 0, 1, \ldots, 15$ are graphed in Fig. 9.1.7; they lie on a damped sinusoid and tend to zero as n tends to infinity.

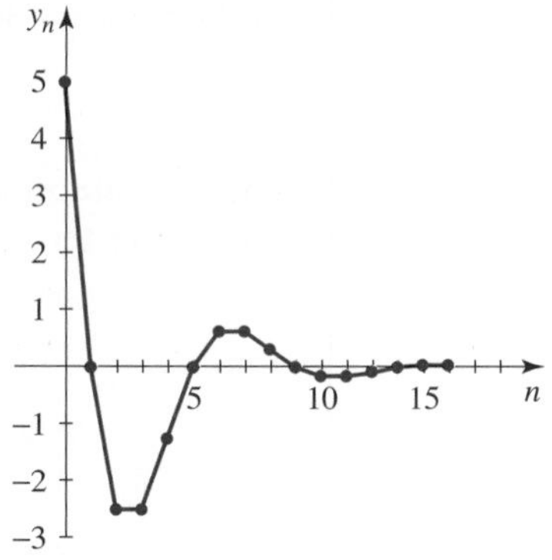

FIGURE 9.1.7 Solution graph for Example 6. $y_e = 0$ is an attracting fixed point.

■

The contrast of Example 6 to the corresponding behavior for a differential equation with these eigenvalues may surprise you. For the differential equation, solutions would grow exponentially, since the real part of λ is positive. For the iterative equation, the significant fact about the eigenvalues is that they have *absolute value less than 1.* The *size* rather than the *sign* is what counts!

EXAMPLE 7 **Spreadsheet Alternative** We can solve the second-order iterative equation of Example 6 using a spreadsheet as follows. Solving for y_{n+2}, we can state the IVP in the form

$$y_{n+2} = y_{n+1} - \frac{1}{2}y_n, \qquad y_0 = 5, \quad y_1 = 0. \tag{13}$$

If we set up a spreadsheet with the initial values y_0 and y_1 in cells B1 and B2, we would enter into cell B3 the formula = B2 − (0.5) ∗ B1. We use column A to generate the index values so that a time series graph can be plotted. Enter 0 into cell A1, and enter the formulas = A1 + 1 into A2 and = A2 + 1 into A3.

The formulas from block $<$ A3 : B3 $>$ can now be copied down through row 16 to obtain the values for $n = 0, 1, 2, \ldots, 15$, as in the previous example. Figure 9.1.8 shows how the spreadsheet is set up, and gives the tabulation through $n = 15$ with the resulting spreadsheet graph; compare this with Fig. 9.1.7.

n	y_n
0	5
1	0
2	−2.5
3	−2.5
4	−1.25
5	0
6	0.625
7	0.625
8	0.3125
9	0
10	−0.15625
11	−0.15625
12	−0.07813
13	0
14	0.039063
15	0.039063

Spreadsheet Formulas		
	A	**B**
1	insert n_0	insert y_{n_0}
2	= A1 + 1	insert y_{n_0+1}
3	= A2 + 1	= B2 − (0.5) ∗ B1

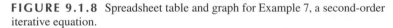

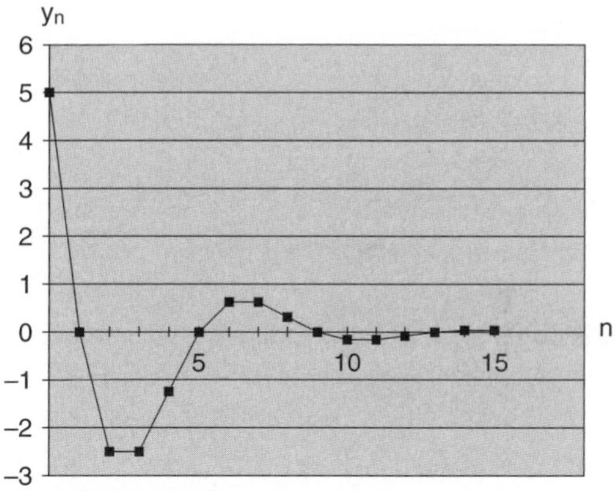

FIGURE 9.1.8 Spreadsheet table and graph for Example 7, a second-order iterative equation. ■

Summary

Iterative equations produce a sequence of values. For linear iterative equations, both first- and second-order, we give explicit solutions (formulas for the nth term of the sequence).

For second-order iterative equations, although the calculation of eigenvalues and the algebraic forms of solutions are superficially similar to those for the corresponding second-order differential equations, the behavior of the solutions can be quite different because *the exponential base is no longer **e** but the eigenvalues themselves.*

9.1 Problems

First-Order Linear Iterative Equations *Problems 1–12 each give a discrete linear IVP.*

(a) *Determine a closed-form solution $y_n = f(n)$ for the problem.*

(b) *Calculate and plot the first ten values of the solution.*

(c) *Describe the long-term behavior of the solution.*

(d) *How much do the initial conditions matter? Describe the differences in behavior of solutions between Problems 1 and 2, 3 and 4, and so on.*

1. $y_{n+1} = \frac{1}{2}y_n + 3, \quad y_0 = 1$

2. $y_{n+1} = \frac{1}{2}y_n + 3, \quad y_0 = -1$

3. $y_{n+1} = -\frac{1}{2}y_n + 3, \quad y_0 = 1$

4. $y_{n+1} = -\frac{1}{2}y_n + 3, \quad y_0 = -1$

5. $y_{n+1} = 2y_n + 3, \quad y_0 = 1$

6. $y_{n+1} = 2y_n + 3, \quad y_0 = -1$

7. $y_{n+1} = -2y_n + 3, \quad y_0 = 1$

8. $y_{n+1} = -2y_n + 3, \quad y_0 = -1$

9. $y_{n+1} = y_n + 3, \quad y_0 = 1$

10. $y_{n+1} = y_n + 3, \quad y_0 = -1$

11. $y_{n+1} = -y_n + 3, \quad y_0 = 1$

12. $y_{n+1} = -y_n + 3, \quad y_0 = -1$

13. Fishing's End Recall the iterative equation constructed in Example 3:

$$y_{n+1} = 1.04y_n - 80{,}000; \qquad y_0 = 1{,}000{,}000.$$

(a) Derive equation (8) from equation (7) for the solution, and give an algebraic argument that extinction is inevitable.

(b) Estimate from Fig. 9.1.2 or from a spreadsheet when extinction will occur.

(c) Calculate from (a) exactly when extinction will occur. Explain any discrepancy between this calculation and your estimate in (b).

Lab Problem: Spreadsheet Predictions *For each of the first-order linear iterative IVPs given in Problems 14–21, do the following:*

(a) *Calculate and graph the solution using a spreadsheet.*

(b) *Calculate the fixed point y_e, and discuss the long-term behavior of the solution with respect to y_e.*

(c) *Explain the impact on long-term behavior of the coefficient of y_n and of the constant term. When it is significant, comment also on the impact of the initial value.*

14. $y_{n+1} = 0.3y_n - 1, \quad y_0 = 0.2$

15. $y_{n+1} = 0.3y_n - 1, \quad y_0 = -1.6$

16. $y_{n+1} = -0.3y_n - 1, \quad y_0 = 0.2$

17. $y_{n+1} = 0.3y_n + 0.5, \quad y_0 = 0.2$

18. $y_{n+1} = 1.3y_n - 1, \quad y_0 = 0.2$

19. $y_{n+1} = -1.3y_n - 1, \quad y_0 = -0.2$

20. $y_{n+1} = 1.3y_n + 0.5, \quad y_0 = -0.2$

21. $y_{n+1} = -1.3y_n - 2, \quad y_0 = 0.2$

Closed-Form Sums *Iterative equations help in developing closed-form expressions for open-form sums (sums containing the ellipsis "..."). For Problems 22–24, do the following:*

(a) *Use the open-form expression to find an iterative equation for s_n.*

(b) *Verify that the closed-form expression given satisfies (a).*

22. $s_n = 1 + 2 + 3 + \cdots + n, \quad s_n = \dfrac{n(n+1)}{2}$

23. $s_n = 1 + 2 + 2^2 + \cdots + 2^n, \quad s_n = 2^{n+1} - 1$

24. $s_n = 1 + 3 + 3^2 + \cdots + 3^n, \quad s_n = \dfrac{1}{2}(3^{n+1} - 1)$

25. Nonhomogeneous Structure For $a \neq 1$ we have written the solution of

$$y_{n+1} = ay_n + b \tag{14}$$

in the form

$$y_n = a^n y_0 + b\left(\frac{a^n - 1}{a - 1}\right).$$

Let

$$u_n = a^n y_0 \quad \text{and} \quad p_n = b\left(\frac{a^n - 1}{a - 1}\right).$$

(a) Show that the u-sequence satisfies the associated homogeneous equation $u_{n+1} - au_n = 0$.

(b) Show that the p-sequence is a particular solution of the nonhomogeneous equation (14).

(c) Show that the general solution y_n of equation (14) is expressible as $y_n = u_n + p_n$, that is, "solution to homogeneous equation plus particular solution of nonhomogeneous equation."

26. Where It Comes From In Problem 22, the solution

$$S_n = \frac{n(n+1)}{2}$$

was given, and you verified that it satisfied the first-order iterative equation

$$S_{n+1} - S_n = n + 1, \quad S_1 = 1. \tag{15}$$

To see where this comes from, use the technique of Problem 25 to solve IVP (15), as follows.

(a) Verify that $h_n \equiv c$ is the general solution of the corresponding *homogeneous* iterative equation.

(b) Let $p_n = An^2 + Bn + C$ be a *particular* solution of (15). Substitute, collect terms, equate like coefficients, and conclude that

$$p_n = \frac{1}{2}n^2 + \frac{1}{2}n.$$

(c) Combine your answers to parts (a) and (b), and use the initial condition to conclude that the solution of (15) is indeed

$$S_n = \frac{n(n+1)}{2}.$$

27. Exceptional Case What happens to solution(s) of the second-order iterative equation (9) in the case that one eigenvalue is zero? Give examples.

28. Nonreal Eigenvalues Verify the form of solutions of (9) for complex conjugate eigenvalues, using Euler's formula $e^{i\theta} = \cos\theta + i\sin\theta$.

Second-Order Linear Iterative Equations *Determine a closed-form solution for each of the iterative equations in Problems 29–32.*

29. $y_{n+2} - 5y_{n+1} + 6y_n = 0$ **30.** $y_{n+2} - 4y_{n+1} + 4y_n = 0$

31. $y_{n+2} + y_{n+1} - 2y_n = 0$ **32.** $y_{n+2} - 4y_{n+1} - 4y_n = 0$

Lab Problem: More Spreadsheets *A second-order discrete IVP is given in each of Problems 33 and 34. You will need to figure out an appropriate adaptation of the procedure given in Example 4. (See Appendix SS.)*

(a) *Calculate and graph the solution using a spreadsheet.*

(b) *Discuss the long-term behavior of the solution.*

33. $y_{n+2} - y_n = 0, \qquad y_0 = 0,\ y_1 = 1$

34. $y_{n+2} + y_n = 0, \qquad y_0 = 0,\ y_1 = 1$

35. Epidemic Model Returning to Example 5, how long would it take for the number of newly infected persons per year to reach 150,000?

36. Rabbits Again Solve the Fibonacci iterative equation (1) with $y_0 = 1$, $y_1 = 1$, using the theory of this section. This closed form solution is called the **Binet formula**.[4]

37. Generalized Fibonacci Sequence: More and More Rabbits Suppose that the adult rabbit pairs studied by Fibonacci have *two* rabbit pairs every month instead of one.

(a) What iterative equation and initial conditions predict the number of rabbit pairs each month?

(b) Solve the discrete initial-value problem in part (a) to find the number of rabbit pairs after the nth month.

(c) Repeat parts (a) and (b), assuming that each adult rabbit pair has k rabbit pairs each month.

38. Probabilistic Fibonacci Sequence Suppose that the adult rabbit pairs studied by Fibonacci lose (on average) half their offspring pairs at birth due to various problems. How many rabbit pairs will we expect after the nth month?

39. Check This with Your Banker Sergei deposits \$1,000 in a bank that pays 8% annual interest compounded daily. How much money will be in Sergei's account after one day? ten days? 365 days?

40. How Much Money Is Enough? Suppose that you win \$200,000 in the tristate lottery, and you decide to retire on your winnings. You deposit your winnings in a bank that pays 8% annual interest, compounded annually, and make yearly withdrawals of \$30,000.

(a) What iterative equation and initial condition describe the future value of your account?

(b) Will you ever run out of money? If so, when?

41. How to Retire as a Millionaire Sheryl wants to become a millionaire, so she takes a discrete mathematics class and learns all about compound interest. Starting with no

[4]Jacques Philippe Binet (1786–1856), a Frenchman with a varied career, made important contributions to mathematics, physics, and astronomy. His work on early matrix theory helped lay the foundation on which Cayley and Sylvester built; another topic on which he wrote was number theory.

capital, she deposits d dollars per year in a bank that pays at an annual interest rate of 8%, compounded annually.

(a) How much money will Sheryl have in the bank at the end of n years?

(b) How much will Sheryl need to deposit every year so that her account will be worth at least a million dollars in 50 years?

42. Amazing But True Wei Chen has just entered college and decided to quit smoking. He deposits the \$25 he saves each week into a savings account that pays 8% annual interest, compounded weekly. (8% annually is (8/52)% weekly.)

(a) What initial-value problem describes the amount of money in Wei Chen's account after n weeks?

(b) How much money will be in Wei Chen's account after n weeks?

(c) How much money will be in Wei Chen's account when he graduates in 208 weeks (four years)?

43. Amortization Problem Amortization is a method for repaying a loan (car loan, home loan) by a series of equal payments. Part of each payment goes toward the outstanding interest, part toward reducing the principal (amount owed). If a person borrows S dollars to buy a house, and if p_n denotes the outstanding principal after the nth payment of d dollars, then p_n satisfies the discrete IVP

$$p_{n+1} = (1+r)p_n - d, \quad p_0 = S,$$

where r is the interest per payment period.

(a) Show that

$$p_n = S(1+r)^n - d\left[\frac{(1+r)^n - 1}{r}\right].$$

(b) Use the equation in part (a) to find the payment d that must be made each period to pay back the loan in N periods.

(c) If you borrow \$200,000 on a house from a bank that charges monthly interest of 1% and you are to repay the loan in 360 monthly payments (30 years), what will be the amount of each payment?

44. Fisheries Management It is estimated that the North Atlantic haddock population would increase by 2% per year under a total ban on fishing. Under new guidelines, fishermen can collectively harvest 1,000 tons of haddock per year. Suppose the current tonnage of haddock present in the North Atlantic is 100,000.

(a) Determine the initial-value problem that describes the future tonnage of North Atlantic haddock (in thousands of tons).

(b) Find the future tonnage under these guidelines.

(c) What would happen if fishing regulations allowed for 5,000 tons to be caught per year?

45. Deer Population A Wyoming deer population has an initial size of 100,000, and with natural growth, increases 10% per year.

(a) Find the future size of the deer population if the state allows 15,000 deer to be killed by hunters each year.

(b) Find the number of deer that can be killed each year to keep the current population constant.

46. Save the Whales Assume that the natural increase in the number of blue whales is 25% per year and that the current population is 1,000. Some countries still harvest blue whales, taking a total of 300 per year.

(a) Find the future blue whale population.

(b) If this trend continues, what will be the fate of the blue whale?

47. Drug Therapy Meena has diabetes, which means that her body does not properly produce insulin. Without taking insulin, the amount of insulin in her body decreases 25% per day. Currently her body has no insulin, but she starts taking shots amounting to 100 grams per day. Assume that her body continues to process insulin at the rate of 25% per day.

(a) What will be the number of grams of insulin in her body after n days?

(b) What will be the long-term amount of insulin in her body if she faithfully continues the treatment regime?

48. Consequence of Periodic Drug Therapy Kaskooli starts to take 100 mg of a given drug to control his asthma. The drug is eliminated from his body at such a rate that it decreases by 25% each day. Let y_n be the amount of drug present in Kaskooli's bloodstream on day n immediately after taking the drug, so $y_0 = 100$.

(a) What initial-value problem describes the amount of the drug present in Kaskooli's bloodstream?

(b) Solve this initial-value problem.

(c) What is the limiting amount of the drug in Kaskooli's body?

(d) What should Kaskooli's daily drug dosage be so that the limiting amount of drug in his body is 800 mg?

49. General Growth Problem Suppose that the growth of given bacteria is such that the increase on a given week is r times the increase on the previous week. If the number last week was y_0 and the number this week is y_1, how many bacteria will be present at week n?

50. Chimes in a Day If a clock chimes the appropriate number of times each hour (and only chimes on the hour), how many times does it chime in a day?

51. Very Interesting If any collection of nonparallel nonconcurrent lines is drawn in the plane, how many distinct regions in the plane are formed? HINT: If y_n denotes the number of regions formed by n nonparallel nonconcurrent lines, then the $(n+1)$st line will intersect the previous n lines at n points and divide $n+1$ previously constructed regions into twice the number. Hence $y_{n+1} = y_n + n + 1$, $y_0 = 1$.

52. Planters Peanuts Puzzle In the early 1950s the makers of Planters Peanuts sponsored a contest and offered a prize to anyone that could determine the number of ways to spell PLANTERS PEANUTS (without the space) by moving either downward, to the left, to the right, or any combination of these directions, in the fifteen-rowed pyramid shown in Fig. 9.1.9. Solve the puzzle as follows.

(a) If T_n denotes the number of ways one can spell an n-lettered word in an n-rowed pyramid, show that T_n satisfies $T_{n+1} = 2T_n + 1$, $T_1 = 1$.

(b) Use the solution of the initial-value problem in part (a) to complete the solution of the puzzle.

```
               P
              PLP
             PLALP
            PLANALP
           PLANTNALP
          PLANTETNALP
         PLANTERETNALP
        PLANTERSRETNALP
       PLANTERSPSRETNALP
      PLANTERSPEPSRETNALP
     PLANTERSPEAEPSRETNALP
    PLANTERSPEANAEPSRETNALP
   PLANTERSPEANUNAEPSRETNALP
  PLANTERSPEANUTUNAEPSRETNALP
 PLANTERSPEANUTSTUNAEPSRETNALP
```

FIGURE 9.1.9 Planters Peanuts puzzle.

53. Suggested Journal Entry A biologist studying a physiological system needs to decide between a continuous model, using differential equations, and a discrete model, using iterative equations. What are some of the factors that might influence her decision?

9.2 Linear Iterative Systems

SYNOPSIS: The study of coupled linear systems of iterative equations leads to analysis in terms of eigenvalues and eigenvectors, as in the case of systems of differential equations, but phase-plane behavior in the discrete case exhibits a greater variety of cases. It turns out that the size of the eigenvalue is at least as important as its sign!

Wild Example

An environmental scientist has been investigating the interdependence of the populations of wolves and moose on a large island in a North American lake. Over a period of several years, she has developed the following model for the transition from wolf population w_n and moose population m_n in the nth year to the corresponding populations in year $n+1$:

$$\begin{aligned} w_{n+1} &= 0.72w_n + 0.24m_n, \\ m_{n+1} &= -0.16w_n + 1.28m_n. \end{aligned} \tag{1}$$

In matrix-vector form, we have

$$\vec{\mathbf{x}}_{n+1} = \mathbf{A}\vec{\mathbf{x}}_n, \tag{2}$$

where

$$\vec{\mathbf{x}}_n = \begin{bmatrix} w_n \\ m_n \end{bmatrix} \quad \text{and} \quad \mathbf{A} = \begin{bmatrix} 0.72 & 0.24 \\ -0.16 & 1.28 \end{bmatrix}.$$

Matrix **A** is called the **transition matrix**.

- The coefficient $a_{11} = 0.72$ is positive but less than 1, and indicates that in the absence of moose the primarily carnivorous wolf population will dwindle. That is, if $m_n \equiv 0$, then $w_{n+1} = 0.72w_n$, and from the previous section we know that $w_n = (0.72)^n w_0$. From this we see that $w_n \to 0$ as n tends to infinity.

Wild Equilibrium:

If our wild example has an equilibrium solution, it will occur when $\vec{\mathbf{x}}_e = \mathbf{A}\vec{\mathbf{x}}_e$. Since $|\mathbf{A}| \neq 0$, equation (2) has an equilibrium solution at

$$\begin{bmatrix} w_e \\ m_e \end{bmatrix} = \begin{bmatrix} 0 \\ 0 \end{bmatrix}.$$

- The coefficient a_{22}, on the other hand, is greater than 1, so that if $w_n \equiv 0$, $m_{n+1} = 1.28m_n$, and the vegetarian moose population will prosper: $m_n = (1.28)^n m_0$.
- The other two coefficients measure the interaction of the two species: $a_{12} > 0$ means that interactions contribute positively to the wolf population, while $a_{21} < 0$ tells us that encounters between moose and wolves are hazardous to the health of moose.

Compare this model to the interpretation of variables in the DE predator-prey model of Sec. 2.5. The difference in this scenario is that the animal populations are only counted once a year.

Eigenstuff Revisited

In studying linear 2×2 systems of differential equations, we sought solutions of the form $e^{\lambda t}\vec{\mathbf{v}}$ and were led to eigenvalues and eigenvectors of the coefficient matrix **A**. The same thing happens if we look for solutions of the iterative equations[1] in the form $\lambda^n \vec{\mathbf{v}}$.

If $\vec{\mathbf{x}}_n = \lambda^n \vec{\mathbf{v}}$, then $\vec{\mathbf{x}}_{n+1} = \lambda^{n+1}\vec{\mathbf{v}}$, and from (2) we obtain

$$\mathbf{A}\lambda^n \vec{\mathbf{v}} - \lambda^{n+1}\vec{\mathbf{v}} = \vec{\mathbf{0}}.$$

Introducing the 2×2 identity matrix **I**, we have

$$\lambda^n \mathbf{A}\vec{\mathbf{v}} - \lambda^{n+1}\mathbf{I}\vec{\mathbf{v}} = \vec{\mathbf{0}},$$

or, in factored form,

$$\lambda^n(\mathbf{A} - \lambda\mathbf{I})\vec{\mathbf{v}} = \vec{\mathbf{0}}. \tag{3}$$

Just as for differential equations, a nonzero λ and the corresponding nonzero $\vec{\mathbf{v}}$ satisfying (3) provide us with an **eigenpair**, such that

$$\mathbf{A}\vec{\mathbf{v}} = \lambda\vec{\mathbf{v}}.$$

EXAMPLE 1 **Wild Eigenstuff** For the introductory wolf/moose example, the characteristic polynomial of **A** is

$$|\mathbf{A} - \lambda\mathbf{I}| = \begin{vmatrix} 0.72 - \lambda & 0.24 \\ -0.16 & 1.28 - \lambda \end{vmatrix} = \lambda^2 - 2\lambda + 0.96,$$

and the eigenvalues, solutions of $\lambda^2 - 2\lambda + 0.96 = (\lambda - 1.2)(\lambda - 0.8) = 0$, are

$$\lambda_1 = 1.2 \quad \text{and} \quad \lambda_2 = 0.8.$$

Eigenvectors are calculated just as in Chapters 5 and 6. For $\lambda_1 = 1.2$, we solve the system $(\mathbf{A} - 1.2\mathbf{I})\vec{\mathbf{v}}_1 = \vec{\mathbf{0}}$; that is, we solve

$$\begin{bmatrix} -0.48 & 0.24 \\ -0.16 & 0.08 \end{bmatrix}\begin{bmatrix} v_1 \\ v_2 \end{bmatrix} = \begin{bmatrix} 0 \\ 0 \end{bmatrix}.$$

[1] We can apply linear algebra techniques to linear iterative systems of any order, but we confine our introductory treatment in this text to 2×2 systems.

Hence, $-0.48v_1 + 0.24v_2 = 0$, which is equivalent to $-2v_1 + v_2 = 0$, so $v_2 = 2v_1$. Thus we can take

$$\vec{\mathbf{v}}_1 = \begin{bmatrix} 1 \\ 2 \end{bmatrix} \quad \text{for } \lambda_1 = 1.2.$$

Similarly, we obtain eigenvector

$$\vec{\mathbf{v}}_2 = \begin{bmatrix} 3 \\ 1 \end{bmatrix} \quad \text{for } \lambda_2 = 0.8.$$

■

Solving the Wolf/Moose System

Suppose that the initial populations of wolves and moose are taken as a multiple of

$$\vec{\mathbf{v}}_1 = \begin{bmatrix} 1 \\ 2 \end{bmatrix}, \quad \text{such as} \quad \vec{\mathbf{x}}_0 = 100\vec{\mathbf{v}}_1 = \begin{bmatrix} 100 \\ 200 \end{bmatrix}.$$

Then, by equation (2),

$$\vec{\mathbf{x}}_1 = \mathbf{A}\vec{\mathbf{x}}_0 = \mathbf{A}(100\vec{\mathbf{v}}_1) = 100(\mathbf{A}\vec{\mathbf{v}}_1) = 100(\lambda_1\vec{\mathbf{v}}_1) = \lambda_1(100\vec{\mathbf{v}}_1) = \lambda_1\vec{\mathbf{x}}_0.$$

From $\vec{\mathbf{x}}_1 = 1.2\vec{\mathbf{x}}_0$, we find $\vec{\mathbf{x}}_2 = (1.2)^2\vec{\mathbf{x}}_0$, and the solution of $\vec{\mathbf{x}}_{n+1} = \mathbf{A}\vec{\mathbf{x}}_n$ is

$$\vec{\mathbf{x}}_n = \lambda_1^n\vec{\mathbf{x}}_0 = (1.2)^n \begin{bmatrix} 100 \\ 200 \end{bmatrix}.$$

For an initial vector that is a multiple of

$$\vec{\mathbf{v}}_2 = \begin{bmatrix} 3 \\ 1 \end{bmatrix}, \quad \text{such as} \quad \vec{\mathbf{x}}_0 = 100\vec{\mathbf{v}}_2 = \begin{bmatrix} 300 \\ 100 \end{bmatrix},$$

we find by similar argument that $\vec{\mathbf{x}}_n = \lambda_2^n\vec{\mathbf{x}}_0 = (0.8)^n \begin{bmatrix} 300 \\ 100 \end{bmatrix}$.

To give the general solution, we only need to express the initial vector $\vec{\mathbf{x}}_0$ in terms of the linearly independent eigenvectors $\vec{\mathbf{v}}_1$ and $\vec{\mathbf{v}}_2$, which are a basis for $\mathbb{R}^2$. If $\vec{\mathbf{x}}_0 = c_1\vec{\mathbf{v}}_1 + c_2\vec{\mathbf{v}}_2$, then

$$\vec{\mathbf{x}}_n = c_1\lambda_1^n\vec{\mathbf{v}}_1 + c_2\lambda_2^n\vec{\mathbf{v}}_2.$$

For the wolf/moose example,

$$\vec{\mathbf{x}}_n = c_1(1.2)^n \begin{bmatrix} 1 \\ 2 \end{bmatrix} + c_2(0.8)^n \begin{bmatrix} 3 \\ 1 \end{bmatrix}.$$

In terms of components, we have the closed-form formulas for w_n and m_n:

$$\begin{aligned} w_n &= c_1(1.2)^n + 3c_2(0.8)^n, \\ m_n &= 2c_1(1.2)^n + c_2(0.8)^n. \end{aligned}$$

Exploring the Wolf/Moose Picture Further

To see the way solutions evolve from step to step, we plot successive population vectors as points in the x_1x_2-plane (or wm-plane for the wolf/moose example). The initial population vector

$$\vec{\mathbf{x}}_0 = \begin{bmatrix} 100 \\ 200 \end{bmatrix} \quad \text{leads to} \quad \vec{\mathbf{x}}_1 = \begin{bmatrix} 120 \\ 240 \end{bmatrix}, \ \vec{\mathbf{x}}_2 = \begin{bmatrix} 144 \\ 288 \end{bmatrix}, \ \vec{\mathbf{x}}_3 = \begin{bmatrix} 173 \\ 346 \end{bmatrix}, \dots.$$

The points all lie on the "eigenline" $x_2 = 2x_1$ and tend to infinity as n tends to infinity. (See Fig. 9.2.1.)

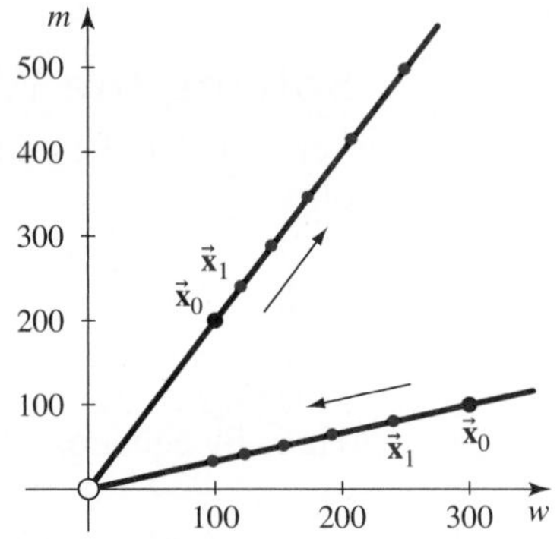

FIGURE 9.2.1 Wolf/moose solutions lying along eigenlines.

On the other hand,

$$\vec{\mathbf{x}}_0 = \begin{bmatrix} 300 \\ 100 \end{bmatrix} \quad \text{iterates to} \quad \vec{\mathbf{x}}_1 = \begin{bmatrix} 240 \\ 80 \end{bmatrix}, \ \vec{\mathbf{x}}_2 = \begin{bmatrix} 192 \\ 64 \end{bmatrix}, \ \vec{\mathbf{x}}_3 = \begin{bmatrix} 154 \\ 51 \end{bmatrix}, \dots.$$

Now successive points lie on the other "eigenline," and they tend to the origin as n tends to infinity (Fig. 9.2.1 again).

For an initial population of 275 wolves and 175 moose, we find that

$$\vec{\mathbf{x}}_0 = 50\vec{\mathbf{v}}_1 + 75\vec{\mathbf{v}}_2,$$

and the solution for this initial state is given by

$$\vec{\mathbf{x}}_n = (1.2)^n \begin{bmatrix} 50 \\ 100 \end{bmatrix} + (0.8)^n \begin{bmatrix} 225 \\ 75 \end{bmatrix}.$$

Several successive iterations are plotted in Fig. 9.2.2, which suggests that the sequence of points will be asymptotic to the eigenline $x_2 = 2x_1$.

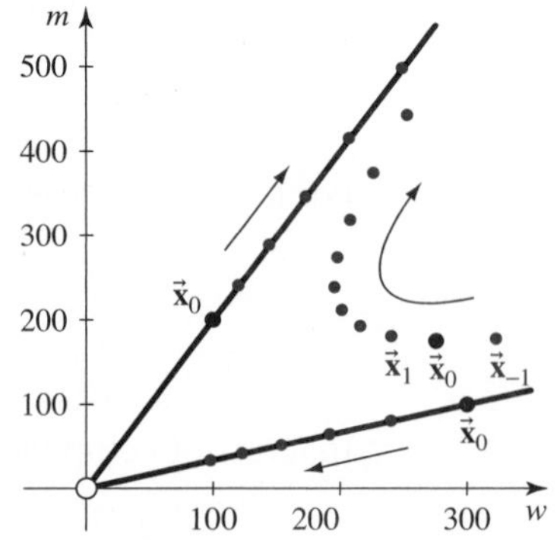

FIGURE 9.2.2 Typical wolf/moose solution for three different initial conditions (in color).

We can attach practical significance to negative values of n in equation (2). If $n = -1$, we write $\vec{\mathbf{x}}_0 = \mathbf{A}\vec{\mathbf{x}}_{-1}$, which says that the population vector would need to have been $\vec{\mathbf{x}}_{-1}$ to give rise to vector $\vec{\mathbf{x}}_0$ a year later. Since $\mathbf{A}$ is invertible, we can obtain $\vec{\mathbf{x}}_{-1}$ from the equation

$$\vec{\mathbf{x}}_{-1} = \mathbf{A}^{-1}\vec{\mathbf{x}}_0 = \begin{bmatrix} 4/3 & -1/4 \\ 1/6 & 3/4 \end{bmatrix}\begin{bmatrix} 275 \\ 175 \end{bmatrix} \approx \begin{bmatrix} 323 \\ 177 \end{bmatrix}.$$

"Projecting" backwards in this way from the initial point $\vec{\mathbf{x}}_0 = [275, 175]$, we obtain a sequence of points tending asymptotically to the eigenline $x_2 = \frac{1}{3}x_1$. (The point corresponding to $\vec{\mathbf{x}}_{-1}$ is plotted in Fig. 9.2.2.)

If we were to extend this analysis beyond the first quadrant (the only values relevant to the practical example from nature), we would find that the equilibrium at the origin is a saddle point for the complete mathematical system.

General Case: A Recursive Solution

Suppose that we now consider the following system of two iterative equations in two discrete variables x_n and y_n:

Linear Iterative System

A **linear iterative system**

$$\begin{aligned} x_{n+1} &= ax_n + by_n, \\ y_{n+1} &= cx_n + dy_n, \end{aligned} \quad \text{or} \quad \vec{\mathbf{x}}_{n+1} = \mathbf{A}\vec{\mathbf{x}}_n = \begin{bmatrix} a & b \\ c & d \end{bmatrix}\begin{bmatrix} x_n \\ y_n \end{bmatrix}, \qquad (4)$$

describes a step-by-step change or transition from state $\vec{\mathbf{x}}_n$ to state $\vec{\mathbf{x}}_{n+1}$, implemented by a **transition matrix A**.

Beginning with initial state $\vec{\mathbf{x}}_0$, we can calculate subsequent state vectors iteratively or recursively:

$$\begin{aligned} \vec{\mathbf{x}}_1 &= \mathbf{A}\vec{\mathbf{x}}_0, \\ \vec{\mathbf{x}}_2 &= \mathbf{A}\vec{\mathbf{x}}_1 = \mathbf{A}(\mathbf{A}\vec{\mathbf{x}}_0) = \mathbf{A}^2\vec{\mathbf{x}}_0, \\ \vec{\mathbf{x}}_3 &= \mathbf{A}\vec{\mathbf{x}}_2 = \mathbf{A}(\mathbf{A}^2\vec{\mathbf{x}}_0) = \mathbf{A}^3\vec{\mathbf{x}}_0. \end{aligned}$$

In general,

$$\vec{\mathbf{x}}_n = \mathbf{A}^n\vec{\mathbf{x}}_0 \qquad (5)$$

provides a recursive solution of (4). Getting a closed form is trickier.

The characteristic polynomial of matrix $\mathbf{A}$ for (4) is

$$\lambda^2 - (a+d)\lambda + (ad - bc) = \lambda^2 - (\operatorname{Tr}\mathbf{A})\lambda + |\mathbf{A}|,$$

where $\operatorname{Tr}\mathbf{A}$ is the trace of $\mathbf{A}$ and $|\mathbf{A}|$ its determinant. If the eigenvalues are denoted by λ_1 and λ_2, we know (from Sec. 6.5, Problem 11) that

$$\operatorname{Tr}\mathbf{A} = \lambda_1 + \lambda_2 \quad \text{and} \quad |\mathbf{A}| = \lambda_1\lambda_2.$$

These eigenvalues determine the behavior of solutions, just as in the differential equations case.

The general solution of system (4)

$$\vec{\mathbf{x}}_{n+1} = \mathbf{A}\vec{\mathbf{x}}_n = \begin{bmatrix} a & b \\ c & d \end{bmatrix} \begin{bmatrix} x_n \\ y_n \end{bmatrix}$$

is given as follows, very much like the box for second-order iterative equations or a system of DEs.[2]

Solution of a Linear Iterative System

Let $\vec{\mathbf{x}}_{n+1} = \mathbf{A}\vec{\mathbf{x}}_n$ be a 2×2 linear iterative system, which has its equilibrium at the origin.

The general solution $\vec{\mathbf{x}}_n$ depends on the eigenvalues of matrix $\mathbf{A}$, as follows:

- For two distinct eigenvalues λ_1 and λ_2 (real or complex), with corresponding eigenvectors $\vec{\mathbf{v}}_1$ and $\vec{\mathbf{v}}_2$,

$$\vec{\mathbf{x}}_n = c_1\lambda_1^n\vec{\mathbf{v}}_1 + c_2\lambda_2^n\vec{\mathbf{v}}_2. \tag{6}$$

- For a repeated eigenvalue λ with two linearly independent eigenvectors,

$$\vec{\mathbf{x}}_n = \lambda^n\vec{\mathbf{x}}_0, \tag{7}$$

for any initial vector $\vec{\mathbf{x}}_0$ (because every vector is an eigenvector for λ).

- For a repeated eigenvalue λ with only one linearly independent eigenvector $\vec{\mathbf{v}}$, and a generalized eigenvector $\vec{\mathbf{u}}$ obtained from $(\mathbf{A} - \lambda\mathbf{I})\vec{\mathbf{u}} = \lambda\vec{\mathbf{v}}$,

$$\vec{\mathbf{x}}_n = c_1\lambda^n\vec{\mathbf{v}} + c_2\lambda^n(n\vec{\mathbf{v}} + \vec{\mathbf{u}}). \tag{8}$$

There is a fundamental difference between the solution of the iterative system (4) and the solution of a system of DEs, caused by the difference in exponential functions (λ^n versus $e^{\lambda t}$). There is now a greater variety of types to classify. As we saw in Sec. 9.1,

> *iterative behavior depends not only on the* ***signs*** *of the eigenvalues but on their* ***sizes*** *as well.*

[2]Samer Habre of Lebanese American University supplied the results for repeated eigenvalues. For a complete exposition, see *Classification of Two-By-Two Iterative Systems*, by S. Habre and J. McDill, 2006 (jmcdill@calpoly.edu).

EXAMPLE 2 **Sink Quartet**[3] The systems $\vec{\mathbf{x}}_{n+1} = \mathbf{A}\vec{\mathbf{x}}$ with the following four transition matrices $\mathbf{A}$ show four types of long-term behavior near the origin:

(a) $\begin{bmatrix} 0.7 & 0 \\ 0 & 0.9 \end{bmatrix}$ (b) $\begin{bmatrix} 0.7 & 0 \\ 0 & -0.9 \end{bmatrix}$

(c) $\begin{bmatrix} -0.7 & 0 \\ 0 & -0.9 \end{bmatrix}$ (d) $\begin{bmatrix} 0 & 0.7 \\ -0.9 & 0 \end{bmatrix}$

Systems (a), (b), and (c) have eigenvalues ± 0.7 or ± 0.9, while the eigenvalues for (d) are approximately $\pm 0.8i$. All are less than 1 in absolute value, and all solutions tend to the origin. We call the origin a **sink** because all the solutions "drain" to $\vec{\mathbf{0}}$ as n tends to infinity.

The plots in Fig. 9.2.3 correspond to the systems (a)–(d), each with an initial point $(-2, 2)$, indicated by a large dot. Arrows help your eye to follow the **solution**, a sequence of points found by iterating the corresponding system. To guide the eye we have drawn segments from each point to its successor.

- Figure 9.2.3(a) depicts an ordinary **sink**, similar to a node from Chapter 6.
- With one of the eigenvalues negative in Fig. 9.2.3(b), there is a dramatic change in behavior; this is called a **flip sink**.
- Two negative eigenvalues produce the more complicated **double-flip sink** in Fig. 9.2.3(c).
- Nonreal eigenvalues produce a **spiral sink**, shown in Fig. 9.2.3(d).

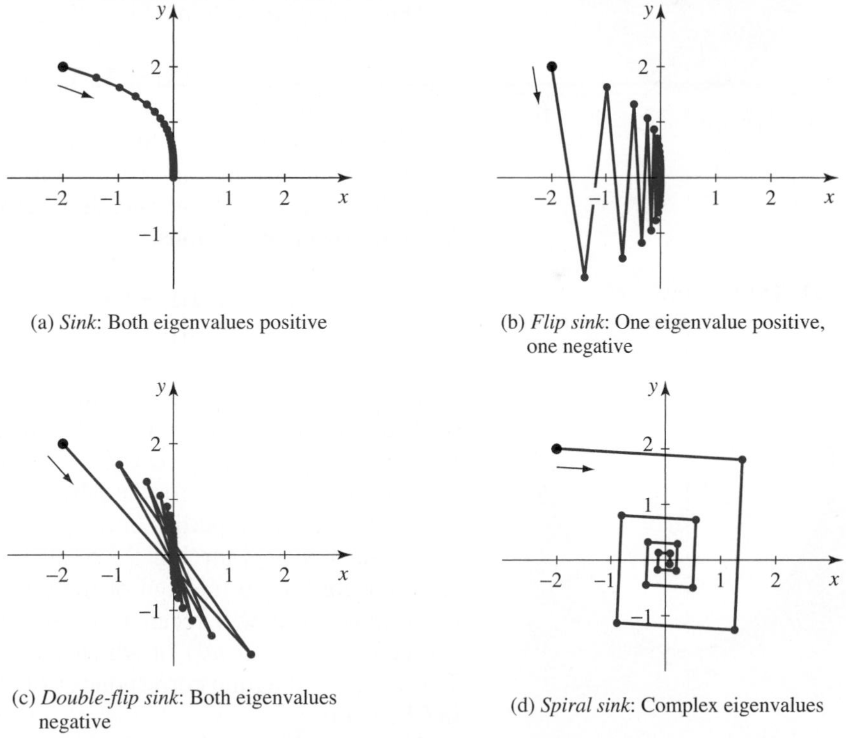
y
2
−2
−1
1
2
x
−1
(a) *Sink*: Both eigenvalues positive
y
2
−2
−1
1
2
x
(b) *Flip sink*: One eigenvalue positive, one negative
y
2
−2
−1
1
2
x
−1
(c) *Double-flip sink*: Both eigenvalues negative
y
2
1
−2
−1
1
2
x
−1
(d) *Spiral sink*: Complex eigenvalues

FIGURE 9.2.3 Example 2: Phase portraits of four sinks. ■

[3] Adapted by permission from Bjørn Felsager's example in the manual for *MacMath*, J. H. Hubbard and B. H. West (NY: Springer-Verlag 1993), 121–127.

Classifying Iterative Behavior with Tr A and |A|

We study the qualitative behavior of solutions of the 2×2 iterative system $\vec{\mathbf{x}}_{n+1} = \mathbf{A}\vec{\mathbf{x}}_n$ by examining either the determinant and trace of $\mathbf{A}$ or by looking at the eigenvalues. If we know the eigenvalues λ_1 and λ_2 of $\mathbf{A}$, we can determine the qualitative nature of the solution by a careful examination of the general solution (6), (7), or (8). On the other hand, the trace and determinant of $\mathbf{A}$ are related to the eigenvalues by

$$\operatorname{Tr}\mathbf{A} = \lambda_1 + \lambda_2 \quad \text{and} \quad |\mathbf{A}| = \lambda_1\lambda_2,$$

so it is possible to determine a useful diagram in the trace-determinant plane that summarizes the entire nature of the solutions. Three diagrams will help us classify the possibilities.

The first diagram (Fig. 9.2.4) distinguishes, as $n \to \infty$,

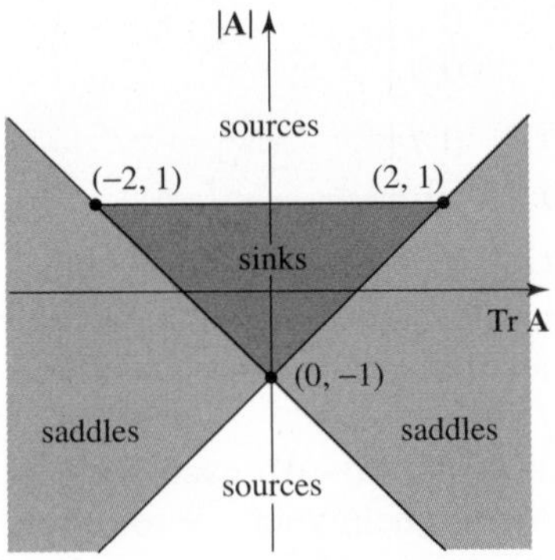

FIGURE 9.2.4 *Sink-source-saddle* classification diagram, based on the *size* of the eigenvalues.

- **sinks**, the case in which solutions tend to the origin;
- **sources**, the case in which solutions move ever further from the origin;
- **saddles**, the case in which solutions move inward in one eigendirection, outward in another.

For a given system (4), we calculate $\operatorname{Tr}\mathbf{A} = \lambda_1 + \lambda_2$ and $|\mathbf{A}| = \lambda_1\lambda_2$, then locate point $(\operatorname{Tr}\mathbf{A}, |\mathbf{A}|)$ in Fig. 9.2.4 to decide whether the origin is a sink, source, or saddle for the system.

Three boundaries divide the sink-source-saddle diagram:

$$|\mathbf{A}| = 1, \quad |\mathbf{A}| = \operatorname{Tr}\mathbf{A} - 1, \quad \text{and} \quad |\mathbf{A}| = -\operatorname{Tr}\mathbf{A} - 1.$$

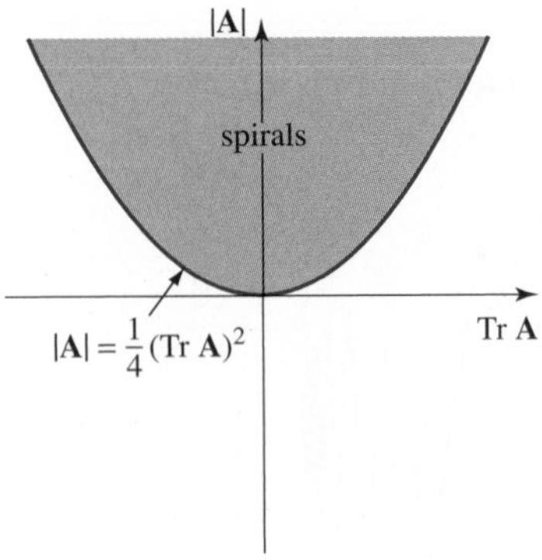

FIGURE 9.2.5 *Spiral* classification diagram, based on real versus complex eigenvalues.

(See Problems 36 and 37.) Since $\operatorname{Tr}\mathbf{A}$ and $|\mathbf{A}|$ depend on the eigenvalues λ_1 and λ_2, we can determine for each region in the figure the *sizes* of λ_1 and λ_2. For example, in the right-hand "saddle" sector (Problem 36), eigenvalues satisfy *one* of the following conditions:

$$\text{(i)} \quad -1 < \lambda_1 < 1 \quad \text{and} \quad \lambda_2 > 1;$$
$$\text{(ii)} \quad -1 < \lambda_2 < 1 \quad \text{and} \quad \lambda_1 > 1.$$

Thus, in this region, powers of one eigenvalue grow in size, while powers of the other diminish. In contrast to the behavior of DE systems, *for a saddle point in iterative systems, both eigenvalues can have the same sign.* These conclusions, and corresponding ones for the other regions of the sink-source-saddle diagram, are derived using properties of inequalities.

The second diagram (Fig. 9.2.5) distinguishes the *nonreal* from the *real* eigenvalues according to the sign of the discriminant $\Delta = \operatorname{Tr}\mathbf{A}^2 - 4|\mathbf{A}|$. Complex eigenvalues, you will recall from Sec. 9.1, correspond to solution terms like $r^n(A\cos n\theta + B\sin n\theta)$ in which the second factor produces spiral behavior. This parabola diagram corresponds exactly to what we found for systems of DEs in Chapter 6.

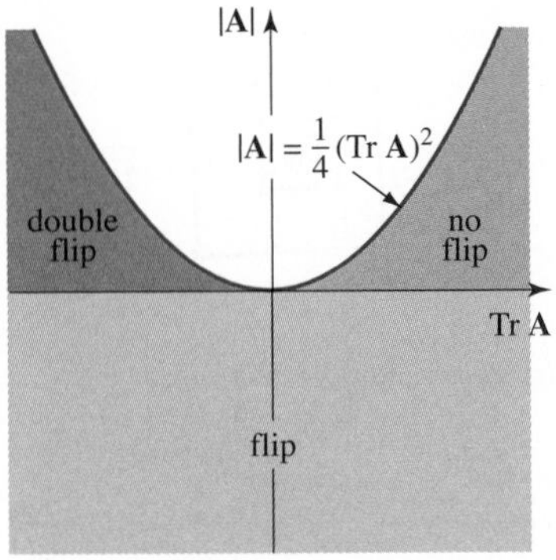

FIGURE 9.2.6 *Flip* classification, based on the *sign* of the eigenvalues.

The third diagram (Fig. 9.2.6) lies behind the geometry of Example 2, based on the combinations of *signs* for the real eigenvalues (where $|\mathbf{A}| < \frac{1}{4}\operatorname{Tr}\mathbf{A}^2$). These results are again consequences of the relationship between the coordinates $(\operatorname{Tr}\mathbf{A}, |\mathbf{A}|)$ and the signs of the eigenvalues. (Compare with Example 2.)

These diagrams and observations illustrate the lessons of equations (6), (7), and (8) for possible general solutions to a linear iterative system.

Iteration Classification

The *size*, not the *sign*, of the eigenvalues determines sink-source-saddle iterative behavior for a discrete system. (The signs introduce flips across axes.)

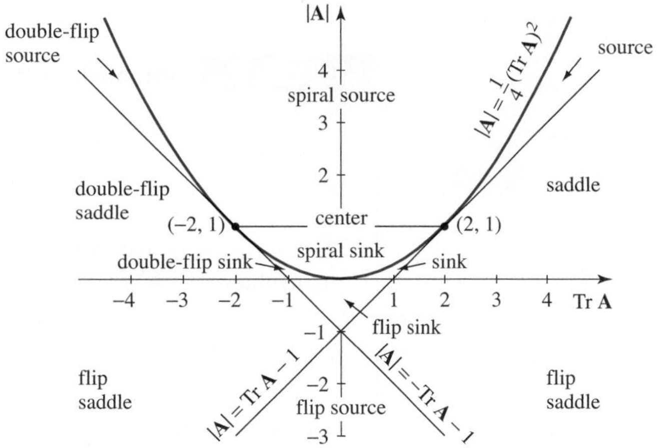

FIGURE 9.2.7 Composite classification diagram for iterative systems.

Combining Figs. 9.2.4–9.2.6 gives the composite classification diagram in Fig. 9.2.7. As you might expect, the line segment $|\mathbf{A}| = 1$ inside the parabola represents **centers**, where spiral sinks bifurcate into spiral sources. (See Problem 37.) For iterative equations a "center" equilibrium means that all iterates in an orbit land on a circular or elliptical locus. Points on other boundaries lead to many interesting behaviors.[4] See Problem 38 to investigate a sample of such a point for yourself.

Recall from Sec. 6.4, Fig. 6.4.18, that only 25% of the trace-determinant plane gives stable behavior to a system of DEs. To make a similar stability analysis for iterative equations, shade in the region of Fig. 9.2.7 that indicates stable long-term behavior. The result is the small inverted triangle surrounding the origin, which is finite and far less than 25% of the infinite trace-determinant plane. So, although iterative systems are easy to analyze solely from the size and sign of their eigenvalues and simple spreadsheets, they are far less likely to be stable.

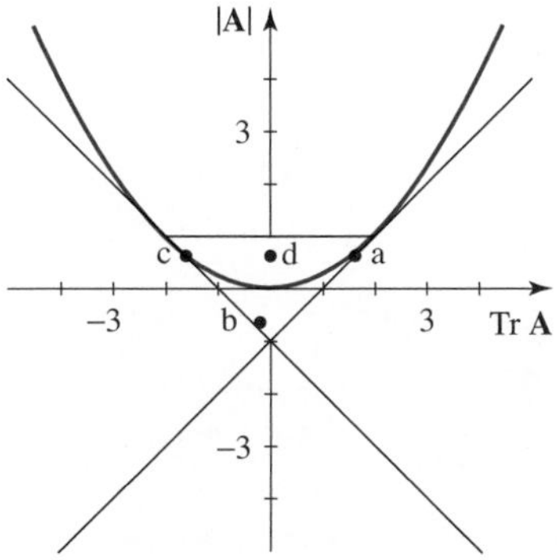

FIGURE 9.2.8 Location of Example 2 systems in the classification diagram.

EXAMPLE 3 **Checking the Diagram** Let us see how the systems from Example 2 correlate with the composite classfication diagram. (See Fig. 9.2.8.)

- For system (a), $\mathrm{Tr}\,\mathbf{A} = \lambda_1 + \lambda_2 = 1.6$ and $|\mathbf{A}| = \lambda_1\lambda_2 = 0.63$. The point $(1.6, 0.63)$ lies near the upper-right corner of the (nonflip) sink region.
- For system (b), $(\mathrm{Tr}\,\mathbf{A}, |\mathbf{A}|) = (-0.2, -0.63)$ in the flip sink region.
- For system (c), $(\mathrm{Tr}\,\mathbf{A}, |\mathbf{A}|) = (-1.6, 0.63)$ in the double-flip sink region.
- For system (d), $(\mathrm{Tr}\,\mathbf{A}, |\mathbf{A}|) = (0, 0.63)$ is also where a spiral sink should be.

■

[4] *Classification of Two-By-Two Iterative Systems*, by S. Habre and J. McDill, 2006 (jmcdill@calpoly.edu).

Our introductory moose and wolves example illustrated that for distinct real eigenvalues, constructing a closed-form solution parallels the differential equations procedure closely. Repeated and complex eigenvalues can be handled in an analogous fashion, but we will omit the details. It should be noted, however, that computers can produce solutions iteratively with ease and speed. One easy way to do this is with a spreadsheet, as we now demonstrate.

EXAMPLE 4 **Spreadsheet Iteration** Let us investigate the system

$$\vec{\mathbf{x}}_{n+1} = \mathbf{A}\vec{\mathbf{x}}_n = \begin{bmatrix} 0 & 1.21 \\ -1 & 0 \end{bmatrix} \vec{\mathbf{x}}_n. \tag{9}$$

The transition matrix **A** has characteristic equation $\lambda^2 + 1.21 = 0$. This gives eigenvalues

$$\lambda_1 = 1.1i \quad \text{and} \quad \lambda_2 = -1.1i,$$

so $\operatorname{Tr}\mathbf{A} = 0$ and $|\mathbf{A}| = 1.21$. The point $(\operatorname{Tr}\mathbf{A}, |\mathbf{A}|) = (0, 1.21)$ is in the spiral source region of Fig. 9.2.7.

To calculate an iterative solution of (9) with $\vec{\mathbf{x}}_0 = [0.5, 0.5]$, we can set up a spreadsheet with 0.5 in cell A1 and 0.5 in cell B1 for the initial state. Since $x_1 = 1.21 y_0$, we enter into cell A2 the formula = 1.21 ∗ B1. In cell B2, where we want to calculate y_1, we enter formula = −A1. (See Fig. 9.2.9.)

We will iterate twelve times, so we need to drag down to copy the formulas from the < A2 : B2 > block through row 14. The resulting *Excel* output and *Excel* graph of these points are shown in Fig. 9.2.9.

x_n	y_n
0.5	0.5
0.605	−0.5
−0.605	−0.605
−0.73205	0.605
0.73205	0.73205
0.885781	−0.73205
−0.88578	−0.88578
−1.07179	0.885781
1.071794	1.071794
1.296871	−1.07179
−1.29687	−0.15625
−1.56921	1.296871
1.569214	1.569214

Spreadsheet Input & Formulas

	A	B
1	insert x_0	insert y_0
2	= 1.21 ∗ B1	= −A1

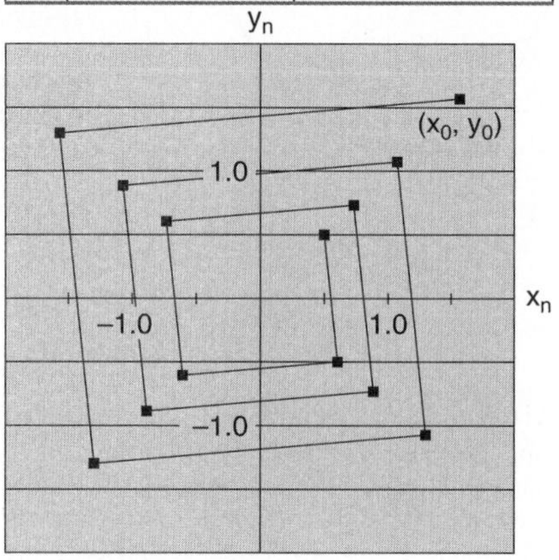

FIGURE 9.2.9 Spreadsheet output for Example 4, a linear iterative system. ■

EXAMPLE 5 **Saddle Spread** For a concluding example, consider the system

$$\vec{\mathbf{x}}_{n+1} = \mathbf{A}\vec{\mathbf{x}}_n = \begin{bmatrix} -1.2 & 0 \\ 0 & -0.8 \end{bmatrix} \vec{\mathbf{x}}_n,$$

for which $\lambda_1 = -1.2$, $\lambda_2 = -0.8$, $\operatorname{Tr}\mathbf{A} = -2$, and $|\mathbf{A}| = 0.96$. We see the point $(\operatorname{Tr}\mathbf{A}, |\mathbf{A}|) = (-2, 0.96)$ is in the double-flip saddle region of Fig. 9.2.7. An *Excel* iteration for $x_0 = 0.1$ and $y_0 = 2$, and the corresponding graph, are shown in Fig. 9.2.10. Notice the characteristic saddle configuration—but we get "two for one" because of the double flip!

x_n	y_n
0.1	2
−0.12	−1.6
0.144	1.28
−0.1728	−1.024
0.20736	0.8192
−0.24883	−0.65536
0.298598	0.524288
−0.35832	−0.41943
0.429982	0.335544
−0.51598	−0.26844
0.619174	0.214748
−0.74301	−0.1718
0.89161	0.137439
−1.06993	−0.10995
1.283918	0.087961
−1.5407	−0.07037
1.848843	0.056295

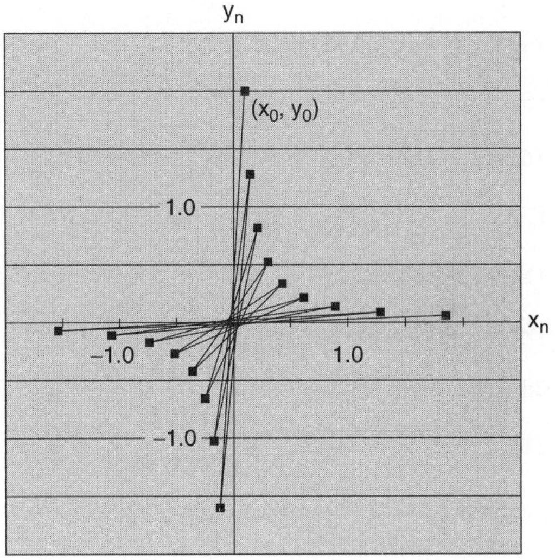

FIGURE 9.2.10 Spreadsheet output for Example 5. ■

Summary

Two-by-two linear iterative systems lead to algebraic solutions and geometric configurations similar to those for systems of differential equations, but with additional variety related to both the size and sign of each eigenvalue. We classify these systems with the classification diagram; we calculate and plot them with a spreadsheet.

9.2 Problems

System Classification *For the iterative system in each of Problems 1–12, classify the behavior of the equilibrium at the origin and sketch a typical solution starting at* $(x_0, y_0) = (3, 2)$. HINT: *Use either the eigenvalues or the trace-determinant plane classification of Fig. 9.2.7.*

1. $x_{n+1} = 0.75x_n$
$y_{n+1} = 0.25y_n$

2. $x_{n+1} = 0.7x_n$
$y_{n+1} = 0.3y_n$

3. $x_{n+1} = y_n$
$y_{n+1} = -x_n$

4. $x_{n+1} = 1.9x_n$
$y_{n+1} = 1.1y_n$

5. $x_{n+1} = -1.9x_n$
$y_{n+1} = -1.1y_n$

6. $x_{n+1} = 1.5x_n$
$y_{n+1} = -0.5y_n$

7. $x_{n+1} = 1.05x_n$
$y_{n+1} = -1.05y_n$

8. $x_{n+1} = -1.5x_n$
$y_{n+1} = 1.1y_n$

9. $x_{n+1} = 0.9y_n$
$y_{n+1} = -1.6x_n$

10. $x_{n+1} = -0.6x_n$
$y_{n+1} = -1.4y_n$

11. $x_{n+1} = -0.8x_n$
$y_{n+1} = -0.6y_n$

12. $x_{n+1} = 1.6y_n$
$y_{n+1} = -0.4x_n$

Classification Using Eigenvalues *For Problems 13–21, consider the* 2×2 *iterative system* $\vec{\mathbf{x}}_{n+1} = \mathbf{A}\vec{\mathbf{x}}_n$ *with given eigenvalues* λ_1 *and* λ_2.

(a) *Classify the equilibrium at* (0, 0).

(b) *Find the solutions* $x_n = f(n)$ *and* $y_n = g(n)$.

(c) *Using as eigenvectors* $\vec{\mathbf{v}}_1 = [1, 0]$ *and* $\vec{\mathbf{v}}_2 = [0, 1]$, *make a rough sketch of the orbit from* (2, 2).

13. $\lambda_1 = 0.5, \quad \lambda_2 = 0.3$

14. $\lambda_1 = -0.5, \quad \lambda_2 = 0.5$

15. $\lambda_1 = -0.5, \quad \lambda_2 = -0.3$

16. $\lambda_1 = 1.5, \quad \lambda_2 = 2$

17. $\lambda_1 = -1.5, \quad \lambda_2 = 2$

18. $\lambda_1 = -1.5, \quad \lambda_2 = -3$

19. $\lambda_1 = 0.5, \quad \lambda_2 = 2$

20. $\lambda_1 = 0.5, \quad \lambda_2 = -2$

21. $\lambda_1 = -0.5, \quad \lambda_2 = -2$

22. Iteration by Rotation From Sec. 5.1, Example 6, we know that the matrix

$$\mathbf{R}_\theta = \begin{bmatrix} \cos\theta & -\sin\theta \\ \sin\theta & \cos\theta \end{bmatrix}$$

rotates a point counterclockwise about the origin through an angle of θ radians.

(a) Show that the eigenvalues of $\mathbf{R}_\theta$ can be written as $\lambda_1, \lambda_2 = e^{\pm i\theta}$.

(b) Show that the nth iterate of

$$\begin{bmatrix} x_{n+1} \\ y_{n+1} \end{bmatrix} = \begin{bmatrix} \cos\theta & -\sin\theta \\ \sin\theta & \cos\theta \end{bmatrix} \begin{bmatrix} x_n \\ y_n \end{bmatrix}$$

can be written as

$$\begin{bmatrix} x_n \\ y_n \end{bmatrix} = \begin{bmatrix} \cos n\theta & -\sin n\theta \\ \sin n\theta & \cos n\theta \end{bmatrix} \begin{bmatrix} x_0 \\ y_0 \end{bmatrix}.$$

Spirals or Circles? *The rotation matrix*

$$\mathbf{R}_\theta = \begin{bmatrix} \cos\theta & -\sin\theta \\ \sin\theta & \cos\theta \end{bmatrix}$$

rotates points in the plane through an angle θ *in the counterclockwise direction. Use this fact to describe the motion of the solution for the iterative systems in Problems 23–25. Then plot a few points with some initial condition using a computer.*

23. $\vec{\mathbf{x}}_{n+1} = \begin{bmatrix} 0 & -1 \\ 1 & 0 \end{bmatrix} \vec{\mathbf{x}}_n$

24. $\vec{\mathbf{x}}_{n+1} = \begin{bmatrix} \sqrt{2} & -\sqrt{2} \\ \sqrt{2} & \sqrt{2} \end{bmatrix} \vec{\mathbf{x}}_n$

25. $\vec{\mathbf{x}}_{n+1} = (0.25) \begin{bmatrix} \sqrt{3} & -1 \\ 1 & \sqrt{3} \end{bmatrix} \vec{\mathbf{x}}_n$

26. Moose Extinction Recall the introductory wolf/moose example.

(a) Suppose that the initial populations are 700 wolves and 100 moose. Iterate to determine when the moose become extinct.

(b) What changes if the initial moose population is raised to 300? Explain.

27. Wolf Extinction If all the moose in our introductory example perish from a deadly liver fluke epidemic, so the initial state is 300 wolves and no moose, what does the model predict for the wolves? Quantify your answer.

28. Moose and Wolves Together If the 1990 populations for our introductory example are 100 wolves and 300 moose, what populations do the model predict for 1995? 2001? What can you say about the relative sizes of these populations in the long run?

System Analysis *For the matrix* $\mathbf{A}$ *in each of Problems 29 and 30, consider the system* $\vec{\mathbf{x}}_{n+1} = \mathbf{A}\vec{\mathbf{x}}_n$.

(a) *Determine eigenvalues* λ_1 *and* λ_2 *and their corresponding eigenvectors.*

(b) *Classify the equilibrium at the origin.*

(c) *Graph the eigenvectors and some typical solution sequences in the phase plane.*

(d) *Obtain closed-form solutions for* x_n *and* y_n.

29. $\mathbf{A} = \begin{bmatrix} -0.92 & -0.36 \\ -0.06 & -0.98 \end{bmatrix}$

30. $\mathbf{A} = \begin{bmatrix} -0.04 & 2.28 \\ 0.38 & 0.34 \end{bmatrix}$

31. Owls and Rats[5] Dusky wood rats provide up to 80% of the diet of spotted owls in California redwood forests. Let O_k denote the population of owls (actual number of individuals) and R_k the population of rats (in thousands) at the kth month of observation. The system has been modeled approximately by the iterative system

$$O_{n+1} = 0.5O_n + 0.4R_n,$$
$$R_{n+1} = -0.1O_n + 1.1R_n.$$

(a) Determine the eigenvalues and eigenvectors, and use them to give closed-form solutions for O_n and R_n.

[5]Problem adapted from David C. Lay, *Linear Algebra and Its Applications* (Reading, MA: Addison-Wesley, 1994), 271–272, 311–312.

(b) Use appropriate software to illustrate the phase portrait for various initial states. Confirm that this portrait agrees with the classification from Fig. 9.2.7.

(c) Find the long-term ratio of owls to rats when both survive.

32. Diabetes Mode[6] Diabetics need to be careful how their bodies metabolize glucose; the natural control mechanism is provided by hormones, chiefly insulin. If x_n represents the deviation of glucose level from the **fasting level** (level approached after many hours of fasting) and y_n the excess hormone level, an iterative model is given by

$$x_{n+1} = 0.978x_n - 0.006y_n,$$
$$y_{n+1} = 0.004x_n + 0.992y_n,$$

where the index n counts one-minute intervals.

(a) Calculate the eigenvalues and eigenvectors to classify behavior of the solution.

(b) After a heavy meal, glucose and hormone levels are $x_0 = 100$ and $y_0 = 0$. Obtain closed formulas for x_n and y_n and graph your results in the phase plane. (See Appendix SS.) Interpret your graph in practical terms.

(c) For the case in part (b), how long does it take for the glucose concentration to fall below fasting level (i.e., become negative)? This is a useful diagnostic measure for diabetes.

33. Conversion Job Convert the 2×2 iterative system

$$x_{n+1} = ax_n + by_n, \qquad (10)$$
$$y_{n+1} = cx_n + dy_n \qquad (11)$$

into an equivalent second-order iterative equation in y_n and show that this equation and the original system have the same eigenvalues. HINT: Use (11) to write an expression for y_{n+2} in terms of x_{n+1} and y_{n+1}, and label it (12). Multiply (10) by c and call this (13). Solve (11) for cx_n and (12) for cx_{n+1}. Substitute these two results into (13) to eliminate x.

34. Decomposition Job Convert the second-order equation

$$px_{n+2} + qx_{n+1} + rx_n = 0, \quad p \neq 0$$

to an equivalent system of two first-order equations. HINT: Let $y_n = x_{n+1}$.

35. The Lilac Bush[7] A lilac bush exhibits the annual growth pattern shown in Fig. 9.2.11. Each new branch simply grows longer in the second year, but in the third it sprouts a pair of new branches from the point where the first year's growth ended. If x_n is the number of new branches grown in year n and y_n is the number of old branches, the stages in the figure result when each old branch grows two new branches in the following year. (Colored branches are new.) Assume that no branches die.

(a) Determine matrix $\mathbf{A}$ so that $\vec{\mathbf{x}}_{n+1} = \mathbf{A}\vec{\mathbf{x}}_n$, where

$$\vec{\mathbf{x}}_n = \begin{bmatrix} x_n \\ y_n \end{bmatrix}.$$

(b) Calculate the eigenvalues and eigenvectors of $\mathbf{A}$.

(c) Determine closed-form solutions for x_n and y_n.

(d) Sketch a diagram of the bush in the sixth year, continuing the color scheme of Fig. 9.2.11, and confirm the values to be calculated from (c).

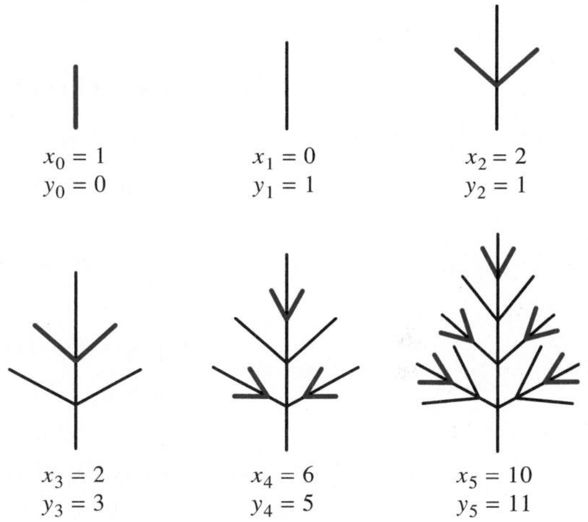

FIGURE 9.2.11 Lilac bush branches for Problem 35.

36. Saddle Regions In Fig. 9.2.4, let region S consist of the right-hand "saddle" region defined by the lines $|\mathbf{A}| = \text{Tr}\,\mathbf{A} - 1$ and $|\mathbf{A}| = -\text{Tr}\,\mathbf{A} - 1$.

(a) Show that $(\text{Tr}\,\mathbf{A}, |\mathbf{A}|)$ is in S when

$$\lambda_1\lambda_2 - \lambda_1 - \lambda_2 + 1 < 0,$$
$$\lambda_1\lambda_2 + \lambda_1 + \lambda_2 + 1 > 0.$$

HINT: $\text{Tr}\,\mathbf{A} = \lambda_1 + \lambda_2$ and $|\mathbf{A}| = \lambda_1\lambda_2$.

(b) Use (a) to show that $(\text{Tr}\,\mathbf{A}, |\mathbf{A}|)$ is in S provided that one or the other of the following conditions holds:

(i) $-1 < \lambda_1 < 1$ and $\lambda_2 > 1$;

(ii) $-1 < \lambda_2 < 1$ and $\lambda_1 > 1$.

HINT: Factor $\lambda_1\lambda_2 - \lambda_1 - \lambda_2 + 1$ into the product of two binomials.

(c) Explain why conditions (i) and (ii) of part (b) make the origin a saddle point.

---

[6]Problem adapted from Otto Bretscher, *Linear Algebra with Applications* (Upper Saddle River, NJ: Prentice-Hall, 1997), 305–306; further information on this model can be found in E. Ackerman, et al., "Blood Glucose Regulation and Diabetes," Chapter 4 of *Concepts and Models of Biomathematics* (NY: Marcel Dekker, 1969).

[7]Problem adapted from Otto Bretscher, *Linear Algebra with Applications* (Upper Saddle River, NJ: Prentice-Hall, 1997), 306.

37. Source and Sink Bifurcation Explain mathematically why the portion of $|\mathbf{A}| = 1$ shown in the trace-determinant plane of Figs. 9.2.4 and 9.2.7 is a bifurcation line between sinks and sources. On this line, what kind of geometric transformation is represented by $\mathbf{A}$?

38. Lab Exercise Pick a point on the boundary of two regions in Fig. 9.2.7 (other than the line segment for centers, which has been discussed).

(a) Devise a system of the form (4) for which $(\mathrm{Tr}\,\mathbf{A}, |\mathbf{A}|)$ corresponds to this point.

(b) Predict what iterative behavior you might expect, given the behaviors on the opposite sides of your boundary.

(c) Explore the behavior of the system using a spreadsheet or other appropriate tool.

(d) Do your answers to (b) and (c) agree? If not, discuss *why* they might not.

39. Suggested Journal Entry For each type of behavior represented by the classification diagram in Fig. 9.2.7, assuming an invertible transition matrix $\mathbf{A}$, discuss the geometric interpretation of the backward iteration $\vec{\mathbf{x}}_{-1} = \mathbf{A}^{-1}\vec{\mathbf{x}}_0$.

9.3 Nonlinear Iterative Equations: Chaos Again

SYNOPSIS: We again meet stable and unstable equilibrium states, and find that near an equilibrium nonlinear functions iterate like their linearizations. The study of nonlinear iterative equations also provides a different introduction to the currently hot topic of chaotic behavior.

The Itchy Finger

A bored student named Pete with an itchy finger entered a number into his calculator and pushed the cosine key. Then he pushed it again and again. Eventually, he realized that the sequence of numbers being displayed had stopped changing: it repeated the number 0.739085 over and over.

Unable to remember the number he started with, Pete entered the number 1 and tried again. (He didn't think he'd used 1 the first time; now he decided to keep track.) The result was the same. With the calculator in radian mode, an initial entry of 1 eventually produced the number 0.739085; Pete just watched it repeat. Next he tried 0.5, and after a while the display got into the same rut: 0.739085 repeatedly.

Thoroughly intrigued, and totally oblivious to the derivatives being calculated by his classmates, Pete tried different starting numbers: 0.1, −1, 2.5; the result was always the same. Maybe his calculator had a hidden bug!

When he stopped to think about it, Pete realized that he had found a number that was equal to its own cosine. That is, within the limits of the calculator's accuracy,

$$\cos(0.739085) = 0.739085.$$

That in itself was kind of neat. But why in the world would you always end up with *that* number even when you started with wildly different values? He tried a few more for good measure: 1000000, 3.14159, 2.718284, −0.222222. Sooner or later, all paths led to 0.739085.

Pete was calculating the successive values of a sequence defined by the *nonlinear iterative equation*

$$x_{n+1} = \cos x_n, \tag{1}$$

and his sequences converged to a limit $L = \cos L = 0.739805$, a *fixed point* or *equilibrium solution*. To visualize this process, we will use a graphical device called a *cobweb diagram*.

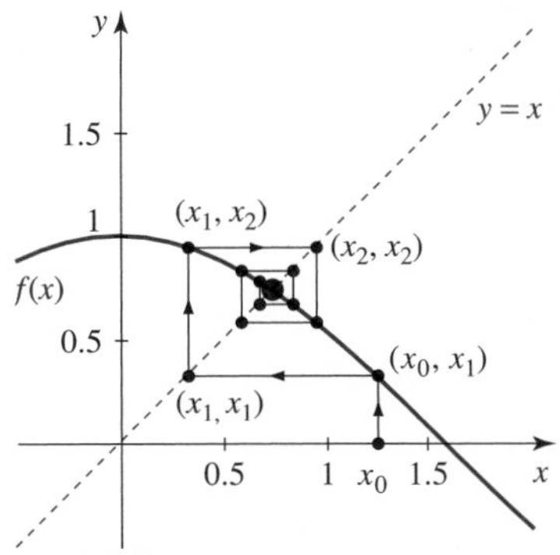

FIGURE 9.3.1 Cobweb diagram for $x_{n+1} = \cos x_n$ with seed $x_0 = 1.25$.

Cobweb Diagrams

For a first-order iterative equation

$$x_{n+1} = f(x_n), \tag{2}$$

f *may* be linear (as in Sec. 9.1), but in Pete's case (1) is clearly *not* linear. Nevertheless, the output value from one step of (2) is the input value for the next, so at each step the role of the value just calculated shifts from output to input, from a "y-value" to an "x-value" in everyday terms.

In Fig. 9.3.1, we have graphed (in color) $f(x) = \cos x$ and $y = x$, the set of points with equal coordinates.

- A **seed** or initial value $x_0 = 1.25$ is plotted on the x-axis.
- By moving vertically from this point $(x_0, 0)$ until we encounter the cosine graph, we reach the point $(x_0, \cos x_0)$, or (x_0, x_1).
- We next want to find $x_2 = \cos x_1$, so we want to locate x_1 on the *horizontal* scale by moving horizontally from (x_0, x_1) to meet line $y = x$ at the point (x_1, x_1). Since the x-coordinate of this point is x_1, we can find $x_2 = \cos x_1$ by moving vertically to meet $y = \cos x$ again. This point will be $(x_1, \cos x_1) = (x_1, x_2)$.

By continuing this process for several additional steps, we see the cobweb path spiraling in to the point $(0.739085, 0.739085)$, where $y = \cos x$ and $y = x$ intersect. (See Fig. 9.3.1.) We have encountered a nonlinear process with a stable equilibrium (sometimes called an **attractor**).

Try this for yourself using the graphs of $f(x) = \cos x$ and $y = x$ in Fig. 9.3.2, and the following mantra:

Vertically to the curve, horizontally to the diagonal.

Repeat, along with Pete, the seeds 0.1, -1, 2.5, and so on.

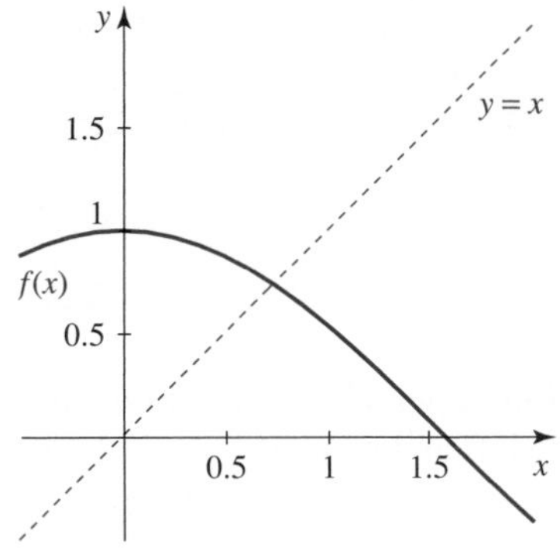

FIGURE 9.3.2 A cobweb diagram of $x_{n+1} = \cos x_n$ starts with a graph of $f(x) = \cos x_n$ and $y = x$.

Cobweb Diagram Algorithm

To construct the diagram for the iterative equation

$$x_{n+1} = f(x_n) \tag{3}$$

with seed x_0, plot $y = f(x)$ and $y = x$ on the same axes. Then:

$\mathbf{x_0}$. Plot $(x_0, 0)$.

$\mathbf{x_1}$. Move vertically to meet the graph of $y = f(x)$ at the point

$$(x_0, f(x_0)) = (x_0, x_1),$$

then horizontally to meet the graph of $y = x$ at (x_1, x_1).

$\mathbf{x_2}$. Move vertically to meet the graph of $y = f(x)$ at the point

$$(x_1, f(x_1)) = (x_1, x_2),$$

then horizontally to meet $y = x$ at (x_2, x_2).

⋮ ⋮

$\mathbf{x_{n+1}}$. In general, from point (x_n, x_n) move vertically to meet $y = f(x)$ at

$$(x_n, f(x_n)) = (x_n, x_{n+1}),$$

then horizontally to meet $y = x$ at (x_{n+1}, x_{n+1}).

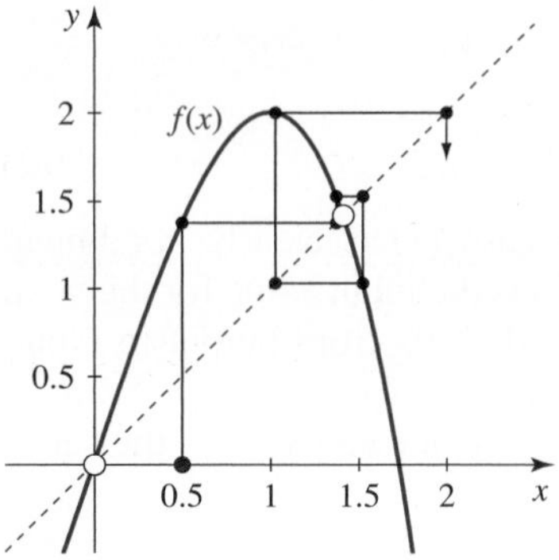

FIGURE 9.3.3 Cobweb diagram for $x_{n+1} = 3x_n - x_n^3$ (Example 1).

EXAMPLE 1 **On the Web** Let us illustrate the algorithm for the iteration

$$x_{n+1} = f(x_n) = 3x_n - x_n^3.$$

For the seed $x_0 = 0.5$, we can easily calculate the following iterates:

$$x_0 = 0.5;$$
$$x_1 = f(x_0) = 1.375;$$
$$x_2 = f(x_1) = 1.52539;$$
$$x_3 = f(x_2) = 1.02686;$$
$$x_4 = f(x_3) = 1.99781.$$

These values are approximated in the cobweb diagram of Fig. 9.3.3. The next iterate will be negative—convergence seems unlikely! ■

Equilibria and Stability

For an iterative equation $x_{n+1} = f(x_n)$, equilibria (fixed points) occur where $x_e = f(x_e)$. In Sec. 9.1, we learned that a *linear* iterative equation has a single equilibrium at the origin, just like a linear DE.

Also as with DEs, *nonlinear* iterative equations can have more than one equilibrium, and the stability of each is determined by the stability of the linear approximation to $f(x)$ at x_e. That is, the stability of an equilibrium depends on the *slope* of the graph of f at the equilibrium point, summarized as follows, which you can explore in Problem 5.

Derivative Test for Stability

For a nonlinear iterative equation $x_{n+1} = f(x_n)$, with fixed points x_e found by solving $x_e = f(x_e)$, the derivative $f'(x_e)$ determines the stability of each equilibrium.

- If $|f'(x_e)| < 1$, then x_e is stable.
- If $|f'(x_e)| > 1$, then x_e is unstable.
- If $|f'(x_e)| = 1$, then x_e is undetermined.

We say undetermined because more information is needed to determine stability. We must know the slopes immediately to the left and right of x_e, as we will demonstrate with the following example.

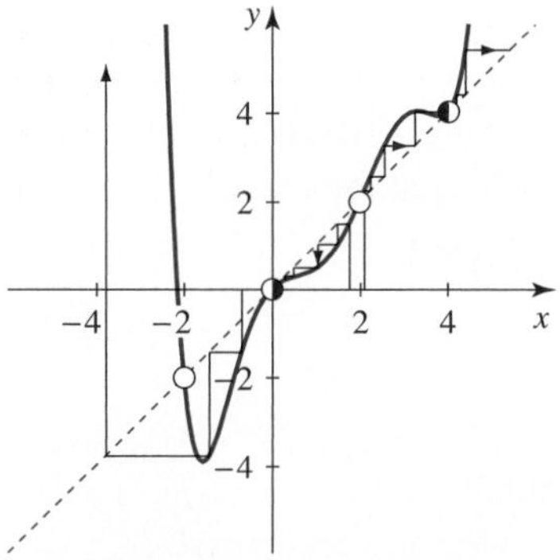

FIGURE 9.3.4 Typical equilibrium points for a nonlinear iterative equation.

EXAMPLE 2 **Stability Demonstration by Cobweb** The function shown in Fig. 9.3.4 has four fixed points. With the graph, and the diagonal of slope 1 for comparison at the equilibria, it is not necessary to know the equation for the function.

The cobwebs drawn from points on either side of each fixed point demonstrate their stabilities.

- $|f'(-2)| > 1$, so $x_e = -2$ is an unstable equilibrium point.
- $|f'(0)| = 1$, so $x_e = 0$ has undetermined stability; however, we can see that $f'(x) > 1$ to the left of 0 and $f'(x) < 1$ to the right of 0, so it is stable from above and unstable from below.

- $|f'(2)| > 1$, so $x_e = 2$ is an unstable equilibrium point.
- $|f'(4)| = 1$, so $x_e = 4$ has undetermined stability; however, we can see that $f'(x) < 1$ to the left of 4 and $f'(x) > 1$ to the right of 4, so it is stable from below and unstable from above. ■

EXAMPLE 3 **Calculating Equilibria and Stability** For the iterative system

$$x_{n+1} = 1.9x_n - x_n^3,$$

the equilibrium points are the points where the graphs of

$$f(x) = 1.9x - x^3 \quad \text{and} \quad y = x$$

intersect, which we find by solving

$$x_e = 1.9x_e - x_e^3,$$

getting $x_e = 0, \pm 0.949$. Evaluating the slopes $f'(x) = 1.9 - 3x^2$ at these points, we get

$$|f'(-0.949)| = 1.9 - 3(-0.949)^2 = -0.799,$$
$$|f'(0)| = 1.9,$$
$$|f'(0.949)| = 1.9 - 3(0.949)^2 = -0.799,$$

hence $x_e = 0$ is unstable and $x_e = \pm 0.949$ are stable. (See Fig. 9.3.5.) ■

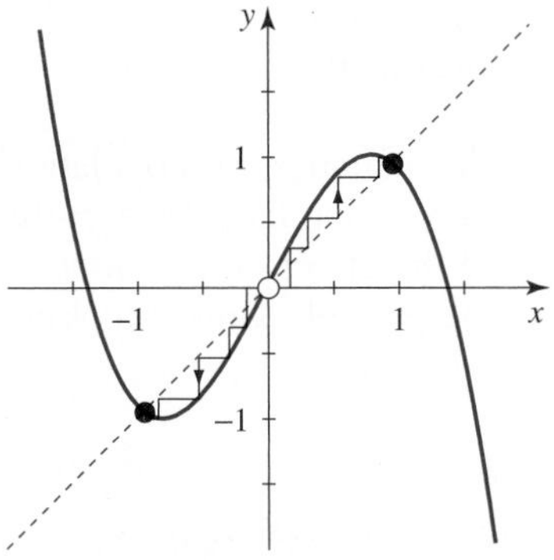

FIGURE 9.3.5 Stable and unstable equilibrium points for $x_{n+1} = 1.9x_n - x_n^3$.

Feedback and Chaotic Behavior

Another example of good (or bad) feedback is Euler's method of numerical approximation for a differential equation, which is actually a discrete iterative scheme. (See Problem 37.)

Iterative equations like (3) model a situation called **feedback**, in which the output is recycled as a new input. Feedback is good when it can be used to improve the data (refining an estimate or correcting a calibration), bad if it magnifies errors or distortions (like the shriek from a public address system).

In recent years a number of iterative systems have been studied because they exhibit **chaotic behavior**: motion with no apparent order or pattern. It is hoped that studying such equations and their solutions will provide better understanding of physical systems in which chaotic behavior is observed. For example, we would like to know more about the transition from the *order* of smooth flow of a fluid to the *chaos* of turbulence.

Of special interest are iterations of the form

$$x_{n+1} = f(x_n, r), \tag{4}$$

where r is a parameter that can be observed, measured, or, in some cases, controlled. The behavior of solutions to equations of this form is often very sensitive to small changes in the parameter r. We will concentrate on one particular equation of this type, the *logistic equation*, because it is relatively easy to study and exhibits many of the characteristics typical for this class of problems.

The Logistic Equation

Logistic Iterative Equation

The **logistic iterative equation** is

$$x_{n+1} = rx_n(1 - x_n), \tag{5}$$

where r is called the **growth parameter**.

- For small values of x_n, we have $x_{n+1} \approx rx_n$, and the pattern is like that of unlimited growth with proportionality constant r.
- As x_n increases, however, $1 - x_n$ decreases and the growth is damped.

The logistic iterative equation (5) is the discrete form of the logistic differential equation studied in Sec. 2.4. However, the variety of possible solution behaviors for (5) is much richer than in the continuous case.

In applications we restrict ourselves to the interval [0, 1], and x_n represents the realized fraction of a potential maximum or whole of 1. To enforce this, we must have $0 \le r \le 4$. (The maximum of the function $f(x) = rx(1-x)$ on [0, 1] is $r/4$ and occurs when $x = 1/2$; thus, $0 \le r \le 4$ assures us that $0 \le f(x) \le 1$. See Problem 26.)

Despite more than a century of investigation of this equation by population biologists, its most interesting properties have been uncovered relatively recently, beginning with the work of Robert May, who studied the effect on solutions of (5) that result from varying the parameter r.[1] This deceptively simple-looking quadratic hides some real surprises.

Quick Survey

For a quick overview of some of the possible solutions of (5) for different r-values, examine Fig. 9.3.6. In all six cases, the seed is $x_0 = 0.5$.

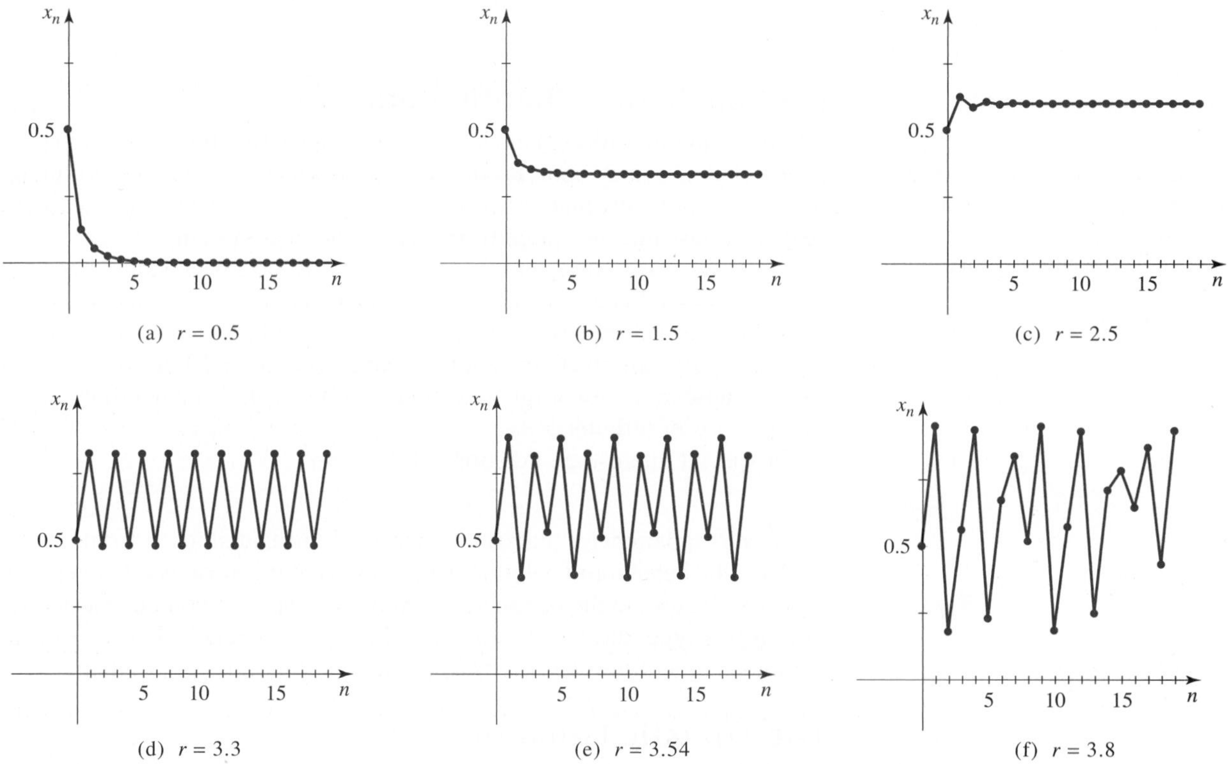

FIGURE 9.3.6 Solutions of $x_{n+1} = rx_n(1 - x_n)$ for six values of parameter r, all with the same seed $x_0 = 0.5$.

[1] R. M. May, "Simple Mathematical Models with Very Complicated Dynamics," *Nature* **261** (1976), 459–467. See also R. M. May and G. F. Oster, "Bifurcation and Dynamic Complexity in Simple Ecological Models," *The American Naturalist* **100** (1976), 573.

- For the smaller r-values, 0.5, 1.5, and 2.5, there is a limiting value (0, $1/3$, and $7/11$, respectively).
- For $r = 3.3$, the solution alternately approaches two different values (**period two**), while for $r = 3.54$ there are four distinct values (**period four**).
- When $r = 3.8$, however, the result is apparently random (and never settles to a cycle even for much larger n).

The chaos exhibited in this last case is not *really* random, however, but **deterministic**, meaning that repeating the iteration with *exactly the same initial value* will give exactly the same "chaotic" orbit. Changing the seed even slightly creates eventually a very different chaotic orbit. The logistic iterative function has the same extreme sensitivity to initial conditions that was discussed for nonlinear differential equations in Sec. 7.4. Of course, not every initial value leads to chaotic orbits. We see again in Fig. 9.3.6 a period-doubling route to chaos, as was discussed for differential equations in Secs. 7.4 and 8.5.[2]

Logistic Behavior Changes as r Increases

As r increases from 0 to 4, the iterative behavior of the logistic equation (5) grows ever more complicated (becoming completely unstable for $r > 4$). We illustrate this growth in complexity with five cases.

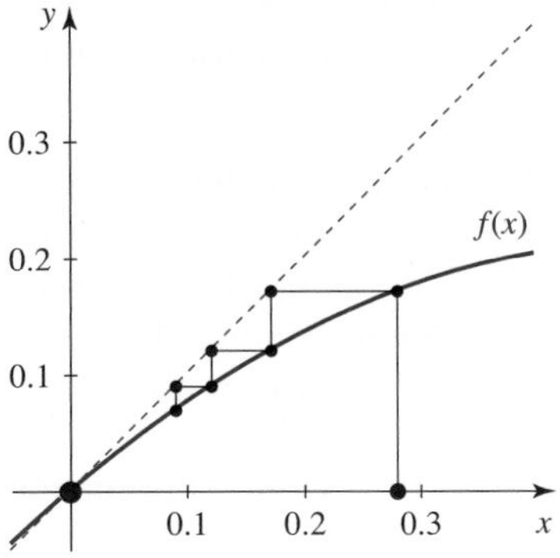

FIGURE 9.3.7 Logistic cobweb: when $0 < r \leq 1$, $x_n \to 0$.

Single Equilibrium (Stable) for $r \in [0, 1]$ As illustrated in Fig. 9.3.7, for $0 < r \leq 1$, the graph of $f(x) = rx(1 - x)$ is strictly below the diagonal on $(0, 1]$; they are equal for $x = 0$. Any seed will result in a cobweb that zigzags down into the funnel between the graphs and tends to zero as a limit. (See also Fig. 9.3.6(a).)

From the graph in Fig. 9.3.7 it is clear that $0 < f(x) < x$ for x-values in $(0, 1]$. Hence, the sequence $\{x_n\}$ is decreasing and bounded below. You are asked to deduce in Problem 27 that this sequence has a limit and that the limit is zero. Thus zero, the only equilibrium point, is asymptotically stable.

Second Equilibrium

For $r > 1$, the parabola $f(x) = rx(1 - x)$ intersects the diagonal $y = x$ *twice*, so there are two equlibria, at 0 and at

$$x_r = \frac{r-1}{r}. \tag{6}$$

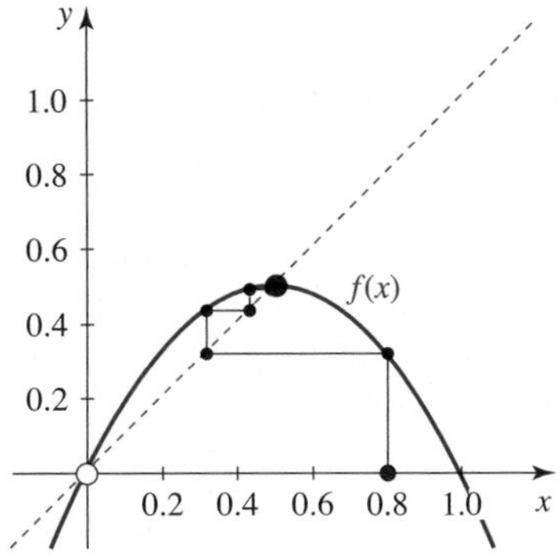

FIGURE 9.3.8 Logistic cobweb: when $r = 2$, $x_n \to 0.5$.

Two Equilibria (One Stable) for $r \in (1, 3]$ For $1 < r \leq 3$, the equilibrium at the origin becomes unstable, while the point x_r is a stable equilibrium. For any seed x_0 in $(0, 1]$, x_n approaches x_r because for these r-values $|f'(x_r)| < 1$. (Compare this with our discussion of eigenvalues for linear iterations in the previous sections.)

The cobweb diagram for $r = 2$ and $x_0 = 0.8$ in Fig. 9.3.8 illustrates this case. The limiting value is $x_2 = (2 - 1)/2 = 1/2$. (See also Fig. 9.3.6(b) and Fig. 9.3.6(c).)

[2]If you did Sec. 7.4, Problem 12, you have already seen an example similar to Fig. 9.3.6. There we set $r = 1$ and varied the step size h; here we have set $h = 1$ and varied r.

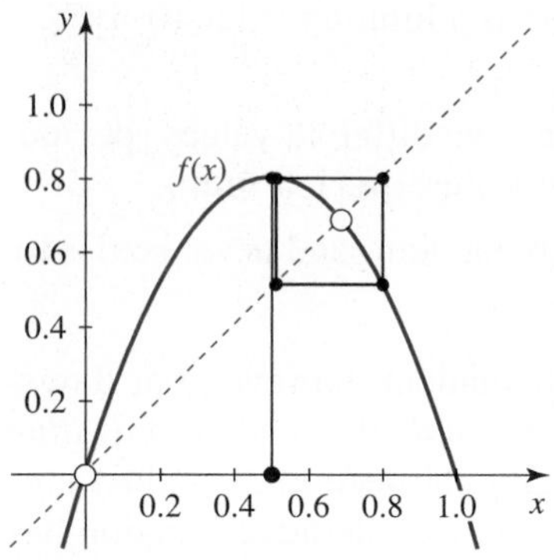

FIGURE 9.3.9 Logistic cobweb, $r = 3.2$; attracting cycle period two.

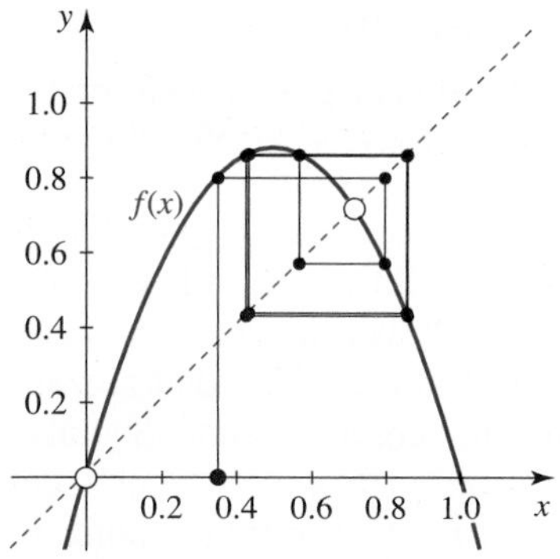

FIGURE 9.3.10 Logistic cobweb, $r = 3.5$; attracting cycle period four.

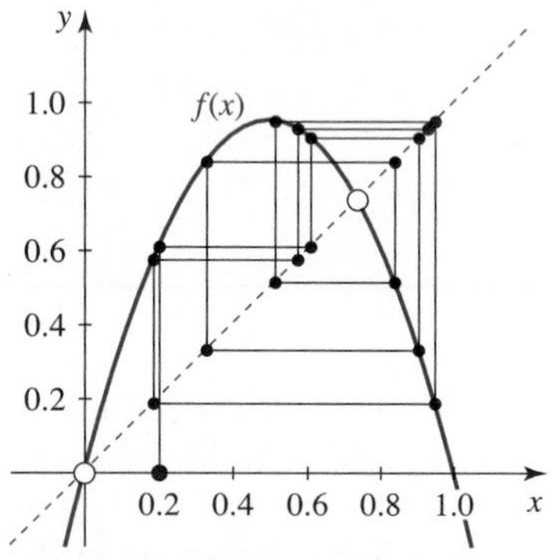

FIGURE 9.3.11 Logistic cobweb, $r = 3.8$; chaos.

Two-Cycle Behavior for $r \in (3, \approx 3.45]$ A qualitative change occurs at $r = 3$. So long as $r \leq 3$, the sequence tends to a limit. As soon as r exceeds 3, however, the "two-cycle" phenomenon we saw in Fig. 9.3.6(d) appears.

For the value $r = 3.2$, so that $f(x) = 3.2x(1 - x)$, the cobweb diagram is shown in Fig. 9.3.9. The two limit values are approximately 0.513 and 0.799. (Problem 30 gives some hints to how these numbers may be calculated.) Neither value 0.513 nor 0.799 is an equilibrium or fixed point, since the solution comes near them only on alternate terms. Together they form a "limiting set" to which the solution is "close" (to one member of the set or the other). For any seed x_0 in $(0, 1]$ that is not an equilibrium, the orbit approaches this same cycle.

We still have two equilibrium points, but both are *unstable*. This is clear for $x_0 = 0$, but less obvious for $x_r = (r - 1)/r$, which is still a fixed point (indicated by the intersection of curve and parabola in Fig. 9.3.9); hence we have $x_{3.2} = 11/16 = 0.6875$. If we use as seed $x_0 = 0.6875$, we will get the constant solution $x_n = 0.6875$. But this fixed point is unstable because the slope of its tangent is now less than -1. As in DEs, nonlinear systems have locally *linear* behavior at the equilibria.

This two-cycle behavior will occur from the first bifurcation point $b_1 = 3$ up to a second bifurcation point $b_2 \approx 3.45$. When r exceeds b_2, the period doubles again.

Four-Cycle Behavior for $r \in (\approx 3.45, \approx 3.54)$ Once beyond the b_2-barrier, say for $r = 3.5$, we have behavior like that in Fig. 9.3.6(e): there are now four values forming a "limiting set," and the solution points cycle among them, getting closer as the process develops. These limiting values are approximately 0.383, 0.501, 0.827, and 0.875 for parameter value $r = 3.5$; the cobweb diagram is shown in Fig. 9.3.10.

The four-cycle behavior continues for r-values up through another bifurcation point $b_3 \approx 3.54$, when yet another doubling occurs.

Further Doubling Leads to Chaos The doubling phenomenon occurs more and more quickly. That is, the successive bifurcation points in the sequence $b_1 = 3$, $b_2 \approx 3.45$, $b_3 \approx 3.54$ (8-cycles begin), $b_4 \approx 3.56$ (16-cycles begin), ..., get closer together and in fact tend to a limit $b_\infty \approx 3.56994$. When r exceeds this "magical" point of accumulation, there are no longer any attractors, and we say the motion is chaotic. This was suggested by Fig. 9.3.6(f) as well as by the cobweb of Fig. 9.3.11 in which $r = 3.8$.

The Orbit Diagram

Further insight into the period-doubling phenomenon is provided by the **orbit diagram**[3] first conceived by Robert May.[4] Parameter r is plotted horizontally, and the limiting value of the solution or the members of the limiting set (for solutions that cycle) are plotted vertically. (See Fig. 9.3.12.) Observe unique limit points as r increases from 0 to 3, then the branching of this curve as the various doublings occur. Finally, at b_∞, the chaotic behavior is indicated by the smearing of the branches into an apparently dense vertical distribution.

---

[3]The orbit diagram is often referred to as the "bifurcation diagram," but strictly speaking it does not qualify for the latter term because the *unstable* fixed points and cyclic points are not included. See Problem 36.

[4]Australian Sir Robert May's (1936–) outstanding career as an eminent mathematical biologist has spanned special professorships at Sydney University (1969–1973), Princeton University (1973–1988), and Oxford University and the Imperial College of London (1989–2000).

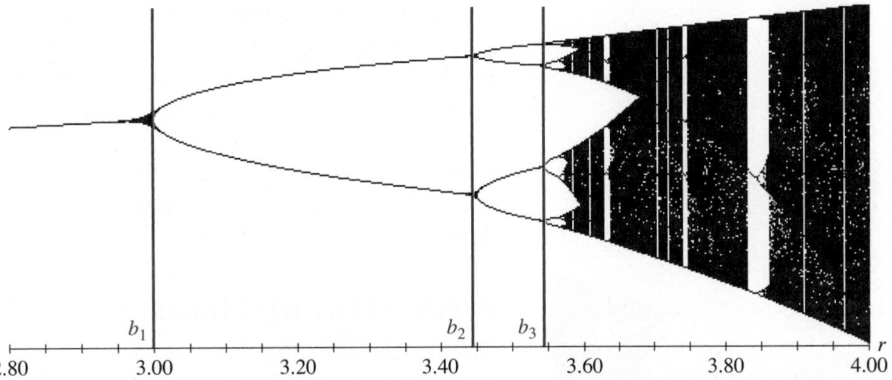

FIGURE 9.3.12 Orbit diagram for the logistic equation, showing bifurcation points b_1, b_2, b_3.

There are "windows of order" in the mostly chaotic region, including three-cycle behavior near $r = 3.84$. (See Problem 33.) Narrower windows occur near parameter values $r = 3.63$ and $r = 3.74$, indicating cycles of periods 6 and 5, respectively. Drawing orbit diagrams for more restricted parameter ranges than that shown in Fig. 9.3.12 will show many other such windows of order for cycles of different periods.

Zooming in on any one of the areas where the lines split, indicating a period-doubling bifurcation, will reveal a smaller "copy" of the entire cascade diagram, distorted but qualitatively identical to the overall bifurcation pattern. (Peek ahead to Fig. 9.3.14 for an example.) Hence, as r increases, a period 6 cycle will soon become period 12 and then period 24; a period 5 cycle will soon split into period 10 and then period 20, and so on. In fact, we will find cycles of many different periods. Where does it end?

Need for Research Approach

To answer the last question, or even to finish asking it, we must explore and experiment. For nonlinear dynamics, mathematicians have had to adopt a non-traditional approach, without formulas. Technology has made such investigation easily available, but thinking about how to use it is a new skill to acquire.

Studying in nonlinear dynamics has to be like research in nonlinear dynamics. It is not difficult to explore, but activity is essential and significant. As we will see in the problems for this section, the initial goal is to seek patterns, to discover what questions to ask and what conjectures to make. Only then can more traditional problem solving and theorem proving begin.

We will close this chapter by scanning some history to illustrate this investigative approach and a few results.

The Biologist's Perspective

Although interest in the logistic iteration and similar nonlinear processes has become quite general, we should not forget its beginning with the population biologists. Logistic recursion as represented by equation (5) might be used to model a gypsy moth population (at one time a topic that brought fear to North American gardeners).

- If the parameter r were found to satisfy $0 < r < 1$, the moths' extinction would be predicted.
- A stable population value of $x_r = (r - 1)/r$ would be expected for $1 \leq r \leq 3$.

- Should it be determined that $3 < r < b_2 \approx 3.4495$, the population would experience a "boom-or-bust" cycle with alternating good and bad years. (Such two-cycle patterns are not uncommon in natural systems.)
- If $b_2 < r < b_3 \approx 3.54$, a four-year cycle would result. (See Fig. 9.3.6(e).)

(One might wonder whether there are systems that could emulate the cycle of the seventeen-year locust; perhaps hidden somewhere in that region of chaos.... Yes—read on.)

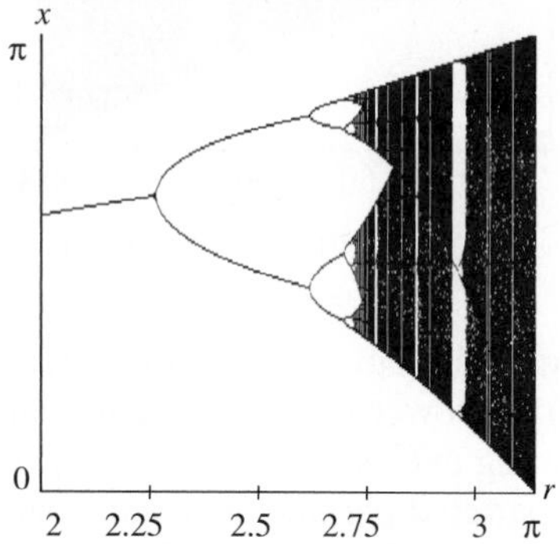

FIGURE 9.3.13 The orbit diagram for iterating $r \sin x$.

Order Out of Chaos

As we scan the orbit diagram of Fig. 9.3.12 from left to right, we see a process that begins in an orderly fashion become increasingly complicated; eventually the order degenerates into chaos. But within such chaotic situations it is also possible to find order. In studying the sequences defined by $x_{n+1} = rx_n(1 - x_n)$, we found a sequence $\{b_n\}$ of parameter values we called bifurcation points:

$$b_1 = 3, \quad b_2 \approx 3.4495, \quad b_3 \approx 3.54, \quad b_4 \approx 3.56, \ldots.$$

This is aptly named the **doubling sequence**.

In the 1970s, Mitchell Feigenbaum of Los Alamos Laboratory studied this doubling sequence.[5] He calculated the ratios of successive differences,

$$q_n = \frac{b_n - b_{n-1}}{b_{n+1} - b_n}, \tag{7}$$

and discovered that the sequence $\{q_n\}$ of ratios tends to a limit $q_\infty \approx 4.6692$. More significantly, this limit of ratios turns out to be the same for a variety of other iterative equations in such diverse fields as electric circuit theory, optical systems, and economics, so Feigenbaum's constant is sometimes referred to as the **universal number**: one of several occurrences of order in the midst of chaos.

Another aspect of the *universality* found by Feigenbaum is that the *same* pictorial structure appears in orbit diagrams for many other functions, such as $r \sin x$, shown in Fig. 9.3.13. (See Problem 35.)

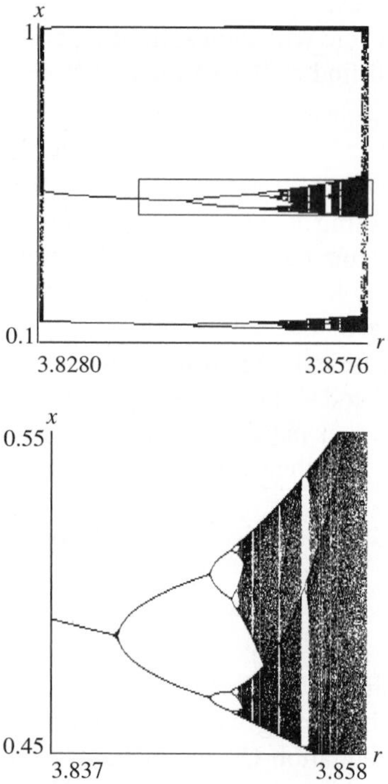

FIGURE 9.3.14 Two successive blowups of the 3-cycle window in the orbit diagram for the logistic $rx(1 - x)$, shown in Fig. 9.3.12.

Furthermore, the orbit diagram is structurally *self-similar*. Figure 9.3.14 shows first a blowup of the 3-cycle window in Fig. 9.3.12, showing that each of the three branching patterns crossing that window has a similar structure to the overall diagram, and second, shows a blowup of the central crossing of the first blowup, which repeats the phenomenon even more clearly. In fact, *every* blowup of *every* window-crossing shows this same structure! The aspect ratio and other scaling details may differ, but all blowups show the same sequence of periodic windows.

There are yet more surprises within the orbit diagram. In 1977 James Yorke of the University of Maryland showed that for any $x_{n+1} = rf(x_n)$ that iterates with a three-cycle, where f is a continuous function, there must be values of a parameter for which n-cycles will occur for *any* integer $n \geq 3$! This fact, which gives an answer to the question of whether there could exist cycles that might model the seventeen-year locust (there are probably several possibilities) startled mathematicians and physicists alike. Not long afterwards, a Russian mathematician A. Sarkovskii told Yorke that he had proved the same result some years earlier. Since that time there has been better communication between the Western and

[5] M. Feigenbaum, "Quantitative Universality for a Class of Nonlinear Transformations," *Journal of Statistical Physics* **19** (1978).

Eastern mathematicians, and between mathematicians and physicists, concerning results in dynamical systems.[6]

Summary

Nonlinear iterative equations can be solved recursively and graphed as time series or cobweb diagrams; in addition to stable and unstable equilibrium points, we find attracting cycles, all of whose members are visited systematically by evolving solutions. These are illustrated by the period-doubling behavior of the logistic iterative equation, which eventually leads to solutions with chaotic behavior; the orbit diagram provides an overview. As with nonlinear DEs, we find these and other patterns show a surprising universality.

9.3 Problems

Attractors and Repellors *Find the equilibrium solutions for the iterative equations in Problems 1–4. Use cobwebs to decide whether each equilibrium is stable (an attractor) or unstable (a repeller).*

1. $x_{n+1} = x_n^2$

2. $x_{n+1} = x_n^2 - 1$

3. $x_{n+1} = x_n^3$

4. $x_{n+1} = -\cos x_n$

5. Hindsight Use cobweb diagrams to illustrate and explain what you learned in Sec. 9.1 about first-order linear iterative equations $y_{n+1} = ay_n + b$. We can look at this as iterating the linear function $ay + b$. The *slope* value a determines the behavior of iterating $ay + b$. (See Fig. 9.3.15.)

(a) Set $b = 0$ and sketch a cobweb example for each of the following cases. Rough but careful handsketches should be fine. For each cobweb example, describe the resulting iteration behavior in terms of equilibrium, convergence and divergence.

(i) $a > 1$ (ii) $a = 1$ (iii) $0 < a < 1$

(iv) $a = 0$ (v) $-1 < a < 0$

(vi) $a = -1$ (vii) $a < -1$

(b) What role does the *size* of a play?

(c) What role does the *sign* of a play?

(d) What role does b play? Experiment with $b \neq 0$.

6. Stability of Fixed Points We have seen that an equilibrium or fixed point x_e of $x_{n+1} = f(x_n)$ is asymptotically stable if $|f'(x_e)| < 1$ and unstable if $|f'(x_e)| > 1$. (No conclusion can be drawn from $|f'(x_e)| = 1$.) Classify the fixed points in the diagrams of Fig. 9.3.16, and confirm with a cobweb.

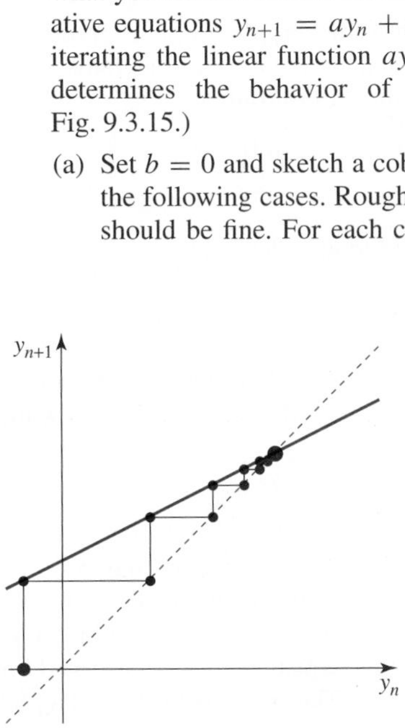

FIGURE 9.3.15 Problem 5: Iterating a straight-line function $ay + b$.

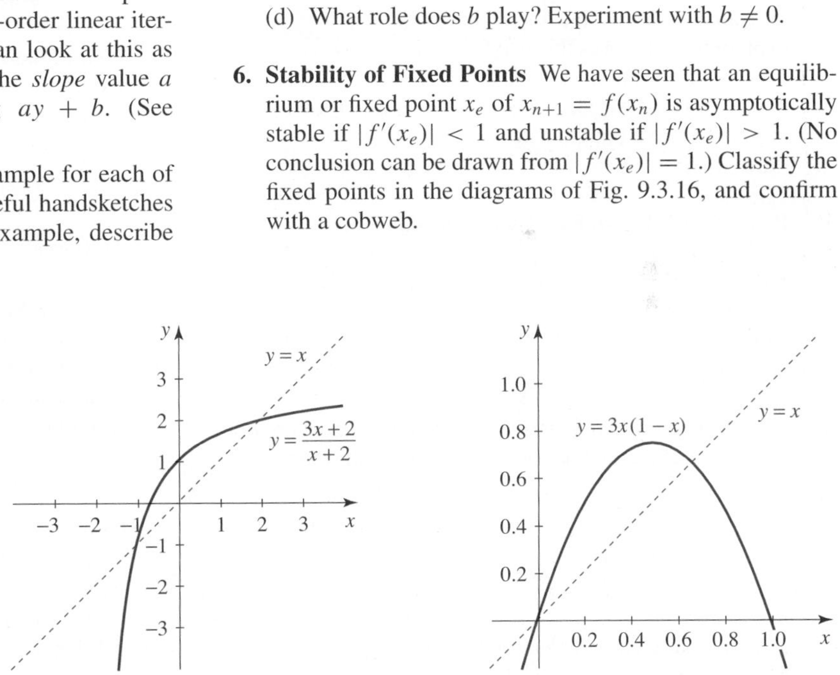

FIGURE 9.3.16 Problem 6: Fixed points for analysis.

---

[6] J. Gleick, *Chaos: Making a New Science* (NY: Viking, 1987), 73–77. See also J. Yorke, "Period Three Implies Chaos," *American Mathematical Monthly* **82** (1975), 985–992; A. N. Sarkovskii, "Coexistence of Cycles of a Continuous Map of a Line into Itself," *Ukrainian Mathematics Journal* **16** (1964), 61.

Analyzing the Data *For the iterative equation in each of Problems 7–16, do the following.*

(a) *Use a calculator or computer and an arbitrary seed to generate a solution sequence. Describe its behavior.*

(b) *Find the fixed points.*

(c) *Use the derivative method to determine the stability of each fixed point.*

7. $x_{n+1} = 0.5x_n(1 - x_n)$

8. $x_{n+1} = 2.8x_n(1 - x_n)$

9. $x_{n+1} = 3.2x_n(1 - x_n)$

10. $x_{n+1} = 4x_n(1 - x_n)$

11. $x_{n+1} = -2\sin x_n$

12. $x_{n+1} = \cos x_n$

13. $x_{n+1} = x_n^2 + 0.1$

14. $x_{n+1} = x_n^2 - 0.1$

15. $x_{n+1} = x_n^2 - 2$

16. $x_{n+1} = x_n^2 - 1$

17. Pete Repeats During a subsequent class of equal dullness, Pete repeats his cosine experiment and is greatly surprised when the constant he arrives at has changed to 0.999848. He verifies this for several different seeds. What is going on? HINT: Consider $x_{n+1} = \cos(\pi x_n/180)$.

18. Repeat Pete's Repeat Now bored out of his mind during an interminable discussion of participles in his English class, Pete unobtrusively takes out his calculator and tries his game with the sine key. What is his result? Can you explain why it works this way?

19. Pete's Parameter Generalize Problem 18 by investigating the iterative equation $x_{n+1} = r\sin x_n$ for $2 \le r \le \pi$. (Take $x_0 = 0.5$.)

20. Pete's Got It Down Pat There is no danger that Pete will suffer from boring discussions any more, even the fine points of oxidation-reduction reactions. Today he's experimenting with iterations on the square root key. Try his game, draw conclusions, and justify your results with cobwebs.

21. The Bernoulli Mapping Make enough cobweb diagrams to decide whether or not the **Bernoulli mapping**

$$x_{n+1} = \begin{cases} 2x_n & \text{if } 0 \le x_n < 0.5, \\ 2x_n - 1 & \text{if } 0.5 \le x_n \le 1 \end{cases}$$

gives rise to chaotic behavior for any $x_0 \in [0, 1]$. HINT: Graph the function and the diagonal; start playing.

Stretch and Fold *The logistic mapping can be interpreted as a process of repeated folding and stretching of the unit interval, resembling the stretching of taffy or the kneading of dough; this is illustrated in Fig. 9.3.17 for the case $r = 4$ (so the range is all of $[0, 1]$).*

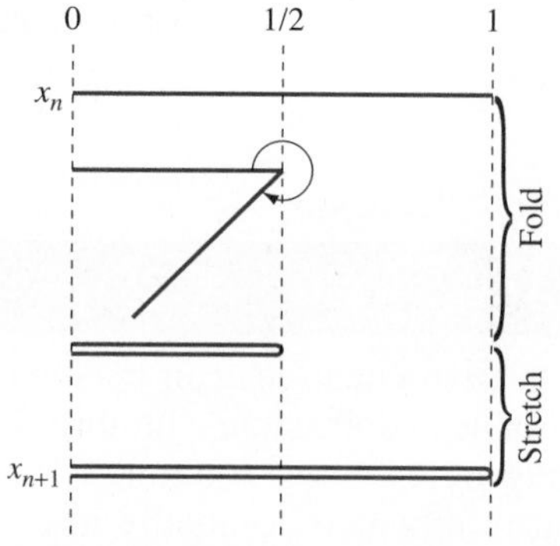

FIGURE 9.3.17 Stretch and fold iteration for $4x(1 - x)$.

Repeating this process gives rise to a chaotic motion of points in the unit interval. For Problems 22 and 23, do the following:

(a) *Show that the mapping displays a "stretch and fold" behavior.*

(b) *Show with a cobweb diagram that iteration will indeed produce a chaotic orbit (e.g., with $x_0 = 0.5$).*

22. $x_{n+1} = 0.98\sin \pi x_n, \quad 0 \le x_n \le 1$

23. $x_{n+1} = \begin{cases} 1.8x_n & \text{if } 0 \le x_n < 0.5, \\ 1.8(1 - x_n) & \text{if } 0.5 \le x_n \le 1 \end{cases}$

24. Orbit Diagram of the Tent Mapping Using appropriate means to automate production of cobweb diagrams, draw the orbit diagram for the **tent mapping**

$$x_{n+1} = \begin{cases} 2rx_n & \text{if } 0 \le x_n < 0.5, \\ 2r(1 - x_n) & \text{if } 0.5 \le x_n \le 1, \end{cases}$$

for $0 \le r \le 1$. HINT: Start all solutions at $x_0 = 0.5$.

25. Chaotic Numerical Iterations An example of the use of iteration in numerical analysis is the computation of a root of the equation $f(x) = 0$ by **Newton's method**. One makes an initial estimate x_0 (the seed), and iterates the formula

$$x_{n+1} = x_n - \frac{f(x_n)}{f'(x_n)}, \tag{8}$$

hoping that the solution sequence will converge to a root. Using suitable means to automate production of cobweb diagrams, draw the orbit diagram for the solution by Newton's method of cubic equation $x^3 + rx + 1 = 0$ for $-1.3 \le r \le -1.25$. HINT: The computed values of x_n lie in the interval $[-1.25, 1.25]$.

26. Extremum Problem Use your calculus and precalculus skills to show that for $f(x) = rx(1 - x)$, restricted to domain $[0, 1]$, we must have $0 \le f(x) \le r \le 4$.

27. Sequential Analysis We noted in the text that for the case $0 \le r \le 1$, the sequence $\{x_n\}$ defined by the logistic equation $x_{n+1} = r(x_n)(1-x_n)$ is decreasing and bounded below.

(a) What result from calculus tells us the sequence has a limit L?

(b) How do we know that $L \geq 0$?

(c) Take limits in the logistic equation to deduce that $L = 0$.

28. Matter of Size

(a) Show that for $f(x) = rx(1-x)$, the fixed point is $x^* = (r-1)/r$.

(b) Show that for $1 < r \leq 3$, $|f'(x^*)| < 1$, so x^* must be a *stable* fixed point.

29. Not Quite a Two-Cycle Consider the logistic function $f(x) = rx(1-x)$. At $r = 3$ we are at a threshold, about to break out into two-cycle behavior for a slightly larger r-value. Analyze what happens, as follows.

(a) Make a time-series plot for an orbit of

$$f(x) = 3x(1-x)$$

to verify that it appears to be possibly settling into a cycle of period 2.

(b) Verify and explain why there is in fact no two-cycle by finding the fixed points of $f(x)$ and of $f(f(x))$, its second iterate.

(c) In light of (b), explain what is really happening in (a).

(d) Make a cobweb diagram and explain how it illustrates all of the above.

30. Finding Two-Cycle Values For $r = 3.2$, the logistic equation has a two-cycle. This means that if you look at alternate terms, those subsequences converge to one of the two-cycle values (one of the numbers in the limiting set). Hence the two-cycle values are *stable* equilibrium points of the *second* iterate. Solve the fourth-degree equation $f(f(x)) = x$, where $f(x) = 3.2x(1-x)$, and verify the values of the two-cycle and their stabilities for the second iterate. HINT: One root is zero; another is $11/16$. Why?

31. Four-Cycle Values Four-cycle behavior for the logistic equation $f(x) = rx(1-x)$ depends on finding x-values in [0, 1] for which $f(f(f(f(x)))) = x$, which are 16th-degree polynomials.

(a) For $r = 3.2$ and then for $r = 3.5$, use a computer algebra system to graph each of these polynomials on [0, 1], and thereby show that the only solutions for the case $r = 3.2$ are the two-cycle points we already knew (plus the fixed points), while for $r = 3.5$ there are eight real solutions (four unstable equilibrium or cyclic points and the four values of the four-cycle).

(b) Estimate the four-cycle values from the graph, and then use one of these four values as a seed to compare with the solution obtained by iteration of $f(x)$.

32. Role of nth Iterates For the logistic function $rx(1-x)$, pick one of the following values of r, which are some of those that yield n-cycles for $4 \leq n \leq 8$: 3.628, 3.702, 3.74, 3.88615, 3.8995, 3.9057, 3.9605, 3.91205.

(a) Find n. A sample figure is shown in Fig. 9.3.18 for $r = 3.5$. HINT: One efficient method is a spreadsheet calculation, continued until a cycle appears, stabilized to two or three decimal places. This might need 100 or more iterates.

(b) Graph the nth iterate function. Then add to your plot a graph of $f(x)$ with a cobweb showing its n-cycle. HINT: Start your iteration at one of the cycle values to get a clean cobweb.

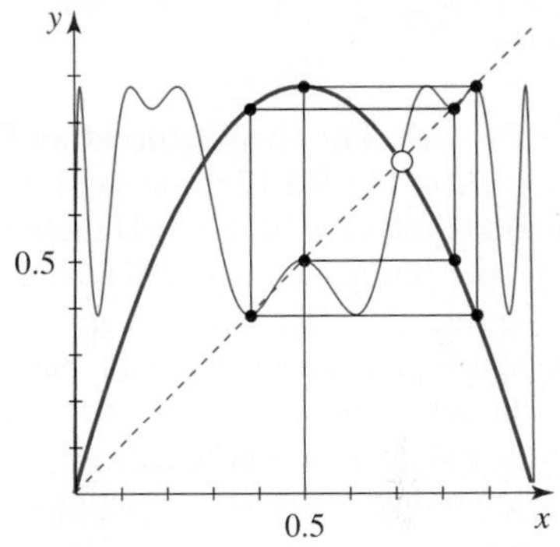

FIGURE 9.3.18 Sample cobweb of n-cycle, showing nth iterate graph as well, for $r = 3.5$ and $n = 4$. Note where and how the nth iterate graph crosses the diagonal.

33. Windows in the Orbit Diagram Figure 9.3.12 shows dark bands where many values occur in an iterated cycle and unshaded bands (windows), like $3.83 < r < 3.845$, in which some three-cycle behavior is indicated because there are only three lines crossing this window for long-term iterates to fall upon.

(a) Compute enough iterations for $r = 3.84$ and $x_0 = 0.5$ to confirm a three-cycle. What are the three-cycle values? Draw a cobweb of the three-cycle. HINT: Use one of the cycle values as x_0 to get a clean cobweb.

(b) Several different regions of the orbit diagram show windows for six-cycles, but you will find that each cycle is in a different pattern.

(i) For $r = 3.628$ and $r = 3.85$ (both visibly occurring within windows in Fig. 9.3.12), make cobweb diagrams to show the iteration pattern for two different six-cycles.

(ii) Compare your two cobwebs and describe their differences.

(c) Find an r-value to create a five-cycle. Give the point values of your cycle and draw its orbit on a cobweb diagram. HINT: You can find a window with five crossings in Fig. 9.3.12.

The Square-and-Add System *The iterative equation*

$$x_{n+1} = x_n^2 + c$$

for real parameter c is another example of an apparently simple process with not so simple results.

34. Investigate the long-term behavior of the solutions of these equations with $x_0 = 0$.

(a) $x_{n+1} = x_n^2 - 1.3$

(b) $x_{n+1} = x_n^2 - 1.755$

(c) $x_{n+1} = x_n^2 - 2$

35. Using appropriate computer assistance, develop an orbit diagram for $x_{n+1} = x_n^2 + c$ for $-2 \le c \le 0.25$. (Take $x_0 = 0$.)

36. Class Project: The Real Bifurcation Diagram The orbit diagram of Fig. 9.3.12 records only the *attracting* fixed points and cycles for the iterated functions at each value of r. A proper bifurcation diagram includes those that repel as well! At every bifurcation, an attracting fixed or cyclic point continues to be a fixed point, but becomes repelling while the new cycle springs up to attract the orbits repelled from the old, as shown schematically in Fig. 9.3.19. The following project guides construction of a proper bifurcation diagram.[7]

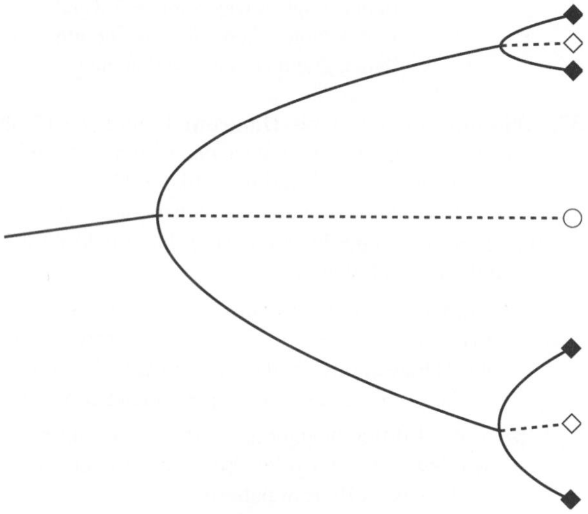

FIGURE 9.3.19 Schematic start to the bifurcation diagram for the logistic equation. Solid curves show attracting fixed points or cycle points; dotted curves show repelling fixed points or cycle points. The right side of the sketch indicates an attracting four-cycle with values separated by a repelling fixed point in the middle and a repelling two-cycle on top and bottom.

For the logistic equation $x_{n+1} = rx_n(1 - x_n)$, assign to each individual a different value of r between 2.8 and 4. Seek r values that show cycles of small period n. To ease calculation, use more values anywhere you suspect cycles of small period n (e.g., between 3.82 and 3.86).

(a) For a given r-value, calculate algebraically the fixed and cyclic points of the first and nth iterate for an n-cycle. That is, express $x_1, x_2, \ldots, x_n$ in terms of x_0, then set $x_n = x_0$. If n is large, you can call on a computer algebra system.

(b) For the r-value used in part (a), determine the stability of each of the fixed and cyclic points.

(c) On a large graph of x versus r, plot your results from (a) and (b) on a vertical line at each r-value, using solid disks for attracting points and open disks for repelling ones. Watch a pattern arise like Fig. 9.3.19!

37. Computer Lab: Double-Well Potential The second-order differential equation

$$\ddot{x} - x + 0.04x^3 = 0, \tag{9}$$

or, in system form,

$$\dot{x} = y,$$
$$\dot{y} = x - 0.04x^3,$$

is a special case of an unforced double-well potential model from Sec. 8.5, Problem 7.

(a) Use a good graphic DE solver to make an accurate phase portrait for solutions to (8).

(b) Recall that solving a DE numerically is an iterative process. Show that, for Euler's method, the formulas would be

$$x_{n+1} = x_n + hy_n,$$
$$y_{n+1} = y_n + h\left(x_n - 0.04x_n^3\right),$$

where h is the step size.

(c) Now set your solver to Euler's method with $h = 0.1$, and plot a phase portrait. Discuss how it fails to capture key characteristics of the real trajectories shown in (a). Will a smaller step size h improve the situation?

(d) Again, using Euler's method, plot solutions $x(t)$ versus t for $0 \le t \le 25$, for three different values of h: 0.05, 0.01, and 0.005.

(e) Describe the changes in your plots from (d) as h decreases. What does this tell you about the discrete iterative approximation model versus a continuous differential equations model for the same system? Before you get too discouraged, consider part (f).

[7] C. Ross and J. Sorensen, "Will the Real Bifurcation Diagram Please Stand Up!" *College Mathematics Journal* **31** (2000), 3–14.

(f) Turn to the Runge-Kutta method and repeat parts (c) and (d). For each of these parts, tell in what respects the Runge-Kutta approximations are better than those obtained by Euler's method.

NOTE: Of course, the Runge-Kutta method is still an iterative process; its formulas are more complicated, but we can see the advantages in this problem. We are fortunate indeed to be able to use computer solvers to eliminate the tedium of the calculations, but we must remain aware of the pitfalls lurking in any numerical approximation to solutions of differential equations.

38. **Suggested Journal Entry** Discuss what happens to the limiting sets for the logistic equation as $b_n \to b_\infty$. Do you think the result is the interval [0, 1]?

CHAPTER 10 Control Theory

Mathematical facts worthy of being studied are those which, by their analogy with other facts, are capable of leading us to the knowledge of a mathematical law, in the same way that experimental facts lead us to the knowledge of a physical law.

—Henri Poincaré

10.1 Feedback Controls

SYNOPSIS: Physical systems can be regulated by feedback controls. We consider three types—proportional, derivative, and integral—and some combinations of these.

What Is Feedback Control?

Consider a physical system modeled by the general linear differential equation $L(\vec{\mathbf{x}}) = f(t)$.[1] A **feedback control** $u(t)$ regulates the system by reducing the difference between actual system output $x(t)$ and a predetermined ideal output x_1. This automatic control measures system output and feeds the results back into the system, where adjustments are made to drive the actual output toward the ideal. The resulting controlled system $L(\vec{\mathbf{x}}) = f(t) + u(t)$ is illustrated schematically in Fig. 10.1.1.

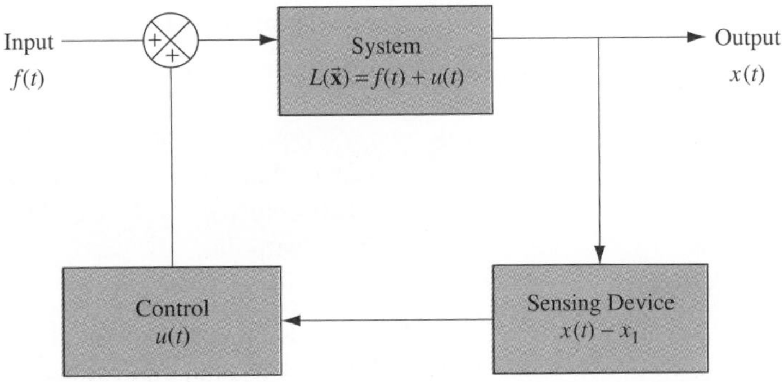

FIGURE 10.1.1 Feedback control.

[1]The examples in this section are all linear DEs with $\vec{\mathbf{x}} = [x, \dot{x}, \ddot{x}]$, the vector formed by $x(t)$ and its derivatives, as in Sec. 2.1 equation (2). However, feedback control is even more common and useful with nonlinear DEs.

Feedback controls are called **closed-loop** controls because they depend on the state of the system. **Open-loop** controls, on the other hand, depend explicitly on time. We will see some examples of open-loop controls when we introduce optimal controls in the next section.

Modern civilization is highly dependent on automatic controls; it is difficult to imagine life without them. Industrial control systems maintain production quality, free workers from routine tasks, reduce waste, and increase efficiency. An airplane autopilot senses changes in wind force and direction, air pressure, and temperature, and makes the necessary adjustment to keep the plane on a given course at a given speed. Another familiar feedback control is the thermostat for a home heating and cooling system.

Natural feedback controls in your own body regulate blood pressure, blood sugar, cell carbon dioxide, and a host of other variables. Few of these controls are well understood, but their importance is obvious.

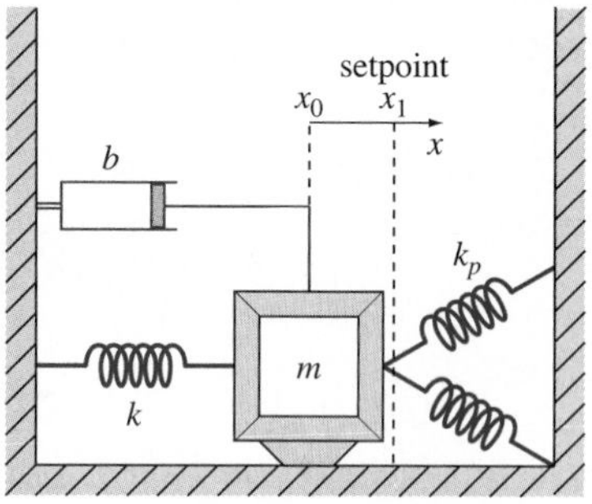

FIGURE 10.1.2 Controlled mass-spring system.

Feedback Control for a Mass–Spring System

We will illustrate different types of feedback control with the familiar mass–spring system, in which a unit mass vibrates horizontally on a flat surface with position measured on the x-axis. (See Fig. 10.1.2.) The unforced motion is modeled by the second-order differential equation

$$\ddot{x} + b\dot{x} + kx = 0, \tag{1}$$

where $b > 0$ is the damping constant and $k > 0$ is the restoring constant of the spring.

The goal is to control the system so that the output $x(t)$ remains near a predetermined value x_1, called the **setpoint**.[2] The controlled system is

$$\ddot{x} + b\dot{x} + kx = u(t), \tag{2}$$

where $u(t)$, which could be $u(x(t))$, is a controlling force.

Proportional Feedback

The proportional control is a very basic type of feedback that applies an external Hooke's Law force, which is proportional to the difference between actual output $x(t)$ and desired output x_1.

Proportional Control

The **proportional control**

$$u_p = k_p(x_1 - x),$$

is a force that opposes the displacement of the system (1). The constant of proportionality $k_p > 0$ is called the **proportional gain** of the control, and x_1 is the setpoint.

When $x > x_1$, a proportional control pushes to the left toward the setpoint x_1, but when $x < x_1$, the control acts to the right.

The controlled system (2) becomes

$$\ddot{x} + b\dot{x} + kx = k_p(x_1 - x),$$

[2]For simplicity we will assume a constant setpoint, but one can also study the situation in which the setpoint varies with time.

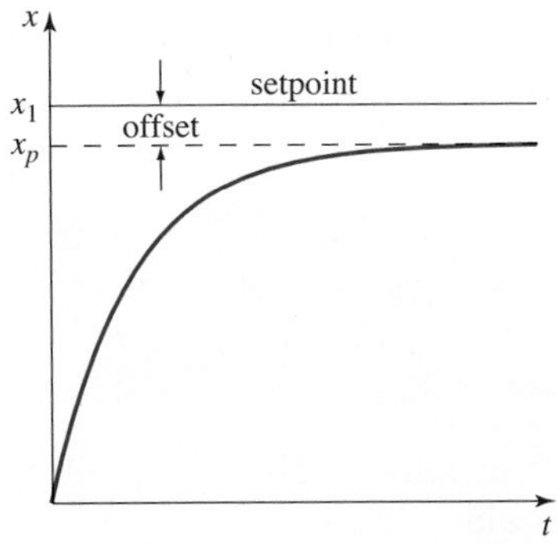

FIGURE 10.1.3 Offset error for a proportional control, starting at $x(0) = 0$, with small gain k_p, $\Delta > 0$.

or, equivalently,

$$\ddot{x} + b\dot{x} + (k + k_p)x = k_p x_1.$$

When the system is in motion, the feedback term drives the mass toward a constant steady-state solution x_p, which is less than the desired setpoint value x_1. (See Fig. 10.1.3.) The difference between x_1 and the steady-state solution x_p is called the **offset error**, exaggerated in the figures to make it visible.

Motion of a Mass-Spring System with Proportional Control

A mass-spring system with proportional control,

$$\ddot{x} + b\dot{x} + (k + k_p)x = k_p x_1, \tag{3}$$

has general solution

$$x(t) = c_1 e^{\lambda_1 t} + c_2 e^{\lambda_2 t} + \underbrace{\left(\frac{k_p}{k + k_p}\right) x_1}_{x_p}, \tag{4}$$

where λ_1, λ_2 are solutions of the characteristic equation with discriminant $\Delta = b^2 - 4(k + k_p)$, and c_1 and c_2 are determined by initial conditions.

Proportional control changes the equilibrium of the system from 0 to x_p. The way in which the system output approaches x_p depends on the gain k_p. (See Fig. 10.1.4.) The design engineer must decide what characteristics are required.

- As the gain k_p increases, x_p approaches x_1, but the system goes from overdamping ($\Delta > 0$) to underdamping ($\Delta < 0$), and oscillations occur.
- If the gain is too small, approach to steady state may be too slow.
- If the gain is too large, undesirable oscillations may be the result. The higher the gain, the more oscillation.

Although there is a tendency to choose a large proportional gain k_p to drive the system to the setpoint as fast as possible, the choice is not always a good one, because large gains cause the system to pass right on through the setpoint and oscillate around it. To rid ourselves of these undesirable oscillations about the setpoint, we add an additional feature to our control.

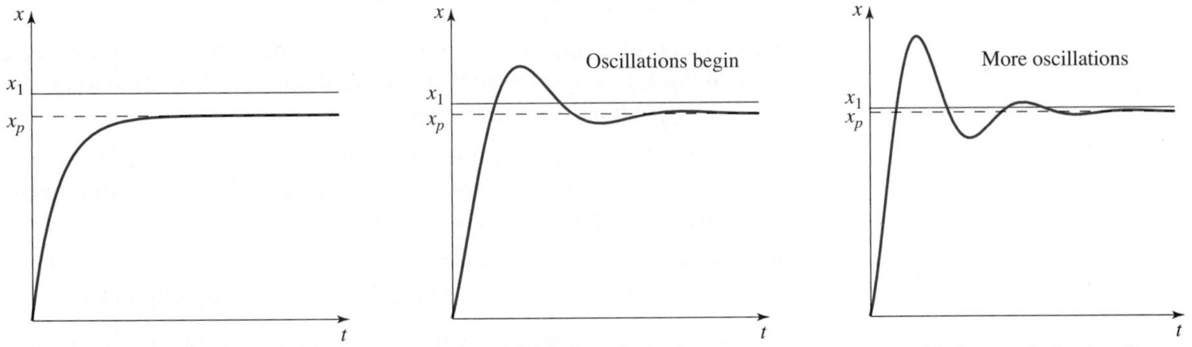

(a) Slightly larger gain k_p, $\Delta = 0$ (b) Larger gain k_p, $\Delta < 0$ (c) Large gain k_p, $\Delta \ll 0$

FIGURE 10.1.4 Effects of proportional gain compared to Fig. 10.1.3. Oscillations begin when Δ becomes negative.

Derivative Feedback

To offset the undesirable oscillations resulting from large proportional gain, the designer may add a damping force.

Derivative Control

The **derivative control,**

$$u_d = -k_d \dot{x},$$

is a force that will oppose the *rate of change of the output*. The constant of proportionality $k_d > 0$ is called the **derivative gain**.

Derivative control effectively modifies the original damping coefficient, which is not useful by itself, but we can obtain a *proportional-plus-derivative* or **PD control**,

$$u_{pd} = k_p(x_1 - x) - k_d \dot{x}, \tag{5}$$

by combining proportional and derivative feedback.

Motion of a Mass-Spring System with PD Control

A mass-spring system with PD control,

$$\ddot{x} + (b + k_d)\dot{x} + (k + k_p)x = k_p x_1, \tag{6}$$

has the same general solution,

$$x(t) = c_1 e^{\lambda_1 t} + c_2 e^{\lambda_2 t} + \underbrace{\left(\frac{k_p}{k + k_p}\right) x_1}_{x_p}, \tag{7}$$

and steady-state solution x_p as system (3) with proportional control alone, but λ_1 and λ_2 are solutions of a different characteristic equation with discriminant

$$\Delta = (b + k_d)^2 - 4(k + k_p),$$

which depends on both the proportional gain k_p and the derivative gain k_d.

Proportional control and PD control have the same steady-state solution for the same proportional gain k_p, but the derivative gain k_d adds to the damping of the system and allows us to choose larger k_p than a system with proportional control alone.

- As the proportional gain k_p increases but the derivative gain k_d remains unchanged, the system goes from overdamping ($\Delta > 0$) to underdamping ($\Delta < 0$), and oscillations occur.
- As the derivative gain k_d increases but the proportional gain k_p remains unchanged, the system goes from underdamping to overdamping, and the oscillations cease.
- A larger derivative gain allows us to choose a large proportional gain, which will move the steady-state solution x_p closer to the setpoint x_1.

One way to choose gain values in a control problem is to require that the eigenvalues have predetermined values. We know that the eigenvalues of a linear differential equation characterize its behavior, and that negative eigenvalues and eigenvalues with negative real parts correlate with stable behavior. This is illustrated in the following example.

EXAMPLE 1 **Eigenvalue Placement by PD Control** We will use PD control to drive the "unstable" system

$$\ddot{x} - \dot{x} + x = 0 \tag{8}$$

(with "negative friction") from some initial point $(x(0), \dot{x}(0))$ toward the setpoint x_1, by choosing negative eigenvalues $\lambda_1 = -2$ and $\lambda_2 = -5$ to control the damping. Given these eigenvalues, we will determine the proportional gain k_p and derivative gain k_d of the PD control.

Equation (8) with control (6) becomes

$$\ddot{x} + (-1 + k_d)\dot{x} + (1 + k_p)x = k_p x_1, \tag{9}$$

which has characteristic polynomial

$$p(\lambda) = \lambda^2 + (-1 + k_d)\lambda + (1 + k_p) = 0. \tag{10}$$

Substituting into (10) our chosen eigenvalues $\lambda_1 = -2$ and $\lambda_2 = -5$, we obtain a system of two equations for k_d and k_p:

$$-2k_d + k_p + 7 = 0,$$
$$-5k_d + k_p + 31 = 0;$$

the solution is $k_d = 8$ and $k_p = 9$. The PD control (6) is

$$u_{pd} = 9(x_1 - x) - 8\dot{x},$$

and the controlled system (9) is modeled by $\ddot{x} + (-1+8)\dot{x} + (1+9)x = 9x_1$, which simplifies to

$$\ddot{x} + 7\dot{x} + 10x = 9x_1. \tag{11}$$

We have stabilized the system solution to

$$x(t) = c_1 e^{-2t} + c_2 e^{-5t} + \frac{9}{10}x_1, \tag{12}$$

the general solution of (11), where c_1 and c_2 depend on the initial values of x and $\dot{x}$. The steady-state solution $x_p = \frac{9}{10}x_1$ results in an offset error of 10%, and the rate at which the actual output tends to x_p is determined by the eigenvalues and the initial conditions. Figure 10.1.5 shows results with initial conditions $x(0) = 0$ and $\dot{x}(0) = 9$ for a setpoint of $x_1 = 1$. ■

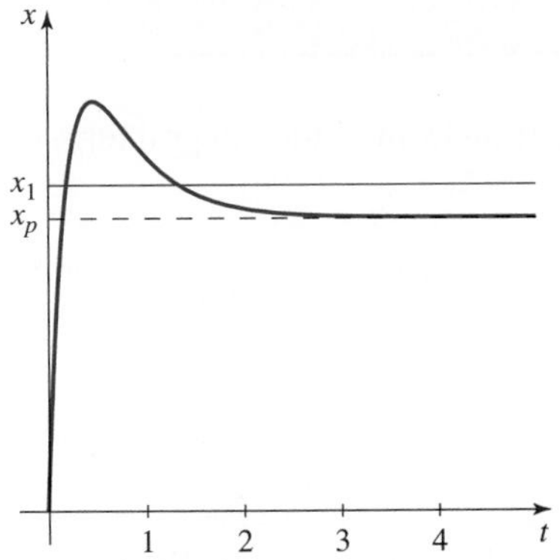

FIGURE 10.1.5 PD feedback with exponential damping, for eigenvalue placement at -2 and -5, with initial conditions $x(0) = 0$ and $\dot{x}(0) = 9$, for a setpoint of $x_1 = 1$.

Integral Feedback

Proportional feedback drives a system closer to its setpoint and derivative feedback limits the oscillations, but neither of these, nor any combination of them, will eliminate the offset error. We need a *control with a memory*.

Integral Control

The **integral control**

$$u_i = k_i \int_0^t [x_1 - x(w)]\,dw \tag{13}$$

keeps track of the history of the relationship between actual output $x(t)$ and the setpoint x_1. The coefficient $k_i > 0$ is called the **integral gain** of the control.

Although integral feedback can be used alone, it is more often combined with proportional and derivative feedback to give the *proportional-plus-integral-plus-derivative* or **PID control**,

$$u_{pid} = k_p(x_1 - x) + k_i \int_0^t [x_1 - x(w)]dw - k_d\dot{x}. \tag{14}$$

The motion is modeled by the **integrodifferential equation**

$$\ddot{x} + (b + k_d)\dot{x} + (k + k_p)x = k_p x_1 + k_i \int_0^t [x_1 - x(w)]dw. \tag{15}$$

We will investigate how PID control is able to eliminate the offset in the mass-spring system using tools from Laplace transform theory (Chapter 8). The following easy consequence of the derivative theorem for Laplace transforms (see Problem 11) allows us to transform the integral terms.

Laplace Transform for an Indefinite Integral
If $\mathcal{L}\{x(t)\} = X(s)$, then

$$\mathcal{L}\left\{\int_0^t x(w)\,dw\right\} = \frac{X(s)}{s}. \tag{16}$$

Using this new tool, we calculate the Laplace transform of the integral term of (15) as follows:

$$\begin{aligned}\mathcal{L}\left\{k_i \int_0^t [x_1 - x(w)]dw\right\} &= \mathcal{L}\left\{k_i x_1 t - k_i \int_0^t x(w)\,dw\right\} \\ &= \frac{k_i x_1}{s^2} - \frac{k_i X(s)}{s}.\end{aligned}$$

The transform of equation (15) is therefore

$$s^2X(s) + (b + k_d)sX(s) + (k + k_p)X(s) = \frac{k_p x_1}{s} + \frac{k_i x_1}{s^2} - \frac{k_i X(s)}{s}.$$

Multiplying through by s and collecting terms involving $X(s)$ gives

$$s^3X(s) + (b + k_d)s^2X(s) + (k + k_p)sX(s) + k_iX(s) = \frac{k_p x_1 s + k_i x_1}{s},$$

and solving for $X(s)$ gives

$$X(s) = x_1\left(\frac{k_p s + k_i}{s}\right)\left(\frac{1}{s^3 + (b + k_d)s^2 + (k + k_p)s + k_i}\right). \tag{17}$$

But to find the steady-state solution x_p, we need a new result from Laplace transform theory. (See Problem 12.)

Final Value Theorem
Suppose that $x(t)$ and $\dot{x}(t)$ have Laplace transforms, that $\lim_{t\to\infty} x(t)$ is finite, and that $\mathcal{L}\{x(t)\} = X(s)$. Then,

$$\lim_{t\to\infty} x(t) = \lim_{s\to 0} sX(s). \tag{18}$$

It follows from (17) that

$$sX(s) = x_1 \frac{k_p s + k_i}{s^3 + (b + k_d)s^2 + (k + k_p)s + k_i},$$

and by the final value theorem (18) that

$$x_p = \lim_{t\to\infty} x(t) = \lim_{s\to 0} sX(s) = x_1.$$

Hence, the addition of integral control has reduced the steady-state error to zero. We summarize these results as follows.

Motion of a Mass-Spring System with PID Control
A mass-spring system with PID control

$$\ddot{x} + (b + k_d)\dot{x} + (k + k_p)x = k_p x_1 + k_i \int_0^t [x_1 - x(w)]dw \qquad (19)$$

has steady-state solution

$$x_p = x_1$$

and general solution

$$x(t) = x_1 \mathcal{L}^{-1}\left\{\frac{k_p s + k_i}{s\left[s^3 + (b + k_d)s^2 + (k + k_p)s + k_i\right]}\right\}.$$

EXAMPLE 2 **Integral Control Alone** A system modeled by the equation

$$\dot{x} + 3x = 0$$

experiences a disturbance to $x(0) = 0$ at reference time zero. The controlled response using integral feedback is described by

$$\dot{x} + 3x = 2\int_0^t [x_1 - x(w)]dw = 2x_1 t - 2\int_0^t x(w)\,dw, \qquad (20)$$

where x_1 is the setpoint. The Laplace transform of (20) is

$$sX(s) + 3X(s) = \frac{2x_1}{s^2} - 2\frac{X(s)}{s}.$$

Solving for $X(s)$, we get

$$X(s) = \frac{2x_1}{s(s^2 + 3s + 2)} = x_1\left(\frac{1}{s} - \frac{2}{s+1} + \frac{1}{s+2}\right),$$

and

$$x(t) = x_1\left(1 - 2e^{-t} + e^{-2t}\right).$$

As expected, $x(t)$ tends to x_1 as $t \to \infty$, and the offset error is zero. When $t = 3$, $x = 0.9x_1$, already within 10% of the setpoint. (See Fig. 10.1.6.) We could achieve faster convergence with the addition of proportional and derivative controls. ■

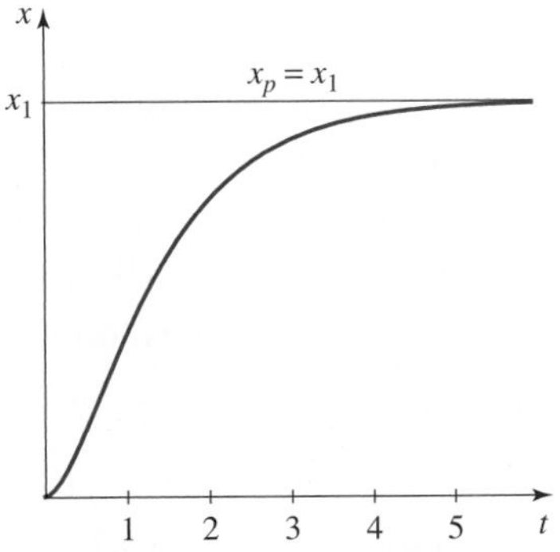

FIGURE 10.1.6 Integral feedback for first-order equation of Example 2, starting at $x(0) = 0$.

Comparison of Controls

While it is easy for the mathematician to dream up a control with the desired effect on a mathematical model, it may be far from easy for the designer to improvise a physical device to implement it. A theoretical ability to control may be difficult to translate into physical reality. Assuming that various controls can be devised, however, it is interesting to see how they stack up against each other.

EXAMPLE 3 **Comparing the Controls** We will compare the effectiveness in driving the system modeled by

$$\ddot{x} + 5\dot{x} + 2x = 0 \tag{21}$$

from $x(0) = x_0$ to setpoint x_1, using three different feedback controls:

(a) P control: $u_p = 18(x_1 - x);$ (22)

(b) PD control: $u_{pd} = 18(x_1 - x) - 4\dot{x};$ (23)

(c) PID control: $u_{pid} = 18(x_1 - x) - 4\dot{x} + 4\int_0^t [x_1 - x(w)]dw.$ (24)

Graphs of the following solutions will be shown in Fig. 10.1.7.

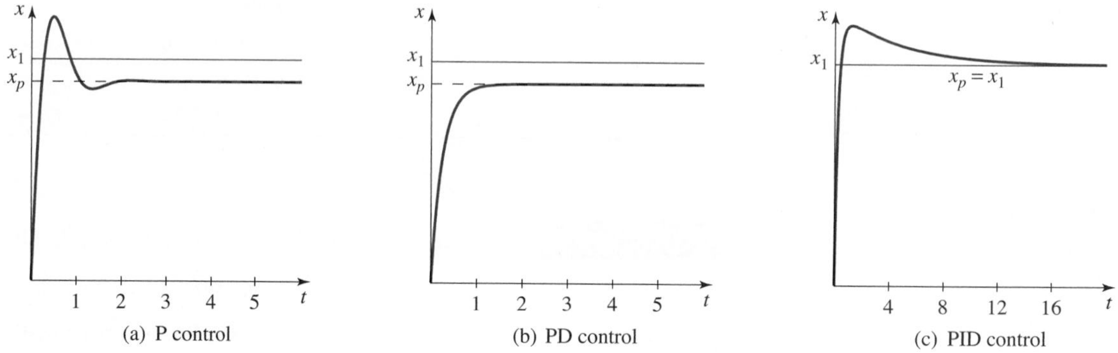

FIGURE 10.1.7 Comparison of controls for Example 3. PID control is better, but it is often far more expensive from an engineering viewpoint and takes a lot longer to get close to x_p in this example. (Note scale change in t.)

(a) The proportional control (22) applied to system (21) yields

$$\ddot{x} + 5\dot{x} + 20x = 18x_1, \tag{25}$$

which has general solution

$$x(t) = e^{-5t/2}\left[c_1 \cos\left(\frac{\sqrt{55}}{2}t\right) + c_2 \sin\left(\frac{\sqrt{55}}{2}t\right)\right] + 0.9x_1.$$

After a disturbance away from the setpoint x_1, system (25) will be driven in an oscillatory fashion toward $x_p = 0.9x_1$ (an offset error of 10%).

(b) If PD feedback from (23) is used instead, the new version of the model is

$$\ddot{x} + 9\dot{x} + 20x = 18x_1, \tag{26}$$

and we find the general solution to be

$$x(t) = c_1e^{-4t} + c_2e^{-5t} + 0.9x_1.$$

The offset error is still 10%, but the system is now overdamped and the approach to $x_p = 0.9x_1$ is no longer oscillatory.

(c) Finally, using the PID control (24) gives the equation

$$\ddot{x} + 9\dot{x} + 20x = 18x_1 + 4\int_0^t [x_1 - x(w)]dw. \quad (27)$$

With the final value theorem (18), we find (Problem 14) that $x_p = x_1$, so the offset error has been eliminated. (See Problem 15 for the explicit general solution.) ■

Summary

Feedback controls permit adjustments of physical systems in operation to achieve predetermined goals. Such controls are essential for stabilizing ships at sea and satellites in space, images on TV and computer screens, nuclear power plants, and computer-controlled brakes.

10.1 Problems

A Matter of Control *For the equations in Problems 1–6, use each of the three controls:*

(a) $u = 2(1 - x)$ (b) $u = -3\dot{x}$ (c) $u = 2(1 - x) - 3\dot{x}$

Determine and compare the response of the uncontrolled equation and the response of each controlled equation. For Problems 1–6, $x(0) = 1$; for Problems 4–6, add $\dot{x}(0) = 0$.

1. $\dot{x} + x = u$

2. $\dot{x} + 2x = u$

3. $2\dot{x} + 3x = u$

4. $\ddot{x} + \dot{x} + x = u$

5. $\ddot{x} + 2\dot{x} + 3x = u$

6. $\ddot{x} + 3\dot{x} + 2x = u$

7. Eigenvalues of a Controlled Equation Show that the eigenvalues of the equation

$$\ddot{x} + 4\dot{x} + x = -k_p x$$

are $-2 \pm \sqrt{3 - k_p}$ if $0 \le k_p \le 3$, $-2 \pm i\sqrt{k_p - 3}$ if $k_p > 3$. Plot these roots as a function of k_p for $k_p \ge 0$, and use the plot to describe the behavior of the solution x as a function of the gain k_p.

8. Eigenvalue Placement As illustrated in Example 1, eigenvalue placement allows us to specify the stability of a system by choosing the feedback gains. Determine the gains k_1 and k_2 in the feedback system $\dot{\vec{\mathbf{x}}} = \mathbf{A}\vec{\mathbf{x}} + \vec{\mathbf{u}}$, where

$$\mathbf{A} = \begin{bmatrix} 0 & 1 \\ -1 & 0 \end{bmatrix} \quad \text{and} \quad \vec{\mathbf{u}} = \begin{bmatrix} -k_1 x_1 \\ -k_2 x_2 \end{bmatrix},$$

that make the eigenvalues of the controlled system -1 and -2.

9. Controlling an Unstable System Consider the unstable system $\ddot{x} - 4\dot{x} + 3x = 0$.

(a) Determine the characteristics of a control of the form

$$u_d = -k_d\dot{x}$$

that will stabilize the system.

(b) For the control

$$u_{pd} = k_p(x_1 - x) - k_d\dot{x},$$

determine the gains k_p and k_d that place both eigenvalues of the controlled system at -2.

10. Burden of Proof Derive equation (16) from the Laplace derivative theorem.

11. Final Value Theorem Derive the final value theorem (18) from the derivative theorem for Laplace transforms.

12. Integral Control Consider the integral feedback system

$$\dot{x} + 4x = 3\int_0^t [1 - x(w)]dw,$$

with setpoint $x_1 = 1$ and integral gain $k_i = 3$.

(a) What is the steady state of the uncontrolled (homogeneous) equation?

(b) Use the Laplace transform to solve the controlled equation with initial condition $x(0) = 0$.

(c) Use the final value theorem to verify that $x(t)$ approaches the setpoint with no offset error.

13. PID Control Show that the steady-state solution of equation (27) is x_1 using the final value theorem for the Laplace transform.

14. Libration Point Control[3] On a line connecting the center of the earth to the center of the moon there is a point L called the **libration point**, where the pulls of the two bodies are equal and opposite.[4] This is an unstable equilibrium point, so if we want to station a space ship there, we need control. (See Fig. 10.1.8.)

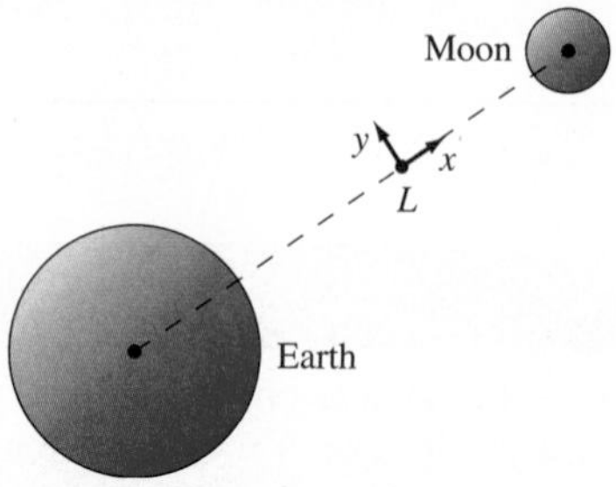

FIGURE 10.1.8 Earth-moon libration point (Problem 14).

The linearized equations that describe the controlled behavior of small deviations x and y away from the libration point are

$$\ddot{x} - 2\omega\dot{y} - 9\omega^2 x = 0,$$

$$\ddot{y} + 2\omega\dot{x} - 4\omega^2 y = u.$$

(a) Show that the origin is an unstable equilibrium point for the uncontrolled system.

(b) Rewrite the two second-order equations as a system of four first-order equations in matrix-vector form $\dot{\vec{\mathbf{x}}} = \mathbf{A}\vec{\mathbf{x}} + \vec{\mathbf{b}}u$, where $\vec{\mathbf{b}} = \vec{\mathbf{e}}_4 = [0, 0, 0, 1]$ and $u = -k_1 x - k_2\dot{x} - k_3 y - k_4\dot{y}$.

15. Out of Control To solve equation (27) without Laplace transforms, do the following:

(a) Differentiate equation (27), using the fundamental theorem of calculus to eliminate the integral.

(b) Use the result of (a) to verify directly that $x_p = x_1$ for part (c) of Example 3.

(c) Use the result of (a) to approximate the general solution of equation (27). (You will need to approximate the eigenvalues from a characteristic equation that is now cubic.)

16. Heating Control In Sec. 2.4, Problem 14, we modeled the temperature T in a building with a broken furnace using Newton's Law of Cooling $\dot{T} = k(M - T)$, where M is the constant temperature of the air surrounding the building. Using control theory, we are now able to incorporate the furnace in our model:

$$\dot{T} = k(M - T) + u(t), \qquad (28)$$

where $u(t)$ is the heat transfer to the building from the furnace. Suppose that the thermostat on the furnace is set so that

$$u(t) = \begin{cases} u_0 & \text{if } T < 65, \\ u_0 & \text{if } 65 \le T \le 75 \text{ and } \dot{T} < 0, \\ 0 & \text{if } 65 \le T \le 75 \text{ and } \dot{T} > 0, \\ 0 & \text{if } T > 75, \end{cases}$$

where u_0 is the heat transfer from furnace to building when the furnace is turned on.

(a) Is this a feedback control? Why?

(b) Is the feedback a linear feedback (that is, a linear function of T)?

(c) Describe in simple language how this control will behave when the furnace is in operation on a cold night.

(d) Use a computer to approximate the solution of (28) when the outside air temperature is $M = 20°$F, the heat loss coefficient is $k = 0.20$, the heat transfer is $u_0 = 1$, and $T(0) = 60$.

Frequency Viewpoint of Feedback *In this section we studied the* ***time-domain*** *approach (i.e., differential equations) for analyzing control systems and linear feedback. An alternate approach is the* ***frequency-domain*** *approach, which focuses on the Laplace transform and is useful for seeing systems pictorially by means of* ***block diagrams****. Problems 17–22 give a glimpse into the frequency-side linear systems.*

17. The transfer function (Sec. 8.4) of a linear system is the ratio of the Laplace transform of the output to the Laplace transform of the input *when the initial conditions are taken as zero*. Find the transfer function of the linear system

$$\ddot{x} + \dot{x} + x = f(t), \qquad x(0) = 0, \qquad \dot{x}(0) = 0. \qquad (29)$$

18. Show that the Laplace transform of the output of the linear system (29) is the product of the transfer function times the Laplace transform of the input. This is illustrated pictorially by the block diagram in Fig. 10.1.9.

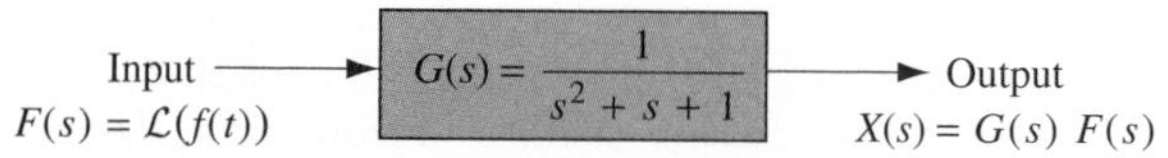

FIGURE 10.1.9 Block diagram for Problem 18.

19. (a) Show that the transfer function of the concatenated system illustrated by the block diagram in Fig. 10.1.10 is

$$\frac{X(s)}{F(s)} = \frac{1}{s(s^2 - 1)}.$$

(b) What differential equation and initial condition are equivalent to this system?

[3] Problem adapted from Thomas Kailath, *Linear Systems* (NY: Prentice-Hall, 1980).

[4] The word *libration* comes from the Latin for "balance."

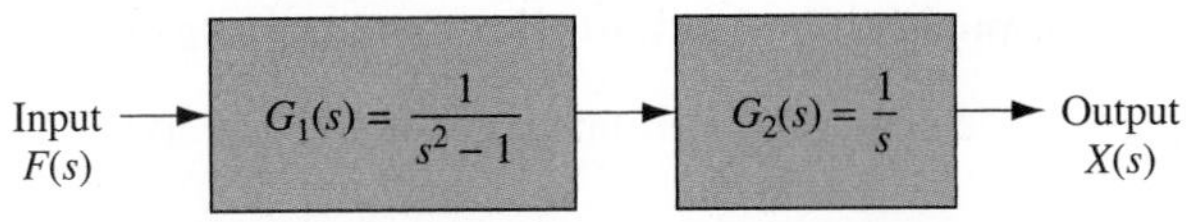

FIGURE 10.1.10 Block diagram for Problem 19. The role of the second block for $G_2(s)$ is to *integrate* the incoming signal $G_1(s)$.

20. Show that

$$\frac{X(s)}{F(s)} = \frac{G(s)}{1 - G(s)H(s)},$$

the transfer function of the feedback control system shown in Fig. 10.1.11. HINT: Use the relationships

$$X(s) = G(s)E(s),$$
$$E(s) = F(s) + B(s),$$
$$B(s) = H(s)X(s).$$

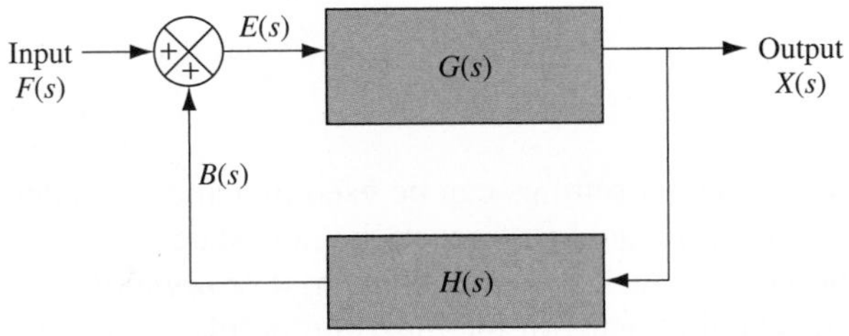

FIGURE 10.1.11 Block diagram for Problem 20.

Here $G(s)$ is the transfer function of the uncontrolled system, and $H(s)$ is the transfer function of the feedback control.

21. Show that the feedback transfer function $H(s)$ is

(a) $H(s) = -k_p$ for proportional control

$$u(x) = -k_p x,$$

(b) $H(s) = -k_d s$ for derivative control

$$u(x) = -k_d \dot{x},$$

(c) $H(s) = -\frac{k_i}{s}$ for integral feedback

$$u(x) = -k_i \int x(s)\,dx.$$

22. Use the results from Problems 20 and 21 to find the transfer function of the feedback systems in (a)–(c).

(a) $\ddot{x} + x = -5x$ (P control)

(b) $\ddot{x} + x = -5\dot{x}$ (D control)

(c) $\ddot{x} + x = -2\int_0^t x(s)\,ds$ (I control)

23. Suggested Journal Entry How many feedback control systems can you list that affect your daily life? Describe the controls and how they operate.

10.2 Introduction to Optimal Control

SYNOPSIS: We introduce the optimal control problem and its basic ingredients: state equations, feasible controls, and the objective function, illustrating these with concrete examples.

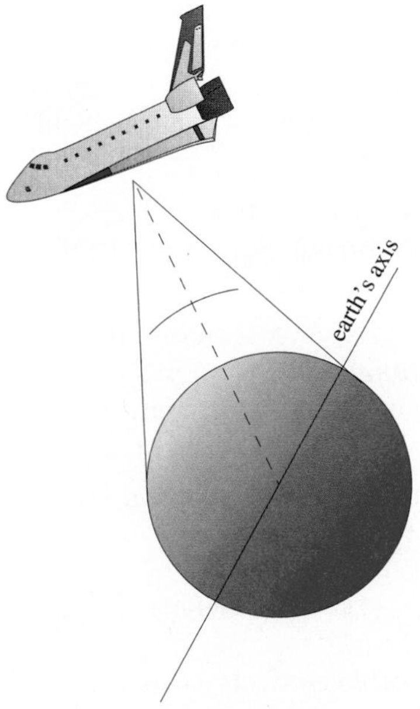

FIGURE 10.2.1 Orbiting satellite.

Who's in Control Here?

Human beings have sought to control natural processes since the dawn of civilization, and to do so in the most efficient, elegant, or economical way. In the twentieth century, mathematical control theory has transformed many of these goals into reality. The control of physical systems now cuts across all scientific, engineering, and technological areas, and is finding application in wider spheres, such as biology and economics.

At the heart of an optimal control problem is a system of equations, often differential equations, which model the evolution of the physical system over time. These equations contain terms called **controls** that represent physical quantities that can be adjusted *during the time the system is evolving* to alter its behavior in order to achieve some goal. The object is to adjust the controls not only to achieve a specific state of the system but to do it so that some measure of system performance is maximized or minimized. Controls that make this possible are called **optimal controls**.

A typical optimal control problem is the orienting of an orbiting satellite. (See Fig. 10.2.1.)

An object in space has both **position** and **orientation** (or **attitude**).

- *Position* is determined by three spatial coordinates relative to the axis and center of the earth.
- *Attitude* is determined by **pitch**, **yaw**, and **roll**, as illustrated in Fig. 10.2.2.

When a signal is received from an on-the-ground or in-flight guidance system advising a change in orientation, that signal must activate changes that will cause the satellite to pitch, yaw, or roll appropriately. In this way the attitude and orbit can be controlled.

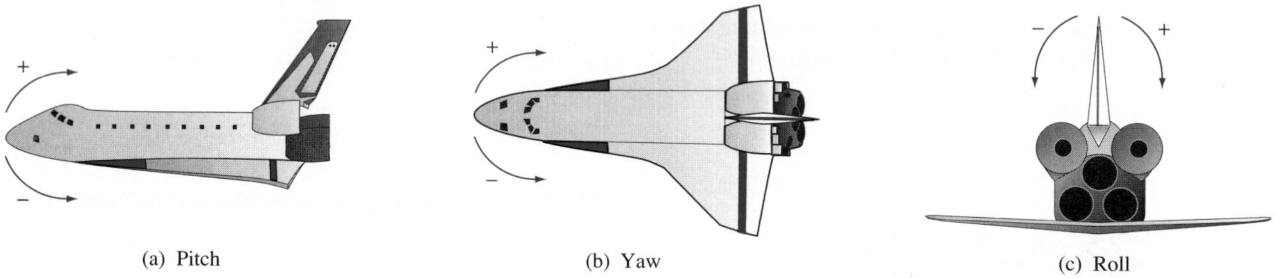

FIGURE 10.2.2 Satellite orientation.

In general, many combinations of settings can be used to achieve the desired change in orbit or orientation. If the goal is to reach the new state of the system in the least time, or using the least fuel, there may be a best or *optimal strategy*. (Compare this with the idea behind many of the max/min problems you studied in calculus.)

Modeling Optimal Control

To formulate our problem mathematically, we need to define the basic ingredients and make assumptions about how they work.

Optimal Control Model

- **Time:** We take time t to vary continuously from initial time 0 to a final time t_f; that is, $0 \leq t \leq t_f$. In some problems, t_f is given in advance; in others, it may be the quantity we wish to minimize.
- **State Variables:** The system is characterized at any time by n real numbers $x_1(t), x_2(t), \ldots, x_n(t)$, called **state variables**, or equivalently by the **state vector** $\vec{\mathbf{x}}(t)$ in $\mathbb{R}^n$.
- **Initial and Terminal States:** The state vector $\vec{\mathbf{x}}$ starts at an **initial point** $\vec{\mathbf{x}}(0) = \vec{\mathbf{x}}_0$ and ends at a **final** or **terminal point** $\vec{\mathbf{x}}(t_f) = \vec{\mathbf{x}}_f$. Generally, these points are specified, but sometimes they are left as unknowns to be determined.
- **Control Function:** We are allowed to choose a **control function** $u(t)$, $0 \leq t \leq t_f$, to make choices about the behavior of the system. We frequently require **bounded controls** satisfying

$$-1 \leq u(t) \leq 1, \tag{1}$$

or restrict attention to a specific class of **feasible controls**, such as continuous or piecewise continuous functions on the interval $[0, t_f]$.

Spaceship Example:

In a spaceship problem, the state vector $\vec{\mathbf{x}}(t)$ has *six* dimensions, so we have three state variables for position and three for orientation.

- **State Equations:** The state vector $\vec{\mathbf{x}}(t)$ is determined by a system of equations

$$\dot{\vec{\mathbf{x}}} = \vec{\mathbf{f}}(\vec{\mathbf{x}}(t), u(t), t), \tag{2}$$

called the **state equations**, which depend in general on the state vector, the control function, and time.[1]

We restrict our discussion to *linear autonomous systems*, $\dot{\vec{\mathbf{x}}} = \mathbf{A}\vec{\mathbf{x}} + \vec{\mathbf{b}}u(t)$, in two dimensions:

$$\begin{bmatrix} \dot{x}_1 \\ \dot{x}_2 \end{bmatrix} = \begin{bmatrix} a_{11} & a_{12} \\ a_{21} & a_{22} \end{bmatrix} \begin{bmatrix} x_1 \\ x_2 \end{bmatrix} + \begin{bmatrix} b_1 \\ b_2 \end{bmatrix} u(t). \tag{3}$$

Given initial conditions $\vec{\mathbf{x}}_0$ and a control function $u(t)$, the state equations (3) can be integrated to determine the state vector $\vec{\mathbf{x}}(t)$.

Our goal will be to *find* a control function $u(t)$ that "sends" the initial state $\vec{\mathbf{x}}_0$ to the final state $\vec{\mathbf{x}}_f$, maximizing or minimizing some auxiliary quantity (the "objective function") along the way.

Most of the ingredients of control theory are new names for familiar objects. The objective function is new; it measures performance and is usually an integral.

Objective Function for an Optimal Control

The **objective function** is an expression of the form

$$J = J(u) = \int_0^{t_f} f_0(\vec{\mathbf{x}}(t), u(t), t)\, dt, \tag{4}$$

where the integrand f_0 is a given function of the state vector, the control function, and the time.[2]

The objective function, which depends explicitly or implicitly on $\vec{\mathbf{x}}$, u, and t over the interval $0 \le t \le t_f$, is chosen to measure some aspect of the system's overall behavior. $J(u)$ is sometimes called the **performance index**.

Using the language and notations just introduced, we can state a fairly general form of the optimal control problem.

Optimal Control Problem

Given the first-order system of *state equations*

$$\dot{\vec{\mathbf{x}}} = \vec{\mathbf{f}}(\vec{\mathbf{x}}(t), u(t), t), \quad \vec{\mathbf{x}}(0) = \vec{\mathbf{x}}_0, \tag{5}$$

determine the *control function* $u(t)$ (belonging to some specified class of functions) that drives the system from $\vec{\mathbf{x}}_0$ to $\vec{\mathbf{x}}_f$, while minimizing (or maximizing) the *objective function*

$$J(u) = \int_0^{t_f} f_0(\vec{\mathbf{x}}(t), u(t), t)\, dt, \tag{6}$$

given f_0 and either the final time t_f or the final state $\vec{\mathbf{x}}_f$.

The following example will illuminate what all this means.

---

[1]In this introduction, the state equations are ordinary first-order differential equations (as in equation (2)); in a broader treatment, they could be iterative equations or partial differential equations.

[2]The function $f_0(\cdots)$ does not always depend explicitly on all three arguments $\vec{\mathbf{x}}$, u, and t. More often than not, it does not depend explicitly on time. Also, the state vector $\vec{\mathbf{x}}$ depends on the control function u through the state equations, so we write the dependence $J = J(u)$.

EXAMPLE 1 **The Hotrod Problem** A hotrod driver sits at an intersection waiting for the light to turn green. She will then race to the next intersection, only to be greeted by another red light. This classic scenario is used by many authors to illustrate basic ideas in control theory. The problem: How to race from one corner to the next in minimum time.

We will first simplify our model. Suppose the hotrod is powered by two rocket engines, fore and aft, which will control the motion of the hotrod through acceleration and deceleration. (See Fig. 10.2.3.) In addition, the design of our hotrod is so efficient that it has no friction, and, for convenience, the mass of the hotrod is 1.

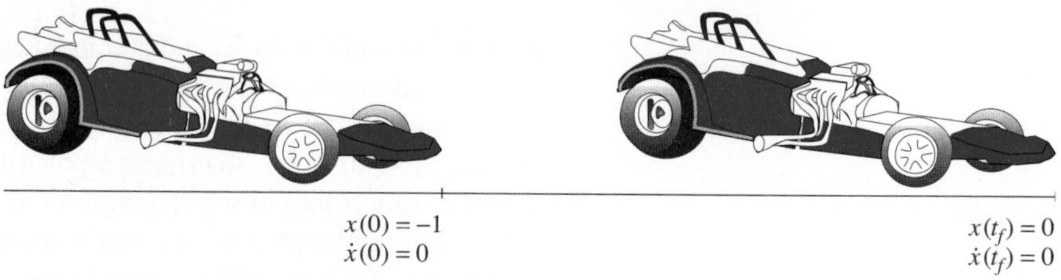

FIGURE 10.2.3 The hotrod problem.

State Vector. Letting x measure the car's position along an axis through the two intersections, and $\dot{x}$ be the car's velocity, the state vector is

$$\vec{\mathbf{x}} = \begin{bmatrix} x \\ \dot{x} \end{bmatrix}.$$

Initial and Final States. We will choose coordinates on the x-axis to set the *initial state* at position -1 with zero velocity, and the *final state* at the origin with zero velocity:

$$\vec{\mathbf{x}}_0 = \begin{bmatrix} -1 \\ 0 \end{bmatrix} \quad \text{and} \quad \vec{\mathbf{x}}_f = \begin{bmatrix} 0 \\ 0 \end{bmatrix}.$$

(See Fig. 10.2.3.) The *final time* t_f remains to be determined.

Control Function. The control is the net thrust $u(t)$ of the rockets, with $u > 0$ in the positive direction along the axis and $u < 0$ in the negative direction. If each engine delivers thrust from zero to one unit, the net thrust in one of the two directions will be in the interval $[-1, 1]$, thus satisfying equation (1). Thrust $+1$ is achieved by using the rear rocket at full power with the front one off, for example. We will restrict attention to a set of *feasible controls*, piecewise continuous functions on the interval $[-1, 1]$.

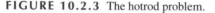

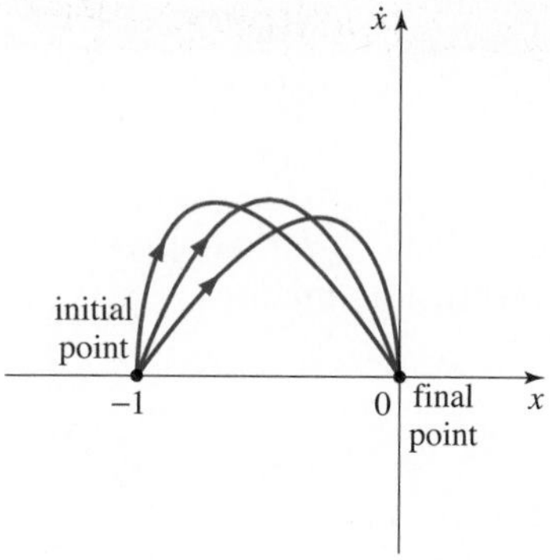

FIGURE 10.2.4 Phase-plane view. Which $x\dot{x}$ path gives minimum time?

State Equation. Assuming zero friction, as stated above, the motion of the hotrod can be modeled using Newton's Second Law of Motion, $F = m\ddot{x}$, where $m = 1$. The car's motion is then governed by

$$\ddot{x} = u(t). \tag{7}$$

Objective Function. Since our goal is to find the control function $u(t)$ that drives the state vector $[x, \dot{x}]$ from $[-1, 0]$ to $[0, 0]$ in minimum time (see Fig. 10.2.4), we have

$$f_0(\vec{\mathbf{x}}, u, t) = 1,$$

so that the objective function to be minimized is

$$J = J(u) = \int_0^{t_f} dt = t_f. \tag{8}$$

Optimal Control. We have two unknowns, the final time t_f and the control function $u(t)$, which will minimize t_f. We show in the next section, using the Pontryagin maximum principle, that the time-optimal control for this problem is the **bang-bang control**,

$$u^*(t) = \begin{cases} 1 & \text{if } 0 \le t < 1, \\ -1 & \text{if } 1 \le t \le 2, \end{cases} \tag{9}$$

where "bang-bang" refers to a control that always takes on one of its extreme values. To find the trajectory of the state vector $[x, \dot{x}]$ using this bang-bang control, we substitute $u^*(t)$ into equation (7), getting

$$\ddot{x} = \begin{cases} 1 & \text{if } 0 \le t < 1, \\ -1 & \text{if } 1 \le t \le 2, \end{cases} \qquad x(0) = -1, \quad \dot{x}(0) = 0. \tag{10}$$

This initial-value problem can be solved easily (Problem 13), getting

$$x(t) = \begin{cases} \frac{1}{2}t^2 - 1 & \text{if } 0 \le t < 1, \\ -\frac{1}{2}t^2 + 2t - 2 & \text{if } 1 \le t \le 2. \end{cases} \tag{11}$$

We plot the trajectory of $(x, \dot{x})$ starting at $(x(0), \dot{x}(0)) = (-1, 0)$, and find (Problem 13) that it ends exactly at $(x(2), \dot{x}(2)) = (0, 0)$, confirming that $t_f = 2$. The trajectory in the $x\dot{x}$-plane (again, see Problem 13) can be found by differentiating x to get $\dot{x}$, then eliminating t from x and $\dot{x}$, to obtain

$$x = \begin{cases} \frac{1}{2}\dot{x}^2 - 1 & \text{if } 0 \le t < 1, \\ -\frac{1}{2}\dot{x}^2 & \text{if } 1 \le t \le 2. \end{cases} \tag{12}$$

We have found the *optimal trajectory*, shown in Fig. 10.2.5, that minimizes our objective function as desired. Keep in mind that the vertical distance on the phase portrait represents the velocity, so the farther away a point is from the x-axis, the faster the hotrod is traveling. The horizontal coordinate represents distance from the final point.

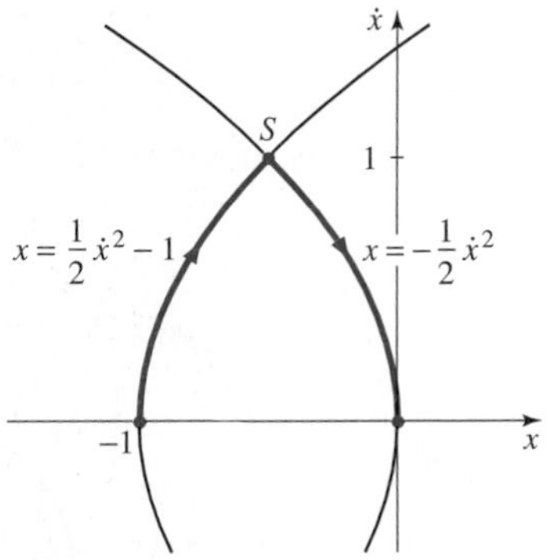

FIGURE 10.2.5 Optimal hotrod phase-plane trajectory (in color).

What this means for the hotrod is that the driver should "floor" ($u = 1$) the hotrod for $0 \le t < 1$, and then fire the retrorocket ($u = -1$) for $1 \le t \le 2$. At $t = 2$ the driver will be sitting at the next red light.

■

The Switching Phenomenon

In Fig. 10.2.5, the point $S = (-1/2, 1)$ is called the "switching point"; it marks the point at which the control function is changed.

We can enlarge the scope of the hotrod problem to include arbitrary initial conditions. For $u(t) = 1$ we obtain a family of parabolas that open to the right, shown in Fig. 10.2.6(a), and for $u(t) = -1$ we obtain a family of parabolas that open to the left (Fig. 10.2.6(b)). (The equations of the trajectories are derived in the more general treatment in Sec. 10.3.)

Switching Point and Boundary

A **switching point** is a point in the phase plane that corresponds to a shift in the control function. The **switching boundary** is the curve made up of all the switching points.

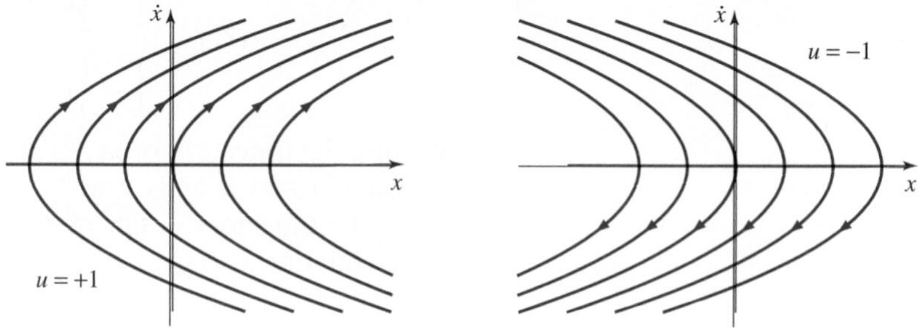

(a) Trajectories $x = \frac{1}{2}\dot{x}^2 + c$ for $u = 1$

(b) Trajectories $x = -\frac{1}{2}\dot{x}^2 + c$ for $u = -1$

FIGURE 10.2.6 Stopping the hotrod with bang-bang controls.

In the hotrod problem, the switching points occur where the control function changes from -1 to 1 or vice versa, and the switching boundary is composed of two parabolic arcs,

$$x = \begin{cases} \dfrac{1}{2}\dot{x}^2 & \text{if } \dot{x} < 0, \\ -\dfrac{1}{2}\dot{x}^2 & \text{if } \dot{x} \geq 0, \end{cases}$$

as shown in Fig. 10.2.7 for the hotrod.

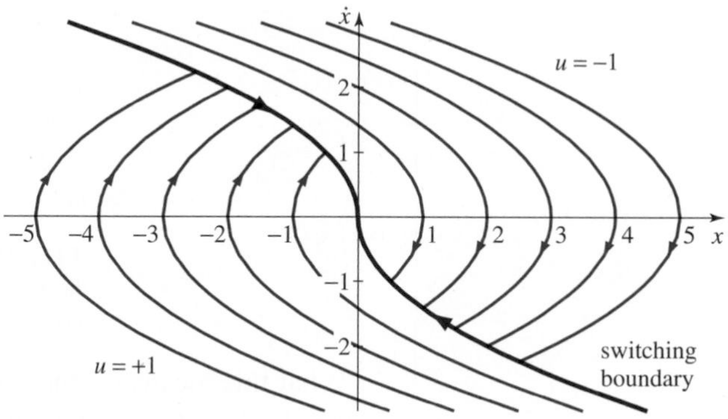

FIGURE 10.2.7 Trajectories and switching boundary for stopping a hotrod.

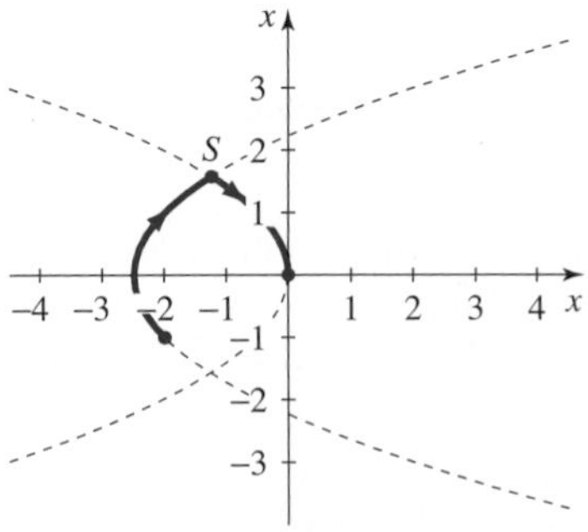

(a) Trajectory from (−2, −1), where the control sequence is 1, −1

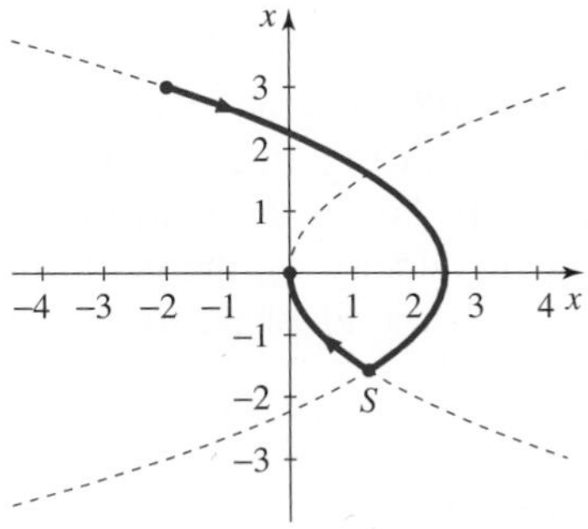

(b) Trajectory from (−2, 3), where the control sequence is −1, 1

FIGURE 10.2.8 Optimal hotrod trajectories from different initial points.

We determine the optimal strategy for moving from initial state $(x_0, \dot{x}_0)$ to final state $(x_f, \dot{x}_f)$ as follows. The control $u(t)$ begins with value $+1$ or -1, according to whether the point $(x_0, \dot{x}_0)$ in the phase plane is above or below the *switching boundary*.

- If $(x_0, \dot{x}_0)$ is below the switching boundary, as in Fig. 10.2.8(a), we take $u = +1$ and switch to $u = -1$ when the trajectory reaches the switching boundary.
- If $(x_0, \dot{x}_0)$ is above the switching boundary, as in Fig. 10.2.8(b), we take $u = -1$ and switch to $u = +1$ when the trajectory reaches the switching boundary.
- If the initial point is *on* the switching boundary, we take $u = -1$ in quadrant II and $u = +1$ in quadrant IV. (See Fig. 10.2.7.)

Choice of the Objective Function

Different aims for the hotrod driver would lead to different objective functions, which would be optimized by different control functions. In Example 1, the hotrod driver wanted to reach her destination as quickly as possible, so she used the **minimum time control**

$$J = \int_0^{t_f} dt = t_f. \tag{13}$$

A more cautious driver may wish to minimize the wear and tear on the vehicle. Physcial arguments show that

$$J = \int_0^{t_f} \dot{x}^2(t)\, dt \tag{14}$$

measures the "energy absorbed" by the hotrod during its transfer, and hence is a measure of wear and tear. The control function that minimizes (14), called a **minimum energy control**, is not concerned with the time of transfer.

A **minimum fuel control** minimizes

$$J = \int_0^{t_f} |u(t)|\, dt, \tag{15}$$

a measure of the amount of fuel used. Minimum fuel controls tend to drive systems very slowly.

In many situations, we need to combine the three controls (13)–(15) to achieve the desired result. A minimum-fuel orbit transfer of a satellite may save on battery usage, but the transfer may take a long time. The control engineer may wish to add a minimum time control. If our hotrod driver discovers that her minimum-time objective is causing too much wear on the tires, she may wish to add minimum energy control.

A general objective function that weighs all three factors for Example 1 is

$$J(u) = \int_0^{t_f} [c_1 + c_2\dot{x}^2(t) + c_3|u(t)|]dt, \tag{16}$$

where c_1, c_2, and c_3 are respective "weights" for the minimum time, minimum energy, and minimum fuel controls.

Controlling the Harmonic Oscillator

Consider the controlled motion of the harmonic oscillator governed by

$$\ddot{x} + x = u(t), \tag{17}$$

where x is the displacement of a spring and u is an external controlling force. We assume the control function u is a piecewise continuous function restricted by $|u(t)| \le 1$, where $u > 0$ in the positive x direction. If we let $y = \dot{x}$, then (17) can be written as the linear control system

$$\begin{aligned} \dot{x} &= y, \\ \dot{y} &= -x + u(t) \end{aligned} \quad \text{or} \quad \begin{bmatrix} \dot{x} \\ \dot{y} \end{bmatrix} = \begin{bmatrix} 0 & 1 \\ -1 & 0 \end{bmatrix} \begin{bmatrix} x \\ y \end{bmatrix} + \begin{bmatrix} 0 \\ 1 \end{bmatrix} u.$$

A typical control problem for the harmonic oscillator involves driving the state vector $[x, y] = [x, \dot{x}]$ of (17) from a starting point $(x(0), \dot{x}(0))$ to rest position $(0, 0)$ in minimum time.

EXAMPLE 2 **How Do We Stop a Moving Elephant?** Imagine an elephant on a platform that slides on a frictionless surface, as in Fig 10.2.9, with the platform attached via a huge spring to a solid wall. When the platform is pulled to the right to position x_0 and released with velocity $\dot{x}_0$, the platform and elephant begin to oscillate. Our goal is to bring the elephant to rest at $(x_f, \dot{x}_f) = (0, 0)$ as quickly as possible.

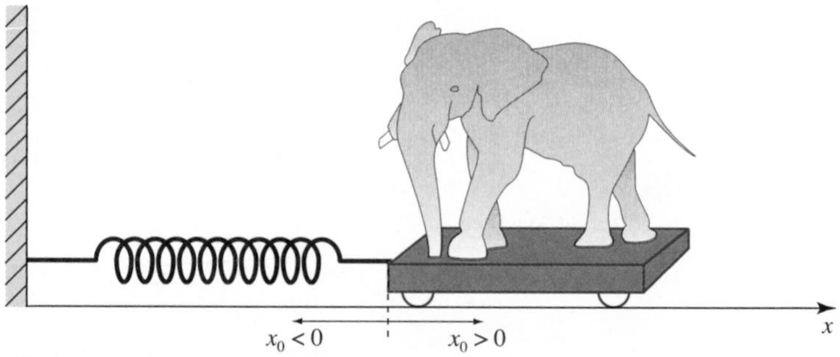

FIGURE 10.2.9 The elephant sits on a platform that is attached to a wall via a huge spring that can be compressed ($x_0 < 0$) or stretched ($x_0 > 0$).

We apply a piecewise continuous control $u(t)$, satisfying a bounded condition $-1 \le u(t) \le 1$, so the elephant's motion is governed by

$$\ddot{x} + x = u(t), \qquad x(0) = x_0, \quad \dot{x}(0) = \dot{x}_0.$$

We will assume that the optimal control function is also a bang-bang control,[3] alternating between the extreme values $+1$ and -1, so a minimum-time trajectory is pieced together from the phase-plane trajectories for

$$\ddot{x} + x = 1 \quad \text{and} \quad \ddot{x} + x = -1,$$

concentric circles centered at $(1, 0)$ and $(-1, 0)$, respectively. (See Fig. 10.2.10 and Problem 16.)

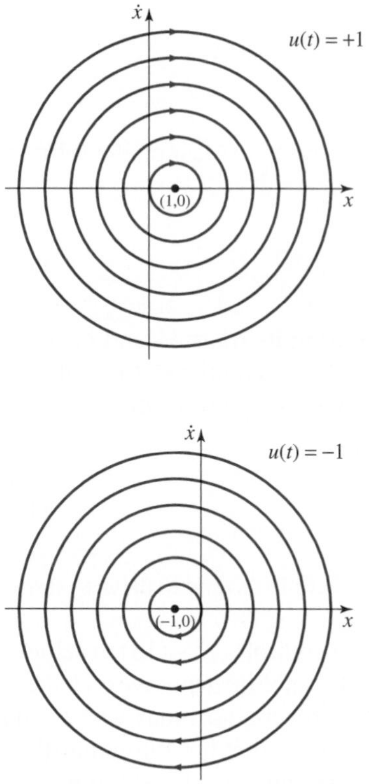

FIGURE 10.2.10 Phase-plane trajectories for the two extremes of the bang-bang control of the harmonic oscillator $\ddot{x} + x = u(t)$.

The solution that *seems* obvious—simply push against the elephant as hard as possible in the direction opposite its motion—is wrong! For small displacement, the control will eventually be *stronger* than the force of the spring, so the elephant will come to rest at a position $(x_f, 0) \neq (0, 0)$. (See Problem 17.)

[3]The proof that the time-optimal control for forcing the harmonic oscillator to rest position is bang-bang is beyond the scope of this text, but it can be found in most advanced books on optimal control theory.

This failed strategy puts the switching curve on the horizontal axis, but the only way to reach $(x, \dot{x}) = (0, 0)$ is by switching the control from 1 to -1 (and back again) on a switching boundary made up of semicircles centered at $(\pm 1, 0)$, $(\pm 3, 0)$, ..., which occur *above* the horizontal axis for *negative* x and *below* for positive x. (See Fig. 10.2.11.)

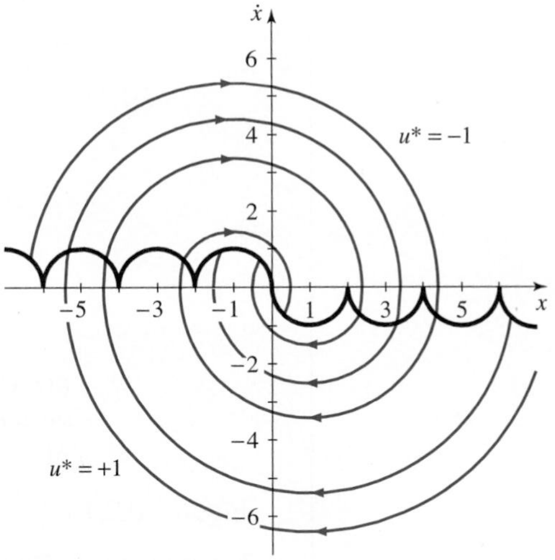

FIGURE 10.2.11 Trajectories and switching boundary for stopping an elephant (on an oscillating platform).

Consider a simple case: The elephant is released at $x_0 = -1$ with no initial velocity. The complete trajectory, shown in Fig. 10.2.12, is pieced together from two circular arcs.

- The first arc is centered at $(1, 0)$ and has radius 2.
- The switching point S occurs at the intersection of the first arc and the switching boundary arc centered at $(-1, 0)$.
- The trajectory then proceeds directly to $(0, 0)$.

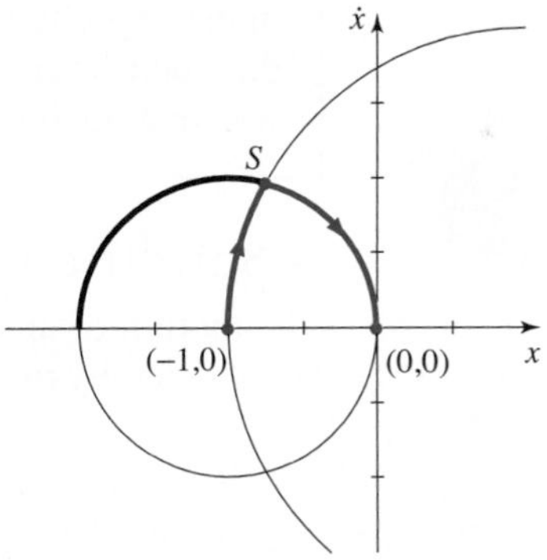

FIGURE 10.2.12 Phase-plane trajectory for stopping an elephant, starting from $x_0 = -1$ with zero velocity.

Any initial point $(x_0, 0)$ for which $-2 < x_0 < 0$ will have a switching point on the semicircle of radius 1 and center at $(-1, 0)$. A more complicated trajectory arises if the initial point is further from the origin. For example:

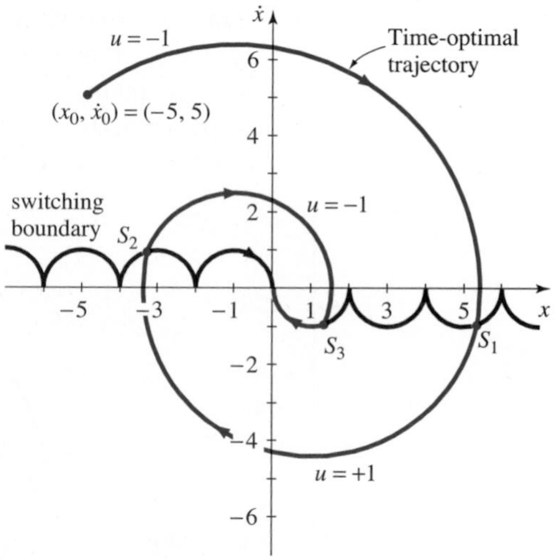

FIGURE 10.2.13 A sample trajectory for stopping an elephant on an oscillating platform with large initial condition $(x_0, \dot{x}_0) = (-5, 5)$.

Figure 10.2.13 illustrates a path with *three* switching points.

- Starting at $(x_0, \dot{x}_0) = (-5, 5)$, we push to the left as hard as we can ($u = -1$), creating a circular trajectory centered at $(-1, 0)$.
- At the first switching point S_1, we stop pushing to the left and push to the right as hard as we can ($u = +1$), creating a new trajectory centered at $(1, 0)$.
- At S_2, we switch back to $u = -1$.
- At S_3, we switch back to $u = +1$ and continue pushing to the right until the elephant comes to a complete stop at $(0, 0)$.

We do not *always* push the elephant in the direction opposite its motion. After the elephant reaches its maximum displacement (either to the left or right), we continue pushing in the same direction momentarily before pushing in the opposite direction. (Look at the direction of motion in Fig. 10.2.13 *after* the x-axis is crossed but *before* the switching boundary is reached.) In this way we are able to make the elephant come to a perfect "soft landing" at $(0, 0)$. ■

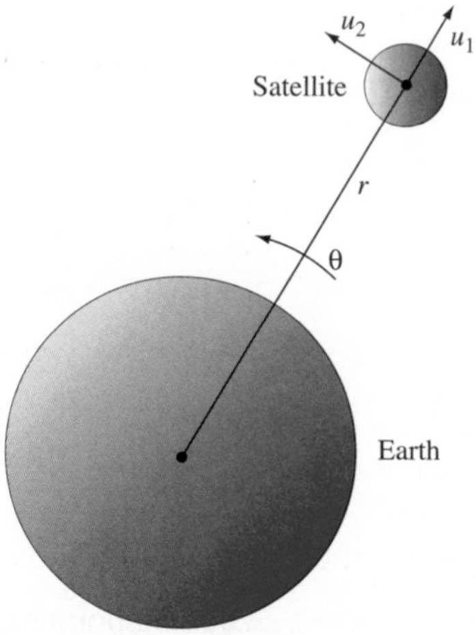

FIGURE 10.2.14 Earth-satellite problem.

A Satellite Control Problem

The motion of an earth satellite in a nearly circular orbit (see Fig. 10.2.14) can be modeled by two second-order differential equations,

$$\begin{aligned} \ddot{r} &= r(t)\dot{\theta}^2(t) - \frac{k}{r^2(t)} + u_1(t), \\ \ddot{\theta} &= -\frac{2\dot{\theta}(t)\dot{r}(t)}{r(t)} + \frac{1}{r(t)}u_2(t), \end{aligned} \tag{18}$$

where

- $r(t)$ is the radial distance of the satellite from Earth's center,
- $\theta(t)$ is the angular position in orbit,
- k is the constant related to earth's mass and the universal gravitational constant,

- $u_1(t)$ is the radial thrust on the satellite,
- $u_2(t)$ is the tangential thrust on the satellite.

Satellite, Earth, and thrust directions are shown schematically in Fig. 10.2.14.

When $u_1(t) = u_2(t) \equiv 0$, system (18) has the circular solution[4]

$$r(t) = r_0 \quad \text{and} \quad \theta(t) = \omega t,$$

where the radius $r_0 > 0$ and the frequency ω are arbitrary constants to be determined from initial conditions.

For the case when $u_1(t)$ and/or $u_2(t)$ are nonzero, let us define

$$x_1 = r - r_0, \quad x_2 = \dot{r}, \quad x_3 = r_0(\theta - \omega t), \quad x_4 = r_0(\dot{\theta} - \omega), \tag{19}$$

where

- x_1 is the difference of altitude from r_0,
- x_3 is the difference in angular velocity from ωt,
- x_2 and x_4 are the corresponding rates of change (velocities).

Then system (18) becomes[5]

$$\begin{aligned}
\dot{x}_1 &= x_2, \\
\dot{x}_2 &= (x_1 + r_0)\left(\omega + \frac{x_4}{r_0}\right)^2 - \frac{r_0^3\omega^2}{(x_1 + r_0)^2} + u_1(t), \\
\dot{x}_3 &= x_4, \\
\dot{x}_4 &= \frac{r_0}{x_1 + r_0}\left[-2x_2\left(\omega + \frac{x_4}{r_0}\right) + u_2(t)\right].
\end{aligned} \tag{20}$$

The nonlinear system (20) of four first-order equations has an equilibrium point at the origin in the *uncontrolled* case $u_1(t) \equiv 0$ and $u_2(t) \equiv 0$. The stability behavior of the system near this equilibrium point is of interest because *no* controls are needed if it is asymptotically stable.[6] The linearization of (20) is the system

$$\vec{\mathbf{x}}' = \begin{bmatrix} 0 & 1 & 0 & 0 \\ 3\omega^2 & 0 & 0 & 2\omega \\ 0 & 0 & 0 & 1 \\ 0 & -2\omega & 0 & 0 \end{bmatrix} \vec{\mathbf{x}}. \tag{21}$$

(See Problem 20.) The eigenvalues of the coefficient matrix in (21) are 0, 0, and $\pm i\omega$. Hence the origin is stable but not asymptotically stable.

Controlling the satellite means finding control functions u_1 and u_2 that will make the origin asymptotically stable for the controlled system

$$\vec{\mathbf{x}}' = \begin{bmatrix} 0 & 1 & 0 & 0 \\ 3\omega^2 & 0 & 0 & 2\omega \\ 0 & 0 & 0 & 1 \\ 0 & -2\omega & 0 & 0 \end{bmatrix} \vec{\mathbf{x}} + \begin{bmatrix} 0 & 0 \\ 1 & 0 \\ 0 & 0 \\ 0 & 1 \end{bmatrix} \begin{bmatrix} u_1 \\ u_2 \end{bmatrix}. \tag{22}$$

The two controls $u_1(t)$ and $u_2(t)$ in the control system (22) represent external thrusters on the satellite, providing forces in the radial and tangential directions,

---

[4]In general, these solutions are elliptical.

[5]For k, we use Kepler's equation $k = r_0^3\omega^2$, named after the great German astronomer Johann Kepler (1571–1630), who first observed the relation.

[6]Recall (from Sec. 6.5) that if an equilibrium solution is stable, then solutions that start close stay close. If, as $t \to \infty$, such solutions also tend to the equilibrium solution, then the equilibrium is *asymptotically* stable.

respectively. Using the Pontryagin Maximum Principle, which will be introduced in Sec. 10.3, it is possible to find controls that will stabilize the satellite's orbit and/or carry out orbital transfers.

Historical Note

The modern mathematical theory of optimal control began shortly after World War II, inspired largely by the seminal work of L. S. Pontryagin and his colleagues in the Soviet Union.[7] A central result growing from this activity is the Pontryagin Maximum Principle, discussed in the next section. Concurrently in the United States a parallel theory, called dynamic programming, was developed by Richard Bellman.[8] Bellman's name for this theory has disappeared because "programming" has acquired a different meaning in the computer age. Nevertheless, the availability and power of contemporary computers has transformed the theory of optimal control, as it has many other specialized areas of differential equations.

Summary

By choosing suitable control functions, which are just forcing terms in a system of equations, we attempt to satisfy both initial conditions and final conditions and, at the same time, extremize an auxiliary quantity that measures system performance; when this objective function is maximized or minimized, the controls are considered to be optimal.

10.2 Problems

Tracking the Hotrod *Trace the optimal trajectory for the hotrod example with the initial conditions given in Problems 1–4. This may require sketching trajectories that are not shown in Fig. 10.2.7, but are in the families of those trajectories shown. The sketches should resemble those in Fig. 10.2.8.*

1. $(x_0, \dot{x}_0) = (3, -1)$

2. $(x_0, \dot{x}_0) = (-3, -1)$

3. $(x_0, \dot{x}_0) = (3, 1)$

4. $(x_0, \dot{x}_0) = (2, 2)$

Conversion and Identification *Write the differential equations given in Problems 5–8 as systems of first-order equations of the form*

$$\dot{\vec{\mathbf{x}}} = \mathbf{A}\vec{\mathbf{x}} + \vec{\mathbf{b}}u,$$

where u is assumed to be a control function $u(t)$. Identify each of the quantities in this formulation.

5. $\ddot{x} + x = u$

6. $\ddot{x} + 2\dot{x} + x = u$

7. $\ddot{x} + b\dot{x} + cx = u$

8. $\ddot{x} + \dot{x} + x + x^3 = u$

More Identity Problems *Write the systems in Problems 9–12 in the form $\dot{\vec{\mathbf{x}}} = \mathbf{A}\vec{\mathbf{x}} + \vec{\mathbf{b}}u$, and identify $\mathbf{A}$ and $\vec{\mathbf{b}}$.*

9. $\dot{x} = y$
$\dot{y} = -x + u$

10. $\dot{x} = y$
$\dot{y} = -x + y + u$

11. $\dot{x} = -x + 2y$
$\dot{y} = -x + u$

12. $\dot{x} = -x + 2y$
$\dot{y} = x - y + u$

---

[7] Lev Semonovich Pontryagin (1908–1988) was one of the world's great topologists before he decided in 1952 to change mathematical fields to work in differential equations and control theory. In 1961 he published *The Mathematical Theory of Optimal Processes* with students V. G. Boltyanskii, R. V. Gamkrelidze, and E. F. Mishchenko, which formulated the Pontryagin Maximum Principle, a cornerstone of optimal control. It is remarkable that although he was blind from the age of 14 (due to an explosion), he was taught mathematics by his mother, who had no training in mathematics.

[8] American applied mathematician Richard Bellman (1920–1984) was a pioneer of control theory in the 1950s and 1960s, and one of the most prolific mathematicians of our time, having authored 44 books and over 600 published research papers. His major lasting contribution is dynamic programming, an approach to optimization problems that decomposes large problems into a number of smaller ones that are easier to solve. He became Professor of Mathematics, Electrical Engineering, Biomedical Engineering, and Medicine at the University of Southern California.

13. Controlling the Hotrod

(a) Solve the IVP

$$\ddot{x} = \begin{cases} +1 & \text{if } 0 \le t < 1, \\ -1 & \text{if } 1 \le t \le 2, \end{cases} \qquad x(0) = -1, \quad \dot{x}(0) = 0$$

by solving it first on the interval [0, 1], and then using the final values from this solution as initial values for a new problem on interval [1, 2].

(b) Show that this control drives the system to the "destination"

$$x(2) = \dot{x}(2) = 0.$$

NOTE: It is not obvious *why* this is optimal. For now, we are just observing the results of analysis from a more advanced treatment.

(c) Derive the phase-plane trajectory equations (12) from your solutions to part (a).

14. Hotrod Revisited Repeat Problem 13 by solving the equation in a single operation using Laplace transforms.

15. Optimal Control Consider the first-order equation

$$\dot{x} = x + u, \quad x(0) = 1$$

with no final condition. It can be shown that the control $u^*(t) = -2e^{-t}$ minimizes the objective

$$J = \frac{1}{2}\int_0^\infty [x^2(t) + u^2(t)]\, dt.$$

(a) Find the optimal trajectory of this system, and evaluate J.

(b) Show that the control $u(t) = -5e^{-4t}$ gives rise to a larger value for J.

16. Elephant Problem Show that the solutions of $\ddot{x} + x = 1$ in the $x\dot{x}$-plane form a family of circles centered at (1, 0), in which the motion of $(x, \dot{x})$ around the circles is clockwise with period 2π. Then show that when the right-hand side of the differential equation is changed to -1, the circles are centered at $(-1, 0)$, but motion is still clockwise.

17. How Not to Stop an Elephant Draw the minimum stopping time trajectory in the $x\dot{x}$-plane for the motion of the elephant discussed in Example 2, using the optimal control function

$$u = \begin{cases} 1 & \text{if } \dot{x} < 0, \\ -1 & \text{if } \dot{x} > 0, \end{cases}$$

starting at $(x_0, \dot{x}_0) = (1, 1)$. For this control, the switching boundary is the x-axis, as shown in Fig. 10.2.15. What happens to the elephant?

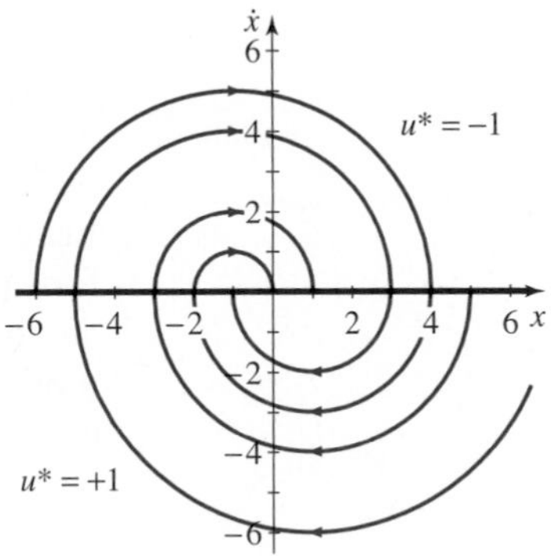

FIGURE 10.2.15 For Problem 17, the switching boundary is the horizontal axis.

18. Satellite System Derive system (20) from system (18), using equations (19).

19. Satellite Equilibrium Show that the system (20) has an equilibrium point at the origin of the **x**-space.

20. Satellite Linearization Linearize the system (20) at the origin and verify that system (21) is the result.

21. Satellite Eigenvalues Show that the characteristic polynomial of the coefficient matrix in (21) is $\lambda^2(\lambda^2+\omega^2)$, thus verifying that the eigenvalues are 0, 0, and $\pm i\omega$. What is the general solution of system (21)?

22. Suggested Journal Entry Describe some optimal control problems in your major field or in another area of interest to you. How difficult would it be to find the optimal controls?

10.3 Pontryagin Maximum Principle

SYNOPSIS: We introduce a form of the Pontryagin Maximum Principle and show how it can be used to determine optimal controls in simple examples.

A New Kind of Calculus

The typical optimization problems of ordinary calculus require us to find the maximum or minimum value of a function of one, two, or three real variables. Problems in optimal control, on the other hand, involve determining the maximum or minimum value of a function of a *function*. In Sec. 10.2, we dealt with objective functions $J(u)$ whose values depended on, among other things, making a critical choice of $u(t)$ from among the admissible control functions.

Functions of functions are sometimes called **functionals**, and the branch of mathematics that deals with their optimization is called the **calculus of variations**, an *old* subject (see the historical note at the end of the section) that is likely to be *new* to you. One of the oldest problems in the calculus of variations is the **brachistochrone problem**. (The name refers to a minimization of time.) It asks for the shape of the curve from a higher point A to a lower point B so that if an object under the influence of gravity slides along the curve with a frictionless motion, the time of descent is a minimum. (See Fig. 10.3.1.) The answer is a curve called the **cycloid**.

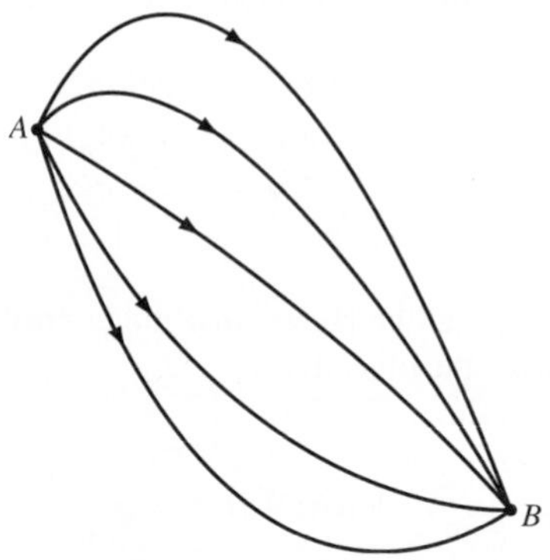

FIGURE 10.3.1 Which path gives quickest descent?

In 1956, Soviet mathematicians Pontryagin, Boltyanskii, and Gamkrelidze formulated and proved what is now known as the **Pontryagin Maximum Principle**, a fundamental advance in the calculus of variations. It allows us to find optimal controls for a large class of both linear and nonlinear problems. The most general statement (as well as the proof) is beyond the scope of this book, but we will use a restricted form of the principle to find controls for some elementary examples.

Pontryagin Maximum Principle

If a differentiable function $f(x)$ of one variable has an extremum at point x^*, then $f'(x^*) = 0$. Pontryagin, Boltyanskii, and Gamkrelidze found that if a control u^* is optimal, a certain function related to the problem, called the **Hamiltonian**, must be maximized.[1] The following statement makes this precise.

[1] Again, this refers to William Rowan Hamilton, discussed in Sec. 3.5, and the Hamiltonian mechanics introduced in Sec. 7.1, Problem 39.

Pontryagin Maximum Principle
If control $u^*(t)$ drives the solution $\vec{\mathbf{x}}$ of the autonomous system

$$\dot{\vec{\mathbf{x}}} = \vec{\mathbf{f}}(\vec{\mathbf{x}}, u) \tag{1}$$

from initial state $\vec{\mathbf{x}}_0$ to final state $\vec{\mathbf{x}}_f = \vec{\mathbf{x}}(t_f)$, where t_f is the final time, and if u^* minimizes the objective function

$$J(u) = \int_0^{t_f} f_0(\vec{\mathbf{x}}, u)\, dt, \tag{2}$$

then this optimal control u^* maximizes the **Hamiltonian function**

$$\begin{aligned} H(u) &= -f_0(\vec{\mathbf{x}}, u) + p_1\dot{x}_1 + p_2\dot{x}_2 + \cdots + p_n\dot{x}_n \\ &= -f_0(\vec{\mathbf{x}}, u) + \vec{\mathbf{p}} \cdot \dot{\vec{\mathbf{x}}} \end{aligned}$$

for each t in the interval $[0, t_f]$, where the **adjoint variables** $p_1, p_2, \ldots, p_n$ are determined by solving the **adjoint equations**

$$\dot{p}_i = -\frac{\partial H}{\partial x_i}, \quad i = 1, 2, \ldots, n.$$

The trajectory $\vec{\mathbf{x}}^*$ corresponding to control u^* is called the **optimal trajectory**.

We use the Pontryagin Maximum Principle to solve optimal control problems, as follows.

Finding Optimal Controls and Trajectories with the Pontryagin Maximum Principle

For the optimal control problem with

- state equations $\dot{\vec{\mathbf{x}}} = \vec{\mathbf{f}}(\vec{\mathbf{x}}, u)$,
- initial state $\vec{\mathbf{x}}(0) = \vec{\mathbf{x}}_0$,
- final state $\vec{\mathbf{x}}(t_f) = \vec{\mathbf{x}}_f$,
- objective function $J(u) = \int_0^{t_f} f_0(\vec{\mathbf{x}}, u)dt$,

the Pontryagin Maximum Principle can be used as follows to find the optimal control $u^*(t)$ that leads to the optimal trajectory $\vec{\mathbf{x}}^*$.

Step 1. Construct the Hamiltonian

$$H(u) = -f_0(\vec{\mathbf{x}}, u) + \vec{\mathbf{p}} \cdot \dot{\vec{\mathbf{x}}} \tag{3}$$

and the adjoint equations

$$\dot{p}_i = -\frac{\partial H}{\partial x_i}, \quad i = 1, 2, \ldots, n. \tag{4}$$

Step 2. Solve the system of adjoint equations (4) to determine the adjoint variables $p_1, p_2, \ldots, p_n$.

Step 3. Maximize the Hamiltonian (3) to find the optimal control $u^*(t)$.

Step 4. Substitute the optimal control into the system of state equations, and solve them to find the optimal trajectory $\vec{\mathbf{x}}^*$.

Equations (3) and (4) may seem to involve circular references, but Step 1 of the following two examples demonstrates how equation (4) indeed gives simple DEs to solve for p_i.

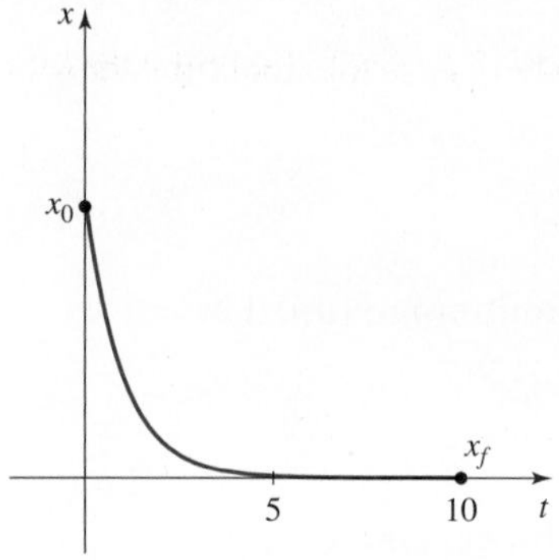

FIGURE 10.3.2 Minimum energy control problem.

EXAMPLE 1 **Minimum Energy Control** We will consider the problem of finding the control $u(t)$ that drives the solution of the first-order equation

$$\dot{x} = x + u(t) \tag{5}$$

from $x(0) = 1$ to $x(10) = 0$ (see Fig. 10.3.2), while minimizing the objective function

$$J(u) = \int_0^{t_f} f_0(x, u)dt = \frac{1}{2}\int_0^{10} u^2(t)dt. \tag{6}$$

(The energy expended by an external controlling force like u is commonly measured by one-half the time integral of the square of the force.) The maximum principle will be our tool.[2]

Step 1. From (3), we can write

$$H(x, p, u) = \underbrace{-\frac{1}{2}u^2}_{-f_0(x,u)} + p\dot{x} = -\frac{1}{2}u^2 + p(x + u). \tag{7}$$

We have omitted subscripts because we are dealing with a first-order equation. We know from (7) that $\partial H/\partial x = p$, so the (single) adjoint variable p is the solution of the adjoint equation

$$\dot{p} = -p. \tag{8}$$

Step 2. The solution of the differential equation (8) is the adjoint variable

$$p(t) = ce^{-t},$$

which involves the arbitrary constant c, to be determined later, because no initial conditions are specified.

Step 3. We maximize the Hamiltonian H as a function of a single variable u, treating $H(u)$ like an ordinary calculus problem.[3] Rewriting (7) by completing the square in u, we obtain

$$\begin{aligned} H(u) &= -\frac{1}{2}u^2 + px + pu \\ &= -\frac{1}{2}(u^2 - 2pu) + px \\ &= -\frac{1}{2}(u^2 - 2pu + p^2) + px + \frac{1}{2}p^2 \\ &= -\frac{1}{2}(u - p)^2 + px + \frac{1}{2}p^2. \end{aligned}$$

[2]The example is adapted from an example in L. M. Hocking, *Optimal Control: An Introduction to the Theory and Applications* (Oxford University Press, 1991).

[3]We can do this because the Pontryagin Maximum Principle ensures that the Hamiltonian is maximized for *each point in time*.

H has a maximum when $u = p$ because

- $\partial H/\partial u = -u + p = 0$ indicates an extremum at $u = p$, and
- the condition $\partial^2 H/\partial u^2 = -1 < 0$ means that the extremum is a maximum.

Combining this result with the solution for p in Step 2, we obtain the optimal control

$$u^* = p = ce^{-t}, \tag{9}$$

with arbitrary constant c to be determined later.

Step 4. The state equation (5) is now

$$\dot{x} = x + ce^{-t},$$

and the solution of this first-order linear equation is

$$x(t) = Ae^t - \frac{c}{2}e^{-t}.$$

The initial state specifies that $x(0) = 1$, so $A - c/2 = 1$; thus, $A = c/2 + 1$. Hence we have

$$x(t) = \left(1 + \frac{c}{2}\right)e^t - \frac{c}{2}e^{-t}.$$

The final state requires that $x(10) = 0$, which means that

$$\left(1 + \frac{c}{2}\right)e^{10} - \frac{c}{2}e^{-10} = 0.$$

Solving for c, we have

$$c = \frac{-2}{1 - e^{-20}} \approx -2.$$

Hence the optimal control and optimal trajectory are *approximated* by

$$u^*(t) = -2e^{-t} \quad \text{and} \quad x^*(t) = e^{-t}, \tag{10}$$

which does not *quite* satisfy $x(10) = 0$. Actually, $e^{-10} \approx 4.5 \times 10^{-5} \neq 0$. The reader is asked to formulate and verify the *exact* $u^*(t)$ and $x^*(t)$ in Problem 10.

The objective function for this control is

$$J(u^*) = \frac{1}{2}\int_0^{10} \left[u^*(t)\right]^2 dt = \frac{1}{2}\int_0^{10} 4e^{-2t}dt = -e^{-2t}\Big|_0^{10} = 1 - e^{-20} \approx 1.$$

Uncontrolled (that is, with $u \equiv 0$), equation (5) models exponential growth. An opposing exponential control is needed to counteract that growth, but its strength continually diminishes. (See Fig. 10.3.3.)

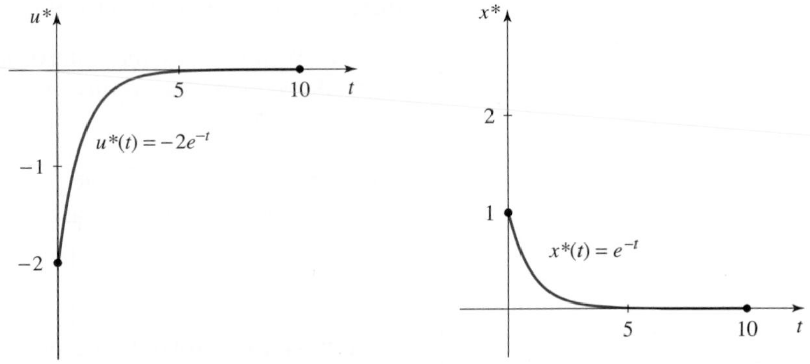

(a) Approximate control $u^*(t) = -2e^{-t}$

(b) Approximate trajectory $x^*(t) = e^{-t}$

FIGURE 10.3.3 Minimum energy control problem.

■

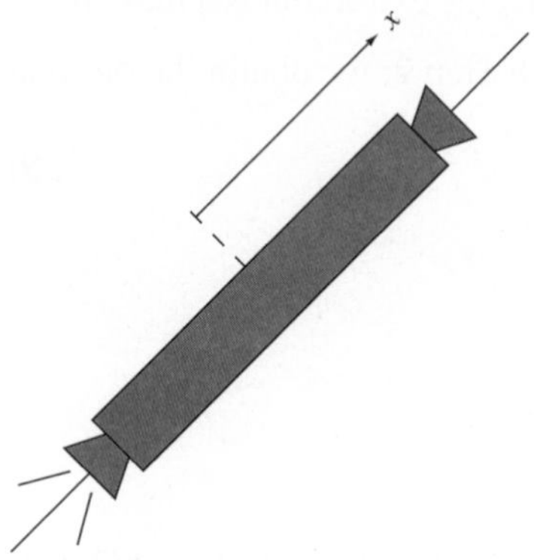

FIGURE 10.3.4 Stopping a space ship.

EXAMPLE 2 **Stopping a Space Ship** The space ship shown in Fig. 10.3.4 moves along a straight line, far from any gravitational field, according to Newton's law

$$\ddot{x} = u(t). \tag{11}$$

(We have taken $m = 1$ for convenience.) The control u represents the net thrust of opposing rocket engines at the two ends of the ship, such that $|u(t)| \leq 1$. The hotrod in Sec. 10.2 and the space ship problem described here have exactly the same equation. There is no point in wasting fuel by running both engines at once, so a positive control refers to firing the left-hand rocket with the right-hand rocket shut down, and a negative control refers to the reverse situation. The control values $+1$ and -1 correspond to full power for the firing engine. The goal is to drive the ship from an arbitrary initial state $x(0) = x_0, \dot{x}(0) = y_0$ to the final state $x = 0, \dot{x} = 0$ in minimum time t_f to be determined.

We will reformulate (11) as a system,

$$\begin{aligned} \dot{x} &= y, \\ \dot{y} &= u(t), \end{aligned} \qquad x(0) = x_0, \quad y(0) = y_0. \tag{12}$$

The objective function $J = t_f$ can be written in integral form as

$$J = \int_0^{t_f} dt,$$

where we identify $f_0(x, y, u) \equiv 1$.

Step 1. The Hamiltonian function is

$$\begin{aligned} H(u) &= -f_0(x, y, u) + p\dot{x} + q\dot{y}, \\ &= -1 + py + qu, \end{aligned} \tag{13}$$

and the adjoint equations are

$$\begin{aligned} \dot{p} &= -\frac{\partial H}{\partial x} = 0, \\ \dot{q} &= -\frac{\partial H}{\partial y} = -p. \end{aligned} \tag{14}$$

Step 2. Solving equations (14), we find the adjoint variables

$$\begin{aligned} p(t) &= a_1, \\ q(t) &= -a_1 t + a_2, \end{aligned} \tag{15}$$

where a_1 and a_2 are constants of integration to be determined.

Step 3. We now seek the control u that maximizes the Hamiltonian $H(u)$, reformulated using the adjoint variables from (15) as

$$H(u) = -1 + a_1 y + (-a_1 t + a_2)u.$$

To maximize H for each t, we require u to be 1 if the coefficient $-a_1 t + a_2$ is positive, -1 in the contrary case, so the optimal control

$$u^*(t) = \begin{cases} 1 & \text{if } -a_1 t + a_2 \geq 0, \\ -1 & \text{if } -a_1 t + a_2 < 0 \end{cases} \tag{16}$$

is a bang-bang control (one engine or the other at full power). Because $-a_1 t + a_2$ is a linear function with straight line graph, $u(t)$ changes sign at most once in the interval $[0, t_f]$ (unless $a_1 = a_2 = 0$, which results in $-a_1 t + a_2 \equiv 0$).[4]

[4]This is the *singular case*; see Problem 11.

Thus, the optimal control $u^*(t)$ takes on the value 1 on one segment of the domain $[0, t_f]$, and -1 on the complementary segment. (One of these segments could have zero length.)

Step 4. We will solve the state equations (12), first with $u(t) = 1$, then with $u(t) = -1$. In the first case, we have

$$\begin{aligned} \dot{x} &= y, \\ \dot{y} &= 1, \end{aligned} \tag{17}$$

which can be integrated without heavy machinery. Integrating the second equation, then the first, the solution of system (17) is given by

$$\begin{aligned} x &= \frac{1}{2}t^2 + b_1 t + b_2, \\ y &= t + b_1. \end{aligned} \tag{18}$$

Solving the second equation for t and substituting into the first equation yields

$$x = \frac{1}{2}y^2 + \left(b_2 - \frac{1}{2}b_1^2\right). \tag{19}$$

Equation (19) represents the one-parameter family of parabolas

$$x = \frac{1}{2}y^2 + K, \tag{20}$$

shown in Fig. 10.3.5(a).

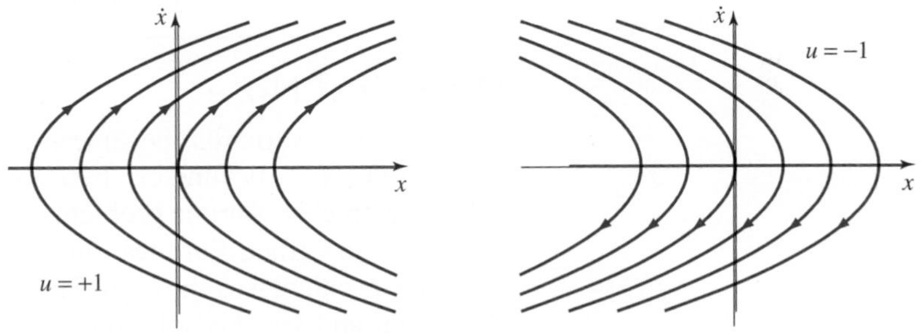

(a) Trajectories $x = \frac{1}{2}\dot{x}^2 + c$ for $u = 1$ (b) Trajectories $x = -\frac{1}{2}\dot{x}^2 + c$ for $u = -1$

FIGURE 10.3.5 Phase-plane trajectories for the two extremes of the bang-bang control of the spaceship equation $\ddot{x} = u(t)$.

We must also solve the state equations for the case $u = -1$; that is,

$$\begin{aligned} \dot{x} &= y, \\ \dot{y} &= -1. \end{aligned} \tag{21}$$

The solution is readily found to be

$$\begin{aligned} x &= -\frac{1}{2}t^2 + c_1 t + c_2, \\ y &= -t + c_1, \end{aligned}$$

and leads to the family of phase-plane parabolas

$$x = -\frac{1}{2}y^2 + \left(c_2 + \frac{1}{2}c_1^2\right) = -\frac{1}{2}y^2 + K,$$

graphed in Fig. 10.3.5(b).

As discussed in Sec. 10.2, the two parabolic arcs can be pieced together to form the switching boundary; see Fig. 10.3.6.

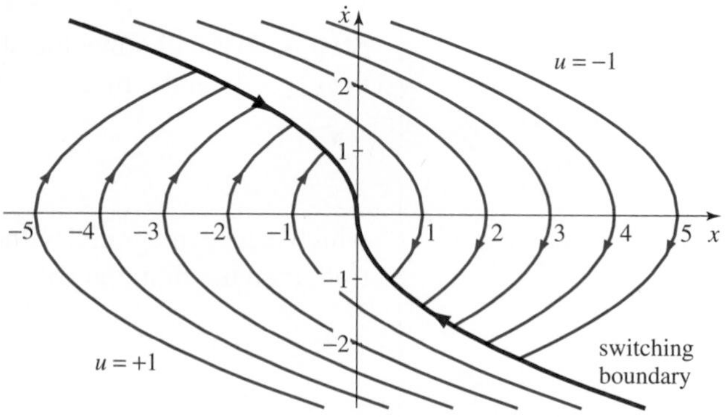

FIGURE 10.3.6 Trajectories and switching boundary for stopping a spaceship.

The time t_f must be computed for any particular initial state because it depends on those initial conditions. (The switching time t_s is also dependent on those values.) The Pontryagin Maximum Principle guarantees, however, that the time t_f *will be the minimum possible time*.

When the control $u^* = u^*(\vec{\mathbf{x}})$ depends on the state of the system, it is called a **closed-loop** control, a desirable circumstance provided that the state of the system can be continuously monitored. (**Open-loop** controls that depend only on time have the drawback that they cannot adapt if something goes wrong.) ■

Historical Note

The calculus of variations became a "hot" topic in the early development of mathematical analysis. In June of 1696, Johann Bernoulli and Gottfried Leibniz proposed the brachistochrone problem to the "acutest mathematicians of the world." The problem baffled European mathematicians for six months. On January 26, 1697, the problem was shown to Sir Isaac Newton, who solved it after dinner that very night and sent his solution anonymously to the Royal Society of London. Despite this anonymity, when Bernoulli saw the solution he is said to have recognized at once that Newton was its author, saying, "Ah, I recognize the lion by his paw."

Close interconnections are to be noted among the calculus of variations, optimization using Lagrange multipliers, and what is called Hamiltonian mechanics. An excellent reference is the classic text on calculus of variations by Weinstock.[5]

Summary

The Pontryagin Maximum Principle facilitates determining the optimal control function in an optimal control problem, using a Hamiltonian energy function and adjoint variables. From a physical point of view, optimality of the control corresponds to maximum total energy, as represented by the Hamiltonian, throughout the motion.

[5]Robert Weinstock, *Calculus of Variations* (NY: McGraw-Hill, 1952).

10.3 Problems

Optimal Control *In each of Problems 1 and 2, determine*

(a) *the Hamiltonian,*

(b) *the adjoint equations, and*

(c) *solutions of the adjoint equations (with constants of integration).*

Be sure to specify the state equations and control function for each of these optimal control problems.

1. The harmonic oscillator with minimum fuel control:

$$\dot{x}_1 = x_2, \quad \dot{x}_2 = -x_1 + u, \qquad J(u) = \int_0^{t_f} u^2(t)\,dt.$$

2. The harmonic oscillator with minimum time control:

$$\dot{x}_1 = x_2, \quad \dot{x}_2 = -x_1 + u, \qquad J(u) = \int_0^{t_f} dt.$$

3. **Controlling a Pure Integrator** The simple first-order *integrator* system is defined by

$$\dot{x} = u, \quad x(0) = 1.$$

The goal is to determine the control function that drives the system from an initial state to the final state $x(2) = 0$, while minimizing

$$J(u) = \frac{1}{2}\int_0^2 u^2 dt.$$

Minimizing J assures that we will minimize the energy expended (the u^2 term in the integral).

(a) Find the Hamiltonian.

(b) Determine and solve the adjoint equations.

(c) Find the optimal control u^*, and use it to find the optimal path $x^*(t)$.

(d) Calculate the value of $J(u)$ for u^* and x^*.

Stopping a Vibrating Spring *The Pontryagin Maximum Principle can be used to show that the optimal control that drives the harmonic oscillator*

$$\begin{aligned} \dot{x} &= y, & x(0) &= x_0, \\ \dot{y} &= -x + u(t), & y(0) &= y_0, \end{aligned}$$

to the origin in minimum time, where $|u| \le 1$, is summarized by the trajectories in Fig. 10.3.7. Starting from any initial position, the control is either $+1$ or -1, depending on the initial condition, until the trajectory reaches a switching boundary, at which time the control switches sign. This process is repeated until the origin is reached. (This is a "bang-bang" control with a finite number of switches.)

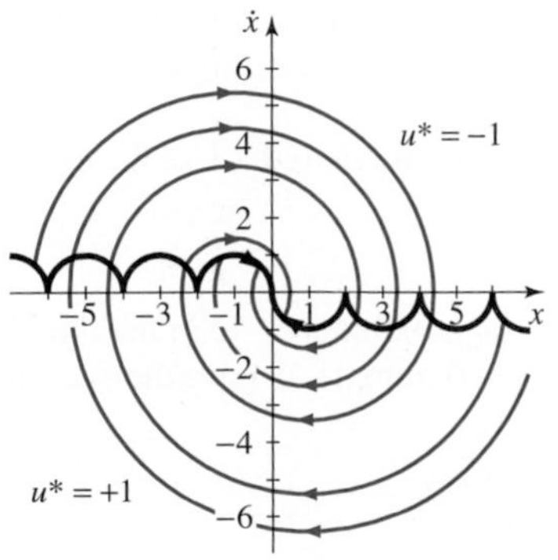

FIGURE 10.3.7 Trajectories and switching curve for Problems 4–7.

Use the initial state specified in each of Problems 4–7 to do the following.

(a) *Describe the optimal control $u^*(t)$. Specify the number of switches, and give the control sequence of positive and negative 1s.*

(b) *Describe the optimal trajectory in the phase plane, illustrating it by the motion of a vibrating spring.*

4. $(x(0), \dot{x}(0)) = (3, 0)$

5. $(x(0), \dot{x}(0)) = (4, 2)$

6. $(x(0), \dot{x}(0)) = (0, -3)$

7. $(x(0), \dot{x}(0)) = (-2, 0)$

8. **Minimum Cost Control** For $\dot{x} = 2x + u$, find the control $u = u(t)$ that drives $x(t)$ from initial condition $x(0) = 1$ to $x(1) = 0$ while minimizing the cost

$$J(u) = \frac{1}{2}\int_0^1 u^2(t)\,dt.$$

9. **Time Optimality and Bang-Bang Control** For the autonomous linear system

$$\dot{\vec{\mathbf{x}}} = \mathbf{A}\vec{\mathbf{x}} + \vec{\mathbf{b}}u,$$

where $|u| \le 1$, $\mathbf{A}$ is an $n \times n$ constant matrix and $\vec{\mathbf{b}}$ is an $n \times 1$ constant vector, show by the following that the time-optimal control $u(t)$ that drives $\vec{\mathbf{x}}$ from $\vec{\mathbf{x}}_0$ to the origin in minimum time is always bang-bang.

(a) Show that the Hamiltonian of the system is

$$H = -1 + \vec{\mathbf{p}} \cdot \mathbf{A}\vec{\mathbf{x}} + \vec{\mathbf{p}} \cdot \vec{\mathbf{b}}u,$$

where $\vec{\mathbf{p}} = [p_1, p_2, \ldots, p_n]$.

(b) Show that $\vec{\mathbf{p}} \cdot \vec{\mathbf{b}}u$ can be written $(\vec{\mathbf{p}} \cdot \vec{\mathbf{b}})u$.

(c) Defining the **switching function** $\sigma(t) = \vec{\mathbf{p}} \cdot \vec{\mathbf{b}}$, show that the optimal control is given by

$$u^*(t) = \begin{cases} 1 & \text{if } \sigma(t) > 0, \\ -1 & \text{if } \sigma(t) < 0. \end{cases}$$

10. Minimum Energy Revisited Use the exact value

$$c = -\frac{2}{1 - e^{-20}}$$

instead of the approximation -2 in the minimum-energy control problem (Example 2). Give the exact formulations for both control and trajectory, and verify the exact value of J.

11. Suggested Journal Entry I Discuss the significance of the "singular case" in the spaceship example.

12. Suggested Journal Entry II Develop an algorithm for determining the extreme values of a piecewise differentiable function of one variable having a finite interval for its domain. What are the parallels with solving optimal control problems?

Appendix CN
Complex Numbers and Complex-Valued Functions

The Complex Plane

Complex Number

By a **complex number** we mean a number of the form

$$z = a + bi. \tag{1}$$

- The **imaginary unit** i is defined by $i = \sqrt{-1}$.
- The real number a is called the **real component** (or **real part**) of z and is often denoted $a = \text{Re}(z)$.
- The real number b is called the **imaginary component** (or **imaginary part**) of z and is often denoted $b = \text{Im}(z)$.

The complex number $a+bi$ can be represented as a point in the **complex plane** with abscissa a and ordinate b, as shown in Fig. CN.1. If the imaginary part of a complex number is zero, then the complex number is simply a real number, such as 3, 5.2, π, -8.53, 0, -1, and so on. If the real part of a complex number is zero and the complex part is nonzero, then we say that the complex number is **purely imaginary**, such as $3i$, $-4i$, and so on. If both the real and imaginary parts of a complex number are zero, then we have the real number $0 = 0 + 0i$.

Two complex numbers are said to be equal if their real parts are equal and their imaginary parts are equal. For example, if

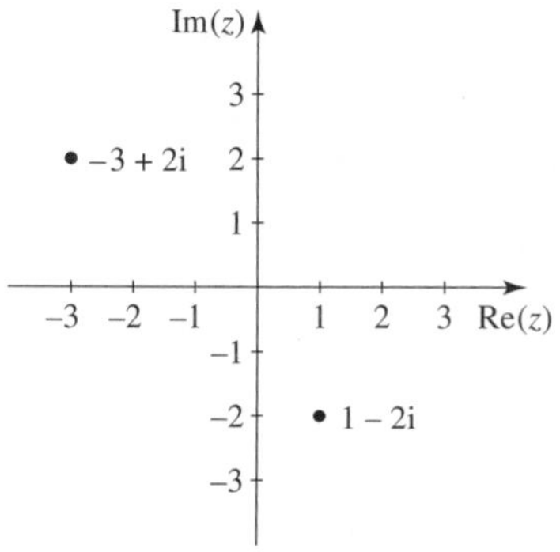

FIGURE CN.1 The complex plane.

$$(x + y - 2) + (x - y)i = 1 - i,$$

then we obtain

$$\begin{aligned} x + y - 2 &= 1, \\ x - y &= -1, \end{aligned}$$

which is true if and only if $x = 1$ and $y = 2$.

Arithmetic of Complex Numbers

The powers of i can be reduced to lowest form according to the following rules.

Powers of i

$$
\begin{aligned}
i^2 &= -1,\\
i^3 &= i^2 i = -i,\\
i^4 &= i^2 i^2 = 1,\\
i^5 &= i^4 i = i,\\
&\vdots
\end{aligned}
$$

With the powers of i defined above, the rules of arithmetic for complex numbers follow naturally from the usual rules for real numbers.

Addition and Subtraction of Complex Numbers

The sum and difference of two complex numbers $a + bi$ and $e + di$ are defined as

$$(a + bi) + (c + di) = (a + c) + (b + d)i, \tag{2}$$

$$(a + bi) - (c + di) = (a - c) + (b - d)i. \tag{3}$$

EXAMPLE 1 Plus or Minus

$$
\begin{aligned}
(3 + 2i) + (1 - i) &= 4 + i,\\
(1 - i) - (2 + 4i) &= -1 - 5i.
\end{aligned}
$$

■

Multiplication of Complex Numbers

The product of two complex numbers $a + bi$ and $c + di$ is defined as

$$
\begin{aligned}
(a + bi)(c + di) &= a(c + di) + bi(c + di)\\
&= ac + adi + bci + bdi^2\\
&= (ac - bd) + (ad + bc)i.
\end{aligned}
\tag{4}
$$

EXAMPLE 2 Multiplication

$$
\begin{aligned}
(4 + i)(2 - 3i) &= 4(2 - 3i) + i(2 - 3i)\\
&= 8 - 12i + 2i - 3i^2\\
&= 11 - 10i.
\end{aligned}
$$

■

Division of Complex Numbers

The quotient of two complex numbers is obtained using a process analogous to rationalizing the denominator and is defined by

$$\begin{aligned}\frac{a+bi}{c+di} &= \frac{a+bi}{c+di}\cdot\frac{c-di}{c-di}\\ &= \frac{a(c-di)+bi(c-di)}{c(c-di)+di(c-di)}\\ &= \frac{ac-adi+bci-bdi^2}{c^2-cdi+cdi-d^2i^2}\\ &= \left(\frac{ac+bd}{c^2+d^2}\right)+\left(\frac{bc-ad}{c^2+d^2}\right)i.\end{aligned} \tag{5}$$

EXAMPLE 3 **Division**

$$\begin{aligned}\frac{1+3i}{3+2i} &= \frac{1+3i}{3+2i}\cdot\frac{3-2i}{3-2i}\\ &= \frac{1(3-2i)+3i(3-2i)}{3(3-2i)+2i(3-2i)}\\ &= \frac{3-2i+9i-6i^2}{9-6i+6i-4i^2}\\ &= \frac{9+7i}{13} = \frac{9}{13}+\frac{7}{13}i.\end{aligned}$$

■

Absolute Value and Polar Angle

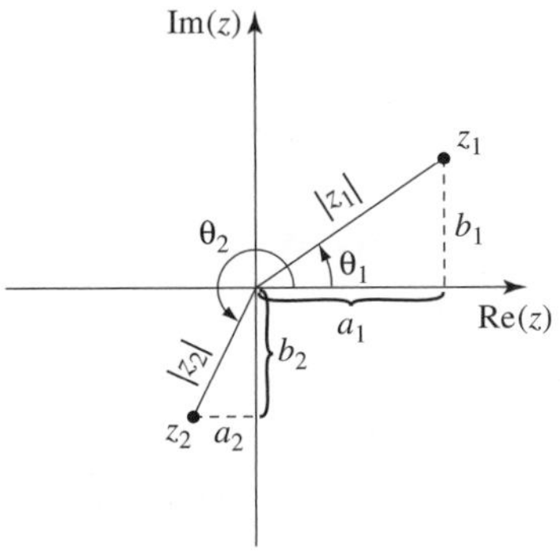

FIGURE CN.2 Absolute value $|z|$ and polar angle θ for $z = a + bi$.

Absolute Value of a Complex Number

The **absolute value** of a complex number $z = a + bi$ is defined by

$$r = |z| = \sqrt{a^2+b^2}, \tag{6}$$

which from the point of view of the complex plane denotes the polar distance from the origin 0 to z.

EXAMPLE 4 **Absolute Value**

$$|3+2i| = \sqrt{3^2+2^2} = \sqrt{13}.$$

■

Polar Angle of a Complex Number

The **polar angle** of a complex number $z = a + bi$, denoted θ, is defined by

$$\theta = \begin{cases} \tan^{-1}\left(\dfrac{b}{a}\right) & \text{if } a > 0, \\ \pi + \tan^{-1}\left(\dfrac{b}{a}\right) & \text{if } a < 0. \end{cases}$$

EXAMPLE 5 **Polar Angle** Complex number $z = 1 + i$ has polar angle

$$\theta = \tan^{-1}(1) = \frac{\pi}{4}.$$

Figure CN.2 shows two other examples. ■

Euler's Formula:

The exponential form $z = re^{i\theta}$ comes from Euler's formula. (See equation (10) below.)

Polar Form of a Complex Number

Using the trigonometric formulas $a = r\cos\theta$ and $b = r\sin\theta$, we can write complex numbers in **polar** or **trigonometric form** as

$$z = a + bi = r\cos\theta + ir\sin\theta = r(\cos\theta + i\sin\theta) = re^{i\theta}.$$

EXAMPLE 6 **Polar Form** The complex number $1 + i$ can be expressed in polar form as

$$\begin{aligned} 1 + i &= r(\cos\theta + i\sin\theta) \\ &= \sqrt{2}\left[\cos(\pi/4) + i\sin(\pi/4)\right]. \end{aligned}$$

■

Powers of a Complex Number

The polar form of a complex number allows us to find **powers** of a complex number with the aid of **De Moivre's formula**

$$z^n = r^n(\cos n\theta + i\sin n\theta), \qquad n = 1, 2, \ldots. \tag{7}$$

Then to find all m of the mth roots of a complex number, we adapt (7) with $n = 1/m$ to read

$$z^{1/m} = r^{1/m}\left(\cos\frac{\theta + 2\pi k}{m} + i\sin\frac{\theta + 2\pi k}{m}\right), \quad k = 0, 1, 2, \ldots, m-1. \tag{8}$$

EXAMPLE 7 **Solving a DE using DeMoivre's Formula** We will solve the homogeneous sixth order DE

$$\frac{d^6y}{dt^6} - y = 0.$$

We write the characteristic equation in terms of z with the expectation that the roots will be complex numbers.

$$z^6 - 1 = 0$$

Then $z^6 = \cos(0) + i\sin(0)$, so that $|z^6| = 1$ and polar angle $\theta = 0$. So then, by equation (8),

$$z = \cos\left(\frac{0 + 2\pi k}{6}\right) + i\sin\left(\frac{0 + 2\pi k}{6}\right), \qquad k = 0, 1, 2, \ldots, 5$$

For

$$\begin{aligned}
k = 0: &\quad z_0 = \cos(0) + i\sin(0) = 1\\
k = 1: &\quad z_1 = \cos\left(\frac{0 + 2\pi}{6}\right) + i\sin\left(\frac{0 + 2\pi}{6}\right) = \frac{1}{2} + i\frac{\sqrt{3}}{2}\\
k = 2: &\quad z_2 = \cos\left(\frac{0 + 4\pi}{6}\right) + i\sin\left(\frac{0 + 4\pi}{6}\right) = -\frac{1}{2} + i\frac{\sqrt{3}}{2}\\
k = 3: &\quad z_3 = \cos\left(\frac{0 + 6\pi}{6}\right) + i\sin\left(\frac{0 + 6\pi}{6}\right) = -1\\
k = 4: &\quad z_4 = \cos\left(\frac{0 + 8\pi}{6}\right) + i\sin\left(\frac{0 + 8\pi}{6}\right) = -\frac{1}{2} - i\frac{\sqrt{3}}{2}\\
k = 5: &\quad z_5 = \cos\left(\frac{0 + 10\pi}{6}\right) + i\sin\left(\frac{0 + 10\pi}{6}\right) = \frac{1}{2} - i\frac{\sqrt{3}}{2}
\end{aligned}$$

Thus the solution of the DE is

$$\begin{aligned}
y(t) = c_1e^t + c_2e^{-t} &+ e^{\frac{1}{2}t}\left[c_3\cos\left(\frac{\sqrt{3}}{2}t\right) + c_4\sin\left(\frac{\sqrt{3}}{2}t\right)\right]\\
&+ e^{-\frac{1}{2}t}\left[c_5\cos\left(\frac{\sqrt{3}}{2}t\right) + c_6\sin\left(\frac{\sqrt{3}}{2}t\right)\right].
\end{aligned}$$

Note that, in this case, an alternative route to the roots could be to factor the characteristic equation,

$$z^6 - 1 = (z^3 - 1)(z^3 + 1) = (z - 1)(z^2 + z + 1)(z + 1)(z^2 - z + 1) = 0,$$

obtaining roots $z = \pm 1, \frac{1}{2} \pm i\frac{\sqrt{3}}{2}, -\frac{1}{2} \pm i\frac{\sqrt{3}}{2}$. ■

When arithmetic operations involve complex numbers, it is not hard to derive various rules relating the absolute values.

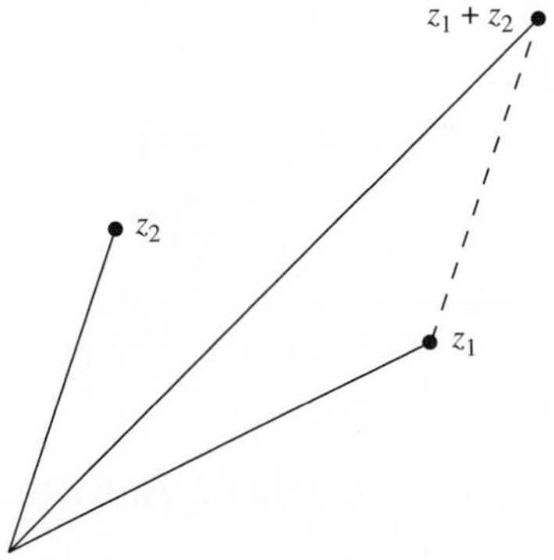

FIGURE CN.3 The triangle inequality.

Properties of the Absolute Value of a Complex Number

A few of the important properties are

$$|z_1z_2| = |z_1||z_2|,$$

$$\left|\frac{z_1}{z_2}\right| = \frac{|z_1|}{|z_2|},$$

$$|z_1 + z_2| \le |z_1| + |z_2|. \quad \text{(triangle inequality)}$$

The triangle inequality is shown in Fig. CN.3.

Complex Conjugates

Complex Conjugates

If two complex numbers differ only in the sign of their imaginary parts, the two complex numbers are called **complex conjugates** (or **conjugate** to each other). The conjugate of a complex number z is usually denoted by $\bar{z}$.

EXAMPLE 8 **Complex Conjugates** The two complex numbers

$$3 + 2i \quad \text{and} \quad 3 - 2i$$

are complex conjugates. ■

EXAMPLE 9 **The Magic of Complex Conjugates** For $z = a + bi$, we can write the magnitude, real part and imaginary part of z in terms of z and its complex conjugate $z = a - bi$.

(a) The magnitude of z is found through multiplication:

$$z\bar{z} = (a + bi)(a - bi) = a^2 + b^2 = |z|^2.$$

Thus, we have

$$|z| = \sqrt{a^2 + b^2} = \sqrt{z\bar{z}}. \tag{9}$$

(b) The real part of z is found through addition:

$$z + \bar{z} = (a + bi) + (a - bi) = 2a = 2\,\mathrm{Re}(z),$$

or

$$\mathrm{Re}(z) = \frac{z + \bar{z}}{2}.$$

(c) The imaginary part of z is found through substraction:

$$z - \bar{z} = (a + bi) - (a - bi) = 2bi = 2i\mathrm{Im}(z),$$

or

$$\mathrm{Im}(z) = \frac{z - \bar{z}}{2i}.$$

■

Complex-Valued Functions

Complex-Valued Functions of a Real Variable

A complex-valued function of a real variable is an expression of the form

$$F(t) = f(t) + ig(t),$$

where f and g are real-valued functions of t, defined on some interval of interest. We define the **derivative** of the complex-valued function $F(t)$ by

$$F'(t) = f'(t) + ig'(t),$$

provided that both f and g are differentiable over the domain of interest. Higher derivatives are defined in the same way; i.e.,
$F''(t) = f''(t) + ig''(t)$, and so on.

EXAMPLE 10 **Complex-Valued Functions** Typical complex-valued functions of a single real variable t are

$$\begin{aligned} F(t) &= \cos 3t + i \sin 3t, \\ G(t) &= t^2 + ite^t, \\ H(t) &= e^{2t} \cos t + ie^{2t} \sin t. \end{aligned}$$

■

EXAMPLE 11 **Differentiating Complex-Valued Functions** For the complex valued function

$$F(t) = \cos 2t + i \sin 2t,$$

we have

$$\begin{aligned} F'(t) &= -2 \sin 2t + 2i \cos 2t, \\ F''(t) &= -4 \cos 2t - 4i \sin 2t, \\ F'''(t) &= 8 \sin 2t - 8i \cos 2t. \end{aligned}$$

■

One of the most important complex-valued functions of a real variable that arises in the study of differential equations is the **complex exponential function**, known as Euler's formula (See Sec. 4.3, Problem 29.)

Euler's Formula

$$e^{it} = \cos t + i \sin t. \tag{10}$$

More generally, we can use the related function

$$e^{(a+bi)t} = e^{at}(\cos bt + i \sin bt), \tag{11}$$

which is also a complex-valued function of the real variable t. When $t = 1$, we have the complex exponential

$$e^{a+bi} = e^a(\cos b + i \sin b).$$

This formula shows how to raise the constant e to a complex number.

EXAMPLE 12 **Complex Powers of e:**

$$\begin{aligned} e^{i\pi/2} &= \cos(\pi/2) + i \sin(\pi/2) = i, \\ e^{3+2\pi i} &= e^3(\cos 2\pi + i \sin 2\pi) = e^3, \\ e^{2\pi i} &= \cos 2\pi + i \sin 2\pi = 1, \\ e^{i\pi/4} &= \cos(\pi/4) + i \sin(\pi/4) = \frac{1}{\sqrt{2}} + i\frac{1}{\sqrt{2}}, \\ e^2 &= e^2\Big[\cos(0) + i \sin(0)\Big] = e^2. \end{aligned}$$

■

As one might suspect, we have the usual rule of exponents for the complex exponential:

$$e^{z_1}e^{z_2} = e^{z_1+z_2}.$$

Hence, we have

$$e^{2+3\pi i} = e^2 e^{3\pi i}.$$

Derivative of a Complex Exponential

From the definition of the derivative of a complex-valued function of a real variable, we can write

$$\begin{aligned}\frac{d}{dt}e^{(a+bi)t} &= \frac{d}{dt}\left[e^{at}(\cos bt + i\sin bt)\right] \\ &= \frac{d}{dt}(e^{at}\cos bt) + i\frac{d}{dt}(e^{at}\sin bt) \\ &= e^{at}(a\cos bt - b\sin bt) + ie^{at}(a\sin bt + b\cos bt) \\ &= (a+bi)e^{(a+ib)t}.\end{aligned}$$

Thus we have proven the important derivative from the calculus of complex-valued functions.

Problems CN

1. **Complex Plane** Plot the following complex numbers in the complex plane.
   (a) $3+3i$ (b) $4i$ (c) 2 (d) $1-i$

2. **Complex Operations** Write the following complex numbers in the form $a+bi$.
   (a) $(2+3i)(4-i)$ (b) $(2+3i)(1+i)$
   (c) $\dfrac{1}{1+i}$ (d) $\dfrac{2+i}{3+i}$

3. **Complex Exponential Numbers** Write each of the following complex exponentials in the form $a+bi$.
   (a) $e^{2\pi i}$ (b) $e^{i\pi/2}$ (c) $e^{-i\pi}$ (d) $e^{(2+\pi i/4)}$

4. **Magnitudes and Angles** Find the absolute value and polar angle of each of the following complex numbers.
   (a) $1+2i$ (b) $-i$ (c) $-1-i$ (d) $-2+3i$
   (e) e^{2i} (f) $\dfrac{2+i}{1+i}$

5. **Complex Verification I** Verify that the two complex numbers $z=-1\pm i$ satisfy the equation $z^2+2z+2=0$.

6. **Complex Verification II** Show that $\dfrac{1+i}{\sqrt{2}}$ satisfies the equation $z^4=-1$.

7. **Real and Complex Parts** If $z=a+bi$, find the following quantities in terms of a and b.
   (a) $\mathrm{Re}(z^2+2z)$ (b) $\mathrm{Im}(z^2+2z)$

8. **Absolute Value Revisited** Use the formula $|z|=\sqrt{z\bar{z}}$ to find the absolute value $|4+2i|$.

9. **Roots of Unity** Find the roots of the following equations.
   (a) $z^2=1$ (b) $z^3=1$ (c) $z^4=1$

10. **Derivatives of Complex Functions** Find the derivatives $F'(t)$ and $F''(t)$ for each of the following complex-valued functions of the real variable t.
    (a) $F(t)=e^{(1-i)t}$
    (b) $F(t)=e^{3it}$
    (c) $F(t)=e^{(2+3i)t}$

11. **Real and Complex Parts of Exponentials** Write each of the following complex numbers in $a+bi$ form.
    (a) $e^{(1+\pi i)}$ (b) $e^{(2+\pi i/2)}$ (c) $e^{\pi i}$ (d) $e^{-\pi i}$

12. **Complex Exponential Functions** Write each of the following complex-valued functions in $a+bi$ form.
    (a) $e^{4\pi it}$ (b) $e^{(-1+2i)t}$

Using DeMoivre's Formula Use formula (8) to find the general solutions for the DE's in Problems 13–14.

13. $\dfrac{d^3y}{dt^3}+y=0$

14. $\dfrac{d^4y}{dt^4}+81y=0$

Appendix LT
Linear Transformations

Linear Transformations

SYNOPSIS: We introduce the coordinate map and use it to prove that for any vector space $\mathbb{V}$ with a basis of n vectors, every basis for $\mathbb{V}$ must have exactly n vectors. We show that all vector spaces of the same finite dimension are isomorphic. We find the matrix associated with a linear transformation $T : \mathbb{V} \to \mathbb{W}$, where $\mathbb{V}$ and $\mathbb{W}$ are finite-dimensional vector spaces. We discover that the matrix depends on the choice of bases for both spaces as well as the linear transformation T.

We will limit our work to vector spaces with finite bases, although these ideas can be extended to infinite-dimensional vector spaces quite naturally.

The Coordinate Map and Dimension

The central idea is that a vector space $\mathbb{V}$ can have many different bases and that a vector $\vec{\mathbf{v}}$ in $\mathbb{V}$ can be expressed as a linear combination of any one of the bases.

In Sec. 5.4, we defined **the coordinate vector for $\vec{\mathbf{v}}$ relative to an ordered basis $B = \{\vec{\mathbf{b}}_1, \vec{\mathbf{b}}_2, \ldots, \vec{\mathbf{b}}_n\}$** to be the column vector

$$\vec{\mathbf{v}}_B = \begin{bmatrix} \beta_1 \\ \beta_2 \\ \vdots \\ \beta_n \end{bmatrix}_B \quad \text{in} \quad \mathbb{R}^n,$$

where $\beta_1, \beta_2, \ldots, \beta_n$ are the coordinates of $\vec{\mathbf{v}}$ relative to B. Now we add the following definition:

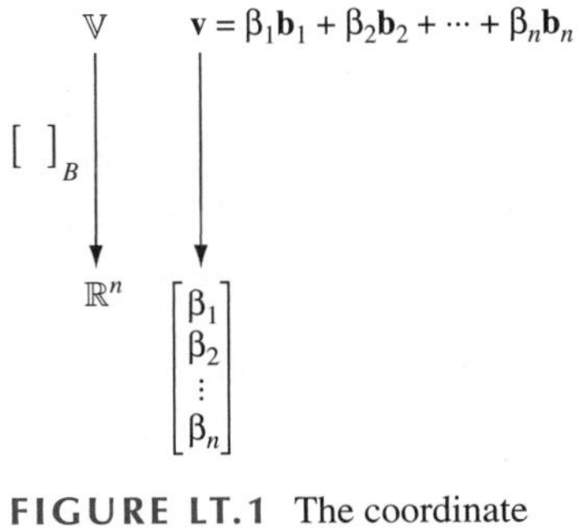

FIGURE LT.1 The coordinate map $[\ ]_B$ for a vector $\vec{\mathbf{v}}$ in $\mathbb{V}$ can be expressed as a linear combination of vectors in an ordered basis

$$B = \{\vec{\mathbf{b}}_1, \vec{\mathbf{b}}_2, \ldots, \vec{\mathbf{b}}_n\},$$

as shown at the top of the diagram.

Coordinate Map

The function $[\ ]_B : \mathbb{V} \to \mathbb{R}^n$ that assigns to each vector $\vec{\mathbf{v}}$ in $\mathbb{V}$ its coordinate vector $\vec{\mathbf{v}}_B$ relative to the ordered basis $B = \{\vec{\mathbf{b}}_1, \vec{\mathbf{b}}_2, \ldots, \vec{\mathbf{b}}_n\}$ is called a **coordinate map**.

In Problem 1 you will be asked to check that the coordinate map from $\mathbb{V}$ to $\mathbb{R}^n$ is a linear transformation. From the uniqueness of the coordinates for a given ordered basis, we can see that coordinate maps are injective. From the fact that a basis is a spanning set, we can see that coordinate maps are surjective. This leads to another definition.

Isomorphism

A linear transformation $T : \mathbb{V} \to \mathbb{W}$ that is both injective and surjective is called an isomorphism, and the vector spaces $\mathbb{V}$ and $\mathbb{W}$ are said to be *isomorphic*, denoted $\mathbb{V} \approx \mathbb{W}$.

If a vector space $\mathbb{V}$ has a basis $B = \{\vec{\mathbf{b}}_1, \vec{\mathbf{b}}_2, \ldots, \vec{\mathbf{b}}_n\}$, then $\mathbb{V} \approx \mathbb{R}^n$ with the isomorphism $[\ ]_B$ so that

Notation

The symbol $\mapsto$ is sometimes used to indicate an isomorphism.

$$\vec{\mathbf{v}} \mapsto \begin{bmatrix} \beta_1 \\ \beta_2 \\ \vdots \\ \beta_n \end{bmatrix}_B ,$$

where $\beta_1, \beta_2, \ldots, \beta_n$ are the coordinates of $\vec{\mathbf{v}}$ relative to B. We have discovered that an n-dimensional vector space $\mathbb{V}$ is isomorphic to $\mathbb{R}^n$.

The Basis Dimension Theorem

Here is the key theorem that allows us to assign a *dimension* to a vector space:

$\dim \mathbb{V} =$ the number of basis vectors.

Basis Dimension Theorem

If a set of n vectors forms a basis for a vector space $\mathbb{V}$, then every basis for $\mathbb{V}$ must have exactly n vectors.

Proof Let $B = \{\vec{\mathbf{b}}_1, \vec{\mathbf{b}}_2, \ldots, \vec{\mathbf{b}}_m\}$ and $C = \{\vec{\mathbf{c}}_1, \vec{\mathbf{c}}_2, \ldots, \vec{\mathbf{c}}_n\}$ be bases for $\mathbb{V}$, where $m \le n$. Consider the coordinate map $[\ ]_B : \mathbb{V} \to \mathbb{R}^m$, which is an isomorphism. Then if $m < n$, the set $\{[\vec{\mathbf{c}}_1]_B, [\vec{\mathbf{c}}_2]_B, \ldots, [\vec{\mathbf{c}}_n]_B\}$ would be linearly dependent in $\mathbb{R}^m$ because this set contains n vectors, which is greater than the number m of entries in each of the vectors, which belong to $\mathbb{R}^m$.

In similar fashion, we can prove that $n \le m$ implies an isomorphism from $\mathbb{V}$ to $\mathbb{R}^n$ and that $n = m$, so B and C must have the same number of vectors. □

Thus our definition in Sec. 3.6 of the dimension of a vector space $\mathbb{V}$ as the number of vectors in a basis is reasonable. Also, we have the following corollary.

Corollary

If a vector space $\mathbb{V}$ has dimension n, then every subset of $\mathbb{V}$ containing more than n vectors must be linearly dependent.

Isomorphic Vector Spaces

The following facts about isomorphisms between vector spaces are easy to prove. (See Problems 4–6.)

Properties of Isomorphisms

An isomorphism $T : \mathbb{V} \to \mathbb{W}$ has the following properties:

- $T^{-1} : \mathbb{W} \to \mathbb{V}$, the inverse map, exists and is also an isomorphism.
- If $\{\vec{\mathbf{b}}_1, \vec{\mathbf{b}}_2, \ldots, \vec{\mathbf{b}}_n\}$ is a basis for vector space $\mathbb{V}$, then $\{T(\vec{\mathbf{b}}_1), T(\vec{\mathbf{b}}_2), \ldots, T(\vec{\mathbf{b}}_n)\}$ is a basis for vector space $\mathbb{W}$.
- If $L : \mathbb{W} \to \mathbb{U}$ is an isomorphism between vector spaces, then the composition $L \circ T : \mathbb{V} \to \mathbb{U}$ is also an isomorphism.

We can use isomorphism properties and the dimension theorem to construct a proof for a quite powerful isomorphism theorem, illustrated in Fig. LT.2.

If $\mathbb{V}$ and $\mathbb{W}$ have dimension n, then $\mathbb{V} \approx \mathbb{W} \approx \mathbb{R}^n$.

Vector Space Isomorphism Theorem
All vector spaces of dimension n are isomorphic.

Proof Since any n-dimensional vector spaces $\mathbb{V}$ and $\mathbb{W}$ must be isomorphic to $\mathbb{R}^n$, they must be isomorphic to each other. □

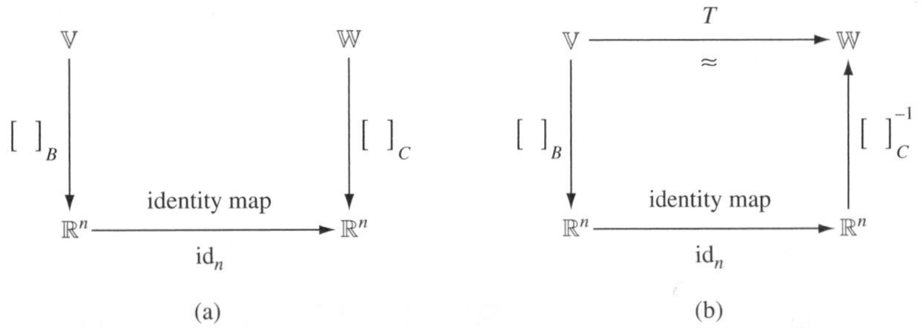

FIGURE LT.2 Diagrams showing the mappings between $\mathbb{V}$, $\mathbb{W}$, and $\mathbb{R}^n$.

A consequence of the Vector Space Isomorphism Theorem is that every n-dimensional vector space "acts like" $\mathbb{R}^n$ in that it has the same number of basis elements and the same vector space structure. Our previous knowledge about $\mathbb{R}^n$ is suddenly applicable to a wide variety of vector spaces.

Matrices Associated with Linear Transformations

At this point, we will use our knowledge of coordinate maps to find the *matrix* associated with a given linear transformation between two finite-dimensional vector spaces with given ordered bases.

Suppose T is a linear transformation with ordered bases $B = \{\vec{\mathbf{b}}_1, \vec{\mathbf{b}}_2, \ldots, \vec{\mathbf{b}}_n\}$ and $C = \{\vec{\mathbf{c}}_1, \vec{\mathbf{c}}_2, \ldots, \vec{\mathbf{c}}_m\}$ for $\mathbb{V}$ and $\mathbb{W}$, respectively. Let $f : \mathbb{R}^n \to \mathbb{R}^m$ be the function defined for each $\vec{\mathbf{v}}$ in $\mathbb{V}$ by

$$f(\vec{\mathbf{v}}_B) = [T(\vec{\mathbf{v}})]_C.$$

With the notation $[\ \ ]_B\vec{\mathbf{v}} = \vec{\mathbf{v}}_B$, expressing $\vec{\mathbf{v}}$ as a vector in B coordinates, we can make a commutative diagram, as shown in Fig. LT.3. We call this diagram

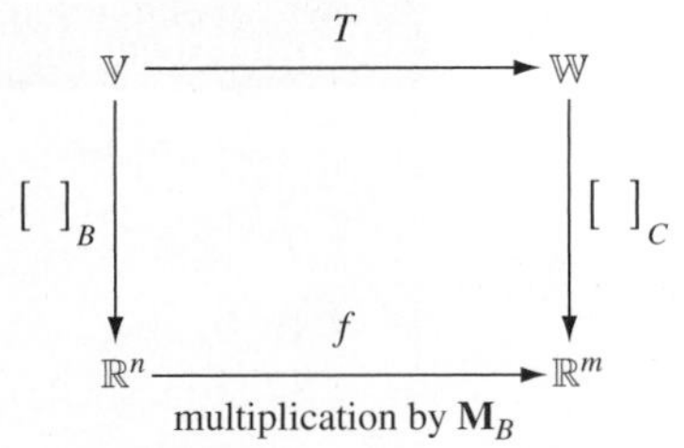

FIGURE LT.3 A *commutative* diagram for $f(\vec{\mathbf{v}}_B) = [T(\vec{\mathbf{v}})]_C$.

commutative because the two routes from upper left to lower right give the same result for any $\vec{\mathbf{v}}$ in $\mathbb{V}$.

Path 1: *downward*: express $\vec{\mathbf{v}}$ in B-coordinates;
right: then apply f.

Path 2: *right*: transform $\vec{\mathbf{v}}$ by T;
downward: express the result in C-coordinates.

Then the fact that f is linear follows directly from the linearity of T, and from Sec. 5.1 we know that there must be an associated matrix $\mathbf{M}_B$ so that

$$\underbrace{\mathbf{M}_B}_{m\times n}\ \underbrace{\vec{\mathbf{v}}_B}_{n\times 1} = \underbrace{[T(\vec{\mathbf{v}})]_C}_{m\times 1}. \tag{1}$$

This matrix $\mathbf{M}_B$ is called **the associated matrix for T from basis B into basis** C. It depends on the transformation T and the two bases B and C, respectively.

We construct $\mathbf{M}_B$ in the following fashion:

$$\left[\left[T(\vec{\mathbf{b}}_1)\right]_C \middle| \left[T(\vec{\mathbf{b}}_2)\right]_C \middle| \cdots \middle| \left[T(\vec{\mathbf{b}}_n)\right]_C\right].$$

Let us try it on a few examples.

EXAMPLE 1 **From 3-Space to $\mathbb{P}_2$** Let $T : \mathbb{R}^3 \to \mathbb{P}_2$ be the linear transformation

$$T(\vec{\mathbf{v}}) = (a + 2b)t^2 + c, \tag{2}$$

defined for each

$$\vec{\mathbf{v}} = \begin{bmatrix} a \\ b \\ c \end{bmatrix} \text{ in } \mathbb{R}^3.$$

We let $C = \{t^2, t, 1\}$ be the standard ordered basis in $\mathbb{P}_2$, and we express $\vec{\mathbf{v}} = \vec{\mathbf{v}}_S$ in terms of the standard ordered basis $S = \{\vec{\mathbf{e}}_1, \vec{\mathbf{e}}_2, \vec{\mathbf{e}}_3\}$, so the associated matrix for T from basis S into basis C is

$$\begin{aligned}
\mathbf{M}_S &= \left[[T(\vec{\mathbf{e}}_1)]_C \middle| [T(\vec{\mathbf{e}}_2)]_C \middle| [T(\vec{\mathbf{e}}_3)]_C \right] \\
&= \left[[t^2]_C \middle| [2t^2]_C \middle| [1]_C \right] \\
&= \left[\begin{bmatrix} 1 \\ 0 \\ 0 \end{bmatrix} \begin{bmatrix} 2 \\ 0 \\ 0 \end{bmatrix} \begin{bmatrix} 0 \\ 0 \\ 1 \end{bmatrix} \right] = \begin{bmatrix} 1 & 2 & 0 \\ 0 & 0 & 0 \\ 0 & 0 & 1 \end{bmatrix}.
\end{aligned}$$

We can check the result shown in Fig. LT.3 by verifying that $\mathbf{M}_B\vec{\mathbf{v}}_B = [T(p)]_C$ for any polynomial $p = at^2 + bt + c$ in $\mathbb{P}_2$:

$$\mathbf{M}_B\vec{\mathbf{v}}_B = \begin{bmatrix} 1 & 2 & 0 \\ 0 & 0 & 0 \\ 0 & 0 & 1 \end{bmatrix} \begin{bmatrix} a \\ b \\ c \end{bmatrix} = \begin{bmatrix} a+2b \\ 0 \\ c \end{bmatrix} = [T(p)]_C.$$

Another way of looking at this transformation is shown in Fig. LT.4. ■

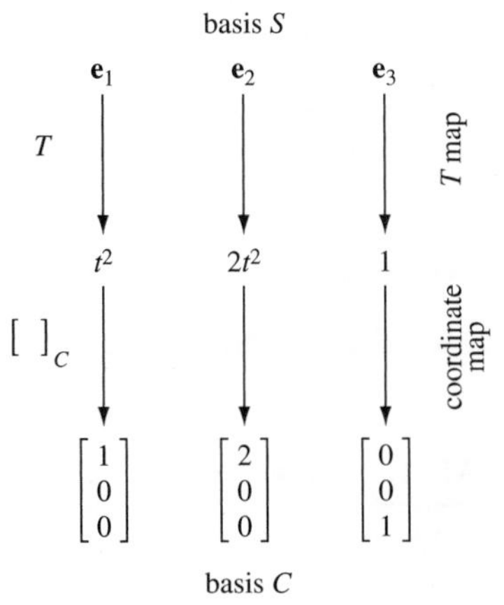

FIGURE LT.4 *T* map and coordinate map for Example 1.

EXAMPLE 2 **The Derivative on Polynomials** Let us find the associated matrix for the linear transformation $D : \mathbb{P}_3 \to \mathbb{P}_2$, which we define as $D(f) = f'$.

Let $B = \{t^3, t^2, t, 1\}$ and $C = \{t^2, t, 1\}$ be the standard ordered bases in $\mathbb{P}_3$ and $\mathbb{P}_2$, respectively. Then

$$\mathbf{M}_B = \Big[[T(t^3)]_C \,\Big|\, [T(t^2)]_C \,\Big|\, [T(t)]_C \,\Big|\, [T(1)]_C \Big]$$

$$= \Big[[3t^2]_C \,\Big|\, [2t]_C \,\Big|\, [1]_C \,\Big|\, [0]_C \Big]$$

$$= \left[\begin{bmatrix} 3 \\ 0 \\ 0 \end{bmatrix} \begin{bmatrix} 0 \\ 2 \\ 0 \end{bmatrix} \begin{bmatrix} 0 \\ 0 \\ 1 \end{bmatrix} \begin{bmatrix} 0 \\ 0 \\ 0 \end{bmatrix} \right] = \begin{bmatrix} 3 & 0 & 0 & 0 \\ 0 & 2 & 0 & 0 \\ 0 & 0 & 1 & 0 \end{bmatrix}$$

is the associated matrix for D from basis B to basis C. See Fig. LT.5. ■

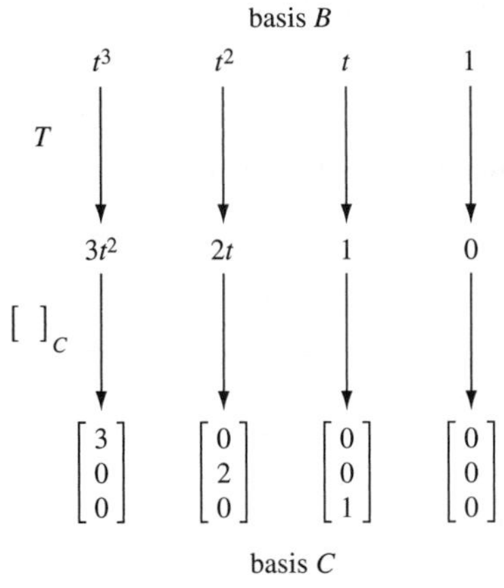

FIGURE LT.5 *T* map and coordinate map for Example 2.

EXAMPLE 3 **The Derivative on a Solution Space** Consider the solution space $\mathbb{V}$ for the differential equation $x'' + 4x = 0$. From Sec. 4.3, we know that $B = \{\cos 2t, \sin 2t\}$ is an ordered basis for $\mathbb{V}$. Let $D : \mathbb{V} \to \mathbb{V}$ be the derivative operator on $\mathbb{V}$ defined by $D(f) = f'$. We will use the basis C, which is the same as basis B, for the codomain $\mathbb{V}$:

$$\mathbf{M}_B = \Big[[D(\cos 2t)]_C \,\Big|\, [D(\sin 2t)]_C \Big]$$

$$= \Big[[-2\sin 2t]_C \,\Big|\, [2\cos 2t]_C \Big]$$

$$= \left[\begin{bmatrix} 0 \\ -2 \end{bmatrix} \begin{bmatrix} 2 \\ 0 \end{bmatrix} \right] = \begin{bmatrix} 0 & 2 \\ -2 & 0 \end{bmatrix}.$$

See Fig. LT.6.

This process means that for an arbitrary solution $x(t) = c_1 \cos 2t + c_2 \sin 2t$ in the solution space S, we can obtain the derivative $\dot{x}(t)$ by matrix multiplication:

$$\dot{x}(t) = [D(x)]_C = \mathbf{M}_B\vec{\mathbf{x}}_B = \begin{bmatrix} 0 & 2 \\ -2 & 0 \end{bmatrix} \begin{bmatrix} c_1 \\ c_2 \end{bmatrix}_B = \begin{bmatrix} 2c_2 \\ -2c_1 \end{bmatrix}_C.$$

■

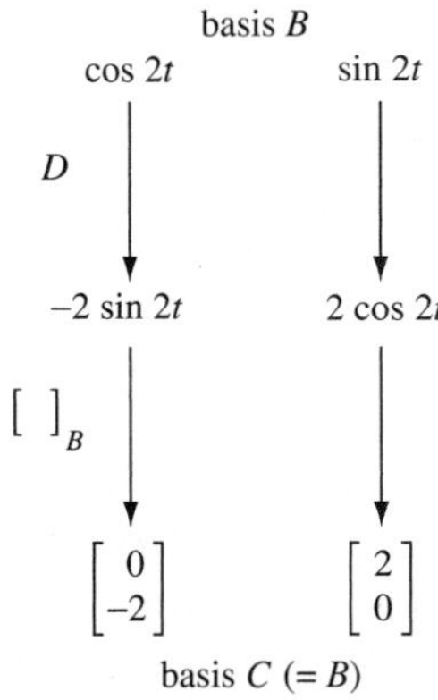

FIGURE LT.6 *D* map and coordinate map for Example 3.

Matrices for a Change of Basis

In Sec. 5.4, we talked about changes of bases in $\mathbb{R}^n$. Now we are going to extend the discussion to changes of bases for arbitrary finite-dimensional vector spaces. At this point, the process is exactly the same as that in the first part of this section, only now our linear transformation is id $(\vec{\mathbf{v}}) = \vec{\mathbf{v}}$ for each $\vec{\mathbf{v}}$ in $\mathbb{V}$. Given two bases B and C for a vector space $\mathbb{V}$ and a vector $\vec{\mathbf{v}}$, we can transform $\vec{\mathbf{v}}_B$ into $\vec{\mathbf{v}}_C$ using matrix multiplication.

$\mathbf{M}_B$ changes bases from B to C.

$\mathbf{M}_C = \mathbf{M}_B^{-1}$ changes bases from C to B.

- $\mathbf{M}_B$ is the matrix associated with the *identity transformation* id *on* $\mathbb{V}$ from basis B into basis C:

$$\mathbf{M}_B \mathbf{v}_B = [\text{id}(\mathbf{v})]_C = \mathbf{v}_C.$$

- $\mathbf{M}_B$ is *not* the identity matrix $\mathbf{I}_n$ as long as the bases B and C are not the same.
- $\mathbf{M}_B$ must be invertible, since for any vector $\mathbf{v}$,

$$\mathbf{M}_C \underbrace{\mathbf{M}_B \mathbf{v}_B}_{\mathbf{v}_C} = \mathbf{I}_n \mathbf{v}_B = \mathbf{v}_B \quad \text{and} \quad \mathbf{M}_B \underbrace{\mathbf{M}_C \mathbf{v}_C}_{\mathbf{v}_B} = \mathbf{I}_n \mathbf{v}_C = \mathbf{v}_C.$$

EXAMPLE 4 **Changes of Bases in $\mathbb{P}_2$** Consider the standard basis for $\mathbb{P}_2$, $B = \{t^2, t, 1\}$, and a new basis $C = \{t, 3 - 2t, t + 3t^2\}$. We can see that

$$\begin{aligned}\mathbf{M}_B &= \left[[\text{id}(t^2)]_C \,\middle|\, [\text{id}(t)]_C \,\middle|\, [\text{id}(1)]_C \right] \\ &= \left[[t^2]_C \,\middle|\, [t]_C \,\middle|\, [1]_C \right].\end{aligned}$$

We must write the elements of basis B in terms of the new basis C vectors. For instance, the basis B vector t^2 can be expressed as

$$t^2 = c_1 t + c_2(3 - 2t) + c_3(t + 3t^2),$$

and equating coefficients of like terms gives $c_1 = -1/3$, $c_2 = 0$, and $c_3 = 1/3$, so

$$[t^2]_C = \begin{bmatrix} -1/3 \\ 0 \\ 1/3 \end{bmatrix}.$$

In similar fashion, we get

$$[t]_C = \begin{bmatrix} 1 \\ 0 \\ 0 \end{bmatrix} \quad \text{and} \quad [1]_C = \begin{bmatrix} 2/3 \\ 1/3 \\ 0 \end{bmatrix},$$

so that

$$\mathbf{M}_B = \left[\begin{bmatrix} -1/3 \\ 0 \\ 1/3 \end{bmatrix} \begin{bmatrix} 1 \\ 0 \\ 0 \end{bmatrix} \begin{bmatrix} 2/3 \\ 1/3 \\ 0 \end{bmatrix} \right] = \begin{bmatrix} -1/3 & 1 & 2/3 \\ 0 & 0 & 1/3 \\ 1/3 & 0 & 0 \end{bmatrix}.$$

■

EXAMPLE 5 **Changing Bases in a Solution Space** Consider the solution space $\mathbb{V}$ for $x'' - x = 0$. From Sec. 4.2, we know that $B = \{e^t, e^{-t}\}$ is a basis for $\mathbb{V}$, and it is easy to verify that $C = \{\cosh t, \sinh t\}$ is also a basis for $\mathbb{V}$. We need to find out what happens to the basis elements of B when mapped by the identity map and then expressed in C-coordinates:

$$[\text{id}(e^t)]_C = [e^t]_C = (\cosh t + \sinh t)_C = \begin{bmatrix} 1 \\ 1 \end{bmatrix},$$

$$[\text{id}(e^{-t})]_C = [e^{-t}]_C = (\cosh t - \sinh t)_C = \begin{bmatrix} 1 \\ -1 \end{bmatrix}.$$

Consequently,

$$\mathbf{M}_B = \left[\begin{bmatrix} 1 \\ 1 \end{bmatrix} \begin{bmatrix} 1 \\ -1 \end{bmatrix} \right] = \begin{bmatrix} 1 & 1 \\ 1 & -1 \end{bmatrix}.$$

Recalling that

$$\cosh t = \frac{e^t + e^{-t}}{2} \quad \text{and} \quad \sinh t = \frac{e^t - e^{-t}}{2},$$

we can see that

$$\mathbf{M}_C = \mathbf{M}_B^{-1} = \begin{bmatrix} 1/2 & 1/2 \\ 1/2 & -1/2 \end{bmatrix}.$$

■

Summary

We used the coordinate map associated with a vector space $\mathbb{V}$ for which there is a finite basis to verify that every basis for $\mathbb{V}$ has the same number of vectors. Then for any linear transformation from one finite-dimensional vector space into another, given a basis for each space, we learned how to construct the associated matrix for the transformation. We applied these concepts to the identity transformation in order to find matrices associated with changes of coordinates. We also proved that all vector spaces of the same finite dimension are isomorphic.

Problems LT

1. **Coordinate Map** Let $\mathbb{V}$ be a vector space with basis $B = \{\vec{\mathbf{b}}_1, \vec{\mathbf{b}}_2, \ldots, \vec{\mathbf{b}}_n\}$. Show that the coordinate map $[\ ]_B : \mathbb{V} \to \mathbb{R}^n$ is an isomorphism. (You need to show three things: linearity, injectivity and surjectivity.)

2. **Isomorphisms** List at least three vector spaces that are isomorphic to $\mathbb{M}_{22}(\mathbb{R})$.

3. **Isomorphism Subtleties** Explain why $\mathbb{M}_{12}(\mathbb{R})$ is not a subspace of $\mathbb{M}_{22}(\mathbb{R})$. Show, however, that it is isomorphic to a subspace of $\mathbb{M}_{22}(\mathbb{R})$ by finding the subspace and the isomorphism.

4. **Isomorphisms Have Inverses** Let $T : \mathbb{V} \to \mathbb{W}$ be an isomorphism. Prove that the inverse map T^{-1} exists and is also an isomorphism. (You must define T^{-1} and show that it is an injective and surjective linear transformation from $\mathbb{W}$ to $\mathbb{V}$.)

5. **Composition of Isomorphisms** Let $T : \mathbb{V} \to \mathbb{W}$ be an isomorphism between vector spaces. Prove that if $L : \mathbb{W} \to \mathbb{U}$ is an isomorphism between vector spaces, then the composition $L \circ T : \mathbb{V} \to \mathbb{U}$ is also an isomorphism.

6. **Isomorphisms and Bases** Let $T : \mathbb{V} \to \mathbb{W}$ be an isomorphism between vector spaces. Use the properties of isomorphisms to prove that if $\{\vec{\mathbf{b}}_1, \vec{\mathbf{b}}_2, \ldots, \vec{\mathbf{b}}_n\}$ is a basis for $\mathbb{V}$, then $\{T(\vec{\mathbf{b}}_1), T(\vec{\mathbf{b}}_2), \ldots, T(\vec{\mathbf{b}}_n)\}$ is a basis for $\mathbb{W}$.

Associated Matrices *In Problems 7–11, find the matrix* $\mathbf{M}_B$ *associated with the linear transformation* $T : \mathbb{V} \to \mathbb{W}$ *from basis* B *for* $\mathbb{V}$ *to basis* C *for* $\mathbb{W}$.

7. $\mathbb{V} = \mathbb{R}^2$, $\mathbb{W} = \mathbb{R}^3$, $T(x, y) = (2x - y, x, y)$, where B and C are the standard bases for $\mathbb{R}^2$ and $\mathbb{R}^3$, respectively.

8. $\mathbb{V} = \mathbb{P}_2$, $\mathbb{W} = \mathbb{R}^3$,

$$T(at^2 + bt + c) = \begin{bmatrix} a - b \\ a \\ 2c \end{bmatrix},$$

where $B = \{t^2, t, 1\}$ and $C = \{\vec{\mathbf{e}}_1, \vec{\mathbf{e}}_2, \vec{\mathbf{e}}_3\}$.

9. $\mathbb{V} = \mathbb{M}_{22}(\mathbb{R})$, $\mathbb{W} = \mathbb{M}_{22}(\mathbb{R})$, $T(\mathbf{A}) = \mathbf{A} + \mathbf{A}^{\mathrm{T}}$ and

$$B = C = \left\{ \begin{bmatrix} 1 & 0 \\ 0 & 0 \end{bmatrix}, \begin{bmatrix} 0 & 1 \\ 0 & 0 \end{bmatrix}, \begin{bmatrix} 0 & 0 \\ 1 & 0 \end{bmatrix}, \begin{bmatrix} 0 & 0 \\ 0 & 1 \end{bmatrix} \right\}.$$

HINT: Note that

$$\begin{bmatrix} a & b \\ c & d \end{bmatrix}_C = \begin{bmatrix} a \\ b \\ c \\ d \end{bmatrix},$$

so $\mathbf{M}_B$ will be a 4×4 matrix.

10. $\mathbb{V} = \mathbb{M}_{22}(\mathbb{R})$, $\mathbb{W} = \mathbb{M}_{22}(\mathbb{R})$, $T(\mathbf{A}) = \begin{bmatrix} \mathrm{Tr}\mathbf{A} & 0 \\ 0 & \mathrm{Tr}\mathbf{A} \end{bmatrix}$ and

$$B = C = \left\{ \begin{bmatrix} 1 & 0 \\ 0 & 0 \end{bmatrix}, \begin{bmatrix} 0 & 1 \\ 0 & 0 \end{bmatrix}, \begin{bmatrix} 0 & 0 \\ 1 & 0 \end{bmatrix}, \begin{bmatrix} 0 & 0 \\ 0 & 1 \end{bmatrix} \right\}.$$

See hint for Problem 9.

11. $\mathbb{V} = \mathbb{W}$ is the solution space for $x'' + 4x' + 4x = 0$, $T(f) = f' - f$ and $B = C = \{e^{-2t}, te^{-2t}\}$.

Changing Bases *In Problems 12–14, determine the matrix associated with a change of bases from basis B to basis C for the given vector space $\mathbb{V}$.*

12. $\mathbb{V} = \mathbb{M}_{21}(\mathbb{R})$,

$$B = \left\{ \begin{bmatrix} 1 \\ 1 \end{bmatrix}, \begin{bmatrix} 3 \\ 0 \end{bmatrix} \right\}, \quad C = \left\{ \begin{bmatrix} -1 \\ 1 \end{bmatrix}, \begin{bmatrix} 0 \\ 2 \end{bmatrix} \right\}.$$

13. $\mathbb{V} = \mathbb{P}_3$, $B = \{t^3, t^2, t, 1\}$, $C = \{2t, t^3, t - t^2, 5\}$.

14. $\mathbb{V} = \mathbb{M}_{22}(\mathbb{R})$,

$$B = \left\{ \begin{bmatrix} 1 & 0 \\ 0 & 0 \end{bmatrix}, \begin{bmatrix} 0 & 1 \\ 0 & 0 \end{bmatrix}, \begin{bmatrix} 0 & 0 \\ 1 & 0 \end{bmatrix}, \begin{bmatrix} 0 & 0 \\ 0 & 1 \end{bmatrix} \right\},$$

$$C = \left\{ \begin{bmatrix} 1 & 0 \\ 0 & 0 \end{bmatrix}, \begin{bmatrix} 1 & 1 \\ 0 & 0 \end{bmatrix}, \begin{bmatrix} 1 & 1 \\ 1 & 0 \end{bmatrix}, \begin{bmatrix} 1 & 1 \\ 1 & 1 \end{bmatrix} \right\}.$$

See hint for Problem 9.

15. Associated Matrix Again Return to Problem 8 but replace the C basis for $\mathbb{R}^3$ by $D = \{\vec{\mathbf{e}}_1, \vec{\mathbf{e}}_1 - \vec{\mathbf{e}}_2, 5\vec{\mathbf{e}}_3 + \vec{\mathbf{e}}_1\}$. Determine the matrix $\mathbf{M}_B^*$ associated with a change from basis B to basis D. HINT: Start with the fact that

$$\begin{aligned} [T(t^2)]_D &= [(1, 1, 0)]_D \\ &= \left[\delta_1 \begin{bmatrix} 1 \\ 0 \\ 0 \end{bmatrix} + \delta_2 \begin{bmatrix} 1 \\ -1 \\ 0 \end{bmatrix} + \delta_3 \begin{bmatrix} 1 \\ 0 \\ 5 \end{bmatrix} \right]_D \\ &= \begin{bmatrix} \delta_1 \\ \delta_2 \\ \delta_3 \end{bmatrix}. \end{aligned}$$

16. Multiplying Associated Matrices Return to Problems 15 and 8, with bases

$$\begin{aligned} B &= \{t^2, t, 1\}, \\ C &= \{\vec{\mathbf{e}}_1, \vec{\mathbf{e}}_2, \vec{\mathbf{e}}_3\}, \\ D &= \{\vec{\mathbf{e}}_1, \vec{\mathbf{e}}_1 - \vec{\mathbf{e}}_2, 5\vec{\mathbf{e}}_3 + \vec{\mathbf{e}}_1\}. \end{aligned}$$

(a) Find $\mathbf{M}_C^*$ for the change of basis from C to D.

(b) Verify that $\mathbf{M}_B^*$ from Problem 15 can be calculated with

$$\mathbf{M}_B^* = \mathbf{M}_C^* \mathbf{M}_B,$$

where $\mathbf{M}_C^*$ is from (a) and $\mathbf{M}_B$ is from Problem 8. Explain why this should be so.

Appendix PF
Partial Fractions

Rational Fractions

The fractions we consider are of the form $p(x)/q(x)$ (**rational** fractions) in lowest terms in which p and q are polynomials in variable x with **real** coefficients and the degree of p is less than the degree of q (the fraction is **proper**). By multiplying numerator and denominator by a suitable constant, we can arrange that the denominator q is **monic** (has leading coefficient 1), and we'll normally assume that this has been done.

It is a consequence of the Fundamental Theorem of Algebra that every real monic polynomial $q(x)$ can be factored (uniquely) into linear factors of the form $x+a$ and irreducible quadratic factors of the form x^2+bx+c with $b^2-4c<0$.[1] Of course, these factors may be repeated; that is, the factorization of q may contain $(x+a)^2$ or $(x^2+bx+c)^3$, and so on.

In analyzing how a rational fraction can be decomposed into a sum of simpler fractions we need to consider four types of factorizations and the "partial fractions" to which they correspond. We assume in each case that monic polynomial $q(x)$ has been factored completely into linear and irreducible quadratic factors.

Case 1: Polynomial $q(x)$ has a **simple** factor

$$x+a;$$

that is, the factor $x+a$ occurs only once in the factorization. To this factor corresponds a partial fraction

$$\frac{A}{x+a},$$

where A is a real constant to be determined.

Case 2: Polynomial $q(x)$ has a linear factor

$$x+a \qquad \text{of multiplicity } k;$$

that is, the factor $x+a$ occurs exactly k times, $k \geq 2$, in the factorization of $q(x)$. To these factors correspond the partial fractions

$$\frac{A_1}{x+a}+\frac{A_2}{(x+a)^2}+\cdots+\frac{A_k}{(x+a)^k},$$

where the real coefficients $A_1, A_2, \ldots, A_k$ are to be determined. (Of course this includes Case 1 if k is allowed to equal 1.)

[1]The condition $b^2-4c<0$ guarantees that x^2+bx+c doesn't have real linear factors; its only factors involve complex numbers.

Case 3: Polynomial $q(x)$ has a simple irreducible quadratic factor

$$x^2 + bx + c;$$

that is, the factor $x^2 + bx + c$ occurs only once in the factorization. To this factor corresponds a partial fraction of the form

$$\frac{Ax + B}{x^2 + bx + c},$$

with real coefficients A and B to be determined.

Case 4: Polynomial $q(x)$ has irreducible quadratic factor

$$x^2 + bx + c \qquad \text{of multiplicity } k \geq 2.$$

To these factors correspond the partial fractions

$$\frac{A_1x + B_1}{x^2 + bx + c} + \frac{A_2x + B_2}{(x^2 + bx + c)^2} + \cdots + \frac{A_kx + B_k}{(x^2 + bx + c)^k},$$

where the real coefficients $A_1, A_2, \cdots, A_k$ and $B_1, B_2, \cdots, B_k$ are to be determined.

EXAMPLE 1 **Patterns** Here are the forms of the partial fraction decompositions for three specific rational fractions:

(a) $$\frac{3x^2 - x}{(x - 1)(x + 2)(x^2 + 4)} = \frac{A}{x - 1} + \frac{B}{x + 2} + \frac{Cx + D}{x^2 + 4}.$$

(b) $$\frac{2x + 3}{(x - 1)^3(x^2 + x + 1)} = \frac{A}{x - 1} + \frac{B}{(x - 1)^2} + \frac{C}{(x - 1)^3} + \frac{Dx + E}{x^2 + x + 1}.$$

(c) $$\frac{3x^3}{(x^2 + 5)^2(x + 5)^2} = \frac{Ax + B}{x^2 + 5} + \frac{Cx + D}{(x^2 + 5)^2} + \frac{E}{x + 5} + \frac{F}{(x + 5)^2}.$$

■

The reason we required q to be monic was to avoid the problem of recognizing that a denominator like $(x - 3)(2x - 6)$ really contains a factor of multiplicity two. Once the correct form of the partial fraction decomposition is determined, it isn't necessary that all leading coefficients be 1, and it may be helpful in some cases to multiply through numerator and denominator by a suitable constant factor to "streamline" the coefficients.

EXAMPLE 2 **Streamlining Coefficients** The rational fraction

$$F(x) = \frac{4x - 1}{2x^2 - x - 1}$$

can be written with monic denominator as

$$F(x) = \frac{2x - 1/2}{x^2 - (1/2)x - 1/2} = \frac{2x - 1/2}{(x - 1)(x + 1/2)},$$

and that denominator has simple linear factors $x - 1$ and $x + 1/2$. Then we can write the partial fraction decomposition as

$$F(x) = \frac{4x - 1}{(x - 1)(2x + 1)} = \frac{A}{x - 1} + \frac{B}{2x + 1}$$

(rather than, say, $A_0/(x - 1) + B_0/(x + 1/2)$).

■

Partial Fraction Decomposition

The determination of the various coefficients in a partial fraction decomposition is made by clearing the general form of fractions (multiplying through by $q(x)$) and, after collection and simplification, equating coefficients of like terms. The result is a system of as many equations as coefficients to be determined. This procedure is illustrated in the examples that follow.

EXAMPLE 3 **Linear Repetition** To resolve

$$\frac{5x^2 - 6x + 4}{x^2(x-1)}$$

into partial fractions, we note in the denominator the simple linear factor $x - 1$ and the linear factor x $(= x - 0)$ of multiplicity two. Hence (for x not equal to 0 or 1),

$$\frac{5x^2 - 6x + 4}{x^2(x-1)} = \frac{A}{x} + \frac{B}{x^2} + \frac{C}{x-1}.$$

Clearing of fractions gives

$$5x^2 - 6x + 4 = Ax(x-1) + B(x-1) + Cx^2,$$

which by continuity must hold for all x. Expanding and collecting terms, we have

$$5x^2 - 6x + 4 = (A + C)x^2 + (-A + B)x - B.$$

Equating coefficients of like terms leads to the system

$$5 = A + C,$$
$$-6 = A + B,$$
$$4 = B,$$

in the three parameters A, B and C, from which we determine

$$B = -4, \quad A = B + 6 = 2 \quad \text{and} \quad C = 5 - A = 3.$$

Therefore,

$$\frac{5x^2 - 6x + 4}{x^2(x-1)} = \frac{2}{x} - \frac{4}{x^2} + \frac{3}{x-1}.$$

EXAMPLE 4 **Quadratic Repetition** The denominator of the rational fraction

$$G(x) = \frac{x^3 - x^2 + 4x}{x^4 + 4x^2 + 4}$$

contains the irreducible quadratic factor $x^2 + 2$ with multiplicity two:

$$G(x) = \frac{x^3 - x^2 + 4x}{(x^2 + 2)^2}.$$

Therefore, by Case 4,

$$G(x) = \frac{x^3 - x^2 + 4x}{(x^2 + 2)^2} = \frac{Ax + B}{x^2 + 2} + \frac{Cx + D}{(x^2 + 2)^2}.$$

Clearing fractions,

$$\begin{aligned}x^3 - x^2 + 4x &= (Ax + B)(x^2 + 2) + (Cx + D)\\ &= Ax^3 + Bx^2 + (2A + C)x + 2B + D.\end{aligned}$$

Therefore,

$$1 = A, \quad -1 = B, \quad 4 = 2A + C \quad \text{and} \quad 0 = 2B + D.$$

Hence $A = 1$, $B = -1$, $C = 4 - 2A = 2$, $D = -2B = 2$ and

$$G(x) = \frac{x - 1}{x^2 + 2} + \frac{2x + 2}{(x^2 + 2)^2}.$$

■

EXAMPLE 5 **Mixed Mode** We resolve

$$\frac{2x^2 - x + 1}{(x^2 + 1)(2x - 1)}$$

into partial fractions, noting that the nonmonic denominator hides no unexpected repetitions. From

$$\frac{2x^2 - x + 1}{(x^2 + 1)(2x - 1)} = \frac{Ax + B}{x^2 + 1} + \frac{C}{2x - 1},$$

we obtain

$$\begin{aligned}2x^2 - x + 1 &= (Ax + B)(2x - 1) + C(x^2 + 1)\\ &= (2A + C)x^2 + (2B - A)x + (C - B).\end{aligned}$$

Therefore,

$$2A + C = 2, \quad 2B - A = -1 \quad \text{and} \quad C - B = 1.$$

This gives the system of equations

$$\begin{aligned}2A \quad\quad + C &= 2,\\ -A + 2B \quad\quad &= -1,\\ -B + C &= 1,\end{aligned}$$

with augmented matrix

$$\left[\begin{array}{rrr|r} 2 & 0 & 1 & 2\\ -1 & 2 & 0 & -1\\ 0 & -1 & 1 & 1\end{array}\right],$$

which has RREF

$$\left[\begin{array}{rrr|r} 1 & 0 & 0 & 3/5\\ 0 & 1 & 0 & -1/5\\ 0 & 0 & 1 & 4/5\end{array}\right].$$

Thus

$$A = \frac{3}{5}, \quad B = -\frac{1}{5} \quad \text{and} \quad C = \frac{4}{5},$$

hence

$$\frac{2x^2 - x + 1}{(x^2 + 1)(2x - 1)} = \frac{(3/5)x - 1/5}{x^2 + 1} + \frac{4/5}{2x - 1} = \frac{3x - 1}{5(x^2 + 1)} + \frac{4}{5(2x - 1)}.$$

■

EXAMPLE 6 **Taking It Literally** To write

$$\frac{-p}{x^2 - 3xp + 2p^2}$$

as a sum of simpler fractions, we factor the denominator as $(x - p)(x - 2p)$ and set up the form

$$\frac{-p}{x^2 - 3xp + 2p^2} = \frac{A}{x - p} + \frac{B}{x - 2p}, \quad x \neq p, x \neq 2p.$$

Clearing of fractions yields

$$-p = A(x - 2p) + B(x - p),$$

valid for all x. Then

$$-p = (A + B)x - 2Ap - Bp,$$

so that by equating constant- and x-coefficients we obtain

$$0 = A + B \quad \text{and} \quad -p = -2Ap - Bp.$$

Hence $A + B = 0$ and $2A + B = 1$, and we find $A = 1$ and $B = -1$. Finally then,

$$\frac{-p}{x^2 - 3xp + 2p^2} = \frac{1}{x - p} - \frac{1}{x - 2p}.$$

Summary

We have shown that the correct form of the partial fraction decomposition of a real proper rational fraction is determined according to the linear and irreducible quadratic factors of its denominator and their multiplicities. The solution for the required coefficients is an exercise in solving systems of linear algebraic equations.

Problems PF

Practice Makes Perfect *Resolve the rational fraction in each of Problems 1–10 into its partial fraction decomposition.*

1. $\dfrac{1}{x(x-1)}$

2. $\dfrac{1}{(x+2)(x-1)}$

3. $\dfrac{x}{(x+1)(x+2)}$

4. $\dfrac{x}{(x^2+1)(x-1)}$

5. $\dfrac{4}{x^2(x^2+4)}$

6. $\dfrac{3}{(x^2+1)(x^2+4)}$

7. $\dfrac{7x-1}{(x+1)(x+2)(x-3)}$

8. $\dfrac{x^2-2}{x(x+7)(x+1)}$

9. $\dfrac{x^2+9x+2}{(x-1)^2(x+3)}$

10. $\dfrac{x^2+1}{x^3-2x^2-8x}$

Appendix SS
Spreadsheets for Systems

Spreadsheet software programs offer an efficient way to organiaze calculation of numerical solutions to initial value problems. This was illustrated earlier for a single differential equations IVP. (See the introduction in Sec. 1.4 to Problems 3–10.)

For *systems* of differential equations (or the iterative equations of Chapter 9), the computations are more tedious and the automation even more welcome.

EXAMPLE 1 **Making Tables** Figure SS.1 shows a sample spreadsheet setup for using Euler's method, with stepsize $h = 0.1$, to solve the IVP

$$\begin{aligned} x' &= xy \\ y' &= y - x^2 + 1 \end{aligned} \qquad x(0) = 0.5,\ y(0) = 0 \tag{1}$$

We entered the following:

Row 1: **column headings**.

Row 2: **initial conditions**, in cells A2, B2, C2; **formulas from the DEs** in cells D2, E2.

Row 3: **formulas** with proper **stepsize** in A3, B3, C3.

Then we selected cells D2:E3 and chose the command to "Fill, Down", which automatically updates the formulas.

◇	A	B	C	D	E
1	t	x	y	xdot	ydot
2	0	0.5	0	=B2*C2↓	=C2-(B2)^2+1 ↓
3	=A2+0.1↓	=B2+0.1*D2↓	=C2+0.1*E2↓	=B3*C3 ↓	=C3-(B3)^2+1 ↓

FIGURE SS.1 Spreadsheet formulas for system (1). Copy Row 3 down as far as you want to go.

Now we are all set. To obtain a complete table of values, we select all of Row 3 and "Fill, Down" as far as we want to go. Figure SS.2 shows the results to $t = 1$.

	A	B	C	D	E
1	t	x	y	xdot	ydot
2	**0.0**	**0.5000**	**0.0000**	0.0000	0.7500
3	**0.1**	**0.5000**	**0.0750**	0.0375	0.8250
4	0.2	0.5038	0.1575	0.0793	0.9037
5	0.3	0.5117	0.2479	0.1268	0.9861
6	0.4	0.5244	0.3465	0.1817	1.0715
7	0.5	0.5425	0.4536	0.2461	1.1593
8	0.6	0.5671	0.5696	0.3230	1.2479
9	0.7	0.5994	0.6943	0.4162	1.3350
10	0.8	0.6411	0.8279	0.5307	1.4169
11	0.9	0.6941	0.9695	0.6730	1.4877
12	**1.0**	**0.7614**	**1.1183**	0.8515	1.5385

FIGURE SS.2 Spreadsheet for system (1), Example 1.

Once a spreadsheet is set up, it is easy to change (e.g., to a different initial condition, or a different stepsize), without having to start all over. Just enter the necessary new data in Rows 2 and 3, then select Row 3 and "Fill, Down" to update the entire spreadsheet.

EXAMPLE 2 **Recycling** To rerun Euler's method for Example 1 with a smaller stepsize, $h = 0.01$, we have only to do the following, shown in Fig. SS.3

- Change the stepsize in cells A3, B3, C3;
- Refill cells D3, E3 from D2, E2 (because their formula entries B3 and C3 have changed).
- Refill down from Row 3 to update all the values, as shown in Fig. SS.4. However, note that we now need more rows (100 steps, not 10) to get to $t = 1$.

	A	B	C	D	E
1	t	x	y	xdot	ydot
2	0	0.5	0	=B2*C2 ↓	=C2-(B2)^2+1 ↓
3	=A2+**0.01** ↓	=B2+**0.01***D2 ↓	=C2+**0.01***E2 ↓	=B3*C3 ↓	=C3-(B3)^2+1 ↓

FIGURE SS.3 Updating a spreadsheet to a new stepsize, $h = 0.01$.

◇	A	B	C	D	E
1	t	x	y	xdot	ydot
2	**0.00**	**0.5000**	**0.0000**	0.0000	0.7500
3	**0.01**	**0.5000**	**0.0075**	0.0038	0.7575
4	0.02	0.5000	0.0151	0.0075	0.7650
5	0.03	0.5001	0.0227	0.0114	0.7726
6	0.04	0.5002	0.0305	0.0152	0.7802
7	0.05	0.5004	0.0383	0.0191	0.7879
8	0.06	0.5006	0.0461	0.0231	0.7956
9	0.07	0.5008	0.0541	0.0271	0.8033
10	0.08	0.5011	0.0621	0.0311	0.8110
11	0.09	0.5014	0.0702	0.0352	0.8188
12	**0.10**	**0.5017**	**0.0784**	0.0393	0.8267
13	0.11	0.5021	0.0867	0.0435	0.8346
14	0.12	0.5026	0.0950	0.0478	0.8425
	:	:	:	:	:
	:	:	:	:	:
98	0.96	0.7893	1.0890	0.8595	1.4660
99	0.97	0.7979	1.1036	0.8806	1.4670
100	0.98	0.8067	1.1183	0.9021	1.4676
101	0.99	0.8157	1.1330	0.9242	1.4676
102	**1.00**	**0.8249**	**1.1477**	0.9468	1.4671

FIGURE SS.4 Updated spreadsheet for Example 2.

As we would expect with Euler's method, Examples 1 and 2, with different stepsizes, give obviously different results. Compare the highlighted values in Figs. SS.2 and SS.4:

- At $t = 0.1$, $h = 0.1$ gives (0.5000, 0.0750),
  $h = 0.01$ gives (0.5017, 0.0784).
- At $t = 1.0$, $h = 0.1$ gives (0.7614, 1.1183),
  $h = 0.01$ gives (0.8249, 1.1477).

Spreadsheets for higher-dimensional systems are made by the same procedures as Examples 1 and 2, by adding columns as necessary.

We are reminded that smaller stepsizes give more accurate approximations, and that error grows as we take more steps from the initial conditions.

Furthermore, if we seek closer accuracy over some distance, we should use a better numerical approximation method, such as Runge-Kutta. This can be done with spreadsheets; it just requires more columns for the formulas to calculate the intermediate slopes. You might consider this Appendix as a peek behind the scenes at exactly what specialized graphical DE solvers are doing—they have simply already built in the formulas for all the columns.

Graphs from Spreadsheets

Phase portraits and time series can be made from spreadsheets by selecting columns and graph type. We shall use our system (1) from Example 1, extended to $t = 3.0$ (for a total of 30 steps), to best illustrate the advantages and disadvantages of spreadsheet graphs. See Fig. SS.5.

◇	A	B	C	D	E
1	t	x	y	xdot	ydot
2	**0.0**	**0.5000**	**0.0000**	0.0000	0.7500
3	**0.1**	**0.5000**	**0.0750**	0.0375	0.8250
4	0.2	0.5038	0.1575	0.0793	0.9037
5	0.3	0.5117	0.2479	0.1268	0.9861
6	0.4	0.5244	0.3465	0.1817	1.0715
7	0.5	0.5425	0.4536	0.2461	1.1593
8	0.6	0.5671	0.5696	0.3230	1.2479
9	0.7	0.5994	0.6943	0.4162	1.3350
10	0.8	0.6411	0.8279	0.5307	1.4169
11	0.9	0.6941	0.9695	0.6730	1.4877
12	**1.0**	**0.7614**	**1.1183**	0.8515	1.5385
13	1.1	0.8466	1.2722	1.0770	1.5554
14	1.2	0.9543	1.4277	1.3625	1.5170
15	1.3	1.0905	1.5794	1.7224	1.3901
16	1.4	1.2628	1.7184	2.1700	1.1238
17	1.5	1.4798	1.8308	2.7092	0.6410
18	1.6	1.7507	1.8949	3.3174	-0.1700
19	1.7	2.0824	1.8779	3.9106	-1.4587
20	1.8	2.4735	1.7320	4.2842	-3.3862
21	1.9	2.9019	1.3934	4.0436	-6.0277
22	2.0	3.3063	0.7906	2.6141	-9.1408
23	2.1	3.5677	-0.1234	-0.4404	-11.8518
24	2.2	3.5236	-1.3086	-4.6111	-12.7247
25	2.3	3.0625	-2.5811	-7.9047	-10.9602
26	2.4	2.2721	-3.6771	-8.3546	-7.8394
27	2.5	1.4366	-4.4611	-6.4088	-5.5249
28	2.6	0.7957	-5.0135	-3.9894	-4.6467
29	2.7	0.3968	-5.4782	-2.1737	-4.6357
30	2.8	0.1794	-5.9418	-1.0661	-4.9740
31	2.9	0.0728	-6.4392	-0.4688	-5.4445
32	3.0	0.0259	-6.9836	-0.1811	-5.9843

FIGURE SS.5 Spreadsheet for system (1), Euler's method, with $h = 0.1$, extended to $t = 3.0$ for Examples 3 and 4. The columns are chosen to plot either phase portrait or time series.

From the spreadsheet in Fig. SS.5 we will create various **Charts**. You may see a code such as

$B$2:$C$32

that represents all the data in adjacent columns B and C, from Rows 2 to 32.

NOTE: The spreadsheet labels **X** and **Y** refer respectively to the horizontal and vertical axes of the graph, which may not coincide with our mathematical spreadsheet labels.

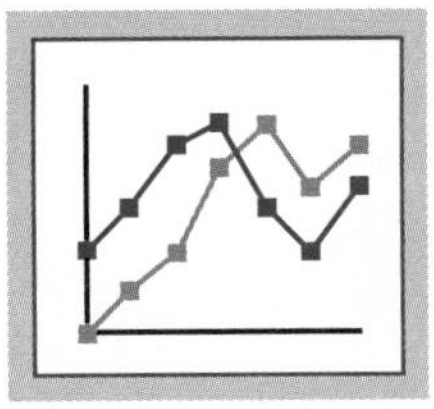

A *phase portrait* is a parametric plot of (x, y) coordinates, so we choose **XY Scatter Plot** in a style that connects the dots in order (with or without individual point highlighting).

EXAMPLE 3 **Phase Portrait** Using the extended spreadsheet of Example 1 as shown in Figure SS.5, we ask for a Chart and choose the XY Scatter option to produce the graph shown in Figure SS.6.

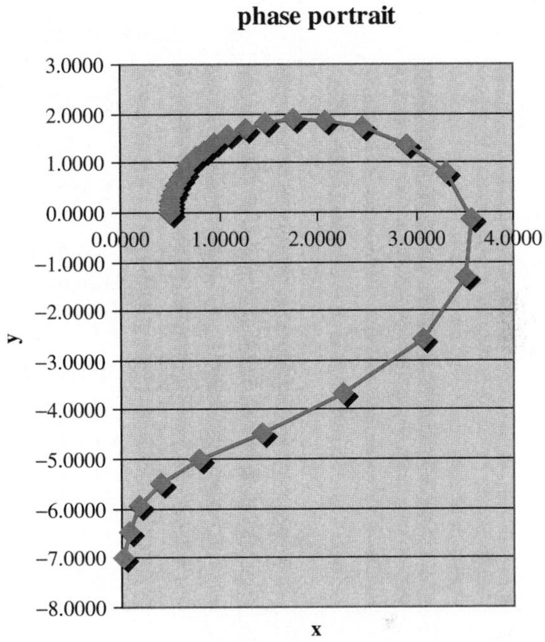

FIGURE SS.6 Phase portrait for system (1), Euler's method, $h = 0.1$, to $t = 3.0$. An arrow is added to show direction in which the trajectory proceeds. ■

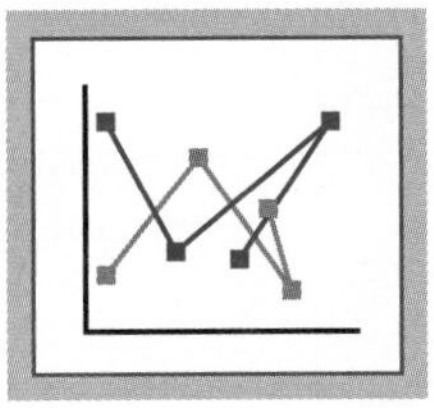

For *time* series, we choose **Line** graphs, which will graph each column vertically against its line number in the spreadsheet. The $x(t)$ graph will be shown with one color and symbol, the $y(t)$ graph with another.

EXAMPLE 4 **Time Series** For system (1) we again use the selected columns B and C of the spreadsheet shown in Fig. SS.5.

However for time series we choose the Line option to obtain the chart shown in Fig. SS.7, which graphs $x(t)$ and $y(t)$ separately. Note that the t axis is labelled in *steps*, rather than in t values.

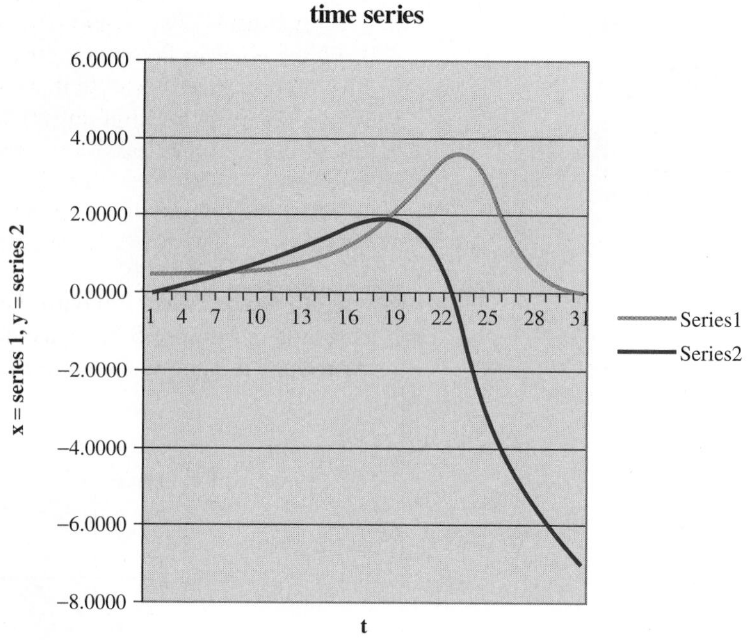

FIGURE SS.7 Time series for system (1), Euler's method, $h = 0.1$, to $t = 3.0$, hence for 30 steps.

EXAMPLE 5 **Three Hundred Steps** We return to system (1), but use the spreadsheet of Example 2, with smaller stepsize $h = 0.01$. We want to make the same charts we made in Examples 3 and 4, and see how they differ. In order to reach $t = 3.0$, we now need 300 steps, but the process is exactly the same and takes no additional work beyond choosing the B and C columns down to Row 302.

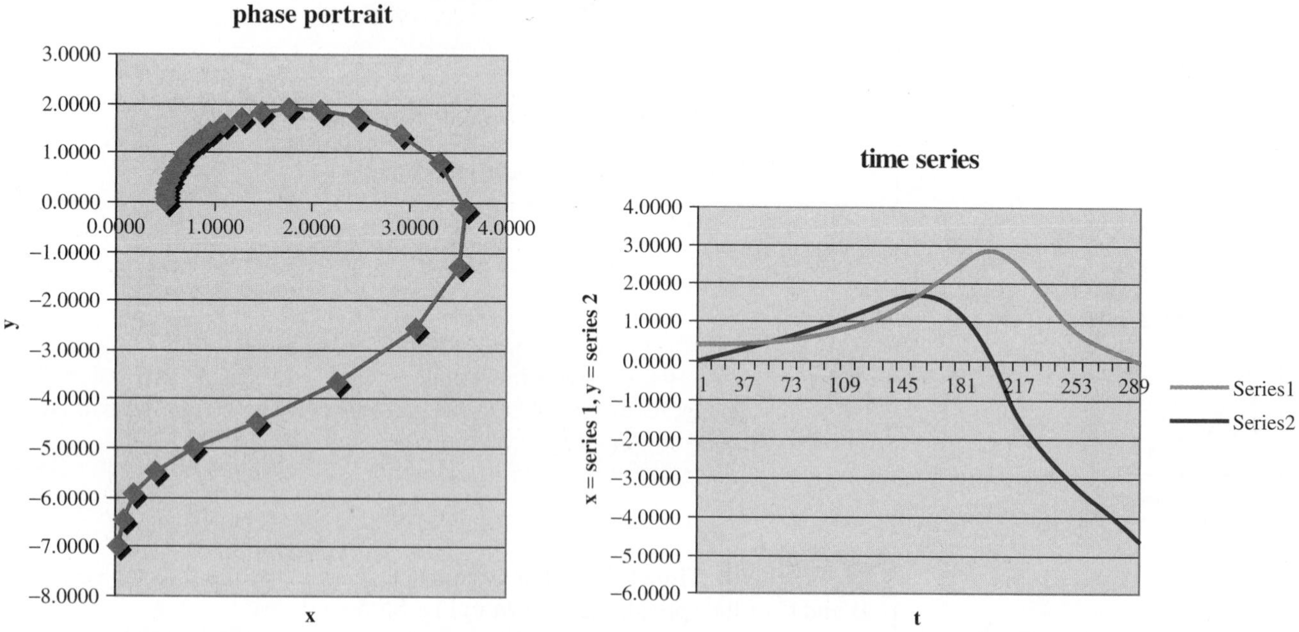

FIGURE SS.8 Charts for system (1), Euler's method, with smaller stepsize $h = 0.01$, to $t = 3.0$, hence for 300 steps.

It is not entirely easy to compare the graphs of Figs. SS.6 and SS.7 with those of Fig. SS.8 that used a smaller stepsize, because the spreadsheet charts adjust their dimensions to the values in the columns, which results in axes on different scales. Nevertheless, the graphs do help to interpret the data in a spreadsheet.

A pictorial comparison of our two approximations with different stepsizes is however easy to obtain with a graphic DE solver, designed to allow us to see many trajectories at once, all on the same scale. See Fig. SS.9.

The lesson, for *differential* equations, is that a *small* stepsize must be used with Euler's method, and the further one wishes to go with t, the smaller will be the stepsize necessary to obtain a good approximation.

However, with *iterative* equations, there is no issue of stepsize, because the discrete step is *fixed* at

$$\Delta n = 1.$$

Hence there is *no* inaccuracy involved in using a spreadsheet to calculate and plot iterative trajectories.

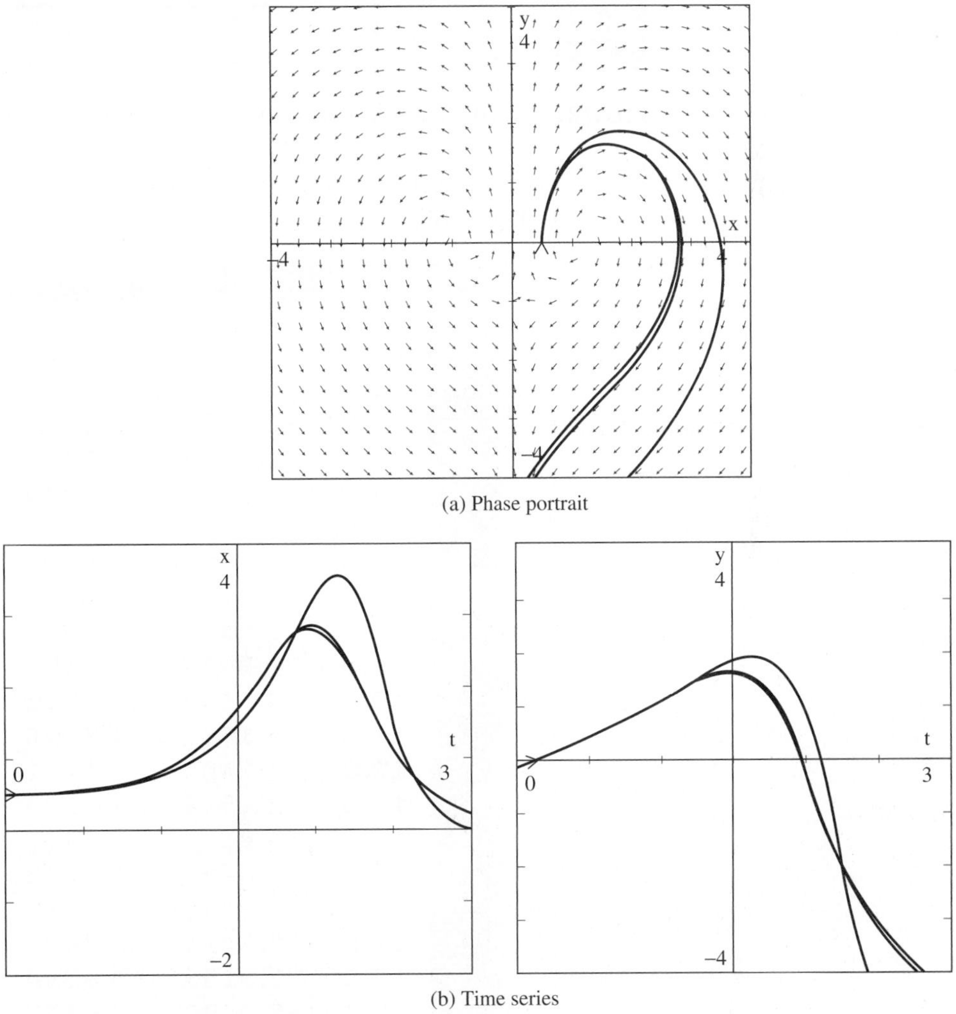

FIGURE SS.9 Comparison, with a graphic DE solver, for system (1) using Euler's method to $t = 3.0$, with stepsizes $h = 0.1$ (outer curve) and $h = 0.01$ (middle curve). The curves in color represent the exact solution, drawn by Runge-Kutta with stepsize $h = 0.1$.

Iterative Systems

See Sec. 9.1, Examples 4 and 7 to set the stage for entering first- and second-order iterative systems. The following example is for a 2-dimensional system of iterative equations.

EXAMPLE 6 **Iterative System** For the iterative IVP

$$\begin{aligned} x_{n+1} &= 1.05y_n \\ y_{n+1} &= -1.1x_n \end{aligned} \qquad x_0 = 2,\ y_0 = 1, \tag{2}$$

we set up a spreadsheet by exactly the same procedure as in Examples 1 and 2.

However, the formulas in B and C are simpler for the iterative system than those for the differential equation system (because there is no stepsize within these cell entries), and the formulas in D and E must be changed to reflect system (2).

	t	x_n	y_n	x_(n+1)	y_(n+1)
2	0	2	1	=1.05*C2	=-1.1*(B2)
3	=A2+1	=D2	=E2	=1.05*C3	=-1.1*(B3)

FIGURE SS.10 Spreadsheet setup for iterative system (2) of Example 5.

The values that result when Row 3 is filled down to Row 22 are shown in Figure SS.11.

◇	A	B	C	D	E
1	t	x_n	y_n	x_(n+1)	y_(n+1)
2	0.00	2.0000	1.0000	1.0500	-2.2000
3	1.0	1.0500	-2.2000	-2.3100	-1.1550
4	2.0	-2.3100	-1.1550	-1.2128	2.5410
5	3.0	-1.2128	2.5410	2.6681	1.3340
6	4.0	2.6681	1.3340	1.4007	-2.9349
7	5.0	1.4007	-2.9349	-3.0816	-1.5408
8	6.0	-3.0816	-1.5408	-1.6178	3.3898
9	7.0	-1.6178	3.3898	3.5592	1.7796
10	8.0	3.5592	1.7796	1.8686	-3.9152
11	9.0	1.8686	-3.9152	-4.1109	-2.0555
12	10.0	-4.1109	-2.0555	-2.1582	4.5220
13	11.0	-2.1582	4.5220	4.7481	2.3741
14	12.0	4.7481	2.3741	2.4928	-5.2229
15	13.0	2.4928	-5.2229	-5.4841	-2.7420
16	14.0	-5.4841	-2.7420	-2.8791	6.0325
17	15.0	-2.8791	6.0325	6.3341	3.1671
18	16.0	6.3341	3.1671	3.3254	-6.9675
19	17.0	3.3254	-6.9675	-7.3159	-3.6580
20	18.0	-7.3159	-3.6580	-3.8408	8.0475
21	19.0	-3.8408	8.0475	8.4499	4.2249
22	20.0	8.4499	4.2249	4.4362	-9.2949

FIGURE SS.11 Spreadsheet values for iterative system (2) of Example 5. ■

Charts for an iterative system are made in the same way as in Examples 3 and 4, using the columns B and C that are highlighted in Fig. SS.11.

- For an *xy*-trajectory, choose XY Scatter Plot, connecting the points to see the order in which they are plotted.
- For time series, choose Line graphs, again connecting the points to clarify the patterns, since both x_n and y_n are plotted on the same chart.

EXAMPLE 7 **Iterative Graphs** From the iterative IVP of Example 5 we obtain the charts shown in Figures SS.12 and SS.13.

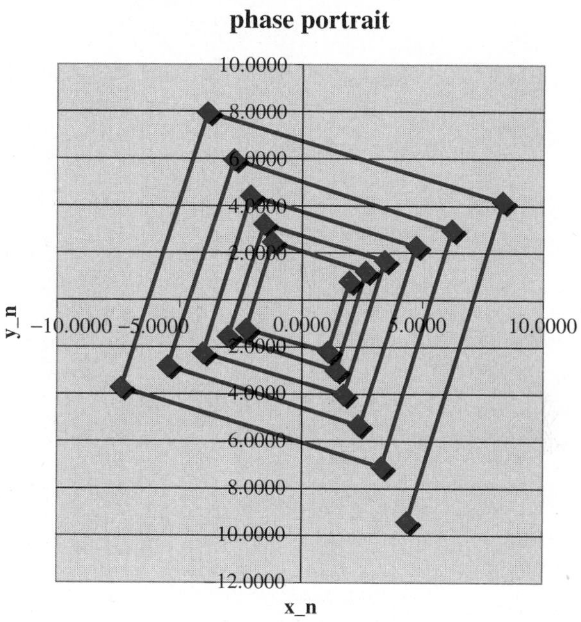

FIGURE SS.12 Trajectory for iterative system (2) of Example 5, for $n = 0$ to 20. An arrow has been added to show that iterates are spiraling outward, away from the origin.

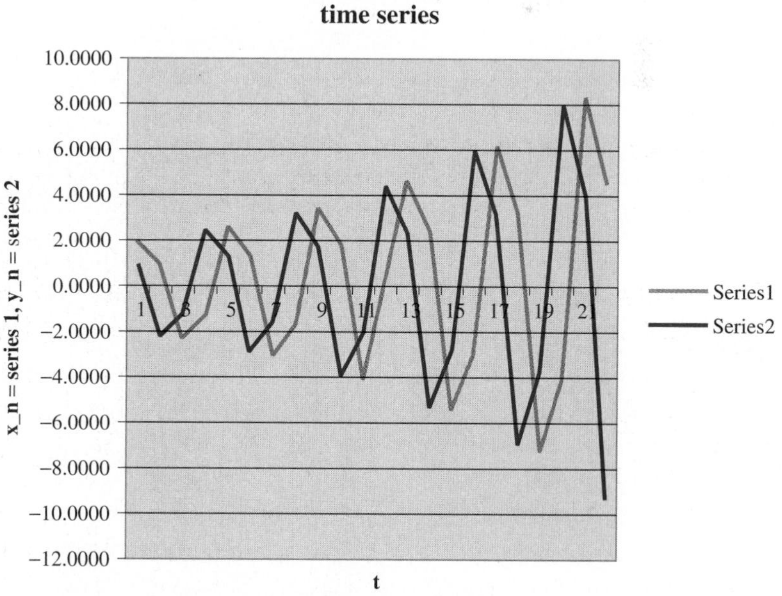

FIGURE SS.13 Time series for iterative system (2) of Example 5, for $n = 0$ to 20. ■

Bibliography

1. E. Ackerman, *et. al.*, "Blood Glucose Regulation and Diabetes," Chapter 4 of *Concepts and Models of Biomathematics* (NY: Marcel Dekker, 1969).
2. Howard Anton and Chris Rorres, *Linear Algebra*, 7th edition (Wiley and Sons, 1994).
3. Martha Boles and Rochelle Newman, *Universal Patterns*, rev. ed. (Bedford, MA: Pythagorean Press, 1990).
4. Robert Borrelli and Courtney Coleman, "Computers, Lies and the Fishing Season," *College Mathematics Journal* **25** no. 5 (1994).
5. Robert Borrelli and Courtney Coleman, *Differential Equations: A Modeling Perspective* (Wiley, 1998).
6. Martin Braun, *Differential Equations and Their Applications* (NY: Springer-Verlag, 1975); 3rd edition (NY: Springer-Verlag, 1983).
7. Otto Bretscher, *Linear Algebra with Applications* (Upper Saddle River, NJ: Prentice-Hall, 1997); 2nd edition (Prentice-Hall, 2001).
8. Roger Brockett, *Finite Dimensional Linear Systems* (NY: Wiley, 1970).
9. R. Bulirsch and J. Stoer, *Introduction to Numerical Analysis* (NY: Springer-Verlag, 1991).
10. Colin J. Campbell and Jean H. Laherrère, "The End of Cheap Oil," *Scientific American* **March** (1998).
11. C-ODE-E Consortium, *ODE Architect*, DE software with a companion book (Wiley, 1998).
12. William C. Dement, MD, *The Promise of Sleep* (Delacorte Press, Random House, 1999).
13. J. R. Dormand and P. J. Prince, "A Family of Embedded Runge-Kutta Formulae," *Journal of Computational and Applied Mathematics* **6** (1980), pp. 19–26.
14. Leah Edelstein-Keshet, *Mathematical Models in Biology* (NY: Random House/Birkhauser, 1988).
15. Mitchell Feigenbaum, "Quantitative Universality for a Class of Nonlinear Transformations," *Journal of Statistical Physics* **19** (1978).
16. Robert E. Gaskell, *Engineering Mathematics* (Dryden Press, 1958).
17. James Gleick, *Chaos: Making a New Science* (Viking, 1987) p. 141.
18. George Green, "An Essay on the Application of Mathematical Analysis to the Theories of Electricity and Magnetism," Nottingham Subscription Library (1828).
19. O. Gurel and O. E. Roessler, *Annals of the New York Academy of Sciences* **316** (1979), p. 376.
20. S. Habre and J. McDill, *Classification of Two-By-Two Iterative Systems*, (2006).
21. J. Higgins, "A Chemical Mechanism for Oscillation of Glycolytic Intermediates in Yeast Cells," *Proceedings of the National Academy of Sciences* (USA) **51** (1964), pp. 989–994.

22. Leslie M. Hocking, *Optimal Control: An Introduction to the Theory and Applications* (Oxford, 1991).
23. Arun V. Holden, ed., *Chaos* (Princeton University Press, 1986).
24. Arun V. Holden and M. A. Muhamad, "A Graphical Zoo of Strange and Peculiar Attractors;" in *Chaos*, ed. A. V. Holden (Princeton University Press, 1986).
25. Philip J. Holmes, "A Nonlinear Oscillator with a Strange Attractor," *Philo. Trans. R. Soc. London A* **292** (1979), pp. 419–448.
26. John H. Hubbard, "What It Means to Understand a Differential Equation," *College Mathematics Journal* **25** no. 5 (1994), pp. 372–384.
27. John H. Hubbard and Beverly H. West, *Differential Equations: A Dynamical Systems Approach, Part 1: Ordinary Differential Equations* (TAM 5, NY: Springer-Verlag, 1991).
28. John H. Hubbard and Beverly H. West, *Differential Equations: A Dynamical Systems Approach, Part 2: Higher Dimensional Systems* (TAM 18, NY: Springer-Verlag, 1995).
29. John H. Hubbard and Beverly H. West, *MacMath* (Springer-Verlag, 1993).
30. Thomas Kailath, *Linear Systems* (NY: Prentice-Hall, 1980).
31. Daniel Kaplan and Leon Glass, *Understanding Nonlinear Dynamics* (NY: Springer-Verlag, 1995).
32. E. M. Landesman and M. R. Hestenes, *Linear Algebra for Mathematics, Science and Engineering* (Prentice-Hall, 1991).
33. David C. Lay, *Linear Algebra and Its Applications* (Reading, MA: Addison-Wesley, 1994).
34. Lengyel, Rabai, and Epstein, *Journal of the American Chemical Society* **112** (1990), p. 9104.
35. Leonardo of Pisa, *Liber Abaci* (1202).
36. Edward Lorenz, "Deterministic Nonperiodic Flow," *Journal of Atmospheric Sciences* **20** (1963), p. 130.
37. Alfred J. Lotka, *Elements of Physical Biology* (reprinted in 1956 by Dover as *Elements of Mathematical Biology*).
38. P. A. Mackowiak, S. S. Wasserman, and M. M. Levine, "A Critical Appraisal of 98.6 Degrees F, the Upper Limit of the Normal Body Temperature, and Other Legacies of Carl Reinhold August Wunderlich," *Journal of the American Medical Association* **268**, **12** (23–30 September 1992), pp. 1578–80.
39. Robert M. May, "Simple Mathematical Models with Very Complicated Dynamics," *Nature* **261** (1976), pp. 459–467.
40. Robert M. May and G. F. Oster, "Bifurcation and Dynamic Complexity in Simple Ecological Models," *The American Naturalist* **100** (1976), p. 573.
41. Z. A. Melzak, *Bypasses: A Simple Approach to Complexity* (NY: Wiley, 1983).
42. Francis C. Moon, *Chaotic Vibrations* (NY: Wiley, 1987).
43. Foster Morrison, *The Art of Modeling Dynamic Systems* (NY: Wiley-Interscience, 1991).
44. James R. Newman, *The World of Mathematics* (NY: Simon & Schuster, 1956).
45. Grégoire Nicolis and Ilya Prigogine, *Exploring Complexity* (San Francisco, CA: Freeman, 1989).
46. Lev Semonovich Pontryagin, V. G. Boltyanskii, R. V. Gamkrelidze, and E. F. Mishchenko, *The Mathematical Theory of Optimal Processes*. Translation

by D. E. Brown (NY: Macmillan, 1964); Translation by K. Trirogoff (NY: Gordan & Breach Science Publishers, 1986).

47. Ilya Prigogine and R. Lefever, "Symmetry Breaking Instabilities in Dissipative Systems," *Journal of Chemistry and Physics* (1968), pp. 1695–1700.
48. Otto E. Roessler, "An Equation for Continuous Chaos," *Physics Letters* **57A** no. 5 (1976), pp. 397–398.
49. Chip Ross and Jody Sorensen, "Will the Real Bifurcation Diagram Please Stand Up!" *College Mathematics Journal* **31** (2000), pp. 3–14.
50. A. N. Sarkovskii, "Coexistence of Cycles of a Continuous Map of a Line into Itself," *Ukrainian Mathematics Journal* **16** (1964), p. 61.
51. J. Maynard Smith, *Mathematical Ideas in Biology* (Cambridge: Cambridge University Press, 1968).
52. Ian Stewart, "The Lorenz Attractor Exists," *Nature* **August 31** (2000), pp. 948–949.
53. Gilbert Strang, *Linear Algebra and its Applications*, 3rd edition (Harcourt Brace Jovanovich, 1988).
54. Steven H. Strogatz, *Nonlinear Dynamics and Chaos* (Reading, MA: Addison-Wesley, 1994; Perseus Press, 2001).
55. Kazuhisa Tomita and Hiroaki Daido, "Possibility of Chaotic Behavior and Multi-basins in Forced Glycolytic Oscillator," *Physics Letters* **79A** (1980), pp. 133–137.
56. Kazuhisa Tomita and Tohru Kai, "Stroboscopic Phase Portrait and Strange Attractors," *Physics Letters* **66A** (1978), pp. 91–93.
57. N. B. Tufillaro and A. M. Albano, "Chaotic Dynamics of a Bouncing Ball," *American Journal of Physics* **54** no. 10 (1986), pp. 939–944.
58. M. Viana, "What's New on Lorenz Strange Attractors?" *Mathematical Intelligencer* **22** no. 3 (2000), pp. 6–9.
59. Andrew Watson, "The Perplexing Puzzle Posed by a Pile of Apples," *New Scientist* **Dec. 14** (1991), p. 19.
60. Robert Weinstock, *Calculus of Variations* (NY: McGraw-Hill, 1952).
61. A. Wolf and T. Bessoir, "Diagnosing Chaos in the Space Circle," *Physica D* **50** (1991), pp. 239–258.
62. James Yorke, "Period Three Implies Chaos," *American Mathematical Monthly* **82** (1975), pp. 985–992.

Answers to Selected Problems

In order to give students quick feedback on their efforts, short (often partial) answers with sample graphs are provided for approximately half of the problems. Exceptions are proofs and open-ended project questions, but in these cases we often provide guidance for at least one problem in a group.

More detail is available in the Student Solutions Manual.

CHAPTER 1

Section 1.1, p. 9

1. $\dfrac{dA}{dt} = kA$

3. $\dfrac{dP}{dt} = kP(20{,}000 - P)$

5. $\dfrac{dG}{dt} = k\dfrac{N}{A}$

6. $d = vt$, where d = distance traveled, v = average velocity and t = time elapsed.

8. (a) Replacing $e^{0.03} \approx 1.03045$ gives $y = 0.9(1.03045)^t$, which increases roughly 3% per year.

10. (a) Population increased exponentially and food supply arithmetically.

(b) The model cannot last forever since the population approaches infinity and reality would produce some limitation; the model does not take under consideration starvation, wars, etc., which slow growth.

(c) A linear growth model for food supply fails to account for technological innovations, such as mechanization, pesticides and genetic engineering.

(d) An exponential model is sometimes reasonable with simple populations over short periods of time.

12. $dy/dt = y(k - cy)$. As y increases, the first factor grows larger, but the second factor becomes smaller as y grows toward k/c.

Section 1.2, p. 20

Sample verification for Problems 1–8:

1. $y = 2\tan t$, so $y' = 4\sec^2 2t$. Substitution into the DE gives a trigonometric identity.

7. $c = 2$

9. Solutions are $y = ce^{2t}$.

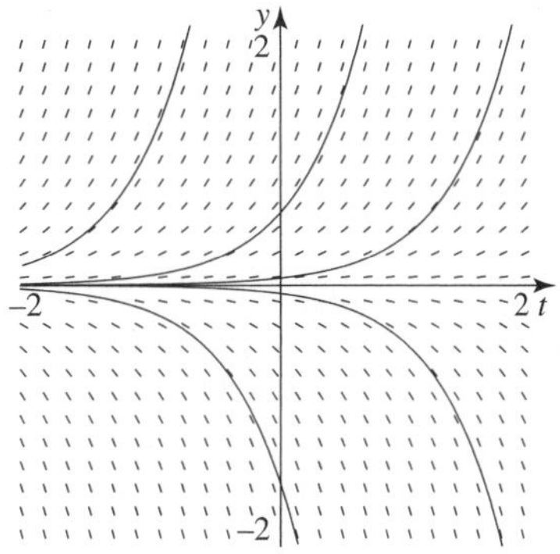

11. Solutions are $y = t - 1 + ce^{-t}$.

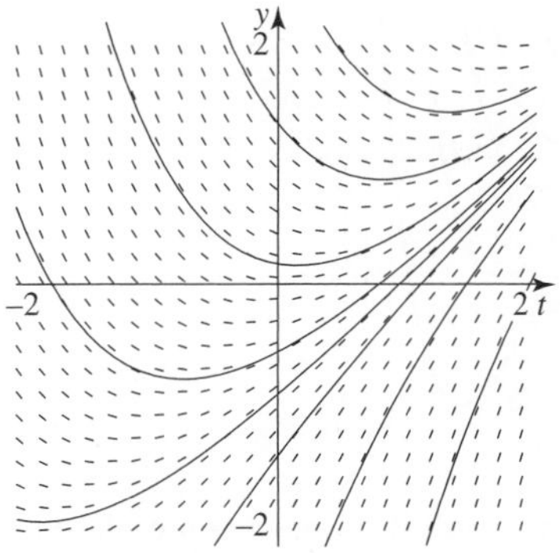

13. $y = 1$ is a *stable* equilibrium.

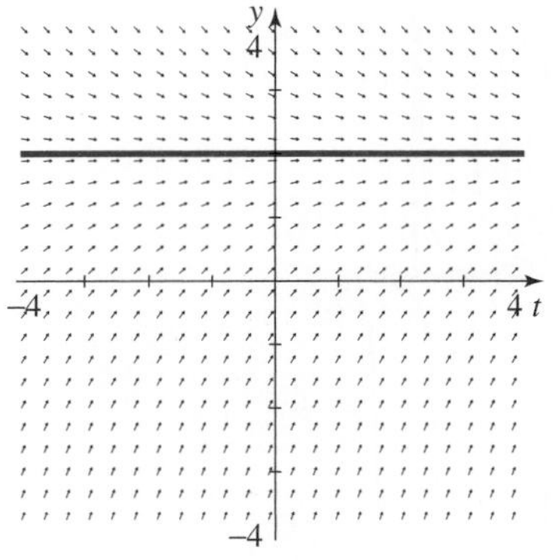

15. $y = 1$ is a *stable* equilibrium.
$y = -1$ is an *unstable* equilibrium.

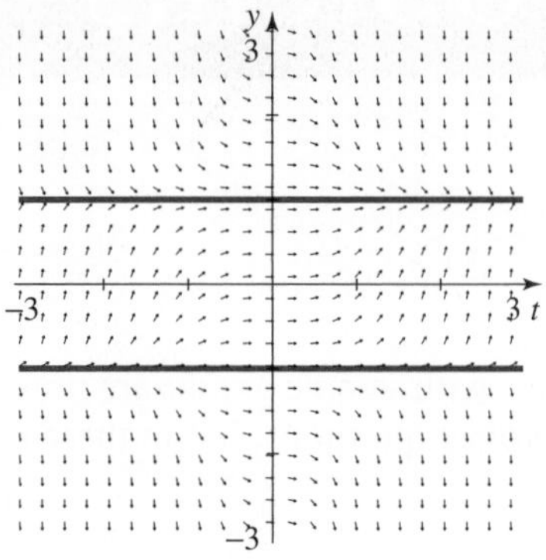

16. (C)

18. (F)

20. (E)

23. $y'' = y + t^2 + 2t$
Inflection points occur on the curve

$$y = -t^2 - 2t.$$

Solutions are concave down below this parabola, in shaded region.

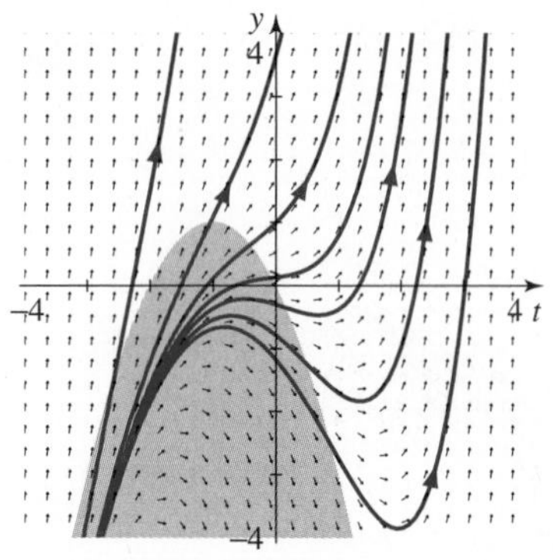

25. We expect a (different) vertical asymptote for each solution

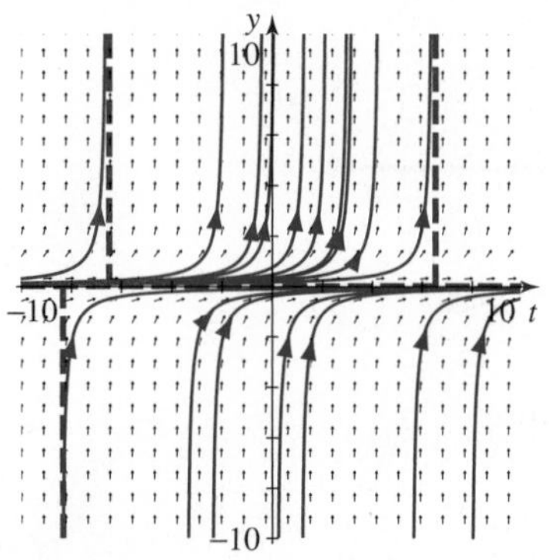

27. There are no vertical asymptotes

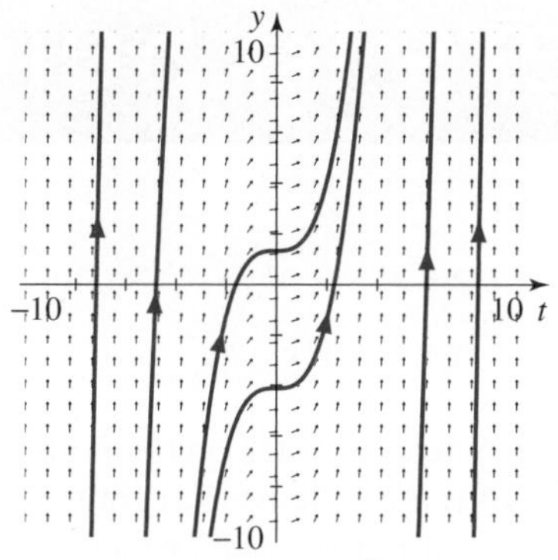

29. We have a horizontal asymptote at $y = \frac{1}{2}$.

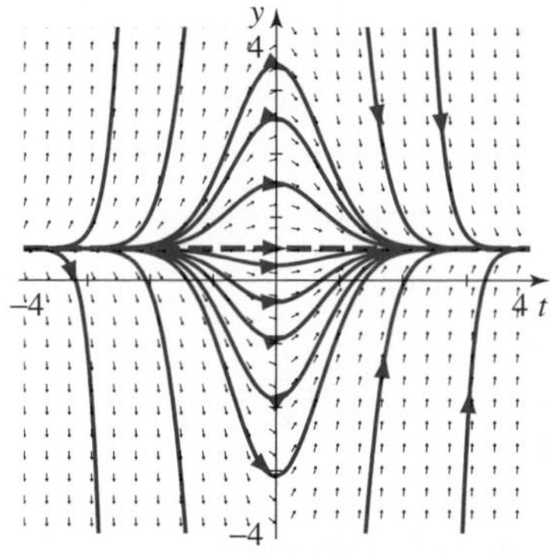

31.

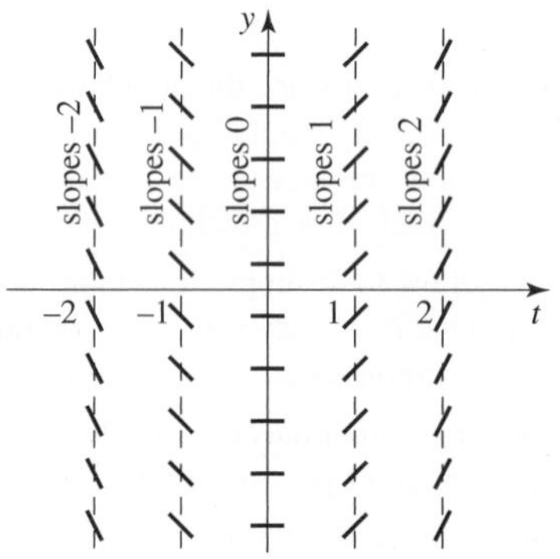

33.

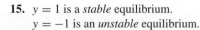

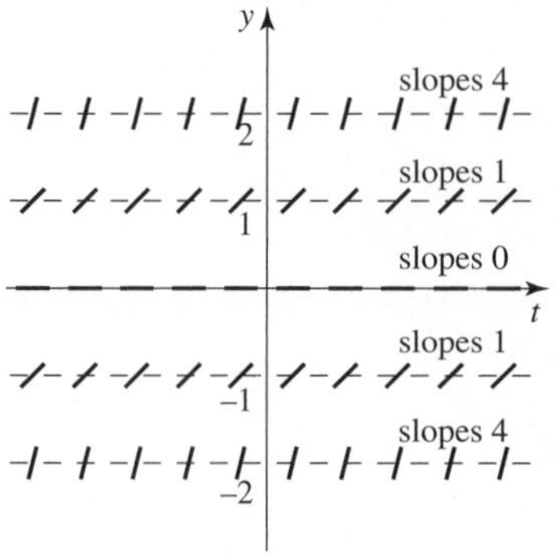

35.

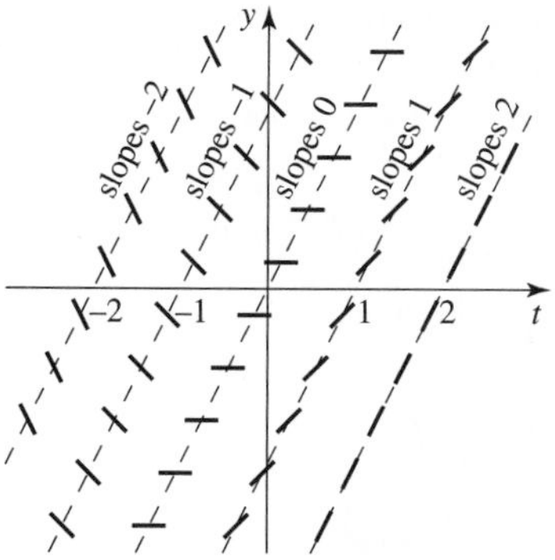

37.

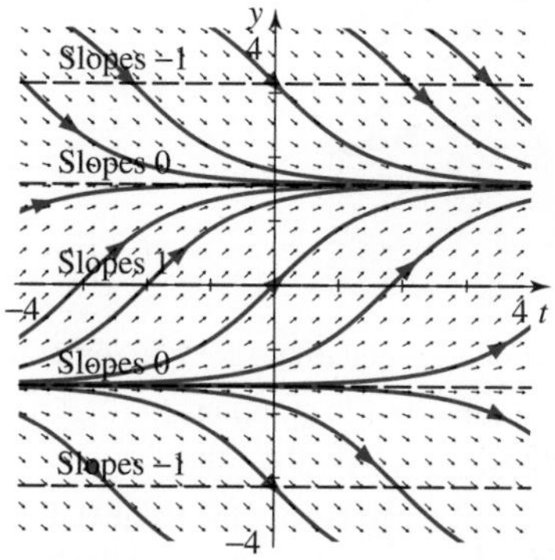

39.

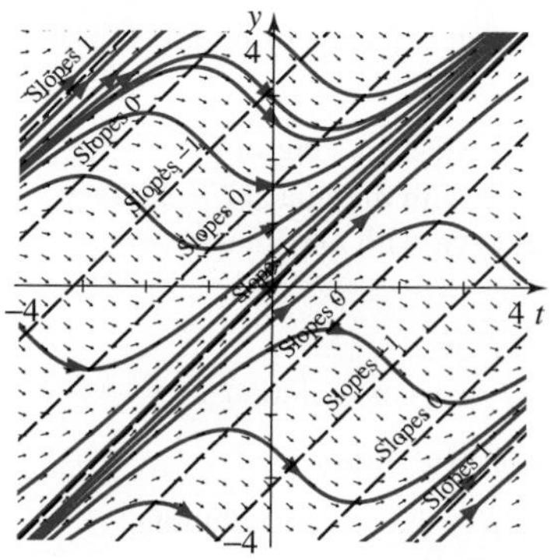

41. Although there is a periodic pattern to the direction field, the *solutions* are *not* periodic.

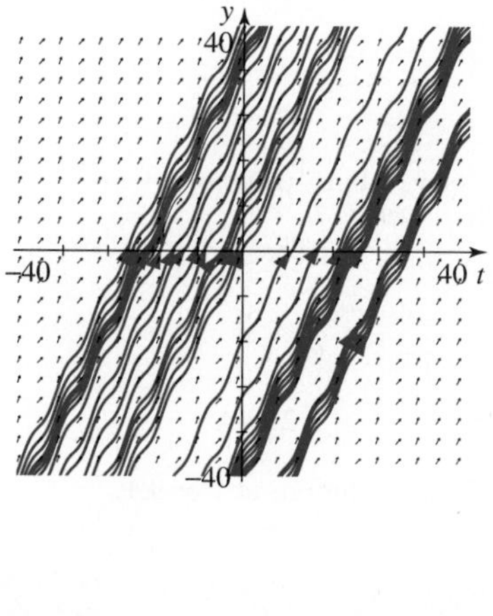

43. The solutions os
ever upward. He

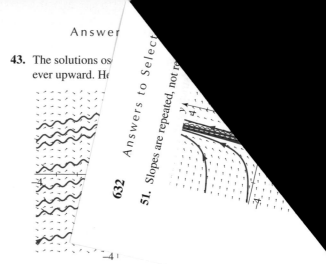

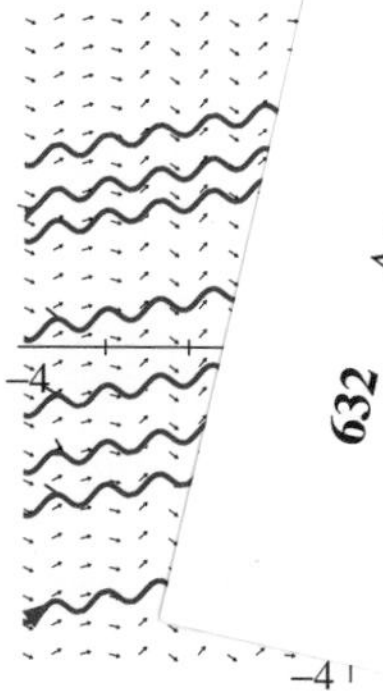

45. The solutions are *not* periodic, despite a
the DE.

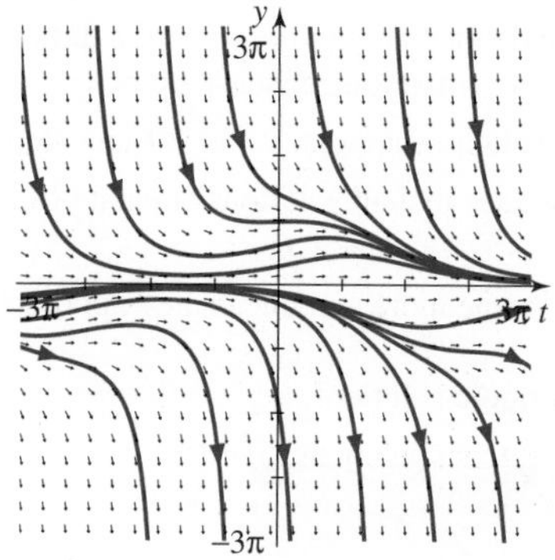

47. Although slope *values* are symmetric about the horizontal axis, the direction field and solutions are not.

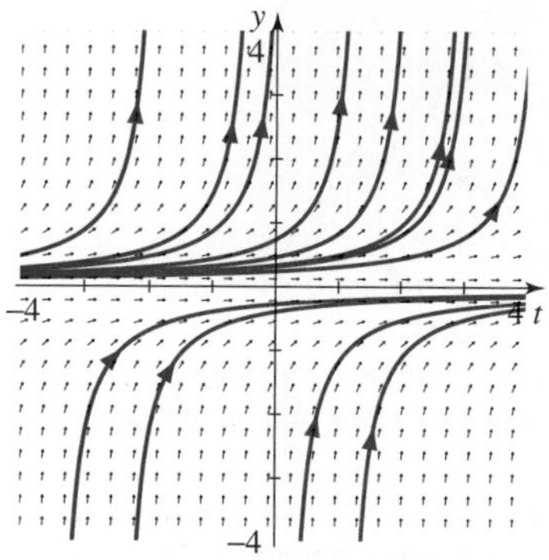

49. The direction field and solutions have pictorial symmetry about the vertical axis.

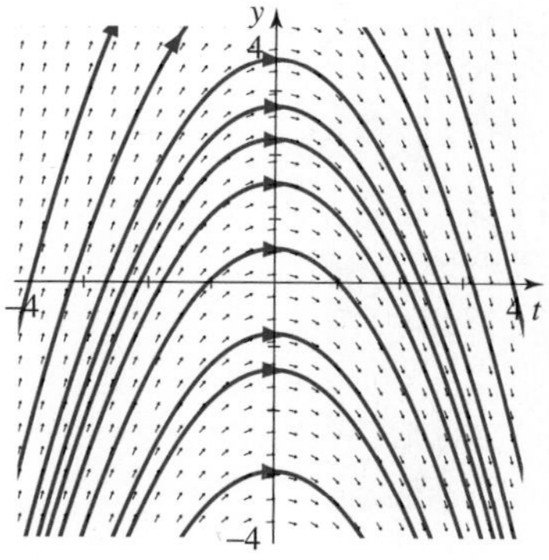

flected, about $t = -1$.

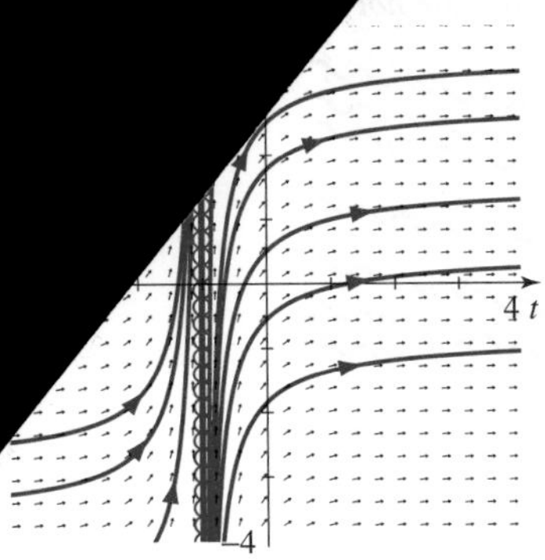

53. (d) $y = -e^{2t} + 3e^{-t}$

55. (a), (c) There are no constant or straight line solutions.

(b), (d) The DE is undefined along $y = -t$, and solutions are concave down above that line, concave up below.

(e) As $t \to \infty$, all solutions approach $y = -t$ and stop.

(f) As $t \to -\infty$, we see that all solutions emanate from $y = -t$.

(g) All solutions become more vertical as they approach $y = -t$.

There are *no* periodic solutions.

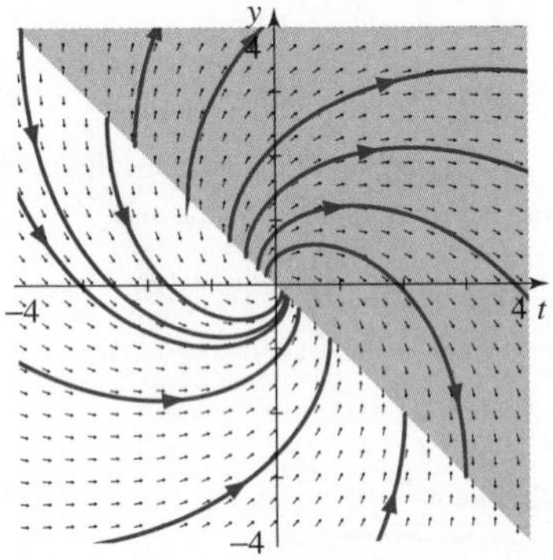

57. The answers include the following observations.

The line $y = t - 1$ is a solution, and an oblique asymptote in backward time.

Along $y = t$ the DE is undefined, and solutions above $y = t - 1$ approach $y = t$ ever more vertically.

Solutions below $y = t - 1$ approach ∞ as $t \to \infty$

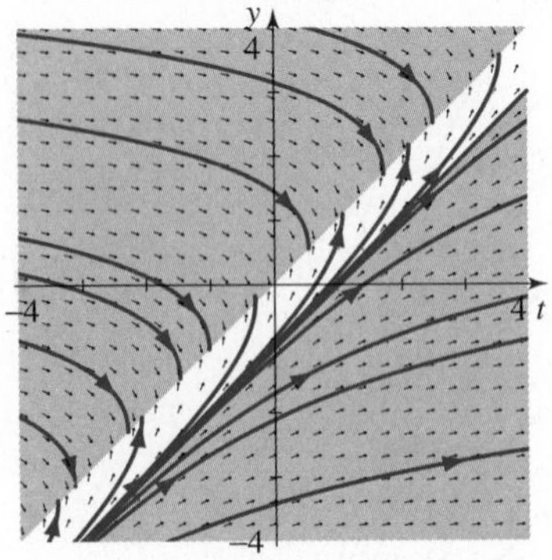

59. Answers include the following observations.

The DE is not defined for $t = 0$.

Concavity changes along the parabola

$$\left(t - \frac{1}{16}\right) = \left(y - \frac{1}{4}\right)^2$$

Solutions are concave down in the shaded portion of the figure.

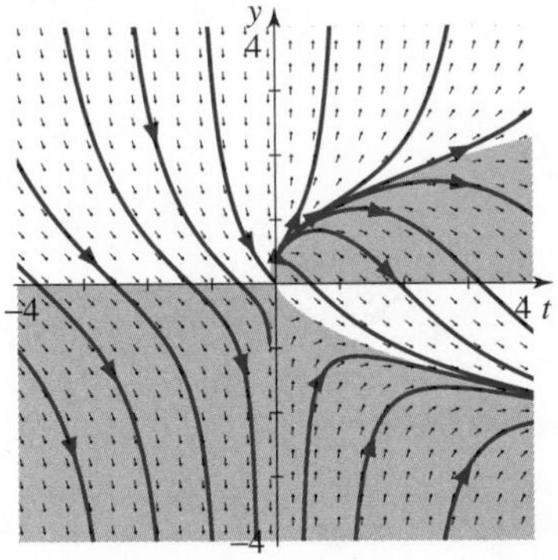

61. (b) Isoclines for autonomous equations are horizontal lines.

63. The basin of attraction for $y \equiv 1$ is all points in $(0, \infty)$. For $y \equiv 0$ the basin of attraction is only the value 0.

65. The basin of attraction for $y \equiv 0$ is only the value 0. For $y \equiv 1$ the basin of attraction is $(0, 2)$. For $y \equiv 2$ the basin of attraction is only the value 2.

Sample description of solution behavior from direction fields for Problems 67–74:

69. For $y' = ty$ there is one constant solution $y \equiv 0$; for $y > 0$ solutions are concave up with minima at $t = 0$; for $y > 0$ solutions are concave down with maxima at $t = 0$.

Section 1.3, p. 29

1. Separable, $\dfrac{dy}{1+y} = dt$; constant solution $y \equiv -1$.

3. Not separable; no constant solutions.

5. Separable, $e^{-y}dy = e^t\,dt$; no constant solutions.

7. Separable, $e^{-y}(y+1)dy = e^t\,dt$; no constant solutions.

9. Not separable; no constant solutions.

11. $\dfrac{1}{2}y^2 = \dfrac{1}{3}t^3 + c$

13. $\dfrac{1}{5}y^5 - y^4 = \dfrac{1}{3}t^3 + 7t + c$

15. $e^{\ln|y|} = e^{c+\sin t}$

17. $y(t) = -\sqrt{-2t^2 + 2t + 4}$

18. $y(t) = \dfrac{2(1 - e^{4t})}{1 + e^{4t}}$

21. $\tan y = t\ln t - t + c$ or $y = \tan^{-1}(t\ln t - t + c)$

23. $e^{-y} = -\dfrac{t^2e^{2t}}{2} + \dfrac{te^{2t}}{2} - \dfrac{e^{2t}}{4} + c$ or

$$y = -\ln\left(-\frac{t^2e^{2t}}{2} + \frac{te^{2t}}{2} - \frac{e^{2t}}{4} + c\right)$$

25. (C) **27.** (E) **29.** (A)

31. $y = \dfrac{ke^{2t} - 1}{ke^{2t} + 1}$; equilibrium solutions at $y = \pm 1$.

33. $y = \pm\sqrt{\dfrac{1}{1+ke^{2t}}}$; equilibrium solutions at $y = 0, \pm 1$.

35. At $(1, 1)$, $y = e^{t-1}$.
At $(1, -1)$, $y = -e^{t+1}$.

37. At $(1, 1)$, $y = \left(3\sqrt{1+t^2} + 1 - 3\sqrt{2}\right)^{1/3}$.
At $(-1, -1)$, $y = \left(3\sqrt{1+t^2} - 1 - 3\sqrt{2}\right)^{1/3}$.

39. At $(1, 1)$, $y - \ln|y+1| = t^2 - \ln 2$.
At $(-1, -1)$, $y = -1$ is the equilibrium solution.

41. $y = t\ln|t| + ct$

43. $\ln|t| = \dfrac{1}{4}\ln\left[\left(\dfrac{y}{t}\right)^4 + 1\right] + c$

45. $y(t) = \tan(t+c) - t$

47. (a) Problems 1, 2 and 17 are autonomous; the others are nonautonomous.
(b) Isoclines of autonomous equations are horizontal lines.

49. $x^2 + 2y^2 = c$, a family of ellipses

51. $y^2 - x^2 = c$, a family of hyperbolas

53.

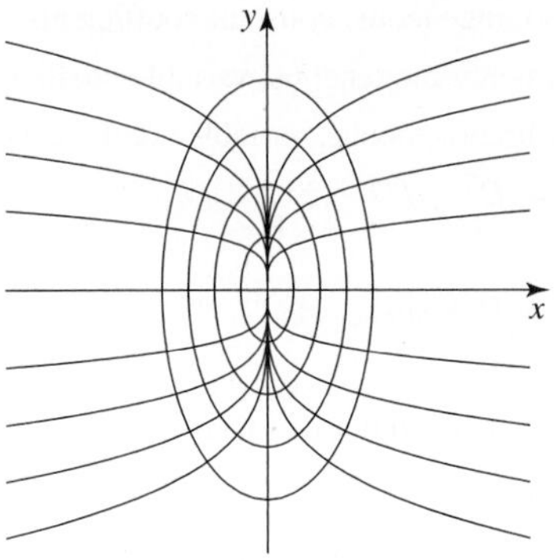

55.

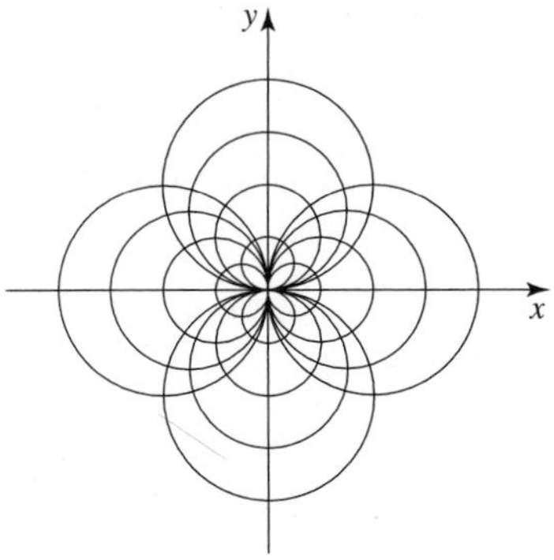

57. (a) $r(t) = -\dfrac{1}{24}t + \dfrac{1}{2}$, $0 \le t \le 12$
(b) One year

59. HINT: Use partial fractions and the facts that

$$\int \frac{dT}{M^2 - T^2} = \frac{1}{2M}\ln\left|\frac{M+T}{M-T}\right|,$$

$$\int \frac{dT}{M^2 + T^2} = \frac{1}{M}\arctan\left(\frac{T}{M}\right).$$

Section 1.4, p. 42

1. (a) Using step size 0.1, $y_3(0.3) \approx 1.0298$.
(b) Using step size 0.05, $y_6(0.3) \approx 1.03698$.
(c) Exact $y(0.2) = 1.0198\ldots$.
Euler approximations are both high.

3. Using step size 0.1 and Euler's method, $y(1) \approx 1.06121$. Smaller steps give higher $y_n(t_n)$.

5. Using step size 0.1 and Euler's method, $y(5) \approx 12.25186$. Smaller steps give higher $y_n(t_n)$.

7. Using step size 0.1 and Euler's method, $y(1) \approx 1.046035$. Smaller steps give higher $y_n(t_n)$.

9. Using step size 0.1 and Euler's method, $y(3) \approx 1.37796$. Smaller steps give lower $y_n(t_n)$.

11. Using step size 0.25 and Euler's method, $T(1) \approx 2.9810$.

13. Using step size 0.1 and Euler's method, $y(1) \approx 2.593742$. The true value of $y(1) = e$ is an irrational number, $2.7182818\ldots$.

15. At $t = 1$, the difference will be εe.
At $t = 10$, the difference will be $\varepsilon e^{10} \approx 22{,}026\varepsilon$.
Accumulative roundoff error grows at an exponential rate.

17. (a) Euler gives $y_1 = 0$.
Second-order Runge-Kutta gives $y_1 = 0.5$.
Fourth-order Runge-Kutta gives $y_1 \approx 0.646$.
(c) The exact solution $y(1) \approx 0.718$.

19. Using step size 0.1 and Runge-Kutta, $y(1) \approx 1.1606$.

21. Using step size 0.1 and Runge-Kutta, $y(1) \approx 0.0488$.

23. (d) $h \le \dfrac{\sqrt{2E}}{M}$

25. At step size 0.1, Richardson's extrapolation on Euler's method gives $y(0.2) \approx 1.2211$. Compare with $e^{0.2} \approx 1.2214$

27. At step size 0.1, Richardson's extrapolation on Euler's method gives $y(0.2) \approx 1.2476$. The exact answer is $y(0.2) = 1.25$; compare your approximation.

Section 1.5, p. 51

1. Unique solution through any initial conditions

3. Unique solution through any initial conditions

5. Unique solution through any initial conditions except $(0, 0)$

7. Unique solution as long as $y_0 \ne 1$

11. Unique solution through A and B for negative t. No unique solution through C, where derivative is not uniquely defined. Unique solution through D for $t > 0$.

13. Unique solution through A, B, C and D. Solutions appear to exist for all t.

15. Unique solution through B, C and D. Solutions exist only for $t > t_A$ or $t < t_A$ because all solutions appear to leave from or go toward A, where there is no unique slope.

17. Unique solution through A, B, C and D. Solutions appear to exist for all t.

19. (a) $f = y^2$, $f_y = 2y$ are continuous, hence Picard's Theorem holds, but does not say how large R can be.

(b)

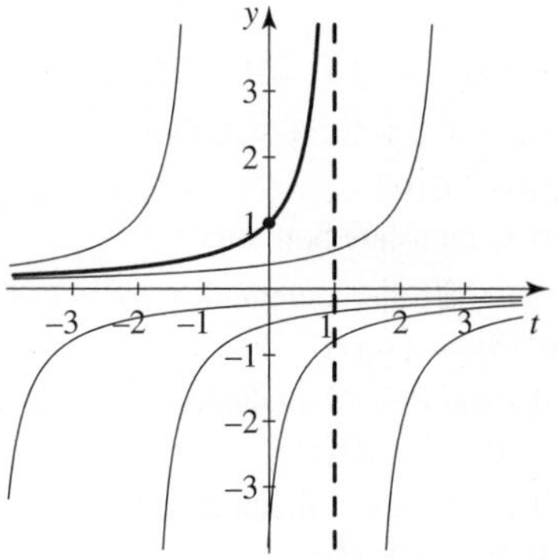

(c) $y(t) = \dfrac{1}{1-t}$, $t < 1$, $y > 0$

(d) $y(t) = \dfrac{-1}{\left(t - t_0 - y_0^{-1}\right)}$ cannot pass through $t = t_0 + \dfrac{1}{y_0}$.

20. $y(t) = \begin{cases} 0, & t < c; \\ \pm\left(\dfrac{2}{3}\right)^{3/2}(t-c)^{3/2}, & t \geq c, \end{cases}$

where c is a real number such that $c \geq 0$. Picard's Theorem tells us that solutions through (0, 0) are not guaranteed to be unique.

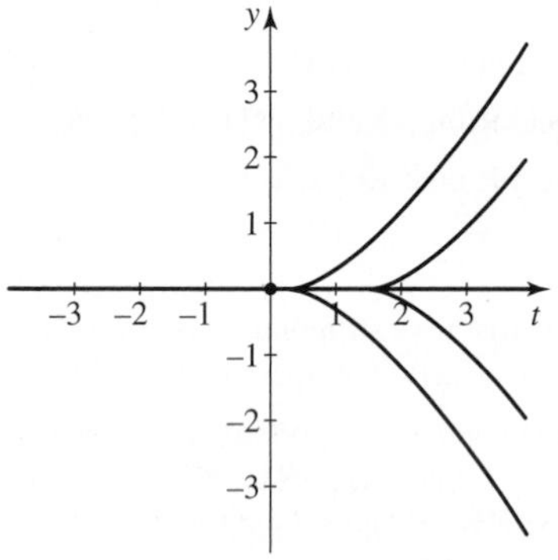

23. (a) The partial derivative $\dfrac{\partial f}{\partial y}$ is not continuous at $y = 0$.

(b) For $\dfrac{dy}{dt} = |y|$, $y(0) = 0$, the general solution is

$$y(t) = \begin{cases} Ce^{-t} & \text{if } y < 0 \\ Ce^{t} & \text{if } y \geq 0 \end{cases}$$

The only solution that satisfies the IVP occurs when $C = 0$ and $y \equiv 0$, so this is a unique solution.

25. (a) $A = \sqrt[3]{36\pi}\, V^{2/3}$

(b) We cannot tell when the snowball melted; the backwards solution is not unique.

(c) $V(t) = \begin{cases} -\left(\dfrac{t-c}{3}\right)^3, & t < t_0; \\ 0, & t \geq t_0, \end{cases}$

where c is an arbitrary constant such that $c \leq t_0$.

(d) $f = -kV^{2/3}$ does not satisfy Picard's Theorem when $V = 0$.

29. $y_0(t) = t - 1$, $y_1(t) = t + 1$,
$y_2(t) = -t + 1$,
$y_3(t) = t^2 - t + 1$

31. $y_0(t) = 1 + t$,
$y_1(t) = 1 - t$, $y_2(t) = t^2 - t + 1$,
$y_3(t) = -\dfrac{1}{3}t^3 + t^2 - t + 1$

33. Existence fails when $y_0 < 0$; uniqueness fails when $y_0 = 0$.

35. Picard's Theorem holds for all (t_0, y_0).

37. Existence holds for all (t_0, y_0); uniqueness fails when $y_0 = t_0$.

CHAPTER 2

Section 2.1, p. 62

1. First-order, nonlinear

3. Second-order, linear, homogeneous, variable coefficients

5. Third-order, linear, homogeneous, constant coefficients

7. Second-order, linear, nonhomogeneous, variable coefficients

9. Second-order, linear, homogeneous, variable coefficients

11. (a) $L(y) = 0$ for $L = D^2 + tD - 3$.

(b) not a linear DE

(c) $L(y) = 1$ for $L = D + \sin t$.

13. Nonlinear

15. Linear (The coefficients need not be linear.)

17. Nonlinear

19. $y(t) = ce^{-t} + 2$

21. $y(t) = ce^{3t} - \dfrac{5}{3}$

23. $y(t) = 2 - e^{-2t}$

Problems 26–31 just require finding derivatives and substituting. The following is a sample of one such line:

27. $y_1 = \sin 2t$, $y_1' = 2\cos 2t$, $y_1'' = -4\sin 2t$, so
$y_1'' + 4y_1 = -4\sin 2t + 4\sin 2t = 0$.

35. $y(t) = ce^{-t} + \dfrac{1}{2}e^t$

37. $y(t) = ce^t + te^t$

39. $y(t) = \dfrac{c}{t} + \dfrac{t^4}{5}$

41. $y(t) = c_1 \sin at + c_2 \cos at$

43. $y(t) = ce^t + 3te^t$

45. $y(t) = ct^2 + t^3$

47. (a) For $y_2 = te^t$, $y_2' = te^t + e^t$, $y_2'' = (te^t + e^t) + e^t$, and $y_2''' = (te^t + e^t) + 2e^t$, so

$$y''' - y'' - y' + y = (te^t + 3e^t) - (te^t + 2e^t) - (te^t + e^t) + te^t = 0.$$

Section 2.2, p. 70

1. $y(t) = ce^{-2t}$

3. $y(t) = ce^{t} + 3te^{t}$

5. $y(t) = ce^{-t} + e^{-t}\ln(1+e^{t})$

7. $y(t) = ce^{-t^3} + \dfrac{1}{3}$

9. $y(t) = \dfrac{c}{t} + t$

11. $y(t) = ct^2 + t^2 \sin t$

13. $y(t) = \dfrac{c}{1+e^{t}}$

15. $y(t) = c\left(\dfrac{e^{-2t}}{t}\right) + \dfrac{1}{2t} + t - 1$

17. $y(t) = \dfrac{1}{2}t^2 - \dfrac{1}{2} + e^{1-t^2}$

19. $y(t) = \dfrac{1}{2}e^{-t^2} + \dfrac{1}{2}$

21.

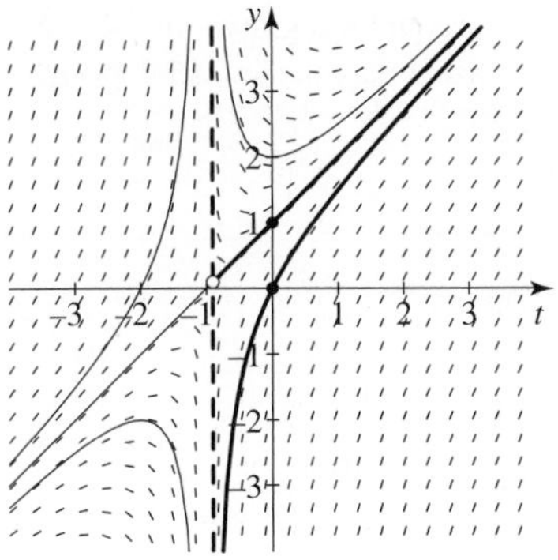

(a) $y(t) = \dfrac{t^2+2t}{t+1}\ (t > -1)$

(b) $y(t) = t + 1\ (t > -1)$

(d) For $t = -1$ the DE is not defined, so $y = t + 1$ is a solution only for $t > -1$.

23. $y(t) = ce^{-2t} + e^{t}$

25. $y(t) = \dfrac{1}{2}(\sin t - \cos t) + ce^{-t}$

27. $y(t) = ce^{-t^2} + \dfrac{1}{2}$

29. $y(t) = c\left(\dfrac{1}{t}\right) + \dfrac{1}{t}\ln t$

31. $y(t) = -t - 1$

33. (a) $y(t) = e^{(a/b + ce^{bt})}$

(b) $y(t) = e^{(1+ce^{t})}$

35. $y(t) = \pm\sqrt{\dfrac{1}{1+ce^{t^2}}}$

37. $y(t) = t^2 \sqrt[3]{\dfrac{5}{c_1 - 9t^5}}$

39. $y(t) = \sqrt[3]{1 + ct^{-3}}$

41. (b) $y(t) = 1 + \dfrac{1}{t+c}$

43. (a) $y' - y = e^{3t}$

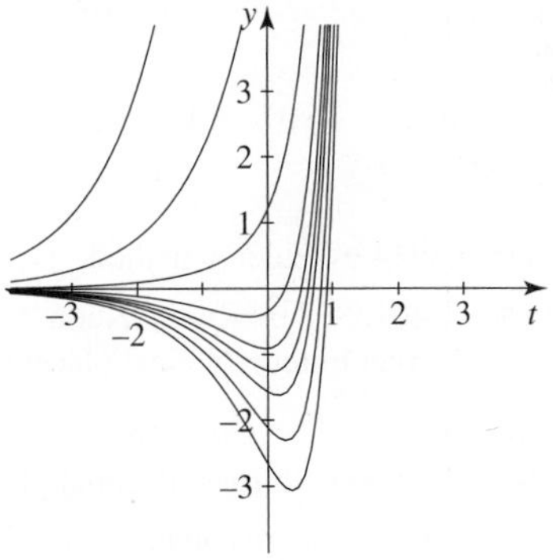

(b) General solution is $y(t) = \underbrace{ce^{t}}_{y_h} + \underbrace{\frac{1}{2}e^{3t}}_{y_p}$.

(c) There is no steady-state solution. Both y_h and y_p go to ∞ as $t \to \infty$.

45. (a) $y' + y = \sin 2t$

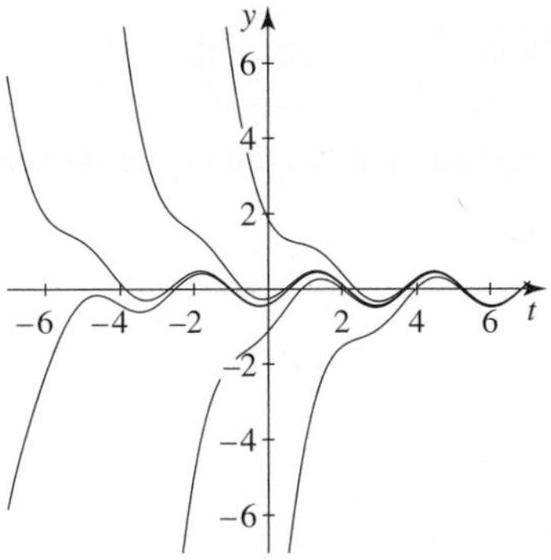

(b) General solution is $y(t) = \underbrace{ce^{-t}}_{y_h} + \underbrace{\dfrac{\sin 2t - 2\cos 2t}{5}}_{y_p}$.

(c) The steady-state solution is y_p, which attracts all other solutions. The transient solution is y_h.

47. (a) $y' + 2ty = 1$

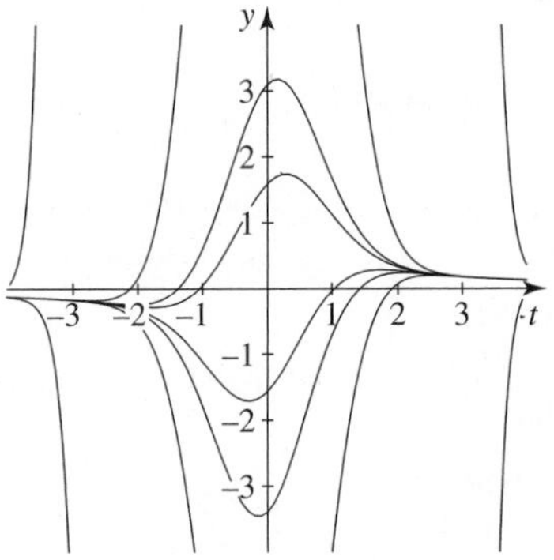

(b) General solution is $y(t) = \underbrace{ce^{-t^2}}_{y_h} + \underbrace{e^{-t^2}\int e^{t^2}\,dt}_{y_p}$.

(c) The steady-state solution is $y(t) = 0$, which is *not* equal to y_p. Both y_h and y_p are transient.

49. Sample analysis:

$$y' - y = e^{3t}, \qquad y(0) = 1, \quad y(1).$$

(a) For step size 0.1,

$$y(1) \approx 9.5944 \text{ by Euler's method,}$$
$$y(1) \approx 11.4018877 \text{ by Runge-Kutta (correct to four decimal places).}$$

(b) Exact solution is $y = 0.5e^t + 0.5e^{3t}$, so y(1)=11.4019090461656 to thirteen decimal places.

(c) The accuracy of Euler's method can be greatly improved by using a smaller step size; but it still is not correct to even one decimal place for step size 0.01,

$$y(1) \approx 11.20206 \text{ by Euler's method.}$$

(d) MORAL: Euler's method converges ever so slowly to the exact answer—clearly a far smaller step would be necessary to approach the accuracy of the Runge-Kutta method.

51. (a) (A) is linear homogeneous; (B) is linear nonhomogeneous; (C) is nonlinear

(c) The sum of any two solutions follows the direction field only in (A).

Section 2.3, p. 77

1. (a) $t_h = -\dfrac{1}{k}\ln 2$

3. If $y(t) = y_0 e^{-kt}$,

$$y\left(\frac{1}{k}\right) = y_0 e^{-1} = y_0(0.3678794\ldots) \approx \frac{y_0}{3}.$$

Hence $\left|\dfrac{1}{k}\right|$ is the time it takes to get to roughly one third of the initial amount.

5. $\dfrac{5\ln 10}{\ln 2} \approx 16.6$ hrs

7. $-\dfrac{5600\ln 0.55}{\ln 2} \approx 4830$ years

9. $\dfrac{1}{2^4} \approx 6.25\%$

11. $\dfrac{258\ln 20}{\ln 2} \approx 1115$ years

13. (a) $P(t) = 0.2e^{(\ln 0.9)t} \approx 0.2e^{-0.105t}$

(b) $-\dfrac{\ln 2}{\ln 0.9} \approx 6.6$ hours

15. 6.16 grams

17. $10\dfrac{\ln 3}{\ln 2} \approx 15.85$ hours

19. $5e^{4\ln 2} = 80$ million

21. $100e^{t\ln(3/2)} \approx 100e^{0.405t}$ cells

23. Account value $\$1 \cdot e^{0.10(10)} \approx \$2.72$

25. $7,382.39

27. 393.6 billion bottles

29. (a) $A(t) = \dfrac{1000}{0.08}\left(e^{0.08t} - 1\right)$

(b) $3,399.55

(c) 9.04%

31. $\dfrac{\ln 5}{0.08} \approx 20.1$ years

33. (a) $S_0 e^{0.08} \approx 1.0832775 S_0$, or 8.33% annual compounding

(c) $r_{\text{daily}} = \left(1 + \dfrac{0.08}{365}\right)^{365} - 1 \approx 0.083287,$

(i.e., 8.3287%) effective annual interest rate

35. After 20 years at 8% the account has grown to $247,064. The interest rate is more important than the annual deposit.

Section 2.4, p. 84

1. $Q(t) = c(t - 100)^2$

3. (a) $Q(t) = 10 + 40e^{-0.04t}$

(b) Concentration is $0.1 - 0.4e^{-0.04t}$.

(c) 10 kg

(d) 0.1 kg/liter

5. Approximately 2.0 lb/gal

7. (a) $dV/dt = 0.004 - 0.4V$, $V(0) = 0.05$ mi^3

(b) $V(t) = 0.01 + 0.04e^{-0.4t}$ cubic miles of pollutant

(c) $2.5\ln 4 \approx 3.5$ years

9. (a) $\dfrac{dx}{dt} = 4 - \dfrac{x}{50}$, $x(0) = 0$

(b) $x_{\text{eq}} = 200$ lb

(c) Resetting time so $t = 0$ when second faucet opens gives

$$\frac{dx}{dt} = 8 - \frac{4x}{200 + 2t}, \quad x(0) = x_{\text{eq}} = 200 \text{ lb.}$$

(d) $t_f = 400$ sec

(e) $x(400) \approx 1330.7$ lb.

11. (c) You should find that maximum $y_n(t)$ occurs when $t = 2n$.

13. $T(t) = M + (T_0 - M)e^{-kt}$

15. (a) 82.9°F

(b) Approximately 1:09 PM

17. $t = \dfrac{\ln 2}{b}$

19. John drinks the hotter coffee.

21. Approximately 5:24 PM

24. $y' + \dfrac{1}{1+t}y = 2,\ y(0) = 0$

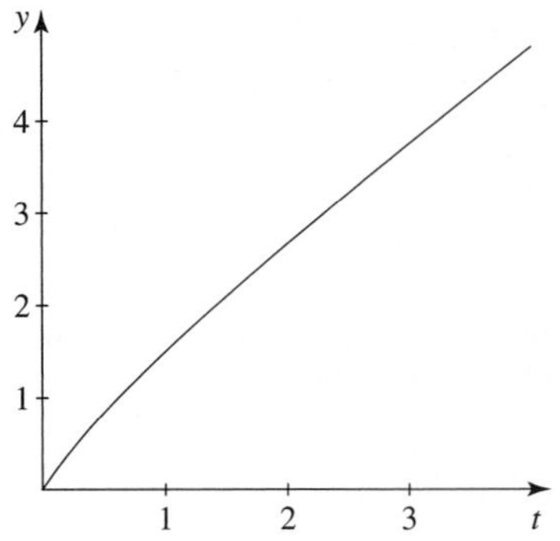

Section 2.5, p. 97

1. Equilibria at $y \equiv 0$ (unstable) and $y \equiv -\dfrac{a}{b}$ (stable). These are also isoclines of horizontal slope.

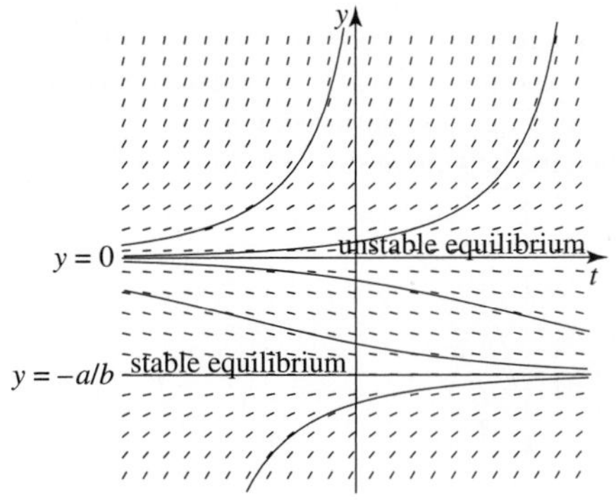

3. Equilibria at $y \equiv 0$ (unstable) and $y \equiv \dfrac{a}{b}$ (unstable). These are also isoclines of horizontal slope.

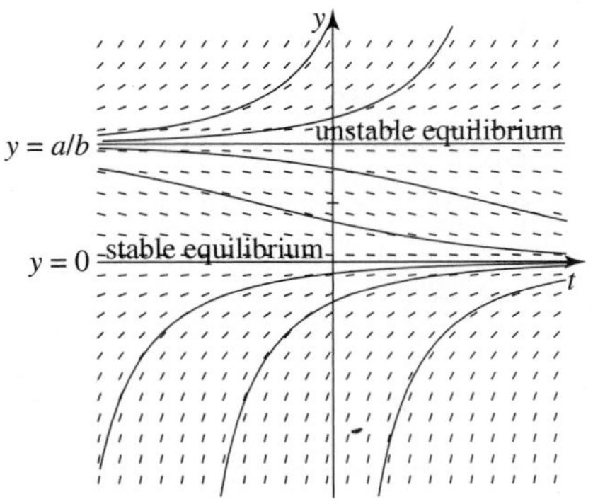

5. Equilibrium at $y \equiv 0$ (unstable), which is also an isocline of horizontal slope.

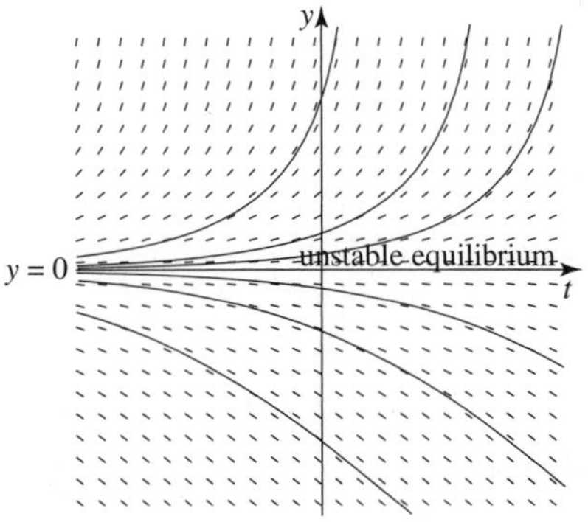

7. Isoclines of horizontal slopes are $y \equiv 0$ and $y = t$; only $y \equiv 0$ is also an equilibrium (unstable for $t < 0$, stable for $t > 0$).

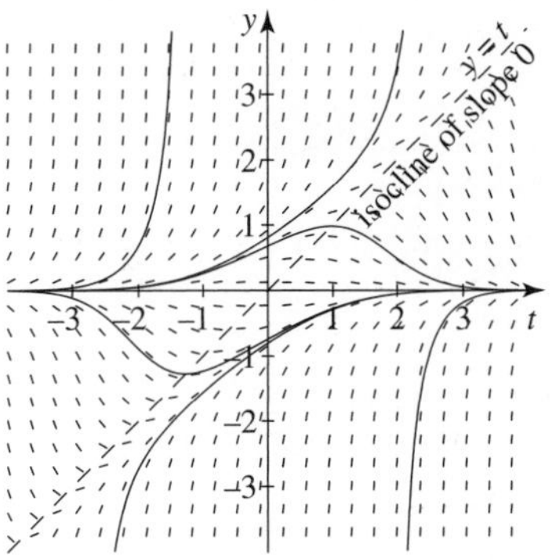

9. Isoclines of horizontal slopes (dashed) are hyperbolas $yt = \pm n\pi$ for $n = 0, 1, 2, \ldots$. Only $y \equiv 0$ is an equilibrium, stable for $t < 0$, unstable for $t > 0$ (note direction field near t-axis).

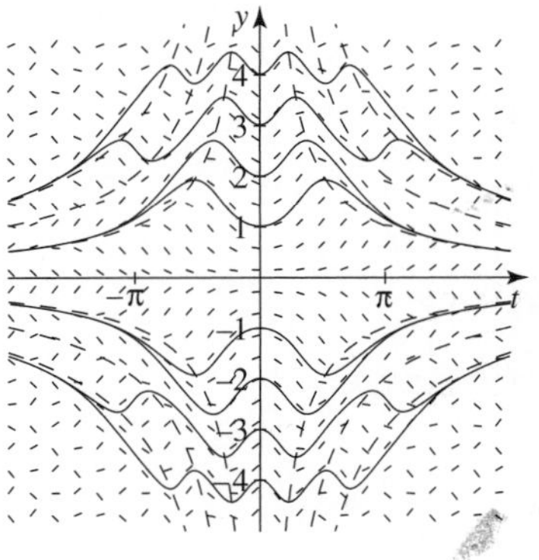

11. Inflection point at $y = T/2$.

13. (d) At t^* the rate is $\dfrac{rL}{4}$.

15. (a) $y(5) = 25{,}348$ cells

(b) $t \approx 6.536$ days

16. (b) Straight logistic $y' = y(1 - y)$

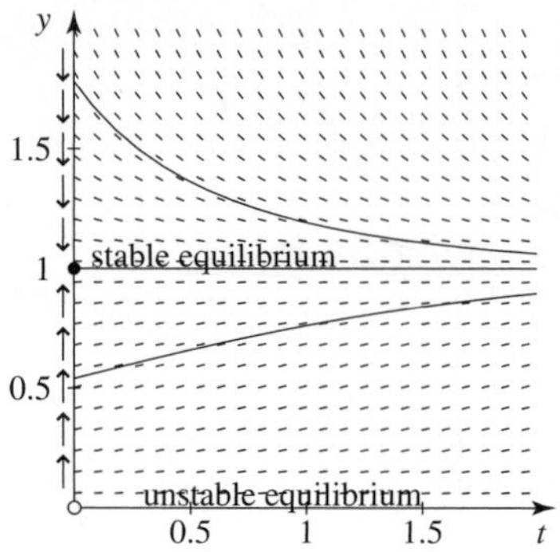

Logistic with harvesting $y' = y(1 - y) - 0.25$

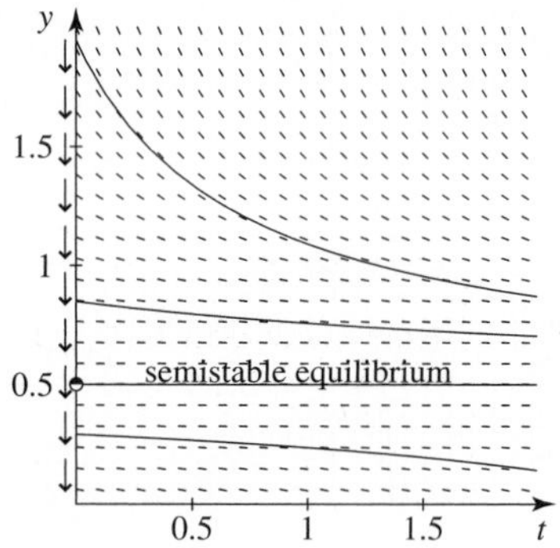

17. $x(t) = \dfrac{80}{1 + 79e^{-2.4235t}}$

20. (b) $y(t) = e^{a/b}e^{ce^{-bt}}$, where $c = \ln y_0 - \dfrac{a}{b}$.

(c) $\lim_{t\to\infty} y(t) = e^{a/b}$ when $b > 0$, $y(t) \to \infty$ when $b < 0$

22. (a)

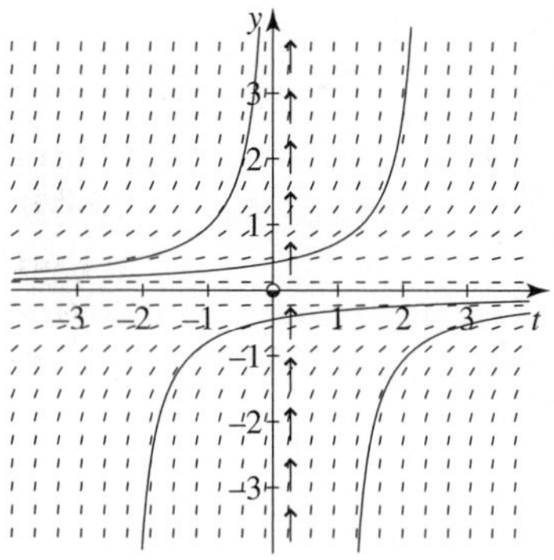

(b) One semistable equilibrium at $y(t) = 0$, stable from below, unstable from above

24. (a)

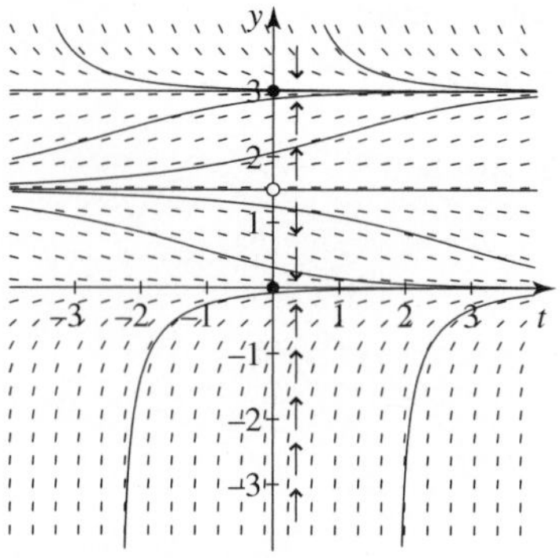

(b) Equilibrium points: $y = 0, L, M$. For $0 < L < M$, 0 and M are stable, L is unstable

26.

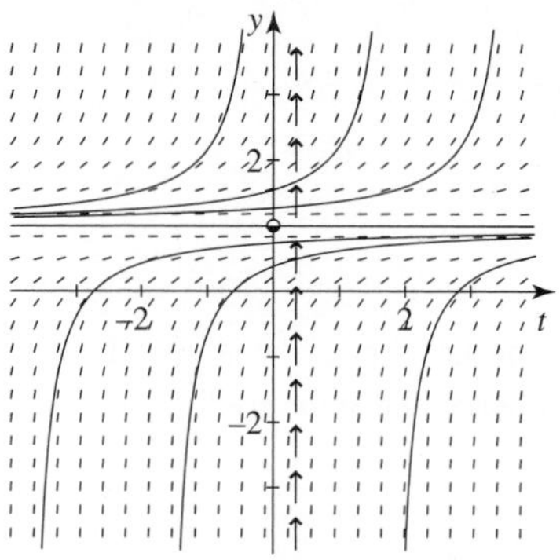

29. (a) From even a hand-sketched logistic curve you can graph its slope y' and find a roughly bell-shaped curve for $y'(t)$. Depending on scales used, it may be steeper or flatter than the bell curve shown in Fig. 1.3.5.

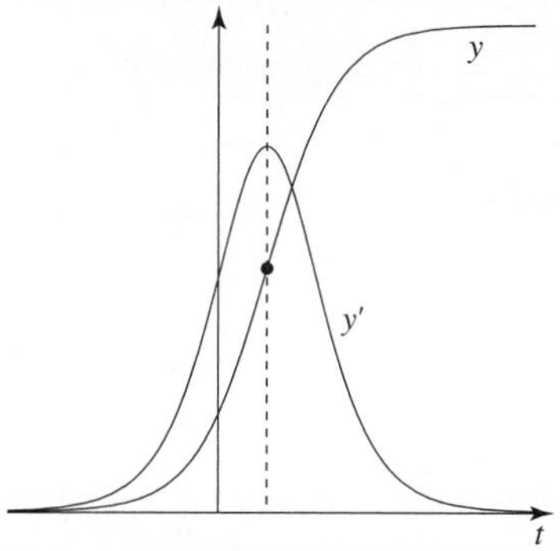

(b) If the inflection point is lower than halfway on an approximately logistic curve, the peak on the y' curve occurs sooner, creating an asymmetric curve for y'.

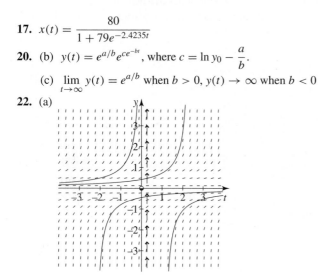

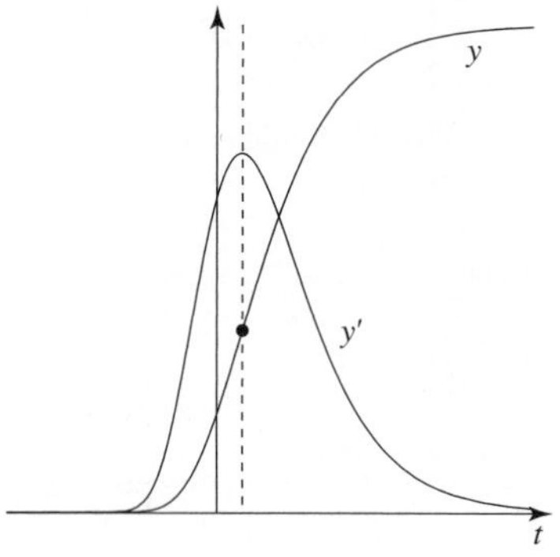

30. $y(t) = \dfrac{1}{1 + (1/c)e^{-kt}}$.

31. The solutions for many initial conditions are shown on the graph below. Conclusions: Any $x(0) > 100$ causes $x(t)$ to increase without bound. On the other hand, for *any* $x(0) \in (0, 100)$ the solution will approach an equilibrium value of 50, which implies the tiniest amount is sufficient to start the reaction. If you are looking for a different scenario, you might consider some other modeling options that appear in Problem 32.

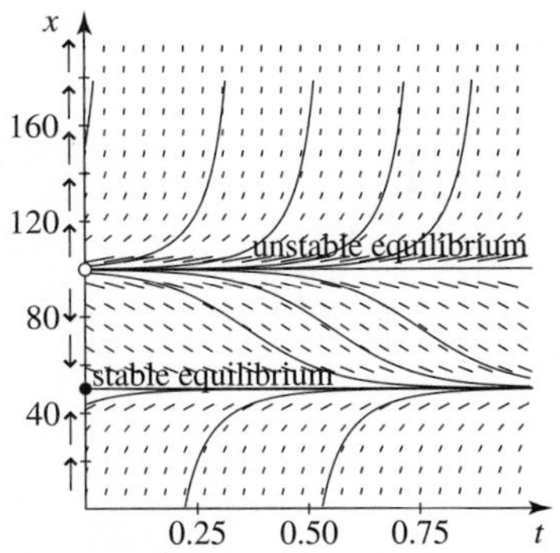

36. (a) An equilibrium occurs when $y' = 0$ for all values of t; i.e., when $y^2 + by + 1 = 0$ or $y = \dfrac{-b \pm \sqrt{b^2 - 4}}{2}$.

(b) Bifurcation points are at $b = \pm 2$.

(e) The above information is summed up in the bifurcation diagram below, showing the equilibrium values of y for each value of b. The solid dots give stable equilibrium values; the open dots give unstable equilibrium values.

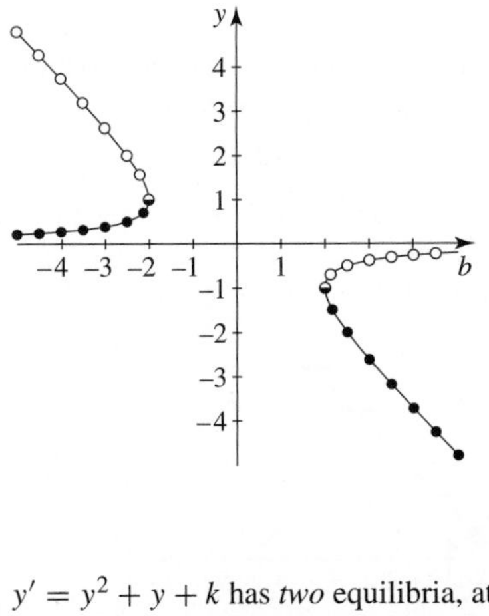

38. (a) $y' = y^2 + y + k$ has *two* equilibria, at $y = \dfrac{-1 \pm \sqrt{1 - 4k}}{2}$, for $k < \dfrac{1}{4}$; *none* for $k > \dfrac{1}{4}$; *one* for $k = \dfrac{1}{4}$. The following phase-plane graphs illustrate the bifurcation.

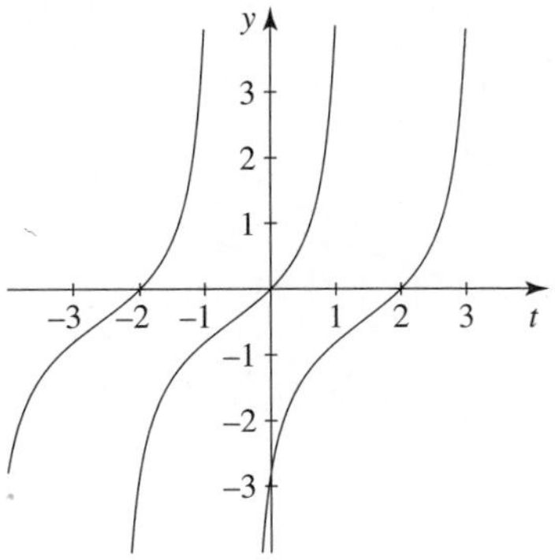

$k = 1$, no equilibria

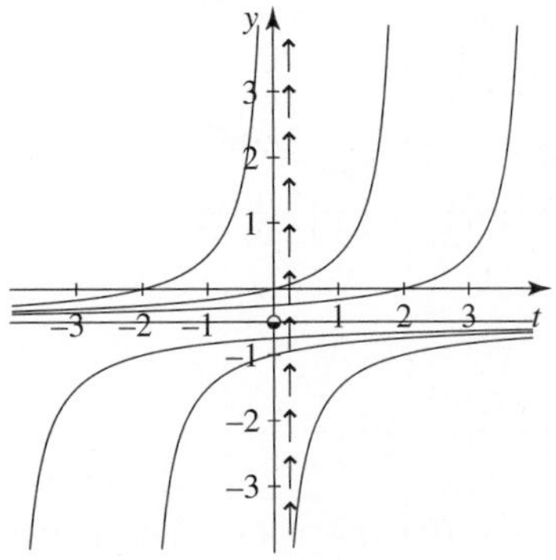

$k = 1/4$, one semistable equilibrium

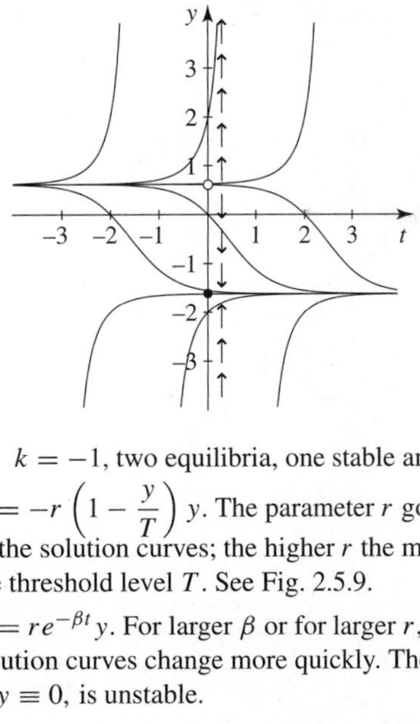

$k = -1$, two equilibria, one stable and one unstable

40. $y' = -r\left(1 - \dfrac{y}{T}\right)y$. The parameter r governs the steepness of the solution curves; the higher r the more steeply y leaves the threshold level T. See Fig. 2.5.9.

42. $y' = re^{-\beta t}y$. For larger β or for larger r, the slopes of solution curves change more quickly. The only equilibrium, at $y \equiv 0$, is unstable.

Section 2.6, p. 112

1. (a) One equilibrium point at the origin;
h-nullcline $x - 3y = 0$;
v-nullcline $y = 0$.

(b)

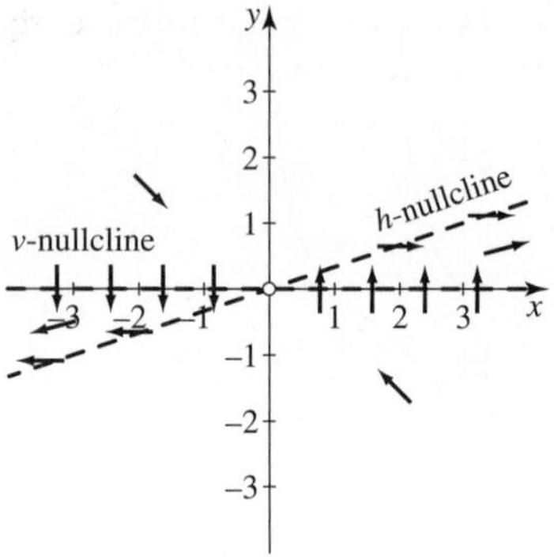

(c)

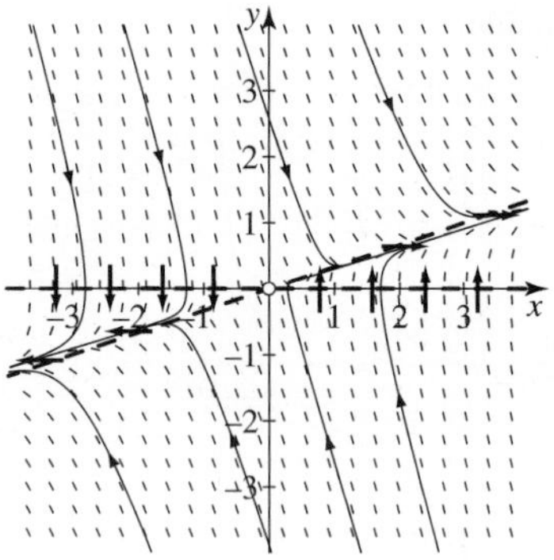

(d) Equilibrium at (0, 0) is unstable—solutions approach a line of small positive slope from above or below then turn away from the origin along that line.

3. (a) Equilibrium points (0, 1), (1, 0);
h-nullcline $x^2 + y^2 = 1$;
v-nullcline $x + y = 1$.

(b)

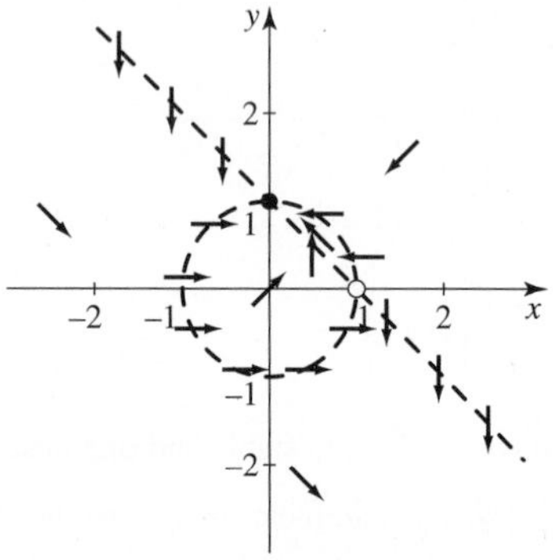

(c)

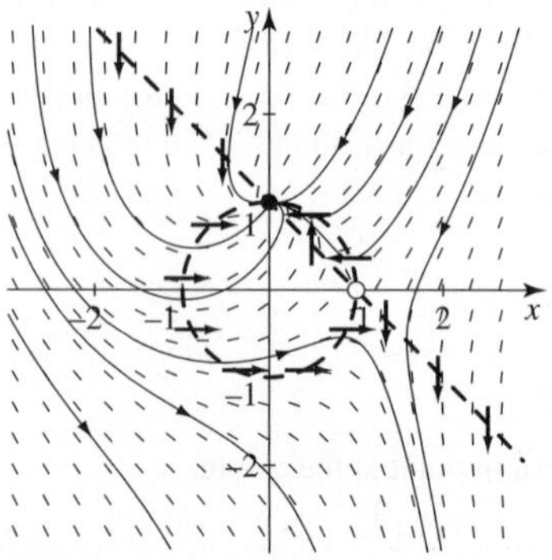

(d) Equilibrium at (1, 0) is unstable; equilibrium at (0, 1) is stable. Most solutions seem to be attracted to stable equilibrium, but those that approach the lower unstable equilibrium turn down toward the lower right.

5. (a) No equilibrium points;
h-nullcline $x^2 + y^2 = 3$;
v-nullcline $y = 4 - x$

(b)

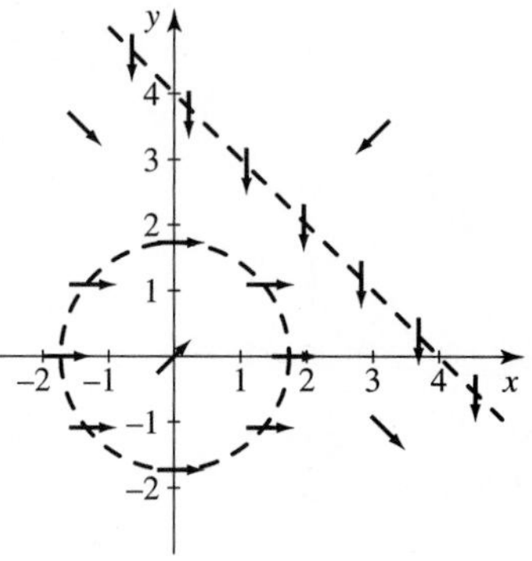

(c)

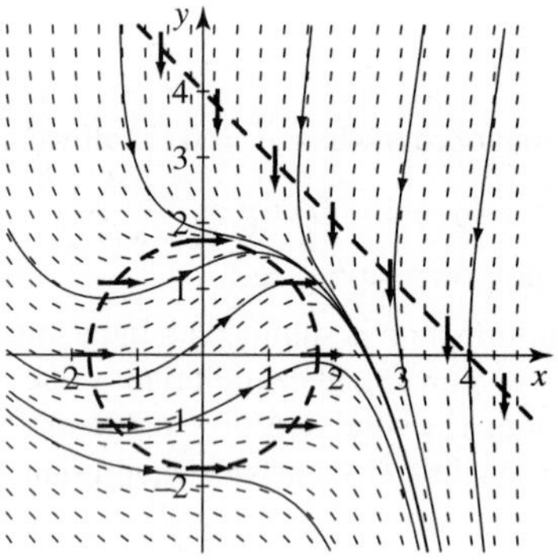

(d) No equilibria—all solutions head down to lower right.

7. (a) Stable equilibrium point (1/2, 1/2);
h-nullcline $|y| = x$;
v-nullcline $y = 1 - x$

(b)

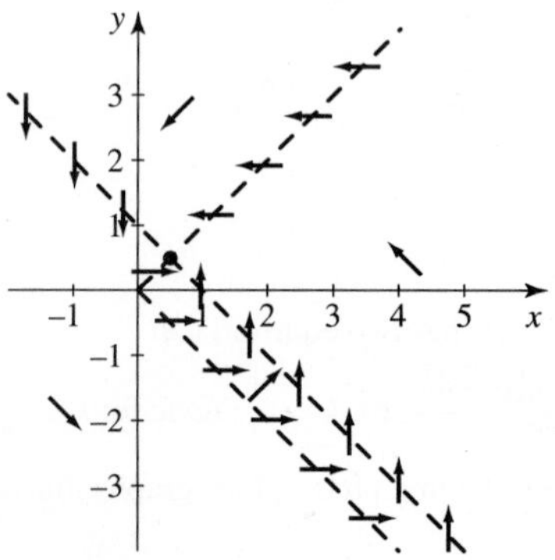

(c)

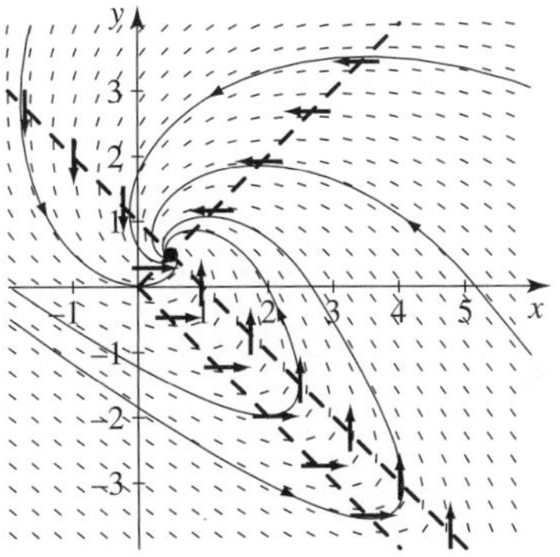

(d) Equilibrium is stable; solutions spiral into it.

9. Equilibrium at (8, 1).

(a)

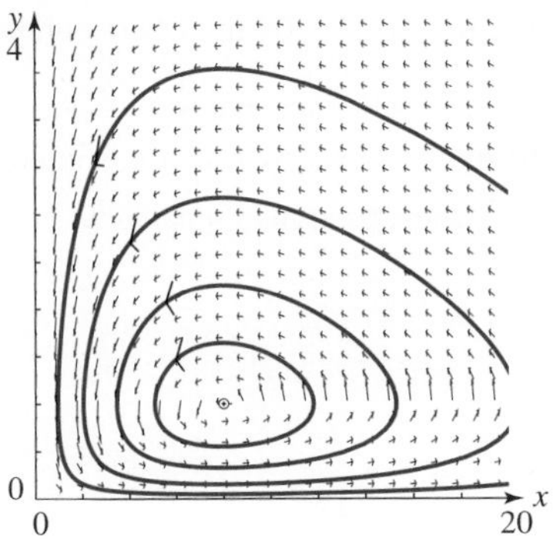

(b) Equilibrium at (11.2, 0.3).

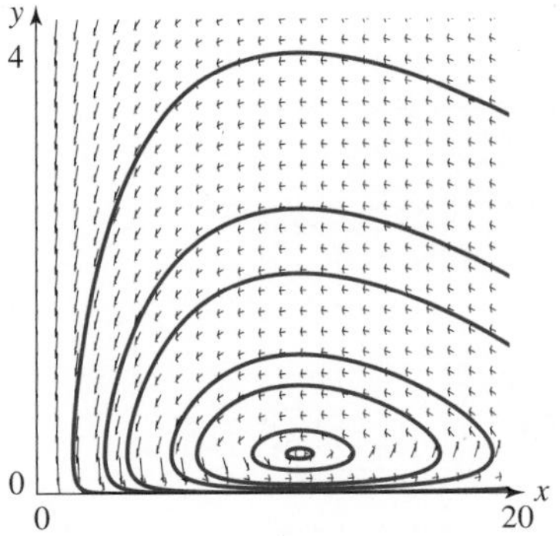

10. (a) $\widetilde{x}_e = \dfrac{c+f}{d}$, $\widetilde{y}_e = \dfrac{a-f}{b}$, where x is prey population and y is predator population.

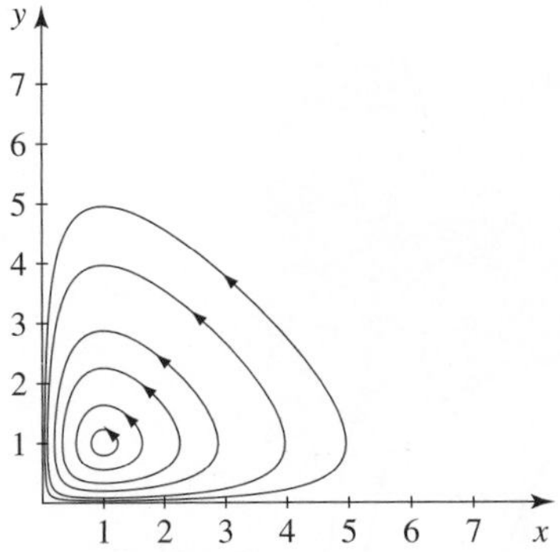

(b) As fishing increases, the equilibrium moves right (more prey) and down (fewer predators).

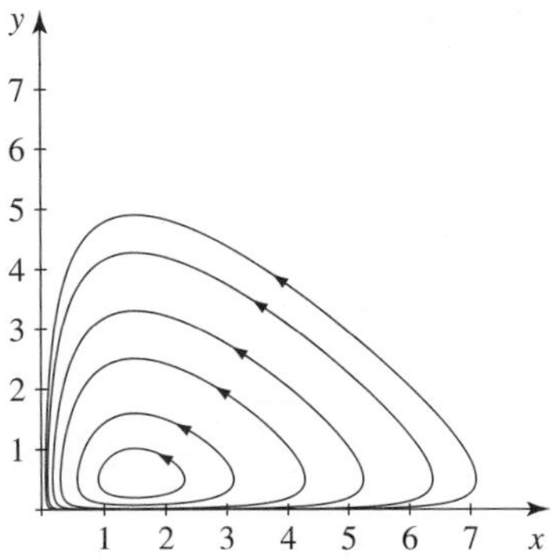

(c) You should fish for sardines when the sardine population is increasing and sharks when the shark population is increasing. In both cases, more fishing tends to move the populations closer to equilibrium while maintaining higher populations in the low parts of the cycle.

(d) Note that with fishing the shark population gets a lot lower each cycle. As fishing increases, predator equilibrium decreases and prey equilibrium increases.

11. Measuring in hundreds, unstable equilibria occur at (0, 0) and (3, 2), stable equilibria at (0, 5) and (6, 0).

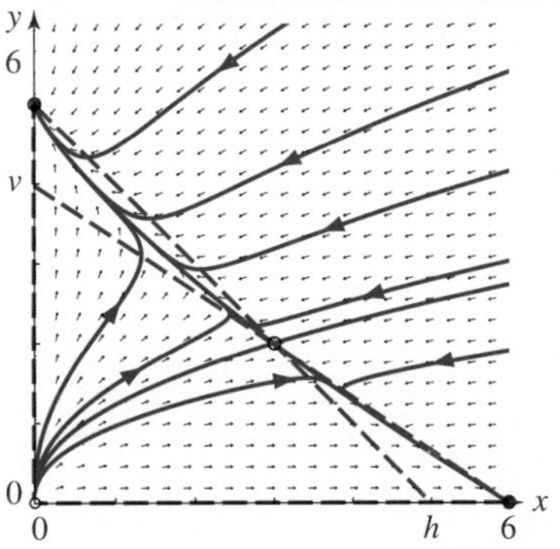

14. $x' = ax + bxy$
$y' = cy - dxy + eyz$
$z' = fz - gz^2 - hyz$

16. (a) One suggested model is

$$H' = aH - c\left(\frac{HP}{1+P}\right),\ P' = -bP + dHP$$

18.

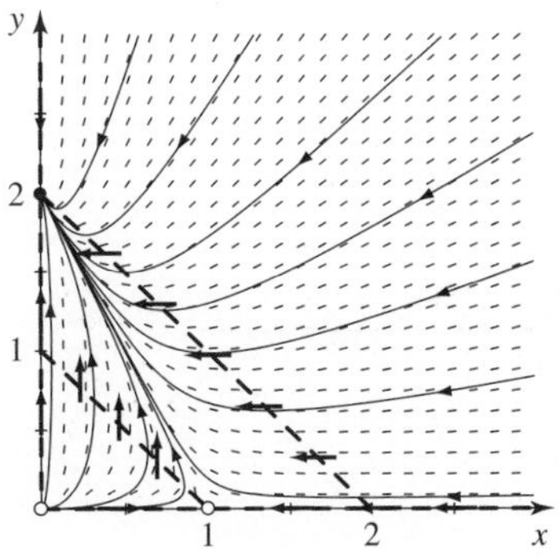

20.

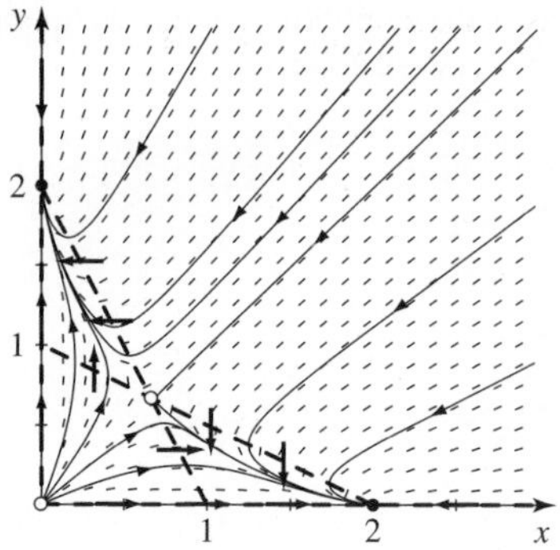

22. As an example,

(a)–(b)

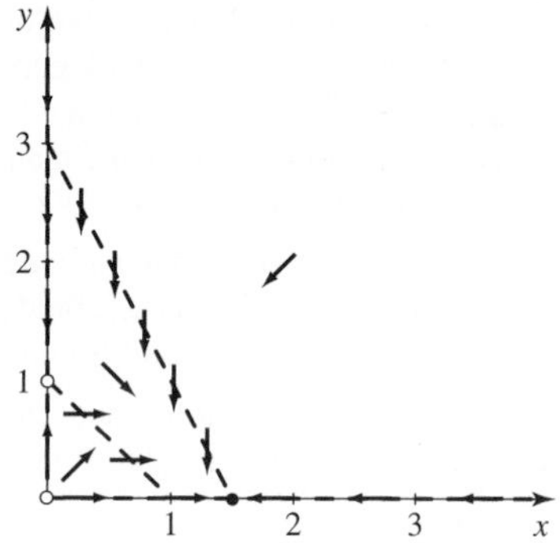

(c)–(d)

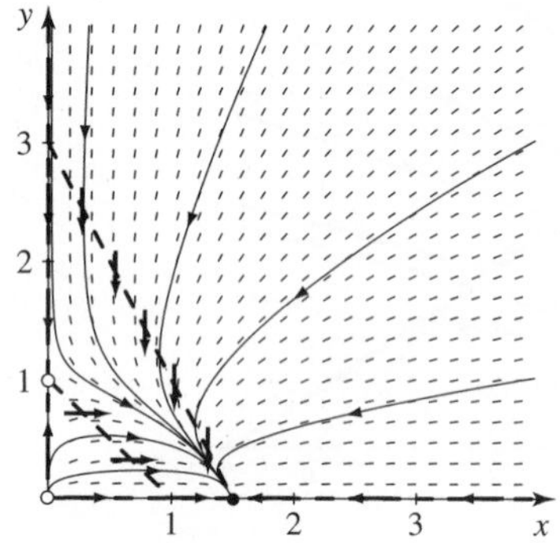

(e) Population x survives.

24. As an example,

(a)–(b)

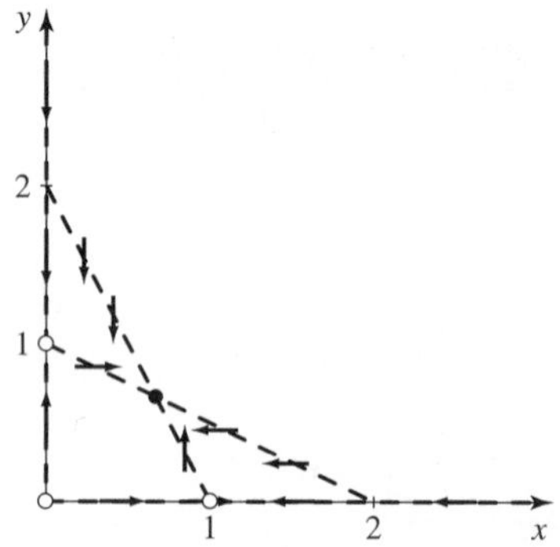

(c)–(d)

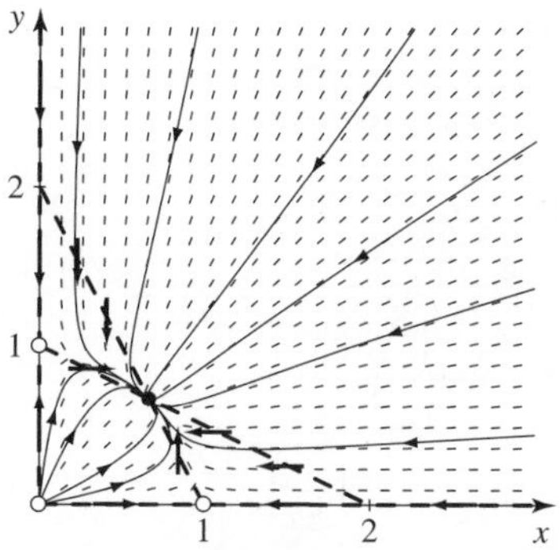

(e) The two populations coexist where $bx + cy = a$ intersects $ex + fy = d$.

26. h-nullclines at $y = 0$ and $x = \frac{d}{e}$; v-nullclines at $x = 0$ and $y = \frac{a}{c} - \frac{ab}{c}x$. Equilibria at (0, 0) and where oblique v-nullcline intersects h-nullclines. For $\frac{1}{b} > \frac{d}{e}$ the information in the figure shows coexistence is impossible. The case for $\frac{1}{b} < \frac{d}{e}$ must be argued separately.

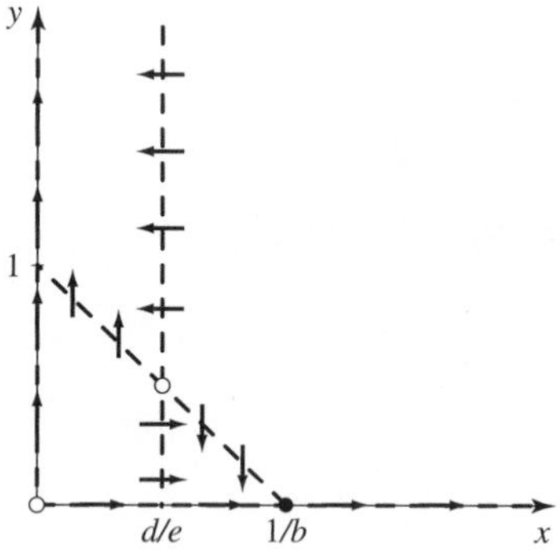

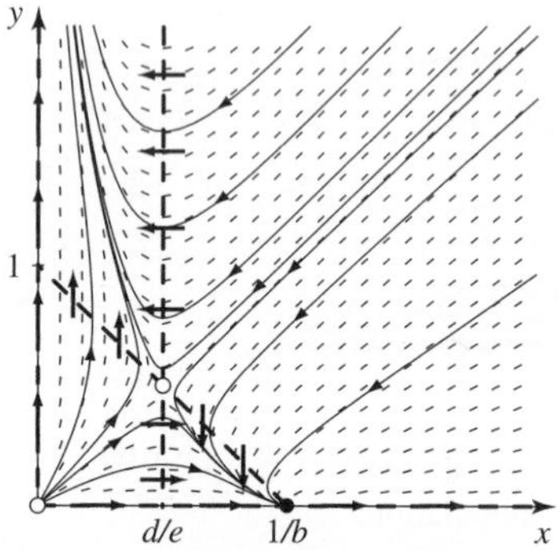

27.

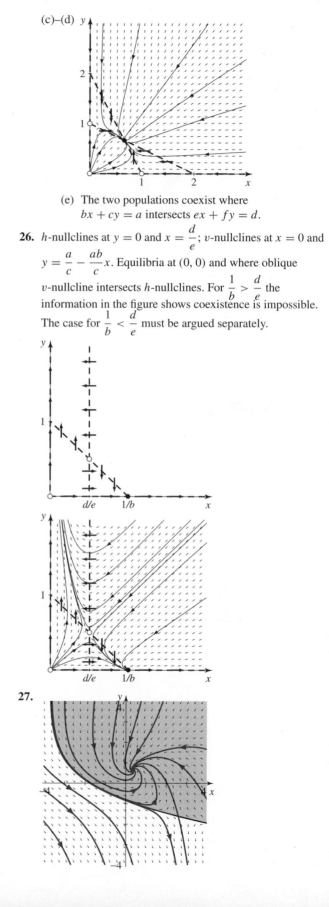

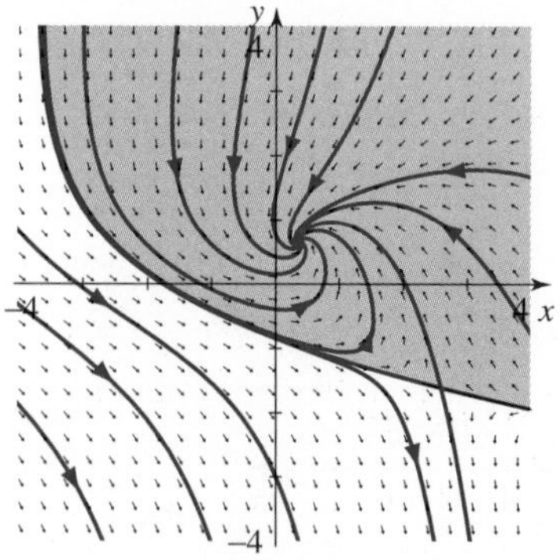

29.

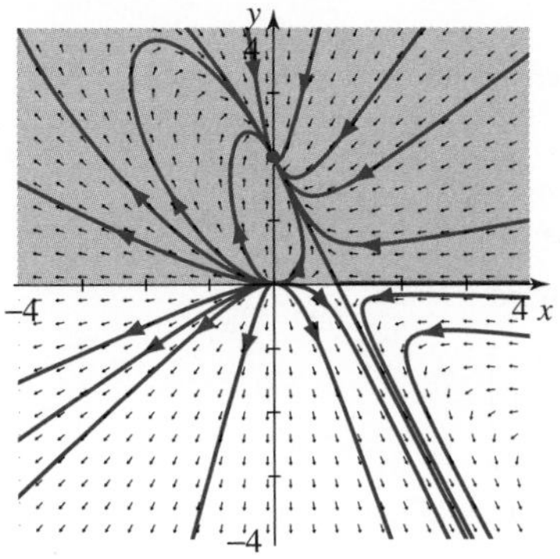

CHAPTER 3

Section 3.1, p. 127

1. $2\mathbf{A} = \begin{bmatrix} -2 & 0 & 6 \\ 4 & 2 & 4 \\ -2 & 0 & 2 \end{bmatrix}$

3. Matrices are not compatible.

5. $\mathbf{BA} = \begin{bmatrix} 5 & 3 & 9 \\ 2 & 1 & 2 \\ -1 & 0 & 1 \end{bmatrix}$

7. $\mathbf{DC} = \begin{bmatrix} 1 & -1 \\ 6 & 7 \end{bmatrix}$

9. Matrices are not compatible.

11. $\mathbf{A}^2 = \begin{bmatrix} -2 & 0 & 0 \\ -2 & 1 & 10 \\ 0 & 0 & -2 \end{bmatrix}$

13. $\mathbf{A} - \mathrm{I}_3 = \begin{bmatrix} -2 & 0 & 3 \\ 2 & 0 & 2 \\ -1 & 0 & 0 \end{bmatrix}$

15. Matrices are not compatible.

17. $\begin{bmatrix} a - 2e \\ b - 2f \end{bmatrix}$

19. $\begin{bmatrix} 1 & 0 \\ 0 & 1 \end{bmatrix}$

21. Not possible

23. (a) 5 columns
    (b) 4 rows
    (c) 6×4

Problems 35–50 are proofs or demonstrations. We give a few sample answers.

25. Not true

27. True

29. $B = \begin{bmatrix} 1 - 3e & -2 - 3f \\ 0 & 1 \\ e & f \end{bmatrix}$ for any real numbers e and f.

31. Every 2×2 matrix

33. Any matrix of the form $\begin{bmatrix} a & b \\ b & a \end{bmatrix}$ with $a, b \in \mathbb{R}$.

37. $$\begin{aligned} (c+d)\mathbf{A} &= \left[(c+d)a_{ij}\right] = \left[ca_{ij} + da_{ij}\right] \\ &= [ca_{ij}] + [da_{ij}] \\ &= c[a_{ij}] + d[a_{ij}] \\ &= c\mathbf{A} + d\mathbf{A} \end{aligned}$$

39. Interchanging rows and columns of a matrix two times reproduces the original matrix.

41. $(k\mathbf{A})^{\mathrm{T}} = k\mathbf{A}^{\mathrm{T}}$. It makes no difference whether you multiply each element of matrix **A** before or after rearranging them to form the transpose.

43. If the matrix $\mathbf{A} = [a_{ij}]$ is symmetric, then $a_{ij} = a_{ji}$. Hence $\mathbf{A}^{\mathrm{T}} = [a_{ji}]$ is symmetric since $a_{ji} = a_{ij}$.

51. $\mathbf{A} + 2\mathbf{B} = \begin{bmatrix} 3+i & 0 \\ 2+4i & 4-i \end{bmatrix}$

53. $\mathbf{BA} = \begin{bmatrix} 1-i & -3 \\ 4i & 1-i \end{bmatrix}$

55. $i\mathbf{A} = \begin{bmatrix} -1+i & -2 \\ 2i & 3+2i \end{bmatrix}$

57. $\mathbf{B}^{\mathrm{T}} = \begin{bmatrix} 1 & 2i \\ -i & 1+i \end{bmatrix}$

59. $\mathbf{A} = \begin{bmatrix} 1+i & 2i \\ 2 & 2-3i \end{bmatrix} = \begin{bmatrix} 1 & 0 \\ 2 & 2 \end{bmatrix} + i \begin{bmatrix} 1 & 2 \\ 0 & -3 \end{bmatrix}.$

    $\mathbf{B} = \begin{bmatrix} 1 & -i \\ 2i & 1+i \end{bmatrix} = \begin{bmatrix} 1 & 0 \\ 0 & 1 \end{bmatrix} + i \begin{bmatrix} 0 & -1 \\ 2 & 1 \end{bmatrix}.$

61. No, $\mathbf{AB} = \mathbf{0}$ does not imply that $\mathbf{A} = \mathbf{0}$ or $\mathbf{B} = \mathbf{0}$. For example, the product $\begin{bmatrix} 1 & 0 \\ 0 & 0 \end{bmatrix} \begin{bmatrix} 0 & 0 \\ 0 & 1 \end{bmatrix}$ is the zero matrix, but neither factor is itself the zero matrix.

67. If $\mathbf{M} = \begin{bmatrix} a & b \\ c & d \end{bmatrix}$ is a square root of $\mathbf{A} = \begin{bmatrix} 0 & 1 \\ 0 & 0 \end{bmatrix}$, then $\mathbf{M}^2 = \mathbf{A}$, which leads to the condition $a^2 = d^2$. Each of the possible cases leads to a contradiction.

    $\begin{bmatrix} 1 & 0 \\ \alpha & -1 \end{bmatrix}$ is a square root of **B** for any α.

69. $k = 0$

71. $k = \pm 1$

73. $\left\{ \begin{bmatrix} a \\ -2a \\ -a \end{bmatrix} : a \in \mathbb{R} \right\}$

75. $\left\{ \begin{bmatrix} a \\ -2a \\ -a \end{bmatrix} : a \in \mathbb{R} \right\}$

77. −6, not orthogonal

79. 0, orthogonal

81. 30, not orthogonal

83. $\mathbf{A}+\mathbf{C}$ lies on the horizontal axis, from 0 to -2.

85. $\mathbf{A}-2\mathbf{B}$ lies on the horizontal axis, from 0 to 7.

87. Right triangle, because dot product is zero

89. Invalid operation, because we ask for the scalar product of a vector and a scalar.

91. True

93. $$\mathbf{T}=\begin{bmatrix}0&1&1&0&1\\0&0&0&1&1\\0&1&0&0&1\\1&0&1&0&1\\0&0&0&0&0\end{bmatrix}$$

Ranking players by the number of games won means summing the elements of each row of $\mathbf{T}$, which in this case gives two ties: 1 and 4, 2 and 3, 5. Players 1 and 4 have each won 3 games. Players 2 and 3 have each won 2 games. Player 5 has won none.
Second-order dominance can be determined from

$$\mathbf{T}^2=\begin{bmatrix}0&1&0&1&2\\1&0&1&0&1\\0&0&0&1&1\\0&2&1&0&2\\0&0&0&0&0\end{bmatrix}.$$

For example, $\mathbf{T}^2$ tells us that Player 1 can dominate Player 5 in two second-order ways (by beating either Player 2 or Player 4, both of whom beat Player 5).
The sum

$$\mathbf{T}+\mathbf{T}^2=\begin{bmatrix}0&2&1&1&3\\1&0&1&1&2\\0&1&0&1&2\\1&2&2&0&3\\0&0&0&0&0\end{bmatrix}$$

gives the number of ways one player has beaten another both directly and indirectly. Reranking players by sums of row elements of $\mathbf{T}+\mathbf{T}^2$ can sometimes break a tie: In this case it does so and ranks the players in order 4, 1, 2, 3, 5.

Section 3.2, p. 143

1. $$\left[\begin{array}{rr|r}1&2&1\\2&-1&0\\3&2&1\end{array}\right]$$

3. $$\left[\begin{array}{rrr|r}1&2&1&1\\1&-3&3&1\\0&4&-5&3\end{array}\right]$$

5. (A)

7. (C)

9. (A)

11. RREF

13. Not RREF (leading nonzero element in row 2 is not 1; nonzero elements above leading one)

15. RREF

17. Not RREF (not all zeroes above leading ones)

19. RREF

21. RREF $\begin{bmatrix}1&1&0&4&5\\0&0&1&1&-1\end{bmatrix}$; pivot columns of the original matrix are first and third.

23. RREF $\begin{bmatrix}1&0&1&-1\\0&1&1&1\\0&0&0&0\end{bmatrix}$;

pivot columns of the original matrix are first and second.

25. RREF $\left[\begin{array}{rr|r}1&0&3\\0&1&6\end{array}\right]$; unique solution; $x=3$, $y=6$.

27. RREF $\left[\begin{array}{rrr|r}1&0&3/7&0\\0&1&-5/7&0\end{array}\right]$; nonunique solutions;

$x=-\frac{3}{7}z$, $y=\frac{5}{7}z$, z is arbitrary.

29. RREF $\left[\begin{array}{rrr|r}1&0&3&4\\0&1&-2&-1\end{array}\right]$; nonunique solutions;

$x_1=4-3x_3$, $x_2=-1+2x_3$, x_3 is arbitrary.

31. RREF $\left[\begin{array}{rrr|r}1&0&0&0\\0&1&0&0\\0&0&1&0\end{array}\right]$; unique solution; $x=y=z=0$.

33. RREF $\left[\begin{array}{rrr|r}1&0&1&1\\0&1&1&0\\0&0&0&0\end{array}\right]$; nonunique solutions;

$x=1-z$, $y=-z$, z is arbitrary.

35. RREF $\left[\begin{array}{rrr|r}1&0&0&24/5\\0&1&0&4/5\\0&0&1&-22/5\\0&0&0&0\end{array}\right]$; consistent system;

unique solution; $x=24/5$, $y=4/5$, $z=-22/5$.

37. $\vec{\mathbf{x}}=\begin{bmatrix}2\\2\end{bmatrix}+\vec{\mathbf{0}}$

39. $\vec{\mathbf{x}}=\begin{bmatrix}-1\\1\\0\end{bmatrix}+r\begin{bmatrix}0\\-1\\1\end{bmatrix}$ for any $r\in\mathbb{R}$

41. $\vec{\mathbf{x}}=\begin{bmatrix}2/3\\1/3\\-1/3\end{bmatrix}+\vec{\mathbf{0}}$

43. $\vec{\mathbf{x}}=\begin{bmatrix}1\\1\\1\end{bmatrix}+\vec{\mathbf{0}}$

45. $\vec{\mathbf{x}}=\vec{\mathbf{0}}+r\begin{bmatrix}-1\\-1\\1\end{bmatrix}$ for any $r\in\mathbb{R}$

47. The system is inconsistent, so there is no $\vec{\mathbf{x}}_p$ and no general solution.

49. $\vec{\mathbf{x}} = \begin{bmatrix} 1 \\ 2 \\ 0 \\ 0 \end{bmatrix} + r \begin{bmatrix} -2 \\ -1 \\ 1 \\ 0 \end{bmatrix} + s \begin{bmatrix} 4 \\ 3 \\ 0 \\ 1 \end{bmatrix}$ for $r, s \in \mathbb{R}$

51. Unique solution; $x = 2$, $y = -1$, $z = -3$.

53. $\begin{bmatrix} w \\ x \\ y \\ z \end{bmatrix} = \begin{bmatrix} 2r - 5s \\ r \\ -2s \\ s \end{bmatrix} = r \begin{bmatrix} 2 \\ 1 \\ 0 \\ 0 \end{bmatrix} + s \begin{bmatrix} -5 \\ 0 \\ -2 \\ 1 \end{bmatrix}$,

r, s any real numbers

55. $\begin{bmatrix} x_1 \\ x_2 \\ x_3 \\ x_4 \end{bmatrix} = \begin{bmatrix} 4r - 3s \\ r \\ s \\ t \end{bmatrix} = r \begin{bmatrix} 4 \\ 1 \\ 0 \\ 0 \end{bmatrix} + s \begin{bmatrix} -3 \\ 0 \\ 1 \\ 0 \end{bmatrix} + t \begin{bmatrix} 0 \\ 0 \\ 0 \\ 1 \end{bmatrix}$

57. Rank is 2; system is inconsistent for all vectors $\begin{bmatrix} a \\ b \\ c \end{bmatrix}$ for which $a - 20b + 5c \neq 0$.

59. Rank is 2; system is inconsistent for any vector $\begin{bmatrix} a \\ b \\ c \end{bmatrix}$ for which $-2a - b + c \neq 0$.

62. Any k will produce a consistent system.

64. The system is inconsistent for all k.

65. The system is consistent if $k = 10$.

67. A system $\mathbf{A}\vec{\mathbf{x}} = \vec{\mathbf{b}}$ of four equations in two unknowns will have a unique solution if the RREF of the augmented matrix has the form

$$\left[\begin{array}{cc|c} 1 & 0 & a \\ 0 & 1 & b \\ 0 & 0 & 0 \\ 0 & 0 & 0 \end{array}\right],$$

where a and b are nonzero real numbers.

69. $R_3 \leftrightarrow R_1$ will undo the operation $R_1 \leftrightarrow R_3$.

$R_1 = \frac{1}{3} R_1$ will undo the operation $R_1 = 3R_1$.

$R_i = R_i - cR_j$ will undo the operation $R_i = R_i + cR_j$.

71. Neither of the last two columns affects the other, so the last two columns will contain the respective solutions.

73. The areas of the two fields are 1200 and 600 square yards.

75. The basic idea is to formalize a strategy like that used in Example 3. The augmented matrix for $\mathbf{Ax} = \mathbf{b}$ is

$$\left[\begin{array}{ccc|c} a_{11} & a_{12} & a_{13} & b_1 \\ a_{21} & a_{22} & a_{23} & b_2 \\ a_{31} & a_{32} & a_{33} & b_3 \end{array}\right].$$

A pseudocode might begin:

(a) To get a one in first place in row 1, multiply every element of row 1 by $1/a_{11}$.

(b) To get a zero in first place in row 2, replace row 2 by

row 2 − a_{21}(row 1).

⋮

77.
$$\begin{aligned} I_1 - I_2 - I_3 &= 0 \\ -I_1 + I_2 + I_3 &= 0 \end{aligned}$$

79.
$$\begin{aligned} I_1 - I_2 - I_3 - I_4 \qquad &= 0 \\ -I_1 + I_2 \qquad\qquad + I_5 &= 0 \\ I_3 + I_4 - I_5 &= 0 \end{aligned}$$

Section 3.3, p. 154

5. $\mathbf{A}^{-1} = \begin{bmatrix} 1/2 & 0 \\ -1/2 & 1 \end{bmatrix}$

7. $\mathbf{A}^{-1} = \begin{bmatrix} 1 & 0 & 1/3 \\ -2 & 1/2 & -5/6 \\ 3 & -1/2 & 5/6 \end{bmatrix}$

9. $\mathbf{A}^{-1} = \begin{bmatrix} 1/k & 0 & 0 \\ 0 & 1 & 0 \\ 0 & 0 & 1 \end{bmatrix}$

11. $\mathbf{A}^{-1} = \begin{bmatrix} 1 & 0 & 0 & 0 \\ 0 & 1 & -k & 0 \\ 0 & 0 & 1 & 0 \\ 0 & 0 & 0 & 1 \end{bmatrix}$

13. $\mathbf{A}^{-1} = \begin{bmatrix} 1 & 0 & 0 & 0 \\ 0 & -1 & 0 & 0 \\ 0 & -1/2 & -1/2 & 0 \\ -1/3 & 1/6 & 1/2 & 1/3 \end{bmatrix}$

16. We seek $\mathbf{A}^{-1} = \begin{bmatrix} a & b \\ c & d \end{bmatrix}$, so we need

$$\begin{bmatrix} a & b \\ c & d \end{bmatrix} \begin{bmatrix} 1 & 3 \\ 1 & 2 \end{bmatrix} = \begin{bmatrix} 1 & 0 \\ 0 & 1 \end{bmatrix}.$$

This gives four equations in the four unknowns. Solving them gives

$$\mathbf{A}^{-1} = \begin{bmatrix} -2 & 3 \\ 1 & -1 \end{bmatrix}.$$

17. Not true

20. $x_1 = 50$, $x_2 = -18$

22. $\vec{\mathbf{x}} = \begin{bmatrix} 3 \\ -2 \\ 1 \end{bmatrix}$.

25. $(\mathbf{AB}^{-1})^{-1} = \mathbf{BA}^{-1}$

27. $\vec{\mathbf{x}} = \mathbf{B}\vec{\mathbf{b}}$

29. $\mathbf{A} + \mathbf{B}$ must be invertible.

31. The key to the proof is to premultiply $\mathbf{AB} = \mathbf{I}$ by $\mathbf{A}^{-1}$.

33. k any real number except ± 1.

37. $\mathbf{A}^{-1}$ does not exist.

41. $\mathbf{E}_{\text{Int}}^{-1} = \begin{bmatrix} 0 & 1 & 0 \\ 1 & 0 & 0 \\ 0 & 0 & 1 \end{bmatrix}$

43. The key is to premultiply **B** by a nonregular matrix **P** and postmultiply by $\mathbf{P}^{-1}$.

48. $x_1 = 11.2,\ x_2 = 12.2$

50. $x_1 = 136.4,\ x_2 = 90.9$

52. (a) $\mathbf{I} - \mathbf{T} = \begin{bmatrix} 0.70 & 0.00 & 0.00 \\ -0.10 & 0.80 & -0.20 \\ -0.05 & -0.10 & 0.98 \end{bmatrix}$

(b) $(\mathbf{I} - \mathbf{T})^{-1} = \begin{bmatrix} 1.43 & 0.00 & 0.00 \\ 0.20 & 1.25 & 0.26 \\ 0.07 & 0.01 & 1.02 \end{bmatrix}$

(c) $\vec{\mathbf{x}} = \begin{bmatrix} \$200{,}200 \\ \$53{,}520 \\ \$12{,}040 \end{bmatrix}$

Section 3.4, p. 164

1. 0

3. 12

5. 0

7. −24

8. Subtract the first row from the second row.

10. Interchange the two rows of the matrix.

12. 0

14. Extending the basketweave hypothesis gives −1, which is not the true answer of 0 (because row 1 equals row 4).

16. −105

18. −8

21. Invertible if $k \neq 0$ and $k \neq 1$.

23. The matrix does not have an inverse because its determinant is zero.

25. The matrix has an inverse because its determinant is nonzero.

27. $\mathbf{AB} = \begin{bmatrix} 3 & 2 \\ 7 & 4 \end{bmatrix}$; $|\mathbf{A}| = -2$, $|\mathbf{B}| = 1$, $|\mathbf{AB}| = -2$.

31. The key to the proof lies in the determinant of a product of matrices.

33. One example is $\mathbf{A} = \begin{bmatrix} 1 & 0 \\ 0 & 1 \end{bmatrix}$, $\mathbf{B} = \begin{bmatrix} -1 & 0 \\ 0 & -1 \end{bmatrix}$.

35. For an $n \times n$ matrix $\mathbf{A}$, $|k\mathbf{A}| = k^n|\mathbf{A}|$.

37. (a) Interchange any two rows of the identity matrix.

(b) The determinant is unchanged, so it is +1.

(c) Multiplying a row by k will give a determinant of k.

39. $x = 10,\ y = -4$

41. All determinants are 3, so $x = 1,\ y = 1,\ z = 1$.

47. $y = 0.62x + 1.68$

49. Least squares plane $y = a + b_1 T + b_2 P$, where a, b_1 and b_2 are determined by solving the system

$$\begin{bmatrix} n & \sum_{i=1}^{n} T_i & \sum_{i=1}^{n} P_i \\ \sum_{i=1}^{n} T_i & \sum_{i=1}^{n} T_i^2 & \sum_{i=1}^{n} T_i P_i \\ \sum_{i=1}^{n} P_i & \sum_{i=1}^{n} T_i P_i & \sum_{i=1}^{n} P_i^2 \end{bmatrix} \begin{bmatrix} a \\ b_1 \\ b_2 \end{bmatrix} = \begin{bmatrix} \sum_{i=1}^{n} y_i \\ \sum_{i=1}^{n} T_i y_i \\ \sum_{i=1}^{n} P_i y_i \end{bmatrix}$$

Section 3.5, p. 175

1. A typical vector is $[x, y]$;
the zero vector is $[0, 0]$;
the negative of $[x, y]$ is $[-x, -y]$.

3. A typical vector is $[a, b, c, d]$;
the zero vector is $[0, 0, 0, 0]$;
the negative of $[a, b, c, d]$ is $[-a, -b, -c, -d]$.

5. A typical vector is $\begin{bmatrix} a & b & c \\ d & e & f \end{bmatrix}$;

the zero vector is $\begin{bmatrix} 0 & 0 & 0 \\ 0 & 0 & 0 \end{bmatrix}$;

the negative of $\begin{bmatrix} a & b & c \\ d & e & f \end{bmatrix}$ is $\begin{bmatrix} -a & -b & -c \\ -d & -e & -f \end{bmatrix}$.

7. A typical vector is a linear function $p(t) = at + b$; the zero vector is $p(t) \equiv 0$; and the negative of $p(t)$ is $-p(t)$.

9. A typical vector is a continuous and differentiable function, such as $f(t) = \sin t$; the zero vector is $f(t) \equiv 0$; and the negative of $f(t)$ is $-f(t)$.

11. Not a vector space; there is no additive inverse.

13. Not a vector space; e.g., the negative of [2, 1] does not lie in the set.

15. Not a vector space; e.g., $x^2 + x$ and $(-1)x^2$ each belongs but their sum $x^2 + x + (-1)x^2 = x$ does not.

17. Not a vector space; the set is not closed under vector addition.

19. Yes, a vector space.

21. Not a vector space; not closed under scalar multiplication; no additive inverse.

23. Yes, a vector space.

25. Not a vector space; not closed under scalar multiplication.

27. Yes, the solution space of the linear homogeneous DE

$$y'' + p(t)y' + q(t)y = 0$$

is indeed a vector space; the linearity properties are sufficient to prove all the vector space properties.

Sample proof for Problems 29–32:

31. $\vec{\mathbf{v}} + 0\vec{\mathbf{v}} = 1\vec{\mathbf{v}} + 0\vec{\mathbf{v}} = (1 + 0)\vec{\mathbf{v}} = 1\vec{\mathbf{v}} = \vec{\mathbf{v}}$, hence $0\vec{\mathbf{v}} = \vec{\mathbf{0}}$, from Problem 30.

33. For $c \neq 0$, $\vec{\mathbf{v}} = 1\vec{\mathbf{v}} = \frac{1}{c}(c\vec{\mathbf{v}}) = \frac{1}{c}(\vec{\mathbf{0}}) = \vec{\mathbf{0}}$.

35. Not a vector space because, for example, the new vector addition is not commutative.

37. Subspace

39. Subspace

41. Subspace

43. Subspace

45. Subspace

47. Not a subspace, because $\vec{\mathbf{x}} = \vec{\mathbf{0}} \notin \mathbb{W}$.

50. Subspace

52. Not a subspace; the last two coordinates are not linear functions of a and b.

54. Not a subspace; no zero vector; not closed under vector addition or scalar multiplication.

56. Not a subspace; e.g., not closed under scalar multiplication. (f may satisfy equation $f' = f^2$, but $2f$ will not, since $2f' \neq 4f^2$.)

58. An example of a set in $\mathbb{R}^2$ that is closed under scalar multiplication but not under vector addition is that of two different lines passing through the origin.

61. The solution space $\mathbb{S} = \left\{ r\begin{bmatrix} -3 \\ -1 \\ 1 \\ 0 \end{bmatrix} : r \in \mathbb{R} \right\}$.

63. The solutions are $y = \dfrac{1}{c-t}$, but the sum of two solutions is not a solution, so the solution set of this nonlinear DE is not a vector space.

65. From the DE we can see that the zero vector is not a solution, so the solution space of the nonlinear DE is not a vector space.

67. The solutions are $y = \dfrac{1}{t-c}$, and the sum of two solutions is not a solution, so the general solution space of this nonlinear DE is not a vector space.

69. Yes, the DE is linear and homogeneous, so the solutions form a vector space. The linearity properties cause all the vector space properties to be satisfied.

Section 3.6, p. 191

1. The given vectors do not span $\mathbb{R}^2$, although they span the one-dimensional subspace $\{k[1, 1] \mid k \in \mathbb{R}\}$.

3. They do not span $\mathbb{R}^3$, because they cannot give any vector $[a, b, c]$ with $b \neq 0$.

5. The given vectors do not span $\mathbb{P}_2$; they only span a one-dimensional subspace of $\mathbb{R}^3$

7. Linearly dependent

9. Linearly independent

11. Linearly independent

13. Linearly independent

15. Linearly independent

17. Linearly independent

19. Linearly independent

21. Linearly independent

23. Linearly independent

25. Linearly dependent

27. Linearly independent

29. Linearly independent

31. Not a basis, because the determinant of the matrix they form is 0.

37. Linearly independent

39. Linearly independent

41. Linearly independent

43. Not a basis because $\{[1, 1]\}$ does not span $\mathbb{R}^2$

45. Not a basis because $[-1, -1]$ and $[1, 1]$ are linearly dependent

47. Not a basis because the vectors are linearly dependent

49. Not a basis because two vectors are not enough to span $\mathbb{R}^3$

51. Not a basis because four vectors must be linearly dependent in $\mathbb{R}^3$

53. Basis

55. Basis

57. The dimension of $\mathbb{W}$ is 2; a basis is $\{[-1, 0, 1], [-1, 1, 0]\}$

59. 2-dimensional

61. 2-dimensional

63. The solution space is one-dimensional, with basis $\left\{\begin{bmatrix} 3 \\ -1 \\ 1 \\ 0 \end{bmatrix}\right\}$.

65. A basis for $\mathbb{P}_{n-1}$ is $\{1, t, t^2, \ldots, t^{n-1}\}$. Dim $\mathbb{P}_{n-1} = n$.

67. A basis $B = \{e^{t^2}\}$, dim $\mathbb{S} = 1$.

69. The solutions $y = \dfrac{1}{t-c}$ do *not* form a vector space.

71. A basis for $\mathbb{W}$ is $\left\{\begin{bmatrix} 1 \\ 0 \\ 0 \\ 1 \end{bmatrix}, \begin{bmatrix} 0 \\ 0 \\ 1 \\ -1 \end{bmatrix}, \begin{bmatrix} 0 \\ 0 \\ 0 \\ 1 \end{bmatrix}\right\}$.

Dim $\mathbb{W} = 3$.

73. A basis for $\mathbb{W}$ is $\left\{\begin{bmatrix} 1 \\ 1 \\ 0 \\ 0 \end{bmatrix}, \begin{bmatrix} 1 \\ 0 \\ 4 \\ 0 \end{bmatrix}\right\}$. Dim $\mathbb{W} = 2$.

74. The given vectors are linearly independent; include the vector $\begin{bmatrix} 0 & 0 \\ 1 & 0 \end{bmatrix}$ to make a basis for $\mathbb{M}_{22}$, which is four-dimensional.

76. The set of four-dimensional vectors

$$\left\{\begin{bmatrix} -3 \\ 1 \\ 0 \\ 0 \end{bmatrix}, \begin{bmatrix} 2 \\ 0 \\ 1 \\ 0 \end{bmatrix}, \begin{bmatrix} -6 \\ 0 \\ 0 \\ 1 \end{bmatrix}\right\}$$

is a basis for the hyperplane.

77. A basis for $\mathbb{W}$ is $\left\{\begin{bmatrix}1&0\\0&0\end{bmatrix},\begin{bmatrix}0&1\\1&0\end{bmatrix},\begin{bmatrix}0&0\\0&1\end{bmatrix}\right\}$.
Dim $\mathbb{W}=3$.

79. A typical different basis is $\left\{\begin{bmatrix}1&0\\0&1\end{bmatrix},\begin{bmatrix}1&0\\0&-1\end{bmatrix}\right\}$, with both elements diagonal and linearly independent.
Dim $\mathbb{D}=2$.

82. (a) True
(b) False
(c) False

85. The coset of [0, 0, 1] in $\mathbb{W}$ is the collection of vectors

$$\{[0,0,1]+\beta[-1,1,0]+\gamma[-1,0,1]\mid \beta,\gamma\in\mathbb{R}\}.$$

Geometrically, this describes a plane passing through (0, 0, 1) and parallel to $x_1+x_2+x_3=0$.

87. The coset through the point $(1,-2,1)$ is given by the points

$$\{(1,-2,1)+t(1,3,2)\}.$$

This describes a line passing through $(1,-2,1)$ parallel to the line $(t,3t,2t)$.

88. The general solution of $y'+2y=e^{-2t}$ is $y(t)=ce^{-2t}+te^{-2t}$. This solution could be considered a "line" in the vector space of solutions, passing through te^{-2t} in the direction of e^{-2t}.

CHAPTER 4

Section 4.1, p. 205

1. $x(t)=\cos t$

3. $x(t)=\cos 3t+\dfrac{1}{3}\sin 3t$

5. $x(t)=-\cos 4t$

7. $x(t)=\dfrac{1}{4}\sin 4\pi t$

9. $A\approx 1.4$; period $=2\pi$; delay $\dfrac{\delta}{\omega_0}\approx 0.8$

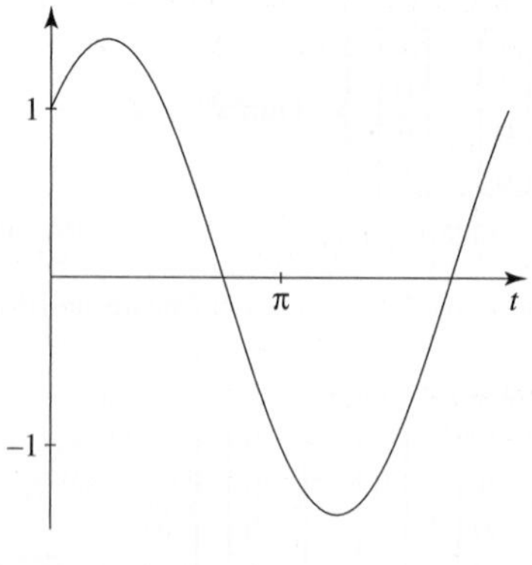

$\cos t+\sin t\approx 1.4\cos(t-0.8)$.

11. $A\approx 5.1$; period $=\dfrac{2\pi}{3}$; delay $\dfrac{\delta}{\omega_0}\approx 0.05$

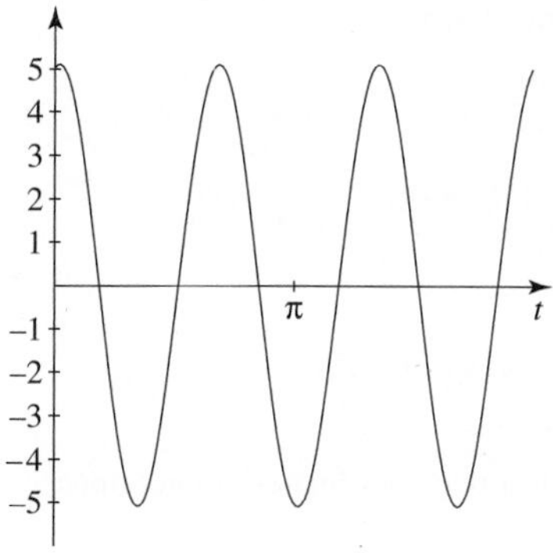

$5\cos 3t+\sin 3t\approx 5.1\cos(3t-0.15)$.

13. $A\approx 2.2$; period $=\dfrac{2\pi}{5}$; delay $\dfrac{\delta}{\omega_0}\approx\dfrac{\pi}{8}$ or 0.4.

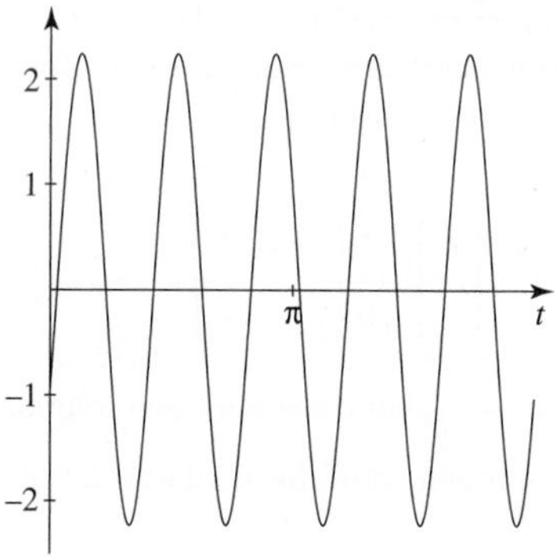

$-\cos 5t+2\sin 5t\approx 2.2\cos(5t-2)$.

15. $\sqrt{2}\cos\left(t-\dfrac{\pi}{4}\right)$ (See answer graph for Problem 9.)

17. $\sqrt{2}\cos\left(t-\dfrac{3\pi}{4}\right)$

19. $-2\cos 2t$

21. $\dfrac{3\sqrt{2}}{2}(\cos t+\sin t)$

23. Amplitude 1;
phase angle $\delta=0$ radians;
period $=2\pi$

25. Amplitude $\dfrac{\sqrt{10}}{3}$;
phase angle $\delta=\tan^{-1}\left(\dfrac{1}{3}\right)\approx 0.3218$ radians;
period $=\dfrac{2\pi}{3}$

27. Amplitude 1;
phase angle $\delta=\pi$ radians;
period $=\dfrac{\pi}{2}$

29. Amplitude $\frac{1}{4}$;

phase angle $\delta = \frac{\pi}{2}$;

period = 8

31. (a) Starting points are where trajectories cross positive $\dot{x}$ axis.

(b) $x(t) = c \sin\left(\frac{t}{2}\right)$

$\dot{x}(t) = \frac{c}{2} \cos\left(\frac{t}{2}\right)$

(c) For $x(t)$, horizontal axis crossing at even multiples of π; for $\dot{x}(t)$, at odd multiples of π.

(d) Amplitudes are approximately $\frac{A}{3}, \frac{A}{2}, \frac{2A}{3}, \frac{5A}{6}$, and A.

32. Trajectories are circles centered at the origin, traversed clockwise.

34. Trajectories are ellipses centered at the origin, each with height thrice its width. Motion is clockwise.

36. Trajectories are ellipses centered at the origin, each with height four times its width. Motion is clockwise.

38. Trajectories are ellipses centered at the origin, each with height twelve times its width. Motion is clockwise.

40. B

42. D

44. (a) $\omega_0 = 0.5$ gives tx curve with lowest frequency (fewest humps); $\omega_0 = 2$ gives the highest frequency (most humps).

(b) $\omega_0 = 0.5$ gives the innermost phase-plane trajectory; as ω_0 increases, the amplitude of $\dot{x}$ increases.

46. (a) $x(t) = \frac{1}{2} \cos \sqrt{8}\, t$

(b) Amplitude $= \frac{1}{2}$ m;

$T = \frac{2\pi}{\omega_0} = \frac{2\pi}{\sqrt{8}}$ sec;

$f = \frac{\sqrt{8}}{2\pi}$

(c) $\frac{\pi}{2\sqrt{8}} \approx 0.56$ seconds; $\dot{x}(0.56) \approx -1.414$ m/sec

48. (a) $\ddot{x} + 64x = 0$, $x(0) = \frac{1}{3}$ ft, $\dot{x}(0) = -4$ ft/sec

(b) $\ddot{x} + 64x = 0$, $x(0) = -\frac{1}{6}$ ft, $\dot{x}(0) = 1$ ft/sec

50. Period is the same and so is the frequency, but the amplitude will be twice that in the first case.

52. Smaller restoring force for larger amplitude vibrations

54. A vibrating mass where the friction starts very big but dies off

56. A vibrating spring with no friction but the restoring force $-tx$ gets stronger as time passes.

58. (a) The charge on the capacitor would oscillate indefinitely.

(b) $L\ddot{Q} + \frac{1}{C}Q = 0$; $Q(0) = 0$, $\dot{Q}(0) = 5$

(c) $Q(t) = 5\dfrac{\sin\left(\sqrt{\frac{1}{LC}}\, t\right)}{\sqrt{\frac{1}{LC}}}$

(d) $Q(t) = \frac{1}{2} \sin 10t$

60. $\dot{x} = y$

$\dot{y} = \frac{1}{4}(-3x + 2y + 17 - \cos t)$

62. $\dot{q} = I$

$\dot{I} = -\frac{1}{50}q - 3I + \cos 3t$

64. $\dot{x} = y$

$\dot{y} = -4x + \sin t$

65. Use $x = r\cos\theta$.

67. The buoy weighs 657 lbs.

68. **(b)** $t_f \approx 42.5$ minutes.

Section 4.2, p. 222

1. $y(t) = c_1 + c_2 t$

3. $y(t) = c_1 e^{3t} + c_2 e^{-3t}$

5. $y(t) = c_1 e^{t} + c_2 e^{2t}$

7. $y(t) = c_1 e^{-t} + c_2 t e^{-t}$

9. $y(t) = c_1 e^{t/2} + c_2 e^{t}$

11. $y(t) = c_1 e^{4t} + c_2 t e^{4t}$

13. $y(t) = e^{-t}\left(c_1 e^{\sqrt{2}t} + c_2 e^{-\sqrt{2}t}\right)$

15. $y(t) = \frac{1}{2}e^{5t} + \frac{1}{2}e^{-5t}$

17. $y(t) = te^{-t}$

19. $y(t) = -te^{3t}$

21. $y = 3 - e^{t}$ **23.** Basis: $\{1, e^{4t}\}$ **25.** Basis: $\{e^{3t}, e^{-t}\}$

27. Show that in each set the elements are solutions, and then that they are linearly independent solutions by calculating the Wronskian.

29. $W = 24 \neq 0$.

31. A basis for the solution space of $y^{(4)} = 0$ must have four linearly independent solutions. The given set has only three solutions, so it cannot be a basis.

33. (a) $x(0) \approx -10$, $\dot{x}(0) \approx 0$

(b) $x(t) = -30e^{-2t} + 20e^{-3t}$

(c) For $t > 0$, each term of $x(t)$ diminishes as t increases; the result remains negative and the solution remains below the t-axis. For $t < 0$, each exponential increases as t decreases. The negative term cancels the positive term when $e^{-t} = 1.5$ or $t \approx -0.405$; the solution graph indeed appears to cross the negative t axis at about that value.

(d) $\dot{x}(t)$ reaches a maximum when $\ddot{x}(t) = 0$, at $t \approx 0.406$. Again, the graph of the solution appears to agree.

For Problems 34–39 see Answer to Problem 33 for sample format of answers to parts (c) and (d).

35. (a) $x(0) \approx 0$, $\dot{x}(0) \approx -8$.
(b) $x(t) = -8e^{-2t} + 8e^{-3t}$.

37. (a) $x(0) \approx 2$, $\dot{x}(0) \approx 0$.
(b) $x(t) = \frac{6}{5}e^{-2t} + \frac{4}{5}e^{3t}$.

39. (a) $x(0) \approx 0$, $\dot{x}(0) \approx -1$.
(b) $x(t) = \frac{1}{5}e^{-2t} - \frac{1}{5}e^{3t}$.

40. (B) **42.** (A)

49. (a) $R^2 - 4\left(\frac{L}{C}\right) < 0$ (underdamped)

$R^2 - 4\left(\frac{L}{C}\right) = 0$ (critically damped)

$R^2 - 4\left(\frac{L}{C}\right) > 0$ (overdamped)

53.

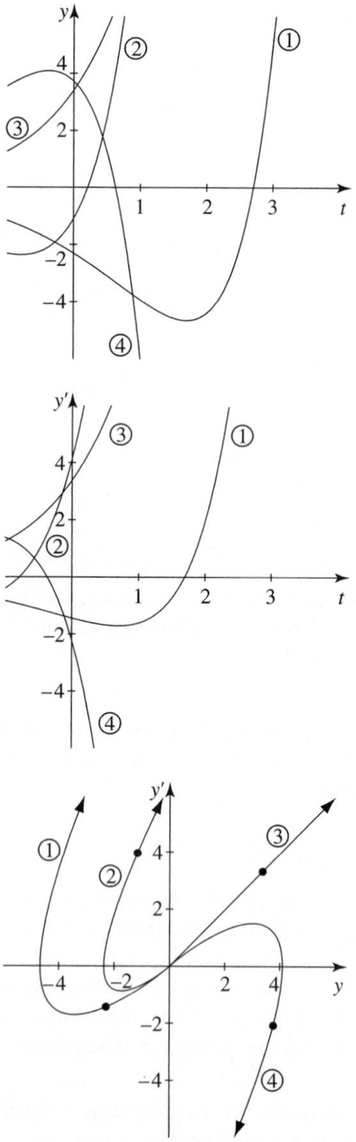

55.

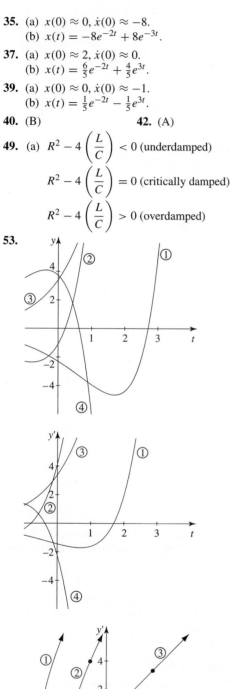

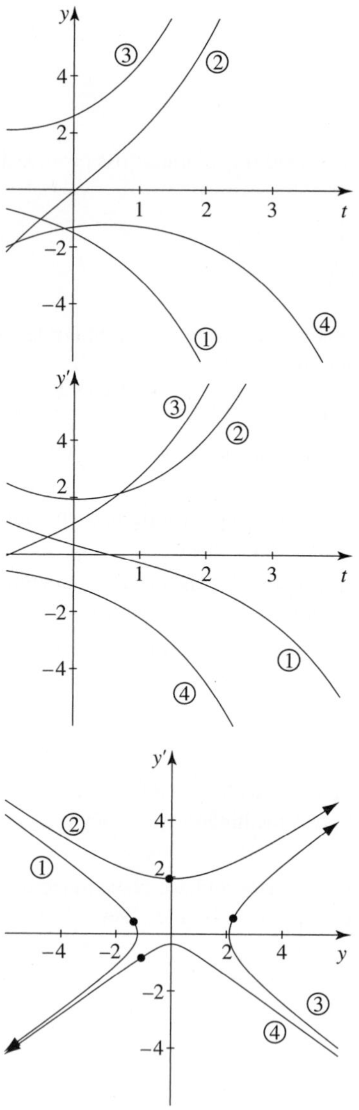

56. $\ddot{x} + 2\dot{x} + x = 0$, $x(0) = 3$ in. $= \frac{1}{4}$ ft, $\dot{x}(0) = 0$ ft/sec.

The solution is $x(t) = \frac{1}{4}e^{-t} + \frac{1}{4}te^{-t}$. This is zero only for $t = -1$, whereas the physical system does not start before $t = 0$.

59. (a) $\ddot{Q} + 15\dot{Q} + 50Q = 0$, $Q(0) = 5$, $\dot{Q}(0) = 0$

(b) $Q(t) = 10e^{-5t} - 5e^{-10t}$

(c) $I(t) = \dot{Q} = -50e^{-5t} - 50e^{-10t}$

(d) As $t \to \infty$, $Q(t) \to 0$ and $I(t) \to 0$.

62. $y(t) = c_1t^{1/2} + c_2t^{-3/2}$

64. $y(t) = c_1t^{1/2} + c_2t^{-1}$

66. $y(t) = c_1t^{-2} + c_2t^{-2}\ln t$ (for $t > 0$)

68. $y(t) = c_1t^{1/3} + c_2t^{1/3}\ln t$, $t > 0$

71. (a) $y'' + 9y' + 8y = 0$

(b) $y(0) = 2$, $y'(0) = -9$

(c)

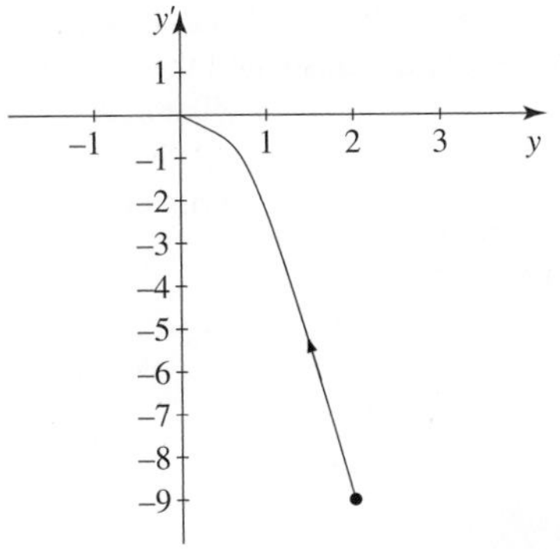

73. (a) $y'' + 2y' + y = 0$

(b) $y(0) = 1,\ y'(0) = 0$

(c)

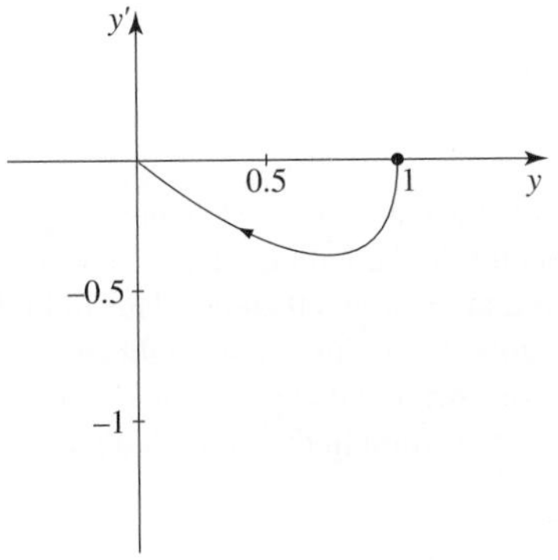

77. $y_2 = te^{2t}$

79. $y_2 = t^2 - 1$

81. $y_2 = t \displaystyle\int \frac{1}{t^2\sqrt{1-t^2}}\,dt$

Section 4.3, p. 238

1. $y(t) = c_1 \cos 3t + c_2 \sin 3t$

3. $y(t) = e^{2t}(c_1 \cos t + c_2 \sin t)$

5. $y(t) = e^{-t}(c_1 \cos \sqrt{3}t + c_2 \sin \sqrt{3}t)$

7. $y(t) = e^{5t}(c_1 \cos t + c_2 \sin t)$

9. $y(t) = e^{t/2}\left(c_1 \cos \dfrac{\sqrt{3}}{2}t + c_2 \sin \dfrac{\sqrt{3}}{2}t\right)$

11. $y(t) = \cos 2t - \dfrac{1}{2}\sin 2t$

13. $y(t) = e^{-t}(\cos t + \sin t)$

15. $y(t) = -\dfrac{1}{3}\sqrt{3}e^{2t}\sin\sqrt{3}t$

17. $y''' - 3y'' + 3y' - y = 0$

19. $y''' - 6y'' + 13y' - 10y = 0$

21. (D) **23.** (A) **25.** (G) **27.** (E)

31. $y(t) = c_1 e^{r_1 t} + c_2$, which approaches c_2 as $t \to \infty$ because $r_1 < 0$.

33. $y(t) = c_1 + c_2 t$ approaches $+\infty$ as $t \to \infty$ when $c_2 > 0$ and $-\infty$ when $c_2 < 0$.

35. $y(t) = c_1 \cos \beta t + c_2 \sin \beta t$ is a periodic function of period $\dfrac{2\pi}{\beta}$ and amplitude $\sqrt{c_1^2 + c_2^2}$.

39. $y(t) = c_1 + c_2 t + c_3 t^2 + c_4 e^{2t} + c_5 t e^{2t}$

41. $y(t) = c_1 + c_2 e^t + c_3 e^{-t} + c_4 \cos t + c_5 \sin t$

43. $y(t) = c_1 e^{-2t} + c_2 t e^{-2t} + c_3 t^2 e^{-2t}$

46.

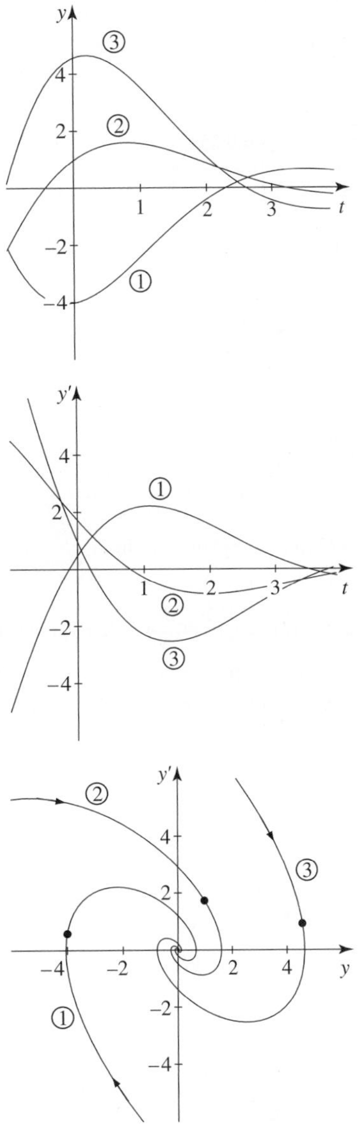

48. (a) For larger k, we have more oscillations.

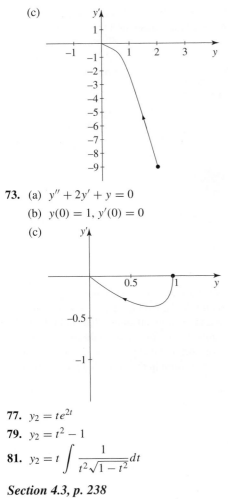

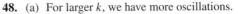

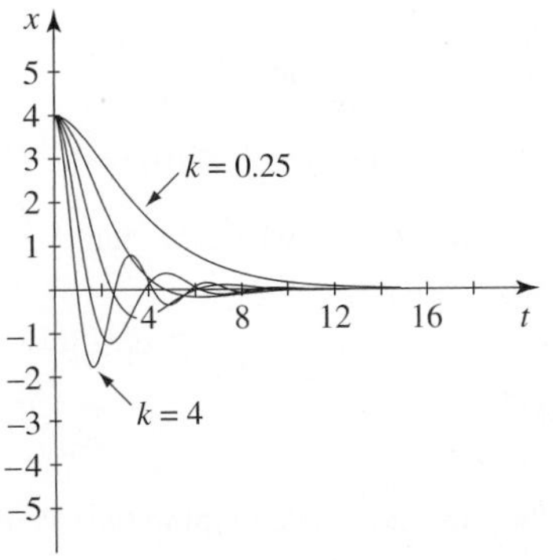

(b) For larger k, since there are more oscillations, the phase-plane trajectory spirals further around the origin.

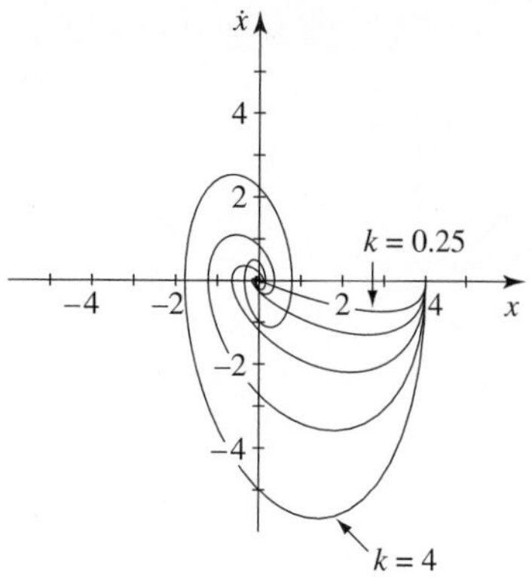

50. Maximum amplitude is $e^{-\pi/\sqrt{2}}$.

52. $y(t) = t^{-1/2}\cos\left(\frac{\sqrt{3}}{2}\ln t\right) + t^{-1/2}\sin\left(\frac{\sqrt{3}}{2}\ln t\right)$

54. $y(t) = t^{-8}(c_1, \cos(4\sqrt{3}\ln t) + c_2\sin(4\sqrt{3}\ln t))$

56. $y(t) = c_1 t + c_2 t^{-1} + c_3 t^2$ for $t > 0$.

58. For $x(0) = 0$, $\dot{x}(0) = 1$, $x(t) = \frac{1}{2}e^t - \frac{1}{2}e^{-t}$. The general solution $x(t) = c_1 e^t + c_2 e^{-t}$ will approach 0 as $t \to \infty$ if $c_1 = 0$. That happens whenever $x(0) = -\dot{x}(0)$.

60. $x(t) = e^{-t}(\cos t + 2\sin t)$

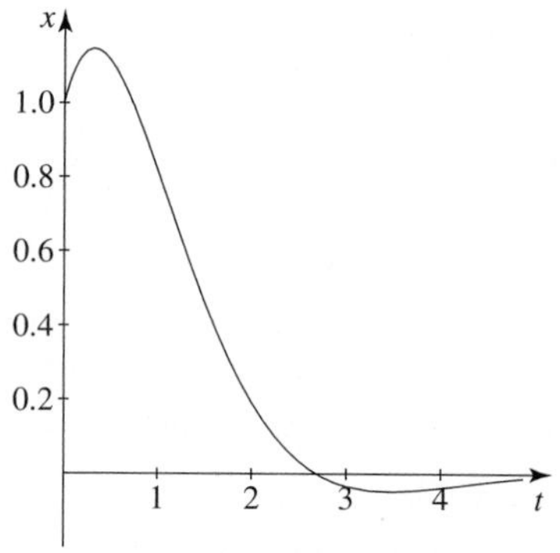

62. $x(t) = e^{-2t}\left(\cos 2\sqrt{3}t + \frac{\sqrt{3}}{2}\sin 2\sqrt{3}t\right)$

64. (a) $\ddot{Q} + 8\dot{Q} + 25Q = 0$, $Q(0) = 1$, $\dot{Q}(0) = 0$

(b) $Q(t) = \frac{5}{3}e^{-4t}\cos(3t - \delta)$, where $\delta = \tan^{-1}\left(\frac{4}{3}\right)$.

(c) $I(t) = -5e^{-4t}\sin(3t - \delta) - \frac{20}{3}e^{-4t}\cos(3t - \delta)$, where $\delta = \tan^{-1}\left(\frac{4}{3}\right)$.

(d) Charge on the capacitor and current in the circuit approach zero as $t \to +\infty$.

68. We expect larger damping initially but as time goes on the damping would steadily diminish until the system behaves more like an undamped harmonic oscillator. An experiment with a graphical DE solver confirms the first expectation but not the second—it cries for more experiment!

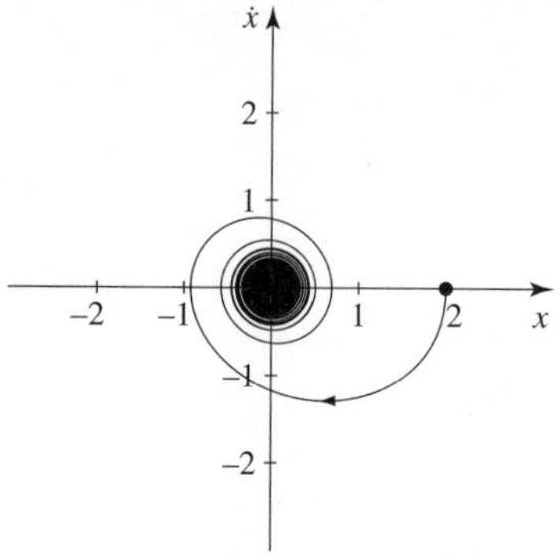

70. Negative friction for $0 < x < 1$, positive damping for $|x| \geq 1$. For a small initial condition near $x = 0$, we might expect the solution to grow and then oscillate around $x = 1$. This would be a good DE to investigate with an open-ended graphical DE solver. An initial experiment confirms all, but shows unexpected distortion in the cyclic long term behavior.

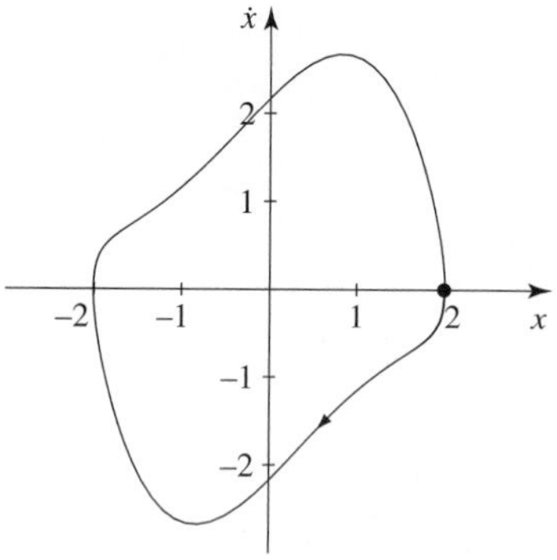

72. Damping starts large but goes to zero for large time; however, the restoring force starts small and becomes larger—which effect wins in the long term? It is difficult to predict; you might explore with an open-ended graphical DE solver. A single experiment produced the following picture.

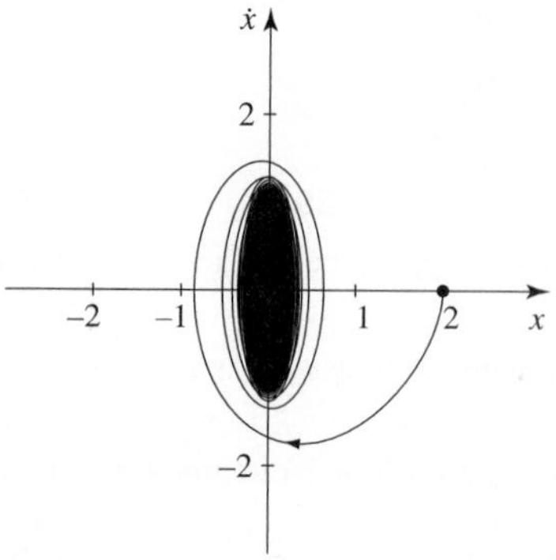

74. $y(t) = 0$

76. No solutions.

78. $y(t) = \frac{c_1}{2}t + \frac{c_2}{t}$

80. $y(t) = \frac{c_1 t + c_2}{t^2 - 2t}$

Section 4.4, p. 253

1. $y_p(t) = -t$

3. $y_p(t) = t^2$

5. $y_p(t) = 2$

7. $y_p(t) = e^t$

9. $y_p(t) = At^3 + Bt^2 + Ct + D$

11. $y_p(t) = A\sin t + B\cos t$

13. $y_p(t) = (At + B)\sin t + (Ct + D)\cos t$

15. $y_p(t) = Ae^{-t} + B\sin t + C\cos t$

17. $y(t) = t + c$

19. $y(t) = ce^{-t} + t - 1$

21. $y(t) = c_1 + c_2e^{-4t} + \frac{1}{4}t$

23. $y(t) = c_1 + c_2e^{-4t} + \frac{1}{8}t^2 - \frac{1}{16}t$

25. $y(t) = c_1\cos t + c_2\sin t + \frac{1}{2}e^t + 3$

27. $y(t) = c_1 + c_2e^{-t} - \frac{3}{5}\cos 2t - \frac{6}{5}\sin 2t$

29. $y(t) = c_1e^{-2t} + c_2te^{-2t} + te^{-t} - 2e^{-t}$

31. $y(t) = c_1\cos t + c_2\sin t - 2\cos 2t + 6$

33. $y(t) = c_1e^{2t} + c_2te^{2t} + \frac{1}{6}t^3e^{2t}$

35. $y(t) = c_1e^t + c_2e^{2t} + \frac{1}{2}e^t(\cos t - \sin t)$

37. $y(t) = c_1 + c_2t + c_3e^{4t} - \frac{1}{4}t^3 - \frac{3}{16}t^2$

39. $y(t) = c_1\cos t + c_2\sin t + c_3e^t + c_4e^{-t} - 10$

41. $y(t) = -\frac{5}{3}e^t + \frac{2}{3}e^{-2t} + 3t$

43. $y(t) = \cos 2t - \frac{5}{8}\sin 2t + \frac{1}{4}t$

45. $y(t) = \frac{16}{15}\cos\left(\frac{t}{2}\right) - \frac{1}{15}\cos 2t$

47. $y(t) = \frac{1}{3}e^{2t} + \frac{2}{3}e^{-t}$

49. $y(t) = \frac{4}{15}e^{-t} + \frac{1}{3}e^{2t} - \frac{3}{5}\cos 2t - \frac{1}{5}\sin 2t$

51. $y(t) = \frac{1}{10}\cos t + \frac{1}{5}\sin t - \frac{1}{4}e^t + \frac{1}{12}e^{-t} + \frac{1}{15}e^{2t}$

53. $y_p(t) = At + B + Ct\cos\left(\frac{t}{2}\right) + Dt\sin\left(\frac{t}{2}\right)$

55. $y_p(t) = A\cos t + B\sin t + e^t(Ct + D)$

57. (a) $y_h(t) = c_1e^{3t} + c_2e^{-2t}$

(b) (i) $y_p(t) = -\frac{1}{6}e^t$

(ii) $y_p(t) = -\frac{1}{4}e^{-t}$

(c) $y_p(t) = -\frac{1}{12}e^t - \frac{1}{8}e^{-t}$

59. $y(t) = \begin{cases} -2 + 2e^{-t} + 2t & 0 \le t < 4 \\ 3 + (2 - e^4)e^{-t} + t & t \ge 4 \end{cases}$

61. Let $y_p = Ae^{it}$. Then $A = i$ and we need $\mathrm{Im}(y_p) = \cos t$ so $y(t) = c_1e^t + c_2te^t + \cos t$.

63. Let $y_p = Ate^{5it}$. Then $A = -2i$ and $\mathrm{Im}(y_p) = -2t\cos 5t$. Thus $y(t) = c_1\cos 5t + c_2\sin 5t - 2t\cos 5t$.

Section 4.5, p. 260

1. $y_p(t) = 2t^2 - 4t + 4$,
$y(t) = c_1 + c_2e^{-t} + 2t^2 - 4t$

3. $y_p(t) = -te^t + te^t\ln t$,
$y(t) = c_1e^t + c_2te^t + te^t\ln t$

5. $y_p(t) = \sin t - t\cos t + \sin t\ln|\sec t|$,
$y(t) = c_1\cos t + c_2\sin t - t\cos t + \sin t\ln|\sec t|$

7. $y_p(t) = (e^t + e^{2t})\ln(1 + e^{-t}) - e^t$,
$y(t) = c_1e^t + c_2e^{2t} + (e^t + e^{2t})\ln(1 + e^{-t})$

9. $y_p = -\frac{1}{4}\cos 2t\,(\ln|\sec 2t + \tan 2t| - \sin 2t) - \frac{1}{4}\sin 2t\cos 2t$
$y(t) = c_1\cos 2t + c_2\sin 2t + y_p$

11. $y(t) = c_1\cos t + c_2\sin t - 1 + \sin t\ln|\sec t + \tan t|$

13. $y(t) = c_1t + c_2t^2 - t\sin t$

15. $y(t) = c_1t + c_2e^t + e^{-t}\left(\frac{1}{2} - t\right)$

18. $y_p(t) = -\frac{1}{2}te^t + \frac{1}{12}e^t - \frac{1}{3}e^t$,
$y(t) = c_1e^t + c_2e^{-t} + c_3e^{2t} - \frac{1}{2}te^t - \frac{1}{4}e^t$

20. $y(t) = c_1 + c_2\cos 3t + c_3\sin 3t - \frac{1}{27}\ln|\cos 3t| + \frac{1}{27}\cos^2 3t + \frac{\sin 3t}{27}(\ln|\sec 3t + \tan 3t - \sin 3t|)$

23. $y_p(t) = \int_0^t \sinh(t - s)f(s)\,ds$

Section 4.6, p. 270

1. $x(t) = c_1e^{-t} + c_2te^{-t} + 3\sin t$
$x_{ss} = 3\sin t = 3\cos(t - \pi/2)$
Amplitude $= 3$, phase shift $= \pi/2$.

3. $x(t) = c_1\cos\sqrt{\frac{3}{2}}t + c_2\sin\sqrt{\frac{3}{2}}t - \frac{4}{125}\cos 8t$
$x_{ss} = -\frac{4}{125}\cos 8t = \frac{4}{125}\cos(8t - \pi)$
Amplitude $= \frac{4}{125}$, phase shift $\frac{\delta}{\beta} = \frac{\pi}{8}$.

5. $x(t) = e^{-t}(c_1 \cos t + c_2 \sin t) + \frac{2}{5}\cos t + \frac{4}{5}\sin t$

$x_{ss} = \frac{2}{\sqrt{5}}\cos(t - 1.1)$

Amplitude $= \frac{2}{\sqrt{5}}$, phase shift $\frac{\delta}{\beta} \approx 1.1$ radians.

7. $x(t) = -\frac{34}{5}e^{-4t} + \frac{23}{5}e^{-6t} + \frac{1}{5}\cos 2t + \frac{1}{5}\sin 2t$

10. (a) $\omega_f = 6.04$ rad/sec

(b) $x_{ss}(t) \approx 0.029\cos\left(7t - \frac{\pi}{2}\right)$

(c) Without damping, $x_{ss}(t) = At\cos 7t + Bt\sin 7t$, or $Ct\cos(7t - \delta)$.

12. $x_{ss}(t) = \frac{3}{2}\sin 6t$

14. $Q(t) = -\frac{5}{18}\cos 5t + \frac{5}{18}\cos 4t$

16. True; $x_{ss}t = C\cos(\omega_f t - \delta)$.

18. $\cos 3t - \cos t = -2\sin 2t \sin t$

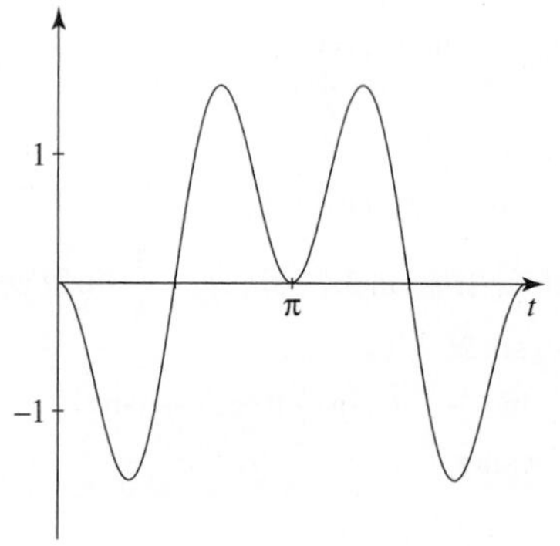

20. $x_{ss}(t) = 0.20\cos(t - 0.93)$

22. $x_{ss}(t) = \frac{4}{\sqrt{73}}\cos(3t - 2.78)$

24. $x(t) = \frac{4\sqrt{3}}{3}t\sin\left(2\sqrt{3}t\right)$

25. (a) $x_{ss}(t) = \frac{49}{80\pi^2}\cos\left(\frac{2\pi t}{7}\right) \approx 0.6\cos\left(\frac{2\pi t}{7}\right)$

(b) The buoy is always at least 0.06 feet above water because the steady-state solution is in phase with the waves that are forcing it.

28. (A)

30. (B)

32. (a) $x_h = 4\cos 4t - 3\sin 4t$

(b) The amplitude of x_h is 5.

(c) The amplitude (time-varying) of x_p is $5t$.

(d) x_p will be unchanged.

(e) $k = 16$ nt/m

(f) Pure resonance

34. (a) $x_h = 3e^{-2t}\cos t - 2e^{-2t}\sin t$

(b) $b = 4$

(c) Underdamped

(d) The amplitude (time-varying) of x_h is $\sqrt{13}e^{-2t}$.

(e) $x_{ss} = x_p = \sqrt{2}\cos(5t - \delta)$

(f) $\omega_f = 5$, $F_0 = 40$ nt

35. Define $\theta = \tan^{-1}\frac{y_0}{x_0}$.

(a) $y_D(t) = v_0 \sin\theta t - \frac{1}{2}gt^2$

$y_T(t) = y_0 - \frac{1}{2}gt^2$

(b) $y_D(t) = y_T(t)$ when $t^* = \frac{y_0}{v_0 \sin\theta}$.
Show that $x_D(t^*) = x_T(t^*)$.

(c) $y_T(t^*) = y_0 - \frac{1}{2}g(t^*)^2 = y_0 - \frac{1}{2}g\left(\frac{x_0^2 + y_0^2}{v_0^2}\right)$.

Section 4.7, p. 281

1. $E = \frac{17}{2}$

3. $E(t) = \frac{1}{2}LI_0^2 + \frac{1}{2C}Q_0^2$

5. $E(t) = e^{-t}(t^2 - 2t + 2)$;

energy loss $= 2 - e^{-t}(t^2 - 2t + 2)$

7. (a) $E(x, \dot{x}) = \frac{1}{2}\dot{x}^2 - \frac{1}{2}x^2 - \frac{1}{4}x^4$

(b) (0, 0) is an unstable equilibrium point.

(c)

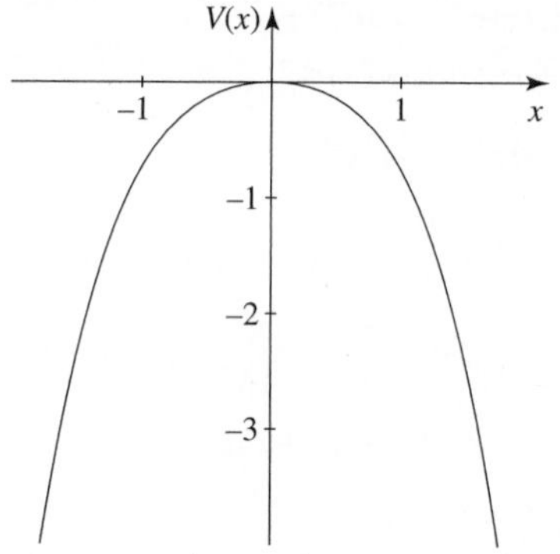

9. (a) $E(x, \dot{x}) = \frac{1}{2}\dot{x}^2 + \frac{1}{3}x^3$

(b) (0, 0) is an unstable (or semistable) equilibrium point.

(c)

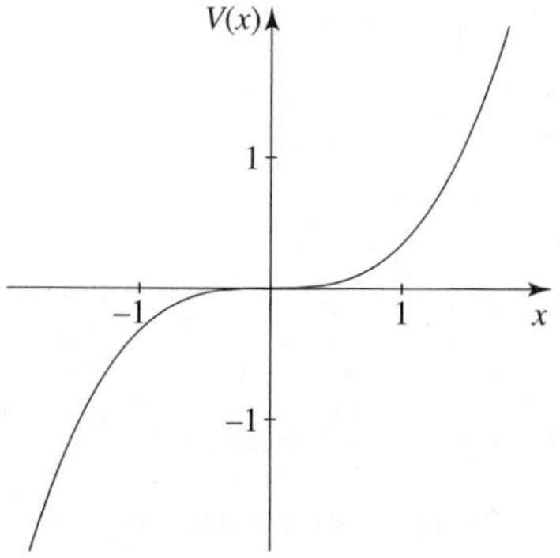

11. (a) $E(x, \dot{x}) = \frac{1}{2}\dot{x}^2 + \frac{1}{3}x^3 - x^2 + x$

(b) $(1, 0)$ is an unstable (or semistable) equilibrium point.

(c)

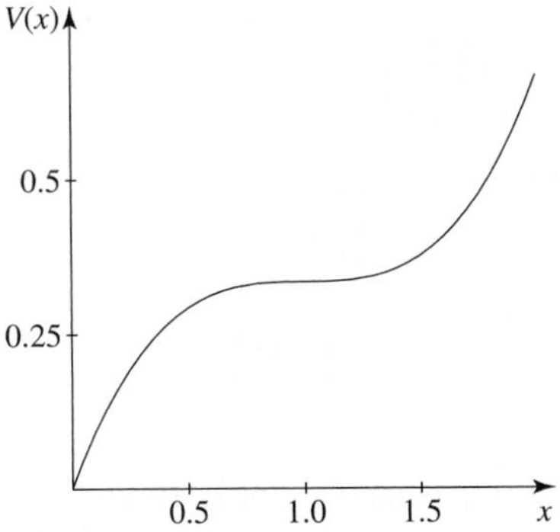

13. (a) $E(x, \dot{x}) = \frac{1}{2}\dot{x}^2 - \frac{1}{3}x^3 + \frac{3}{2}x^2 - 2x$

(b) $(1, 0)$ is a stable equilibrium point; $(2, 0)$ is an unstable equilibrium point.

(c)

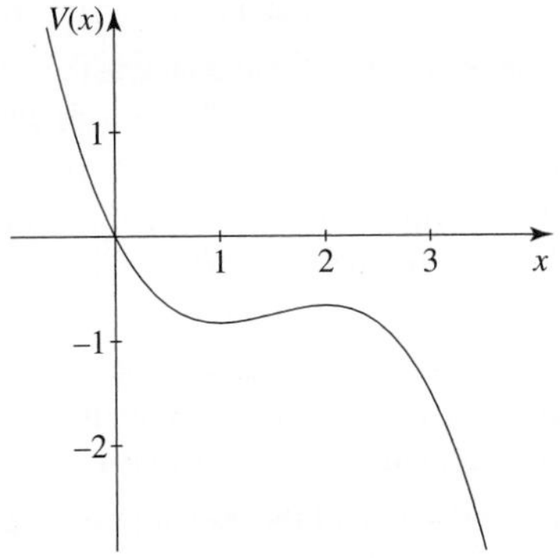

15. Conservative, trajectories are ellipses each with height $\sqrt{K}$ times its width.

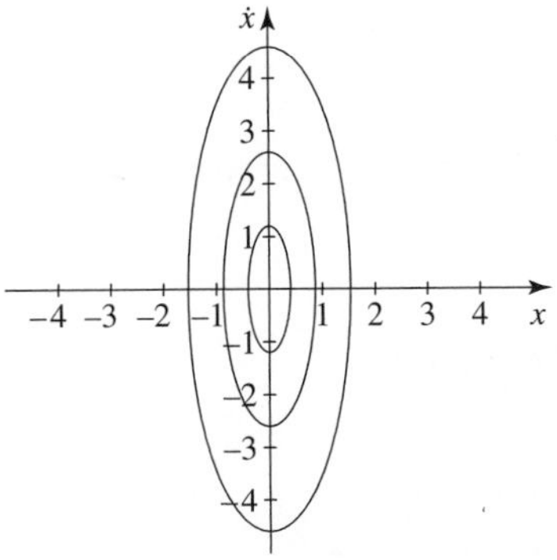

17. Conservative, trajectories are drawn below.

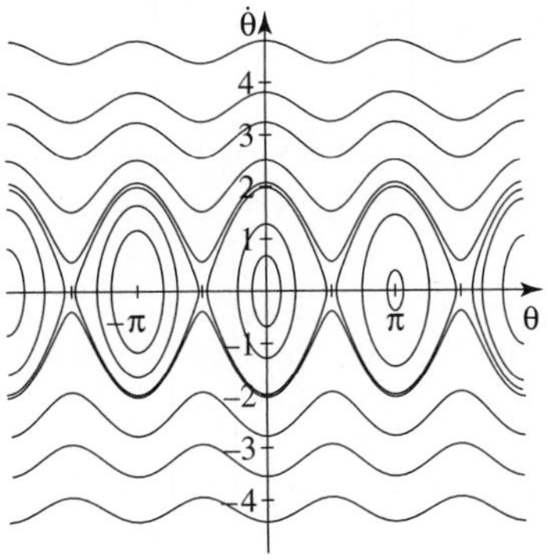

19. Not conservative; trajectories cannot be level curves for any surface. See graph.

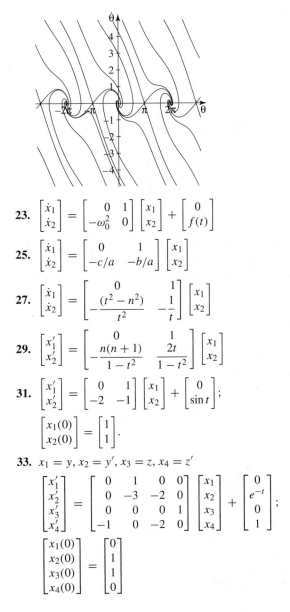

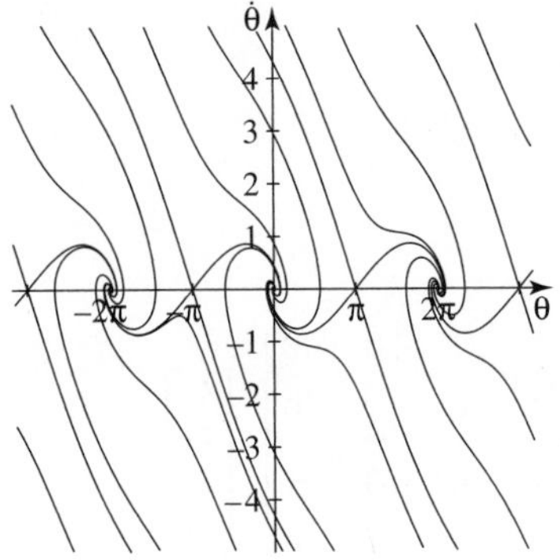

23. $\begin{bmatrix} \dot{x}_1 \\ \dot{x}_2 \end{bmatrix} = \begin{bmatrix} 0 & 1 \\ -\omega_0^2 & 0 \end{bmatrix} \begin{bmatrix} x_1 \\ x_2 \end{bmatrix} + \begin{bmatrix} 0 \\ f(t) \end{bmatrix}$

25. $\begin{bmatrix} \dot{x}_1 \\ \dot{x}_2 \end{bmatrix} = \begin{bmatrix} 0 & 1 \\ -c/a & -b/a \end{bmatrix} \begin{bmatrix} x_1 \\ x_2 \end{bmatrix}$

27. $\begin{bmatrix} \dot{x}_1 \\ \dot{x}_2 \end{bmatrix} = \begin{bmatrix} 0 & 1 \\ -\dfrac{(t^2 - n^2)}{t^2} & -\dfrac{1}{t} \end{bmatrix} \begin{bmatrix} x_1 \\ x_2 \end{bmatrix}$

29. $\begin{bmatrix} x_1' \\ x_2' \end{bmatrix} = \begin{bmatrix} 0 & 1 \\ -\dfrac{n(n+1)}{1 - t^2} & \dfrac{2t}{1 - t^2} \end{bmatrix} \begin{bmatrix} x_1 \\ x_2 \end{bmatrix}$

31. $\begin{bmatrix} x_1' \\ x_2' \end{bmatrix} = \begin{bmatrix} 0 & 1 \\ -2 & -1 \end{bmatrix} \begin{bmatrix} x_1 \\ x_2 \end{bmatrix} + \begin{bmatrix} 0 \\ \sin t \end{bmatrix}$;

$\begin{bmatrix} x_1(0) \\ x_2(0) \end{bmatrix} = \begin{bmatrix} 1 \\ 1 \end{bmatrix}$.

33. $x_1 = y, x_2 = y', x_3 = z, x_4 = z'$

$\begin{bmatrix} x_1' \\ x_2' \\ x_3' \\ x_4' \end{bmatrix} = \begin{bmatrix} 0 & 1 & 0 & 0 \\ 0 & -3 & -2 & 0 \\ 0 & 0 & 0 & 1 \\ -1 & 0 & -2 & 0 \end{bmatrix} \begin{bmatrix} x_1 \\ x_2 \\ x_3 \\ x_4 \end{bmatrix} + \begin{bmatrix} 0 \\ e^{-t} \\ 0 \\ 1 \end{bmatrix}$;

$\begin{bmatrix} x_1(0) \\ x_2(0) \\ x_3(0) \\ x_4(0) \end{bmatrix} = \begin{bmatrix} 0 \\ 1 \\ 1 \\ 0 \end{bmatrix}$

35. $z_1 = x_1, z_2 = \dot{x}_1, z_3 = x_2, z_4 = \dot{x}_2$

$$\begin{bmatrix} \dot{z}_1 \\ \dot{z}_2 \\ \dot{z}_3 \\ \dot{z}_4 \end{bmatrix} = \begin{bmatrix} 0 & 1 & 0 & 0 \\ -1 & 0 & -2 & 0 \\ 0 & 0 & 0 & 1 \\ 0 & 0 & -2 & 0 \end{bmatrix} \begin{bmatrix} z_1 \\ z_2 \\ z_3 \\ z_4 \end{bmatrix} + \begin{bmatrix} 0 \\ e^{-t} \\ 0 \\ 0 \end{bmatrix}$$

37. $z_1 = x_1, z_2 = \dot{x}_1, z_3 = x_2, z_4 = \dot{x}_2, z_5 = x_3, z_6 = \dot{x}_3$

$$\begin{bmatrix} \dot{z}_1 \\ \dot{z}_2 \\ \dot{z}_3 \\ \dot{z}_4 \\ \dot{z}_5 \\ \dot{z}_6 \end{bmatrix} = \begin{bmatrix} 0 & 1 & 0 & 0 & 0 & 0 \\ a_{11} & 0 & a_{12} & 0 & a_{13} & 0 \\ 0 & 0 & 0 & 1 & 0 & 0 \\ a_{21} & 0 & a_{22} & 0 & a_{23} & 0 \\ 0 & 0 & 0 & 0 & 0 & 1 \\ a_{31} & 0 & a_{32} & 0 & a_{33} & 0 \end{bmatrix} \begin{bmatrix} z_1 \\ z_2 \\ z_3 \\ z_4 \\ z_5 \\ z_6 \end{bmatrix}$$

39. $x_1 = c_1e^{-t} + c_2e^{2t}$

$x_2 = 2c_1e^{-t} + \frac{1}{2}c_2e^{2t}$

41. $x_1 = c_1e^{t} + c_2e^{2t} - \frac{3}{2}t + \frac{3}{4}$

$x_2 = c_1e^{t} + 2c_2e^{2t} - t - \frac{3}{2}$

43. $x_1 = e^{-t}$

$x_2 = -e^{-t}$

45. $z_1 = x_1, z_2 = \dot{x}_1, z_3 = x_2, z_4 = \dot{x}_2$

$$\begin{bmatrix} \dot{z}_1 \\ \dot{z}_2 \\ \dot{z}_3 \\ \dot{z}_4 \end{bmatrix} = \begin{bmatrix} 0 & 1 & 0 & 0 \\ -\frac{(k_1+k_2)}{m} & 0 & \frac{k_2}{m} & 0 \\ 0 & 0 & 0 & 1 \\ \frac{k_2}{m} & 0 & -\frac{k_2}{m} & 0 \end{bmatrix} \begin{bmatrix} z_1 \\ z_2 \\ z_3 \\ z_4 \end{bmatrix}$$

47. $x_1 = \theta_1, x_2 = \dot{\theta}_1, x_3 = \theta_2, x_4 = \dot{\theta}_2$

$$\begin{bmatrix} \dot{x}_1 \\ \dot{x}_2 \\ \dot{x}_3 \\ \dot{x}_4 \end{bmatrix} = \begin{bmatrix} 0 & 1 & 0 & 0 \\ mg+1 & 0 & mg & 0 \\ 0 & 0 & 0 & 1 \\ mg & 0 & mg+1 & 0 \end{bmatrix} \begin{bmatrix} x_1 \\ x_2 \\ x_3 \\ x_4 \end{bmatrix} + \begin{bmatrix} 0 \\ -u(t) \\ 0 \\ -u(t) \end{bmatrix}$$

CHAPTER 5

Section 5.1, p. 294

1. Not linear; $T(\mathbf{u}+\mathbf{v}) \neq T(\mathbf{u}) + T(\mathbf{v})$
3. Not linear; $cT(\mathbf{u}) \neq T(c\mathbf{u})$
5. Linear
7. Linear
9. Linear
11. Linear
13. Linear
15. Linear
21. Not linear; $T(k\mathbf{x}) \neq kT(\mathbf{x})$
23. Not linear; e.g., $T(2+3) \neq T(2) + T(3)$
25. Linear

29. $\begin{bmatrix} 1 & 0 \\ 0 & -1 \end{bmatrix}$; reflects points about the x-axis

31. $\begin{bmatrix} 1 & 0 \\ 1 & 0 \end{bmatrix}$; projects points vertically onto the 45-degree line $y = x$

33. $T(x, y) = \begin{bmatrix} 1 & 2 \end{bmatrix} \begin{bmatrix} x \\ y \end{bmatrix}$

35. $T(x, y) = \begin{bmatrix} 1 & 2 \\ 1 & -2 \end{bmatrix} \begin{bmatrix} x \\ y \end{bmatrix}$

37. $T(x, y, z) = \begin{bmatrix} 1 & 2 & 0 \\ 1 & -2 & 0 \\ 1 & 1 & -2 \end{bmatrix} \begin{bmatrix} x \\ y \\ z \end{bmatrix}$

39. $T(v_1, v_2, v_3) = \begin{bmatrix} 1 & 2 & 0 \\ 0 & 0 & 1 \\ -1 & 4 & 3 \end{bmatrix} \begin{bmatrix} v_1 \\ v_2 \\ v_3 \end{bmatrix}$

41. $T(0, 0) = (0, 0)$. The point $(0, 0)$ maps into $(0, 0)$.

43. $T(0, 1, 2) = (0, 3)$. Any point $(1, 2 - \alpha, \alpha)$ maps into $(1, 2)$ where α is any real number. These points form a line in $\mathbb{R}^3$.

45. $T(1, 1) = (1, 2, 0)$. No points map into $(1, 1, 0)$.

47. $T(1, 1, 1) = (2, 0)$. The line $\{(-\alpha, \alpha, \alpha) \mid \alpha \in \mathbb{R}\}$ in $\mathbb{R}^3$ maps into $(0, 0)$.

49. The original square has area 1; the image is the parallelogram with vertices $(0, 0)$, $(1, 2)$, $(0, 3)$ and $(-1, 1)$ and area 3.

51. The original rectangle has area 2; the image is the parallelogram with vertices $(0, 0)$, $(1, 2)$, $(-1, 4)$ and $(-2, 2)$ and area 6.

53. For the square of Problem 49 under the transformation defined by $\mathbf{B}$, the image is a parallelogram with vertices $(0, 0)$, $(2, -4)$, $(1, -1)$ and $(-1, 3)$ and area 2.

For the rectangle of Problem 51 under the transformation defined by $\mathbf{B}$, the image is a parallelogram with vertices $(0, 0)$, $(2, -4)$, $(0, -2)$ and $(-2, 6)$ and area 4.

$|\mathbf{B}| = 2$; in each case the area of the transformed image is twice the area of the original figure. The determinant is a scale factor for the area.

55. **J** describes (C).

57. **L** describes (G).

59. **N** describes (A).

61.

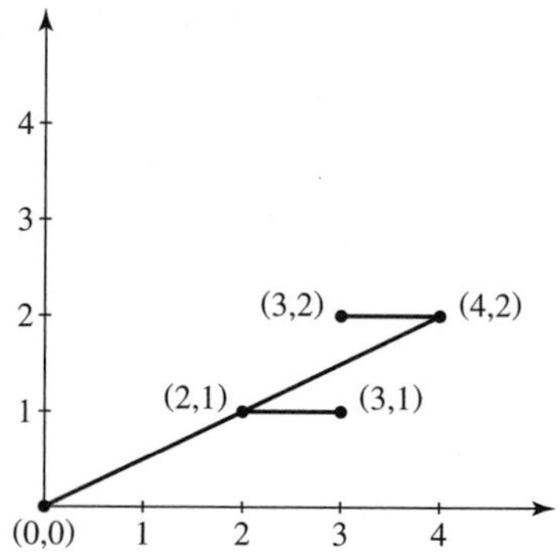

63. (a) A negative shear of 1 in the y direction is $\begin{bmatrix} 1 & 0 \\ -1 & 1 \end{bmatrix}$.

Twelve rotations of 30 degrees will give the identity matrix.

(b) $(\mathbf{R}_{30^\circ})^n = \mathbf{I}$ only when n is a multiple of 12.

65. $\frac{1}{2}\begin{bmatrix} \sqrt{3}-1 & -1 \\ \sqrt{3}+1 & \sqrt{3} \end{bmatrix}$

(0, 0)

67. (a) $(DI)(f) = f(x)$

(b) $(ID)(f) = f(x) - f(a)$

(c) They commute if $f(a) = 0$.

69. (a) $\alpha\begin{bmatrix} 1 \\ -1 \\ 0 \end{bmatrix}$, α any real number

(b) $\begin{bmatrix} 2 \\ 0 \\ 1 \end{bmatrix} + \alpha\begin{bmatrix} 1 \\ -1 \\ 0 \end{bmatrix}$, α any real number.

(c) The image of T is all of $\mathbb{R}^2$.

71. Not a linear functional

73. Not a linear functional

77. $\mathbb{W}$ is the xy-plane in $\mathbb{R}^3$.

79. $\mathbb{W}$ is the line spanned by $\begin{bmatrix} -1 \\ 0 \\ 3 \end{bmatrix}$. T is not a projection because T does not reduce to the identity on $\mathbb{W}$.

83. (a) Only $\begin{bmatrix} 0 \\ 0 \end{bmatrix}$

(b) None

(c) All

(d) Only $\begin{bmatrix} 0 \\ 0 \end{bmatrix}$

(e) $\begin{bmatrix} 0 \\ 1 \end{bmatrix}$

(f) $\mathbb{R}^2$

85. (a) Only vectors in the direction of $\begin{bmatrix} 1 \\ 1 \end{bmatrix}$

(b) Only vectors in the direction of $\begin{bmatrix} 1 \\ 1 \end{bmatrix}$

(c) All

(d) Only $\begin{bmatrix} 0 \\ 0 \end{bmatrix}$

(e) $\begin{bmatrix} 0 \\ 1 \end{bmatrix}$

(f) $\mathbb{R}^2$

87. (a) Only $\begin{bmatrix} 0 \\ 0 \end{bmatrix}$

(b) Only vectors in the direction of $\begin{bmatrix} 1 \\ 0 \end{bmatrix}$ or $\begin{bmatrix} 0 \\ 1 \end{bmatrix}$

(c) Only $\begin{bmatrix} 0 \\ 0 \end{bmatrix}$

(d) Only $\begin{bmatrix} 0 \\ 0 \end{bmatrix}$

(e) $\begin{bmatrix} 1/2 \\ 0 \end{bmatrix}$

(f) $\mathbb{R}^2$

Section 5.2, p. 309

1. $\{(0, 0)\}$

3. Line $\{(0, 0, \alpha) \mid \alpha \in \mathbb{R}\}$; i.e., the z axis in $\mathbb{R}^3$

5. Family of constant functions $f(t) = c$

7. Family of solutions $y(t) = ce^{-\int p(t)dt}$

9. $\left\{\begin{bmatrix} 0 & 0 & 0 \\ 0 & 0 & 0 \end{bmatrix}\right\}$

11. All functions $p(t)$ for which $\int_0^x p(t)dt = 0$

13. All polynomials in $\mathbb{P}_2$ of the form $p(t) = bt + c$

15. All polynomials in $\mathbb{P}_3$ of the form $p(t) = ct + d$

19. $y(t) = \cos t - \sin t + t^2 - 2$

21. The kernel consists of all points in $\mathbb{R}^2$. The dim Ker (T) is 2. The image contains only the zero vector. The dim Im(T) is 0. T is neither injective nor surjective.

23. The kernel is $\{[0, \alpha] \mid \alpha \in \mathbb{R}\}$. The dim Ker$(T)$ is 1. The image is $\{[\beta, 0] \mid \beta \in \mathbb{R}\}$. The dim Im$(T)$ is 1. T is neither injective nor surjective.

25. The kernel is $\{[x, y] \mid x + 2y = 0\}$. The dim Ker$(T)$ is 1. The image is the line in $\mathbb{R}^2$ spanned by $\begin{bmatrix} 1 \\ 2 \end{bmatrix}$. The dim Im$(T)$ is 1. T is neither injective nor surjective.

27. The kernel is the line $\{[-\alpha, 0, \alpha] \mid \alpha \in \mathbb{R}\}$ in $\mathbb{R}^3$. The dim Ker(T) is 1. The image is all vectors of the form $(x+z)\begin{bmatrix} 1 \\ 1 \end{bmatrix} + y\begin{bmatrix} 1 \\ 2 \end{bmatrix}$. The dim Im$(T)$ is 2. T is surjective but not injective.

29. The kernel is the line $\{[-\alpha, 0, \alpha] \mid \alpha \in \mathbb{R}\}$ in $\mathbb{R}^3$. The dim Ker(T) is 1. The image is $\mathbb{R}^2$. The dim Im(T) is 2. T is surjective but not injective.

31. The kernel contains only $\begin{bmatrix}0\\0\end{bmatrix}$. The dim Ker($T$) is 0. The image is all vectors of the form $x\begin{bmatrix}1\\1\\1\end{bmatrix}+y\begin{bmatrix}1\\2\\1\end{bmatrix}$.
The dim Im(T) is 2. T is injective but not surjective.

33. The kernel is all of $\mathbb{R}^2$. The dim Ker(T) is 2. The image contains only $\begin{bmatrix}0\\0\\0\end{bmatrix}$. The dim Im($T$) is 0. T is neither injective nor surjective.

35. The kernel contains only $\begin{bmatrix}0\\0\\0\end{bmatrix}$. The dim Ker($T$) is 0. The image is all vectors of the form $x\begin{bmatrix}1\\0\\0\end{bmatrix}+y\begin{bmatrix}2\\1\\0\end{bmatrix}+z\begin{bmatrix}1\\1\\1\end{bmatrix}$.
The dim Im(T) is 3. T is both injective and surjective.

37. The dim Ker(T) is 0. T is both injective and surjective. The dim Im(T) is 3.

39. The kernel contains only $\begin{bmatrix}0\\0\\0\end{bmatrix}$. The dim Ker($T$) is 0. The image is $\mathbb{R}^3$. The dim Im(T) is 3. T is both injective and surjective.

45. (b) A basis for Ker(T) is $\{t^2 - t\}$.
(c) A basis for Im(T) is $\left\{\begin{bmatrix}1\\0\end{bmatrix}, \begin{bmatrix}0\\1\end{bmatrix}\right\}$

47. Ker $(T) = \{d \mid d \in \mathbb{R}\}$
Im $(T) = \{qx^2 + rx + s \mid q, r, s \in \mathbb{R}\}$

49. Ker $(T) = \left\{\begin{bmatrix}-b & b\\-d & d\end{bmatrix} \middle| b, d \in \mathbb{R}\right\}$
Im $(T) = \mathbb{R}^2$

51. Ker $(T) = \{\vec{\mathbf{0}}\}$
Im $(T) = \left\{x\begin{bmatrix}1\\0\\1\end{bmatrix}+y\begin{bmatrix}1\\0\\-1\end{bmatrix} \middle| x, y \in \mathbb{R}\right\}$

53. $\mathbf{A} = \begin{bmatrix}2 & 0 & 0\\0 & 0 & 0\\0 & 0 & 0\end{bmatrix}$, or any matrix for which a_{11} is the only nonzero element.

55. False

57. True

59. False

61. The dim Ker(T) is 2. The dim Im(T) is 2. T is neither injective nor surjective.

63. The dim Ker(T) is 0. The dim Im(T) is 3. T is injective but not surjective.

64. The dim Ker(T) is 2. The dim Im(T) is 1. T is neither injective nor surjective.

66. $\frac{1}{2}\begin{bmatrix}1\\1\end{bmatrix}+c\begin{bmatrix}-1\\1\end{bmatrix}$

68. $\frac{1}{3}\begin{bmatrix}2\\2\end{bmatrix}$

70. $\begin{bmatrix}0\\3\\0\end{bmatrix}+c\begin{bmatrix}-1\\1\\1\end{bmatrix}$

72. $y(t) = ce^t - 3$

74. $y(t) = \frac{c}{t} + 1$

76. $y(t) = c_1e^{-t^3/3} + 3$

78. $y(t) = c_1e^t + c_2e^{-2t} - t + 1$

80. $y(t) = c_1e^t + c_2te^t + t - 1$

Section 5.3, p. 324

1. $\lambda_1 = 1, \lambda_2 = 2; \vec{\mathbf{v}}_1 = c\begin{bmatrix}0\\1\end{bmatrix}, \vec{\mathbf{v}}_2 = c\begin{bmatrix}1\\0\end{bmatrix}, c \in \mathbb{R}$.

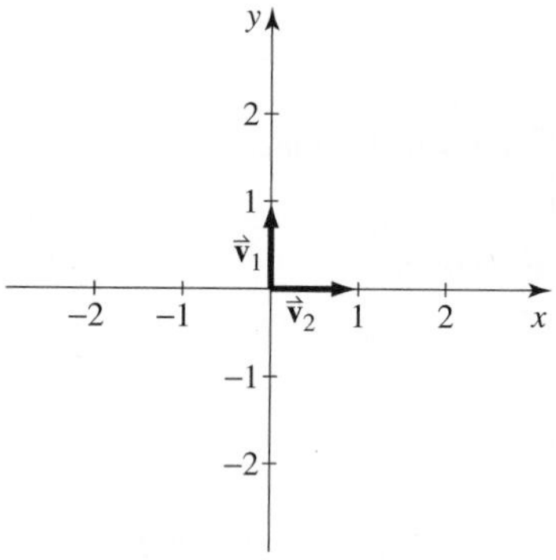

3. $\lambda_1 = 0, \lambda_2 = 3; \vec{\mathbf{v}}_1 = c\begin{bmatrix}2\\-1\end{bmatrix}, \vec{\mathbf{v}}_2 = c\begin{bmatrix}1\\1\end{bmatrix}, c \in \mathbb{R}$.

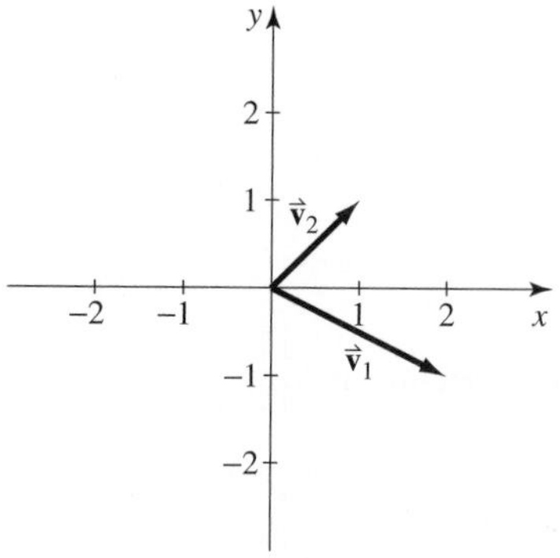

5. $\lambda_1 = 0, \lambda_2 = 4; \vec{\mathbf{v}}_1 = c\begin{bmatrix}-3\\1\end{bmatrix}, \vec{\mathbf{v}}_2 = c\begin{bmatrix}1\\1\end{bmatrix}, c \in \mathbb{R}.$

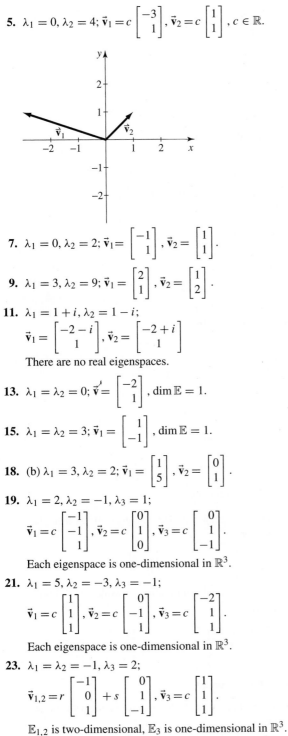

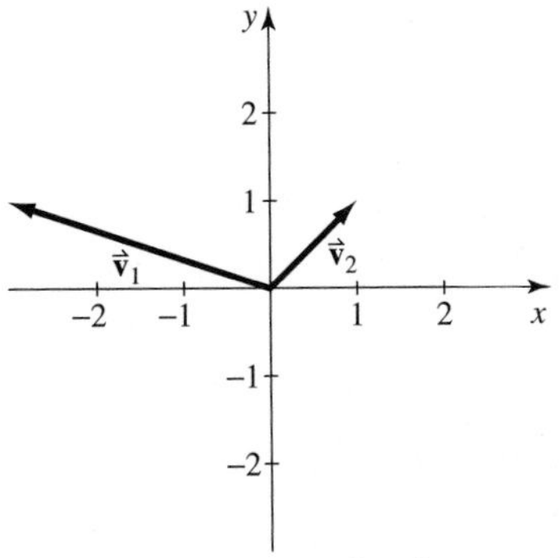

7. $\lambda_1 = 0, \lambda_2 = 2; \vec{\mathbf{v}}_1 = \begin{bmatrix}-1\\1\end{bmatrix}, \vec{\mathbf{v}}_2 = \begin{bmatrix}1\\1\end{bmatrix}.$

9. $\lambda_1 = 3, \lambda_2 = 9; \vec{\mathbf{v}}_1 = \begin{bmatrix}2\\1\end{bmatrix}, \vec{\mathbf{v}}_2 = \begin{bmatrix}1\\2\end{bmatrix}.$

11. $\lambda_1 = 1 + i, \lambda_2 = 1 - i;$

$$\vec{\mathbf{v}}_1 = \begin{bmatrix}-2-i\\1\end{bmatrix}, \vec{\mathbf{v}}_2 = \begin{bmatrix}-2+i\\1\end{bmatrix}$$

There are no real eigenspaces.

13. $\lambda_1 = \lambda_2 = 0; \vec{\mathbf{v}} = \begin{bmatrix}-2\\1\end{bmatrix}, \dim \mathbb{E} = 1.$

15. $\lambda_1 = \lambda_2 = 3; \vec{\mathbf{v}}_1 = \begin{bmatrix}1\\-1\end{bmatrix}, \dim \mathbb{E} = 1.$

18. (b) $\lambda_1 = 3, \lambda_2 = 2; \vec{\mathbf{v}}_1 = \begin{bmatrix}1\\5\end{bmatrix}, \vec{\mathbf{v}}_2 = \begin{bmatrix}0\\1\end{bmatrix}.$

19. $\lambda_1 = 2, \lambda_2 = -1, \lambda_3 = 1;$

$$\vec{\mathbf{v}}_1 = c\begin{bmatrix}-1\\-1\\1\end{bmatrix}, \vec{\mathbf{v}}_2 = c\begin{bmatrix}0\\1\\0\end{bmatrix}, \vec{\mathbf{v}}_3 = c\begin{bmatrix}0\\1\\-1\end{bmatrix}.$$

Each eigenspace is one-dimensional in $\mathbb{R}^3$.

21. $\lambda_1 = 5, \lambda_2 = -3, \lambda_3 = -1;$

$$\vec{\mathbf{v}}_1 = c\begin{bmatrix}1\\1\\1\end{bmatrix}, \vec{\mathbf{v}}_2 = c\begin{bmatrix}0\\-1\\1\end{bmatrix}, \vec{\mathbf{v}}_3 = c\begin{bmatrix}-2\\1\\1\end{bmatrix}.$$

Each eigenspace is one-dimensional in $\mathbb{R}^3$.

23. $\lambda_1 = \lambda_2 = -1, \lambda_3 = 2;$

$$\vec{\mathbf{v}}_{1,2} = r\begin{bmatrix}-1\\0\\1\end{bmatrix} + s\begin{bmatrix}0\\1\\-1\end{bmatrix}, \vec{\mathbf{v}}_3 = c\begin{bmatrix}1\\1\\1\end{bmatrix}.$$

$\mathbb{E}_{1,2}$ is two-dimensional, $\mathbb{E}_3$ is one-dimensional in $\mathbb{R}^3$.

25. $\lambda_1 = 2, \lambda_2 = -\frac{1}{2} + \frac{1}{2}i\sqrt{3}, \lambda_3 = -\frac{1}{2} - \frac{1}{2}i\sqrt{3};$

$$\vec{\mathbf{v}}_1 = \begin{bmatrix}1\\1\\3\end{bmatrix}, \vec{\mathbf{v}}_2 = \begin{bmatrix}\frac{7}{2} - \frac{1}{2}i\sqrt{3}\\1\\\frac{5}{2} + \frac{3}{2}i\sqrt{3}\end{bmatrix}, \vec{\mathbf{v}}_3 = \begin{bmatrix}\frac{7}{2} + \frac{1}{2}i\sqrt{3}\\1\\\frac{5}{2} - \frac{3}{2}i\sqrt{3}\end{bmatrix}.$$

$\mathbb{E}_1$ is one-dimensional in $\mathbb{R}^3$; λ_2 and λ_3 have no real eigenspaces.

27. $\lambda_1 = \lambda_2 = 1; \lambda_2 = 3;$

$$\mathbb{E}_{1,2} = \text{span}\left\{\begin{bmatrix}1\\2\\0\end{bmatrix}, \begin{bmatrix}0\\0\\1\end{bmatrix}\right\}, \vec{\mathbf{v}}_3 = \begin{bmatrix}0\\1\\1\end{bmatrix}.$$

29. $\lambda_1 = 1, \lambda_2 = 2, \lambda_3 = 3;$

$$\vec{\mathbf{v}}_1 = \begin{bmatrix}1\\1\\0\end{bmatrix}, \vec{\mathbf{v}}_2 = \begin{bmatrix}2\\1\\1\end{bmatrix}, \vec{\mathbf{v}}_3 = \begin{bmatrix}0\\0\\1\end{bmatrix}.$$

31. $\lambda_1 = 2, \lambda_2 = \lambda_3 = 4, \lambda_4 = 6;$

$$\vec{\mathbf{v}}_1 = \begin{bmatrix}1\\0\\0\\0\end{bmatrix}, \mathbb{E}_{2,3} = \text{span}\left\{\begin{bmatrix}1\\2\\0\\0\end{bmatrix}, \begin{bmatrix}-1\\0\\0\\2\end{bmatrix}\right\}, \vec{\mathbf{v}}_4 = \begin{bmatrix}2\\0\\1\\0\end{bmatrix}.$$

33. $\lambda_1 = \lambda_2 = 2, \lambda_3 = 4, \lambda_4 = 6;$

$$\mathbb{E}_{1,2} = \text{span}\left\{\begin{bmatrix}1\\0\\0\\0\end{bmatrix}, \begin{bmatrix}0\\1\\0\\0\end{bmatrix}\right\},$$

$$\vec{\mathbf{v}}_3 = \begin{bmatrix}1\\0\\0\\1\end{bmatrix}, \vec{\mathbf{v}}_4 = \begin{bmatrix}1\\0\\2\\1\end{bmatrix}.$$

39. An example is $\mathbf{A} = \begin{bmatrix}2 & 0\\0 & 3\end{bmatrix}$ with eigenvalues 2, 3;

$\mathbf{A}^{-1} = \begin{bmatrix}\frac{1}{2} & 0\\0 & \frac{1}{3}\end{bmatrix}$ with eigenvalues 1/2, 1/3.

41. For $\mathbb{I}_2$, $\lambda_1 = \lambda_2 = 1$ and every vector in $\mathbb{R}^2$ is an eigenvector. For $\mathbb{I}_n$, we have a repeated eigenvalue 1 with multiplicity n and n linearly independent eigenvectors that span $\mathbb{R}^n$.

42. The eigenvalues of $\mathbf{A}$ and $\mathbf{A}^{-1}$ are reciprocals.

50. (a) $\lambda_1 = 0, \lambda_2 = 5; \vec{\mathbf{v}}_1 = [-2, 1], \vec{\mathbf{v}}_2 = [1, 2].$

54. All matrices of the form $\begin{bmatrix}\lambda - b & b\\\lambda - d & d\end{bmatrix}$ have eigenvector $\begin{bmatrix}1\\1\end{bmatrix}$ with eigenvalue λ.

56. $\begin{bmatrix}1 & -\frac{1}{2}\\0 & 2\end{bmatrix}$ has the given eigenvalues and eigenvectors.

58. $\begin{bmatrix}1 & 0\\0 & -1\end{bmatrix}; \lambda_1 = 1, \lambda_2 = -1; \vec{\mathbf{v}}_1 = \begin{bmatrix}1\\0\end{bmatrix}, \vec{\mathbf{v}}_2 = \begin{bmatrix}0\\1\end{bmatrix}.$

60. $\mathbf{A} = \frac{\sqrt{2}}{2}\begin{bmatrix}1 & 1\\-1 & 1\end{bmatrix}; \lambda = \frac{\sqrt{2}}{2}(1 \pm i); \vec{\mathbf{v}} = \begin{bmatrix}1\\\mp i\end{bmatrix}.$

62. $\mathbf{A} = \begin{bmatrix}1 & 0\\2 & 1\end{bmatrix}; \lambda = 1, 1; \vec{\mathbf{v}} = \begin{bmatrix}0\\1\end{bmatrix}.$

68. $\mathbf{A}^{-1} = \frac{1}{6}(\mathbf{A}^2 - 6\mathbf{A} + 11\mathbf{I}) = \frac{1}{6}\begin{bmatrix} 4 & -6 & 2 \\ -1 & 9 & -2 \\ -4 & 12 & -2 \end{bmatrix} = \begin{bmatrix} 2/3 & -1 & 1/3 \\ -1/6 & 3/2 & -1/3 \\ -2/3 & 2 & -1/3 \end{bmatrix}$

71. $\lambda^3 - (\text{Tr}\mathbf{A})\lambda^2 + \left[\cdots\right]\lambda - |\mathbf{A}|$, where $\left[\cdots\right] = \left[(a_{11}a_{22} - a_{12}a_{21}) + (a_{11}a_{33} - a_{13}a_{31}) + (a_{22}a_{33} - a_{23}a_{32})\right]$.

73. $\begin{bmatrix} y_1' \\ y_2' \end{bmatrix} = \begin{bmatrix} 0 & 1 \\ -5 & 2 \end{bmatrix}\begin{bmatrix} y_1 \\ y_2 \end{bmatrix}$

75. $\begin{bmatrix} y_1' \\ y_2' \\ y_3' \end{bmatrix} = \begin{bmatrix} 0 & 1 & 0 \\ 0 & 0 & 1 \\ -6 & 5 & 2 \end{bmatrix}\begin{bmatrix} y_1 \\ y_2 \\ y_3 \end{bmatrix}$

77. There are nonzero solutions if $\lambda = \left(\frac{2n+1}{2}\right)^2$, for n an integer.

Section 5.4, p. 338

1. $\mathbf{M}_B = \begin{bmatrix} 3 & -4 \\ -2 & 3 \end{bmatrix}$; $\mathbf{M}_B^{-1} = \begin{bmatrix} 3 & 4 \\ 2 & 3 \end{bmatrix}$

3. $\begin{bmatrix} 13 \\ -9 \end{bmatrix}, \begin{bmatrix} -2 \\ 2 \end{bmatrix}, \begin{bmatrix} 3 \\ -2 \end{bmatrix}$

5. $\begin{bmatrix} 5 \\ 4 \end{bmatrix}, \begin{bmatrix} -1 \\ 0 \end{bmatrix}, \begin{bmatrix} 13 \\ 9 \end{bmatrix}$

7. $\mathbf{M}_B = \begin{bmatrix} 1 & 1 & 1 \\ 0 & 1 & 1 \\ 0 & 0 & 1 \end{bmatrix}, \mathbf{M}_B^{-1} = \begin{bmatrix} 1 & -1 & 0 \\ 0 & 1 & -1 \\ 0 & 0 & 1 \end{bmatrix}$

9. $\begin{bmatrix} 0 \\ -1 \\ -1 \end{bmatrix}, \begin{bmatrix} 5 \\ 4 \\ 3 \end{bmatrix}, \begin{bmatrix} 0 \\ 2 \\ 1 \end{bmatrix}$

11. $\begin{bmatrix} 5 \\ -1 \\ -2 \end{bmatrix}, \begin{bmatrix} -3 \\ 1 \\ 1 \end{bmatrix}, \begin{bmatrix} 3 \\ 2 \\ 0 \end{bmatrix}$

13. $\mathbf{M}_N = \begin{bmatrix} 2 & 1 & 1 \\ -1 & 0 & 0 \\ 0 & 0 & 1 \end{bmatrix}, \mathbf{M}_N^{-1} = \begin{bmatrix} 0 & -1 & 0 \\ 1 & 2 & -1 \\ 0 & 0 & 1 \end{bmatrix}$

15. $u(x) = 4x^2 - x + 2$, $v(x) = x^2 + 2x + 3$, $w(x) = -3x^2 + x$

17. $\vec{\mathbf{p}}_Q = [1, 0, -1, 3]$, $\vec{\mathbf{q}}_Q = [1, -1, 3, -2]$, $\vec{\mathbf{r}}_Q = [1, 0, -1, 1]$

19. $\mathbf{M}_B = \begin{bmatrix} 0 & 0 & 0 & 0 & 0 \\ 0 & 0 & 0 & 0 & 0 \\ 12 & 0 & 0 & 0 & 0 \\ 0 & 6 & 0 & 0 & 0 \\ 0 & 0 & 2 & 0 & 0 \end{bmatrix}$

(a) $[0, 0, 12, -6, 2]$

(b) $[0, 0, 12, 0, 4]$

(c) $[0, 0, -48, 18, 0]$

(d) $[0, 0, 12, 0, -16]$

21. $\mathbf{M}_B = \begin{bmatrix} 0 & 0 & 0 & 0 & 0 \\ 0 & 0 & 0 & 0 & 0 \\ 0 & 0 & 0 & 0 & 0 \\ 24 & 0 & 0 & 0 & 0 \\ 0 & 6 & 0 & 0 & 0 \end{bmatrix}$

(a) $[0, 0, 0, 24, -6]$

(b) $[0, 0, 0, 24, 0]$

(c) $[0, 0, 0, -96, 18]$

(d) $[0, 0, 0, 24, 0]$

23. $\mathbf{M}_B = \begin{bmatrix} -2 & 0 & 0 & 0 & 0 \\ 4 & -2 & 0 & 0 & 0 \\ 0 & 3 & -2 & 0 & 0 \\ 0 & 0 & 2 & -2 & 0 \\ 0 & 0 & 0 & 1 & -2 \end{bmatrix}$

(a) $[-2, 6, -5, 4, -3]$

(b) $[-2, 4, -4, 4, -8]$

(c) $[8, -22, 9, 0, 0]$

(d) $[-2, 4, 16, -16, -32]$

25. $\mathbf{P} = \begin{bmatrix} -\frac{1}{2}\sqrt{5} - \frac{3}{2} & \frac{1}{2}\sqrt{5} - \frac{3}{2} \\ 1 & 1 \end{bmatrix}$;

$\mathbf{P}^{-1}\mathbf{A}\mathbf{P} = \begin{bmatrix} \sqrt{5} & 0 \\ 0 & -\sqrt{5} \end{bmatrix}$

NOTE: **P** is not unique.

27. $\mathbf{P} = \begin{bmatrix} -1 & 1 \\ 1 & 1 \end{bmatrix}$; $\mathbf{P}^{-1}\mathbf{A}\mathbf{P} = \begin{bmatrix} -1 & 0 \\ 0 & 3 \end{bmatrix}$

NOTE: **P** is not unique.

29. Cannot be diagonalized (double eigenvalue with single eigenvector)

31. $\mathbf{P} = \begin{bmatrix} 3 & 2 \\ 5 & 3 \end{bmatrix}$; $\mathbf{P}^{-1}\mathbf{A}\mathbf{P} = \begin{bmatrix} 2 & 0 \\ 0 & 3 \end{bmatrix}$

33. Cannot be diagonalized (double eigenvalue with single eigenvector).

35. $\mathbf{P} = \begin{bmatrix} -1 & 1 & 0 \\ 0 & 0 & 1 \\ 1 & 1 & 0 \end{bmatrix}$; $\mathbf{P}^{-1}\mathbf{A}\mathbf{P} = \begin{bmatrix} 0 & 0 & 0 \\ 0 & 2 & 0 \\ 0 & 0 & 1 \end{bmatrix}$

NOTE: **P** is not unique.

37. Cannot be diagonalized (double eigenvalue with single eigenvector)

39. Cannot be diagonalized (only two linearly independent eigenvectors)

41. Cannot be diagonalized (double eigenvalue with single eigenvector)

43. $\mathbf{P} = \begin{bmatrix} 1 & 0 & 0 \\ 2 & 0 & 1 \\ 0 & 1 & 1 \end{bmatrix}$; $\mathbf{P}^{-1}\mathbf{A}\mathbf{P} = \begin{bmatrix} 1 & 0 & 0 \\ 0 & 1 & 0 \\ 0 & 0 & 3 \end{bmatrix}$

45. $\mathbf{P} = \begin{bmatrix} 0 & 2 & 1 \\ 1 & 0 & 1 \\ 0 & 1 & 1 \end{bmatrix}$; $\mathbf{P}^{-1}\mathbf{AP} = \begin{bmatrix} 1 & 0 & 0 \\ 0 & 1 & 0 \\ 0 & 0 & 2 \end{bmatrix}$

47. $\mathbf{P} = \begin{bmatrix} -2 & 4 & 1 & 0 \\ 1 & 0 & 0 & 0 \\ 0 & 0 & 1 & 0 \\ 0 & 1 & 0 & 1 \end{bmatrix}$; $\mathbf{P}^{-1}\mathbf{AP} = \begin{bmatrix} 4 & 0 & 0 & 0 \\ 0 & 4 & 0 & 0 \\ 0 & 0 & 8 & 0 \\ 0 & 0 & 0 & 8 \end{bmatrix}$

49. (b) $\mathbf{A}^{50} = \frac{1}{4}\begin{bmatrix} 2(3^{50}) + 2(-1)^{50} & (3^{50}) - (-1)^{50} \\ 4(3^{50}) - 4(-1)^{50} & 2(3^{50}) + 2(-1)^{50} \end{bmatrix}$

$$= \frac{(3)^{50}}{4}\begin{bmatrix} 2 & 1 \\ 4 & 2 \end{bmatrix} + \frac{1}{4}\begin{bmatrix} 2 & -1 \\ -4 & 2 \end{bmatrix}$$

(c) See Sec. 3.1, Problem 48.

(d) Yes

52. An example is $\begin{bmatrix} 0 & 0 \\ 0 & 1 \end{bmatrix}$.

54. (a) Cannot be diagonalized (only one eigenvector)

(b) $\begin{bmatrix} 0 & 0 & 0 & 0 & 0 \\ 0 & -3 & 0 & 0 & 0 \\ 0 & 0 & -3 & 0 & 0 \\ 0 & 0 & 0 & 3 & 0 \\ 0 & 0 & 0 & 0 & -1 \end{bmatrix}$

(c) $\begin{bmatrix} 3 & 0 & 0 & 0 \\ 0 & 0 & 0 & 0 \\ 0 & 0 & 2 & 0 \\ 0 & 0 & 0 & 5 \end{bmatrix}$

56. (a) Use $\mathbf{A} \sim \begin{bmatrix} 2 & 0 \\ 0 & 3 \end{bmatrix} \sim \mathbf{B}$.

61. For $\mathbf{Q} = \begin{bmatrix} 1 & 1 \\ -2 & 0 \end{bmatrix}$, $\mathbf{Q}^{-1}\mathbf{AQ} = \begin{bmatrix} 4 & -2 \\ 0 & 4 \end{bmatrix}$.

CHAPTER 6

Section 6.1, p. 356

1. $x_1' = x_1 + 2x_2$
$x_2' = 4x_1 - x_2$

3. $x_1' = 4x_1 + 3x_2 + e^{-t}$
$x_2' = -x_1 - x_2$

5. $\vec{\mathbf{x}}(t) = c_1 \begin{bmatrix} e^{4t} \\ e^{4t} \end{bmatrix} + c_2 \begin{bmatrix} e^{-2t} \\ -e^{-2t} \end{bmatrix}$

7. $\vec{\mathbf{x}}(t) = c_1 \begin{bmatrix} e^{-t} \\ -2e^{-t} \end{bmatrix} + c_2 \begin{bmatrix} e^{3t} \\ 2e^{3t} \end{bmatrix}$

9. We have drawn three distinct trajectories for six initial conditions
$(x(0), y(0)) = (1, 0), (2, 0), (3, 0), (0, 1), (0, 2), (0, 3)$.

Note that although trajectories may (and do) coincide if one starts at a point lying on another, they never *cross* each other.

However, if we plot $x = x(t)$ or $y = y(t)$ for these same six initial conditions, we get the six intersecting curves shown below.

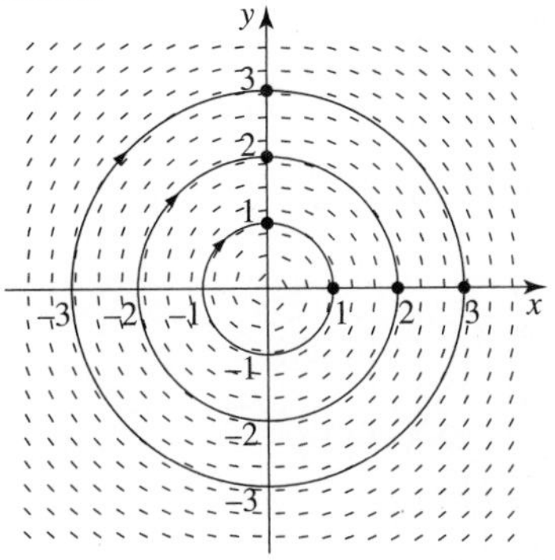

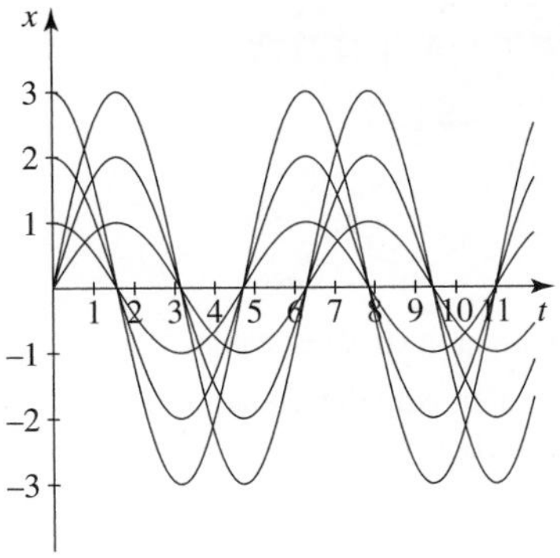

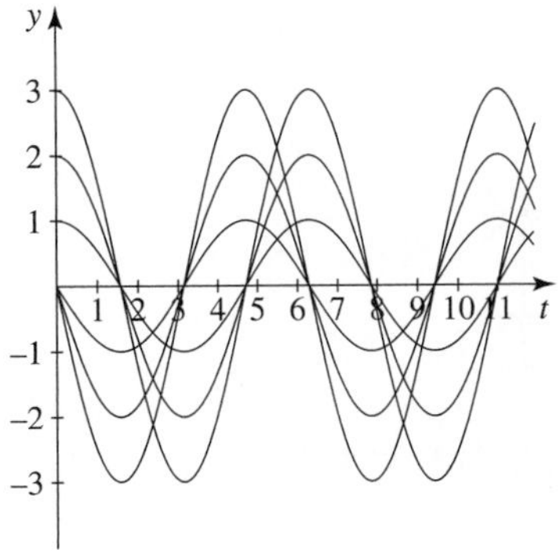

13. $y(t) = c|x(t)|$. As one leaves the origin on any trajectory, speed would keep increasing.

15. (a) The computer phase portrait for Problem 13 is as follows:

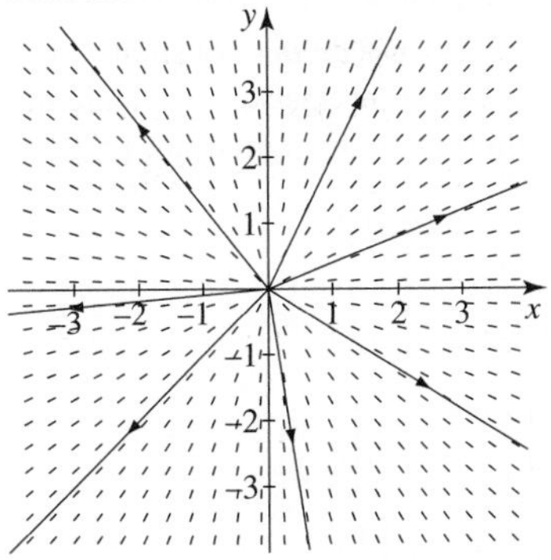

17. $W = -e^{-3t} \neq 0$, so the vectors form a fundamental set.

19. $W = -e^{2t} \neq 0$, so the vectors form a fundamental set.

21. $W = e^{2t} \neq 0$, so the vectors form a fundamental set.

Section 6.2, p. 368

1. (a) $x' = y$
$y' = -x - y$

(b) Equilibrium at $(0, 0)$

(c) h-nullcline: $x + y = 0$
v-nullcline: $y = 0$

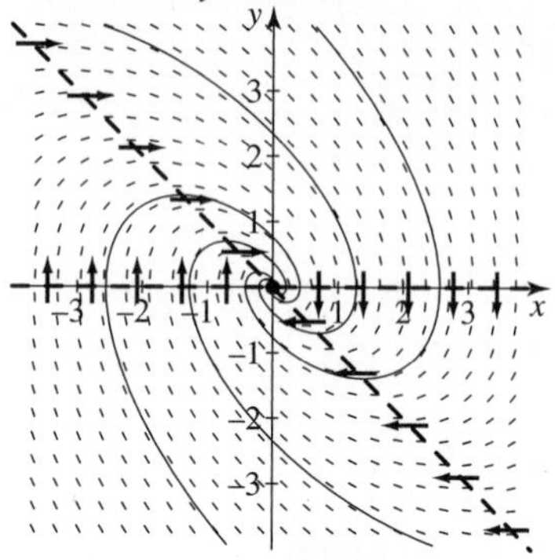

(d) The equilibrium point at $(0, 0)$ is stable.

(e) A mass-spring system with this equation shows damped oscillatory motion about $x(t) \equiv 0$.

3. (a) $x' = y$
$y' = -x - 1$

(b) Equilibrium at $(1, 0)$

(c) h-nullcline: $x = 1$
v-nullcline: $y = 0$

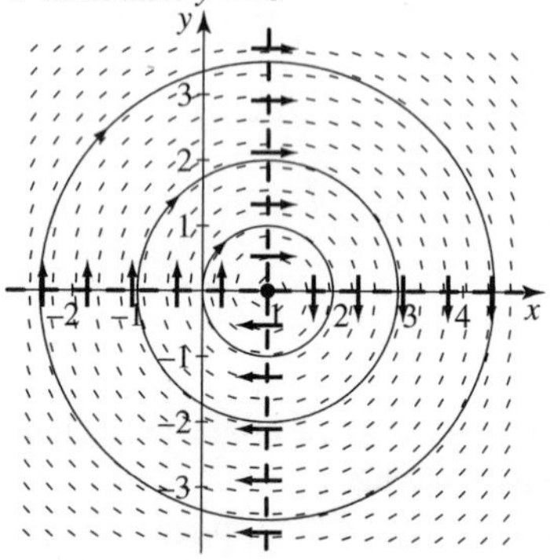

(d) The equilibrium point at $(1, 0)$ is stable.

(e) A mass-spring system with this equation shows no damping and steady forcing, hence periodic motion about an equilibrium to the right of the origin.

5. (A)

7. (D)

9. $\mathbf{x}(t) = c_1 \begin{bmatrix} 1 \\ 2 \end{bmatrix} + c_2 e^{-5t} \begin{bmatrix} -2 \\ 1 \end{bmatrix}$

11. $\mathbf{x}(t) = c_1 e^{2t} \begin{bmatrix} 1 \\ -1 \end{bmatrix} + c_2 e^{3t} \begin{bmatrix} 1 \\ -2 \end{bmatrix}$

13. $\mathbf{x}(t) = c_1 e^{2t} \begin{bmatrix} 1 \\ 3 \end{bmatrix} + c_2 e^{4t} \begin{bmatrix} 1 \\ 1 \end{bmatrix}$

15. $\mathbf{x}(t) = c_1 e^{t} \begin{bmatrix} 1 \\ 2 \end{bmatrix} + c_2 e^{2t} \begin{bmatrix} 0 \\ 1 \end{bmatrix}$

17. $\mathbf{x}(t) = c_1 e^{-t} \begin{bmatrix} 1 \\ 2 \end{bmatrix} + c_2 e^{2t} \begin{bmatrix} 2 \\ 1 \end{bmatrix}$

19. $\mathbf{x}(t) = c_1 e^{-2t} \begin{bmatrix} 2 \\ 3 \end{bmatrix} + c_2 e^{-t} \begin{bmatrix} 1 \\ 1 \end{bmatrix}$

21. $\mathbf{x}(t) = c_1 \begin{bmatrix} 3 \\ 4 \end{bmatrix} + c_2 e^{-2t} \begin{bmatrix} 1 \\ 2 \end{bmatrix}$

23. $\mathbf{x}(t) = c_1 e^{t} \begin{bmatrix} 1 \\ 2 \end{bmatrix} + c_2 \left\{ te^{t} \begin{bmatrix} 1 \\ 2 \end{bmatrix} + e^{t} \begin{bmatrix} 0 \\ 1 \end{bmatrix} \right\}$

25. $\mathbf{x}(t) = \frac{1}{2} e^{3t} \begin{bmatrix} 1 \\ 5 \end{bmatrix} + \frac{1}{2} e^{-t} \begin{bmatrix} 1 \\ 1 \end{bmatrix}$

27. $\mathbf{x}(t) = 5e^{2t} \begin{bmatrix} 1 \\ 0 \end{bmatrix} + 4e^{3t} \begin{bmatrix} 0 \\ 1 \end{bmatrix} = \begin{bmatrix} 5e^{2t} \\ 4e^{3t} \end{bmatrix}$

29. $\mathbf{x}(t) = \frac{1}{2} \begin{bmatrix} -1 \\ 1 \end{bmatrix} + \frac{5}{2} e^{2t} \begin{bmatrix} 1 \\ 1 \end{bmatrix}$

31. $\mathbf{x}(t) = \begin{bmatrix} 2 \\ 4 \end{bmatrix}$

33. $\mathbf{x}(t) = -2e^{2t} \begin{bmatrix} -1 \\ 1 \end{bmatrix} - e^{3t} \begin{bmatrix} 1 \\ -2 \end{bmatrix}$

35. The following are examples:

(a) $\mathbf{A} = \begin{bmatrix} a & 1 & 0 \\ 0 & a & 0 \\ 0 & 0 & b \end{bmatrix}$

(b) $\mathbf{A} = \begin{bmatrix} a & 1 & 0 \\ 0 & a & 0 \\ 0 & 0 & a \end{bmatrix}$

38. (c) $\vec{\mathbf{x}}(t) = c_1 e^{t} \begin{bmatrix} 1 \\ 0 \\ 0 \end{bmatrix} + c_2 e^{t} \left(t \begin{bmatrix} 1 \\ 0 \\ 0 \end{bmatrix} + \begin{bmatrix} 0 \\ 1 \\ 0 \end{bmatrix} \right)$
$+ c_3 e^{t} \left(\frac{1}{2} t^2 \begin{bmatrix} 1 \\ 0 \\ 0 \end{bmatrix} + t \begin{bmatrix} 0 \\ 1 \\ 0 \end{bmatrix} + \begin{bmatrix} 0 \\ -1 \\ 1 \end{bmatrix} \right)$

39. (b) $\vec{\mathbf{x}}_1(t) = ce^{t} \begin{bmatrix} 1 \\ -2 \\ 1 \end{bmatrix}$

(c) $\vec{\mathbf{x}}_2(t) = te^{t} \begin{bmatrix} 1 \\ -2 \\ 1 \end{bmatrix} + e^{t} \begin{bmatrix} -1 \\ 1 \\ 0 \end{bmatrix}$

(d) $\vec{\mathbf{x}}_3(t) = \frac{1}{2} t^2 e^{t} \begin{bmatrix} 1 \\ -2 \\ 1 \end{bmatrix} + te^{t} \begin{bmatrix} -1 \\ 1 \\ 0 \end{bmatrix} + e^{t} \begin{bmatrix} 1 \\ 0 \\ 0 \end{bmatrix}$

40. $\mathbf{x}(t) = c_1 e^{t} \begin{bmatrix} 1 \\ 0 \\ -1 \end{bmatrix} + c_2 e^{2t} \begin{bmatrix} -2 \\ 1 \\ 0 \end{bmatrix} + c_3 e^{3t} \begin{bmatrix} 0 \\ 1 \\ -1 \end{bmatrix}$

42. $\mathbf{x}(t) = \frac{1}{4} \begin{bmatrix} 3 \\ 3 \\ 1 \end{bmatrix} + \frac{3}{8} e^{2t} \begin{bmatrix} -1 \\ 1 \\ 1 \end{bmatrix} + \frac{3}{8} e^{-2t} \begin{bmatrix} -1 \\ -3 \\ 1 \end{bmatrix}$

45. (a) The adjoint system is $\vec{\mathbf{w}}' = \begin{bmatrix} 0 & -1 \\ -1 & 0 \end{bmatrix} \vec{\mathbf{w}}$.

(c) $\vec{\mathbf{x}}(t) = \dfrac{1}{2}e^{t}\begin{bmatrix} 1 \\ 1 \end{bmatrix} + \dfrac{1}{2}e^{-t}\begin{bmatrix} 1 \\ -1 \end{bmatrix}$

(d) $\vec{\mathbf{w}}(t) = \dfrac{1}{2}e^{t}\begin{bmatrix} 1 \\ -1 \end{bmatrix} - \dfrac{1}{2}e^{-t}\begin{bmatrix} 1 \\ 1 \end{bmatrix}$

(e) Trajectories are orthogonal.

47. $x(t) = c_1 e^{t}$,
$y(t) = c_2 e^{-t}$;

$y = c/x$.

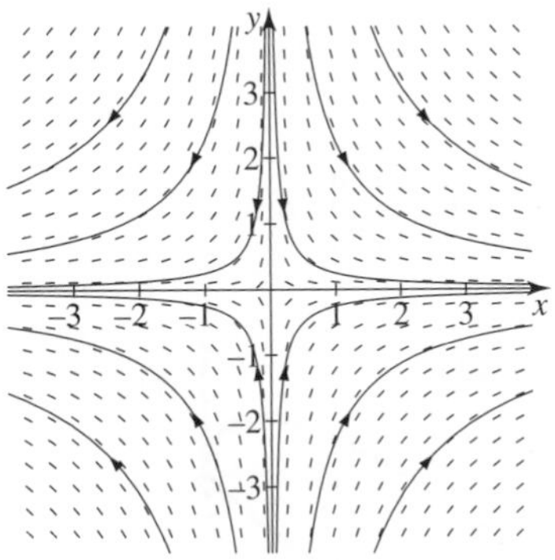

49. $x = y + c$, because $x' = y'$;
$x = -y$ is a line of equilibrium points.

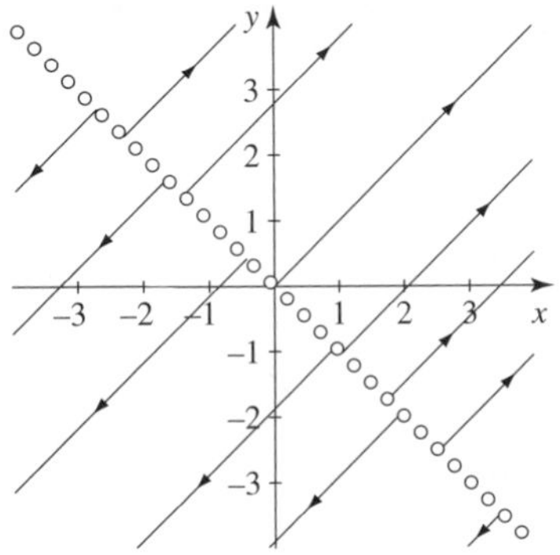

52. (a) $\begin{bmatrix} x_A \\ x_B \end{bmatrix} = 12.5\left(e^{-0.025t}\begin{bmatrix} 1 \\ \sqrt{3} \end{bmatrix} + e^{-0.095t}\begin{bmatrix} 1 \\ -\sqrt{3} \end{bmatrix}\right)$

(b)

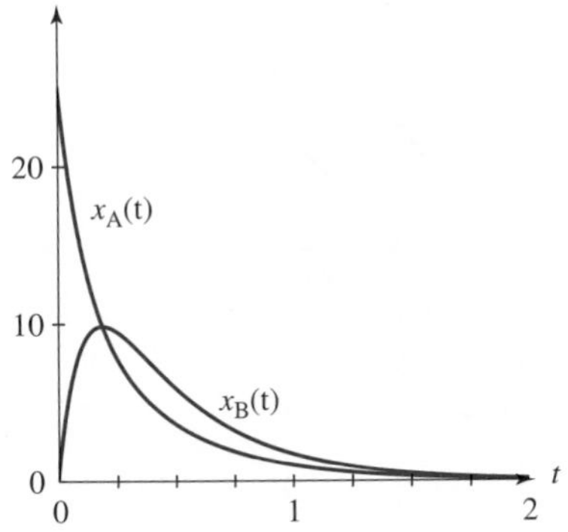

(c) x_B never exceeds x_A at any t.

(d) $x_B \to 0$; $x_A \to 0$.

55. $\vec{\mathbf{x}}' = \begin{bmatrix} -.10 & .10 & 0 \\ .06 & -.11 & .05 \\ .04 & .01 & -.05 \end{bmatrix} \vec{\mathbf{x}}$

57. $\vec{\mathbf{I}}(t) = \begin{bmatrix} I_1 \\ I_2 \end{bmatrix} = c_1 e^{-t}\begin{bmatrix} 2 \\ 3 \end{bmatrix} + c_2 e^{-12t}\begin{bmatrix} 3 \\ -1 \end{bmatrix}$;
$I_3 = I_1 - I_2$.

Section 6.3, p. 381

1. $\vec{\mathbf{x}}(t) = c_1\begin{bmatrix} \cos t \\ -\sin t \end{bmatrix} + c_2\begin{bmatrix} \sin t \\ \cos t \end{bmatrix}$

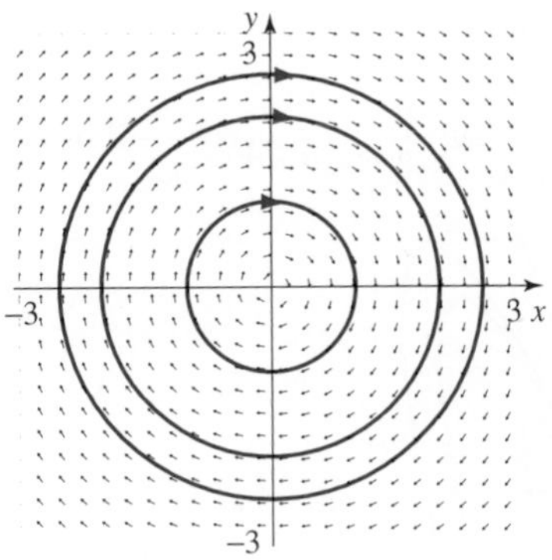

3. $\vec{\mathbf{x}}(t) = c_1 e^{t}\begin{bmatrix} \cos 2t \\ -\sin 2t \end{bmatrix} + c_2 e^{t}\begin{bmatrix} \sin 2t \\ \cos 2t \end{bmatrix}$

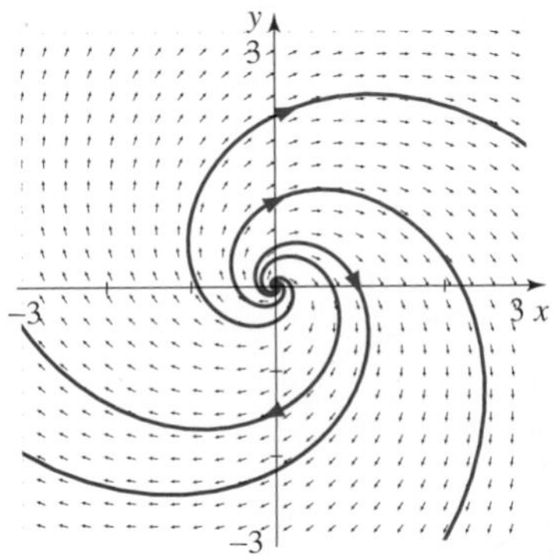

5. $\vec{\mathbf{x}}(t) = c_1\begin{bmatrix} \cos t \\ -\cos t - \sin t \end{bmatrix} + c_2\begin{bmatrix} \sin t \\ \cos t - \sin t \end{bmatrix}$

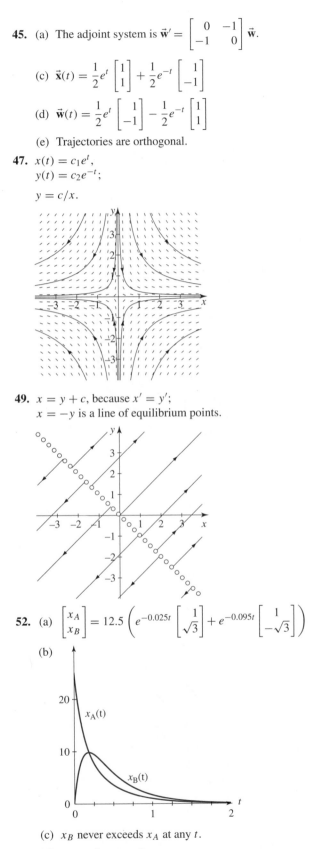

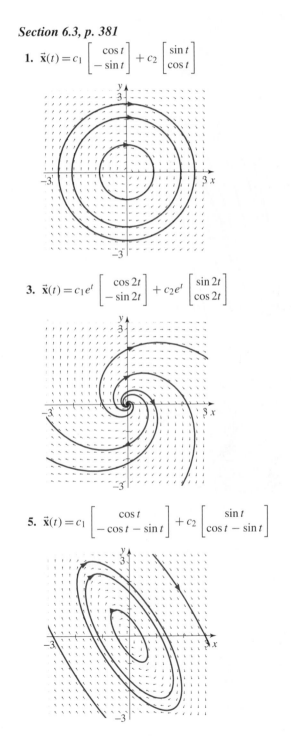

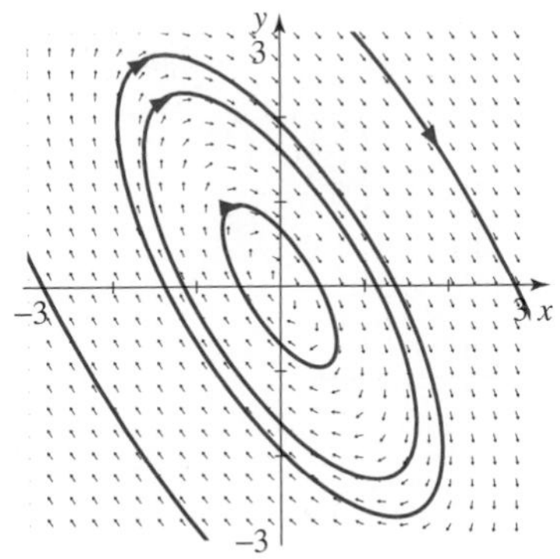

7. $\vec{\mathbf{x}}(t) = c_1 e^t \begin{bmatrix} \cos 2t \\ \cos 2t + \sin 2t \end{bmatrix} + c_2 e^t \begin{bmatrix} \sin 2t \\ -\cos 2t + \sin 2t \end{bmatrix}$

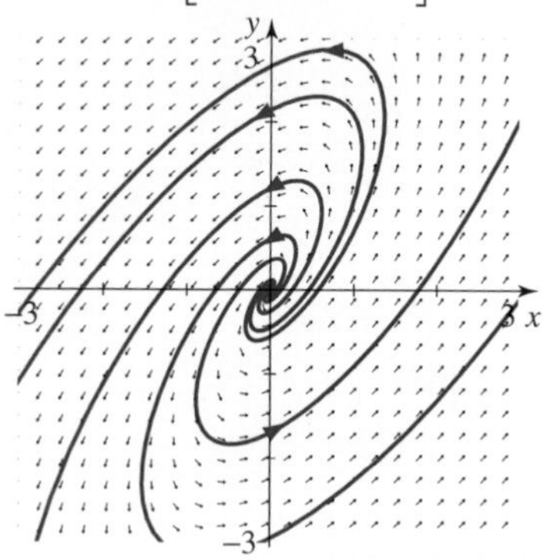

9. $\vec{\mathbf{x}}(t) = c_1 e^{-t} \begin{bmatrix} 2\cos t - \sin t \\ 5\cos t \end{bmatrix} + c_2 e^{-t} \begin{bmatrix} \cos t + 2\sin t \\ 5\sin t \end{bmatrix}$

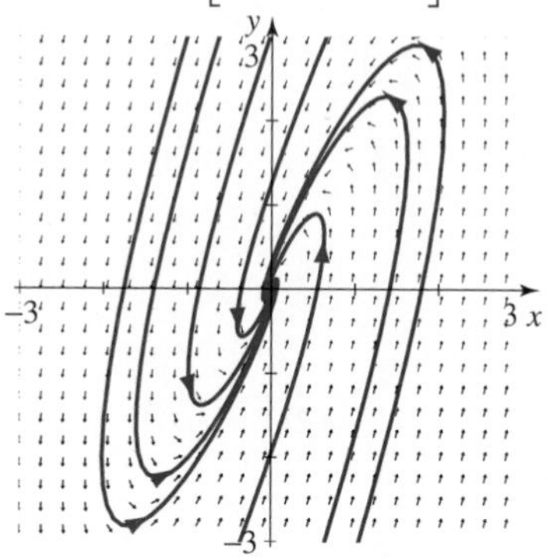

11. $\vec{\mathbf{x}}(t) = c_1 e^{-2t} \begin{bmatrix} -\cos t \\ \cos t - \sin t \end{bmatrix} + c_2 e^{-2t} \begin{bmatrix} -\sin t \\ \cos t + \sin t \end{bmatrix}$

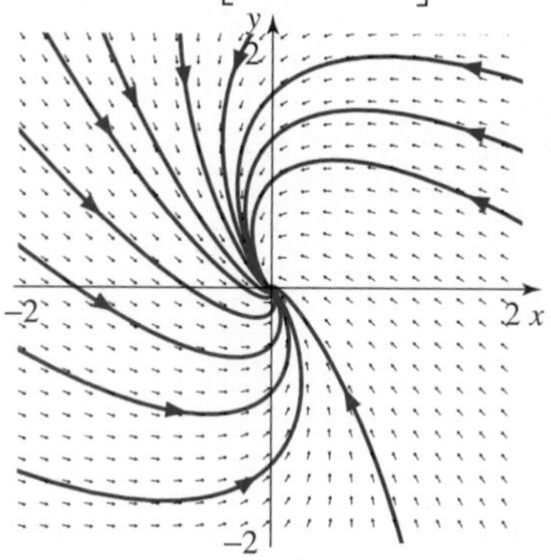

13. $\vec{\mathbf{x}}(t) = e^t \begin{bmatrix} -\sin t - \cos t \\ \cos t - \sin t \end{bmatrix}$

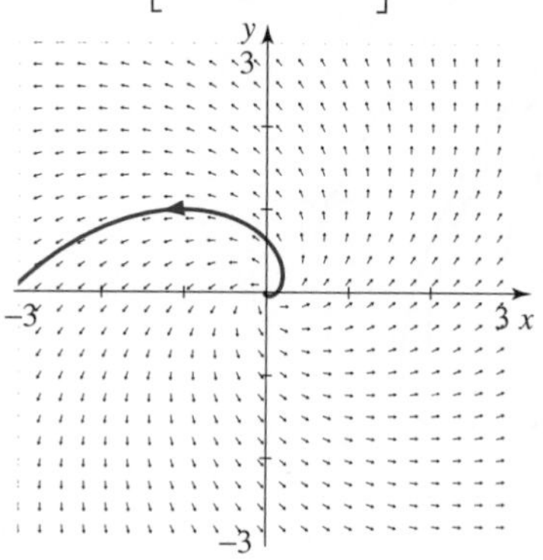

15. $\vec{\mathbf{x}}(t) = e^{-2t} \begin{bmatrix} \cos t + \sin t \\ \cos t \end{bmatrix}$

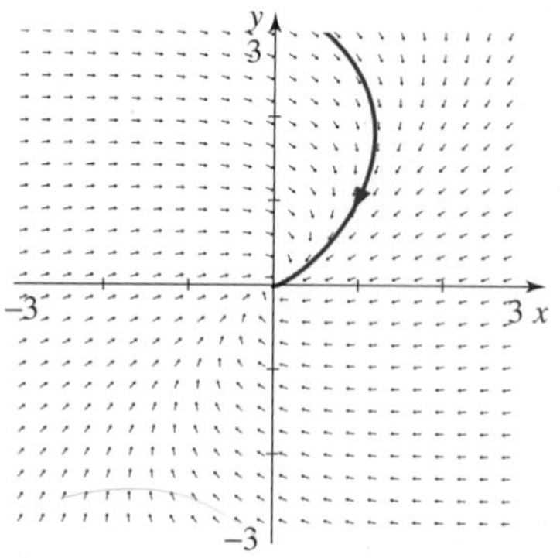

21. $\lambda = \pm 3i$, $\vec{\mathbf{v}} = \begin{bmatrix} 5 \\ 4 \mp 3i \end{bmatrix}$, $\vec{\mathbf{p}} = \begin{bmatrix} 5 \\ 4 \end{bmatrix}$, $\vec{\mathbf{q}} = \begin{bmatrix} 0 \\ -3 \end{bmatrix}$.

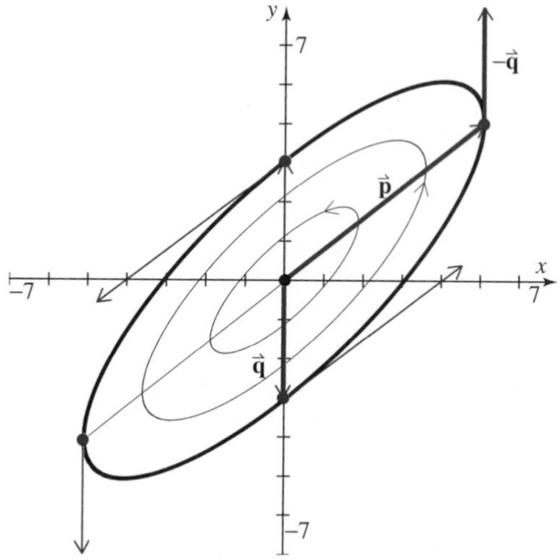

23. $\lambda = \pm i$, $\vec{\mathbf{v}} = \begin{bmatrix} 1 \\ 1 \mp i \end{bmatrix}$, $\vec{\mathbf{p}} = \begin{bmatrix} 1 \\ 1 \end{bmatrix}$, $\vec{\mathbf{q}} = \begin{bmatrix} 0 \\ -1 \end{bmatrix}$.

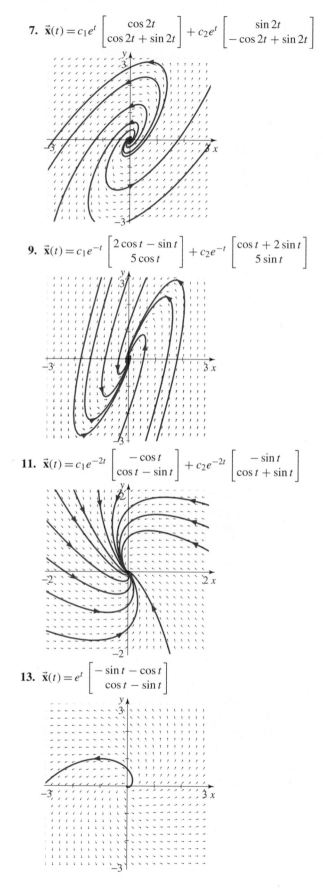

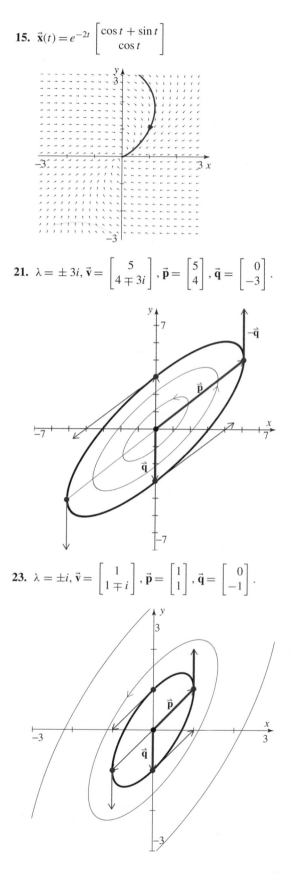

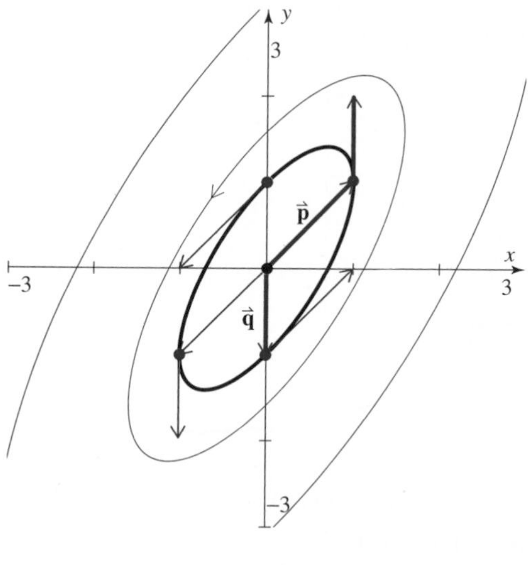

27. $\lambda = \pm 2i$, $\vec{\mathbf{v}} = \begin{bmatrix} 1 \\ -1 \mp 2i \end{bmatrix}$, $\vec{\mathbf{p}} = \begin{bmatrix} 1 \\ -1 \end{bmatrix}$, $\vec{\mathbf{q}} = \begin{bmatrix} 0 \\ -2 \end{bmatrix}$.

The parameter $\beta t^* = \frac{1}{2}\tan^{-1} 2 \approx 1.11$ radians or 4.25 radians. Endpoints of the ellipse axes occur at approximately (.85, .19) and (−.52, 2.22).

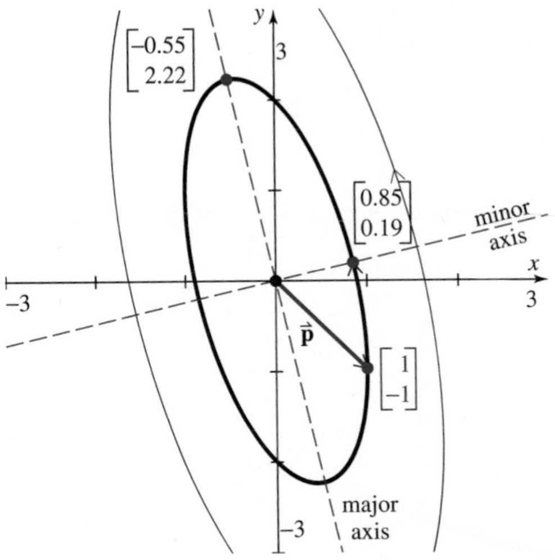

29. $\lambda = \pm i$, $\vec{\mathbf{v}} = \begin{bmatrix} -1 \\ -1 \mp i \end{bmatrix}$, $\vec{\mathbf{p}} = \begin{bmatrix} -1 \\ -1 \end{bmatrix}$, $\vec{\mathbf{q}} = \begin{bmatrix} 0 \\ -1 \end{bmatrix}$.

The parameter $\beta t^* = \frac{1}{2}\tan^{-1}(-2) \approx -.55$ radians or 1.02 radians. Endpoints of the ellipse axes occur approximately at (−.85, −1.37) and (−.52, .33).

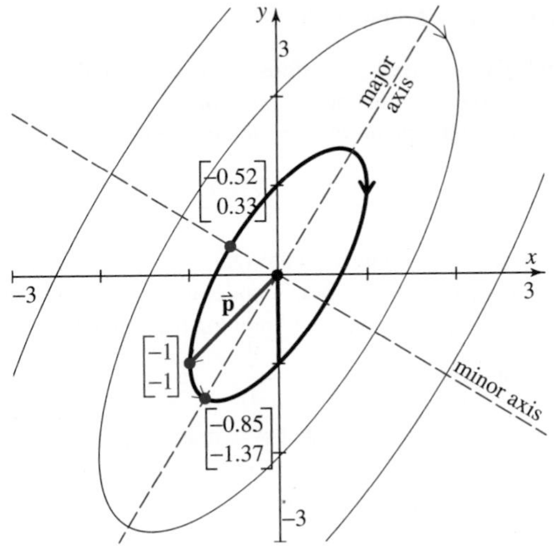

30. (e) $\vec{\mathbf{x}}(t) = \begin{bmatrix} e^{-t} \\ \sin 2t \\ \cos 2t \end{bmatrix}$

(f)

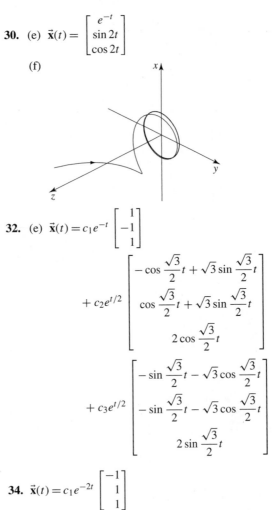

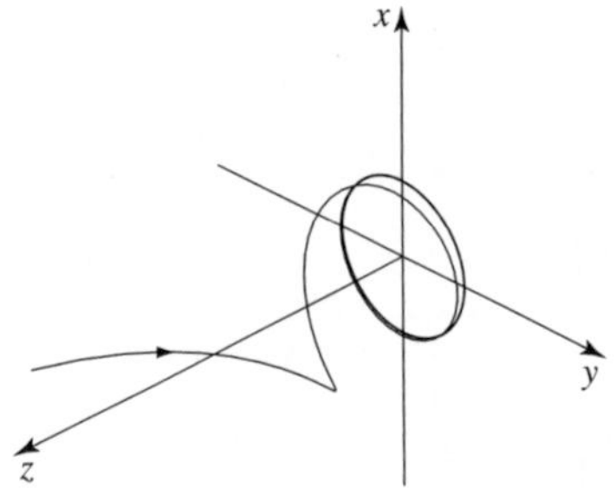

32. (e) $\vec{\mathbf{x}}(t) = c_1 e^{-t}\begin{bmatrix} 1 \\ -1 \\ 1 \end{bmatrix}$

$$+ c_2 e^{t/2}\begin{bmatrix} -\cos\frac{\sqrt{3}}{2}t + \sqrt{3}\sin\frac{\sqrt{3}}{2}t \\ \cos\frac{\sqrt{3}}{2}t + \sqrt{3}\sin\frac{\sqrt{3}}{2}t \\ 2\cos\frac{\sqrt{3}}{2}t \end{bmatrix}$$

$$+ c_3 e^{t/2}\begin{bmatrix} -\sin\frac{\sqrt{3}}{2}t - \sqrt{3}\cos\frac{\sqrt{3}}{2}t \\ -\sin\frac{\sqrt{3}}{2}t - \sqrt{3}\cos\frac{\sqrt{3}}{2}t \\ 2\sin\frac{\sqrt{3}}{2}t \end{bmatrix}$$

34. $\vec{\mathbf{x}}(t) = c_1 e^{-2t}\begin{bmatrix} -1 \\ 1 \\ 1 \end{bmatrix}$

$$+ c_2 e^{-t}\begin{bmatrix} 2\cos\sqrt{2}t - \sqrt{2}\sin\sqrt{2}t \\ 2\cos\sqrt{2}t \\ 2\sqrt{2}\sin\sqrt{2}t \end{bmatrix}$$

$$+ c_3 e^{-t}\begin{bmatrix} \sqrt{2}\cos\sqrt{2}t + 2\sin\sqrt{2}t \\ 2\sin\sqrt{2}t \\ -2\sqrt{2}\cos\sqrt{2}t \end{bmatrix}$$

36. $\vec{\mathbf{x}}(t) =$

$$\frac{1}{2}\begin{bmatrix} -1 \\ 0 \\ 1 \end{bmatrix} + \frac{1}{2}\cos\sqrt{2}t\begin{bmatrix} 1 \\ \sqrt{2} \\ 1 \end{bmatrix} + \sin\sqrt{2}t\begin{bmatrix} 1 \\ -\sqrt{2} \\ 1 \end{bmatrix}$$

38. $\vec{\mathbf{x}}(t) = c_1\begin{bmatrix} \cos kt \\ -\sin kt \end{bmatrix} + c_2\begin{bmatrix} \sin kt \\ \cos kt \end{bmatrix}$

39. $x(t) = \cos t - \cos\sqrt{3}t$
$y(t) = \cos t + \cos\sqrt{3}t$

40. $\vec{\mathbf{x}}' = \begin{bmatrix} 0 & 1 \\ -5 & -2 \end{bmatrix}, \vec{\mathbf{x}(0)} = \begin{bmatrix} 2 \\ 2 \end{bmatrix}$

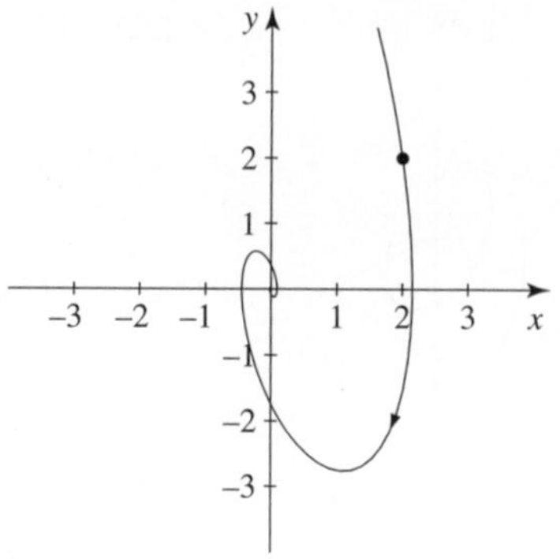

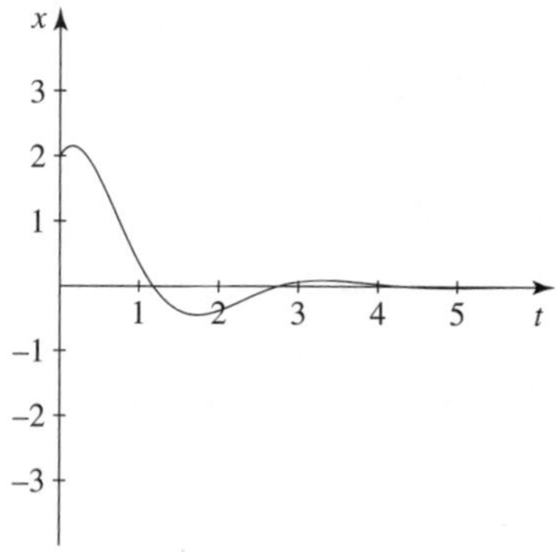

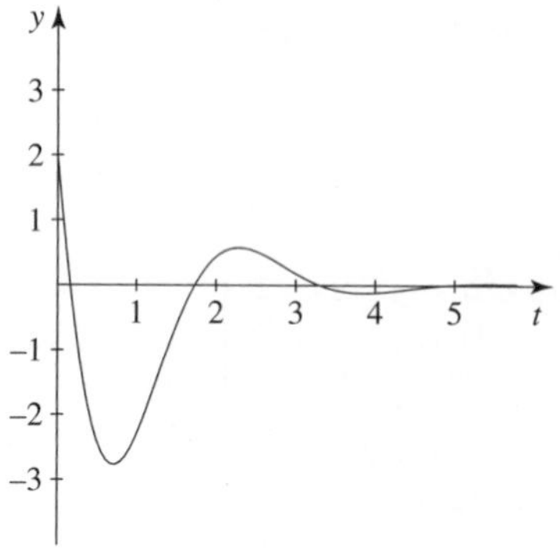

Section 6.4, p. 394

1. Eigenvalues -3 and 2 give a saddle point.
3. Eigenvalues -2 and -2 give an asymptotically stable star node.
5. Eigenvalues 1 and 5 give an unstable node.
7. The origin (0, 0) is a center point and thus neutrally stable.
11. There is a double eigenvalue, $\lambda = 0$, with only one linearly independent eigenvector, $\vec{\mathbf{v}} = \begin{bmatrix} 1 \\ 0 \end{bmatrix}$.

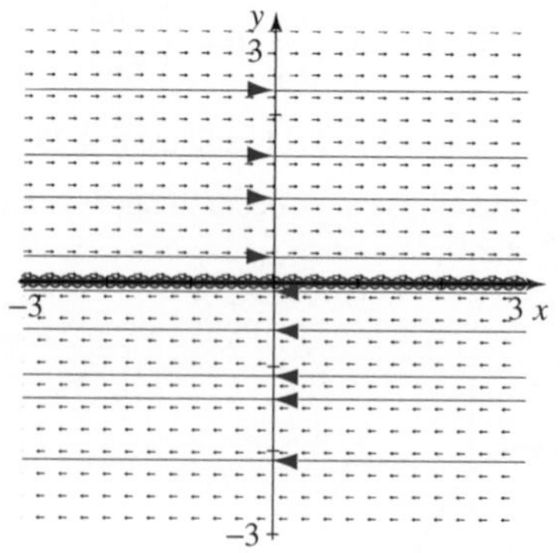

13. There is a double eigenvalue, $\lambda = 0$, with only one linearly independent eigenvector, $\vec{\mathbf{v}} = \begin{bmatrix} 3 \\ 1 \end{bmatrix}$.

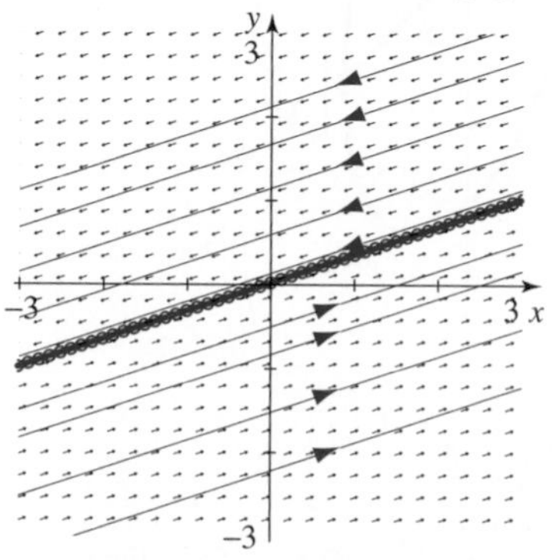

16. Nothing moves; all trajectories are points.
18. The bifurcation values are $k = \pm 2$.

Sample argument for Problems 20–27:

20. Graphically, the given conditions place us above the horizontal axis and outside the parabola in the trace-determinant plane. Algebraically, from

$$\lambda_1, \lambda_2 = \frac{\operatorname{Tr}\mathbf{A} \pm \sqrt{(\operatorname{Tr}\mathbf{A})^2 - 4|\mathbf{A}|}}{2},$$

we know that the eigenvalues are real, unequal and of the same sign. Hence the origin is a node; we can't tell if it is attracting or repelling without knowing the sign of Tr **A**.

Section 6.5, p. 401

1. $\mathbf{P} = \begin{bmatrix} 1 & 2 \\ -2 & 1 \end{bmatrix}$. The decoupled system is

$$w_1' = 3w_1,\ w_2' = -2w_2,$$

which has solutions

$$w_1(t) = c_1 e^{3t},\ w_2(t) = c_2 e^{-2t}.$$

The original system has solution

$$\vec{\mathbf{x}}(t) = \mathbf{P}\vec{\mathbf{w}}(t) = c_1 e^{3t} \begin{bmatrix} 1 \\ -2 \end{bmatrix} + c_2 e^{-2t} \begin{bmatrix} 2 \\ 1 \end{bmatrix}.$$

3. $\mathbf{P} = \begin{bmatrix} -1 & 1 \\ 1 & 1 \end{bmatrix}$. The decoupled system is

$$w_1' = w_1,\ w_2' = -w_2,$$

which has solutions

$$w_1(t) = c_1 e^t,\ w_2(t) = c_2 e^{-t}.$$

The original system has solution

$$\vec{\mathbf{x}}(t) = \mathbf{P}\vec{\mathbf{w}}(t) = c_1 e^t \begin{bmatrix} -1 \\ 1 \end{bmatrix} + c_2 e^{-t} \begin{bmatrix} 1 \\ 1 \end{bmatrix}.$$

5. $\mathbf{P} = \begin{bmatrix} 3 & 1 \\ 1 & 2 \end{bmatrix}$. The decoupled system is

$$w_1' = w_1,\ w_2' = -4w_2,$$

which has solutions

$$w_1(t) = c_1 e^t,\ w_2(t) = c_2 e^{-4t}.$$

The original system has solution

$$\vec{\mathbf{x}}(t) = \mathbf{P}\vec{\mathbf{w}}(t) = c_1 e^t \begin{bmatrix} 3 \\ 1 \end{bmatrix} + c_2 e^{-4t} \begin{bmatrix} 1 \\ 2 \end{bmatrix}.$$

7. $\mathbf{P} = \begin{bmatrix} 1 & -1 & -1 \\ 1 & 1 & 0 \\ 1 & 0 & 1 \end{bmatrix}$. The decoupled system is

$$w_1' = 3w_1,\ w_2' = 0,\ w_3' = 0,$$

with solution

$$w_1(t) = c_1 e^{3t},\ w_2(t) = c_2,\ w_3(t) = c_3.$$

The solution of the original system is

$$\vec{\mathbf{x}}(t) = \mathbf{P}\vec{\mathbf{w}}(t) = c_1 e^{3t} \begin{bmatrix} 1 \\ 1 \\ 1 \end{bmatrix} + c_2 \begin{bmatrix} -1 \\ 1 \\ 0 \end{bmatrix} + c_3 \begin{bmatrix} -1 \\ 0 \\ 1 \end{bmatrix}.$$

9. $\mathbf{P} = \begin{bmatrix} 0 & 0 & \frac{1}{2} \\ 1 & 0 & 1 \\ 1 & 1 & 0 \end{bmatrix}$

$$\vec{\mathbf{x}}(t) = c_1 e^{3t} \begin{bmatrix} 0 \\ 1 \\ 1 \end{bmatrix} + c_2 e^t \begin{bmatrix} 0 \\ 0 \\ 1 \end{bmatrix} + c_3 e^t \begin{bmatrix} 1 \\ 2 \\ 0 \end{bmatrix}$$

11. $\mathbf{P} = \begin{bmatrix} 1 & -1 \\ 1 & 1 \end{bmatrix}$. The decoupled system is

$$w_1' = w_1 + 1,\ w_2' = -w_2,$$

with solution

$$w_1(t) = c_1 e^t - 1,\ w_2(t) = c_2 e^{-t}.$$

The solution of the original system is

$$\vec{\mathbf{x}}(t) = \mathbf{P}\vec{\mathbf{w}}(t) = c_1 e^t \begin{bmatrix} 1 \\ 1 \end{bmatrix} + c_2 e^{-t} \begin{bmatrix} -1 \\ 1 \end{bmatrix} + \begin{bmatrix} -1 \\ -1 \end{bmatrix}.$$

13. $\mathbf{P} = \begin{bmatrix} 1 & 1 \\ -1 & 1 \end{bmatrix}$. The decoupled system is

$$w_1' = \frac{1}{2}(t-1),\ w_2' = 2w_2 + \frac{1}{2}(t+1),$$

with solution

$$w_1(t) = \frac{1}{4}t^2 - \frac{t}{2} + c_1,\ w_2(t) = c_2 e^{2t} - \frac{t}{4} - \frac{3}{8}.$$

The solution of the original system is

$$\begin{aligned} \vec{\mathbf{x}}(t) &= \mathbf{P}\vec{\mathbf{w}}(t) \\ &= c_1 \begin{bmatrix} 1 \\ -1 \end{bmatrix} + c_2 e^{2t} \begin{bmatrix} 1 \\ 1 \end{bmatrix} + \begin{bmatrix} \frac{t^2}{4} - \frac{3t}{4} - \frac{3}{8} \\ -\frac{t^2}{4} + \frac{t}{4} - \frac{3}{8} \end{bmatrix}. \end{aligned}$$

15. $\mathbf{P} = \begin{bmatrix} -1 & \frac{1}{2} \\ 1 & 1 \end{bmatrix}$

$$\vec{\mathbf{x}}(t) = c_1 e^{-t} \begin{bmatrix} -1 \\ 1 \end{bmatrix} + c_2 e^{5t} \begin{bmatrix} 1 \\ 2 \end{bmatrix}$$

17. See Problem 9. Add $\vec{\mathbf{x}}_p(t) = \begin{bmatrix} -1 \\ -\frac{4}{3} \\ -\frac{7}{3} \end{bmatrix}$.

19. $\mathbf{P} = \begin{bmatrix} 1 & 0 & 1 \\ 2 & 1 & 0 \\ 1 & 1 & 1 \end{bmatrix}$. The decoupled system is

$$\vec{\mathbf{w}}' = \begin{bmatrix} 5 & 0 & 0 \\ 0 & 3 & 0 \\ 0 & 0 & 3 \end{bmatrix} \vec{\mathbf{w}} + \frac{1}{2} \begin{bmatrix} -t^2 + t + 1 \\ 2t^2 - 2 \\ t^2 - t - 1 \end{bmatrix}$$

with solutions

$$\begin{aligned} w_1(t) &= c_1 e^{5t} + \left(\frac{t^2}{10} - \frac{3t}{50} - \frac{14}{125} \right), \\ w_2(t) &= c_2 e^{3t} + \left(-\frac{t^2}{3} - \frac{2t}{9} + \frac{7}{27} \right), \\ w_3(t) &= c_3 e^{3t} + \left(-\frac{t^2}{6} + \frac{t}{18} - \frac{4}{27} \right). \end{aligned}$$

The solution of the original system is

$$\begin{aligned} \vec{\mathbf{x}}(t) &= \mathbf{P}\vec{\mathbf{w}}(t) \\ &= \begin{bmatrix} c_1 e^{5t} + c_3 e^{3t} - \left(\frac{t^2}{15} + \frac{t}{225} + \frac{878}{3375} \right) \\ 2c_1 e^{5t} + c_2 e^{3t} - \left(\frac{2t^2}{15} + \frac{77t}{225} - \frac{119}{3375} \right) \\ c_1 e^{5t} + (c_2 + c_3) e^{3t} - \left(\frac{2}{5}t^2 + \frac{17}{75}t + \frac{1}{1175} \right) \end{bmatrix}. \end{aligned}$$

21. $\mathbf{A} = \begin{bmatrix} 3 & -2 \\ 4 & -3 \end{bmatrix}$

Section 6.6, p. 410

1. $e^{\mathbf{A}t} = \begin{bmatrix} e^t & 0 \\ 0 & e^{-t} \end{bmatrix}$

3. $e^{\mathbf{A}t} = \begin{bmatrix} e^t & 0 \\ e^t - 1 & 1 \end{bmatrix}$

5. $e^{\mathbf{A}t} = \begin{bmatrix} e^t & 0 & 0 \\ 0 & e^{2t} & 0 \\ 0 & 0 & e^{3t} \end{bmatrix}$

7. $e^{\mathbf{A}t} = \begin{bmatrix} e^t & 0 \\ 0 & e^t \end{bmatrix}$, $\vec{\mathbf{x}}(t) = \begin{bmatrix} c_1 e^t \\ c_2 e^t \end{bmatrix}$.

9. $e^{\mathbf{A}t} = \begin{bmatrix} e^t & te^t \\ 0 & e^t \end{bmatrix}$; $\vec{\mathbf{x}}(t) = \begin{bmatrix} e^t & te^t \\ 0 & e^t \end{bmatrix} \begin{bmatrix} c_1 \\ c_2 \end{bmatrix}$

11. $e^{\mathbf{A}(t)} = \begin{bmatrix} e^{-t} & 0 \\ 0 & e^{2t} \end{bmatrix}$;

$\vec{\mathbf{x}}(t) = \begin{bmatrix} e^{-t} & 0 \\ 0 & e^{2t} \end{bmatrix} \begin{bmatrix} c_1 \\ c_2 \end{bmatrix} + \begin{bmatrix} 1 - e^{-t} \\ 0 \end{bmatrix}$.

13. $e^{\mathbf{A}t} = \begin{bmatrix} \cosh t & \sinh t \\ \sinh t & \cosh t \end{bmatrix}$;

$\vec{\mathbf{x}}(t) = \begin{bmatrix} \cosh t & \sinh t \\ \sinh t & \cosh t \end{bmatrix} \begin{bmatrix} c_1 \\ c_2 \end{bmatrix} + \begin{bmatrix} -1 + e^t \\ -1 + e^t \end{bmatrix}$.

15. (a) $e^{\mathbf{A}t} = \begin{bmatrix} 1 & t \\ 0 & 1 \end{bmatrix}$, $e^{\mathbf{B}t} = \begin{bmatrix} 1 & t \\ 0 & 1 \end{bmatrix}$

(c) No.

19. (c) $\vec{\mathbf{x}}(t) = \begin{bmatrix} \cosh t & \sinh t \\ \sinh t & \cosh t \end{bmatrix} \begin{bmatrix} c_1 \\ c_2 \end{bmatrix}$

21. $e^{\mathbf{A}t} = \begin{bmatrix} e^t & te^t \\ 0 & e^t \end{bmatrix}$

23. $e^{\mathbf{A}t} = \frac{1}{4} \begin{bmatrix} 2e^{-t} + 2e^{3t} & -e^{-t} + e^{3t} \\ -4e^{-t} + 4e^{3t} & 2e^{-t} + 2e^{3t} \end{bmatrix}$

25. $e^{\mathbf{A}t} = \begin{bmatrix} \cos t & 0 & 0 & \sin t \\ 0 & \cos t & -\sin t & 0 \\ 0 & \sin t & \cos t & 0 \\ -\sin t & 0 & 0 & \cos t \end{bmatrix}$

27. $\vec{\mathbf{x}}(t) = \begin{bmatrix} \cos 3t + 2\sin 3t \\ -\sin 3t + \cos 3t \end{bmatrix}$

29. $\vec{\mathbf{x}}(t) = \begin{bmatrix} e^{3t} \\ 0 \\ 0 \end{bmatrix}$

Section 6.7, p. 418

1. $\vec{\mathbf{x}}_p = \begin{bmatrix} -\frac{1}{2}e^t + 2 \\ -\frac{1}{2}e^t - 2 \end{bmatrix}$

3. $\vec{\mathbf{x}}_p = c_1 e^{3t} \begin{bmatrix} 1 \\ 2 \end{bmatrix} + c_2 e^{-t} \begin{bmatrix} 1 \\ -2 \end{bmatrix} + \begin{bmatrix} e^t - t \\ 1 - e^t \end{bmatrix}$

5. $\vec{\mathbf{x}}_p = c_1 e^{3t} \begin{bmatrix} 2 \\ 1 \end{bmatrix} + c_2 e^{-t} \begin{bmatrix} 2 \\ -1 \end{bmatrix} + t \begin{bmatrix} -12 \\ 3 \end{bmatrix} + \begin{bmatrix} 8 \\ -5 \end{bmatrix}$

7. $\vec{\mathbf{x}}_p = c_1 e^{3t} \begin{bmatrix} 2 \\ 1 \end{bmatrix} + c_2 e^{-t} \begin{bmatrix} 2 \\ -1 \end{bmatrix} + \cos t \begin{bmatrix} 4 \\ -3 \end{bmatrix} + \sin t \begin{bmatrix} -8 \\ 1 \end{bmatrix}$

9. $\vec{\mathbf{x}}(t) = c_1 e^{3t} \begin{bmatrix} 1 \\ 2 \end{bmatrix} + c_2 e^{-t} \begin{bmatrix} 1 \\ -2 \end{bmatrix} + \begin{bmatrix} 2 \\ 1 \end{bmatrix}$

11. $\vec{\mathbf{x}}(t) = c_1 e^{t} \begin{bmatrix} 1 \\ -1 \end{bmatrix} + c_2 e^{3t} \begin{bmatrix} 1 \\ -3 \end{bmatrix} + t \begin{bmatrix} -4 \\ 3 \end{bmatrix} + \begin{bmatrix} -\frac{22}{3} \\ 4 \end{bmatrix}$

13. $\vec{\mathbf{x}}(t) = c_1 e^{t} \begin{bmatrix} -2 \\ 1 \end{bmatrix} + c_2 e^{4t} \begin{bmatrix} 1 \\ 1 \end{bmatrix} + \frac{1}{2}t \begin{bmatrix} -1 \\ 1 \end{bmatrix} + \begin{bmatrix} -\frac{11}{8} \\ \frac{5}{8} \end{bmatrix}$

15. $\vec{\mathbf{x}}(t) = c_1 \begin{bmatrix} 1 \\ 2 \end{bmatrix} + c_2 \begin{bmatrix} t \\ 2t - \frac{1}{2} \end{bmatrix}$

$+ \frac{1}{2t^2} \begin{bmatrix} -1 + 4t - 4t^2(\ln t + 1) \\ 10t + 8t^2(-\ln t - 1) \end{bmatrix}$

17. (c) The equilibrium solution is $\vec{\mathbf{x}}(t) = \begin{bmatrix} 100 \\ 100 \end{bmatrix}$.

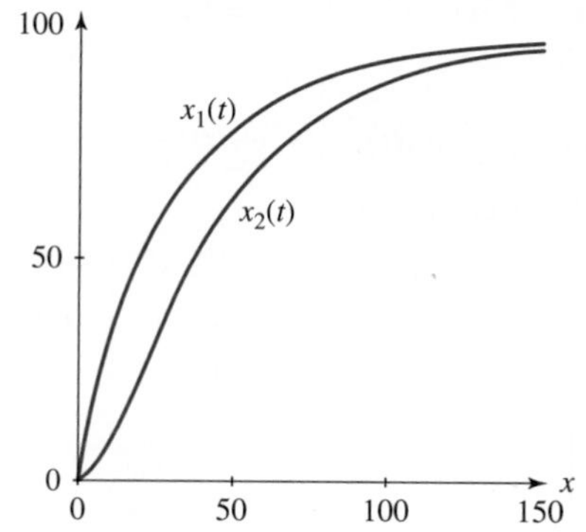

19.

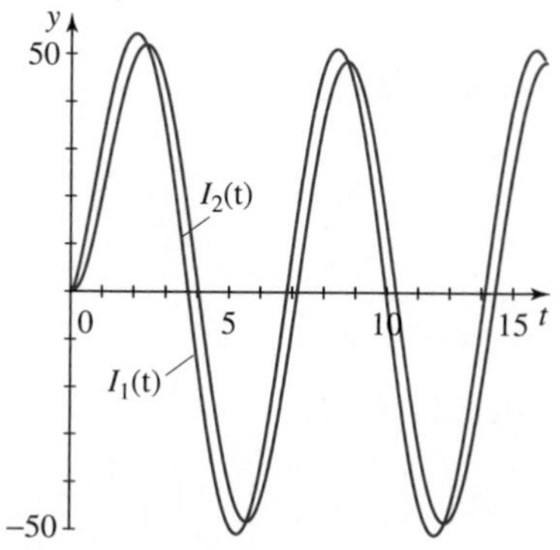

Phase portrait:

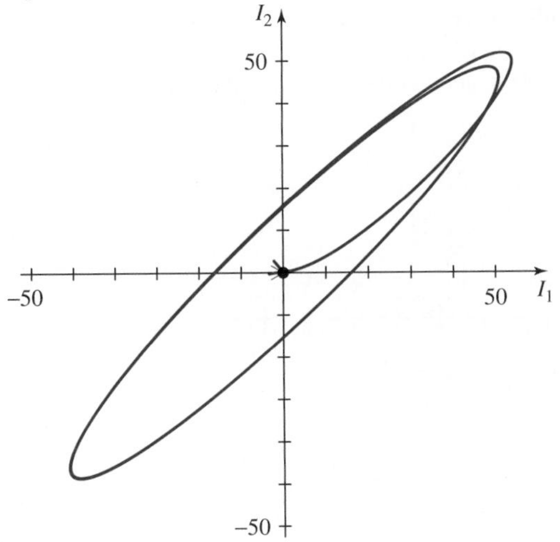

CHAPTER 7

Section 7.1, p. 428

1. Dependent variables: x, y
Parameter: γ
Nonautonomous linear system
Nonhomogeneous ($\gamma \sin t$)

3. Dependent variables: x_1, x_2
Parameter: κ
Autonomous nonlinear ($\sin x_1$) system

5. Dependent variables: R, S, I
Parameters: r, γ
Autonomous nonlinear (SI) system

Sample for Problems 6–9:

7. Substituting the given x, y into the two differential equations, we get

$$\cos t = \cos t,$$
$$-\sin t = -\sin t.$$

11. h-nullcline: $y = 3\cos x$
v-nullclines: the x- and y-axes
The equilibrium points are located at the points $(0, 3)$, $(\pm\pi/2, 0)$, $(\pm 3\pi/2, 0)$, $(\pm 5\pi/2, 0), \ldots$. The equilibrium at $(0, 3)$ is unstable. The x-intercepts at $\pm\pi/2$, $\pm 5\pi/2$, $\pm 9\pi/2, \ldots$ are unstable saddle points. The x-intercepts at $\pm 3\pi/2$, $\pm 7\pi/2, \ldots$ are either centers or spirals; the phase portrait shows they are unstable spirals.

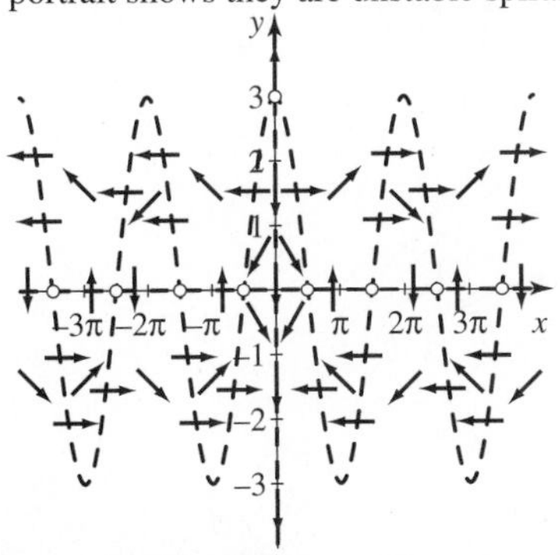

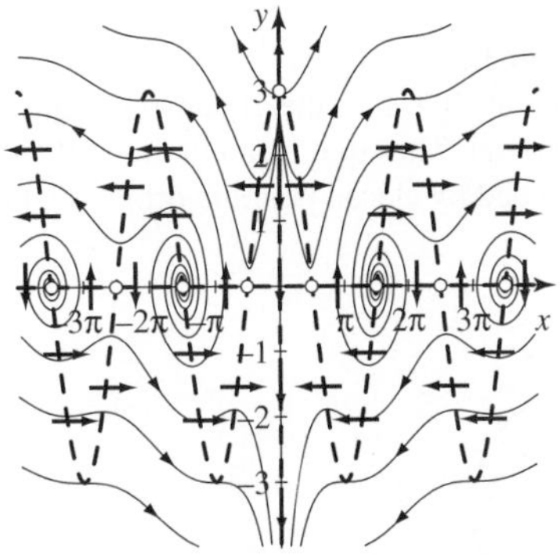

13. h-nullcline: $y = 6x$
v-nullcline: $y = -x^2 - 1$
The equilibrium point that shows in Fig. 7.1.1 is located at the intersection of the nullclines near $(-0.2, -1.2)$; it is either a center point or an unstable spiral.

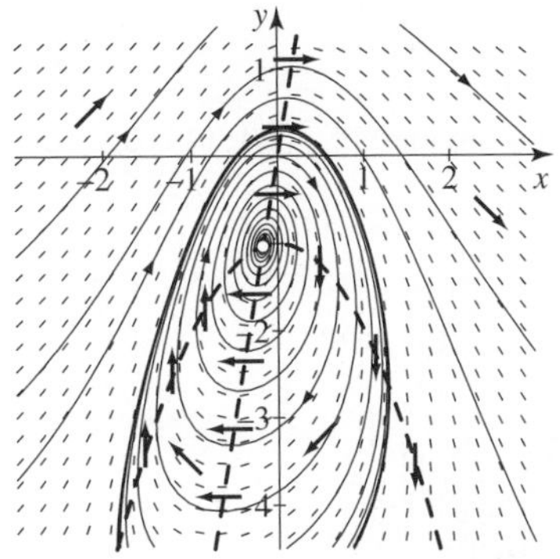

The phase portrait indicates that this equilibrium point is an *unstable* spiral point from which nearby solutions seem to be attracted to a *limit cycle*. There is a *second* equilibrium point, which the reader should find.

15. h-nullcline: $x = \ln|y|$ or $y = \pm e^x$
v-nullcline: $y = \ln|x|$ or $x = \pm e^y$
There are saddle points at approximately $(-1.31, 0.27)$ and $(0.27, -1.31)$ and an unstable node at approximately $(-0.57, -0.57)$. Solutions do not cross either axis because the DEs are not defined for $x = 0$ or $y = 0$.

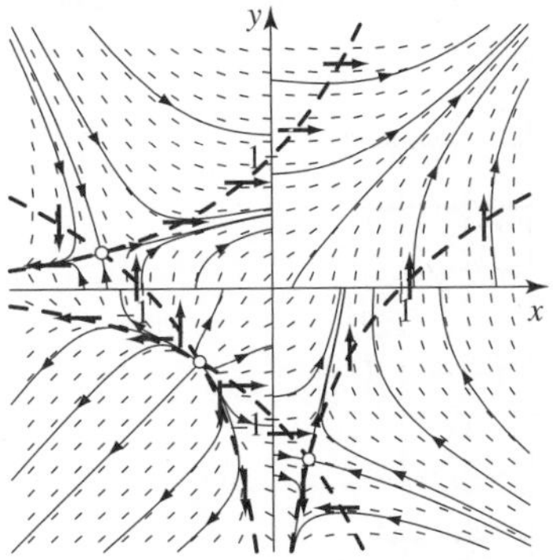

17. h-nullcline: y-axis
v-nullcline: $x^2 + y^2 = 1$
The equilibrium point at $(0, -1)$ is clearly unstable and looks locally like a saddle point. The equilibrium point at $(0, 1)$ is a stable center, surrounded by closed periodic orbits (that are not limit cycles).

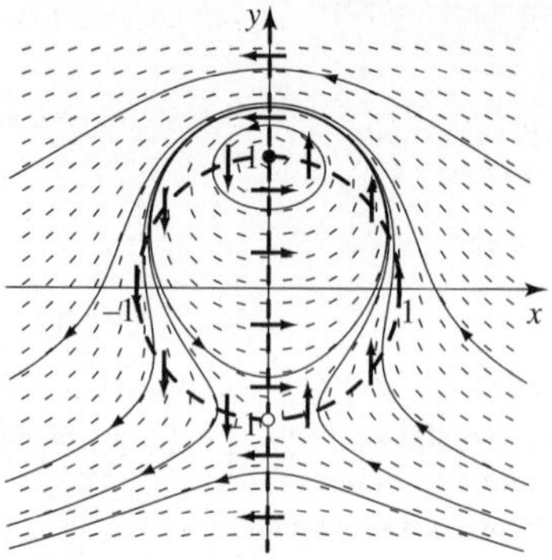

19. h-nullcline: $y = -|x| + 1$
v-nullcline: $y = |x| - 1$
Both equilibrium points are unstable; $(-1, 0)$ is a saddle; $(1, 0)$ is a spiral source.

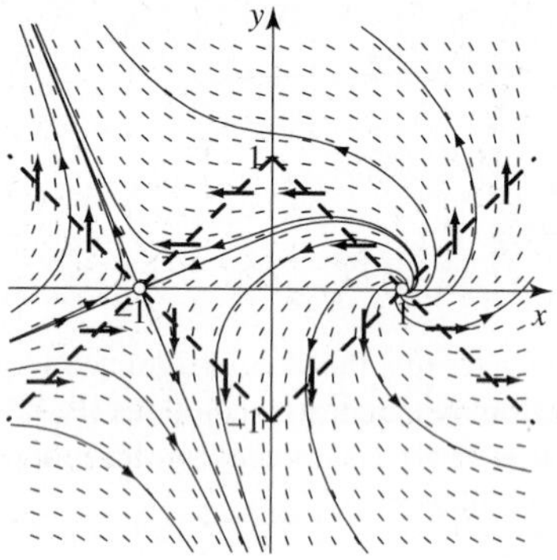

21. h-nullclines: $\theta = (n\pi, 0)$, $n = 0, \pm 1, \pm 2, \ldots$
v-nullcline: $y = 0$ (θ-axis)
From the phase plane we see that the points $(0, 0)$, $(\pm 2\pi, 0)$, $(\pm 4\pi, 0), \ldots$ are center points and hence stable, and the points $(\pm\pi, 0)$, $(\pm 3\pi, 0), \ldots$ are saddle points and hence unstable.

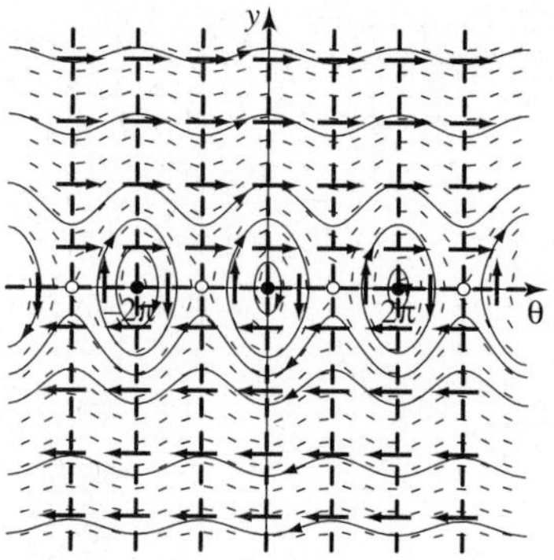

23. h-nullcline: $x^2 + y^2 = 0$ (the origin)
v-nullcline: $y = 0$ (x-axis)
The origin $(0, 0)$ is unstable. Although one trajectory heads towards it, another heads away. All others pass it by.

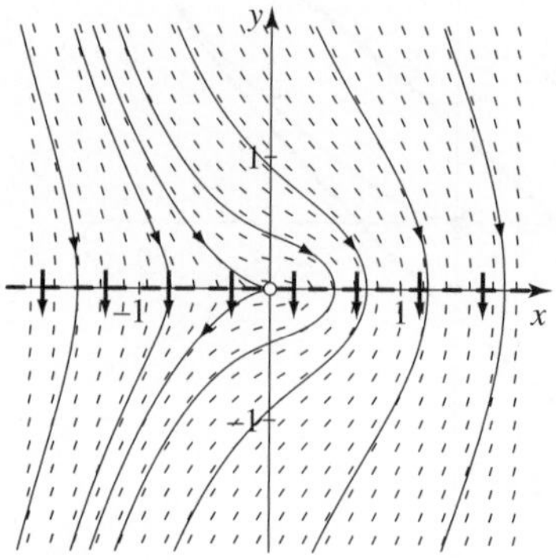

25. h-nullcline: $x + (y^2 - 1)y = 0$
v-nullcline: $y = 0$ (x-axis)
The origin $(0, 0)$ is unstable. Hence $x(t) \equiv 0$ is an unstable solution of the second-order equation. The phase portrait shows also an attracting limit cycle surrounding the equilibrium.

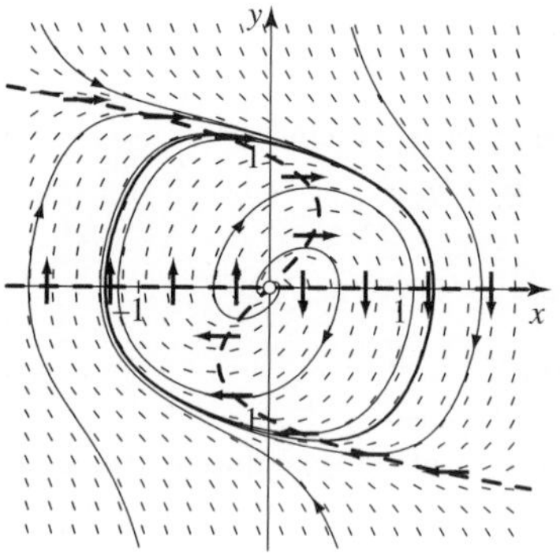

27. $x^2 - y^2 = c$

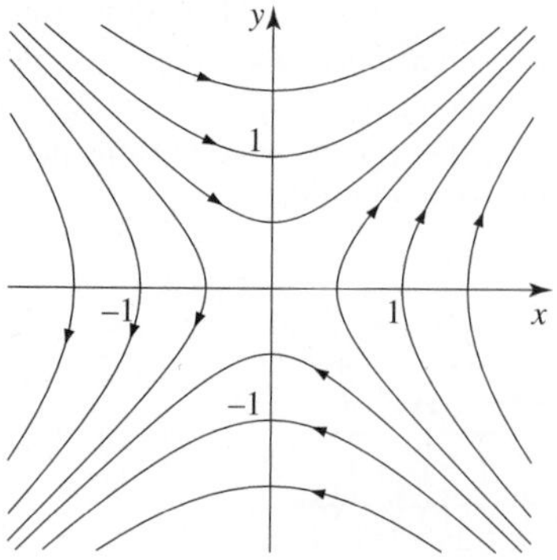

29. $y = c(x^2 + 1)$

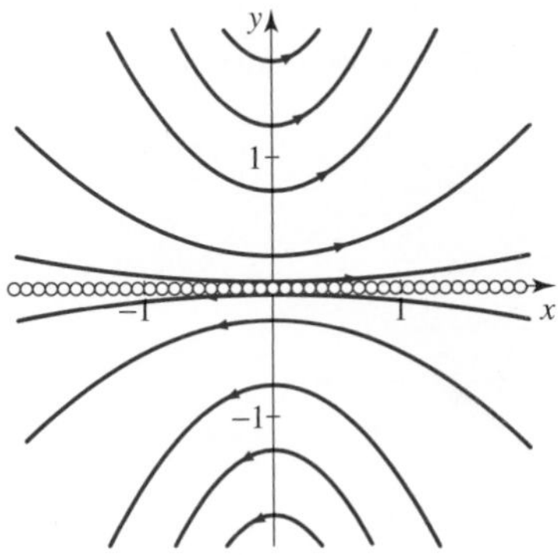

Note that the x axis consists entirely of equilibrium points for this system; both x' and y' are zero.

31. Trajectories move on elliptical paths from an unstable equilibrium at $(0, 1)$ to a stable equilibrium at $(0, -1)$.

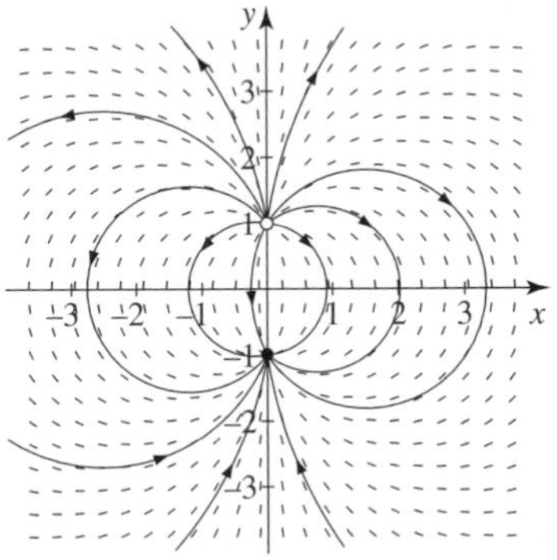

33. There is one equilibrium at $(0, 0)$, which is stable, but solutions start to spiral around the points $(\pm 1, 0)$ and whenever $-1 \le x \le 1$ they "chatter" because the friction force $(\operatorname{sgn} x)$ opposes the direction of motion of the spring.

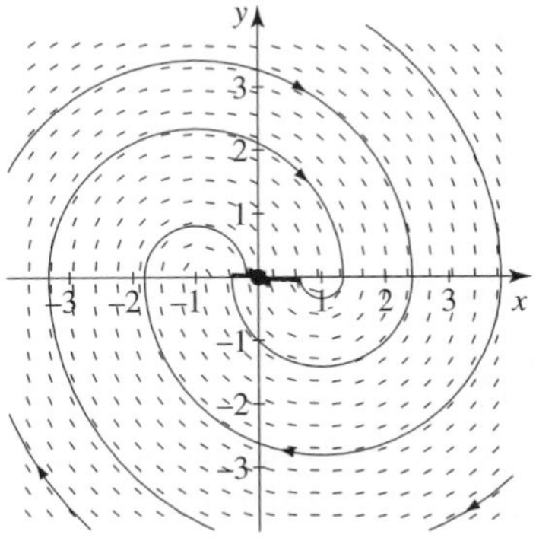

35. The equation $\dot{\theta} = 1$ tells us that trajectories rotate around the origin at constant angular velocity (1 radian per unit time) in the counterclockwise direction, so $r = 1$ is a limit cycle, stable on the inside and unstable on the outside.

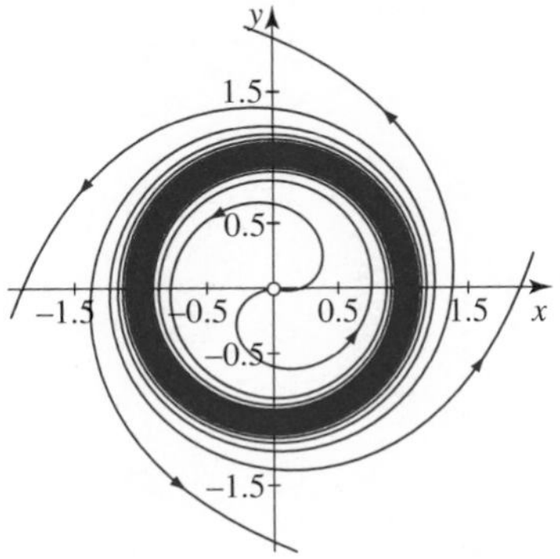

37. The origin is an unstable equilibrium, and there are limit cycles at $r = 1$ (stable) and $r = 2$ (unstable).

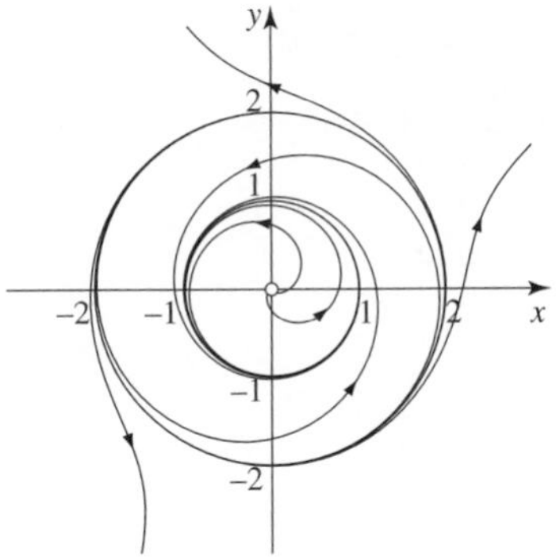

39. (a) To start, we expect no DE or direction field for $y < 0$, where $\sqrt{y}$ is not real.

(b) We note that $y = 0$ is OK for existence, but not for uniqueness because $\partial y'/\partial y$ has a factor of $1/\sqrt{y}$. This shows up in our phase portrait as many solutions seem to melt into the x-axis. For $y > 0$ solutions *are* unique, which means they cannot cross the line of equilibria at $x = -1$.

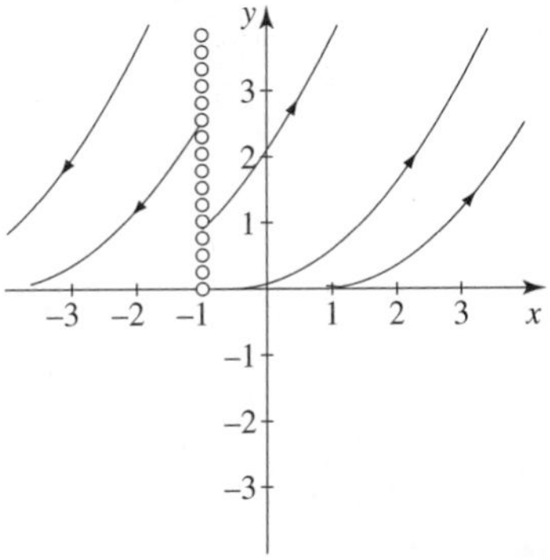

43. Equilibrium points are at $(0, 0)$ (unstable) and $(1, 0)$ (stable). The direction field doesn't detect any periodic solutions, which would appear as closed-loop trajectories. The long-term behavior of this system depends on the initial conditions. For $x > 0$, trajectories move toward the stable equilibrium at $(1, 0)$. For $x < 0$, trajectories approach the x-axis and go off to $-\infty$.

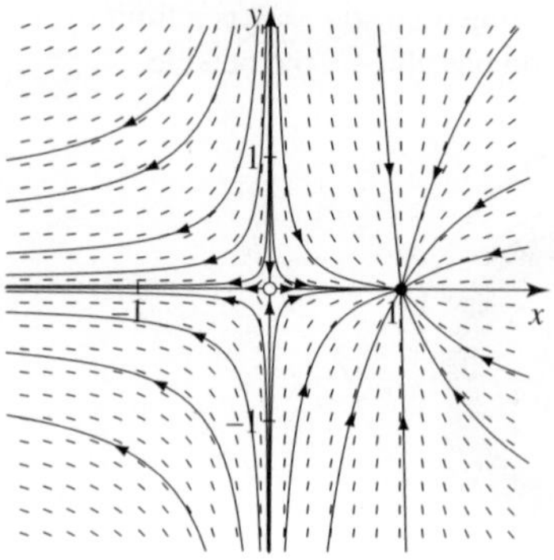

45.

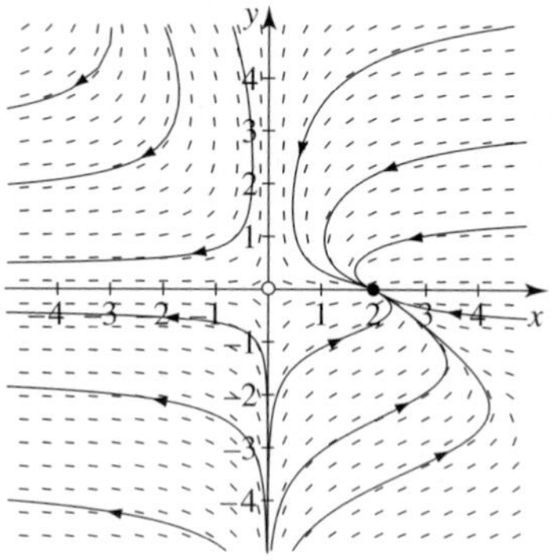

47.

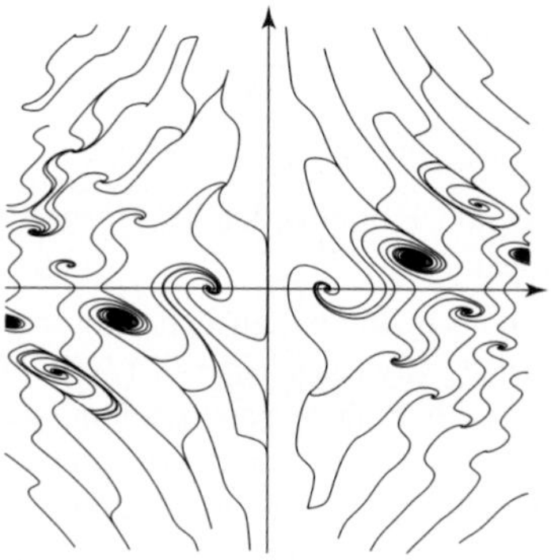

Section 7.2, p. 439

1. At $(0, 0)$, $\mathbf{J} = \begin{bmatrix} -2 & 3 \\ -1 & 1 \end{bmatrix}$,

$\lambda = -\frac{1}{2} \pm \frac{1}{2}\sqrt{3}i$, stable spiral.

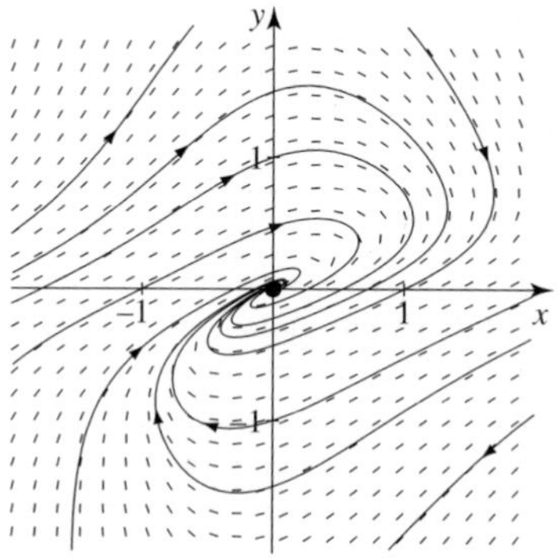

3. At $(0, 0)$, $\mathbf{J} = \begin{bmatrix} 1 & 1 \\ -2 & 1 \end{bmatrix}$,

$\lambda_1, \lambda_2 = 1 \pm \sqrt{2}i$, unstable spiral.

Note that a second equilibrium point exists at $(-1, -1)$.

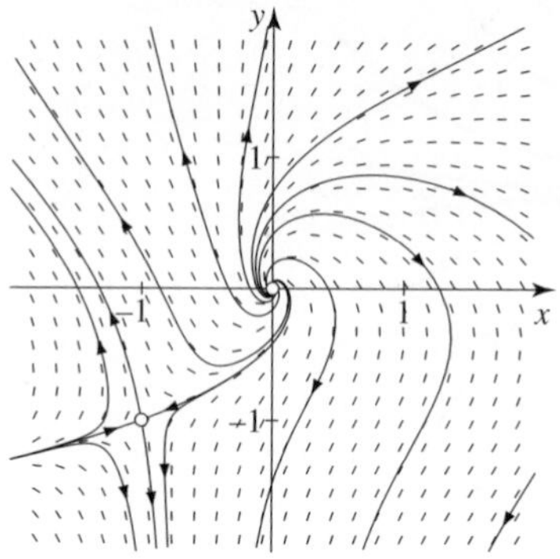

5. At $(0, 0)$, $\mathbf{J} = \begin{bmatrix} 1 & 0 \\ 0 & 0 \end{bmatrix}$,

$\lambda_1 = 1, \lambda_2 = 0$, unstable (half saddle, half node).

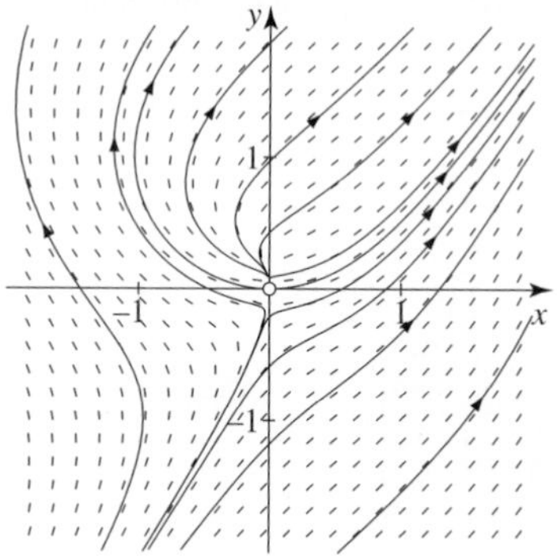

7. At $(1, 1)$, $\mathbf{J} = \begin{bmatrix} -1 & -1 \\ 1 & -3 \end{bmatrix}$,

$\lambda = -2$ (double eigenvalue), stable degenerate node.

At $(-1, -1)$, $\mathbf{J} = \begin{bmatrix} 1 & 1 \\ 1 & -3 \end{bmatrix}$,

$\lambda_1, \lambda_2 = -1 \pm \sqrt{5}$, saddle.

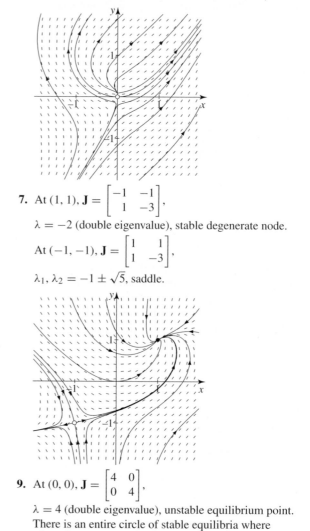

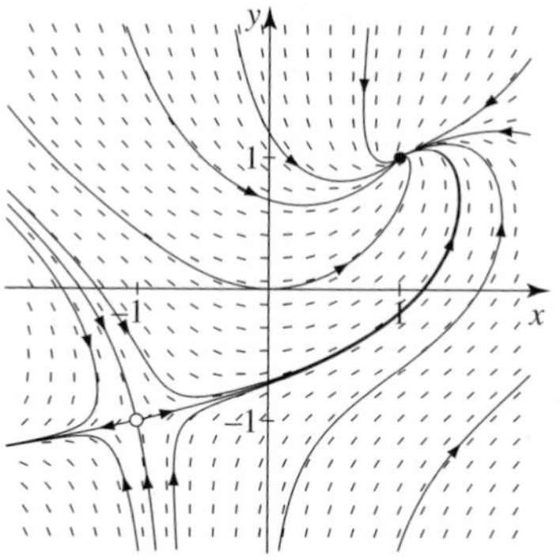

9. At $(0, 0)$, $\mathbf{J} = \begin{bmatrix} 4 & 0 \\ 0 & 4 \end{bmatrix}$,

$\lambda = 4$ (double eigenvalue), unstable equilibrium point. There is an entire circle of stable equilibria where $x^2 + y^2 = 4$,

$$\mathbf{J} = \begin{bmatrix} -2x^2 & -2xy \\ -2xy & -2y^2 \end{bmatrix}.$$

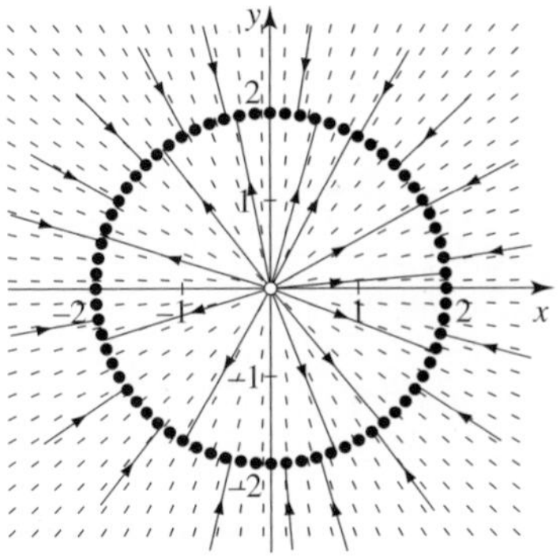

11. The equilibrium at (0, 0) is stable.

15. The equilibrium at $\left(\frac{c}{d}, \frac{a}{b}\right)$ is a center point *or* a spiral point of unknown stability. A phase portrait with trajectories shows it is a center. (See the answer for Problem 10 in Sec. 2.6.)

18. The equilibrium at (0, 0) is an asymptotically stable spiral point. The phase portrait also shows a periodic solution (the limit cycle) that is unstable.

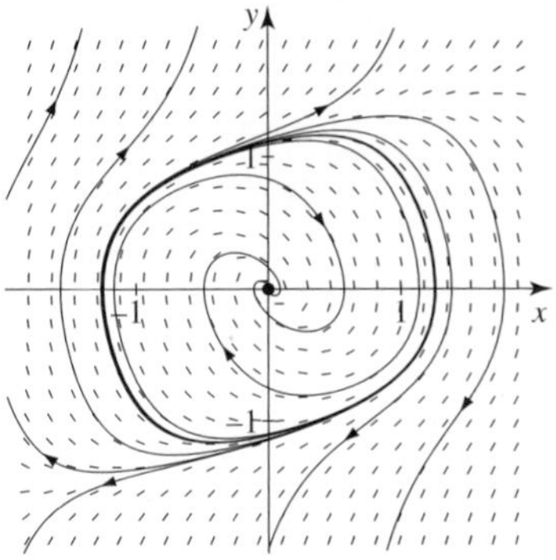

20. The system has negative damping so we suspect the origin is unstable. The eigenvalues of the system around (0, 0) are $\frac{1}{2} \pm i\frac{\sqrt{3}}{2}$, which confirms that the equilibrium point is an unstable spiral point.

21. Liapunov's method verifies $L(x, y) = 2x^2 + y^2$ is positive definite and $dL/dt = -(8x^4 + 6y^6)$ is negative definite.

23. (a) For $k < 0$ there are *two* equilibria.

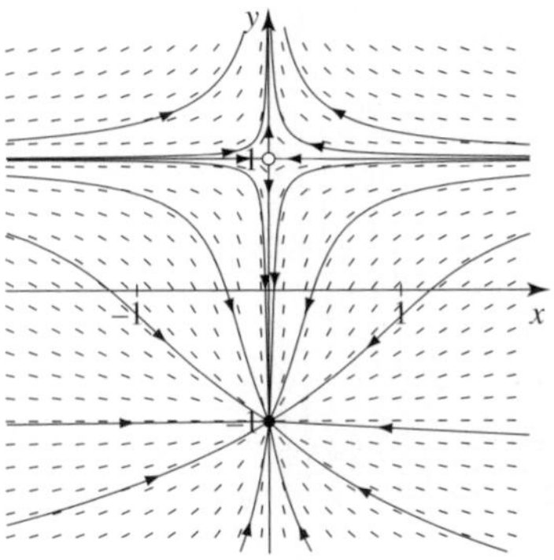

(b) For $k = 0$ there is *one* equilibrium.

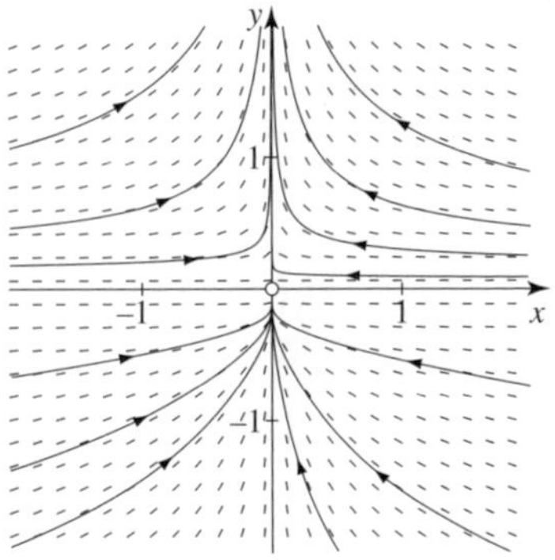

(c) For $k > 0$ there are *no* equilibria.

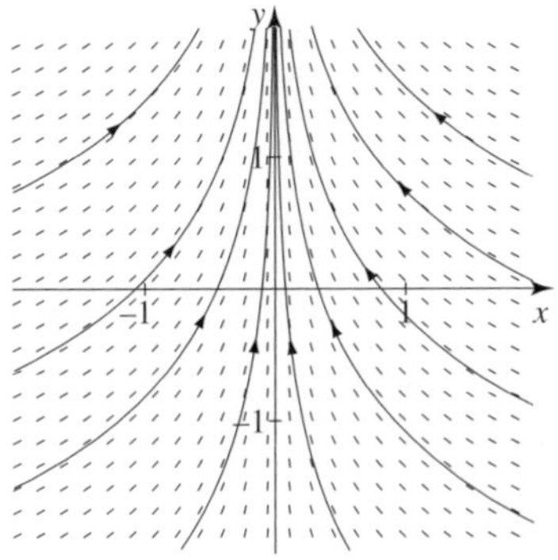

(d) When $k = 0$ the linearized system at (0, 0) is $\dot{x} = -x$, $\dot{y} = 0$, and there is a line of stable equilibria along the vertical axis for the linearized system.

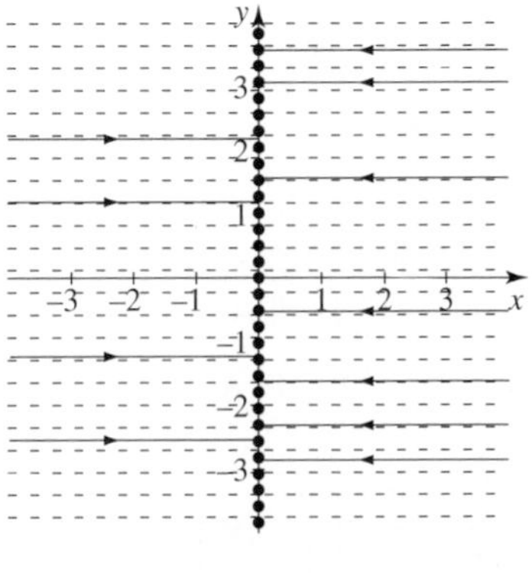

25.

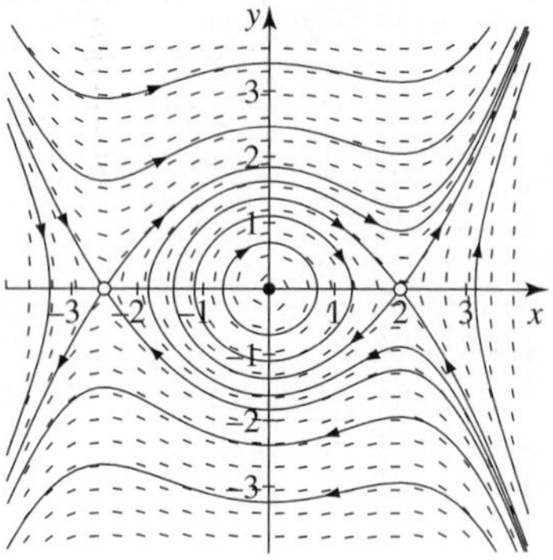

27.

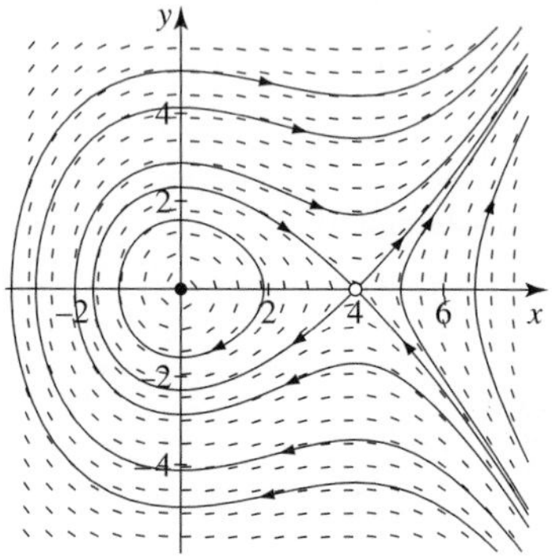

Section 7.3, p. 447

1.

Method	h	$x(1)$	$y(1)$
Euler	0.1	1.9596	1.6559
Euler	0.05	2.0126	1.8071

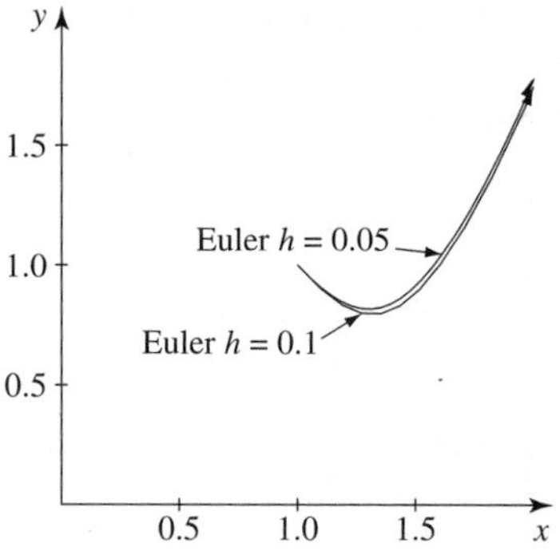

3.

Method	h	$x(1)$	$y(1)$
Euler	0.1	1.3064	−0.4467
Euler	0.05	1.2864	−0.4197

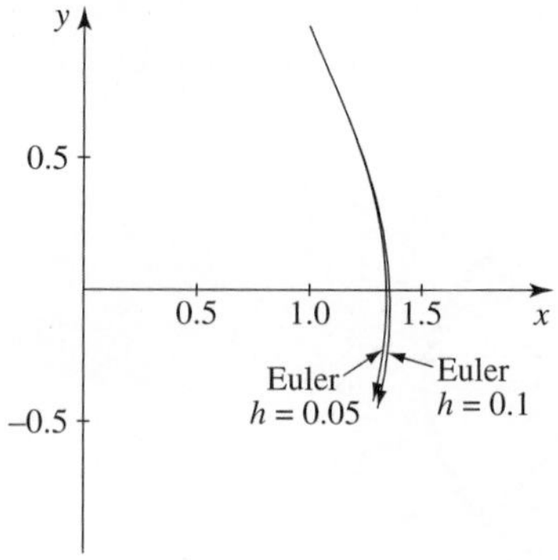

5. Speed $= \sqrt{y^4 + x^4}$ increases with distance from the origin.

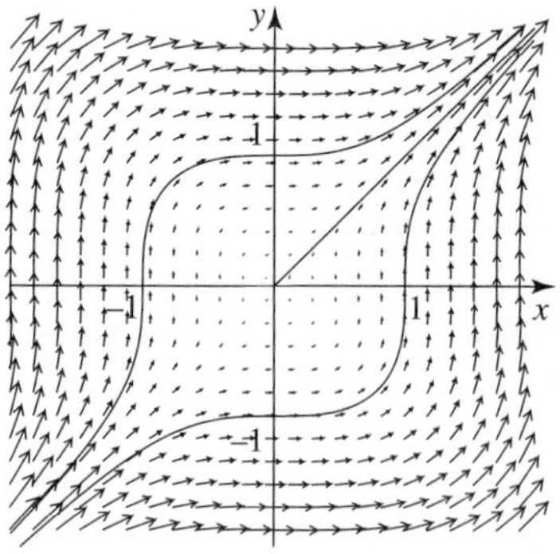

7. Speed $= \sqrt{y^2 + x^2}$ increases directly with distance from the origin.

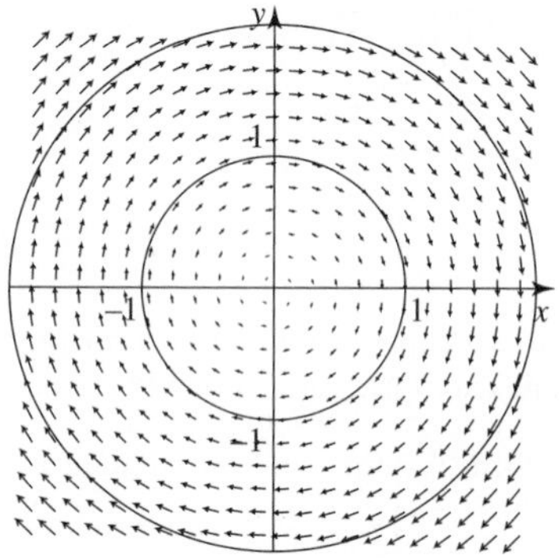

9. Speed $= \sqrt{y^2 + x^2 + 2x^4 + x^6}$ increases faster than distance from the origin.

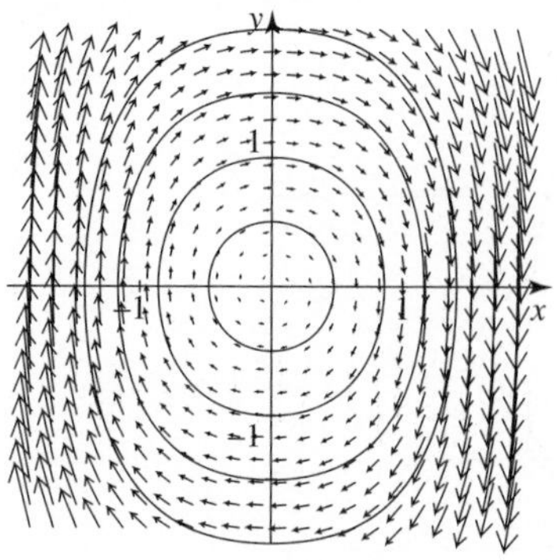

11. A sample sequence follows. Note the "drastic differences" in location/type of equilibria between each pair of phase portraits.

$\varepsilon = -3$

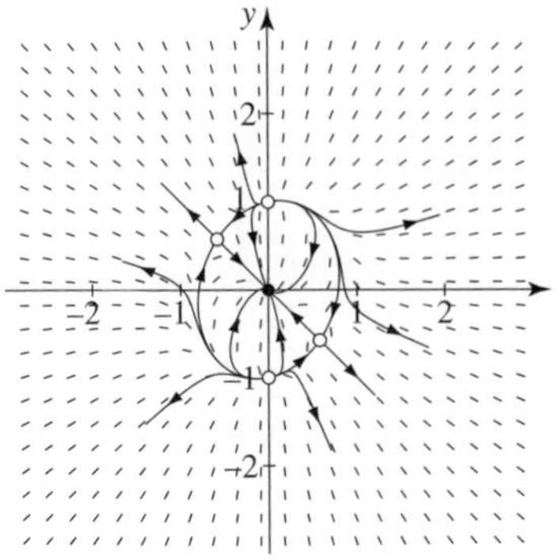

$\varepsilon = -0.3$

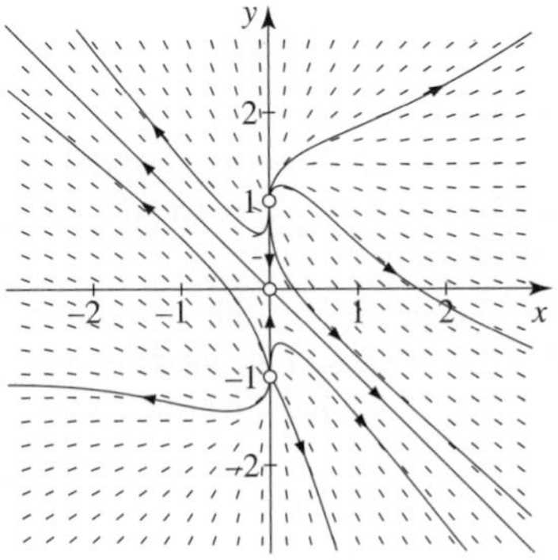

$\varepsilon = 0$

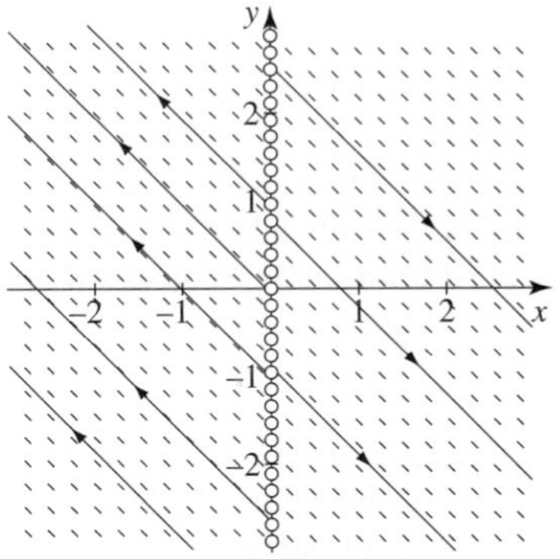

$\varepsilon = 0.3$

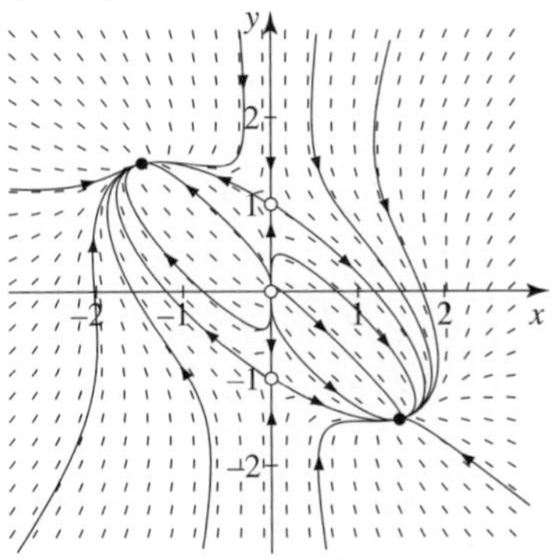

13.

Method	h	$x(1)$	$y(1)$	$z(1)$
Euler	0.1	5.88	4.68	1.40
Euler	0.05	6.25	4.79	1.19
Exact	—	6.66	4.87	0.93

15.

Method	h	$x(1)$	$y(1)$	$z(1)$
Euler	0.1	5.90	1.75	20.84
Euler	0.05	6.18	2.50	23.05

17. For initial conditions in the region plotted, all epidemics rise from I_0 to a maximum I not too much higher then descend until $I = 0$ for a much smaller S (e.g., for $S_0 = 930$, $I_0 = 30$).

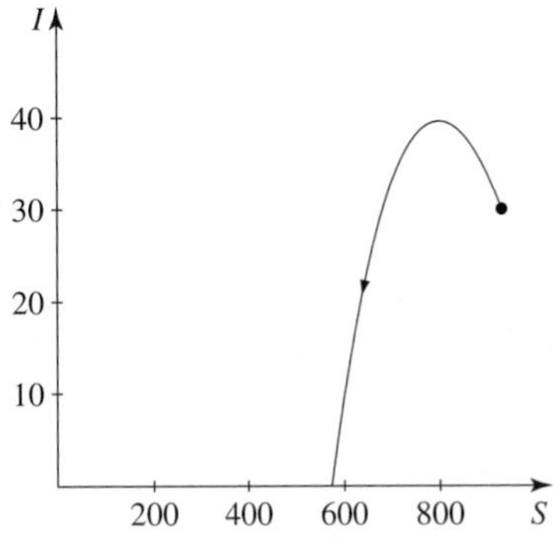

20.

Method	h	$x(1)$	$y(1)$
Euler	0.1	0.7502	−1.1894
Euler	0.05	0.7454	−1.0679
Runge-Kutta	0.05	0.7451	−0.9649

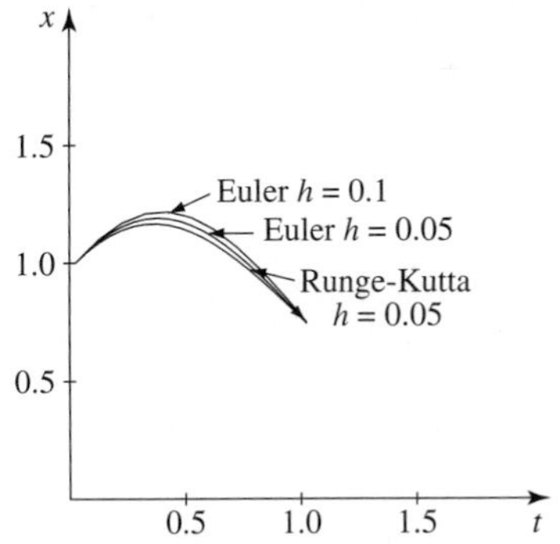

Note on the tx graph how close the approximations become at $t = 1$.

22.

Method	h	$x(1)$	$y(1)$
Euler	0.1	1.1565	−0.4885
Euler	0.05	1.1484	−0.4525
Runge-Kutta	0.05	1.1397	−0.4188

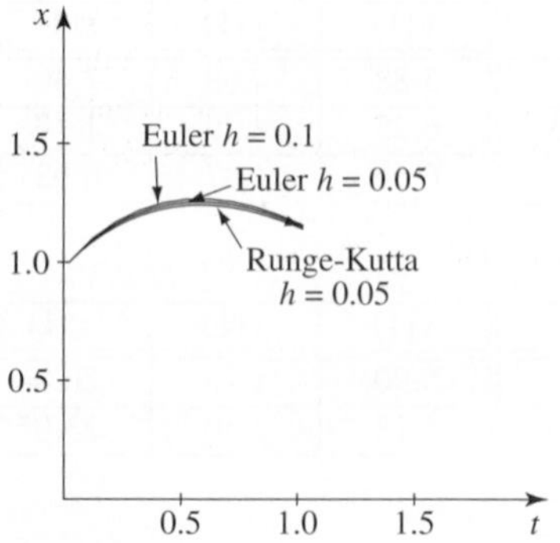

24.

Method	h	$x(1)$	$y(1)$
Euler	0.1	6.1917	6.1917
Euler	0.05	6.7275	6.7275
Runge-Kutta	0.05	7.3890	7.3890

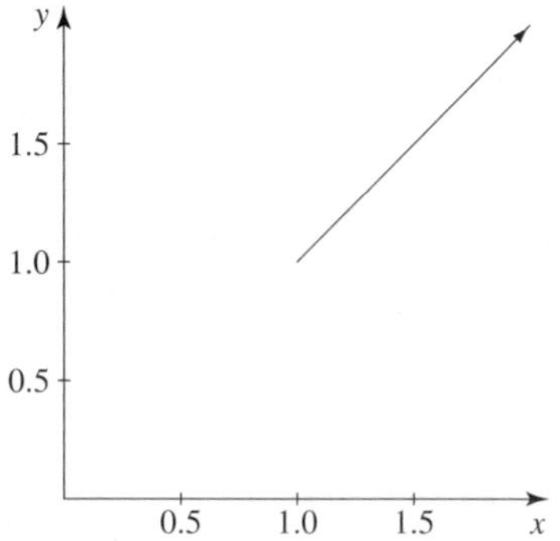

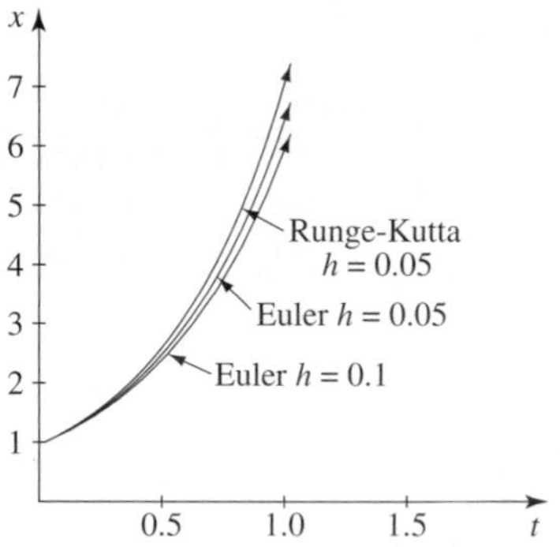

Because x and y have the same time series, the xy trajectory lies along the same diagonal path in all approximations, but the tx graphs shows how the trajectory $x = y$ is traced even faster as t increases, and the approximations give different results.

25.

Method	h	$x(2\pi)$	$y(2\pi)$
Euler	0.1	1.3657	0.0283
Euler	0.01	1.0319	0.0002
Euler	0.001	1.003	−0.0000026
Runge-Kutta	0.1	0.99999957	0.00000049

Note that Euler approximations always return *outside* the unit circle and Runge-Kutta approximations always return *inside* the unit circle. Note also that the latter returns closer to $(1, 0)$ with $h = 0.1$ than Euler does with $h = 0.001$.

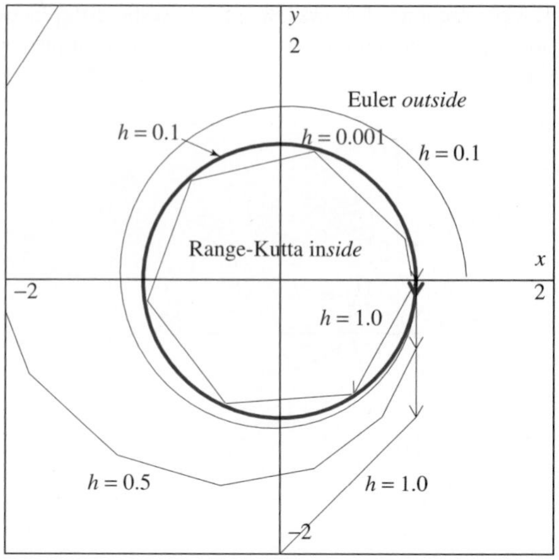

Section 7.4, p. 455

1. Equilibria at $(0, 0, 0)$, $(6\sqrt{2}, 6\sqrt{2}, 27)$ and $(-6\sqrt{2}, -6\sqrt{2}, 27)$. All three equilibria are unstable, which helps explain why an orbit can never settle and the system could exhibit a strange attractor.

3. Figure 7.4.2 provides a sample set of graphs—with longer time series there will be fewer gaps in the phase portraits so they tend to look more similar, but the time series never do.

5. (c) The linearized Lorenz system has a single equilibrium at the origin, which is unstable. The solutions to this linearized system will simply go off to infinity in the unstable direction, so chaos cannot result.

7. (a) There are two equilibrium points, at $(0.0070, -0.0351, 0.0351)$ and $(5.69297, -28.4648, 28.4648)$; both are unstable.

11. You should get a series of plots like Fig. 7.4.5. Some examples are an 8-cycle at $r = 4.13$ and a 16-cycle emerging at $r = 4.15$. At $r = 5$ an outward spiral becomes a chaotic orbit caught in a strange attractor.

Section 7.5, p. 463

1. (a) Equilibrium points are $\pm n\pi$, $n = 0, 1, 2, \ldots$.
Odd multiples of π give saddles;
even multiples of π give stable spirals.
The trajectory from $\theta(0) = 3$, $\dot{\theta}(0) = 1$ is shown below.

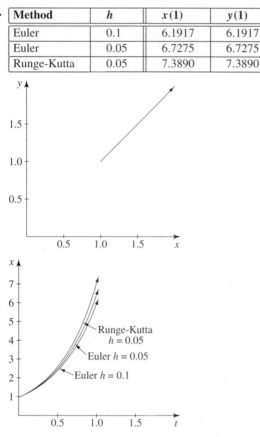

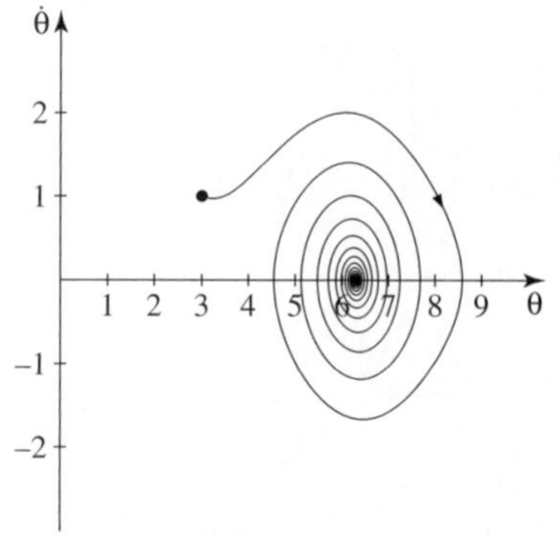

(b) Adding forcing changes the phase portrait.

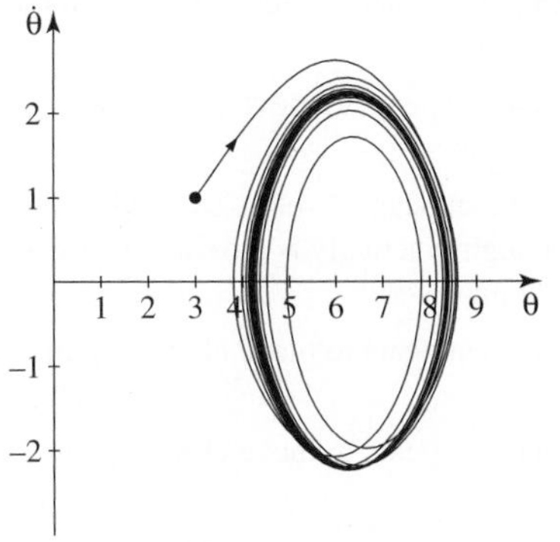

(c) In part (a) an unforced damped pendulum eventually comes to rest at $\theta = 2\pi$. The forced damped pendulum of (b) simply swings back and forth around $\theta = 2\pi$.

3. (a) $\theta(t) = e^{-0.05t}(-3.28\cos t + 0.836\sin t) + 2\pi$

(b) $\theta(t) = e^{-0.05t}(-3.28\cos t - 9.16\sin t) + 2\pi + 10\sin t$

The factor of 10 gives a dramatic difference in scale, which shows up in the phase portrait for Problem 2(b).

5. (a) If the pendulum starts at (3, 1), it goes 360° over the top before settling down to periodic motion about $(2\pi, 0)$. If the pendulum starts at $(-1, 1)$, it settles directly into periodic motion about (0, 0).
The graphs of θ versus t show this difference in the height of the cycles.

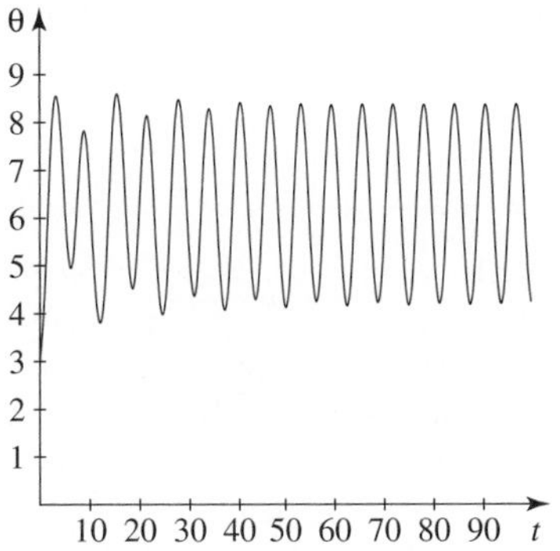

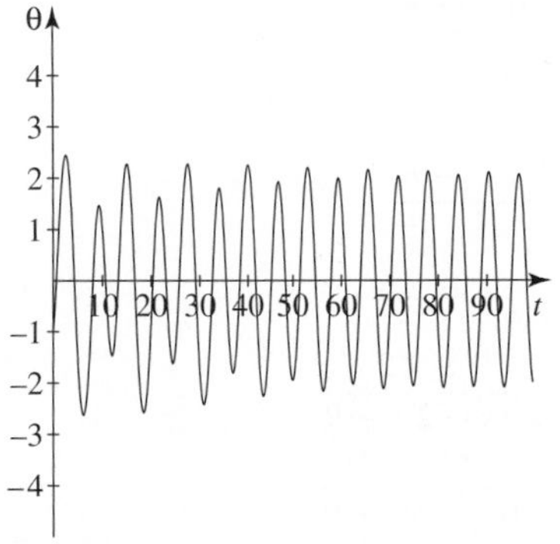

(b) The eleven initial conditions will give $\theta(t)$ graphs that settle out at *several* different levels, representing different numbers of swings "over the top." There appears to be no pattern that would allow prediction of level for a given initial condition, and different computer software will give results that differ in detail.

7. (a) For the damped unforced pendulum there is an unstable equilibrium at (0, 0) and stable equilibria at $(\pm 1, 0)$.

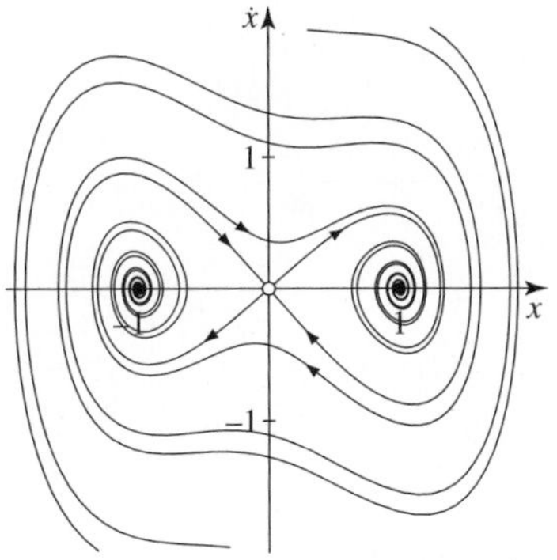

$\ddot{x} + 0.25\dot{x} - x + x^3 = 0$

(b) For this damped forced pendulum the trajectory seems to settle down into a nice closed triple loop (a periodic result en route to chaos, similar to that seen for the Lorenz attractor in Fig. 7.4.6).

(c) For the undamped and unforced pendulum the stable equilibria seen in (a) are now centers instead of spirals.

(d) For this undamped and unforced oscillator the trajectory winds chaotically, similar to Fig. 7.5.6(c).

11. Sample graphs:

(a)

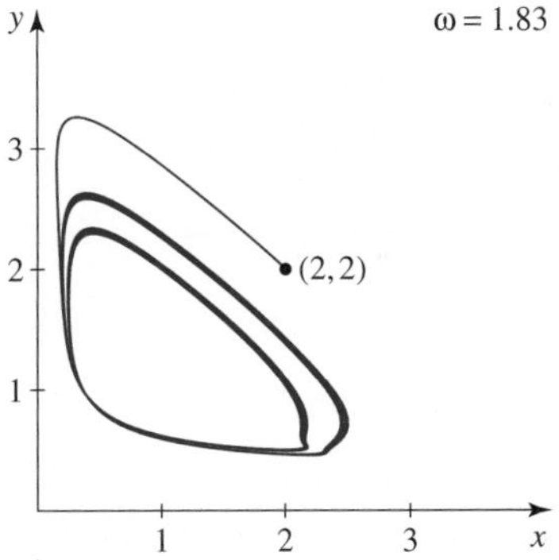

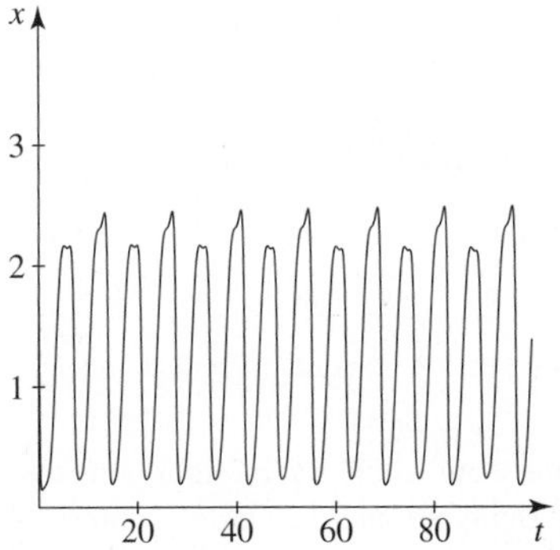

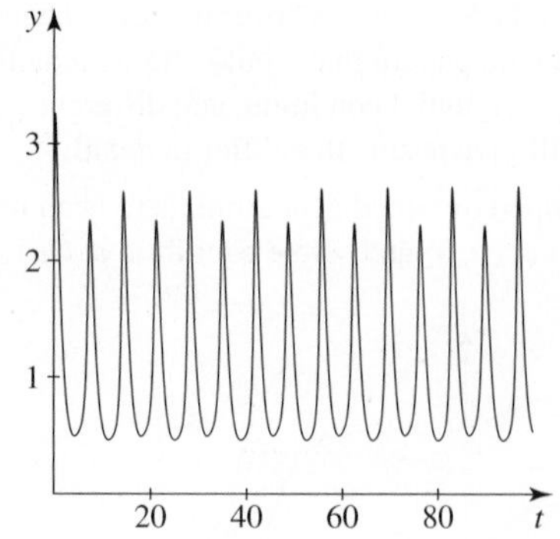

(b)
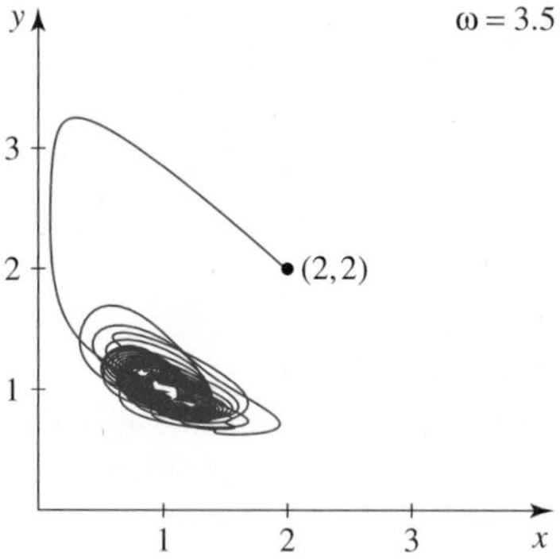

(b)

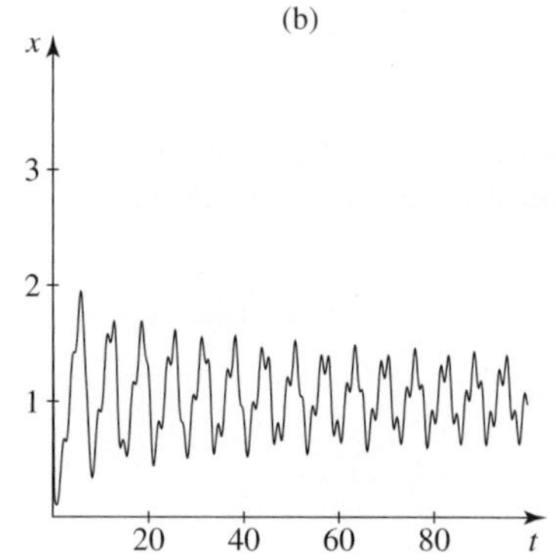

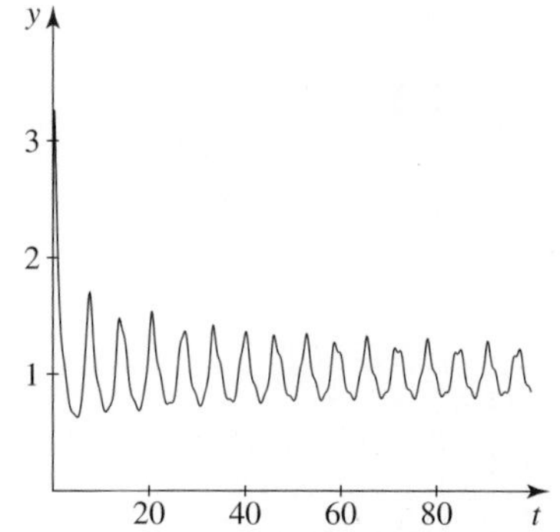

13. (a) The Poincaré section consists of two points (0, 1) and (0, −1).

(b) The Poincaré section consists of eight points, starting at (0, 1) and moving around the circle counterclockwise in jumps of 45°.

15. (a) The clockhand appears unmoved.

(c) The clockhand appears to alternate between top and bottom positions.

(d) The clockhand appears to move clockwise in 15-second jumps.

(f) The clockhand appears to rotate clockwise in 20-second jumps, although it actually is moving clockwise in 80-second jumps.

17. (a) The clockhand appears to move clockwise in 24-second jumps.

(b) The clockhand appears to move clockwise in 36-second jumps.

(c) The clockhand moves clockwise 22.5 seconds with each strobe flash; *apparent* motion is erratic.

(d) The clockhand appears to move *counterclockwise* in 5-second jumps.

CHAPTER 8

Section 8.1, p. 474

1. $\mathcal{L}\{5\} = \dfrac{5}{s}$

3. $\mathcal{L}\{e^{2t}\} = \dfrac{1}{s-2}$

5. $\mathcal{L}\{\sin 2t\} = \dfrac{2}{s^2+4}$

7. $\mathcal{L}\{t^2\} = \dfrac{2}{s^3}$

9. $\mathcal{L}\{f(t)\} = \dfrac{1}{s^2}(1-e^{-2s}) - \dfrac{1}{s}$

11. $\mathcal{L}\{a+bt+ct^2\} = \dfrac{a}{s} + \dfrac{b}{s^2} + \dfrac{2c}{s^3}$

13. $\mathcal{L}\{e^{2t}+e^{-2t}\} = \dfrac{1}{s-2} + \dfrac{1}{s+2} = \dfrac{2s}{s^2-4}$

15. $\mathcal{L}\{e^{-t}(t+3t^2)\} = \dfrac{1}{(s+1)^2} + \dfrac{6}{(s+1)^3}$

17. $\mathcal{L}\{2e^{at}-e^{-at}\} = \dfrac{2}{(s-a)} - \dfrac{1}{(s+a)} = \dfrac{s+3a}{s^2-a^2}$

23. (a) $\dfrac{1}{s}$ (b) $\dfrac{1}{(s-b)^3} + \dfrac{1}{(s+b)^3}$

27. $\mathcal{L}\{t\sin 3t\} = \dfrac{6s}{(s^2+9)^2}$

29. $\mathcal{L}\{3t\cos at\} = 3\dfrac{s^2-a^2}{(s^2+a^2)^2}$

33. $\mathcal{L}\{e^t\sin 2t\} = \dfrac{2}{(s-1)^2+4}$

35. $\mathcal{L}\{e^{2t}\cosh 3t\} = \dfrac{s-2}{(s-2)^2-9}$

37. $\mathcal{L}\{te^{2t}\sin 3t\} = \dfrac{6(s-2)}{[(s-2)^2+9]^2}$

41. $2+3e^t+\dfrac{7}{2}t^2$

43. $3e^{3t}+4e^{-3t}$

45. $e^{-t}\cos 3t$

47. $e^{3t}\left(3\cos 4t + \dfrac{7}{2}\sin 4t\right)$

49. $-e^{-3t} + e^{2t}$

51. $\dfrac{7}{\sqrt{3}}e^{-2t}\sin\sqrt{3}t$

53. $t - \dfrac{1}{2}\sin 2t$

Section 8.2, p. 483

1. $y(t) = t + 1$

3. $y(t) = te^t + e^t$

5. $y(t) = -e^{2t} + 2e^t$

7. $y(t) = 2e^{-t} - 2\cos 3t + \sin 3t$

9. $y(t) = 3 - 4e^{-t} + e^{-2t}$

11. $y(t) =$
$$\sin t - \cos t + \frac{1}{\sqrt{3}}e^{-t/2}\left[\sin\left(\frac{\sqrt{3}}{2}t\right) + \cos\left(\frac{\sqrt{3}}{2}t\right)\right].$$

14. $y(t) = (2A + B)e^{-t} + (-A - B)e^{-2t}$

16. $y(t) = \cos t$

17. The solution $y(t)$ to the first equation, $y' = ky$, gives $y(0) = 1$ if $k \geq 1$.

19. (c) $y(t) = t^6 - 30t^4 + 360t^2 - 720$

Section 8.3, p. 496

1. $f(t) = a\ \text{step}(t) + (b - a)\ \text{step}(t - 1) + (c - b)\ \text{step}(t - 2)$

3. $f(t) = 1 + (4t - t^2 - 1)\ \text{step}(t - 1) + (1 - 4t + t^2)\ \text{step}(t - 4)$

5. $\mathcal{L}\{f(t)\} = \dfrac{1}{s}\left(\dfrac{e^{-s}}{1 - e^{-s}}\right)$

7. $\mathcal{L}\{f(t)\} = \dfrac{2(1 - e^{-s})}{s(2 - e^{-s})}$

9. $\mathcal{L}\{f(t)\} = \dfrac{1}{s} - \dfrac{2e^{-s}}{s}\left(\dfrac{1}{1 + e^{-s}}\right)$

11. $\mathcal{L}\{f(t)\} = \dfrac{1}{s} - \dfrac{1}{2s^2}e^{-s} + \dfrac{1}{2s^2}e^{-3s}$

13. $\mathcal{L}\{f(t)\} = \dfrac{b(\pi/a)}{s^2 + (\pi/a)^2}(1 + e^{-as})$

15. $\mathcal{L}\{f(t)\} = \dfrac{\pi}{s^2 + \pi^2}(1 + 3e^{-s} + 2e^{-2s})$

17. $\mathcal{L}\{f(t)\} = 1 - 2e^{-\pi s} + e^{-2\pi s}$

19. $\mathcal{L}\{f(t)\} = 1 + e^{-\pi s} + e^{-2\pi s} + \cdots$

20. $\dfrac{1}{s} - \dfrac{e^{-s}}{s}$

21. $\dfrac{1}{s} - \dfrac{2e^{-s}}{s} + \dfrac{e^{-2s}}{s}$

22. $\dfrac{e^{-s}}{s^2}$

24. $\dfrac{e^{(3-3s)}}{s - 1}$

26. $e^{-2s}\left(\dfrac{2}{s^3} + \dfrac{4}{s^2} + \dfrac{4}{s}\right)$

28. $(t - 1)\ \text{step}(t - 1)$

30. $e^{-4(t-4)}\ \text{step}(t - 4)$

32. $\text{step}(t - 1) - 2\ \text{step}(t - 2) + 2\ \text{step}(t - 3) - \text{step}(t - 4)$

34. $x(t) = t - 2(t - 1)\ \text{step}(t - 1) + (t - 2)\ \text{step}(t - 2)$

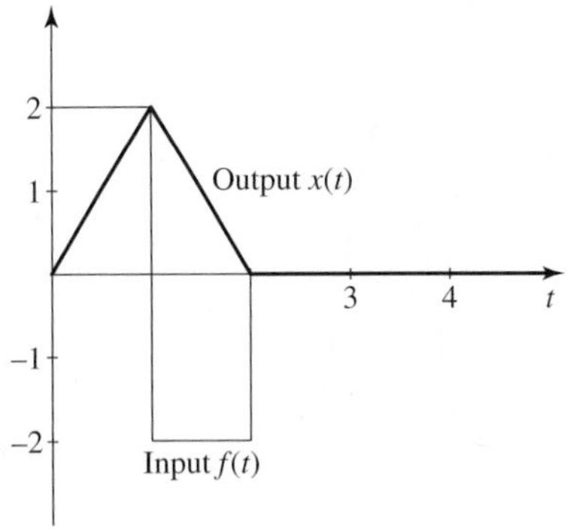

36. $x(t) = \sin t + (1 + \cos t)\ \text{step}(t - \pi) - (1 - \cos t)\ \text{step}(t - 2\pi)$

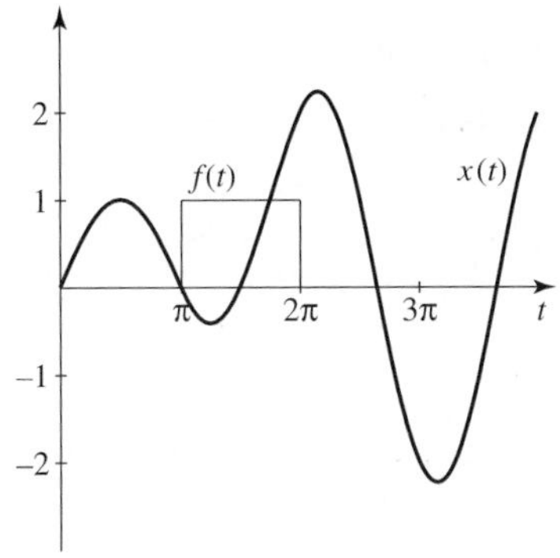

38. $\mathcal{L}\{f(t)\} = \dfrac{1}{1 - e^{-3s}}\left(\dfrac{1 - 2e^{-s} + e^{-2s}}{s^2}\right)$

40. $\mathcal{L}\{f(t)\} = \dfrac{1}{s^2 + 1}\left(\dfrac{1 + e^{-\pi s}}{1 - e^{-\pi s}}\right)$

42. $\mathcal{L}\{f(t)\} = \dfrac{1}{1 - e^{-4s}}\left[\dfrac{\left(\frac{\pi}{2} - se^{-s}\right)}{s^2 + \left(\frac{\pi}{2}\right)^2} + \dfrac{e^{-s}}{s} - \dfrac{e^{-2s}}{s^2}\right]$

44. $\mathcal{L}\{f(t)\} = \left(\dfrac{1}{1 - e^{-2s}}\right)\left(\dfrac{\pi}{s^2 + \pi^2}\right)\left(1 + 3e^{-s} + 2e^{-2s}\right)$

46. $x(t) = t - 1 + e^{-t} - \left[1 - e^{-(t-1)}\right]\text{step}(t - 1)$
$-\left[1 - e^{-(t-2)}\right]\text{step}(t - 2) - \cdots$

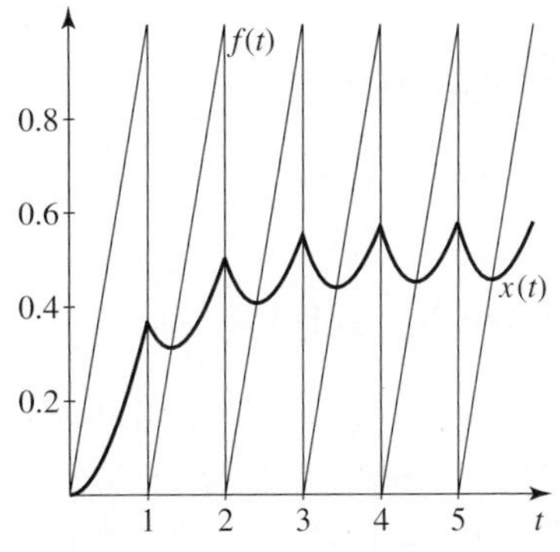

48. $x(t) = 1$

50. $x(t) = \sin(t - 2\pi)\ \text{step}(t - 2\pi)$; its graph is the sine function starting at $t = 2\pi$.

52. $x(t) = \cos t + \sin t\ \text{step}(t - 2\pi)$

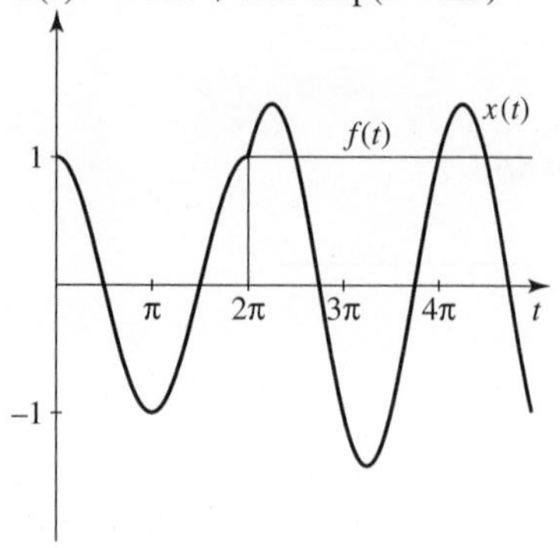

Section 8.4, p. 507

2. Prove by writing out the integrals.

5. $\frac{1}{2}t^2$ **7.** $\frac{1}{6}t^3$ **8.** $\sinh t$

11. $a * \left(\frac{b}{a}\right) = \int_0^t (a)\left(\frac{b}{a}\right) dw = bw\Big]_0^t = bt$

13. $x(t) = 1 + 1 * f(t)$
$= 1 + \int_0^t f(\tau)d\tau$

15. $x(t) = e^{-t} + e^{-t} * f(t)$
$= e^{-t} + \int_0^t e^{-(t-\tau)} f(\tau)d\tau$

17. $x(t) = (e^{-t} + e^{-2t}) * f(t)$
$= \int_0^t \left(e^{-(t-\tau)} - e^{-2(t-\tau)}\right) f(\tau)d\tau$

19. The transfer function is $\frac{1}{s+a}$, impulse response $I(t) = e^{-at}$,
$x(t) = I(t) * f(t)$
$= \int_0^t e^{-a(t-\tau)} f(\tau)d\tau$

21. The transfer function is $\frac{1}{(s+2)^2+1}$, impulse response $I(t) = e^{-2t}\sin t$,
$x(t) = I(t) * f(t)$
$= \int_0^t e^{-2(t-\tau)} \sin(t-\tau) f(\tau)d\tau$

22. t **24.** $1 - e^{-t}$ **26.** $-\sin t + t$

28. $\frac{1}{2k}\sin kt - \frac{1}{2}t\cos kt.$

31. $x(t) = h(t) * 4\cos t + \cos 2t - \frac{1}{2}\sin 2t$

32. (a) $I_{1/2}(1) = \frac{2\sqrt{t}}{\sqrt{\pi}}$

(c) $I_{1/2}(at^2 + bt + c) = \frac{\sqrt{t}}{\sqrt{\pi}}\left[\frac{16}{15}at^2 + \frac{4}{3}bt + 2c\right]$

34. (a) $\frac{d^{1/2}}{dt^{1/2}}(1) = \frac{1}{\sqrt{\pi}\sqrt{t}}$ (b) $\frac{d^{1/2}}{dt^{1/2}}(t) = \frac{2\sqrt{t}}{\sqrt{\pi}}$

36. \$334,712

38. $\int_0^t e^{-0.1(t-w)} 2e^{0.05w}dw = \frac{40}{3}\left[e^{0.05t} - e^{-0.1t}\right]$

40. $y(t) = e^{-t}$ **42.** $y(t) = t^3 + \frac{1}{20}t^5$ **44.** $y(t) = 1$

46. (b) $I(s) = \frac{12}{\left(5 + \frac{5}{2}\right)^2 + \frac{75}{4}}(1 - 2e^{-s})$

(c) $i(t) = \frac{24}{5\sqrt{2}}e^{-(5/2)t}\sin\frac{5\sqrt{2}}{2}t$
$- \text{step}(t-1)\frac{48}{5\sqrt{2}}e^{-(5/2)(t-1)}\sin\frac{5\sqrt{2}}{2}(t-1)$

48. (b) $I(s) = \frac{1}{L}\left(\frac{1}{s + \frac{R}{L}}\right)V(s)$

(c) $i(t) = \frac{1}{L}e^{-(R/L)t} * V(t)$

51. $y(t) = (e^t + e^{-t} - 2) * f(t)$

Section 8.5, p. 516

1. $\vec{\mathbf{x}}(t) = \begin{bmatrix} \sin t \\ \cos t \end{bmatrix}$

3. $\vec{\mathbf{x}}(t) = \begin{bmatrix} 5e^t - 6e^{2t} + 2e^{4t} \\ 5e^t - 12e^{2t} + 8e^{4t} \end{bmatrix}$

5. $\vec{\mathbf{x}}(t) = \begin{bmatrix} -e^t + e^{2t} \\ -e^t + 2e^{2t} - e^{3t} \end{bmatrix}$

7. $\vec{\mathbf{x}}(t) = 3e^{-t}\begin{bmatrix} 2 \\ -1 \end{bmatrix} - \frac{2}{3}e^{3t}\begin{bmatrix} 2 \\ 1 \end{bmatrix} + \frac{1}{3}\begin{bmatrix} -14 \\ 11 \end{bmatrix}$

9. $\vec{\mathbf{x}}(t) = e^{-3t}\begin{bmatrix} 1 \\ 1 \end{bmatrix} + e^{5t}\begin{bmatrix} 1 \\ 2 \end{bmatrix}$

11. $\vec{\mathbf{x}}(t) = \begin{bmatrix} \frac{1}{2}e^{-t} + \frac{1}{2}e^{3t} & -\frac{1}{4}e^{-t} + \frac{1}{4}e^{3t} \\ -e^{-t} + e^{3t} & \frac{1}{2}e^{-t} + \frac{1}{2}e^{3t} \end{bmatrix}\begin{bmatrix} x_0 \\ y_0 \end{bmatrix}$

13. $\begin{bmatrix} 1 & 1 \\ 1 & 0 \end{bmatrix}\dot{\vec{\mathbf{x}}} + \begin{bmatrix} 4 & 0 \\ -2 & 1 \end{bmatrix}\vec{\mathbf{x}} = \begin{bmatrix} 0 \\ 0 \end{bmatrix}, \vec{\mathbf{x}}(0) = \begin{bmatrix} 2 \\ -1 \end{bmatrix}.$

$\vec{\mathbf{x}}(t) = \begin{bmatrix} \frac{1}{5}e^{-t} - \frac{1}{5}e^{4t} \\ \frac{3}{5}e^{-t} + \frac{2}{5}e^{4t} \end{bmatrix}$

17. $\vec{\mathbf{x}}(t) = \begin{bmatrix} \frac{1}{2}\sqrt{\frac{2}{3}}\sin\left(\sqrt{\frac{3}{2}}t\right) + \frac{\sqrt{2}}{2}\sin\left(\sqrt{\frac{1}{2}}t\right) \\ -\frac{1}{2}\sqrt{\frac{2}{3}}\sin\left(\sqrt{\frac{3}{2}}t\right) + \frac{\sqrt{2}}{2}\sin\left(\sqrt{\frac{1}{2}}t\right) \end{bmatrix}$

19. $\frac{dx_2}{dt} = -s + k_{12}x_1 - k_{21}x_2 - k_{23}x_2 + k_{32}x_3$
$\frac{dx_3}{dt} = k_{23}x_2 - k_{32}x_3$

20. (a) $\begin{array}{ll} \ddot{x}_1 = -4x_1 - 2(x_1 - x_2), & x_1(0) = \dot{x}_1(0) = 0 \\ 2\ddot{x}_2 = 2(x_1 - x_2), & x_2(0) = 1, \dot{x}_2(0) = 0 \end{array}$

22. $\vec{\mathbf{x}}(t) = \begin{bmatrix} 0 \\ 2e^{3t} - 2 \end{bmatrix}$.

CHAPTER 9

Section 9.1, p. 530

1. (a) $y_n = -5\left(\frac{1}{2}\right)^n + 6$

(b)

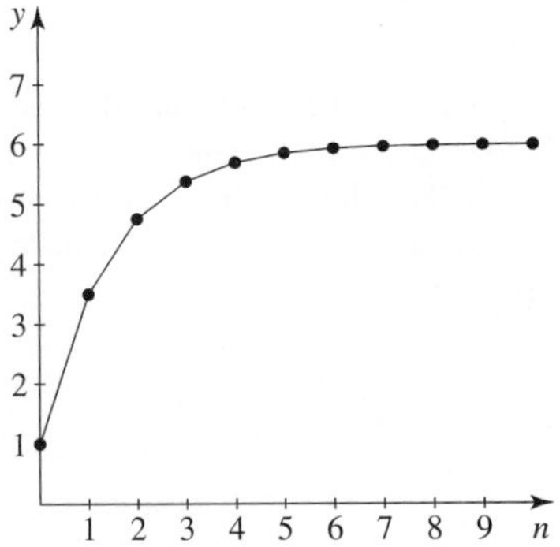

(c) $y_n \to 6 = y_e$

3. (a) $y_n = -\left(-\frac{1}{2}\right)^n + 2$

(b)

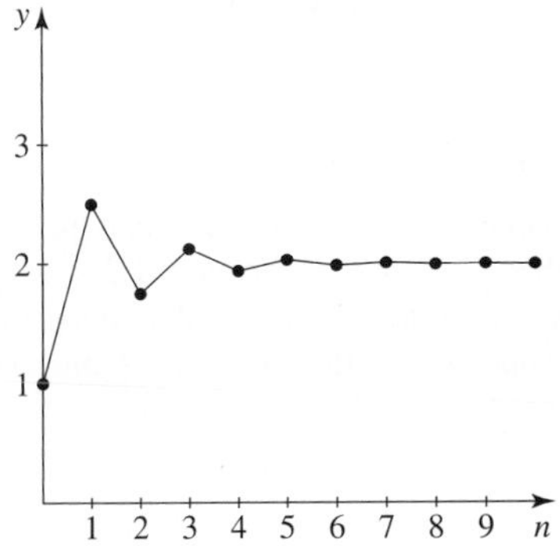

(c) $y_n \to 2 = y_e$

5. (a) $y_n = 4(2^n) - 3$

(b)

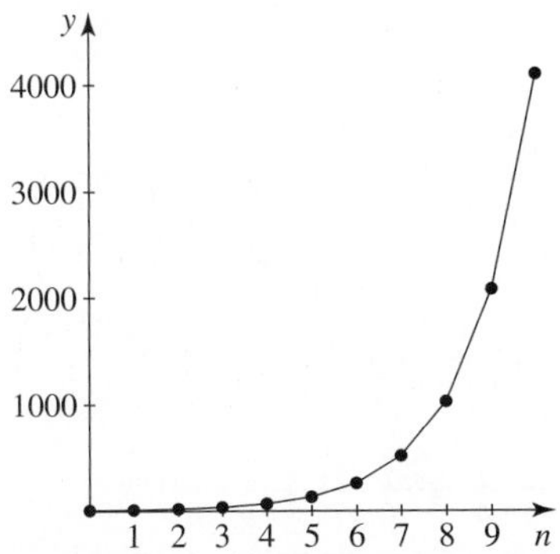

(c) The solution grows without bound.

7. (a) $y_n = 1$ for all n.

(b)

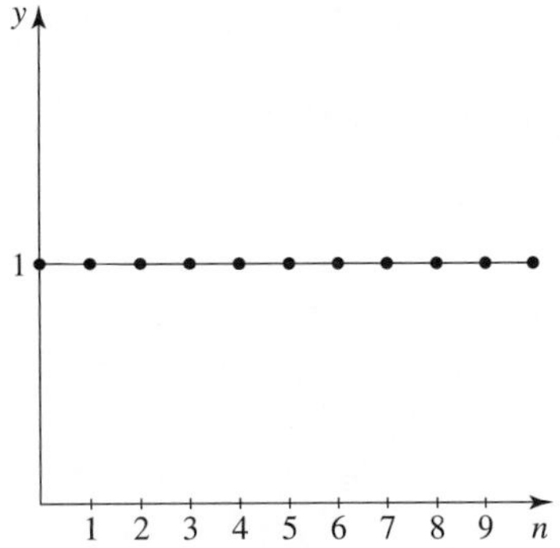

(c) The orbit starts at $y_0 = 1$ and remains there, because this is a fixed point. It is, however, a *repelling* fixed point, so at any other value for y_0 the solution is unbounded. See Problem 8.

9. (a) $y_n = 1 + 3n$

(b)

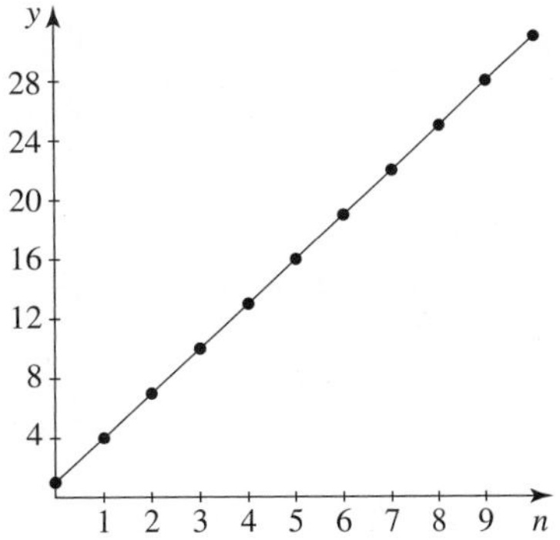

(c) Values grow without bound.

11. (a) $y_n = -\frac{1}{2}(-1)^n + \frac{3}{2}$

(b)

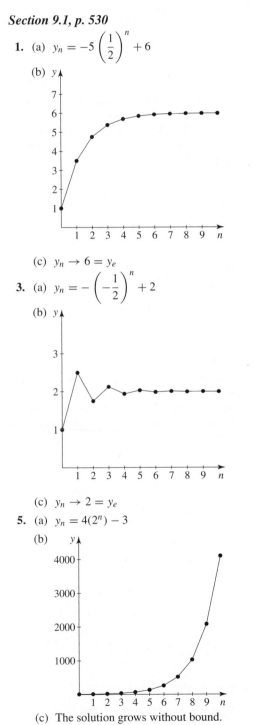

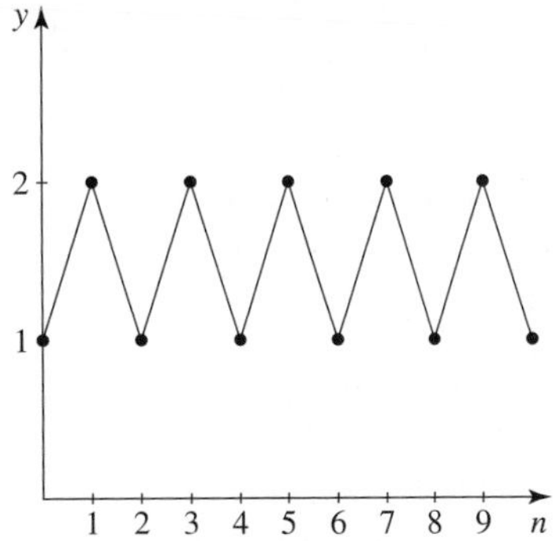

(c) No steady state; iterates cycle between 1 and 2.

13. (a) $y_n = 1{,}000{,}000\left[2 - (1.04)^n\right]$. The negative term will grow until it exceeds the positive term, so extinction will occur.

(b)

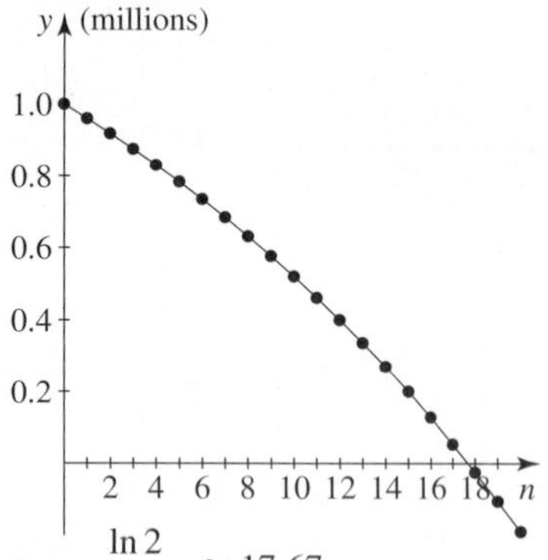

(c) $n = \dfrac{\ln 2}{\ln 1.04} \approx 17.67$

15. (a) $y_{n+1} = 0.3y_n - 1;\ y_0 = -1.6$

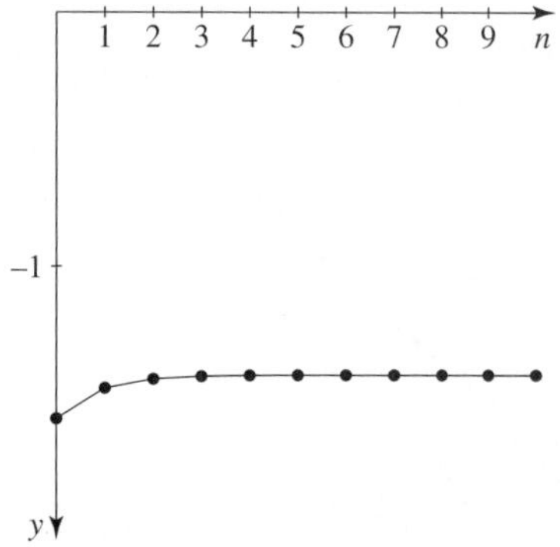

(b) $y_e = -1.428\ldots$

(c) The initial value $y_0 = -1.6$ determines that the solution approaches y_e from below; the coefficient $0.3 < 1$ causes the solution to approach rather than diverge from y_e; the constant -1 affects the level of y_e.

17. (a) $y_{n+1} = 0.3y_n + 0.5;\ y_0 = 0.2$

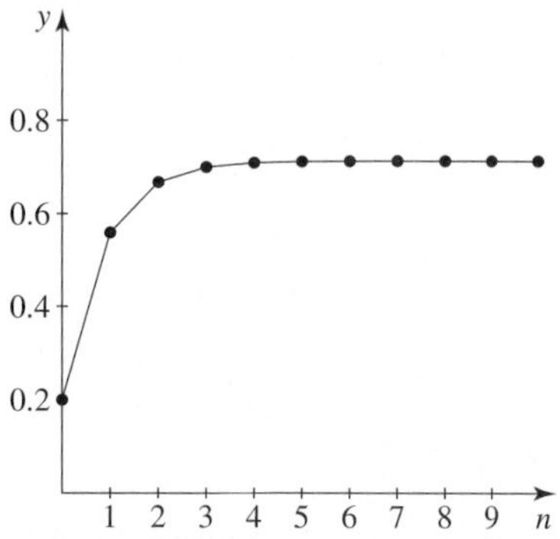

(b) $y_e = 0.714\ldots$

(c) The coefficient of y_n is small, so we are not surprised to see little effect as n becomes large. The constant 0.5 contributes to the steady-state level.

19. (a) $y_{n+1} = -1.3y_n - 1;\ y_0 = -0.2$

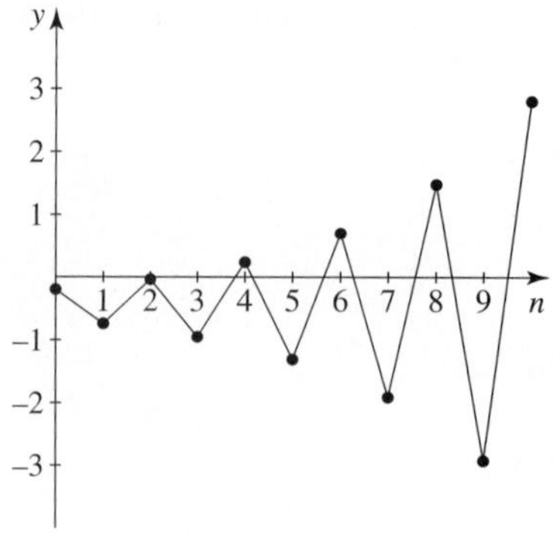

(b) $y_e = -0.4348\ldots$

(c) The initial value $y_0 = -0.2$ determines that the solution starts from below y_e; the coefficient $-1.3 < -1$ causes the solution to oscillate about y_e and to diverge from rather than approach y_e; the constant -1 affects the level of y_e.

21. (a) $y_{n+1} = -1.3y_n - 2;\ y_0 = 0.2$

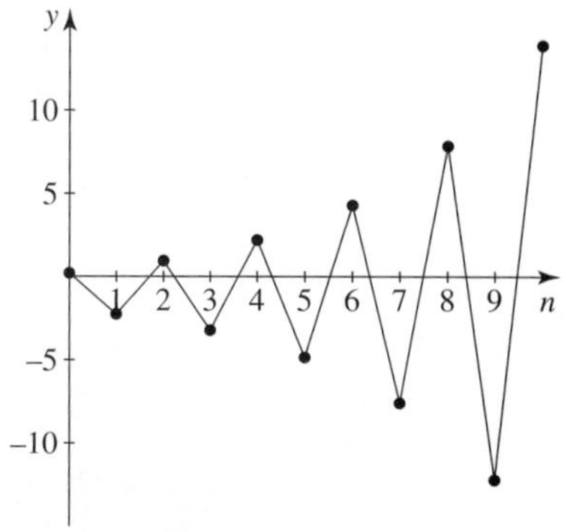

(b) $y_e = -0.8696\ldots$

(c) The initial value $y_0 = 0.2$ determines that the solution starts from below y_e; the coefficient $-1.3 < -1$ causes the solution to oscillate about y_e and to diverge from rather than approach y_e; the constant -2 affects the level of y_e.

23. (a) $s_{n+1} = s_n + 2^{n+1};\ s_0 = 1$

27. $y_n = c_1 + c_2\left(-\dfrac{b}{a}\right)^n$

29. $y_n = c_1 2^n + c_2 3^n$ **31.** $y_n = c_1(-2)^n + c_2$

33. (a)

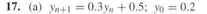

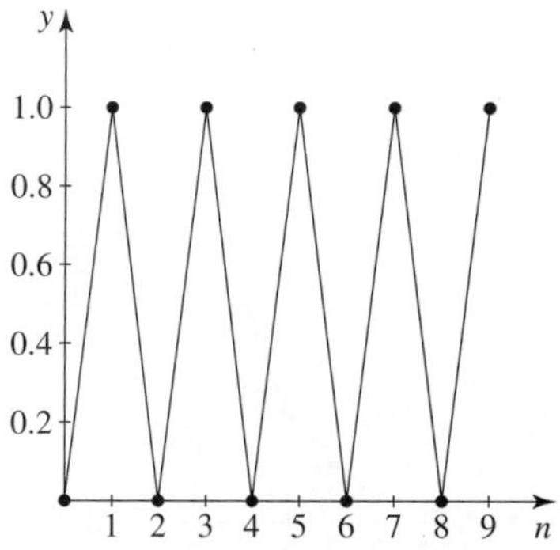

(b) Long-term behaviors cycle between 0 and 1.

35. $n = \dfrac{\log 3000}{\log 1.1} \approx 84$ years

37. (a) $y_{n+2} = y_{n+1} + 2y_n;\ y_0 = 1,\ y_1 = 1$

(b) $y_n = \dfrac{1}{3}\left[(-1)^n + 2^{n+1}\right]$

(c) $y_{n+2} - y_{n+1} - ky_n = 0;\ y_0 = 1,\ y_1 = 1$

$$y = c_1\left(\frac{1}{2} + \frac{\sqrt{1+4k}}{2}\right)^n + c_2\left(\frac{1}{2} - \frac{\sqrt{1+4k}}{2}\right)^n$$

39. $A_1 = \$1{,}000.22$, $A_{10} = \$1{,}002.19$, $A_{365} = \$1{,}083.28$

41. (a) $y_n = 12.5d\left[(1.08)^n - 1\right]$ (b) $d \approx \$1{,}742.86$

43. (b) $d = S\left[\dfrac{r}{1-(1+r)^{-N}}\right]$ (c) \$1,028.61 per month

45. (a) $y_n = -50(1.10)^n + 150$ thousands of deer

(b) 10,000 deer/year

47. (a) $y_n = 400\left[1 - (0.75)^n\right]$ grams of insulin

(b) 400 grams

49. (a) $y_n = \dfrac{1}{r-1}\left[(y_1 - y_0)r^n + (y_0 r - y_1)\right]$

51. (a) $y_n = \dfrac{n^2 + 2n + 2}{2}$ distinct regions

Section 9.2, p. 543

1. $\mathbf{A} = \begin{bmatrix} 0.75 & 0 \\ 0 & 0.25 \end{bmatrix}$, $\text{Tr}\,\mathbf{A} = 1$, $|\mathbf{A}| = 0.1875$;

both eigenvalues less than 1; sink

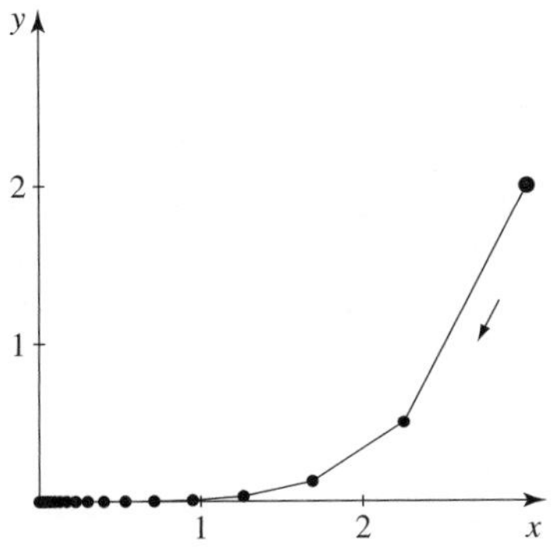

3. $\mathbf{A} = \begin{bmatrix} 0 & 1 \\ -1 & 0 \end{bmatrix}$, $\text{Tr}\,\mathbf{A} = 0$, $|\mathbf{A}| = 1$;

eigenvalues $\pm i$; center

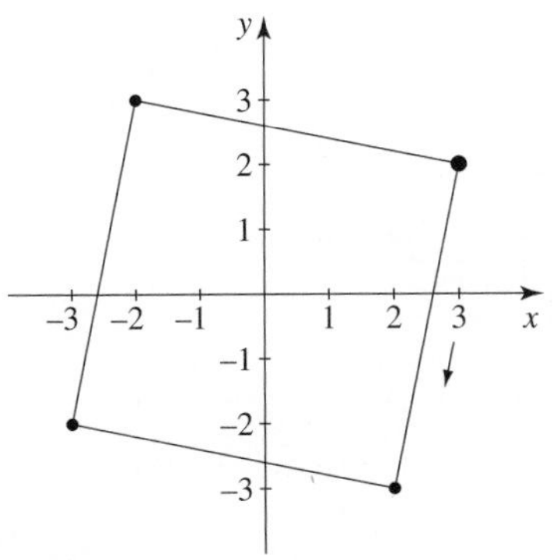

5. $\mathbf{A} = \begin{bmatrix} -1.9 & 0 \\ 0 & -1.1 \end{bmatrix}$, $\text{Tr}\,\mathbf{A} = -3$, $|\mathbf{A}| = 2.09$;

both eigenvalues greater than 1 in absolute value, both negative; double flip source

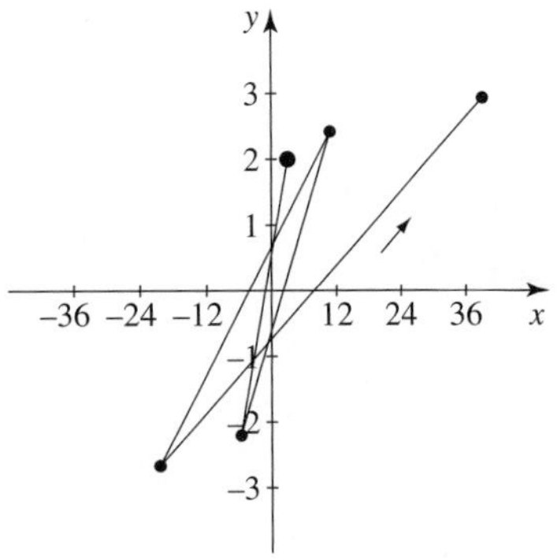

7. $\mathbf{A} = \begin{bmatrix} 1.05 & 0 \\ 0 & -1.05 \end{bmatrix}$, $\text{Tr}\,\mathbf{A} = 0$, $|\mathbf{A}| = -1.1025$;

both eigenvalues greater than one in absolute value (source), one negative (flip), one positive; flip source

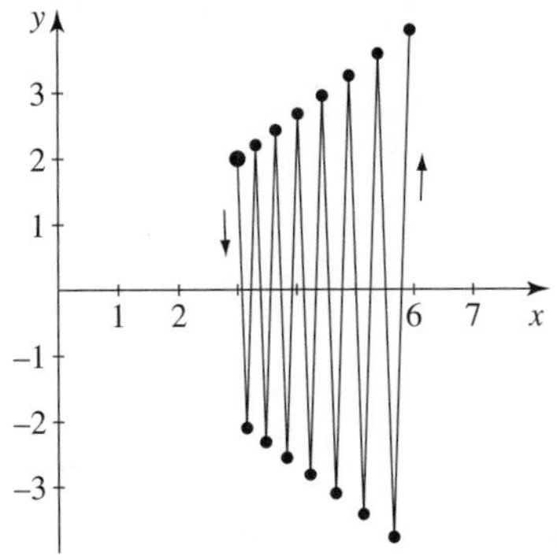

9. $\mathbf{A} = \begin{bmatrix} 0 & 0.9 \\ -1.6 & 0 \end{bmatrix}$, $\text{Tr}\,\mathbf{A} = 0$, $|\mathbf{A}| = 1.44$;

eigenvalues $\pm 1.2i$; spiral source

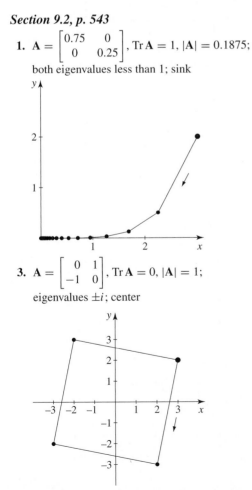

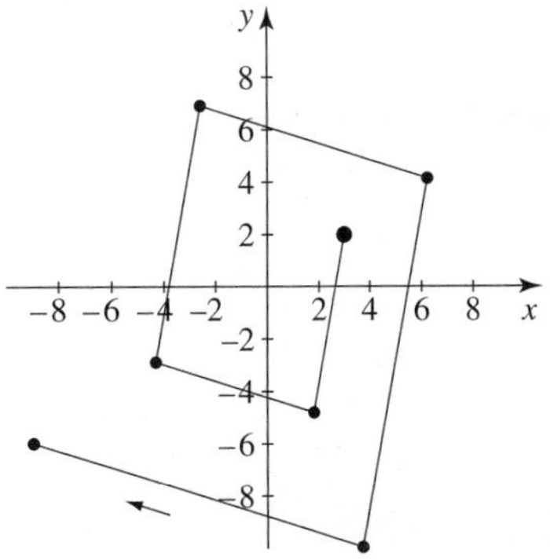

11. $\mathbf{A} = \begin{bmatrix} -0.8 & 0 \\ 0 & -0.6 \end{bmatrix}$, $\mathrm{Tr}\,\mathbf{A} = -1.4$, $|\mathbf{A}| = 0.48$; both eigenvalues negative but less than 1 in absolute value; double flip sink

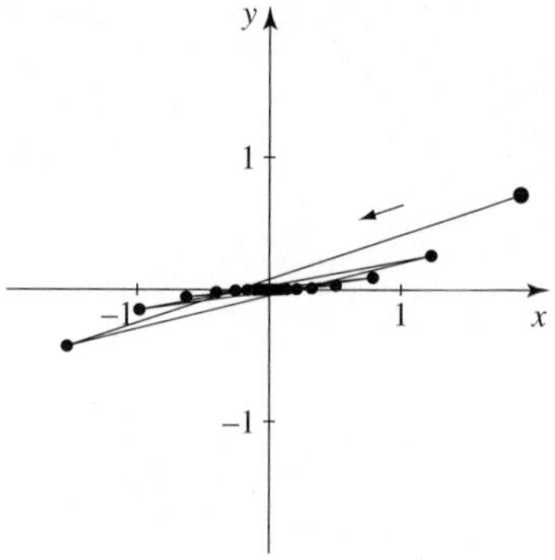

13. (a) Sink

(b) Solution is $x_n = x_0(0.5)^n$
$y_n = y_0(0.3)^n$

(c)

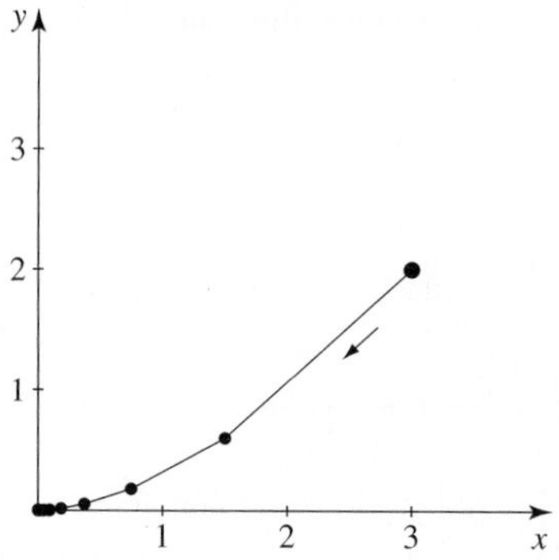

15. (a) Double flip sink

(b) Solution is $x_n = x_0(-0.5)^n$
$y_n = y_0(-0.3)^n$

(c)

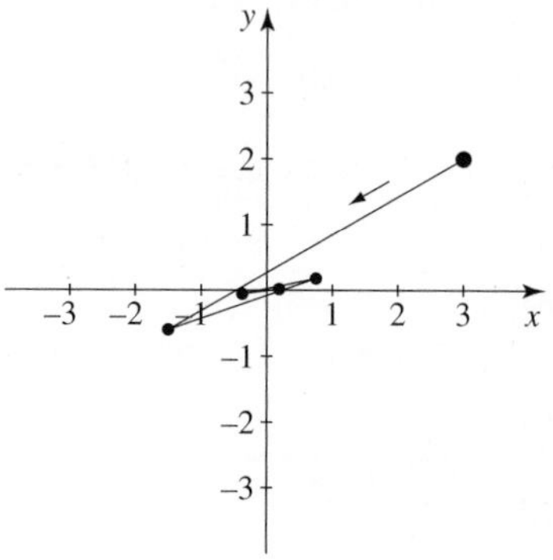

17. (a) Flip source

(b) Solution is $x_n = x_0(-1.5)^n$
$y_n = y_0(2)^n$

(c)

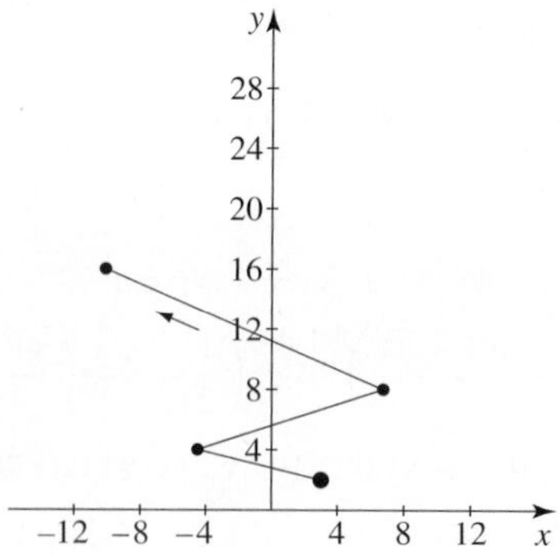

19. (a) Saddle

(b) Solution is $x_n = x_0(0.5)^n$
$y_n = y_0(2)^n$

(c)

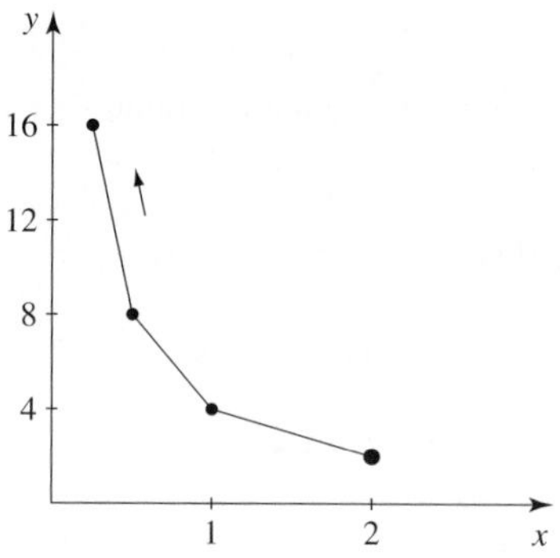

21. (a) Double flip sink

(b) Solution is $x_n = x_0(-0.5)^n$
$y_n = y_0(-2)^n$

(c)

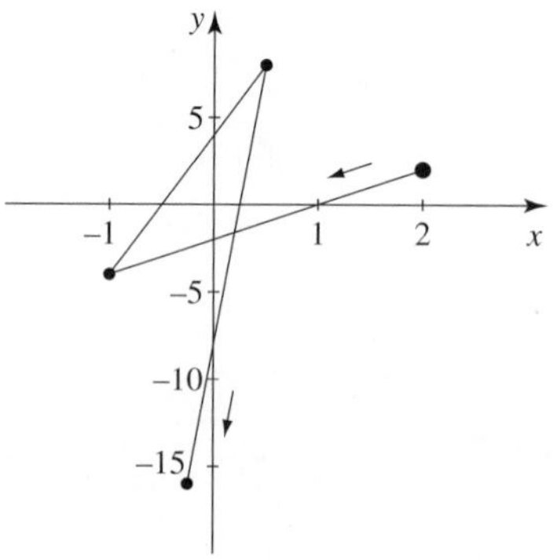

23. Counterclockwise 90° rotation

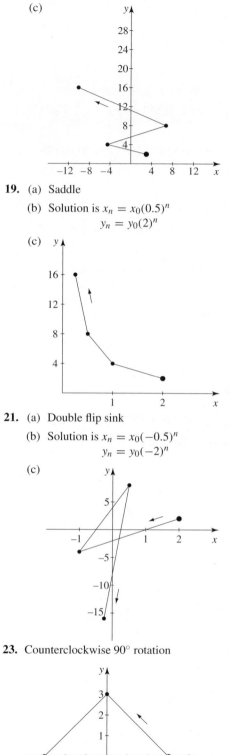

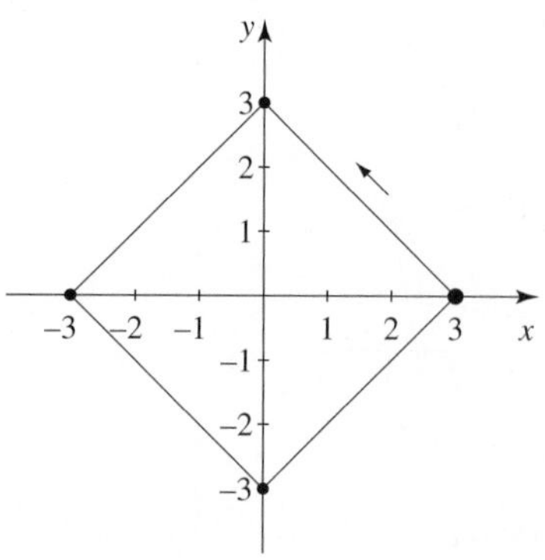

25. Counterclockwise inward spiral

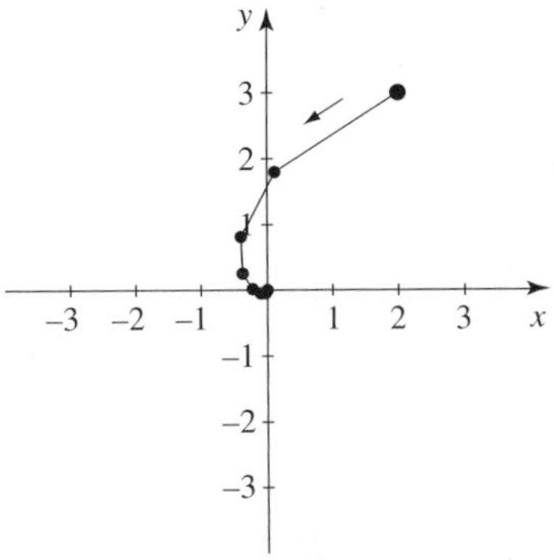

27. $w_n = 300(0.72)^n$ implies extinction.

29. (a) $\lambda_1 = -0.8, \lambda_2 = -1.1;$

$$\vec{\mathbf{v}}_1 = \begin{bmatrix} 0.95 \\ -0.32 \end{bmatrix}, \vec{\mathbf{v}}_2 = \begin{bmatrix} 1.3 \\ 0.75 \end{bmatrix}$$

(b) Double flip saddle

(c)

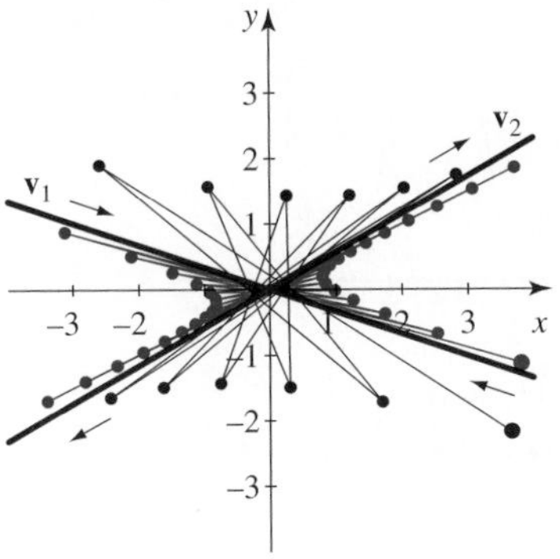

(d) $\vec{\mathbf{x}}_n = c_1(-0.8)^n \begin{bmatrix} 0.95 \\ -0.32 \end{bmatrix} + c_2(-1.1)^n \begin{bmatrix} 1.3 \\ 0.65 \end{bmatrix}$

31. (a) $\lambda_1 = 0.58, \lambda_2 = 1.02;$

$$\vec{\mathbf{v}}_1 = \begin{bmatrix} 0.98 \\ 0.19 \end{bmatrix}, \vec{\mathbf{v}}_2 = \begin{bmatrix} 0.61 \\ 0.79 \end{bmatrix}$$

$$\vec{\mathbf{x}}_n = c_1(0.58)^n \begin{bmatrix} 0.98 \\ 0.19 \end{bmatrix} + c_2(1.02)^n \begin{bmatrix} 0.61 \\ 0.79 \end{bmatrix}$$

(b)

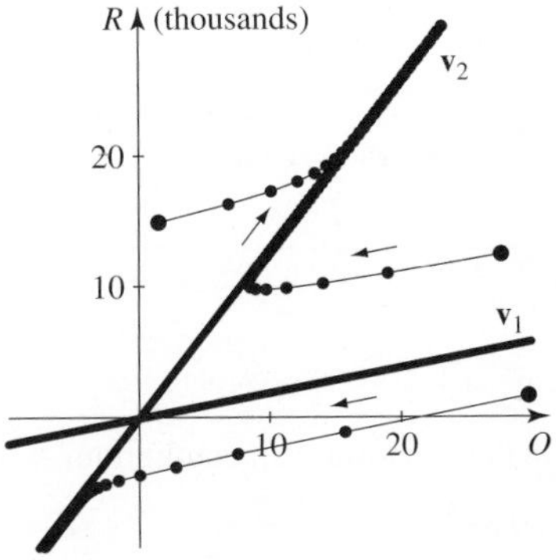

(c) 61 owls to 79,000 rats

33. (a) $\mathbf{A} = \begin{bmatrix} 0 & 2 \\ 1 & 1 \end{bmatrix}$

(b) $\lambda_1 = 2, \lambda_2 = -1;$

$$\vec{\mathbf{v}}_1 = \begin{bmatrix} 1 \\ 1 \end{bmatrix}, \vec{\mathbf{v}}_2 = \begin{bmatrix} 2 \\ -1 \end{bmatrix}$$

(c) $\vec{\mathbf{x}}_n = c_1 2^n \begin{bmatrix} 1 \\ 1 \end{bmatrix} + c_2(-1)^n \begin{bmatrix} 2 \\ -1 \end{bmatrix}$

(d) $\vec{\mathbf{x}}_6 = \begin{bmatrix} 22 \\ 21 \end{bmatrix}$

35.
$$x_{n+1} = y_n$$
$$y_{n+1} = -\frac{r}{p}x_n - \frac{q}{p}y_n$$

Section 9.3, p. 555

1. Equilibria at 0 (attracting) and 1 (repelling)

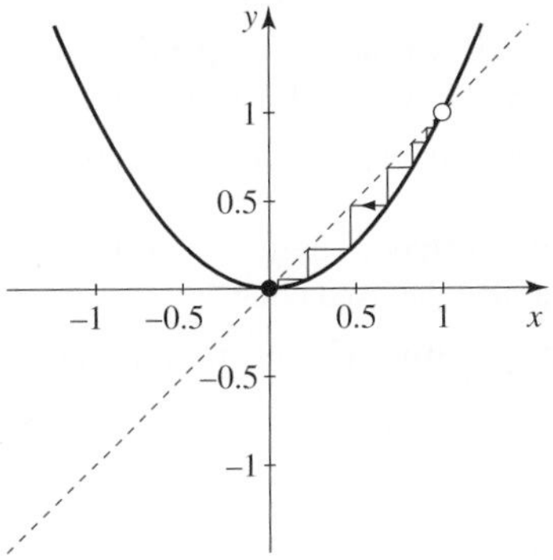

3. Attracting fixed point at 0, repelling fixed points at ±1

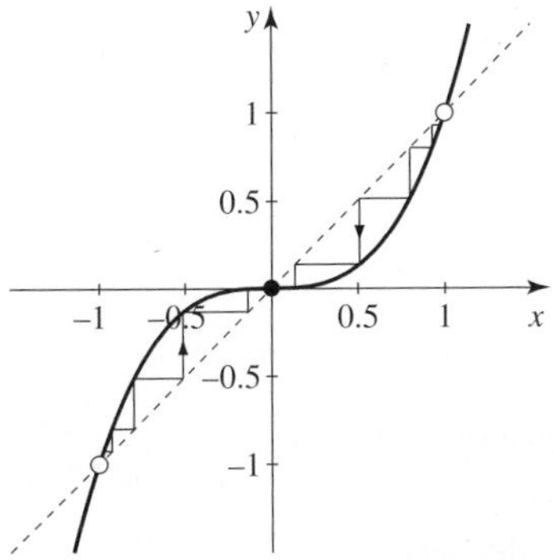

5. (a) (i) For $a > 1$, orbits head monotonically toward $\pm\infty$.

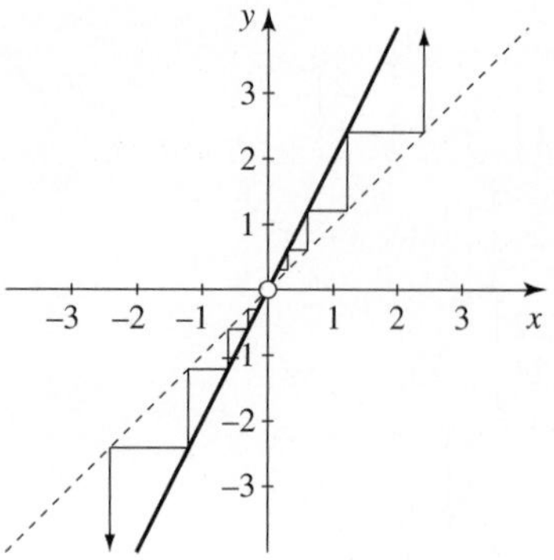

(vii) For $a < -1$, iterates alternate positive and negative and diverge from the starting point.

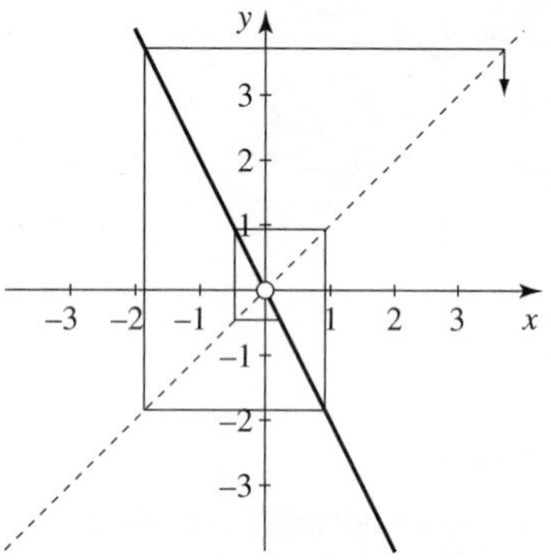

(b) When $|a| < 1$, $y = 0$ is a stable fixed point.
When $|a| > 1$, $y = 0$ is an unstable fixed point.

(c) When a is positive, iteration is monotonic; cobwebs look like stairsteps.
When a is negative, iterations oscillate in value; cobwebs wind around the fixed point.

(d) The value of b plays a role in locating the fixed point.
$x_e = \dfrac{b}{1-a}$.

7. (a)

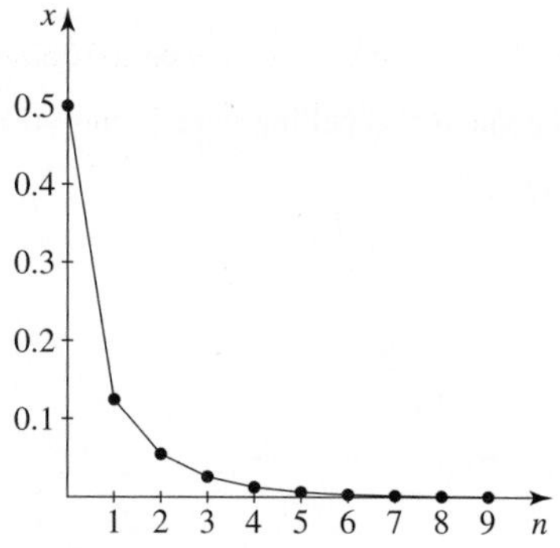

The function iterates toward zero.

(b) $x_e = 0, -1$

(c) 0 is stable, -1 is unstable.

9. (a) The function iterates toward a cycle.

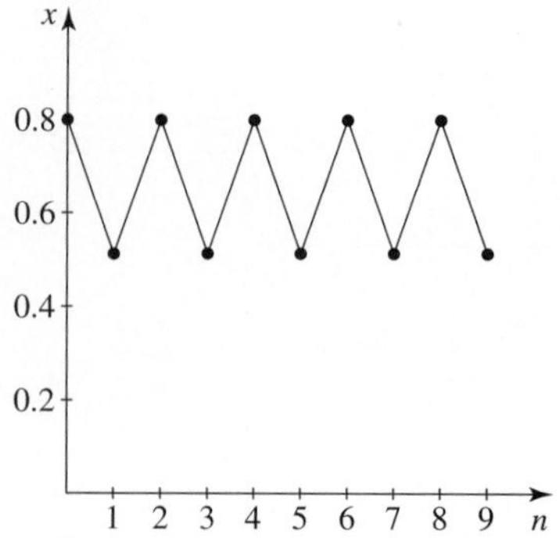

(b) $x_e = 0, 0.6875$

(c) Both fixed points are unstable or repelling, with an attracting *cycle* passing between them.

11. (a)

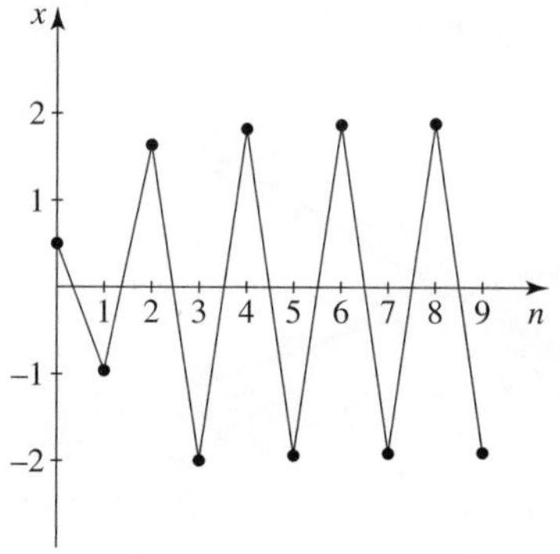

(b) $x_e = 0$

(c) 0 is unstable, repelling toward a cycle period 2.

13. (a)

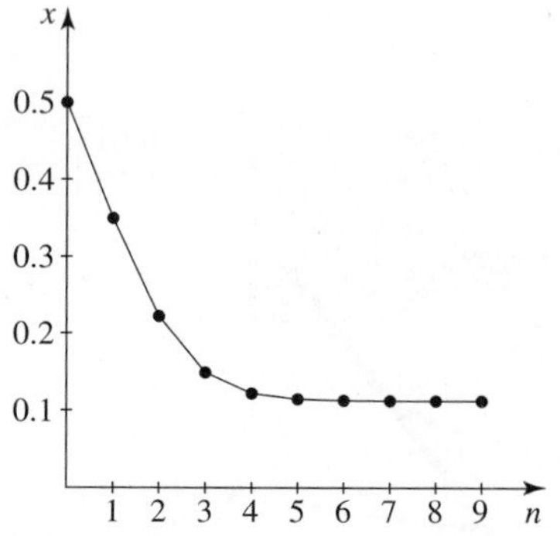

The function iterates toward an equilibrium ≈ 0.1.

(b) $x_e = 0.113, 0.887$

(c) 0.113 is stable; 0.887 is unstable.

15. (a)

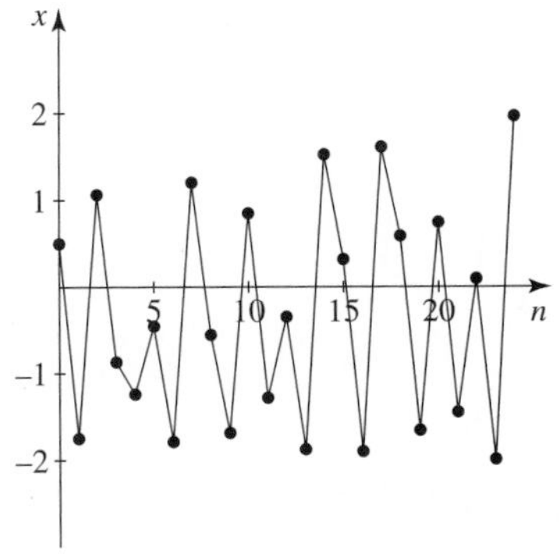

The iterative behavior seems chaotic.

(b) $x_e = -1, 2$

(c) Both fixed points are unstable, with no evidence of anything attracting between. This gives rise to the chaos we see in the time series.

17. Pete is using degrees instead of radian measure. That means $x_e = \cos \dfrac{\pi x_e}{180} \approx 0.999848$.

19. Your results should agree with Fig. 9.3.13.

21.

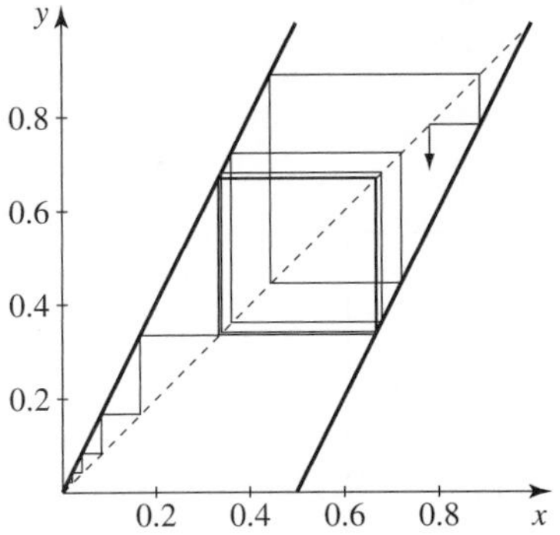

Chaotic orbits arise for any seeds except 0, 0.5, 1.

23. (b)

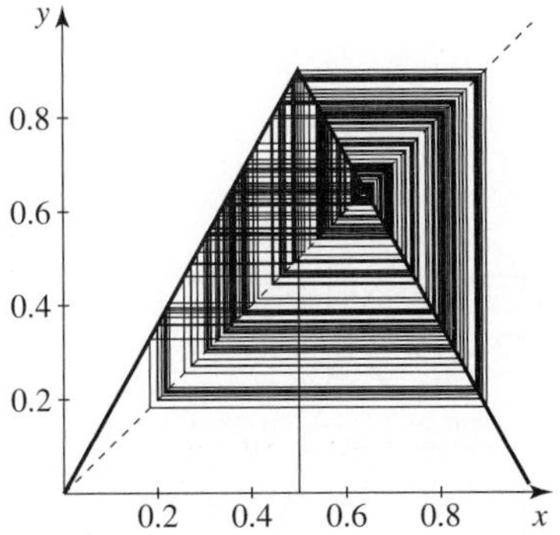

25. Newton's method gives the iterative equation

$$x_{n+1} = x_n - \frac{x_n^3 + rx_n + 1}{3x_n^2 + r}.$$

We used the following BASIC program to make our picture.

```
10  REM ORBIT DIAGRAM
20  REM N = ITERATIONS FOR
    EACH R
30  REM MINR = MINIMUM R
40  REM MAXR = MAXIMUM R
50  REM NSTEP = # OF R VALUES
60  SCREEN2
70  WINDOW (-1.3, -0.5) -
    (-1.25, 2)
80  LET N = 30
90  LET MINR = -1.30
100 LET MAXR = -1.25
110 LET NSTEP = 101
120 LET D = (MAXR -
    MINR)/(NSTEP - 1)
130 FOR I = 1 TO NSTEP
140   LET X = 0.5
150   LET R = MINR + (I - 1) * D
160   FOR J = 1 TO N
170     LET X = X - (X^3 +
        R * X + 1)/(3 * X^2 + R)
180     IF J < 10 THEN GO TO 200
190     CIRCLE (R, X), 0
200   NEXT J
210 NEXT I
220 END
```

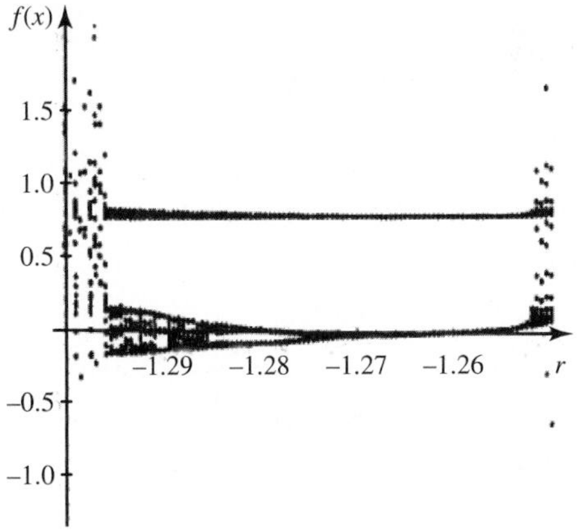

27. (a) A monotone decreasing sequence bounded below by 0 converges to its greatest lower bound L.

29. (a)

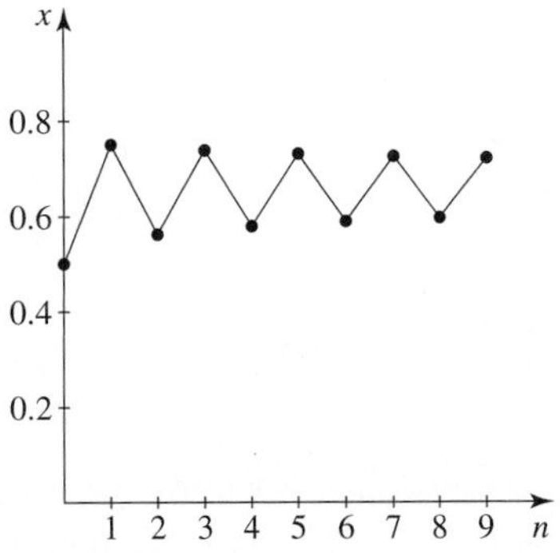

(b) Fixed points for $f(x)$ are $x_e = 0, \frac{2}{3}$. The first at 0 is repelling, but for the second at $\frac{2}{3}$ the derivative test $f'\left(\frac{2}{3}\right) = -1$ is inconclusive since $f'\left(\frac{2}{3}\right) = 1$. The second iterate function $f(f(x))$ has only two fixed points, 0 and $\frac{2}{3}$ (The root $\frac{2}{3}$ has multiplicity 3.), so there is not in fact a 2-cycle.

(c) The lack of a 2-cycle means the time series in (a) is in fact *converging*, extremely slowly, to the fixed point $\frac{2}{3}$, between the highs and lows of the orbit shown in (a).

(d)

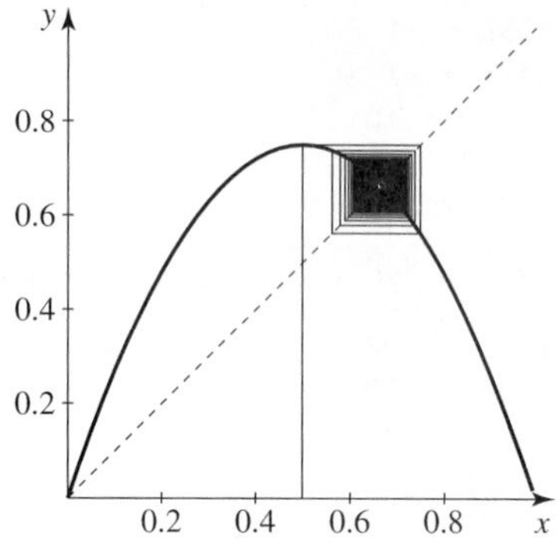

The extremely slow convergence to the fixed point $\frac{2}{3}$ causes the cobweb to appear solid black in its vicinity.

31. (a)

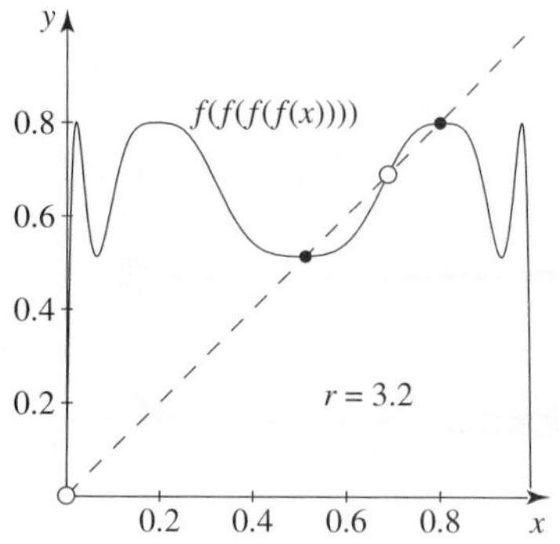

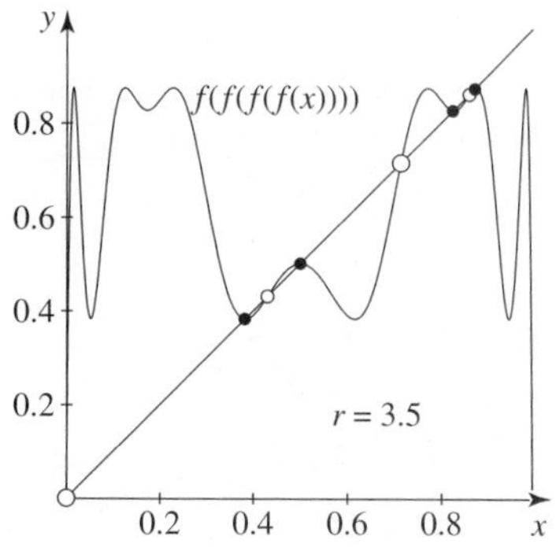

(b) The four-cycle values are 0.383, 0.501, 0.827, 0.874, and we can see the cycle emerging in a short time series.

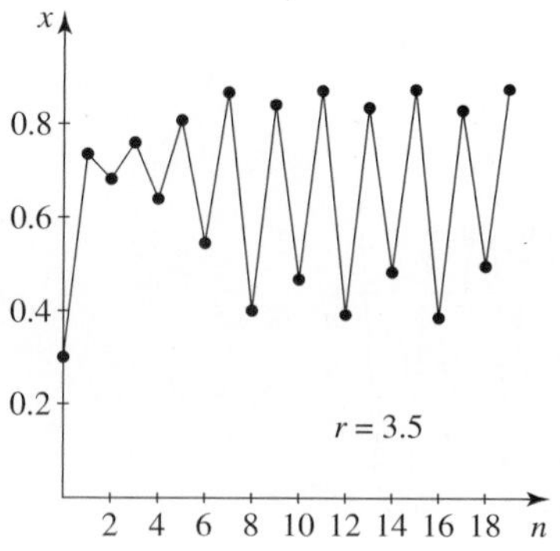

33. (a)

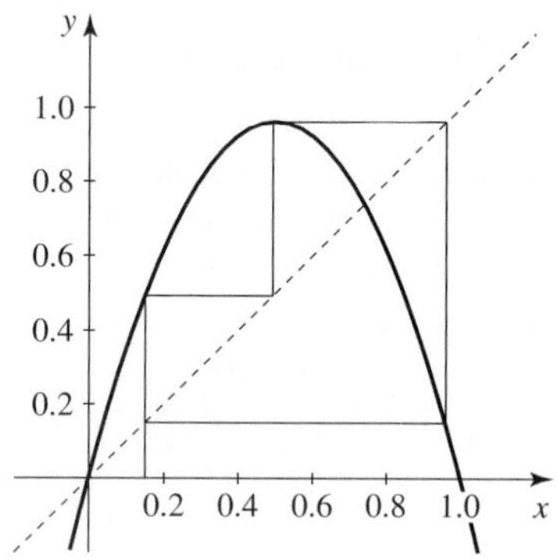

(b)

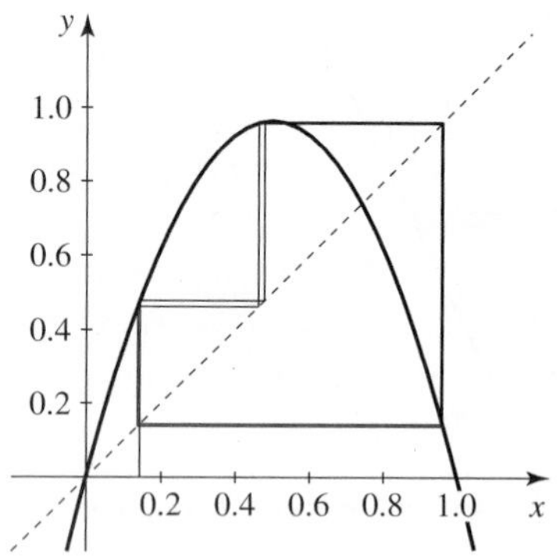

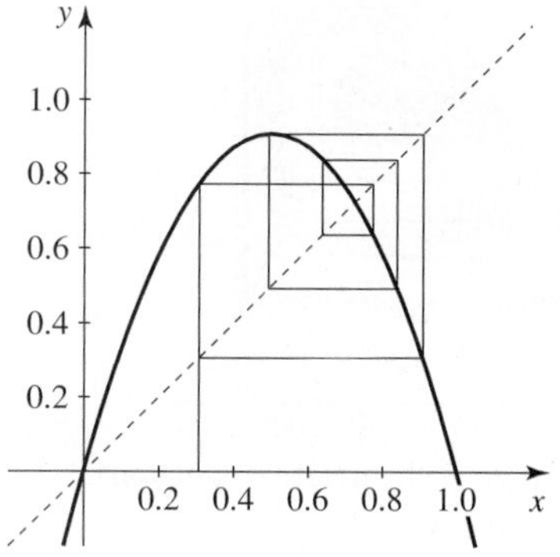

(c)

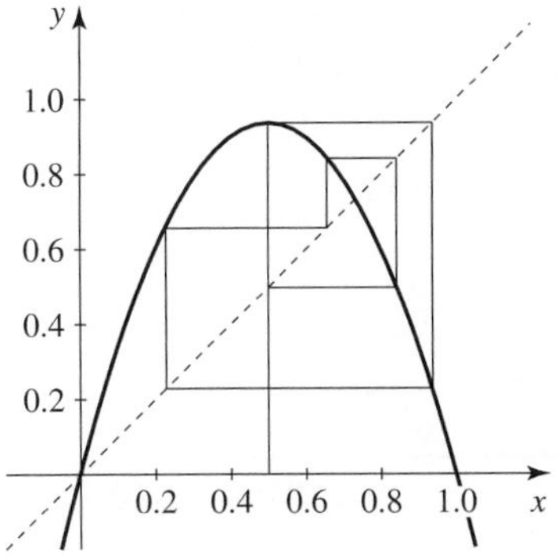

35. We adapt the BASIC program from Problem 25 to draw the orbit diagram for $x^2 + c$.

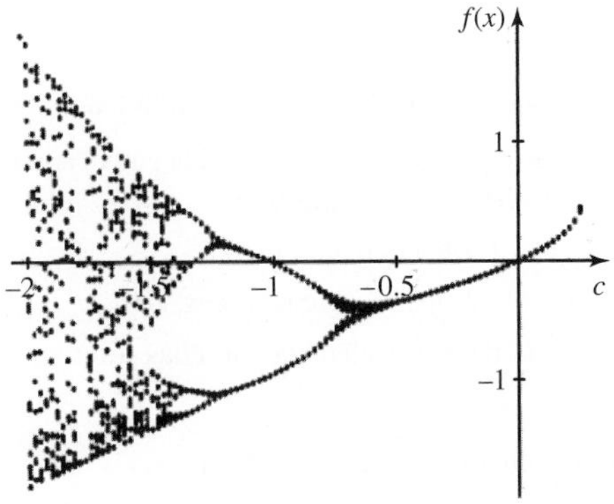

37. (a)

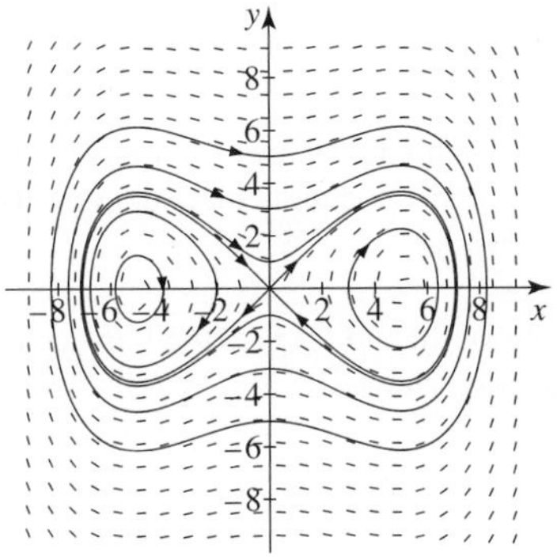

(c) Euler's method (three orbits) with $h = 0.1$:

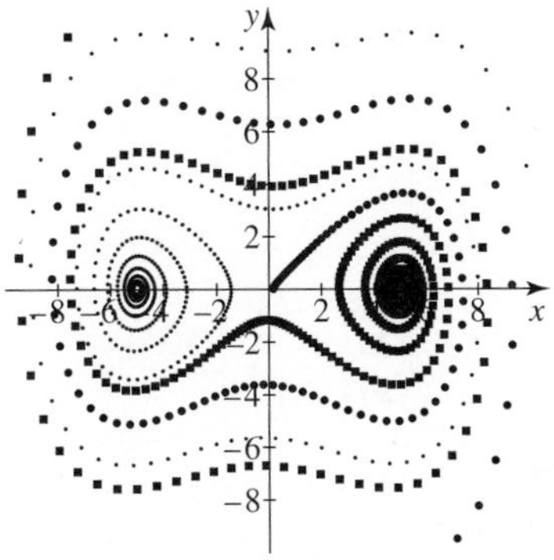

(d) Euler's method with $h = 0.05, 0.01, 0.005$:

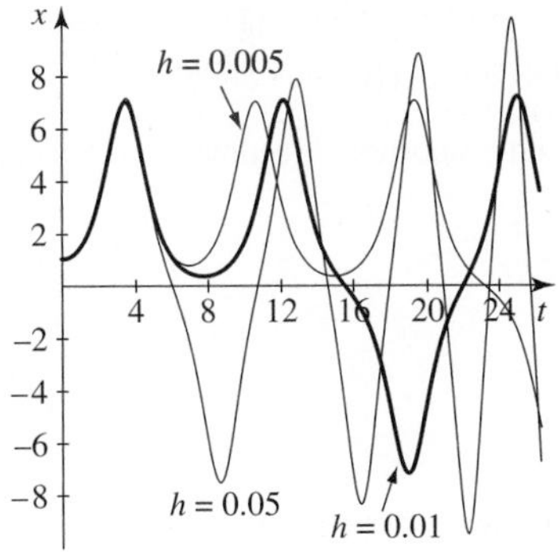

(e) The Runge-Kutta method has the same curve for $h = 0.05, 0.01, 0.005$:

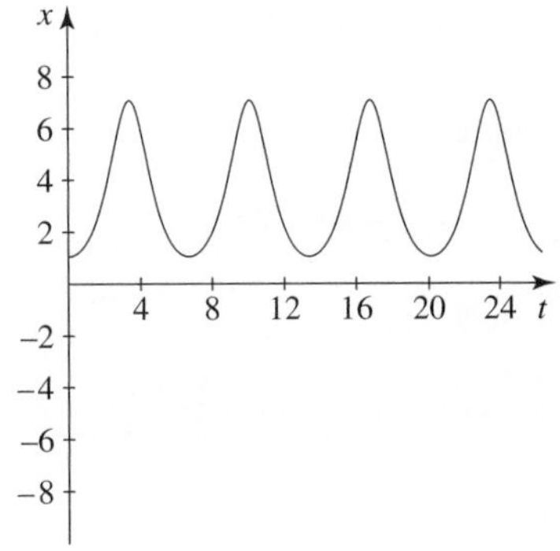

CHAPTER 10

Section 10.1, p. 569

1. Uncontrolled response: $x(t) = e^{-t}$.

(a) Proportional feedback: $x(t) = \frac{1}{3}e^{-3t} + \frac{2}{3}$.

(b) Derivative feedback: $x(t) = e^{-t/4}$.

(c) Derivative and proportional feedback: $x(t) = \frac{1}{3}e^{-3t/4} + \frac{2}{3}$.

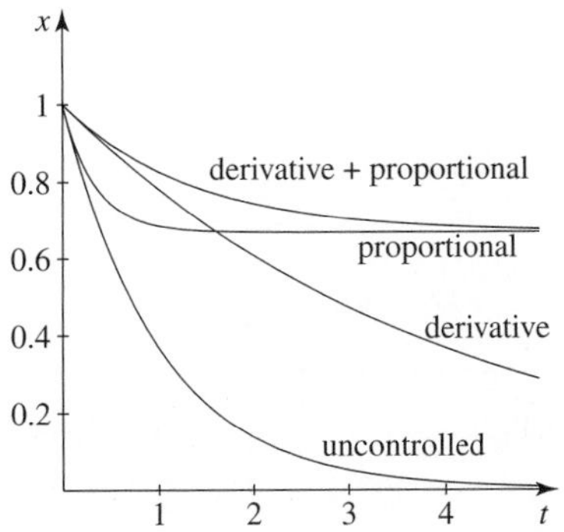

3. Uncontrolled response: $x(t) = e^{-3t/2}$.

(a) Proportional feedback: $x(t) = \frac{3}{5}e^{-5t/2} + \frac{2}{5}$.

(b) Derivative feedback: $x(t) = e^{-3t/5}$.

(c) Derivative and proportional feedback: $x(t) = \frac{3}{5}e^{-t} + \frac{2}{5}$.

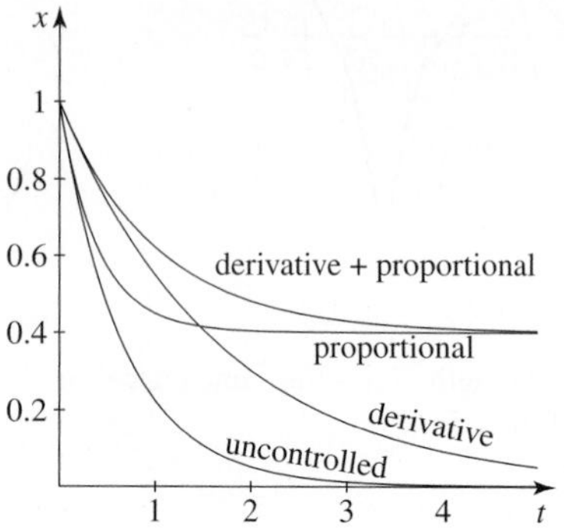

5. Uncontrolled response:

$$x(t) = e^{-t}\left(\cos\sqrt{2}t + \frac{\sqrt{2}}{2}\sin\sqrt{2}t\right).$$

(a) Proportional feedback:

$$x(t) = e^{-t}\left(\frac{3}{5}\cos 2t + \frac{3}{10}\sin 2t\right) + \frac{2}{5}.$$

(b) Derivative feedback:

$$x(t) = \left(\frac{1}{2} + \frac{5}{26}\sqrt{13}\right)e^{(-5+\sqrt{13})t/2} + \left(\frac{1}{2} - \frac{5}{26}\sqrt{13}\right)\sqrt{13}e^{(-5-\sqrt{13})t/2}.$$

(c) Derivative and proportional feedback:

$$x(t) = \left(\frac{3}{10} + \frac{3}{10}\sqrt{5}\right)e^{(-5+\sqrt{5})t/2} + \left(\frac{3}{10} - \frac{3}{10}\sqrt{5}\right)e^{(-5-\sqrt{5})t/2} + \frac{2}{5}.$$

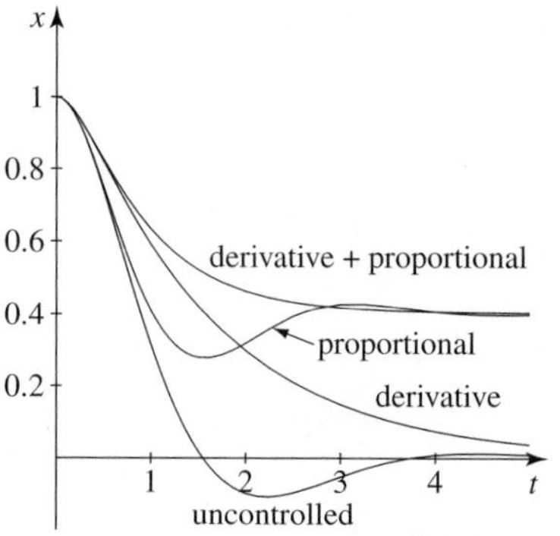

7. For $0 \le k_p \le 3$, $\lambda_1, \lambda_2 = -2 \pm \sqrt{3 - k_p}$.
For $k_p > 3$, $\lambda_1, \lambda_2 = -2 \pm i\sqrt{k_p - 3}$.

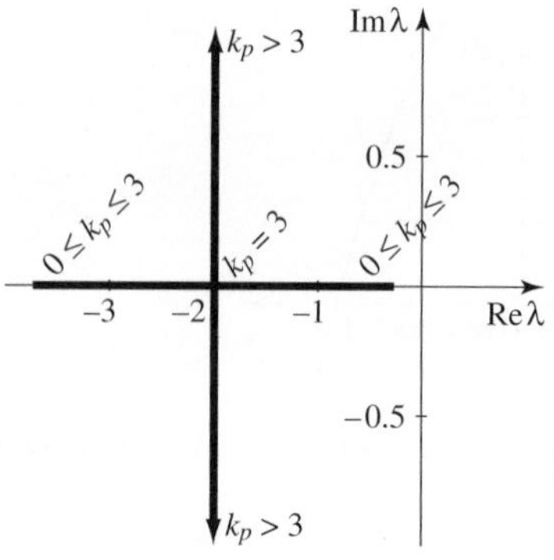

Location of eigenvalues (bold) as a function of gain k_p.

9. (a) If $k_d > 4$ the eigenvalues are real and negative and hence the solutions have stable equilibria.

(b) $k_d = 8$; $k_p = 1$; offset error $= 0.75x_1$.

12. (a) The uncontrolled steady-state response is 0.

(b) For $x(0) = 0$, the controlled equation has solution $x(t) = 1 + \frac{1}{2}e^{-3t} - \frac{3}{2}e^{-t}$.

15. The roots of the characteristic polynomial

$$\lambda^3 + 9\lambda^2 + 20\lambda + 4 = 0$$

are approximately -0.2215, -3.2892 and -5.4893. They are all negative; hence the homogeneous solutions go to zero, and all solutions approach the particular solution $x(t) \equiv x_1$.

17. $G(s) = \dfrac{1}{s^2 + s + 1}$

21. HINT: For each of (a), (b) and (c) take the Laplace transform of the control equation and set

$$H(s) = \frac{U(s)}{X(s)}.$$

Section 10.2, p. 582

1.

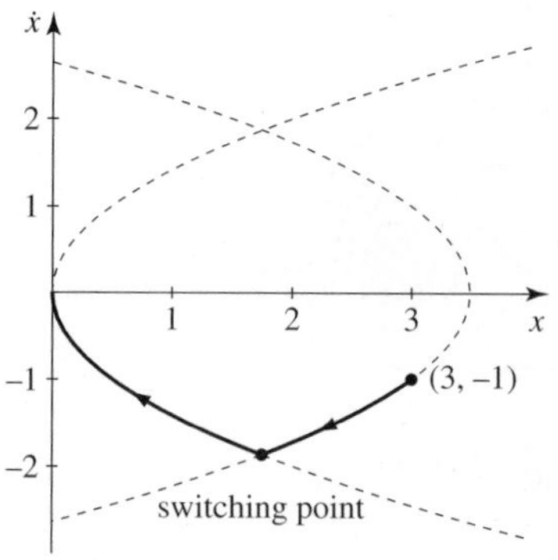

3.

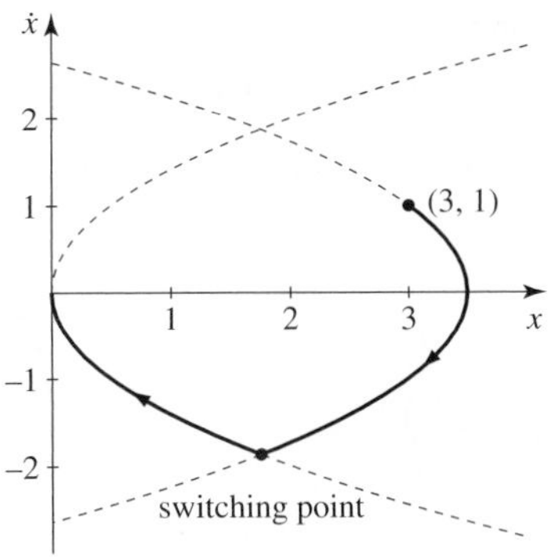

5. $\begin{bmatrix} \dot{x}_1 \\ \dot{x}_2 \end{bmatrix} = \begin{bmatrix} 0 & 1 \\ -1 & 0 \end{bmatrix} \begin{bmatrix} x_1 \\ x_2 \end{bmatrix} + \begin{bmatrix} 0 \\ u(t) \end{bmatrix}.$

7. $\begin{bmatrix} \dot{x}_1 \\ \dot{x}_2 \end{bmatrix} = \begin{bmatrix} 0 & 1 \\ -c & -b \end{bmatrix} \begin{bmatrix} x_1 \\ x_2 \end{bmatrix} + \begin{bmatrix} 0 \\ u(t) \end{bmatrix}.$

9. $\begin{bmatrix} \dot{x} \\ \dot{y} \end{bmatrix} = \begin{bmatrix} 0 & 1 \\ -1 & 0 \end{bmatrix} \begin{bmatrix} x \\ y \end{bmatrix} + \begin{bmatrix} 0 \\ u(t) \end{bmatrix}.$

11. $\begin{bmatrix} \dot{x} \\ \dot{y} \end{bmatrix} = \begin{bmatrix} -1 & 2 \\ -1 & 0 \end{bmatrix} \begin{bmatrix} x \\ y \end{bmatrix} + \begin{bmatrix} 0 \\ u(t) \end{bmatrix}.$

13. (a) $x(t) = \begin{cases} \frac{1}{2}t^2 - 1, & 0 \le t < 1; \\ -\frac{1}{2}t^2 + 2t - 2, & 1 \le t \le 2. \end{cases}$

15. (a) $J = 1.25$

(b) $J = 1.625$

17.

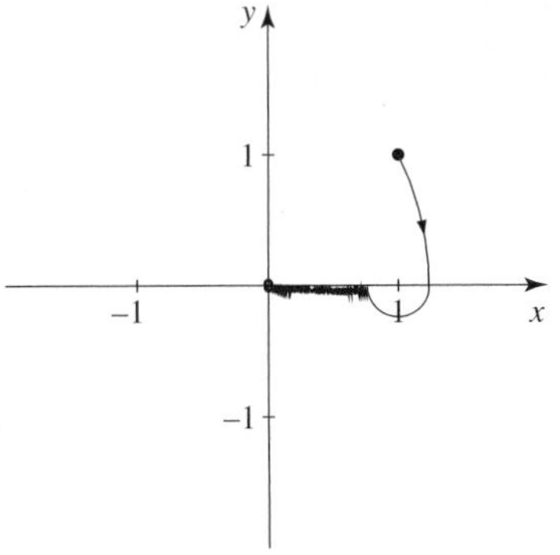

The initial control is $u = -1$ but changes to 1 once the trajectory crosses below the x-axis. The trajectory then follows the bottom half of the semicircle centered at $(1, 0)$ and wraps around until it comes back up and hits the x-axis near $(0.7, 0)$ at which time the control changes back to 1, thus turning the trajectory back down into minus y territory, where the control changes back to 1. The net result is that when the trajectory reaches $(0.7, 0)$ the control starts to *chatter* back and forth between 1 and -1.

Section 10.3, p. 590

1. (a) $H(u) = -f_0 + p_1\dot{x}_1 + p_2\dot{x}_2$
$= -u^2 + p_1x_2 + p_2(-x_1 + u)$

(b) $\dot{p}_1 = -\partial H/\partial x_1 = p_2$
$\dot{p}_2 = -\partial H/\partial x_2 = -p_1$

(c) $\begin{bmatrix} p_1 \\ p_2 \end{bmatrix} = c_1 \begin{bmatrix} \cos t \\ -\sin t \end{bmatrix} + c_2 \begin{bmatrix} \sin t \\ \cos t \end{bmatrix}$

3. (a) $H(u) = -\frac{1}{2}u^2 + pu$

(b) $\dot{p} = -\partial H/\partial x = 0$

$p(t) = c_1$ (c_1 a constant)

(c) $u^*(t) = -\frac{1}{2}$, $x^*(t) = -\frac{1}{2}t + 1$

(d) $J(u) = \frac{1}{4}$

5. (a) From the trajectories in Fig. 10.3.5, we see that the control function starts with -1 and has three switches. In other words the control sequence is $-1, 1, -1, 1$.

(b) Physically, the spring is initially stretched to the right 4 units and is moving to the right with velocity 2. We begin by pushing to the left in the direction of equilibrium, and then make three switches until finally we push to the right against the motion of the spring ($u = +1$) and make a soft landing at $(0, 0)$.

7. (a) From the trajectories in Fig. 10.3.5, we see that the control function is always -1 with no switches.

(b) Physically, the spring is initially compressed to the right 2 units. We then simply push to the left against the motion of the spring until it makes a soft landing at $(0, 0)$.

Appendix CN

1. The complex numbers $3 + 3i$, $4i$, 2 and $1 - i$ are plotted as the respective points $(3, 3)$, $(0, 4)$, $(2, 0)$ and $(1, -1)$ in the complex plane.

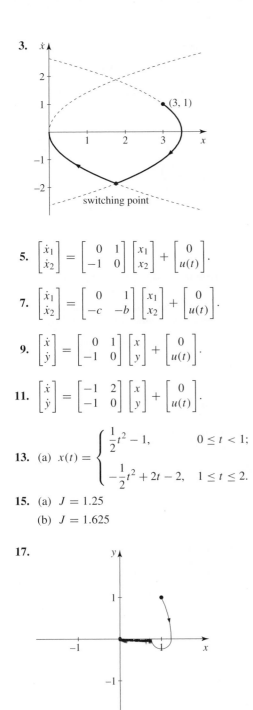

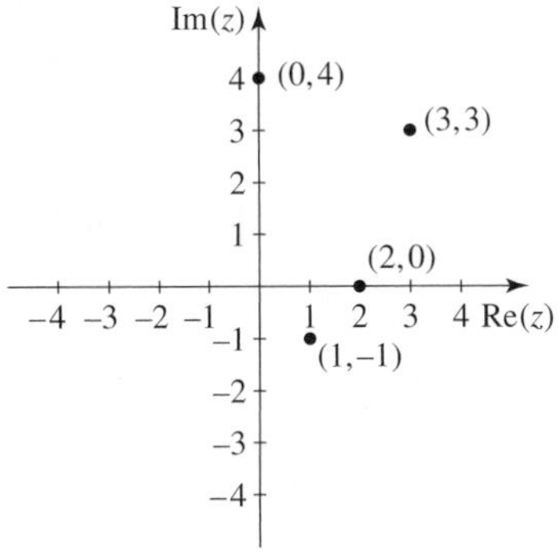

3. (a) $e^{2\pi i} = 1$

(b) $e^{i\pi/2} = i$

(c) $e^{-i\pi} = -1$

(d) Using the property $e^{a+b} = e^a e^b$ and Euler's formula, we write

$$e^{(2+\pi i/4)} = e^2\frac{\sqrt{2}}{2} + ie^2\frac{\sqrt{2}}{2}.$$

5. By direct substitution, $(-1+i)^2 + 2(-1+i) + 2 = 0$.

7. Calling the complex number $z = a + ib$, we write

$$\begin{aligned} z^2 + 2z &= (a+ib)^2 + 2(a+ib) \\ &= (a^2 - b^2 + 2iab) + 2(a+ib) \\ &= (a^2 - b^2 + 2a) + i(2ab + 2b). \end{aligned}$$

Hence

(a) $\operatorname{Re}(z^2 + 2z) = a^2 - b^2 + 2a$,

(b) $\operatorname{Im}(z^2 + 2z) = 2b(a+1)$.

9. The m roots of $z^m = 1$ (called the roots of unity) are the m values

$$z = \cos\frac{2\pi k}{m} + i\sin\frac{2\pi k}{m} = 1^{1/m}, \quad k = 0, 1, \ldots, m-1.$$

Note that for $z = 1$ we have polar angle $\theta = 0$ for the above formula.

(a) $z^2 = 1$ has two roots:

$$\begin{aligned} z_k &= \cos\left(\frac{2\pi k}{2}\right) + i\sin\left(\frac{2\pi k}{2}\right) \\ &= \cos\pi k + i\sin\pi k, \quad k = 0, 1 \end{aligned}$$

or

$$z = \pm 1.$$

(b) $z^3 = 1$ has three roots:

$$z_k = \cos\left(\frac{2\pi k}{3}\right) + i\sin\left(\frac{2\pi k}{3}\right), \quad k = 0, 1, 2$$

or

$$\begin{aligned} z_0 &= 1, \\ z_1 &= \cos\left(\frac{2\pi}{3}\right) + i\sin\left(\frac{2\pi}{3}\right) = -\frac{1}{2} + \frac{\sqrt{3}}{2}i, \\ z_2 &= \cos\left(\frac{4\pi}{3}\right) + i\sin\left(\frac{4\pi}{3}\right) = -\frac{1}{2} - \frac{\sqrt{3}}{2}i. \end{aligned}$$

(c) $z^4 = 1$ has four roots:

$$z_k = \cos\left(\frac{2\pi k}{4}\right) + i\sin\left(\frac{2\pi k}{4}\right), \quad k = 0, 1, 2, 3$$

or

$$\begin{aligned} z_0 &= 1, \\ z_1 &= i, \\ z_2 &= -1, \\ z_3 &= -i. \end{aligned}$$

11. (a) $e^{(1+\pi i)} = e^1(\cos\pi + i\sin\pi) = -e$

(b) $e^{(2+\pi i/2)} = e^2\left(\cos\frac{\pi}{2} + i\sin\frac{\pi}{2}\right) = ie^2$

(c) $e^{\pi i} = \cos\pi + i\sin\pi = -1$

(d) $e^{-\pi i} = \cos(-\pi) + i\sin(-\pi) = -1$

13. $y(t) = c_1e^{(-t)} + c_2e^{(t/2)}\sin\left(\frac{\sqrt{3}t}{2}\right) + c_3e^{(t/2)}\cos\left(\frac{\sqrt{3}t}{2}\right)$

14. $y(t) =$

$$-c_1e^{\left(-3\sqrt{2}t/2\right)}\sin\left(\frac{3\sqrt{2}t}{2}\right) - c_2e^{\left(3\sqrt{2}t/2\right)}\sin\left(\frac{3\sqrt{2}t}{2}\right) + c_3e^{\left(-3\sqrt{2}t/2\right)}\cos\left(\frac{3\sqrt{2}t}{2}\right) + c_4e^{\left(3\sqrt{2}t/2\right)}\cos\left(\frac{3\sqrt{2}t}{2}\right)$$

Appendix LT

Sample proof for Problems 1–6:

5. Suppose $\mathbf{w} \in \mathbb{W}$. Then if T is surjective, $T(\mathbf{v}) = \mathbf{v}$ for some $\mathbf{v} \in \mathbb{V}$. And if L is surjective, $L(\mathbf{u}) = \mathbf{v}$ for some $\mathbf{u} \in \mathbb{U}$. Therefore $T \circ L(\mathbf{u}) = T(L(\mathbf{u})) = T(\mathbf{v}) = \mathbf{w}$, and we have proved that the composition of surjective functions is surjective.

7. $\mathbf{M}_B = \begin{bmatrix} 2 & -1 \\ 1 & 0 \\ 0 & 1 \end{bmatrix}$

9. $\mathbf{M}_B = \begin{bmatrix} 2 & 0 & 0 & 0 \\ 0 & 1 & 1 & 0 \\ 0 & 2 & 1 & 0 \\ 0 & 0 & 0 & 2 \end{bmatrix}$

11. $\mathbf{M}_B = \begin{bmatrix} -3 & 1 \\ 0 & -3 \end{bmatrix}$

13. $\mathbf{M}_B = \begin{bmatrix} 0 & \frac{1}{2} & \frac{1}{2} & 0 \\ 1 & 0 & 0 & 0 \\ 0 & -1 & 0 & 0 \\ 0 & 0 & 0 & \frac{1}{5} \end{bmatrix}$

15. $\mathbf{M}_B^* = \begin{bmatrix} 2 & -1 & -\frac{2}{5} \\ -1 & 0 & 0 \\ 0 & 0 & \frac{2}{5} \end{bmatrix}$

Appendix PF

1. $\dfrac{1}{x(x-1)} = -\dfrac{1}{x} + \dfrac{1}{x-1}$

3. $\dfrac{x}{(x+1)(x+2)} = -\dfrac{1}{x+1} + \dfrac{2}{x+2}$

5. $\dfrac{4}{x^2(x^2+4)} = \dfrac{1}{x^2} - \dfrac{1}{x^2+4}$

7. $\dfrac{7x-1}{(x+1)(x+2)(x-3)} = \dfrac{2}{x+1} - \dfrac{3}{x+2} + \dfrac{1}{x-3}$

9. $\dfrac{x^2+9x+2}{(x-1)^2(x+3)} = \dfrac{2}{x-1} + \dfrac{3}{(x-1)^2} - \dfrac{1}{x+3}$

Differential Equations & Linear Algebra

Index

A

B

C

M

N

Q

R

S

T

U

Table of Laplace Transforms

$$\mathcal{L}\{f(t)\} = F(s) \equiv \int_0^{\infty} e^{-st} f(t)dt$$

By linearity, $\mathcal{L}\{af(t) + bg(t)\} = aF(s) + bG(s)$.

1. $\mathcal{L}\{1\} = \dfrac{1}{s}, \quad s > 0$

2. $\mathcal{L}\{t^n\} = \dfrac{n!}{s^{n+1}}$, n a positive integer, $s > 0$

3. $\mathcal{L}\{e^{at}\} = \dfrac{1}{s-a}, \quad s > a$

4. $\mathcal{L}\{t^n e^{at}\} = \dfrac{n!}{(s-a)^{n+1}}$, n a positive integer, $s > a$

5. $\mathcal{L}\{\sin bt\} = \dfrac{b}{s^2 + b^2}, \quad s > 0$

6. $\mathcal{L}\{\cos bt\} = \dfrac{s}{s^2 + b^2}, \quad s > 0$

7. $\mathcal{L}\{e^{at} \sin bt\} = \dfrac{b}{(s-a)^2 + b^2}, \quad s > a$

8. $\mathcal{L}\{e^{at} \cos bt\} = \dfrac{s-a}{(s-a)^2 + b^2}, \quad s > a$

9. $\mathcal{L}\{\sinh bt\} = \dfrac{b}{s^2 - b^2}, \quad s > |b|$

10. $\mathcal{L}\{\cosh bt\} = \dfrac{s}{s^2 - b^2}, \quad s > |b|$

11. $\mathcal{L}\{tf(t)\} = -\dfrac{d}{ds}F(s), \quad s > 0$

12. $\mathcal{L}\{e^{at} f(t)\} = F(s-a), \quad s > a$

13.* $\mathcal{L}\{f'(t)\} = sF(s) - f(0), \quad s > \alpha$

14.* $\mathcal{L}\{f''(t)\} = s^2 F(s) - sf(0) - f'(0), \quad s > \alpha$

15.* $\mathcal{L}\{f^{(n)}(t)\} = s^n F(s) - s^{n-1} f(0) - s^{n-2} f'(0) - \cdots - f^{n-1}(0), \quad s > \alpha$ (nth derivative)

16.* $\mathcal{L}\{ay'' + by' + cy\} = \mathcal{L}\{f(t)\} \Rightarrow Y(s) = \dfrac{F(s) + asy(0) + by'(0)}{as^2 + bs + c}, \quad s > \alpha$

17. $\mathcal{L}\{\text{step}(t-a)\} = \dfrac{e^{-as}}{s}, \quad s > a$

18. $\mathcal{L}\{f(t-a)\, \text{step}(t-a)\} = e^{-as} F(s), \quad s > a$

19. $\mathcal{L}\{f(t)\, \text{step}(t-a)\} = e^{-as} \mathcal{L}\{f(t+a)\}, \quad s > a$

20. $\mathcal{L}\{\delta(t)\} = 1, \quad s > 0$

21. $\mathcal{L}\{\delta(t-a)\} = e^{-as}, \quad s > a$

22. $\mathcal{L}\{f(t+P)\} = \mathcal{L}\{f(t)\} = \dfrac{1}{1 - e^{-sP}} \displaystyle\int_0^P e^{-st} f(t)dt$ (f is periodic, with period P)

For $F(s) = \dfrac{p(s)}{q(s)}$, where p and q are polynomials, use partial fractions to rewrite, if possible, in terms of simple denominators listed above; then use linearity to find $\mathcal{L}^{-1}\{F(s)\}$.

*α is the exponential order of $f(t)$ and its derivatives. [See Sec. 8.3 equation (4).]